Diagnosis, Management and Modeling of Neurodevelopmental Disorders

The Neuroscience Of Development

Diagnosis, Management and Modeling of Neurodevelopmental Disorders

The Neuroscience of Development

Edited by

Colin R. Martin
Institute for Clinical and Applied Health Research (ICAHR)
University of Hull
Hull, United Kingdom

Victor R. Preedy
King's College London
London
United Kingdom

Rajkumar Rajendram
Department of Medicine, King Abdulaziz Medical City,
Ministry of National Guard Health Affairs,
Riyadh, Saudi Arabia

ELSEVIER

ACADEMIC PRESS
An imprint of Elsevier

Academic Press is an imprint of Elsevier
125 London Wall, London EC2Y 5AS, United Kingdom
525 B Street, Suite 1650, San Diego, CA 92101, United States
50 Hampshire Street, 5th Floor, Cambridge, MA 02139, United States
The Boulevard, Langford Lane, Kidlington, Oxford OX5 1GB, United Kingdom

Library of Congress Cataloging-in-Publication Data
A catalog record for this book is available from the Library of Congress

British Library Cataloguing-in-Publication Data
A catalogue record for this book is available from the British Library

ISBN: 978-0-12-818371-7 (Set)
ISBN: 978-0-12-817988-8

For information on all Academic Press publications visit our website at https://www.elsevier.com/books-and-journals

Publisher: Nikki Levy
Acquisitions Editor: Natalie Farra
Editorial Project Manager: Timothy Bennett
Production Project Manager: Paul Prasad Chandramohan
Cover Designer: Matthew Limbert

Typeset by TNQ Technologies

Contents

Part II
Impairments and diseases

9. Pediatric brain tumors

Sherif G. Shaaban and Stephanie A. Terezakis

10. Neurofibromatosis type 1: from cellular phenotypes to human brain function

Inês Bernardino, Joana Gonçalves and Miguel Castelo-Branco

11. Current insights into neural crest cell development and pathologies

Rebekah M. Charney, Maneeshi S. Prasad and Martín I. García-Castro

Part III
Biomarkers, screening, methods and diagnosis

28. Neurodevelopment and the Ages and Stages Questionnaire, third edition (ASQ-3)

N. Granana and A.M. Romero Otalvaro

29. Attention deficit hyperactivity disorder screening in childhood and adolescence

Steven J. Korzeniewski

30. The Mini-Mental State Pediatric Examination: a cognitive screening test for Italian children

Valeria Peviani, Pina Scarpa and Gabriella Bottini

31. Prediction of neurodevelopment in preterm infants: from the neonatal intensive care unit into adulthood

Pilar Medina–Alva

Part V
Models and modelling

44. Microplatforms as a model for neurological conditions

George Deeb, Zeina Bou Diab, Jessica Saliba, Arij Daou and Rami Mhanna

45. Disease models in neurodevelopmental disorders

Pietro Fazzari, Angela Rodríguez-Prieto and Máximo Ibo Galindo

46. Developmental neurotoxicity of endosulfan: modeling with zebrafish

Karina Dale, Josef Daniel Rasinger and Ståle Ellingsen

47. Neurodevelopment of the zebrafish spinal serotonin system

Mark A. Masino and Jacob E. Montgomery

48. Inbred mouse model of brain development and intestinal microbiota

Reinaldo B. Oriá, Daniel V. Pinto, Ronaldo P. Dias, Ramon S. Raposo, Patricia L. Foley, João O. Malva and Richard L. Guerrant

49. The myelin mutant *taiep* rat as a model for developmental brain disorders

Jose R. Eguibar, Carmen Cortes, Juan C. Ahumada-Juárez, Valeria Piazza and Victor Hugo Hernandez Gonzalez

50. The MAM-E17 neurodevelopmental model of schizophrenia

Marzena Maćkowiak

Part VI
Resources

Contributors

Carmen Cortes, Behavioral Neurophysiological Laboratory, Institute of Physiology, Benemérita Universidad Autónoma de Puebla, Puebla, Pue, México

Mohamed Ahmed AboEllail, Department of Perinatology and Gynecology, Kagawa University, Graduate School of Medicine, Miki, Kagawa, Japan

Janiel Ahkin Chin Tai, School of Health Sciences, Purdue University, West Lafayette, IN, United States

Juan C. Ahumada-Juárez, Behavioral Neurophysiological Laboratory, Institute of Physiology, Benemérita Universidad Autónoma de Puebla, Puebla, Pue, México

Peter J. Anderson, Turner Institute for Brain and Mental Health, School of Psychological Sciences, Monash University, Clayton, VIC, Australia

Vicente Andreu- Fernández, Grup de Recerca Infancia i Entorn (GRIE), Institut d'investigacions Biomèdiques August Pi i Sunyer (IDIBAPS), Barcelona, Spain; Valencian International University (VIU), Valencia, Spain

Panagiotis Antsaklis, Department of Obstetrics and Gynecology, Department of Fetal Maternal Medicine, Alexandra Maternity Hospital, University of Athens, Athens, Greece

R.A. Armstrong, Vision Sciences, Aston University, Birmingham, United Kingdom

Eva Ausó, Department of Optics, Pharmacology and Anatomy, Universidad de Alicante, Alicante, Spain

Laura Avagliano, Department of Health Sciences, University of Milan, Milano, Italy

Tandy Aye, Division of Pediatric Endocrinology, Stanford University School of Medicine, Stanford, CA, United States

Silvia Bassani, Neuroscience Institute, Consiglio Nazionale delle Ricerche (CNR), Milan, Italy

Inês Bernardino, FMUC, University of Coimbra, Coimbra, Portugal; ICNAS/CIBIT, University of Coimbra, Coimbra, Portugal

Reid Blanchett, Institute of Quantitative Health Science and Engineering, Bioengineering Facility, Michigan State University, East Lansing, MI, United States

Sarah J. Blossom, Department of Pharmaceutical Sciences, College of Pharmacy, University of New Mexico, New Mexico, United States

Gabriella Bottini, Cognitive Neuropsychology Center, ASST Grande Ospedale Metropolitano Niguarda, Milan, Italy; Department of Brain and Behavioural Sciences, University of Pavia, Pavia, Italy; NeuroMI, Milan Center for Neuroscience, Italy

Zeina Bou Diab, Biomedical Engineering Program, Maroun Semaan Faculty of Engineering and Architecture, American University of Beirut, Beirut, Lebanon

Kristen R. Breit, Department of Psychology, West Chester University, West Chester, PA, United States

Soledad Bárez-López, Translational Health Sciences, University of Bristol, Bristol, United Kingdom

Gaetano P. Bulfamante, Department of Health Sciences, University of Milan, Milano, Italy

Miguel Castelo-Branco, ICNAS/CIBIT, University of Coimbra, Coimbra, Portugal

Rebekah M. Charney, Division of Biomedical Sciences, University of California Riverside, Riverside, CA, United States

Karina Dale, University of Bergen, Department of Biological Sciences, Bergen, Norway

Arij Daou, Biomedical Engineering Program, Maroun Semaan Faculty of Engineering and Architecture, American University of Beirut, Beirut, Lebanon

George Daskalakis, Department of Obstetrics and Gynecology, Department of Fetal Maternal Medicine, Alexandra Maternity Hospital, University of Athens, Athens, Greece

Josepheen De Asis-Cruz, Developing Brain Institute, Diagnostic Imaging and Radiology, Children's National Health System, Washington, D.C., United States

George Deeb, Biomedical Engineering Program, Maroun Semaan Faculty of Engineering and Architecture, American University of Beirut, Beirut, Lebanon

Dana DeMaster, Department of Pediatrics, University of Texas Health Science Center at Houston, Children's Learning Institute, Houston, TX, United States

Briana R. De Miranda, Center for Neurodegeneration and Experimental Therapeutics, Department of Neurology, College of Medicine, University of Alabama at Birmingham, Birmingham, AL, United States

Ronaldo P. Dias, Laboratory of Biology of Tissue Healing, Federal University of Ceara, Faculty of Medicine, Fortaleza, Ceará, Brazil

Maja Drobnič Radobuljac, Unit for Intensive Child and Adolescent Psychiatry, Center for Mental Health, University Psychiatric Clinic Ljubljana, Slovenia; Department of Psychiatry, Medical Faculty, University of Ljubljana, Ljubljana, Slovenia

Jose R. Eguibar, Behavioral Neurophysiological Laboratory, Institute of Physiology, Benemérita Universidad Autónoma de Puebla, Puebla, Pue, México

Lior M. Elkaim, Division of Neurosurgery, McGill University, McGill University Hospital Center, Montreal, QC, Canada

Ståle Ellingsen, University of Bergen, Department of Biological Sciences, Bergen, Norway

Pietro Fazzari, Laboratory of Cortical Circuits, Centro de Investigacion Príncipe Felipe, Valencia, Spain

Marlena S. Fejzo, Department of Obstetrics and Gynecology, Keck School of Medicine, University of Southern California, Los Angeles, CA, United States

Sacri R. Ferrón, Department of Cell Biology/ERI BiotecMed, University of Valencia, Burjassot, Spain

Richard H. Finnell, Center for Precision Environmental Health, Department of Molecular and Cellular Biology and Medicine, Baylor College of Medicine, Houston, TX, United States

Cecilia Flores, McGill University, Department of Psychiatry, Douglas Mental Health University Institute, Montreal, QC, Canada

Patricia L. Foley, Division of Comparative Medicine, Georgetown University, Washington, DC, United States

Jennifer L. Freeman, School of Health Sciences, Purdue University, West Lafayette, IN, United States

Máximo Ibo Galindo, Instituto Interuniversitario de Investigacion de Reconocimiento Molecular y Desarrollo Tecnologico (IDM), Universitat Politecnica de Valencia, Universitat de Valencia, Valencia, Spain; UPV-CIPF Joint Research Unit "Disease Mechanisms and Nanomedicine", Centro de Investigacion Príncipe Felipe, Valencia, Spain

Oscar García-Algar, Neonatal Unit, Hospital Clinic-Maternitat, ICGON, BCNatal-Barcelona Center for Maternal-Fetal and Neonatal Medicine, Grup de Recerca Infancia i Entorn (GRIE), Institut d'investigacions Biomèdiques August Pi i Sunyer (IDIBAPS), Barcelona, Spain

Martín I. García-Castro, Division of Biomedical Sciences, University of California Riverside, Riverside, CA, United States

Timothy M. George, Department of Neurosurgery, The University of Texas Dell Medical School, Austin, TX, United States

Robert Gerlai, Department of Psychology, University of Toronto Mississauga, Mississauga, ON, Canada

Gabrielle Gloston, Menninger Department of Psychiatry and Behavioral Sciences, Baylor College of Medicine, Houston, TX, United States

Joana Gonçalves, ICNAS/CIBIT, University of Coimbra, Coimbra, Portugal

N. Granana, Department of Pediatrics, Hospital C Durand, Buenos Aires, Argentina

Paolo Grazioli, Department of Health Sciences, University of Milan, Milano, Italy

Carmen Grijota-Martínez, Department of Endocrine and Nervous System Pathophysiology, Instituto de Investigaciones Biomédicas Alberto Sols, Consejo Superior de Investigaciones Científicas-Universidad Autónoma de Madrid, Madrid, Spain; Department of Cellular Biology, Universidad Complutense de Madrid, Madrid, Spain

Ana Guadaño-Ferraz, Department of Endocrine and Nervous System Pathophysiology, Instituto de Investigaciones Biomédicas Alberto Sols, Consejo Superior de Investigaciones Científicas-Universidad Autónoma de Madrid, Madrid, Spain

Richard L. Guerrant, Center for Global Health, Division of Infectious Diseases and International Health, University of Virginia, School of Medicine, Charlottesville, VA, United States

Toshiyuki Hata, Department of Perinatology and Gynecology, Kagawa University, Graduate School of Medicine, Miki, Kagawa, Japan; Department of Obstetrics and Gynecology, Miyake Clinic, Okayama, Japan

Victor Hugo Hernandez Gonzalez, Division of Sciences and Engineering, Department of Chemical, Electronic and Biomedical Engineering, Loma del Bosque, León, Mexico

Grayson N. Holmbeck, Loyola University Chicago, Department of Psychology, Chicago, IL, United States

George M. Ibrahim, Division of Neurosurgery, Department of Surgery, University of Toronto, Toronto, ON, Canada

F. Klamt, Laboratory of Cellular Biochemistry, ICBS/UFRGS, Porto Alegre, Rio Grande do Sul, Brazil; Programa de Pós-Graduação em Ciências Biológicas: Bioquímica, ICBS/UFRGS, Porto Alegre, Rio Grande do Sul, Brazil; National Institutes of Science and Technology—Translational Medicine (INCT-TM), Porto Alegre, Rio Grande do Sul, Brazil

Rebecca C. Knickmeyer, Institute of Quantitative Health Science and Engineering, Bioengineering Facility, Michigan State University, East Lansing, MI, United States

Kristi A. Kohlmeier, Department of Drug Design and Pharmacology, Faculty of Health and Medical Sciences, University of Copenhagen, Copenhagen, Denmark

Munekazu Komada, Laboratory of Mammalian Embryology, Department of Life Science, Faculty of Science and Engineering, Kindai University, Higashiosaka, Osaka, Japan

Steven J. Korzeniewski, Department of Family Medicine and Public Health Sciences, Wayne State University School of Medicine, Detroit, MI, United States

Asim Kurjak, Department of Obstetrics and Gynecology, Medical School University of Zagreb, Clinical Hospital Sveti Duh, Dubrovnik International University, Zagreb, Croatia

Danica Limon, Menninger Department of Psychiatry and Behavioral Sciences, Baylor College of Medicine, Houston, TX, United States

Catherine Limperopoulos, Developing Brain Institute, Diagnostic Imaging and Radiology, Children's National Health System, Washington, D.C., United States

Anna Lozano-Ureña, Department of Cell Biology/ERI BiotecMed, University of Valencia, Burjassot, Spain

Inés López del Castillo, Department of Cell Biology/ERI BiotecMed, University of Valencia, Burjassot, Spain

João O. Malva, Coimbra Institute for Clinical and Biomedical Research (iCBR), Faculty of Medicine, University of Coimbra, Coimbra, Portugal

Louis N. Manganas, Epilepsy Fellow, Department of Neurology, Stony Brook University Hospital, Stony Brook, NY, United States

Carla Marini, Pediatric Neurology Unit and Laboratories, Children's Hospital A. Meyer, Florence, Italy

Mark A. Masino, Department of Neuroscience, University of Minnesota, Minneapolis, MN, United States

Valentina Massa, Department of Health Sciences, University of Milan, Milano, Italy

Marzena Maćkowiak, Laboratory of Pharmacology and Brain Biostructure, Department of Pharmacology, Maj Institute of Pharmacology, Polish Academy of Sciences, Kraków, Poland

Pilar Medina—Alva, Department of Neonatology, Instituto Materno Perinatal, Department of Pediatrics, Universidad Peruana Cayetano Heredia, Lima, Peru

Rami Mhanna, Biomedical Engineering Program, Maroun Semaan Faculty of Engineering and Architecture, American University of Beirut, Beirut, Lebanon

Ana Montero-Pedrazuela, Department of Endocrine and Nervous System Pathophysiology, Instituto de Investigaciones Biomédicas Alberto Sols, Consejo Superior de Investigaciones Científicas-Universidad Autónoma de Madrid, Madrid, Spain

Jacob E. Montgomery, Department of Neuroscience, University of Minnesota, Minneapolis, MN, United States

Bartlett D. Moore, IV, Department of Neurosurgery, University of Texas Health Science Center at Houston, McGovern Medical School, Houston, TX, United States; Program in Cognitive Sciences, School of Social Science, Rice University, Houston, Texas, United States

Patrick M. Mullin, Department of Obstetrics and Gynecology, Keck School of Medicine, University of Southern California, Los Angeles, CA, United States

Laura M. Nicholson, Loyola University Chicago, Department of Psychology, Chicago, IL, United States

Yoshihiro Noda, Multidisciplinary Translational Research Lab, Department of Neuropsychiatry, Keio University School of Medicine, Shinjuku-ku, Tokyo, Japan

Kateryna Nohejlová, Department of Pathophysiology, Third Faculty of Medicine, Charles University, Prague, Czech Republic

Diana M. Ohanian, Loyola University Chicago, Department of Psychology, Chicago, IL, United States

Reinaldo B. Oriá, Laboratory of Biology of Tissue Healing, Federal University of Ceara, Faculty of Medicine, Fortaleza, Ceará, Brazil

Cynthia Ortinau, Department of Pediatrics, Washington University in St. Louis School of Medicine, St. Louis, MO, United States

Chiara Parodi, Department of Health Sciences, University of Milan, Milano, Italy

R.B. Parsons, Institute of Pharmaceutical Science, School of Cancer and Pharmaceutical Studies, King's College London (KCL), London, United Kingdom

Leona Pascoe, Turner Institute for Brain and Mental Health, School of Psychological Sciences, Monash University, Clayton, VIC, Australia

Maria Passafaro, Neuroscience Institute, Consiglio Nazionale delle Ricerche (CNR), Milan, Italy

Mayra I. Perez, Menninger Department of Psychiatry and Behavioral Sciences, Baylor College of Medicine, Houston, TX, United States

Valeria Peviani, Cognitive Neuropsychology Center, ASST Grande Ospedale Metropolitano Niguarda, Milan, Italy; Department of Brain and Behavioural Sciences, University of Pavia, Pavia, Italy; Department of Neuroscience, Max Planck Institute for Empirical Aesthetics, Frankfurt am Main, Germany

Valeria Piazza, Center for Research in Optics, Loma Del Bosque, Leon, Mexico

Daniel V. Pinto, Laboratory of Biology of Tissue Healing, Federal University of Ceara, Faculty of Medicine, Fortaleza, Ceará, Brazil

Barbara Plemeniti Tololeski, Unit for Adolescent Psychiatry, Center for Mental Health, University Psychiatric Clinic Ljubljana, Ljubljana, Slovenia

Filip S. Polli, Department of Drug Design and Pharmacology, Faculty of Health and Medical Sciences, University of Copenhagen, Copenhagen, Denmark

Shannon Pollock, Epilepsy Fellow, Department of Neurology, Stony Brook University Hospital, Stony Brook, NY, United States

Maneeshi S. Prasad, Division of Biomedical Sciences, University of California Riverside, Riverside, CA, United States

Victor R. Preedy, Diabetes and Nutritional Sciences Research Division, Faculty of Life Science and Medicine, King's College London, London, United Kingdom

Rajkumar Rajendram, College of Medicine, King Saud Bin Abdulaziz University for Health Sciences, Riyadh, Central Region, Saudi Arabia; Department of Medicine, King Abdulaziz Medical City, King Abdullah International Medical Research Center, Ministry of National Guard Health Affairs, Riyadh, Central Region, Saudi Arabia; Diabetes and Nutritional Sciences Research Division, Faculty of Life Science and Medicine, King's College London, London, United Kingdom

Ana C. Ramirez, Menninger Department of Psychiatry and Behavioral Sciences, Baylor College of Medicine, Houston, TX, United States

Ramon S. Raposo, Experimental Biology Core, University of Fortaleza, Fortaleza, Ceará, Brazil

Josef Daniel Rasinger, Institute of Marine Research, Bergen, Norway

Lauren M. Reynolds, McGill University, Integrated Program in Neuroscience, Montreal, QC, Canada; Sorbonne Université, CNRS UMR 8246, INSERM U1130, Neuroscience Paris Seine - Institut de Biologie Paris Seine, Paris, France; McGill University, Department of Psychiatry, Douglas Mental Health University Institute, Montreal, QC, Canada

Angela Rodríguez-Prieto, Laboratory of Cortical Circuits, Centro de Investigacion Príncipe Felipe, Valencia, Spain

A.M. Romero Otalvaro, Universidad Pontificia Bolivariana, Monteria, Cordoba, Colombia

Sebastian Sailer, Neonatal Unit, Hospital Clinic-Maternitat, ICGON, BCNatal-Barcelona Center for Maternal-Fetal and Neonatal Medicine, Grup de Recerca Infancia i Entorn (GRIE), Institut d'investigacions Biomèdiques August Pi i Sunyer (IDIBAPS), Barcelona, Spain

Jessica Saliba, Department of Biology, Faculty of Science, Lebanese University, Beirut, Lebanon

Pina Scarpa, Cognitive Neuropsychology Center, ASST Grande Ospedale Metropolitano Niguarda, Milan, Italy; NeuroMI, Milan Center for Neuroscience, Italy

Sophie C. Schneider, Menninger Department of Psychiatry and Behavioral Sciences, Baylor College of Medicine, Houston, TX, United States

P. Schonhofen, Laboratory of Cellular Biochemistry, ICBS/UFRGS, Porto Alegre, Rio Grande do Sul, Brazil; Programa de Pós-Graduação em Ciências Biológicas: Bioquímica, ICBS/UFRGS, Porto Alegre, Rio Grande do Sul, Brazil; National Institutes of Science and Technology—Translational Medicine (INCT-TM), Porto Alegre, Rio Grande do Sul, Brazil

Giorgia Sebastiani, Neonatal Unit, Hospital Clinic-Maternitat, ICGON, BCNatal-Barcelona Center for Maternal-Fetal and Neonatal Medicine, Grup de Recerca Infancia i Entorn (GRIE), Institut d'investigacions Biomèdiques August Pi i Sunyer (IDIBAPS), Barcelona, Spain

Sherif G. Shaaban, Department of Radiation Oncology, University of Minnesota, Minneapolis, MN, United States

Merina Shrestha, Department of Child Health, Tribhuvan University Teaching Hospital, Kathmandu, Nepal

Eric A. Storch, Menninger Department of Psychiatry and Behavioral Sciences, Baylor College of Medicine, Houston, TX, United States

Michael A. Sustaíta, Menninger Department of Psychiatry and Behavioral Sciences, Baylor College of Medicine, Houston, TX, United States

Stephanie A. Terezakis, Department of Radiation Oncology, University of Minnesota, Minneapolis, MN, United States

Romana Šlamberová, Department of Physiology, Third Faculty of Medicine, Charles University, Prague, Czech Republic

D.M. Vargas, Laboratory of Cellular Biochemistry, ICBS/UFRGS, Porto Alegre, Rio Grande do Sul, Brazil; Programa de Pós-Graduação em Ciências Biológicas: Bioquímica, ICBS/UFRGS, Porto Alegre, Rio Grande do Sul, Brazil; National Institutes of Science and Technology−Translational Medicine (INCT-TM), Porto Alegre, Rio Grande do Sul, Brazil

Kelly A. Vaughn, Department of Pediatrics, University of Texas Health Science Center at Houston, Children's Learning Institute, Houston, TX, United States

Andres G. Viana, Department of Psychology, University of Houston, Houston, TX, United States

Saira A. Weinzimmer, Menninger Department of Psychiatry and Behavioral Sciences, Baylor College of Medicine, Houston, TX, United States

Eva Widerstrom-Noga, Miami Project to Cure Paralysis University of Miami, Miller School of Medicine, Miami, FL, United States

Adrien M. Winning, Loyola University Chicago, Department of Psychology, Chicago, IL, United States

Marlon L. Wong, Department of Physical Therapy, Miller School of Medicine University of Miami, Miami, FL, United States

Walter Zegarra, Division of Pediatric Endocrinology, Stanford University School of Medicine, Stanford, CA, United States

Shao Jia Zhou, School of Agriculture, Food & Wine, Robinson Research Institute, University of Adelaide, Adelaide, SA, Australia

Jill G. Zwicker, Department of Occupational Science & Occupational Therapy, University of British Columbia, Vancouver, BC, Canada; Brain, Behaviour, & Development, BC Children's Hospital Research Institute, Vancouver, BC, Canada

Preface

The effects of abnormal neurological development can have drastic consequences for the individual and also for the family unit. In the United States, about 10% of hospitalized children have a neurological diagnosis. In children, neurological conditions may include cerebral palsy, genetic abnormalities (e.g., Down syndrome), epilepsy and seizures, neural tube defects, and pathologies due to malnutrition, toxic substances, environmental factors, cancers, infections, and disturbed physiology. Such conditions will perturb intellectual and behavioural development. There are many other conditions and some have a global impact. For example, iodine deficiency causes neurological damage, and over 300 million children worldwide are at risk. Iodine deficiency is argued to be the most preventable cause of mental retardation in the world. Some neurological conditions will arise *in utero* and others will occur during postnatal growth. Maternal malnutrition will impact on fetal brain development. Postnatal malnutrition will impact on neurological development, which may persist into adulthood. A prelude to understanding these conditions is the normative processes of development. In other words, there is a need to understand both normal and abnormal processes and how they are interlinked. There is also a need to address the fact that experts in one field of neuroscience may not be experts in other areas.

Hitherto such material on the neuroscience of normal and abnormal development has been published in different scientific domains and/or written for the expert who specializes in a narrow and focused area. Such specialized text omits the fact that some processes, techniques, or abnormalities may be common across fields of neuroscience or have translational applicability. For example, methods for the utilization of biomarkers in one area of neuroscience may be applicable to others. Understanding the science of epigenetics *in utero* may be applicable to a variety of fields of postnatal neurodevelopment. There are many other example of cross-transferable knowledge, concepts, and methods, but a single source of material is necessary to address this. This is achieved in **The Neuroscience of Development**, which is divided into two books:

Diagnosis, Management, and Modeling of Neurodevelopmental Disorders: The Neuroscience of Development
Factors Affecting Neurodevelopment: Genetics, Neurology, Behavior, and Diet

In this book **Diagnosis, Management, and Modeling of Neurodevelopmental Disorders: The Neuroscience of Development**, there are over 50 chapters in 6 sections:

[1] Introductory Chapters: Setting the Scene for the Neuroscience of Development
[2] Impairments and Diseases
[3] Biomarkers, Screening, Methods, and Diagnosis
[4] Management and Treatments
[5] Models and Modeling
[6] Resources

It has been difficult to ascribe some chapters to specific sections as they could equally be located in more than one part of the book. However, this is addressed by the excellent indexing system provided by the publishers.

Each chapter has the following:

- An Abstract (published online)
- Key Facts
- Mini-Dictionary of terms
- Applications to Other Areas of Development
- Summary Points

The sections ***Key Facts***, ***Mini-Dictionary of Terms***, and ***Applications to Other Areas of Development*** enable the reader to cross the transintellectual and transdisciplinary divides. The section ***Summary Points*** encapsulates each chapter in a succinct way.

Diagnosis, Management, and Modeling of Neurodevelopmental Disorders: The Neuroscience of Development is designed for research and teaching purposes. It is suitable for pediatricians, neurologists, psychologists, developmental biologists, health scientists, public health workers, doctors, pharmacologists, and research scientists. It is valuable as a personal reference book and also for academic libraries that cover the domains of development, neurology, and health sciences. Contributions are from leading national and international experts including those from world-renowned institutions. It is suitable for undergraduates, postgraduate, lecturers, and academic professors.

The Editors

Part I

Introductory chapters: Setting the scene for the neuroscience of development

Chapter 1

The structure of the human brain as revealed in six histological sections

R.A. Armstrong
Vision Sciences, Aston University, Birmingham, United Kingdom

List of abbreviations

B17 Brodmann area 17: Primary visual cortex
B18 Brodmann area 18: Association visual cortex
B19 Brodmann area 19: Association visual cortex
BG Band of Gennari
CA (1−4) Cornu ammonis sectors
CAq Cerebral aqueduct
Cl Claustrum
CN Caudate nucleus
CSF Corticospinal fibers
DG Dentate gyrus
DN Dentate nucleus
EC External capsule
EnC Entorhinal cortex
FPF Frontopontine fibers
GCL Granule cell layer of cerebellum
GM Gray matter
GP Globus pallidus
HC Hippocampus
IC Internal capsule
ICP Inferior cerebellar peduncle
ION Inferior olivary nucleus
IVv Fourth ventricle
LGN Lateral geniculate nucleus
LHE Hematoxylin and Eosin/Luxol fast blue
LOT Lateral occipitotemporal gyrus
MCP Middle cerebellar peduncle
MGB Medial geniculate body
ML Medial lemniscus
MLA Molecular layer of cerebellum
MLF Medial longitudinal fasciculus
ON Oculomotor nucleus
ONv Oculomotor nerve
PAG Periaqueductal gray matter
PCL Purkinje cell layer
PHG Parahippocampal gyrus
PUT Putamen
RB Restiform body (ICP)
RN Red nucleus

Diagnosis, Management and Modeling of Neurodevelopmental Disorders. https://doi.org/10.1016/B978-0-12-817988-8.00001-4

SC Superior colliculus
SCP Superior cerebellar peduncle
SNc Substantia nigra compacta
SNr Substantia nigra reticulate
SUB Subiculum
TPF Temporopontine fibers
VN Vagus nerve
WM White matter

Mini-dictionary of terms

Association cortex Regions of cerebral cortex providing more complex processing of sensory information
Basal ganglia A group of subcortical nuclei, acting as a single functional unit, and with extensive connections to the *cerebral cortex*, thalamus, and *cerebellum*.
Cerebellum A major component of the hindbrain involved in the control of movement but may also contribute to attention, language, and the control of emotions
Corticocortical projections Fibers connecting adjacent and remote gyri of the cerebral cortex and form a significant part of the *external capsule*.
Gray Matter Regions that contain the majority of neurons
Hippocampus A region of the temporal lobe involved in transfer of information from short to long-term memory, spatial memory, and in the control of behavior
Medulla A region of the hindbrain involved in a number of vital bodily functions such as the cardiac cycle and respiration
Neural tube A structure derived from surface ectoderm during development, which gives rise to the nervous system
Primary sensory area Regions of cerebral cortex that carry out initial processing of sensory information received from the sense organs
White matter Regions that contain the fiber tracts

Introduction

The human brain develops from a layer of ectoderm located on the surface of the developing embryo. Invagination of this cell layer forms a neural tube, which becomes differentiated anteriorly into forebrain, midbrain, and hindbrain. The brain is composed of many billions of neurons, each with the capability of communicating with a large number of other cells (Nauta & Feirtag, 1986). In association with white matter fiber tracts, networks of neurons are organized anatomically into pathways, which are involved in sensory, motor, and cognitive function (Diamond, Scheibel, & Elson, 1985; Haines, 1987; Moyer, 1980, sections 3). This chapter provides a basic guide to the structure of the brain based on six key histological sections: (1) occipital cortex, (2) temporal cortex, (3) basal ganglia, (4) midbrain, (5) medulla, and (6) cerebellum. Throughout the text, regions and fiber tracts illustrated in the sections are shown in italics.

Section 1: occipital cortex

Fig. 1.1 shows the typical appearance of the cerebral cortex in a section of occipital cortex with its characteristic folds (gyri) and fissures (sulci) and external gray matter and internal white matter. The cerebral cortex comprises six layers (I–VI), which vary in prominence from region to region. Information from a sense organ is conveyed first via a region of thalamus to a "primary" sensory area for initial processing. Hence, in the visual system shown in Fig. 1.1, for example, information arrives via the lateral geniculate nucleus (LGN) to the *primary visual cortex (B17)* in layer IV, marked by the distinctive *"Band of Gennari" (BG)* (Singer, 1979), and is then conveyed to layers II and III and to a lesser extent layer V. The output from B17 is to *"association cortex" B18 and B19* where more complex visual processing takes place in layers II/III (Zeki, 1971; DeLacoste & White, 1993). Subsequently, neurons in the lower layers provide the output to subcortical visual areas, e.g., from lamina VI to the dorsal and ventral LGN and from lamina V to the *superior colliculus (SC)* and the pretectal area (Parnavelos & McDonald, 1983). Similar processing takes place in unique regions of cortex for each sensory modality.

Section 2: temporal cortex

Ultimately, all types of sensory information are integrated by the cerebral cortex into an image of the "state of the world," a process that involves the parietal lobe. This information is transferred to many other regions of the brain and is used in more complex cognitive processing. In particular, the temporal cortex (Fig. 1.2) is involved in a variety of cognitive

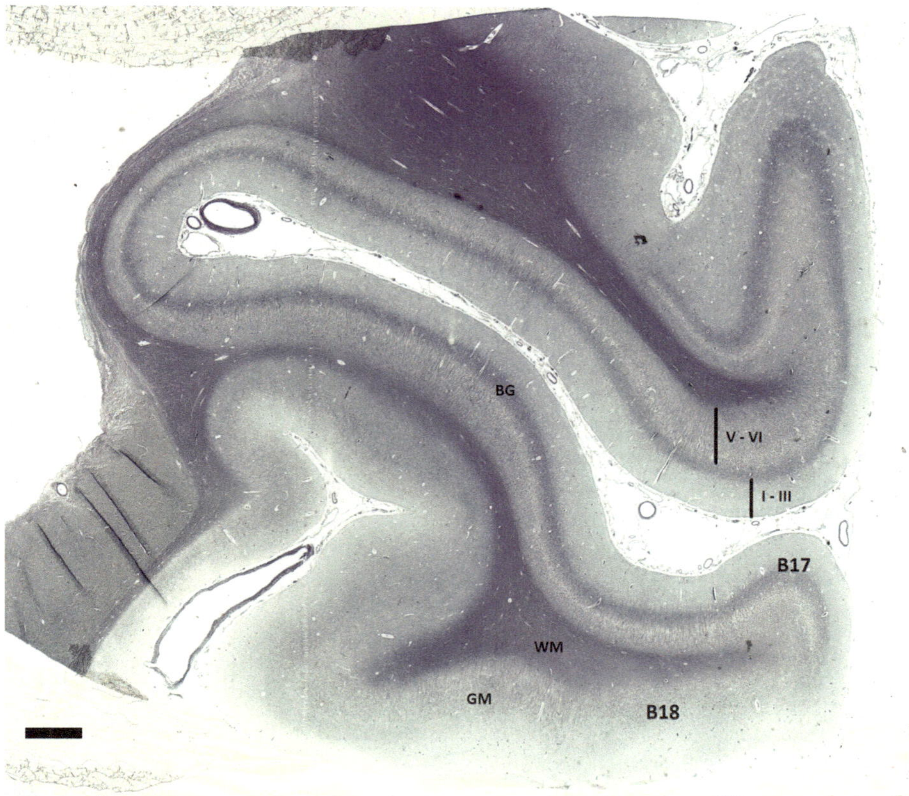

FIGURE 1.1 Occipital lobe through visual cortex. The figure shows a section of the human occipital lobe through the visual cortex stained with a silver impregnation method (*B17*, Primary visual cortex; *B18*, Association visual cortex; *BG*, Band of Gennari; *GM*, Gray matter; *WM*, White matter; *I–VI*, Cortical layers; Unpublished).

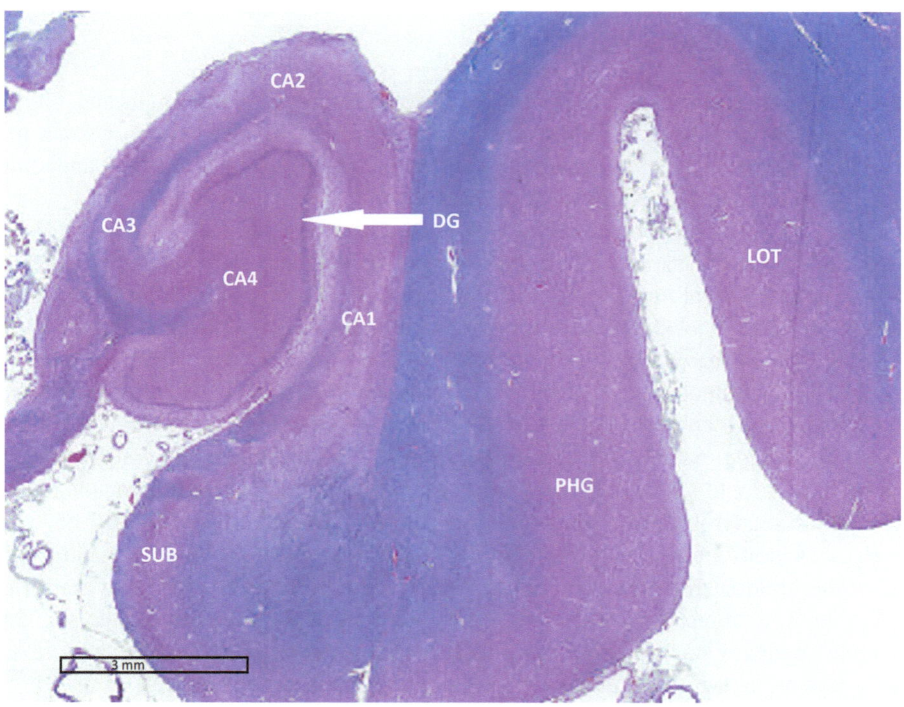

FIGURE 1.2 Temporal lobe and hippocampus at the level of the lateral geniculate body. The figure shows a section of the human temporal lobe through the hippocampus (HC) stained with combination of H/E and luxol fast blue (LHE) (*CA1-4*, Sectors of cornu ammonis; *DG*, Dentate gyrus; *LOT*, Lateral-occipitotemporal gyrus; *PHG*, Parahippocampal gyrus; *SUB*, Subiculum; Unpublished).

functions in humans (Brodal, 1981). Hence, the left temporal lobe is involved in comprehension, naming of objects, and in verbal memory. The ventral surface of the temporal lobe and especially the *parahippocampal gyrus (PHG)*, which includes the more anterior entorhinal cortex (EnC), are involved in more complex visual processing including facial recognition and analysis of the visual scene. Areas of medial temporal lobe are also involved in the processing of olfactory information and in memory including episodic/declarative memory.

Hippocampus (HC)

The *hippocampus (HC)* is involved in various cognitive functions including transfer of information from short to long-term memory, spatial memory, and in the control of behavior (Gray & McNaughton, 2000; O'Keefe & Nadel, 1978). Projections from the EnC take the perforant path and terminate in the molecular layer of the *dentate gyrus (DG)*. From the *DG*, mossy fibers travel to *cornu ammonis (CA) sectors CA3/4*, while sector *CA3* fibers have collaterals, which travel to sector *CA1*. Sector *CA1* axons connect with the *subiculum (SUB)* representing the major output pathway of the *HC* via the fornix. In addition, there are direct projections from the EnC to sectors *CA3* and *CA1* and a return pathway from sector *CA3* to the *DG*. Within this circuit, sector *CA3* is an important "nodal" point at which many major projections converge. The *HC* may act as a "comparator" comparing current sensory events with expected or predicted events (Gray & McNaughton, 2000; Vinogradova, 2001). If an expectation continues to be verified by information derived from sensory analysis, then no alteration of current behavior may be required. If, however, a discrepancy is detected, the *HC* may intervene by initiating appropriate action and directly controlling behavior by active inhibition of current motor activity and initiation of new sensory data gathering.

Dentate gyrus

The *DG* is a narrow notched band of cortex that lies on the dorsal surface of the *PHG* covered by the fimbria. It has a characteristic "horse-shoe" appearance in coronal section (Fig. 1.2) and consists of a dense band of granule cells with dendrites, which extend into the molecular layer. This region receives a major excitatory input from the EnC via the perforant path. The axons of the granule cells, namely the mossy fibers, then pursue a course along the hippocampal pyramidal cells making synaptic contacts with the basal parts of their apical dendrites via the "mossy tufts," which are enlarged, synaptic terminals (Brodal, 1981). The *DG* is believed to be involved in a variety of brain functions including episodic and spatial memory and the exploration of novel environments.

Section 3. the basal ganglia

The result of cognitive processing is a motor response involving many areas of the brain and ultimately, a change in behavior. Motor action is initiated by the precentral gyrus ("motor cortex") and adjacent regions, but it is the ventral horn cells of the spinal cord that directly connect to the muscles and initiate their movement. Many anatomical regions exert a controlling influence over movement, one of the most important being the *basal ganglia*.

The *basal ganglia* are a group of subcortical nuclei with extensive connections to the *cerebral cortex*, thalamus, and *cerebellum*. They exert control over a variety of functions including voluntary motor control, procedural learning, and motivation. Fig. 1.3 shows the *caudate nucleus (CN)* and *putamen (PUT)*, which constitute the "striatum," and various adjacent structures. Additional areas include the globus pallidus (GP), subthalamic nucleus, and the *substantia nigra (SN)*, the latter visible in Fig. 1.4. The *CN* has a relatively large anterior "head," which is directly connected to the *PUT*. The rest of the *CN* then forms an arch continuously diminishing in size as it curves around laterally toward the more posteriorly located thalamus. The slender tail then proceeds in a rostral direction terminating in the amygdala. There are two fiber bundles traversing the *basal ganglia*, namely the *internal capsule (IC)*, which principally contains the *corticospinal fibers (CSF)* connecting the motor cortex to spinal cord, and the *external capsule (EC)*, which contains the cortical association fibers and cholinergic projections from basal forebrain to the cortex. The GP ("pale globe" or "pallidum") regulates voluntary movement so that actions are carried out "smoothly." Its primary action is inhibitory and balances the excitatory effects of other areas of the brain such as the *cerebral cortex* and *cerebellum*. There are two main functional pathways involving the *basal ganglia*. The "direct" pathway exerts an excitatory effect on the cerebral cortex, especially the motor cortex, while the "indirect" pathway has an inhibitory effect resulting ultimately in reduced muscle activity. Hence, the direct and indirect pathways are antagonistic in their overall control of muscle movement.

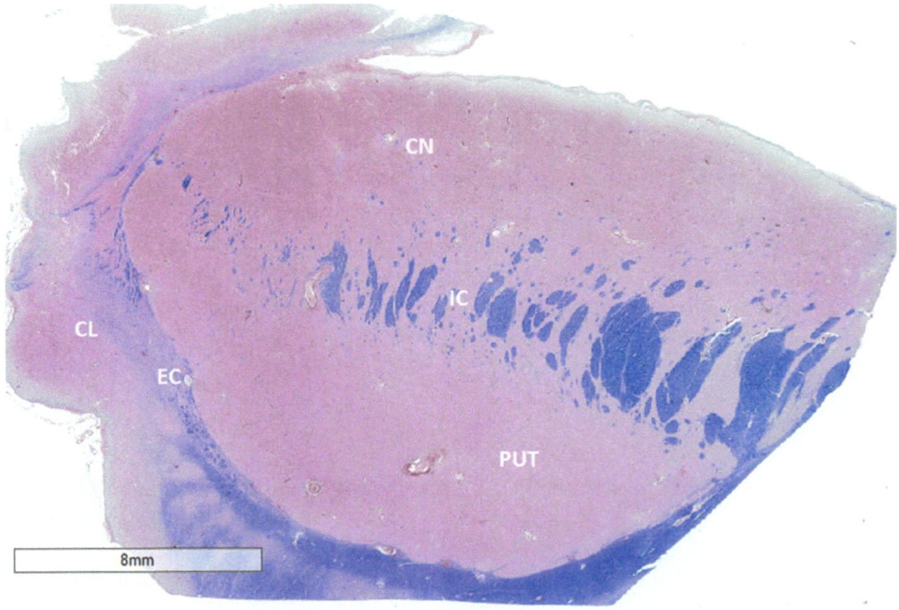

FIGURE 1.3 The basal ganglia through the anterior portion of the striatum. The figure shows the basal ganglia in coronal section stained with combination of H/E and luxol fast blue (LHE) (*Cl*, Claustrum; *CN*, Head of caudate nucleus; *EC*, External capsule; *IC*, Internal capsule; *PUT*, Putamen; Unpublished).

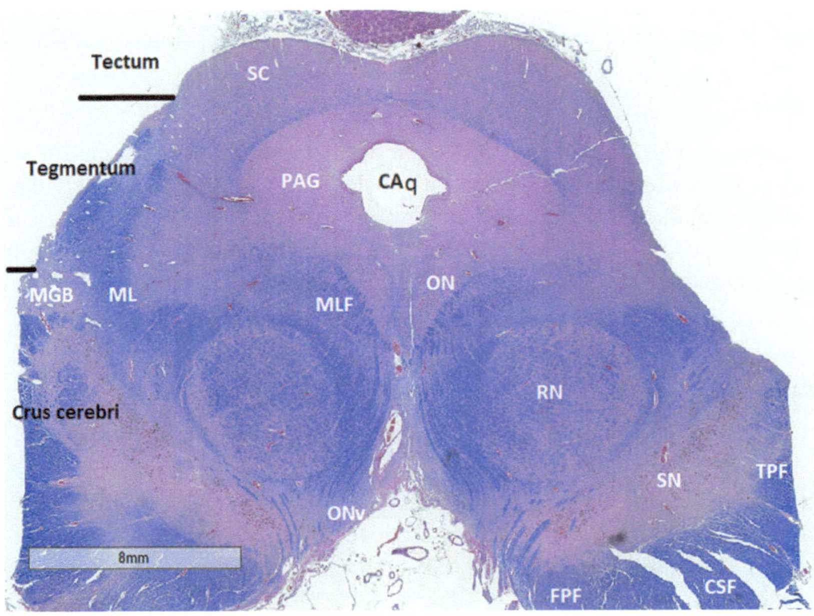

FIGURE 1.4 The midbrain through the superior colliculus. The figure shows a section of midbrain stained with combination of H/E and luxol fast blue (LHE) (*CA*, Cerebral aqueduct; *CSF*, Corticospinal fibers; *FPF*, Frontopontine fibers; *MGB*, Medial geniculate body; *ML*, Medial lemniscus; *MLF*, Medial longitudinal fasciculus; *ON*, Oculomotor nucleus; *ONv*, Oculomotor nerve; *RN*, Red nucleus; *PAG*, Periaqueductal gray; *SC*, Superior colliculus; *SN*, Substantia nigra; *TPF*, Temperopontine fibers; Unpublished).

Section 4: the midbrain

Fig. 1.4 shows a number of areas located in the midbrain that are largely involved in the control of movement.

The superior colliculus (SC)

The *SC* is prominent in the upper part of the tectum adjacent to the inferior colliculus and is involved in directing a behavioral response, via eye movements, toward a specific point or object (Klier, Wang, & Crawford, 2003; Kustov & Robinson, 1996; Sprague, 1996). It consists of seven layers, with alternating fibre-rich and cell-rich bands, the superficial layers (I/II/III) being sensory in function and receiving input from the eyes and other sensory systems, while the deeper layers (VI/VII) are motor-related and involved in the control of eye movements. The intermediate layers (IV/V) are involved in both sensory and motor function. The inferior colliculus has a similar function with regard to the auditory system.

The oculomotor nucleus and nerve

The *oculomotor nerve (III)* is visible in this section and, together with the trochlear (IV) and abducens (VI) nerves, enters the orbit to supply the eye muscles. There are 12 pairs of cranial nerves, which either have an exclusively sensory or motor function or have elements of both (Diamond et al., 1985).

Red nucleus (RN)

The *red nucleus (RN)* is located in the rostral part of the tegmentum adjacent to the *SN*. It receives inputs from the *cerebellum* and motor cortex and its outputs are mainly to the *cerebellum* via the *medial tegmental tract.*

Substantia nigra (SN)

The *SN* is divided into two parts, namely the *SN reticulata (SNr)* and the *SN compacta (SNc)*. The *SNr* in combination with the GP acts to inhibit the thalamus. By contrast, the *SNc* provides a major projection to the striatum, the "striatonigral pathway," which regulates the activity of the striatum as a whole and uses dopamine as a neurotransmitter.

The cerebral peduncles

The *cerebral peduncles ('crus cerebri')* are a large collection of fiber bundles in the ventral midbrain, which originate in the cerebral cortex. Many of these fibers connect either with nuclei in the pons and ultimately the *cerebellum* or the spinal cord. Fibers are arranged topographically from their point of origin and include: (1) *the frontopontine fibers (FPF)*, which originate in the frontal lobe and travel via the anterior limb of the *IC* to terminate in the nuclei of the pons; (2) the *CSF*, which enervates the motor neurons of the spinal cord; and (3) the *temperopontine fibers (TPF)* originating largely in the temporal lobe.

Section 5. the medulla

The *medulla* (Fig. 1.5) is located beneath the pons and is concerned with several vital functions including those involved in cardiac and respiratory function. The *inferior cerebellar peduncle (ICP)* is prominent in this section, and the olivocerebellar fibers enter this region from the *inferior olivary nucleus (ION)*. The *ION* coordinates signals received from the spinal cord and sends them to the *cerebellum* where they are involved in the process of motor coordination and learning. The vagus cranial nerve (X) also emerges at this level, has both sensory and motor functions, and supplies the pharynx, larynx, and viscera of the chest and abdomen.

Section 6. the cerebellum

The *cerebellum* (Fig. 1.6) is a major component of the hindbrain appearing as a distinctive region located beneath the *occipital lobes* (Wright et al., 2016). Its most significant functional role is the control of movement, but it may also be involved in attention, language, and the control of emotions. The *cerebellum* does not directly evoke a movement but receives information from many brain regions, which is then integrated to provide fine-tuning of motor function.

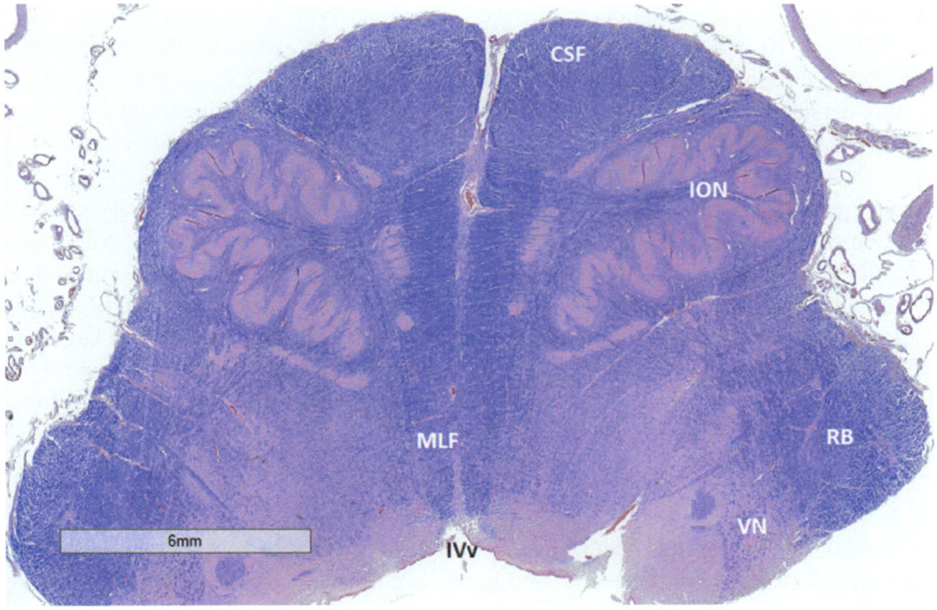

FIGURE 1.5 The lower medulla through the inferior olivary nucleus. The figure shows a section of the medulla stained with combination of H/E and luxol fast blue (LHE) (*CSF*, Corticospinal fibers; *ION*, Inferior olivary nucleus, *IVv*, Fourth ventricle; *MLF*, Medial longitudinal fasciculus, *RB*, Restiform body; *VN*, Vagus nerve; Unpublished).

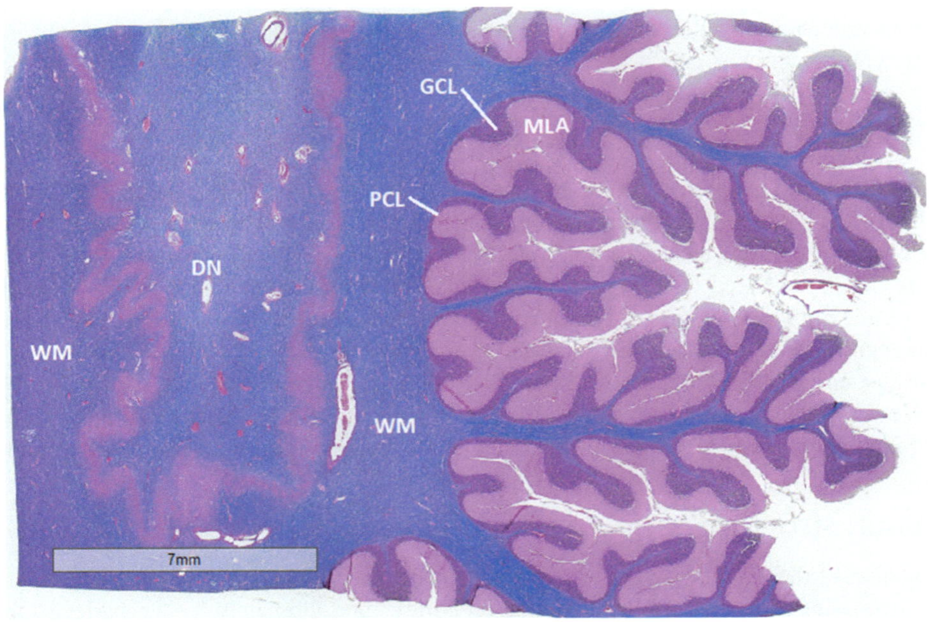

FIGURE 1.6 The cerebellum through the cerebellar cortex and dentate nucleus. The figure shows a section of the cerebellar hemispheres stained with combination of H/E and Luxol fast blue (LHE) (*DN*, Dentate nucleus; *GCL*, Granule cell layer; *MLA*, Molecular layer; *PLC*, Purkinje cell layer, *WM*, White matter; Unpublished).

Cerebellar cortex

The cerebellar cortex consists of two hemispheres connected by the vermis. Each hemisphere comprises a layer of gray matter highly folded into gyri or folia with a thick layer of white matter beneath connecting the *cerebellum*, *cerebral cortex,* and other regions (Fig. 1.6). Embedded within the white matter are four cerebellar nuclei, namely the *dentate*, emboliform, globose, and fastigial nuclei, which are largely composed of gray matter. The cerebellar cortex is made of three distinct layers, namely the *molecular layer (MLA)*, *Purkinje cell layer (PCL)*, and *granule cell layer (GCL)*. The *PCL* controls the output of the cerebellar cortex and is influenced by the activity of many other neurons. The *GCL* comprises a dense layer of approximately 10 billion granule cells, the axons of which ascend to the *MLA*, each dividing into two branches that run parallel to the surface of the folia and transfer impulses to the *PCL*. In addition, the *GCL* contains Golgi type II cells whose long dendrites ascend to the *MLA* terminating in complex three-dimensional dendritic trees while their short dendrites make many synaptic contacts with the *GCL*. Within the *MLA*, the basket cells and stellate cells make inhibitory synaptic contacts with *PCL* dendrites.

There are two major excitatory inputs to the *cerebellum*. First, the large-diameter rapidly conducting mossy fibers represent the termination of several major tracts, which have their origin in the spinal cord, pons, and vestibular nuclei. Second, the climbing fibers represent the termination of fibers from the pons and reticular formation, and each connects with a *PCL* by coiling around its axon. The climbing fibers exert a powerful excitatory influence on the *PCL*. Virtually all of the output from the cerebellar cortex is via inhibitory connections with the deep nuclei. The largest of these nuclei, the *dentate nucleus (DN)*, connects exclusively with the lateral sector of the cerebellar cortex. The majority of neurons in these nuclei consist of large cell bodies with more or less spherical dendritic trees. Subsequently, axons from cells in the deep nuclei enter the brain stem.

White matter

Three white matter bundles (the "cerebellar peduncles") connect the *cerebellum* to the brain stem. First, the *ICP* carries proprioceptive information principally from the head and limbs, lateral vestibular nucleus, the contralateral inferior olive, and from the lateral reticular nucleus and projects widely to the cerebellar cortex. In addition, there is a single output pathway from the *PCL*, which directly inhibits the vestibular nuclei but without first synapsing in the cerebellar nuclei. Second, the middle cerebellar peduncle (MCP), the largest of the three bundles, carries sensory/motor information, largely involving the body's extremities, from the *cerebral cortex* to the *cerebellum*. Third, the superior cerebellar peduncle (SCP) carries information to the *cerebellum* from the spinal cord to the anterior lobe via the anterior *CSF* and outputs to the *RN*, thalamus, and medulla and comprises the "cerebro-ponto-cerebellar thalamic loop," a feedback mechanism for fine-tuning the output from the cerebral cortex.

Applications to other areas of development

In this chapter, the structure and main functions of the human brain are revealed via six histological sections taken through the forebrain, midbrain, and hindbrain; the three major regions that develop from the anterior segment of the original neural tube. Much of the *cerebral cortex*, which develops from the forebrain, carries out a sensory analysis of the world using information supplied from specialized sense organs and provides the brain with an image of the "state of the world." This information is used in cognitive processing by the *hippocampus* and adjacent areas, which results in appropriate motor action. Many of the regions in the midbrain and hindbrain also receive this sensory information and use it to provide fine-tuning of the motor response. Ultimately, motor responses result in a change of behavior dictated by changes in sensory information. Many of the developmental problems described in this book will impinge directly on various aspects of sensory analysis of incoming information and/or the ability of the brain to make an appropriate response.

Key facts of brain structure

- The primary function of the cerebral cortex is sensory analysis using information provided by the sense organs.
- More complex processing is carried out by "association cortex" and ultimately by the parietal lobe to construct an "image of the world."
- The entorhinal cortex receives information from all sensory systems via the *temporal lobe*, the latter acting as a data storage region.
- This information is transmitted to the *dentate gyrus* via the perforant path and also to the *hippocampus* and *subiculum*.
- Cognitive processing by the *hippocampus* and adjacent regions results in motor action initiated by the motor cortex and transferred to the spinal cord via the *corticospinal tract*.
- Many of the regions in the midbrain and hindbrain function to fine-tune the motor response resulting ultimately in a "behavioral" response to changing sensory data.

Summary points

- The nervous system is the most complex body system comprising a very large number of interconnected neurons.
- In association with white matter fiber tracts, networks of neurons are organized into anatomical pathways, which are involved in sensory, motor, and cognitive functions.
- A large part of the brain is concerned with the reception and analysis of sensory data and in the control of subsequent motor responses.
- The parietal lobe of the brain integrates data from the different senses and provides an image of the "state of the world."
- Data concerning the state of the world are used in more complex cognitive processing in the temporal lobe resulting ultimately in motor action and a change in behavior.
- Many areas in the midbrain and hindbrain are involved in the control of the motor response.

References

Brodal, A. (1981). *Neurological anatomy* (3rd ed.). Oxford: Oxford University Press (Chapters 2–12).

De Lacoste, M., & White, C. L. (1993). The role of cortical connectivity in alzheimer's disease pathogenesis: A review and model system. *Neurobiology of Aging, 14*, 1–16.

Diamond, M. C., Scheibel, A. B., & Elson, L. M. (1985). *The human brain coloring book*. New York: Barnes & Noble Books (Unit 6).

Gray, J. A., & McNaughton, N. (2000). *The neuropsychology of anxiety: An enquiry into the functions of the septo-hippocampal system*. New York: Oxford University Press (Chapters 7–10).

Haines, D. E. (1987). *Neuroanatomy: An atlas of structures, sections, and systems* (2nd ed.). Baltimore & Munich: Urban & Schwarzenberg (Chapter 5).

Klier, E., Wang, H., & Crawford, D. (2003). Three-dimensional eye-head coordination is implimented downstream from the superior colliculus. *Journal of neurophysiology, 89*, 2839–2853.

Kustov, A., & Robinson, D. (1996). Shared neural control of attentional shifts and eye movements. *Nature, 384*, 74–77.

Moyer, K. E. (1980). *Neuroanatomy*. New York: Harper & Row.

Nauta, W. J. H., & Feirtag, M. (1986). *Fundamental neuroanatomy*. New York: WH Freeman & Co (Chapter 2).

O'Keefe, J., & Nadel, L. (1978). *The Hippocampus as a cognitive map*. Oxford: Oxford University Press (Chapters 2,3,6–8,14).

Parnavelas, J. G., & McDonald, J. K. (1983). The cerebral cortex. In P. C. Emson (Ed.), *Chemical neuroanatomy* (pp. 337–358). New York: Raven Press.

Singer, W. (1979). Central-core control of visual cortex functions. In F. O. Schmitt, & F. G. Worden (Eds.), *The neurosciences: 4th study program* (pp. 1093–1110). Cambridge: MIT Press.

Sprague, J. M. (1996). Neural mechanisms of visual orienting responses. *Progress in Brain Research, 112*, 1–15.

Vinogradova, O. S. (2001). Hippocampus as comparator: Role of the two input and two output systems of the hippocampus in selection and registration of information. *Hippocampus, 11*, 578–598.

Wright, M., Skaggs, W., Nielsen, F. A., et al. (2016). The cerebellum. *WikiJournal of Medicine*, 3. https://doi.org/10.15347/wjm/2016.001.

Zeki, S. M. (1971). Cortical projections from two striate areas in the monkey. *Brain Research, 34*, 19–35.

Chapter 2

Neural tube defects: embryonic origin, cell survival equilibrium impact, and clinical features

Chiara Parodi[1,a], Paolo Grazioli[1,a], Laura Avagliano[1], Timothy M. George[2], Gaetano P. Bulfamante[1], Richard H. Finnell[3] and Valentina Massa[1]

[1]*Department of Health Sciences, University of Milan, Milano, Italy;* [2]*Department of Neurosurgery, The University of Texas Dell Medical School, Austin, TX, United States;* [3]*Center for Precision Environmental Health, Department of Molecular and Cellular Biology and Medicine, Baylor College of Medicine, Houston, TX, United States*

List of abbreviations

CM-II Chiari II malformation
CNS Central nervous system
NTDs Neural tube defects

Mini-dictionary of terms

Anencephaly: the absence of a major portion of the brain, skull, and scalp.
Anoikis: form of programmed cell death that happens following loss of cell contact (with neighboring cells or extracellular matrix).
Craniorachischisis: most severe NTD by which the entire spinal cord and brain are exposed to the amniotic environment.
Encephalocele: protrusion of the cranial contents beyond the normal confines of the skull through a defect in the calvarium.
Exencephaly: exposure of the brain outside of the skull.
Iniencephaly: NTD characterized by extreme retroflection of the head associated with severe spinal distortion.
Meningocele: developmental defect allows the meninges to herniate between the vertebrae while the nervous system remains intact.
Myelomeningocele: birth defect due to impaired closure of the posterior region of the neural tube with a protruding cystic mass through vertebral arches.
Neural tube defect: congenital malformation due to a failure of neurulation.
Primary neurulation: formation of neural tube during the third and fourth gestational week, in humans.
Secondary neurulation: process necessary for the sacral and caudal formation of the spinal cord by cavitation.
Spina bifida occulta: common birth defect characterized by protrusion of the meninges forming a sack covered by skin in which cerebrospinal fluid is present.

Introduction

This chapter aims at describing the most common neural tube defects (NTDs) and their embryonic origins. Such birth defects may result from abnormalities occurring during primary or secondary neurulation in early phases of development. Depending on the type and site of lesion, the defect might lead to devastating neurological consequences, including death.

[a] These authors contributed equally to this work.

Diagnosis, Management and Modeling of Neurodevelopmental Disorders. https://doi.org/10.1016/B978-0-12-817988-8.00002-6

Main text

Neural tube defects: embryonic origin

Neurulation in mammals

In humans, the development of the central nervous system (CNS) is a long and complicated process that starts very early during pregnancy (Fig. 2.1). Twenty-one days postfertilization the embryo, referred to as neurula, is ready to initiate shaping the CNS. This early process is divided in two consecutive and coordinated steps: *primary* and *secondary* neurulation. Primary neurulation shapes the neural tube of the future brain and majority of the spinal cord, while secondary neurulation determines the formation of the most posterior spinal cord. Primary and secondary neurulation processes are conserved in mammals, although there are some differences in the precise sequence of neural tube closure events (Copp, Greene, & Murdoch, 2003).

Primary and secondary neurulation

The formation of the brain and the spinal cord begins with the development of the neural tube, through the process of embryonic neurulation. The neural plate originates from thickening of the dorsal surface ectoderm that folds and merges into the midline to create the neural tube. In mammals, closure of the neural tube starts sequentially at different levels of the body axis (multisite closure).

Primary neurulation can be divided into four main steps: (a) formation, (b) shaping and (c) bending of the neural plate, (d) closure of the neural tube.

(a) The formation of the neural tube starts from the ectoderm, one of the three embryonic layers, that upon inductive molecular cues thickens to become the typical pseudostratified epithelium.
(b) Cells of the neural plate change their shape from cuboidal to columnar.
(c) Formation of the neural groove; elevation of the lateral walls; formation of dorsolateral hinge points.
(d) These steps are followed by a remodeling process leading to a closed neural tube covered by the surface ectoderm.

The closure of the primary neural tube is a discontinuous process that starts at different points along the rostrum–caudal axis. The fusion spreads bidirectionally in the brain and back along the spinal region. The fusion progression ("zipping") proceeds along the spine, culminating in the final closure of the posterior neuropore, at the level of the second sacral segment. This completes the process of primary neurulation (by neural folding) (Copp & Greene, 2010) (Fig. 2.2A–G).

In the posterior part of the growing embryos, the neural tube formation occurs via a different process that is not fully elucidated (Catala, 2002). The most accepted mechanism is through cavitation. Undifferentiated mesenchymal cells change their shape to form a neural epithelium by mesenchymal-to-epithelial transition (Shimokita & Takahashi, 2011) (Fig. 2.2H–N).

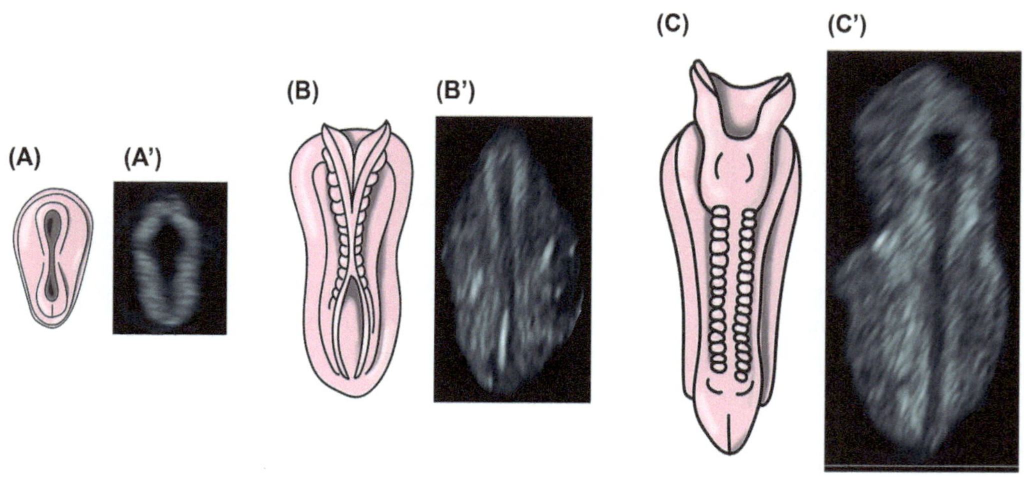

FIGURE 2.1 **Development of the human neural tube.** Schematic drawings (A–C) and ultrasounds scans (A′–C′) of the neural tube formation and closure, from early (third to fourth gestational week) to late (fifth to sixth gestational week) embryonic stages, are shown.

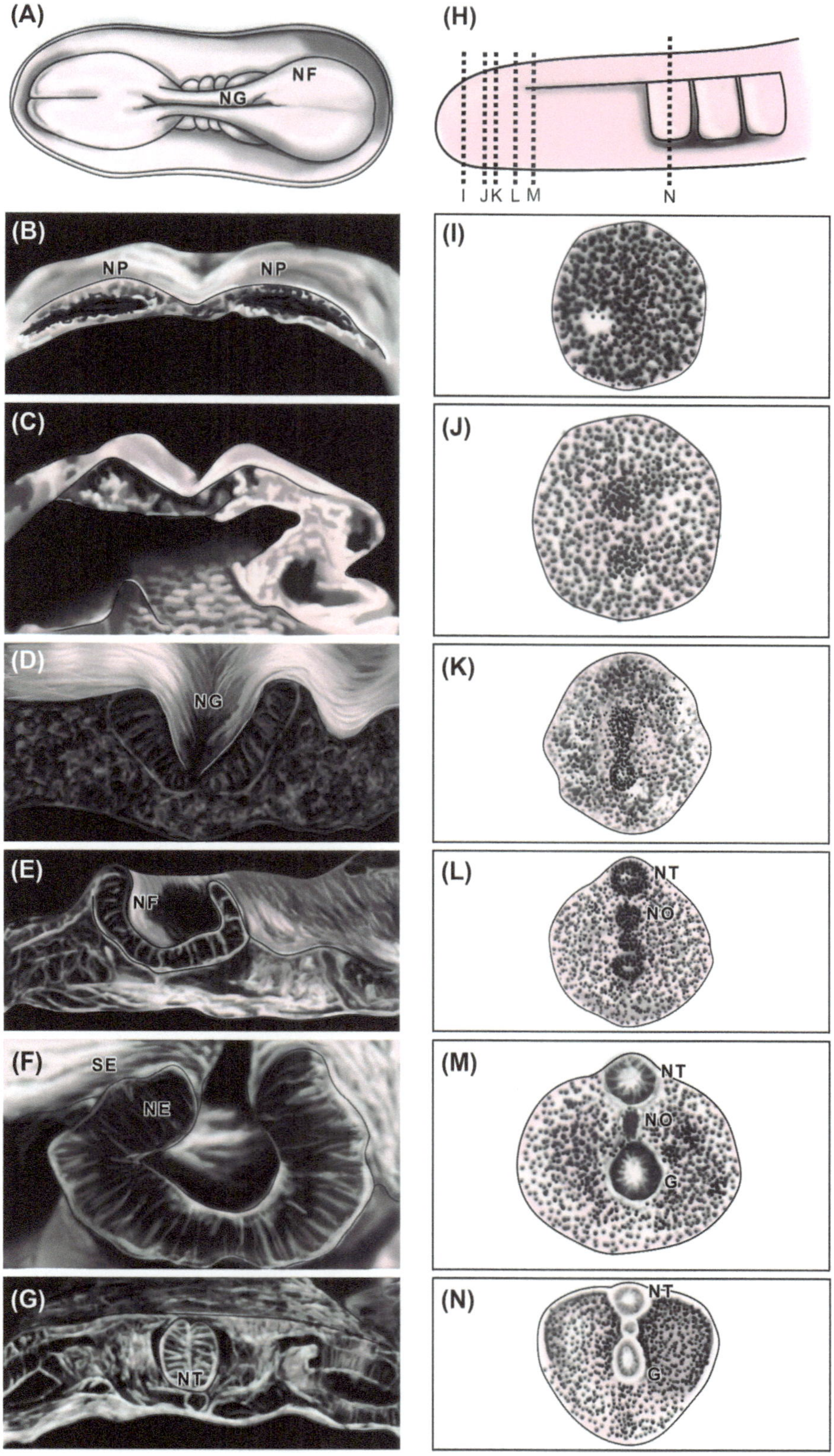

FIGURE 2.2 **Schematic representation of primary and secondary neurulation.** Schematic drawings of the primary neurulation (A: caudal to the left; B–G: ventral to the bottom) and secondary neurulation (H: caudal to the left; I–N: ventral to the bottom). At the end of the third gestational week, the neural folds form at the lateral extremes of the neural plate (A), elevate (B–D) and converge toward the dorsal midline forming the neural groove (D–F), and fuse at their dorsal tips to form the closed neural tube (G). Secondary neurulation happens by cavitation of a solid mass of cells, the tail bud. A rod-like condensation of mesenchymal cells forms beneath the dorsal ectoderm of the tail bud (I–L), these mesenchymal cells then transform into an epithelium and a lumen develops to form the neural tube (M–N). (*G*, gut; *NE*, neuroepithelium; *NF*, neural fold; *NG*, neural groove; *NO*, notochord; *NP*, neural plate; *NT*, neural tube; *SE*, surface ectoderm).

The secondary neurulation, or *canalization*, in which a multipotential cellular population in the caudal eminence differentiates into future neuronal cells, permits the formation of the spinal cord at the lower sacral and caudal level. Therefore, the formation of a sacred caudal lumen occurs, which will eventually join with the newly established primary lumen.

Neural tube defects: cell survival equilibrium impact

During the primary neurulation process, the embryo grows rapidly, and cells proliferate and differentiate in a highly dynamic way. One of the processes that has been observed during neurulation process and neural tube closure in several experimental models involves programmed cell death (Nikolopoulou, Galea, Rolo, Greene, & Copp, 2017) (Fig. 2.3A–C). This morphogenetic process is known to occur during several embryonic milestones (i.e., cell removal in the developing interdigital region in mammals) (Suzanne & Steller, 2013), and it is known to be of fundamental importance during nervous system development during the later stages of embryonic development (Yamaguchi & Miura, 2015).

Such cell death is thought to be a peculiar type: "anoikis" (Greek for homelessness). For both surviving and proliferating, cells require a continuous cross talk with the extracellular matrix (ECM) or neighboring cells. During the highly dynamic process of neurulation, some cells become "homeless," associated with marked changes in shape and loss of contact and association with ECM components. During the elevation and folding of the neuroepithelium, some cells at the fold tips will alter their contacts with the adjacent cells and/or with ECM (Gilmore, 2005) (Fig. 2.3D–E).

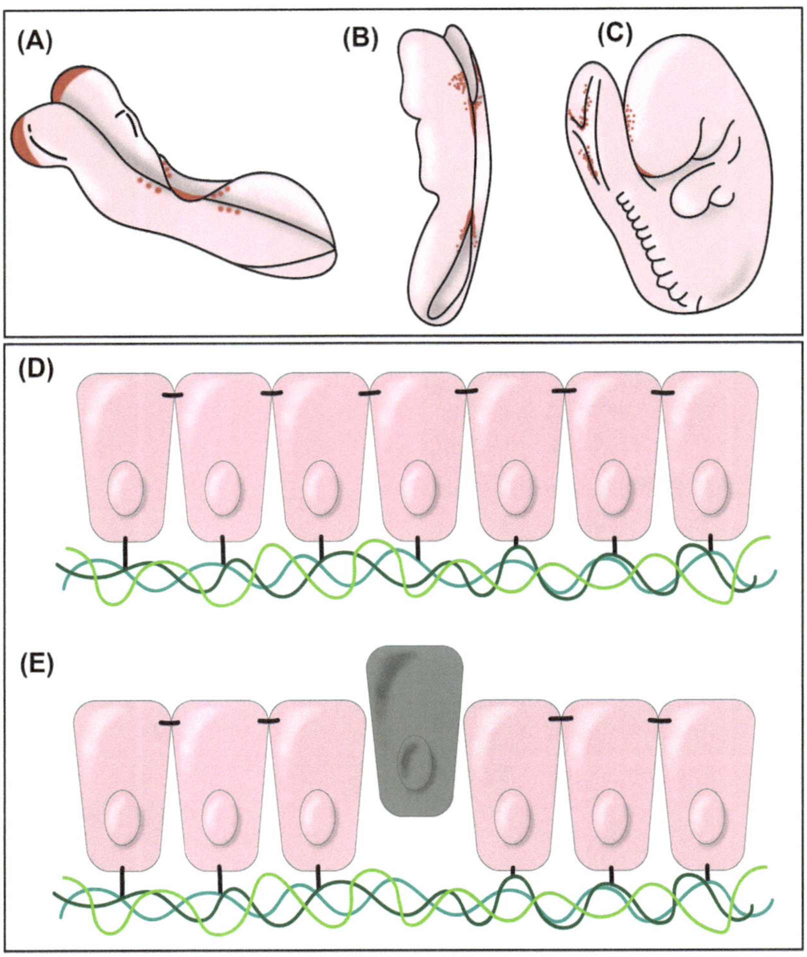

FIGURE 2.3 **Cell death during neurulation.** Schematic drawings of the cells undergoing death in mouse embryo (A: E8.5; B: E9.0 C: E9.5). In red are shown the main sites of apoptosis during neural tube formation. Schematic drawings of the anoikis process in the neuroepithelium (D–E). The neuroepithelium is composed of a pseudostratified layer of epithelial cells with basal nuclei, in which cells are in contact with surrounding cells laterally and with the extracellular matrix on basal surfaces through junctional complexes (D). When a cell loses contact with the extracellular matrix and/or the neighbor cells, *anoikis* happens. This event happens during the dynamic process of neurulation, and it is due to the cellular reorganization that helps the fusion process during the closure of neural tube. (*E*, embryonic day).

In this context, an important cross talk occurs between integrins and their receptors. Integrins regulate signaling pathways that control apoptosis in growth-factor-mediated survival, DNA damage responses, and death-receptor-mediated apoptosis (Boudreau, Sympson, Werb, & Bissell, 1995; Hynes, 2002; Pullan et al., 1996). The role of integrin receptor signaling in cell survival has been demonstrated through a number of experimental evidences (Salmela et al., 2017), as well as the ECM—cytoskeleton intimate relationship in this frame (Chen, Mrksich, Huang, Whitesides, & Ingber, 1997).

Moreover, it is undoubted that embryonic tissue fusion events, including proper neural tube closure, are associated with abundant apoptosis. However, it has been shown that this is not an essential step for the completion of the process, but more likely a secondary step due to the cellular reorganization that helps the fusion process during the closure of neural tube (Massa, Savery, et al., 2009) and/or as a consequence of excessive cells, the removal of which is a necessary embryonic event, for guaranteeing the success of a fundamental embryonic step as the formation of CNS primordium (Massa, Greene, & Copp, 2009).

Neural tube defects: clinical features

Failure of primary neurulation leads to NTDs, which are common congenital anomalies in humans Despite the growing understanding of many molecular and cellular events through the use of different experimental models (Butler & Wallingford, 2017; Colas & Schoenwolf, 2001; Greene & Copp, 2014; Wallingford, Niswander, Shaw, & Finnell, 2013), the elucidation of the etiopathogenesis of NTDs is yet to be defined.

A strict classification of human NTDs is still lacking consensus, especially with regard to clinical effects and embryonic origin. However, NTDs can be divided in two main subgroups: open, whereby neural tissue is directly exposed to the environment, and closed, in which neural tissue is covered.

Epidemiology

NTDs are the second most common congenital malformation (Committee on Practice Bulletins-Obstetrics, 2017). While their prevalence is believed to be underestimated because of terminations of pregnancy and spontaneous abortions, the incidence of new cases has greatly decreased following antenatal folic acid maternal supplementation (Kondo et al., 2017).

The prevalence of NTDs varies considerably among different geographic areas: in the United States and in many European countries, it is estimated between 0.5 and 0.8/1000 births (Castilla, Botto, & Bakker, 2011), while in China it is estimated 20 times higher (Li et al., 2006). Further, areas with higher prevalence of NTDs show higher frequencies of rarer subtypes such as craniorachischisis and iniencephaly (Copp et al., 2015; Moore et al., 1997). Globally, it is estimated that approximately 300,000 babies are born each year with NTDs worldwide (Zaganjor et al., 2016).

There is also evidence that in twins, the concordance for NTDs is higher between same-sex twin pairs than with twins of the opposite sex (Copp & Greene, 2013). In addition, in humans, anencephaly affects females more often than males, suggesting either a genetic relationship or a sex-related epigenetic impact. Conversely, less than 10% of NTDs cases are syndromic (Juriloff & Harris, 2012; Rampersaud et al., 2005; Wen et al., 2010).

Worldwide differences in prevalence may be related to the complexity of the pathogenesis of NTDs. In fact, NTDs are an archetype of complex birth defects caused by gene—environment interactions (Wallingford et al., 2013). In support of this causative interaction, several distinct lines of evidence have been reported in the literature. For example, the prevalence varies among different countries. Such variation persists when considering different ethnic groups independently of countries of origin supporting a greater impact of genetics compared to the environment (Leck, 1974; Shaw, Velie, & Wasserman, 1997). Despite numerous genetic studies, a single causative gene that can explain the population burden for NTDs has not been found. Several studies have reported an increased risk of NTD outcome in association with select gene variants (Ross, Mason, & Finnell, 2017). Among them, an association has been identified between the presence of mutations in *VANGL1* (MIM # 610132) and the appearance of NTDs. These mutations affect highly conserved residues in the Vangl family (Kibar et al., 2007; Lei et al., 2010). However, mutations in *VANGL1* have only been found in a small proportion of patients with NTDs. This could be due to several factors, for example, *VANGL1* mutations in NTDs could be present in noncoding regulatory regions or DNA methylation changes in *VANGL1* could be associated with NTDs. This methylation hypothesis is particularly interesting and suggests how folic acid protects against NTDs by stimulating cell methylation reactions (Blom, Shaw, den Heijer, & Finnell, 2006).

Accordingly, most of the sporadic cases seem to be related to a deficiency of folate (Blom et al., 2006). Folate deficiency may be secondary to a reduced intestinal intake or malabsorption or consumption of drugs that interfere with its metabolism, known as folic acid antagonists (Hernández-Díaz, Werler, Walker, & Mitchell, 2000; Robert & Guibaud, 1982). Some studies have shown that the maintenance of homeostasis of folate plays a role for many cellular reactions, including the biosynthesis of purines (for the synthesis of DNA and RNA) and the production of methyl donor S-adenosyl-methionine, which is essential in methylation processes. Folate could therefore play a role in the methylation of important cytoskeletal proteins and ciliogenesis necessary for proper neural tube closure (Toriyama, Toriyama, Wallingford, & Finnell, 2017). A robust maternal folate status is not sufficient to prevent all NTD cases. Recently another related compound, inositol (Vitamin B8), has been suggested as a possible player in the supplementation studies (Greene et al., 2016; Greene, Leung, & Copp, 2017). It is interesting to note that both folic acid and inositol, despite molecularly acting at different levels (i.e., one-carbon metabolism and intracellular kinases cascade), are known to be pivotal in cell cycle regulation promoting and supporting cell proliferation (Kibar et al., 2007; Noventa et al., 2016; Wong et al., 2008), hence participating in the finely tuned cell survival equilibrium during development.

NTDs are commonly subdivided into open and closed defects. Craniorachischisis, exencephaly–anencephaly, and myelomeningoceles are considered to be open and are characterized by exposure of neural tissue. Closed defects are encephalocele, meningocele, and spina bifida occulta, which present an epithelium covering the neural tissue. Open NTDs are clinically more severe than closed ones (Copp & Greene, 2013; McComb, 2015). NTDs incidence/prevalence, clinical outcome and management are summarized in Table 2.1.

Open NTDs

Exencephaly/anencephaly

When the main defect implies a failure of the cranial neural tube closure, the defect is named exencephaly (compare Fig. 2.4B–A and D–C). Upon degeneration of the brain-neural tissues, due to destructive exposure of the brain to the intraamniotic environment, the exencephaly becomes an anencephaly.

The bones of the face appear structurally normal, although the eyes often seem to protrude due to the superficiality of the orbits and the forehead is absent or shortened.

Some cerebral remnants are present, called cerebrovasculosa area appearing as a dark brown undifferentiated mass (Anand, Javia, & Lakhani, 2015). Residual brain tissue appears as an irregular mass containing vascular tissue, islands of nervous tissue including interspersed glial cells, nerve cells, and cavities surrounded by the epithelium. Despite a severe abnormality of the rostral neural tube, the spinal cord in anencephalic fetuses appears to be structurally normal (Golden & Harding, 2004).

Anencephaly is a lethal condition (Fig. 2.4E). Detection of anencephaly during early gestation is generally followed by legal termination of the pregnancy. In those cases of pregnancies that come to term, most anencephalic newborns die in the first day or two postpartum (The Medical Task Force of Anencepahly, 1990).

It is estimated that approximately three out of 10,000 pregnancies in the United States will present with anencephaly every year. This means that approximately 1206 pregnancies are affected by these conditions every year in the United States (Williams et al., 2015).

Prenatal diagnosis is mainly based on the absence of a normal cranial format and the brain above the orbital line. The cerebrovascular area can be detected as fluctuating echogenic tissue. The differential diagnosis between anencephaly and other causes of an absence of the cranial vault is very important for family counseling, as the presence of anencephaly increases the risk of the onset of other NTDs in any subsequent pregnancy (Goldstein & Filly, 1988; Keeling & Kjaer, 1994). Nowadays, the approaches used for NTD screening are: biochemical tests of maternal blood for alpha-fetoprotein or the use of traditional 2D ultrasounds. Some screening programs combine the two techniques. The development of more sophisticated imaging techniques (for example, 3D ultrasound) could improve the understanding of the injury allowing a better definition of structural anomalies. This information is vital when informing prospective parents about the likely prognosis for their child (Cameron & Moran, 2009).

Open spina bifida: myelomeningocele/myelocele

The failure to close the posterior regions of the neural tube, or the rupture of the posterior neuropore, leads to the formation of spina bifida. Spina bifida has a greater or lesser severity depending on the site of lesion, with more rostral lesions being more clinically severe.

The developmental defect involves the inability to close the posterior spinal portion of the neural tube, shape and size of the lesion can vary significantly, the lumbar portion is the most frequently observed; a cystic mass that protrudes although a bony defect in the vertebral arches is detectable; this condition is called myelomeningocele. In myelomeningocele, the fetus presents with a cystic meningeal sac that contains cerebrospinal fluid and nervous tissue, including nerve roots and spinal cord (Figs. 2.5A and 2.6B). The histological sections show the protrusion of the marrow through the bone defect of

TABLE 2.1 Features of neural tube defects.

	Type	Incidence/prevalence	Clinical outcome	Clinical management
Open NTDs	Exencephaly/ anencephaly	Three every 10,000 pregnancies in the United States (Williams et al., 2015)	Lethal (The Medical Task Force of Anencepahly, 1990)	There are no neurosurgical management options and almost all live children die shortly after birth. Considering this poor prognosis, most pregnancies are interrupted shortly after diagnosis. (Weprin & Oakes, 2000)
	Myelomeningocele/ myelocele/open spina bifida (or rachischisis)	1.8 per 10,000 live births in the United States (Zaganjor et al., 2016)	Motor and sensorial deficit below the level of the spinal lesion. Often bladder and rectal incontinence. Hydrocephaly. Intellectual disability is relatively infrequent (20%–25% of cases) and related to hydrocephalus. Deformity of the lower extremities.	*In utero* surgical repair might be considered for open spina bifida, and this procedure has shown significant benefits for the newborns, leading to a significant improvement in spinal neurological function. (Adzick et al., 2011). In order to prevent infections, swelling, and damage to the spinal cord, an early surgery is often suggested.
	Craniorachischisis	Different geographic distribution prevalence. In northern China, 0.01 per 1000 newborns (Moore et al., 1997) and in a Texas–Mexico border population, the prevalence was 0.51 per 10,000 live births (Johnson et al., 2004).	Lethal	Such defect is not compatible with life. No neurosurgical management options.
Closed NTDs	Encephalocele	One in 12,200 live births in the United States each year (Parker et al., 2010)	Clinical characteristics depend on the site and extent of the lesion. Often cognitive, neurodevelopmental delay and sometimes hydrocephalus.	Encephaloceles always require surgery correction under general anesthetic. The timing of surgery will vary depending on whether the sac is covered with skin or a thin membrane.
	Meningocele	Generally unknown. In one study the incidence of meningocele was 2.9 per 1000 in a nigerian population (Uba et al., 2004)	Generally normal neurologic examination without deformity of the lower extremities or sphincter abnormality.	In order to prevent infections, swelling, and damage to the spinal cord, an early surgery is often suggested.
	Spina bifida occulta	Common defect, estimated in one out of 10 people (Eubanks & Cheruvu, 2009)	Asymptomatic in 10% of cases. Possible: Motor and sensorial deficit below the level of the spinal lesion. Bowel and/or bladder dysfunction, scoliosis, foot or leg length discrepancies, deformity of the lower extremities.	Neurosurgical intervention is the therapy chosen in cases of symptomatic patients with neurological deterioration.
	Iniencephaly	The incidence ranges from 0.1 to 10/10,000 and most of the patients are females (Aytar et al., 2007)	Most iniencephalic children who are born alive die shortly after birth (Khatami et al., 2015).	No neurosurgical management options.

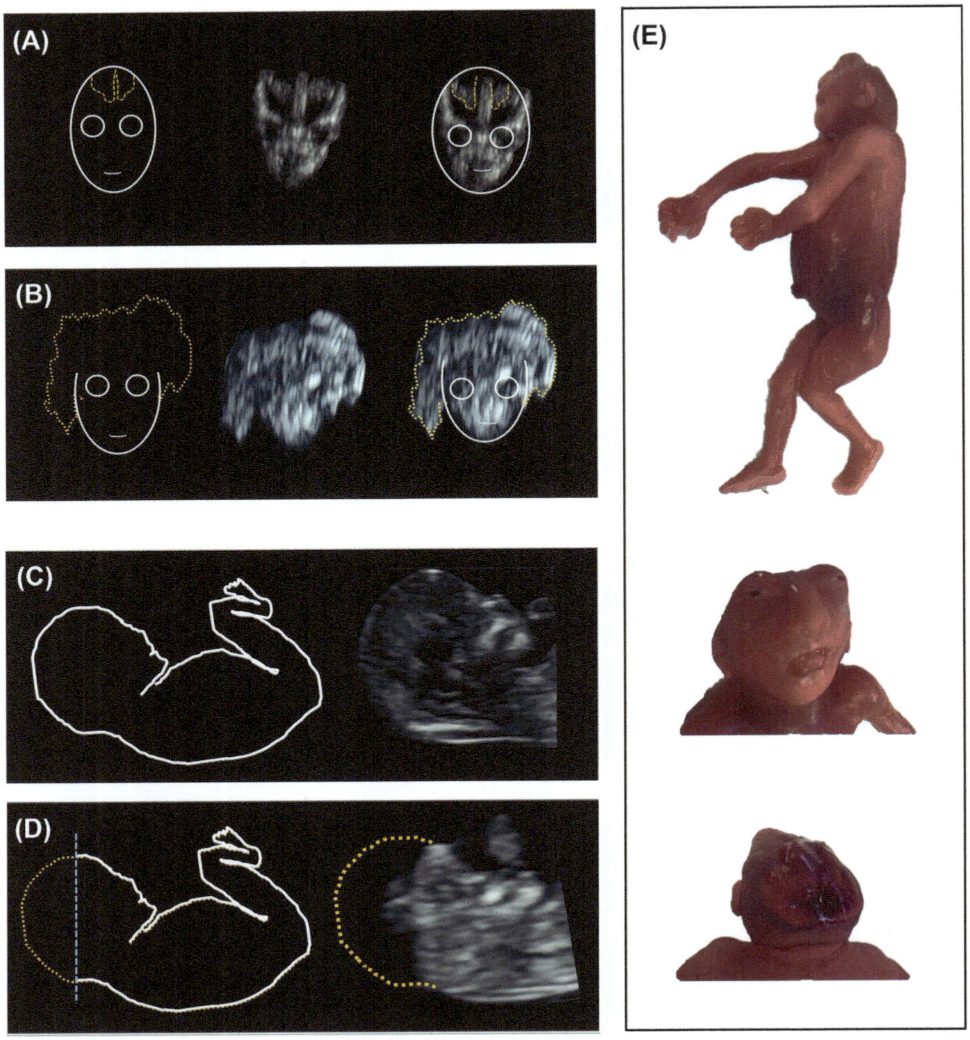

FIGURE 2.4 **Exencephaly/anencephaly.** Comparison between normal (A) and abnormal (B) frontal view of human embryo affected by exencephaly. The developmental defect of cranial portion was observed at 11 weeks of gestation. The defect involves the cranial portion of the neural tube. The calvarium is absent and brain tissues are externally exposed and float in the amniotic fluid (B: *orange dotted line*), whereas the face is normally structured. Comparison between normal (C) and abnormal (D) sagittal view of human embryo affected by anencephaly. The absence of the upper portion of the cranial vault was detected at 13 weeks of gestation, without external appearance of brain remnants (D: *orange dotted line*). Macroscopic aspect of anencephalic fetus is shown (E; from top to bottom: sagittal, front, and posterior view). Legal interruption of pregnancy was performed at 18 weeks of gestation. The face of the fetus is normal, although the eyes are protruded because of the shallow of the orbits, the forehead and the calvarium are absent. Posteriorly, the macroscopic aspect of the brain remnant appears as a dark brown undifferentiated mass (area cerebrovasculosa).

the vertebral arches. Even the meninges are herniated forming the cystic mass. Both the spinal cord and the meninges are damaged. The spinal cord is generally hypervascularized and in some cases the spinal cord could be closed with a dilated central channel, while in other cases the spinal cord appears open as a flat mass (Golden & Harding, 2004).

According to the Center for Disease Control and Prevention, myelomeningocele incidence is estimated to occur with a prevalence of 1.8 per 10,000 live births in United States (Zaganjor et al., 2016).

In some cases, the open defect is present but without protrusion of the meningeal sac. These cases are referred to as myelocele. In infants with a myelocele, it is possible to detect the neural tissue through the vertebral cleft (Emery & Lendon, 1973; Naik & Emery, 2008).

In myelomeningocele and myelocele, the cerebral trunk is stretched as a consequence of the lower open spinal cord, and this leads to their association with Chiari Type II malformation (CM-II), also known as Arnold–Chiari malformation. CM-II is characterized by a downward shift of the cerebellar vermis, tonsils and presents a mesencephalon malformation. The stretching of the structures into the medullary canal alters the normal cerebrospinal flow through the ventricles, leading to

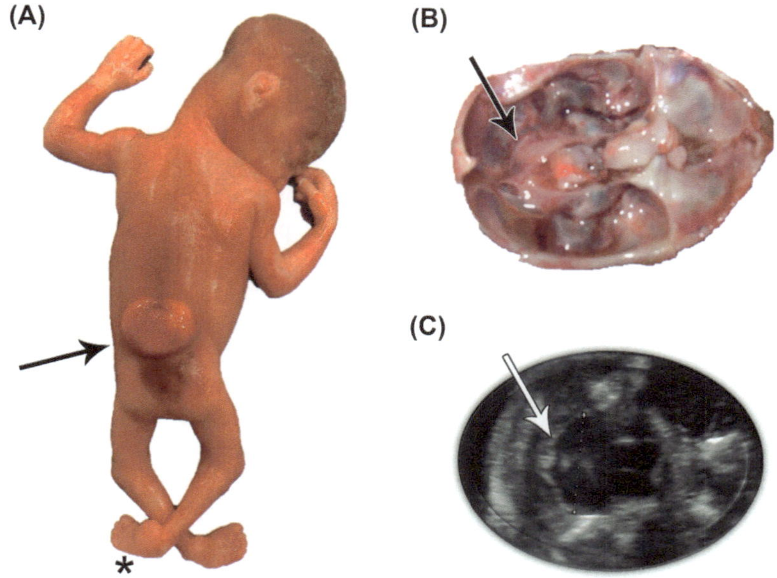

FIGURE 2.5 **Myelomeningocele.** Posterior sac-like protrusion from the spine is shown (A: *arrow*) in an embryo affected by myelomeningocele. The cystic lesion herniates from a cleft in vertebral arch and contains meninges, cerebrospinal fluid, nerve roots, and spinal cord. The abnormal foot position in which the foot is internally rotated (*asterisk*) is called "clubfoot." Generally, it is due to the damage of the nerve roots and virtually always involves both feet. The abnormal shape of the cerebellum (called "banana sign") is shown (B, C), at both macroscopic and ultrasound evaluation of the posterior cranial fossa (*arrows*). The banana sign is due to downward traction on the cerebellum related to tethering of the spine though the spinal opening.

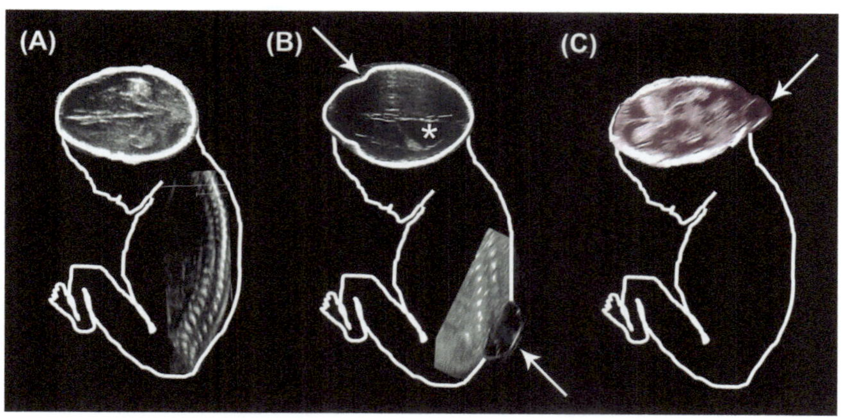

FIGURE 2.6 **Myelomeningocele/encephaloceles.** Normal ultrasonographic aspects of transverse cephalic scan and sagittal spinal scan are shown (A). Fetus affected by myelomeningocele (B) displays vertebral dysraphism with an associated cystic mass protruding from the lumbar portion of the spine (*lower arrow*) and containing tissues and internal septations. Cranial signs are also associated: abnormal shape of the skull in transverse section is present, displaying the so called "lemon sign" related to scalloping of the frontal bones (*upper arrow*); the cerebrospinal fluid through the ventricles is increased, resulting in secondary hydrocephalus (*asterisk*). Fetus affected by cephalocele (C) displays skull defect with a paracranial anechogenic sac. The posterior shape of choroid plexus is abnormal and brain tissues are herniated in the sac that protrudes through the opening of the skull (*arrow*).

progressive hydrocephalus (Sarnat, 2008) (Figs. 2.5B–C and 2.6B). CM-II is usually detected prenatally or at birth. Manifestations during childhood can include dysphagia, weakness in the arms, and episodes of apnea and aspiration.

In the absence of other serious congenital malformations, newborns with myelomeningocele or myelocele survive albeit presenting clinical features of motor and sensory deficits below the spinal cord injury and, in some cases, exhibit varying degrees of intellectual disability (related to the presence of hydrocephalus) (Copp et al., 2015; Thompson, 2009). *In utero* surgical repair might be considered for open spina bifida, and this procedure has shown significant benefits for the newborns, leading to a significant improvement in spinal neurological function despite the obvious risks associated to the surgical intervention (Adzick et al., 2011).

Craniorachischisis

The craniorachischisis represents the most severe NTD resulting in open brain and spine. In this condition, both anencephaly and open spina bifida are present. Such a malformation is not compatible with life. Typically, the posterior telencephalon and spinal cord tissue are exposed externally. The fetuses lack a cranial vault, with the defect extending with various degrees to the vertebrae. In craniorachischisis, both the brain and the spinal cord are exposed to the intraamniotic environment, leading to the destruction of nervous tissue due to the toxic effect of amniotic fluid. Craniorachischisis is a lethal condition for which there is no cure or possible surgery. The prevalence differs geographically. For example, in Northern China, the prevalence is 0.01 per 1000 newborns (Moore et al., 1997) and in a Texas−Mexico border population, the prevalence was 0.51 per 10,000 live births (Johnson, Suarez, Felkner, & Hendricks, 2004).

Closed NTDs

Closed NTDs form a less defined group of defects. Embryologically, there is scant experimental evidence to clearly establish the origin of the closed spinal lesions. Caudal closed defects result from alterations in secondary neurulation and are less severe birth defects. In fact, these defects typically occur at the sacrococcygeal level (i.e., at the level of secondary neurulation) and do not open to the external environment. The spinal cord remains bound to adjacent tissues, as expected for a defective separation of tissues during secondary neurulation (Copp & Greene, 2013). As a whole, closed defects typically have less severe loss of neurological function. Rostral closed defects result from alterations confined to the calvarium and may have more severe consequences.

Rostral closed defects: encephalocele

Encephalocele is the protrusion of the cranial contents beyond the normal confines of the skull through a defect in the calvarium (Raja, Qureshi, Memon, Ali, & Dev, 2008) (compare Fig. 2.6C−A).

It is classified according to the site: anterior, parietal, occipital, basal. Depending on the type of tissue involved in the herniation, cephaloceles are classified as meningocele (herniation of meninges), encephalomeningocele (herniation of meninges and brain), and encephalomeningocystocele (herniation of meninges, brain, and ventricle) (Pilu, Buyukkurt, Youssef, & Tonni, 2014). Moreover, cystic variants also exist, containing only cerebrospinal fluid (Gao, Massimi, Rogerio, Raybaud, & Di Rocco, 2014). According to the localization, the herniated mass can contain various types of tissues including the cerebral and cerebellar part, the ventricles with choroid plexus, gray matter heterotopia. Brain tissues involved in the hernia are generally nonfunctional. Meningeal inflammation was also detected (Ivashchuk, Loukas, Blount, Tubbs, & Oakes, 2015). Encephaloceles are one of the most common NTDs, and each year in the United States 1 out of 12,200 babies is born with this condition. Encephaloceles always require surgery correction under general anesthetic. The timing of surgery will vary depending on whether the sac is covered with skin or a thin membrane. If it is only covered with a thin membrane, surgery soon after birth will be required to prevent infection and damage/drying of the exposed tissue. If there is a covering of skin, surgery can be delayed for a month or two to allow the baby to grow and develop.

Caudal closed defects: Meningocele

Meningocele is a congenital malformation characterized by failed closure of the spine in its posterior part, generally at last lumbar vertebrae level (Copp & Greene, 2013). The skin is often dysplastic but present to some degree. Nervous disorders in the lower limbs are sometimes associated with this malformation. Posterior meningocele is the less common form of spina bifida. In this subtype, the nervous system remains intact, but a single developmental defect allows the meninges to herniate between the vertebrae. Therefore, meningocele is characterized by a protrusion of the meninges.

The incidence of this condition is unknown. As for the earlier described NTDs, in order to prevent infections, swelling, and damage to the spinal cord, an early surgery is often suggested.

Caudal closed defects: Closed spina bifida

Closed spina bifida represents a range of anomalies of the caudal spinal cord and has many variants. Spina bifida occulta is one form and likely the most common defect. Most of the cases are undiagnosed and without clinical manifestations or consequences. The prevalence is estimated in 1 out of 10 people (Eubanks & Cheruvu, 2009; Fidas et al., 1987). Closed spina bifida also includes lipomyelomeningocele, low conus, and thickened or fibrolipoma of the filum terminal. Midline cutaneous stigmata are often associated to closed spina bifida. These skin markers include nevi, nodules, depigmented region, subcutaneous lipomas, capillary hemangiomas, tufts of hair (localized hypertrichosis), and dermal sinus tract (Hertzler, DePowell, Stevenson, & Mangano, 2010; Sewell, Chiu, & Drolet, 2015). Neurosurgical intervention is the therapy chosen in cases of symptomatic patients with neurological deterioration.

Mixed defect: iniencephaly

Iniencephaly is an uncommon and fatal NTD characterized by an occipital bone defect, open spina bifida of the cervical vertebra, and an increase in cervical lordosis. The retroflexion of the head and the severe distortion of the spine are related to the localization of the defects that involve occiput, inion, and cervical and thoracic spine. Fetuses with iniencephaly are rarely live born, but at least one live birth has been reported (Khatami, Hasanzadeh, Norouzi, Esfandiari, & Mehrafarin, 2015). The prevalence of iniencephaly ranges from 0.1 to 10/10,000, and most of the patients are females (Aytar et al., 2007).

Conclusions

NTDs are the second most common congenital malformations in humans, and despite their potentially devastating consequences on patients and families, they represent an example of possible successful preventable birth defect. Numerous studies using an array of experimental models have been reported in the literature, leading to a better understanding of molecular and cellular events (Butler & Wallingford, 2017; Greene & Copp, 2014; Wallingford et al., 2013); however, NTD pathogenesis is not yet fully elucidated. It is possible to hypothesize that genetic predisposition and/or environmental factors impact on cell survival equilibrium, which must be finely tuned in this pivotal embryonic process.

Application to other areas of development

Cellular and molecular mechanisms regulating neurulation process are shared among several other pivotal developmental events. Indeed, anoikis has been suggested to intervene also in palatal shelves fusion and trachea–esophagus separation. In this context, it is possible to hypothesize that key embryonic steps are facilitated by abundant cells that obviously must be removed in subsequent developmental phases.

Key facts

Key facts on neurulation:

- Neurulation is a multistep process leading to the development of the central nervous system.
- It starts at 21-day postfertilization in humans.
- Neurulation is a process divided into primary neurulation and secondary neurulation.
- Primary neurulation is a discontinuous process that starts at different points along the rostrum–caudal axis necessary for the neural tube closure.
- The secondary neurulation permits the formation of the spinal cord at the lower sacral and caudal level.

Key facts on cell survival:

- Programmed cell death has been observed during neurulation and neural tube closure in several models.
- Cell death/survival occurs during the morphogenetic process of embryo development, and it is known to be of fundamental importance during nervous system formation.
- During neurulation, some cells become "homeless" because of the loss of contact and association with neighboring cells and/or ECM components (anoikis).
- Anoikis is not necessary, but more likely a secondary step due to the cellular reorganization that facilitates the fusion processes.
- It is possible that several core molecular cascades converge in controlling such cellular events.

Key facts on neural tube defects:

- Neural tube defects can be grouped as closed or open.
- Open NTDs result from failure of primary neurulation.
- Closed NTDs result from impairment of secondary neurulation.
- NTDs might result in mild clinical manifestation or in extremely severe, including death.
- NTDs represent an archetype of complex birth defect (i.e., of both genetic and environmental origin).

Summary points

- This chapter focuses on neural tube defects (NTDs), which are common congenital anomalies arising from a failure neural tube formation (early neurulation).
- Primary neurulation is the process that shapes the neural plate in the anterior part and secondary neurulation happens in the caudal region.
- During neural tube formation, cells undergo apoptosis, probably as anoikis.
- NTDs are complex birth defects: both genetic and environmental factors participate in the pathogenesis.
- Despite the pathogenesis of NTDs being not fully elucidated, maternal folic acid supplementation significantly reduces the incidence.
- NTDs are classified as *open* (characterized by exposure of neural tissue) and *closed* (presenting an epithelium covering the neural tissue).
- Open NTDs are clinically more severe than closed ones.

Acknowledgments

The authors would like to express their deep gratitude to Ms. Susanna Brusa for graphic support.

References

Adzick, N. S., Thom, E. A., Spong, C. Y., Brock, J. W., Burrows, P. K., Johnson, M. P., … Farmer, D. L. (2011). A randomized trial of prenatal versus postnatal repair of myelomeningocele. *New England Journal of Medicine, 364*(11), 993–1004. https://doi.org/10.1056/NEJMoa1014379

Anand, M. K., Javia, M. D., & Lakhani, C. J. (2015). Development of brain and spinal cord in anencephalic human fetuses. *Anatomy, 9*(2), 60–65. https://doi.org/10.2399/ana.15.009

Aytar, M. H., Doğulu, F., Cemil, B., Ergün, E., Kurt, G., & Baykaner, K. (2007). Iniencephaly and long-term survival: A rare case report. *Child's Nervous System· ChNS : Official Journal of the International Society for Pediatric Neurosurgery, 23*(6), 719–721. https://doi.org/10.1007/s00381-007-0309-6

Blom, H. J., Shaw, G. M., den Heijer, M., & Finnell, R. H. (2006). Neural tube defects and folate: Case far from closed. *Nature Reviews Neuroscience, 7*(9), 724–731. https://doi.org/10.1038/nrn1986

Boudreau, N., Sympson, C., Werb, Z., & Bissell, M. (1995). Suppression of ICE and apoptosis in mammary epithelial cells by extracellular matrix. *Science, 267*(5199), 891–893. https://doi.org/10.1126/science.7531366

Butler, M. T., & Wallingford, J. B. (2017). Planar cell polarity in development and disease. *Nature Reviews Molecular Cell Biology, 18*(6), 375. https://doi.org/10.1038/NRM.2017.11

Cameron, M., & Moran, P. (2009). Prenatal screening and diagnosis of neural tube defects. *Prenatal Diagnosis, 29*(4), 402–411. https://doi.org/10.1002/pd.2250

Castilla, E. E., Botto, L. D., & Bakker, M. K. (2011). Annual report 2011, with data for 2009. *International clearinghouse for birth defects surveillance and research.* Roma, Italia: The International Centre on Birth Defects – ICBDSR Centre. Retrieved from: www.icbdsr.org.

Catala, M. (2002). Genetic control of caudal development. *Clinical Genetics, 61*(2), 89–96. https://doi.org/10.1034/j.1399-0004.2002.610202.x

Chen, C. S., Mrksich, M., Huang, S., Whitesides, G. M., & Ingber, D. E. (1997). Geometric control of cell life and death. *Science, 276*(5317), 1425–1428. https://doi.org/10.1126/science.276.5317.1425

Colas, J.-F., & Schoenwolf, G. C. (2001). Towards a cellular and molecular understanding of neurulation. *Developmental Dynamics, 221*(2), 117–145. https://doi.org/10.1002/dvdy.1144

Committee on Practice Bulletins-Obstetrics. (2017). Practice bulletin No. 187. *Obstetrics and Gynecology, 130*(6), e279–e290. https://doi.org/10.1097/AOG.0000000000002412

Copp, A. J., Adzick, N. S., Chitty, L. S., Fletcher, J. M., Holmbeck, G. N., & Shaw, G. M. (2015). Spina bifida. *Nature Reviews Disease Primers, 1*(1), 15007. https://doi.org/10.1038/nrdp.2015.7

Copp, A. J., & Greene, N. D. E. (2010). Genetics and development of neural tube defects. *The Journal of Pathology, 220*(2), 217–230. https://doi.org/10.1002/path.2643

Copp, A. J., & Greene, N. D. E. (2013). Neural tube defects-disorders of neurulation and related embryonic processes. *Wiley Interdisciplinary Reviews: Developmental Biology, 2*(2), 213–227. https://doi.org/10.1002/wdev.71

Copp, A. J., Greene, N. D. E., & Murdoch, J. N. (2003). The genetic basis of mammalian neurulation. *Nature Reviews Genetics, 4*(10), 784–793. https://doi.org/10.1038/nrg1181

Emery, J. L., & Lendon, R. G. (1973). The local cord lesion in neurospinal dysraphism (meningomyelocele). *The Journal of Pathology, 110*(1), 83–96. https://doi.org/10.1002/path.1711100110

Eubanks, J. D., & Cheruvu, V. K. (2009). Prevalence of sacral spina bifida occulta and its relationship to age, sex, race, and the sacral table angle: An anatomic, osteologic study of three thousand one hundred specimens. *Spine, 34*(15), 1539–1543. https://doi.org/10.1097/BRS.0b013e3181a98560

Fidas, A., MacDonald, H. L., Elton, R. A., Wild, S. R., Chisholm, G. D., & Scott, R. (1987). Prevalence and patterns of spina bifida occulta in 2707 normal adults. *Clinical Radiology, 38*(5), 537–542. https://doi.org/10.1016/s0009-9260(87)80150-2

Gao, Z., Massimi, L., Rogerio, S., Raybaud, C., & Di Rocco, C. (2014). Vertex cephaloceles: A review. *Child's Nervous System, 30*(1), 65–72. https://doi.org/10.1007/s00381-013-2249-7

Gilmore, A. P. (2005). Anoikis. *Cell Death and Differentiation, 12*(S2), 1473–1477. https://doi.org/10.1038/sj.cdd.4401723

Golden, J. A., & Harding, B. N. (2004). *Pathology and genetics: Developmental neuropathology*. Los Angeles: ISN Neuropathlogy Press.

Goldstein, R., & Filly, R. (1988). Prenatal diagnosis of anencephaly: Spectrum of sonographic appearances and distinction from the amniotic band syndrome. *American Journal of Roentgenology, 151*(3), 547–550. https://doi.org/10.2214/ajr.151.3.547

Greene, N. D. E., & Copp, A. J. (2014). Neural tube defects. *Annual Review of Neuroscience, 37*, 221–242. https://doi.org/10.1146/annurev-neuro-062012-170354

Greene, N. D. E., Leung, K., & Copp, A. J. (2017). Inositol, neural tube closure and the prevention of neural tube defects. *Birth Defects Research, 109*(2), 68. https://doi.org/10.1002/BDRA.23533

Greene, N. D. E., Leung, K.-Y., Gay, V., Burren, K., Mills, K., Chitty, L. S., & Copp, A. J. (2016). Inositol for the prevention of neural tube defects: A pilot randomised controlled trial. *British Journal of Nutrition, 115*(6), 974. https://doi.org/10.1017/S0007114515005322

Hernández-Díaz, S., Werler, M. M., Walker, A. M., & Mitchell, A. A. (2000). Folic acid antagonists during pregnancy and the risk of birth defects. *New England Journal of Medicine, 343*(22), 1608–1614. https://doi.org/10.1056/NEJM200011303432204

Hertzler, D. A., DePowell, J. J., Stevenson, C. B., & Mangano, F. T. (2010). Tethered cord syndrome: A review of the literature from embryology to adult presentation. *Neurosurgical Focus, 29*(1), E1. https://doi.org/10.3171/2010.3.FOCUS1079

Hynes, R. O. (2002). Integrins: Bidirectional, allosteric signaling machines. *Cell, 110*(6), 673–687. https://doi.org/10.1016/s0092-8674(02)00971-6

Ivashchuk, G., Loukas, M., Blount, J. P., Tubbs, R. S., & Oakes, W. J. (2015). Chiari III malformation: A comprehensive review of this enigmatic anomaly. *Child's Nervous System: ChNS : Official Journal of the International Society for Pediatric Neurosurgery, 31*(11), 2035–2040. https://doi.org/10.1007/s00381-015-2853-9

Johnson, K. M. K., Suarez, L., Felkner, M. M., & Hendricks, K. (2004). Prevalence of craniorachischisis in a Texas-Mexico border population. *Birth Defects Research Part A: Clinical and Molecular Teratology, 70*(2), 92–94. https://doi.org/10.1002/bdra.10143

Juriloff, D. M., & Harris, M. J. (2012). A consideration of the evidence that genetic defects in planar cell polarity contribute to the etiology of human neural tube defects. *Birth Defects Research Part A: Clinical and Molecular Teratology, 94*(10), 824–840. https://doi org/10.1002/bdra.23079

Keeling, J. W., & Kjaer, I. (1994). Diagnostic distinction between anencephaly and amnion rupture sequence based on skeletal analysis. *Journal of Medical Genetics, 31*(11), 823–829. https://doi.org/10.1136/jmg.31.11.823

Khatami, A., Hasanzadeh, M., Norouzi, H., Esfandiari, E., & Mehrafarin, M. (2015). Iniencephaly clausus: A new case with clinical and imaging findings. *Iranian Journal of Radiology: A Quarterly Journal Published by the Iranian Radiological Society, 12*(3), e4790. https://doi.org/10.5812/iranjradiol.4790v2

Kibar, Z., Torban, E., McDearmid, J. R., Reynolds, A., Berghout, J., Mathieu, M., … Gros, P. (2007). Mutations in *VANGL1* associated with neural-tube defects. *New England Journal of Medicine, 356*(14), 1432–1437. https://doi.org/10.1056/NEJMoa060651

Kondo, A., Matsuo, T., Morota, N., Kondo, A. S., Okai, I., & Fukuda, H. (2017). Neural tube defects: Risk factors and preventive measures. *Congenital Anomalies, 57*(5), 150–156. https://doi.org/10.1111/cga.12227

Leck, I. (1974). Letter: Neural-tube defects and twinning. *Lancet, 1*(7849), 178. https://doi.org/10.1016/s0140-6736(74)92487-8

Lei, Y.-P., Zhang, T., Li, H., Wu, B.-L., Jin, L., & Wang, H.-Y. (2010). *VANGL2* mutations in human cranial neural-tube defects. *New England Journal of Medicine, 362*(23), 2232–2235. https://doi.org/10.1056/NEJMc0910820

Li, Z., Ren, A., Zhang, L., Ye, R., Li, S., Zheng, J., … Li, Z. (2006). Extremely high prevalence of neural tube defects in a 4-county area in Shanxi Province, China. *Birth Defects Research Part A: Clinical and Molecular Teratology, 76*(4), 237–240. https://doi.org/10.1002/bdra.20248

Massa, V., Greene, N. D. E., & Copp, A. J. (2009). Do cells become homeless during neural tube closure? *Cell Cycle*. https://doi.org/10.4161/cc.8.16.9272

Massa, V., Savery, D., Ybot-Gonzalez, P., Ferraro, E., Rongvaux, A., Cecconi, F., … Copp, A. J. (2009). Apoptosis is not required for mammalian neural tube closure. *Proceedings of the National Academy of Sciences, 106*(20), 8233–8238. https://doi.org/10.1073/pnas.0900333106

McComb, J. G. (2015). A practical clinical classification of spinal neural tube defects. *Child's Nervous System, 31*(10), 1641–1657. https://doi.org/10.1007/s00381-015-2845-9

Moore, C. A., Li, S., Li, Z., Hong, S. X., Gu, H. Q., Berry, R. J., … Erickson, J. D. (1997). Elevated rates of severe neural tube defects in a high-prevalence area in northern China. *American Journal of Medical Genetics, 73*(2), 113–118. Retrieved from: http://www.ncbi.nlm.nih.gov/pubmed/9409858.

Naik, D. R., & Emery, J. L. (2008). The position of the spinal cord segments related to the vertebral bodies in children with meningomyelocele and hydrocephalus. *Developmental Medicine and Child Neurology, 10*, 62–68. https://doi.org/10.1111/j.1469-8749.1968.tb04848.x

Nikolopoulou, E., Galea, G. L., Rolo, A., Greene, N. D. E., & Copp, A. J. (2017). Neural tube closure: Cellular, molecular and biomechanical mechanisms. *Development, 144*(4), 552–566. https://doi.org/10.1242/dev.145904

Noventa, M., Vitagliano, A., Quaranta, M., Borgato, S., Abdulrahim, B., & Gizzo, S. (2016). Preventive and therapeutic role of dietary inositol supplementation in periconceptional period and during pregnancy. *Reproductive Sciences, 23*(3), 278–288. https://doi.org/10.1177/1933719115594018

Parker, S. E., Mai, C. T., Canfield, M. A., Rickard, R., Wang, Y., Meyer, R. E., … Streuli, C. H. (2010). Updated national birth prevalence estimates for selected birth defects in the United States, 2004-2006. *Birth Defects Research Part A - Clinical and Molecular Teratology, 88*(12), 1008–1016. https://doi.org/10.1002/bdra.20735

Pilu, G., Buyukkurt, S., Youssef, A., & Tonni, G. (2014). *Visual encyclopedia of ultrasound in obstetrics and gynecology*. VISUOG.

Pullan, S., Wilson, J., Metcalfe, A., Edwards, G. M., Goberdhan, N., Tilly, J., … Streuli, C. H. (1996). Requirement of basement membrane for the suppression of programmed cell death in mammary epithelium. *Journal of Cell Science, 109*(Pt 3), 631–642. Retrieved from: http://www.ncbi.nlm.nih.gov/pubmed/8907708.

Raja, R. A., Qureshi, A. A., Memon, A. R., Ali, H., & Dev, V. (2008). Pattern of encephaloceles: A case series. *Journal of Ayub Medical College, Abbottabad: JAMC, 20*(1), 125–128. Retrieved from: http://www.ncbi.nlm.nih.gov/pubmed/19024205.

Rampersaud, E., Bassuk, A. G., Enterline, D. S., George, T. M., Siegel, D. G., Melvin, E. C., … Speer, M. C. (2005). Whole genomewide linkage screen for neural tube defects reveals regions of interest on chromosomes 7 and 10. *Journal of Medical Genetics, 42*(12), 940–946. https://doi.org/10.1136/jmg.2005.031658

Robert, E., & Guibaud, P. (1982). Maternal valproic acid and congenital neural tube defects. *Lancet, 2*(8304), 937. https://doi.org/10.1016/s0140-6736(82)90908-4

Ross, M. E., Mason, C. E., & Finnell, R. H. (2017). Genomic approaches to the assessment of human spina bifida risk. *Birth Defects Research, 109*(2), 120–128. https://doi.org/10.1002/bdra.23592

Salmela, M., Jokinen, J., Tiitta, S., Rappu, P., Cheng, R. H., & Heino, J. (2017). Integrin α2β1 in nonactivated conformation can induce focal adhesion kinase signaling. *Scientific Reports, 7*(1), 3414. https://doi.org/10.1038/s41598-017-03640-w

Sarnat, H. B. (2008). Disorders of segmentation of the neural tube: Chiari malformations. In *Handbook of clinical neurology* (Vol. 87, pp. 89–103). https://doi.org/10.1016/S0072-9752(07)87006-0

Sewell, M. J., Chiu, Y. E., & Drolet, B. A. (2015). Neural tube dysraphism: Review of cutaneous markers and imaging. *Pediatric Dermatology, 32*(2), 161–170. https://doi.org/10.1111/pde.12485

Shaw, G. M., Velie, E. M., & Wasserman, C. R. (1997). Risk for neural tube defect-affected pregnancies among women of Mexican descent and white women in California. *American Journal of Public Health, 87*(9), 1467–1471. https://doi.org/10.2105/ajph.87.9.1467

Shimokita, E., & Takahashi, Y. (2011). Secondary neurulation: Fate-mapping and gene manipulation of the neural tube in tail bud. *Development Growth and Differentiation, 53*(3), 401–410. https://doi.org/10.1111/j.1440-169X.2011.01260.x

Suzanne, M., & Steller, H. (2013). Shaping organisms with apoptosis. *Cell Death and Differentiation, 20*(5), 669–675. https://doi.org/10.1038/cdd.2013.11

The Medical Task Force of Anencepahly. (1990). The infant with anencephaly. *New England Journal of Medicine, 322*(10), 669–674. https://doi.org/10.1056/NEJM199003083221006

Thompson, D. N. P. (2009). Postnatal management and outcome for neural tube defects including spina bifida and encephalocoeles. *Prenatal Diagnosis, 29*(4), 412–419. https://doi.org/10.1002/pd.2199

Toriyama, M., Toriyama, M., Wallingford, J. B., & Finnell, R. H. (2017). Folate-dependent methylation of septins governs ciliogenesis during neural tube closure. *The FASEB Journal: Official Publication of the Federation of American Societies for Experimental Biology, 31*(8), 3622–3635. https://doi.org/10.1096/fj.201700092R

Uba, A., Isamade, E., Chirdan, L., Edino, S., Ogbe, M. E., & Igun, G. (2004). Epidemiology of neural tube defects in North Central Nigeria title: Neural tube defects in Nigeria. *Child Health International African Journal of Paediatric Surgery, 1*. Retrieved from: http://www.afrjpaedsurg.org.

Wallingford, J. B., Niswander, L. A., Shaw, G. M., & Finnell, R. H. (2013). The continuing challenge of understanding, preventing, and treating neural tube defects. *Science, 339*(6123). https://doi.org/10.1126/science.1222002, 1222002–1222002.

Wen, S., Zhu, H., Lu, W., Mitchell, L. E., Shaw, G. M., Lammer, E. J., & Finnell, R. H. (2010). Planar cell polarity pathway genes and risk for spina bifida. *American Journal of Medical Genetics, Part A, 152A*(2), 299–304. https://doi.org/10.1002/ajmg.a.33230

Weprin, B., & Oakes, W. (2000). Occult spinal dysraphism: The clinical presentation and diagnosis. *Operative Techniques in Plastic and Reconstructive Surgery, 7*, 39.

Williams, J., Mai, C. T., Mulinare, J., Isenburg, J., Flood, T. J., Ethen, M., & Centers for Disease Control and Prevention. (2015). Updated estimates of neural tube defects prevented by mandatory folic acid fortification - United States, 1995–2011. *MMWR. Morbidity and Mortality Weekly Report, 64*(1), 1–5. Retrieved from: http://www.ncbi.nlm.nih.gov/pubmed/25590678.

Wong, R. L. Y., Wlodarczyk, B. J., Min, K. S., Scott, M. L., Kartiko, S., Yu, W., … Finnell, R. H. (2008). Mouse Fkbp8 activity is required to inhibit cell death and establish dorso-ventral patterning in the posterior neural tube. *Human Molecular Genetics, 17*(4), 587–601. https://doi.org/10.1093/hmg/ddm333

Yamaguchi, Y., & Miura, M. (2015). *Developmental cell review programmed cell death in neurodevelopment*. https://doi.org/10.1016/j.devcel.2015.01.019

Zaganjor, I., Sekkarie, A., Tsang, B. L., Williams, J., Razzaghi, H., Mulinare, J., … Rosenthal, J. (2016). Describing the prevalence of neural tube defects worldwide: A systematic literature review. *PLoS One, 11*(4), e0151586. https://doi.org/10.1371/journal.pone.0151586

Chapter 3

Chemical substances affecting neurodevelopment

Munekazu Komada
Laboratory of Mammalian Embryology, Department of Life Science, Faculty of Science and Engineering, Kindai University, Higashiosaka, Osaka, Japan

List of abbreviations

ACE acetamiprid
BDNF brain-derived neurotrophic factor
BPA bisphenol A
CNS central nervous systems
DCX doublecortin
DEHP di(2-ethylhexyl) phthalate
DFC dorsal frontal cortex
DG dentate gyrus
ER estrogen receptor
GPER1 G-protein-coupled estrogen receptor 1
HIP hippocampus
IGF1 insulin-like growth factor 1
IL 10 interleukin 10
IL-12 interleukin 12
IL-1β interleukin 1 beta
IL-4 interleukin 4
IL-6 interleukin 6
IMI imidacloprid
SGZ subgranular zone
TGF-β transforming growth factor beta
TNFα tumor necrosis factor alpha
VFC ventral frontal cortex
VZ ventricular zone

Mini-dictionary of terms

Blood–brain barrier A structure that regulates material exchange between the blood and tissue fluid in the capillaries of the brain. It is composed of capillary endothelial cells and astrocytes.
Cell cycle The process of producing two daughter cells from one cell. The cell cycle is divided into the G1, S, G2, and M phase.
Central nervous system In vertebrates, the brain and the spinal cord.
Cerebellum Localized below the brain. It integrates perception and motor function and is involved in the regulation of voluntary muscle movements.
Compartmentalization The areas of the neocortex are divided according to their functions, for example, the visual cortex, the auditory cortex, and the association cortex.
Cortical plate The primordium of the neocortex formed during morphogenesis of the neocortex.
Corticogenesis The process by which the neocortex is formed during the development of the nervous system.

Diagnosis, Management and Modeling of Neurodevelopmental Disorders. https://doi.org/10.1016/B978-0-12-817988-8.00003-8

Cytoarchitecture A structure in which cells with certain characteristics form a group.
Dentate gyrus A part of the hippocampus with a layered structure.
Developmental disorder A disorder of brain function that usually manifests at a younger age. Includes autism, Asperger's syndrome, and attention deficit hyperactivity disorder.
DNA methylation Chemical reaction in which a methyl group is added to a carbon atom of a nucleotide in, for example, a CpG island in a DNA sequence. It is involved in epigenetic regulation of gene expression.
Dopaminergic neurons Neurons that release dopamine as a neurotransmitter. They are abundant in the substantia nigra and the ventral tegmental area.
Dorsal telencephalon The dorsal part of the brain formed most rostral to the neural tube. As development progresses, the dorsal telencephalon becomes the neocortex.
Endocrine disruptor A chemical substance that causes or inhibits hormonal effects in the living body.
Estrogen A type of steroid hormone. It is produced by the gonads and is involved in sexual differentiation of the brain and body and maintenance of pregnancy.
Glial cells Generic term for cells other than neurons in the nervous system. This includes astrocytes, oligodendrocytes, microglia, and Schwann cells.
Hippocampus Part of the limbic system. The hippocampus has a layered structure and is involved in memory and spatial learning.
Hypospadias A birth defect in male external genitalia. Due to the incomplete fusion of the left and right urethral folds, the urethra opens on the ventral side of the penis.
Hypothalamus Part of the diencephalon. It is the center of the endocrine and autonomic nervous systems.
Lamination Formation of a layer structure. Neurons born at the same time and having the same characteristics are often found in the same layer.
Molecular marker A molecule expressed in a specific cell.
Morphogenesis In development, the process by which the form of an organism follows the growth and differentiation of cells.
Neocortex A layer of gray matter that extends to the surface of the cerebrum. Neuronal cell bodies are localized here. The neocortex is involved in higher brain functions such as perception, thinking, and memory.
Neurogenesis The division of one neural stem cell into one neural stem cell and one neuron by asymmetric division.
Neuronal migration During the development of the neocortex, juvenile neurons born in the ventricular surface move toward the pial surface via translocation and locomotion.
Neuronal projection A neuron extends an axon toward a target neuron. Also, forming a synapse with the dendrites of a target neuron.
Phagocytosis A mechanism by which cells such as monocytes and macrophages take up and degrade foreign substances by endocytosis.
Plasticity Change in synaptic transmission efficiency due to changes in synapse activity by neuronal activity.
Proliferation The expansion of one neural stem cell into two cells by symmetric division.
Pyramidal neuron The major excitatory neuron type in the neocortex and the hippocampus.
Subplate During formation of the neocortex, when the cortical plate is formed from the preplate, the subplate is formed together with the marginal zone.
Synapse Contact structure formed between an output neuron that transmits information to an input neuron.
Thyroid hormone A hormone secreted by the thyroid gland. Increases cell respiration and energy production, thereby promoting basal metabolism.

Application in other areas of development

In this chapter, we have reviewed the adverse effects of endocrine disruptors on the morphogenesis of neocortex (Table 3.1). Endocrine disruptors with estrogenic effects cause morphogenetic abnormalities in other areas of the brain. For example, studies have reported abnormalities in the hypothalamus—a region associated with sexual dimorphism—in response to sex hormones, as well as in the hippocampus (HIP) where neurogenesis continues even after maturation. The sexually dimorphic nucleus in the hypothalamus regulates the onset of sex-specific behaviors. Endocrine disruptors not only impede estrogen signaling but also disrupt other signal transduction pathways, such as thyroid hormone signaling. Studies have reported that epigenetic modifications (e.g., histone methylation) induce abnormal gene expression (Table 3.2). Epigenetic abnormalities affect the expression of genes involved in proliferation, neurogenesis, and migration of neural stem cells and juvenile neurons.

Neuro-inflammation in the brain due to prenatal exposure to chemical substances has been reported for alcohol, as well as for endocrine disruptors. Prenatal exposure to alcohol causes fetal alcohol syndrome. Alcohol induces cell death in neurons and neural stem cells due to cytotoxicity. In addition, it induces neuroinflammation and abnormal activation of microglia, adversely affecting morphogenesis, organization, and corticogenesis of the neocortex. Neuroinflammation in the embryonic brain due to prenatal exposure to chemical substances might be a widespread phenomenon (Table 3.3).

TABLE 3.1 Key facts of neurogenesis during embryogenesis.

- In the rodent dorsal telencephalon, neurogenesis begins at approximately embryonic day 10.
- Neural stem cells undergo self-renewal through symmetric division and neurogenesis through asymmetric division.
- The differentiated juvenile neurons move to the upper layer of the dorsal telencephalon along the radial fiber of the neural stem cell, also called the radial grail cells.
- In the developing dorsal telencephalon, late-born neurons move to the upper layers, overtaking the early-born neurons (inside-out pattern).
- Neural stem cells and juvenile neurons exist in a dynamic state, and therefore, are vulnerable to the adverse influence of environmental factors.
- Prenatal exposure to chemical substances affects the cell cycle of neural stem cells and induces cell death.
- Decreased proliferation and increased apoptosis of neural stem cells induce abnormalities in neocortical tissue organization and neuronal projection.

This table lists the key facts of dorsal telencephalon corticogenesis, including proliferation, differentiation, and neuronal migration. Chemical substances affect these processes in neural stem cells and juvenile neurons.

TABLE 3.2 Chemical substances that induce inflammation in the rodents brain during prenatal exposure.

Chemical substances	Regions	Phenotypes	References
Bisphenol A	Dorsal telencephalon hypothalamus	Microglia, abnormal activation, increase. Inflammatory factors, signal transducing molecules, and neurotrophic factor were altered	Takahashi M. et al. (2018)
Acetamiprid (neonicotinoid)	Dorsal telencephalon	Increased Iba1 -immunoreactive microglia	Kagawa & Nagao (2018)
	Dorsal telencephalon	Microglia, abnormal activation, increase. Inflammatory factors, signal transducing molecules, and neurotrophic factor were altered	Komada M. et al. (2017)
Ethanol	Hippocampus Cerebellum Cerebral cortex Cerebellum	IL-1β, TNFα, ccl expression was increased	Drew P.D. et al. (2015)
Valproic acid	Prefrontal cortex brainstem	Inflammation (increased in IL-6, TNFα and decreased in IL-10)	Mirza & Sharma (2019)

Morphogenesis in the neocortex

The complex neuronal organization and structure of the neocortex are responsible for higher brain functions in mammals. The compartmentalization and lamination of the neocortex are dictated by neuronal distribution, which is defined before birth, and their characteristics (including specific gene expression and neuronal input and output projections). The mammalian neocortex is structurally organized into six layers, identifiable with specific molecular markers and localized neuron types. The layered neocortical structure is established during the prenatal period based on the "inside-out" pattern, meaning that the neurons that produced later in the prenatal period will be distributed within the upper layers, on top of the lower layers formed by neurons that developed earlier in the prenatal period. The initial cytoarchitecture provides the basis needed to guide the accurate formation of neuronal projections and synapse formation. The "inside-out" pattern is considered to have been an evolutionary trigger for the development of the large and highly developed neocortex of mammals. Mechanistically, the inside-out pattern of the neocortex is coordinated by proliferation, neurogenesis, neuronal migration, and cell death, as well as responses to extracellular cues and the regulation of intracellular signaling pathways. Structural anomalies in neocortex formation are a cause of developmental and psychiatric disorders, which can be induced by prenatal exposure to certain chemicals. The developmental stage of neocortical formation is highly sensitive to the disruptive effects of chemical exposure. These chemical substances, including endocrine disruptors with estrogenic signaling and activity, can be neurotoxic during prenatal development and the early neonatal period. Endocrine disruptors can modulate hormonal responses; therefore, there is concern about their physiological effects, even from low dose exposure.

TABLE 3.3 Mechanisms of prenatal exposure to chemical substances that adversely affect neural stem cells.

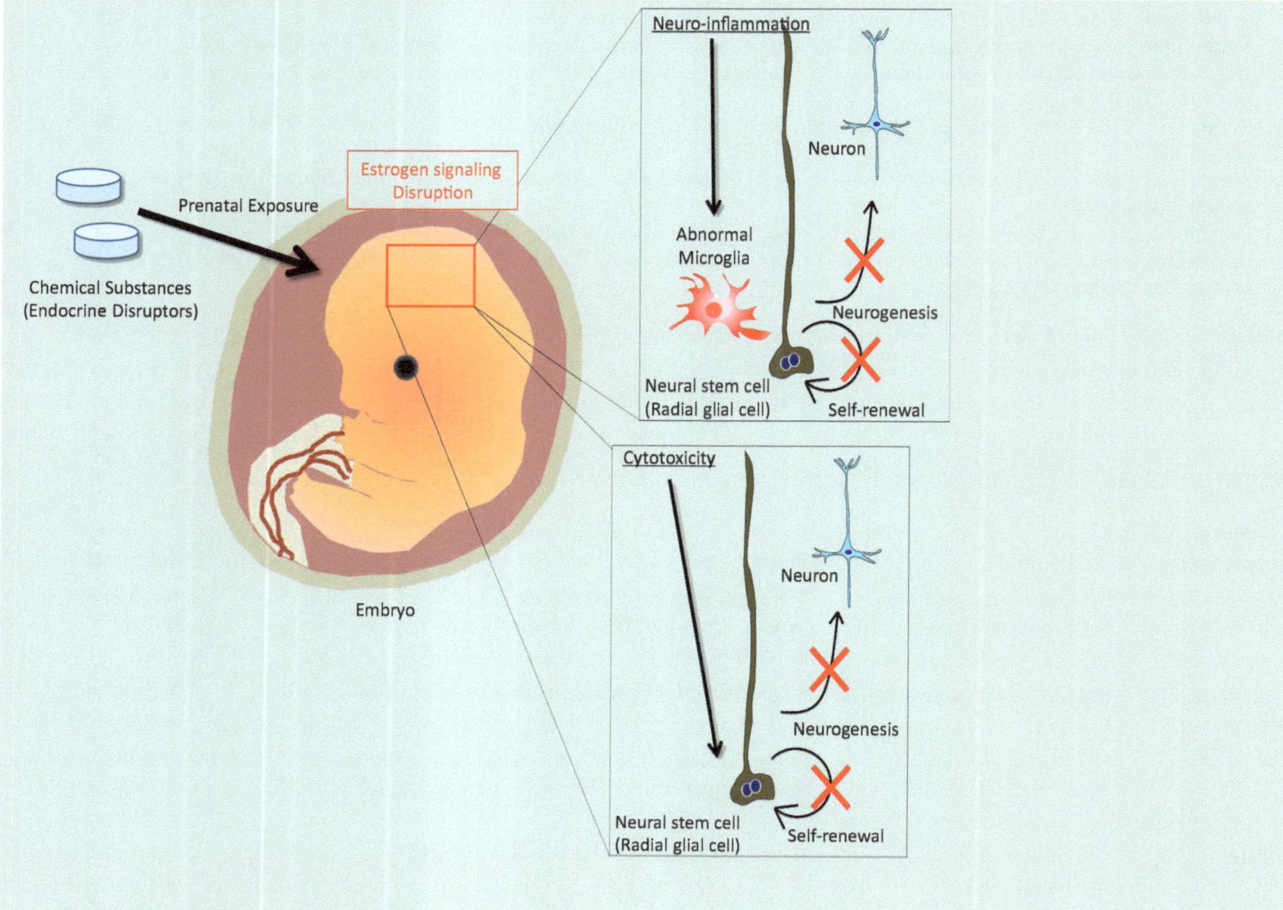

Endocrine disruptors ingested during pregnancy pass through the placenta into the embryos. Disruption of estrogen signaling by endocrine disruptors causes abnormalities in neural stem cell proliferation and differentiation in the following two ways: First, endocrine disruptor-induced "cytotoxicity" directly affects the self-renewal and neurogenesis of neural stem cells. In contrast, neuroinflammation induces microglial abnormalities, which adversely affect neural stem cells.

The roles of estrogen signaling in the developing neocortex

Estrogens are a class of steroid hormones, of which estradiol is the main biologically active form (Blaustein, 2008). This class of steroids has long been known to exert effects on central nervous system (CNS) development, neuronal functional activity, and plasticity. During the prenatal periods, estrogen signaling may also play an important role in hippocampal and neocortical morphogenesis (Srivastava, Woolfrey, & Penzes, 2013).

Aromatase, the key enzyme that converts androgens into estrogens, is highly expressed in the CNS, including in the hypothalamus, HIP, and neocortex in humans and other mammals (Azcoitia, Yague, & Garcia-Segura, 2011). Aromatase is highly expressed in pyramidal neurons as well as glial cells, and there is increasing evidence that estradiol is produced within the brain. Importantly, estradiol is expressed in the developing neocortex during the prenatal periods of both male and female rats and mice (Komada, Takahashi, & Ikeda, 2015; Konkle & McCarthy, 2011). These findings suggest a functional role of aromatase activity in the developing neocortex and HIP during the prenatal periods, and aromatase activity might be the main source of estradiol biosynthesis in the developing neocortex, which could affect neocortical morphogenesis.

Estrogen receptors (ERs) have an important functional role during morphogenesis in the developing neocortex. The pyramidal neurons of the HIP and the neocortical layers II-IV express ER mRNA and translated proteins (Shughrue & Merchenthaler, 2000). During neocortical morphogenesis in the prenatal period, extranuclear ERα (an ER subtype) is expressed within dendritic spines and astrocytes (Milner, 2000). ERβ mRNA transcripts and protein have also been detected in the rodent neocortex (Shughrue & Merchenthaler, 2001). These data suggest that ERs are expressed during periods of cortical morphogenesis and laminar development in rodents. It is also important to note that the three major ER subtypes, ERα, ERβ, and G-protein-coupled estrogen receptor 1 (GPER1), are expressed during the developmental of several brain regions, including the neocortex (Milner et al., 2005); all three are also expressed during the prenatal period in the dorsal frontal cortex (DFC), ventral frontal cortex (VFC), and HIP (this review). ERα expression increases over the prenatal development period, but ERβ is highly expressed during early corticogenesis (this review). These data indicate that the molecular machinery required for estrogen signaling is expressed throughout the development period of the neocortex and that estrogen signaling may play an important regulatory role during corticogenesis and laminar development (Tables 3.1–3.3).

Endocrine disruptors that affect neocortical development: bisphenol A

Further evidence that estrogen signaling plays an important role in the corticogenesis and laminar organization of the neocortex comes from studies examining the detrimental effects of endocrine disruptors with estrogenic activity. For example, bisphenol A (BPA) is a compound widely used in the food industry in the manufacturing of polycarbonate bottles and aluminum can linings, as well as in dentistry as a hardener for sealants and for tooth lacquering (Table 3.2). BPA has estrogenic activity (Gould et al., 1998) and disrupts methylation processes (Kundakovic et al., 2013). ERs and ARs are also expressed in many areas of the developing brain of rodents (Matsuda, Sakamoto, & Kawata, 2008), where estrogen signaling has neurotrophic effects and promotes cellular differentiation. Both ERα and ERβ serve crucial roles during the critical period of brain development (Beyer, 1999). ERβ knockout experiments in mice have demonstrated that this receptor subtype serves important roles in neuronal migration, differentiation, and survival of neurons and neural stem cells in the developing neocortex (Bromer, Zhou, Taylor, Doherty, & Taylor, 2010). In mouse embryos, prenatal BPA exposure induces abnormal, accelerated neurogenesis and hyperplasia of the cortical plate in the dorsal telencephalon (Itoh, Yaoi, & Fushiki, 2012; Komada et al., 2012; Nakamura et al., 2006), leading to disruption of the cell cycle in neural stem/progenitor cells. In addition, accelerated neurogenesis induced by prenatal exposure to BPA leads to hyperplasia of layer 6b and abnormal layer formation in the neocortex of newborn mice. Layer 6 is formed in the subplate of the dorsal telencephalon and has important roles in guiding the projections of dopaminergic neurons to the neocortex and in regulating neuronal morphology and neocortical physiology (Briggs, 2010). Similarly, in adult mice, BPA exposure induces morphological anomalies in the neocortex (Nakamura, Itoh, Sugimoto, & Fushiki, 2007). In humans, BPA interacts with the human ER (Krishnan, Stathis, Permuth, Tokes, & Feldman, 1993), acts as an antagonist for the human androgen receptor (Xu et al., 2005), and strongly binds to human estrogen-related receptor gamma (Takayanagi et al., 2006).

Endocrine disruptors that affect neocortical development: phthalates

Phthalates are industrial chemicals used primarily to increase the flexibility and durability of plastic products (Table 3.2). Their worldwide consumption exceeds 3 million metric tons annually, with the majority of these chemicals used in common household products, including food storage containers, children's toys, pharmaceutical products, and cosmetics (Lyche et al., 2009). Phthalates are not chemically bound to plastic polymers and are easily released into the environment (Heudorf, Mersch-Sundermann, & Angerer, 2007). It was assumed that upon entry into the human body, phthalates are converted into metabolites that are rapidly excreted in the urine, but biomonitoring studies that have measured urine metabolites in humans have shown widespread phthalate exposure. Phthalates have been identified as potential endocrine disruptors, as they interfere with hormonal regulation in the reproductive systems of humans and animals. The reproductive toxicity resulting from phthalate exposure early in life has been well documented, and recent toxicological studies in animals have indicated that pre- and postnatal exposure to phthalates results in severe disruptions in the developing male reproductive system, including defects in the external genitalia, hypospadias, and other reproductive tract malformations (Swan, 2008).

Recently, there has been mounting public concern about the possibility that phthalates, including di(2-ethylhexyl) phthalate (DEHP), may affect neurodevelopment and neuronal functioning. DEHP exposure in utero and during the lactation period impairs neurons of the sexually dimorphic CNS in male rats (Moore, Rudy, Lin, Ko, & Peterson, 2001) and causes neurodegeneration in the rat brain (Dhanya, Indu, Deepadevi, & Kurup, 2003). DEHP inhibits cell proliferation and induces apoptosis via the activation of caspase-3 in a neuroblastoma cell line, Neuro-2a (Lin, Chen, Chen, Hsiao, & Yang, 2011), and increases the number of apoptotic cells in the dorsal telencephalon. The disruption of cell cycle kinetics affects the neurogenesis and proliferation of neuronal stem cells. Both mono(2-ethylhexyl) phthalate (MEHP) and DEHP induce G2/M phase cell cycle arrest in PC12 cells (Chen, Yang, Li, Chen, & Xu, 2011), and prenatal DEHP exposure decreases neurogenesis and differentiation caused by the prolongation of the cell cycle of neural stem cells in mouse embryos (Komada, Gendai, Kagawa, & Nagao, 2016). In addition, the acute postnatal exposure of DEHP adversely affects hippocampal development in male rats; a decreased cell density of both mature and immature neurons is observed in the dentate gyrus and CA3, respectively. These abnormalities disrupt the organization and neuronal distribution of the neocortex. Therefore, phthalates distributed in the environment can disrupt the developing brain, especially development of the neocortex.

Prenatal abnormalities induced by DEHP exposure depend on estrogen signaling, which regulates neurogenesis, differentiation, and neuronal migration of neural stem cells in the development of the neocortex (Komada et al., 2016). The ability of phthalates to bind to ERs in human breast cancer cells and rat uterine cells has been reported. Using MCF-7 (Michigan Cancer Foundation-7), human breast cancer cells, ER-mediated transcriptional activation was increased by dibutyl phthalate, though diethylthexyl phthalate had no effect (Jobling, Reynolds, White, Parker, & Sumpter, 1995). In an estrogenicity study using the proliferation ability of MCF-7, dibutyl phthalate, dimethyl isophthalate, dimethyl terephthalate, and dinonyl phthalate were not estrogenic. However, benzylbutyl phthalate is reported to have a slight estrogenic effect (Sato & Koshiro, 1995). In ER-binding assays in tissue from the rat uterus, the half maximal inhibitory concentration (IC50) of estradiol, used as a positive control, was 9—10 M, whereas dibutyl phthalate and diisooctyl phthalate had IC50s of 9% and 15% at 3—10 M, respectively. However, no effect was seen with diisononyl phthalate, diisodecyl phthalate, or dioctyl phthalate.

Prenatal DEHP exposure might disrupt estrogen signaling and induce morphological abnormalities of radial fibers in the ventricular zone (VZ); these fibers act as footholds for the immature neurons during migration and form the ventricle, which is directed toward the pial surface. Disrupted estrogen signaling induces morphological abnormalities of these radial fibers, leading to migration defects in immature neurons during the development of the dorsal telencephalon (Komada et al., 2012; Wang, Andersson, Warner, & Gustafsson, 2003). This abnormal neuronal distribution in the dorsal telencephalon has been found in P7 (postnatal day 7) newborns exposed prenatally to DEHP (Komada et al., 2016).

Endocrine disruptors that affect neocortex development: neonicotinoids

Neonicotinoids are an important class of synthetic insecticides used to control insect populations, both in agriculture and to provide protection for companion animals (Table 3.2). Neonicotinoids are registered in 120 countries and their use has steadily increased (Jeschke, Nauen, Schindler, & Elbert, 2011), particularly in the farming of corn, canola, soybeans, and most fruits and vegetables. Unlike pesticides that remain on plant surfaces, neonicotinoids are taken up by plants and accumulate in their root structures, stems, leaves, flowers, pollen, and nectar. There is evidence that the concentrations of potentially neurotoxic chemicals in our environment have increased and that these chemicals may disrupt the development of the CNS in humans (Julvez et al., 2016). In this context, neonicotinoid pesticides have attracted considerable attention (Sano et al., 2016). With a base chemical structure similar to nicotine (NIC), neonicotinoids exhibit agonist activity at insect nicotinic acetylcholine receptors (nAChRs), though their precise effects on mammalian nAChRs are unclear. Given the importance of nAChRs for mammalian brain development, and the fact that the developing brain is more susceptible to injury caused by toxic agents than the adult brain in both humans (Weiss, 2000) and animals, a thorough investigation of neonicotinoid exposure in a developmental context is essential for protecting pediatric populations and for promoting the healthy development of human children.

Acetamiprid (ACE) is a globally utilized neonicotinoid that has higher affinities for rodent nAChRs than other neonicotinoids (Tomizawa & Casida, 1999). In adult rats, ACE is rapidly and extensively absorbed after repeated oral administration at a dose of 1 mg/kg (Kimura-Kuroda, Komuta, Kuroda, Hayashi, & Kawano, 2012). ACE bioaccumulates in various brain regions, including the cortex and HIP of mature mice, and ACE and its metabolic products have been observed in the brain, liver, plasma, and urine of ACE-treated rats. Because of its lipid solubility, ACE readily crosses the blood—brain barrier and may come into contact with fetuses after repeated oral administration in pregnant dams. The reproductive toxicity of ACE has been extensively investigated (Bal et al., 2012; Kong et al., 2017). ACE, imidacloprid (IMI), and NIC exert similar excitatory effects on mammalian nAChRs in primary cultures of cerebellar neurons from neonatal rats, suggesting that neonicotinoids may adversely affect human health, particularly the developing brain (Kimura-Kuroda et al., 2016). NIC is thought to interfere with neuronal development by exerting adverse effects on various brain regions including the neocortex, HIP, and cerebellum (Dwyer, McQuown, & Leslie, 2009). Oral administration of NIC during the early postnatal period increases high-affinity binding in the HIP, cortex, and hypothalamus (Huang, Abbott, & Winzer-Serhan, 2007). In addition, studies have shown that repeated prenatal ACE exposure adversely affects the neurogenesis and neuronal distribution in the dorsal telencephalon in embryos and neonatal mice. In a study by Nakayama et al. (2019), postnatal mice were orally administered ACE (5 mg/kg/day) or IMI (5 mg/kg/day); a reduction in neurogenesis was observed in the hippocampal dentate gyrus of neonates following ACE and IMI administration. Other abnormalities included cortical plate hypoplasia and an altered neuronal distribution in the neocortex of ACE-exposed mice. These finding suggest that prenatal exposure to ACE induces histological alterations in the dorsal telencephalon and affects corticogenesis in the neocortex. Neuronal disruption might result from morphological abnormalities in the fetal radial fibers of radial glial cells induced by ACE exposure, resulting in migration defects in immature neurons during the development of the dorsal telencephalon (Komada et al., 2012).

A growing number of studies describe the endocrine-disrupting potential of neonicotinoids (Bhaskar, Mishra, & Mohanty, 2017; Sekeroglu, Sekeroglu, & Demirhan, 2014). Neonicotinoids induce aromatase expression in a promoter-specific manner in vitro (Caron-Beaudoin, Viau, Hudon-Thibeault, Vaillancourt, & Sanderson, 2017). Prenatal ACE exposure disrupts estrogen signaling to produce morphological abnormalities in aforementioned radial fibers, resulting in an abnormal neuronal distribution in the developing neocortex. In addition, oral administration of the neonicotinoids ACE and IMI during the early postnatal period reduces cell proliferation, the number of immature neurons, and neurogenesis by altering the cell cycle exit point in the hippocampal dentate gyrus (DG) of mice.

Doublecortin (DCX) is expressed in type 2b and type 3 progenitor cells and in immature granule cells (Kempermann, Jessberger, Steiner, & Kronenberg, 2004). Low-intensity, long-term exposure to tobacco smoke reduces the number of DCX-positive immature neurons in the hippocampal DG of mice. Therefore, oral dosing of neonicotinoids in neonatal mice inhibits neurogenesis in the developing hippocampal DG. A reduction in the number of NeuN- (a cell marker of postmitotic neurons) positive cells is observed in neonicotinoid- and NIC-treated neonates. Thus, neonatal neonicotinoid exposure might affect neuronal cells in the hippocampal DG, possibly due to promotion of neuronal cell death in the SGZ and granule cell layer of the hippocampal DG. Neonicotinoid toxicity induces apoptosis in several mammalian species (Wang et al., 2016), and neonicotinoids are associated with decreased structural plasticity in the HIP.

Effects of brain inflammation and microglial abnormalities on neocortical development

Microglia originate from immature erythromyeloid progenitor cells and perform phagocytic functions in the CNS (Ginhoux, Lim, Hoeffel, Low, & Huber, 2013). In the developing mouse neocortex, microglia are morphologically ameboid in shape, but transition to a ramified morphology in the early postnatal period (Cunningham, Martinez-Cerdeno, & Noctor, 2013). Microglia act as the principle immune cells in the brain; they phagocytize apoptotic cells, help mediate signaling during cell death, and are important for proper neocortical development (Bilimoria & Stevens, 2015; Ginhoux et al., 2013). In addition, microglia promote proliferation and differentiation of neural stem cells during CNS development (Bilimoria & Stevens, 2015; Cunningham et al., 2013). CNS insult by neurotoxic chemical exposure causes microglia to shift from a ramified, quiescent state to an activated phenotype, characterized by an ameboid morphology, with a round cell body and short, thick, retracted processes (Long-Smith, Sullivan, & Nolan, 2009). Microglia are activated by a variety of inflammatory factors, each of which may induce a different pattern of response. M1 "pro-inflammatory" microglial activation plays a vital role in the defense against pathogens by producing proinflammatory cytokines such as interleukin 1 beta (IL-1β), tumor necrosis factor alpha (TNFα), interleukin 6 (IL-6), and interleukin 12 (IL-12) (Wang, Tan, Yu, & Tan, 2015). In contrast, M2 "anti-inflammatory" microglial activation promotes tissue remodeling and repair and releases antiinflammatory cytokines such as interleukin 10 (IL-10), interleukin 4 (IL-4), and transforming growth factor beta (TGF-β) (Wang et al., 2015). In addition, M2 microglia produce numerous neuronal protective and trophic factors, such as insulin-like growth factor 1 (IGF1) and brain-derived neurotrophic factor (BDNF) (Wang et al., 2015). Activated M2 microglia secrete IGF1 to support neuronal survival, neurogenesis, and axonal growth during development of the neocortex (Ueno et al., 2013). M1 and M2 microglia react differentially in response to pathological alterations in the injured brain. In addition, CX3CL1 (also known as fractalkine or chemokine ligand 1) signaling can exert both neuroprotective (antiinflammatory responses) and proinflammatory responses, associated with M1 or M2 microglia, respectively (Lauro, Catalano, Trettel, & Limatola, 2015).

During neocortical maturation, microglia play an important role in the formation of neuronal networks by controlling neuronal migration, synaptogenesis, and neurotransmission, particularly in cortical layers 2/3 (Bilimoria & Stevens, 2015). Furthermore, microglia release growth factors to promote neuronal survival, neurogenesis, and axonal growth by providing IGF1 (Ueno et al., 2013). Microglia preferentially accumulate in specific regions surrounding dopaminergic axonal tracts and affect axonal extension. Abnormal microglial distribution, function, and activation have been extensively documented in various developmental and mental disorders, including autism, schizophrenia, and Rett syndrome (Bilimoria & Stevens, 2015), and may be associated with problems related to social (Zhan et al., 2014).

ACE-exposed mice exhibit increased numbers of total and ameboid-type microglia in the dorsal telencephalon. IMI- and NIC-exposed mice exhibit increases in the number of ameboid-type microglia in the developing hippocampal DG. In addition, prenatal exposure to ACE, IMI, and NIC increases M1 activation (CD11b expression) but not M2 activation (CD206/mannose receptor expression) in the newborn mouse neocortex and hippocampal DG. Prenatal and postnatal exposure of these neonicotinoids promotes M1 activation. Prenatal ACE exposure stimulates microglial activation and associated expression of microglial inflammatory factors in the developing brain. Similarly, prenatal neonicotinoid exposure affects the expression of various microglial inflammatory factors (TNFα, IL-6, and IL-1β) in the developing hippocampal DG and might be related to dysplasia and dysfunction of the HIP. Accordingly, alterations in microglia and neuroinflammatory processes may represent sensitive endpoints for assessing neurodevelopmental toxicity after exposure to environmental chemicals.

Association of chemicals exposure with brain inflammation and microglial abnormalities

Ethanol (EtOH) exposure induces neuroinflammation as well as excessive microglial activation and apoptosis in the developing rodent neocortex (Drew, Johnson, Douglas, Phelan, & Kane, 2015). Prenatal EtOH exposure adversely affects neocortical development by inducing abnormal distribution of neurons across neocortical layers, hypoplasia in layers 2/3, and reductions in dopaminergic afferents. Prenatal treatment with antiinflammatory agents helps prevent morphological abnormalities in the formation of layer 2/3, reduces ethanol's effects on dopaminergic neuronal projections within the neocortex, and attenuates EtOH-induced activation of microglia and the associated release of proinflammatory cytokines and other inflammatory mediators. Various environmental or industrial chemicals that affect the developing brain have the potential to induce neurodevelopmental or mental disorders associated with neuroinflammation, but antiinflammatory drugs can help attenuate inflammation, microglial activation, and histological anomalies in the neocortex resulting from maternal chemical exposure during pregnancy. Antiinflammatory agents might be effective in the treatment of neurodevelopmental disabilities resulting from prenatal exposure to industrial chemicals.

Peroxisome proliferator-activated receptor gamma (PPARγ) is a member of the nuclear receptor family for steroid/thyroid hormones and is expressed in neurons, astrocytes, oligodendrocytes, and microglia of the CNS (Bernardo & Minghetti, 2008). PPARγ activation inhibits gliosis in microglia and the associated production of proinflammatory factors (Diab et al., 2002). Inhibition of PPARγ attenuates abnormalities in microglial development as well as changes in the M1/M2 ratio in mice prenatally exposed to EtOH, leading to a reduction in neuroinflammatory processes in the brain.

Prenatal exposure to EtOH excessively and specifically promotes M1 at the expense of M2 activation in the newborn neocortex. CX3CL1 is expressed in neurons and is the exclusive ligand for the chemokine receptor CX3CR1, which is exclusively expressed in microglia in the brain. Notably, in transgenic mice lacking CX3CR1, M1 microglial activation is significantly increased compared to wild-type animals (Cardona et al., 2006). CX3CL1-CX3CR1 complex signaling suppresses M1 differentiation and induces the activation of M2 microglia, promoting the release of neurotrophic protective factors (Lauro et al., 2015). CX3CL1 is a candidate effector of abnormal microglial differentiation and activation associated with prenatal EtOH-induced cytotoxicity. Neuronal expression of CX3CL1 decreases following prenatal exposure to EtOH, contributing to the abnormal pattern of increased M1 activation and decreased M2 activation. Neuronal degradation and abnormal microglial differentiation result in morphological abnormalities in the neocortex of neonates.

The expression levels of M1 microglial markers, TNF-α, and CX3CR1/CX3CR1 are increased in the dorsal telencephalon and hypothalamus of BPA-exposed embryos. However, it is difficult to classify embryonic microglia as M1 and M2 phenotypes because marker expression and function differ from those in the adult brain. Microglia in embryonic brains basally express both M1 and M2 markers under noninflammatory physiological conditions (Durafourt et al., 2012), and the expression levels change throughout development (Crain, Nikodemova, & Watters, 2013). At embryonic day 17, maternal administration of polyinosinic:polycytidylic acid (Poly C), an immunostimulant, increased gene expression levels of both M1 and M2 markers at 6 h after maternal administration compared to controls (Meyer et al., 2006). M1 and M2 factors expressed in the adult brain, in the dorsal telencephalon and hypothalamus of embryos, are also increased following prenatal BPA exposure. BPA also competitively inhibits the binding of protein disulfide isomerase to thyroid hormone 3,3′,5-triiodo-L-thyronine (Hiroi, Okada, Imaoka, Osada, & Funae, 2006). Throughout pregnancy, the embryo is largely dependent on the maternal thyroid hormone supply, and proper hypothalamic development requires functional thyroid hormone signaling (Alkemade, 2015), which regulates the morphology and density of microglia in the cortex during the prenatal and postnatal periods in mice (Lima et al., 2001). Thus, interaction of BPA with these hormone systems might be a mechanism of inducing changes in microglial number and activity detected in the two embryonic brain regions.

ACE-exposed mice exhibit increased numbers of total and ameboid-type microglia in the dorsal telencephalon. IMI- and NIC-exposed mice exhibit increases in the number of ameboid-type microglia in the developing hippocampal DG. In addition, prenatal exposure to ACE, IMI, and NIC increases M1 activation (CD11b expression) but not M2 activation (CD206/mannose receptor expression) in the newborn mouse neocortex and hippocampal DG. Prenatal and postnatal exposure of these neonicotinoids promotes M1 activation. Prenatal ACE exposure stimulates microglial activation and associated expression of microglial inflammatory factors in the developing brain. Similarly, prenatal neonicotinoid exposure affects the expression of various microglial inflammatory factors (TNFα, IL-6, and IL-1β) in the developing hippocampal DG and might be related to dysplasia and dysfunction of the HIP. Accordingly, alterations in microglia and neuroinflammatory processes may represent sensitive endpoints for assessing neurodevelopmental toxicity after exposure to environmental chemicals.

Summary points

- The neocortex controls higher brain functions.
- Complex organization of the neocortical tissue ensures strict regulation of proliferation, differentiation, and migration of neocortical neurons during the prenatal periods.
- Estrogen, a steroid hormone, plays an important role in neocortical morphogenesis.
- Prenatal exposure to endocrine disruptors with estrogen-like effects (Bisphenol A, phthalates, and neonicotinoid) affects neocortical morphogenesis.
- These chemical substances adversely affect the proliferation and neurogenesis of neural stem cells and migration of juvenile neurons.
- They induce neuroinflammation in the embryonic brain, causing abnormalities in microglial activity and numbers.
- These morphological and functional abnormalities in the developing neocortex might be one of the pathogenetic mechanisms underlying developmental disorders.

References

Alkemade, A. (2015). Thyroid hormone and the developing hypothalamus. *Frontiers in Neuroanatomy, 9*, 15.

Azcoitia, I., Yague, J. G., & Garcia-Segura, L. M. (2011). Estradiol synthesis within the human brain. *Neuroscience, 191*, 139–147.

Bal, R., Naziroglu, M., Turk, G., Yilmaz, O., Kuloglu, T., Etem, E., & Baydas, G. (2012). Insecticide imidacloprid induces morphological and DNA damage through oxidative toxicity on the reproductive organs of developing male rats. *Cell Biochemistry and Function, 30*(6), 492–499.

Bernardo, A., & Minghetti, L. (2008). Regulation of glial cell functions by PPAR-gamma natural and synthetic agonists. *PPAR Research, 2008*, 864140.

Beyer, C. (1999). Estrogen and the developing mammalian brain. *Anatomy and Embryology, 199*(5), 379–390.

Bhaskar, R., Mishra, A. K., & Mohanty, B. (2017). Neonatal exposure to endocrine disrupting chemicals impairs learning behaviour by disrupting hippocampal organization in male Swiss Albino mice. *Basic and Clinical Pharmacology and Toxicology, 121*(1), 44–52.

Bilimoria, P. M., & Stevens, B. (2015). Microglia function during brain development: New insights from animal models. *Brain Research, 1617*, 7–17.

Blaustein, J. D. (2008). An estrogen by any other name. *Endocrinology, 149*(6), 2697–2698.

Briggs, F. (2010). Organizing principles of cortical layer 6. *Frontiers in Neural Circuits, 4*, 3.

Bromer, J. G., Zhou, Y., Taylor, M. B., Doherty, L., & Taylor, H. S. (2010). Bisphenol-A exposure in utero leads to epigenetic alterations in the developmental programming of uterine estrogen response. *The FASEB Journal, 24*(7), 2273–2280.

Cardona, A. E., Pioro, E. P., Sasse, M. E., Kostenko, V., Cardona, S. M., Dijkstra, I. M., ... Ransohoff, R. M. (2006). Control of microglial neurotoxicity by the fractalkine receptor. *Nature Neuroscience, 9*(7), 917–924.

Caron-Beaudoin, E., Viau, R., Hudon-Thibeault, A. A., Vaillancourt, C., & Sanderson, J. T. (2017). The use of a unique co-culture model of fetoplacental steroidogenesis as a screening tool for endocrine disruptors: The effects of neonicotinoids on aromatase activity and hormone production. *Toxicology and Applied Pharmacology, 332*, 15–24.

Chen, T., Yang, W., Li, Y., Chen, X., & Xu, S. (2011). Mono-(2-ethylhexyl) phthalate impairs neurodevelopment: Inhibition of proliferation and promotion of differentiation in PC12 cells. *Toxicology Letters, 201*(1), 34–41.

Crain, J. M., Nikodemova, M., & Watters, J. J. (2013). Microglia express distinct M1 and M2 phenotypic markers in the postnatal and adult central nervous system in male and female mice. *Journal of Neuroscience Research, 91*(9), 1143–1151.

Cunningham, C. L., Martinez-Cerdeno, V., & Noctor, S. C. (2013). Microglia regulate the number of neural precursor cells in the developing cerebral cortex. *Journal of Neuroscience, 33*(10), 4216–4233.

Dhanya, C. R., Indu, A. R., Deepadevi, K. V., & Kurup, P. A. (2003). Inhibition of membrane Na(+)-K+ Atpase of the brain, liver and RBC in rats administered di(2-ethyl hexyl) phthalate (DEHP) a plasticizer used in polyvinyl chloride (PVC) blood storage bags. *Indian Journal of Experimental Biology, 41*(8), 814–820.

Diab, A., Deng, C., Smith, J. D., Hussain, R. Z., Phanavanh, B., Lovett-Racke, A. E., … Racke, M. K. (2002). Peroxisome proliferator-activated receptor-gamma agonist 15-deoxy-delta(12,14)-prostaglandin J(2) ameliorates experimental autoimmune encephalomyelitis. *The Journal of Immunology, 168*(5), 2508–2515.

Drew, P. D., Johnson, J. W., Douglas, J. C., Phelan, K. D., & Kane, C. J. (2015). Pioglitazone blocks ethanol induction of microglial activation and immune responses in the hippocampus, cerebellum, and cerebral cortex in a mouse model of fetal alcohol spectrum disorders. *Alcoholism: Clinical and Experimental Research, 39*(3), 445–454.

Durafourt, B. A., Moore, C. S., Zammit, D. A., Johnson, T. A., Zaguia, F., Guiot, M. C., … Antel, J. P. (2012). Comparison of polarization properties of human adult microglia and blood-derived macrophages. *Glia, 60*(5), 717–727.

Dwyer, J. B., McQuown, S. C., & Leslie, F. M. (2009). The dynamic effects of nicotine on the developing brain. *Pharmacology and Therapeutics, 122*(2), 125–139.

Ginhoux, F., Lim, S., Hoeffel, G., Low, D., & Huber, T. (2013). Origin and differentiation of microglia. *Frontiers in Cellular Neuroscience, 7*, 45.

Gould, J. C., Leonard, L. S., Maness, S. C., Wagner, B. L., Conner, K., Zacharewski, T., … Gaido, K. W. (1998). Bisphenol A interacts with the estrogen receptor alpha in a distinct manner from estradiol. *Molecular and Cellular Endocrinology, 142*(1–2), 203–214.

Heudorf, U., Mersch-Sundermann, V., & Angerer, J. (2007). Phthalates: Toxicology and exposure. *International Journal of Hygiene and Environmental Health, 210*(5), 623–634.

Hiroi, T., Okada, K., Imaoka, S., Osada, M., & Funae, Y. (2006). Bisphenol A binds to protein disulfide isomerase and inhibits its enzymatic and hormone-binding activities. *Endocrinology, 147*(6), 2773–2780.

Huang, L. Z , Abbott, L. C., & Winzer-Serhan, U. H. (2007). Effects of chronic neonatal nicotine exposure on nicotinic acetylcholine receptor binding, cell death and morphology in hippocampus and cerebellum. *Neuroscience, 146*(4), 1854–1868.

Itoh, K., Yaoi, T., & Fushiki, S. (2012). Bisphenol A, an endocrine-disrupting chemical, and brain development. *Neuropathology, 32*(4), 447–457.

Jeschke, P., Nauen, R., Schindler, M., & Elbert, A. (2011). Overview of the status and global strategy for neonicotinoids. *Journal of Agricultural and Food Chemistry, 59*(7), 2897–2908.

Jobling, S., Reynolds, T., White, R., Parker, M. G., & Sumpter, J. P. (1995). A variety of environmentally persistent chemicals, including some phthalate plasticizers, are weakly estrogenic. *Environmental Health Perspectives, 103*(6), 582–587.

Julvez, J., Paus, T., Bellinger, D., Eskenazi, B., Tiemeier, H., Pearce, N., … Sunyer, J. (2016). Environment and brain development: Challenges in the global context. *Neuroepidemiology, 46*(2), 79–82.

Kagawa, N., & Nagao, T. (2018). Neurodevelopmental toxicity in the mouse neocortex following prenatal exposure to acetamiprid. *Journal of Applied Toxicology, 38*(12), 1521–1528.

Kempermann, G., Jessberger, S., Steiner, B., & Kronenberg, G. (2004). Milestones of neuronal development in the adult hippocampus. *Trends in Neurosciences, 27*(8), 447–452.

Kimura-Kuroda, J., Komuta, Y., Kuroda, Y., Hayashi, M., & Kawano, H. (2012). Nicotine-like effects of the neonicotinoid insecticides acetamiprid and imidacloprid on cerebellar neurons from neonatal rats. *PLoS One, 7*(2), e32432.

Kimura-Kuroda, J., Nishito, Y., Yanagisawa, H., Kuroda, Y., Komuta, Y., Kawano, H., & Hayashi, M. (2016). Neonicotinoid insecticides alter the gene expression profile of neuron-enriched cultures from neonatal rat cerebellum. *International Journal of Environmental Research and Public Health, 13*(10).

Komada, M., Asai, Y., Morii, M., Matsuki, M., Sato, M., & Nagao, T. (2012). Maternal bisphenol A oral dosing relates to the acceleration of neurogenesis in the developing neocortex of mouse fetuses. *Toxicology, 295*(1–3), 31–38.

Komada, M., Gendai, Y., Kagawa, N., & Nagao, T. (2016). Prenatal exposure to di(2-ethylhexyl) phthalate impairs development of the mouse neocortex. *Toxicology Letters, 259*, 69–79.

Komada, M., Hara, N., Kawachi, S., Kawachi, K., Kagawa, N., Nagao, T., & Ikeda, Y. (2017). Mechanisms underlying neuro-inflammation and neurodevelopmental toxicity in the mouse neocortex following prenatal exposure to ethanol. *Scientific Reports, 7*(1), 4934. https://doi.org/10.1038/s41598-017-04289-1.

Komada, M., Takahashi, M., & Ikeda, Y. (2015). Involvement of SF-1 in neurogenesis and neuronal migration in the developing neocortex. *Neuroscience Letters, 600*, 85–90.

Kong, D., Zhang, J., Hou, X., Zhang, S., Tan, J., Chen, Y., … Cai, R. (2017). Acetamiprid inhibits testosterone synthesis by affecting the mitochondrial function and cytoplasmic adenosine triphosphate production in rat Leydig cells. *Biology of Reproduction, 96*(1), 254–265.

Konkle, A. T., & McCarthy, M. M. (2011). Developmental time course of estradiol, testosterone, and dihydrotestosterone levels in discrete regions of male and female rat brain. *Endocrinology, 152*(1), 223–235.

Krishnan, A. V., Stathis, P., Permuth, S. F., Tokes, L., & Feldman, D. (1993). Bisphenol-A: An estrogenic substance is released from polycarbonate flasks during autoclaving. *Endocrinology, 132*(6), 2279–2286.

Kundakovic, M., Gudsnuk, K., Franks, B., Madrid, J., Miller, R. L., Perera, F. P., & Champagne, F. A. (2013). Sex-specific epigenetic disruption and behavioral changes following low-dose in utero bisphenol A exposure. *Proceedings of the National Academy of Sciences of the United States of America, 110*(24), 9956–9961.

Lauro, C., Catalano, M., Trettel, F., & Limatola, C. (2015). Fractalkine in the nervous system: Neuroprotective or neurotoxic molecule? *Annals of the New York Academy of Sciences, 1351*, 141–148.

Lima, F. R., Gervais, A., Colin, C., Izembart, M., Neto, V. M., & Mallat, M. (2001). Regulation of microglial development: A novel role for thyroid hormone. *Journal of Neuroscience, 21*(6), 2028–2038.

Lin, C. H., Chen, T. J., Chen, S. S., Hsiao, P. C., & Yang, R. C. (2011). Activation of Trim17 by PPARgamma is involved in di(2-ethylhexyl) phthalate (DEHP)-induced apoptosis on Neuro-2a cells. *Toxicology Letters, 206*(3), 245–251.

Long-Smith, C. M., Sullivan, A. M., & Nolan, Y. M. (2009). The influence of microglia on the pathogenesis of Parkinson's disease. *Progress in Neurobiology, 89*(3), 277–287.

Lyche, J. L., Gutleb, A. C., Bergman, A., Eriksen, G. S., Murk, A. J., Ropstad, E., … Skaare, J. U. (2009). Reproductive and developmental toxicity of phthalates. *Journal of Toxicology and Environmental Health Part B: Critical Reviews, 12*(4), 225–249.

Matsuda, K., Sakamoto, H., & Kawata, M. (2008). Androgen action in the brain and spinal cord for the regulation of male sexual behaviors. *Current Opinion in Pharmacology, 8*(6), 747–751.

Meyer, U., Nyffeler, M., Engler, A., Urwyler, A., Schedlowski, M., Knuesel, I., … Feldon, J. (2006). The time of prenatal immune challenge determines the specificity of inflammation-mediated brain and behavioral pathology. *Journal of Neuroscience, 26*(18), 4752–4762.

Milner, G. R. (2000). Conquistadors, excavators, or rodents: What damaged the king site skeletons? *American Antiquity, 65*(2), 355–363.

Milner, T. A., Ayoola, K., Drake, C. T., Herrick, S. P., Tabori, N. E., McEwen, B. S., … Alves, S. E. (2005). Ultrastructural localization of estrogen receptor beta immunoreactivity in the rat hippocampal formation. *The Journal of Comparative Neurology, 491*(2), 81–95.

Mirza, R., & Sharma, B. (2019). A selective peroxisome proliferator-activated receptor-γ agonist benefited propionic acid induced autism-like behavioral phenotypes in rats by attenuation of neuroinflammation and oxidative stress. *Chemico-Biological Interactions, 25*(311), 108758. https://doi.org/10.1016/j.cbi.2019.108758.

Moore, R. W., Rudy, T. A., Lin, T. M., Ko, K., & Peterson, R. E. (2001). Abnormalities of sexual development in male rats with in utero and lactational exposure to the antiandrogenic plasticizer di(2-ethylhexyl) phthalate. *Environmental Health Perspectives, 109*(3), 229–237.

Nakamura, K., Itoh, K., Sugimoto, T., & Fushiki, S. (2007). Prenatal exposure to bisphenol A affects adult murine neocortical structure. *Neuroscience Letters, 420*(2), 100–105.

Nakamura, K., Itoh, K., Yaoi, T., Fujiwara, Y., Sugimoto, T., & Fushiki, S. (2006). Murine neocortical histogenesis is perturbed by prenatal exposure to low doses of Bisphenol A. *Journal of Neuroscience Research, 84*(6), 1197–1205.

Nakayama, A., Yoshida, M., Kagawa, N., & Nagao, T. (2019). The neonicotinoids acetamiprid and imidacloprid impair neurogenesis and alter the microglial profile in the hippocampal dentate gyrus of mouse neonates. *Journal of Applied Toxicology, 39*(6), 877–887. https://doi.org/10.1002/jat.3776. In press.

Sano, K., Isobe, T., Yang, J., Win-Shwe, T. T., Yoshikane, M., Nakayama, S. F., … Maekawa, F. (2016). In utero and lactational exposure to acetamiprid induces abnormalities in socio-sexual and anxiety-related behaviors of male mice. *Frontiers in Neuroscience, 10*, 228.

Sato, S., & Koshiro, A. (1995). Protein binding of chlorpromazine in vivo and in vitro: Effect of chlorpromazine metabolite on chlorpromazine protein binding in rat. *Biological and Pharmaceutical Bulletin, 18*(4), 586–592.

Sekeroglu, V., Sekeroglu, Z. A., & Demirhan, E. (2014). Effects of commercial formulations of deltamethrin and/or thiacloprid on thyroid hormone levels in rat serum. *Toxicology and Industrial Health, 30*(1), 40–46.

Shughrue, P. J., & Merchenthaler, I. (2000). Estrogen is more than just a "sex hormone": Novel sites for estrogen action in the hippocampus and cerebral cortex. *Frontiers in Neuroendocrinology, 21*(1), 95–101.

Shughrue, P. J., & Merchenthaler, I. (2001). Distribution of estrogen receptor beta immunoreactivity in the rat central nervous system. *The Journal of Comparative Neurology, 436*(1), 64–81.

Srivastava, D. P., Woolfrey, K. M., & Penzes, P. (2013). Insights into rapid modulation of neuroplasticity by brain estrogens. *Pharmacological Reviews, 65*(4), 1318–1350.

Swan, S. H. (2008). Environmental phthalate exposure in relation to reproductive outcomes and other health endpoints in humans. *Environmental Research, 108*(2), 177–184.

Takahashi, M., Komada, M., Miyazawa, K., Goto, S., & Ikeda, Y. (2018). Bisphenol A exposure induces increased microglia and microglial related factors in the murine embryonic dorsal telencephalon and hypothalamus. *Toxicology Letters, 284*, 113–119. https://doi.org/10.1016/j.toxlet.2017.12.010.

Takayanagi, S., Tokunaga, T., Liu, X., Okada, H., Matsushima, A., & Shimohigashi, Y. (2006). Endocrine disruptor bisphenol A strongly binds to human estrogen-related receptor gamma (ERRgamma) with high constitutive activity. *Toxicology Letters, 167*(2), 95–105.

Tomizawa, M., & Casida, J. E. (1999). Minor structural changes in nicotinoid insecticides confer differential subtype selectivity for mammalian nicotinic acetylcholine receptors. *British Journal of Pharmacology, 127*(1), 115–122.

Ueno, M., Fujita, Y., Tanaka, T., Nakamura, Y., Kikuta, J., Ishii, M., & Yamashita, T. (2013). Layer V cortical neurons require microglial support for survival during postnatal development. *Nature Neuroscience, 16*(5), 543–551.

Wang, L., Andersson, S., Warner, M., & Gustafsson, J. A. (2003). Estrogen receptor (ER)beta knockout mice reveal a role for ERbeta in migration of cortical neurons in the developing brain. *Proceedings of the National Academy of Sciences of the United States of America, 100*(2), 703–708.

Wang, W. Y., Tan, M. S., Yu, J. T., & Tan, L. (2015). Role of pro-inflammatory cytokines released from microglia in Alzheimer's disease. *Annals of Translational Medicine, 3*(10), 136.

Wang, C. J., Wang, G., Wang, X. Y., Liu, M., Chuai, M., Lee, K. K., … Yang, X. (2016). Imidacloprid exposure suppresses neural crest cells generation during early chick embryo development. *Journal of Agricultural and Food Chemistry, 64*(23), 4705–4715.

Weiss, B. (2000). Vulnerability of children and the developing brain to neurotoxic hazards. *Environmental Health Perspectives, 108*(Suppl. 3), 375–381.

Xu, L. C., Sun, H., Chen, J. F., Bian, Q., Qian, J., Song, L., & Wang, X. R. (2005). Evaluation of androgen receptor transcriptional activities of bisphenol A, octylphenol and nonylphenol in vitro. *Toxicology, 216*(2–3), 197–203.

Zhan, Y., Paolicelli, R. C., Sforazzini, F., Weinhard, L., Bolasco, G., Pagani, F., … Gross, C. T. (2014). Deficient neuron-microglia signaling results in impaired functional brain connectivity and social behavior. *Nature Neuroscience, 17*(3), 400–406.

Chapter 4

High-risk babies and neurodevelopmental outcome

Merina Shrestha
Department of Child Health, Tribhuvan University Teaching Hospital, Kathmandu, Nepal

List of abbreviations

EI Early intervention
IQ Intelligence quotient
KC Kangaroo care
LMIC Low- and middle-income countries
NICU Neonatal intensive care unit

Mini-dictionary of terms

Birth asphyxia: Failure to initiate and sustain normal breathing at birth.
Early intervention: Identifying and providing effective support to all the children who are at risk of poor outcomes
Multidisciplinary: Involving people from different disciplines or professional specializations to work together, in an approach to a problem.
Neurodevelopment: Brain's ability to develop according to different ages for the normal functioning to learn, focus, develop memories and social skills, etc.
Preterm babies: Babies born before 37 weeks of gestation.

Introduction

The high-risk infants refer to newborns exposed to high-risk factors in the prenatal, natal, or postnatal period or who require added observation and care during neonatal as well post neonatal period. Of all newborn babies, about 10%–20% require intensive care. Table 4.1 shows different causes of high risk for any newborn baby. The listed factors are associated with numerous immediate complications and mortalities. Even if babies survive these complications, they might be at risk of many long-term neurodevelopmental morbidities.

For ease of follow-up care of these high-risk babies, American Academy of Pediatrics in their policy statement has categorized babies into four broad categories; (1) the preterm infant; (2) the infant with special health care needs or dependence on technology; (3) the infant at risk because of family issues; and (4) the infant with anticipated early death (American Academics of American Academy of Pediatrics, 2004).

Babies who are discharged from neonatal care are always at risk of not attaining optimal development, and they require more regular follow-up care and assessments. In this chapter we are more focusing on preterm babies and babies with birth asphyxia as these conditions are more common causes affecting neurodevelopmental outcome of a newborn baby.

Preterm babies

Preterm babies are babies born before 37 weeks of gestation. Worldwide, it was estimated that about 10.6% of live births were preterm in 2014. Of these 15 million babies, more than 80% were born in South Asia and sub-Saharan Africa. The global trend also showed preterm birth rate in increasing from 9.8% (8.3–10.9) in 2000 to 10.6% (9.0–12.0) in 2014 (Lawn, Cousens, Zupan, & Team, 2005). The incidence of preterm deliveries was found to be the lowest in Europe and

Diagnosis, Management and Modeling of Neurodevelopmental Disorders. https://doi.org/10.1016/B978-0-12-817988-8.00004-X

TABLE 4.1 Causes of high-risk babies.

Biologic risk factor	Interventions	Social/environmental
Low-birth-weight babies <1500 g Seizure, hydrocephalus Necrotizing enterocolitis Chronic lung disease	Resuscitation Postnatal steroids	Low maternal education and teen mothers and single marital status Drugs/alcohol, smoking, substance abuse No prenatal care Environmental stress
Complex medical problems	High-frequency ventilation Prolonged ventilation 7 days	
Higher-order multiples	Total parenteral nutrition	
Twin–twin transfusion	Prolonged oxygen requirements	
Complex congenital anomalies	Nutritional therapies	
Recurrent apnea and bradycardia Hyperbilirubinemia requiring exchange transfusion Failure to grow in the NICU	Other medications Surgical interventions for NEC, patent ductus arteriosus, shunt	
Sepsis, meningitis, nosocomial infections Multiparity		
Abnormal neurologic exam or encephalopathy at discharge		

Adapted from American Academic of Pediatrics. (2004). Follow-up care of high-risk infants. *Pediatrics*, 114(Suppl. 5), 1377–1397.

North America. Of these preterm babies, only about 4.3 million (28%) had access to intensive neonatal care, 2.5 million (17%) to special care, and 8.2 million (55%) to basic/limited care only. There seems to be a huge gap in survival of these preterm babies; in high-income countries, the mortality is as less as 1% as compared to 75% in South Asian and sub-Saharan region. In countries such as Sweden, the survival of extremely preterm babies has even increased significantly over the past few years (Fischer, Steurer, Adams, & Berger, 2009).

Ideally, babies who are less than 28 weeks require management in neonatal intensive care unit (NICU), and babies between 28 and 32 weeks will need special newborn care at a minimum. However, these neonatal care facilities are very scarce in low- and middle-income countries (LMICs). Of the preterm babies who survive, many may develop consequences described in Table 4.2. Of them, impaired neurodevelopmental outcome is one of the important comorbidities. The neurodevelopment impairment has direct relation to gestational age of the baby. Babies born less than 28 weeks of gestation have the highest risk of neurodevelopmental impairment. About half of the babies born less than 28 weeks and one-third between 28 and 31 weeks develop impairments. Only about 5% of those born at 32–36 weeks are estimated to have some degree of neurodevelopmental impairment. It is seen that the babies even born near term i.e., between 34 and 36 weeks, are three times at risk of developing neurodevelopmental impairments such as developmental delay and cerebral palsy as compared to babies born at term (Petrini et al., 2009). The preterm survivors are also at risk of learning impairment with academic underachievement and behavioral problems (Aarnoudse-Moens, Weisglas-Kuperus, van Goudoever, & Oosterlaan, 2009).

All these impairments exert a heavy burden on families, societies, and the health system. Hence, global burden of disease analysis has recognized preterm birth as one the largest single conditions resulting in a high mortality and the considerable risk of lifelong impairment (Murray et al., 2012).

Birth asphyxia

Another important cause for future neurodevelopmental impairment in children is birth asphyxia. Birth asphyxia is a failure to initiate and sustain normal breathing at birth. The immediate outcomes of birth asphyxia are hypoxic encephalopathy, seizure, and death. Worldwide, birth asphyxia contributes about 10% of all deaths in children less than 5 years of age

TABLE 4.2 **Long-term outcomes in babies with birth asphyxia.**

Long-term outcomes		Examples:	Frequency in survivors:
Specific physical effects	Visual impairment	• Blindness or high myopia after retinopathy of prematurity • Increased hypermetropia and myopia	Around 25% of all extremely preterm affected Also risk in moderately preterm babies especially if poorly monitored oxygen therapy
	Hearing impairment		Up to 5%–10% of extremely preterm
	Chronic lung disease of prematurity	• From reduced exercise tolerance to requirement for home oxygen • Increased hospital admissions in childhood for LRTI	Up to 40% of extremely preterm
	Long-term cardiovascular ill-health and noncommunicable disease	• Increased blood pressure • Reduced lung function • Increased rates of asthma • Growth failure in infancy, accelerated weight gain in adolescence	Full extent of burden still to be quantified
Neurodevelopmental/ behavioral effects	Mild Disorders of executive functioning	• specific learning impairments, dyslexia, reduced academic achievement	
	Moderate to severe global developmental delay	• Moderate/severe cognitive impairment • motor impairment • Cerebral palsy	Affected by gestational age and quality of care dependent
	Psychiatric/behavioral sequelae	• Attention deficit hyperactivity disorder • Increased anxiety and depression	
Family, economic, and societal effects	Impact on family Impact on health service Intergenerational	• Psychosocial, emotional, and economic • Cost of care – acute and ongoing • Risk of preterm birth in offspring	Common varying with medical risk factors, disability, socioeconomic status

Adapted from Blencowe, H., Cousens, S., Chou, D., Oestergaard, M., Say, L., Moller, A. B., et al. (2013). Born too soon: the global epidemiology of 15 million preterm births. *Reproductive Health*, 10(1), S2.

(Bryce, Boschi-Pinto, Shibuya, Black, & WHO Child Health Epidemiology Reference Group, 2005; Jones et al., 2003). Babies who survive are at risk of developing adverse neurodevelopmental outcomes. Cerebral palsy, mental retardation, and cognitive disabilities are few of the consequences in babies who survive the birth asphyxia (Simon, 1999; Robertson and Finer, 1993; Cioni et al., 1992). Even with mild to moderate birth asphyxia, infants may develop permanent severe neurologic impairments (Ellis et al., 1999). It is seen that when newborns have seizure either due to birth asphyxia or other vascular events during neonatal period, there is added risk of adverse neurologic outcome including increased risk of postnatal seizure and developmental delay (De Haan, Langeslag, van der Lee, & van Kaam, 2018; Shrestha et al., 2017).

The adverse outcome of asphyxia is summarized in Table 4.3.

To minimize the adverse outcomes in high-risk babies, early intervention (EI) plays a critical role.

TABLE 4.3 Long-term outcome of perinatal asphyxia (Ahearne, Boylan & Murray, 2016).

Motor	Cerebral palsy
Sensory impairment	Hearing loss, visual
Cognitive	Attention/executive function impairments Lower working memory, reading accuracy
Educational	Increased support requirements, Lower test scores
Behavioral	Inattention, explosiveness, and irritability
Neuropsychiatric	Psychotic symptoms
Neurodevelopmental	Autistic spectrum

Early intervention (EI) for high-risk babies

The basis of EI lies in the neuroplasticity of the young brain. Neuroplasticity is the ability of the brain to change structurally and functionally to different trainings and experiences (Mundkur, 2005; Johnston, 2009). EI promotes child's health and well-being. Thus, EI is very important for all newborns whether they are at risk or not, of developing developmental disorders in later life. There is added benefit of EI, specially in high-risk babies to minimize developmental delay, remediate existing or emerging disabilities, and prevent functional deterioration. EI consists of multidisciplinary services that start from birth and also promote adaptive parenting and family functioning (Bieleninik and Gold, 2014). The services aim at treating medical causes, minimizing the impact of the disability with appropriate and suitable program for the child and the family.

EI should start from birth even when they are in neonatal care. The continuous sounds of monitors and other instruments, unnecessary lights, and handling quite often for cleaning or medical interventions can cause overstimulation. There is little chance to be held or touched by parents when babies are in incubators or multiple monitoring devices on their bodies. This lack of opportunity of social interaction may lead to disabilities in motor functioning, hearing, eyesight, eating, and abnormal behavior or psychological disorders.

It is thus recommended to avoid overstimulation of the newborn babies by avoiding direct light over babies, covering the top of the incubators to support sleep cycles and also decreasing the noise level to 50 dB or less while babies are cared in NICU (VandenBerg, 2007). For preterm babies another beneficial intervention is kangaroo care (KC). Apart from decreased infection and better weight gain, KC has significant positive effects on child's behavior states, including less crying and more quiet sleep (Dodd, 2005; Messmer et al., 1997). KC also helps for significant positive impact on infants' perceptual, cognitive, and motor development at 6 months (Feldman and Eidelman, 2003). Music therapy even during early life is found to be beneficial on children's behavioral states (Loewy, Stewart, Dassler, Telsey, & Homel, 2013; Standley, 2012).

EI not only involves the baby but also takes into account the parents and family involvement as a whole. It is seen that when parents are themselves stressed, it also affects the care of the child. Parental anxiety increases parent hostility and decreases child responsiveness with poor quantity and quality of stimulation for their infants (Bettes, 1988; Field, Healy, Goldstein, & Guthertz, 1990). Thus, EI program ideally includes three domains: (1) psychosocial support to parents for responsive parenting and to alleviate parental stress, anxiety, and depressive symptoms; (2) educational interventions for parents including knowledge about normal growth and development, taking care of babies; and (3) therapeutic developmental intervention (Benzies, Magill-Evans, Hayden, & Ballantyne, 2013). The comprehensive EI can also be adopted in low-income settings to improve child's development (Chomba et al., 2011). Even home-based EI implemented by parents improves developmental outcomes (Wallander et al., 2014). In home-based therapy, the caretaker, mostly mother, primarily takes a role of "Developmental Therapist." Physiotherapy, occupational therapy, cognitive stimulation, and speech and language therapy are incorporated together in developmental therapy. This low-cost holistic approach of developmental therapy when applied in preterm and birth-asphyxiated babies, there is improved outcome in motor, mental, and behavior functions (Ehsanul, Rahman, & Fatema, 2018). Along with EI or developmental therapy, strict follow-up of high-risk babies is advised for early detection of delays and timely more specific and focused interventions.

When to do follow-up?

If developmental delays are identified early in infants, interventions could be started early. Developmental delay can be predicted as early as 3 months of age. Absence of milestones such as social smile at 3 months, unable to sit upon pulling, or unable to transfer objects and voluntary grasp at 6 months can be the early signs of developmental delay (Godbole, Barve, & Chaudhari, 1997).

Ideally, all the high-risk babies should be assessed in each hospital visits for any delay. Close monitoring of development, together with coordination of treatment for any emerging problems, is often referred to as developmental follow-up.

Follow-up care of high-risk infants

Regular follow-up visits help to identify the developmental problems early so that interventions could be started early.

For the developmental follow-up for preterm babies, corrected age should be taken. Corrected age is the sum of chronologic age in weeks minus the difference between gestational age at birth and 40 weeks' gestation. This corrected age is considered till the child becomes 18–24 months, then chronological age is taken.

Schedule for follow-up

Different developmental issues emerge at different age, as shown in the figure. Thus, the assessment should also include the tests that address the given issue. Developmental assessments are not very useful before 6 months of corrected age. Commonly, complete neurodevelopmental assessments should be performed at 4–6 months, 8 months, 12 months, and 18–24 months corrected. After 18 months, 6 monthly follow-up until 3 years and annually thereafter is required. This has to be continued until 2 years after school entry. Sometimes extended neurodevelopmental follow-up is required to detect early but low severity disabilities and disorders as many of these are evident only after infancy. For example, in preterm babies, there might be some transient neurologic abnormalities before 1 year of age, which might improve over time (Chaudhari et al., 2010); however, other neurologic manifestations in high-risk babies such as asphyxia might emerge over time (Drillien, 1972).

When babies are discharged, detailed discharge summary must be sent with a schedule for neurodevelopmental follow-up. A local pediatrician can perform simple developmental screening during the visits. Screening tests cannot be relied upon completely in all cases. When screening test suggests of delay or presence of other abnormal neurologic findings or there are concerns about a child's development, motor tone, or movement patterns, the child should be referred for elaborative neurodevelopmental assessments.

In such cases, the multidisciplinary approach is ideal. The multidisciplinary team should include pediatrician, orthopedician, psychologist, ophthalmologist, audiologist, speech therapist, physiotherapist, occupational therapist, etc.

6 months' corrected age

By this age medical problems such as issues of chronic lung disease are usually resolved, primitive reflexes should have disappeared and the child's motor development and tone and use of hands for reach and grasp can be evaluated. Sensorineural impairments must be ruled out by appropriate tests and referrals.

12 months' corrected age

At this age, the cognitive and language development starts emerging. The environmental factors are less influential during this time. Different motor and cognitive functions are intertwined. There is variability of time for developmental acquisition. Some transient neurological abnormalities in the first year of life may improve later.

18–24 months' corrected age

At this age, the test results on cognitive and motor abilities are influenced by the environmental factors. The early school-age performance can even be predicted as language and reasoning skills are developing at this stage.

3–4 years

At 3–4 years, preacademic skills that are early indicators of executive functions and visual-motor integrative abilities can be assessed. At this age, verbal and nonverbal skills can be better differentiated. The test results are more influenced by socioeconomic status and social support beyond 3 years of age.

6 years

By this attention problems and school achievements can be well tested.

8 years

At 8 years or when children are at grade 3, intelligence, neuropsychological function, learning disabilities, school performance, and behavioral adjustment can be adequately assessed. The intelligence quotient (IQ) measured at 8 years predicts later IQ more accurately than at earlier ages.

Summary points

- The high-risk infants refer to newborns exposed to high-risk factors during pregnancy, delivery, or after birth.
- Almost 80% of high-risk preterm babies are born in low- and middle-income countries of South-East Asian and sub-Saharan African regions.
- About 10% of babies born asphyxiated will die before 5 years of age.
- High-risk babies are at risk of adverse neurodevelopmental outcome such as developmental delay, seizure, hearing and visual impairments, poor academic achievements, etc.
- Regular and detailed follow-up is mandatory for any high-risk babies for early identification of any delay and intervention at the earliest to minimize the adverse outcome.

References

Aarnoudse-Moens, C. S. H., Weisglas-Kuperus, N., van Goudoever, J. B., & Oosterlaan, J. (2009). *Meta-analysis of neurobehavioral outcomes in 728.*

Ahearne, C. E., Boylan, G. B., & Murray, D. M. (2016). Short and long term prognosis in perinatal asphyxia: An update. *World Journal of Clinical Pediatrics, 5*(1), 67.

American Academy of Pediatrics. (2004). Follow-up care of high-risk infants. *Pediatrics, 114*(Suppl. 5), 1377–1397.

Benzies, K. M., Magill-Evans, J. E., Hayden, K. A., & Ballantyne, M. (2013). Infants and their parents: a systematic review and meta-analysis. *BMC Pregnancy and Childbirth, 13*(1), S10.

Bettes, B. A. (1988). *Maternal depression and motherese: Temporal and intonational features* (pp. 1089–1096). Child development.

Bieleninik, Ł., & Gold, C. (2014). Early intervention for premature infants in neonatal intensive care unit. *Acta Neuropsychologica, 12*(2).

Blencowe, H., Cousens, S., Chou, D., Oestergaard, M., Say, L., Moller, A.-B., & Lawn, J. (2013). Born too soon: The global epidemiology of 15 million preterm births. *Reproductive Health, 10*(1), S2.

Bryce, J., Boschi-Pinto, C., Shibuya, K., Black, R. E., & WHO Child Health Epidemiology Reference Group. (2005). WHO estimates of the causes of death in children. *The Lancet, 365*(9465), 1147–1152.

Chaudhari, S., Bhalerao, M., Chitale, A., Patil, B., Pandit, A., & Hoge, M. (2010). Transient tone abnormalities in "high risk" infants and cognitive outcome at five years. *Indian Pediatrics, 47*(11), 931–935.

Chomba, E., Carlo, W. A., McClure, E. M., Basini, F., Wright, L., Mpabalwani, E., & Wallander, J. L. (2011). Feasibility of implementing an early intervention program in an urban low- income setting to improve neurodevelopmental outcome in survivors following birth asphyxia. Field Actions Science Reports *The Journal of Field Actions, 5*.

Cioni, G., Bartalena, L., Biagioni, E., Boldrini, A., & Canapicchi, R. (1992). Neuroimaging and functional outcome of neonatal leukomalacia. *Behavioural Brain Research, 49*(1), 7–19.

De Haan, T., Langeslag, J., van der Lee, J., & van Kaam, A. (2018). A systematic review comparing neurodevelopmental outcome in term infants with hypoxic and vascular brain injury with and without seizures. *BMC Pediatrics, 18*(1), 147.

Dodd, V. L. (2005). Implications of kangaroo care for growth and development in preterm infants. *Journal of Obstetric, Gynecologic, and Neonatal Nursing, 34*(2), 218–232.

Drillien, C. (1972). Abnormal neurologic signs in the first year of life in low-birthweight infants: Possible prognostic significance. *Developmental Medicine and Child Neurology, 14*(5), 575–584.

Ehsanul, K., Rahman, M., & Fatema, K. (2018). Effectiveness of developmental therapy in children with cerebral palsy. *Asia-Pacific Journal of Intellectual Disabilities, 5*(2), 64–74.

Ellis, M., Manandhar, N., Shrestha, P. S., Shrestha, L., Manadhar, D. S., & de L Costello, A. M. (1999). Outcome at 1 year of neonatal encephalopathy in Kathmandu, Nepal. *Developmental Medicine and Child Neurology, 41*, 689–695 [PubMed: 10587046].

Feldman, R., & Eidelman, A. I. (2003). Direct and indirect effects of breast milk on the neurobehavioral and cognitive development of premature infants. *Developmental Psychobiology: The Journal of the International Society for Developmental Psychobiology, 43*(2), 109–119.

Field, T., Healy, B. T., Goldstein, S., & Guthertz, M. (1990). Behavior-state matching and synchrony in mother-infant interactions of nondepressed versus depressed dyads. *Developmental Psychology, 26*(1), 7.

Fischer, N., Steurer, M. A., Adams, M., & Berger, T. M. (2009). Survival rates of extremely preterm infants (gestational age< 26 weeks) in Switzerland: Impact of the Swiss guidelines for the care of infants born at the limit of viability. *Archives of Disease in Childhood - Fetal and Neonatal Edition, 94*(6), F407–F413.

Godbole, K., Barve, S., & Chaudhari, S. (1997). Early predictors of neurodevelopmental outcome in high risk infants. *Indian Pediatrics, 34*, 491–496.

Johnston, M. V. (2009). Plasticity in the developing brain: Implications for rehabilitation. *Developmental disabilities research reviews, 15*(2), 94–101.

Jones, G., Steketee, R. W., Black, R. E., Bhutta, Z. A., Morris, S. S., & Bellagio Child Survival Study Group. (2003). How many child deaths can we prevent this year? *The Lancet, 362*(9377), 65–71.

Lawn, J. E., Cousens, S., Zupan, J., & Team, L. N. S. S. (2005). 4 million neonatal deaths: When? Where? Why? *The Lancet, 365*(9462), 891–900.

Loewy, J., Stewart, K., Dassler, A.-M., Telsey, A., & Homel, P. (2013). The effects of music therapy on vital signs, feeding, and sleep in premature infants. *Pediatrics, 131*(5), 902–918.

Messmer, P. R., Rodriguez, S., Adams, J., Wells-Gentry, J., Washburn, K., Zabaleta, I., & Abreu, S. (1997). Effect of kangaroo care on sleep time for neonates. *Pediatric Nursing, 23*(4), 408–415.

Mundkur, N. (2005). Neuroplasticity in children. *Indian Journal of Pediatrics, 72*(10), 855–857.

Murray, C. J., Vos, T., Lozano, R., Naghavi, M., Flaxman, A. D., Michaud, C., & Abdalla, S. (2012). Disability-adjusted life years (DALYs) for 291 diseases and injuries in 21 regions, 1990–2010: A systematic analysis for the global burden of disease study 2010. *The Lancet, 380*(9859), 2197–2223.

Petrini, J. R., Dias, T., McCormick, M. C., Massolo, M. L., Green, N. S., & Escobar, G. J. (2009). Increased risk of adverse neurological development for late preterm infants. *The Journal of Pediatrics, 154*(2), 169–176. e163.

Robertson, C. M., & Finer, N. N. (1993). Long-term follow-up of term neonates with perinatal asphyxia. *Clinics in Perinatology, 20*(2), 483–499.

Simon, N. P. (1999). Long-term neurodevelopmental outcome of asphyxiated newborns. *Clinics in Perinatology, 26*(3), 767–778.

Shrestha, M., Bajracharya, L., & Shrestha, L. (2017). Neurodevelopmental outcome of high risk babies at one year of age born in a Tertiary Centre. *Journal of Nepal Paediatric Society, 37*(1).

Standley, J. (2012). Music therapy research in the NICU: An updated meta-analysis. *Neonatal Network, 31*(5), 311–316.

VandenBerg, K. A. (2007). Individualized developmental care for high risk newborns in the NICU: A practice guideline. *Early Human Development, 83*(7), 433–442.

Wallander, J. L., Bann, C. M., Biasini, F. J., Goudar, S. S., Pasha, O., Chomba, E., & Carlo, W. A. (2014). Development of children at risk for adverse outcomes participating in early intervention in developing countries: A randomized controlled trial. *Journal of Child Psychology and Psychiatry, 55*(11), 1251–1259.

Chapter 5

Genomic imprinting and neurodevelopment

Inés López del Castillo, Anna Lozano-Ureña and Sacri R. Ferrón
Department of Cell Biology/ERI BiotecMed, University of Valencia, Burjassot, Spain

List of abbreviations

5hmC 5′hydroxymethylcytosine
5mC 5′methylcytosine
adSVZ adult subventricular zone
AS Angelman syndrome
ASD autism spectrum disorder
CSC cancer stem cell
DMR differentially DNA-methylated region
ICR imprinting control region
IPC intermediate progenitor cell
LOI loss of imprinting
mdDA midbrain dopamine
NB neuroblasts
NE neuroepithelial cells
NSC neural stem cell
OB olfactory bulb
PGC primordial germ cell
PWS Prader−Willi syndrome
RG radial glia
RMS rostral migratory stream
SGZ subgranular zone
SVZ subventricular zone
VZ ventricular zone

Mini-dictionary of terms

Biallelic when both copies of a particular gene are expressed. Most genes are biallelic.
Epigenetics refers to the molecular processes that can affect gene expression independently of DNA sequence.
Genomic imprinting epigenetic process that reversibly marks one of the two alleles of a gene during development causing them to be expressed depending on their parental origin. Genomic imprinting is implicated in the control of gene dosage.
Human imprinting syndromes Developmental human diseases involving alterations of genomic imprinting.
Methylation process by which a methyl group is added to the cytosine in the DNA. DNA methylation participates in the regulation of gene expression.
Methyltransferase enzyme that catalyses the transfer of a methyl group to DNA.
Monoallelic when only one copy of a particular gene is expressed and the other is silenced or repressed. Only a set of genes known as imprinted genes are monoallelically expressed.

Diagnosis, Management and Modeling of Neurodevelopmental Disorders. https://doi.org/10.1016/B978-0-12-817988-8.00005-1

Neural stem cell self-renewing and multipotent cells that during the embryonic development generate radial glial progenitors to give rise to neurons and glia of the nervous system and that in adult give rise to neurons, oligodendrocytes, and astrocytes during the entire life of the organisms.

Neurogenesis process by which new neurons are formed in the brain. Neurogenesis is crucial during brain development but also continues in certain brain regions after birth and throughout life span.

Reprogramming process of erasure of epigenetic marks, such as DNA methylation and histones modifications, during mammalian development.

Introduction

The development of the mammalian brain is a highly complex process that begins during the embryonic period and extends throughout adulthood. Neurons and glia arise from neural stem cells (NSCs) and undergo further differentiation and migration to acquire their final fate. These processes of neurogenesis and gliogenesis require a precise control of gene transcription (Obernier & Alvarez-Buylla, 2019). For that, epigenetic regulation modulates gene expression and gene dosage control through several mechanisms, including genomic imprinting (Yao & Jin, 2014). Genomic imprinting is an epigenetic phenomenon that restricts gene expression in a parent-of origin manner, silencing either the maternally or the paternally inherited allele and resulting in monoallelic expression of the gene (Fig. 5.1) (Yao & Jin, 2014). Genomic imprinting is tissue and developmental-stage specific, and selective loss of imprinting has also been observed, suggesting a key role in gene dosage fine-tuning (Ferron et al., 2011, 2015). Expression of imprinted genes is highly prevalent in the brain, where they show a critical function in the regulation of pre- and postnatal development (Perez, Rubinstein, & Dulac, 2016). Therefore, abnormalities in genomic imprinting are associated with the development of neurological disorders including imprinting syndromes, such as Prader–Willi (PWS) and Angelman (AS), and some human cancer types, such as glioma (Lim & Maher, 2010).

Here, we provide a comprehensive review of the bases of neurodevelopment and genomic imprinting, discussing how imprints are established and maintained and their dynamic regulation. We also gather the current knowledge about the contribution of imprinted genes to differentiation and maintenance of the NSCs pool, in both the embryonic and adult neurogenic niches. Finally, we also review the pathological phenotypes associated with genetic mutations or loss of imprinting at several loci in order to elucidate the impact of imprinted genes in health and disease of the human brain.

Genomic imprinting and gene dosage control

Diploid organisms have two copies of each gene, one homologue allele from each parent. During mammalian development, the vast majority of genes are expressed or repressed from both alleles. Nonetheless, few genes, known as *imprinted genes*, are differently expressed or silenced from either the maternal or the paternal homologue, resulting functionally monoallelic (Ferguson-Smith, 2011) (Fig. 5.1). This process, known as *genomic imprinting*, is crucial for normal mammalian development pre- and postnatally. Until now, around 150 imprinted genes have been identified in the mouse genome (listed in http://www.mousebook.org/mousebook-catalogs/imprinting-resource) and even though there are some species-specific imprinted genes, most of known imprinted genes are conserved between mouse and human (Babak et al., 2015; Tucci, Isles, Kelsey, & Ferguson-Smith, 2019). The establishment and maintenance of genomic imprints are not

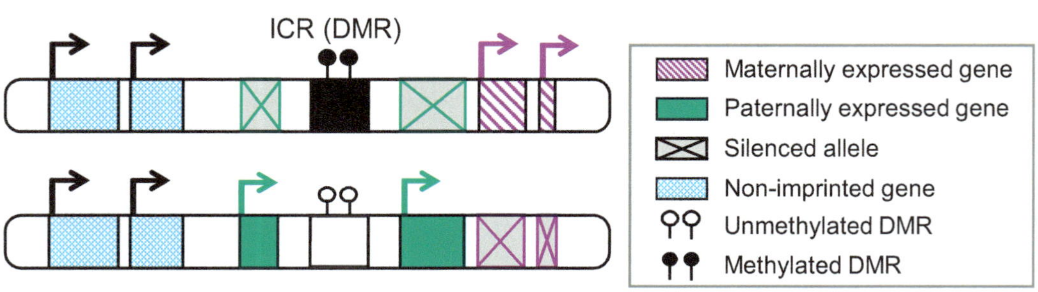

FIGURE 5.1 **Epigenetic control of imprinted genes.** Maternally and paternally imprinted genes are generally located together in clusters. Parental-specific expression of each imprinted gene is achieved by differentially DNA-methylated regions (DMRs) on both homologues named as imprinting control regions (ICR). While nonimprinted genes show the same epigenetic modifications on both chromosomes and are expressed or silenced biallelically (blue boxes), imprinted genes (purple and green boxes) are monoallelically expressed due to DMRs (black and white boxes). Black circles show methylation and white circles show unmethylation.

based on changes on the DNA sequence, but on inheritable epigenetic marks. Epigenetics consists of regulatory mechanisms that are essential to maintain a specific gene expression in each cell type during development (Bird, 2007). These mechanisms include chemical modifications to DNA such as methylation, posttranslational modifications of histones (the proteins around which the DNA is wrapped), and noncoding RNAs (ncRNAs). As a result of these epigenetic modifications, the appearance and exposure of the DNA to transcription factors and other DNA-binding proteins vary, regulating gene expression. The two parental alleles of a nonimprinted gene show the same epigenetic landscape, being both homologues equally expressed or repressed. However, monoallelic expression of imprinted genes is achieved by differential epigenetic regulation between the maternal and the paternal homologues (Adalsteinsson & Ferguson-Smith, 2014).

Imprinted genes are not uniformly distributed across the genome but grouped together in clusters, even though some singleton imprinted genes have been described (Adalsteinsson & Ferguson-Smith, 2014). These clusters commonly include maternally and paternally expressed and silenced genes and, at least, one ncRNA. Parental-specific pattern of expression is defined by differentially DNA-methylated regions (DMRs) on both parental homologues. Each cluster of imprinted genes has several DMRs, which are coregulated by a cis-acting imprinting control region (ICR) (Adalsteinsson & Ferguson-Smith, 2014) (Fig. 5.1). ICRs differential methylation results in establishing and controlling the imprints of the entire cluster (Adalsteinsson & Ferguson-Smith, 2014).

Establishment of imprints during development

The establishment of imprints takes place in the germline, and they are maintained throughout development and adult life (Ferguson-Smith, 2011). However, some DMRs acquire their differential methylation pattern after fertilization, somatic DMRs, being directly dependent on the methylation of ICRs established in the germline (Fig. 5.2). Previous to gametogenesis, during proliferation and migration of the primordial germ cells (PGCs) toward the genital ridge, an erasure of DNA methylation at a genome-wide scale takes place (Zeng & Chen, 2019). This phenomenon is known as epigenetic reprogramming and allows the establishment of new DNA-methylation patterns, including genomic imprints (Zeng & Chen, 2019). DNA-methylation marks at the ICRs must be reestablished to ensure the sex-dependent imprint during fertilization. Both active and passive demethylation have been shown as mediators in this process of erasure (Kawasaki et al., 2014). Ten-eleven translocation (TET) enzymes are implicated in passive demethylation by oxidation of

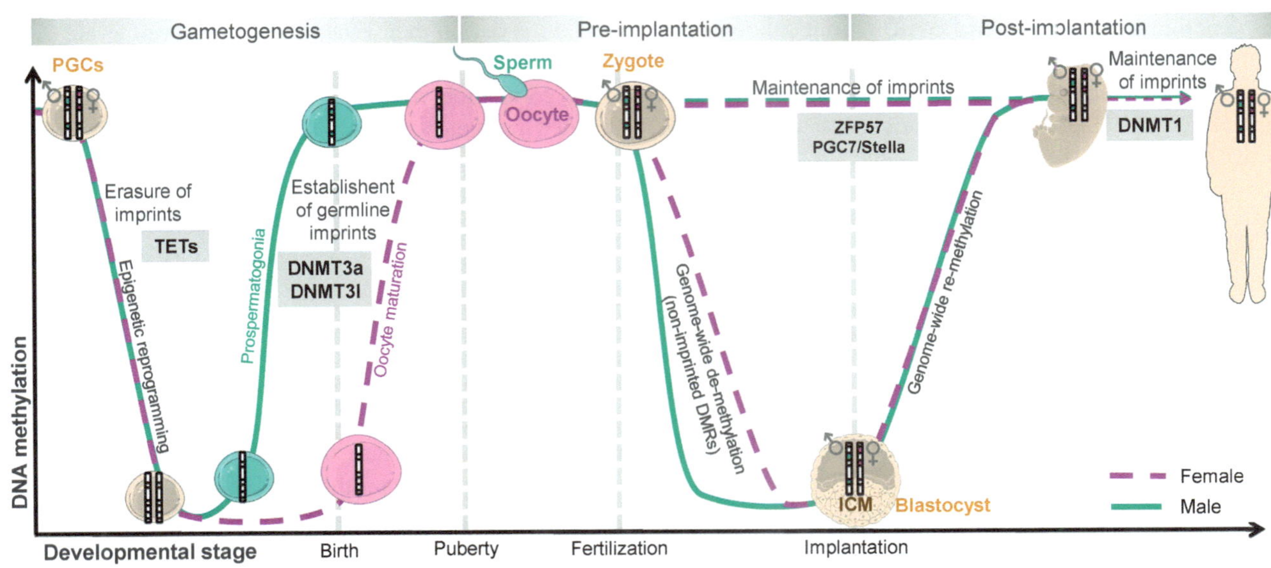

FIGURE 5.2 **DNA methylation dynamics throughout development and life span.** Before gametogenesis, during proliferation and migration of primordial germ cells (PGCs), DNA methylation marks are erased on a genome-wide scale. This process is called epigenetic reprogramming. New DNA-methylation patterns, including genomic imprints, are established at different time points in male and female developing embryos. In males, this occurs during prospermatogonia and completes before birth; while in females, it takes place postnatally, during oocyte growth. The DNA methyltransferases DNMT3a and DNMT3l are essential for the establishment of genomic imprints. Maternal and paternal imprinted genomes are passed to the zygote after fertilization reconstituting asymmetrical methylation of ICRs. There is another epigenetic reprogramming postfertilization. However, the factors ZFP57 and PGC7/Stella protect imprinted DMRs from this second round of demethylation. The methyltransferase DNMT1 plays an important role in imprint maintenance and heritability. ICM: Inner Cell Mass. Females genome in pink and male genome in green.

5-methylcytosine (5mC) to 5-hydroxymethylcytosine (5hmC). TET family enzymes include three members, TET1, TET2, and TET3, being *Tet1* and *Tet2* expressed in PGCs (Vincent et al., 2013). The protein TET1 has been reported to play a critical role in the erasure of genomic imprints, by removing remaining methylation at a late reprogramming stage (Yamaguchi, Shen, Liu, Sendler, & Zhang, 2013). However, further research is required to fully understand the demethylation process of imprinted DMRs.

De novo DNA-methylation at the ICRs occurs at different time points in male and female developing embryos (Fig. 5.2). In males, it takes place during prospermatogonia, before meiosis, and completes before birth. On the contrary, in oogenesis, new methylation patterns are established postnatally, during oocyte growth (Smallwood & Kelsey, 2012). Furthermore, the location of paternal and maternal germline DMRs is also different; DMRs methylated during oogenesis are located in promoters of protein-coding or noncoding genes, while those methylated during prospermatogonia are found in intergenic areas. In mammals, DNA methylation occurs in CpG dinucleotides, and it is mediated by three functional methyltransferases: DNMT3a, DNMT3b, and DNMT3l (Fig. 5.2). While DNMT3a and DNMT3b are essential for de novo methylation and normal mammalian development, only DNMT3a seems to be indispensable in the establishment of maternal and paternal genomic imprints (Kaneda et al., 2004). DNMT3l, which is similar in sequence to DNMT3a and DNMT3b, lacks some of the catalytic motifs. However, it colocalizes and interacts with DNMT3a and DNMT3b in the nucleus, being essential for the establishment and appropriate expression of maternally imprinted genes (Hata, Okano, Lei, & Li, 2002). After fertilization, maternal and paternal imprinted genomes are passed to the zygote reconstituting asymmetrical methylation of ICRs (Hata et al., 2002).

A second round of genome-wide DNA demethylation takes place upon fertilization; in the zygote, the paternal genome is actively demethylated, while the maternal genome demethylates passively over cell division (Albert & Peters, 2009). Nonetheless, imprinted DMRs are resistant to this event thanks to the factors KRAB-zinc finger protein 57 (ZFP57) and PGC7/Stella, which is required to protect maternal and paternal imprints from demethylation after fertilization (Li et al., 2008) (Fig. 5.2). As most members of the KRAB-ZFP family, it functions as a transcriptional repressor by interacting with the KAP-1/TIF1β complex through its KRAB domain. In turn, KAP-1 recruits epigenetic repressive factors, such as DNA-methylation factors, histone deacetylases, and histone methyltransferases (Schultz, Ayyanathan, Negorev, Maul, & Rauscher, 2002).

A third critical factor that participates in somatic maintenance and heritability of genomic imprints is DNMT1. Due to its strong affinity for hemimethylated DNA molecules, DNMT1 is able to recognize and methylate the new-synthesized DNA strand (Hirasawa et al., 2008). An oocyte-specific promoter of *Dnmt1* expresses an isoform named DNMT1o in oocytes and preimplantation embryos. It is located upstream the somatic and spermatocyte promoter, and its removal results critical in the maintenance of imprints, though no effect is observed on their establishment (Howell et al., 2001). Differential distribution of histone modifications between the two parental alleles has been reported at imprinted regions. Interestingly, specific combinations of histone modifications show correlation with different types of repression. While H3K9me3 and H4K20me3 are enriched at differentially methylated promoters of imprinted genes, the enrichment of H3K27me3 and H3K4me3 on the same allele is correlated with developmental repression (McEwen & Ferguson-Smith, 2010). Notably, all ICRs show a specific epigenetic signature comprising a histone "tri-mark": H3K4me3, H3K9me3, and H4K20me3 (McEwen & Ferguson-Smith, 2010). Nevertheless, there are many unanswered questions concerning the role of histone modifications in imprinting and whether they are a cause or a consequence of monoallelic expression.

Neural stem cells and neurodevelopment

The development of the mammalian telencephalon commences in early embryo stages with the formation of the neural plate that closes on the midline forming the neural tube. After gastrulation and the formation of the three primordial layers, the first neural progenitors appear, and shortly after, neurogenesis begins. Neurogenesis is the process by which new neurons are generated by NSCs or progenitor cells, and it takes place both pre- and postnatally (Stiles & Jernigan, 2010). The first neural progenitors in the neural tube are called neuroepithelial cells (NEs). In order to increase the size of the pool of neural progenitors, NEs divide symmetrically originating two identical progenitors in every round of division (Rakic, 1995). Around embryonic day (E) 10 (E10) in mouse development (E42 in humans) (Bystron, Blakemore, & Rakic, 2008), NEs start dividing asymmetrically, producing one progenitor and one neuron, and initiating the process of neurogenesis (Kriegstein & Alvarez-Buylla, 2009) (Fig. 5.3). At this time point, they start expressing the intermediate filament Nestin and the antigen recognized by the mouse-specific antibody RC2 (Edwards, Yamamoto, & Caviness, 1990). Shortly after the beginning of neurogenesis, NE cells suffer a transition in their cellular and molecular identities acquiring an astroglial phenotype (Kriegstein & Alvarez-Buylla, 2009). At this developmental stage, cells named as radial glial cells (RGs) are the primary neuron and glia progenitors in the developing brain (Alvarez-Buylla, Garcia-Verdugo, & Tramontin, 2001).

Importantly, apicobasal bipolar morphology is conserved in both NE and RG, which stretch from the ventricular lumen to the pial surface and have their soma lying on the ventricular zone (VZ) (Alvarez-Buylla et al., 2001) (Fig. 5.3). In addition, both cell types show interkinetic nuclear migration, a movement of the nucleus between the basal and the apical surface, which is synchronized with the phases of the cell cycle (Del Bene, 2011). Glial molecular markers, such as GLAST (astrocyte-specific glutamate transporter), brain lipid-binding protein (BLBP), and Tenascin C (TN-C), start to appear at E12 in the developing mouse telencephalon (Kriegstein & Alvarez-Buylla, 2009).

RGs are a heterogeneous population comprising multipotent stem cells and different lineage-restricted progenitors. Its progeny not only includes neurons, but also astrocytes, oligodendrocytes, postnatal ependymocytes, and adult NSCs (Fuentealba et al., 2015; Spassky et al., 2005). In each round of division, RGs self-renew producing one RG and also an immature neuron or an intermediate progenitor cell (IPC). Thus, RGs originate neurons both directly in the VZ and indirectly through IPCs, which migrate and divide in the subventricular zone (SVZ). The SVZ zone is a proliferative region, which overlaps the basal VZ and is mainly populated by IPCs (Kriegstein & Alvarez-Buylla, 2009), multipolar cells that do not show interkinetic nuclear migration and mainly divide symmetrically to produce two neurons or two new IPCs (Kriegstein & Alvarez-Buylla, 2009). RGs not only function as stem cells, their long pial process serves as scaffold to guide migrating neurons to the developing cortex (Hatten, 1999). Importantly, RGs transform postnatally into astrocytes, after neurogenesis and neuronal migration when they lose their ventricular attachment, develop branched processes, and migrate toward the cortical plate acquiring the unipolar organization of astrocytes (Culican, Baumrind, Yamamoto, & Pearlman, 1990).

Between E13.5 and E15.5 of murine development, a subset of RGs, which is VCAM-1 positive (Fuentealba et al., 2015), slows down their division rate until acquiring a quiescent state (Fuentealba et al., 2015). These quiescent cells will get reactivated at postnatal stages, serving as adult NSCs. Moreover, a second population of RGs, which slows its proliferation rate at E17.5 and shares molecular identity with quiescent adult NSCs, has been identified (Alvarez-Buylla et al., 2001). Thus, adult NSCs show astroglial morphology (Doetsch, Caille, Lim, Garcia-Verdugo, & Alvarez-Buylla, 1999) and express the same set of astroglial markers as RGs (Alvarez-Buylla et al., 2001). This is consistent with the idea that quiescence of adult NSCs is necessary for their long-term maintenance, preventing exhaustion due to excess division. It makes sense that the embryonic origin of adult NSCs would lie in cells that become quiescent early during development to reduce the number of times they divide.

Imprinted genes regulation during brain development

Genomic imprinting is essential for normal mammalian growth and shows a tissue and developmental-stage specific pattern (Ferguson-Smith, 2011), suggesting a key role in gene dosage fine-tuning during development. Moreover, the prevalence of genomic imprinting is higher in the brain than in other organs, and many genes with a crucial role in neurodevelopment are expressed in a maternal or paternal specific manner (Perez et al., 2016). For instance, the *MiR379-410* cluster, located on the *Dlk1-Dio3* region, contains more than 40 microRNAs (miRs) genes, all of them maternally expressed in mouse (Kircher, Bock, & Paulsen, 2008), and is regulated by an intergenic DMR (IG-DMR) ~200 kb upstream the cluster (Seitz et al., 2004) (Fig. 5.3). During early embryonic development, RGs are interconnected by N-cadherin-mediated adherens junctions, which mediates calcium-dependent cell adhesion and inhibits RG differentiation by activating β-catenin-signalling (Zhang et al., 2010). The maternally expressed miRs, miR369-3p, miR496, and miR543 of the *MiR379-410* cluster (Seitz et al., 2004) bind to the 3′-untranslated region (UTR) of N-cadherin and repress its expression promoting cell differentiation and migration of NSCs (Rago, Beattie, Taylor, & Winter, 2014). In addition, miR134, from this same imprinting cluster, promotes cell differentiation and reduces cell migration and dendritogenesis in vivo (Gaughwin, Ciesla, Yang, Lim, & Brundin, 2011) (Fig. 5.3).

The maternally expressed ZFP *Zac1* (also known as *Plagl1*) is located at an imprinted region on mouse chromosome 10 (human chromosome 6q24). ZAC1 is highly present in RG and colocalizes with Nestin and GFAP (Valente, Junyent, & Auladell, 2005); it regulates a wide cohort of imprinted genes. For example, elevated in vivo levels of this gene promote cell cycle exit through induction of the maternally expressed *Cdkn1c* gene in the brain (Rraklli, Sodersten, Nyman, Hagey, & Holmberg, 2016) (Fig. 5.3). This is accompanied by hindered neurogenesis and neuronal differentiation, provoking a partial switch from neural to mesodermal/myogenic lineage with additional expression of pluripotency genes (Rraklli et al., 2016). On the contrary, loss or lower levels of *Zac1* expression affect migration of cortical neurons (Adnani et al., 2015). In turn, *Cdkn1c*, which is absolutely required for embryo and early life survival (Yan, Frisén, Lee, Massagué, & Barbacid, 1997), promotes NSC cell cycle arrest and proglial differentiation by repressing proneural factors, such as *Ascl1* (achaete-scute complex homolog 1, also known as *Mash1*) (Joseph et al., 2009). CDKN1C (also known as p57) and the maternally biased Neurabin-I (encoded by PPP1R9A) are also required for actin polymerization, which is critical for cell motility within the cortical plate (Tury, Mairet-Coello, & DiCicco-Bloom, 2012) (Fig. 5.3).

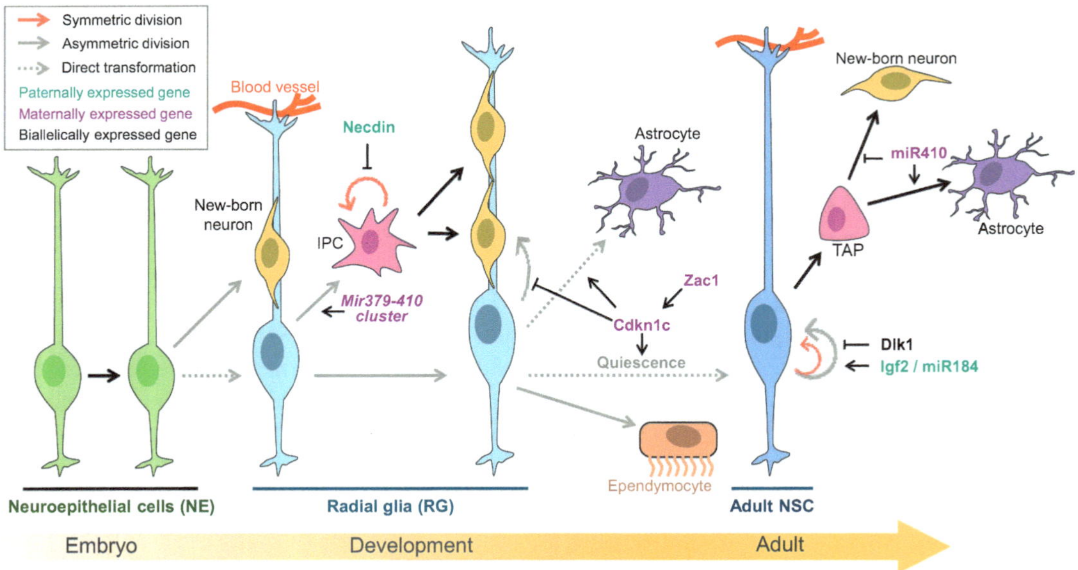

FIGURE 5.3 **Imprinted genes are essential for neurodevelopment and neural stem cell maintenance.** Neuroepithelial cells (NEs) divide symmetrically increasing the pool of neural progenitors. Around embryonic day (E) 10 (E10), NEs start dividing asymmetrically, initiating neurogenesis. Shortly after, NEs acquire astroglial phenotype and become radial glia (RG). RG divide asymmetrically to self-renew and give rise to intermediate progenitors (IPCs), neurons, astrocytes, oligodendrocytes, ependymocytes, and adult neural stem cells (NSCs). Two subsets of RG enter quiescence around E13.5-15.5 and E17.5 respectively, until adulthood when they get reactivated serving as adult NSCs. The maternally expressed *Cdkn1c* regulates quiescence entry by promoting cell cycle arrest. RG also serve as scaffold for migrating neurons and, finally, transform into astrocytes postnatally. Several imprinted genes regulate differentiation and proliferation of neural progenitors pre- and postnatally. The paternally expressed genes *Igf2* and *Dlk1* lose their imprints, acquiring biallelic expression, in order to fine-tune gene dosage in adult NSC niches.

Similarly, *Grb10* (*Meg1*, maternally expressed gene 1) is a downstream effector of LMX1A, a very potent inducer of stem cells toward dopaminergic neurons; thus, playing an important role in the differentiation of mdDA neurons (Hoekstra et al., 2013). *Grb10* is a tissue-specific imprinted gene. In most mouse tissues, it is transcribed from a maternally expressed major promoter; however, in the brain it is also, and predominantly, expressed from a paternally expressed neuronal-specific promoter located downstream the major promoter. Interestingly, in humans *GRB10* is biallelically expressed in most tissues, but, in the fetal brain, it is paternally expressed from the same neuronal-specific promoter as in mouse (Hikichi, Kohda, Kaneko-Ishino, & Ishino, 2003). In addition, the overgrowth observed in most tissues in heterozygous *Grb10* mutant mice that inherited the paternal deletion is not observed in the central nervous system (Hikichi et al., 2003). This suggests a *Grb10* gene dose fine-tuning through imprinting modulation in the brain, although its neuronal function is not fully understood.

Finally, the paternally expressed *Necdin* (*Ndn*) is highly expressed in postmitotic neurons and moderately in the VZ of the developing murine neocortex (Kuwajima, Hasegawa, & Yoshikawa, 2010). It suppresses transcriptional activation of Cdk1 and serves as a suppressor of proliferation of intermediate progenitors in the embryonic neocortex. In addition, Necdin promotes neocortical GABAergic interneurons tangential migration during development (Kuwajima et al., 2010). Thus, selective regulation of imprinting seems to be a normal mechanism for modulating gene dosage to control stem cell potential in brain development.

Genomic imprinting in postnatal neurogenesis

Neurogenesis continues postnatally in discrete regions of the mammalian brain. The subgranular zone (SGZ) of the hippocampus that generates excitatory neurons, which integrate into the granule cell layer (Kempermann, Song, & Gage, 2015), and the adult subventricular zone (adSVZ) of the lateral ventricles, which generates inhibitory neurons that migrate through the rostral migratory stream (RMS) to the olfactory bulb (OB) where they differentiate and integrate into the existing circuitry (Lim & Alvarez-Buylla, 2016). Neurogenesis has also been suggested in hippocampus of the adult human brain (Moreno-Jimenez et al., 2019); however, it remains a controversial area of research among neuroscientists.

In the adSVZ, the process of neurogenesis is continually sustained by the activity of NSCs, also known as B1 cells, which are characterized by their ability to balance self-renewal with multipotential differentiation into astrocytes, oligodendrocytes, and neurons (Lim & Alvarez-Buylla, 2016). B1 cells have astroglial features and retain expression of Nestin,

BLBP, GLAST, and the transcription factor Sox2 (Sry-related HMG box 2) markers that were also expressed in RG (Obernier & Alvarez-Buylla, 2019). Indeed, B1 cells have a long basal process contacting blood vessels (Tavazoie et al., 2008) and extend an apical process ending in a primary cilium that protrudes into the ventricle; cytoarchitecture that has been suggested to be an epithelial apicobasal organization reminiscent of RG morphology (Mirzadeh, Merkle, Soriano-Navarro, Garcia-Verdugo, & Alvarez-Buylla, 2008), consistently with the embryonic origin of adult NSCs. Activated and quiescent NSCs appear to coexist in the adult SVZ (Chaker, Codega, & Doetsch, 2016). Once activated, NSCs give rise to transit-amplifying progenitors or type C cells, which in turn generate neuroblasts (NB) or type A cells that migrate anteriorly via the RMS to the OB, where they mature into functional neurons involved in olfactory discrimination (Bond, Ming, & Song, 2015). Although less frequently, B1 cells also generate a small number of oligodendrocyte precursor cells and mature oligodendrocytes that migrate to the *corpus callosum* and striatum where they differentiate into myelinating and nonmyelinating oligodendrocytes (Menn et al., 2006). NSCs in the SGZ are also called type I progenitors, show a long radial process, which extends through the granular layer of the hippocampus, and are positive for Sox2 and Nestin. Type I cells originate type II cells, which express the proneural gene *Ascl1* (achaete-scute complex homolog 1, also known as Mash1) and the T-box gene-2 (*Tbr2*), a molecular marker for progenitor cells. Type II cells, in turn, give rise to neuronal precursors (type III cells), which are positive for doublecortin (DCX), a marker for immature migrating neurons (Goncalves, Schafer, & Gage, 2016).

Recent evidence suggests that the genomic imprinting process can be selectively lost or "switched off" in these adult neurogenic niches to activate the allele that is usually repressed (Ferron et al., 2011, 2015). For example, in the adult SVZ, a soluble isoform of the atypical NOTCH ligand delta-like homologue 1 (DLK1) is secreted by niche astrocytes and signals through a membrane-tethered alternative isoform of DLK1, expressed by NSCs, to maintain NSC self-renewal and regulate postnatal neurogenesis (Ferron et al., 2011). *Dlk1* is a single gene located on mouse chromosome 12 (human chromosome 14), which encodes both, soluble and membrane, isoforms. Notably, *Dlk1*, which is a canonically imprinted gene elsewhere in the brain, shows a selective absence of imprinting in adult NSCs, requiring expression from both maternally and paternally inherited chromosomes. This is associated with postnatal acquisition of DNA methylation at the IG-DMR of the *Dlk1-Dio3* imprinted cluster and seems to be a postnatal and gene-specific imprinting relaxation (Ferron et al., 2011).

Similarly, the modulation of insulin-like growth factor 2 (*Igf2*) imprinted gene is crucial for the regulation of NSCs in the adult SVZ and SGZ niches. SGZ NSCs rely on monoallelic expression from the paternally inherited allele to maintain IGF2 concentration at the site of action in autocrine signaling. However, biallelic expression of *Igf2* in the choroid plexus, the leptomeninges, and the vasculature of the adult brain is required to maintain a sufficiently high concentration of IGF2 at the target SVZ NSCs (Ferron et al., 2015). As in the embryonic brain, biallelic expression regulates NSC self-renewal and homeostasis (Ferron et al., 2015). This demonstrates that *Igf2*, as *Dlk1*, is highly dosage-sensitive and that genomic imprinting is critical in gene dosage control in particular developmental contexts and distinct regulatory mechanisms (Fig. 5.3).

Loss of imprinting and developmental syndromes

The monoallelic expression of imprinted genes makes them very vulnerable to mutation or to loss of imprinting (LOI). Thus, alterations of these genes have been associated with several human pathologies. Imprinted genes are highly and widely expressed during embryonic and fetal development, but largely downregulated after birth, playing an essential role in mammalian development (Perez et al., 2016). Therefore, abnormalities in imprinting are associated with perturbations in growth, placental function, postnatal adaptations, and neurobehavioural processes, causing several human congenital imprinting syndromes (Cleaton, Edwards, & Ferguson-Smith, 2014). Considering that genomic imprinting is highly prevalent in the brain even at postnatal stages, some of these disorders are characterized by cognitive or behavioral impairments (Perez et al., 2016).

AS and PWS syndromes are two distinct complex neurodevelopmental disorders, which map to the same cluster of imprinted genes on human chromosome 15q (Eggermann et al., 2015). AS arises from disruptions that lead to a lack of paternally imprinted *Ube3a* expression and is characterized by severe learning difficulties, ataxia, and abnormal "happy" disposition with inappropriate bouts of laughter (Mabb, Judson, Zylka, & Philpot, 2011). *Ube3a* encodes an E3 ubiquitin ligase, and it is maternally expressed in the cortex, thalamus, olfactory bulb, the hippocampus, and the cerebellum, but biallelically elsewhere in the rest of the brain and other body tissues. Normal silencing of the paternal allele of *Ube3a* is mediated by a large antisense RNA expressed in a paternal-specific manner. The imprinting of *Ube3a* in mice has allowed the generation of several disease models, which recapitulate many AS patient phenotypes. These reveal multiple potential targets of UBE3A, including proteins that regulate neuronal outgrowth, survival, differentiation, synapse, and spine dendrite density (Mabb et al., 2011). On the contrary, PWS results from disruptions on paternally expressed transcripts, including the protein coding *Snrpn* (small nuclear ribonucleoprotein polypeptide N) gene and *Necdin* (Schaaf et al., 2013). However, the precise contribution of these genes to PWS etiology is unknown. PWS is characterized by early hypotonia followed by hyperphagia, mild mental retardation, and hypogonadism. Notably, maternal duplication of the PWS/AS locus is one of the few characterized persistent genetic abnormalities associated with autism spectrum disorder (ASD) (Mabb et al., 2011). Furthermore, truncating mutations in the paternally expressed gene *Magel2* located in PWS/AS cluster have been reported to cause Schaaf–Yang syndrome, a Prader–Willi-like disease that manifests together with ASD (Schaaf et al., 2013). Other imprinted syndromes such as Temple and Kagami–Ogawa, associated with the altered dosage of imprinted genes on human chromosome 14, also exhibit neurological and behavioral impairments (Eggermann et al., 2015; Ioannides, Lokulo-Sodipe, Mackay, Davies, & Temple, 2014). Patients who suffer from pseudohypoparathyroidism type 1A (PHP1A) also show mental retardation. This syndrome is caused by genetic or epigenetic alterations at the imprinted *Gnas* locus, at human chromosome 20q (Eggermann et al., 2015). Beckwith–Wiedemann syndrome (BWS) is another classical human imprinting disorder characterized by prenatal and postnatal overgrowth and variable developmental anomalies and is associated with a hypomethylation defect at the imprinting center 2 (KvDMR1) (Pandita, Gupta, & Gupta, 2018).

Aberrations of imprinting and cancer

Importantly, several epigenetic marks, especially DNA methylation, become altered in tumor progression and lead to changes in gene expression. Thus, biallelic expression of oncogenes or repression of tumor suppressor genes favors the development of cancer. Indeed LOI is highly frequent in a large variety of human tumors being considered a valuable tool for cancer diagnosis and treatment, as it is the most abundant early alteration observed in cancer (Fenner, 2017). It includes activation of the imprinted copy of a gene and silencing of the normally expressed allele. For instance, *IGF2* and *CDKN1C* have been observed to lose their imprints in different human cancers (Fenner, 2017). The fact that LOI most likely precedes tumor formation suggests that it might occur originally in the "cancer stem cell" (CSC) population or part of it which then gives rise to the growing tumor (Galli et al., 2004). Indeed, a CSC theory of gliomas has emerged based on the discovery of NSCs in the adult mammalian brain and the identification of stem-like cells in several brain tumors. This theory states that gliomas originate from the malignant transformation of cells deriving from adult NSCs or lineage-restricted progenitors, instead of from dedifferentiation of mature cells (Galli et al., 2004). For example, *Dlk1*, which is essential for SVZ NSC maintenance, is highly expressed in human glioma and promotes proliferation of glioblastoma cell lines suggesting a role for DLK1 in glioma formation and/or progression (Yin et al., 2006). Besides, paternally expressed gene 3 (*Peg3*), located on human chromosome 19q, is abnormally regulated in glioma cell lines (Otsuka et al., 2009) also suggesting a relevant role for PEG3 in tumorigenesis. Therefore, a deeper comprehension of the regulatory mechanisms, which are common to healthy and cancer neural stem cells and the epigenetic modulation underneath them, including alterations in imprinting, will help to elucidate which changes contribute to the development of gliomas.

Applications to other areas of development

In this chapter we have reviewed the role of imprinted genes on the regulation of cell identity, differentiation, migration, and communication of neural stem cells (NSC) and newborn neurons in the developing and adult brain. Dynamic modulation of genomic imprinting is used as a gene dosage fine-tuning mechanism in NSCs before and after birth, increasing the sophistication of gene expression control during neurodevelopment. The importance of imprinted genes contribution to normal brain function is evidenced by the pathological conditions resulting from mutations or loss of imprinting in these loci; the high prevalence and the precocity of LOI in human cancer are remarkable. Therefore, it is now a priority to unravel the epigenetic profile and mechanisms that control imprints at ICRs and how they are modulated by specific environments, such as neurogenic niches. Understanding the similarities and differences in the regulation of genomic imprints in NSCs between health and disease will shed light on the normal NSC function and maintenance, but also in the tumorigenic process and whether cancer stem cells in the brain originate from adult NSCs as a result of aberrations on epigenetic regulation, including genomic imprinting.

Key facts

Key facts of genomic imprinting

- Epigenetic process that causes genes to be expressed in a parent-of-origin specific manner.

- It is an epigenetic process that involves DNA methylation and histone methylation without altering the genetic sequence.
- Cis-acting mechanism that affects gene expression.
- Imprinting genes are mostly clustered together with a noncoding RNA
- Imprints are epigenetic modifications acquired by one parental gamete.
- Imprints are maintained after fertilization
- Aberrations or loss-of-genomic imprinting can lead to delayed development and intellectual disability.

Key facts of neurogenesis

- Process by which new neurons are formed in the brain.
- Neurogenesis is most active during embryonic development and is responsible for producing all types of neurons in the organism.
- New neurons are formed in prenatal and neonatal brain.
- Adult neurogenesis occurs throughout life in two areas of the mammalian brain: the subventricular zone (SVZ) and the subgranular zone (SGZ).
- Neural stem cells (NSCs) divide indefinitely to produce more stem cells or differentiate to give rise to neural progenitor cells.

Summary points

- Genomic imprinting is an epigenetic process that restricts gene expression in a parent-of-origin manner, silencing either the maternally or the paternally inherited allele.
- The establishment of imprints takes place in the germline, and they are maintained throughout development and adult life.
- Neurogenesis is the process by which new neurons are generated by NSCs or progenitor cells, and it takes place both pre- and postnatally
- Selective regulation of imprinting is a normal mechanism for modulating gene dosage in brain development.
- Imprinted genes makes them very vulnerable to mutation and loss of imprinting (LOI) has been associated to several human pathologies.

References

Adalsteinsson, B. T., & Ferguson-Smith, A. C. (2014). Epigenetic control of the genome-lessons from genomic imprinting. *Genes (Basel), 5*(3), 635–655.

Adnani, L., Langevin, L. M., Gautier, E., Dixit, R., Parsons, K., Li, S., … Schuurmans, C. (2015). Zac1 regulates the differentiation and migration of neocortical neurons via Pac1. *Journal of Neuroscience, 35*(39), 13430–13447.

Albert, M., & Peters, A. H. (2009). Genetic and epigenetic control of early mouse development. *Current Opinion in Genetics and Development, 19*(2), 113–121.

Alvarez-Buylla, A., Garcia-Verdugo, J. M., & Tramontin, A. D. (2001). A unified hypothesis on the lineage of neural stem cells. *Nature Reviews Neuroscience, 2*(4), 287–293.

Babak, T., DeVeale, B., Tsang, E. K., Zhou, Y., Li, X., Smith, K. S., Kukurba, K. R., Zhang, R., Li, J. B., van der Kooy, D., et al. (2015). Genetic conflict reflected in tissue-specific maps of genomic imprinting in human and mouse. *Nature Genetics, 47*(5), 544–549.

Bird, A. (2007). Perceptions of epigenetics. *Nature, 447*(7143), 396–398.

Bond, A. M., Ming, G. L., & Song, H. (2015). Adult mammalian neural stem cells and neurogenesis: Five decades later. *Cell Stem Cell, 17*(4), 385–395.

Bystron, I., Blakemore, C., & Rakic, P. (2008). Development of the human cerebral cortex: Boulder Committee revisited. *Nature Reviews Neuroscience, 9*(2), 110–122.

Chaker, Z., Codega, P., & Doetsch, F. (2016). A mosaic world: Puzzles revealed by adult neural stem cell heterogeneity. *Wiley Interdisciplinary Reviews: Developmental Biology, 5*(6), 640–658.

Cleaton, M. A., Edwards, C. A., & Ferguson-Smith, A. C. (2014). Phenotypic outcomes of imprinted gene models in mice: Elucidation of pre- and postnatal functions of imprinted genes. *Annual Review of Genomics and Human Genetics, 15*, 93–126.

Culican, S. M., Baumrind, N. L., Yamamoto, M., & Pearlman, A. L. (1990). Cortical radial glia: Identification in tissue culture and evidence for their transformation to astrocytes. *Journal of Neuroscience, 10*(2), 684–692.

Del Bene, F. (2011). Interkinetic nuclear migration: Cell cycle on the move. *The EMBO Journal, 30*(9), 1676–1677.

Doetsch, F., Caille, I., Lim, D. A., Garcia-Verdugo, J. M., & Alvarez-Buylla, A. (1999). Subventricular zone astrocytes are neural stem cells in the adult mammalian brain. *Cell, 97*(6), 703–716. https://doi.org/10.1016/S0092-8674(00)80783-7.

Edwards, M. A., Yamamoto, M., & Caviness, V. S. (1990). Organization of radial glia and related cells in the developing murine CNS. An analysis based upon a new monoclonal antibody marker. *Neuroscience, 36*(1), 121–144. https://doi.org/10.1016/0306-4522(90)90356-9.

Eggermann, T., Perez de Nanclares, G., Maher, E. R., Temple, I. K., Tumer, Z., Monk, D., Mackay, D. J., Gronskov, K., Riccio, A., Linglart, A., et al. (2015). Imprinting disorders: A group of congenital disorders with overlapping patterns of molecular changes affecting imprinted loci. *Clinical Epigenetics, 7*, 123.

Fenner, A. (2017). Prostate cancer: IGF2 imprinting loss promotes cancer. *Nature Reviews Urology, 14*(10), 583.

Ferguson-Smith, A. C. (2011). Genomic imprinting: The emergence of an epigenetic paradigm. *Nature Reviews Genetics, 12*(8), 565–575.

Ferron, S. R., Charalambous, M., Radford, E., McEwen, K., Wildner, H., Hind, E., Morante-Redolat, J. M., Laborda, J., Guillemot, F., Bauer, S. R., et al. (2011). Postnatal loss of Dlk1 imprinting in stem cells and niche astrocytes regulates neurogenesis. *Nature, 475*(7356), 381–385.

Ferron, S. R., Radford, E. J., Domingo-Muelas, A., Kleine, I., Ramme, A., Gray, D., Sandovici, I., Constancia, M., Ward, A., Menheniott, T. R., et al. (2015). Differential genomic imprinting regulates paracrine and autocrine roles of IGF2 in mouse adult neurogenesis. *Nature Communications, 6*, 8265.

Fuentealba, L. C., Rompani, S. B., Parraguez, J. I., Obernier, K., Romero, R., Cepko, C. L., & Alvarez-Buylla, A. (2015). Embryonic origin of postnatal neural stem cells. *Cell, 161*(7), 1644–1655.

Galli, R., Binda, E., Orfanelli, U., Cipelletti, B., Gritti, A., De Vitis, S., Fiocco, R., Foroni, C., Dimeco, F., & Vescovi, A. (2004). Isolation and characterization of tumorigenic, stem-like neural precursors from human glioblastoma. *Cancer Research, 64*(19), 7011–7021.

Gaughwin, P., Ciesla, M., Yang, H., Lim, B., & Brundin, P. (2011). Stage-specific modulation of cortical neuronal development by Mmu-miR-134. *Cerebral Cortex, 21*(8), 1857–1869.

Goncalves, J. T., Schafer, S. T., & Gage, F. H. (2016). Adult neurogenesis in the hippocampus: From stem cells to behavior. *Cell, 167*(4), 897–914.

Hata, K., Okano, M., Lei, H., & Li, E. (2002). Dnmt3L cooperates with the Dnmt3 family of de novo DNA methyltransferases to establish maternal imprints in mice. *Development, 129*(8), 1983–1993.

Hatten, M. E. (1999). Central nervous system neuronal migration. *Annual Review of Neuroscience, 22*, 511–539.

Hikichi, T., Kohda, T., Kaneko-Ishino, T., & Ishino, F. (2003). Imprinting regulation of the murine Meg1/Grb10 and human GRB10 genes; roles of brain-specific promoters and mouse-specific CTCF-binding sites. *Nucleic Acids Research, 31*(5), 1398–1406.

Hirasawa, R., Chiba, H., Kaneda, M., Tajima, S., Li, E., Jaenisch, R., & Sasaki, H. (2008). Maternal and zygotic Dnmt1 are necessary and sufficient for the maintenance of DNA methylation imprints during preimplantation development. *Genes and Development, 22*(12), 1607–1616.

Hoekstra, E. J., von Oerthel, L., van der Heide, L. P., Kouwenhoven, W. M., Veenvliet, J. V., Wever, I., Jin, Y. R., Yoon, J. K., van der Linden, A. J., Holstege, F. C., et al. (2013). Lmx1a encodes a rostral set of mesodiencephalic dopaminergic neurons marked by the Wnt/B-catenin signaling activator R-spondin 2. *PLoS One, 8*(9), Article e74049.

Howell, C. Y., Bestor, T. H., Ding, F., Latham, K. E., Mertineit, C., Trasler, J. M., & Chaillet, J. R. (2001). Genomic imprinting disrupted by a maternal effect mutation in the Dnmt1 gene. *Cell, 104*(6), 829–838.

Ioannides, Y., Lokulo-Sodipe, K., Mackay, D. J., Davies, J. H., & Temple, I. K. (2014). Temple syndrome: Improving the recognition of an underdiagnosed chromosome 14 imprinting disorder: An analysis of 51 published cases. *Journal of Medical Genetics, 51*(8), 495–501.

Joseph, B., Andersson, E. R., Vlachos, P., Södersten, E., Liu, L., Teixeira, A. I., & Hermanson, O. (2009). p57Kip2 is a repressor of Mash1 activity and neuronal differentiation in neural stem cells. *Cell Death and Differentiation, 16*(9), 1256–1265. https://doi.org/10.1038/cdd.2009.72.

Kaneda, M., Okano, M., Hata, K., Sado, T., Tsujimoto, N., Li, E., & Sasaki, H. (2004). Essential role for de novo DNA methyltransferase Dnmt3a in paternal and maternal imprinting. *Nature, 429*(6994), 900–903.

Kawasaki, Y., Lee, J., Matsuzawa, A., Kohda, T., Kaneko-Ishino, T., & Ishino, F. (2014). Active DNA demethylation is required for complete imprint erasure in primordial germ cells. *Scientific Reports, 4*, 3658.

Kempermann, G., Song, H., & Gage, F. H. (2015). Neurogenesis in the adult Hippocampus. *Cold Spring Harbor Perspectives in Biology, 7*(9). a018812.

Kircher, M., Bock, C., & Paulsen, M. (2008). Structural conservation versus functional divergence of maternally expressed microRNAs in the Dlk1/Gtl2 imprinting region. *BMC Genomics, 9*, 346.

Kriegstein, A., & Alvarez-Buylla, A. (2009). The glial nature of embryonic and adult neural stem cells. *Annual Review of Neuroscience, 32*, 149–184.

Kuwajima, T., Hasegawa, K., & Yoshikawa, K. (2010). Necdin promotes tangential migration of neocortical interneurons from basal forebrain. *Journal of Neuroscience, 30*(10), 3709–3714.

Li, X., Ito, M., Zhou, F., Youngson, N., Zuo, X., Leder, P., & Ferguson-Smith, A. C. (2008). A maternal-zygotic effect gene, Zfp57, maintains both maternal and paternal imprints. *Developmental Cell, 15*(4), 547–557.

Lim, D. A., & Alvarez-Buylla, A. (2016). The adult ventricular-subventricular zone (V-SVZ) and olfactory bulb (OB) neurogenesis. *Cold Spring Harbor Perspectives in Biology, 8*(5).

Lim, D. H., & Maher, E. R. (2010). Genomic imprinting syndromes and cancer. *Advances in Genetics, 70*, 145–175.

Mabb, A. M., Judson, M. C., Zylka, M. J., & Philpot, B. D. (2011). Angelman syndrome: Insights into genomic imprinting and neurodevelopmental phenotypes. *Trends in Neurosciences, 34*(6), 293–303.

McEwen, K. R., & Ferguson-Smith, A. C. (2010). Distinguishing epigenetic marks of developmental and imprinting regulation. *Epigenetics and Chromatin, 3*(1), 2.

Menn, B., Garcia-Verdugo, J. M., Yaschine, C., Gonzalez-Perez, O., Rowitch, D., & Alvarez-Buylla, A. (2006). Origin of oligodendrocytes in the subventricular zone of the adult brain. *Journal of Neuroscience, 26*(30), 7907–7918.

Mirzadeh, Z., Merkle, F. T., Soriano-Navarro, M., Garcia-Verdugo, J. M., & Alvarez-Buylla, A. (2008). Neural stem cells confer unique pinwheel architecture to the ventricular surface in neurogenic regions of the adult brain. *Cell Stem Cell, 3*(3), 265–278.

Moreno-Jimenez, E. P., Flor-Garcia, M., Terreros-Roncal, J., Rabano, A., Cafini, F., Pallas-Bazarra, N., Avila, J., & Llorens-Martin, M. (2019). Adult hippocampal neurogenesis is abundant in neurologically healthy subjects and drops sharply in patients with Alzheimer's disease. *Nature Medicine, 25*(4), 554–560.

Obernier, K., & Alvarez-Buylla, A. (2019). Neural stem cells: Origin, heterogeneity and regulation in the adult mammalian brain. *Development, 146*(4).

Otsuka, S., Maegawa, S., Takamura, A., Kamitani, H., Watanabe, T., Oshimura, M., & Nanba, E. (2009). Aberrant promoter methylation and expression of the imprinted PEG3 gene in glioma. *Proceedings of the Japan Academy Series B Physical and Biological Sciences, 85*(4), 157–165.

Pandita, A., Gupta, S., Gupta, G., et al. (2018). Beckwith-weidemann syndrome with IC2 (KvDMR1) hypomethylation defect: A novel mutation. *Case Reports.*

Perez, J. D., Rubinstein, N. D., & Dulac, C. (2016). New perspectives on genomic imprinting, an essential and multifaceted mode of epigenetic control in the developing and adult brain. *Annual Review of Neuroscience, 39*, 347–384.

Rago, L., Beattie, R., Taylor, V., & Winter, J. (2014). miR379-410 cluster miRNAs regulate neurogenesis and neuronal migration by fine-tuning N-cadherin. *The EMBO Journal, 33*(8), 906–920.

Rakic, P. (1995). A small step for the cell, a giant leap for mankind: A hypothesis of neocortical expansion during evolution. *Trends in Neurosciences, 18*(9), 383–388.

Rraklli, V., Sodersten, E., Nyman, U., Hagey, D. W., & Holmberg, J. (2016). Elevated levels of ZAC1 disrupt neurogenesis and promote rapid in vivo reprogramming. *Stem Cell Research, 16*(1), 1–9.

Schaaf, C. P., Gonzalez-Garay, M. L., Xia, F., Potocki, L., Gripp, K. W., Zhang, B., Peters, B. A., McElwain, M. A., Drmanac, R., Beaudet, A. L., et al. (2013). Truncating mutations of MAGEL2 cause Prader-Willi phenotypes and autism. *Nature Genetics, 45*(11), 1405–1408.

Schultz, D. C., Ayyanathan, K., Negorev, D., Maul, G. G., & Rauscher, F. J., 3rd (2002). SETDB1: A novel KAP-1-associated histone H3, lysine 9-specific methyltransferase that contributes to HP1-mediated silencing of euchromatic genes by KRAB zinc-finger proteins. *Genes and Development, 16*(8), 919–932.

Seitz, H., Royo, H., Bortolin, M. L., Lin, S. P., Ferguson-Smith, A. C., & Cavaille, J. (2004). A large imprinted microRNA gene cluster at the mouse Dlk1-Gtl2 domain. *Genome Research, 14*(9), 1741–1748.

Smallwood, S. A., & Kelsey, G. (2012). De novo DNA methylation: A germ cell perspective. *Trends in Genetics, 28*(1), 33–42.

Spassky, N., Merkle, F. T., Flames, N., Tramontin, A. D., Garcia-Verdugo, J. M., & Alvarez-Buylla, A. (2005). Adult ependymal cells are postmitotic and are derived from radial glial cells during embryogenesis. *Journal of Neuroscience, 25*(1), 10–18.

Stiles, J., & Jernigan, T. L. (2010). The basics of brain development. *Neuropsychology Review, 20*(4), 327–348.

Tavazoie, M., Van der Veken, L., Silva-Vargas, V., Louissaint, M., Colonna, L., Zaidi, B., Garcia-Verdugo, J. M., & Doetsch, F. (2008). A specialized vascular niche for adult neural stem cells. *Cell Stem Cell, 3*(3), 279–288.

Tucci, V., Isles, A. R., Kelsey, G., Ferguson-Smith, A. C., & Erice Imprinting Group. (2019). Genomic imprinting and physiological processes in mammals. *Cell, 176*(5), 952–965. https://doi.org/10.1016/j.cell.2019.01.043.

Tury, A., Mairet-Coello, G., & DiCicco-Bloom, E. (2012). The multiple roles of the cyclin-dependent kinase inhibitory protein p57(KIP2) in cerebral cortical neurogenesis. *Developmental Neurobiology, 72*(6), 821–842.

Valente, T., Junyent, F., & Auladell, C. (2005). Zac1 is expressed in progenitor/stem cells of the neuroectoderm and mesoderm during embryogenesis: Differential phenotype of the Zac1-expressing cells during development. *Developmental Dynamics, 233*(2), 667–679.

Vincent, J. J., Huang, Y., Chen, P. Y., Feng, S., Calvopina, J. H., Nee, K., Lee, S. A., Le, T., Yoon, A. J., Faull, K., et al. (2013). Stage-specific roles for tet1 and tet2 in DNA demethylation in primordial germ cells. *Cell Stem Cell, 12*(4), 470–478.

Yamaguchi, S., Shen, L., Liu, Y., Sendler, D., & Zhang, Y. (2013). Role of Tet1 in erasure of genomic imprinting. *Nature, 504*(7480), 460–464.

Yan, Y., Frisén, J., Lee, M. H., Massagué, J., & Barbacid, M. (1997). Ablation of the CDK inhibitor p57(Kip2) results in increased apoptosis and delayed differentiation during mouse development. *Genes and Development, 11*(8), 973–983. https://doi.org/10.1101/gad.11.8.973.

Yao, B., & Jin, P. (2014). Unlocking epigenetic codes in neurogenesis. *Genes and Development, 28*(12), 1253–1271.

Yin, D., Xie, D., Sakajiri, S., Miller, C. W., Zhu, H., Popoviciu, M. L., Said, J. W., Black, K. L., & Koeffler, H. P. (2006). DLK1: Increased expression in gliomas and associated with oncogenic activities. *Oncogene, 25*(13), 1852–1861.

Zeng, Y., & Chen, T. (2019). DNA methylation reprogramming during mammalian development. *Genes (Basel), 10*(4).

Zhang, J., Woodhead, G. J., Swaminathan, S. K., Noles, S. R., McQuinn, E. R., Pisarek, A. J., Stocker, A. M., Mutch, C. A., Funatsu, N., & Chenn, A. (2010). Cortical neural precursors inhibit their own differentiation via N-cadherin maintenance of beta-catenin signaling. *Developmental Cell, 18*(3), 472–479.

Chapter 6

Insights from model systems: alcohol, neurodevelopment, and zebrafish

Robert Gerlai
Department of Psychology, University of Toronto Mississauga, Mississauga, ON, Canada

List of abbreviations

5HIAA – 5-hydroxyindoleacetic acid
BDNF – brain-derived neurotrophic factor
CNS – central nervous system
DOPAC – 3,4-dihydroxyphenylacetic acid
dpf – days postfertilization
FASD – fetal alcohol spectrum disorder
GABA – gamma-aminobutyric acid
hpf – hours postfertilization
HPLC – high-precision liquid chromatography
NCAM – neuronal cell adhesion molecule
SEM – Standard error of the mean
TUNEL staining – Terminal deoxynucleotidyl transferase dUTP nick end labeling staining

Mini-dictionary of terms

Anxiety – is defined as a set of fear-like behavioral responses that appear for a prolonged period of time, does not require the appearance or clear presence or the appearance of aversive stimuli, and is considered pathological.

Comparative approach – is a scientific method that emphasizes the use of multiple species and the comparison of these species in order to advance our knowledge of biology, evolution, and/or medicine.

DNA microarray – or gene chip is a method with which the experimenter may test the mRNA expression (transcription) of tens of thousands of genes. The technology is based upon binding short nucleotide sequences (oligonucleotides) complementary to mRNA sequences on glass chips and using fluorescent probes to detect binding of mRNA to these sequences.

Dopamine – is a neurotransmitter released from the presynaptic terminal of dopaminergic synapses and is involved in a number of behavioral functions including motor function, motivation, and learning and memory.

Event recording – is an observation-based behavior quantification method with which the human observer, usually with the aid of a computer, records the time when and for how long particular motor responses were emitted by the subject, animal or human.

Evolutionarily conserved features – are characteristics that appear across multiple species and have the same evolutionary origin. The common evolutionary origin also means that biological mechanisms underlying such characteristics are also similar.

Fear – is defined as a set of adaptive behavioral responses induced by the appearance or clear presence of aversive stimuli.

Forward genetics – is a genetic approach whose goal is to discover novel genes. In forward genetics, the experimenter induces mutations randomly (most often using a chemical mutagen), then searches for mutants based upon their phenotypical characteristics, and subsequently identifies the gene carrying the mutation using linkage analysis-based positional cloning.

Gene expression – refers to a multistep process. The two most important steps of this process are, one, transcription, i.e., mRNA synthesis, and two, translation, i.e., protein synthesis.

Pharmacological profile – is a set of features of a drug that may include, for example, the targets with which the drug interacts or binds to and the biochemical interactions/pathways via which the drug alters the functioning of biological systems.

Diagnosis, Management and Modeling of Neurodevelopmental Disorders. https://doi.org/10.1016/B978-0-12-817988-8.00006-3

Pharmacology – is the discipline that is focused on studying drugs, relatively simple compounds and molecules that may interact with biological systems.
Phene – is the "unit" of the phenotype, i.e., it is a phenotypical characteristic one can quantify.
Phenotype – is the collection of characteristics one can observe and measure.
Reverse genetics – is a genetic approach in which the experimenter already knows the identity of the gene, i.e., its nucleotide sequence, and experimentally manipulates the gene (for example, knocks it out or overexpresses it) in order to tests its phenotypical function.
Schooling – is a form of group forming behavior in which group members move in a synchronized manner, i.e., the direction of movement of subject within the group is similar or identical.
Serotonin – is a neurotransmitter released from the presynaptic terminal of serotoninergic synapses and in the brain it is involved in a number of behavioral functions including mood and emotion.
Shoaling – is a form of group-forming behavior in which group members do not synchronize their motion paths but stay close to each other. Shoaling is often observed in fish, but similar group-forming behaviors have been described in numerous bird and mammalian species as well.
Synaptic plasticity – is the ability of synapses to become strengthened or weakened. Strengthening or weakening of synaptic connections means increased or decreased probability of action potential generation in the postsynaptic neuron, changes that are believed to underly memory formation.
Transcriptome – is the set of genes that transcribe (express) mRNA. The term is usually used in the context of alteration of gene expression induced by treatment, disease, or underlying biological function.
Video tracking – is a computer-based movement analysis system with which the experimenter monitors and quantifies patterns of locomotory paths or motion/movement of the experimental subject, animal or human.

Introduction

Fetal alcohol spectrum disorders (FASDs) arise as a result of pregnant women drinking and exposing their developing children to alcohol (ethanol, ethyl alcohol, or EtOH) in utero. FASD represents the most prevalent preventable form of mental retardation (Lange et al., 2017). Despite being fully preventable by abstinence from alcohol, FASD continues to be a major problem leading to lifelong suffering for the patient and a substantial cost to the society (Popova, Stade, Bekmuradov, Lange, & Rehm, 2011). FASD patients present a variety of symptoms. On the most extreme and severe end of the spectrum, the patient may have significant anatomical abnormalities such as facial dysmorphia, on the other end of the spectrum, the patient may not even be diagnosed, but may exhibit milder abnormalities in behavior, including abnormal or inappropriate social behavior and/or learning disabilities (Chudley, 2018). Although clearly much less debilitating, the milder FASD cases represent an important issue as the prevalence of this form of FASD is the highest (Lange, Shield, Rehm, & Popova, 2013). Furthermore, these patients are often undiagnosed or misdiagnosed and as a result, receive no appropriate treatment. However, even when diagnosed (for example, because recorded information about the mother's drinking during pregnancy is available), treatment options are limited for children suffering from FASD, because we still do not understand the mechanism of this disease.

Analysis of such mechanisms is hampered by two important issues. One, alcohol is a complex drug from a pharmacological standpoint as it interacts with a large number of molecular targets in the brain directly and with an even larger number of biochemical pathways indirectly (Abrahao, Salinas, & Lovinger, 2017; Most, Ferguson, & Harris, 2014). Coupled with this is the fact that such interactions may lead to a cascade of developmental changes if alcohol reaches the brain during the ontogenesis of this organ. In summary, the manner in which alcohol may alter brain development is expected to be rather complicated. Two, addressing this complexity is difficult in humans, as the concentration of alcohol to which the developing human embryo is exposed, the timing, the frequency, and length of this exposure all vary from patient to patient and quite often are unknown. Furthermore, the genotype and the health status of both the mother and of the fetus are unknown and uncontrolled, so are many other environmental factors that potentially influence the effects of alcohol.

To alleviate these issues, investigators proposed numerous animal models where the aforementioned exemplified factors could be experimentally controlled and precisely manipulated. Most of these attempts have been made using rodents (Barron, Hawkey, Fields, & Littleton, 2016). Nevertheless, more recently, the zebrafish has also been proposed for this purpose (Gerlai, 2015; Lovely, Fernandes, & Eberhart, 2016; Seguin & Gerlai, 2018). Systematic comparison between the rodent and the zebrafish models or a thorough review of the rodent literature is beyond the scope of this chapter. Instead, we will focus on the literature pertinent to the use of zebrafish in the modeling and analysis of FASD, mainly from the perspectives of research conducted in the laboratory of the author of this chapter.

The zebrafish in FASD research: advantages, disadvantages, and consideration of the comparative approach

It is customary to emphasize the advantages of one's preferred animal model. This is also how we will start our discussion here. However, we note that there is another overarching reason why zebrafish should be employed, a point to which we will return later: the importance of the comparative approach (Gerlai, 2014). But first let us talk about the advantages associated with this organism. These advantages can be summarized in one sentence. The zebrafish, arguably, represents an optimal compromise between system complexity and practical simplicity. Briefly, it is complex enough to be translationally relevant, but at the same time it is much simpler, cheaper, and easier to keep and study in the laboratory than mammals. The zebrafish is a vertebrate whose basic body plan, including the major structural components of its brain, is similar to that of higher-order vertebrates. It possesses numerous evolutionarily conserved features at several levels of its biological organization, from the nucleotide sequence of its genes through the neurotransmitter systems of its brain to its behavior (Pickart & Klee, 2014; Brennan, 2011). Yet, it has only a few million neurons as opposed to billions like in our own brain. It develops to a fully formed little fish within 5 days from a single cell zygote, a process throughout which it mostly remains transparent. It is only 4 cm long when fully grown, and because of its shoaling (group forming) nature, it can be housed in large numbers in small fish tanks. Furthermore, numerous sophisticated forward and reverse genetic methods as well as neuroscience techniques have been developed for this species. From the perspective of FASD research, the zebrafish also enjoys some advantages. Perhaps the most important of them is that these fish are externally fertilized and thus their eggs develop outside the mother. As a result, alcohol administration is simple and precise. The dose of alcohol, the developmental stage at which this substance is administered, as well as the length of time of the exposure can be precisely controlled (Fernandes & Gerlai, 2009). Furthermore, external development also means that indirect effects of alcohol altering the mother's physiology do not have influence. Lastly, as zebrafish do not exhibit parental care for their young, alcohol-induced alterations in behavior of parental care are also not a confounding factor, unlike in mammalian species. However, some of these advantages may be viewed as cons. Due to the differences between the externally developing fish embryo and the in utero developing mammalian embryo, the fish model cannot recapitulate the mammalian situation. Nevertheless, from the perspective of the analysis of alcohol's direct effects on the vertebrate embryo's brain development and brain function, the zebrafish is a clear winner.

There is another, overarching, reason we eluded to at the beginning, for which zebrafish should be chosen in FASD research (and any other research field too) irrespective of whether this particular species is actually better or not in the laboratory compared to others. This reason is the power of the comparative approach. Science usually evolves a bit like solar systems. Once a critical mass has been reached, celestial bodies start to coalesce. Once scientists have obtained enough information on a particular species, and once numerous methods have been developed for this particular species, it is easier to work with this species. It is also easier to connect the pieces of information discovered across multiple laboratories and form a coherent and cohesive understanding of the studied biological phenomena if one uses the same species as others. So, why is it good to start working on a different species like the zebrafish? A different species allows one to discover common phenomena and mechanisms underlying the target of investigation. For example, while some of the biological mechanisms mediating the effects of alcohol in mice may be similar to those underlying alcohol effects in the human, it is not easy to tell which ones are evolutionarily conserved (evolutionary homology) and which ones may represent spurious analogies, or which ones are unique to the studied species. Adding at least one more species to the comparative approach will aid the investigator to identify mechanisms, features, and phenomena common to all three species, human, mouse, and fish in this case. These common features represent aspects of the biology of these species that have been conserved for at least 400 million years (the first common ancestor of fish and humans lived about 400 million years ago). For this reason, these common mechanisms allow one to tell what is most fundamentally important for the biological phenomenon in question, alcohol-induced abnormalities in brain development in our case. Fundamental importance does not just mean evolutionarily old, it likely also represents functional fundamentality too. In summary, adding the zebrafish to the analysis of FASD will likely enhance translatability of animal research to human and thus, will likely lead to better understanding of the human disease.

Zebrafish in FASD research: the alcohol exposure method

One of the reasons why the zebrafish has become popular in alcohol research is the ease and precision with which these animals can be exposed to alcohol. Unlike with other model organisms, zerbrafish can be immersed in alcohol solution from which the fish take up the substance through their gills and body surface (Gerlai, Lahav, Guo, & Rosenthal, 2000). Immersion allows one to "clamp" the concentration of alcohol, and thus, ADME (absorption, distribution, metabolism, and excretion)-related processes are less likely to influence the effects of this drug, unlike in the case of bolus-type delivery methods such as intraperitoneal injection or oral administration. Immersion also allows precise determination of the concentration and length of exposure to the substance. These alcohol delivery advantages also apply to studies where alcohol is administered to the developing embryo. Most studies of the latter kind employ intact eggs (i.e., do not dechorionate). The chorion (egg shell) is protective and does not allow much alcohol to infuse inside the egg. Depending on the length of immersion, and the method of quantification of alcohol concentration, about 1/4th to 1/25th of the external bath concentration of alcohol has been found inside the zebrafish egg after immersing the eggs into the alcohol bath (Fernandes & Gerlai, 2009; Mahabir, Chatterjee, & Gerla, 2014; Lovely, Nobles, & Eberhart, 2014). Importantly, the internal alcohol concentrations have been found to be linearly correlated with the external bath concentration (Fernandes & Gerlai, 2009; Mahabir et al., 2014).

Earlier attempts to investigate the effects of alcohol on embryonic development of zebrafish employed higher concentrations of and/or prolonged exposure to alcohol (Bilotta, Barnett, Hancock, & Saszik, 2004; Loucks & Carvan, 2004; Arenzana et al., 2006). These studies demonstrated robust structural abnormalities arising from alcohol exposure that recapitulated several aspects of human fetal alcohol syndrome (FAS), i.e., the most severe form of FASD. Subsequently, we proposed a different model, one with which we wanted to mimic the milder forms of FASD (Fernandes & Gerlai, 2009). In this model, we exposed the developing zebrafish embryo to only a low dose of alcohol (up to 1% vol/vol external bath) and for only a short period of time (2 h). Knowing about the relatively low prevalence of the severe FAS cases, we argued that this model was more realistic, as most pregnant women would not consume large quantities of alcohol and for prolonged periods of time. We were hoping to establish a model of the milder and most prevalent form of FASD. To accomplish this, we immersed zebrafish eggs into alcohol solutions at 0%, 0.25%, 0.50%, 0.75%, or 1.00% concentration and measured the concentration of alcohol inside the egg using the Analox alcohol analyzer. Our results showed that the internal concentration was about 1/16th of that of the external bath. That is, in case of the highest (1.00% alcohol concentration), it translated to about 0.06%, which would be about the legal limit of blood alcohol concentration for driving in North America.

Characterization of phenotypical effects of embryonic alcohol exposure: starting the analysis with behavioral tests

Given the complex pharmacological profile of alcohol, it is not easy to forecast what changes one may expect resulting from exposure to this substance during embryonic development. Most scientists, grant funding agencies, as well as peer-reviewed scientific journals and their referees prefer what we call "hypothesis driven" studies. Hypothesis driven means that we have enough knowledge to form reasonably good working hypotheses that we may accept or reject after the empirical study has been performed. However, hypothesis drivenness can limit one's scope and can lead too narrow focus. In other words, postulating what results one may expect too early may lead to inability to make proper discoveries. An alternative to the hypothesis-driven approach is comprehensive screening. In this approach, the experimenter characterizes many features, phenes, and only after thorough phenotyping and substantial accumulation of data she/he attempts to interpret results. One may argue that analysis of alcohol's effects on embryonic development and the ensuing changes should fall under this latter category of research.

But how should one start the phenotypical characterization of embryonic alcohol exposure induced changes in zebrafish? Given how complex the brain of even this relatively simple organism is and how many different ways alcohol may influence the development of this organ, this is not a trivial question. We have argued that perhaps the best way to start such phenotyping effort is to measure behavior (Gerlai, 2010; 2002). Briefly, behavioral analysis may be viewed as a first pass, comprehensive, and unbiased method that may reveal altered functioning of the brain without having any assumptions about the anatomical locale or neurobiological mechanism underlying the changes induced, in this case, by embryonic alcohol exposure.

Following this logic, we started phenotyping potential embryonic alcohol-induced changes using a behavioral test battery. We observed no anatomical alterations induced by exposing zebrafish eggs to low concentrations of alcohol for 2 h (Fernandes & Gerlai, 2009). The embryos appeared to develop normally. They hatched on time, and mortality did not differ between alcohol-exposed and control embryos. The larval fish exposed to alcohol during their embryonic development grew like control fish and reached the same size within the same amount of time as their control, alcohol-unexposed, counterparts. Superficial observation of the fish when fully grown also did not reveal any abnormalities. All previously alcohol-exposed zebrafish looked healthy and appeared to behave just like their control counterparts (Fernandes & Gerlai, 2009). Subsequently, we decided to analyze the behavior of adult zebrafish exposed to alcohol during their embryonic development. Adult fish exhibit a sophisticated behavioral repertoire and do not suffer from fast developmental change related temporal confounds. In other words, they do lots of things and they are quite stable across time. First, we have quantified motor function of the fish by measuring numerous parameters of their swim paths using video tracking. These included, for example, swim speed, the fish's location in the tank (distance to bottom, for example), angular velocity (speed and degree of turning), immobility (Fernandes & Gerlai, 2009). The analysis was extended to the mean of these parameters, the bout length (the length of time for which a behavior occurs once it starts), as well as the intraindividual (temporal) variability of these behaviors, which quantifies how consistently the behavior was exhibited by the given individual. In addition to the video-tracking measures, we also quantified complex motor/posture patterns using event recording. This technique requires a human observer to watch and identify the appearance of particular complex behaviors. Software applications developed to help one measure such behavioral responses essentially turn the computer into a sophisticated multichannel stopwatch with which the experimenter can measure the duration and the frequency (number of occurrence) of particular behaviors. We quantified numerous motor and posture patterns, including thrashing at the glass, erratic movement, and jumping or leaping. Analysis of all these behaviors showed no change in zebrafish exposed to alcohol during their embryonic development. We also tested the response of zebrafish to the appearance of an animated (moving) image of a sympatric predator of zebrafish, the Indian leaf fish (*Nandus nandus*). Previously, we found zebrafish to show robust avoidance, antipredatory responses toward this aversive stimulus (Ahmed, Seguin, & Gerlai, 2011; Bass & Gerlai, 2008). However, control fish and the fish exposed to alcohol during their embryonic development did not differ in their avoidance reactions suggesting that the alcohol-exposed fish exhibit normal motor function, fear, and can also see as well as their control counterparts (Fernandes & Gerlai, 2009). However, we did find a significant abnormality in the behavior of the embryonic alcohol-exposed fish.

When presented with animated (moving) images of conspecifics, single zebrafish exposed to alcohol during their development showed significant reduction of shoaling responses (Fernandes & Gerlai, 2009). Zebrafish are social. They form groups in nature and in the laboratory (Miller & Gerlai, 2011; 2007). This group-forming behavior, called shoaling, can be induced and measured in multiple ways. One of them is to measure interindividual distance (the distance among the focal fish and all other fish around it) or the nearest neighbor distance (the distance between the focal fish and the fish closest to it) when the fish swim together with other fish. Another way to measure shoaling is to place a single fish into the test tank and measure its distance to a stimulus tank that contains conspecifics. Qin, Wong, Seguin, and Gerlai (2014) demonstrated, however, that one does not have to show live stimulus fish, or even video recordings of live stimulus fish, as computer animated images of conspecifics also induce a shoaling response (reduction of distance to the stimulus upon its appearance) equivalent to what live conspecifics would induce. It was this reduction of distance upon stimulus delivery that was impaired by embryonic alcohol exposure (Fernandes & Gerlai, 2009). Importantly, the impairment was linearly dose dependent (Fig. 6.1), with fish that were administered 0.25% alcohol during their embryonic development showing only a small (but still significant) impairment, and the 1% alcohol—treated ones the strongest. In fact, the 1% alcohol—treated fish were so impaired that their distances from the stimulus screen before, during, and after the stimulus presentation were statistically indistinguishable. In other words, 1% alcohol administered during embryonic development of the fish abolished their shoaling response at their adult stage.

How can we interpret this result? It is tempting to jump to conclusion that, just like in the milder forms of human FASD, zebrafish too may suffer from abnormal social behavior as a result of embryonic alcohol exposure. However, impaired response to social stimuli may have a number of causes unrelated to social behavior itself. For example, if embryonic alcohol exposure impaired visual perception, the exposed fish may not be able to properly see the computer animated images and as a result respond to them less or not at all. Embryonic alcohol exposure is known to disrupt eye development, leading, for example, to fusion of eyes (cyclopia) in zebrafish when the concentration of alcohol employed was high and/or when the substance is administered for prolonged periods of time (e.g., Arenzana et al., 2006). In our alcohol administration procedure, however, zebrafish were exposed to alcohol concentrations not exceeding 1% (vol/vol) and the exposure period was only 2 h. Furthermore, structural abnormalities of any sort were not observed, e.g., the eyes looked normal. Nevertheless, in principle, microstructural or functional alterations in the eye could have remained undetected leading to impaired vision. This possibility, however, was excluded after perusal of the results obtained by Fernandes and Gerlai (2009). These results showed that during the presentation of the animated stimulus fish, two behavioral parameters changed. One, as mentioned before, test fish moved closer to the stimulus presentation screen, and two, their swim speed decreased. Importantly, this speed decrease was observable in fish of all groups, control and embryonic alcohol-exposed alike. Thus, all fish were able to see and respond to the stimulus fish, and only the reduction of distance to stimulus was affected by embryonic alcohol treatment. Another possible factor that may result in altered response to the social stimulus is motor function. If the alcohol-exposed fish could not swim properly, they may not have

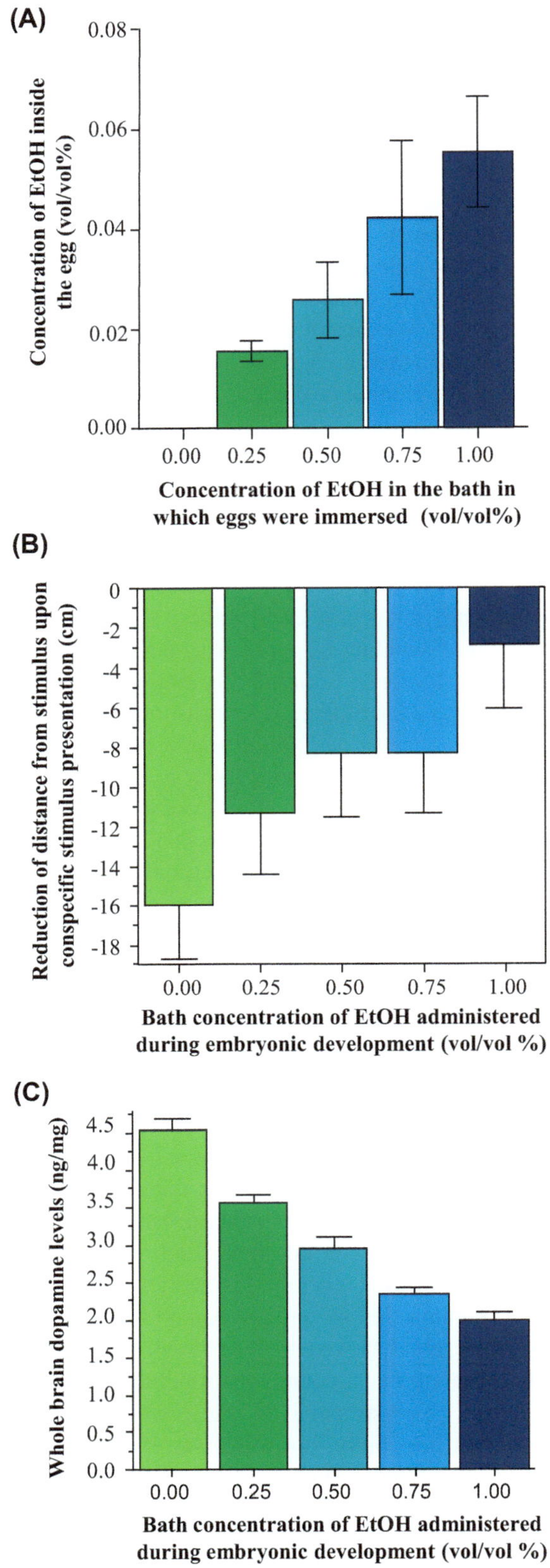

FIGURE 6.1 Alcohol administered via immersion of eggs at 24th hour postfertilization (hpf) for 2 h enters the eggs in a bath concentration-dependent manner (A), leads to significant and linearly concentration-dependent disruption of shoaling (responding to animated images of conspecifics) when tested at the adult stage of the fish (B), and leads to significant dose-dependent reduction of levels of neurochemicals, including dopamine (C). Mean + SEM are shown. The concentration of the alcohol bath is indicated on the X axis. Note that concentration inside the egg does not necessarily equal, although likely is close to, the concentration of alcohol that actually reaches the brain of the developing embryo. Note that the reduction of distance from stimulus is calculated as the distance between test zebrafish and the stimulus presentation screen during stimulus presentation minus the distance before the stimulus was turned on. Note that dopamine levels are expressed relative to total brain protein weight. *Modified from Fernandes, Y., Gerlai, R. (2009). Long-term behavioral changes in response to early developmental exposure to ethanol in zebrafish.* Alcoholism: Clinical and Experimental Research, 33, *601-609; Mahabir, S., Chatterjee, D., & Gerlai, R. (2014). Strain dependent neurochemical changes induced by embryonic alcohol exposure in zebrafish.* Neurotoxicology and Teratology, 41, *1–7.*

been able to move closer to the stimulus. However, as mentioned before, detailed analysis of the parameters of their swim path revealed no alterations, excluding the possibility that motor function could explain the abnormal response to social stimuli.

Confirmation of the aforementioned came from an independent study in which the same alcohol dosing procedure was employed (0%–1% alcohol administered for 2 h at 24th hpf), but social behavior of the fish was quantified using freely moving shoals (Buske & Gerlai, 2011). Shoal cohesion in this latter study was quantified by measuring interindividual distance among fish in 10-member groups. Embryonic alcohol exposure was found to increase interindividual distance, i.e., reduce shoal cohesion. It is notable that in freely moving fish, shoal cohesion does not solely depend upon vision. Blind cave tetras and blinded shoaling fish of other species will form shoals appropriately even without being able to see as the shoal members can rely upon their lateral line, a sensory system that can detect low-frequency waves in the water. Motor function was also found unaltered in the embryonic alcohol-treated fish compared to control (Buske & Gerlai, 2011). Thus, again, impaired vision or motor function could be excluded as possible explanations for the impaired shoaling response. A third independent study (Seguin unpublished) also confirmed the aforementioned. In this study the investigator observed increased distance between pairs of fish housed singly in transparent acrylic tanks adjacent to each other.

Does the aforementioned mean that shoaling, a species-typical social behavior of zebrafish, was impaired by embryonic alcohol administration? It seems to be the most parsimonious explanation, except that there is another factor that influences performance in all behavioral tasks. So far, we have considered perception and motor function. The third factor is motivation. One of the functions of shoaling from an adaptive evolutionary fitness standpoint has been shown to minimize predation risk (Hamilton, 1971). Fish forming a shoal reduce the likelihood of being caught by a predator as many eyes, olfactory systems, and lateral lines can detect the approaching predator better. Being in a large group also reduces the chance of being caught for the single individual. Fish escaping in many directions in response to a predatory attack can confuse the predator making it difficult for the predator to focus on a single target. For all these experimentally proven reasons, being in a group is a good thing for small fish that are preyed upon by several piscivorous fish sympatric with zebrafish and also by several fishing birds. In summary, one important adaptive function of shoaling is reduction of predatory threat. Indeed, shoaling has been shown to be influenced by fear-inducing stimuli in zebrafish (Speedie & Gerlai, 2008; Miller & Gerlai, 2007). In response to general aversive conditions, shoal cohesion increases (Speedie & Gerlai, 2008). In response to imminent threat, e.g., an approaching predator, the shoal abruptly breaks up and subsequently reassembles into a tighter group (Miller & Gerlai, 2007). In other words, experimental manipulations that alter the level of anxiety and/or fear in zebrafish are expected to influence shoaling as well. Thus, could the impaired shoaling response seen in the embryonic alcohol-exposed fish be due to reduced fear or anxiety?

Seguin, Shams, and Gerlai (2016) conducted a study to address this question. The authors tested behavior of zebrafish in a novel tank as well as in response to presentation of an animated image of a sympatric predator of the zebrafish, the clown knife fish. Novelty has been shown to induce anxiety-like responses in zebrafish, and presentation of sympatric predators, including the knife fish, has been shown to elicit fear responses (Ahmed, Fernandes, & Gerlai, 2012). Here, we note that anxiety-like and fear responses are actually the same in terms of their appearance. What distinguishes them is the inducing stimuli. Anxiety is induced by diffuse and chronically present aversive conditions, whereas fear is induced by the appearance of a clearly present aversive stimulus with a precise onset and offset of presentation (Gerlai, 2013). Seguin et al. (2016) found embryonic alcohol-exposed zebrafish to respond to anxiety and fear-inducing stimuli in a manner that was statistically indistinguishable from responses of control, alcohol-unexposed fish. Thus, the authors concluded that embryonic alcohol exposure likely disrupted brain function responsible for shoaling, a form of social behavior in zebrafish.

Mechanisms underlying embryonic alcohol-induced impairment of the shoaling response

What neurobiological mechanisms could explain the earlier described impairment of shoaling behavior? How could we start unraveling such mechanisms? This is a potentially complex question. Ideally, one would like to conduct comprehensive and systematic analyses, for example, assaying the transcriptome. DNA microarrays, or high-throughput deep sequencing, could reveal a number of genes whose expression is modified as a result of embryonic alcohol treatment. Discovering such transcriptome level changes may aid one to obtain a comprehensive picture of possible biochemical alterations resulting from this treatment. Similarly, systematic analysis of neuronal activity by, for example, measuring immediate early gene expression (e.g., c-fos) in the brain, may allow one to identify brain regions whose activity is reduced or increased as a result of embryonic alcohol treatment. When combined with double labeling, subpopulations of neurons may be identified as particularly affected or resistant. Similarly, comprehensive mutation screens could identify genes that alter, enhance, or blunt the effects of embryonic alcohol exposure, thereby illuminating possible biochemical pathways through which this drug acts throughout the development of the brain. A similar idea would be to use drugs with possible embryonic alcohol effects altering properties. All of these approaches, however, require significant resources and time. They have not been performed for zebrafish FADS models. Instead, more hypothesis-driven, proof-of-principle studies have been conducted as a start to address mechanistic questions. The first of these was the analysis of the dopaminergic system.

Dopamine is known to be involved in a number of important brain functions, including motivation, motor function, and learning and memory (Iversen, Iversen, Dunnett, & Björklund, 2010). Interestingly, significant increase of levels of dopamine as well as of DOPAC (dopamine's metabolite) has been observed in response to presentation of conspecific images in zebrafish (Saif, Chatterjee, Buske, & Gerlai, 2013). The method used for this analysis was high-precision liquid chromatography (HPLC), a technique that was adapted specifically for zebrafish to measure neurochemicals from whole brain extracts (Chatterjee & Gerlai, 2009). The results showed that in response to the appearance of conspecific images, levels of dopamine and DOPAC significantly rose, while other neurochemical levels, including that of serotonin and 5HIAA, remained unchanged (Said et al., 2013). It appeared as if the presentation of conspecific images engaged the reward pathway. A similar conclusion was drawn from another study in which a D1 dopamine receptor antagonist was administered acutely before presentation of conspecific images (Scerbina, Chatterjee, & Gerlai, 2012). The drug dose dependently reduced shoaling without altering motor function and, at its highest dose employed, actually completely abolished the shoaling response (Scerbina et al., 2012). Thus, the authors concluded that the dopaminergic system is involved in shoaling.

Did the dopaminergic system get altered by embryonic alcohol exposure? This question was answered by Mahabir et al. (2014), who studied the levels of dopamine, DOPAC, serotonin, and 5HIAA in fish exposed to alcohol for 2 h at their 24th hpf developmental stage. The authors investigated potential changes in these neurochemicals not only in adult zebrafish but also in younger fish, i.e., they conducted a cross-sectional developmental study. They found adult fish to exhibit a linear dose-dependent reduction of levels of these neurochemicals. This is noteworthy because just like in case of the impaired shoaling response, the neurochemical changes resulting from embryonic alcohol treatment were observed in adult fish, i.e., demonstrated a lasting alteration that correlated precisely with the behavioral impairment. Also important was the finding that showed that the impairments became progressively more robust as the fish matured (Mahabir et al., 2014). For example, at the youngest age tested, i.e., at 15 days postfertilization (dpf), the levels of dopamine and DOPAC relative to the total brain protein amount were found low, indicating an immature neurotransmitter system. At this age, no alcohol effect was observed. At the next age studied, however, i.e., at 40 dpf, the embryonic alcohol-induced impairments started to manifest, and when tested at later ages (70 and 102 dpf), it became increasingly more pronounced (Mahabir et al., 2014). Thus, embryonic alcohol was found to dose dependently impair shoaling and the dopaminergic and serotoninergic systems.

It is also important to note that without being stimulated, i.e., at baseline conditions, dopamine and DOPAC levels in the brain of adult fish that were exposed to alcohol during their embryonic development were statistically indistinguishable from levels seen in control, alcohol-unexposed fish. For example, Fernandes, Rampersad, and Gerlai (2015) found no difference between embryonic alcohol-treated and control fish in dopamine and DOPAC levels as long as the fish were isolated from their shoal mates. However, upon presentation of shoaling stimuli (animated images of zebrafish), control fish responded with a robust increase of dopamine and DOPAC levels, as also shown before (Saif et al., 2013), but embryonic alcohol-exposed fish did not, effectively mimicking the lack of shoaling behavior response (Fig. 6.2). Thus, it is possible that impaired behavioral response to social stimuli is the result of impaired dopaminergic function. It is also notable that although the dopaminergic system is known to be involved in motivation in general and also motor function, motor responses were unaltered and general motivation to explore novelty, and/or to seek out food, for example, was also unaltered in embryonic alcohol-treated fish when tested at their adult age (e.g. Fernandes, Tran, Abraham, & Gerlai, 2014). Thus, it appears that the impaired serotoninergic and dopaminergic responses were fairly specific to social behavior.

Interestingly, analysis of whole-brain levels of numerous neurotransmitters, including glutamate, aspartate, GABA, glycine, and taurine suggested that these other neurotransmitter systems functioned normally in adult zebrafish exposed to embryonic alcohol treatment (Mahabir, Chatterjee & Gerlai, 2018a), results that support at least some specificity of alcohol treatment on the serotoninergic and dopaminergic systems. What could explain this apparent specificity? Dopamine and serotonin are monoaminergic neurotransmitters. All monoamines are derived from aromatic amino acids by the action of aromatic amino acid decarboxylase enzymes. These neurotransmitters are deactivated by monoamine oxidases. It is possible that embryonic alcohol treatment interfered with neuronal functions associated with the synthesis and/or metabolism of monoamines. Furthermore, monoaminergic neurons mediate secretion and production of neurotrophin-3 by astrocytes (Mele, Carman-Krzan, & Juric, 2010), a neurotrophic factor that in turn affects a variety of neuronal functions including neurite growth and apoptotic cell death during brain development and synaptic

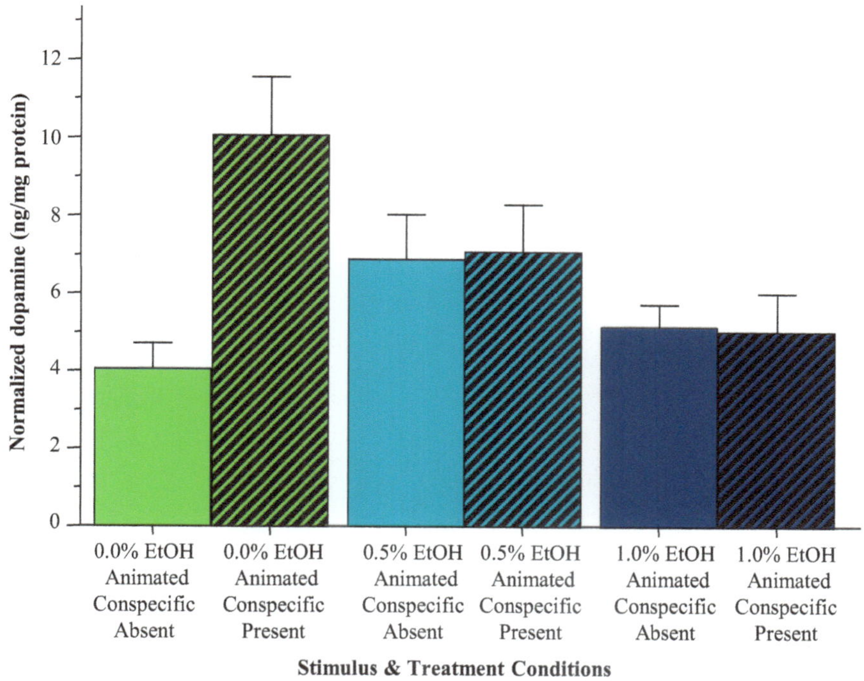

FIGURE 6.2 Alcohol administered at 24th hpf for 2 h does not affect baseline dopamine levels but abolishes the response induced by presentation of animated conspecific images. Mean + SEM are shown. Note that all fish were tested after a 24 h isolation period. Note the robust increase of dopamine level in response to the conspecific stimulus in control, but not in the alcohol treated fish. Note that the treatment was conducted at the embryonic stage and the behavioral test was administered at the adult stage of the fish. *Modified from Fernandes, Y., Rampersad, M., & Gerlai, R. (2015). Embryonic alcohol exposure impairs the dopaminergic system and social behavioural responses in adult zebrafish.* The International Journal of Neuropsychopharmacology, 18(6), *1–8. https://doi.org/10.1093/ijnp/pyu089. pii: pyu089.*

plasticity in the adult brain (Bernd, 2008). It is possible that embryonic alcohol affected the development and functioning of monoaminergic neurons leading to abnormal astrocyte–neuron communication, a possibility to which we return later.

The dopaminergic and serotoninergic neurotransmitter systems are well mapped in the zebrafish brain, but no one has tested whether structurally or in terms of connectivity these neurons are specifically or uniquely affected by embryonic alcohol treatment. Even gross morphology of the brain of embryonic alcohol-treated fish has not been analyzed. Nevertheless, given that DOPAC and dopamine levels appear normal under baseline, unstimulated conditions, one possible working hypothesis is that these monoaminergic neurons themselves may be unaffected, but connections to them from other neurons or brain areas may be abnormal. For example, if social stimulus–induced excitatory inputs do not properly reach the dopaminergic neuron, it will not release dopamine at the dopaminergic synapse, and thus, it will not increase DOPAC levels in response to social stimuli. It will also not increase dopamine synthesis in response to such stimulation. What connections or brain areas may be involved in this impairment are unknown at this point, but numerous authors have uncovered that an extended network of neurons termed the social behavior neuronal network they argue is specifically involved in processing and responding to social stimuli (Albers, 2012). Thus, it is possible that embryonic alcohol treatment damaged some parts of this network, its neurons, and/or the connectome among them.

Based on the earlier arguments, one may expect changes in numerous brain regions. To investigate this possibility, Mahabir, Chatterjee, Misquitte, Chatterjee, and Gerlai (2018b) analyzed protein expression of three genes, BDNF (brain-derived neurotrophic factor), NCAM (neuronal cell adhesion molecule), and synaptophysin (a gene that encodes a synaptic protein). These genes and their proteins have been known to play fundamental roles in brain development as well as synaptic plasticity. Molecular mechanisms underlying synaptic plasticity have been known to be engaged by drugs of abuse, including alcohol. Importantly, antibodies developed against the aforementioned molecular targets in mammals have been found to work well in zebrafish too, suggesting strong evolutionary conservation of these targets. Using these antibodies for immunohistochemical detection, Mahabir et al. (2018b) found significant impairments in the brain of embryonic alcohol-treated fish. In almost all brain areas studied, the levels of the aforementioned proteins were significantly reduced in the embryonic alcohol-treated fish compared to control, suggesting a widespread alcohol effect (for example, see Fig. 6.3 for BDNF). How this global alcohol effect is associated with the earlier discussed fairly specific

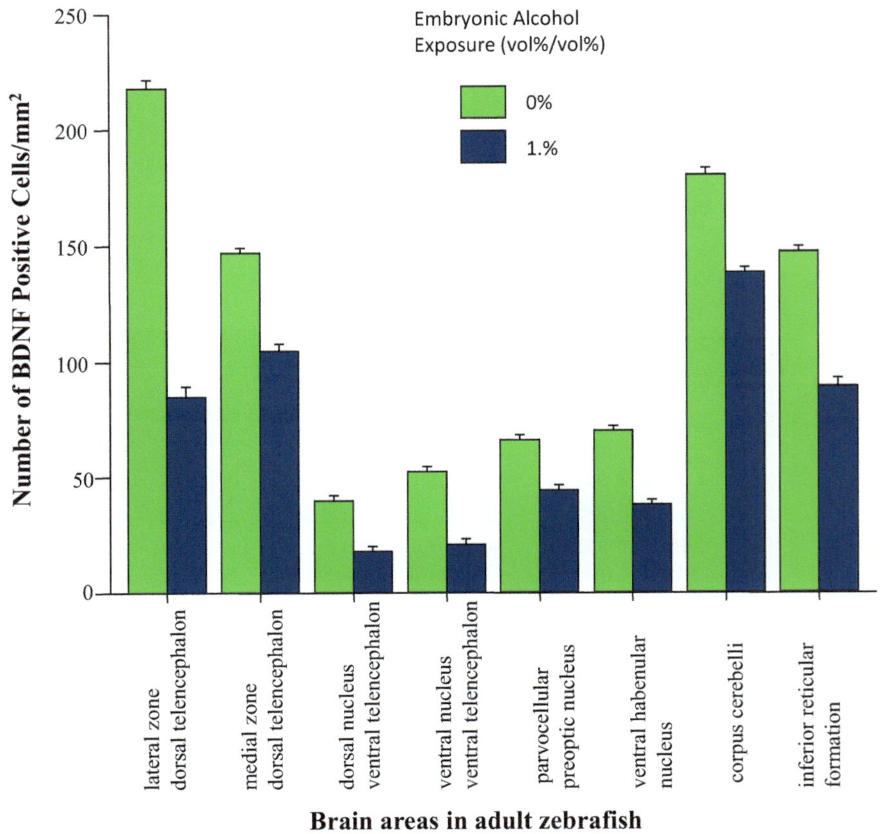

FIGURE 6.3 Number of BDNF positive cells as detected by immunohistochemical analysis in the brain of adult zebrafish exposed to alcohol for 2 h at their age of 24 hpf as compared to control. Mean + SEM are shown. Brain areas analyzed are specified under the X-axis. The treatment condition (0% alcohol, control, and 1% alcohol) is indicated by the legend. Note that in all brain areas analyzed, the alcohol exposure led to significant reduction in the number of BDNF positive neurons (dark blue bars) compared to control (green bars). *Modified from Mahabir, S., Chatterjee, D., Misquitte, K., Chatterjee, D., & Gerlai, R. (2018b). Lasting changes induced by mild alcohol exposure during embryonic development in brain derived neurotrophic factor, neuronal cell adhesion molecule and synaptophysin positive neurons quantified in adult zebrafish.* European Journal of Neuroscience, 47, *1457–1473.*

social behavioral and correlated monoaminergic system-related impairment is unknown at this point. Two possible explanations may be considered. One, some methods employed so far for the phenotypical analysis of embryonic alcohol-induced changes may be biased, i.e., may have differential sensitivity. For example, it is possible that the shoaling paradigm is more sensitive to detect behavioral impairment than other behavioral tasks we employed. Such differential sensitivity of test paradigms may lead to biased conclusions. Another possibility is that embryonic alcohol treatment although may affect numerous developmental and functional aspects of the brain, e.g., the social behavioral network and the monoaminergic neurotransmitter system are particularly sensitive to the deleterious effects of embryonic alcohol treatment. In other words, it is possible that what we observed is not an artifact resulting from our biased methods, but represents reality.

Unfortunately, at this point, we cannot distinguish between these two possibilities, but we do know that a fundamental molecular mechanism is altered soon after alcohol is administered to the embryo. Alcohol is known to enhance apoptotic cell death (Granato & Dering, 2018). Apoptosis is a natural process that is important during brain development, and it is thought, for example, to allow pruning of neurons that were overproduced and not properly integrated into the connectome. Mahabir et al. (unpublished data) found significant overexpression of the proapoptotic protein bax while they found no change in the level of the antiapoptotic protein bcl-2 when tested 2 days after embryonic alcohol exposure. Furthermore, the number of apoptotic cells was also increased, as quantified using TUNEL staining, in embryonic alcohol versus control zebrafish. Thus, it is possible that apoptotic cell death enhanced by embryonic alcohol expression contributes, or underlies, the observed reduction of protein levels associated with neuronal plasticity as well as the impaired dopaminergic and behavioral responses to social stimuli.

Until now, we focused our attention on neuronal function–related phenotypes, and we only mentioned the possibility of altered glial cell (astrocytic) function briefly. The majority of cells of the brain are not neurons, but glial cells, yet most neuroscientists studying CNS disorders focus on neurons. Nevertheless, embryonic alcohol is known to affect glial cells, and thus, behavioral as well as neuronal abnormalities resulting from exposure to this substance may arise due to altered glial cell functions. Glial cells play multiple distinct roles in the CNS. Reviewing these would be beyond the scope of this chapter. Briefly, the three main functions glial cells serve also distinguish the three main classes of these cells. Oligodendrocytes form myelin sheets around neurites and serve as insulators, allowing efficient transmission of action potentials. Astrocytes by now we know perform almost all functions a neuron does except they cannot generate action potentials. Lastly, microglia are important part of the immune system. Analysis of numerous phenotypical aspects of these glial cell types has revealed robust and widespread changes induced by embryonic alcohol exposure in the zebrafish (Chatterjee, Mahabir, Chatterjee, & Gerlai, 2020). For example, these authors found the number of astrocytes reduced, myelin sheet damaged, the expression of galactocerebrosides and myelin basic protein (major lipid components of the myelin sheet) reduced. The authors also found embryonic alcohol treatment to lead to activation of microglia along with enhanced caspase 3/7 activity in the brain of embryonic alcohol-exposed zebrafish suggesting the presence of an inflammatory response (Chatterjee et al., unpublished). It is thus likely that both the neuronal phenotypes and the behavioral responses found altered by embryonic alcohol treatment in zebrafish are not isolated examples of impairment but likely are associated with the aforementioned glial cell—related abnormalities.

Developmental stage of exposure

The earlier discussions focused on results obtained using an alcohol exposure paradigm in which administration of low concentrations of alcohol was conducted at a single developmental stage, i.e., at 24th hpf. However, alcohol likely affects ontogenesis and subsequent functioning of the brain at a developmental stage—specific manner. Thus, both the phenotypical manifestation and the molecular or neurobiological mechanisms underlying the effects of embryonic alcohol exposure should change with the developmental stage at which this substance reached the brain of the embryo. Testing such developmental stage dependency may allow one to discover developmentally sensitive periods and may also allow the investigation of mechanistic aspects of how alcohol alters brain development. The first steps toward such analyses have already been made. The earlier described alcohol exposure paradigm employed with zebrafish is particularly appropriate for such analysis because the short, 2 h long, alcohol exposure window allows high temporal resolution mapping of possible developmental stage—dependent effects of this substance. For example, Fernandes and Gerlai (unpublished results) found significant developmental stage dependency of alcohol effects on social behavioral responses with the 24th hpf exposure group showing the strongest impairment and fish exposed to alcohol earlier and later showing decreasing impairments. A somewhat similar finding was obtained by Facciol, Bailleul, Nguyen, & Gerlai (2020), who studied neurochemical changes induced by embryonic alcohol exposure at different stages of development and quantified dopamine and DOPAC at the age of 30 dpf of the exposed fish. These authors found that alcohol administered at or after 32 dpf had a less deleterious effect than when it was administered during earlier developmental stages, with the most robust effects being found for the 16th and 24th hpf administration period. Abozaid et al. (2019) also found significant alcohol exposure x developmental stage interaction. These authors studied the behavior of larval fish and could detect changes in motor responses induced by alcohol as early as by the eighth dpf. Importantly, the alcohol effects were found dependent upon the developmental stage at which the substance was administered with particularly strong effects detected when this substance was administered at or around 24 hpf.

Concluding remarks

Short exposure to low alcohol concentration in zebrafish has been found to lead to significant and lasting behavioral and neurobiological changes. Some of these changes mimic what is reported for the milder cases of human FASD. The zebrafish shares numerous evolutionarily conserved features with other vertebrates including even humans. From the nucleotide sequence of its genes, through the general anatomical layout of its brain, its neurotransmitter systems, numerous molecular and biochemical mechanisms, this species has been found to exhibit similarities to other vertebrates. It is a small animal that breeds well in captivity, and it is easy to keep in large numbers in the laboratory. Thus, the zebrafish has been suggested as a potentially excellent, and certainly the cheapest, vertebrate animal model alternative to more traditional mammalian laboratory species, including the house mouse or the rat. The previous pages demonstrate that the first steps toward building the zebrafish FASD model have been made. These pages, however, also show how little we know and how much more we need to do to make this species translationally relevant for the analysis of FASD. Nevertheless, although the zebrafish is a relative newcomer in the field, the fast progress it enjoyed in alcohol research in general, and in the analysis of FASD-related phenomena in particular, is noteworthy and suggests a bright future.

Applications to other areas of development

The utility of zebrafish in the modeling and analysis of human fetal alcohol spectrum disorders is increasingly appreciated as data accumulate about the behavioral, neurobiological, and molecular effects of this substance on brain development. In this chapter, we focused on findings obtained in the author's laboratory and reviewed the behavioral effects of embryonic alcohol exposure on zebrafish, along with changes in neurochemistry, neuronal protein expression, as well as glial cell—related phenotypes. We note that our discussions were centered on how one can utilize this small vertebrate to model FASD and how such studies may advance our understanding of the mechanisms underlying this devastating brain disorder.

However, one can easily view these approaches from another angle too. Alcohol is a complex drug, whose psychopharmacological profile depends upon the concentration, the length and timing of exposure, and the manner in which it is administered. Alcohol is known to interact directly with a large number of molecular targets and may influence an even larger number of biochemical mechanisms and thus additional molecular players. This complex molecular/biochemistry profile is further complicated by the potential cascade of changes resulting from the molecular level effects of this substance during development. From the perspective of modeling and analysis of human FASD, all the aforementioned represent a significant challenge. However, one can view these complications also as an opportunity. By challenging brain development using alcohol, one may be able to uncover fundamentally important mechanisms underlying ontogenesis of the vertebrate brain, and given the well-established translational relevance of the zebrafish, such discoveries are likely to be extendable to other vertebrates including our own.

Key facts

About fetal alcohol spectrum disorders (FASD):

- FASD is the most prevalent preventable cause of mental disability in the Western world.
- More than 10% of mothers admit to consuming alcohol during pregnancy.
- FASD is an umbrella term that describes a range of symptoms with differing severity.
- The most severe forms of FASD present debilitating symptoms including facial dysmorphia, severe cognitive and behavioral deficits such as social dysfunction, drug abuse, sleep disorders, and memory deficits.
- The milder forms of FASD are more frequent with prevalence estimates up to 2%—3%.
- The milder forms of FASD still represent lifelong suffering for the patients, who often exhibit impaired learning and memory and abnormal social behaviors.
- Biomarkers aiding diagnosis of FASD are not available.
- Proper treatment of FASD is not available.

About zebrafish:

- The zebrafish is the simplest vertebrate species employed in biomedical research.
- The zebrafish is an often-preferred subject of developmental biologists and geneticists.
- A single female zebrafish can produce 200 eggs multiple times a week.
- The zebrafish takes about 5 days to develop from fertilization to free swimming stage.
- The zebrafish remains transparent throughout its embryonic development.
- The zebrafish lives for about 5 years.
- The zebrafish has a sophisticated behavioral repertoire.
- Numerous recombinant DNA methods (forward and reverse genetic techniques) have been developed for the zebrafish.
- Numerous neurobiology techniques have been developed for the zebrafish.

Summary points

- The timing of alcohol exposure can be precisely determined in zebrafish due to external development of the embryo.
- Short- and low-dose alcohol exposure of zebrafish embryos leads to lasting changes that mimic several aspects of the prevalent milder forms of human FASD.
- Embryonic alcohol exposure was found to lead to impaired ability to respond to social stimuli in zebrafish.
- Embryonic alcohol exposure was found to lead to reduced levels of monoaminergic neurotransmitters dopamine and serotonin and their metabolites.

- Embryonic alcohol exposure was found to reduce expression of neuronal markers involved in synaptic plasticity and communication, including BDNF, NCAM, and synaptophysin
- Embryonic alcohol exposure was found to impair a variety of glial cell functions and phenotypes.

References

Abozaid, A., Trzuskot, L., Najmi, Z., Paul, I., Tsang, B., & Gerlai, R. (2019). Developmental stage and genotype dependent behavioral effects of embryonic alcohol exposure in zebrafish larvae. *Progress in Neuropsychopharmacology and Biological Psychiatry, 97*, 109774. https://doi.org/10.1016/j.pnpbp.2019.109774

Abrahao, K. P., Salinas, A. G., & Lovinger, D. M. (2017). Alcohol and the brain: Neuronal molecular targets, synapses, and circuits. *Neuron, 96*, 1223–1238.

Ahmed, T. S., Fernandes, Y., & Gerlai, R. (2012). Effects of animated images of sympatric predators and abstract shapes on fear responses in zebrafish. *Behaviour, 149*, 1125–1153.

Ahmed, O., Seguin, D., & Gerlai, R. (2011). An automated predator avoidance task in zebrafish. *Behavioural Brain Research, 216*, 166–171.

Albers, H. E. (2012). The regulation of social recognition, social communication and aggression: Vasopressin in the social behavior neural network. *Hormones and Behavior, 61*, 283–292.

Arenzana, F. J., Carvan, M. J., III, Aijon, J., Sanchez-Gonzalez, R., Arevalo, R., & Porteros, A. (2006). Teratogenic effects of ethanol exposure on zebrafish visual system development. *Neurotoxicology and Teratology, 28*, 342–348.

Barron, S., Hawkey, A., Fields, L., & Littleton, J. M. (2016). Animal models for medication development and application to treat fetal alcohol effects. *International Review of Neurobiology, 126*, 423–440.

Bass, S. L. S., & Gerlai, R. (2008). Zebrafish (*Danio rerio*) responds differentially to stimulus fish: The effects of sympatric and allopatric predators and harmless fish. *Behavioural Brain Research, 186*, 107–117.

Bernd, P. (2008). The role of neurotrophins during early development. *Gene Expression, 14*, 241–250.

Bilotta, J., Barnett, J. A., Hancock, L., & Saszik, S. (2004). Ethanol exposure alters zebrafish development: A novel model of fetal alcohol syndrome. *Neurotoxicology and Teratology, 26*, 737–743.

Brennan, C. H. (2011). Zebrafish behavioural assays of translational relevance for the study of psychiatric disease. *Reviews in the Neurosciences, 22*, 37–48.

Buske, C., & Gerlai, R. (2011). Early embryonic ethanol exposure impairs shoaling and the dopaminergic and serotoninergic systems in adult zebrafish. *Neurotoxicology and Teratology, 33*, 698–707.

Chatterjee, D., & Gerlai, R. (2009). High precision liquid chromatography analysis of dopaminergic and serotoninergic responses to acute alcohol exposure in zebrafish. *Behavioural Brain Research, 200*, 208–213.

Chatterjee, D., Mahabir, S., Chatterjee, D., & Gerlai, R. (2020). *Lasting alterations induced in glial cell phenotypes by short exposure to low alcohol dose during embryonic development in zebrafish.* Addiction Biology (under revision).

Chudley, A. E. (2018). Diagnosis of fetal alcohol spectrum disorder: Current practices and future considerations. *Biochemistry and Cell Biology, 96*, 231–236.

Facciol, A., Bailleul, C., Nguyen, S., & Gerlai, R. (2020). Developmental stage-dependent deficits induced by embryonic ethanol exposure in zebrafish: A neurochemical analysis. *Progress in Neuropsychopharmacology and Biological Psychiatry*. https://doi.org/10.1016/j.pnpbp.2020.109859 (in press).

Fernandes, Y., & Gerlai, R. (2009). Long-term behavioral changes in response to early developmental exposure to ethanol in zebrafish. *Alcoholism: Clinical and Experimental Research, 33*, 601–609.

Fernandes, Y., Rampersad, M., & Gerlai, R. (2015). Embryonic alcohol exposure impairs the dopaminergic system and social behavioural responses in adult zebrafish. *The International Journal of Neuropsychopharmacology, 18*(6), 1–8. https://doi.org/10.1093/ijnp/pyu089. pii: pyu089.

Fernandes, Y., Tran, S., Abraham, E., & Gerlai, R. (2014). Embryonic alcohol exposure impairs associative learning performance in adult zebrafish. *Behavioural Brain Research, 265*, 181–187.

Gerlai, R. (2002). Phenomics: Fiction or the future? *Trends in Neurosciences, 25*, 506–509.

Gerlai, R. (2010). High-throughput behavioral screens: The first step towards finding genes involved in vertebrate brain function using zebrafish. *Molecules, 15*, 2609–2622.

Gerlai, R. (2013). Antipredatory behavior of zebrafish: Adaptive function and a tool for translational research. *Evolutionary Psychology, 11*, 1–15.

Gerlai, R. (2014). Fish in behavior research: Unique tools with a great promise! *Journal of Neuroscience Methods, 234*, 54–58.

Gerlai, R. (2015). Embryonic alcohol exposure: Towards the development of a zebrafish model of fetal alcohol spectrum disorders. *Developmental Psychobiology, 57*, 787–798.

Gerlai, R., Lahav, M., Guo, S., & Rosenthal, A. (2000). Drinks like a fish: Zebra fish (*Danio rerio*) as a behavior genetic model to study alcohol effects. *Pharmacology Biochemistry and Behavior, 67*, 773–782.

Granato, A., & Dering, B. (2018). Alcohol and the developing brain: Why neurons die and how survivors change. *International Journal of Molecular Sciences, 19*. https://doi.org/10.3390/ijms19102992. pii: E2992.

Hamilton, W. D. (1971). Geometry for the selfish herd. *Journal of Theoretical Biology, 31*, 295–311.

Iversen, L. L., Iversen, S. D., Dunnett, S. B., & Björklund, A. (Eds.). (2010). *Dopamine handbook* (p. 614). Oxford: Oxford University Press.

Lange, S., Probst, C., Gmel, G., Rehm, J., Burd, L., & Popova, S. (2017). Global prevalence of fetal alcohol spectrum disorder among children and youth: A systematic review and meta-analysis. *JAMA Pediatrics, 171*, 948–956.

Lange, S., Shield, K., Rehm, J., & Popova, S. (2013). Prevalence of fetal alcohol spectrum disorders in child care settings: A meta-analysis. *Pediatrics, 132*, e980–e995. https://doi.org/10.1542/peds.2013-0066

Loucks, E., & Carvan, M. J., 3rd (2004). Strain-dependent effects of developmental ethanol exposure in zebrafish. *Neurotoxicology and Teratology, 26*, 745–755.

Lovely, C. B., Fernandes, Y., & Eberhart, J. K. (2016). Fishing for fetal alcohol spectrum disorders: Zebrafish as a model for ethanol teratogenesis. *Zebrafish, 13*, 391–398.

Lovely, C. B., Nobles, R. D., & Eberhart, J. K. (2014). Developmental age strengthens barriers to ethanol accumulation in zebrafish. *Alcohol, 48*, 595–602.

Mahabir, S., Chatterjee, D., & Gerlai, R. (2014). Strain dependent neurochemical changes induced by embryonic alcohol exposure in zebrafish. *Neurotoxicology and Teratology, 41*, 1–7.

Mahabir, S., Chatterjee, D., & Gerlai, R. (2018a). Short exposure to low concentrations of alcohol during embryonic development has only subtle and strain- dependent effect on the levels of five amino acid neurotransmitters in zebrafish. *Neurotoxicology and Teratology, 68*, 91–96.

Mahabir, S., Chatterjee, D., Misquitte, K., Chatterjee, D., & Gerlai, R. (2018b). Lasting changes induced by mild alcohol exposure during embryonic development in brain derived neurotrophic factor, neuronal cell adhesion molecule and synaptophysin positive neurons quantified in adult zebrafish. *European Journal of Neuroscience, 47*, 1457–1473.

Mele, T., Carman-Krzan, M., & Juric, D. M. (2010). Regulatory role of monoamine neurotransmitters in astrocytic NT-3 synthesis. *International Journal of Developmental Neuroscience, 28*, 13–19.

Miller, N., & Gerlai, R. (2007). Quantification of shoaling behaviour in zebrafish (*Danio rerio*). *Behavioural Brain Research, 184*, 157–166.

Miller, N., & Gerlai, R. (2011). Shoaling in zebrafish: What we don't know. *Reviews in the Neurosciences, 22*, 17–25.

Most, D., Ferguson, L., & Harris, R. A. (2014). Molecular basis of alcoholism. *Handbook of Clinical Neurology, 125*, 89–111.

Pickart, M. A., & Klee, E. W. (2014). Zebrafish approaches enhance the translational research tackle box. *Translational Research, 163*, 65–78.

Popova, S., Stade, B., Bekmuradov, D., Lange, S., & Rehm, J. (2011). What do we know about the economic impact of fetal alcohol spectrum disorder? A systematic literature review. *Alcohol, 46*, 490–497.

Qin, M., Wong, A., Seguin, D., & Gerlai, R. (2014). Induction of social behaviour in zebrafish: Live versus computer animated fish as stimuli. *Zebrafish, 11*, 185–197.

Saif, M., Chatterjee, D., Buske, C., & Gerlai, R. (2013). Sight of conspecific images induces changes in neurochemistry in zebrafish. *Behavioural Brain Research, 243*, 294–299.

Scerbina, T., Chatterjee, D., & Gerlai, R. (2012). Dopamine receptor antagonism disrupts social preference in zebrafish, a strain comparison study. *Amino Acids, 43*, 2059–2072.

Seguin, D., & Gerlai, R. (2018). Fetal Alcohol Spectrum Disorders: Zebrafish in the analysis of the milder and more prevalent form of the disease. *Behavioural Brain Research, 352*, 125–132.

Seguin, D., Shams, S., & Gerlai, R. (2016). Behavioural responses to novelty or to a predator stimulus are not altered in adult zebrafish by early embryonic alcohol exposure. *Alcoholism: Clinical and Experimental Research, 40*, 2667–2675.

Speedie, N., & Gerlai, R. (2008). Alarm substance induced behavioral responses in zebrafish (*Danio rerio*) *Behav. Brain Research, 188*, 168–177.

Chapter 7

Maternal iodine nutrition and neurodevelopment in the offspring

Shao Jia Zhou
School of Agriculture, Food & Wine, Robinson Research Institute, University of Adelaide, Adelaide, SA, Australia

List of abbreviations

IDD Iodine deficiency disorders
IQ Intelligence quotient
NIS Sodium-iodide symporter
RCT Randomized controlled trial
RDA Recommended dietary allowance
RDI Recommended dietary intake
T3 Triiodothyronine
T4 Thyroxin
TRH Thyrotropin-releasing hormone
TSH Thyroid-stimulating hormone
UIC Urinary iodine concentration
UIE Urinary iodine excretion
UL Upper level of intake
UNICEF United Nations Children's Fund
USI universal salt iodization
WHO World Health Organization

Mini-dictionary of terms

Endemic cretinism is the main effect of severe fetal iodine deficiency. There are two types of endemic cretinism: the neurological cretinism characterized by severe mental retardation, deaf mutism, and spastic diplegia, and the less common myxedematous cretinism, which is characterized by hypothyroidism and dwarfism.

Iodine deficiency disorders (IDD) describes a spectrum of iodine deficiency symptoms ranging from goiter, hypothyroidism, subtle impairments in cognitive function, cretinism, to congenital anomalies and adverse pregnancy outcomes.

Wolff−Chaikoff effect acute temporary inhibition of the organification of iodide in the thyroid to prevent the synthesis of large quantities of thyroid hormone in response to excess iodide intake. In healthy adults without thyroid dysfunction, the acute Wolff−Chaikoff effect generally lasts for a few days after which normal synthesis of thyroid hormone resumes.

Introduction

Nutrition in all stages of life is important for health and well-being. The World Health Organization (WHO) has designated the first 1000 days (pregnancy and the first 2 years of life) as the "window of opportunity" because appropriate nutrition during these early stages of life is critical for normal growth, development, and long-term health. Malnutrition in all forms during this critical period can result in irreversible adverse effects on the growth and development of offspring. Pregnant women are at risk of deficiencies in key micronutrients such as iron, folate, and iodine due to increased requirements, which can affect nutrient availability to the growing fetus. Micronutrient deficiency is often viewed as being more of an

Diagnosis, Management and Modeling of Neurodevelopmental Disorders. https://doi.org/10.1016/B978-0-12-817988-8.00007-5

issue in developing countries; however, iodine deficiency in particular has become a public health concern in developed countries with the reemergence of iodine deficiency in a number of countries such as the United States, United Kingdom, and Australia. Maternal iodine deficiency has been linked to impaired neurodevelopment in the offspring.

Iodine and its role in brain development

Iodine is an essential micronutrient that needs to be obtained from the diet and has a unique function that cannot be replaced or compensated for by other nutrients. Iodine is a substrate required for thyroid hormone, thyroxin (T4), and triiodothyronine (T3) synthesis in the thyroid gland. When the level of thyroid hormone in the blood is low, the hypothalamus releases thyrotropin-releasing hormone (TRH). TRH in turn stimulates the secretion of thyroid-stimulating hormone (TSH) from the pituitary, which signals to the thyroid gland to release thyroid hormone into the circulation. This process is regulated through a negative feedback mechanism in such a way that when the circulating level of thyroid hormone reaches a threshold, the release of TRH, TSH, and thyroid hormone is inhibited under normal physiological conditions to maintain homeostasis (see Fig. 7.1).

Thyroid hormone exerts its action primarily through binding to thyroid hormone receptors on target cells, thereby regulating gene expression (Brent, 1994). The pattern of gene expression determines which proteins in the target cells will be synthesized and at what level. Thyroid hormone receptors are present in many tissues and organs including the developing brain (Bernal & Pekonen, 1984), thus the actions of thyroid hormone are wide ranging. One of the key roles of thyroid hormone is supporting growth and brain development, which is particularly important during fetal development.

Brain development during the fetal stage

Brain development follows a specific sequence in which different parts of the brain develop and mature at different time points. While brain development continues throughout the fetal growth period and into the postnatal period, the embryonic stage (from 2 to 8 weeks postconception) is a stage of rapid fetal brain development. The neural tube forms during the third week of gestation, cortical neurogenesis (the generation of neurons) begins in humans from week 5 (Stenzel & Huttner, 2013) or week 6 of gestation (Stiles & Jernigan, 2010) and is largely completed by midgestation. By the end of week 8 gestation, the basic structure of the brain and the central and peripheral nervous systems are established (Stiles & Jernigan, 2010). The mature brain is composed of more than 100 billion neurons differing in shape, size, and function (Stiles & Jernigan, 2010). Most neurons are produced in the center of the brain and gradually migrate out to the neocortex. Neurons are connected with each other by short- and long-distance connecting fibers called dendrites and axons, and the point of

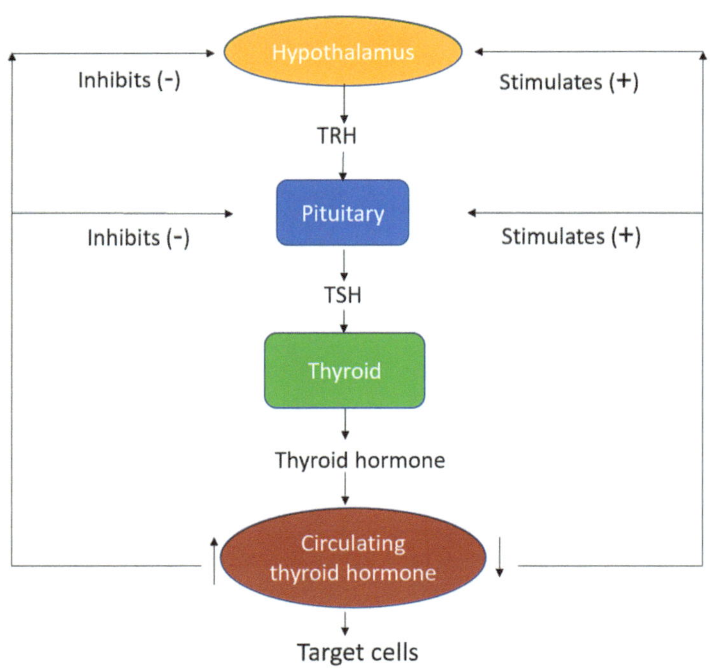

FIGURE 7.1 Regulation of thyroid hormone synthesis and secretion via negative feedback.

connection between two neurons is called a synapse (Stiles & Jernigan, 2010). The connection between neurons results in a neural network for information processing in the brain that controls all biological processes. Most axons are surrounded by a fatty membranous sheath called myelin and the process of encasing the nerve cells with myelin is referred to as myelination, which begins at around 16 weeks gestation and continues after birth. Myelinated nerve cells transmit signals more efficiently compared with unmyelinated nerve cells resulting in faster information processing.

The role of thyroid hormone in neurodevelopment

Extensive evidence from animal experiments shows that thyroid hormone is involved in the regulation of neurogenesis, neuron differentiation, migration, myelination, and synapse formation (Bernal, 2002; Stenzel & Huttner 2013; Thompson & Potter, 2000). Thyroid hormone is needed at all stages of brain development, but the fetal thyroid gland is not mature enough to synthesize its own thyroid hormone until approximately 13 weeks of gestation, and even then the level of thyroid hormone produced is insufficient and the thyroid reserve is low (Williams, 2008). The presence of thyroid hormone (de Escobar, Obregon, & del Rey, 2004) before the onset of fetal thyroid hormone production indicates the importance of thyroid hormone in fetal brain development and the dependency on the maternal supply of thyroid hormone during the early stage of fetal development. Perturbations in maternal thyroid hormone levels before the onset of fetal thyroid function affect gene expression patterns in a dose-dependent and brain region specific manner (Dowling, Martz, Leonard, & Zoeller, 2000; Royland, Parker, & Gilbert, 2008). The impact of thyroid hormone deficiency on brain development is dependent on the severity and timing of the deficiency.

Iodine requirements and the consequences of iodine deficiency

In food, iodine is found in several chemical forms with the inorganic iodide being the most common form. Iodide is easily absorbed from the gastrointestinal tract into the blood circulation and is actively transported to the thyroid by sodium-iodide symporter (NIS) for thyroid hormone synthesis. Other forms of iodine in food are reduced to iodide before absorption. The amount of iodine being absorbed is regulated by the activity of NIS. Low iodine intake or iodine deficiency increases the NIS expression and iodine absorption while excess iodine intake reduces its expression. Iodine intake in excess of requirements is excreted primarily through urine with a small amount in feces. The total iodine store of a healthy iodine-replete adult is approximately 15–20 mg, with 70%–80% of this stored in the thyroid gland (Zimmermann, Jooste, & Pandav, 2008).

Iodine requirements and the recommended dietary intake for iodine

Iodine requirements vary depending on gender and stages of the life cycle. A daily intake of 150 μg/day of iodine for adults and adolescents is the commonly recommended intake (Table 7.1). This recommendation is based on the amount of iodine required for the prevention of goiter, which is approximately 50–75 μg/day, with an additional buffer to account for increased requirements in certain conditions and goitrogens that may be present in the diet (Underwood, 1977, pp. 271–301; Food and Nutrition Board, 1989). Goitrogens are compounds in food or drugs that either compete for the transport of iodide into the thyroid or inhibit the synthesis of thyroid hormone. Pregnant women have increased requirements to support the growing fetus and the recommended intake for pregnant women ranges from 200 to 220 μg/day (Table 7.1). Interestingly the recommendation for the upper intake level (UL) varies significantly (Table 7.1), reflecting the uncertainty regarding the safe UL for iodine. The American Thyroid Association recommended that sustained iodine intake exceeding 500 μg daily (including supplements) should be avoided during pregnancy, due to concerns about the potential for fetal thyroid dysfunction as the fetal thyroid is sensitive to excessive iodine levels (Alexander et al., 2017).

The consequences of iodine deficiency

The consequences of iodine deficiency consist of a spectrum of disorders termed iodine deficiency disorders (IDD). The signs of IDD range from goiter, hypothyroidism, subtle impairments in cognitive function, to neurological cretinism characterized by severe mental retardation, deaf mutism, and spastic diplegia (Hetzel, 1983). IDD can also affect pregnancy outcomes and result in stillbirth, spontaneous abortion, and congenital anomalies (Hetzel, 1983). Decades of public health campaigns worldwide to eliminate iodine deficiency have shifted the epidemiology of iodine deficiency from severe (endemic cretinism) to a milder degree. Whether mild to moderate iodine deficiency in pregnancy affects the neurodevelopment of children is less certain and has been a focus of recent research.

TABLE 7.1 Recommended iodine intake for pregnant women versus nonpregnant adults.

Age	Recommended iodine intake (μg/day)				Recommended upper intake level (UL) (μg/day)			
	Australia: RDI (National Health & Medical Research Council and Aging, 2006)	EFSA: AI (European Food Safety Authority, 2014)	USA: RDA (Institute of Medicine, 2001)	WHO (World Health Organization and Food and Agriculture Organization of the United Nations, 2004)	Australia (National Health & Medical Research Council and Aging. 2006)	ECSCF(European Commission Health & Consumer Protection Directorate-General, 2002)	USA (Institute of Medicine, 2001)	WHO (World Health Organization and Food and Agriculture Organization of the United Nations, 2004)
14–18 years	150	120–130[a]	150	150	900	450–500[b]	300–900[c]	30 (μg/kg/day)
Adults (19+ years)	150	150[d]	150	150	1100	600	1100	30 (μg/kg/day)
Pregnant women	220	200	220	200	1100[e]	600	1100	40 (μg/kg/day)
Lactating	270	200	290	200	1100[e]	600	1100	40 (μg/kg/day)

AI, Adequate Intake; *EFSA*, European Food Safety Authority; *EUSCF*, The European Commission Scientific Committee on Food, *RDA*, recommended dietary allowance; *RDI*, recommended dietary intake, *WHO*, World Health Organization.

[a]*The recommended AI is 120 μg/day for children aged 4–14 years and 130 μg/day for aged 15–17 years.*

[b]*The recommended UL is 450 μg/day for aged 11–14 years and 500 μg/day for aged 15–17 years.*

[c]*The recommended UL is 300 μg/day for children aged 4–14 years and 900 μg/day for aged 15–17 years.*

[d]*The recommended AI is for adults ≥18 years.*

[e]*The UL is 900 μg/day for pregnant and lactating women aged 14–18 years and 1100 μg/day for aged ≥19 years.*

Assessment of iodine status

Central to the monitoring and control of iodine deficiency is the ability to accurate assess iodine status. Table 7.2 outlines the criteria to define iodine status of populations by different indicators (World Health Organization, 2007). WHO recommends the use of the median urinary iodine concentration (UIC) as the preferred indicator for assessing the iodine status of a population because in healthy iodine-replete individuals, approximately 90% of iodine intake is excreted in the urine. It is a simple, convenient, and inexpensive method of classifying the iodine status of populations. Goiter prevalence in school-age children is also used as an indicator of population iodine status, but its use in monitoring iodine status in populations with recent iodine interventions is not recommended as goiter reflects chronic inadequate iodine intake and is not sensitive to acute changes in iodine intake (Ristic-Medic et al., 2009). Blood biomarkers of thyroid function including T3, T4, and TSH may be affected by IDD, but T4 and T3 are not recommended for monitoring IDD as they are not sensitive unless IDD is more severe. WHO recommends the use of neonatal TSH concentration to monitor the iodine status of populations in countries where newborn screening is routinely practiced (World Health Organization, 2007). Newborn TSH concentration can be affected by factors such as exposure to iodine containing antiseptics during delivery and the timing of the newborn's blood sampling due to a physiological surge in TSH concentration soon after birth. These factors need to be considered when using newborn TSH data to monitor iodine status. A recent systematic review of 57 studies assessing the agreement between the different indicators of population iodine status showed that the TSH marker has a better agreement with goiter prevalence than median UIC when classifying the iodine status of populations (Wassie, Middleton, & Zhou, 2019). There is no single gold standard marker for population iodine status. Use of multiple indicators to monitor iodine status of populations over time may provide a better indication of population iodine status.

UIC is not recommended as an indicator to assess individual iodine status due to its large individual day-to-day as well as within-day variation. It has been shown that 10–12 repeat spot urine collections are needed to more accurately assess individuals' iodine status (Andersen, Pedersen, Pedersen, & Laurberg, 2001; König, Andersson, Hotz, Aeberli, & Zimmermann, 2011). Adjusting UIC by urine creatinine as UIC in μg per g of creatinine or using 24-hour urinary excretion (UIE) has been suggested as a better marker of individual iodine status. However, 24-hour UIE does not address the issue of day-to-day variation in iodine intake and excretion, and creatinine excretion is largely influenced by protein intake and other factors such as gender and body weight. Currently there is a lack of biomarker indicator to assess iodine status of individuals. Although newborn TSH is considered a sensitive marker of iodine status in neonates, there is currently no established criteria to define iodine status of individuals. Further research to develop sensitive and inexpensive indicators suitable to assess individual iodine status is warranted.

Maternal iodine nutrition and neurodevelopment of offspring

Research investigating the impact of maternal iodine nutrition in pregnancy on the neurodevelopment of the offspring in humans can be categorized into three areas: (1) observational studies that examine the association between maternal hypothyroxinemia or thyroid dysfunction and the neurodevelopmental outcome of the offspring; (2) observational studies that examine the association between maternal iodine status and the neurodevelopmental outcome of offspring; (3) maternal iodine supplementation in pregnancy and neurodevelopmental outcome of offspring.

TABLE 7.2 World Health Organization's criteria for assessing iodine status of populations (World Health Organization, 2007).

		Degree of iodine deficiency			Iodine sufficiency
	Iodine deficiency	Mild	Moderate	Severe	
Median UIC in (μg/L)					
Children and nonpregnant adults	<100	50–99	20–49	<20	≥100[a]
Pregnant women	<150	NA	NA	NA	≥150[b]
Proportion of neonatal TSH >5 mIU/L (%)	≥3	3–19	20–39	≥40	<3
Goiter prevalence (%)	≥5	5–19	20–29	≥30	<5

NA, not applicable; *TSH*, thyroid stimulating hormone, *UIC*, urinary iodine concentration.
[a]*UIC 100–199 μg/L indicates adequate intake, UIC 200–299 μg/L indicates intake above requirements, UIC ≥300 μg/L indicates excessive iodine intake and a potential risk of adverse health consequences.*
[b]*UIC 150–249 μg/L indicates adequate intake, UIC 250–499 μg/L indicates intake above requirements, UIC ≥500 μg/L indicates excessive iodine intake a potential risk of adverse health consequences.*

Maternal thyroid function in pregnancy and neurodevelopment of offspring

The growing fetus is completely dependent on the supply of maternal thyroid hormone for normal brain development before the onset of fetal thyroid function and is still partially reliant on the maternal thyroid hormone until the fetal thyroid becomes fully mature at birth (Williams, 2008; Stenzel & Huttner, 2013). Numerous studies have reported an association between maternal hypothyroxinaemia (T4 below the normal range) in pregnancy with lower developmental scores and a higher risk of developmental delay in the offspring (see Table 7.3). Subclinical hypothyroidism in pregnancy, defined as elevated TSH but normal T3 and T4 concentration, was associated with increased risk of intellectual and motor developmental delay in a meta-analysis of 15 cohort studies (Liu, Chen, Jing, & Li, 2018). Elevated TSH in pregnancy in the absence of thyroid disease may indicate maternal iodine deficiency. However, a Cochrane systematic review reported that screening and subsequent treatment for subclinical hypothyroidism or hypothyroxinaemia in pregnancy does not have a clear impact on maternal or infant health outcomes including neurodevelopment (Spencer, Bubner, Bain, & Middleton, 2015). Residual confounding effect in observational studies may partly explain the contrasting findings between randomized control trials (RCTs) and observational studies. Findings from the Cochrane review should be interpreted with caution as only one (Lazarus et al., 2012) of the two included trials in the Cochrane review reported on neurodevelopmental outcome in children. The study is a large RCT ($n = 21,846$) comparing universal screening and treatment for hypothyroidism versus no treatment on the IQ of children at 3 years of age. There were no differences in the IQ score or risk of developmental delay between children whose mothers were treated for hypothyroxinaemia identified at screening ($\sim$12 weeks' gestation) compared with children whose mothers' hypothyroxinaemia was not identified until after birth and thus hypothyroxinaemia during pregnancy was not treated. The authors suggested that treatment of hypothyroidism commencing at 13 weeks' gestation may be too late to impact the developing brain (Lazarus et al., 2012).

Maternal iodine intake or status in pregnancy and neurodevelopment in the offspring

Two earlier studies have been frequently cited as evidence supporting the view that mild iodine deficiency in pregnancy can impair intellectual ability in the offspring. These two retrospective cohort studies reported an association between mild iodine deficiency in pregnancy, defined as UIC <150 μg/L (Hynes, Otahal, Hay, & Burgess, 2013) or 150 μg/g creatinine (Bath, Steer, Golding, Emmett, & Rayman, 2013), and lower intelligence quotient (IQ) or poorer educational performance in children aged 8–9 years (Bath et al., 2013; Hynes et al., 2013). In contrast, more recent large prospective cohort studies (>1500 children) in populations of mild iodine deficiency (Rebagliato et al., 2013) or iodine sufficiency (Ghassabian, Steenweg-de Graaff, et al., 2014b) did not observe an association between maternal UIC <150 μg/L or 150 μg/g creatinine in early pregnancy and the IQ of children at 6 years or Bayley score at 30 months of age. While the degree and timing of iodine deficiency during pregnancy may partly explain the contrasting findings; however, UIC or UIC adjusted for creatinine is not recommended to be used to classify the iodine status of individuals. In the absence of more suitable indictors of an individual's iodine status, assessment of maternal habitual iodine intake as the exposure provides a better indication of the adequacy of maternal iodine nutrition in pregnancy.

Evidence of the relationship between maternal iodine intake in pregnancy and neurodevelopment in the offspring is limited. In an Australian study where fortification of bread with iodine is mandatory and the iodine status of the population is sufficient, both low (<220 μg/day) and high (>390 μg/d) iodine intake in pregnancy was associated with a lower Bayley scores and higher risk of developmental delay at 18 months of age in 696 children assessed (Zhou et al., 2019). Iodine intakes at both low and high levels were also associated with language delay and behavioral problems at 3 years of age in the Norwegian Mother and Child Cohort Study, a population of iodine sufficiency (Abel et al., 2017).

It is difficult to adjust for all social and environmental confounders influencing childhood development in observational studies. RCTs investigating the effects of iodine supplementation in pregnancy on the developmental outcomes of children are needed to establish a causal relationship.

Iodine supplementation in pregnancy and neurodevelopment in the offspring

As discussed earlier, the impact of iodine deficiency in pregnancy on brain development is dependent on the severity and the timing of the deficiency. The effect of iodine supplementation in pregnancy on neurodevelopment in offspring should be examined according to the iodine status of the study populations.

Historical RTCs in regions of severe iodine deficiency demonstrate a clear benefit of perinatal iodine supplementation in reducing the risk of cretinism in the offspring, but there was little benefit to childhood intelligence or gross development (See Table 7.4) (Zhou, Anderson, Gibson, & Makrides, 2013).

TABLE 7.3 Association between maternal hypothyroxinaemia in pregnancy and neurodevelopment in the offspring.

Country and reference	Design and participants (N)	Timing of fT4 assessment and definition of hypo-thyroxinaemia (HT)	Developmental outcome assessment & finding
The Netherlands (Pop et al., 1999)	Prospective cohort study, N = 220.	fT4: 12 & 32 weeks' GA • HT (n = 22): fT4 <10th percentile. • CG (n = 198): fT4 ?10th percentile.	Bayley at 10 months of age: • Score: HT < CG. • Risk of developmental delay: HT > CG. • fT4 evel at 32 weeks' GA: not associated with development.
The Netherlands (Pop et al., 2003)	A 3-year follow-up of a prospective cohort study. Women were recruited from an antenatal clinic, N = 125	fT4: 12, 24 & 32 weeks' GA • HT (n = 63 at 12 weeks' GA): fT4 <10th percentile. • CG (n = 62 at 12 weeks' GA): fT4: 50–90th percentiles.	Bayley at 1 and 2 years of age: • Score: HT < CG (8–10 points lower). • fT4 level at 24 and 32 weeks' GA: was not associated with bayley score.
The Netherlands (Kooistra, Crawford, van Baar, Brouwers, & Pop, 2006)	Prospective cohort study. Women were recruited from an antenatal clinic N = 204.	fT4: 12 weeks GA • HT (n = 108 at 12 weeks' GA): fT4 <10th percentile. • CG (n = 96 at 12 weeks' GA): fT4: 50–90th percentiles.	Behaviors at 3 weeks of age using NBAS: • Compared to the CG, HT had poorer behaviors orientation (1of 6 clusters of the NBAS scale).
The Netherlands: 18 and 30 months' assessment (Henrichs et al., 2010) and 6 years' assessment (Ghassabian, El Marroun, et al., 2014a)	Population-based cohort study (The generation R study), N = 3659.	fT4: <18 weeks' GA • Mild HT (n = NR): fT4 <10th percentile. • Severe HT (n = NR): fT4 < 5th percentile. • CG (n = NR): fT4 > 10th percentile	At 18 months of age (n = 2736): • Verbal (expressive vocabulary) cognitive using MCDI – risk of verbal cognitive delay[a]: severe HT > CG. At 30 months of age: • Verbal (n = 2225) and nonverbal (n = 2086) cognitive – risk of verbal (using LDS) and nonverbal (using PRCA) cognitive delay[a]: HT > CG. At 6 years of age (n = 3727): • nonverbal IQ: severe HT < CG (~5 points lower).
USA (Chevrier et al., 2011)	Population-based cohort study of latino, N = 287	fT4: 27 weeks GA • HT (n = NR): fT4 < 10th percentile. • CG (n = NR): fT4 ?10th & ? 90th percentile	At 6 (n = 271), 12 (n = 258), and 24 (n = 240) months: • Bayley II preschool language scale: ND At 60 months (n = 207): • Performance IQ using WPPSI-III and other scales: ND
USA (Craig et al., 2012)	Population-based cohort study (antenatal screening for Down syndrome), N = 4150.	fT4: in second trimester • HT (n = 99): fT4 ? 3rd percentile. • Matched CG (n = 99): fT4 ?10th & ? 90th percentile.	Bayley at 2 years: • Score: ND • Risk of developmental delay: ND

Continued

TABLE 7.3 Association between maternal hypothyroxinaemia in pregnancy and neurodevelopment in the offspring.—cont'd

Country and reference	Design and participants (N)	Timing of fT4 assessment and definition of hypo-thyroxinaemia (HT)	Developmental outcome assessment & finding
Spain (Julvez et al., 2013)	Population-based cohort study, N = 1761	fT4: 13 weeks GA • HT (n = NR): fT4 <5th percentile. • CG (n = NR): fT4 ?10th percentile	Bayley at 14 months • Mental score: HT < CG • Psychomotor score: ND

Bayley, Bayley Scales of Infant Development; *CG*, control group; *HT*, hypothyroxinaemia; *LDS*, Language Development Survey; *MCDI*, MacArthur Communicative Development Inventory; *NBAS*, Neonatal Behavioral Assessment Scale; *ND*, no difference; *PRCA*, Parent Report of Children's Abilities; *SVA*, spatial visualization abilities; *WPPSI-III*, Wechsler Preschool and Primary Scale of Intelligence third edition.
[a]*Verbal and nonverbal cognitive delay: defined as a score <15th percentile for age and gender-specific centile.*

TABLE 7.4 Randomized controlled trials of perinatal iodine supplementation and neurodevelopmental outcomes of children.

Country (reference)	Design, setting, and N (participants)	Intervention	Developmental outcome assessment and finding
Peru (Pretell, Torres, Zenteno, & Cornejo, 1972)	• Quasi-RCT (alternate families) in three villages. • Women of childbearing age randomized: N = 747 (Pretell, Palacois, Tello, Utiger, & Stanbury, 1974) and the number of newborns registered: N = 456 (Pretell et al., 1972). • UIE of the study population: 17 ?g/24 hr	• Intervention (IG): (n = 1992): 1) IG: 2 mL iodized oil containing 960 mg iodine if > 6 years, no nodular goiter; 2) IG: 0.2 mL iodized oil containing 96 mg iodine if nodular goiter present. • Control (CG): (n = 1191) poppy seed oil without iodine • Duration: single injection in 1966, followed by a reinjection 3 years later when a new group of participants were injected for the first time	At 5-year follow-up: • Gross development: ND[a] • IQ score (n = 81): ND[b] • Mental or language or hearing deficit (%): ND[b]
Papua New Guinea (Pharoah, Buttfield, & Hetzel, 1971; Connolly, Pharoah, & Hetzel, 1979; Pharoah & Connolly, 1987)	Quasi-RCT (alternate individuals) in 27 villages, N: ~16,500.	• Intervention (IG): Iodized oil containing approximately 1600 mg of iodine (4 mL, if aged >12 years) • Control (CG): saline solution • Duration: single IM injection in 1966	At 4-year follow-up (in 16/27 villages randomised) • Cretinism (%): IG < CG (Pharoah et al., 1971). At 10–16-year follow-up (in 5/27 villages with highest % of cretinism) • Developmental score < 10th centile: IG < CG (Pharoah & Connolly, 1987). • Grip strength, speed of movement, and unimanual accuracy: ND (Connolly et al., 1979). • Bimanual accuracy: IG > CG (Connolly et al., 1979).

Continued

TABLE 7.4 Randomized controlled trials of perinatal iodine supplementation and neurodevelopmental outcomes of children.—cont'd

Country (reference)	Design, setting, and N (participants)	Intervention	Developmental outcome assessment and finding
Thailand and India (Gowachirapant et al., 2017)	RCT. Women were recruited from an antenatal clinic in each country, N = 832	• Intervention: 200 ?g/d Iodine as KI • Control: placebo • Duration: from ? 14 weeks GA to delivery	IQ at 5–6 years using WPPSI-III: ND
Spain (Santiago et al., 2013)	Women were recruited from an antenatal clinic, N = 131	• Intervention: (1) KI 200 ?g/d + IS (n = 55); (2) KI 300 ?g/d + IS (n = 38) • Control: IS (n = 38) • Duration: from ?10 weeks' GA to delivery	Bayley-III at 1 year of age: ND
Australia (Zhou et al., 2015)	RCT. Women were recruited from antenatal clinics in two tertiary teaching hospitals N = 53	• Intervention: KI 150 ?g/d • Control: Placebo • Duration: < 20 weeks' GA to delivery	Bayley-III at 18 months of age: ND

Bayley-III, Bayley Scales of Infant and Toddler Development (third edition); *IG*, Iodine intervention group; *CG*, Control group; *IS*, iodized salt replaced noniodized salt in cooking and on table; *GA*, gestational age; *KI*, potassium iodide; *RCT*, randomized controlled trial; *ND*, no difference; *UIE*, urinary iodine excretion; *WPPSI-III*, Wechsler Preschool and Primary Scale of Intelligence Third Edition.
[a]*Gross development were assessed in ~70% of children.*
[b]*Assessed in one village only and children whose mothers were pregnant at the time of intervention were excluded from the analysis (eight in the IG and three in the CG).*

Although there are several RCTs of iodine supplementation in pregnancy, conducted in regions with less severe iodine deficiency, demonstrating that iodine supplements at dosages between 100 and 230 μg/day increased UIC (Zhou et al., 2013), only two RCTs examined the developmental outcome of children (see Table 7.4). Both RCTs reported no difference in the developmental outcomes of children between the iodine supplemented and control groups; however, both trials have methodological limitations due to small samples size (n = 131) (Santiago et al., 2013) and a high attrition rate with only 40% (330 out of 832 pregnant women randomized) children assessed for IQ at 5.4 years (mean value) of age (Gowachirapant et al., 2017).

There is a lack of RCTs conducted in regions of iodine sufficiency except one small RCT conducted in Australia. Unfortunately, the trial was prematurely terminated following the withdrawal of funding support, which resulted in a small sample size of 58 participants and no firm conclusion could be drawn from this small study (see Table 7.4). The efficacy and safety of maternal iodine supplementation in iodine sufficient populations on child development remain uncertain. In a large cohort (>1500 mother–infant pairs) in iodine-sufficient and mildly iodine-deficient areas of Spain, children whose mothers took iodine supplements in pregnancy at the current recommended level of 150 μg/day or above had five points lower score in the Bayley-II and 1.8-fold increase in developmental delay compared to children whose mothers took iodine supplements < 100 μg/day or no supplements (Murcia et al., 2011; Rebagliato et al., 2013).

These emerging data suggest that both inadequate and excess maternal iodine intake in pregnancy may impair the neurodevelopment of children as both situations can lead to inhibition of thyroid hormone production. The inhibitory effect of excess iodine is well recognized as the acute Wolff–Chaikoff effect (Wolff, Chaikoff et al., 1949), which is a protective mechanism against exposure to harmful effect of high thyroid hormone concentration by the temporary blocking of thyroid hormone production. Healthy adults are able to "escape" from the acute Wolff–Chaikoff effect after a few days and resume thyroid hormone synthesis. The fetus is more vulnerable to excess iodine as the ability to "escape" from the acute Wolff–Chaikoff effect is limited. Although the level of maternal iodine intake that triggers the Wolff–Chaikoff effect in fetus is unknown, the available evidence suggests that it could be in the range achievable in populations with sustained universal salt iodization or iodine fortification of staple foods and concurrent iodine supplementation at the commonly recommended dosages. This raises questions as to whether the current recommended safe upper limit for iodine intake in pregnancy of 1100 μg/day by most authorities may be too high, and whether universal iodine supplementation in pregnancy in populations of mild iodine-deficient or iodine-sufficient is warranted.

Conclusion

The most severe consequence of iodine deficiency in pregnancy is cretinism in offspring, but of greater public health relevance is the more common mild deficiency, which has emerged in a number of developed countries. Universal iodine supplementation in pregnancy is now widely recommended and practiced due to increasing concerns that mild iodine deficiency in pregnancy may lead to subtle intellectual impairment and behavior disorders in the offspring. However, the benefits and safety of universal iodine supplementation in pregnancy in mild iodine-deficient or iodine-sufficient populations remain unclear due to a lack of quality evidence from RCTs. Emerging evidence suggests that excess iodine intake in pregnancy (within the current recommended safe upper intake limit) may impairs neurodevelopment in the offspring. Reevaluation of the safe upper intake level in pregnancy and the establishment of more accurate indicators for classifying individual iodine status are warranted. RCTs to evaluate the safety and efficacy of universal iodine supplementation in pregnancy in mild iodine-deficient and iodine-sufficient populations are also needed to ensure that public health policy and clinical practice guidelines are evidence-based and safe.

Applications to other areas of development

Thyroid dysfunction has been linked to other health outcomes in adults including but not limited to diabetes, cardiovascular diseases, and neurodegenerative diseases such as Alzheimer's and dementia. Whether iodine nutrition in early life influences the development of these noncommunicable diseases and neurodegenerative diseases in human has not been investigated. Due to the critical role of iodine and thyroid hormone in brain development, neurodegenerative diseases may have a fetal origin related to iodine nutrition in pregnancy. Further research investigating the impact of early life iodine nutrition on the development of neurodegenerative diseases and other neurological disorders may potentially contribute to strategies for reducing the risk of developing Alzheimer's and dementia later in life.

Key facts of universal salt iodization (USI)

- USI was recommended as a safe and cost-effective strategy for the prevention and control of IDD by the WHO and UNICEF Joint Committee on Health Policy in 1993.
- USI requires that all salt for human and animal consumption is fortified with iodine and iodized salt should be consumed by > 90% of households.
- The level of iodine fortification was recommended to be at 20 mg of iodine per kg of salt, which aimed to provide 150 μg of iodine per person per day based on the average salt intake of 10 g/day and 30% iodine losses from manufacturing to household.
- In 2012, WHO recommended a reduction to salt intake to <5 g/d for adults as a public health intervention to reduce the risk of hypertension, cardiovascular disease, and their associated morbidity and mortality.
- As a result of the public health measure to reduce salt intake in the population, the USI guideline established in 1993 was updated in 2014 to reflect the lower salt intake target. The updated guideline recommends that the level of iodine fortification should be based on the estimated daily consumption of salt in the populations, ranging from 20 mg iodine per kg of salt for salt intake of 10 g/d to 39 mg/kg for salt intake of 5 g/d or 65 mg/kg for salt intake of 3 g/d.

Summary points

- Iodine is an essential nutrient required for thyroid hormone synthesis.
- Thyroid hormone regulates gene expression involving in neurogenesis, neuron differentiation, migration, myelination, and synapse formation.
- The fetus is completely dependent on the maternal supply of thyroid hormone for brain development before the onset of its own thyroid function in the second trimester of pregnancy.
- Severe iodine deficiency in pregnancy can cause irreversible intellectual impairment in the offspring and neurological cretinism in severe cases, but the effects of mild to moderate iodine deficiency in pregnancy on neurodevelopment are less clear.
- Perinatal iodine supplementation in regions of severe iodine deficiency is effective in reducing the incidence of cretinism but its efficacy and safety in mild iodine deficient or iodine sufficient populations are uncertain.
- Emerging evidence suggests that in iodine-sufficient populations, high maternal iodine intake from foods and supplements may adversely impair neurodevelopment in the offspring.
- Further research to investigate the safe upper intake limit of iodine in pregnancy and the efficacy and safety of universal iodine supplementation in pregnancy in mild iodine-deficient or iodine-sufficient populations is warranted to ensure optimal growth and development in the offspring.

References

Abel, M. H., Caspersen, I. H., Meltzer, H. M., Haugen, M., Brandlistuen, R. E., Aase, H., … Brantsaeter, A. L. (2017). Suboptimal maternal iodine intake is associated with impaired child neurodevelopment at 3 years of age in the Norwegian mother and child cohort study. *Journal of Nutrition, 147*, 1314−1324.

Alexander, E. K., Pearce, E. N., Brent, G. A., Brown, R. S., Chen, H., Dosiou, C., … Sullivan, S. (2017). Guidelines of the American thyroid association for the diagnosis and management of thyroid disease during pregnancy and the postpartum. *Thyroid, 27*, 315−389, 2017.

Andersen, S., Pedersen, K. M., Pedersen, I. B., & Laurberg, P. (2001). Variations in urinary iodine excretion and thyroid function. A 1-year study in healthy men. *European Journal of Endocrinology, 144*, 461−465.

Bath, S. C., Steer, C. D., Golding, J., Emmett, P., & Rayman, M. P. (2013). Effect of inadequate iodine status in UK pregnant women on cognitive outcomes in their children: Results from the avon longitudinal study of parents and children (ALSPAC). *Lancet, 382*, 331−337.

Bernal, J. (2002). Action of thyroid hormone in brain. *Journal of Endocrinological Investigation, 25*, 268−288.

Bernal, J., & Pekonen, F. (1984). Ontogenesis of the nuclear 3,5,3'-triiodothyronine receptor in the human fetal brain. *Endocrinology, 114*, 677−679.

Brent, G. A. (1994). The molecular basis of thyroid hormone action. *New England Journal of Medicine, 331*, 847−853.

Chevrier, J., Harley, K. G., Kogut, K., Holland, N., Johnson, C., & Eskenazi, B. (2011). Maternal thyroid function during the second half of pregnancy and child neurodevelopment at 6, 12, 24, and 60 months of age. *Journal of Thyroid Research*, 426427, 2011.

Connolly, K. J., Pharoah, P. O., & Hetzel, B. S. (1979). Fetal iodine deficiency and motor performance during childhood. *Lancet, 2*, 1149−1151.

Craig, W. Y., Allan, W. C., Kloza, E. M., Pulkkinen, A. J., Waisbren, S., Spratt, D. I., … Haddow, J. E. (2012). Mid-gestational maternal free thyroxine concentration and offspring neurocognitive development at age two years. *The Journal of Cinical Endocrinology and Metabolism, 97*, E22−E28.

Dowling, A. L., Martz, G. U., Leonard, J. L., & Zoeller, R. T. (2000). Acute changes in maternal thyroid hormone induce rapid and transient changes in gene expression in fetal rat brain. *Journal of Neuroscience: The Official Journal of the Society for Neuroscience, 20*, 2255−2265.

de Escobar, G. M., Obregon, M. J., & del Rey, F. E. (2004). Maternal thyroid hormones early in pregnancy and fetal brain development. *Best Practice and Research Clinical Endocrinology and Metabolism, 18*, 225−248.

European Commission Health & Consumer Protection Directorate-General. (2002). *Opinion of the scientific committee on food on the tolerable upper level of intake of iodine*. Brussels: European Commission.

European Food Safety Authority. (2014). Scientific opinion on dietary reference values for iodine. *EFSA Journal, 12*, 3660.

Food and Nutrition Board. (1989). *Recommended dietary allowances: Section 10. Trace elements*. Washington, D.C.USA): National Research Council.

Ghassabian, A., El Marroun, H., Peeters, R. P., Jaddoe, V. W., Hofman, A., Verhulst, F. C., … White, T. (2014a). Downstream effects of maternal hypothyroxinemia in early pregnancy: Nonverbal IQ and brain morphology in school-age children. *The Journal of Cinical Endocrinology and Metabolism, 99*, 2383−2390.

Ghassabian, A., Steenweg-de Graaff, J., Peeters, R. P., Ross, H. A., Jaddoe, V. W., Hofman, A., … Tiemeier, H. (2014b). Maternal urinary iodine concentration in pregnancy and children's cognition: Results from a population-based birth cohort in an iodine-sufficient area. *BMJ Open, 4*, e005520.

Gowachirapant, S., Jaiswal, N., Melse-Boonstra, A., Galetti, V., Stinca, S., Mackenzie, I., … Zimmermann, M. B. (2017). Effect of iodine supplementation in pregnant women on child neurodevelopment: A randomised, double-blind, placebo-controlled trial. *Lancet Diabetes Endocrinology, 5*, 853−863.

Henrichs, J., Bongers-Schokking, J. J., Schenk, J. J., Ghassabian, A., Schmidt, H. G., Visser, T. J., … Tiemeier, H. (2010). Maternal thyroid function during early pregnancy and cognitive functioning in early childhood: the generation R study. *The Journal of Cinical Endocrinology and Metabolism, 95*, 4227−4234.

Hetzel, B. S. (1983). Iodine deficiency disorders (IDD) and their eradication. *Lancet, 2*, 1126−1129.

Hynes, K. L., Otahal, P., Hay, I., & Burgess, J. R. (2013). Mild iodine deficiency during pregnancy is associated with reduced educational outcomes in the offspring: 9-year follow-up of the gestational iodine cohort. *The Journal of Cinical Endocrinology and Metabolism, 98*, 1954−1962.

Institute of Medicine. (2001). *Dietary reference intakes for vitamin A, vitamin K, arsenic, boron, chromium, copper, iodine, iron, manganese, molybdenum, nickel, silicon, vanadium, and zinc. Washington DC.*

Julvez, J., Alvarez-Pedrerol, M., Rebagliato, M., Murcia, M., Forns, J., Garcia-Esteban, R., … Sunyer, J. (2013). Thyroxine levels during pregnancy in healthy women and early child neurodevelopment. *Epidemiology, 24*, 150−157.

König, F., Andersson, M., Hotz, K., Aeberli, I., & Zimmermann, M. B. (2011). Ten repeat collections for urinary iodine from spot samples or 24-hour samples are needed to reliably estimate individual iodine status in women. *Journal of Nutrition, 141*, 2049−2054.

Kooistra, L., Crawford, S., van Baar, A. L., Brouwers, E. P., & Pop, V. J. (2006). Neonatal effects of maternal hypothyroxinemia during early pregnancy. *Pediatrics, 117*, 161−167.

Lazarus, J. H., Bestwick, J. P., Channon, S., Paradice, R., Maina, A., Rees, R., & Wald, N. J. (2012). Antenatal thyroid screening and childhood cognitive function. *New England Journal of Medicine, 366*(6), 493−501.

Liu, Y., Chen, H., Jing, C., & Li, F. (2018). The association between maternal subclinical hypothyroidism and growth, development, and childhood intelligence: A meta-analysis. *Journal of Clinical Research in Pediatric Endocrinology, 10*, 153−161.

Murcia, M., Rebagliato, M., Iniguez, C., Lopez-Espinosa, M. J., Estarlich, M., Plaza, B., … Ballester, F. (2011). Effect of iodine supplementation during pregnancy on infant neurodevelopment at 1 year of age. *American Journal of Epidemiology, 173*, 804−812.

National Health & Medical Research Council. (2006). *Nutrient references for Australia and New Zealand. Canberra.*

Pharoah, P. O., Buttfield, I. H., & Hetzel, B. S. (1971). Neurological damage to the fetus resulting from severe iodine deficiency during pregnancy. *Lancet, 1*, 308−310.

Pharoah, P. O., & Connolly, K. J. (1987). A controlled trial of iodinated oil for the prevention of endemic cretinism: A long-term follow-up. *International Journal of Epidemiology, 16*, 68–73.

Pop, V. J., Brouwers, E. P., Vader, H. L., Vulsma, T., van Baar, A. L., & de Vijlder, J. J. (2003). Maternal hypothyroxinaemia during early pregnancy and subsequent child development: A 3-year follow-up study. *Clinical Endocrinology, 59*, 282–288.

Pop, V. J., Kuijpens, J. L., van Baar, A. L., Verkerk, G., van Son, M. M., de Vijlder, J. J., … Vader, H. L. (1999). Low maternal free thyroxine concentrations during early pregnancy are associated with impaired psychomotor development in infancy. *Clinical Endocrinology, 50*, 149–155.

Pretell, E. A., Palacois, P., Tello, L., Wan, M., Utiger, R. D., & Stanbury, J. B. (1974). Iodine deficiency and the maternal/fetal relationship. *Nutritional Biochemistry and Pathology*, 143–154. Endemic goiter and cretinism: Continuing Threats to World Health D. JG and G. Medeiros-Neto. Washington, PAHO: 1974.

Pretell, E. A., Torres, T., Zenteno, V., & Cornejo, M. (1972). Prophylaxis of endemic goiter with iodized oil in rural Peru: Preliminary report of the effect on the development of newborn children. *Human development and the thyroid gland*, 246–265 (J. B. Stanbury and R. L. Kroc. New York, Plenum Publishing Corporation).

Rebagliato, M., Murcia, M., Alvarez-Pedrerol, M., Espada, M., Fernandez-Somoano, A., Lertxundi, N., … Ballester, F. (2013). Iodine supplementation during pregnancy and infant neuropsychological development: INMA mother and child cohort study. *American Journal of Epidemiology, 177*(9), 944–953.

Ristic-Medic, D., Piskackova, Z., Hooper, L., Ruprich, J., Casgrain, A., Ashton, K., … Glibetic, M. (2009). Methods of assessment of iodine status in humans: A systematic review. *American Journal of Clinical Nutrition, 89*, 2052S–2069S.

Royland, J. E., Parker, J. S., & Gilbert, M. E. (2008). A genomic analysis of subclinical hypothyroidism in hippocampus and neocortex of the developing rat brain. *Journal of Neuroendocrinology, 20*, 1319–1338.

Santiago, P., Velasco, I., Muela, J. A., Sanchez, B., Martinez, J., Rodriguez, A., … Soriguer, F. (2013). Infant neurocognitive development is independent of the use of iodised salt or iodine supplements given during pregnancy. *British Journal of Nutrition, 110*, 831–839.

Spencer, L., Bubner, T., Bain, E., & Middleton, P. (2015). Screening and subsequent management for thyroid dysfunction pre-pregnancy and during pregnancy for improving maternal and infant health. *Cochrane Database of Systematic Reviews, 21*(9), CD011263.

Stenzel, D., & Huttner, W. B. (2013). Role of maternal thyroid hormones in the developing neocortex and during human evolution. *Frontiers in Neuroanatomy, 7*, 19.

Stiles, J., & Jernigan, T. L. (2010). The basics of brain development. *Neuropsychology Review, 20*, 327–348.

Thompson, C. C., & Potter, G. B. (2000). Thyroid hormone action in neural development. *Cerebral Cortex, 10*, 939–945.

Underwood, E. J. (1977). *11 - iodine. Trace elements in human and animal nutrition* (4th ed.). Academic Press. E. J. Underwood.

Wassie, M. M., Middleton, P., & Zhou, S. J. (2019). Agreement between markers of population iodine status in classifying iodine status of populations: A systematic review. *American Journal of Clinical Nutrition, 110*, 949–958.

Williams, G. R. (2008). Neurodevelopmental and neurophysiological actions of thyroid hormone. *Journal of Neuroendocrinology, 20*, 784–794.

Wolff, J., Chaikoff, I. L., et al. (1949). The temporary nature of the inhibitory action of excess iodine on organic iodine synthesis in the normal thyroid. *Endocrinology, 45*, 504–513 (illust).

World Health Organization. (2007). *Assessment of iodine deficiency disorders and monitoring their elimination-a guide for programme managers Geneva, Switzerland.*

World Health Organization and Food and Agriculture Organization of the United Nations. (2004). *Vitamin and mineral requirements in human nutrition.* Geneva: Switzerland World Health Organization.

Zhou, S. J., Anderson, A. J., Gibson, R. A., & Makrides, M. (2013). Effect of iodine supplementation in pregnancy on child development and other clinical outcomes: A systematic review of randomized controlled trials. *American Journal of Clinical Nutrition, 98*, 1241–1254.

Zhou, S. J., Condo, D., Ryan, P., Skeaff, S. A., Howell, S., Anderson, P. J., … Makrides, M. (2019). Association between maternal iodine intake in pregnancy and childhood neurodevelopment at age 18 months. *American Journal of Epidemiology, 188*, 332–338.

Zhou, S. J., Skeaff, S. A., Ryan, P., Doyle, L. W., Anderson, P. J., Kornman, L., … Makrides, M. (2015). The effect of iodine supplementation in pregnancy on early childhood neurodevelopment and clinical outcomes: Results of an aborted randomised placebo-controlled trial. *Trials, 16*, 563.

Zimmermann, M. B., Jooste, P. L., & Pandav, C. S. (2008). Iodine-deficiency disorders. *Lancet, 372*, 1251–1262.

Chapter 8

Endocrine aspects of development. Thyroid hormone actions in neurological processes during brain development

Ana Montero-Pedrazuela[1], Carmen Grijota-Martínez[1,2], Eva Ausó[3], Soledad Bárez-López[4] and Ana Guadaño-Ferraz[1]

[1]Department of Endocrine and Nervous System Pathophysiology, Instituto de Investigaciones Biomédicas Alberto Sols, Consejo Superior de Investigaciones Científicas-Universidad Autónoma de Madrid, Madrid, Spain; [2]Department of Cellular Biology, Universidad Complutense de Madrid, Madrid, Spain; [3]Department of Optics, Pharmacology and Anatomy, Universidad de Alicante, Alicante, Spain; [4]Translational Health Sciences, University of Bristol, Bristol, United Kingdom

List of abbreviations

CNS central nervous system
DIO deiodinase
E embryonic day
EDC endocrine-disrupting chemical
GW gestational week
MCT8 Monocarboxylate transporter 8
NSC neural stem cell
OPC oligodendrocyte progenitor cell
SVZ subventricular zone
T3 3-5-3′-triiodothyronine
T4 thyroxine or 3-5-3′-5′-tetraiodothyronine
TH thyroid hormones
TR thyroid hormone receptor
VZ ventricular zone

Mini-dictionary of terms

Hormones Chemical messengers produced by the endocrine glands that integrate the function of different organs.
Hypothyroidism Disease characterized by low T4 and T3 levels and high TSH levels.
Hypothyroxinemia Disease characterized by low T4 levels and normal TSH values. It is frequently associated to dietary alterations as iodine deficiency and/or food goitrogens (mainly vegetables).
Myelination Covering axons with multiple layers of oligodendrocyte cell's membrane forming a sheath, which allows a faster transmission of the electric impulse.
Neurite outgrowth Generation of axons and dendrites in neurons by the elongation, sprouting and branching of the cell membrane.
Neurogenesis Process of generation of new differentiated neurons from neural stem cells.
Neuronal migration Movement of neuronal-committed cells through the brain tissue, from the place of birth to their final position, following two main pathways: radial and tangential migration.
Neurotransmission Transmission of electric impulses across synapsis between neurons, delivering information to/from the nervous system.
Synaptogenesis Process of generation of synaptic contacts between neurons, which is essential for their function and survival.
Thyroid hormones Hormones synthetized in the thyroid gland derived from the amino acid tyrosine, by adding 3 (T3) or 4 (T4) iodine atoms.

Diagnosis, Management and Modeling of Neurodevelopmental Disorders. https://doi.org/10.1016/B978-0-12-817988-8.00008-7

Hormones in brain development

Hormones are messengers produced by the endocrine glands that integrate the function of different organs. In the central nervous system (CNS), hormones regulate plastic changes in the brain along the life span: from the neural plate forming the neural tube, throughout the fetal and early postnatal brain development, until the posterior plasticity events in the puberty, adulthood, and aging (Garcia-Segura, 2009). There is a cross talk between the endocrine and nervous systems contributing during development to the maturation of both of them.

The main role of the hormones reaching the brain during development is to provide environmental information leading to plastic modulation of the CNS, that is, adaptation. The hormonal imprinting during neurodevelopment will be decisive for adult functions, namely neuroendocrine, neuroimmune, behavioral, affective, and cognitive. Hence, altered hormonal levels during development may induce permanent impairments of brain function.

The actions of hormones on the CNS include, but are not restricted to, the regulation of cell generation and death, migration, growth and retraction of cellular processes, synaptic contacts, and establishment of neuronal circuits. Thus, hormonal action involves different cell types (neurons subtypes, diverse glial cells, cells forming blood vessels, and other brain barriers) under a specific regional and temporal pattern of activity.

This chapter will be mainly focused on thyroid hormones' (TH) actions as they play a major role as indispensable factors for the normal development of the CNS, regulating different events in a timely and regionally specific manner.

Other important hormones for neurodevelopment are gonadal and stress hormones, as well as different peptide hormones including ghrelin, peptide YY, leptin, growth hormone, and IGF-1 (Fig. 8.1).

Gonadal hormones, mainly estradiol and testosterone, induce sex differences in the organization of certain brain areas, generating female or male traits. These sexually dimorphic structures (which include neocortex, hippocampus, hypothalamus, or amygdala) show different neuroendocrine regulation leading to dimorphic behaviors and plasticity in the adult (Schwarz & McCarthy, 2008).

Stress hormones as glucocorticoids cross the placenta. They regulate myelin maturation, hippocampal function, and plasticity, as well as the maturation of the hypothalamus–pituitary–adrenal axis. Prenatal or perinatal stress is an important

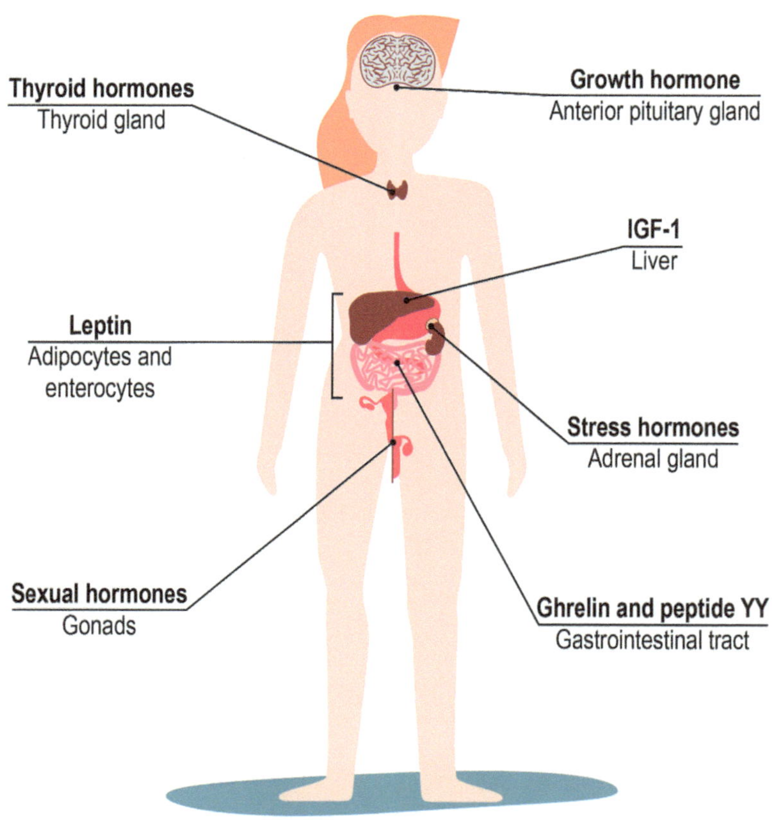

FIGURE 8.1 **Hormones involved in CNS development.** The figure shows the main hormones influencing the development of CNS in humans and their main place of production.

risk factor for psychopathological alterations, including anxiety, aggression, hyperactivity, and cognitive problems in the adult. Alterations in perinatal stress hormone levels induce inadequate responses to stress in later life (van den Bergh, Dahnke, & Mennes, 2018).

Ghrelin and peptide YY are involved in the development of the neural tube (Yuzuriha et al., 2007). Leptin regulates the maturation of hypothalamic neuronal circuits that control energy homeostasis and food intake, and although its actions in the developing brain are not well known, they may affect later body functions (Bouret, 2010). Growth hormone has a role in neuronal proliferation and survival, synaptogenesis, and myelination (Scheepens, Moderscheim, & Gluckman, 2005). IGF-1 affects the production of neurons and oligodendrocytes from neural stem cells (NSC) and regulates axon outgrowth, neuronal migration, synaptogenesis, and myelination (Dyer, Vahdatpour, Sanfeliu, & Tropea, 2016).

As described, different hormones regulate the same events during brain development. Moreover, these hormones interplay between them. For example, stress hormones interact with TH, IGF-1, or leptin to modulate brain plasticity; and the differential effects of prenatal stress in male and female offspring rely on gonadal hormones-related brain dimorphism.

In the fetus, the main hormonal source is of maternal origin, as the fetal hormonal secretions and their control are still immature. However, some hormones such as estradiol are also produced locally, acting as autocrine or paracrine factors (McCarthy, 2008).

Thyroid hormones have an essential role in brain development

TH carry out pleiotropic actions during brain development, with region and time-specific patterns. The availability of TH to the developing brain relies on maternal hormones crossing the placenta until the genesis of the fetal thyroid gland, which takes place at 16th–20th gestational weeks (GW) in humans.

The thyroid gland synthetizes TH, mainly T4 (thyroxine or 3-5-3′-5′-tetraiodothyronine) and a small amount of T3 (3-5-3′-triiodothyronine). In the adult, the nuclear active form of TH, T3, reaches the brain from the blood, together with the more abundant T4 through specific transporter proteins such as monocarboxylate transporter 8 (MCT8; Bernal, Guadaño-Ferraz, & Morte, 2015). However, at fetal stages, T4 is the main form reaching fetal tissues (Morreale de Escobar, Obregon, & Escobar del Rey, 2004). The activation of T4 to T3 and degradation of them both are achieved by deiodinase enzymes (DIO2 and DIO3, respectively), which control the region and timing for TH action. For more details on TH availability and metabolism during neurodevelopment, refer to chapter "Availability and metabolism of thyroid hormones in the developing brain." The main action of T3 is the modulation of gene expression through its binding to nuclear receptors (TRs) in the target cells (Bernal, 2005). There are two TR genes: THRA and THRB, encoding three functional receptors, TRα1, TRβ1, and TRβ2.

Due to the pleiotropic features of TH activity and the unique cellular diversity of the brain, TH actions show a characteristic complexity (Fig. 8.2). Direct and indirect actions as well as genomic and nongenomic activity of both T3 and T4 have been described. In addition to TH activation and inactivation by deiodinases, the differential expressions of transporter proteins and transducers of TH actions in different cell populations and regions with different temporal patterns have to be taken into account. These transducers (nuclear receptors, coactivators, or corepressors that induce the expression of different target genes) frequently show multiple levels of interaction. Much of this molecular complexity is currently unknown.

TH activity affects several processes throughout brain development, including cell proliferation, migration, differentiation, synaptogenesis, and myelination. Altered TH levels during brain development induce mild to devastating impairments on the neurological outcome of the offspring depending on the severity of the alteration on TH levels, its duration, and the time window affected, as different neural processes are taking place at different time points (Fig. 8.3).

Much of the current knowledge on TH actions during brain development has been obtained using experimental animals, especially rodents. It is essential to consider the big differences among ontogenic events between these murine models and humans. For example, although neurogenesis is a phenomenon that extends into the adulthood, neocorticogenesis in humans takes place from 5th to 20th GW (Marin-Padilla, 1978) and in mice at embryonic days E11–17 (Levers, Edgar, & Price, 2001).

In humans, thyroid function begins at midgestation, around 16–20 GW. Developmental processes that occur before fetal TH production, such as proliferation of neural progenitors and early neuronal migration in the cerebral cortex and medial ganglionic eminence, are exclusively dependent on maternal TH. After the onset of fetal thyroid function, processes such as ongoing neurogenesis and neuronal migration, axonal and dendritic growth, synaptogenesis, and onset of gliogenesis and myelination depend on fetal and maternal TH contribution. The late migration of granular cells in the hippocampus and granular and Purkinje cells in the cerebellum and the later events of gliogenesis and myelination occur in perinatal and postnatal periods and are therefore controlled mainly by neonatal TH.

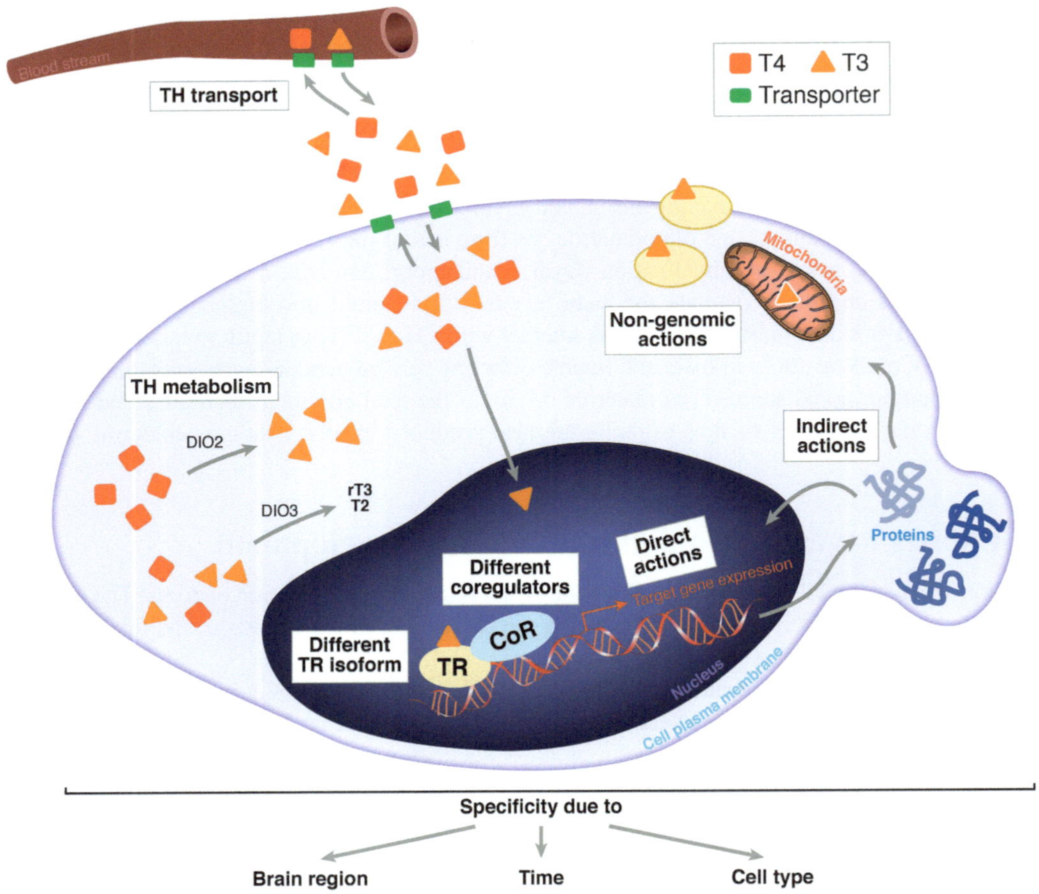

FIGURE 8.2 **Complexity of thyroid hormone actions in neural cells.** Different transporter proteins are involved in TH transport from/to the blood, crossing the blood–brain barrier, and between the different neural cells. TH are mainly metabolized by deiodination by DIO2 (activation of T4) and DIO3 (inactivation of both T4 and T3). TH exert genomic actions by binding of T3 to nuclear receptors (TR). These nuclear receptors are transcription factors that modulate the expression of target genes. There are different isoforms encoded in two genes, THRA (TRα) and THRB (TRβ). The binding of T3 to these receptors promotes the recruitment of different coregulators (CoR), modulating the expression of different genes, which mediate direct or indirect actions. TH extranuclear nongenomic actions are carried out through receptors at the plasma membrane, cytosol, or mitochondria. The expression of the different proteins involved in these processes (transporters, deiodinases, TR isoforms, coregulators) and therefore, the final actions, depend on cell type, brain region, and the time of development.

We describe further in some detail the main events of neurodevelopment influenced by TH. Table 8.1 summarizes these events and the associated alterations secondary to TH deficiency during neurodevelopment.

Role of thyroid hormones in neurogenesis

Corticogenesis is an early event during neurodevelopment, whereas the development of the hippocampus and cerebellum is mainly perinatal, making these regions vulnerable to TH deficiency at different moments.

Extremely summarized, in the development of the vertebrate CNS, the neuronal and glial cells derive from NSC populating the neuroepithelium along the anteroposterior axis of the CNS primordium. After several asymmetric cell divisions of NSC in the ventricular zone (VZ) and subventricular zone (SVZ) germinal layers, the final number of cells is fixed, determining brain size (Rakic, 1972). In mammals, TH are one of the main regulators of neurogenesis, in both the developing and adult brain. The main evidence of the influence of TH on progenitor cells at embryonic stages has been obtained from mammalian models. Imbalances in TH levels result in a decrease in the proliferating progenitor cell pool size (Mohan et al., 2012). In vitro studies revealed that T3 promotes the neuronal differentiation of embryonic mouse NSC, suggesting that T3 contributes to early neurogenesis. Silencing the TRα1 protein expression abolishes the effects of T3 on embryonic NSC (Chen et al., 2012). It has been shown in rats that the cellular progenitors of the VZ expressing TRα1, the TH transporter MCT8, and the enzyme DIO2 are sensitive to TH actions (Mohan et al., 2012). Indeed, maternal

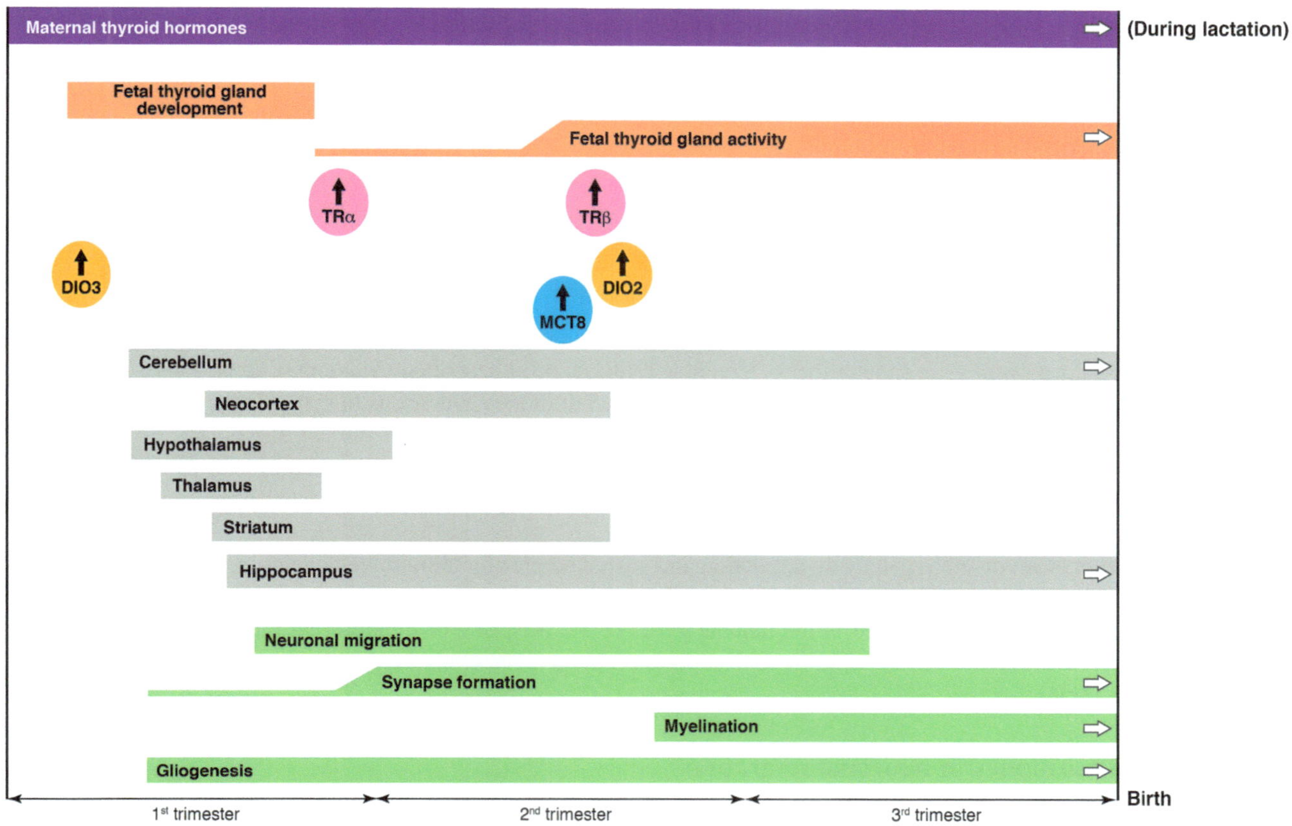

FIGURE 8.3 **Chronology of human brain development and the influence of thyroid hormones from conception to birth.** Fetal thyroid gland starts functioning around the 11th gestational week (GW) and completes its maturation at 16–20th GW. Before fetal gland functioning, the unique source of TH is from maternal origin. The expression of the specific TH transporter MCT8 peaks at the 18th GW. Thyroid hormone receptors (TRα) are detected already in the 10th GW. In addition to the control by transporter proteins, deiodination by DIO2 and DIO3 enzymes is the main mechanism of control of T3 availability inside the neural cells. The development of the different brain regions is represented by *gray rectangles*. The rest of the indicated processes (*green rectangles*) take place at different moments depending on the brain regions. *Horizontal arrows* indicate processes that continue after birth. *Vertical arrows* indicate a peak in the expression of the indicated proteins. *Figure adapted from Howdeshell KL (2002) with permission. Data for TRs, MCT8 and deiodinases peak expression have been obtained from http://evodevoapp.kaessmannlab.org, developed by Cardoso-Moreira M et al. Nature (2019).*

TABLE 8.1 Main alterations related to thyroid hormones deficiency on neurodevelopment.

Event	Alterations
Neurogenesis	Reduced proliferation of progenitor cells Altered corticogenesis: reduced number of cortical neurons Downregulation of cell cycle genes
Neuronal migration	Delayed neuronal radial migration Alteration in neuronal tangential migration directionality Delay in radial glial cells maturation Downregulation of genes related to neuronal migration
Neuritogenesis	Impaired neurite outgrowth and dendrite and axonal branching Altered regulation of cytoskeleton, growth cone, and axon guidance proteins
Synaptogenesis	Altered distribution of dendritic spines and synaptic connections Downregulation of proteins related to synaptogenesis Reduced synaptic transmission and plasticity
Myelination	Impaired oligodendrocyte precursor cells differentiation Reduced number of oligodendrocytes Downregulation of myelin proteins Reduction in myelin deposition and delayed myelination
Neurotransmission	Delayed maturation of the cholinergic system Altered development of the inhibitory GABAergic system

Sequential neurodevelopment events and alterations due to TH deficiency.

hypothyroidism during gestation causes cell cycle disruption, increases apoptosis, decreases the proliferation and the number of neural progenitors, and produces defects in neuronal differentiation. The result is a reduction in cortical thickness and a decrease in the number of neurons, especially in the upper layers, and disrupted learning and memory capacities in the adulthood (Mohan et al., 2012).

Similar effects are found on cell proliferation and survival in other areas of the brain such as the hippocampus, which leads to a moderate reduced brain weight and hippocampal volume, resulting in abnormal brain function (Gilbert, Goodman, Gomez, Johnstone, & Ramos, 2017). In addition, a reduction in the synaptic transmission and impairment of short-term synaptic plasticity and long-term potentiation in the dentate gyrus, as well as altered spatial learning in the progeny of hypothyroid mothers throughout gestation (Gilbert & Sui, 2006) have been found.

Perinatal TH deficiency also induces a decrease in hippocampal cell proliferation and survival that results in a significant reduction of its area (Farahvar & Meisami, 2007). Similar effects are found in the dentate gyrus, where a reduction in its volume as well as in the number of granular cells has been described (Madeira, Cadete-Leite, Andrade, & Paula-Barbosa, 1991). Perinatal TH deficiency has also been shown to induce alterations in synaptic functions in CA1 area of the hippocampus (Gilbert, 2004).

TH deficiency at perinatal stages also has a profound effect on cerebellar neurogenesis, leading to a persistent external granular layer due to delayed proliferation and migration of granular cells, as well as to a severe impairment of Purkinje cells development (Horn & Heuer, 2010).

TH upregulate genes involved in cell cycle regulation and sustained proliferation in the developing cortex (Morte et al., 2010). TH deficiency induces shifts in the expression of genes related to neurogenesis, such as the gene encoding cyclooxygenase 2 that was downregulated in the dentate gyrus of the offspring of hypothyroid rats (Shiraki et al., 2014). Even mild hypothyroxinemia induces changes in gene expression in the developing hippocampus and neocortex (Royland, Parker, & Gilbert, 2008).

Role of thyroid hormones in neuronal migration

In cortical development, neurons generated in the SVZ follow different patterns of migration. Glutamatergic neurons migrate radially, following the processes of radial glial cells, reaching their position in the cortex. On the other hand, interneurons forming inhibitory cortical circuits follow a pattern of tangential migration. These migratory processes are necessary to form the six cortical layers characteristic of the mammalian brain. Therefore, alterations in migration produce changes in the structure and morphology of the brain as well as in its functions. TH play a crucial role in neuronal migration since it regulates the expression of genes related to this process such as *Reelin* and *Dab1* (Alvarez-Dolado et al., 1999), *Laminin* (Farwell, Tranter, & Leonard, 1995), or *Prostaglandin D2* (García-Fernández et al., 1997), expression of which is reduced in the brain of hypothyroid animals. The deficiency of TH produces alterations in the cytoarchitecture and functions of the brain. At the structural level, in rodents, TH deficiency during the second half of gestation produces alterations in the cerebral cortex morphology, with a reduction in the cortical thickness, absence of cortical lamination and heterotopic cells in the subcortical white matter (Berbel, Ausó, Garcia-Velasco, Molina, & Camacho, 2001). Moreover, hypothyroidism alters the development of radial glial cells in the neocortex (Martinez-Galan, Escobar del Rey, Morreale de Escobar, Santacana, & Ruiz-Marcos, 2004). A situation of moderate and transitory maternal hypothyroxinemia produces the same structural alterations in the cerebral cortex and in the hippocampus (Lavado-Autric et al., 2003). In these two regions, there is an evident delay in the radial migration of glutamatergic excitatory neurons along the radial glia. On the other hand, although the timing of the tangential migration of GABAergic inhibitory neurons is not altered, the correct direction of migration is lost (Ausó et al., 2004; Cuevas et al., 2005), suggesting that TH modulate genes directly involved not only in the migration of neurons, but also in their migratory guidance. Indeed, recent studies in the prenatal human brain point to the radial glia as an important source of T3 for migrating neural cells (López-Espíndola et al., 2019).

An inadequate neuronal migration due to TH deficiency also leads to alterations in callosal and anterior commissure connectivity (Guadaño-Ferraz, Escobar del Rey, Morreale de Escobar, Innocenti, & Berbel, 1994; Lucio et al., 1997) and in thalamocortical connections (Ausó et al., 2001).

Effects of thyroid hormones in astroglial cells maturation

Although the study of TH actions in the brain has historically been focused on neurons and oligodendrocyte precursor cells (OPCs) maturation (see further), lately it has been expanded to other glial cells such as astrocytes.

Astrocytes follow a maturation program, milestoned by the transition in the expression of the microfilaments nestin–vimentin–glial fibrillary acidic protein (GFAP), from the most immature to the mature state. Radial glial cells, which guide the neuronal radial migration in the cerebral cortex, are the precursors of the astrocytic lineage.

TH deficiency is shown to delay the maturation of radial glial cells of the hippocampus and neocortex (Martinez-Galan et al., 2004; Rami & Rabie, 1988). Besides, TH regulate astrocyte number and maturation of Bergmann glia in the cerebellum, and a delay in vimentin–GFAP transition due to hypothyroidism has also been found in the rat cerebellum inducing retarded astrocyte differentiation (Das, Ghosh, Gharami, & Das, 2018). This effect could be mediated by T3 directly (if TH receptors are expressed in radial glial cells) or indirectly, most probably through regulation of the extracellular matrix protein REELIN (Alvarez-Dolado et al., 1999).

Moreover, astroglial cells secrete different proteins, including growth factors and regulators of extracellular matrix proteins that induce neuronal differentiation and axonal elongation, whose release is under the influence of T3 (Das et al., 2018).

Involvement of thyroid hormones in later events of neurodevelopment

During brain development, different processes take place following a well-established spatiotemporal sequence. Thus, alterations in some of the early events can affect posterior events, making it difficult to differentiate between the direct effects of TH deficiency during late gestation and those arising from alterations during early neurodevelopment. For example, altered proliferation and maturation of neurons affect their migration, which in turn affects their positioning. This misplacing will alter the patterns of connectivity between neural cells, which will additionally influence the neuronal arborization, affecting the synaptic activity, the establishment of neuronal networks, and in general, the neuronal wiring. Developmental TH deficiency alters OPCs differentiation (see further). In addition, the impairment of axonal maturation will also affect myelination. The alteration in other glial cell types, such as radial glial cells or microglia, has a great impact on neuronal survival, migration, synaptic transmission, etc. Nevertheless, specific effects of TH deficiency have also been described for events that take place at later developmental stages. These studies have been carried out mainly in rodents and have focused on regions known to be especially sensitive to TH deprivation, as neocortex, hippocampus, and cerebellum.

These later events of brain development include axon and dendrite outgrowth, synaptogenesis, and myelination. These processes are essential for the efficiency of cell-to-cell communication and consequently for the functional maturation of the neural circuits. Finally, TH are also involved in the maturation of neurotransmission systems.

Neurite outgrowth

After migration, the neurons reach their final position and begin to extend processes to make synaptic contacts. In the cerebellum, TH insufficiency causes altered branching of the dendritic tree in Purkinje cells (Legrand, 1984). This effect is accompanied by shorter parallel fibers of the granule cells (Lauder, 1978), which appears to be caused by their slower migration. Hence, an important TH action in the cerebellum would be to synchronize differentiation events on Purkinje and granule cells. As Purkinje and granule cells establish synaptic contacts, this reduction in their respective neurites would probably induce a decrease in the number of contacts.

A parallel mechanism has also been proposed for pyramidal and granule cells of the hippocampus (Koromilas et al., 2010), where developmentally hypothyroid animals show altered dendritic arborization in pyramidal and granule cells (Rami, Patel, & Rabie, 1986). Therefore, a major effect of TH would be to synchronize and/or coordinate the differentiation programs of different neuronal cell types.

Neuronal process outgrowth relies highly on cytoskeleton proteins regulation along the axon and in the growth cone. One possible mechanism of TH influence on neurite outgrowth is by regulating the expression of microtubule-associated proteins and some tubulin isotypes (Nunez, Couchie, Aniello, & Bridoux, 1991). In fact, it has been described that hypothyroidism reduces the number of microtubules in the dendrites of the Purkinje cells of the cerebellum (Faivre, Legrand, & Rabié, 1983). Besides, hypothyroidism alters the expression of growth cone proteins and molecules involved in axon guidance (Wong & Leung, 2001).

In cerebellum, an additional mechanism of TH action involves a T3-stimulated loop of growth factors secretion between Purkinje and granule cells, which is thought to be an important mechanism for the organization of this region (Flamant, Gauthier, & Richard, 2017).

Synaptogenesis

Synaptogenesis is intimately related to neurite outgrowth. After elongation, axons and dendrites make contacts and establish the initial synapses. This synaptic activity itself modulates the stabilization or elimination of these contacts, as well as the maturation of the synaptic terminals. TH deficiency downregulates the expression of several proteins involved in synaptogenesis as the receptor EphA5 and its ligand ephrin-A5, NMDAR-1, PSD95, and CaMKII in hippocampal neurons (Suo et al., 2018). This could explain the alterations in the distribution of dendritic spines and synaptic connections shown in the pyramidal cells of the neocortex and hippocampus due to developmental TH deficiency (Rami et al., 1986; Ruiz-Marcos, Sanchez-Toscano, Obregon, Escobar del Rey, & Morreale de Escobar, 1982). Additionally, synaptic plasticity alterations may contribute to the neurological deficits associated with TH deficiency. In this sense, hypothyroidism decreases the expression of neurogranin, a postsynaptic protein involved in synaptic plasticity (Iñiguez et al., 1996).

Myelination

In humans, the process of myelination begins around mid-gestation (Fig. 8.3). Myelination in the CNS is dependent on oligodendrocytes, which derive from OPCs. Two populations of OPCs (early and late) have been described in murine models, and it has been postulated that they respond differently to TH (Wong & Leung, 2001). The initial myelination relies on early OPCs, differentiation of which is indirectly affected by T3 that regulates the secretion of neurotrophins by other cell types. Later OPCs are responsible for myelin repair and oligodendrocyte renewal and respond directly to T3 (Flamant et al., 2017). Developmental hypothyroidism in rodents decreases the number of oligodendrocytes, due to an impaired differentiation of OPCs, and delays myelin deposition (Billon, Tokumoto, Forrest, & Raff, 2001). This delay is normalized with age. T3 induces the expression of myelin genes during OPCs differentiation, directly, or through the regulation of enzymes involved in myelin synthesis. This way, prenatal TH deficiency reduces the expression of the myelin basic protein, proteolipid protein, myelin-associated glycoprotein, and 2′,3′-cyclic nucleotide 3′phosphodiesterase (Ibarrola & Rodriguez-Pena, 1997). Besides these effects on oligodendrocytes, other factors associated to TH deficiency during development may affect myelination, as the lower size of axonal diameter associated to impaired axonal maturation that may hinder myelination.

All these hypothyroidism-related alterations involve a reduction in myelin deposition and induce a decrease in the number of myelinated axons in the corpus callosum and anterior commissure, which impair interhemispheric connections (Berbel, Guadaño-Ferraz, Angulo, & Ramon Cerezo, 1994; Guadaño-Ferraz et al., 1994).

In humans, a few studies have used imaging techniques to study brain morphological changes in children born to mothers diagnosed with hypothyroidism (Samadi, Skocic, & Rovet, 2015; Willoughby, McAndrews, & Rovet, 2014a,b). These works have found alterations in the morphology of the corpus callosum and reduced hippocampal volumes, associated with memory deficits. Due to the lack of a consistent body of evidence, it would be important to take advantage of the now affordability of these imaging techniques to deepen into the morphological alterations of children due to TH deficiency during neurodevelopment.

Neurotransmission

TH are essential in the process of formation and maturation of several neurotransmission systems (as cholinergic and GABAergic systems), influencing neuronal electric properties and general brain connectivity and wiring.

Developmental TH deficiency induces an important delay in the maturation of the cholinergic system in several brain regions. It induces a reduction in the expression of the muscarinic receptor 1, as well as a reduced activity of choline acetyltransferase and an increased activity of acetylcholinesterase (Domingues et al., 2018; Koromilas et al., 2010). These alterations decrease acetylcholine and increase choline levels, modifying the homeostasis of the cholinergic system during neurodevelopment.

TH play a critical role also in the maturation of the inhibitory GABAergic system in different brain regions. In the cerebellum, hypothyroidism reduces the proliferation of GABAergic interneurons and delays the maturation of their precursors, which produces a decrease in the number of mature GABAergic neurons and terminals (Manzano, Bernal, & Morte, 2007). In the neocortex and hippocampus, it reduces the expression of the calcium-binding protein parvalbumin, a marker of a subpopulation of GABAergic interneurons (Berbel, Marco, Cerezo, & DeFelipe, 1996; Gilbert et al., 2007). This alteration in parvalbumin expression has been associated with functional deficits in inhibitory synaptic transmission in the dentate gyrus (Gilbert et al., 2007). The expression of the presynaptic protein glutamic acid decarboxylase 65, which produces GABA, is also reduced in both cells and processes due to TH deficiency during development (Sawano, Takahashi, Negishi, & Tashiro, 2013). Furthermore, TH regulate the expression of the sodium-chloride cotransporter KCC2, an essential postsynaptic component involved in the maturation of the GABAergic system (Sawano et al., 2013). The increase of this transporter during postnatal mice development is critical for the switching of GABAergic transmission from excitatory to inhibitory and hypothyroidism induces a delay in this switch, altering the maturation of the GABAergic system. It has been suggested that these effects of TH action on GABAergic neurotransmission could be partly mediated by TH regulation of neurotrophin expression.

TH also affect other neurotransmitter systems. Hypothyroid condition during brain development delays the maturation of the dopaminergic system (Vaccari et al., 1990), particularly related to a reduction in striatal D1 receptors and altered dopamine metabolism, and affects glutamatergic neurotransmission (Domingues et al., 2018), mainly by alterations in glutamine–glutamate metabolism in astroglial cells.

All these effects in the maturation of the different neurotransmitter systems will probably interfere with later synaptic transmission and plasticity and would be one of the most important causes of the neurological deficits associated with TH deficiency during critical periods of development.

Pathologies related to altered thyroid hormone levels during brain development

Severe deficiency of TH (fetal and maternal) during gestation and perinatal ages causes irreversible damage to the neurological development of the fetus (Morreale de Escobar, Obregon, & Escobar del Rey, 2000). The extent of the neurological impairments related to TH deficiency during development depends on the duration, severity, and time window of the deficiency (Table 8.2). These conditions can be prevented by routine blood analyses during pregnancy and subsequent treatment with L-T4 if necessary in cases of hypothyroidism and with supplemental iodine programs during pregnancy to avoid hypothyroxinemia.

On the other hand, in situations of congenital hypothyroidism, with a fetal gland that does not produce or produces insufficient quantities of TH, maternal TH can compensate this deficiency during the gestational period allowing a correct neurological development of the fetus (Obregon, Calvo, Escobar Del Rey, & Morreale de Escobar, 2007; Vulsma, Gons, & de Vijlder, 1989). Routine screening programs are implemented in hospitals to detect this problem right after delivery.

TABLE 8.2 Main origin of thyroid hormones deficiency during fetal development.

		Aetiology	Effects in newborn
Maternal TH deficiency	Hypothyroidism	Various, frequently autoimmune thyroiditis	Alterations related to timing and severity of the deficiency: from milder to severe neurological alterations. Milder forms: impaired cognition, altered psychomotor and language development, behavioral problems. Most severe: cretinism (profound mental retardation and neurological impairments).
	Hypothyroxinemia	Iodine deficiency (due to low iodine intake and/or goitrogens)	
Congenital hypothyroidism		Defects in fetal gland development or TH production (maternal TH protects the before birth; TH treatment must be administered lifelong after birth)	At birth: hypotonia, feeding problems, constipation. In child if untreated: mental retardation, deficits in memory, language, fine motor abilities, lack of attention, delay in bone maturation and growth.
Genetic resistance to TRα (RTHα)		Mutations in TRα	Neurodevelopmental delay, bradycardia, skeletal dysplasia
Genetic resistance to TRβ (RTHβ)		Mutations in TRβ	Goiter, attention deficit disorder, tachycardia
Allan–Herndon–Dudley syndrome (AHDS)		Mutations in the TH transporter MCT8	Global developmental delay, profound mental retardation, lack of speech, hypotonia, very low IQ
Other causes		Exposure to endocrine disruptor chemicals	Potentially affect neurodevelopment (epidemiological studies are needed)
		Preterm birth (transient hyothyroxinemia due to hypothalamic immaturity)	Related to increased risk of poor cognitive development and of cerebral palsy

The table classifies different situations that lead to thyroid hormone deficiency during fetal development, indicating their origin and consequences on the newborn.

In addition to the lack of TH directly related to the maternal/fetal production, other problems of genetic origin such as mutations in specific TH receptors or transporters can be the cause of an impaired TH action, causing severe neurological damage during development (Table 8.2).

Applications to other areas of development

Iodine deficiency is the most frequent preventable cause of reproductive failure, mental impairments, and learning disabilities in the child. Nowadays, due to screening and iodination programs, severe TH deficiency is almost eradicated in developed countries. However, it has been demonstrated that mild degrees of iodine deficiency during the development, with a large prevalence across developed and developing countries, are potentially dangerous for a correct brain development.

Beyond iodine deficiency, a more unknown factor that interferes with the proper functioning of the thyroid system is the exposure to environmental endocrine disrupting chemicals (EDCs). EDCs comprise a heterogeneous group of substances capable of altering the synthesis or action of the natural hormone in the organism. These compounds can affect most of the steps in the pathway of action of the TH (Mughal, Fini, & Demeneix, 2018). EDCs can alter the function of the hypothalamus−pituitary−thyroid axis, as well as modify the uptake of iodine by the thyroid gland. EDCs can also impair TH interaction with serum transporter proteins or disrupt TH metabolism, affecting the activity of deiodinases and other enzymes. They can also interfere with genomic TH action (agonizing or antagonizing TR activity). Different epidemiological studies have shown a correlation between EDCs exposure and alterations of children neurodevelopment. Further studies are needed to assess the influence of EDCs during neurodevelopment, also in combination with iodine deficiency.

Key facts on neurodevelopmental actions of thyroid hormones

- Due to their involvement in diverse events during neurodevelopment, TH could be considered as master regulators on brain development and later functioning.
- Until the functioning of the fetal thyroid gland, the maternal supply of TH is essential for a correct neurodevelopment.
- The main way of action of TH is by binding of T3 to nuclear receptors in the nucleus of the target cells, modulating gene expression patterns.
- The amount of T3 that reaches the nucleus is mainly controlled by plasma membrane transporters and the activity of deiodinase enzymes.
- The lack of TH during development leads to diverse neurological damage that depends on the severity and the time of the deficiency.

Summary points

- Hormones are key factors throughout neurodevelopment.
- Thyroid hormones are essential for the correct development and functioning of the nervous system.
- Thyroid hormones regulate essential processes such as cell proliferation, differentiation and migration, development of cellular processes, synapsis establishment, and myelination.
- The main action of thyroid hormones is conducted through interaction of T3 with nuclear receptors, modulating the expression of target genes.
- Among the target genes regulated by thyroid hormones, there are cell cycle proteins, growth factors and neurotrophins, cytoskeleton proteins, axon guidance and synaptogenesis-related proteins, myelin proteins, and enzymes related to neurotransmitter metabolism.
- The actions of T3 in brain vary across cell types, regions, and time.
- In the fetus, the maternal supply of thyroid hormones is essential for a correct neurodevelopment until the fetal gland is active.
- TH deficiency during brain development induces mild to devastating impairments on the neurological outcome of the offspring depending on the severity of the deficiency, its duration, and the time window affected.

References

Alvarez-Dolado, M., Ruiz, M., Del Río, J. A., Alcántara, S., Burgaya, F., Sheldon, M., Nakajima, K., Bernal, J., Howell, B. W , Curran, T., Soriano, E., & Muñoz, A. (1999). Thyroid hormone regulates reelin and dab1 expression during brain development. *Journal of Neuroscience, 19*(16), 6979–6993.

Ausó, E., Cases, O., Fouquet, C., Camacho, M., Garcia-Velasco, J. V., Gaspar, P., & Berbel, P. (2001). Protracted expression of serotonin transporter and altered thalamocortical projections in the barrelfield of hypothyroid rats. *European Journal of Neuroscience, 14*(12), 1968–1980.

Ausó, E., Lavado-Autric, R., Cuevas, E., Del Rey, F. E., Morreale De Escobar, G., & Berbel, P. (2004). A moderate and transient deficiency of maternal thyroid function at the beginning of fetal neocorticogenesis alters neuronal migration. *Endocrinology, 145*(9), 4037–4047.

Berbel, P., Ausó, E., Garcia-Velasco, J. V., Molina, M. L., & Camacho, M. (2001). Role of thyroid hormones in the maturation and organisation of rat barrel cortex. *Neuroscience, 107*(3), 383–394.

Berbel, P., Guadaño-Ferraz, A., Angulo, A., & Ramon Cerezo, J. (1994). Role of thyroid hormones in the maturation of interhemispheric connections in rats. *Behavioural Brain Research, 64*(1–2), 9–14.

Berbel, P., Marco, P., Cerezo, J. R., & DeFelipe, J. (1996). Distribution of parvalbumin immunoreactivity in the neocortex of hypothyroid adult rats. *Neuroscience Letters, 204*(1–2), 65–68.

van den Bergh, B. R. H., Dahnke, R., & Mennes, M. (2018). Prenatal stress and the developing brain: Risks for neurodevelopmental disorders. *Development and Psychopathology, 30*(3), 743–762.

Bernal, J. (2005). Thyroid hormones and brain development. *Vitamines and Hormones, 71*, 95–122.

Bernal, J., Guadaño-Ferraz, A., & Morte, B. (2015). Thyroid hormone transporters–functions and clinical implications. *Nature Reviews Endocrinology, 11*(7), 406–417.

Billon, N., Tokumoto, Y., Forrest, D., & Raff, M. (2001). Role of thyroid hormone receptors in timing oligodendrocyte differentiation. *Developmental Biology, 235*(1), 110–120.

Bouret, S. G. (2010). Neurodevelopmental actions of leptin. *Brain Research, 1350*, 2–9.

Cardoso-Moreira, M., Halbert, J., Valloton, D., Velten, B., Chen, C., Shao, Y., … Kaessmann, H. (2019). Gene expression across mammalian organ development. *Nature, 571*(7766), 505–509.

Chen, C., Zhou, Z., Zhong, M., Zhang, Y., Li, M., Zhang, L., Qu, M., Yang, J., Wang, Y., & Yu, Z. (2012). Thyroid hormone promotes neuronal differentiation of embryonic neural stem cells by inhibiting STAT3 signaling through TRalpha1. *Stem Cells and Development, 21*(14), 2667–2681.

Cuevas, E., Ausó, E., Telefont, M., Morreale de Escobar, G., Sotelo, C., & Berbel, P. (2005). Transient maternal hypothyroxinemia at onset of corticogenesis alters tangential migration of medial ganglionic eminence-derived neurons. *European Journal of Neuroscience, 22*(3), 541–551.

Das, M., Ghosh, M., Gharami, K., & Das, S. (2018). Thyroid hormone and astrocyte differentiation. *Vitamins and Hormones, 106*, 283–312.

Domingues, J. T., Wajima, C. S., Cesconetto, P. A., Parisotto, E. B., Winkelmann-Duarte, E., Santos, K. D., Saleh, N., Filippin-Monteiro, F. B., Razzera, G., Mena Barreto Silva, F. R., Pessoa-Pureur, R., & Zamoner, A. (2018). Experimentally-induced maternal hypothyroidism alters enzyme activities and the sensorimotor cortex of the offspring rats. *Molecular and Cellular Endocrinology, 478*, 62–76.

Dyer, A. H., Vahdatpour, C., Sanfeliu, A., & Tropea, D. (2016). The role of insulin-like growth factor 1 (IGF-1) in brain development, maturation and neuroplasticity. *Neuroscience, 325*, 89–99.

Faivre, C., Legrand, Ch, & Rabié, A. (1983). Effects of thyroid deficiency and corrective effects of thyroxine on microtubules and mitochondria in cerebellar Purkinje cell dendrites of developing rats. *Developmental Brain Research, 8*(1), 21–30.

Farahvar, A., & Meisami, E. (2007). Novel two-dimensional morphometric maps and quantitative analysis reveal marked growth and structural recovery of the rat hippocampal regions from early hypothyroid retardation. *Experimental Neurology, 204*(2), 541–555.

Farwell, A. P., Tranter, M. P., & Leonard, J. L. (1995). Thyroxine-dependent regulation of integrin-laminin interactions in astrocytes. *Endocrinology, 136*(9), 3909–3915.

Flamant, F., Gauthier, K., & Richard, S. (2017). Genetic investigation of thyroid hormone receptor function in the developing and adult brain. *Current Topics in Developmental Biology, 125*, 303–335.

García-Fernández, L. F., Rausell, E., Urade, Y., Hayaishi, O., Bernal, J., & Munoz, A. (1997). Hypothyroidism alters the expression of prostaglandin D2 synthase/beta trace in specific areas of the developing rat brain. *European Journal of Neuroscience, 9*(8), 1566–1573.

Garcia-Segura, L. M. (2009). *Hormones and brain plasticity*. Oxford, NY: Oxford University Press.

Gilbert, M. E. (2004). Alterations in synaptic transmission and plasticity in area CA1 of adult hippocampus following developmental hypothyroidism. *Developmental Brain Research, 148*(1), 11–18.

Gilbert, M. E., Goodman, J. H., Gomez, J., Johnstone, A. F. M., & Ramos, R. L. (2017). Adult hippocampal neurogenesis is impaired by transient and moderate developmental thyroid hormone disruption. *Neurotoxicology, 59*, 9–21.

Gilbert, M. E., & Sui, L. (2006). Dose-dependent reductions in spatial learning and synaptic function in the dentate gyrus of adult rats following developmental thyroid hormone insufficiency. *Brain Research, 1069*(1), 10–22.

Gilbert, M. E., Sui, L., Walker, M. J., Anderson, W., Thomas, S., Smoller, S. N., Schon, J. P., Phani, S., & Goodman, J. H. (2007). Thyroid hormone insufficiency during brain development reduces parvalbumin immunoreactivity and inhibitory function in the hippocampus. *Endocrinology, 148*(1), 92–102.

Guadaño-Ferraz, A., Escobar del Rey, F., Morreale de Escobar, G., Innocenti, G. M., & Berbel, P. (1994). The development of the anterior commissure in normal and hypothyroid rats. *Developmental Brain Research, 81*(2), 293–308.

Howdeshell, K. L. (2002). A model of the development of the brain as a construct of the thyroid system. *Environmental Health Perspectives, 110*(Suppl 3), 337–348.

Horn, S., & Heuer, H. (2010). Thyroid hormone action during brain development: More questions than answers. *Molecular and Cellular Endocrinology, 315*(1–2), 19–26.

Ibarrola, N., & Rodriguez-Pena, A. (1997). Hypothyroidism coordinately and transiently affects myelin protein gene expression in most rat brain regions during postnatal development. *Brain Research, 752*(1–2), 285–293.

Iñiguez, M. A., De Lecea, L., Guadaño-Ferraz, A., Morte, B., Gerendasy, D., Sutcliffe, J. G., & Bernal, J. (1996). Cell-specific effects of thyroid hormone on RC3/neurogranin expression in rat brain. *Endocrinology, 137*(3), 1032–1041.

Koromilas, C., Liapi, C., Schulpis, K. H., Kalafatakis, K., Zarros, A., & Tsakiris, S. (2010). Structural and functional alterations in the hippocampus due to hypothyroidism. *Metabolic Brain Disease, 25*(3), 339–354.

Lauder, J. M. (1978). Effects of early hypo- and hyperthyroidism on development of rat cerebellar cortex. IV. The parallel fibers. *Brain Research, 142*(1), 25–39.

Lavado-Autric, R., Ausó, E., Garcia-Velasco, J. V., Arufe Mdel, C., Escobar del Rey, F., Berbel, P., & Morreale de Escobar, G. (2003). Early maternal hypothyroxinemia alters histogenesis and cerebral cortex cytoarchitecture of the progeny. *Journal of Clinical Investigation, 111*(7), 1073–1082.

Legrand, J. (1984). Effects of thyroid hormones on central nervous system. In J. Yanai (Ed.), *Neurobehavioral teratology* (pp. 331–363). Amsterdam: Elsevier

Levers, T. E., Edgar, J. M., & Price, D. J. (2001). The fates of cells generated at the end of neurogenesis in developing mouse cortex. *Journal of Neurobiology, 48*(4), 265–277.

López-Espíndola, D., García-Aldea, A., Gómez de la Riva, I., Rodríguez-García, A. M., Salvatore, D., Visser, T. J., Bernal, J., & Guadaño-Ferraz, A. (2019). Thyroid hormone availability in the human fetal brain: novel entry pathways and role of radial glia. *Brain Structure and Function, 224*(6), 2103–2119.

Lucio, R. A., Garcia, J. V., Ramon Cerezo, J., Pacheco, P., Innocenti, G. M., & Berbel, P. (1997). The development of auditory callosal connections in normal and hypothyroid rats. *Cerebral Cortex, 7*(4), 303–316.

Madeira, M. D., Cadete-Leite, A., Andrade, J. P., & Paula-Barbosa, M. M. (1991). Effects of hypothyroidism upon the granular layer of the dentate gyrus in male and female adult rats: A morphometric study. *Journal of Comparative Neurology, 314*(1), 171–186.

Manzano, J., Bernal, J., & Morte, B. (2007). Influence of thyroid hormones on maturation of rat cerebellar astrocytes. *International Journal of Developmental Neuroscience, 25*(3), 171–179.

Marin-Padilla, M. (1978). Dual origin of the mammalian neocortex and evolution of the cortical plate. *Anatomy and Embryology, 152*(2), 109–126.

Martinez-Galan, J. R., Escobar del Rey, F., Morreale de Escobar, G., Santacana, M., & Ruiz-Marcos, A. (2004). Hypothyroidism alters the development of radial glial cells in the term fetal and postnatal neocortex of the rat. *Developmental Brain Research, 153*(1), 109–114.

McCarthy, M. M. (2008). Estradiol and the developing brain. *Physiological Reviews, 88*(1), 91–124.

Mohan, V., Sinha, R. A., Pathak, A., Rastogi, L., Kumar, P., Pal, A., & Godbole, M. M. (2012). Maternal thyroid hormone deficiency affects the fetal neocorticogenesis by reducing the proliferating pool, rate of neurogenesis and indirect neurogenesis. *Experimental Neurology, 237*(2), 477–488.

Morreale de Escobar, G., Obregon, M. J., & Escobar del Rey, F. (2000). Is neuropsychological development related to maternal hypothyroidism or to maternal hypothyroxinemia? *The Journal of Cinical Endocrinology and Metabolism, 85*(11), 3975–3987.

Morreale de Escobar, G., Obregon, M. J., & Escobar del Rey, F. (2004). Role of thyroid hormone during early brain development. *European Journal of Endocrinology, 151*(Suppl. 3), U25–U37.

Morte, B., Díez, D., Ausó, E., Belinchón, M. M., Gil-Ibáñez, P., Grijota-Martínez, C., Navarro, D., Morreale de Escobar, G., Berbel, P., & Bernal, J. (2010). Thyroid hormone regulation of gene expression in the developing rat fetal cerebral cortex: prominent role of the Ca^{2+}/calmodulin-dependent protein kinase IV pathway. *Endocrinology, 151*(2), 810–820.

Mughal, B. B., Fini, J. B., & Demeneix, B. A. (2018). Thyroid-disrupting chemicals and brain development: An update. *Endocrine Connections, 7*(4), R160–R186.

Nunez, J., Couchie, D., Aniello, F., & Bridoux, A. M. (1991). Regulation by thyroid hormone of microtubule assembly and neuronal differentiation. *Neurochemical Research, 16*(9), 975–982.

Obregon, M. J., Calvo, R. M., Escobar Del Rey, F., & Morreale de Escobar, G. (2007). Ontogenesis of thyroid function and interactions with maternal function. *Endocrine Development, 10*, 86–98.

Rakic, P. (1972). Mode of cell migration to the superficial layers of fetal monkey neocortex. *Journal of Comparative Neurology, 145*(1), 61–83.

Rami, A., Patel, A. J., & Rabie, A. (1986). Thyroid hormone and development of the rat hippocampus: Morphological alterations in granule and pyramidal cells. *Neuroscience, 19*(4), 1217–1226.

Rami, A., & Rabie, A. (1988). Effect of thyroid deficiency on the development of glia in the hippocampal formation of the rat: An immunocytochemical study. *Glia, 1*(5), 337–345.

Royland, J. E., Parker, J. S., & Gilbert, M. E. (2008). A genomic analysis of subclinical hypothyroidism in hippocampus and neocortex of the developing rat brain. *Journal of Neuroendocrinology, 20*(12), 1319–1338.

Ruiz-Marcos, A., Sanchez-Toscano, F., Obregon, M. J., Escobar del Rey, F., & Morreale de Escobar, G. (1982). Thyroxine treatment and recovery of hypothyroidism-induced pyramidal cell damage. *Brain Research, 239*(2), 559–574.

Samadi, A., Skocic, J., & Rovet, J. F. (2015). Children born to women treated for hypothyroidism during pregnancy show abnormal corpus callosum development. *Thyroid, 25*(5), 494–502.

Sawano, E., Takahashi, M., Negishi, T., & Tashiro, T. (2013). Thyroid hormone-dependent development of the GABAergic pre- and post-synaptic components in the rat hippocampus. *International Journal of Developmental Neuroscience, 31*(8), 751–761.

Scheepens, A., Moderscheim, T. A., & Gluckman, P. D. (2005). The role of growth hormone in neural development. *Hormone Research, 64*(Suppl. 3), 66–72.

Schwarz, J. M., & McCarthy, M. M. (2008). Steroid-induced sexual differentiation of the developing brain: Multiple pathways, one goal. *Journal of Neurochemistry, 105*(5), 1561–1572.

Shiraki, A., Saito, F., Akane, H., Takeyoshi, M., Imatanaka, N., Itahashi, M., Yoshida, T., & Shibutani, M. (2014). Expression alterations of genes on both neuronal and glial development in rats after developmental exposure to 6-propyl-2-thiouracil. *Toxicology Letters, 228*(3), 225–234.

Suo, G., Shen, F., Sun, B., Song, H., Xu, M., & Wu, Y. (2018). Abnormal expression of ephrin-A5 affects brain development of congenital hypothyroidism rats. *NeuroReport, 29*(11), 877–882.

Vaccari, A., Rossetti, Z. L., de Montis, G., Stefanini, E., Martino, E., & Gessa, G. L. (1990). Neonatal hypothyroidism induces striatal dopaminergic dysfunction. *Neuroscience, 35*(3), 699–706.

Vulsma, T., Gons, M. H., & de Vijlder, J. J. (1989). Maternal-fetal transfer of thyroxine in congenital hypothyroidism due to a total organification defect or thyroid agenesis. *New England Journal of Medicine, 321*(1), 13–16.

Willoughby, K. A., McAndrews, M. P., & Rovet, J. F. (2014a). Accuracy of episodic autobiographical memory in children with early thyroid hormone deficiency using a staged event. *Developmental Cognitive Neuroscience, 9*, 1–11.

Willoughby, K. A., McAndrews, M. P., & Rovet, J. F. (2014b). Effects of maternal hypothyroidism on offspring hippocampus and memory. *Thyroid, 24*(3), 576–584.

Wong, C. C., & Leung, M. S. (2001). Effects of neonatal hypothyroidism on the expressions of growth cone proteins and axon guidance molecules related genes in the hippocampus. *Molecular and Cellular Endocrinology, 184*(1–2), 143–150.

Yuzuriha, H., Inui, A., Asakawa, A., Ueno, N., Kasuga, M., Meguid, M. M., Miyazaki, J., Ninomiya, M., Herzog, H., & Fujimiya, M. (2007). Gastrointestinal hormones (anorexigenic peptide YY and orexigenic ghrelin) influence neural tube development. *FASEB Journal, 21*(9), 2108–2112.

Part II

Impairments and diseases

Chapter 9

Pediatric brain tumors

Sherif G. Shaaban and Stephanie A. Terezakis
Department of Radiation Oncology, University of Minnesota, Minneapolis, MN, United States

List of abbreviations

AFP Alpha fetoprotein
AT/RT Atypical teratoid rhabdoid tumors
CCSS Childhood Cancer Survivors Study
CEA Carcinoembryonic antigen
CNS Central nervous system
CSF Cerebrospinal fluid
CSI Craniospinal irradiation
DFCI Dana-Farber Cancer Institute
DFS Disease-free survival
DIPG: Diffuse intrinsic pontine glioma
EFS Event-free survival
EPN Ependymoma
FAP Familial adenomatous polyposis
FCCS Familial colon cancer syndrome
FSIQ Full-scale intelligence quotient
GBM Glioblastoma multiforme
GHD Growth hormone deficiency
GTR Gross total resection
HCG Human chorionic gonadotropin
HCPs Hereditary cancer predisposition syndromes
HGG High-grade glioma
HPA Hypothalamic–pituitary axis
IMRT Intensity-modulated radiation therapy
LFS Li–Fraumeni syndrome
LGG Low-grade glioma
MMR: Mismatch repair
NFI Neurofibromatosis type I
NGGCT Nongerminomatous germ cell tumors
OPG Optic pathway glioma
OS Overall survival
PE Cisplatin and etoposide
PFS Progression-free survival
PT Proton therapy
RT Radiation therapy
SJMB 96 St. Jude Medulloblastoma-96 study
WHO World Health Organization

Diagnosis, Management and Modeling of Neurodevelopmental Disorders. https://doi.org/10.1016/B978-0-12-817988-8.00009-9

Mini-dictionary of terms

Adjuvant Therapy Therapy delivered after the definitive treatment.

BRAF A gene belonging to the oncogene class. Its mutation is found in some types of cancer including melanoma and colorectal cancer.

FSIQ Used to test intelligence and cognitive capacity. The Wechsler Intelligence Scale for Children (WISC) is the most commonly used.

Gray (Gy) The unit of radiation dose expressed in terms of absorbed energy per unit mass of tissue.

IMRT Intensity-modulated radiotherapy is an advanced technique of high-precision radiotherapy that uses computer-controlled linear accelerators to deliver certain dose to the tumor.

Neoadjuvant Therapy Therapy delivered before the definitive treatment.

Proton Therapy Type of particle therapy uses protons to treat different types of cancer while sparing the surrounding critical structures due to its physical properties (Bragg Peak).

SHH Sonic hedgehog molecular subgroup of medulloblastoma. Associated with mutation in protein patched homolog 1 (PTCH1) gene.

TP53 The guardian of the genome that codes for a protein that regulates the cell cycle and functions as a tumor suppressor.

WNT Molecular subgroup of medulloblastoma. Associated with mutation in CTNNB1 gene. Carries best prognosis.

Introduction

Pediatric brain tumors represent the second most common pediatric malignancy, following hematological malignancy. These tumors include embryonal tumors, gliomas, germ cell tumors, ependymoma, and craniopharyngioma. This chapter will discuss the contemporary multidisciplinary management of each type including surgery, chemotherapy, and radiotherapy as well as treatment-related toxicity.

Epidemiology

Childhood brain tumors are the most common type of pediatric solid tumors representing the leading cause of cancer death among children and adolescents, accounting for 3 out of 10 cancer deaths in 2014 in the United States and Canada. In the United States, pediatric brain tumors are generally more common in males, non-Hispanic whites, and Asian-Pacific Islanders (Johnson et al., 2014).

Risk factors

Genetic risk factors

Hereditary cancer predisposition syndromes (HCPSs) are associated with an increased incidence of childhood brain tumors 10–100-fold compared to the general population (Strahm & Malkin, 2006). HCPSs are classified into autosomal dominant or autosomal recessive. Autosomal dominant syndromes include neurocutaneous syndromes, familial colon syndrome (Turcot), and exclusively cancer predisposing syndrome (Li–Fraumeni).

Neurofibromatosis type I (NF1)

Affected individuals present with multiple café-au-lait spots, cutaneous neurofibromas, iris lisch nodules, skeletal abnormalities, and axillary and groin freckling. The most common associated tumor is optic gliomas (Fisher et al., 2012).

Turcot syndrome

Turcot syndrome is characterized by development of brain tumors in association with familial colon cancer syndrome (FCCS), which occurs as a result of germline mutations of the APC gene, which causes familial adenomatous polyposis (FAP) and mismatch repair (MMR) gene aberrations. The most common brain tumors reported in this setting are medulloblastoma and glioblastoma (Barrow et al., 2009).

Li–Fraumeni syndrome (LFS)

LFS is a rare condition with TP53 mutation associated with a variety of different cancers including brain tumors. The most frequent primary brain tumors are high-grade glioma (HGG), choroid plexus carcinoma, and medulloblastoma (Tabori et al., 2010).

Environmental risk factors

Exposure to radiation during treatment of head and neck cancers is a known risk factor for childhood brain tumors. Children diagnosed with acute lymphoblastic leukemia who received prophylactic cranial irradiation are at higher risk of developing brain tumors particularly gliomas and PNETs (Ohgaki & Kleihues, 2005).

Embryonal brain tumors

Medulloblastoma

Epidemiology

Medulloblastoma is a primary brain tumor that originates in the cerebellum. It accounts for about 20% of brain and 40% of all posterior fossa tumors in children. The median age at diagnosis is 6 years. Medulloblastoma is more common in white males (Ostrom et al., 2014).

Clinical presentation

Children present with signs and symptoms of cerebellar involvement including ataxia, gait disturbances, lack of coordination in fine motor movements, and cranial nerve palsies. They also present with a clinical picture of increased intracranial pressure including headaches, lethargy, and morning vomiting. Patients with spinal dissemination can present with back pain, focal neurological deficits, and rarely spinal cord compression.

Histology and molecular subgroups

Medulloblastoma has four histologic subtypes including nodular desmoplastic "classic," medulloblastoma with extensive nodularity (MBEN), anaplastic medulloblastoma, and large cell medulloblastoma. The 2016 World Health Organization (WHO) classification of central nervous system (CNS) tumors included four subtypes of medulloblastoma; WNT-activated, SHH-activated and TP53-mutant, SHH-activated and TP53 wild-type, non-WNT/non-SHH, which includes group 3 and group 4 patients (Louis et al., 2016).

Workup

All children should initially undergo a preoperative MRI of the brain with and without gadolinium. Postoperatively, another MRI should be performed within 48 h to evaluate extent of resection. Neuro-axis evaluation by MRI of the spine and lumbar puncture is required to assess for spinal dissemination. Spinal imaging should be delayed 10−14 days for an accurate reading due to false-positive results secondary to tumor debris and blood products postsurgery. Lumbar puncture is performed 10−14 days after surgery.

Treatment

Surgery Maximal safe resection is required for all patients. Children presenting with hydrocephalus might need a ventriculoperitoneal shunt after resection of primary tumor (Lin & Riva-Cambrin, 2015). Potential complications of surgery include posterior fossa syndrome, pseudomeningocele, aseptic meningitis, and cervical spine instability. The posterior fossa or cerebellar mutism syndrome is typically seen in 10%−30% of medulloblastoma patients. Cerebellar mutism syndrome is characterized by mutism, truncal ataxia, personality changes, emotional lability, and cranial nerve palsies (Avula, Mallucci, Kumar, & Pizer, 2015).

Adjuvant treatment

Standard risk disease. The standard treatment of average-risk patients includes concurrent vincristine with a reduced dose of craniospinal radiation (23.4 Gy) and 54 Gy to the postsurgical tumor bed. Craniospinal irradiation (CSI) is considered the standard treatment for all medulloblastoma patients except for children (<3 years) and infants where CSI might result in significant neurocognitive and musculoskeletal morbidities (Ashley et al., 2012). This technique encompasses the entire central nervous system including the brain and thecal sac (Carrie et al., 1999).

Following CSI, additional radiation is delivered to the operative bed. With advances in imaging, improved radiation treatment delivery, and concerns regarding toxicity of RT, only the tumor bed and any residual tumor can be boosted with improved neurocognitive outcome (Moxon-Emre et al., 2014).

High-risk disease. The outcomes for high-risk disease still remain inferior with significant room for improvement. The current standard of care involves concurrent chemotherapy with high-dose CSI and posterior fossa boost along with boosts to metastatic deposits followed by adjuvant chemotherapy.

The St. Jude Medulloblastoma-96 (SJMB96) study treated high-risk patients with 36–39.6 Gy CSI followed by tumor bed boost to 55.8 Gy followed by four cycles of high-dose chemotherapy (cisplatin, vincristine, cyclophosphamide, Mesna) each followed by stem-cell rescue. The 5-year (OS) and disease-free survival (DFS) were both 70% (Gajjar et al., 2006). The German HIT-SKK′92 study included 43 children under the age of 3 years with medulloblastoma. All children were treated with maximal surgical resection followed by chemotherapy including cyclophosphamide, vincristine, and intraventricular and intravenous methotrexate. Five-year progression-free survival (PFS) and OS were 58 ± 9% and 66 ± 7%, respectively (Rutkowski et al., 2005).

Atypical teratoid rhabdoid tumors (AT/RT)

Epidemiology

These are rare tumors accounting for 0.33 per 100,000 with a peak age group 0–4 years. SMARCB1 gene mutation is one of the most common genetic risk factors responsible for one-third of cases (Eaton, Tooke, Wainwright, Judkins, & Biegel, 2011).

Clinical picture

Clinical presentation varies depending on the site of the tumor. The most common location of the tumor is the posterior fossa. Patients usually present with ataxia, balance problems, and hydrocephalus-related symptoms.

Workup

Initial imaging including CT and MRI of the brain is required for determination of tumor extent. The presence of a wavy, band-like enhancing zone surrounding an area of necrosis is an imaging marker for AT/RT (Warmuth-Metz et al., 2008). MRI of the spine is used for staging the entire neuraxis.

Treatment

Surgery

Maximal safe surgical resection is standard as part of the treatment approach. The reported median survival of patients undergoing gross total resection (GTR) was 20 versus 15 months for those treated with partial resection or biopsy (Hilden et al., 2004).

Adjuvant therapy

Different chemotherapeutic regimens have been investigated. While the Dana-Farber Cancer Institute (DFCI) protocol included temozolomide, intrathecal methotrexate, cytarabine, and hydrocortisone, the 2-year PFS was 53% and OS was 70% (Chi et al., 2009).

For children younger than 3 years, multiple trials studied the use of conventional and intensified chemotherapy with stem-cell rescue to avoid use of RT. The 3-year DFS ranged from 20% to 40% (Gardner et al., 2008; Zaky et al., 2014).

ACNS0333 study is addressing the impact of multimodality treatment on the survival of AT/RT patients. Following surgery, children received two cycles of induction chemotherapy (methotrexate, vincristine, cyclophosphamide, etoposide, and cisplatin). Patients without evidence of tumor progression received three cycles of high-dose thiotepa and carboplatin with autologous stem-cell rescue and involved field radiation therapy.

Pediatric glioma

Low-grade glioma

Epidemiology

Low-grade glioma (LGG) is the most common CNS tumor in pediatric patients accounting for approximately 26% of childhood CNS tumors in the United States (Ostrom et al., 2013). According to WHO classification of CNS tumors,

LGGs are considered grade I and II. WHO grade I tumors include pilocytic astrocytoma, subependymal giant cell astrocytoma, dysembryoplastic neuroepithelial tumor, and desmoplastic infantile ganglioglioma while grade II tumors include pilomyxoid astrocytoma, pleomorphic xanthoastrocytoma, oligodendroglioma, oligoastrocytoma, and diffuse astrocytoma (Louis et al., 2007).

Clinical picture

Owing to the low proliferative potential of these tumors, clinical signs and symptoms develop gradually over time. Symptoms and signs of increased intracranial pressure as well as focal neurological signs depending on the location of the tumor are usually present.

Workup

Brain imaging with MRI is required for assessment of tumor extension. LGGs appear as abnormal areas of increased intensity on T2-weighted MR images with mild or no enhancement. While pilocytic astrocytomas have enhanced centrally cystic lesions.

Treatment

Surgery Maximal safe surgical resection is the main goal of treatment. Scope of resection has an impact on prognosis and survival. A large retrospective study included 351 patients with LGG. Multivariate analysis showed improved PFS with GTR and adjuvant radiotherapy, while overall survival was improved with GTR and pilocytic histology (Youland et al., 2013).

Chemotherapy Chemotherapy is often used in younger children to delay the onset of radiotherapy. The most commonly used agents are carboplatin and vincristine. Use of single agent carboplatin has a similar efficacy compared with combined chemotherapy (Dodgshun, Maixner, Heath, Sullivan, & Hansford, 2016).

Targeted therapy acting on activation of the RAS-RAF-MEK-ERK signaling pathway represents an emerging treatment for LGGs. BRAF inhibitors (Sorafenib) have been investigated and found to interfere with tyrosine kinase activity leading to unpredicted acceleration of tumor growth in LGGs (Karajannis et al., 2014). These agents warrant further study.

Radiation therapy Radiation therapy is typically indicated after incomplete surgical resection to enhance local tumor control and functional outcomes. Use of IMRT technique helps to provide optimal coverage of the tumor while sparing the surrounding structures. The Children's Oncology Group Study (ACNS0221) investigated the impact of radiotherapy on survival of LGG patients. Patients younger than 10 years were required to receive at least one cycle of chemotherapy. All children received radiotherapy 54Gy in 30 fractions. The 5-year PFS was $71 \pm 6\%$ and OS was $93 \pm 4\%$. OS was poorly associated with male sex, large tumor size, and nonpilocytic astrocytoma histology (Cherlow et al., 2019).

High-grade glioma

Epidemiology

Childhood HGGs are grade III or IV tumors that originate from glial tissue. They represent 6%–15% of all pediatric CNS tumors. Grade III tumors include anaplastic astrocytoma and anaplastic oligodendroglioma, while grade IV includes glioblastoma (GBM) (Broniscer & Gajjar, 2004).

Clinical picture

Presenting symptoms are typically due to increased intracranial pressure, including persistent headaches, nausea, vomiting, diplopia, and papilledema, or site-dependent symptoms secondary to compression or invasion of adjacent structures.

Workup

MR imaging is used for tumor localization and relation to surrounding structures. Although these tumors have a low incidence of leptomeningeal spread, spinal evaluation should be obtained if there are concerning symptoms (Fangusaro, 2012).

Treatment

Surgery

Maximal safe resection represents the standard of care in the treatment of HGG. Extent of resection has prognostic significance. The 5-year (DFS) when extent of resection exceeds 90% is 35% and only 17% with a lesser extent (Wisoff et al., 1998).

Adjuvant therapy

Chemotherapy is combined with surgery and radiotherapy. ACNS0126 study reported 3-year EFS of 11 ± 3% and OS of 22 ± 5% (Cohen et al., 2011). The current standard RT dose for HGG is 54–60 Gy delivered over 6 weeks. Involved-field radiotherapy is the standard treatment volume.

Optic pathway glioma

Epidemiology

Optic pathway gliomas (OPGs) are uncommon tumors accounting for approximately 5% of all pediatric intracranial tumors. There is a strong association of (NF-1) and development of OPGs in approximately 30% of cases (Bikowska-Opalach & Jackowska, 2013).

Clinical picture

Patients commonly present with loss of visual acuity in one or both eyes and proptosis. Hydrocephalus, developmental delay, and precocious puberty may also be present.

Workup

High-resolution MRI of the brain at 1–2 mm thickness with and without contrast is preferred to define tumor extent along the optic pathway. OPGs are predominantly solid but can have a cystic component particularly in non-NF-1 patients (Kornreich et al., 2001). Calcifications are usually present. All patients should undergo complete ophthalmologic and endocrine evaluation before starting treatment.

Treatment

The role of surgery is limited to localized optic nerve lesions. Debulking surgery is indicated to provide immediate decompression to preserve vision or relieve obstructive hydrocephalus. Radiation therapy is more commonly recommended for older children particularly when surgery is not feasible and after patients have progressed through systemic therapy. The standard radiation dose is 50.4–54 Gy (Fig. 9.1). The role of chemotherapy is controversial. Its application is mainly to delay use of radiotherapy in young children.

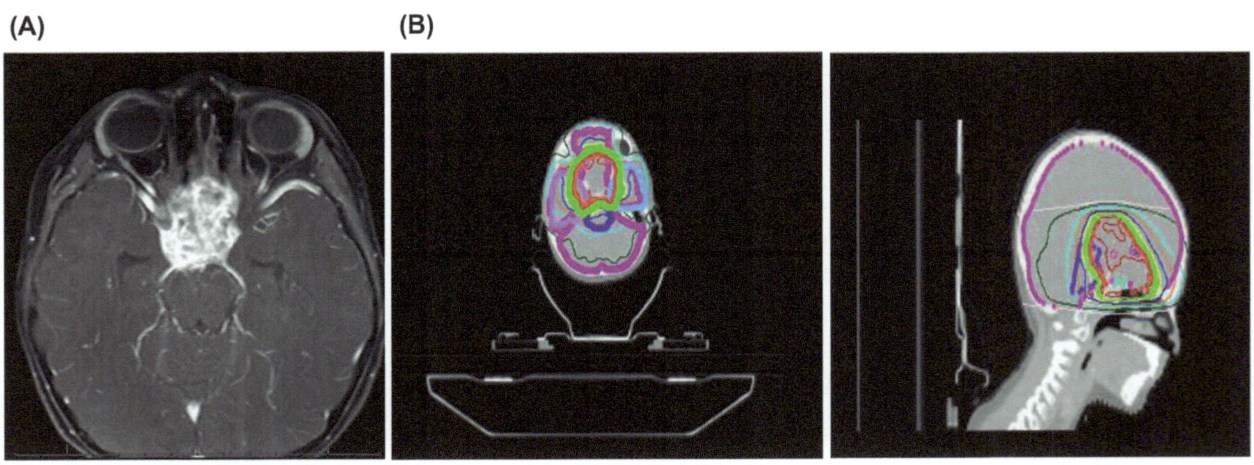

FIGURE 9.1 **Treatment of optic pathway glioma.** A 6-year-old female with optic pathway glioma (A). She received chemotherapy followed by radiotherapy 54Gy in 30 fractions (B).

Diffuse intrinsic pontine gliomas (DIPGs)

Epidemiology

Diffuse intrinsic pontine gliomas (DIPGs) represent approximately 50% of tumors arising in the brainstem in children. The median age of diagnosis is 7 years. Males and females are equally affected.

Clinical picture

The most common symptoms and signs are bilateral cranial nerve deficits and ataxia. Symptoms usually develop within 6 months or shorter.

Workup

MRI of the brain is the primary imaging modality for determination of tumor extension. On MRI, DIPG causes diffuse enlargement of the brainstem particularly the pons. The degree of enhancement is usually variable (Poussaint et al., 2011).

Treatment

Surgery

The role of surgery is limited in DIPG due to high surgical morbidity given the critical location of the tumor and difficulty in obtaining maximal safe resection. Historically, the diagnosis of DIPG was made based on radiographic characteristics of the tumor. More commonly, biopsy is performed when morbidity is considered low to obtain molecular information about the tumor.

Adjuvant therapy

Chemotherapy has a limited role in DIPG as it has failed to show any survival benefit in both preradiation and postradiation settings. Radiation treatment represents the standard of care for treatment of DIPG. The radiation treatment dose is typically 54 Gy over 6 weeks. Hypofractionated radiotherapy with shorter treatment time was shown to be as safe and effective as conventional radiotherapy. There was no significant difference in OS (9.0 vs. 9.4 months) and DFS (5.0 vs. 7.6 months) compared to conventional radiotherapy (Janssens et al., 2013).

Germ cell tumors

Central nervous system germinoma

Epidemiology

Intracranial germ cell tumors account for 0.1%–2% of childhood CNS tumors in Western countries and up to 6% in East Asia (Keene et al., 2007). Germinomas occur more frequently among male teenagers.

Clinical picture

Clinical symptoms vary according to tumor location. In patients with pineal germinomas, Parinaud syndrome (upward gaze palsy, convergence retraction nystagmus, and light-near dissociation) is typically seen with large tumors. In germinomas that present in the suprasellar region, children often present with diabetes insipidus.

Workup

Primary brain imaging and tumor markers including β-human chorionic gonadotropin (HCG), α-fetoprotein (AFP), and carcinoembryonic antigen (CEA), in the serum and/or cerebrospinal fluid (CSF) are obtained. β-HCG level is used to differentiate between pure or mixed nongerminomatous germ cell tumors with levels higher than 200 ng/mL and pure germinoma with lower levels below 10 mIU/mL (Suzuki, Konishi, Kondo, Fujita, & Ishikawa, 1985), whereas elevation of AFP or CEA levels supports the diagnosis of nongerminomatous germ cell tumors (NGGCT) (Echevarria, Fangusaro, & Goldman, 2008).

Treatment

Surgery Surgery has a limited role in the management of germinomas. It is typically indicated to obtain a biopsy and aggressive surgical resection should be avoided due to the risk of associated serious surgical morbidity.

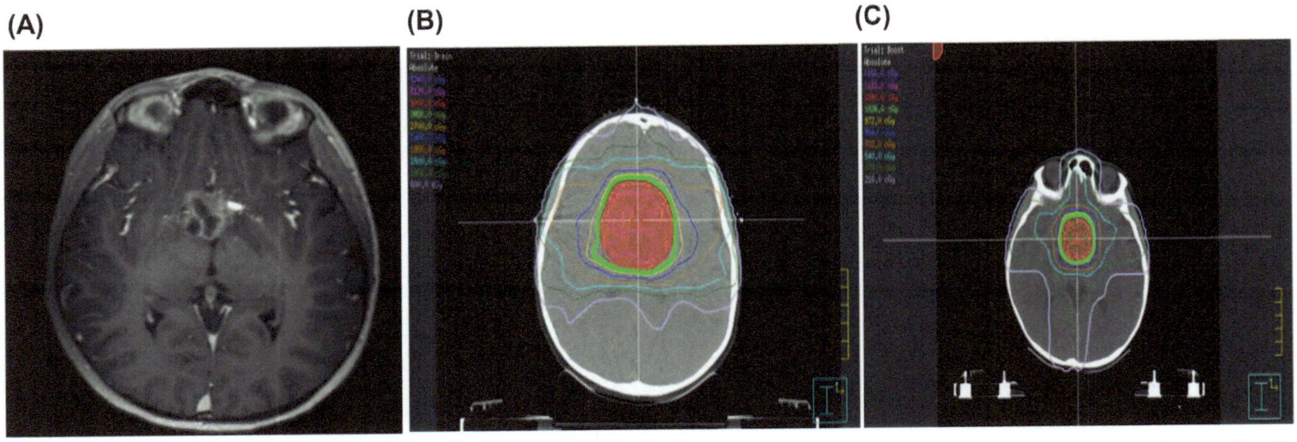

FIGURE 9.2 **Treatment of suprasellar germinoma.** A 12-year-old female with multicystic heterogeneous mass suprasellar germinoma (A). She received neoadjuvant chemotherapy followed by radiotherapy including whole ventricular radiotherapy 2520 cGy (B) followed by a boost 1080 cGy (C).

Adjuvant therapy These tumors are chemoradiotherapy sensitive. The treatment options include radiation alone to the whole ventricular system with a boost or neoadjuvant chemotherapy followed by reduced dose radiotherapy. The standard radiation dose when treating with radiation alone is 24 Gy to the whole ventricular system followed by a 45 Gy boost to the primary tumor delivered in 1.8–2 Gy per fraction (Fig. 9.2). When chemotherapy is used neoadjuvantly, tumor response is evaluated. For patients that achieve a complete response or near complete response, they are eligible to receive a reduced dose of radiotherapy. A multicentric study conducted by Matsutani and colleagues utilized three courses of carboplatin and etoposide followed by focal radiation field or extended local field encompassing the tumor site, the third and lateral ventricles, and the sellar and pineal regions, delivering 24 Gy. The recurrence rate was 28% in patient received focal field and 6% with extended field patients. It was recommended that treatment volume should include the whole ventricular system (Matsutani, 2009). Currently, radiation to the entire craniospinal axis is reserved for CSF dissemination cases.

Central nervous system nongerminomatous germ cell tumors (NGGCT)

Clinical picture

The clinical presentation of germinomas and NGGCTs are very similar. Symptoms and signs are related to tumor location and size.

Workup

The workup is also similar to germinomas. Radiologically, NGGCTs show a large mass with heterogenous signal. Teratomas exhibit high signal intensity of the fat tissue in T1-weighted MRI images (Reis et al., 2006). Mature teratomas are well circumscribed with a mixed cystic and solid components, while immature teratoma tends to have fewer cysts with a characteristic “honeycomb appearance” (Kim et al., 2011).

Treatment

Surgery

The role of surgery is reserved for tissue biopsy or primary resection. Second resection is indicated in management of residual tumors after neoadjuvant chemotherapy (Weiner et al., 2002). GTR alone is the curable treatment option for mature teratomas. In other types of NGGCTs, surgery is typically combined with chemotherapy and radiotherapy to achieve satisfactory outcomes (Goldman et al., 2015).

Chemotherapy

Systemic chemotherapy represents an integral part of the multimodality treatment of NGGCTs except for mature teratomas. Use of combined chemoradiation in the neoadjuvant setting has been associated with better local control and survival outcomes (Goldman et al., 2015). Cisplatin and etoposide (PE) and carboplatin and etoposide are frequently used regimens.

Radiation therapy

Patients with localized NGGCT can be treated with combined chemotherapy and radiotherapy while patients with disseminated disease should receive 30 Gy CSI with 24 Gy boost to the gross tumor, as demonstrated in the SIOP trial (Calaminus et al., 2013). ACNS1123 tested the effectiveness of reduced radiotherapy to 30.6 Gy to the whole ventricular field and a tumor-bed boost to a total dose of 54 Gy for localized NGGCT in children with complete and partial response to chemotherapy.

Ependymoma

Epidemiology

Ependymomas (EPNs) represent 6%–10% of childhood brain tumors. Approximately 90% of pediatric EPNs are intracranial with the most common location in the posterior fossa (Vaidya, Smee, & Williams, 2012).

Treatment

The standard of care for EPN involves maximal safe surgical resection, followed by adjuvant localized focal radiotherapy. The role of chemotherapy is controversial with little evidence supporting its efficacy (Vaidya et al., 2012). Adjuvant radiation treatment has been associated with local control and survival benefit. The current standard RT dose after GTR is 54–59.4 Gy to the involved field (Reni, Gatta, Mazza, & Vecht, 2007). CSI was utilized historically but leptomeningeal failure rates were found to be low at 3%–10% (Merchant, Happersett, Finlay, & Leibel, 1999). Therefore, localized, involved field RT has become the current standard treatment volume for intracranial EPN (Chan & McMullen, 2012).

Childhood craniopharyngioma

Epidemiology

In the United States, craniopharyngiomas account for 1%–3% of all brain tumors (Jane & Laws, 2006). In children, the peak incidence is between 5 and 14 years.

Clinical picture

Craniopharyngiomas are slow-growing tumors that can frequently reach large sizes before diagnosis. Symptoms are widely variable according to tumor location and compression on nearby structures. Headaches are the most common presenting symptoms. Visual symptoms are frequently reported and include diplopia, blurred vision, and reduced visual acuity (Müller, 2013). Endocrinopathies commonly develop due to disruption of the hypothalamic–pituitary axis. The most common hormonal deficiency is growth hormone, followed by gonadotropins, adrenocorticotropic hormone, and thyroid-stimulating hormone.

Workup

Neuroimaging evaluation by obtaining CT and MRI scan of the brain is the initial step. Features such as calcifications and cystic areas often help in diagnosis. A baseline visual acuity exam as well as comprehensive endocrine assessment is crucial prior to surgery.

Treatment

Surgery The role of surgery is limited to obtain a biopsy for diagnosis, decompression, or maximal safe resection (Van Effenterre & Boch, 2002). Microsurgery, endoscopic assistance, and minimally invasive approaches are currently used to improve the quality of tumor resection while minimizing morbidity.

Radiation therapy Radiation therapy is typically indicated for patients with subtotal resection due to the high rate of local failure (Clark et al., 2013). For fractionated radiotherapy, the current standard radiation dose ranges from 50.4 to 54 Gy. Patients with small tumors (<3 cm) could potentially be treated by radiosurgery or hypofractionated stereotactic radiotherapy (Lee et al., 2014).

Treatment-related toxicity

The toxicity profile of radiation treatment differs based on the treatment fields, dose, and age of the patient. Acute side effects include fatigue, nausea, vomiting, headaches, alopecia, skin reaction, seizure, and myelosuppression. Long-term effects can be significant and include cognitive disability, organ dysfunction, neurologic deficits, psychosocial difficulties, and secondary malignancy.

Neurocognitive dysfunction

Neurocognitive impairment is one of the most devastating and feared sequelae in survivors of childhood brain tumors. The severity of impairment depends on age at treatment, volume of irradiated brain, and radiation dose. The most common deficits are attention and concentration problems, executive function, memory, and processing speed. Neurocognitive status is measured using full-scale intelligence quotient (FSIQ). However, cognitive decline can be limited with certain radiotherapy approaches for particular tumor sites. For example, the cognitive outcomes of LGG patients treated with an involved radiotherapy field were evaluated and showed no change of FSIQ score at 60 months from the baseline (Merchant, Conklin, Wu, Lustig, & Xiong, 2009).

Ototoxicity

Sensorineural hearing loss may result from radiation to the cochlea. It was found that risk of ototoxicity is dose-dependent. High-frequency hearing loss rate was 37% at 60 Gy while doses less than 35 Gy were rarely associated with ototoxicity (Hua, Bass, Khan, Kun, & Merchant, 2008).

Hearing deficits are more common following combined cisplatin-based chemotherapy and radiotherapy. Use of advanced radiotherapy techniques such as IMRT and proton therapy allows for cochlear sparing and reduces the risk of ototoxicity. In medulloblastoma, use of smaller boost fields to treat the tumor bed was associated with a reduced grade 3–4 ototoxicity rate of 13% compared to 65% when treating the entire posterior fossa (Huang et al., 2002).

Endocrine dysfunction

Childhood brain tumor survivors are at risk for a number of endocrinopathies including diabetes insipidus, hypothyroidism, growth hormone deficiency (GHD), precocious puberty, gonadotropin deficiency, adrenocorticotropic hormone deficiency, and panhypopituitarism. GHD is the most common endocrine deficit to develop after cranial irradiation. It has been noted that nearly all children treated with doses higher than 35Gy to the hypothalamic–pituitary axis (HPA) will develop GHD.

In a German study, the cumulative risk of GHD 2.5 years after radiotherapy for children receiving more than and less than 37.5 Gy to the HPA was 87% and 33%, respectively (Schmiegelow et al., 1999). Stubberfield and colleagues evaluated growth rates in children with acute lymphoblastic leukemia after cranial irradiation at 18 and 24 Gy. Growth rates in both groups did not differ for the first 5 years after treatment. On longer follow-up, however, a significant height decrease was observed in children who had received 24 Gy compared to the group of patients that received 18 Gy (Stubberfield, Byrne, & Jones, 1995).

Secondary malignancy

Development of second tumors represents a major concern following treatment of pediatric population. It takes several decades after treatment to occur. The Childhood Cancer Survivors Study (CCSS) found 76 subsequent neoplasms among 1877 CNS tumor survivors at a median of 16 years from diagnosis. Meningioma was the most common second malignancy followed by gliomas (Armstrong, 2010).

Proton therapy

Proton therapy (PT) has become more widely used in the treatment of childhood brain tumors owing to its unique physical properties that eliminate dose delivery beyond the target (Bragg Peak) and thus reduce the dose to critical structures (Fig. 9.3).

Multiple studies have compared protons versus photon therapy in treatment of pediatric CNS tumors. They found that both have comparable control and survival outcomes. However, PT was associated with reduced doses to normal brain, cochlea, hippocampus, and HPA leading to decreased incidence of long-term sequelae including neurocognitive function, neuroendocrine disorder, and growth impairment. Pulsifer et al. analyzed the cognitive function among 60 children after treatment with PT. They found that (FSIQ) was stable at mean follow-up of 2.5 years compared to progressive cognitive decline at 1–2 years after photon therapy (Pulsifer et al., 2015).

Key facts of proton therapy

- Advanced form of radiotherapy that uses charged particles (protons) to damage DNA and kill cancer cells.
- There are over 75 particle therapy facilities worldwide. The first treatments were carried out at Berkeley Radiation Laboratory (USA) in 1954 and at Uppsala in Sweden in 1957.

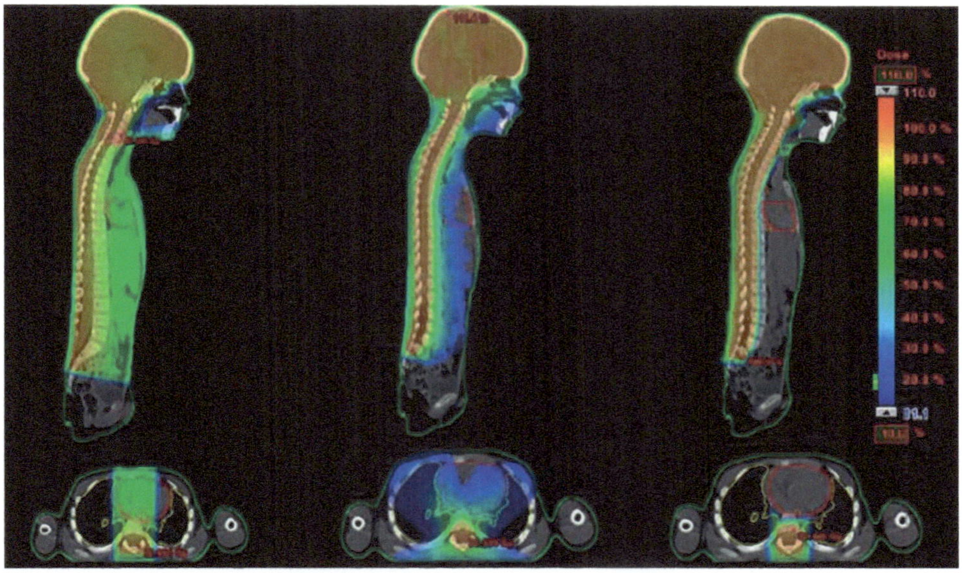

FIGURE 9.3 **Dose distribution of different radiation therapy modalities.** Craniospinal irradiation dose distribution of three dimensional—conformal, intensity-modulated radiation therapy, and proton therapy. *Permission from Brodin, N. P., Munck Af Rosenschold, P., Aznar, M. C., Kiil-Berthelsen, A., Vogelius, I. R., Nilsson, P., … Bjork-Eriksson, T. (2011). Radiobiological risk estimates of adverse events and secondary cancer for proton and photon radiation therapy of pediatric medulloblastoma.* Acta Oncologica, 50*(6), 806–816. https://doi:10.3109/0284186x.2011.582514.*

- Characterized by ability to release most of their energy within the tumor itself. The point where the highest energy deposition occurs called (Bragg Peak).
- After energy delivery to the tumor, the protons stop and don't go beyond the tumor allowing for sparing of surrounding normal tissues.
- This helps to deliver more conformal treatment to the tumor and avoid injury of surrounding normal structures achieving higher tumor control and fewer harmful side effects.
- Its utilization has substantially increased over the last decade in treatment of various types of cancers with great success particularly in pediatric tumors.

Summary points

- Maximal safe resection is the standard surgical approach for treatment of the majority of pediatric brain tumors.
- Treatment of average-risk medulloblastoma includes maximal safe resection followed by concurrent vincristine with craniospinal radiation followed by a boost to the surgical bed and adjuvant chemotherapy
- Treatment of germinomas is definitive radiotherapy ± chemotherapy.
- The most common cognitive deficits are attention and concentration problems, executive function, memory, and processing speed.
- Meningioma is the most commonly reported second malignancy after treatment of childhood brain tumors followed by gliomas

References

Armstrong, G. T. (2010). Long-term survivors of childhood central nervous system malignancies: The experience of the childhood cancer survivor study. *European Journal of Paediatric Neurology, 14*(4), 298–303. https://doi.org/10.1016/j.ejpn.2009.12.006.

Ashley, D. M., Merchant, T. E., Strother, D., Zhou, T., Duffner, P., Burger, P. C., … Pollack, I. F. (2012). Induction chemotherapy and conformal radiation therapy for very young children with nonmetastatic medulloblastoma: Children's Oncology Group study P9934. *Journal of Clinical Oncology, 30*(26), 3181–3186. https://doi.org/10.1200/JCO.2010.34.4341.

Avula, S., Mallucci, C., Kumar, R., & Pizer, B. (2015). Posterior fossa syndrome following brain tumour resection: Review of pathophysiology and a new hypothesis on its pathogenesis. *Child's Nervous System, 31*(10), 1859–1867. https://doi.org/10.1007/s00381-015-2797-0.

Barrow, E., Robinson, L., Alduaij, W., Shenton, A., Clancy, T., Lalloo, F., … Evans, D. G. (2009). Cumulative lifetime incidence of extracolonic cancers in Lynch syndrome: A report of 121 families with proven mutations. *Clinical Genetics, 75*(2), 141–149. https://doi.org/10.1111/j.1399-0004.2008.01125.x.

Bikowska-Opalach, B., & Jackowska, T. (2013). Neurofibromatosis type 1 - description of clinical features and molecular mechanism of the disease. *Medycyna Wieku Rozwojowego, 17*(4), 334–340.

Brodin, N. P., Munck Af Rosenschold, P., Aznar, M. C., Kiil-Berthelsen, A., Vogelius, I. R., Nilsson, P., … Bjork-Eriksson, T. (2011). Radiobiological risk estimates of adverse events and secondary cancer for proton and photon radiation therapy of pediatric medulloblastoma. *Acta Oncologica, 50*(6), 806–816. https://doi.org/10.3109/0284186x.2011.582514.

Broniscer, A., & Gajjar, A. (2004). Supratentorial high-grade astrocytoma and diffuse brainstem glioma: Two challenges for the pediatric oncologist. *The Oncologist, 9*(2), 197–206. https://doi.org/10.1634/theoncologist.9-2-197.

Calaminus, G., Kortmann, R., Worch, J., Nicholson, J. C., Alapetite, C., Garrè, M. L., … Frappaz, D. (2013). SIOP CNS GCT 96: Final report of outcome of a prospective, multinational nonrandomized trial for children and adults with intracranial germinoma, comparing craniospinal irradiation alone with chemotherapy followed by focal primary site irradiation for patients with localized disease. *Neuro-Oncology, 15*(6), 788–796. https://doi.org/10.1093/neuonc/not019.

Carrie, C., Hoffstetter, S., Gomez, F., Moncho, V., Doz, F., Alapetite, C., … Habrand, J. L. (1999). Impact of targeting deviations on outcome in medulloblastoma: Study of the French society of pediatric Oncology (SFOP). *International Journal of Radiation Oncology, Biology, Physics, 45*(2), 435–439. https://doi.org/10.1016/s0360-3016(99)00200-x.

Chan, M. D., & McMullen, K. P. (2012). Multidisciplinary management of intracranial ependymoma. *Current Problems in Cancer, 36*(1), 6–19. https://doi.org/10.1016/j.currproblcancer.2011.10.013.

Cherlow, J. M., Shaw, D. W. W., Margraf, L. R., Bowers, D. C., Huang, J., Fouladi, M., … Merchant, T. E. (2019). Conformal radiation therapy for pediatric patients with low-grade glioma: Results from the children's Oncology group phase 2 study ACNS0221. *International Journal of Radiation Oncology, Biology, Physics, 103*(4), 861–868. https://doi.org/10.1016/j.ijrobp.2018.11.004.

Chi, S. N., Zimmerman, M. A., Yao, X., Cohen, K. J., Burger, P., Biegel, J. A., … Kieran, M. W. (2009). Intensive multimodality treatment for children with newly diagnosed CNS atypical teratoid rhabdoid tumor. *Journal of Clinical Oncology, 27*(3), 385–389. https://doi.org/10.1200/JCO.2008.18.7724.

Clark, A. J., Cage, T. A., Aranda, D., Parsa, A. T., Sun, P. P., Auguste, K. I., & Gupta, N. (2013). A systematic review of the results of surgery and radiotherapy on tumor control for pediatric craniopharyngioma. *Child's Nervous System, 29*(2), 231–238. https://doi.org/10.1007/s00381-012-1926-2.

Cohen, K. J., Heideman, R. L., Zhou, T., Holmes, E. J., Lavey, R. S., Bouffet, E., & Pollack, I. F. (2011). Temozolomide in the treatment of children with newly diagnosed diffuse intrinsic pontine gliomas: A report from the children's Oncology group. *Neuro-Oncology, 13*(4), 410–416. https://doi.org/10.1093/neuonc/noq205.

Dodgshun, A. J., Maixner, W. J., Heath, J. A., Sullivan, M. J., & Hansford, J. R. (2016). Single agent carboplatin for pediatric low-grade glioma: A retrospective analysis shows equivalent efficacy to multiagent chemotherapy. *International Journal of Cancer, 138*(2), 481–488. https://doi.org/10.1002/ijc.29711.

Eaton, K. W., Tooke, L. S., Wainwright, L. M., Judkins, A. R., & Biegel, J. A. (2011). Spectrum of SMARCB1/INI1 mutations in familial and sporadic rhabdoid tumors. *Pediatric Blood and Cancer, 56*(1), 7–15. https://doi.org/10.1002/pbc.22831.

Echevarria, M. E., Fangusaro, J., & Goldman, S. (2008). Pediatric central nervous system germ cell tumors: A review. *The Oncologist, 13*(6), 690–699. https://doi.org/10.1634/theoncologist.2008-0037.

Fangusaro, J. (2012). Pediatric high grade glioma: A review and update on tumor clinical characteristics and biology. *Frontiers in Oncology, 2*, 105. https://doi.org/10.3389/fonc.2012.00105.

Fisher, M. J., Loguidice, M., Gutmann, D. H., Listernick, R., Ferner, R. E., Ullrich, N. J., … Liu, G. T. (2012). Visual outcomes in children with neurofibromatosis type 1-associated optic pathway glioma following chemotherapy: A multicenter retrospective analysis. *Neuro-Oncology, 14*(6), 790–797. https://doi.org/10.1093/neuonc/nos076.

Gajjar, A., Chintagumpala, M., Ashley, D., Kellie, S., Kun, L. E., Merchant, T. E., … Gilbertson, R. J. (2006). Risk-adapted craniospinal radiotherapy followed by high-dose chemotherapy and stem-cell rescue in children with newly diagnosed medulloblastoma (St Jude medulloblastoma-96): Long-term results from a prospective, multicentre trial. *The Lancet Oncology, 7*(10), 813–820. https://doi.org/10.1016/S1470-2045(06)70867-1.

Gardner, S. L., Asgharzadeh, S., Green, A., Horn, B., McCowage, G., & Finlay, J. (2008). Intensive induction chemotherapy followed by high dose chemotherapy with autologous hematopoietic progenitor cell rescue in young children newly diagnosed with central nervous system atypical teratoid rhabdoid tumors. *Pediatric Blood and Cancer, 51*(2), 235–240. https://doi.org/10.1002/pbc.21578.

Goldman, S., Bouffet, E., Fisher, P. G., Allen, J. C., Robertson, P. L., Chuba, P. J., … Pollack, I. F. (2015). Phase II trial assessing the ability of neoadjuvant chemotherapy with or without second-look surgery to eliminate measurable disease for nongerminomatous germ cell tumors: A children's Oncology group study. *Journal of Clinical Oncology, 33*(22), 2464–2471. https://doi.org/10.1200/jco.2014.59.5132.

Hilden, J. M., Meerbaum, S., Burger, P., Finlay, J., Janss, A., Scheithauer, B. W., … Biegel, J. A. (2004). Central nervous system atypical teratoid/rhabdoid tumor: Results of therapy in children enrolled in a registry. *Journal of Clinical Oncology, 22*(14), 2877–2884. https://doi.org/10.1200/JCO.2004.07.073.

Hua, C., Bass, J. K., Khan, R., Kun, L. E., & Merchant, T. E. (2008). Hearing loss after radiotherapy for pediatric brain tumors: Effect of cochlear dose. *International Journal of Radiation Oncology, Biology, Physics, 72*(3), 892–899. https://doi.org/10.1016/j.ijrobp.2008.01.050.

Huang, E., Teh, B. S., Strother, D. R., Davis, Q. G., Chiu, J. K., Lu, H. H., … Woo, S. Y. (2002). Intensity-modulated radiation therapy for pediatric medulloblastoma: Early report on the reduction of ototoxicity. *International Journal of Radiation Oncology, Biology, Physics, 52*(3), 599–605. https://doi.org/10.1016/s0360-3016(01)02641-4.

Jane, J. A., Jr., & Laws, E. R. (2006). Craniopharyngioma. *Pituitary, 9*(4), 323–326. https://doi.org/10.1007/s11102-006-0413-8.

Janssens, G. O., Jansen, M. H., Lauwers, S. J., Nowak, P. J., Oldenburger, F. R., Bouffet, E., ... Hargrave, D. (2013). Hypofractionation vs conventional radiation therapy for newly diagnosed diffuse intrinsic pontine glioma: A matched-cohort analysis. *International Journal of Radiation Oncology, Biology, Physics, 85*(2), 315–320. https://doi.org/10.1016/j.ijrobp.2012.04.006.

Johnson, K. J., Cullen, J., Barnholtz-Sloan, J. S., Ostrom, Q. T., Langer, C. E., Turner, M. C., ... Scheurer, M. E. (2014). Childhood brain tumor epidemiology: A brain tumor epidemiology consortium review. *Cancer Epidemiology Biomarkers and Prevention, 23*(12), 2716–2736. https://doi.org/10.1158/1055-9965.EPI-14-0207.

Karajannis, M. A., Legault, G., Fisher, M. J., Milla, S. S., Cohen, K. J., Wisoff, J. H., ... Allen, J. C. (2014). Phase II study of sorafenib in children with recurrent or progressive low-grade astrocytomas. *Neuro-Oncology, 16*(10), 1408–1416. https://doi.org/10.1093/neuonc/nou059.

Keene, D., Johnston, D., Strother, D., Fryer, C., Carret, A. S., Crooks, B., ... Consortium, C. P. B. T. (2007). Epidemiological survey of central nervous system germ cell tumors in Canadian children. *Journal of Neuro-Oncology, 82*(3), 289–295. https://doi.org/10.1007/s11060-006-9282-2.

Kim, C. Y., Choi, J. W., Lee, J. Y., Kim, S. K., Wang, K. C., Park, S. H., ... Cho, B. K. (2011). Intracranial growing teratoma syndrome: Clinical characteristics and treatment strategy. *Journal of Neuro-Oncology, 101*(1), 109–115. https://doi.org/10.1007/s11060-010-0238-1.

Kornreich, L., Blaser, S., Schwarz, M., Shuper, A., Vishne, T. H., Cohen, I. J., ... Horev, G. (2001). Optic pathway glioma: Correlation of imaging findings with the presence of neurofibromatosis. *American Journal of Neuroradiology, 22*(10), 1963–1969.

Lee, C. C., Yang, H. C., Chen, C. J., Hung, Y. C., Wu, H. M., Shiau, C. Y., ... Liu, K. D. (2014). Gamma knife surgery for craniopharyngioma: Report on a 20-year experience. *Journal of Neurosurgery, 121*(Suppl. 1), 167–178. https://doi.org/10.3171/2014.8.GKS141411.

Lin, C. T., & Riva-Cambrin, J. K. (2015). Management of posterior fossa tumors and hydrocephalus in children: A review. *Child's Nervous System, 31*(10), 1781–1789. https://doi.org/10.1007/s00381-015-2781-8.

Louis, D. N., Ohgaki, H., Wiestler, O. D., Cavenee, W. K., Burger, P. C., Jouvet, A., ... Kleihues, P. (2007). The 2007 WHO classification of tumours of the central nervous system. *Acta Neuropathologica, 114*(2), 97–109. https://doi.org/10.1007/s00401-007-0243-4.

Louis, D. N., Perry, A., Reifenberger, G., von Deimling, A., Figarella-Branger, D., Cavenee, W. K., ... Ellison, D. W. (2016). The 2016 World Health Organization classification of tumors of the central nervous system: A summary. In *Acta neuropathol* (Vol. 131, pp. 803–820) (Germany).

Matsutani, M. (2009). Pineal germ cell tumors. *Progress in Neurological Surgery, 23*, 76–85. https://doi.org/10.1159/000210054.

Merchant, T. E., Conklin, H. M., Wu, S., Lustig, R. H., & Xiong, X. (2009). Late effects of conformal radiation therapy for pediatric patients with low-grade glioma: Prospective evaluation of cognitive, endocrine, and hearing deficits. *Journal of Clinical Oncology, 27*(22), 3691–3697. https://doi.org/10.1200/jco.2008.21.2738.

Merchant, T. E., Happersett, L., Finlay, J. L., & Leibel, S. A. (1999). Preliminary results of conformal radiation therapy for medulloblastoma. *Neuro-Oncology, 1*(3), 177–187. https://doi.org/10.1093/neuonc/1.3.177.

Moxon-Emre, I., Bouffet, E., Taylor, M. D., Laperriere, N., Scantlebury, N., Law, N., ... Mabbott, D. (2014). Impact of craniospinal dose, boost volume, and neurologic complications on intellectual outcome in patients with medulloblastoma. *Journal of Clinical Oncology, 32*(17), 1760–1768. https://doi.org/10.1200/JCO.2013.52.3290.

Müller, H. L. (2013). Childhood craniopharyngioma. *Pituitary, 16*(1), 56–67. https://doi.org/10.1007/s11102-012-0401-0.

Ohgaki, H., & Kleihues, P. (2005). Population-based studies on incidence, survival rates, and genetic alterations in astrocytic and oligodendroglial gliomas. *Journal of Neuropathology and Experimental Neurology, 64*(6), 479–489. https://doi.org/10.1093/jnen/64.6.479.

Ostrom, Q. T., Gittleman, H., Farah, P., Ondracek, A., Chen, Y., Wolinsky, Y., ... Barnholtz-Sloan, J. S. (2013). CBTRUS statistical report: Primary brain and central nervous system tumors diagnosed in the United States in 2006-2010. *Neuro-Oncology, 15*(Suppl. 2), ii1–56. https://doi.org/10.1093/neuonc/not151.

Ostrom, Q. T., Gittleman, H., Liao, P., Rouse, C., Chen, Y., Dowling, J., ... Barnholtz-Sloan, J. (2014). CBTRUS statistical report: Primary brain and central nervous system tumors diagnosed in the United States in 2007-2011. *Neuro-Oncology, 16*(Suppl. 4), iv1–63. https://doi.org/10.1093/neuonc/nou223.

Poussaint, T. Y., Kocak, M., Vajapeyam, S., Packer, R. I., Robertson, R. L., Geyer, R., ... Kun, L. E. (2011). MRI as a central component of clinical trials analysis in brainstem glioma: A report from the pediatric brain tumor consortium (PBTC). *Neuro-Oncology, 13*(4), 417–427. https://doi.org/10.1093/neuonc/noq200.

Pulsifer, M. B., Sethi, R. V., Kuhlthau, K. A., MacDonald, S. M., Tarbell, N. J., & Yock, T. I. (2015). Early cognitive outcomes following proton radiation in pediatric patients with brain and central nervous system tumors. *International Journal of Radiation Oncology, Biology, Physics, 93*(2), 400–407. https://doi.org/10.1016/j.ijrobp.2015.06.012.

Reis, F., Faria, A. V., Zanardi, V. A., Menezes, J. R., Cendes, F., & Queiroz, L. S. (2006). Neuroimaging in pineal tumors. *Journal of Neuroimaging, 16*(1), 52–58. https://doi.org/10.1177/1051228405001514.

Reni, M., Gatta, G., Mazza, E., & Vecht, C. (2007). Ependymoma. *Critical Reviews in Oncology, 63*(1), 81–89. https://doi.org/10.1016/j.critrevonc.2007.03.004.

Rutkowski, S., Bode, U., Deinlein, F., Ottensmeier, H., Warmuth-Metz, M., Soerensen, N., ... Kuehl, J. (2005). Treatment of early childhood medulloblastoma by postoperative chemotherapy alone. *New England Journal of Medicine, 352*(10), 978–986. https://doi.org/10.1056/NEJMoa042176.

Schmiegelow, M., Lassen, S., Weber, L., Poulsen, H. S., Hertz, H., & Muller, J. (1999). Dosimetry and growth hormone deficiency following cranial irradiation of childhood brain tumors. *Medical and Pediatric Oncology, 33*(6), 564–571.

Strahm, B., & Malkin, D. (2006). Hereditary cancer predisposition in children: Genetic basis and clinical implications. *International Journal of Cancer, 119*(9), 2001–2006. https://doi.org/10.1002/ijc.21962.

Stubberfield, T. G., Byrne, G. C., & Jones, T. W. (1995). Growth and growth hormone secretion after treatment for acute lymphoblastic leukemia in childhood. 18-Gy versus 24-Gy cranial irradiation. *Journal of Pediatric Hematology, 17*(2), 167–171.

Suzuki, N., Konishi, E., Kondo, K., Fujita, I., & Ishikawa, E. (1985). A highly sensitive colorimetric sandwich-enzyme immunoassay for human chorionic gonadotropin using specific antibodies against the carboxy-terminal portion of the beta-subunit. *Clinica Chimica Acta, 150*(3), 247–253. https://doi.org/10.1016/0009-8981(85)90251-7.

Tabori, U., Shlien, A., Baskin, B., Levitt, S., Ray, P., Alon, N., … Malkin, D. (2010). TP53 alterations determine clinical subgroups and survival of patients with choroid plexus tumors. *Journal of Clinical Oncology, 28*(12), 1995–2001. https://doi.org/10.1200/JCO.2009.26.8169.

Vaidya, K., Smee, R., & Williams, J. R. (2012). Prognostic factors and treatment options for paediatric ependymomas. *Journal of Clinical Neuroscience, 19*(9), 1228–1235. https://doi.org/10.1016/j.jocn.2012.02.006.

Van Effenterre, R., & Boch, A. L. (2002). Craniopharyngioma in adults and children: A study of 122 surgical cases. *Journal of Neurosurgery, 97*(1), 3–11. https://doi.org/10.3171/jns.2002.97.1.0003.

Warmuth-Metz, M., Bison, B., Dannemann-Stern, E., Kortmann, R., Rutkowski, S., & Pietsch, T. (2008). CT and MR imaging in atypical teratoid/rhabdoid tumors of the central nervous system. *Neuroradiology, 50*(5), 447–452. https://doi.org/10.1007/s00234-008-0369-7.

Weiner, H. L., Lichtenbaum, R. A., Wisoff, J. H., Snow, R. B., Souweidane, M. M., Bruce, J. N., & Finlay, J. L. (2002). Delayed surgical resection of central nervous system germ cell tumors. *Neurosurgery, 50*(4), 727–733. https://doi.org/10.1097/00006123-200204000-00008. discussion 733-724.

Wisoff, J. H., Boyett, J. M., Berger, M. S., Brant, C., Li, H., Yates, A. J., … Finlay, J. L. (1998). Current neurosurgical management and the impact of the extent of resection in the treatment of malignant gliomas of childhood: A report of the children's cancer group trial no. CCG-945. *Journal of Neurosurgery, 89*(1), 52–59. https://doi.org/10.3171/jns.1998.89.1.0052.

Youland, R. S., Khwaja, S. S., Schomas, D. A., Keating, G. F., Wetjen, N. M., & Laack, N. N. (2013). Prognostic factors and survival patterns in pediatric low-grade gliomas over 4 decades. *Journal of Pediatric Hematology, 35*(3), 197–205. https://doi.org/10.1097/MPH.0b013e3182678bf8.

Zaky, W., Dhall, G., Ji, L., Haley, K., Allen, J., Atlas, M., … Finlay, J. L. (2014). Intensive induction chemotherapy followed by myeloablative chemotherapy with autologous hematopoietic progenitor cell rescue for young children newly-diagnosed with central nervous system atypical teratoid/rhabdoid tumors: The head start III experience. *Pediatric Blood and Cancer, 61*(1), 95–101. https://doi.org/10.1002/pbc.24648.

Chapter 10

Neurofibromatosis type 1: from cellular phenotypes to human brain function

Inês Bernardino[1,2], Joana Gonçalves[2] and Miguel Castelo-Branco[2]
[1]FMUC, University of Coimbra, Coimbra, Portugal; [2]ICNAS/CIBIT, University of Coimbra, Coimbra, Portugal

List of abbreviations

ADHD Attention deficit hyperactivity disorder
ASD Autism spectrum disorder
BOLD Blood oxygen level–dependent
DMN Default mode network
DTI Diffusion tensor imaging
E/I Excitatory/inhibitory
FEF Frontal eye fields
fMRI Functional magnetic resonance imaging
GABA γ-Aminobutyric acid
MRS Magnetic resonance spectroscopy
NF1 Neurofibromatosis type 1
OCD Obsessive–compulsive disorder
PET Positron-emission tomography
PFC Prefrontal cortex
SVM Support vector machine
VBM Voxel-based morphometry

Mini-dictionary of terms

Autism spectrum disorder A pervasive neurodevelopmental condition characterized by marked impairments in social interaction and communication in the presence of repetitive stereotyped behavior.
Default mode network A set of brain regions that deactivate during performance of cognitively demanding tasks and are active during task-irrelevant thoughts.
Excitation/inhibition balance Contributions of excitatory and inhibitory synaptic inputs corresponding to some neuronal event.
Frontal eye fields (FEF) Area of the frontal cortex involved in the control of several types of eye movements such as saccades, fixation, smooth pursuit; it is also part of visuospatial networks and is important in distinct aspects of attentional and visual cognition.
GABA Main inhibitory neurotransmitter in the brain that contributes to brain function regulation.
Inhibitory control Executive function that allows the suppression of thoughts or inappropriate actions in order to select a more appropriate behavior to achieve a given goal.
Magnetic resonance spectroscopy A noninvasive brain imaging technique that allows in vivo quantification of biochemical changes in localized brain regions.
Neural oscillations Rhythmic and/or repetitive electrical activity generated by neural tissue in the central nervous system in response to stimuli.
Neurofibromatosis type 1 One of the most common inherited neurological disorders caused by a single mutation of the Nf1 gene, characterized by altered skin pigmentation, increased tumor predisposition, and cognitive deficits.
Neurofibromin GTPase-activating protein that negatively regulates the RAS/MAPK pathway.
Phenotype Observable features present in an individual that result from the interaction of genetic and environmental factors.

Diagnosis, Management and Modeling of Neurodevelopmental Disorders. https://doi.org/10.1016/B978-0-12-817988-8.00010-5

Visual dorsal stream Part of visual system that projects from V1 to the parietal cortex and is involved in the guidance of actions and spatial information processing.

Introduction

Neurofibromatosis type 1 (NF1) is a neurodevelopmental monogenetic disorder, characterized by a high incidence of cognitive deficits and tumorigenesis (Gutmann et al., 2017). Moderate to severe impairments, in one or more areas of the cognitive functioning, are present in 80% of children with NF1 (Hyman, Shores, & North, 2005) and occur even in the absence of visible pathology. This suggests the existence of molecular mechanisms that contribute to the systems phenotype, irrespective of gross brain alterations. A long-standing hypothesis to explain the neurobehavioral deficits in NF1 is based on neurochemical imbalance in the γ-aminobutyric acid (GABA) system versus glutamatergic neurotransmission. Indeed, NF1 belongs to a cluster of neurodevelopmental conditions, including autism spectrum disorder (ASD), for which alterations in GABA were proposed as the underlying mechanism of the observed cognitive difficulties (Ramamoorthi & Lin, 2011). However, the nature of such imbalance across brain regions and different disease models has remained poorly understood.

The GABAergic dysfunction theory emerged from the initial groundbreaking work in the mouse model of the disorder ($Nf1^{+/-}$), which established a connection between RAS regulation by the *NF1* gene, GABAergic signaling, and learning deficits (Cui et al., 2008). In fact, a successful mouse model of NF1 has been generated based on the high homology between mouse and human neurofibromin protein sequence. The first mouse strain carrying a germline mutation in *NF1* gene was described by Jacks and colleagues (Jacks et al., 1994), who demonstrated that heterozygous animals developed a variety of tumors. Later, it was shown that $Nf1^{+/-}$ mice have hippocampal learning and memory deficits that resemble those present in the NF1 phenotype (Silva et al., 1997). Since then, several studies have identified altered behaviors in NF1 mice, most of these alterations being shared by other neurodevelopmental disorders, such as ASD. Hence, learning deficits (Cui et al., 2008; Molosh et al., 2014; Silva et al., 1997), working memory impairments (Shilyansky et al., 2010), visuospatial attention impairment (Li et al., 2005), sensorimotor gating deficits (Li et al., 2005; Molosh et al., 2014), and exploratory behavior dysfunction (Brown et al., 2010) have been identified in the mice models. Studies employing $Nf1^{+/-}$ mice had a pivotal role in unveiling potential neurobiological mechanisms underlying cognitive and behavioral alterations found in patients with NF1. Cellular and molecular studies in the $Nf1^{+/-}$ model indicated an imbalance in excitatory/inhibitory (E/I) leading to a sort of push–pull disequilibrium as a consequence of increased GABA-mediated inhibitory neurotransmission (Gonçalves et al., 2017).

The effort to translate the GABA hypothesis to the human disorder was performed by using in vivo magnetic resonance spectroscopy (MRS), the only brain imaging tool suited for noninvasively obtaining brain metabolites concentrations. Importantly, GABA in the human visual cortex was early on shown to be correlated with performance in cognitive tasks (Edden, Muthukumaraswamy, Freeman, & Singh, 2009). This supports the hypothesis that alterations in the push–pull mechanism between physiological excitation and inhibition might underlie cognitive deficits in some neurodevelopmental disorders and particularly in NF1 given that a similar association was previously reported in the animal model. Studies in children and adults with NF1 using MRS have shown systematically reduced GABA levels, across different brain regions (Ribeiro, Violante, Bernardino, Edden, & Castelo-Branco, 2015; Violante et al., 2016; Violante, Ribeiro, Edden, et al., 2013). These results may be conciliated with the evidence of increased GABA release found in the $Nf1^{+/-}$ by hypothesizing that the low GABA concentration found in patients with NF1 might reflect a compensatory mechanism. Such a mechanism might work if one takes into account that a dynamic modulation of GABA levels likely occurs in response to variations in the network activity. Combined MRS and positron emission tomography (PET) studies, to explore both pre- and postsynaptic involvement, represent a challenging but crucial step to deepen our understanding regarding homeostatic mechanisms governing GABAergic function in NF1 (Violante et al., 2016). A comprehensive view of these mechanisms, taking advantage of multimodal approaches, is an important step to uncover fundamental mechanisms underlying NF1 pathogenesis and to identify potential therapeutic strategies for targeted disease mechanisms.

The need to simultaneously understand pre- and postsynaptic phenotypes in NF1

The puzzling findings from both animal and human studies led to the hypothesis that, in a homeostatic system, low GABA levels lead to increased number of GABA receptors in order to maintain efficient neurotransmission. The converse might also actually hold true. The reduced concentration of GABA in patients could instead be a result of regulatory mechanisms to counterbalance pre- and postsynaptically the effects of increased GABA release related to reduced neurofibromin expression. To understand the implications of reduced GABA levels in NF1, it is important to know whether this is accompanied by a change in the number of GABA receptors in both human and animal models and in which brain regions.

In order to investigate this question and to comprehensively characterize the GABAergic system in adults with NF1, we previously investigated the distribution of $GABA_A$ receptors using [^{11}C]-Flumazenil PET while concomitantly measuring GABA concentration, in the occipital cortex and in the frontal eye fields (FEF), using MRS (Violante et al., 2016). Occipital cortex has particular relevance, given our prior finding showing reduced GABA levels in this area (Violante, Ribeiro, Edden, et al., 2013) in addition to altered visual sensory processing in children with NF1(Ribeiro et al., 2012). The FEF region, on the other hand, is linked to numerous cortical inhibitory circuits and its oculomotor task induced neural activation, which can be used as a metric to investigate relationships between blood oxygen level−dependent (BOLD) and GABA levels. In that study, an evidence for a global deficiency of the GABA system in NF1 was found. An overall reduction of GABA concentration in both occipital and FEF areas was reported in addition to reduced expression of $GABA_A$ receptors in other brain regions (parieto-occipital cortex, midbrain, and thalamus), suggesting both pre- and postsynaptic involvement. These findings are at odds with a simple homeostatic mechanism.

To probe the same question in the rodent model, and in order to link preclinical and clinical models, pre- and postsynaptic phenotypes were investigated in *Nf1*$^{+/-}$ mice (Gonçalves et al., 2017). Although our data did provide support for the hypothesis of E/I imbalance in NF1, it also showed, similarly to the human model, that pre- and postsynaptic changes are region-specific (Fig. 10.1). Cortical and striatal GABA/glutamate ratios proved to increase. At the postsynaptic level, very high $GABA_A$ receptor expression was found specifically in the hippocampus, disproportionately to the small reduction in GABA levels. GABAergic tone (either by receptor levels change or GABA/glutamate ratios) seemed therefore to be enhanced in all regions, although by a different mechanism. The findings are consistent with physiological evidence of increased inhibitory tone.

Taken together, these findings reveal that E/I imbalance may depend on distinct pre- and postsynaptic push−pull mechanisms that might be region-dependent, in both animals and humans. This is highly relevant for the development of therapeutic approaches, which may need to focus on targets where convergent physiological mechanisms can be found.

The visual system as a model system to study inhibitory phenotypes in NF1

Contrast processing and inhibition

The role of inhibition in the visual system is widely recognized, in particular in which concerns contrast sensitivity and perception. We therefore predicted that if the GABA hypothesis is true in NF1, then such processes should be impaired. Neurophysiological alterations in low-level visual processing that could be relevant to explain the cognitive phenotype are nevertheless still poorly understood.

In a previous psychophysical study, we found abnormalities in low-level sensory processing visual in individuals with NF1 (Ribeiro et al., 2012). Contrast sensitivity deficits were found for peripheral magnocellular (assessed with low-spatial/high-temporal frequency probes) and both central and peripheral parvocellular (assessed with intermediate-spatial/low-temporal frequency and chromatic red-green tests) retinocortical pathways. In contrast, central magnocellular and koniocellular (chromatic blue-yellow test) function was relatively spared. Interestingly, the same pattern of deficits using this indirect visual marker of inhibition was found in both children and adults with NF1, suggesting that these deficits persist in adulthood. The neural substrates of these deficits in low-level sensory processing were explored in a subsequent functional magnetic resonance imaging (fMRI) study employing stimuli differing in contrast as well as spatial and temporal frequencies to provide a relative bias regarding both magnocellular and parvocellular frequency channels (Violante et al., 2012). Hemodynamic responses in the early cortical visual system, V1, V2, and V3 were investigated through retinotopic mapping. Patients with NF1 showed deficient activation of the low-level visual cortex for both types of visual stimulation in accordance with the previous evidence suggesting impaired contrast sensitivity in both visual pathways. Importantly, this study corroborates the idea that low-level visual processing deficits do not ameliorate with age, since these impairments were consistently found in children, adolescents, and adults with NF1.

Interestingly, NF1 patients also showed a failure in deactivating anterior and posterior midline regions of the default mode network (DMN) during the performance of a cognitively demanding task recruiting the magnocellular visual pathway, which may be related with highly prevalent attentional impairments in this pathology (Violante, Ribeiro, Edden, et al., 2013). Taken together, these data may provide insightful information that can help to further understand the high-level visual cognitive profile of patients with NF1, particularly visuospatial and attentional impairments. This intriguing link between impaired magnocellular, peripheral attention, and DMN processing may generalize to neuropsychiatric disorders where such deficits have been separately identified, such as ASD and schizophrenia (Pomarol-Clotet et al., 2008).

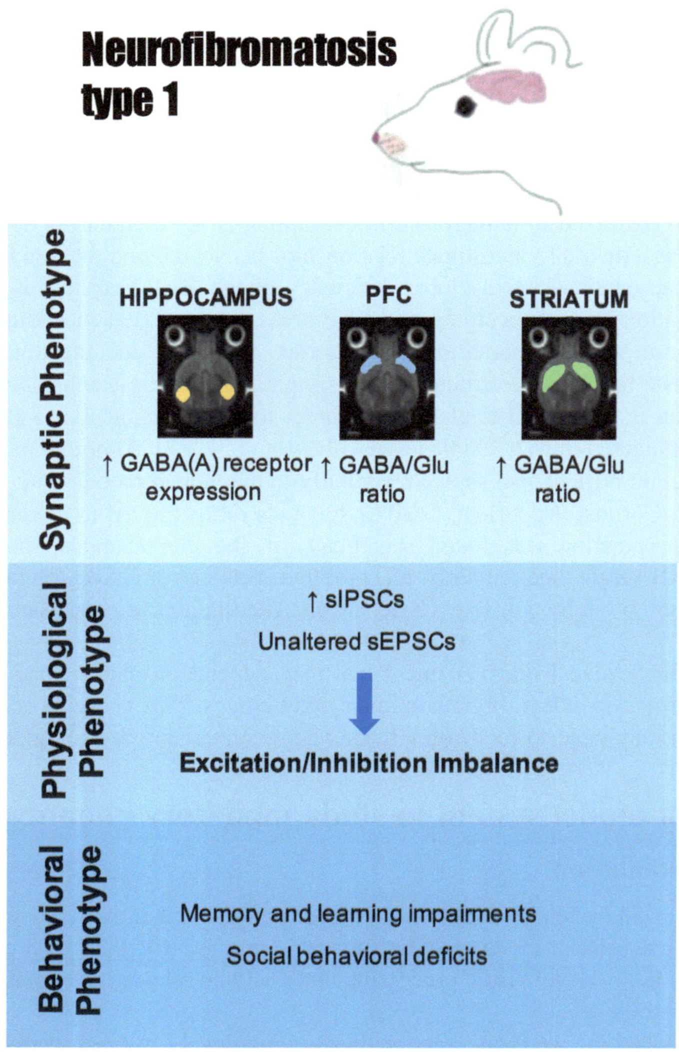

FIGURE 10.1 **Regional phenotypes of excitation/inhibition imbalance in the mouse model of NF1.** The $Nf1^{+/-}$ model shows GABAergic changes in a region-dependent way. The synaptic phenotype in hippocampus is characterized by $GABA_A$ receptor disproportionate increases while in the prefrontal cortex and striatum it is characterized by GABA/glutamate ratio increases. In spite of the differences in synaptic phenotype in these brain regions, the physiological phenotype is similar with converging increases of sIPSCs relative to sEPSCs resulting in increased inhibitory drive. Accordingly, behavioral phenotype is always characterized by memory and learning impairments and social behavioral deficits. *GABA*, γ-aminobutyric acid; *GLU*, glutamate; *PFC*, prefrontal cortex; *sEPSCs*, spontaneous excitatory postsynaptic currents; *sIPSCs*, spontaneous inhibitory postsynaptic currents.

Later on, a subsequent fMRI study of the visual cortex indicated that the BOLD signal was negatively correlated with GABA levels in both patients and control subjects (Violante, Ribeiro, Edden, et al., 2013). Moreover, there were differences in curve slopes of NF1 patients and controls indicating that changes in BOLD activity are larger for the same unity change in GABA levels in patients. This may suggest a more sensitive push–pull mechanism between physiological excitation and inhibition in NF1.

Neural oscillations in visual cortex—a new functional probe in NF1?

Neural oscillations also represent a hallmark of inhibitory processing, given the role of inhibitory processing to sculpt the frequency of such oscillations. This makes NF1 a useful model to study the role of inhibition in shaping brain oscillations A specific enhancement of alpha brain oscillations' amplitudes triggered by a low-level visual task, as well as deficits in later stages of visual processing, was observed in children and adolescents with NF1 supporting the existence of deficits in basic sensory processing and difficulties in attention allocation (Ribeiro et al., 2014). Dysregulation of alpha oscillations was subsequently corroborated using a covert attention task (Silva et al., 2016). Interestingly, in the latter study patients with NF1 demonstrated surprisingly excessive alpha oscillations desynchronization along with high accuracy, which suggests a compensatory mechanism to achieve normal levels of performance. Considering the link between alpha oscillations, visual processing, and attention processes, these results indicate that abnormal increases or decreases in alpha oscillations are task and performance-dependent and result from a dysregulatory mechanism in disorders affecting attention, such as NF1.

Beyond the visual system, inhibition is also involved in the control of frequency and stability of motor rhythms. For that reason, we explored the link between behavioral motor control, brain rhythms, and brain activity in NF1, through the application of complementary quantitative measures, based on electroencephalography and fMRI. We found impaired precision in rhythmic pacing behavior in NF1 along with matched oscillatory patterns at the beta subband, which are linked to the behavior of inhibitory interneurons and have been associated with E/I balance (Whittington, Traub, Kopell, Ermentrout, & Buhl, 2000). Additionally, a decreased recruitment of the cerebellum, a structure where inhibitory interneurons are essential regulators of rhythmic synchronization, was found as well as in deep brain regions pivotally involved in motor pacing.

Applications to other areas of development with a focus on ASD

What is the interest of this disorder in the general context of neurodevelopmental disorders?

NF1 is a single gene disorder, which enables to better understand nature–nurture interactions during development. This may provide clues to deeply investigate other neurodevelopmental disorders that exhibit high comorbidity with NF1 (Table 10.1). Hence, this pathology can be used as a model to understand the neural basis of impulsivity and obsessive–compulsive disorder (OCD) and also attention deficit hyperactivity disorder (ADHD), which are highly prevalent in the NF1 population. Importantly, NF1 can be used as a model to study impaired social cognition, given its autism comorbidity. Indeed, there is robust evidence regarding the presence of ASD-like traits and behaviors in the NF1phenotype, with a prevalence that is substantially higher than that found in the general population (Chisholm et al., 2018; Eijk et al., 2018; Garg et al., 2013). Interestingly, a common pathophysiological basis associated with dysfunction of inhibitory GABAergic circuits has been reported in both NF1 and ASD (Ramamoorthi & Lin, 2011). However, the much higher heterogeneity found in the latter renders these findings much more difficult to replicate (Carvalho Pereira, Violante, Mouga, Oliveira, & Castelo-Branco, 2018).

NF1 represents, therefore, a monogenic model for several neurobehavioral features of ASD, which are discussed further. Moreover, it provides a way to identify putative-mechanism-based treatments, which may be more difficult in the broader context of ASD (Molosh & Shekhar, 2018). Heterogeneity of ASD remains a challenge, and studies of monogenic syndromic forms of ASD, such as NF1, may be very helpful in this context.

TABLE 10.1 Neurodevelopmental disorders affecting E/I imbalance.

Idiopathic and syndromic autism (tuberous sclerosis)	Low brain GABA levels in ASD. Significant increases in GABA concentrations in tuberous sclerosis.
Down syndrome	Evidence for enhanced GABAergic inhibition.
Schizophrenia	Despite previous evidence of altered GABAergic system in schizophrenia, current literature does not support consistent alterations and GABA levels were only significantly lower compared with controls after averaging GABA levels across multiple brain regions.
Tourette syndrome	Inconsistent results regarding GABA neurotransmission.
Epilepsy	Low GABA concentrations in the cerebrospinal fluid are consistently observed in a variety of epileptic syndromes.
Fragile X syndrome	Altered GABAergic circuit function.

The table lists neurodevelopmental disorders where there is evidence of altered E/I imbalance. In some cases, results are not consistent across studies and seem to be regionally and temporally specific. Methodological aspects may also contribute to this variability (Egerton, Modinos, Ferrera, & McGuire, 2017; Gao et al., 2018; Mori et al., 2012; Schür et al., 2016). *ASD*, autism spectrum disorder; *GABA*, γ-aminobutyric acid.

Heterogeneity in ASD and NF1

It is quite remarkable that despite being a monogenetic disorder, NF1 shows quite high heterogeneity in terms of its clinical and cognitive phenotype (Monroe, Dahiya, & Gutmann, 2017). To date, it remains a challenge to establish genotype−phenotype correlations in NF1 because of the heterogeneous disease expression observed in humans. A notable exception concerning the impact in neurotransmission is the study by Violante and colleagues (Violante, Ribeiro, Edden, et al., 2013).

The heterogeneity described in both the human and animal models is present for core clinical and cognitive features and relevant comorbidities. Indeed, genomic background, sex, and the germline NF1 gene mutation interact in predicting development and progression of NF1 symptoms as well as response to treatments. There are studies demonstrating that the *NF1* mutation type significantly predicted severity of plexiform neurofibromas and levels of GABA in the cortex (Li et al., 2016; Violante, Ribeiro, Edden, et al., 2013) (Fig. 10.2). Importantly, NF1 heterogeneity can be present also at the within-subject level. We have shown that E/I imbalance in NF1 is brain-region-specific with distinct pre- and postsynaptic mechanisms, suggesting the need for region-tailored therapies directed to different phenotypes of ASD (Gonçalves et al., 2017).

Heterogeneity of volumetric findings in ASD is quite often quoted. Volumetric changes have been consistently found in NF1 and seem to be more consistent than in ASD. Subcortical analysis revealed disproportionately larger thalami, right caudate, and middle corpus callosum in patients with NF1. Cortical analyses on volume, thickness, and surface area were less remarkable (Violante, Ribeiro, Silva, & Castelo-Branco, 2013), but machine learning approaches could detect distributed abnormalities even at this level (Duarte et al., 2014). Interestingly, patients with NF1 had significantly lower gyrification indices than typically developing children primarily in the frontal and temporal lobes, but also affecting the insula, cingulate cortex, parietal and occipital regions. Duarte et al. (2014) performed voxel-based morphometry (VBM) analysis, which revealed structural differences, which were further corroborated by the support vector machine (SVM) weight maps representing the most relevant brain regions for group discrimination. These included the hippocampus, basal ganglia, thalamus, and visual cortex.

Morphometric alterations were also observed in $Nf1^{+/-}$ mice, namely significant increase in the volume of prefrontal cortex (PFC), caudate−putamen, and hippocampus (Petrella et al., 2016). Moreover, diffusion tensor imaging (DTI) analysis determined that mean diffusivity maps were higher in $Nf1^{+/-}$ mice in cortical structures, septum, striatum, globus pallidus, hippocampus, thalamus, periaqueductal gray, and midbrain (Petrella et al., 2016). However, the relations between such volumetric changes and behavioral impairments remain poorly understood in both animal models and the human disease. The association between neurobehavioral phenotypes and the mutation type also needs to be established. This might also illuminate the morphometric heterogeneity found in ASD.

Visual dorsal stream deficits

Visual dorsal stream dysfunction, related with visuospatial impairments, is one of the most common manifestations in NF1 and is also described in other neurodevelopmental disorders such as ASD and Williams syndrome (Bernardino, Castelhano, Farivar, Silva, & Castelo-Branco, 2013; Bernardino, Rebola, Farivar, Silva, & Castelo-Branco, 2014). Performance in the judgment of line orientation test is suggestive of visual dorsal stream deficits (Ribeiro et al., 2012). However, it remains to be understood how these impairments relate to the ones described in ASD, which we found to be milder than in Williams syndrome, in particular concerning visuoconstructive abilities and motion coherence perception and integration (Bernardino et al., 2012; Castelo-Branco et al., 2007; Mendes et al., 2005).

Impulsivity and inhibitory control

Individuals with NF1 showed impaired impulse control and reduced EEG correlates of early visual processing (parieto-occipital P1) and inhibitory control (frontal P3) (Ribeiro et al., 2015). These findings could be related with MRS data revealing a reduction in medial frontal GABA+/tCr (total Creatine) levels in the NF1 patients, simultaneously with the replication of reduced occipital GABA levels. In contrast, glutamate + glutamine/tCr levels were normal. Surprisingly, the relationship between inhibitory control and medial frontal GABA was reversed in NF1: higher GABA was associated with a faster and more erroneous response style, whereas in controls it was related to a cautious strategy. Abnormal GABAergic physiology appears, thus, as an important factor underlying impaired cognition in NF1, in a level and region-dependent manner. These reversed correlation patterns are quite intriguing and suggest that the relation between genes, GABA levels, and behavior maybe more complex than previously thought.

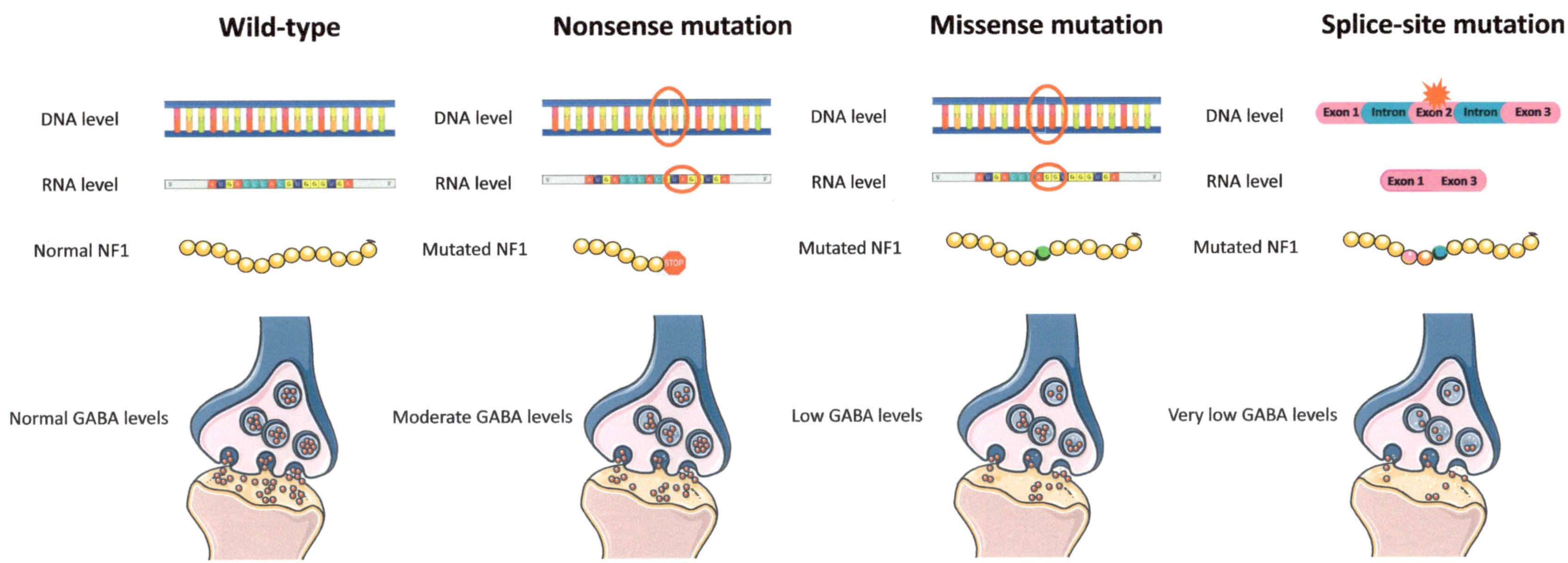

FIGURE 10.2 **Genotype-phenotype correlation in NF1. Schematic representation of genotypic impact on GABA levels.** By ordering mutations by type, it was found that abnormal neurofibromin proteins resulting from nonsense mutations do not necessarily predict low GABA levels. However, missense or splice-site mutations have significantly lower GABA levels than unaffected individuals, resulting in a more severe phenotype than nonsense mutations. *DNA*, deoxyribonucleic acid; *GABA*, γ-aminobutyric acid; *RNA*, ribonucleic acid.

Social cognition deficits

Alterations in social cognition, a pivotal feature in the ASD cognitive phenotype, were reported in the NF1 animal model. Significant structure—function correlations were identified concerning social recognition performance and PFC volumes in *Nf1*$^{+/-}$ mice (Petrella et al., 2016). Voluntary initiation of social interactions and discrimination of social novelty are ASD-relevant behaviors (Peça et al., 2011). Using three-chamber social arena, *Nf1*$^{+/-}$ mice displayed a lower index of preference for social interactions and for novel stimuli (Fig. 10.3). Similar findings were observed in another animal model of ASD (Edfawy et al., 2019; Peça et al., 2011). These types of impairments are being increasingly described also in the human disorder, reinforcing the notion that NF1 is a tenable model for ASD, in spite of the heterogeneity found in both conditions.

Implications for therapy

A comprehensive knowledge of the pathophysiology of NF1 (Fig. 10.4) opens a window for the identification of targeted therapeutic strategies. In both NF1 and ASD, there is intensive research trying to identify biomarkers that can predict treatment response, as well as therapeutic targets. It remains an open question whether modulation of GABA levels, which can be achieved using drugs or neurostimulation approaches, will be useful in any of these conditions. It is important to take into account the relatively small effect size of their changes and the ongoing debate on whether these changes reflect a homeostatic or a maladaptive alteration.

Interestingly, treatment of *Nf1*$^{+/-}$ mice with $GABA_A$ receptor antagonist, picrotoxin, improved spatial learning, and decreased GABA inhibition (Cui et al., 2008). Currently there are ongoing clinical trials investigating effects of drugs that modulate GABAergic neurotransmission on social or cognitive impairments in children and adults with ASD (ClinicalTrials.gov Identifier: *NCT01288716*) (Veenstra-Vanderweele et al., 2017). Similar types of clinical trials have been attempted in NF1 using statins but results have been so far inclusive (Mainberger et al., 2013; Payne et al., 2016; Stivaros et al., 2018). The great advantage in NF1 relates to the fact that RAS pathway has been causally described in the animal model of this developmental disorder, and the same pathway is known to be altered in the human disease and in a widely distributed set of regions of the central nervous system. Hence, the main challenge in the future remains to identify the best approaches to modulate this pathway.

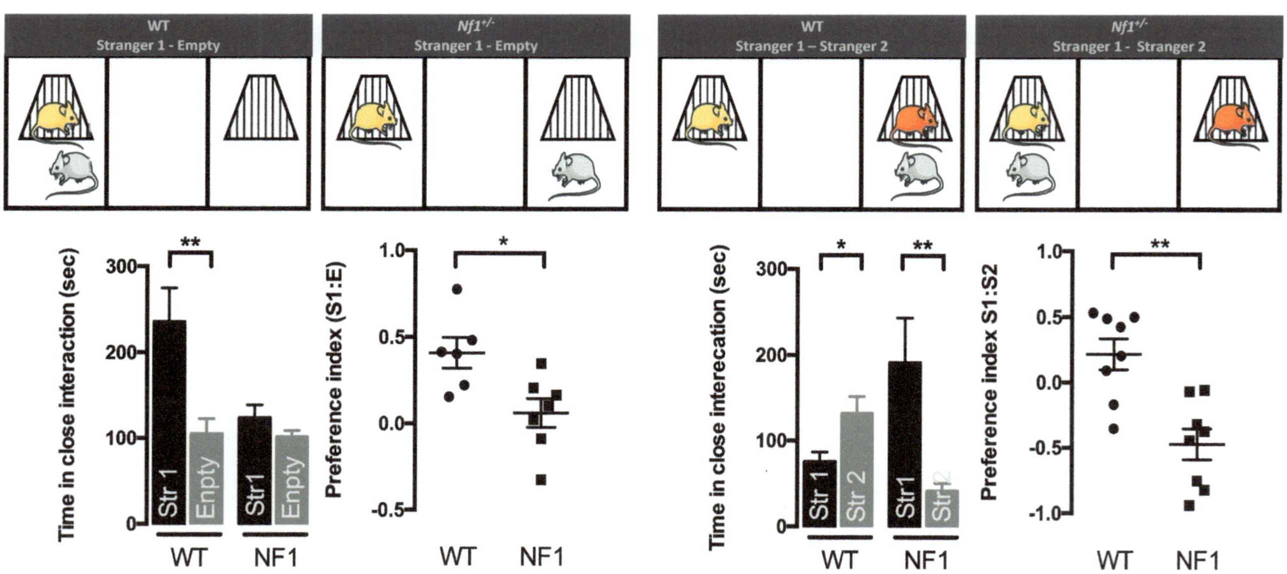

FIGURE 10.3 **Adult *Nf1*$^{+/-}$ mice exhibit social behavioral alterations.** Illustration of the social behavior of *Nf1*$^{+/-}$ mice and littermate controls in three-chamber paradigm tests at day 1 (boxes "Stranger 1 — Empty") and day 2 (boxes "Stranger 1 — Stranger 2"). Below representative images, graphs show that mutated mice have difficulty in interacting with "Stranger" stimulus mice displaying reduced preference for social interaction when compared with wild-type mice in day 1. At day 2, NF1 mice shown an increase in total time interacting with "Stranger1" and a reduced preference index for "Stranger 2." All data are presented as mean ± SEM (n = 8—10). $^{*}P < .05$, $^{**}P < .01$, statistical significance using two-tailed *t*-test. *Str 1*, Stranger 1; *Str 2*, Stranger 2; *WT*, wild-type.

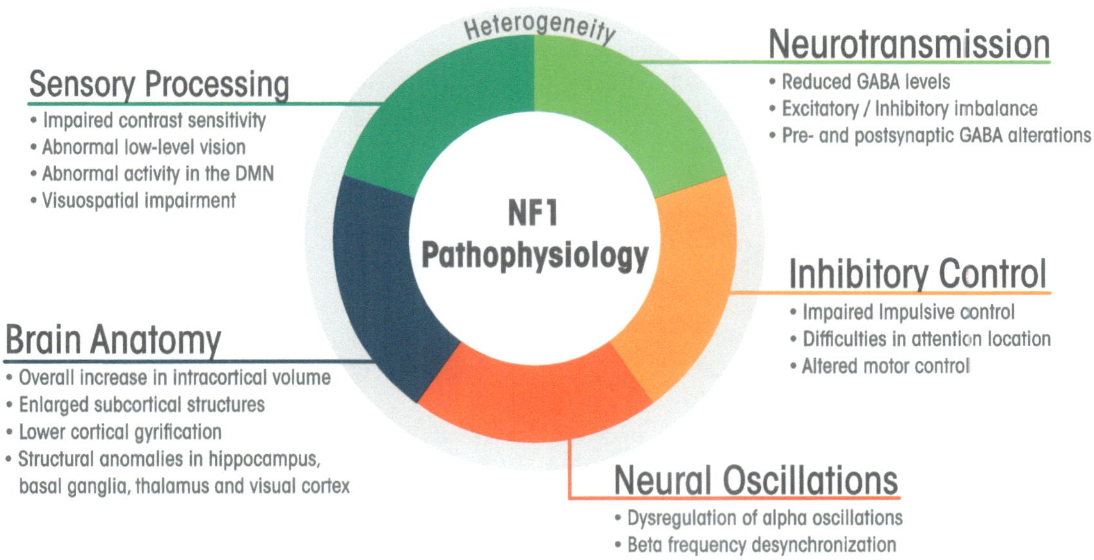

FIGURE 10.4 **Pathophysiology of NF1.** NF1 phenotype is characterized by alterations in several domains that encompass sensory processing, brain anatomy, neural oscillations, inhibitory control, and GABAergic neurotransmission. *NF1*, Neurofibromatosis Type 1; *DMN*, default mode network, *GABA*, γ-aminobutyric acid.

Key facts of ASD prevalence in NF1

- There is evidence of increased ASD symptomatology in NF1, with considerably higher prevalence (11%–25%) than in the general population.
- Reduced social skills and prosocial behaviors, as well as social difficulties including isolation and reduced peer acceptance, are observed in patients with NF1.
- Social behavior impairments in NF1 patients are not explained simply by cognitive deficits.
- Emotion perception deficits are found in patients with NF1, particularly regarding the identification of negative emotions such as anger and fear.
- A higher incidence of ASD traits is often found in studies with both children and adults with NF1.

Key facts of comorbid phenotypes in NF1

- Approximately 40% of children with NF1 meet the diagnostic criteria for ADHD symptomatology.
- About 25% of children with NF1 meet criteria for both ASD and ADHD.
- Patients with NF1 show a vulnerability to symptoms of social anxiety, which is related to a predisposition for anxiety disorder.
- Epilepsy symptomatology is present in approximately 14% of NF1 patients.
- Other neurological comorbidities include migraine, optic nerve glioma, brain tumor.

Key facts of GABA in NF1

- The GABAergic system is altered in NF1 and may contribute to the pathophysiology of the disease.
- Patients with NF1 have significantly lower GABA levels in the brain.
- The NF1 mutation type significantly predicts cortical GABA.
- GABA is altered in NF1 both pre- and postsynaptically.
- Alterations in GABA concentrations in NF1 are region-dependent.

Key facts of cognitive function in NF1

- Intellectual difficulties are frequent in NF1 with the full-scale intellectual quotient being within the average to low average range (80–90).

- Learning disabilities are reported in 30%–60% of the patients with NF1.
- Impairments in mathematical ability and visuospatial abilities are one of the hallmark features of the NF1 cognitive profile.
- Executive function is affected in patients with NF1, including functions such as problem-solving, attentional shifting, planning, impulsivity control, mental flexibility, monitoring, and changing behaviors.
- NF1 patients show motor impairments, even in simple motor tasks such as finger tapping, and deficits in visuomotor function.

Key facts of the mouse model in NF1

- The mouse model of NF1 has an abnormal upregulation of RAS activity that induces synaptic network deficits.
- *Nf1*$^{+/-}$ mice exhibit hippocampal learning and memory deficits that resemble those seen in NF1 patients.
- Large volumes in regions involved in social cognition, such as prefrontal cortex and *caudate–putamen*, are observed in the NF1 mouse.
- MRS analysis confirms E/I imbalance in *Nf1*$^{+/-}$ mice with increased GABA/glutamate ratios in cortex and striatum and high expression of $GABA_A$ receptors in the hippocampus.
- The NF1 mouse model represents a promising approach to isolate mechanisms underlying ASD-related cognitive disabilities.

Summary points

- **Summary Point 1:** NF1 provides a monogenic model to understand a broad range of manifestations that also occur in ASD.
- **Summary Point 2:** Inhibitory neurotransmission is impaired in both human and animal models of NF1.
- **Summary Point 3:** Inhibitory phenotypes in NF1 are changed at pre- and postsynaptic levels.
- **Summary Point 4:** Inhibitory phenotypes in NF1 are brain-region-dependent.
- **Summary Point 5:** RAS is the affected pathway in NF1 and can be modulated by pharmacological approaches, such as the use of statins.

References

Bernardino, I., Castelhano, J., Farivar, R., Silva, E. D., & Castelo-Branco, M. (2013). Neural correlates of visual integration in Williams syndrome: Gamma oscillation patterns in a model of impaired coherence. *Neuropsychologia, 51*(7), 1287–1295.

Bernardino, I., Mouga, S., Almeida, J., van Asselen, M., Oliveira, G., & Castelo-Branco, M. (2012). A direct comparison of local-global integration in autism and other developmental disorders: Implications for the central coherence hypothesis. *PLoS One, 7*(6).

Bernardino, I., Rebola, J., Farivar, R., Silva, E., & Castelo-Branco, M. (2014). Functional reorganization of the visual dorsal stream as probed by 3-D visual coherence in Williams syndrome. *Journal of Cognitive Neuroscience, 26*(11), 2624–2636.

Brown, J. A., Emnett, R. J., White, C. R., Yuede, C. M., Conyers, S. B., O'Malley, K. L., & Gutmann, D. H. (2010). Reduced striatal dopamine underlies the attention system dysfunction in neurofibromatosis-1 mutant mice. *Human Molecular Genetics, 19*(22), 4515–4528.

Carvalho Pereira, A., Violante, I. R., Mouga, S., Oliveira, G., & Castelo-Branco, M. (2018). Medial frontal lobe neurochemistry in autism spectrum disorder is marked by reduced N-acetylaspartate and unchanged gamma-aminobutyric acid and glutamate + glutamine levels. *Journal of Autism and Developmental Disorders, 48*(5), 1467–1482.

Castelo-Branco, M., Mendes, M., Sebastião, A. R., Reis, A., Soares, M., Saraiva, J., & Silva, E. (2007). Visual phenotype in Williams-Beuren syndrome challenges magnocellular theories explaining human neurodevelopmental visual cortical disorders. *The Journal of clinical investigation, 117*(12), 3720–3729.

Chisholm, A. K., Anderson, V. A., Pride, N. A., Malarbi, S., North, K. N., & Payne, J. M. (2018). Social function and autism spectrum disorder in children and adults with neurofibromatosis type 1: A systematic review and meta-analysis. *Neuropsychology Review, 28*(3), 317–340.

Cui, Y., Costa, R. M., Murphy, G. G., Elgersma, Y., Zhu, Y., Gutmann, D. H., & Silva, A. J. (2008). Neurofibromin regulation of ERK signaling modulates GABA release and learning. *Cell, 135*(3), 549–560.

Duarte, J. V., Ribeiro, M. J., Violante, I. R., Cunha, G., Silva, E., & Castelo-Branco, M. (2014). Multivariate pattern analysis reveals subtle brain anomalies relevant to the cognitive phenotype in neurofibromatosis type 1. *Human Brain Mapping, 35*(1), 89–106.

Edden, R. A. E., Muthukumaraswamy, S. D., Freeman, T. C. A., & Singh, K. D. (2009). Orientation discrimination performance is predicted by GABA concentration and gamma oscillation frequency in human primary visual cortex. *Journal of Neuroscience, 29*(50), 15721–15726.

Edfawy, M., Guedes, J. R., Pereira, M. I., Laranjo, M., Carvalho, M. J., Gao, X., & Peça, J. (2019). Abnormal mGluR-mediated synaptic plasticity and autism-like behaviours in Gprasp2 mutant mice. *Nature Communications, 10*, 1431–1446.

Egerton, A., Modinos, G., Ferrera, D., & McGuire, P. (2017). Neuroimaging studies of GABA in schizophrenia: A systematic review with meta-analysis. *Translational Psychiatry, 7*(6). e1147-10.

Eijk, S., Mous, S. E., Dieleman, G. C., Dierckx, B., Rietman, A. B., de Nijs, P. F. A., & Legerstee, J. S. (2018). Autism spectrum disorder in an unselected cohort of children with neurofibromatosis type 1 (NF1). *Journal of Autism and Developmental Disorders, 48*(7), 2278–2285.

Gao, F., Qi, L., Yang, Z., Yang, T., Zhang, Y., Xu, H., & Zhao, H. (2018). Impaired GABA neural circuits are critical for fragile X syndrome. *Neural Plasticity*, 1–7, 2018.

Garg, S., Green, J., Leadbitter, K., Emsley, R., Lehtonen, A., Evans, D. G., & Huson, S. M. (2013). Neurofibromatosis type 1 and autism spectrum disorder. *Pediatrics, 132*(6), e1642–e1648.

Gonçalves, J., Violante, I. R., Sereno, J., Leitão, R. A., Cai, Y., Abrunhosa, A., & Castelo-Branco, M. (2017). Testing the excitation/inhibition imbalance hypothesis in a mouse model of the autism spectrum disorder: In vivo neurospectroscopy and molecular evidence for regional phenotypes. *Molecular Autism, 8*(47), 1–8.

Gutmann, D. H., Ferner, R. E., Listernick, R. H., Korf, B. R., Wolters, P., & Johnson, K. J. (2017). Neurofibromatosis type 1. *Nature Reviews, Disease Primers, 3*, 1–17.

Hyman, S. L., Shores, E. A., & North, K. N. (2005). The nature of frequency of cognitive deficits in children with neurofibromatosis type 1. *Neurology, 65*(7), 1037–1044.

Jacks, T., Shih, T. S., Schmitt, E. M., Bronson, R. T., Bernards, A., & Weinberg, R. A. (1994). Tumour predisposition in mice heterozygous for a targeted mutation in Nf1. *Nature Genetics, 7*(3), 353–361.

Li, W., Cui, Y., Kushner, S. A., Brown, R. A. M., Jentsch, J. D., Frankland, P. W., & Silva, A. J. (2005). The HMG-CoA reductase inhibitor lovastatin reverses the learning and attention deficits in a mouse model of neurofibromatosis type 1. *Current Biology, 15*(21), 1961–1967.

Li, K., Turner, A. N., Chen, M., Brosius, S. N., Schoeb, T. R., Messiaen, L. M., & Kesterson, R. A. (2016). Mice with missense and nonsense NF1 mutations display divergent phenotypes compared with human neurofibromatosis type I. *Disease Models and Mechanisms, 9*(7), 759–767.

Mainberger, F., Jung, N. H., Zenker, M., Wahlländer, U., Freudenberg, L., Langer, S., & Mall, V. (2013). Lovastatin improves impaired synaptic plasticity and phasic alertness in patients with neurofibromatosis type 1. *BMC Neurology, 13*(131), 1–12.

Mendes, M., Silva, F., Simões, L., Jorge, M., Saraiva, J., & Castelo-Branco, M. (2005). Visual magnocellular and structure from motion perceptual deficits in a neurodevelopmental model of dorsal stream function. *Cognitive Brain Research, 25*(3), 788–798.

Molosh, A. I., Johnson, P. L., Spence, J. P., Arendt, D., Federici, L. M., Bernabe, C., & Shekhar, A. (2014). Social learning and amygdala disruptions in Nf1 mice are rescued by blocking p21-activated kinase. *Nature Neuroscience, 17*, 1583–1590.

Molosh, A. I., & Shekhar, A. (2018). Neurofibromatosis type 1 as a model system to study molecular mechanisms of autism spectrum disorder symptoms. *Progress in Brain Research*, 37–62.

Monroe, C. L., Dahiya, S., & Gutmann, D. H. (2017). Dissecting clinical heterogeneity in neurofibromatosis type 1. *Annual Review of Pathology: Mechanisms of Disease, 12*(1), 53–74.

Mori, K., Mori, T., Toda, Y., Fujii, E., Miyazaki, M., Harada, M., & Kagami, S. (2012). Decreased benzodiazepine receptor and increased GABA level in cortical tubers in tuberous sclerosis complex. *Brain and Development, 34*(6), 478–486.

Payne, J. M., Barton, B., Ullrich, N. J., Cantor, A., Hearps, S. J. C., Cutter, G., & North, K. N. (2016). Randomized placebo-controlled study of lovastatin in children with neurofibromatosis type 1. *Neurology, 87*(24), 2575–2584.

Petrella, L. I., Cai, Y., Sereno, J. V., Gonçalves, S. I., Silva, A. J., & Castelo-Branco, M. (2016). Brain and behaviour phenotyping of a mouse model of neurofibromatosis type-1: An MRI/DTI study on social cognition. *Genes, Brain and Behavior, 15*(7), 637–646.

Peça, J., Feliciano, C., Ting, J. T., Wang, W., Wells, M. F., Venkatraman, T. N., & Feng, G. (2011). Shank3 mutant mice display autistic-like behaviours and striatal dysfunction. *Nature, 472*(7344), 437–442.

Pomarol-Clotet, E., Salvador, R., Sarró, S., Gomar, J., Vila, F., Martínez, Á., & McKenna, P. (2008). Failure to deactivate in the prefrontal cortex in schizophrenia: Dysfunction of the default mode network? *Psychological Medicine, 38*(8), 1185–1193.

Ramamoorthi, K., & Lin, Y. (2011). The contribution of GABAergic dysfunction to neurodevelopmental disorders. *Trends in Molecular Medicine, 17*(8), 452–462.

Ribeiro, M. J., D'Almeida, O. C., Ramos, F., Saraiva, J., Silva, E. D., & Castelo-Branco, M. (2014). Abnormal late visual responses and alpha oscillations in neurofibromatosis type 1: A link to visual and attention deficits. *Journal of Neurodevelopmental Disorders, 6*(4), 1–19.

Ribeiro, M. J., Violante, I. R., Bernardino, I., Edden, R. A. E., & Castelo-Branco, M. (2015). Abnormal relationship between GABA, neurophysiology and impulsive behavior in neurofibromatosis type 1. *Cortex, 64*, 194–208.

Ribeiro, M. J., Violante, I. R., Bernardino, I., Ramos, F., Saraiva, J., Reviriego, P., & Castelo-Branco, M. (2012). Abnormal achromatic and chromatic contrast sensitivity in neurofibromatosis type 1. *Investigative Ophthalmology and Visual Science, 53*(1), 287–293.

Schür, R. R., Draisma, L. W. R., Wijnen, J. P., Boks, M. P., Koevoets, M. G. J. C., Joëls, M., & Vinkers, C. H. (2016). Brain GABA levels across psychiatric disorders: A systematic literature review and meta-analysis of 1H-MRS studies. *Human Brain Mapping, 37*(9), 3337–3352.

Shilyansky, C., Karlsgodt, K. H., Cummings, D. M., Sidiropoulou, K., Hardt, M., James, A. S., & Silva, A. J. (2010). Neurofibromin regulates corticostriatal inhibitory networks during working memory performance. *Proceedings of the National Academy of Sciences United States of America, 107*(29), 13141–13146.

Silva, A. J., Frankland, P. W., Marowitz, Z., Friedman, E., Lazlo, G., Cioffi, D., & Bourtchuladze, R. (1997). A mouse model for the learning and memory deficits associated with neurofibromatosis type I. *Nature Genetics, 15*(3), 281–284.

Silva, G., Ribeiro, M., Costa, G., Violante, I., Ramos, F., Saraiva, J., & Castelo-Branco, M. (2016). Peripheral attentional targets under covert attention lead to paradoxically enhanced alpha desynchronization in neurofibromatosis type 1. *PLoS One, 11*(2), 1–15.

Stivaros, S., Garg, S., Tziraki, M., Cai, Y., Thomas, O., Mellor, J., & Steiger, C. (2018). Randomised controlled trial of simvastatin treatment for autism in young children with neurofibromatosis type 1 (SANTA). *Molecular Autism, 9*(12), 1–13.

Veenstra-Vanderweele, J., Cook, E. H., King, B. H., Zarevics, P., Cherubini, M., Walton-Bowen, K., & Carpenter, R. L. (2017). Arbaclofen in children and adolescents with autism spectrum disorder: A randomized, controlled, phase 2 trial. *Neuropsychopharmacology, 42*(7), 1390–1398.

Violante, I. R., Patricio, M., Bernardino, I., Rebola, J., Abrunhosa, A. J., Ferreira, N., & Castelo-Branco, M. (2016). GABA deficiency in NF1. *Neurology, 1–9.*

Violante, I. R., Ribeiro, M. J., Cunha, G., Bernardino, I., Duarte, J. V., Ramos, F., & Castelo-Branco, M. (2012). Abnormal brain activation in neurofibromatosis type 1: A link between visual processing and the default mode network. *PLoS One, 7*(6), 1–11.

Violante, I. R., Ribeiro, M. J., Edden, R. A. E., Guimarães, P., Bernardino, I., Rebola, J., & Castelo-Branco, M. (2013). GABA deficit in the visual cortex of patients with neurofibromatosis type 1: Genotype-phenotype correlations and functional impact. *Brain*, 1–8.

Violante, I. R., Ribeiro, M. J., Silva, E. D., & Castelo-Branco, M. (2013). Gyrification, cortical and subcortical morphometry in neurofibromatosis type 1: An uneven profile of developmental abnormalities. *Journal of Neurodevelopmental Disorders, 5*(1), 1–13.

Whittington, M. A., Traub, R. D., Kopell, N., Ermentrout, B., & Buhl, E. H. (2000). Inhibition-based rhythms: Experimental and mathematical observations on network dynamics. *International Journal of Psychophysiology, 38*(3), 315–336.

Chapter 11

Current insights into neural crest cell development and pathologies

Rebekah M. Charney[a], Maneeshi S. Prasad[a] and Martín I. García-Castro
Division of Biomedical Sciences, University of California Riverside, Riverside, CA, United States

List of abbreviations

BMP Bone morphogenetic protein
EMT Epithelial-to-mesenchymal transition
ENS Enteric nervous system
FGF Fibroblast growth factor
GRN Gene regulatory network
NC Neural crest
NPB Neural plate border
PNS Peripheral nervous system
PSCs Pluripotent stem cells
TFs Transcription factors

Mini-dictionary of terms

Ectoderm: The outer layer of the three primary germ layers in triploblastic embryos which gives rise to the nervous system and epidermis.
Induction: The interaction of two groups of cells whereby one group elicits a signal that directs the development of the receiving group.
Mesoderm: The middle layer of the three primary germ layers. The mesoderm gives rise to the connective tissues and muscle.
Multipotent: A progenitor cell population with the capacity to give rise to multiple cell types. Multipotent cells exhibit a narrower differentiation potential than pluripotent cells.
Neural crest: Vertebrate specific multipotent cells which migrate extensively and contribute to numerous tissues and organs.
Neural plate border: The region in between the neural plate and the nonneural ectoderm containing the NC and preplacodal cells.
Neurocristopathy: A human pathology that arises from abnormal development, migration, and/or differentiation of NC cells.
Pluripotent: A cell with the capacity to give rise to any cell type in the body derived from each of the three germ layers.
Specification: The initiation of a differentiation program which will continue under permissive conditions. This program can be derailed in the presence of nonpermissive cues.

Introduction

The neural crest (NC) is an embryonic cell population which contributes widely to the tissues and organs of the body. NC cells contribute to vertebrate-specific features such as the jaw, a large brain enclosure, and sense organs (Northcutt, 2005). First identified in chick embryos by Wilhelm His, NC cells arise at the dorsal aspect of the neural tube, undergo an epithelial-to-mesenchymal transition (EMT), delaminate, and migrate extensively to give rise to numerous derivatives.

NC cells arise along the embryonic anterior–posterior axis and exhibit differentiation capabilities according to their axial identity (Le Douarin et al., 2004) (Fig. 11.1). Cranial NC cells migrate along a dorsolateral pathway and give rise to parasympathetic and sensory ganglia, pigment cells, and ectomesenchymal derivatives including craniofacial bone and

a Authors contributed equally.

Diagnosis, Management and Modeling of Neurodevelopmental Disorders. https://doi.org/10.1016/B978-0-12-817988-8.00011-7

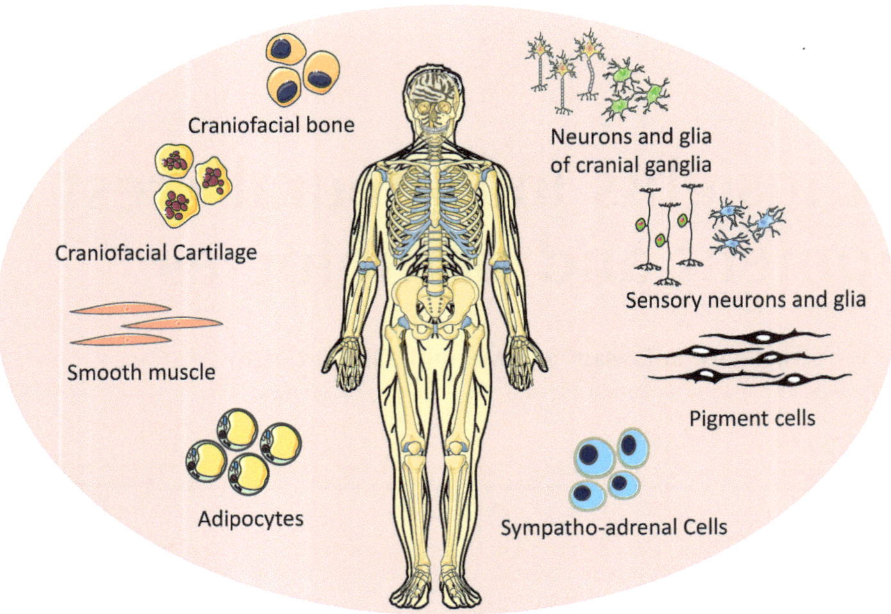

FIGURE 11.1 Contributions of neural crest cells to derivatives throughout the vertebrate body. NC cells are a transient multipotent cell population which arises at multiple axial levels and contributes to numerous cell types throughout the vertebrate body. NC derivatives include cartilage, bone, adipocytes, pigment cells, sensory neurons, and glia (cranial, vagal, and trunk NC); smooth muscle cells, sensory and enteric neurons/glia (vagal and trunk NC); sympathoadrenal cells (trunk NC). *This figure was created using images from Servier Medical Art (http://smart.servier.com), which are licensed under a Creative Commons Attribution 3.0 Unported License.*

cartilage, adipocytes, and odontoblasts. Vagal NC gives rise to the enteric nervous system (ENS), pigment cells, and contributes to the cardiac outflow tract. Trunk NC differentiates into chromaffin cells of the adrenal medulla, sympathetic ganglia, dorsal root ganglia, and pigment cells. Finally, sacral NC gives rise to the ganglia that innervate the hindgut and the ENS.

Early neural crest cell development

NC development is a progressive process that occurs over multiple developmental stages. There is significant evidence for the role of Wnt, FGF, and BMP signaling pathways in stimulating NC formation at a region between the developing neural plate and the nonneural ectoderm termed the neural plate border (NPB). Interestingly, recent work has suggested that NC cell specification begins earlier than previously thought.

Tissue interactions during neural crest cell induction

NC development has been studied in many model systems, but work performed in chick and amphibian embryos—which are particularly amenable to grafting and coculture assays—have yielded enormous insight. NC marker expression is initially described at the NPB, which lies in between the medial neural plate and the lateral nonneural ectoderm and in direct contact with underlying mesoderm. Experiments using grafts and ex-vivo explants demonstrated that interactions between the neural and nonneural ectoderm elicit the formation of NC cells (Labonne & Bronner-Fraser, 1998; Selleck & Bronner-Fraser, 1995). In amphibians, the mesoderm is also thought to provide NC inducing signals (Mayor et al., 1995). In the *Xenopus* gastrula, prospective NC cells are located above the prospective paraxial mesoderm (Steventon et al., 2009), and coculturing the dorsolateral marginal zone with naive animal caps triggers the expression of NC marker genes (Mancilla & Mayor, 1996). However, the mesoderm appears to be dispensable for NC induction in zebrafish (Ragland & Raible, 2004), and explants generated from blastula and gastrula stage chick embryos can generate NC cells in the absence of mesodermal markers (Basch et al., 2006; Prasad et al., 2020a). Similar results were obtained from gastrula stage explants generated from the prospective NC territory in rabbit embryos (Betters et al., 2018). Further, recent work using a model of human cranial NC based on human pluripotent stem cells (hPSC) suggests that human NC forms without apparent contribution from mesodermal tissue (Leung et al., 2016). However, the role of tissue induction in NC formation has not been demonstrated in mammalian embryonic tissue.

Early specification of neural crest and lineage segregation

The neural and nonneural ectodermal contribution to the NC is consistent with NC ectodermal derivatives such as pigment cells and the PNS. However, cranial NC forms bone, cartilage, and fat cells, which in other parts of the body are mesoderm derived. The ability of the NC to contribute to derivatives of two different germ layers has been an important topic of discussion (Hall, 2018).

Studies examining the origin of chick NC have provided insights into this topic. This work uncovered evidence of NC specification in early stages of development prior to the formation of the neural plate (Basch et al., 2006; Patthey et al., 2009; Prasad et al., 2020a). In the earliest embryonic stage examined, a territory containing specified NC cells was identified in the blastula chick epiblast Prasad et al., 2020a). In vivo lineage tracing validated that cells from this territory contribute to Pax7+ NC cells. This study also revealed that the expression of NC markers in low-density explant cultures is independent of contact-mediated induction and Wnt ligand induced signaling, suggesting that the initiation of the NC specification program began even earlier. These results in chick have been supported by studies in mammalian NC including rabbit and human PSC models (Betters et al., 2018; Leung et al., 2016) and point to a model of blastula stage specification of amniote NC.

Studies in *Xenopus* and chick have eluded to some early factors that might be involved during the pregastrula specification of NC. A study in *Xenopus* revealed the expression of some NC genes along with pluripotency markers in blastula stage embryos (Buitrago-Delgado et al., 2015). In chick, NC specifier genes are coexpressed with pluripotency factors at neurula stages (Lignell et al., 2017), and a study in zebrafish has suggested the presence of a network functioning during NC formation that is paralogous to the epiblast pluripotency regulatory network (Lukoseviciute et al., 2018). Models of human NC have also revealed coexpression of pluripotency and NC factors during early NC specification (Leung et al., 2016; Prasad et al., 2020b). However, recent findings from a human NC model revealed distinct stoichiometric changes in the expression of pluripotency factors in the prospective NC cell state compared to the pluripotent ES cell state (Prasad et al., 2020b). Further, prospective NC cells, unlike pluripotent stem cells, were found to have a restricted stem cell potential (Prasad et al., 2020b). As stemness factors have been implicated in the lineage segregation of other cell fates, a similar role could be suggested during NC development that does not confer pluripotency to NC cells but rather supports the specification and differentiation of this multipotent stem cell population. These findings suggest that there is an operational gene regulatory network (GRN) for NC specification prior to gastrulation that shares factors with, but is distinct from, the pluripotency GRN.

Signaling pathways governing neural crest cell formation

Wnt, BMP, and FGF signaling pathways play critical roles during the formation, migration, and differentiation of NC (Fig. 11.2). In the late gastrula, Wnt and FGF signaling are required for the formation of the NPB (García-Castro et al., 2002; Mayor et al., 1997; Monsoro-Burq et al., 2003). Next, during neurulation, Wnt and BMP pathways activate and maintain the expression of NC specifier transcription factors (TFs), which go on to activate specifiers of EMT and migration (Liem et al., 1995; Steventon et al., 2009) (Fig. 11.2).

WNT signaling

Wnt signaling is required for NC induction in fish, frog, chick, and human PSC-derived NC cells. Studies in *Xenopus* and chick have demonstrated that the expression of Wnt ligands, along with modulation of BMP signaling, is required for the induction of the NPB and NC marker expression (García-Castro et al., 2002; Labonne & Bronner-Fraser, 1998; Saint-Jeannet et al., 1997). Wnt6 in chick and Wnt3a and Wnt8 in both chick and *Xenopus* have been implicated in NC induction, and in zebrafish there is a requirement for Wnt8.1 (García-Castro et al., 2002; Hong et al., 2008; Lewis et al., 2004; Patthey et al., 2009). Consistent with the importance of Wnt in NC induction, many hPSC-based models of NC rely on the activation of WNT signaling, and loss of function has validated the requirement for β-catenin (Leung et al., 2016; Prasad et al., 2020b). A recent report has further exposed the role of WNT signaling in modulating the axial identity of human NC, with higher levels of WNT triggering posterior NC (Gomez et al., 2019b). In mouse embryos, it is clear that Wnt signaling functions in controlling NC cell migration and differentiation, but there is currently no direct evidence for the requirement of Wnt signaling during NC induction (Ikeya et al., 1997; van Amerongen & Berns, 2006).

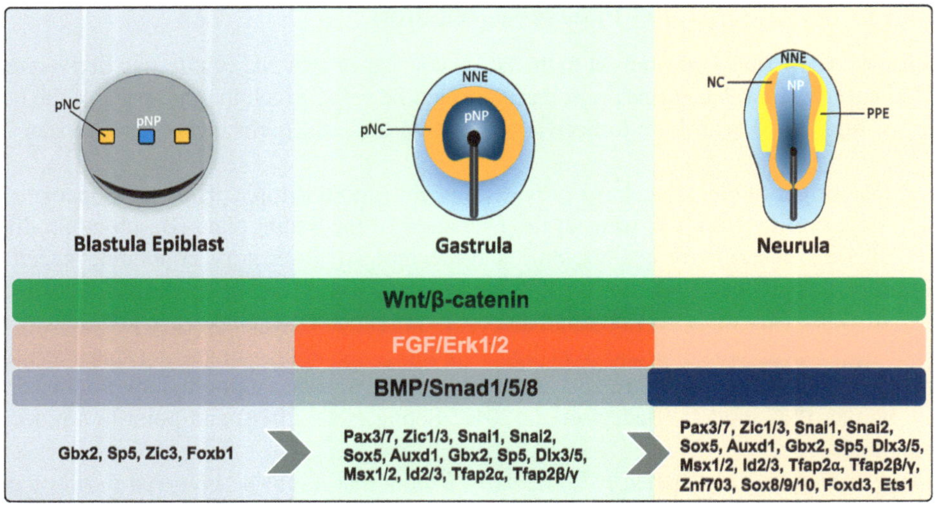

FIGURE 11.2 Signaling and transcription factor contributions to neural crest cell formation. Formation of NC cells begins in the blastula embryo and continues through neurula stages. A population of specified NC cells has been revealed in the chick blastula epiblast. Wnt/β-catenin signaling is required for NC formation at all stages. FGF signaling regulates NC cell formation during gastrulation but is not required at later stages. BMP signaling is attenuated during gastrula stage NC formation, with higher levels required during neurulation. The roles of FGF and BMP signaling during blastula stage NC specification are unknown. Key transcription factors involved in NC cell development at each stage are noted. Schematics of gastrula and neurula chick embryos are adapted from Stuhlmiller and García-Castro (2012a). *pNC*, prospective NC; *pNP*, prospective neural plate; *NNE*, nonneural ectoderm; *NP*, neural plate; *NC*, neural crest; *PPE*, preplacodal ectoderm.

Bone morphogenetic protein signaling

BMP signaling is necessary for NC induction with different requirements for the level of signal during early NC formation and later NC maintenance. An intermediate or attenuated level of BMP signaling induces NC at the NPB, with this reduced signal providing competence for NC induction by Wnt and FGF signals (LaBonne & Bronner-Fraser, 1998; Steventon et al., 2009). The intermediate level of BMP signaling at the NPB is generated by antagonistic interactions between the high levels of BMP from the nonneural ectoderm and BMP signaling inhibitors secreted from the neural plate. This is then followed by the reactivation of BMP signaling at the NPB to promote the expression of NC genes. Importantly, intermediate levels of BMP signaling alone cannot induce NC in any vertebrate species (García-Castro et al., 2002; Labonne & Bronner-Fraser, 1998).

The role of BMP signaling in NC induction has also been explored in mammalian systems. In the mouse, BMP receptor inducible knockouts result in a downregulation of NC marker gene expression and severe craniofacial, pharyngeal, and cardiac defects (Dudas et al., 2004; Jia et al., 2007; Stottmann & Klingensmith, 2011). A PSC model of human NC has demonstrated that BMP is required for NC induction, but that high levels are inhibitory (Leung et al., 2016).

Fibroblast growth factor signaling

FGF signaling is required for NC induction in most model systems. In *Xenopus*, the source of the FGF signal appears to be the mesoderm underlying the prospective NPB (Monsoro-Burq et al., 2003). In chick, FGFR1/4 are expressed in the prospective NPB but not in the underlying mesoderm, and inhibition of FGF signaling in the prospective NPB region of the gastrula epiblast results in a loss of NC markers (Lunn et al., 2007; Stuhlmiller & García-Castro, 2012b). The role of FGF4 in chick NC induction has also been examined by implantation of FGF4-soaked beads in the nonneural ectoderm, revealing an upregulation of early NC markers without apparent mesoderm contribution (Yardley & García-Castro, 2012). In rabbit embryos, mesoderm-independent NC specification appears dependent on FGF signaling (Betters et al., 2018). Recent work has also revealed a requirement for FGF in the expression of NPB markers in zebrafish (Garnett et al., 2012). In mouse, FGF8 null mice exhibit defects in NC terminal differentiation in the pharyngeal arches (Frank et al., 2002), but a role for FGF and FGFR in early NC induction has not been revealed. Finally, FGF signaling is required for human NC induction. The loss of FGF reduces cranial NC induction (Leung et al., 2016), and FGF appears to play a role in posterior NC induction by regulating proliferation and modulating the degree of posteriorization by WNT (Gomez et al., 2019b; Hackland et al., 2019).

Gene regulatory network governing neural crest specification and formation

The actions of WNT, BMP, and FGF signaling activate and maintain the GRN controlling the establishment of the NPB and the formation of the NC. The NPB expresses TFs whose expression results in the establishment of the NC cell state and subsequently, migration and differentiation networks. Recent high-throughput technologies, including transcriptomic and epigenomic analysis of single cells, are allowing for the expansion of the NC-GRN (Williams et al., 2019). We provide an overview of crucial transcriptional interactions that govern the NPB and NC states of the NC-GRN (Fig. 11.2).

Neural plate border specifiers

During NC development, the prospective NPB cells receive inducing signals resulting in the expression of TFs known as NPB specifiers. These include Zic1, Msx1/2, Pax3/7, Dlx5, Gbx2, AP2α, and SP5, which are currently considered to be the earliest players in NC cell formation. In *Xenopus*, expression of Msx1, Pax3, and Zic1 is induced by BMP, Wnt, and FGF signals (Monsoro-Burq et al., 2005; Sato et al., 2005; Hong & Saint-Jeannet, 2007), while lower levels of BMP signaling regulate Dlx5 expression (Luo et al., 2001). Gbx2 is a direct target of Wnt signaling and in turn activates the expression of NPB specifier genes such as Msx1 and Pax3 (Li et al., 2009). Further, BMP signaling directly activates Msx2 through a *cis*-regulatory element conserved between mouse and *Drosophila* (Brugger et al., 2004).

Pax3 and Pax7 are expressed at the NPB in most vertebrates and have been the subject of much focus. Pax7 appears to be the earliest marker of the NC in chick and is necessary for its formation (Basch et al., 2006). Studies have validated the role of FGF signaling in activating Pax7 expression (Stuhlmiller & García-Castro, 2012b). Early expression of Pax7 in the NPB is also found in rabbit embryos (Betters et al., 2018). In *Xenopus*, Pax3 plays a prominent role in NC formation through its expression in the ectoderm (Maczkowiak et al., 2010). Additional studies in *Xenopus* have further elaborated on the GRN activated by Pax3 and Zic1 (Bae et al., 2014; Plouhinec et al., 2014). In mouse, it is suggested that Pax3 and Pax7 play redundant roles during NC formation and differentiation (Relaix et al., 2004).

Neural crest specifiers

TFs expressed at the NPB activate and maintain the expression of NC specifier genes, which include Sox8/9/10, Snai1/2, Foxd3, cMyc, and the Id genes. Continued expression of NPB specifiers and the activation of NC specifiers regulate the decision toward the premigratory and migratory NC states.

SoxE (Sox8/9/10) and SoxD (Sox5) TFs are critical for NC formation and migration. They are expressed at the NPB following NC induction, with the exception of the blastula stage expression of Sox5 in *Xenopus* (Buitrago-Delgado et al., 2015). In chick, Sox10 is transiently expressed in migratory cranial NC populating the pharyngeal arches, but is not found in ectomesenchymal derivatives (Cheng et al., 2000). By contrast, the Sox9-positive NC cells that migrate into the pharyngeal arches differentiate into cardiac derivatives and cranial skeletal elements (Cheung & Briscoe, 2003; Montero et al., 2002). SOX9 is also expressed in both premigratory and migratory NC cells in human and rabbit embryos (Betters et al., 2018, 2010). The overlapping expression of Sox8 with other SoxE factors suggests their functional redundancy in NC development (Cheung & Briscoe, 2003).

The function and regulation of SoxE/D factors have been extensively studied during NC development. During early NC development, Sox5 regulates the expression of NPB and NC specifiers by controlling BMP signaling, and modulates the expression of Sox10 and RhoB during EMT (Nordin & LaBonne, 2014; Perez-Alcala et al., 2004). Sox9 is regulated by Ap2α and Gbx2, while Sox10 expression is regulated by Snai2, Sox9, Ets1, and cMyb (Aoki et al., 2003; Betancur et al., 2010; Li et al., 2009; Luo et al., 2001). During early NC induction, SoxE factors are involved in inhibiting differentiation, while at later stages SoxE factors play a crucial role in NC differentiation, e.g., Sox10 in melanocyte and glia (Aoki et al., 2003; Stolt et al., 2008) and Sox9 in ectomesenchymal differentiation (Lefebvre et al., 1997).

Snai TFs also play crucial roles in NC development. The expression of Snai1 or Snai2 is observed in premigratory NC of *Xenopus*, zebrafish, mouse and chick, while a recent study has suggested an earlier blastula stage expression of Snai1 in *Xenopus* (Buitrago-Delgado et al., 2015; Locascio et al., 2002). In premigratory NC, Snai2 plays a key role in delamination of NC cells from the dorsal neural tube. Given its role in inducing EMT of NC as well as in cancer metastasis, the regulation of Snai2 has been well studied. Wnt signaling and BMP signaling, along with Zic1, Msx1, and Pax3/7, have all been shown to regulate Snai2 expression in *Xenopus* and chick embryos (Conacci-Sorrell et al., 2003; Sato et al., 2005).

In addition, numerous other TFs are required for the proper development of NC cells. Myc genes function during early NC formation in part by regulating the Id genes, which are also direct targets of BMP signaling. Together, Id and Myc regulate proliferation and differentiation in the multipotent NC state (Light et al., 2005; Nikitina et al., 2008). The TF

Foxd3 is involved in maintenance of NC multipotency by preventing early differentiation (Lister et al., 2006). Recently, a study in zebrafish has characterized the bimodal role of Foxd3 during NC development first as an activator during early NC formation and later as a repressor of other cell fates to promote NC development (Lukoseviciute et al., 2018). The expression of Foxd3 is regulated by Pax7, Msx1/2, and Ets1 at all axial levels (Sato et al., 2005; Simões-Costa et al., 2012; Wettstein et al., 1997). Other key TFs that are also involved in setting up the NC gene regulatory network are Myb, Twist1, and Ets1. Myb regulates Pax7, Snai2, and Twist expression in chick (Betancur, 2014; Vadasz et al., 2013), Twist1 modulates later NC development, and Ets1 is a downstream effector of FGF/Erk signaling and can transduce FGF/Erk signaling effects (Nelson et al., 2010). Studies continue to identify new factors involved in defining the NC-GRN in different species.

Relationship to human pathologies

Owing to their extensive contribution to derivatives throughout the body, a large number of human health conditions are associated with improper NC development and differentiation (Table 11.1). These pathologies are collectively known as neurocristopathies and include craniofacial malformations such as cleft lip and palate, cancers such as neuroblastoma, and rare diseases such as Waardenburg syndrome and Hirschsprung disease (Bolande, 1974; Watt & Trainor, 2014). These and other neurocristopathies can be categorized based on the developmental stage and axial level of affected NC. Here, we discuss select neurocristopathies pertaining to the PNS contribution of NC.

Hirschsprung disease (HSCR)

Migrating vagal NC cells enter the foregut during early gestation and migrate in a rostral-caudal direction as enteric NC-derived cells to sequentially colonize the foregut, midgut, and hindgut. HSCR is caused by the failure of enteric NC-derived cells to colonize the terminal regions of the gut, leading to intestinal obstruction and megacolon (Heanue & Pachnis, 2007). Occurring in approximately 1 in 5,000 births, HSCR is classified as short-segment or long-segment depending on the length of the aganglionic bowl, and can be either sporadic or familial. Mutations in a number of genes have been associated with HSCR, including the RET receptor tyrosine kinase and the SIP1 TF (Pichel et al., 1996; Sánchez et al., 1996; Schuchardt et al., 1994)—which are expressed in migrating and differentiating NC cells—and GDNF, which is expressed in the developing gut (Moore et al., 1996), among others. If untreated, HSCR can be fatal following enterocolitis-induced sepsis. Children born with HSCR typically undergo surgical treatment, which includes resection of the bowel to remove or circumvent the region lacking the ENS.

TABLE 11.1 Neurocristopathies related to neural crest cell induction, migration, and differentiation.

Neurocristopathy	Affected NC Stage	Genes	Prevalence
Craniosynostosis	Differentiation	TWIST1, MSX2, FGFR1-3, EFNB1	1:2,500
DiGeorge syndrome (22q11.2 deletion)	Migration	TBX1	1:3,000
Familial dysautonomia	Migration, Differentiation	ELP1/IKBKAP	1:3,700 (Ashkenazi Jewish population)
Hirschsprung disease	Migration, Differentiation	GDNF, RET, SOX10, SIP1, HAND2, PHOX2B	1:5,000
Neuroblastoma	Migration, Differentiation	MYCN, ALK, PHOX2B	1:100,000
Treacher Collins syndrome	Induction	TCOF1, POLR1C, POLR1D	1:50,000
Waardenburg syndrome	Induction, Migration, Differentiation	PAX3, SNAI2, SOX10	1:40,000

Selected neurocristopathies are listed along with their associated genetic defects, affected stage of neural crest cell formation, and prevalence in live births.

Neuroblastoma

Neuroblastoma is the most common extracranial tumor in children and accounts for up to 15% of all pediatric cancer deaths (Louis & Shohet, 2015). An embryonal malignancy of the sympathetic nervous system, tumors most commonly arise in the adrenal medulla and the paraspinal sympathetic ganglia of the abdomen. The median age of neuroblastoma diagnosis is 18 months. Clinical presentation is extremely variable and ranges from spontaneous tumor regression to an aggressive metastatic disease that often develops resistance to current therapies. Nearly half of new neuroblastoma cases are metastatic. Approximately 25% of neuroblastoma patients have tumors with MYCN amplification, which is associated with an aggressive tumor and poor prognosis. MYCN is a member of the MYC proto-oncogene family and encodes a TF that functions extensively in developmental processes where it plays multifaceted roles in stemness, proliferation, and migration.

It is well accepted that the embryonic origins of neuroblastoma are trunk NC cells, which give rise to adrenal chromaffin cells or sympathetic ganglia through a common sympathoadrenal precursor (Johnsen et al., 2019). Interestingly, evidence has also suggested that peripheral Schwann cell precursors, which are also NC-derived, are a major source of adrenal chromaffin cells (Furlan et al., 2017). As hallmarks of cancer cells, including EMT and migration, are similar to NC, it will be crucial to examine commonalities between neuroblastoma and NC (Gonzalez Malagon et al., 2018; Rabadán et al., 2013).

Familial dysautonomia

FD is a rare, congenital sensory neuropathy that occurs in individuals of Ashkenazi Jewish descent. Affected individuals experience widespread sensory and autonomic dysfunction, resulting in severe symptoms that include tachycardia, diminished pain and temperature sensation, and uncontrolled swings in blood pressure, among many others (Norcliffe-Kaufmann et al., 2017). FD is caused by a founder mutation in ELP1/IKBKAP, and tissue-specific expression of a splice defect accounts for almost all cases (Slaugenhaupt et al., 2001). ELP1 functions broadly in the cell during transcription, cell migration and adhesion, and tRNA modification, among other functions (Rubin & Anderson, 2017). As FD results from impaired development and survival of sensory and sympathetic neurons, this pathology can be considered to result, at least in part, from the NC. Recently, transcriptional changes and a decreased migratory capacity were revealed in human NC derived from FD patient cells (Kim et al., 2014).

Stem cell models of neural crest development

The human pathologies associated with NC cells underscore the need to investigate human NC cell formation and differentiation. Human PSC-based models have become a powerful tool to interrogate human NC biology (Fig. 11.3). A

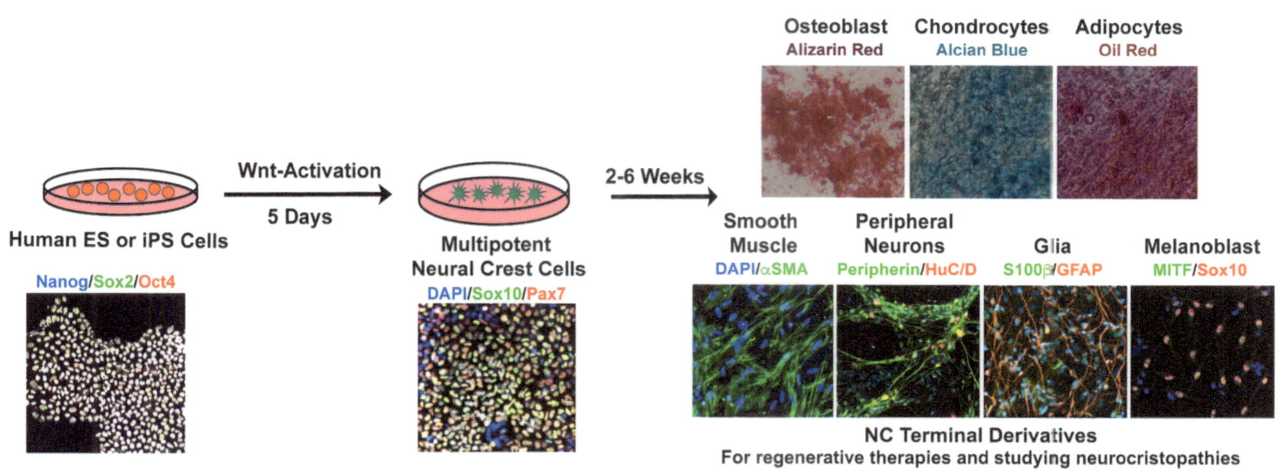

FIGURE 11.3 Pluripotent stem-cell-based model to investigate human neural crest cells and neurocristopathies. Human PSCs can be induced by Wnt activation to form NC cells in five days (Gomez et al., 2019a; Leung et al., 2016). Human NC cells can be subsequently differentiated into NC terminal derivatives over the course of 2–6 weeks. Models of human NC based on PSCs can be used to investigate cell type specification and neurocristopathies.

number of methods have been established that allow for the induction of human PSCs into NC (Chambers et al., 2009; Gomez et al., 2019a; Menendez et al., 2011), and current studies are investigating the transcriptomic and epigenetic events during human NC development (Leung et al., 2016; Rada-Iglesias et al., 2012; Prasad et al., 2020b). Further, the human NC model can be used to explore the underlying molecular mechanisms and novel therapeutics for neurocristopathies. Currently, pathologies including Hirschsprung disease (Fattahi et al., 2016), familial dysautonomia (Kim et al., 2014), and Hermansky–Pudlak syndrome (Mica et al., 2013) are being investigated. A major benefit of using the human NC model is the unprecedented access to unlimited biological material that can be readily manipulated (e.g., CRISPR/Cas9 genome editing) and enable "omic" scale studies to advance both basic and translational research.

Application to other areas of development

The NC gives rise to numerous cell types that contribute to tissues and organs throughout the body, and as such, impaired development, migration, and differentiation of the NC can result in human pathologies. In this chapter, we have highlighted neurocristopathies related to NC contributions to the PNS; however, NC contributions to other cell types underscore additional neurocristopathies. NC contributions to craniofacial development result in numerous malformations including cleft lip and cleft palate and craniosynostosis. NC cells also give rise to melanocytes throughout the body. Waardenburg syndrome is a rare disease resulting from abnormal NC differentiation into melanocytes and is characterized by sensorineural hearing loss, partial albinism, and bright blue eyes or heterochromia iridis. NC cells are also associated with melanoma, an aggressive cancer, which accounts for the vast majority of skin cancer deaths. Finally, due to the reiterative role many factors play during development, syndromes often arise as a result of multiple cell lineages. For example, Mowat–Wilson syndrome is caused by mutations in the transcription factor ZEB2, and the phenotype of patients is consistent with the role of ZEB2 in NC cells and the central nervous system.

Key facts of neural crest cells

- NC cells are multipotent and differentiate into a wide variety of cell types.
- NC cells are specific to vertebrates.
- WNT, FGF, and BMP signaling are required for proper NC cell formation.
- Proper NC development relies on transcription factors including PAX3/7 and SOX family, among others.
- Abnormal development of NC cells results in human pathologies known as neurocristopathies.

Summary points

- NC cells are multipotent and migrate extensively to contribute to numerous tissues and organs throughout the body.
- NC cells form at the neural plate border, but recent evidence suggests that they are specified earlier in the blastula epiblast.
- WNT and FGF signaling function in the early specification of NC, while WNT and BMP signaling function in later NC formation.
- A transcriptional gene regulatory network has been proposed for NC cell formation based on the initial involvement of neural plate border specifiers which activate later NC specifiers.
- Abnormal NC cell development and differentiation lead to neurocristopathies affecting the peripheral nervous system including Hirschsprung disease and neuroblastoma.
- Pluripotent stem cell models can be used to understand human NC development and pathologies.

Acknowledgements

We apologize to researchers whose work we were unable to include due to space limitations. This work is supported by the National Institute of Dental & Craniofacial Research of the National Institutes of Health Award Numbers R01DE017914 and R21DE028112 to M.I.G.C., the National Institute of Dental & Craniofacial Research of the National Institutes of Health Award Numbers F32DE027862 and K99DE029878 to R.M.C., and a UC Riverside School of Medicine Dean's Postdoc to Faculty fellowship to M.S.P.

References

Aoki, Y., et al. (2003). Sox10 regulates the development of neural crest-derived melanocytes in *Xenopus*. *Developmental Biology, 259*(1), 19–33.

Bae, C.-J., et al. (2014). Identification of Pax3 and Zic1 targets in the developing neural crest. *Developmental Biology, 386*(2), 473–483.

Basch, M. L., et al. (2006). Specification of the neural crest occurs during gastrulation and requires Pax7. *Nature, 441*(7090), 218–222.

Betancur, P., et al. (2010). Genomic code for Sox10 activation reveals a key regulatory enhancer for cranial neural crest. *Proceedings of the National Academy of Sciences of the United States of America, 107*(8), 3570–3575.

Betancur, P., et al. (2014). Expression and function of transcription factor cMyb during cranial neural crest development. *Mechanisms of Development, 132*, 38–43.

Betters, E., et al. (2010). Analysis of early human neural crest development. *Developmental Biology, 344*(2), 578–592.

Betters, E., et al. (2018). Early specification and development of rabbit neural crest cells. *Developmental Biology, 444*(Suppl. 1), S181–S192.

Bolande, R. (1974). The neurocristopathies: A unifying concept of disease arising in neural crest maldevelopment. *Human Pathology, 5*(4), 409–429.

Brugger, S. M., et al. (2004). A phylogenetically conserved cis-regulatory module in the Msx2 promoter is sufficient for BMP-dependent transcription in murine and Drosophila embryos. *Development, 131*(20), 5153–5165.

Buitrago-Delgado, E., et al. (2015). Shared regulatory programs suggest retention of blastula-stage potential in neural crest cells. *Science, 348*(6241), 1332–1335.

Chambers, S. M., et al. (2009). Highly efficient neural conversion of human ES and iPS cells by dual inhibition of SMAD signaling. *Nature Biotechnology, 27*(3), 275–280.

Cheng, Y.-C., et al. (2000). Chick Sox10, a transcription factor expressed in both early neural crest cells and central nervous system. *Developmental Brain Research, 121*(2), 233–241.

Cheung, M., & Briscoe, J. (2003). Neural crest development is regulated by the transcription factor Sox9. *Development, 130*(23), 5681–5693.

Conacci-Sorrell, M., et al. (2003). Autoregulation of E-cadherin expression by cadherin-cadherin interactions: The roles of β-catenin signaling, Slug, and MAPK. *The Journal of Cell Biology, 163*(4), 847–857.

Dudas, M., et al. (2004). Craniofacial defects in mice lacking BMP type I receptor Alk2 in neural crest cells. *Mechanisms of Development, 121*(2), 173–182.

Fattahi, F., et al. (2016). Deriving human ENS lineages for cell therapy and drug discovery in Hirschsprung disease. *Nature, 531*(7592), 105–109.

Frank, D. U., et al. (2002). An Fgf8 mouse mutant phenocopies human 22q11 deletion syndrome. *Development, 129*(19), 4591–4603.

Furlan, A., et al. (2017). Multipotent peripheral glial cells generate neuroendocrine cells of the adrenal medulla. *Science, 357*(6346), eaal3753.

García-Castro, M. I., et al. (2002). Ectodermal Wnt function as a neural crest inducer. *Science, 297*(5582), 848–851.

Garnett, A. T., et al. (2012). BMP, Wnt and FGF signals are integrated through evolutionarily conserved enhancers to achieve robust expression of Pax3 and Zic genes at the zebrafish neural plate border. *Development, 139*(22), 4220–4231.

Gomez, G. A., et al. (2019a). Human neural crest induction by temporal modulation of WNT activation. *Developmental Biology, 449*(2), 99–106.

Gomez, G. A., et al. (2019b). WNT/β-catenin modulates the axial identity of embryonic stem cell-derived human neural crest. *Development, 146*(16).

Gonzalez Malagon, S. G., et al. (2018). Glycogen synthase kinase 3 controls migration of the neural crest lineage in mouse and Xenopus. *Nature Communications, 9*(1), 1126–15.

Hackland, J. O. S., et al. (2019). FGF modulates the axial identity of trunk hPSC-derived neural crest but not the cranial-trunk decision. *Stem Cell Reports, 12*(5), 920–933.

Hall, B. K. (2018). Germ layers, the neural crest and emergent organization in development and evolution. *Genesis, 56*(6–7), e23103.

Heanue, T. A., & Pachnis, V. (2007). Enteric nervous system development and hirschsprung's disease: Advances in genetic and stem cell studies. *Nature Reviews Neuroscience, 8*(6), 466–479.

Hong, C. S., et al. (2008). Fgf8a induces neural crest indirectly through the activation of Wnt8 in the paraxial mesoderm. *Development, 135*(23), 3903–3910.

Hong, C. S., & Saint-Jeannet, J.-P. (2007). The activity of Pax3 and Zic1 regulates three distinct cell fates at the neural plate border. *Molecular Biology of the Cell, 18*(6), 2192–2202.

Ikeya, M., et al. (1997). Wnt signalling required for expansion of neural crest and CNS progenitors. *Nature, 389*(6654), 966–970.

Jia, Q., et al. (2007). Smad signaling in the neural crest regulates cardiac outflow tract remodeling through cell autonomous and non-cell autonomous effects. *Developmental Biology, 311*(1), 172–184.

Johnsen, J. I., et al. (2019). Neuroblastoma—a neural crest derived embryonal malignancy. *Frontiers in Molecular Neuroscience, 12*, 450–11.

Kim, Y. J., et al. (2014). Generation of multipotent induced neural crest by direct reprogramming of human postnatal fibroblasts with a single transcription factor. *Stem Cells, 15*(4), 497–506.

Labonne, C., & Bronner-Fraser, M. (1998). Neural crest induction in Xenopus: Evidence for a two-signal model. *Development, 125*(13), 2403–2414.

Le Douarin, N. M., et al. (2004). Neural crest cell plasticity and its limits. *Development, 131*(19), 4637–4650.

Lefebvre, V., et al. (1997). SOX9 is a potent activator of the chondrocyte-specific enhancer of the pro alpha1(II) collagen gene. *Molecular and Cellular Biology, 17*(4), 2336–2346.

Leung, A. W., et al. (2016). WNT/β-catenin signaling mediates human neural crest induction via a pre-neural border intermediate. *Development, 143*(3), 398–410.

Lewis, J. L., et al. (2004). Reiterated Wnt signaling during zebrafish neural crest development. *Development, 131*(6), 1299–1308.

Li, B., et al. (2009). The posteriorizing gene Gbx2 is a direct target of Wnt signalling and the earliest factor in neural crest induction. *Development, 136*(19), 3267–3278.

Liem, K. F., et al. (1995). Dorsal differentiation of neural plate cells induced by Bmp-mediated signals from epidermal ectoderm. *Cell, 82*(6), 969–979.

Light, W., et al. (2005). Xenopus Id3 is required downstream of Myc for the formation of multipotent neural crest progenitor cells. *Development, 132*(8), 1831–1841.

Lignell, A., et al. (2017). Identification of a neural crest stem cell niche by Spatial Genomic Analysis. *Nature Communications, 8*(1), 1830.

Lister, J. A., et al. (2006). Zebrafish Foxd3 is required for development of a subset of neural crest derivatives. *Developmental Biology, 290*(1), 92–104.

Locascio, A., et al. (2002). Modularity and reshuffling of Snail and Slug expression during vertebrate evolution. *Proceedings of the National Academy of Sciences, 99*(26), 16841–16846.

Louis, C. U., & Shohet, J. M. (2015). Neuroblastoma: Molecular pathogenesis and therapy. *Annual Review of Medicine, 66*, 49–63.

Lukoseviciute, M., et al. (2018). From pioneer to repressor: Bimodal foxd3 activity dynamically remodels neural crest regulatory landscape in vivo. *Developmental Cell, 47*(5), 608–628.e6.

Lunn, J. S., et al. (2007). A spatial and temporal map of FGF/Erk1/2 activity and response repertoires in the early chick embryo. *Developmental Biology, 302*(2), 536–552.

Luo, T., et al. (2001). Distinct roles for Distal-less genes Dlx3 and Dlx5 in regulating ectodermal development in Xenopus. *Molecular Reproduction and Development, 60*(3), 331–337.

Maczkowiak, F., et al. (2010). The Pax3 and Pax7 paralogs cooperate in neural and neural crest patterning using distinct molecular mechanisms in *Xenopus laevis* embryos. *Developmental Biology, 340*(2), 381–396.

Mancilla, A., & Mayor, R. (1996). Neural crest formation in *Xenopus laevis*: Mechanisms of Xslug induction. *Developmental Biology, 177*(2), 580–589.

Mayor, R., et al. (1995). Induction of the prospective neural crest of Xenopus. *Development, 121*(3), 767–777.

Mayor, R., et al. (1997). Role of FGF and Noggin in neural crest induction. *Developmental Biology, 189*(1), 1–12.

Menendez, L., et al. (2011). Wnt signaling and a Smad pathway blockade direct the differentiation of human pluripotent stem cells to multipotent neural crest cells. *Proceedings of the National Academy of Sciences of the United States of America, 108*(48), 19240–19245.

Mica, Y., et al. (2013). Modeling neural crest induction, melanocyte specification, and disease-related pigmentation defects in hESCs and patient-specific iPSCs. *Cell Reports, 3*(4), 1140–1152.

Monsoro-Burq, A.-H., et al. (2003). Neural crest induction by paraxial mesoderm in Xenopus embryos requires FGF signals. *Development, 130*(14), 3111–3124.

Monsoro-Burq, A.-H., et al. (2005). Msx1 and Pax3 cooperate to mediate FGF8 and WNT signals during Xenopus neural crest induction. *Developmental Cell, 8*(2), 167–178.

Montero, J. A., et al. (2002). Expression of Sox8, Sox9 and Sox10 in the developing valves and autonomic nerves of the embryonic heart. *Mechanisms of Development, 118*(1–2), 199–202.

Moore, M. W., et al. (1996). Renal and neuronal abnormalities in mice lacking GDNF. *Nature, 382*(6586), 76–79.

Nelson, M. L., et al. (2010). Ras signaling requires dynamic properties of Ets1 for phosphorylation-enhanced binding to coactivator CBP. *Proceedings of the National Academy of Sciences of the United States of America, 107*(22), 10026–10031.

Nikitina, N., et al. (2008). Dissecting early regulatory relationships in the lamprey neural crest gene network. *Proceedings of the National Academy of Sciences of the United States of America, 105*(51), 20083–20088.

Norcliffe-Kaufmann, L., et al. (2017). Familial dysautonomia: History, genotype, phenotype and translational research. *Progress in Neurobiology, 152*, 131–148.

Nordin, K., & LaBonne, C. (2014). Sox5 Is a DNA-binding cofactor for BMP R-Smads that directs target specificity during patterning of the early ectoderm. *Developmental Cell, 31*(3), 374–382.

Northcutt, R. G. (2005). The new head hypothesis revisited. In *Journal of experimental zoology Part B: Molecular and developmental evolution* (pp. 274–297).

Patthey, C., et al. (2009). Wnt-regulated temporal control of BMP exposure directs the choice between neural plate border and epidermal fate. *Development, 136*(1), 73–83.

Perez-Alcala, S., et al. (2004). LSox5 regulates RhoB expression in the neural tube and promotes generation of the neural crest. *Development, 131*(18), 4455–4465.

Pichel, J. G., et al. (1996). Defects in enteric innervation and kidney development in mice lacking GDNF. *Nature, 382*(6586), 73–76.

Plouhinec, J.-L., et al. (2014). Pax3 and Zic1 trigger the early neural crest gene regulatory network by the direct activation of multiple key neural crest specifiers. *Developmental Biology, 386*(2), 461–472.

Prasad, M. S., et al. (2020a). Blastula stage specification of avian neural crest. *Developmental Biology, 458*(1), 64–74.

Prasad, M. S., et al. (2020b). Distinct molecular profile and restricted stem cell potential defines the prospective human cranial neural crest from embryonic stem cell state. *Stem Cell Research, 49*.

Rabadán, M. A., et al. (2013). Identification of a putative transcriptome signature common to neuroblastoma and neural crest cells. *Developmental Neurobiology, 73*(11), 815–827.

Rada-Iglesias, A., et al. (2012). Epigenomic annotation of enhancers predicts transcriptional regulators of human neural crest. *Cell Stem Cell, 11*(5), 633–648.

Ragland, J. W., & Raible, D. W. (2004). Signals derived from the underlying mesoderm are dispensable for zebrafish neural crest induction. *Developmental Biology, 276*(1), 16–30.

Relaix, F., et al. (2004). Divergent functions of murine Pax3 and Pax7 in limb muscle development. *Genes and Development, 18*(9), 1088–1105.

Rubin, B. Y., & Anderson, S. L. (2017). IKBKAP/ELP1 gene mutations: Mechanisms of familial dysautonomia and gene-targeting therapies. *The Application of Clinical Genetics, 10*, 95–103.

Saint-Jeannet, J. P., et al. (1997). Regulation of dorsal fate in the neuraxis by Wnt-1 and Wnt-3a. *Proceedings of the National Academy of Sciences, 94*(25), 13713–13718.

Sato, T., et al. (2005). Neural crest determination by co-activation of Pax3 and Zic1 genes in Xenopus ectoderm. *Development, 132*(10), 2355–2363.

Schuchardt, A., et al. (1994). Defects in the kidney and enteric nervous system of mice lacking the tyrosine kinase receptor Ret. *Nature, 367*(6461), 380–383.

Selleck, M. A., & Bronner-Fraser, M. (1995). Origins of the avian neural crest: The role of neural plate-epidermal interactions. *Development, 121*(2), 525–538.

Simões-Costa, M. S., et al. (2012). Dynamic and differential regulation of stem cell factor FoxD3 in the neural crest is encrypted in the genome. *PLoS Genetics, 8*(12), e1003142.

Slaugenhaupt, S. A., et al. (2001). Tissue-specific expression of a splicing mutation in the gene causes familial dysautonomia. *The American Journal of Human Genetics, 68*(3), 598–605.

Steventon, B., et al. (2009). Differential requirements of BMP and Wnt signalling during gastrulation and neurulation define two steps in neural crest induction. *Development, 136*(5), 771–779.

Sánchez, M. P., et al. (1996). Renal agenesis and the absence of enteric neurons in mice lacking GDNF. *Nature, 382*(6586), 70–73.

Stolt, C. C., et al. (2008). The transcription factor Sox5 modulates Sox10 function during melanocyte development. *Nucleic Acids Research, 36*(17), 5427–5440.

Stottmann, R. W., & Klingensmith, J. (2011). Bone morphogenetic protein signaling is required in the dorsal neural folds before neurulation for the induction of spinal neural crest cells and dorsal neurons. *Developmental Dynamics, 240*(4), 755–765.

Stuhlmiller, T. J., & García-Castro, M. I. (2012a). Current perspectives of the signaling pathways directing neural crest induction. *Cellular and Molecular Life Sciences, 69*(22), 3715–3737.

Stuhlmiller, T. J., & García-Castro, M. I. (2012b). FGF/MAPK signaling is required in the gastrula epiblast for avian neural crest induction. *Development, 139*(2), 289–300.

Vadasz, S., et al. (2013). Pax7 is regulated by cMyb during early neural crest development through a novel enhancer. *Development, 140*(17), 3691–3702.

van Amerongen, R., & Berns, A. (2006). Knockout mouse models to study Wnt signal transduction. *Trends in Genetics, 22*(12), 678–689.

Watt, K., & Trainor, P. A. (2014). Neurocristopathies: The etiology and pathogenesis of disorders arising from defects in neural crest cell development. In *Neural crest cells* (pp. 361–394). Academic Press.

Wettstein, D. A., et al. (1997). The Xenopus homolog of Drosophila Suppressor of Hairless mediates Notch signaling during primary neurogenesis. *Development, 124*(3), 693–702.

Williams, R. M., et al. (2019). Reconstruction of the global neural crest gene regulatory network in vivo. *Developmental Cell, 51*(2), 255–276.e7.

Yardley, N., & García-Castro, M. I. (2012). FGF signaling transforms non-neural ectoderm into neural crest. *Developmental Biology, 372*(2), 166–177.

Chapter 12

Spina bifida: a biopsychosocial perspective

Diana M. Ohanian, Adrien M. Winning, Laura M. Nicholson and Grayson N. Holmbeck
Loyola University Chicago, Department of Psychology, Chicago, IL, United States

List of abbreviations

ADHD-I Attention-deficit/hyperactivity disorder-inattentive type
CC Corpus callosum
CII Chiari II malformation
CNS Central nervous system
CSF Cerebrospinal fluid
EF Executive functioning
NTD Neural tube defect
NVLD Nonverbal learning disability
SES Socioeconomic status
SLD Specific learning disability
TD Typically developing

Mini-dictionary of terms

Assembled processing: constructing and integrating information (e.g., mental rotations, applying previously acquired knowledge to tasks of reading comprehension).
Associative processing: the formation of associations and categories (e.g., decoding familiar words, recognizing faces).
Cerebral spinal shunt: a hollow tube with a valve that is surgically inserted in the brain to drain cerebral spinal fluid.
Executive functioning: higher-order attentional and organizational skills, such as working memory, shifting, sequencing, and planning.
Nonverbal learning disability: a cognitive presentation, which includes deficits in visuospatial intelligence, fine-motor skills, math, and socioemotional skills.
Self-management: management of condition-related physical and medical needs to maintain one's health.
Typically developing: individuals who do not have cognitive or physical disabilities and who acquire the same skills as individuals their own age, at the same pace.

Spina bifida is the most common congenital birth defect of the central nervous system (CNS), which occurs during the first month of pregnancy when the embryonic neural tube fails to completely close (Copp et al., 2015). Individuals with spina bifida often live with numerous physical, neurocognitive, and psychosocial impairments associated with this prenatal neural insult (Fig. 12.1).

Diagnosis

Spina bifida can be diagnosed prenatally or postnatally. During the prenatal period, spina bifida is diagnosed via ultrasound and/or biochemical screening. With regard to biochemical screening, elevated concentrations of alpha fetoprotein in amniotic fluid or serum samples may indicate the presence of a neural tube defect, such as spina bifida. Spina bifida is diagnosed postnatally using X-ray, MRI, or tomography.

There are four types of spina bifida, which vary in severity (Table 12.1):

Diagnosis, Management and Modeling of Neurodevelopmental Disorders. https://doi.org/10.1016/B978-0-12-817988-8.00012-9

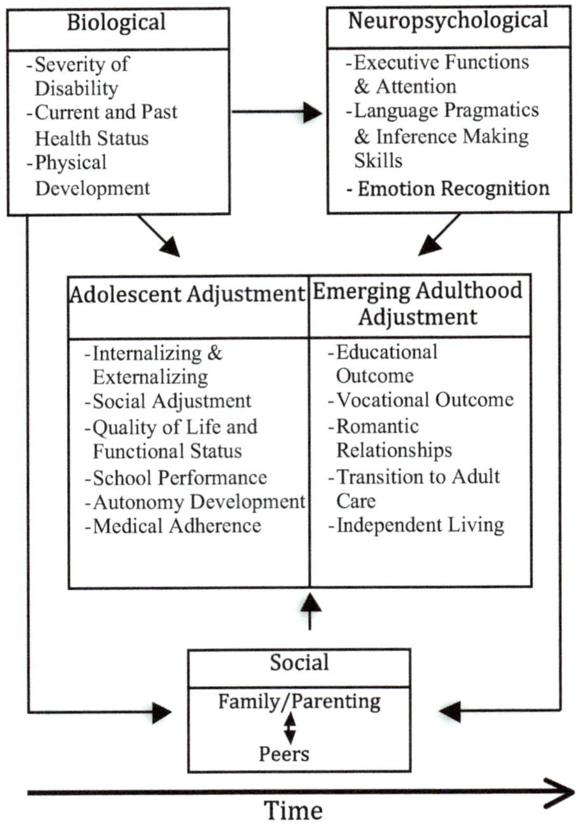

FIGURE 12.1 Bioneuropsychosocial model of psychosocial adjustment in children, adolescents, and emerging adults with spina bifida. This figure reviews important domains to consider for healthy adjustment in individuals with spina bifida, as well as the interaction of these domains throughout development. *From Holmbeck, G. N., & Devine, K. A. (2010). Psychosocial and family functioning in spina bifida.* Developmental Disabilities Research Reviews, *16(1), 40–46. Copyright 2010 by Wiley-Blackwell. Reprinted with permission.*

TABLE 12.1 Types of spina bifida.

Types of spina bifida	Severity	Vertebrae	Spinal fluid	Meninges	Spinal cord	Lesion site
Occulta	Mild	Gap (small)	Intact	Intact	Intact	Marked by difference in skin color, possible presence of fatty lump or indentation
Lipomeningocele	Mild	Gap	Intact	Intact	Intact	Marked by the presence of fatty lump
Meningocele	Moderate	Gap	Exits vertebrae	Exits vertebrae	Intact	Marked by a fluid filled sack on the back of spine
Myelomeningocele	Severe	Gap	Exits vertebrae	Exits vertebrae	Exits vertebrae	Marked by exposed spinal cord and membranes or covered by fluid-filled sack

This table reviews the four types of spina bifida. Types vary in severity, vertebrae gap, location of lesion, and if spinal fluid, meninges, and spinal cord remain within or exit the vertebrae.

- **Occulta** is the mildest form of spina bifida and occurs when vertebrae are malformed with no associated damage to the spinal cord or nerves. Because there is no or minimal damage, disabilities rarely occur.
- **Lipomeningocele** is another mild form of spina bifida characterized by minimal nerve damage and the presence of a fatty tumor that covers the spine. Mild forms of disability may be present, such as bladder and bowel dysfunction.

- **Meningocele** develops when the spinal fluid and meninges exit through a vertebral opening into a skin-covered sack. The spinal cord remains intact; however, malformations of the spinal fluid and meninges may cause mild disabilities (e.g., loss of motor function, bladder and bowel dysfunction).
- **Myelomeningocele** is the most common form of spina bifida and occurs when the spinal cord and associated membranes exit the malformed vertebrae opening and are either exposed or covered by a fluid-filled sack. This type of lesion may present in the middle or lower part of the spinal cord and is generally reflective of the level of disability experienced (e.g., loss of motor function, neurological deficits, orthopedic issues, scoliosis, bowel and bladder dysfunction, etc.).

Epidemiology

Spina bifida occurs in approximately 3 out of every 10,000 live births in the United States (Centers for Disease Control and Prevention, 2011). Medical advancements and the use of cerebrospinal fluid shunts have dramatically increased life expectancy for individuals with spina bifida. Nevertheless, the mortality rate is still significantly higher than in the general population (about 1% between the ages of 5 and 30; Oakeshott, Hunt, Poulton, & Reid, 2010). Spina bifida is more common in certain geographic locations and ethnic groups; rates are higher in Mexico, Northern China, and South America, and Hispanics and Caucasians are the more affected compared to Blacks and Asians. Lifetime costs have been estimated at $600,000 due to medical bills, as well as special education, assistive technology, caregiver costs, and loss of future earnings (Copp et al., 2015).

Causation

The exact cause of spina bifida remains unknown; current literature suggests that spina bifida is associated with multiple nongenetic and genetic factors. Regarding nongenetic factors, higher levels of folic acid intake during pregnancy have been associated with a lower prevalence of spina bifida. In fact, research has found that increased folic acid consumption reduced the risk of NTD by 50%–70% (Volcik et al., 2000). Nevertheless, the reason why folic acid is associated with the development of spina bifida remains unclear (Yeates, Fletcher, & Dennis, 2008). Certain maternal health characteristics have also been found to be significant risk factors for spina bifida, including obesity, illness, hyperthermia, and teratogens (Yeates et al., 2008). Research has suggested that maternal obesity increases risk for NTD development by 1.5–3 times (Copp et al., 2015; Stothard, Tennant, Bell, & Rankin, 2009). Multiple mechanisms, including problems with glycemic control (e.g., diabetes, hyperglycemia), nutritional deficiencies, oxidative stress, and metabolic syndrome, have been suggested to explain the link between obesity and spina bifida (Stothard et al., 2009).

With regard to genetics, twin and family studies have found that heritability is a significant risk factor. Indeed, the risk of recurrence is substantially higher for siblings (Copp et al., 2015). Research has not yet determined which genes cause spina bifida; however, certain genes have been implicated in the etiology of spina bifida. Specifically, mutations of the 5,10 methylene-tetrahydrofolate reductase (MTHFR) gene have been associated with decreased plasma folate levels (Copp et al., 2015).

Prevention and surgical intervention

Prevention efforts have largely targeted nongenetic and potentially modifiable factors (Copp et al., 2015). Many countries (e.g., United States, Canada, South Africa, Costa Rica) have implemented policies, such as food fortification programs, to improve maternal folic acid intake, resulting in notable declines in the overall prevalence of NTDs (Copp et al., 2015). Preventative efforts aimed at improving maternal health and diet may also be important to reduce risk of spina-bifida-affected pregnancies, and policies to address increasing obesity rates have been implemented internationally (Chriqui, 2013).

Finally, early surgical intervention can mitigate complications associated with spina bifida. Prenatal surgery is often performed to repair the spinal lesion and has resulted in a reduction in hindbrain herniation, reducing the rate of shunting and improving neuromotor function (Adzick et al., 2011). Postnatal surgery is typically performed within 48 hours after birth to close the spinal lesion, thereby preventing risk for infection and meningitis.

Neurological presentation

A distinct set of neurological features accompany spina bifida, including the Chiari II malformation (CII), brain abnormalities, hydrocephalus, and seizures (Table 12.2). CII is a structural abnormality in which the cerebellum extends into the cervical spinal canal via the foramen magnum. CII is associated with abnormalities in the midbrain, hindbrain, and the cervical spinal cord. Common abnormalities include posterior and interior stretching of the mesencephalic tectum, often referred to as "tectal beaking" on magnetic resonance imagining scans (MRI). CII is also associated with abnormal development of the corpus callosum (CC); research has found associations between lesion location and location of CC agenesis (Juranek & Salman, 2010).

These abnormalities can impede flow of cerebrospinal fluid (CSF) and cause an array of neurological consequences, including hydrocephalus (Yeates et al., 2008, Fig. 12.2). Characterized by the buildup of excess CSF in the ventricles, hydrocephalus and its associated treatment can cause further injury to the brain in individuals with spina bifida (e.g., demyelination, stretching of white matter, dysplasia of the CC, and thinning of posterior cortical regions; Del Bigio, 1993; Juranek, 2008). Treatment for hydrocephalus typically includes shunting, whereby a hollow tube with a valve is

TABLE 12.2 Neurological presentations.

Neurological features	CNS abnormalities
Chiari II malformation (CII)	Cerebellum
	Tectum
	Corpus callosum
Hydrocephalus	Ventricles
	White matter
	Corpus callosum
	Various cortical regions
Seizures	Electrical stimulation
Tethered cord	Spinal cord

This table reviews neurological features found in spina bifida and associated abnormalities of the central nervous system.

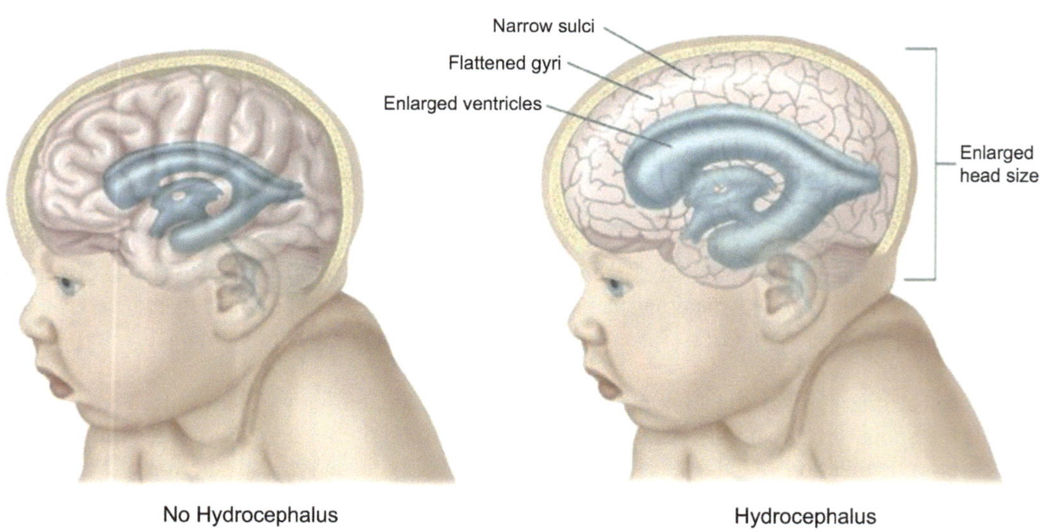

FIGURE 12.2 Hydrocephalus. This figure shows the impact of hydrocephalus or the buildup of excess cerebral spinal fluid, on the structure of the brain. Hydrocephalus results in enlarged ventricles, flattened gyri, and narrow sulci, as well as an enlarged head size. Individuals with spina bifida often require ventricular shunt placement to drain excess cerebral spinal fluid and release pressure on the brain. *This image was provided by the Centers for Disease Control and Prevention, National Center on Birth Defects and Developmental Disabilities.*

surgically inserted in the brain to drain CSF (Copp et al., 2015). While shunting has improved survival rates in spina bifida, it is also associated with multiple risks (e.g., malfunction, infection, surgical replacement) and can negatively affect cognitive functioning (Copp et al., 2015). An additional CNS anomaly is tethered cord. In spina bifida, the abnormal spinal cord can get tethered to scar tissue or bone deformities, which can restrict blood flow and result in tissue damage (Sandler, 2010).

In addition to the Chiari II malformation, hydrocephalus, and tethered cord, individuals with spina bifida are also at an increased risk for seizures. Indeed, the incidence of seizures among those with spina bifida is notably higher than in the general population (i.e., about 5%—15%; Yeates et al., 2008). Individuals with ventriculitis and more shunt revisions are at greatest risk for seizures (Yeates et al., 2008). Tonic—clonic seizures are the most common form of seizure seen in this population (Sandler, 2010).

Physical presentation

Physical disabilities are also common in spina bifida and involve the bowel/bladder, musculoskeletal system, and skin. Lower spinal cord damage can result in neurogenic bladder and bowel functioning. Clean intermittent catheterization (CIC), or the insertion of a catheter several times per day to urinate, is often used to prevent the consequences of bladder dysfunction, including urinary tract and kidney infections. Bowel functioning may be managed with dietary changes and in some instances surgical intervention (Copp et al., 2015).

Musculoskeletal deformities and orthopedic problems found in spina bifida include rotational deformities, contracture of the hip and knee, hip subluxation and dislocation, paralysis, spinal curvature (e.g., scoliosis), and nonplantigrade foot. The severity of secondary condition parameters often corresponds to the neurological level of the lesion. Orthopedic interventions, including bracing, serial casting, and surgery, are employed early in development to encourage independent ambulation and reduce the progression of possible spinal deformities. Finally, skin-related concerns (e.g., pressure injury) commonly occur in this population (26%) and similarly correspond with one's lesion level and wheelchair use, among other factors (Kim et al., 2015).

Cognitive presentation

While cognitive presentation can vary significantly in this population, a specific cognitive phenotype is commonly observed (Summarized in Table 12.3). Cognitive deficits are frequently present in spina bifida, but, unlike other conditions, deficits are thought to be process-specific rather than domain-specific (Fletcher, Ostermaier, Cirino, & Dennis, 2008). Individuals with spina bifida have relative strengths in "associative processing" and weaknesses in "assembled processing" (Dennis, Landry, Barnes, & Fletcher, 2006; Fletcher et al., 2008). Associative processing involves the formation of associations and categories (e.g., decoding familiar words, recognizing faces), whereas assembled processing involves

TABLE 12.3 Cognitive strengths and weaknesses in spina bifida.

Domain	Relative strengths	Relative weaknesses
Intelligence	Verbal IQ	Nonverbal and performance IQ
Visual-spatial	Object-based (e.g., object and facial recognition)	Action-based (e.g., mental rotation, drawing)
Attention and executive functioning	Sustaining attention	Shifting and focusing attention Planning, organization, self-monitoring
Memory	Rote and implicit	Encoding/retrieval, contextualized information, prospective memory
General academic	Verbal achievement (e.g., reading language arts)	Nonverbal achievement (e.g., mathematics)
Reading	Word recognition and decoding	Comprehension
Mathematics	Number fact retrieval	Procedural math and fluency

This table demonstrates the unique cognitive phenotype commonly found in spina bifida and details specific cognitive strengths and weaknesses. Please see Dennis, Fletcher, Rogers, Hetherington, & Francis, 2002; Dennis, Landry, Barnes, & Fletcher, 2006; Dennis & Barnes, 2010 for more detail.

constructing and integrating information (e.g., mental rotations, applying previously acquired knowledge to tasks of reading comprehension). The following sections will further elaborate on the specific strengths and deficits commonly found in this population (Table 12.3).

Intellectual functioning

Intellectual functioning in spina bifida often falls within the low average to average range. Illness severity (e.g., lesion level and hydrocephalus) is a strong predictor of intellectual functioning, with 20%–25% of individuals with spina bifida myelomeningocele and hydrocephalus estimated to have an intellectual disability (Copp et al., 2015).

The cognitive profile of spina bifida is characterized by stronger verbal abilities as compared to both nonverbal abilities and overall performance IQ (Dennis & Barnes, 2010). This pattern of strengths and weaknesses has been likened to a nonverbal learning disability (NVLD), a syndrome of deficits thought to be attributable to white matter damage in the right hemisphere (Harnadek & Rourke, 1994). NVLD is a cluster of symptoms including deficits in visuospatial intelligence, fine-motor skills, math, and socioemotional skills (Mammarella & Cornoldi, 2014). Notably, the presence of NVLD is often seen within the context of hydrocephalus, and recent work has identified associations between NVLD and posterior regions of the CC, areas that are also implicated in spina bifida (Rissman, 2011). Therefore, the NVLD profile is a useful framework for conceptualizing the cognitive and intellectual presentation of spina bifida, particularly when hydrocephalus is present.

Visual–spatial

Visual–spatial deficits are common in spina bifida, as many individuals may have fewer opportunities to develop visual–spatial skills in infancy and early childhood due to condition-related motor and visual impairments. Additionally, abnormal brain development (i.e., thinner posterior cortex, agenesis/hypoplasia of the CC) is associated with greater impairment in visual–perceptual, visual–motor integration, and nonverbal abilities (Fletcher et al., 1996). Prior work has considered the role of two visual processing pathways involved in visual perception (i.e., ventral and dorsal) to characterize visual spatial perception deficits among youth with spina bifida. Findings demonstrate that performances are lower on action-based tasks (relevant to the dorsal pathway) as compared to object-based tasks (relevant to the ventral pathway). Specifically, Dennis and colleagues revealed that, in the context of spina bifida with hydrocephalus, youth experience deficits in tasks that require representations of visual space and visually guided action (2002). As such, aspects of visual spatial perception may be reduced as a result of developmental and/or other secondary consequences of spina bifida (e.g., hydrocephalus).

Attention

Spina bifida is characterized by a unique attentional profile. The incidence rate of ADHD in spina bifida is significantly higher than in the general population (31% vs. 17%; Burmeister et al., 2005). However, contrary to the typical clinical manifestation of ADHD, youth with spina bifida struggle with focusing and shifting attention rather than sustaining attention (Burmeister et al., 2005; Rose & Holmbeck, 2007). This difference reflects relative strengths in the top-down attentional network (e.g., anterior system) and relative weaknesses in the bottom-up attentional network (e.g., posterior system; Ramsundhar & Donald, 2014). Further, individuals with spina bifida who receive an ADHD diagnosis are typically diagnosed with ADHD-predominately inattentive (as opposed to the hyperactive or combined) subtype, which may indicate that the pathophysiology of ADHD in spina bifida is different from that of ADHD in the general population (Ramsundhar & Donald, 2014).

In infants and young children, attention deficits typically manifest as problems orienting and shifting from one stimulus to another (Fletcher et al., 1996). In middle childhood and adolescence, difficulties in filtering irrelevant information (e.g., selective attention) arise (Caspersen & Habekost, 2013). Brain abnormalities common in spina bifida, particularly midbrain dysmorphology, may be related to these attentional deficits (Kulesz et al., 2015). Individuals with spina bifida have also demonstrated activation deficits during selective attention and response inhibition tasks, including decreased frontal and superior parietal activation and low involvement of the left brain hemisphere (Ou, Snow, Byerley, Hall, & Glasier, 2013). Finally, medical factors, such as hydrocephalus and shunt surgeries, can also contribute to attention difficulties (Kulesz et al., 2015).

Executive functioning

Individuals with spina bifida also exhibit deficits in executive functioning (EF) or higher-order attentional, organizational, and goal-directed skills. These include problems with cognitive flexibility, abstract reasoning, sequencing, and switching (Rose & Holmbeck, 2007). Further, metacognitive skills, such as working memory, initiation, planning, organizing, and self-monitoring, are also weaker in this population (Zabel et al., 2011). Several CNS anomalies have been linked to executive dysfunction in spina bifida, including reduced superior parietal cortical volume, increased frontal cortical thickness, decreased overall cortical surface area, reduced cerebral white matter, and abnormalities in the dorsolateral prefrontal cortex circuitry (Juranek et al., 2010; Kulesz et al., 2015; Ware et al., 2016). In addition to brain abnormalities, EF appears to be worse in those with shunted hydrocephalus, a higher frequency of shunt-related surgeries, and seizures (Brown et al., 2008).

Memory

While rote and implicit memory appear to be intact in individuals with spina bifida (Yeates & Enrile, 2005), they often demonstrate deficits in both the encoding and retrieval of verbal and nonverbal memory. Specifically, they exhibit profound recency effects and perform poorly on delayed memory tasks. Further, memory of contextualized information is also reduced, indicating that individuals with spina bifida do not appear to benefit from increased structure when encoding information (Dennis et al., 2007). Research has also shown that individuals with spina bifida often struggle with prospective memory or memory involving planned actions. This deficit has been linked to reduced hippocampal volume (Treble-Barna et al., 2015). Difficulties with prospective memory may also be related to the aforementioned problems with attention and EF. Shunt revision is also associated with memory deficits in spina bifida (Zebracki, Zaccariello, Zelko, & Holmbeck, 2010).

Academic functioning

Academic performance in spina bifida is typically consistent with the intellectual profile described earlier. Individuals with spina bifida show relative strengths in verbal achievement (e.g., reading, language arts) and weaknesses in nonverbal achievement (e.g., mathematics). Nevertheless, academic functioning is also characterized by specific strengths and weaknesses within these domains. Specifically, with regard to verbal academic skills, word and sentence-level processing (e.g., word recognition and decoding) typically emerges as a relative strength. In contrast, complex verbal skills such as verbal fluency and reading comprehension are significantly weaker (Barnes et al., 2014; Dennis et al., 2006). With regard to nonverbal academic skills, math fact retrieval, numeration, and exact calculations are generally found to be intact. With regard to more complex mathematics, individuals with spina bifida often struggle with procedural math, demonstrate deficits in math fluency, and often use inefficient math strategies (e.g., counting instead of using multiplication in a word problem; Dennis et al., 2006). Consequently, specific learning disabilities (SLDs) are quite prevalent in this population, with SLD in mathematics being the most common (Fletcher, Dennis, Northrup, & Francis, 2004). While most children with spina bifida start off school in a general education classroom, many will move to a special education classroom as the demands of school increase.

Research has identified several illness and demographic factors related to academic achievement in this population. As seen in typically developing youth, lower socioeconomic status (SES) has been found to negatively impact academic performance in youth with spina bifida (Swartwout, Garnaat, Myszka, Fletcher, & Dennis, 2010). Shunt status, hydrocephalus, and seizure history are also related to academic achievement in this population (Wasserman & Holmbeck, 2016). With regard to academic outcomes, most youth with spina bifida graduate high school (85%) but they have significantly lower rates of college attendance (14.6% vs. 69.7%). These statistics are important to note as education has been found to be the best predictor of work participation, above and beyond lesion level, hydrocephalus, IQ, functional independence, and ambulation (van Melchen et al., 2008).

Psychological functioning

Individuals with spina bifida are at an increased risk for poor psychosocial adjustment (Table 12.4). Specifically, youth with spina bifida are at risk for higher internalizing symptoms (e.g., anxiety and depressive symptoms; Holmbeck et al., 2010), lower self-concept (Shields, Taylor, & Dodd, 2008), and poor quality of life (Murray et al., 2015). The current literature has identified several factors associated with poor psychological functioning, including parenting behaviors

TABLE 12.4 Psychosocial and functional adjustment in spina bifida.

Domain	Common difficulties
Psychological	• Internalizing symptoms (e.g., anxiety and depressive symptoms) • Lower self-concept
Social	• Social immaturity and passivity • Less adaptive social behaviors (e.g., verbosity) • Fewer friendships and romantic partners
Self-management	• Difficulties in managing bladder and bowel programs, skin checks, medical appointments/adherence, and other self-care tasks
Independence	• Lower employment rates • Difficulties in mastering daily living skills (e.g., cooking and cleaning) and navigating transportation • Many individuals continue to live with their parents into adulthood

This table reviews domains of psychosocial and functional adjustment, in which individuals with spina bifida can experience difficulties.

(Holmbeck, Coakley, Hommeyer, Shapera, & Westhoven, 2002), condition severity (Cate, Kennedy, & Stevenson, 2002), maternal depression (Schellinger, Holmbeck, Essner, & Alvarez, 2012), neuropsychological functioning (Stern, Driscoll, Ohanian, & Holmbeck, 2018), pain, and SES (Bellin et al., 2013). Furthermore, research has also found that reduced social experiences and reduced positive social contexts are also associated with poor psychological functioning, specifically increased depressive symptoms (Essner & Holmbeck, 2010).

Social functioning

Given the high prevalence of hydrocephalus and associated neurocognitive deficits, youth with spina bifida often exhibit impairments in social functioning (Table 12.4). Research consistently indicates that youth with spina bifida have lower social competence, including greater social immaturity and passivity, less adaptive social behavior, and inappropriate sociability and verbosity (Burmeister et al., 2005; Holmbeck et al., 2003). Such impairments in social functioning persist throughout development, which may partially explain why youth with spina bifida have been shown to have fewer social contacts, friends, and romantic partners as compared to typically developing youth and even compared to those with other chronic health conditions (Holmbeck et al., 2010; Zukerman, Devine, & Holmbeck, 2011). Contrary to previous findings, a longitudinal examination of friendship revealed that although youth with spina bifida may have fewer relationships, they describe comparable levels of friendship quality as typically developing peers over the course of childhood to emerging adulthood (Stiles-Shields, Bechtel Driscoll, Rausch, & Holmbeck, 2019). As youth with spina bifida transition to young adulthood, there is less participation in romantic relationships as compared to same-age peers (Zukerman et al., 2011). This may be due, in part, to a lack of participation in social activities, as well as the degree to which their disease impacts aspects of functioning (Copp et al., 2015).

Family functioning

Spina bifida impacts not only affected individuals, but also their caregivers and family members. Caring for a child with complex medical needs can result in higher levels of caregiver stress, feelings of incompetence, and isolation (Holmbeck & Devine, 2010), as well as disrupt normative family functioning. Indeed, research has found that spina bifida can negatively impact mothers' and fathers' quality of parenting (Holmbeck & Devine, 2010) and that families of these children demonstrate less cohesion during preadolescence as compared to the families of TD children (Holmbeck et al., 2002). Families that manage additional stressors (e.g., lower SES) may be especially vulnerable to experiencing difficulties (Holmbeck et al., 2002). Despite these findings, a resilience-disruption model of family functioning is generally supported, in which most families demonstrate considerable resilience despite having a child with spina bifida (Holmbeck & Devine, 2010). Unlike families of typically developing children, evidence indicates that families of children with spina bifida also do not experience normative increased family conflict during pubertal development in adolescence (Coakley, Holmbeck, Friedman, Greenley, & Thrill, 2002).

Family factors, such as environment and parenting practices, can impact individuals with spina bifida. Evidence suggests that a family environment characterized by cohesion and encouragement of independence/achievement is positively related to youth's use of adaptive coping strategies, level of independence, and social engagement (Loomis, Javornisky, Monahan, Burke, & Lindsay, 1997; McKernon et al., 2001). Parents of children with spina bifida have been found to exhibit high levels of overprotectiveness and psychological control, which can result in lower levels of autonomy development and poor psychosocial adjustment (Holmbeck & Devine, 2010; Holmbeck et al., 2002). Conversely, family factors may be able to buffer (i.e., moderate) the impact of spina bifida on youth psychosocial adjustment. For example, one study found that parental behavioral and psychological control moderated the relationship between EF deficits and poor medical adherence in youth with spina bifida (O'Hara & Holmbeck, 2013). Through parenting, caregivers may be able to help children with spina bifida compensate for cognitive difficulties and develop skills to better manage deficits associated with this complex condition (e.g., Landry, Taylor, Swank, Barnes, & Juranek, 2013).

Self-management and independent functioning

Given the complexity of spina bifida and the potential life-threatening complications associated with the condition (e.g., urinary tract infections, pressure sores, shunt malfunctions; Dicianno & Wilson, 2010, Table 12.4), self-management and medical adherence are particularly important skills for children with spina bifida to acquire during development (Fig. 12.1). From birth through childhood, caregivers typically manage the extensive treatment regimen associated with spina bifida, including bladder and bowel programs, skin checks, medical appointments/adherence, and various other self-care tasks (Roberts & Steele, 2017, Table 12.4). However, as children with spina bifida transition into adolescence and adulthood, they often become progressively more responsible for managing their spina bifida and its sequelae, which can negatively impact their medical adherence and, in turn, their health (Psihogios, Kolbuck, & Holmbeck, 2015). Self-management requires individuals with spina bifida to not only understand their condition and associated self-management tasks, but also acquire and effectively implement the skills necessary for these tasks (Binks, Barden, Burke & Young, 2007).

Successfully gaining independence requires individuals to also master daily living skills (e.g., cooking, cleaning), navigate transportation, and obtain employment. The cognitive and physical impairments that often accompany spina bifida can ultimately delay or prevent independent functioning for some individuals, especially for those with higher lesion levels or shunts (Zebracki et al., 2010). Emerging adults with spina bifida are less likely to go to college and have lower employment rates than their typically developing peers, and many continue to live with their parents (e.g., 43% in one study; Copp et al., 2015).

Interventions

Interventions for neuropsychological and psychosocial vulnerabilities in this population are still in their infancy. Further, few interventions address all of the areas of vulnerability reviewed in this chapter (Fig. 12.1). However, there are several interventions that can improve functioning for individuals with spina bifida that should be noted; some of these interventions are condition-specific and some are well-established treatments that can be implemented in this population. Selected domains for intervention are reviewed further (summarized in Table 12.5).

Attention and Executive Functioning

While many youth with spina bifida and ADHD are prescribed stimulant medication, it has been suggested that the unique attentional deficits found in this population may not respond solely to stimulant medication (Fletcher et al., 2008); therefore, behavioral interventions should be a first-line treatment for ADHD and EF deficits in spina bifida. There are several well-established interventions for ADHD; those specifically geared toward ADHD, inattentive (ADHD-I) subtype focus on organization, planning, study skills, and social skills (Evans et al., 2016). Most of these interventions include parent training, child skills training, and teacher involvement. One study specifically aimed at EF skills in adults with spina bifida demonstrated significant improvements in EF through the use of a goal-management training program (Stubberud, Langenbahn, Levine, Stanghelle, & Schanke, 2013). Furthermore, assistive technology, such as smartphone apps focused on organizational skills, may also provide extra support for individuals with spina bifida with EF deficits (Dicianno et al., 2008).

TABLE 12.5 Areas for intervention and treatment.

Domain	Treatment
Attention and executive functioning	• ADHD-I treatment components: parent training, child skills training, teacher involvement • ADHD-I treatment skills training: organization, planning, study skills, social skills • Executive functioning goal-management training (Stubberud et al., 2013) • Assistive technology
Psychological	• Cognitive behavioral therapy • Social skills training • Family-based interventions: parenting skills, communication skills, problem-solving skills
Academic	• Individualized education program • Cognitive strategy training program (Coughlin & Montague, 2011) • Increased time/instruction for lessons • Skill remediation • Training for teachers regarding teaching youth with chronic conditions • Vocational training
Self-management and independence	• Family-based interventions: see above • Camp-based intervention • Behavioral intervention technologies

This table reviews treatments for psychosocial and adjustment difficulties associated with spina bifida; this includes treatments designed specifically for individuals with spina bifida as well as common evidence-based treatments.

Psychosocial interventions

ADHD-I, EF deficits, anxiety, and depression are highly comorbid. Therefore, it is no surprise that the aforementioned goal-management training intervention aimed at improving EF in adults with spina bifida also improved coping skills, health-related quality of life, and symptoms of depression and anxiety (Stubberud, Langenbahn, Levine, Stanghelle, & Schanke, 2015). Furthermore, well-established interventions, such as cognitive behavioral therapy (CBT), are often the first-line treatment for depression and anxiety; for those with cognitive deficits, many interventions, including CBT, have been appropriately adapted and have yielded significant treatment gains (Hartley et al., 2015).

There are few interventions focused on improving social and family functioning (Holmbeck, Greenley, Coakley, Greco, & Hagstrom, 2006), as well as autonomy and self-management, in spina bifida. Though targeted family-based interventions have yet to be developed for this population, research with other populations with neurological conditions suggests that interventions that teach parenting skills (e.g., behavior management), problem-solving, and communication skills to families may be beneficial (Roberts & Steele, 2017). With regard to social functioning, autonomy, and self-management, Holbein et al. (2013) reported significant improvement in youth's skills across all of these domains after attending a camp-based intervention targeting independence and social skills in youth with spina bifida. Relatedly, some health care organizations in the United States have created structured, comprehensive programs to facilitate a smooth transition from pediatric to adult spina bifida care (Hopson et al., 2018), yet there remains a lack of continuity in preparing patients for this transition (Kelly et al., 2017). Finally, behavioral intervention technologies ([BITs], e.g., mobile apps, websites) may help to promote self-management in adolescents and young adults with spina bifida. Currently, one mobile app has been piloted to improve self-management skills in adults with spina bifida and has demonstrated feasibility and skill improvement (Dicianno et al., 2016). A recent systematic review of BITs for other chronic health populations has further demonstrated the importance of developing interventions in this domain for youth with spina bifida (Stiles-Shields et al., 2018).

Academic interventions

With regard to academics, youth with spina bifida should be monitored by a developmental pediatrician to determine if/what accommodations are necessary, including altering their classroom (e.g., special education vs. regular education classroom; Ramsundhar & Donald, 2014). Academic interventions can be put into place informally or through a formal individualized education program. There are only a few studies that have assessed spina-bifida-specific interventions for academic performance. Coughlin and Montague (2011) found evidence for the efficacy of a cognitive strategy training program aimed at bettering math performance. This study also suggested that increased instruction time in mathematics is crucial in this population. Providing skill remediation or vocational training may also be beneficial for older youth with spina bifida who have learning difficulties or cognitive impairments. Finally, one study found that both special education and mainstream teachers were not confident in their ability to meet the academic needs of children with chronic medical conditions (Nabors, Little, Akin-Little, & Lobst, 2008). Therefore, teachers should receive information about how best to support and educate children with chronic medical conditions, such as spina bifida.

Burgeoning research

Sleep and pain difficulties are not often considered part of condition management in spina bifida; however, recent work has demonstrated the importance of assessing and treating sleep problems and pain symptoms in this population.

Sleep

Research suggests that children, adolescents, and young adults with spina bifida are at increased risk for sleep-related difficulties as compared to TD peers (e.g., lower sleep quality, shorter sleep duration; Murray et al., 2016). There is also a high incidence of obstructive sleep apnea among individuals with spina bifida (Patel et al., 2015), which can go unidentified and result in preventable deaths (Kirk, Morielli, & Brouillette, 1999). As such, sleep quality should be routinely evaluated by providers as well as considered a critical agenda item for future clinical and research endeavors.

Pain

There is currently limited research on pain in spina bifida; however, existing research suggests that it is an important parameter to consider in relation to health and psychosocial outcomes. Youth with spina bifida typically experience musculoskeletal pain, abdominal pain, joint pain, and headache, with headache being the most common form of pain in spina bifida (Stellman-Ward, Bannister, Lewis, & Shaw, 1997). Pain in spina bifida may be related to shunt malfunction, ambulation method and muscle overuse, constipation, and scoliosis (Clancy, McGrath, & Oddson, 2005). Pain in spina bifida is significantly more prevalent than in the general population (Clancy et al., 2005) and has been associated with poor quality of life (Bellin et al., 2013), increased depressive symptoms (Oddson, Clancy, & McGrath, 2006), and poor social activity involvement and social competence (Essner, Murray, & Holmbeck, 2014). Therefore, it is essential that clinicians assess for pain symptoms when working with individuals with spina bifida.

Cultural considerations

While spina bifida can affect multiple domains of functioning (e.g., cognitive, psychological, family, social, physical, etc.), it does not have a monolithic effect on individuals. In addition to individual differences in clinical manifestation, spina bifida can have a differential effect on individuals and families from different ethnicities and cultures. In recent years, researchers have begun to recognize the importance of examining diversity characteristics in spina bifida. Indeed, the current literature has examined psychosocial, cognitive, and health-related outcomes across ethnicity (e.g., Chowanadisai et al., 2013; Devine, Holbein, Psihogios, Amaro, & Holmbeck, 2012; Papadakis et al., 2018; Swartout et al., 2010). There is some evidence that Hispanic youth with spina bifida demonstrate lower social competence, but also have fewer externalizing symptoms compared to non-Hispanic White youth with spina bifida (Papadakis et al., 2018). Further, with regard to family functioning, Hispanic families have been found to exhibit less conflict compared to non-Hispanic White families; however, Hispanic mothers of children with spina bifida may also be more vulnerable to lower levels of parental satisfaction and competence and higher levels of stress compared to non-Hispanic White mothers (Devine et al., 2012). Recently, a longitudinal case study demonstrated the importance of considering collectivist cultural values when developing and implementing interventions for medical and nonmedical autonomy for individuals with spina bifida (Ohanian et al., 2018). More research is necessary in order to create culturally sensitive interventions that respect the individual, family, and communities affected by spina bifida.

Applications to other areas of development

In this chapter, we have provided an overview of spina bifida from a biopsychosocial perspective, including the neurological, physical, cognitive, and psychosocial presentation of spina bifida. We have noted that spina bifida is a unique condition with significant variability in presentation, including variability in areas of vulnerability and intervention. Other conditions share similar areas of vulnerability, such as chronic health conditions, physical and motor disabilities, and intellectual disabilities (Stiles-Shields et al., 2018). This chapter will be informative for researchers examining conditions involving CNS insults, complex medical regimen, and deficits in cognitive, social, and independent functioning, as well as for clinicians working with these types of populations. Further, this chapter describes this condition within both a developmental and ecological framework, which should serve as an example for others examining similar conditions or conducting research in related areas. An individual with spina bifida interacts with multiple environments and systems; the degree to which certain vulnerabilities affect an individual with spina bifida is often a product of these interactions. Finally, we also review new areas of research, examining domains such as sleep and pain that are often overlooked in spina bifida and other chronic conditions but can have a significant impact on overall functioning throughout development.

Key facts of spina bifida

- Spina bifida is a congenital birth defect that occurs in the first month of pregnancy when the neural tube fails to close completely.
- The exact etiology of spina bifida has yet to be determined; however, folic acid intake is one major cause of spina bifida.
- There is a wide range of illness severity, including varying degrees of neurological problems, physical disability, cognitive deficits, and psychosocial difficulties.
- Brain anomalies are common in spina bifida.
- Spina bifida is associated with a complex medical regimen.
- Psychosocial and neuropsychological interventions are necessary to treat the many needs of individuals with spina bifida.

Summary points

- Spina bifida is a congenital condition and is associated with multiple neurological, physical, cognitive, and psychosocial difficulties.
- Myelomeningocele is the most prevalent and severe form of spina bifida.
- Brain dysmorphology in spina bifida is often associated with specific neurocognitive deficits (e.g., visual spatial skills), in the context of low average to average intellectual functioning.
- Spina bifida is also associated with deficits in attention and executive functioning, as well as academic difficulties.
- Social and family functioning is vulnerable in this population and should be the target of interventions.
- Existing interventions in this populations are generally limited; current interventions target socioemotional functioning and self-management.
- It is important to consider culture when treating individuals with spina bifida.

References

Adzick, N. S., Thom, E. A., Spong, C. Y., Brock, J. W., 3rd, Burrows, P. K., Johnson, M. P., … MOMS Investigators. (2011). A randomized trial of prenatal versus postnatal repair of myelomeningocele. *New England Journal of Medicine, 364*(11), 993–1004.

Barnes, M., Raghubar, K., English, L., Williams, J., Taylor, H., & Landry, S. (2014). Longitudinal mediators of achievement in mathematics and reading in typical and atypical development. *Journal of Experimental Child Psychology, 119*, 1–16.

Bellin, M. H., Dicianno, B. E., Osteen, P., Dosa, N., Aparicio, E., Braun, P., & Zabel, T. A. (2013). Family satisfaction, pain, and quality-of-life in emerging adults with spina bifida: A longitudinal analysis. *American Journal of Physical Medicine and Rehabilitation, 92*(8), 641–655.

Binks, Barden, Burke, & Young. (2007). What do we really know about the transition to adult-centered health care? A focus on cerebral palsy and spina bifida. *Archives of Physical Medicine and Rehabilitation, 88*(8), 1064–1073.

Brown, T., Ris, M., Beebe, D., Ammerman, R., Oppenheimer, S., Yeates, K., & Enrile, B. (2008). Factors of biological risk and reserve associated with executive behaviors in children and adolescents with spina bifida myelomeningocele. *Child Neuropsychology, 14*(2), 118–134.

Burmeister, R., Hannay, H. J., Copeland, K., Fletcher, J. M., Boudousquie, A., & Dennis, M. (2005). Attention problems and executive functions in children with spina bifida and hydrocephalus. *Child Neuropsychology: A Journal on Normal and Abnormal Development in Childhood and Adolescence, 11*(3), 265–283.

Caspersen, I. D., & Habekost, T. (2013). Selective and sustained attention in children with spina bifida myelomeningocele. *Child Neuropsychology, 19*(1), 55–77.

Cate, I. M. P., Kennedy, C., & Stevenson, J. (2002). Disability and quality of life in spina bifida and hydrocephalus. *Developmental Medicine and Child Neurology, 44*(5), 317–322.

Center for Disease Control and Prevention [CDC]. (2011). *Spina bifida*. Retrieved from cdc.gov/ncbddd/spinabifida/facts.html 2015.

Chowanadisai, M., de la Rosa Perez, D. L., Weitzenkamp, D. A., Wilcoz, D. T., Clayton, G. H., & Wilson, P. E. (2013). The role of ethnicity and culture on functional status in children with spina bifida. *Journal of Pediatric Rehabilitation Medicine, 6*(4), 203–213.

Chriqui, J. F. (2013). Obesity prevention policies in US states and localities: Lessons from the field. *Current Obesity Reports, 2*(3), 200–210.

Clancy, C. A., McGrath, P. J., & Oddson, B. E. (2005). Pain in children and adolescents with spina bifida. *Developmental Medicine and Child Neurology, 47*(1), 27–34.

Coakley, R. M., Holmbeck, G. N., Friedman, D., Greenley, R. N., & Thill, A. W. (2002). A longitudinal study of pubertal timing, parent-child conflict, and cohesion in families of young adolescents with spina bifida. *Journal of Pediatric Psychology, 27*(5), 461–473.

Copp, A. J., Adzick, N. S., Chitty, L. S., Fletcher, J. M., Holmbeck, G. N., & Shaw, G. M. (2015). Spina bifida. *Nature Review, Disease Primers, 1*, 1–18.

Coughlin, J., & Montague, M. (2011). The effects of cognitive strategy instruction on the mathematical problem solving of adolescents with spina bifida. *The Journal of Special Education, 45*(3), 171–183.

Del Bigio, M. R. (1993). Neuropathological changes caused by hydrocephalus. *Acta Neuropathologica, 85*(6), 573–585.

Dennis, M., & Barnes, M. A. (2010). The cognitive phenotype of spina bifida meningomyelocele. *Developmental Disabilities Research Reviews, 16*(1), 31–39. https://doi.org/10.1002/ddrr.89

Dennis, M., Fletcher, J. M., Rogers, T., Hetherington, R., & Francis, D. J. (2002). Object-based and action-based visual perception in children with spina bifida and hydrocephalus. *Journal of the International Neuropsychological Society, 8*(1), 95–106.

Dennis, M., Jewell, D., Drake, J., Misakyan, T., Spiegler, B., Hetherington, R., ... Barnes, M. (2007). Prospective, declarative, and nondeclarative memory in young adults with spina bifida. *Journal of the International Neuropsychological Society, 13*(2), 312–323.

Dennis, M., Landry, S. H., Barnes, M., & Fletcher, J. M. (2006). A model of neurocognitive function in spina bifida over the life span. *Journal of the International Neuropsychological Society, 12*, 285–296.

Devine, K. A., Holbein, C. E., Psihogios, A. M., Amaro, C. M., & Holmbeck, G. N. (2012). Individual adjustment, parental functioning, and perceived social support in Hispanic and non-Hispanic white mothers and fathers of children with spina bifida. *Journal of Pediatric Psychology, 37*(7), 769–778.

Dicianno, B., Fairman, A., McCue, M., Parmanto, B., Yih, E., McCoy, A., ... Brienza, D. (2016). Feasibility of using mobile health to promote self-management in spina bifida. *American Journal of Physical Medicine and Rehabilitation, 95*(6), 425–437.

Dicianno, B., Kurowski, B., Yang, J., Chancellor, M., Bejjani, G., Fairman, A., ... Sotirake, J. (2008). Rehabilitation and medical management of the adult with spina bifida. *American Journal of Physical Rehabilitation, 87*(12), 1027–1050.

Dicianno, B. E., & Wilson, R. (2010). Hospitalizations of adults with spina bifida and congenital spinal cord anomalies. *Archives of Physical Medicine and Rehabilitation, 91*, 529–535.

Essner, B. S., & Holmbeck, G. N. (2010). The impact of family, peer, and school contexts on depressive symptoms in adolescents with spina bifida. *Rehabilitation Psychology, 55*(4), 340–350.

Essner, B. S., Murray, C. B., & Holmbeck, G. N. (2014). The influence of condition parameters and internalizing symptoms on social outcomes in youth with spina bifida. *Journal of Pediatric Psychology, 39*(7), 718–734.

Evans, S. W., Langberg, J. M., Schultz, B. K., Vaughn, A., Altaye, M., Marshall, S. A., & Zoromski, A. K. (2016). Evaluation of a school-based treatment program for young adolescents with ADHD. *Journal of Consulting and Clinical Psychology, 84*(1), 15–30.

Fletcher, J. M., Bohan, T. P., Brandt, M. E., Kramer, L. A., Brookshire, B. L., Thorstad, K., ... Baumgartner, J. E. (1996). Morphometric evaluation of the hydrocephalic brain: Relationships with cognitive development. *Children's Nervous System, 12*(4), 192–199.

Fletcher, J. M., Dennis, M., Northrup, H., & Francis, D. J. (2004). Spina bifida: Genes, brain, and development. *International Review of Research in Mental Retardation, 29*, 63–117.

Fletcher, J. M., Ostermaier, K. K., Cirino, P. T., & Dennis, M. (2008). Neurobehavioral outcomes in spina bifida: Processes versus outcomes. *Journal of Pediatric Rehabilitation Medicine, 1*(4), 311–324.

Harnadek, M. C. S., & Rourke, B. P. (1994). Principal identifying features of the syndrome of nonverbal learning disabilities in children. *Journal of Learning Disabilities, 27*, 144–154.

Hartley, S. L., Esbensen, A. J., Shalev, R., Vincent, L. B., Mihaila, I., & Bussanich, P. (2015). Cognitive behavioral therapy for depressed adults with mild intellectual disability: A pilot study. *Journal of Mental Health Research in Intellectual Disabilities, 8*(2), 72–97.

Holbein, C. E., Murray, C. B., Psihogios, A. M., Wasserman, R. M., Essner, B. S., O'Hara, L. K., & Holmbeck, G. N. (2013). A camp-based psychosocialIntervention to promote independence and social function in individuals with spina bifida: Moderators of treatment effectiveness. *Journal of Pediatric Psychology, 38*(4), 412–424.

Holmbeck, G. N., Coakley, R. M., Hommeyer, J. S., Shapera, W. E., & Westhoven, V. C. (2002). Observed and perceived dyadic and systemic functioning in families of preadolescents with spina bifida. *Journal of Pediatric Psychology, 27*(2), 177–189.

Holmbeck, G. N., DeLucia, C., Essner, B., Kelly, L., Zebracki, K., & Friedman, D. (2010). Trajectories of psychosocial adjustment in adolescents with spina bifida: A 6-year, four-wave longitudinal follow-up. *Journal of Consulting and Clinical Psychology, 78*(4), 511–525.

Holmbeck, G. N., & Devine, K. A. (2010). Psychosocial and family functioning in spina bifida. *Developmental Disabilities Research Reviews, 16*(1), 40–46.

Holmbeck, G., Greenley, R., Coakley, R., Greco, J., & Hagstrom, J. (2006). Family functioning in children and adolescents with spina bifida: An evidence-based review of research and interventions. *Journal of Developmental and Behavioral Pediatrics, 27*(3), 249–277.

Holmbeck, G. N., Westhoven, V. C., Phillips, W. S., Bowers, R., Gruse, C., Nikolopoulos, T., … Davison, K. (2003). A multimethod, multi-informant, and multidimensional perspective on psychosocial adjustment in preadolescents with spina bifida. *Journal of Consulting and Clinical Psychology, 71*(4), 782–796.

Hopson, B., Rocque, B. G., Joseph, D. B., Powell, D., McLain, A., Davis, R. D., … Blount, J. P. (2018). The development of a lifetime care model in comprehensive spina bifida care. *Journal of pediatric rehabilitation medicine, 11*(4), 323–334. https://doi.org/10.3233/PRM-180548

Juranek, J., Fletcher, J. M., Hasan, K. M., Breir, J. I., Cirino, P. T., Pazo-Alvarez, P., … Papanicolaou, A. C. (2008). Neocortical reorganization in spina bifida. *NeuroImage, 40*(4), 1516–1522.

Juranek, J., & Salman, M. S. (2010). Anomalous development of brain structure and function in spina bifida myelomeningocele. *Developmental Disabilities Research Reviews, 16*(1), 23–30.

Kelly, M. S., Thibadeau, J., Struwe, S., Ramen, L., Ouyang, L., & Routh, J. (2017). Evaluation of spina bifida transitional care practices in the United States. *Journal of Pediatric Rehabilitation Medicine, 10*(3–4), 275–281.

Kim, S., Ward, E., Dicianno, B. E., Clayton, G. H., Sawin, K. J., Beierwaltes, P., … Chinarian, J. (2015). Factors associated with pressure ulcers in individuals with spina bifida. *Archives of Physical Medicine and Rehabilitation, 96*(8), 1435–1441.

Kirk, V., Morielli, A., & Brouillette, R. (1999). Sleep-disordered breathing in patients with myelomeningocele: The missed diagnosis. *Developmental Medicine and Child Neurology, 41*(1), 40–43.

Kulesz, P. A., Treble-Barna, A., Williams, V. J., Juranek, J., Cirino, P. T., Dennis, M., & Fletcher, J. M. (2015). Attention in spina bifida myelomeningocele: Relations with brain volume and integrity. *NeuroImage Clinical, 8*, 72–78.

Landry, S. H., Taylor, H. B., Swank, P. R., Barnes, M., & Juranek, J. (2013). Longitudinal mediators of social problem solving in spina bifida and typical development. *Rehabilitation Psychology, 58*(2), 196.

Loomis, J. W., Javornisky, J. G., Monahan, J. J., Burke, G., & Lindsay, A. (1997). Relations between family environment and adjustment outcomes in young adults with spina bifida. *Developmental Medicine and Child Neurology, 39*(9), 620–627.

Mammarella, I. C., & Cornoldi, C. (2014). An analysis of the criteria used to diagnose children with Nonverbal Learning Disability (NLD). *Child Neuropsychology, 20*(3), 255–280.

McKernon, W. L., Holmbeck, G. N., Colder, C. R., Hommeyer, J. S., Shapera, W., & Westhoven, V. (2001). Longitudinal study of observed and perceived family influences on problem-focused coping behaviors of preadolescents with spina bifida. *Journal of Pediatric Psychology, 26*(1), 41–54.

van Mechelen, M. C., Verhoef, M., van Asbeck, F. W. A., & Post, M. W. M. (2008). Work participation among young adults with spina bifida in The Netherlands. *Developmental Medicine and Child Neurology, 50*(10), 772–777.

Murray, C. B., Holmbeck, G. N., Ros, A. M., Flores, D. M., Mir, S. A., & Varni, J. W. (2015). A longitudinal examination of health-related quality of life in children and adolescents with spina bifida. *Journal of Pediatric Psychology, 40*(4), 419–430.

Murray, C. B., Kirsch, A. C., Palermo, T. M., Holmbeck, G. N., Kolbuck, V., Psihogios, A., & Pigott, T. (2016). Developmental course and determinants of sleep disturbances in adolescents with spina bifida. *Journal of Pediatric Psychology, 41*(6), 631–642.

Nabors, L. A., Little, S. G., Akin-Little, A., & Iobst, E. A. (2008). Teacher knowledge of and confidence in meeting the needs of children with chronic medical conditions: Pediatric psychology's contribution to education. *Psychology in the Schools, 45*, 217–226.

Oakeshott, P., Hunt, G. M., Poulton, A., & Reid, F. (2010). Expectation of life and unexpected death in open spina bifida: A 40-year complete, non-selective, longitudinal cohort study. *Developmental Medicine and Child Neurology, 52*(8), 749–753.

Oddson, B. E., Clancy, C. A., & McGrath, P. J. (2006). The role of pain in reduced quality of life and depressive symptomology in children with spina bifida. *The Clinical Journal of Pain, 22*(9), 784–789.

Ohanian, D. M., Stiles-Shields, C., Afzal, K. I., Driscoll, C. F. B., Papadakis, J. L., Stern, A., … Holmbeck, G. N. (2018). Cultural considerations for autonomy and medical adherence in a young Palestinian American Muslim female with spina bifida: A longitudinal case study in a research context. *Clinical Practice in Pediatric Psychology, 6*(4), 386–397.

Ou, X., Snow, J., Byerley, A., Hall, J., & Glasier, C. (2013). Decreased activation and increased lateralization in brain functioning for selective attention and response inhibition in adolescents with spina bifida. *Child Neuropsychology, 19*(1), 23–36.

O'Hara, L. K., & Holmbeck, G. N. (2013). Executive functions and parenting behaviors in association with medical adherence and autonomy among youth with spina bifida. *Journal of Pediatric Psychology, 38*(6), 675–687.

Papadakis, J. L., Acevedo, L., Ramirez, S., Stern, A., Driscoll, C. F., & Holmbeck, G. N. (2018). Featured article: Psychosocial and family functioning among Latino youth with spina bifida. *Journal of Pediatric Psychology, 43*(2), 105–119.

Patel, D. M., Rocque, B. G., Hopson, B., Arynchyna, A., Ralee'Bishop, E., Lozano, D., & Blount, J. P. (2015). Sleep-disordered breathing in patients with myelomeningocele. *Journal of Neurosurgery: Pediatrics, 16*(1), 30–35.

Psihogios, A. M., Kolbuck, V., & Holmbeck, G. N. (2015). Condition self-management in pediatric spina bifida: A longitudinal investigation of medical adherence, responsibility-sharing, and independence skills. *Journal of Pediatric Psychology, 40*(8), 790–803.

Ramsundhar, N., & Donald, K. (2014). An approach to the developmental and cognitive profile of the child with spina bifida. *South African Medical Journal, 104*(3), 221.

Rissman, B. (2011). Nonverbal learning disability explained: The link to shunted hydrocephalus. *British Journal of Learning Disabilites, 39*(3), 209–215.

Roberts, M. C., & Steele, R. G. (Eds.). (2017). *Handbook of pediatric psychology*. Guilford Press.

Rose, B. M., & Holmbeck, G. N. (2007). Attention and executive functions in adolescents with spina bifida. *Journal of Pediatric Psychology, 32*(8), 983–994.

Sandler, A. D. (2010). Children with spina bifida: Key clinical issues. *Pediatric Clinics of North America, 57*, 879–892.

Schellinger, K. B., Holmbeck, G. N., Essner, B. S., & Alvarez, R. (2012). Mother and child depressive symptoms in youth with spina bifida: Additive, moderator, and mediator models. *Journal of Child and Family Studies, 21*(2), 281–292.

Shields, N., Taylor, N. F., & Dodd, K. J. (2008). Self-concept in children with spina bifida compared with typically developing children. *Developmental Medicine and Child Neurology, 50*(10), 733–743.

Stellman-Ward, G. R., Bannister, C. M., Lewis, M. A., & Shaw, J. (1997). The incidence of chronic headache in children with shunted hydrocephalus. *European Journal of Pediatric Surgery, 7*(Suppl. 1), 12–14.

Stern, A., Driscoll, C. F. B., Ohanian, D., & Holmbeck, G. (2018). A longitudinal study of depressive symptoms, neuropsychological functioning, and medical responsibility in youth with spina bifida: Examining direct and mediating pathways. *Journal of Pediatric Psychology, 43*(8), 895–905.

Stiles-Shields, C., Crowe, A. N., Driscoll, C. F. B., Ohanian, D. M., Stern, A., Wartman, E., … Holmbeck, G. N. (2018). A systematic review of behavioral intervention technologies for youth with chronic health conditions and physical and intellectual disabilities: Implications for adolescents and young adults with spina bifida. *Journal of Pediatric Psychology, 44*(3), 349–362.

Stiles-Shields, C., Driscoll, C. F. B., Rausch, J. R., & Holmbeck, G. N. (2019). Friendship quality over time in youth with spina bifida compared to peers. *Journal of Pediatric Psychology, 44*(5), 601–610.

Stothard, K. J., Tennant, P. W., Bell, R., & Rankin, J. (2009). Maternal overweight and obesity and the risk of congenital anomalies: A systematic review and meta-analysis. *JAMA, 301*(6), 636–650.

Stubberud, J., Langenbahn, D., Levine, B., Stanghelle, J., & Schanke, A. (2013). Goal management training of executive functions in patients with spina bifida: A randomized controlled trial. *Journal of the International Neuropsychological Society, 19*(6), 672–685.

Stubberud, J., Langenbahn, D., Levine, B., Stanghelle, J., & Schanke, A. K. (2015). Emotional health and coping in spina bifida after goal management training: a randomized controlled trial. *Rehabilitation psychology, 60*(1), 1–16. https://doi.org/10.1037/rep0000018

Swartwout, M., Garnaat, S., Myszka, K., Fletcher, J., & Dennis, M. (2010). Associations of ethnicity and SES with IQ and achievement in spina bifida meningomyelocele. *Journal of Pediatric Psychology, 35*(9), 927–936.

Treble-Barna, A., Juranek, J., Stuebing, K. K., Cirino, P. T., Dennis, M., & Fletcher, J. M. (2015). Prospective and episodic memory in relation to hippocampal volume in adults with spina bifida myelomeningocele. *Neuropsychology, 29*(1), 92–101.

Volcik, K. A., Blanton, S. H., Tyerman, G. H., Jong, S. T., Rott, E. J., Page, T. Z., … Northrup, H. (2000). Methylenetetrahydrofolate reductase and spina bifida: Evaluation of level of defect and maternal genotypic risk in Hispanics. *American Journal of Medical Genetics, 95*(1), 21–27.

Ware, A., Kulesz, P., Williams, V., Juranek, J., Cirino, P., & Fletcher, J. (2016). Gray matter integrity within regions of the dorsolateral prefrontal cortical-subcortical network predicts executive function and fine motor dexterity in spina bifida. *Neuropsychology, 30*(4), 492–501.

Wasserman, R., & Holmbeck, G. (2016). Profiles of neuropsychological functioning in children and adolescents with spina bifida: Associations with biopsychosocial predictors and functional outcomes. *Journal of the International Neuropsychological Society, 22*(8), 804–815.

Yeates, K. O., & Enrile, B. G. (2005). Implicit and explicit memory in children with congenital and acquired brain disorder. *Neuropsychology, 19*(5), 618–628.

Yeates, K. O., Fletcher, J. M., & Dennis, M. (2008). Spina bifida and hydrocephalus. In J. E. Morgan, & J. H. Ricker (Eds.), *Textbook of clinical neuropsychology* (pp. 128–148). New York, NY: Psychology Press.

Zabel, T. A., Jacobson, L., Zachik, C., Levey, E., Kinsman, S., & Mahone, E. (2011). Parent- and self-ratings of executive functions in adolescents and young adults with spina bifida. *The Clinical Neuropsychologist, 25*(6), 926–941.

Zebracki, K., Zaccariello, M., Zelko, F., & Holmbeck, G. (2010). Adolescence and emerging adulthood in individuals with spina bifida: A developmental neuropsychological perspective. In J. Donders, & S. J. Hunter (Eds.), *Principles and practice of lifespan developmental neuropsychology*. Cambridge, MA: Cambridge University Press.

Zukerman, J. M., Devine, K. A., & Holmbeck, G. N. (2011). Adolescent predictors of emerging adulthood milestones in youth with spina bifida. *Journal of Pediatric Psychology, 36*(3), 265–276.

Chapter 13

Diabetic ketoacidosis and neurodevelopment

Walter Zegarra and Tandy Aye
Division of Pediatric Endocrinology, Stanford University School of Medicine, Stanford, CA, United States

List of abbreviations

CMS Children's Memory Scale
CPT Continuous Performance Test II
DKA Diabetic ketoacidosis
DTI Diffusion Tensor Imaging
IQ Intelligence Quotient
MRI Magnetic Resonance Imaging
MRS Magnetic Resonance Spectroscopy
NAA N-acetylaspartate
T1DM Type 1 Diabetes Mellitus
VBM Voxel-Based Morphometry
WASI Wechsler Adult Scale of Intelligence
WPPSI3 Wechsler Preschool and Primary Scale of Intelligence, third edition

Mini-dictionary of terms

Glucosuria: presence of glucose in the urine
Glycogenolysis: the breakdown of glycogen into glucose
Ketogenesis: the formation of ketones
Ketonemia: presence of ketones in the blood
Ketonuria: presence of ketones in the urine
Kussmaul respirations: deep and labored breathing used for compensation for acidosis seen in diabetic ketoacidosis
Lipolysis: the breakdown of fats and lipids into fatty acids
Polyuria: increased urination
Proteolysis: the breakdown of proteins into amino acids

Introduction

Diabetic ketoacidosis (DKA) is the most common cause of morbidity and mortality in children with type 1 diabetes mellitus (T1DM) (Edge, Ford-Adams, & Dunger, 1999; Scibilia, Finegold, Dorman, Becker, & Drash, 1986; Warner, McKinney, Law, & Bodansky, 1998). In this chapter we will define the pathophysiology, prevalence, and presentation of DKA in children and address the effect of DKA on the brain and its impact on neurodevelopment. Comparison of overall full scale intelligence quotient (IQ) between children with recently diagnosed T1DM and controls does not show any differences (Kovacs, Goldston, & Iyengar, 1992). However, longitudinal research shows that children with T1DM have deficits in IQ and academic performance years after the diagnosis (Nadebaum, Scratch, Northam, & Cameron, 2012; Northam et al., 2009). The exact mechanism of these deficits is not fully understood and is likely related to specific complications of T1DM rather than any factor intrinsic to T1DM (Ghetti, Lee, Sims, Demaster, & Glaser, 2010). As the

Diagnosis, Management and Modeling of Neurodevelopmental Disorders. https://doi.org/10.1016/B978-0-12-817988-8.00013-0

most common complication of T1DM, especially in younger children, it is important to understand the acute and chronic neurologic complications that can arise from experiencing an episode of DKA and its possible effect in neurodevelopment.

Definition and pathophysiology of DKA

DKA is characterized by hyperglycemia (blood glucose >11 mmol/L [>200 mg/dL]), acidosis (venous pH < 7.3 or serum bicarbonate <15 mmol/L), and ketonemia or ketonuria (Dunger et al., 2004; Kitabchi et al., 2003; Wolfsdorf et al., 2018) (See Table 13.1). However, there are instances in which blood glucose levels may be lower than 11 mmol/L (200 mg/dL), such as in very young children, those consuming very low amounts of carbohydrate, or those whose hyperglycemia was partially treated with insulin; this is often referred to as "euglycemic ketoacidosis" (Baş, Uytun, & Torun, 2015; Peters et al., 2015; Wolfsdorf et al., 2018). DKA is further divided into severity based on the degree of acidosis. Mild DKA shows a venous pH < 7.3 or serum bicarbonate <15 mmol/L, moderate DKA is defined by a venous pH < 7.2 or serum bicarbonate <10 mmol/L, and severe DKA by a venous pH < 7.1 or serum bicarbonate <5 mmol/L (Chase, Garg, & Jelley, 1990; Wolfsdorf et al., 2018). The severity of DKA has important implications on the acute management of DKA.

DKA occurs when there is an absolute insulin deficiency or increased demand (due to stress or infection) leading to insufficient insulin (See Fig. 13.1). This leads to an increase in the counterregulatory hormones: glucagon, cortisol, growth hormone, and catecholamines. The increase in these hormones leads to the key components for DKA: hyperglycemia,

TABLE 13.1 Definitions of diabetic ketoacidosis.

D	Hyperglycemia (Diabetes)	Blood glucose >11 mmol/L (200 mg/dL)		
K	Ketosis	Ketonuria (moderate-large, >2+, or >40 mg/dL) or Ketonemia (beta-hydroxybutyrate level ≥3 mmol/L)		
A	Acidosis	**Mild:** pH < 7.3 or serum bicarbonate <15 mmol/L	**Moderate:** pH < 7.2 or serum bicarbonate <10 mmol/L	**Severe:** pH < 7.1 or serum bicarbonate <5 mmol/L

Definition and severity of diabetes ketoacidosis based on blood glucose levels, presence of ketones, and level of acidosis seen in blood.
Sources: Based on Chase, H. P., Garg, S. K., & Jelley, D. H. (1990). Diabetic ketoacidosis in children and the role of outpatient management. *Pediatrics in Review, 11*(10), 297–304. doi:10.1542/pir.11-10-297 and Wolfsdorf, J. I., Glaser, N., Agus, M., Fritsch, M., Hanas, R., Rewers, A. ... Codner, E. (2018). ISPAD Clinical Practice Consensus Guidelines 2018: Diabetic ketoacidosis and the hyperglycemic hyperosmolar state. *Pediatric Diabetes, 19*(S27), 155–177. doi:10.1111/pedi.12701.

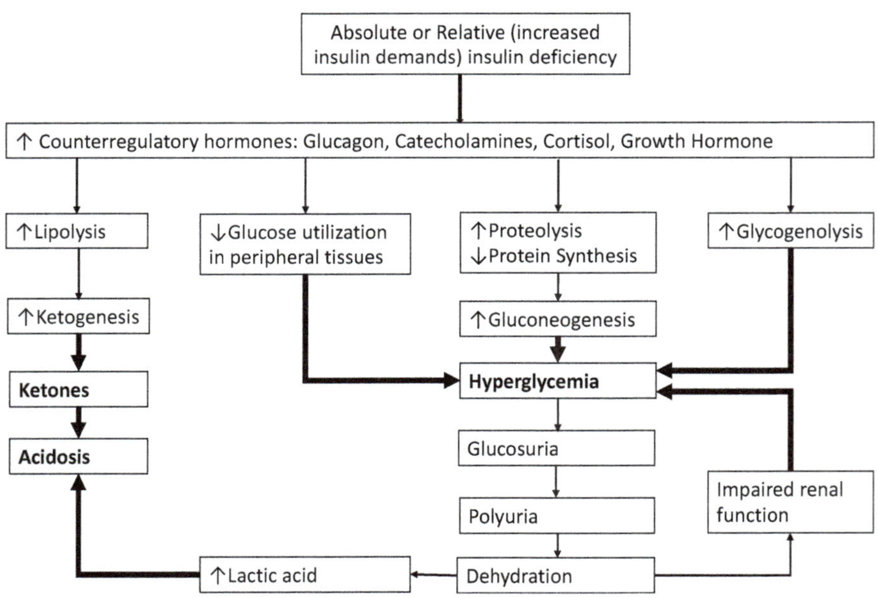

FIGURE 13.1 Pathophysiology of diabetic ketoacidosis. Pathophysiology of diabetic ketoacidosis (DKA), showing the mechanism behind the key components seen in DKA: hyperglycemia, ketosis, and acidosis. *Adapted from Sperling's Pediatric Endocrinolgy (Sperling, 2014).* Pediatric endocrinology *(4th ed.). Philadelphia, PA: ELSEVIER.).*

acidosis, and ketosis. Hyperglycemia occurs due to the counterregulatory hormones increasing glycogenolysis in the liver, impairing glucose utilization in peripheral tissues, and inducing proteolysis and decreased protein synthesis leading to gluconeogenesis. Acidosis is formed as the counterregulatory hormones increase lipolysis, which leads to ketogenesis in the liver. Acidosis is further worsened by the glucosuria leading to dehydration and hypoperfusion of body tissues, creating lactic acid. Dehydration further impairs kidney function leading to worsened hyperglycemia and hyperosmolarity (Cox et al., 2012; Krane, 1987; Wolfsdorf, Glaser, & Sperling, 2006; Wolfsdorf et al., 2018).

Frequency of DKA

The frequency of DKA at onset of diabetes in children varies greatly depending on the regional incidence of T1DM. Frequencies range from approximately 15% to 70% in Europe and North America (Dabelea et al., 2014; Dunger et al., 2004; Fritsch et al., 2013; Lévy-Marchal et al., 1992; Lévy-Marchal, Patterson, & Green, 2001; Usher-Smith, Thompson, Ercole, & Walter, 2012; Wolfsdorf et al., 2018; Zucchini et al., 2016). DKA presenting at diagnosis of T1DM is more common in younger children and infants, those of ethnic minority groups, those without a first degree relative with T1DM, and those with lower access to medical care due to socioeconomic reasons (Bui, To, Stein, Fung, & Daneman, 2010; Dunger et al., 2004; Quinn, Fleischman, Rosner, Nigrin, & Wolfsdorf, 2006; Szypowska & Skórka, 2011; Wolfsdorf et al., 2018).

In children with known T1DM, the risk of DKA ranges from 1% to 10% per patient per year (Cengiz et al., 2013; Hanas, Lindgren, & Lindblad, 2009; Maahs et al., 2015; Morris et al., 1997; Rewers et al., 2002; Smith, Firth, Bennett, Howard, & Chisholm, 1998). The risk of children with T1DM developing DKA increases in children who omit insulin, have poor metabolic control, have a history of previous DKA, have acute gastroenteritis with inability to take fluids, have a history of psychiatric disorders (including eating disorders), have unstable family circumstances, are near or in puberty (particularly females), binge on alcohol, and/or have limited access to medical services (Dunger et al., 2004; Morris et al., 1997; Rewers et al., 2002; Smith et al., 1998) (See Table 13.2). Since DKA is caused by an absolute insulin deficiency or insufficient insulin for increased demands, insulin omission, insulin pump failures, and poor adherence to sick-day management account for nearly all episodes of recurrent DKA.

Diagnosing DKA in patients with and without known T1DM remains critical. The lack of prompt recognition of DKA can lead to misdiagnosis and delay in appropriate treatment, thus increasing the morbidity and mortality of T1DM. Physicians, especially those in the acute care setting, must be familiar with the signs and symptoms of DKA in order to prevent any further complications from T1DM and DKA.

Clinical manifestations of DKA

When a patient with known T1DM presents with any symptoms of illness, providers consider DKA high on the list of differential diagnoses. However, when presented with a previously well child, there are many signs that can allude to the diagnosis. The most common signs include symptoms consistent with dehydration (sunken eyes, dry-cracked lips, poor

TABLE 13.2 Risk factors for diabetic ketoacidosis in children with known type 1 diabetes mellitus.

- Insulin omission
  - Skipping insulin injections
  - Not following sick-day management
  - Not reacting to insulin pump failures
- Poor metabolic control or previous episodes of diabetic ketoacidosis
- Gastroenteritis with persistent vomiting and inability to maintain hydration
- Psychiatric disorders (including eating disorders)
- Unstable family circumstances
- Peripuberty and adolescence in girls
- Binge alcohol consumption
- Limited access to medical services

List of risk factors that have been associated with a higher risk for developing diabetic ketoacidosis in children with a known diagnosis of type 1 diabetes mellitus.
Adapted from Wolfsdorf, J. I., Glaser, N., Agus, M., Fritsch, M., Hanas, R., Rewers, A. ... Codner, E. (2018). ISPAD Clinical Practice Consensus Guidelines 2018: Diabetic ketoacidosis and the hyperglycemic hyperosmolar state. *Pediatric Diabetes, 19*(S27), 155–177. doi:10.1111/pedi.12701.

skin turgor, decreased capillary reflex). Gastrointestinal complaints are also common due to the presence of ketones and include nausea, vomiting, and abdominal pain, which often mimic an acute abdomen. In moderate to severe cases, confusion, drowsiness, progressive obtundation, and even loss of consciousness may occur. Additionally, respiratory rate increases to lower acidosis, which can cause Kussmaul respirations (deep labored breathing). The presence of any of these signs should increase clinical suspicion of DKA. Early identification of DKA and prompt treatment can significantly reduce further complications related to DKA.

Morbidity and mortality of DKA

Mortality from DKA remains very low, ranging from 0.15% to 0.30%, and increased awareness is helping further decrease mortality rates (Edge et al., 1999; Saydah, Imperatore, Geiss, & Gregg, 2012; Wolfsdorf et al., 2018). However, DKA remains the leading cause of death for children with T1DM diagnosed before 15 years of age (Morgan et al., 2018). Cerebral injury is the major cause of mortality and morbidity and cerebral edema accounts for 60%–90% of all deaths due to DKA (Daneman, 2001; Edge et al., 1999; Edge, Hawkins, Winter, & Dunger, 2001; Glaser et al., 2001). Other rare causes of morbidity and mortality in DKA include: electrolyte imbalances (hypokalemia, hypocalcemia, hypomagnesemia, and severe hypophosphatemia), hypochloremic alkalosis, hypoglycemia, venous thrombosis, pulmonary embolism, sepsis, rhinocerebral or pulmonary mucormycosis, aspiration pneumonia, pulmonary edema, adult respiratory distress syndrome, pneumothorax, pneumomediastinum and subcutaneous emphysema, rhabdomyolysis, ischemic bowel necrosis, acute kidney injury including renal failure, acute pancreatitis, and other central nervous system complications including dural sinus thrombosis, basilar artery thrombosis, intracranial hemorrhage, and cerebral infarction (Wolfsdorf et al., 2018).

Neurocognitive complications due to DKA

While clinically noted cerebral edema is the most worrisome complication from DKA, studies have shown that there may still be subtle evidence of brain injury in subclinical cases (Ghetti et al., 2010; Krane, Rockoff, Wallman, & Wolfsdorf, 1985). Clinical and subclinical cerebral injury can have many neurocognitive sequelae on the developing brain. Few studies have been done to observe the direct effect of DKA on neurocognitions. Table 13.3 summarizes the main findings of these studies. Ghetti et al. (2010) showed decreases in memory function by color and spatial position task tests in children with a history of DKA when compared to matched peers with T1DM but no history of DKA. Similarly, Cameron et al. (2014), showed that children who presented in DKA had poorer scores in memory and attention on the Wechsler Memory Scale Paired Associate Learning subtest and the Test of Everyday Attention for Children respectively. These changes were noted 6 months after an episode of DKA showing long-lasting sequela in children presenting in DKA, compared to those who presented without DKA at diagnosis of T1DM. Another prospective study done by Semenkovich et al. (2016) that compares patients with newly diagnosed T1DM and their siblings without T1DM showed that patients who presented in DKA had poorer long delay spatial memory scores on the Spatial Delayed Response Task, than sibling controls without T1DM. The most recent study performed by Aye et al. (2019) evaluated the effect of the severity of DKA in neurocognitive function and concluded that children who presented in moderate or severe DKA scored lower in full-scale IQ tests, attention tests (detectability and commission on Conner's Continuous Performance Test II—CPT2), and memory tests (Immediate and Delayed Dots Location in the Children's Memory Scale—CMS) when compared to children who presented in mild or no DKA.

The mechanism through which T1DM and DKA affect brain development is not fully understood, though some theories can be drawn from its pathophysiology and clinical effects on the brain. DKA is more common in younger children and previous studies (Gaudieri, Chen, Greer, & Holmes, 2008; Naguib, Kulinskaya, Lomax, & Garralda, 2009) have shown that patients with an early diagnosis of T1DM have lower cognitive performance, making DKA and its severity the likely culprit for the neurocognitive deficits seen in children with T1DM. However, more studies looking at the effect of DKA in neurocognitive performance in children need to be performed to further confirm these findings and understand the mechanism through which DKA affects the developing brain.

Brain structural differences in DKA

Brain structure itself is also different upon presentation with T1DM and even more so in patients presenting with DKA. Only a limited number of pediatric studies in T1DM have examined the brain structure of the developing brain using magnetic resonance imaging (MRI). Cameron et al. found that at 6 months after diagnosis of T1DM, children who presented in DKA had a decrease in total white matter volume on brain MRI and an initially decreased level of N-

TABLE 13.3 Neurocognitive and brain structural studies in children with diabetic ketoacidosis.

				Findings	
First Author	Year	Journal	Cohort	Brain Imaging	Neurocognition
Ghetti	2010	The Journal of Pediatrics	Ages 6–18 Patients with a history of DKA versus T1DM patients without a history of DKA	N/A	Patients with a history of DKA showed lower scores in memory tests (Color and Spatial Position task tests)
Cameron	2014	Diabetes Care	Ages 7–16 Patients in DKA versus T1DM patients not in DKA	MRI: Total white matter volume decreased over the first 6 months in those with DKA DTI: Higher mean diffusivity in frontal, temporal, and parietal white matter MRS: Initial decrease in N-acetylaspartate (NAA) in frontal gray matter and basal ganglia	Patients with DKA had poorer scores in delayed memory recall (Wechsler Memory Scale Paired Associate Learning subtests) and sustained/divided attention (Test of Everyday Attention for Children) at 6 months after DKA episode
Semenkovich	2016	Pediatric Diabetes	Ages 7–17 Patients in DKA at diagnosis versus siblings without T1DM (controls)	N/A	Patients who presented in DKA at diagnosis performed worse on spatial memory tasks (The Spatial Delayed Response Task) when compared to sibling controls
Siller	2017	Pediatric Diabetes	Ages 7–17 Patients in DKA at diagnosis versus siblings without T1DM (controls)	DWI: DKA at presentation was associated with lower radial, axial, and mean diffusivity throughout major white matter tracts	N/A
Aye	2019	Diabetes Care	Ages 4 to <10 Patients in Mild/No DKA versus patients in Moderate/Severe DKA	VBM: Patients in moderate/severe DKA gained more total and regional white matter and gray matter volume after DKA. DTI: No differences between the moderate/severe DKA and mild/none DKA group	Patients in moderate/severe DKA scored lower full-scale IQ scores, lower attention test scores (Conners' CPT2: detectability and commission), and lower memory scores (CMS: Immediate and Delayed Dots Location)

Summary of the findings of the five studies that have been performed to study the effects of diabetic ketoacidosis in neurocognition and/or brain structure in children with type 1 diabetes mellitus. *CMS*, Children's Memory Scale; *CPT2*, Continuous Performance Test II; *DKA*, diabetic ketoacidosis; *DTI*, diffusion tensor imaging; *DWI*, diffusion weighted imaging; *IQ*, intelligence quotient; *MRI*, magnetic resonance imaging; *MRS*, magnetic resonance spectroscopy; *T1DM*, type 1 diabetes mellitus; *VBM* = voxel-based morphometry.
Based on review of the following studies: Aye, T., Mazaika, P. K., Mauras, N., Marzelli, M. J., Shen, H., Hershey, T., ... Reiss, A. L. (2019). Impact of early diabetic ketoacidosis on the developing brain. *Diabetes Care, 42*(3), 443–449. doi:10.2337/dc18-1405; Cameron, F. J., Scratch, S. E., Nadebaum, C., Northam, E. A., Koves, I., Jennings, J., ... Inder, T. E. (2014). Neurological consequences of diabetic ketoacidosis at initial presentation of type 1 diabetes in a prospective cohort study of children. *Diabetes Care, 37*(6), 1554–1562. doi:10.2337/dc13-1904; Ghetti, S., Lee, J. K., Sims, C. E., Demaster, D. M., & Glaser, N. S. (2010). Diabetic ketoacidosis and memory dysfunction in children with type 1 diabetes. *The Journal of Pediatrics, 156*(1), 109–114. doi:10.1016/j.jpeds.2009.07.054; Semenkovich, K., Bischoff, A., Doty, T., Nelson, S., Siller, A. F., Hershey, T., & Arbelaez, A. M. (2016). Clinical presentation and memory function in youth with type 1 diabetes. *Pediatric Diabetes, 17*(7), 492–499. doi:10.1111/pedi.12314 and Siller, A. F., Lugar, H., Rutlin, J., Koller, J. M., Semenkovich, K., White, N. H. ... Hershey, T. (2017). Severity of clinical presentation in youth with type 1 diabetes is associated with differences in brain structure. *Pediatric Diabetes, 18*(8), 686–695. doi:10.1111/pedi.12420.

acetylaspartate (NAA) by magnetic resonance spectroscopy (MRS) in frontal gray matter and basal ganglia. NAA is considered a marker of neuronal density and thus a temporary decrease in neuronal density likely represents acute brain injury seen in patients presenting with DKA. The study further looked at mean diffusivity by diffusion tension imaging (DTI), which showed an increase in mean diffusivity in frontal, temporal, and parietal white matter in children presenting in DKA. Increased mean diffusivity values can reflect increased tissue fluid content and/or more subtle cellular injury or loss, while a decreased mead diffusivity value is associated with acute and subacute brain injury (Cameron et al., 2014). Siller et al. (2017) looked at the same population previously studied by Semenkovich et al. (2016) using brain MRI diffused weighted imaging (DWI) and showed that patients who presented in DKA had lower radial, axial, and mean diffusivity throughout major white matter tracts. While Cameron et al. and Siller et al., have different trends on mean diffusivity, both studies are looking at different times in point in the course of DKA and possibly represent different mechanisms of cerebral injury and recovery in children who have experienced DKA. Nevertheless, more studies need to be performed to elucidate these findings. The study by Aye et al. (2019) also used DTI but did not find any significant difference between children with moderate/severe DKA and those with mild/no DKA. However, their voxel-based morphometry (VBM) MRI modality did show that children with moderate or severe DKA gained more total and regional white matter and gray matter volume after the DKA episode (see Figs. 13.1 and 13.2).

The sustained brain structural changes seen in these studies could represent brain injury and/or neuro-developmental compensation at different points throughout the natural history of DKA. How DKA causes these brain structural changes is not fully understood. In cases of clinically evident cerebral edema, it was thought to be due to rapid fluid administration with abrupt changes in serum osmolality. Conversely, recent studies have shown that dehydration and cerebral hypoperfusion may be the cause (Glaser et al., 2004; Tasker & Acerini, 2014). In this theory, there are intrinsic factors in DKA causing brain injury and these could be worsened during treatment. This same concept can apply for cases of subclinical cerebral edema or brain injury. The metabolic disturbances seen in DKA may cause acute ischemia and reperfusion of the brain (Glaser et al., 2004) or an increase in inflammatory factors and cytokines (Levin, 2008), which in turn can lead to permanent brain changes in the developing brain of children who experience DKA. More studies devoted specifically to DKA and its effect on brain structure are needed in order to understand the pathophysiology behind the structural differences seen patients who present with DKA, even in those without clinically evident brain injury.

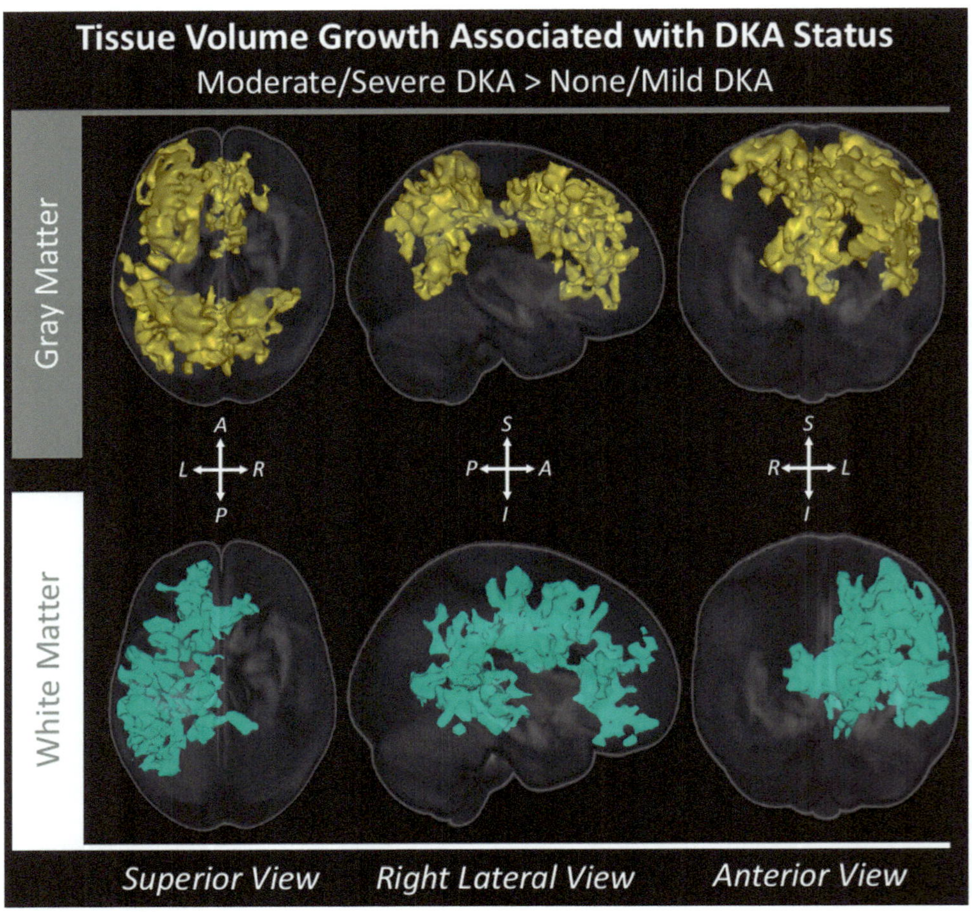

FIGURE 13.2 White matter and gray matter volumes in patients with diabetic ketoacidosis. Brain magnetic resonance imaging (MRI) reconstructions showing total tissue white matter and gray matter volumes studied in children who presented with moderate or severe diabetic ketoacidosis (DKA) versus those who presented with mild or no DKA. *Based on Aye, T., Mazaika, P. K., Mauras, N., Marzelli, M. J., Shen, H., Hershey, T. … Reiss, A. L. (2019). Impact of early diabetic ketoacidosis on the developing brain.* Diabetes Care, *42(3), 443–449. doi:10.2337/dc18-1405.*

Clinical implications

While more research and studies need to be performed to fully understand the effect of DKA and neurodevelopment, evidence thus far points to clinically significant changes in brain structure and neurocognitive function. Thus, early identification of new onset T1DM before the development of DKA and prompt diagnosis and treatment of DKA in patients with known T1DM are instrumental in preventing short- and long-term neurologic complications of DKA. As memory is greatly affected in the short term, initial patient and family education on diabetes management should be reviewed again with the child or adolescent on subsequent visits, particularly if the child presented with moderate or severe DKA upon diagnosis. With more understanding of T1DM and DKA on neurodevelopment, accommodations in school for children with T1DM and history of DKA should be considered in order to meet the education needs for each individual.

Applications to other areas of development

In this chapter we have reviewed the effects of DKA on brain structure and neurocognition. Based on the studies done up to date on DKA, it is evident that DKA causes structural and metabolic changes on the brain (Aye et al., 2019; Cameron et al., 2014; Siller et al., 2017). These changes are likely to represent brain injury and neurologic compensation after an episode of DKA. Having an episode of DKA during key years of neuroplasticity can have a significant impact in neurodevelopment. At this time research points to neurocognitive deficits particularly in attention, memory, and full-scale IQ in children who have an episode of DKA (Aye et al., 2019; Cameron et al., 2014; Ghetti et al., 2010; Semenkovich et al., 2016). The memory problems seen shortly after DKA closely resemble memory deficits seen in cerebral hypoxia or ischemia in conditions such as living in high altitudes or children with obstructive sleep-disordered breathing (Suratt et al., 2006; Virues-Ortega, Garrido, Javierre, & Kloezeman, 2006). Understanding changes in brain structure leading to neurocognitive deficits in the setting of illnesses such as DKA can help us further understand the impact of similar injuries caused by chronic conditions in children. While DKA is episodic and the changes seen in brain structure are transient or evolving the neurocognitive effects are long-lasting. Understanding the mechanism of neuronal injury can aid in the advancement of treatments and prevention of further damage, not only in DKA but in other chronic conditions that may show a similar mechanism of injury.

Key facts about diabetic ketoacidosis

- Diabetic ketoacidosis (DKA) is characterized by hyperglycemia, acidosis, and the presence of ketones in blood or urine.
- The frequency of DKA at onset of diabetes in children ranges from approximately 15% to 70% in Europe and North America.
- In children with known type 1 diabetes mellitus, the risk of DKA ranges from 1% to 10% per patient per year.
- DKA occurs when there is an absolute insulin deficiency or an increased demand in insulin due to stress or infection.
- Insulin omission, insulin pump failures, and not following sick-day management account for nearly all episodes of recurrent DKA.
- Mortality from DKA ranges from 0.15% to 0.30% and increased awareness is helping to decrease mortality rates.
- Cerebral injury is the major cause of mortality and morbidity and cerebral edema accounts for 60%–90% of all deaths due to DKA in children.

Summary points

- Diabetic ketoacidosis (DKA) is the most common cause of morbidity and mortality in children with type 1 diabetes mellitus (T1DM)
- Cerebral injury is the major cause of mortality and morbidity and cerebral edema accounts for 60%–90% of all deaths due to DKA in children
- Several studies on DKA and neurocognitive function have found deficits in memory, attention, and/or full-scale IQ in children with T1DM who present with DKA
- Brain magnetic resonance imaging (MRI) studies using advanced modalities such as voxel-based morphometry have shown both increased and decreased volumes in total or regional white and gray matter in children with a history of DKA

- One study on brain magnetic resonance spectroscopy (MRS) showed an initial decrease in a metabolic marker of neuronal density in specific brain regions soon after an episode of DKA
- Diffusion-weighted and diffusion tensor imaging (DWI/DTI) studies in children with DKA have shown increases and decreases in brain mean diffusivity, which can be associated with subtle cellular injury and acute/subacute brain injury
- Mores studies devoted specifically to DKA and its effect on brain structure and neurocognitive deficits are required to fully understand the effect of DKA in neurodevelopment
- With more understanding of T1DM and DKA on neurodevelopment, accommodations in school for children with T1DM and history of DKA should be considered to meet the education needs for each individual

References

Aye, T., Mazaika, P. K., Mauras, N., Marzelli, M. J., Shen, H., Hershey, T., … Reiss, A. L. (2019). Impact of early diabetic ketoacidosis on the developing brain. *Diabetes Care, 42*(3), 443–449. https://doi.org/10.2337/dc18-1405.

Baş, V. N., Uytun, S., & Torun, Y. A. (2015). Diabetic euglycemic ketoacidosis in newly diagnosed type 1 diabetes mellitus during Ramadan fasting. *Journal of Pediatric Endocrinology and Metabolism, 28*(3–4), 333–335. https://doi.org/10.1515/jpem-2013-0497.

Bui, H., To, T., Stein, R., Fung, K., & Daneman, D. (2010). Is diabetic ketoacidosis at disease onset a result of missed diagnosis? *The Journal of Pediatrics, 156*(3), 472–477. https://doi.org/10.1016/j.jpeds.2009.10.001.

Cameron, F. J., Scratch, S. E., Nadebaum, C., Northam, E. A., Koves, I., Jennings, J., … Inder, T. E. (2014). Neurological consequences of diabetic ketoacidosis at initial presentation of type 1 diabetes in a prospective cohort study of children. *Diabetes Care, 37*(6), 1554–1562. https://doi.org/10.2337/dc13-1904.

Cengiz, E., Xing, D., Wong, J. C., Wolfsdorf, J. I., Haymond, M. W., Rewers, A., … Beck, R. W. (2013). Severe hypoglycemia and diabetic ketoacidosis among youth with type 1 diabetes in the T1D Exchange clinic registry. *Pediatric Diabetes, 14*(6), 447–454. https://doi.org/10.1111/pedi.12030.

Chase, H. P., Garg, S. K., & Jelley, D. H. (1990). Diabetic ketoacidosis in children and the role of outpatient management. *Pediatrics in Review, 11*(10), 297–304. https://doi.org/10.1542/pir.11-10-297.

Cox, K., Cocchi, M. N., Salciccioli, J. D., Carney, E., Howell, M., & Donnino, M. W. (2012). Prevalence and significance of lactic acidosis in diabetic ketoacidosis. *Journal of Critical Care, 27*(2), 132–137. https://doi.org/10.1016/j.jcrc.2011.07.071.

Dabelea, D., Rewers, A., Stafford, J. M., Standiford, D. A., Lawrence, J. M., Saydah, S., … Pihoker, C. (2014). Trends in the prevalence of ketoacidosis at diabetes diagnosis: The SEARCH for diabetes in youth study. *Pediatrics, 133*(4), e938–945. https://doi.org/10.1542/peds.2013-2795.

Daneman, D. (2001). Diabetes-related mortality. A pediatrician's view. *Diabetes Care, 24*(5), 801–802. https://doi.org/10.2337/diacare.24.5.801.

Dunger, D. B., Sperling, M. A., Acerini, C. L., Bohn, D. J., Daneman, D., Danne, T. P., … Wolfsdorf, J. I. (2004). ESPE/LWPES consensus statement on diabetic ketoacidosis in children and adolescents. *Archives of Disease in Childhood, 89*(2), 188–194. https://doi.org/10.1136/adc.2003.044875.

Edge, J. A., Ford-Adams, M. E., & Dunger, D. B. (1999). Causes of death in children with insulin dependent diabetes 1990–96. *Archives of Disease in Childhood, 81*(4), 318–323. https://doi.org/10.1136/adc.81.4.318.

Edge, J. A., Hawkins, M. M., Winter, D. L., & Dunger, D. B. (2001). The risk and outcome of cerebral oedema developing during diabetic ketoacidosis. *Archives of Disease in Childhood, 85*(1), 16–22. https://doi.org/10.1136/adc.85.1.16.

Fritsch, M., Schober, E., Rami-Merhar, B., Hofer, S., Fröhlich-Reiterer, E., & Waldhoer, T. (2013). Diabetic ketoacidosis at diagnosis in Austrian children: A population-based analysis, 1989–2011. *The Journal of Pediatrics, 163*(5), 1484–1488. https://doi.org/10.1016/j.jpeds.2013.06.033. e1481.

Gaudieri, P. A., Chen, R., Greer, T. F., & Holmes, C. S. (2008). Cognitive function in children with type 1 diabetes: A meta-analysis. *Diabetes Care, 31*(9), 1892–1897. https://doi.org/10.2337/dc07-2132.

Ghetti, S., Lee, J. K., Sims, C. E., Demaster, D. M., & Glaser, N. S. (2010). Diabetic ketoacidosis and memory dysfunction in children with type 1 diabetes. *The Journal of Pediatrics, 156*(1), 109–114. https://doi.org/10.1016/j.jpeds.2009.07.054.

Glaser, N., Barnett, P., McCaslin, I., Nelson, D., Trainor, J., Louie, J., … Kuppermann, N. (2001). Risk factors for cerebral edema in children with diabetic ketoacidosis. The pediatric emergency medicine collaborative research committee of the American academy of pediatrics. *New England Journal of Medicine, 344*(4), 264–269. https://doi.org/10.1056/nejm200101253440404.

Glaser, N. S., Wootton-Gorges, S. L., Marcin, J. P., Buonocore, M. H., Dicarlo, J., Neely, E. K., … Kuppermann, N. (2004). Mechanism of cerebral edema in children with diabetic ketoacidosis. *The Journal of Pediatrics, 145*(2), 164–171. https://doi.org/10.1016/j.jpeds.2004.03.045.

Hanas, R., Lindgren, F., & Lindblad, B. (2009). A 2-yr national population study of pediatric ketoacidosis in Sweden: Predisposing conditions and insulin pump use. *Pediatric Diabetes, 10*(1), 33–37. https://doi.org/10.1111/j.1399-5448.2008.00441.x.

Kitabchi, A. E., Umpierrez, G. E., Murphy, M. B., Barrett, E. J., Kreisberg, R. A., Malone, J. I., & Wall, B. M. (2003). Hyperglycemic crises in patients with diabetes mellitus. *Diabetes Care, 26*(Suppl. 1), S109–S117. https://doi.org/10.2337/diacare.26.2007.s109.

Kovacs, M., Goldston, D., & Iyengar, S. (1992). Intellectual development and academic performance of children with insulin-dependent diabetes mellitus: A longitudinal study. *Developmental Psychology, 28*(4), 676–684. https://doi.org/10.1037/0012-1649.28.4.676.

Krane, E. J. (1987). Diabetic ketoacidosis. Biochemistry, physiology, treatment, and prevention. *Pediatric Clinics of North America, 34*(4), 935–960. https://doi.org/10.1016/s0031-3955(16)36296-4.

Krane, E. J., Rockoff, M. A., Wallman, J. K., & Wolfsdorf, J. I. (1985). Subclinical brain swelling in children during treatment of diabetic ketoacidosis. *New England Journal of Medicine, 312*(18), 1147–1151. https://doi.org/10.1056/nejm198505023121803.

Levin, D. L. (2008). Cerebral edema in diabetic ketoacidosis. *Pediatric Critical Care Medicine, 9*(3), 320–329. https://doi.org/10.1097/PCC.0b013e31816c7082.

Lévy-Marchal, C., Papoz, L., de Beaufort, C., Doutreix, J., Froment, V., Voirin, J., & Czernichow, P. (1992). Clinical and laboratory features of type 1 diabetic children at the time of diagnosis. *Diabetic Medicine, 9*(3), 279–284. https://doi.org/10.1111/j.1464-5491.1992.tb01777.x.

Lévy-Marchal, C., Patterson, C. C., & Green, A. (2001). Geographical variation of presentation at diagnosis of type I diabetes in children: The EURODIAB study. European and dibetes. *Diabetologia, 44*(Suppl. 3), B75–B80. https://doi.org/10.1007/pl00002958.

Maahs, D. M., Hermann, J. M., Holman, N., Foster, N. C., Kapellen, T. M., Allgrove, J., … Holl, R. W. (2015). Rates of diabetic ketoacidosis: International comparison with 49,859 pediatric patients with type 1 diabetes from England, Wales, the U.S., Austria, and Germany. *Diabetes Care, 38*(10), 1876–1882. https://doi.org/10.2337/dc15-0780.

Morgan, E., Black, C. R., Abid, N., Cardwell, C. R., McCance, D. R., & Patterson, C. C. (2018). Mortality in type 1 diabetes diagnosed in childhood in Northern Ireland during 1989–2012: A population-based cohort study. *Pediatric Diabetes, 19*(1), 166–170. https://doi.org/10.1111/pedi.12539.

Morris, A. D., Boyle, D. I., McMahon, A. D., Greene, S. A., MacDonald, T. M., & Newton, R. W. (1997). Adherence to insulin treatment, glycaemic control, and ketoacidosis in insulin-dependent diabetes mellitus. The DARTS/MEMO Collaboration. Diabetes Audit and Research in Tayside Scotland. Medicines Monitoring Unit. *Lancet, 350*(9090), 1505–1510. https://doi.org/10.1016/s0140-6736(97)06234-x.

Nadebaum, C., Scratch, S. E., Northam, E. A., & Cameron, F. J. (2012). Clinical utility of mental state screening as a predictor of intellectual outcomes 6 months after diagnosis of type 1 diabetes. *Pediatric Diabetes, 13*(8), 632–637. https://doi.org/10.1111/j.1399-5448.2012.00870.x.

Naguib, J. M., Kulinskaya, E., Lomax, C. L., & Garralda, M. E. (2009). Neuro-cognitive performance in children with type 1 diabetes—a meta-analysis. *Journal of Pediatric Psychology, 34*(3), 271–282. https://doi.org/10.1093/jpepsy/jsn074.

Northam, E. A., Rankins, D., Lin, A., Wellard, R. M., Pell, G. S., Finch, S. J., … Cameron, F. J. (2009). Central nervous system function in youth with type 1 diabetes 12 years after disease onset. *Diabetes Care, 32*(3), 445–450. https://doi.org/10.2337/dc08-1657.

Peters, A. L., Buschur, E. O., Buse, J. B., Cohan, P., Diner, J. C., & Hirsch, I. B. (2015). Euglycemic diabetic ketoacidosis: A potential complication of treatment with sodium-glucose cotransporter 2 inhibition. *Diabetes Care, 38*(9), 1687–1693. https://doi.org/10.2337/dc15-0843.

Quinn, M., Fleischman, A., Rosner, B., Nigrin, D. J., & Wolfsdorf, J. I. (2006). Characteristics at diagnosis of type 1 diabetes in children younger than 6 years. *The Journal of Pediatrics, 148*(3), 366–371. https://doi.org/10.1016/j.jpeds.2005.10.029.

Rewers, A., Chase, H. P., Mackenzie, T., Walravens, P., Roback, M., Rewers, M., … Klingensmith, G. (2002). Predictors of acute complications in children with type 1 diabetes. *Jama, 287*(19), 2511–2518. https://doi.org/10.1001/jama.287.19.2511.

Saydah, S. H., Imperatore, G., Geiss, L. S., & Gregg, E. (2012). Diabetes death rates among youths aged </= 19 years—United States, 1968–2009. *Morbidity and Mortality Weekly Report, 61*(43), 869–872.

Scibilia, J., Finegold, D., Dorman, J., Becker, D., & Drash, A. (1986). Why do children with diabetes die? *Acta Endocrinologica – Supplementum, 279*, 326–333. https://doi.org/10.1530/acta.0.112s326.

Semenkovich, K., Bischoff, A., Doty, T., Nelson, S., Siller, A. F., Hershey, T., & Arbelaez, A. M. (2016). Clinical presentation and memory function in youth with type 1 diabetes. *Pediatric Diabetes, 17*(7), 492–499. https://doi.org/10.1111/pedi.12314.

Siller, A. F., Lugar, H., Rutlin, J., Koller, J. M., Semenkovich, K., White, N. H., … Hershey, T. (2017). Severity of clinical presentation in youth with type 1 diabetes is associated with differences in brain structure. *Pediatric Diabetes, 18*(8), 686–695. https://doi.org/10.1111/pedi.12420.

Smith, C. P., Firth, D., Bennett, S., Howard, C., & Chisholm, P. (1998). Ketoacidosis occurring in newly diagnosed and established diabetic children. *Acta Paediatrica, 87*(5), 537–541. https://doi.org/10.1080/08035259850158245.

Sperling, M. A. (2014). In *Pediatric Endocrinology* (4th ed.). Philadelphia, PA: ELSEVIER.

Suratt, P. M., Peruggia, M., D'Andrea, L., Diamond, R., Barth, J. T., Nikova, M., … Johnson, M. L. (2006). Cognitive function and behavior of children with adenotonsillar hypertrophy suspected of having obstructive sleep-disordered breathing. *Pediatrics, 118*(3), e771–781. https://doi.org/10.1542/peds.2006-0173.

Szypowska, A., & Skórka, A. (2011). The risk factors of ketoacidosis in children with newly diagnosed type 1 diabetes mellitus. *Pediatric Diabetes, 12*(4 Pt 1), 302–306. https://doi.org/10.1111/j.1399-5448.2010.00689.x.

Tasker, R. C., & Acerini, C. L. (2014). Cerebral edema in children with diabetic ketoacidosis: Vasogenic rather than cellular? *Pediatric Diabetes, 15*(4), 261–270. https://doi.org/10.1111/pedi.12153.

Usher-Smith, J. A., Thompson, M., Ercole, A., & Walter, F. M. (2012). Variation between countries in the frequency of diabetic ketoacidosis at first presentation of type 1 diabetes in children: A systematic review. *Diabetologia, 55*(11), 2878–2894. https://doi.org/10.1007/s00125-012-2690-2.

Virues-Ortega, J., Garrido, E., Javierre, C., & Kloezeman, K. C. (2006). Human behaviour and development under high-altitude conditions. *Developmental Science, 9*(4), 400–410. https://doi.org/10.1111/j.1467-7687.2006.00505.x.

Warner, D. P., McKinney, P. A., Law, G. R., & Bodansky, H. J. (1998). Mortality and diabetes from a population based register in Yorkshire 1978–93. *Archives of Disease in Childhood, 78*(5), 435–438. https://doi.org/10.1136/adc.78.5.435.

Wolfsdorf, J. I., Glaser, N., Agus, M., Fritsch, M., Hanas, R., Rewers, A., … Codner, E. (2018). ISPAD Clinical Practice Consensus Guidelines 2018: Diabetic ketoacidosis and the hyperglycemic hyperosmolar state. *Pediatric Diabetes, 19*(S27), 155–177. https://doi.org/10.1111/pedi.12701.

Wolfsdorf, J., Glaser, N., & Sperling, M. A. (2006). Diabetic ketoacidosis in infants, children, and adolescents: A consensus statement from the American Diabetes Association. *Diabetes Care, 29*(5), 1150–1159. https://doi.org/10.2337/diacare.2951150.

Zucchini, S., Scaramuzza, A. E., Bonfanti, R., Buono, P., Cardella, F., Cauvin, V., … Rabbone, I. (2016). A multicenter retrospective survey regarding diabetic ketoacidosis management in Italian children with type 1 diabetes. *Journal of Diabetes Research, 2016*, 5719470. https://doi.org/10.1155/2016/5719470.

Chapter 14

Four-dimensional features of fetal brain: applications to diabetes

Panagiotis Antsaklis[1], George Daskalakis[1] and Asim Kurjak[2]

[1]*Department of Obstetrics and Gynecology, Department of Fetal Maternal Medicine, Alexandra Maternity Hospital, University of Athens, Athens, Greece;* [2]*Department of Obstetrics and Gynecology, Medical School University of Zagreb, Clinical Hospital Sveti Duh, Dubrovnik International University, Zagreb, Croatia*

Introduction

The study of fetal neurology is one of the most interesting areas in fetal medicine, but at the same time one of the less studied ones. Particularly assessment of fetal behavior in utero has been one of the greatest obstetrical challenges, with small progress until recently. The method that allowed a better assessment of the fetal behavior in real time, with direct observation of fetal reactions, fetal movements, and activities, including detailed characteristics such as facial expressions and motoric activity or even detailed finger movements, was the introduction of four-dimensional (4D) ultrasound (Joseph, 1999; Rees & Harding, 2004; Yigiter & Kavak, 2006). 4D ultrasound technology allowed the study of fetal behavior in a systematic way, it helped in the understanding of the pattern of normal fetal behavior, which corresponds to each trimester of pregnancy and also in the detection of abnormal patterns of fetal behavior (Antsaklis & Antsaklis, 2012; Kurjak et al., 2004). Kurjak's antenatal neurodevelopmental test (KANET) is a recognized method that by application of 4D ultrasound assesses a combination of parameters of fetal behavior and forms a scoring system that assesses the fetus in a comprehensive and systematic approach, in the same way that neonatologists perform a neurological assessment in newborns, during the first days of their life, in order to determine their neurological status (Kurjak et al., 2008). KANET has succeeded to define which behavior is expected for each stage of pregnancy, which corresponds to the brain maturation of that stage and to distinguish between normal and abnormal fetal behavior. The development and maturation of the fetal brain follow a very strict and complicated process, but at the same time a very sensitive procedure that makes it susceptible to many genetic or epigenetic factors that could be expressed at any time of in utero life.

One of the most common medical conditions that we are called to manage during pregnancy and could affect not only the growth but also the behavior and finally the overall outcome of the fetus and later the neonate is gestational diabetes. Gestational diabetes (GD) includes the so-called true GD, which is defined as abnormal glucose metabolism that is identified during the present pregnancy and is related to insulin resistance that is expressed during the second or third trimester of pregnancy, in a pregnant women without a history of abnormal glucose levels and previous normal glycemic control and secondly the preexisting diabetes that includes women with known type-I or type-II diabetes or pregnant women that are diagnosed at the beginning of their pregnancy (first trimester). The incidence of GD in pregnancy varies from 5% to 15% of all pregnancies, with the incidence increasing continuously the last few years, an increase that is in agreement with the overall increase of diabetes mellitus type-II in young people and the increase of body mass index (BMI). From the GD cases, 87% of cases are due to true GD and 13% due to preexisting DM. Diabetes is a well-known factor of obstetrical and fetal complications, and that shows how GD can affect the in utero but also the future life of a fetus. Studies show that the earlier GD is diagnosed in pregnancy and that the better control of preexisting or true GD leads to better fetal and maternal outcome with decreased incidence of complications for both of them. A great challenge in fetal medicine is to define how different epigenetic factors affect the fetal brain development and which could be expressed through fetal behavior. Especially for the effect of glucose levels and how a good glycemic control affects the fetal brain development would be of great importance, as GD is a complication that we already know that affects the growth of the fetus and becomes more and more frequent in clinical practice.

Diagnosis, Management and Modeling of Neurodevelopmental Disorders. https://doi.org/10.1016/B978-0-12-817988-8.00014-2

Assessment of fetal neurobehavior in diabetic patients

The assessment of fetal neurobehavior has not been assessed in diabetic pregnancies with 4D ultrasound technology and application of KANET. We recruited 80 women with singleton pregnancies, with gestational age that ranged according to the KANET protocol from 28 to 38 weeks. These 80 women included two different types of populations, the women with diabetes and the control women group, which were at low risk. Specifically the first group, which was the so-called diabetic group, included 40 pregnancies with known diagnosed preexisting diabetes mellitus or gestational diabetes mellitus (GDM), but who all required treatment with insulin. The second group, which was the control group, or nondiabetic group, consisted of 40 women, normoglycemic and with normal oral glucose tolerance test (OGTT). This study was a collaboration of two university hospitals: (1) the department of fetal and maternal medicine of "Alexandra" Maternity hospital of the University of Athens and (2) the department of obstetrics and gynecology of Sveti Duh clinical hospital, Zagreb, Croatia. The duration of the study for the recruitment of the patients, the performance of the ultrasound examinations, and glucose level monitoring, plus the recording of the pregnancy outcome was 2 years. The research protocol was approved by the ethics committee of both hospitals, and a written consent form was obtained from all pregnant women who took part in the study. The two populations that participated in the study had similar demographic characteristics, and these characteristics are shown in Table 14.1.

TABLE 14.1 Demographic characteristics.

Characteristics	Mean Value or percentage (%)
• Maternal age	
Mean value	**30.2 (15–45)**
Median value	**20**
• *Parity*	
Primigravida	**57 (71%)**
Multigravida	**23 (29%)**
• Nationality	
Caucasian	**78 (97%)**
African-American	**2 (3%)**
• *Smoking*	
No	**75 (94%)**
Yes	**5 (6%)**
• *Conception*	
Normal	**77 (96%)**
ART	**3 (4%)**
• *BMI (kg/m^2)*	**24.07**
• *Height (cm)/Weight (kg)*	**164/64.5**
• Previous adverse pregnancy outcome	**2 (3%)**
CP	**1 (1.25%)**
Metabolic disorder	**1 (1.25%)**
• *First trimester bleeding*	
Yes No	**18 (22%)** **62 (78%)**
• *Gestational age at delivery*	
≥37 weeks	**74 (92%)**
<37 weeks	**6 (8%)**

For the diabetic group, all pregnant women were assessed by the endocrinology department and were confirmed cases of diabetes, with preexisting diabetes, and these were cases that were monitored by the endocrinology department even before pregnancy (both type-I and type-II), and during pregnancy were all treated with insulin from early pregnancy (either were on insulin before pregnancy or converted to insulin as soon as pregnancy was diagnosed), or cases of true GDM, with normal glucose levels at the beginning of the pregnancy but abnormal OGTT at or after the second trimester of pregnancy and were treated with insulin at the time of enrollment to the study (at 28 weeks), mainly because dietary changes failed to control glucose levels adequately. As mentioned earlier for the control group, all women had no known history of diabetes, had normal glucose levels at early pregnancy and normal oral glucose tolerance test of 75 gr at 24–28 weeks as per protocol, and were also assessed by the endocrinology department. From the study, the following cases were excluded: multiple pregnancies, known fetal chromosomal or anatomical anomalies, hypertension or preeclampsia diagnosed at the time or before the recruitment to the study, and oligohydramnios. The 80 cases that were included in the study were followed up by the fetal medicine department of each university hospital and all had dating scans, first trimester scans (nuchal translucency), anomaly scans, and all follow-up growth scans and KANET were performed in these departments.

The ultrasound equipment used consisted of two machines: GE Voluson 730 Pro (4D Wide Band Convex probe) and GE E8 Expert (RAB4-8-D Wide Band Convex Volume Probe). KANET test was applied to all patients every 2–4 weeks and after 28 weeks. All tests were performed transabdominally and lasted from 15 to 20 min, they were performed by two examiners and were reviewed by one of them. Performance of KANET was according to the OSAKA consensus statement, consisting of eight parameters: isolated head anteflexion, cranial sutures and head circumference, isolated eye blinking/facial alteration (grimace or tongue expulsion)/mouth opening (yawning or mouthing), isolated hand movement, isolated leg movement, hand to face movement, fingers movement, gestalt perception of general fetal movements. The interpretation of KANET score is abnormal for 0–5, borderline for 6–13, and normal for more than 14.

Statistical analysis was carried out using Statistical Package for Social Sciences (SPSS). Mean and medians were calculated for all quantitative variables, and the measures of dispersion (standard deviation and standard error) were calculated. Normality of data was checked by measures of Kolmogorov–Smirnov tests of normality. For normally distributed data, means were compared by using *t*-test. For skewed data and scores, Mann–Whitney test was applied. Qualitative or categorical variables were described as frequencies and proportions. Proportions were compared by using Chi-squared test or Fisher's-exact test, depending on which was more appropriate. All statistical tests were two-sided. Statistical significance was reached at $P < 0{,}5$.

The differences in fetal behavior between diabetic and nondiabetic patients

Although the number of cases studied was limited, the study had strict criteria and not big differences at least regarding the demographic characteristics and the pregnancy characteristics of the two groups, so that the comparison between them would be more objective. For example, regarding the age of the pregnant women of the two groups, there were no statistically significant differences (30.5 ± 5.1 years for the diabetic group compared to 29.8 ± 6.2 years for the nondiabetic group) and respectively regarding the gestational age at which the examinations were performed (32 ± 1.6 weeks for the diabetic group compared to 33 ± 1.2 weeks for the nondiabetic group). The ultrasound examinations and KANET were performed at the fetal medicine department working hours from 8.30 a.m. to 20.00 p.m., and this is mentioned because the diurnal rhythm of the fetus and the time of performance of KANET can affect the result as mentioned by previous studies (Table 14.2). The only limitation that should be mentioned is that at least for the women that belonged to the diabetic group, ultrasound examination and KANET were performed mostly in the morning hours, and that was due to the fact that these women had their endocrinology appointment scheduled early in the morning and in order to avoid transportation and delays were seen also in the morning hours. Data analysis of all KANET scores for both groups (diabetic and nondiabetic) showed that KANET scores as an average were higher in the low risk (nondiabetic) group. Specifically in the low-risk nondiabetic group, there were no abnormal scores recorded, while the border line scores were 15 (11%) and the normal scores were 120 (89%). For the diabetic group, the scores of KANET noted were: 2 abnormal (1.3%), 23 borderline (14.7%), and 131 (84%) normal (Table 14.3). Another interesting part of the study was to try and identify which parameters of fetal behavior differ between the two groups, and it was shown that the biggest differences were identified for the following parameters: isolated eye blinking, facial alterations, and finger movements (Fig. 14.1), and finally, a difference of less degree was noted for isolated hand movements, without having a difference that had a statistical significance. The details of these findings are shown in Table 14.4. What is also mentioned is the so-called success rate for each parameter of KANET, which means that for each parameter of KANET what was the possibility of achieving all parameters.

TABLE 14.2 KANET study group.

Methods	Numbers (n)
• Prospective longitudinal cohort study	
• Gestational age	**28–38 weeks**
• Singleton pregnancies	*(early dating US)*
• January 2013–January 2015	**2 years**
• Fetal & Maternal Medicine Unit, "ALEXANDRA" Maternity Hospital, University of Athens/Sveti Duh Clinical Hospital, Zagreb	
• Study approved by the ethics committee/consent form	
• Two operators (*all reviewed by one*)	
• **Number participants**	**80**
• **Number of KANET performed**	**291**
• Neonatal Follow-up	**1–3 days of life III**
• Time of performance of KANET	**08.30––20.00** *(mean 15.00)*
• GE Voluson 730 Pro/GE E8	

TABLE 14.3 Comparison of KANET scores between the two groups.

Population	Study group (Diabetic)	Control (Non-Diabetic)	P value
Number of patients	**N = 40**	**N = 40**	
KANET performed	**n = 156**	**n = 135**	**Total = 291**
GA at KANET (*weeks*)	**32 ± 1.6**	**33 ± 1.2**	
N of KANET/Patient	**3.9**	**3.3**	
Total score			
• median	**14**	**16**	*<0.001*
• Range	**4–16**	**6–16**	
Interpretation			
• abnormal	**2** *(1.3%)*	**0** *(0%)*	*<0.001*
• Borderline	**23** *(14.7%)*	**15** *(11%)*	
• Normal	**131** *(84%)*	**120** *(89%)*	

Fetal behavior in diabetes: differences and factors that affect it

The study of fetal neurology, the development of fetal neurological status, and how the development and maturation of fetal brain and central nervous system (CNS) are reflected through fetal behavior is an area of great interest and still has many unanswered questions (Amiel-Tison, Gosselin, & Kurjak, 2006; Kurjak, Talic, Honemeyer, Stanojevic, & Zalud, 2013). As it happens with neonates, the behavioral patterns of a fetus could be a mirror of its nervous system integrity, and by assessing the fetal behavior in utero, the developmental and maturational degree of fetal central nervous system could be examined (de Vries & Fong, 2007; Greenwood, Newman, Impey, & Johnson, 2003; Strijbis, Oudman, van Essen, & MacLennan, 2006). Neonatal assessment for the examination of the integrity of the nervous system is a routine examination, and since it is known that there is a continuity from fetal to neonatal neurobehavior, assessment of fetal behavior could provide useful information for the neurological status and possibly at an earlier stage, when the window of opportunity for a prompt intervention would be ideal. Fetal brain development is a very detailed and complex process that goes through specific stages, and these are reflected via the fetal behavior. The fetal behavior that corresponds to each trimester of pregnancy for normal neurologic development has been defined up to a point and that helps to distinguish

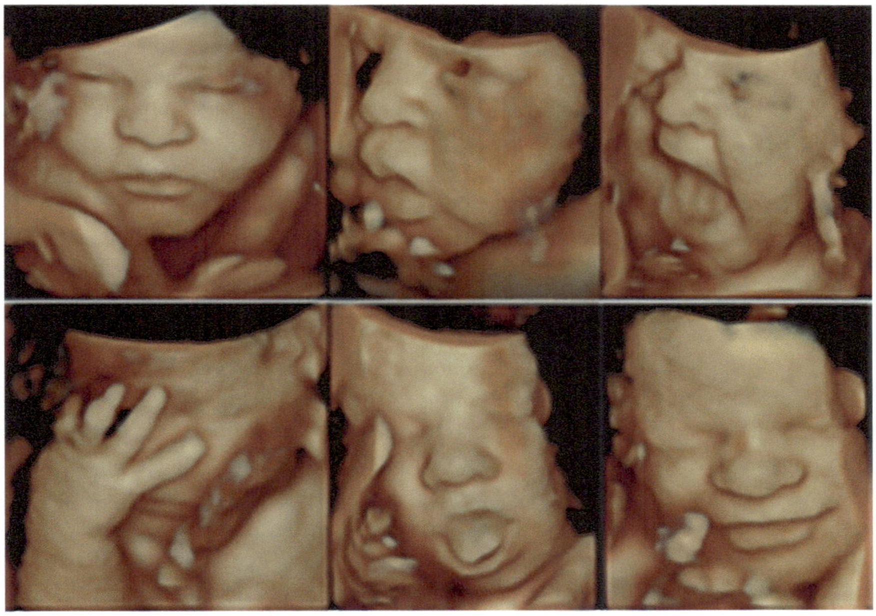

FIGURE 14.1 For the diabetic group, the parameters of KANET that were decreased were isolated eye blinking, facial alterations, and finger movements. In the other parameters, there were no differences.

TABLE 14.4 Comparison of KANET scores between the two groups.

Sign	Success rate	Difference in the 2 groups
1. Isolated head anteflexion	94%	–
2. Cranial sutures and HC	98%	–
3. Isolated eye Blinking	88%	↓
4. Facial alteration	98%	↓
5. Mouth opening	100%	–
6. Isolated leg movement	100%	–
7. Isolated hand movement	96%	–
8. Hand to face movement	96%	–
9. Finger movements	96%	↓

between normal and abnormal behavior and may be detect abnormal conditions. If abnormal neurological conditions can be detected in utero, then possibly it could be determined if this abnormal behavior is related to an anatomical or chromosomal finding or whether fetal behavior may be affected by external factors or obstetrical complications, such as: chorioamnionitis, GD, intrauterine growth restriction, and preterm contractions or preterm rupture of membranes, can affect a fetus (de Vries et al., 1985). GD at the form of preexisting or true GD is a condition that is becoming more frequent in antenatal care. It is well known that good control of diabetes in pregnancy decreases the risk of complications for the mother and the fetus, without being clear however how poor glycemic control can affect the fetal behavior and neuronal development both in utero and long term. Studies have shown that a maternal environment of continuous hyperglycemic state affects the fetal behavior, but in an opposite way that would be expected, as although the glucose levels are increased, the fetal movements are decreased (Edelberg, Dierker, Kalhan, & Rosen, 1987). It has also been stated that the high glucose levels in diabetic patients may affect the development of the fetal sensory-motor response system, and this is why in diabetic mothers with poor glycemic control, although due to the increased glucose levels, it would be expected that the fetus would be more active, actually the fetal activity is decreased (Cosmi, Anceschi, Cosmi, Piazze, & La Torre, 2003). Eldelberget et al. and Allen and Kisilevsky suggested in their studies with diabetic pregnancies who had elevated blood glucose levels that the immaturity of the sensory-motor response system was responsible for the decreased fetal activity

(Allen & Kisilevsky, 1999; Cosmi et al., 2003). Regarding fetal behavioral state including not only body, but also breathing movements in case of pregnancies complicated by diabetes, it appears to be less vigorous and less mature compared to pregnancies without diabetes, although at least from a frequency point of view the breathing movements appear to be more, without though this difference in the number of the breathing movements being at a statistical significant level (Mulder, O'Brien, Lems, Visser, & Prechtl, 1990). Differences in fetal behavior and perception of fetal movements between fetuses of diabetic and nondiabetic pregnancies have been recorded even in cases of good glycemic control, with the main differences being in the number of fetal breathing movements and fetal heart rate rhythm (they are higher in the diabetic group) and the fetal body movements (they were lower in the diabetic group) (Devoe, Youssef, Castillo, & Croom, 1994). Especially for cases of type-I diabetes, if glucose is poorly controlled and the diabetes optimality score is low, it has been documented that the fetal movements after 20 weeks have an abnormal pattern proving the effect that diabetes, especially hyperglycemic state or nonstable levels of glucose for long periods, can affect fetal behavior (Kainer, Prechtl, Engele, & Einspieler, 1997). As mentioned earlier, it is not very well understood how diabetes affects fetal behavior, by affecting the maturation of the nervous system of the fetus, as it has been noticed that in diabetic pregnancies hypoglycemia rather than hyperglycemia can increase fetal activity, and in contrast, hyperglycemia is related with a decrease in the frequency of fetal movements (Holden, Jovanovic, Druzin, & Peterson, 1984). There are studies with contradictory results, which showed that diabetic pregnancies may be related with an increase of fetal activity and fetal body and breathing movements. However, the inconsistency of these studies could be explained by the fact that these studies do not have a uniform and objective method of assessing fetal behavior and activity (Yeoshoua, Goldstein, Zlozover, & Wiener, 2012). The method that could fill in this gap is KANET, which a standardized way of assessing the fetal behavior and activity. The current study is the only study that has used KANET and 4D technology for the assessment of fetal behavior in diabetic pregnancies. KANET appears to be the most complete method so far that utilizes 4D ultrasound technology in order to examine the fetus in utero by assessing detailed parameters of fetal activity and not just general movements, in a similar way that neonatologists assess a neonate postpartum during the first days of their life, in order to determine their neurological status. With the application of these parameters and the development of the scoring system, KANET offers the opportunity for an objective and systematic assessment that could be used in many studies and for the better collection of data (Antsaklis, Porovic, Daskalakis, & Kurjak, 2017; Kurjak et al., 2007, 2008). With the application of KANET on a very strictly designed study and very selective inclusion criteria of diabetic patients, the difference in fetal behavior between diabetic and nondiabetic fetuses was confirmed, but it also showed which specific parameters—movements could be different between the two groups, and possibly with more results of other studies, to identify which nervous centers are affected by these conditions during fetal life. Of course, in order to draw safe conclusions, a larger number of examined patients is needed and also some details should be looked at more carefully, such as the time that KANET was performed, since it has been documented that fetal diurnal rhythm could affect fetal activity and finally that a better correlation of the lower KANET results with glucose levels, especially hypo- or hyperglycemia. However, KANET by application of 4D ultrasound and real-time assessment of the fetus appears to offer the most complete method of assessing fetal behavior in contrast to the classical way of assessing simply the fetal body or breathing movements with 2D ultrasound, as almost all studies that examine fetal activity behavior in pregnancies complicated by diabetes have done. It is important that the two groups (diabetic and nondiabetic pregnancies) were strictly divided and have clear inclusion criteria, all diabetic pregnancies had similar way of treatment as all were under treatment with insulin and the nondiabetic were excluded from having GD as all of them had a timely OGTT. For these two groups, the difference in fetal activity was demonstrated and specified accordingly for each parameter of fetal activity. The development of fetal nervous system and the assessment of fetal neurobehavior, which reflects the integrity of the fetal nervous system, is an area of great interest and research in fetal and perinatal medicine. Especially the pathway that different epigenetic factors and medical conditions, such as glucose levels and GD, which is becoming more and more common, affect the development of fetal neurological development is still to be determined. Many studies with large number of patients are needed in order to define normal and abnormal fetal neurobehavior and identify which parameters and how they may affect the fetus. KANET offers a structured way of performing this assessment and for the management of diabetic pregnancies could add useful information on how a better glycemic control can improve the fetal neurological development.

References

Allen, C. L., & Kisilevsky, B. S. (July 1999). Fetal behavior in diabetic and nondiabetic pregnant women: An exploratory study. *Developmental Psychobiology, 35*(1), 69–80.

Amiel-Tison, C., Gosselin, J., & Kurjak, A. (2006). Neurosonography in the second half of fetal life: A neonatologist's point of view. *Journal of Perinatal Medicine, 34*, 437–446.

Antsaklis, P., & Antsaklis, A. (October 2012). The assessment of fetal neurobehavior with four-dimensional ultrasound: The Kurjak antenatal neurodevelopmental test. *Donald School Journal of Ultrasound in Obstetrics and Gynecology, 6*(4), 362–375.

Antsaklis, P., Porovic, S., Daskalakis, G., & Kurjak, A. (August 28, 2017). 4D assessment of fetal brain function in diabetic patients. *Journal of Perinatal Medicine, 45*(6), 711–715.

Cosmi, E. V., Anceschi, M. M., Cosmi, E., Piazze, J. J., & La Torre, R. (March 2003). Ultrasonographic patterns of fetal breathing movements in normal pregnancy. *International Journal of Gynaecology & Obstetrics, 80*(3), 285–290.

Devoe, L. D., Youssef, A. A., Castillo, R. A., & Croom, C. S. (August 1994). Fetal biophysical activities in third-trimester pregnancies complicated by diabetes mellitus. *American Journal of Obstetrics and Gynecology, 171*(2), 298–303. discussion 303-5.

Edelberg, S. C., Dierker, L., Kalhan, S., & Rosen, M. G. (May 1987). Decreased fetal movements with sustained maternal hyperglycemia using the glucose clamp technique. *American Journal of Obstetrics and Gynecology, 156*(5), 1101–1105.

Greenwood, C., Newman, S., Impey, L., & Johnson, A. (2003). Cerebral palsy and clinical negligence litigation: A cohort study. *BJOG, 110*, 6–11.

Holden, K. P., Jovanovic, L., Druzin, M. L., & Peterson, C. M. (January 1984). Increased fetal activity with low maternal blood glucose levels in pregnancies complicated by diabetes. *American Journal of Perinatology, 1*(2), 161–164.

Joseph, R. (1999). Fetal brain and cognitive development. *Developmental Review, 20*, 81–98.

Kainer, F., Prechtl, H. F., Engele, H., & Einspieler, C. (November 24, 1997). Assessment of the quality of general movements in fetuses and infants of women with type-I diabetes mellitus. *Early Human Development, 50*(1), 13–25.

Kurjak, A., Carrera, J. M., Stanojevic, M., et al. (2004). The role of 4D sonography in the neurological assessment of early human development. *The Ultrasound Review of Obstetrics and Gynecology, 4*, 148–159.

Kurjak, A., Miskovic, B., Andonotopo, W., Stanojevic, M., Azumendi, G., & Vrcic, H. (2007). How useful is 3D and 4D ultrasound in perinatal medicine? *Journal of Perinatal Medicine, 35*, 10–27.

Kurjak, A., Talic, A., Honemeyer, U., Stanojevic, M., & Zalud, I. (2013). Comparison between antenatal neurodevelopmental test and fetal Doppler in the assessment of fetal well being. *Journal of Perinatal Medicine, 41*, 107–114.

Kurjak, A., Tikvica, A., Stanojevic, M., et al. (2008). The assessment of fetal neurobehavior by three-dimensional and four-dimensional ultrasound. *Journal of Maternal-Fetal and Neonatal Medicine, 21*, 675–684.

Mulder, E. J., O'Brien, M. J., Lems, Y. L., Visser, G. H., & Prechtl, H. F. (1990). Body and breathing movements in near-term fetuses and newborn infants of type-1 diabetic women. *Early Human Development, 24*(2), 131–152.

Rees, S., & Harding, R. (2004). Brain development during fetal life: Influences of the intra-uterine environment. *Neuroscience Letters, 361*, 111–114.

Strijbis, E. M., Oudman, I., van Essen, P., & MacLennan, A. H. (2006). Cerebral palsy and the application of the international criteria for acute intrapartum hypoxia. *Obstetrics and Gynecology, 107*, 1357–1365.

de Vries, J. I., & Fong, B. F. (2007). Changes in fetal motility as a result of congenital disorders: An overview. *Ultrasound in Obstetrics and Gynecology, 29*, 590–599.

de Vries, J. I., Visser, G. H., & Prechtl, H. F. (1985). The emergence of fetal behaviour. II. Quantitative aspects. *Early Human Development, 12*, 99–120.

Yeoshoua, E., Goldstein, I., Zlozover, M., & Wiener, Z. (June 2012). Sonographic study of the relationship between gestational diabetes mellitus and fetal activity. *Journal of Maternal-Fetal and Neonatal Medicine, 25*(6), 623–626.

Yigiter, A. B., & Kavak, Z. N. (2006). Normal standards of fetal behavior assessed by four-dimensional sonography. *Journal of Maternal-Fetal and Neonatal Medicine, 19*, 707–721.

Further reading

Kurjak, A., Antsaklis, P., Stanojevic, M., & Porovic, S. (July 31, 2017). Fetal behavior assessed by four-dimensional sonography. In *Donald school textbook of ultrasound in obstetrics and gynecology 4th edition* (4 edition, pp. 642–663). Jaypee Brothers Medical Pub.

Chapter 15

Developmental effects of nicotine on cognitive, motivated, and executive behaviors

Kristi A. Kohlmeier and Filip S. Polli[a]

Department of Drug Design and Pharmacology, Faculty of Health and Medical Sciences, University of Copenhagen, Copenhagen, Denmark

Mini-dictionary of terms

ADHD Attention deficit/hyperactive disorder, manifested by the expression or combination of behaviors including inattention, impulsivity, and hyperactivity.
nAChR The nicotinic acetylcholine receptor (nAChR) at which nicotine binds.
NRTs Nicotine replacement therapies, which are clinically employed to support smoking cessation.
PNE Prenatal nicotine exposure in rodent animal models
Teratogens Chemical substances that can compromise fetal development by exerting negative effects

Nicotine exposure and replacement therapies during pregnancy

Nicotine is powerfully addictive. Together with biological and social–environmental contexts, this property might explain why approximately 8% of women continue to smoke during pregnancy in certain global regions, such as Europe (Lange, Probst, Rehm, & Popova, 2018), despite warnings from health care professionals and public health entities that smoking while pregnant is one of the most preventable risk factors for a variety of adverse outcomes to the offspring. Discerningly, when levels of nicotine are assayed to determine the rate of smoking while pregnant, rather than self-reporting, a gap appears between those who self-report smoking and the numbers who exhibit levels of nicotine in their blood, suggesting underreporting or secondhand exposures. Therefore, the number of pregnancies associated with nicotine exposure could actually be higher than that estimated in epidemiological studies (Hall, Wexelblatt, & Greenberg, 2016). Despite this discrepancy, it is believed that maternal smoking has truly been on the decline in certain regions of the world thanks to wider public awareness of the risks. However, the increase in popularity of vaping through the usage of electronic cigarettes (e-cigarettes) threatens to not only slow tobacco quitting rates, but also to result in a stagnation, or more alarmingly, a rise in the current level of the number of individuals exposed to nicotine during gestation.

In early patent applications in the 1960s for e-cigarettes, the technology was suggested as useful as a nicotine replacement therapy (NRT) to reduce smoking and smoking harm. Since their wide-commercial appearance in 2003, use of e-cigarettes to "vape" has been especially attractive to youth, leading to calls of alarm from regulatory agencies and public health officials of a "Juul epidemic" running rampant among adolescents. While there is some harm reduction when compared to the use of cigarettes, in the past years, it has become apparent that e-cigarettes are far from being harmless, as they can serve as a gateway to cigarette smoking or to the development of a dual use pattern, which has been seen in adolescents (Farsalinos, 2018). Further, in adults, those who used e-cigarettes as abstinence tools were reported to be 28% less likely to give up smoking, supporting the lack of efficacy of this approach in smoking cessation (Kalkhoran & Glantz, 2016). Young women have embraced these devices, and rates of nicotine users have systematically increased in this

[a] **Present Address:** Department of Neuroscience, The Scripps Research Institute, La Jolla, CA, United States.

Diagnosis, Management and Modeling of Neurodevelopmental Disorders. https://doi.org/10.1016/B978-0-12-817988-8.00015-4

population. An increase is also supported by indications that women do not perceive e-cigarettes as harmful as tobacco smoking (Adkison et al., 2013). The ongoing development of pleasant flavors and more stylish, attractive devices will likely result in an increase in the percentages of female users. The predicted increase in women who utilize nicotine-containing products in the form of e-cigarettes or who transition to combustible tobacco cigarettes and the high addiction potential of nicotine raise the concern that a larger percentage of women will become pregnant while dependent on nicotine.

Once pregnant, women dependent on nicotine in traditional cigarettes often make a well-intentioned effort to eliminate exposure to the chemicals present in cigarettes while avoiding abstinence syndromes and switch during their pregnancies to NRTs. The most commonly used NRT strategy nowadays is the nicotine patch, which engenders a constant nicotine absorption into maternal blood, thereby reducing withdrawal symptoms and precluding euphoric effects of nicotine, should other nicotine-containing product be utilized while under NRT management. However, the patch results in a constant exposure of the fetus to a steady level of nicotine during its development, which could account for an even higher nicotine exposure relative to that seen in day-time period smokers (Slotkin, 2008). More and more women are choosing to utilize e-cigarettes as a strategy to reduce harm during pregnancy believing that e-cigarettes are less harmful (Baeza-Loya et al., 2014). Despite evidence that their efficacy in successfully increasing quitting rates during pregnancy has been questioned (Baeza-Loya et al., 2014), health care professionals have promoted their use during pregnancy (Bar-Zeev, Lim, Bonevski, Gruppetta, & Gould, 2018). In addition, some studies have suggested that e-cigarette usage during pregnancy might be a preferential choice over FDA-approved NRTs (Oncken et al., 2017). However, this can result in unprecedented levels of nicotine exposure to the developing fetus as e-cigarettes can provide nicotine at levels even higher than other NRTs or traditional, commercially available cigarettes.

Nicotine is teratogenic

Nicotine is not only addictive; it is also a teratogen (Fig. 15.1). Functional nicotine acetylcholine receptors (nAChRs) are present before neurulation, which suggests that nicotine's teratogenic effects could start during the first trimester of pregnancy in humans (Atluri et al., 2001). Developmentally inappropriate activation of nAChRs is expected to alter the trajectory of development of almost all facets of the fetal brain, which can result in altered tissue structure affecting networks, circuits, and neuroactive substances, which end up developing along an entirely different path than they would otherwise have in the absence of this out of sequence cue resulting in overstimulation of cholinergic systems. Human studies have been conducted of the outcomes on the offspring of smoking during pregnancy (Fig. 15.2). The multiplicity of adverse physiological actions includes preterm delivery, lower birth weight, reduced lung function, sudden death syndrome, childhood high blood pressure, and indirect data have suggested that gestational exposure to nicotine could be related to later-life development of cardiovascular disease, obesity, and Type 2 diabetes (Bakker & Jaddoe, 2011). Although combustible tobacco-based cigarettes expose the smoker to over 4000 chemicals, the well-known teratogenic effects on development of nicotine likely underlie the higher risk seen in negative maladaptive behavioral outcomes for the individuals exposed during gestation to maternal smoking and, in

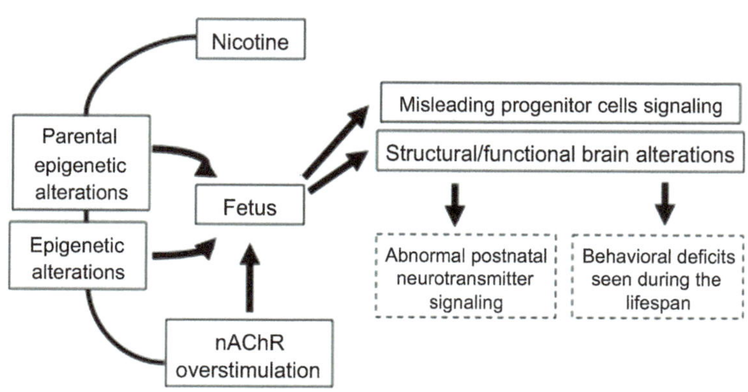

FIGURE 15.1 **Nicotine has actions on progenitor cell development if present during gestation.** Although it is not completely understood how nicotine exerts its teratogenic effects, which could be exacerbated by epigenetic actions through DNA methylation, nicotine in the amniotic fluid can interact with nAChRs in the fetal brain and body, which are expressed at very early embryonic ages, and play a very important role in brain development and maturation in fetuses and later in infants up to adolescence (Dwyer, McQuown, & Leslie, 2009). This role includes induction of gene expression, cell proliferation, differentiation, apoptosis, and cell death (Smith, Dwoskin, & Pauly, 2010).

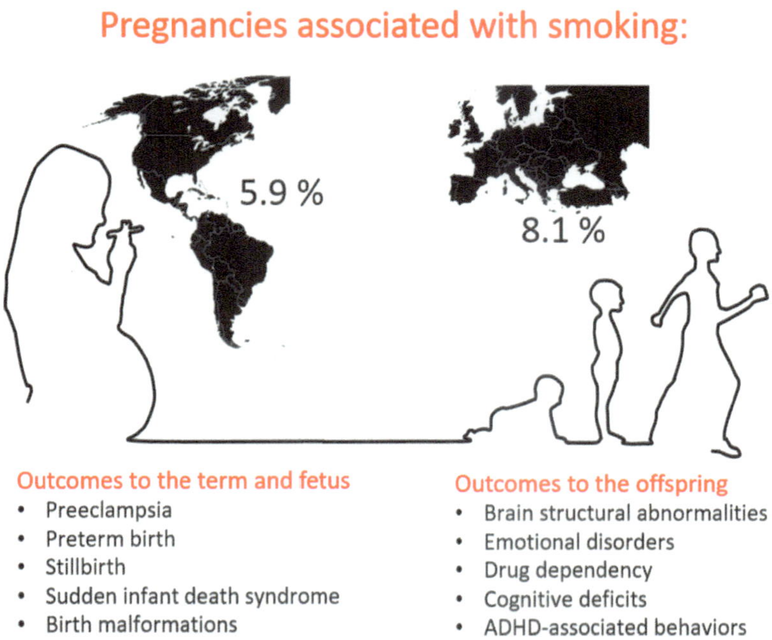

FIGURE 15.2 **Rates for smoking while pregnant vary geographically. Smoking while pregnant is associated with negative physiological and behavioral fetal outcomes.** When a pregnant woman ingests nicotine, it enters her bloodstream where it is able to cross the placenta and accumulate in the amniotic fluid. Due to the poor clearance properties of the placenta, the dwell time of nicotine is actually longer in the fetal compartment than in the mother's blood. Also contributing to prolonging fetal exposure to nicotine, the effects of this drug on the mother include an increase in blood pressure and heart rate, while concurrently there is a decrease in uterine blood flow, resulting in an enhancement of retention of nicotine in the amniotic fluid. Consequently, levels in the fetal compartment are significantly higher than are maternal levels (Jauniaux, Gulbis, Acharya, Thiry, & Rodeck, 1999), and nicotine remains present for significantly longer durations of time than in the mother, almost ensuring a constant exposure of the developing fetal brain and body to this teratogenic compound (Lambers & Clark, 1996). Teratogenic effects of nicotine certainly play a role in negative outcomes for the pregnancy and the offspring.

the few studies that have been conducted, for individuals prenatally exposed to nicotine in NRTs. Negative behavioral and mental health outcomes on the individual include a higher risk for the development of attention-deficit/hyperactivity disorder (ADHD) and ADHD-related behaviors, such as conduct problems including increased externalized behavior. Other associated outcomes are a higher risk of depression, anxiety, aggression, obsessive–compulsive disorders, and drug dependency. Importantly, these associations are also present in studies in which prenatally exposed individuals were adopted and in which parental postnatal smoking and other potential confounders were controlled (Daseking, Petermann, Tischler, & Waldmann, 2015; Fitzpatrick, Barnett, & Pagani, 2014; Schwenke et al., 2018; Thakur et al., 2013). Further, general cognitive impairments, assessed in reading performance, language ability, and other skills scored by a variety of tests, have also been associated with prenatal exposures (Abbott & Winzer-Serhan, 2012; Cho, Frijters, Zhang, Miller, & Gruen, 2013; Clifford, Lang, & Chen, 2012). Moreover, prenatal exposure appears to be associated with increased drug experimentation during adolescence, particularly compulsive abuse of nicotine, which is an association also seen in studies that have corrected for confounds such as maternal cigarette consumption after birth (Buka, Shenassa, & Niaura, 2003; De Genna, Goldschmidt, Day & Cornelius, 2017; Lotfipour et al., 2014). This association can persist during adulthood stages, in which as young adults, prenatally exposed individuals displayed significantly higher rates of cigarette smoking and nicotine dependence, as shown by independent studies (O'Callaghan et al., 2009; Cornelius, Goldschmidt, & Day, 2012). An increased risk for nicotine dependence was also found in sibling pairs discordant for prenatal tobacco exposure and in which postnatal maternal smoking was controlled, strongly linking prenatal cigarette exposure with increased liability for nicotine addiction (Shenassa, Papandonatos, Rogers, & Buka, 2015). Unfortunately, it is not just a higher risk for later-life development of dependence to nicotine, which is associated with gestational exposure. Prenatal exposure has been associated with a higher likelihood for other drug abuses, particularly during the adolescent time when exploratory and risk-taking behaviors are developmentally appropriate (Lotfipour et al., 2009).

Although few epidemiological studies have examined the effects of sex, or included females in their study populations at all, from available data, effects of prenatal smoking exposure on cognitively related behaviors appear to be sex-sensitive.

Sensory processing systems are affected by gestational exposure in a sex-based manner, as deficits in auditory and visual attention accuracy in exposed female adolescents have been detected, while effects were restricted to auditory attentional performance in males (Jacobsen, Slotkin, Mencl, Frost, & Pugh, 2007). Females also exhibit a several-fold higher risk than males for development of substance abuse following prenatal exposure (Brennan, Grekin, Mortensen, & Mednick, 2002; Weissman, Warner, Wickramaratne, & Kandel, 1999). Hyperactivity associated with gestational smoking exposure has a higher incidence in children, and many studies have indicated a particular vulnerability in male offspring during adolescence (Motlagh et al., 2011; Schwenke et al., 2018; Thakur et al., 2013). As nicotine is a known teratogen, these and other studies support the conclusion that nicotine is altering the brain during development within regions that control behavior and that these alterations may occur in a sex-specific manner, which is not surprising, given the sexually dimorphic development of the brain in utero (Brennan et al., 2002; Weissman et al., 1999) (for review, see (Slotkin et al., 2007)), which would be expected to respond differentially to perturbations during development and reflect differences in behavioral and cognitive outcomes.

Prenatal smoking exposure is associated with structural changes in the human brain

In order to provide a mechanistic basis for the negative behavioral outcomes in prenatally exposed individuals, as it is known that these are neurally controlled behaviors and therefore likely reflect alterations in neural networks and circuits, many studies have been conducted in regions of the brain known to control, or to be associated with, cognitive behaviors to determine whether alterations in development in those regions have occurred at different points in ontogeny (Table 15.1). Structural changes were noted in prenatally smoking-exposed human infants or during the early postnatal period, including reduced volumes of the cerebellum, lateral ventricular system and reduced sizes of the frontal lobe, and during childhood and adolescence, the gray matter in the cerebral cortex, volume of the corpus callosum, and thickness of the frontal temporal and parietal regions were found to be reduced (Bublitz & Stroud, 2012). A reduction in function and structural changes that resembled alterations seen in ADHD patients (Knoch et al., 2006) was observed in prenatally exposed young adults within the inferior frontal cortex, which is a region of the brain shown to be involved in choices with risky outcomes and in response inhibition (Holz et al., 2014; Knoch et al., 2006). In a study investigating the effects of gestational smoke exposure on the corpus callosum, a smaller size was noted in adolescents up to 18 years of age; however, this difference was only found in females, which indicates sex-based differences in prenatal exposure-mediated effects on brain structure, likely mechanistically underlying sex-dependent behavioral outcomes (Paus et al., 2008). This conclusion is further supported by observations that the

TABLE 15.1 Structural changes in human brain regions associated with tobacco smoke exposure during pregnancy.

	Human infant and early postnatal	
Brain region	**PNE versus controls**	**References**
Cerebellum	Reduced volume	Bublitz & Stroud, 2012
Latera ventricles	Reduced volume	Bublitz & Stroud. 2012
Frontal lobe	Reduced size	Bublitz & Stroud, 2012
	Childhood, adolescence, and adult	
Brain region	**Gestational exposure to tobacco smoke versus controls**	**References**
Cerebral cortex	Reduction in gray matter	Bublitz & Stroud, 2012
Corpus callosum	Reduced volume	Bublitz & Stroud, 2012
Frontal lobe	Reduced thickness	Bublitz & Stroud, 2012
Inferior cortex	Reduction in function, and presence of features characteristic of ADHD	Knoch et al., 2006
Orbitofrontal, mediofrontal, and parahippocampal cortices	Thinned to a greater extent in females than males	Toro et al., 2008
Corpus callosum	Smaller size in females only	Paus et al., 2008

Regions of the human brain that have been found to be altered at different stages of ontogeny in association with prenatal tobacco smoke exposure.

thinning of the orbitofrontal, mediofrontal, and parahippocampal cortices was more pronounced in exposed female adolescents than that seen in age-matched, prenatally exposed males (Toro et al., 2008).

Animal models of prenatal nicotine exposure

Studies on effects of gestational smoking on behavior have been conducted using rodent models of prenatal nicotine exposure (PNE), and PNE models have shown alterations in behavior. However, PNE models vary, which complicates comparisons of effects of PNE on behavior across studies, as well as the translational significance of the findings relative to exposure effects seen in humans (Polli & Kohlmeier, 2020). However, with these caveats in mind, behavioral testing of PNE rodents (Fig. 15.3) has mostly revealed associations with deficits in executive control, hyperactivity, and emotional disruptions, which can also exhibit sex-dependent outcome differences.

Prenatal nicotine exposure is associated with alterations in cognitive- and emotional-related behaviors

In Sprague-Dawley rats, PNE was associated with anxiogenic-like behavior in male and female offspring when evaluated in the elevated plus maze, with females presenting a more severe phenotype than males (Vaglenova, Birru, Pandiella, & Breese, 2004). Conversely, using the same PNE model, in which nicotine exposure occurred via osmotic minipumps, anxiolytic-like behavior was detected in both sexes in the same behavioral task (Sobrian, Marr, & Ressman, 2003). Such an opposite effect using the same PNE animal model could be explained by the different ages of the offspring employed in

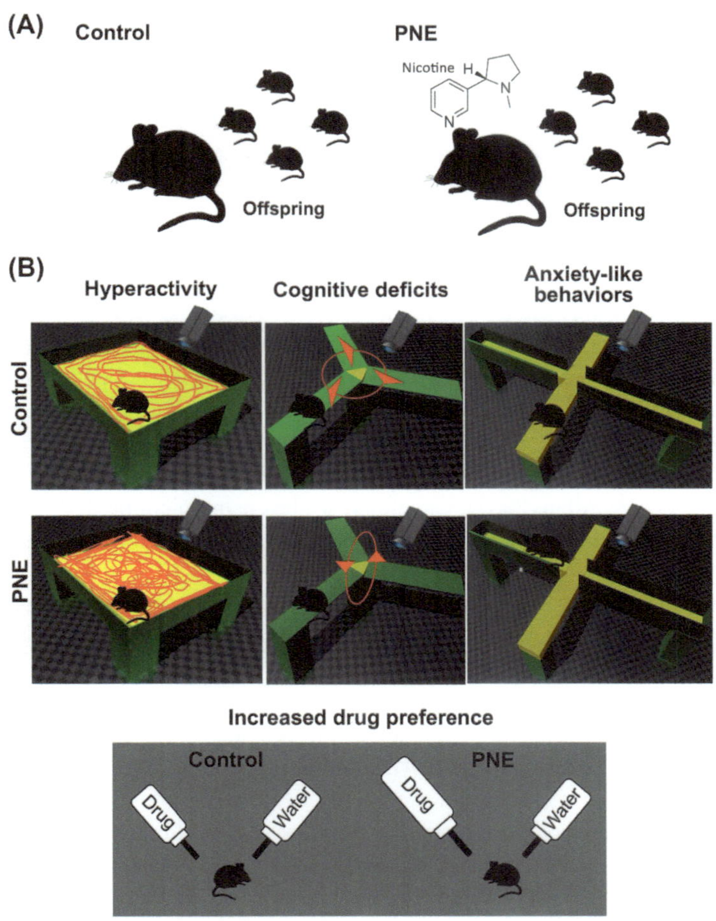

FIGURE 15.3 **Behavioral tests are employed to evaluate deficits associated with prenatal nicotine exposure (PNE).** (A) Offspring from mice who received nicotine during pregnancy have served as animal models of PNE. (B) Among other tests, the open arena, Y-maze, elevated plus maze, and drug preference have been utilized to evaluate hyperactivity, cognitive deficits, anxiety, and alterations in motivated behaviors in FNE offspring at different developmental ages.

these studies, or it could be explained by the heightened impulsive, risk-taking behavior apparent in the PNE offspring, thereby counteracting or masking anxious or emotionally based cautionary behaviors assessed in spatial-based tasks (Sobrian et al., 2003). Using another model of PNE in which nicotine was present in the drinking water during late pregnancy in Sprague-Dawley rats, offspring exhibited anxiogenic behavior when evaluated in the light/dark box test (Lee, Chung, & Noh, 2016).

PNE studies in mice have also reported inconsistent results in the association of PNE with anxiogenic-like behaviors. In C57/BlB6 mice in which nicotine was introduced in the drinking water late in pregnancy, offspring spent a reduced time in the light compartment in the light/dark box test; however, this anxiogenic behavior was only seen in males (Alkam, Kim, Hiramatsu, et al., 2013a), whereas e-cigarette vapor exposure of pregnant dams at similar gestational periods resulted in offspring that exhibited only a trend toward reduced latency to enter the light room (Smith et al., 2015). Moreover, by employing the elevated zero maze task, the later study found no differences in time spent in the open maze quadrants (Smith et al., 2015). Interestingly, PNE male, but not female offspring, also displayed anxiogenic-like behaviors when an outbred mouse model was used (Polli et al., 2020a). These findings likely vary due to the later time of nicotine exposure during pregnancy, since a study investigating the PNE effects when nicotine was administered at different gestational periods demonstrated lower anxiety-related outcomes when nicotine exposure was restricted to late pregnancy, whereas anxiety-like behaviors in the elevated plus maze and marble burying tests were seen in both males and female offspring exposed during the entire pregnancy (Alkam et al., 2013a).

Further, only a few studies have assessed social behavior in rodents after gestational nicotine exposure, but both males and female PNE mice were found to show heighted avoidance of social interaction with a stranger mouse in a social novelty preference paradigm (Alkam et al., 2013a), and PNE was associated in male offspring with an increased aggression behavior in a resident-intruder task by displaying decreased latency to, and an enhanced number of, attacks (Yochum et al., 2014).

Due to the clinical associations of maternal smoking with a heightened risk of development of ADHD, several PNE experimental models have assessed locomotor behavior in rodents, in which conflicting results have been reported. PNE treatment using minipumps was shown to increase motor activity in juvenile rats, an effect only seen in male offspring (Schlumpf, Gahwiler, Ribary, & Lichtensteiger, 1988). However, in subsequent studies, juvenile PNE rats of both sexes tested at postnatal day (PND) 14–21 displayed increased spontaneous locomotor activity (Fung, 1988; Tizabi, Russell, Nespor, Perry, & Grunberg, 2000). When assessing PNE effects on rats at PND25, a sex-based difference was detected, in that hyperactivity was only associated with males (Vaglenova et al., 2004). In a PNE model in which pregnant mice were exposed to nicotine via the drinking water, male PNE offspring displayed increased horizontal activity and stereotypy behavior at PND40 and PND60, while hypolocomotion was seen in PNE female mice at PND20 (Pauly, Sparks, Hauser, & Pauly, 2004). At adulthood (PND90), this same PNE model was not associated with locomotor changes or differences in anxiety as assayed by the elevated plus maze in males or females (Zhang, Spencer, Biederman, & Bhide, 2018). Furthermore, deficits in attention and impulsivity were found in PNE male offspring using an outbred mouse model in the rodent continuous performance task (Polli et al., 2020b). Increased locomotor activity was seen in adult PNE mice in both sexes in which nicotine solutions were given to pregnant dams together with nicotine-free water, which suggests that nicotine effects during pregnancy are not driven by any putative stress due to forced nicotine intake in this PNE model (Paz, Barsness, Martenson, Tanner, & Allan, 2007). Generally, studies using nicotine delivery via maternal drinking water have reported a more consistent association of PNE with hyperactivity relative to other methods of gestational nicotine exposure.

In addition, effects of prenatal cigarette smoke exposure on behavior have been examined, which is interesting as this exposure mimics human exposure, but interpretation of the findings is confounded by the failure in the ability of isolating the effects of nicotine itself, since several teratogenic compounds are present in inhaled tobacco smoke. Smoke exposure was associated with increased locomotor activity only in male offspring when assessed at both adolescence and adulthood, which are similar findings in rats and mice with studies exposing pregnant mice to nicotine via the drinking water (Paz et al., 2007; Schneider et al., 2002; Yochum et al., 2014). Finally, in an attempt to isolate nicotine effects from those of vapor during development, exposure to nicotine and nicotine-free vapor was conducted in male PNE offspring, which attributed heightened locomotor activity including more distance traveled and higher rates of rearing behaviors to nicotine, but the nicotine exposure was delayed until gestational day (GD) 15 and prolonged through postnatal day (PND) 16 in that investigation (Smith et al., 2015). Overall, while data from the literature do contain some confounds, the majority of PNE studies conclude that nicotine during gestation is associated with hyperlocomotion, particularly in males.

As deficiencies in cognitive function are one of the most prominent hallmarks found in ADHD patients, PNE studies have also been conducted to understand the association of early-life nicotine exposure in memory, learning, and attentional performances by using different rodent cognitive tasks. Nicotine delivery through the drinking water of pregnant rats detected impairment of correct choices in an 8-radial-arm maze in both male and female PNE young adult rats (Sorenson, Raskin, & Suh, 1991). Decreased performance in the radial-arm maze and water maze was also found in PNE young adult mice in which nicotine was delivered by subcutaneous injection to pregnant dams (Yanai, Pick, Rogel-Fuchs, & Zahalka, 1992). However, conflicting findings have been reported using this model, as only minor PNE effects on radial-arm maze performance have been reported in other studies, leading to the conclusion that age variability in subjects could explain differential outcomes (Levin, Briggs, Christopher, & Rose, 1993, 1996). Cognitive performance in spatial working memory and attention as assayed by the Y maze and object attention tasks were decreased in PNE males exposed via maternal drinking water, whereas no effects were seen in these parameters in female offspring from the same litters, although sex-dependent differences in response to relatively low nicotine levels could exist (Zhang, Spencer, Biederman, & Bhide, 2018). In the 5-choice serial reaction time task (5-CSRTT), young adult rats prenatally exposed to nicotine via maternal consumption of nicotine-containing drinking water exhibited an increased number of anticipatory responses, omission errors, more variable response times, and lower accuracy when the stimulus display time was reduced, and these effects were observed in both sexes, associating PNE treatment with impulsive-like behavior and attentional deficits (Schneider et al., 2011, Schneider, Bizarro, Asherson, & Stolerman, 2012). The time window during which nicotine is present during gestation can alter the outcome on later-life behavior. In C57BL/6J mice, cognitive performance was assessed in the Y-maze task by examination of spontaneous alternate behavior (SAB) in PNE cohorts who differed only in the time during gestation at which exposure to nicotine occurred via the drinking water (Alkam, Kim, Mamiya, et al., 2013b). In this study, impaired performance in SAB in the Y-maze was found in young PNE offspring of both sexes only when nicotine was present during GD 0-PND 0 or GD14-PND0, whereas nicotine exposure from GD0-PND13 did not impair performance, suggesting different mechanisms by which nicotine alters working memory according to the time window of its presence during the pre- or postnatal periods (Alkam et al., 2013a,b).

A limited number of experimental investigations have studied PNE effects on the susceptibility to later drug dependency. A study using nicotine exposure through maternal drinking water showed that an increased nicotine preference was exhibited by young adult PNE mice (Klein, Stine, Pfaff, & Vandenbergh, 2003). Adolescent PNE mice were shown to display increased voluntary nicotine intake and to exhibit higher nicotine preference than controls (Buck et al., 2019). Shockingly, but powerfully indicative of the epigenetic effects of nicotine, nicotine preference extended across multiple generations as offspring of first-generation PNE females, who were not exposed otherwise to nicotine, also exhibited nicotine preference (Buck et al., 2019). Using the intravenous (*i.v.*) route of nicotine delivery in pregnant rats, PNE was associated with increased nicotine self-administration, but this effect was only seen following a nicotine-abstinence period (Levin et al., 2006). Further, using nicotine minipumps during GD4-18, PNE rats tested on PND32 for operant responding with *i.v.* cocaine injections exhibited a larger cocaine intake at the highest dose, with a concurrently reduced intake of sucrose pellets, leading to the suggestion that PNE could elicit changes in both natural and drug reward pathways (Franke, Park, Belluzzi, & Leslie, 2008). Increased cocaine place preference, a behavioral test indicative of motivation for experience of drug, was also exhibited by both sexes of PNE mice using the model in which nicotine was delivered via maternal drinking water and offered together with nicotine-free water in order to eliminate any stress induced by forced nicotine choice (Paz et al., 2007). Finally, a study using *i.v.* nicotine delivery in pregnant rats from G8-G21 found that both male and female PNE offspring during adulthood exhibited more methamphetamine-maintained responses and self-infusions at high doses, indicative of a higher motivation to self-administer methamphetamine (Lacy, Brown, Morgan, Mactutus, & Harrod, 2016). While there are many inconsistencies in the animal data, likely due to differences in methods or in the models utilized, when taken together, data from animal studies corroborate associations seen in clinical investigations, and the majority of literature suggests that PNE leads to changes in development of regions within locomotive, mood control, and cognitive centers of the brain, which are importantly involved in executive control, and these changes would predict the negative behavioral outcomes seen in individuals following gestational nicotine exposure.

Conclusion

Exposure of a fetus to nicotine has been associated with negative outcomes, including maladaptive behaviors, which can be sex-specific, and are suggestive of alterations in development within regions of the brain important in executive function. This is supported by structural and functional imaging studies in infants that differences in brain development are apparent and that these differences can occur in a gender-dependent fashion. While interpretation of data from animal PNE studies is complicated by differences in the animal models and experimental designs, nicotine does appear to be associated with changes in behaviors similar to those seen in humans gestationally exposed to tobacco smoke and data support the hypothesis of sex-dependent PNE effects in the offspring. Clinical studies have associated maternal tobacco smoking during pregnancy with increased chances of development of alterations in motivated behaviors and ADHD-related behaviors late in life, and despite some conflicting findings, most rodent investigations have associated PNE with this increased susceptibility, placing nicotine, among other teratogenic compounds present in tobacco smoke, as an important substance underlying these risks. Future studies will have to be conducted to determine the association of alteration in behaviors with nicotine changes of the trajectory of neural development; however, based on available human and animal data, targeted efforts should be implemented to sway women from utilizing any nicotine-containing product during pregnancy.

Applications to other areas of development

In this chapter, we have reviewed the effects of the presence of nicotine in the fetal compartment during development in both animal models and epidemiological studies using outcomes in humans exposed to this compound during gestation. In animal studies, prenatal presence of nicotine was associated with alterations in brain areas and circuits implicated in cognitive and executive functioning, which likely play a role in differences seen in behavior and cognitive functioning such as hyperlocomotion, anxiety, cognitive flexibility, and motivation. We have compared these outcomes to those seen in human subjects exposed prenatally to nicotine, mostly via maternal smoking of combustible cigarettes. Outcomes in humans recapitulate to large degree the results seen in animal models and effects suggest gender-based differences as well as exhibition of a particular vulnerability in the young to the negative outcomes associated with early-life exposures. In animal models, nicotine was shown to exert teratogenic effects irrespective of the application method, suggesting that in nicotine-dependent pregnant women, nicotine replacement strategies to avoid withdrawal effects such as the patch, gum, or vaping are not an advisable strategy to avoid negative neural and behavioral outcomes associated with exposure to cigarettes during pregnancy, as there is a high likelihood that negative outcomes are due to nicotine teratogenicity.

Key facts

- Nicotine acetylcholine receptors (nAChRs) in the fetal brain appear early during gestational periods
- nAChRs mediate intercellular signals between progenitor cells associated with cellular growth, migration, and differentiation.
- nAChRs are stimulated endogenously by nicotine in the uterine compartment if the mother is exposed to nicotine.
- Early-life exposure to nicotine alters neural development.
- Early-life exposure to nicotine is associated with attention deficient hyperactivity disorder–related behaviors and alterations in motivated behaviors.

Summary points

- Nicotine exposure is present in a high number of pregnancies.
- Nicotine is a teratogen likely involved in the negative physiological and behavioral outcomes following prenatal tobacco smoke exposure.
- Prenatal nicotine exposure is associated with structural changes in the human brain.
- Prenatal nicotine exposure is associated with alterations in cognitive behaviors and executive function in animal models.
- Nicotine replacement therapy is not recommended to replace smoking during pregnancy.

Acknowledgments

The authors would like to acknowledge Astrid Sabina Kristensen for original artwork contributions to Fig. 15.3. The Philip Morris External Research Project and the Department of Drug Design and Pharmacology at the University of Copenhagen are acknowledged for funding received.

References

Abbott, L. C., & Winzer-Serhan, U. H. (2012). Smoking during pregnancy: Lessons learned from epidemiological studies and experimental studies using animal models. *Critical Reviews in Toxicology, 42*, 279–303.

Adkison, S. E., O'Connor, R. J., Bansal-Travers, M., Hyland, A., Borland, R., Yong, H. H., … Fong, G. T. (2013). Electronic nicotine delivery systems: International tobacco control four-country survey. *American Journal of Preventive Medicine, 44*, 207–215.

Alkam, T., Kim, H. C., Hiramatsu, M., Mamiya, T., Aoyama, Y., Nitta, A., … Nabeshima, T. (2013a). Evaluation of emotional behaviors in young offspring of C57BL/6J mice after gestational and/or perinatal exposure to nicotine in six different time-windows. *Behavioural Brain Research, 239*, 80–89.

Alkam, T., Kim, H. C., Mamiya, T., Yamada, K., Hiramatsu, M., & Nabeshima, T. (2013b). Evaluation of cognitive behaviors in young offspring of C57BL/6J mice after gestational nicotine exposure during different time-windows. *Psychopharmacology, 230*, 451–463.

Atluri, P., Fleck, M. W., Shen, Q., Mah, S. J., Stadfelt, D., Barnes, W., … Schneider, A. S. (2001). Functional nicotinic acetylcholine receptor expression in stem and progenitor cells of the early embryonic mouse cerebral cortex. *Developmental Biology, 240*, 143–156.

Baeza-Loya, S., Viswanath, H., Carter, A., Molfese, D. L., Velasquez, K. M., Baldwin, P. R., … Salas, R. (2014). Perceptions about e-cigarette safety may lead to e-smoking during pregnancy. *Bulletin of the Menninger Clinic, 78*, 243–252.

Bakker, H., & Jaddoe, V. W. (2011). Cardiovascular and metabolic influences of fetal smoke exposure. *European Journal of Epidemiology, 26*, 763–770.

Bar-Zeev, Y., Lim, L. L., Bonevski, B., Gruppetta, M., & Gould, G. S. (2018). Nicotine replacement therapy for smoking cessation during pregnancy. *Medical Journal of Australia, 208*, 46–51.

Brennan, P. A., Grekin, E. R., Mortensen, E. L., & Mednick, S. A. (2002). Relationship of maternal smoking during pregnancy with criminal arrest and hospitalization for substance abuse in male and female adult offspring. *American Journal of Psychiatry, 159*, 48–54.

Bublitz, M. H., & Stroud, L. R. (2012). Maternal smoking during pregnancy and offspring brain structure and function: Review and agenda for future research. *Nicotine and Tobacco Research, 14*, 388–397.

Buck, J. M., Sanders, K. N., Wageman, C. R., Knopik, V. S., Stitzel, J. A., & O'Neill, H. C. (2019). Developmental nicotine exposure precipitates multigenerational maternal transmission of nicotine preference and ADHD-like behavioral, rhythmometric, neuropharmacological, and epigenetic anomalies in adolescent mice. *Neuropharmacology, 149*, 66–82.

Buka, S. L., Shenassa, E. D., & Niaura, R. (2003). Elevated risk of tobacco dependence among offspring of mothers who smoked during pregnancy: A 30-year prospective study. *American Journal of Psychiatry, 160*, 1978–1984.

Cho, K., Frijters, J. C., Zhang, H., Miller, L. L., & Gruen, J. R. (2013). Prenatal exposure to nicotine and impaired reading performance. *The Journal of Pediatrics, 162*, 713, 718 e712.

Clifford, A., Lang, L., & Chen, R. (2012). Effects of maternal cigarette smoking during pregnancy on cognitive parameters of children and young adults: A literature review. *Neurotoxicology and Teratology, 34*, 560–570.

Cornelius, M. D., Goldschmidt, L., & Day, N. L. (2012). Prenatal cigarette smoking: Long-term effects on young adult behavior problems and smoking behavior. *Neurotoxicology and Teratology, 34*, 554–559.

Daseking, M., Petermann, F., Tischler, T., & Waldmann, H. C. (2015). Smoking during pregnancy is a risk factor for executive function deficits in preschool-aged children. *Geburtshilfe und Frauenheilkunde, 75*, 64–71.

De Genna, N. M., Goldschmidt, L., Day, N. L., & Cornelius, M. D. (2017). Prenatal tobacco exposure, maternal postnatal nicotine dependence and adolescent risk for nicotine dependence: Birth cohort study. *Neurotoxicology and Teratology, 61*, 128–132.

Dwyer, J. B., McQuown, S. C., & Leslie, F. M. (2009). The dynamic effects of nicotine on the developing brain. *Pharmacology and Therapeutics, 122*, 125–139.

Farsalinos, K. (2018). Electronic cigarettes: An aid in smoking cessation, or a new health hazard? *Therapeutic Advances in Respiratory Disease, 12*, 1753465817744960.

Fitzpatrick, C., Barnett, T. A., & Pagani, L. S. (2014). Parental bad habits breed bad behaviors in youth: Exposure to gestational smoke and child impulsivity. *International Journal of Psychophysiology, 93*, 17–21.

Franke, R. M., Park, M., Belluzzi, J. D., & Leslie, F. M. (2008). Prenatal nicotine exposure changes natural and drug-induced reinforcement in adolescent male rats. *European Journal of Neuroscience, 27*, 2952–2961.

Fung, Y. K. (1988). Postnatal behavioural effects of maternal nicotine exposure in rats. *Journal of Pharmacy and Pharmacology, 40*, 870–872.

Hall, E. S., Wexelblatt, S. L., & Greenberg, J. M. (2016). Self-reported and laboratory evaluation of late pregnancy nicotine exposure and drugs of abuse. *Journal of Perinatology, 36*, 814–818.

Holz, N. E., Boecker, R., Baumeister, S., Hohm, E., Zohsel, K., Buchmann, A. F., … Laucht, M. (2014). Effect of prenatal exposure to tobacco smoke on inhibitory control: Neuroimaging results from a 25-year prospective study. *JAMA Psychiatry, 71*, 786–796.

Jacobsen, L. K., Slotkin, T. A., Mencl, W. E., Frost, S. J., & Pugh, K. R. (2007). Gender-specific effects of prenatal and adolescent exposure to tobacco smoke on auditory and visual attention. *Neuropsychopharmacology, 32*, 2453–2464.

Jauniaux, E., Gulbis, B., Acharya, G., Thiry, P., & Rodeck, C. (1999). Maternal tobacco exposure and cotinine levels in fetal fluids in the first half of pregnancy. *Obstetrics and Gynecology, 93*, 25–29.

Kalkhoran, S., & Glantz, S. A. (2016). E-cigarettes and smoking cessation in real-world and clinical settings: A systematic review and meta-analysis. *Lancet Respiratory Medicine, 4*, 116–128.

Klein, L. C., Stine, M. M., Pfaff, D. W., & Vandenbergh, D. J. (2003). Maternal nicotine exposure increases nicotine preference in periadolescent male but not female C57B1/6J mice. *Nicotine and Tobacco Research, 5*, 117–124.

Knoch, D., Gianotti, L. R., Pascual-Leone, A., Treyer, V., Regard, M., Hohmann, M., & Brugger, P. (2006). Disruption of right prefrontal cortex by low-frequency repetitive transcranial magnetic stimulation induces risk-taking behavior. *Journal of Neuroscience, 26*, 6469–6472.

Lacy, R. T., Brown, R. W., Morgan, A. J., Mactutus, C. F., & Harrod, S. B. (2016). Intravenous prenatal nicotine exposure alters METH-induced hyperactivity, conditioned hyperactivity, and BDNF in adult rat offspring. *Developmental Neuroscience, 38*, 171–185.

Lambers, D. S., & Clark, K. E. (1996). The maternal and fetal physiologic effects of nicotine. *Seminars in Perinatology, 20*, 115–126.

Lange, S., Probst, C., Rehm, J., & Popova, S. (2018). National, regional, and global prevalence of smoking during pregnancy in the general population: A systematic review and meta-analysis. *Lancet Global Health, 6*, e769–e776.

Lee, H., Chung, S., & Noh, J. (2016). Maternal nicotine exposure during late gestation and lactation increases anxiety-like and impulsive decision-making behavior in adolescent offspring of rat. *Toxicology Research, 32*, 275–280.

Levin, E. D., Briggs, S. J., Christopher, N. C., & Rose, J. E. (1993). Prenatal nicotine exposure and cognitive performance in rats. *Neurotoxicology and Teratology, 15*, 251–260.

Levin, E. D., Conners, C. K., Sparrow, E., Hinton, S. C., Erhardt, D., Meck, W. H., … March, J. (1996). Nicotine effects on adults with attention-deficit/hyperactivity disorder. *Psychopharmacology, 123*, 55–63.

Levin, E. D., Lawrence, S., Petro, A., Horton, K., Seidler, F. J., & Slotkin, T. A. (2006). Increased nicotine self-administration following prenatal exposure in female rats. *Pharmacology Biochemistry and Behavior, 85*, 669–674.

Lotfipour, S., Ferguson, E., Leonard, G., Miettunen, J., Perron, M., Pike, G. B., … Paus, T. (2014). Maternal cigarette smoking during pregnancy predicts drug use via externalizing behavior in two community-based samples of adolescents. *Addiction, 109*, 1718–1729.

Lotfipour, S., Ferguson, E., Leonard, G., Perron, M., Pike, B., Richer, L., … Paus, T. (2009). Orbitofrontal cortex and drug use during adolescence: Role of prenatal exposure to maternal smoking and BDNF genotype. *Archives of General Psychiatry, 66*, 1244–1252.

Motlagh, M. G., Sukhodolsky, D. G., Landeros-Weisenberger, A., Katsovich, L., Thompson, N., Scahill, L., … Leckman, J. F. (2011). Adverse effects of heavy prenatal maternal smoking on attentional control in children with ADHD. *Journal of Attention Disorders, 15*, 593–603.

O'Callaghan, F. V., Al Mamun, A., O'Callaghan, M., Alati, R., Najman, J. M., Williams, G. M., & Bor, W. (2009). Maternal smoking during pregnancy predicts nicotine disorder (dependence or withdrawal) in young adults - a birth cohort study. *Australian and New Zealand Journal of Public Health, 33*, 371–377.

Oncken, C., Ricci, K. A., Kuo, C. L., Dornelas, E., Kranzler, H. R., & Sankey, H. Z. (2017). Correlates of electronic cigarettes use before and during pregnancy. *Nicotine and Tobacco Research, 19*, 585–590.

Pauly, J. R., Sparks, J. A., Hauser, K. F., & Pauly, T. H. (2004). In utero nicotine exposure causes persistent, gender-dependant changes in locomotor activity and sensitivity to nicotine in C57Bl/6 mice. *International Journal of Developmental Neuroscience, 22*, 329–337.

Paus, T., Nawazkhan, I., Leonard, G., Perron, M., Pike, G. B., Pitiot, A., … Pausova, Z. (2008). Corpus callosum in adolescent offspring exposed prenatally to maternal cigarette smoking. *NeuroImage, 40*, 435–441.

Paz, R., Barsness, B., Martenson, T., Tanner, D., & Allan, A. M. (2007). Behavioral teratogenicity induced by nonforced maternal nicotine consumption. *Neuropsychopharmacology, 32*, 693–699.

Polli, F. S., & Kohlmeier, K. A. (2020). Prenatal nicotine exposure in rodents: why are there so many variations in behavioral outcomes? *Nicotine and Tobacco Research, 22*(10), 1694–1710. https://doi.org/10.1093/ntr/ntz196

Polli, F. S., Scharff, M. B., Ipsen, T. H., Aznar, S., Kohlmeier, K. A., & Andreasen, J. T. (2020a). Prenatal nicotine exposure in mice induces sex-dependent anxiety-like behavior, cognitive deficits, hyperactivity, and changes in the expression of glutamate receptor associated-genes in the prefrontal cortex. *Pharmacology Biochemistry and Behavior*. https://doi.org/10.1016/j.pbb.2020.172951. In press.

Polli, F. S., Ipsen, T. H., Caballero-Puntiverio, M., Østerbøg, T. B., Aznar, S., Andreasen, J. T., & Kohlmeier, K. A. (2020b). Cellular and molecular chnages in hippocampal glutamate signaling and alterations in learning, attention, and impulsivity following prenatal nicotine exposure. *Molecular Neurobiology, 57*(4), 2002–2020. https://doi.org/10.1007/s12035-019-01854-9

Schlumpf, M., Gahwiler, M., Ribary, U., & Lichtensteiger, W. (1988). A new device for monitoring early motor development: Prenatal nicotine-induced changes. *Pharmacology Biochemistry and Behavior, 30*, 199–203.

Schneider, A. S., Atluri, P., Shen, Q., Barnes, W., Mah, S. J., Stadfelt, D., … Fleck, M. W. (2002). Functional nicotinic acetylcholine receptor expression on stem and progenitor cells of the early embryonic nervous system. *Annals of the New York Academy of Sciences, 971*, 135–138.

Schneider, T., Bizarro, L., Asherson, P. J., & Stolerman, I. P. (2012). Hyperactivity, increased nicotine consumption and impaired performance in the five-choice serial reaction time task in adolescent rats prenatally exposed to nicotine. *Psychopharmacology, 223*, 401–415.

Schneider, T., Ilott, N., Brolese, G., Bizarro, L., Asherson, P. J., & Stolerman, I. P. (2011). Prenatal exposure to nicotine impairs performance of the 5-choice serial reaction time task in adult rats. *Neuropsychopharmacology, 36*, 1114–1125.

Schwenke, E., Fasching, P. A., Faschingbauer, F., Pretscher, J., Kehl, S., Peretz, R., … Schneider, M. (2018). Predicting attention deficit hyperactivity disorder using pregnancy and birth characteristics. *Archives of Gynecology and Obstetrics, 298*, 889–895.

Shenassa, E. D., Papandonatos, G. D., Rogers, M. L., & Buka, S. L. (2015). Elevated risk of nicotine dependence among sib-pairs discordant for maternal smoking during pregnancy: Evidence from a 40-year longitudinal study. *Epidemiology, 26*, 441–447.

Slotkin, T. A. (2008). If nicotine is a developmental neurotoxicant in animal studies, dare we recommend nicotine replacement therapy in pregnant women and adolescents? *Neurotoxicology and Teratology, 30*, 1–19.

Slotkin, T. A., MacKillop, E. A., Rudder, C. L., Ryde, I. T., Tate, C. A., & Seidler, F. J. (2007). Permanent, sex-selective effects of prenatal or adolescent nicotine exposure, separately or sequentially, in rat brain regions: Indices of cholinergic and serotonergic synaptic function, cell signaling, and neural cell number and size at 6 months of age. *Neuropsychopharmacology, 32*, 1082–1097.

Smith, D., Aherrera, A., Lopez, A., Neptune, E., Winickoff, J. P., Klein, J. D., … McGrath-Morrow, S. A. (2015). Adult behavior in male mice exposed to E-cigarette nicotine vapors during late prenatal and early postnatal life. *PloS One, 10*. e0137953.

Smith, A. M., Dwoskin, L. P., & Pauly, J. R. (2010). Early exposure to nicotine during critical periods of brain development: Mechanisms and consequences. *Journal of Pediatric Biochemistry, 1*, 125–141.

Sobrian, S. K., Marr, L., & Ressman, K. (2003). Prenatal cocaine and/or nicotine exposure produces depression and anxiety in aging rats. *Progress In Neuro-Psychopharmacology and Biological Psychiatry, 27*, 501–518.

Sorenson, C. A., Raskin, L. A., & Suh, Y. (1991). The effects of prenatal nicotine on radial-arm maze performance in rats. *Pharmacology Biochemistry and Behavior, 40*, 991–993.

Thakur, G. A., Sengupta, S. M., Grizenko, N., Schmitz, N., Page, V., & Joober, R. (2013). Maternal smoking during pregnancy and ADHD: A comprehensive clinical and neurocognitive characterization. *Nicotine and Tobacco Research, 15*, 149–157.

Tizabi, Y., Russell, L. T., Nespor, S. M., Perry, D. C., & Grunberg, N. E. (2000). Prenatal nicotine exposure: Effects on locomotor activity and central [125I]alpha-BT binding in rats. *Pharmacology Biochemistry and Behavior, 66*, 495–500.

Toro, R., Leonard, G., Lerner, J. V., Lerner, R. M., Perron, M., Pike, G. B., … Paus, T. (2008). Prenatal exposure to maternal cigarette smoking and the adolescent cerebral cortex. *Neuropsychopharmacology, 33*, 1019–1027.

Vaglenova, J., Birru, S., Pandiella, N. M., & Breese, C. R. (2004). An assessment of the long-term developmental and behavioral teratogenicity of prenatal nicotine exposure. *Behavioural Brain Research, 150*, 159–170.

Weissman, M. M., Warner, V., Wickramaratne, P. J., & Kandel, D. B. (1999). Maternal smoking during pregnancy and psychopathology in offspring followed to adulthood. *Journal of the American Academy of Child & Adolescent Psychiatry, 38*, 892–899.

Yanai, J., Pick, C. G., Rogel-Fuchs, Y., & Zahalka, E. A. (1992). Alterations in hippocampal cholinergic receptors and hippocampal behaviors after early exposure to nicotine. *Brain Research Bulletin, 29*, 363–368.

Yochum, C., Doherty-Lyon, S., Hoffman, C., Hossain, M. M., Zelikoff, J. T., & Richardson, J. R. (2014). Prenatal cigarette smoke exposure causes hyperactivity and aggressive behavior: Role of altered catecholamines and BDNF. *Experimental Neurology, 254*, 145–152.

Zhang, L., Spencer, T. J., Biederman, J., & Bhide, P. G. (2018). Attention and working memory deficits in a perinatal nicotine exposure mouse model. *PloS One, 13*. e0198064.

Chapter 16

Maternal methamphetamine and impact on the developing brain

Romana Šlamberová[1] and Kateryna Nohejlová[2]

[1]*Department of Physiology, Third Faculty of Medicine, Charles University, Prague, Czech Republic;* [2]*Department of Pathophysiology, Third Faculty of Medicine, Charles University, Prague, Czech Republic*

List of abbreviations

CNS - central nervous system
DNA - deoxyribonucleic acid
GD - gestational day
MA - methamphetamine
MDMA - 3,4-methylenedioxymethamphetamine
mRNA - messenger ribonucleic acid
PD - postnatal day

Mini-dictionary of terms

Critical developmental periods Maturational stages of fetus during which developmental processes are especially sensitive to environmental stimuli.
Epigenetic factors External factors that lead to DNA modifications without changing the DNA sequence but turn genes "on" or "off" and affect how cells "read" genes.
Methamphetamine Synthetic highly addictive psychostimulant drug.
Prenatal drug exposure Fetal exposure to maternal drug abuse.
Prenatal sensitization Condition when prenatal exposure to one drug leads to an increased sensitivity to the same or another drug in later life.
Psychoactive drugs Substances of abuse that affect mental processes, as cognition or emotions.

Introduction

One of the most severe social problems of the last decades is the abuse of drugs by women during pregnancy. Since psychoactive drugs can easily cross the blood—brain barrier, the placental barrier offers little impediment to these drugs. Moreover, many drugs can pass into a nursing mother's breast milk and expose a developing infant to drugs.

Women, especially during pregnancy, are more sensitive to drugs. This is due to changes in sex hormones, primarily due to elevated levels of progesterone and estrogen (Plessinger, 1998). In this period, smaller doses of the drug are sufficient to induce euphoria. Animal studies showed gender-related behavioral differences in female rats, which showed increased locomotor activity and stereotyped amphetamine behaviors (Robinson & Becker, 1986). These are primarily explained by faster drug metabolism in males compared to females (thus more locomotor activity in females) and lower drug levels in brain structures of males (Meyerson & Malmnas, 1978). Moreover, the increased sensitivity of females is directly dependent on the phase of the estrous cycle (Becker, Robinson, & Lorenz, 1982), which can also be induced by steroid hormones.

Methamphetamine (MA) is one of the more commonly used addictive drugs in the world and of particular importance is that nearly half of MA users are women of reproductive age (Williams et al., 2003). MA has been shown to be a favorite drug among pregnant drug addicts because of its stimulant and anorectic effects, i.e., it gives them energy and prevents

Diagnosis, Management and Modeling of Neurodevelopmental Disorders. https://doi.org/10.1016/B978-0-12-817988-8.00016-6

weight gain (Klee, 1997). Statistical studies have found that only 17% of drug addicts in the United States used MA as their primary drug of abuse, but this increased to 38% during pregnancy. This is worrisome since MA can cross the placental barrier and reach the fetus. The concentration of MA in the fetus usually reaches 50% of the drug concentration circulating in the mother. However, fetal concentrations may increase (even exceeding the concentration in the mother's body) and persist longer due to slower MA degradation in the fetal liver. Therefore, pregnant women abusing MA expose not only themselves but also their developing offspring to the risks associated with MA (Marwick, 2000). MA affects fetal development, both directly and indirectly. The indirect effect is associated with placental and umbilical abnormalities that decrease the blood supply to the fetus, which can lead to hypoxia and malnutrition and can cause intrauterine growth retardation of the fetus (Vavřínková, Binder, Vítková, & Živný, 2001a). Additionally, MA use can lead to a decline in prenatal care and an increase in prenatal stress, both of which can impact fetal development.

Clinical studies

Effects of MA exposure during critical prenatal periods have been studied and recorded in clinical studies for decades. The most remarkable and descriptive findings are summarized in Table 16.1.

Clinical studies, however, have some limitations. It is usually not possible to run prospective studies for long periods of time, due to the problems associated with tracking a sufficiently large group of children (from drug-dependent mothers) from birth into adulthood. Additional problems are related to drug-abusing mothers being less than conscientious regarding prenatal and postnatal care of their children often neglecting regular visits to their physician (Vavřínková, Binder, & Živný, 2001b). In addition, no clear conclusions can be drawn since pregnant mothers often use or abuse other substances such as alcohol and nicotine as well as MA. The abuse of multiple dependence-inducing substances is a common problem in monitoring drug effects in humans since it is virtually impossible to distinguish the effects of individual addictive substances. Moreover, drug-induced changes in the placenta can lead to decreased placental perfusion, due to vasoconstriction,

TABLE 16.1 Summary of prenatal methamphetamine exposure adverse effects in humans.

Manifestations	Description of abnormality	Clinical study
Altered general development	Lower birth weight, length, and head circumference	(Oro & Dixon, 1987)
	Cardiac defects, cleft lip, and biliary atresia	(Plessinger, 1998)
Structural CNS abnormalities	Intraventricular hemorrhages and necroses in the basal ganglia and frontal lobe	(Dixon & Bejar, 1989) *abnormalities were found in 35% of cases compared to only 5% in healthy newborns*
	Lower putamen, pallidum, and hippocampus volumes	(Chang et al., 2004)
Functional CNS deficits and altered patterns of behavior	Abnormal sleep, milk rejection, tremor, and hypertension in neonates	(Oro & Dixon, 1987)
	Impaired visual memory test results	(Hansen, Struthers, & Gospe, 1993)
	1st months of life – increased sleepiness Within 1st year – symptoms of autism At the age of 4 – average IQ is lower At the age of 8 – aggressive behavior and problems with peers At the age of 14 – lagging behind at school due to difficulties with mathematics and language; difficulties with physical activities	(Cernerud, Eriksson, Jonsson, Steneroth, & Zetterstrom, 1996)
	Signs of increased stress, increased lethargy, and impaired locomotion in neonates	(Smith et al., 2008)
Other, related to mother	Increased incidence of premature births and stillbirth	(Eriksson, Larsson, Winbladh, & Zetterstrom, 1978)

Main adverse effects of exposure to methamphetamine during prenatal period are classified into several groups according to clinical studies.

and poor nutrition (Vavřínková et al., 2001a), which can also negatively affect newborns. Regardless of the mechanism, MA abuse during pregnancy is a risk to the fetus, although, but unlike fetal alcohol or cocaine syndrome, no specific syndrome associated with MA abuse in the prenatal period has been described. At present, scientific research in humans is mostly limited to statistical comparisons.

Experimental studies

There are studies showing the pharmacokinetics of MA in rodents (Cappon & Vorhees, 2001) and in humans (Segal & Kuczenski, 2006). The 5 and 10 mg/kg dosages of MA for female laboratory mice or rats were determined as an experimental model for the potential risks of MA exposure during the prenatal period. Drug concentrations in the brains of these females correspond to those found in maternal-dependent fetuses (Acuff-Smith, Schilling, Fisher, & Vorhees, 1996; Cho, Lyu, Lee, Kim, & Chin, 1991; Martin, Martin, Radow, & Sigman, 1976).

In a high-performance liquid chromatography study by Rambousek et al., (2014), gender differences in the pharmacokinetics of MA (5 mg/kg) in male and female rats were examined. Chronic exposure of human prenatal infants to MA was compared to the brain and plasma levels of MA and amphetamine in dams and pups. Exposure of MA in milk during breastfeeding was simulated by repeated administration of MA to rat mothers, then the amounts MA as well as amphetamine (a metabolite of MA), via breast milk, in rat pups were analyzed. Results showed significantly higher levels of MA and amphetamine in the plasma and brain of female rats compared to male rats, indicating a greater risk of addiction and toxicity in females. Additionally, brain concentrations of MA and amphetamine in pups exposed to MA prenatally were equivalent to 62.13% and 37.78% relative to dams, respectively. Plasma concentrations of amphetamine were equal to 100% of the concentration in MA-treated dams, while MA was equal to only 36.98%.

The experimental model of prenatal MA exposure (5 mg/kg) showed results similar to clinical studies that examined the children of MA-addicted mothers (Table 16.2).

The effect of methamphetamine on maternal care

Experimental administration of MA (5 mg/kg) to female rats during gestation was shown to impair maternal behavior (Šlamberová, 2012). Acuff-Smith et al. (1996) studied the effect of MA on the weight of mothers and pups and found lower weights for both, compared to controls fed ad libitum as well as a group with the same food intake as the drug-exposed animals. Thus, the lower body weight was not caused only by lower nutritional intake and decreased appetite.

TABLE 16.2 Summary of prenatal methamphetamine exposure adverse effects in animal models.

Manifestations	Description of abnormality	Animal model study
Altered general development	Increased pup mortality, reduced pup weight gain, delayed development, and slower reflexes	(Acuff-Smith et al.. 1996; Cho et al., 1991; Martin et al., 1976)
	Affected the second generation of animals	(Šlamberová, Pometlová, & Rokyta, 2007)
Functional CNS deficits and altered patterns of behavior	Persistent behavioral modifications through a deterioration in learning and memory	(Cho et al., 1991)
	Changes in behavior of adult male and female rats associated with stress and anxiety, learning and memory, nociceptive sensitivity, and susceptibility to epileptic seizures	(Rokyta et al., 2008)
Other, related to mother	Shorter gestation periods, less weight gain during pregnancy, and fewer pups per litter than in control mothers	(Martin et al., 1976)
	Increased risk of spontaneous abortion and maternal death	(Acuff-Smith, George, Lorens, & Vorhees, 1992)

Main adverse effects of exposure to methamphetamine during prenatal period have been studied in laboratory rat models. Rat mothers received repeated subcutaneous methamphetamine through pregnancy and/or during lactation periods.

One of the mechanisms that might explain the effect on fetal weight is the vasoconstrictor effect on the uteroplacental blood supply, which reduces the supply of nutrients to the fetus. To differentiate the effect of vasoconstriction, anorexia, and maternal hypoxia on the developing fetus, Yamamoto, Yamamoto, Fukui, and Kurishita (1992) isolated 10.5-day-old rat embryos and cultivated them in the presence of different concentrations of MA. They found that the drug had a dose-related effect on yolk sac diameter and the incidence of malformations and growth disorders. Other studies have found that high doses (50 mg/kg) of prenatal MA lead to eye disorders such as anophthalmia, microphthalmia, and retinal detachment (Acuff-Smith et al., 1996).

When talking about delayed behavioral development, sensorimotor functions have been shown to be most affected (Šlamberová, 2012). Studies on this topic have shown delayed development using the Righting reflex, the Negative Geotaxis test, which evaluates the vestibular and proprioceptive functions, and the Rotarod and Bar-holding test, which assesses the development of cerebellar coordination (Acuff-Smith et al., 1996; Šlamberová, 2012).

There are several epigenetic factors that may influence the long-term changes induced by early MA exposure (Table 16.3). The impairing effect of maternal MA exposure on pup development can be partially suppressed by cross-fostering to control dams, which shows the positive impact of good postnatal foster care (Hrubá, Schutová, Šlamberová, Pometlová, & Rokyta, 2009). Specifically, animals that were raised by MA-exposed mothers during lactation, i.e., had postnatal exposure to MA via breast milk, also displayed delayed development. On the other hand, animals that were exposed to MA prenatally, but were raised by control mothers scored better on tests than their maternal siblings with postnatal MA exposure via breastmilk. The results of studies listed earlier and in Table 16.3 support the hypothesis that maternal care can fundamentally affect functional changes in prenatally MA-exposed animals, both in the near and long term, and could serve as a mechanism for a nongenomic, behavioral, mode of transmission of traits.

On the other hand, maternal separation and maternal physical stress aggravate cognitive impairment and response to stressors in the exposed offspring (Table 16.3). For most species, the mother represents the prime source of food, warmth, protection, and education essential for the development of normal social skills (Kuhn & Schanberg, 1998). There are

TABLE 16.3 Factors affecting fetal and pups development aggravated by methamphetamine.

Effectors	Examples	Possible mechanisms	Study
Epigenetic factors	Postnatal foster care	Maternal licking and grooming, as tactile stimulation affect somatic growth and neural development, may be absent in rat mothers exposed to MA	(Levine, 1994)
	Nutrition	Deficiency of nutrients essential for adequate development	(de Souza, Fernandes, & do Carmo, 2011)
	Environmental conditions	Decrease physical activity in neglected pups from MA-exposed mothers plays role in cognitive disturbances	(van Praag, 2009)
Pre-, peri-, or early postnatal stress	Maternal separation and maternal physical stress	Social and maternal physical stress aggravates cognitive impairment and response to stressors in the exposed offspring	(Holubová et al., 2017)
	Exposure to stress	It may cause delayed development of the pups directly	(Mesquita et al., 2007)
		Increased activity of the hypothalamus–pituitary–adrenal axis over the long term; behavioral changes; increased potential for drug addiction in adulthood	(Plotsky et al., 2005)
	Maternal separation	Prolonged lack of contact between a mother and her infants soon after birth results in significant deficits in maternal attachment	(Kuhn & Schanberg, 1998)
		The lack of maternal care means almost certain death by predation or starvation for the newborns	

Maternal behavior changes with repeated administration of methamphetamine. This has direct and indirect effects on fetal and pups development summarized in the table. *MA*, methamphetamine.

studies showing that the combination of perinatal stress and prenatal MA can have a detrimental effect on maternal behavior as well as on sensorimotor development in pups. However, MA exposure during pregnancy seems to be a decisive factor regarding impairment (Holubová et al., 2017). Thus, postnatal care, as well as pre-, peri-, and postnatal stress, plays an essential role in the healthy development of an individual.

Critical developmental period

In rats, neurogenesis occurs mostly in the cortical and subcortical regions between prenatal (gestational) day (GD) 9.5 and postnatal day (PD) 15, in humans, cortical neurogenesis continues up to 2.5 years of age (Semple, Blomgren, Gimlin, Ferriero, & Noble-Haeusslein, 2013). The striatum and basal ganglia, structures important for motor activity, begin to form during GD 12–13 in rats and at 5 weeks of gestation in humans. Neurogenesis of limbic system (hippocampus, amygdala) structures, important for learning, memory, emotion, and behavior processes, occur in rats between GD 14–17, and in humans, it culminates around the eighth week of gestation (Rice & Barone, 2000). At about that same time, the first neurotransmitter cells gradually begin to appear (Rice & Barone, 2000). The main neurotransmitter systems associated with the effects of psychostimulant drugs are the catecholamines (noradrenaline, dopamine) and indoles (serotonin).

Brain development continues with nerve cell maturation. Cells aggregate and create synaptic connections relatively quickly, but initially, lack the functions typical of mature synapses. The peak of neurogenesis is estimated to occur at 40 weeks of gestation in humans and at PD 7–14 in rats (first and second week postnatally) (Clancy, Finlay, Darlington, & Anand, 2007). A study by Bockhorst et al. (2008) suggests that the most significant structural changes in the gray matter of rats occur during the first 5 days of postnatal development.

One possibility is to compare the development of human and rat for individual parts of the CNS (as aforementioned studies), and the second possibility is to generalize the comparison of rat development with human trimesters. Della Grotta et al. (2010) reported that 84.3% of pregnant, drug-addicted women use MA during the first trimester, 56.0% in the second trimester, and 42.4% continue to use MA during the third trimester. From experimental animal studies examining the impact of psychostimulants on the development of offspring of drug-exposed mothers, it is clear that the timing of MA exposure during gestation and/or lactation significantly affects the degree of drug effect on the offspring of these mothers. Amphetamines, including MA, affect neuronal endings during CNS development and interfere with the maturation of brain structures, which begins during the second half of the rat gestation period and continues until weaning (Andersen, 2005). Therefore, applications during the early stages of pregnancy may not affect fetal brain development or cause functional changes in structure; however, application during the later stages of prenatal or postnatal development, i.e., during development and maturation of neurotransmitter systems, may be harmful since these stages are more sensitive to drug effects (Rice & Barone, 2000; Williams, Moran, & Vorhees, 2003b).

Influence on future generations

Often the question is raised as to whether the mother's drug addiction can affect future generations. Thus, whether, for example, the offspring of drug-addicted mothers are more prone to drugs and addiction, i.e., whether the "like mother-like daughter" rule applies and whether the second generation of the offspring of a drug addict, i.e., her grandchildren, will be affected. Another critical question is whether the main factor is a change in the genome or if it is epigenetic factors that play the main role.

There are studies (Becker & Randall, 1987; Lam, Homewood, Taylor, & Mazurski, 2000; Šlamberová, Pometlová, & Rokyta, 2007) showing that drug addiction during gestation can also affect the second generation of offspring, i.e., the generation that was not exposed to the drug at all. For psychostimulants, data shows that second-generation rat pups of MA-treated mothers have impaired sensorimotor development based on righting reflexes and the rotarod and bar-holding tests (Šlamberová et al., 2007). These findings are supported by data of other studies showing impairment of the righting reflex in the second generation of mothers that received morphine or ethanol (Lam et al., 2000; Šlamberová, Riley, & Vathy, 2005). It appears that both genetic and epigenetic factors may play a role in these long-term effects. Group of Meaney (Champagne & Meaney, 2001) repeatedly showed that there is a transgenerational rule "like mother-like daughter." They found that female pups of rat mothers that displayed increased maternal care were also "better mothers" in their adulthood and vice versa. However, other studies (Šlamberová et al., 2007) cast doubt upon this idea.

As for changes in genetic information, a study by Lam et al. (2000) already found neuromotor deficits in second-generation rats that were induced by prenatal alcohol exposure, which suggests a role for altered oocyte mitochondrial

TABLE 16.4 Methamphetamine and other psycho stimulants induce damage or changes in DNA.

	Effect on DNA	Study
MA (at a single dose of 20 or 40 mg/kg) injected into pregnant mice	Induction or oxidative damage to the DNA in embryonic and fetal brains of offspring	(Jeng, Wong, Ting, & Wells, 2005)
MA and other psychostimulants induce changes in genes	Genes for serotonin, dopamine, and vesicular monoamine transporters	(Biezonski & Meyer, 2010)
	Genes for dopamine developmental factors, *Shh*, *Lmx1b*, and *Nurr1*	(Krasnova, Ladenheim, Hodges, Volkow, & Cadet, 2011)
	Molecules related to sensitization, such as the MA-inducible gene *mrt1*	(Fujiyama, Kajii, Hiraoka, & Nishikawa, 2003)
	Plasticity-related genes *arc* and *homer1a*	(Martin, 2001); (Thomas, 2002)
	Prodeath and antideath *Bcl-2* genes	(Jayanthi, Deng, Bordelon, McCoy, & Cadet, 2001)
	Small heat shock protein gene *HspB2*, and brain-derived neurotrophic factor (BDNF) gene	(Cadet et al., 2009)

Prenatal psychostimulant exposure induces mutagenic effects observed in the series of studies. These effects may then lead to impaired neurodevelopment in the second generation by so-called transgenerational rule "like mother-like daughter." *MA*, methamphetamine.

DNA. Because mitochondrial DNA is inherited exclusively from the mother and since oocyte maturation begins early in fetal life, prenatal drug administration may induce mutagenic effects in the oocytes of a developing, drug-exposed fetus. These effects may then lead to impaired neurodevelopment in the second generation. This hypothesis is supported by the recent works summarized in Table 16.4. Moreover, the most recent study from Zoubková, Tomášková, Nohejlová, Černá, & Šlamberová (2019) identified susceptible genes candidates for a study of impairment related to MA addiction in specific regions of the brain.

The effect of prenatal sensitization

Sensitization or increased sensitivity is defined as the progressive response of an organism to repeated drug administration at the same or even lower doses that manifests as an increased expression of a specific type or pattern of behavior (Suzuki, Fukuoka, Mori, Miyatake, & Narita, 2004). Various forms of behavior have been described in animals as manifestations of increased sensitivity to the drug after repeated intermittent drug administration. In the case of rats, it mainly appears as an increase in locomotion and exploratory activity or the development of stereotyped behaviors (Malanga & Kosofsky, 2003). Once animals are sensitized, the increased susceptibility to increased psychomotor activity lasts from months to years (Robinson & Becker, 1986; Suzuki et al., 2004).

The neuronal background of increased susceptibility to drugs seems to have phases of development: the first phase is associated with short-term changes at the molecular and cellular level in the ventral tegmental area, while sensitization represents permanent adaptations at the level of motivational pathways. From a neurochemical and neuroanatomical point of view, the aforementioned mesolimbic, mesocortical, and nigrostriatal dopamine systems play key roles (Robinson & Becker, 1986).

Research on increased drug sensitivity has revealed that the use of one drug can lead to an increased sensitivity to another drug. In this case, we are talking about cross-sensitivity or cross-sensitization to a drug (Shuster, Yu, & Bates, 1977). So far, cross-sensitivity has been found not only for related drugs, but also for unrelated drugs (Table 16.5).

A similar phenomenon may also be induced by prenatal MA exposure, which also means that one can talk about prenatal sensitization. Prenatal MA exposure can sensitize to the effect of MA in adulthood (prenatal sensitization) or to different drugs (prenatal cross-sensitization). In contrast to sensitivity testing for passive, involuntary, drug administration (Table 16.6), in studies investigating the effect of MA on active drug-seeking, using tests such as conditioned place preference and intravenous self-administration, there were no differences between prenatally MA-exposed rats and their controls (Šlamberová, Schutová, Hrubá, & Pometlová, 2011). Thus, compared to studies testing the effect of prenatal cocaine exposure (Estelles, Rodriguez-Arias, Maldonado, Aguilar, & Minarro, 2006), prenatal MA exposure appears to induce weaker sensitization to drugs in adulthood.

TABLE 16.5 Cross-sensitivity.

	Combination of drugs	Study
Cross-sensitivity for related drugs	Amphetamine and cocaine	(Horger, Giles & Schenk, 1992)
	Amphetamines and methylphenidate	(Valvassori et al., 2007)
Cross-sensitivity for unrelated drugs	Opioids and cocaine	(Leri, Flores, Rajabi, & Stewart, 2003)
	Endocannabinoids and opioids	(Vela et al., 1998)

An exposure and later increased sensitivity to one drug can lead to an increased sensitivity to another drug from both related and unrelated group. Such combinations and sensitization effect were studied thoroughly.

TABLE 16.6 Cross-sensitivity of prenatal methamphetamine.

	Combination	Studied effect	Study
Sensitization of MA	MA and MA	Increased dopamine levels in the nucleus accumbens with increased locomotion	(Bubeníková-Valešová et al., 2009)
		Increased dopamine levels in the hippocampus with changes in the cognitive function	(Fujáková-Lipski et al., 2017)
Cross-sensitivity for related drugs	MA and amphetamine	Laboras test and the Open field test – increased locomotor and exploratory activity	(Šlamberová et al., 2013)
	MA and MDMA (ecstasy) and cocaine	Sex differences relative to acute drug administration	(Šlamberová et al., 2013)
Cross-sensitivity for unrelated drugs	MA and morphine or THC	No changes were found in the sensitivity	(Šlamberová et al., 2013)

Prenatal methamphetamine exposure leads to sensitization of the tested subject to drugs applied acutely later in adulthood. The results of mentioned studies suggest that sensitization is mainly associated with the same drug in adulthood or with drugs having a similar mechanism of action. *MA*, methamphetamine; *MDMA* – 3,4, Methylenedioxymethamphetamine; *THC* – *Delta-9*, tetrahydrocannabinol.

Conclusion

The number of MA addicts is increasing all over the world, and this increase also includes pregnant women (Marwick, 2000). The effect of MA on the fetus is either through direct placental transmission and/or maternal breastmilk or indirectly through changes in uteroplacental circulation and homeostasis in the mother's body and/or epigenetic factors. Studies suggest that, inter alia, the final effect depends on the dose, the route of administration, and the gestational period during which the drug was administered. However, when comparing the results of experimental animal studies and the real impact on humans, we must always remember that the role of prenatal drug exposure is often very ambiguous due to the potential impact of the postnatal environmental on the offspring. Even though these effects are multifactorial, it does not mean that the influence of the drug as such should be taken lightly.

In summary, animal studies to date have shown that drugs administered to mothers affect maternal behavior and have a negative effect on the development of pups, which can persist into adulthood. Moreover, experimental data suggest that the negative effects of prenatal MA administration are at least partially remedied by normal postnatal care.

Applications to other areas of development

We live in hectic times, and maximum performance, as well as nearly perfect work, is expected without hesitation and regardless of our current physical and mental status. Each of us acts according to a learned pattern of behavior, and each of us has inherent dispositions. However, our social environment (i.e., parents, school, and the people around us) also strongly impacts our actions. Each person is unique in terms of the demands placed on them as well as their ability to respond to these demands. For some, the response to increased stress is depression, anxiety, and illness, while others may turn to alcohol, drugs, or other addictive substances as coping mechanisms. In today's modern, stressful environment, substance abuse and addiction are likely to be problems that will increase and expand. Because pregnancy and newborn childcare are well-established stressors, this chapter describes the mechanism of action and potential harm associated with MA administration during pregnancy and postnatal care.

Surveys and statistics have shown that drug-dependent pregnant women often switch from other hard drugs to MA because of its anorectic and stimulant effects. The issue of drug addiction and abuse is a serious problem in our society. It is necessary to realize that these future mothers not only harm themselves with their substance abuse, but they also endanger the health and development of their children. The direct effect of high-risk drugs, as well as drug-induced postpartum behavior changes, can critically affect the development of their offspring. Similar effects can also be caused by other drugs or substances that influence the psychological reward system and the CNS neurotransmitter systems (especially noradrenergic, dopaminergic, and serotoninergic).

In addition, we must not forget the epigenetic factors that can affect, not only postnatally, but also prenatally, the development of the individual. Preclinical research on this issue is essential to know the potential extent of this damage and to find new ways to mitigate it, although the best solution is to prevent drug addiction in the first place.

Key facts of methamphetamine

- Methamphetamine (a.k.a. meth, speed, cristal) is a psychostimulant abuse substance that is often ingested orally, smoked, snorted, or applied by injection.
- It is a derivative of amphetamine, which was medication of choice for treatment of depression and obesity in the United States in the 1950 and 1960s.
- Health effects of methamphetamine abuse include CNS disorders as anxiety, insomnia, paranoia, hallucinations, and structural changes to the brain.
- Use of methamphetamine during pregnancy can cause developmental disorders and growth retardation of the fetus and development of cognitive deficits in its later life.

Key acts of critical developmental period

- The term "critical developmental period" was used in experimental psychology for the first time by Charles R. Stockard in 1921.
- It refers to certain stages of embryogenesis when a temporary interruption of the developmental process can have serious consequences.
- The author of the early theory further argued that the importance of the rate of development is not limited to embryogenesis, but also occurs in postnatal ontogenesis.
- J. P. Scott in 1962 expanded on this hypothesis such that postnatal development and growth depend on organized processes that can only be influenced while they are occurring, i.e., during critical development periods.
- At last, C. Grobstein in 1970 defined critical developmental periods as the existence of stages in which the development process is somehow sensitive to environmental factors.

Summary points

- Methamphetamine is one of the most commonly used addictive drugs in the world and a favorite among pregnant drug addicts.
- Pregnant women abusing methamphetamine not only expose themselves to the effect of the drug but also their developing offspring.
- Animal studies have shown that the late pregnancy and early postnatal periods are the most critical periods for development relative to the behavioral effects of methamphetamine (both long-term and short-term).
- Maternal methamphetamine abuse can sensitize the offspring to the effects of the same or similar drugs later in life.
- Epigenetic factors can influence the consequence of maternal methamphetamine use on the developing offspring during the pre-, peri-, and early postnatal periods.

References

Acuff-Smith, K. D., George, M., Lorens, S. A., & Vorhees, C. V. (1992). Preliminary evidence for methamphetamine-induced behavioral and ocular effects in rat offspring following exposure during early organogenesis. *Psychopharmacology, 109*(3), 255–263.

Acuff-Smith, K. D., Schilling, M. A., Fisher, J. E., & Vorhees, C. V. (1996). Stage-specific effects of prenatal d-methamphetamine exposure on behavioral and eye development in rats. *Neurotoxicology and Teratology, 18*(2), 199–215.

Andersen, S. L. (2005). Stimulants and the developing brain. *Trends in Pharmacological Sciences, 26*(5), 237–243.

Becker, H. C., & Randall, C. L. (1987). Two generations of maternal alcohol consumption in mice: Effect on pregnancy outcome. *Alcoholism: Clinical and Experimental Research, 11*(3), 240–242.

Becker, J. B., Robinson, T. E., & Lorenz, K. A. (1982). Sex differences and estrous cycle variations in amphetamine-elicited rotational behavior. *European Journal of Pharmacology, 80*(1), 65–72.

Biezonski, D. K., & Meyer, J. S. (2010). Effects of 3,4-methylenedioxymethamphetamine (MDMA) on serotonin transporter and vesicular monoamine transporter 2 protein and gene expression in rats: Implications for MDMA neurotoxicity. *Journal of Neurochemistry, 112*(4), 951–962.

Bockhorst, K. H., Narayana, P. A., Liu, R., Ahobila-Vijjula, P., Ramu, J., Kamel, M., et al. (2008). Early postnatal development of rat brain: In vivo diffusion tensor imaging. *Journal of Neuroscience Research, 86*(7), 1520–1528.

Bubeníková-Valešová, V., Kačer, P., Syslová, K., Rambousek, L., Janovský, M., Schutová, B., et al. (2009). Prenatal methamphetamine exposure affects the mesolimbic dopaminergic system and behavior in adult offspring. *International Journal of Developmental Neuroscience, 27*(6), 525–530.

Cadet, J. L., McCoy, M. T., Cai, N. S., Krasnova, I. N., Ladenheim, B., Beauvais, G., et al. (2009). Methamphetamine preconditioning alters midbrain transcriptional responses to methamphetamine-induced injury in the rat striatum. *PloS One, 4*(11). e7812.

Cappon, G. D., & Vorhees, C. V. (2001). Plasma and brain methamphetamine concentrations in neonatal rats. *Neurotoxicology and Teratology, 23*(1), 81–88.

Cernerud, L., Eriksson, M., Jonsson, B., Steneroth, G., & Zetterstrom, R. (1996). Amphetamine addiction during pregnancy: 14-year follow-up of growth and school performance. *Acta Paediatrica, 85*(2), 204–208.

Champagne, F., & Meaney, M. J. (2001). Like mother, like daughter: Evidence for non-genomic transmission of parental behavior and stress responsivity. *Progress in Brain Research, 133*, 287–302.

Chang, L., Smith, L. M., LoPresti, C., Yonekura, M. L., Kuo, J., Walot, I., et al. (2004). Smaller subcortical volumes and cognitive deficits in children with prenatal methamphetamine exposure. *Psychiatry Research, 132*(2), 95–106.

Cho, D. H., Lyu, H. M., Lee, H. B., Kim, P. Y., & Chin, K. (1991). Behavioral teratogenicity of methamphetamine. *Journal of Toxicological Sciences, 16*(Suppl. 1), 37–49.

Clancy, B., Finlay, B. L., Darlington, R. B., & Anand, K. J. (2007). Extrapolating brain development from experimental species to humans. *Neurotoxicology, 28*(5), 931–937.

Della Grotta, S., LaGasse, L. L., Arria, A. M., Derauf, C., Grant, P., Smith, L. M., et al. (2010). Patterns of methamphetamine use during pregnancy: Results from the infant development, environment, and lifestyle (IDEAL) study. *Maternal and Child Health Journal, 14*(4), 519–527.

Dixon, S. D., & Bejar, R. (1989). Echoencephalographic findings in neonates associated with maternal cocaine and methamphetamine use: Incidence and clinical correlates. *The Journal of Pediatrics, 115*(5 Pt 1), 770–778.

Eriksson, M., Larsson, G., Winbladh, B., & Zetterstrom, R. (1978). The influence of amphetamine addiction on pregnancy and the newborn infant. *Acta Paediatrica Scandinavica, 67*(1), 95–99.

Estelles, J., Rodriguez-Arias, M., Maldonado, C., Aguilar, M. A., & Minarro, J. (2006). Gestational exposure to cocaine alters cocaine reward. *Behavioural Pharmacology, 17*(5–6), 509–515.

Fujáková-Lipski, M., Kaping, D., Šírová, J., Horáček, J., Páleníček, T., Zach, P., et al. (2017). Trans-generational neurochemical modulation of methamphetamine in the adult brain of the Wistar rat. *Archives of Toxicology, 91*(10), 3373–3384.

Fujiyama, K., Kajii, Y., Hiraoka, S., & Nishikawa, T. (2003). Differential regulation by stimulants of neocortical expression of mrt1, arc, and homer1a mRNA in the rats treated with repeated methamphetamine. *Synapse, 49*(3), 143–149.

Hansen, R. L., Struthers, J. M., & Gospe, S. M., Jr. (1993). Visual evoked potentials and visual processing in stimulant drug-exposed infants. *Developmental Medicine and Child Neurology, 35*(9), 798–805.

Holubová, A., Ševčíková, M., Macúchová, E., Hrebíčková, I., Pometlová, M., & Šlamberová, R. (2017). Effects of perinatal stress and drug abuse on maternal behavior and sensorimotor development of affected progeny. *Physiological Research, 66*(Suppl. um 4), S481–S491.

Horger, B. A., Giles, M. K., & Schenk, S. (1992). Preexposure to amphetamine and nicotine predisposes rats to self-administer a low dose of cocaine. *Psychopharmacology, 107*(2–3), 271–276.

Hrubá, L., Schutová, B., Šlamberová, R., Pometlová, M., & Rokyta, R. (2009). Effect of methamphetamine exposure and cross-fostering on sensorimotor development of male and female rat pups. *Developmental Psychobiology, 51*(1), 73–83.

Jayanthi, S., Deng, X., Bordelon, M., McCoy, M. T., & Cadet, J. L. (2001). Methamphetamine causes differential regulation of pro-death and anti-death Bcl-2 genes in the mouse neocortex. *The FASEB Journal, 15*(10), 1745–1752.

Jeng, W., Wong, A. W., Ting, A. K. R., & Wells, P. G. (2005). Methamphetamine-enhanced embryonic oxidative DNA damage and neurodevelopmental deficits. *Free Radical Biology and Medicine, 39*(3), 317–326.

Klee, H. (1997). A typolgy of amphetamine users in the United Kingdom. In H. Klee (Ed.), *Amphetamine misuse*. Reading: Harwood Academic.

Krasnova, I. N., Ladenheim, B., Hodges, A. B., Volkow, N. D., & Cadet, J. L. (2011). Chronic methamphetamine administration causes differential regulation of transcription factors in the rat midbrain. *PloS One, 6*(4). e19179.

Kuhn, C. M., & Schanberg, S. M. (1998). Responses to maternal separation: Mechanisms and mediators. *International Journal of Developmental Neuroscience, 16*(3–4), 261–270.

Lam, M. K., Homewood, J., Taylor, A. J., & Mazurski, E. J. (2000). Second generation effects of maternal alcohol consumption during pregnancy in rats. *Progress In Neuro-Psychopharmacology & Biological Psychiatry, 24*(4), 619–631.

Leri, F., Flores, J., Rajabi, H., & Stewart, J. (2003). Effects of cocaine in rats exposed to heroin. *Neuropsychopharmacology, 28*(12), 2102–2116.

Levine, S. (1994). The ontogeny of the hypothalamic-pituitary-adrenal axis. The influence of maternal factors. *Annals of the New York Academy of Sciences, 746*, 275–288. discussion 289-293.

Malanga, C. J., & Kosofsky, B. E. (2003). Does drug abuse beget drug abuse? Behavioral analysis of addiction liability in animal models of prenatal drug exposure. *Brain Res. Dev. Brain Res., 147*(1–2), 47–57.

Martin, K. C. (2001). Arc mRNA dynamics: Return to sender–the NMDA receptor provides the targeting address for arc mRNA. *Trends in Neurosciences, 24*(11), 621–623.

Martin, J. C., Martin, D. C., Radow, B., & Sigman, G. (1976). Growth, development and activity in rat offspring following maternal drug exposure. *Experimental Aging Research, 2*(3), 235–251.

Marwick, C. (2000). NIDA seeking data on effect of fetal exposure to methamphetamine. *Journal of the American Medical Association, 283*(17), 2225–2226.

Mesquita, A. R., Pego, J. M., Summavielle, T., Maciel, P., Almeida, O. F., & Sousa, N. (2007). Neurodevelopment milestone abnormalities in rats exposed to stress in early life. *Neuroscience, 147*(4), 1022–1033.

Meyerson, B., & Malmnas, C. (1978). Brain monoamines and sexual behaviour. In J. Hutchinson (Ed.), *Biological determinants of sexual behavior*. New York: John Wiley and Sons, 1965-1191.

Oro, A. S., & Dixon, S. D. (1987). Perinatal cocaine and methamphetamine exposure: Maternal and neonatal correlates. *The Journal of Pediatrics, 111*(4), 571–578.

Plessinger, M. A. (1998). Prenatal exposure to amphetamines. Risks and adverse outcomes in pregnancy. *Obstetrics and Gynecology Clinics of North America, 25*(1), 119–138.

Plotsky, P. M., Thrivikraman, K. V., Nemeroff, C. B., Caldji, C., Sharma, S., & Meaney, M. J. (2005). Long-term consequences of neonatal rearing on central corticotropin-releasing factor systems in adult male rat offspring. *Neuropsychopharmacology, 30*(12), 2192–2204.

van Praag, H. (2009). Exercise and the brain: Something to chew on. *Trends in Neurosciences, 32*(5), 283–290.

Rambousek, L., Kačer, P., Syslová, K., Bumba, J., Bubeníková-Valešová, V., & Šlamberová, R. (2014). Sex differences in methamphetamine pharmacokinetics in adult rats and its transfer to pups through the placental membrane and breast milk. *Drug and Alcohol Dependence, 139*, 138–144.

Rice, D., & Barone, S., Jr. (2000). Critical periods of vulnerability for the developing nervous system: Evidence from humans and animal models. *Environmental Health Perspectives, 108*(Suppl. 3), 511–533.

Robinson, T. E., & Becker, J. B. (1986). Enduring changes in brain and behavior produced by chronic amphetamine administration: A review and evaluation of animal models of amphetamine psychosis. *Brain Research, 396*(2), 157–198.

Rokyta, R., Yamamotová, A., Šlamberová, R., Franěk, M., Vaculín, S., Hrubá, L., et al. (2008). Prenatal and perinatal factors influencing nociception, addiction and behavior during ontogenetic development. *Physiological Research, 57*(Suppl. 3), S79–S88.

Segal, D. S., & Kuczenski, R. (2006). Human methamphetamine pharmacokinetics simulated in the rat: Single daily intravenous administration reveals elements of sensitization and tolerance. *Neuropsychopharmacology, 31*(5), 941–955.

Semple, B. D., Blomgren, K., Gimlin, K., Ferriero, D. M., & Noble-Haeusslein, L. J. (2013). Brain development in rodents and humans: Identifying benchmarks of maturation and vulnerability to injury across species. *Progress in Neurobiology, 106–107*, 1–16.

Shuster, L., Yu, G., & Bates, A. (1977). Sensitization to cocaine stimulation in mice. *Psychopharmacology, 52*(2), 185–190.

Šlamberová, R. (2012). Drugs in pregnancy: The effects on mother and her progeny. *Physiological Research, 61*(Suppl. 1), S123–S135.

Šlamberová, R., Macúchová, E., Nohejlová-Deykun, K., Schutová, B., Hrubá, L., & Rokyta, R. (2013). Gender differences in the effect of prenatal methamphetamine exposure and challenge dose of other drugs on behavior of adult rats. *Physiological Research, 62*(Suppl. 1), S99–S108.

Šlamberová, R., Pometlová, M., & Rokyta, R. (2007). Effect of methamphetamine exposure during prenatal and preweaning periods lasts for generations in rats. *Developmental Psychobiology, 49*(3), 312–322.

Šlamberová, R., Riley, M. A., & Vathy, I. (2005). Cross-generational effect of prenatal morphine exposure on neurobehavioral development of rat pups. *Physiological Research, 54*(6), 655–660.

Šlamberová, R., Schutová, B., Hrubá, L., & Pometlová, M. (2011). Does prenatal methamphetamine exposure affect the drug-seeking behavior of adult male rats? *Behavioural Brain Research, 224*, 50–86.

Smith, L. M., Lagasse, L. L., Derauf, C., Grant, P., Shah, R., Arria, A., et al. (2008). Prenatal methamphetamine use and neonatal neurobehavioral outcome. *Neurotoxicology and Teratology, 30*, 20–28.

de Souza, A. S., Fernandes, F. S., & do Carmo, M. (2011). Effects of maternal malnutrition and postnatal nutritional rehabilitation on brain fatty acids, learning, and memory. *Nutrition Reviews, 69*(3), 132–144.

Suzuki, T., Fukuoka, Y., Mori, T., Miyatake, M., & Narita, M. (2004). Behavioral sensitization to the discriminative stimulus effects of methamphetamine in rats. *European Journal of Pharmacology, 498*(1–3), 157–161.

Thomas, U. (2002). Modulation of synaptic signalling complexes by Homer proteins. *Journal of Neurochemistry, 81*(3), 407–413.

Valvassori, S. S., Frey, B. N., Martins, M. R., Reus, G. Z., Schimidtz, F., Inacio, C. G., et al. (2007). Sensitization and cross-sensitization after chronic treatment with methylphenidate in adolescent Wistar rats. *Behavioural Pharmacology, 18*(3), 205–212.

Vavřínková, B., Binder, T., Vítková, I., & Živný, J. (2001a). Placental and umbilical cord changes in drug-addicted women. *Ceská Gynekologie, 66*(5), 345–349.

Vavřínková, B., Binder, T., & Živný, J. (2001b). Characteristics of a population of drug dependent pregnant women in the Czech Republic. *Ceská Gynekologie, 66*(4), 285–291.

Vela, G., Martin, S., Garcia-Gil, L., Crespo, J. A., Ruiz-Gayo, M., Javier Fernandez-Ruiz, J., et al. (1998). Maternal exposure to delta9-tetrahydrocannabinol facilitates morphine self-administration behavior and changes regional binding to central mu opioid receptors in adult offspring female rats. *Brain Research, 807*(1–2), 101–109.

Williams, M. T., Blankenmeyer, T. L., Schaefer, T. L., Brown, C. A., Gudelsky, G. A., & Vorhees, C. V. (2003a). Long-term effects of neonatal methamphetamine exposure in rats on spatial learning in the Barnes maze and on cliff avoidance, corticosterone release, and neurotoxicity in adulthood. *Brain Research Development, 147*(1–2), 163–175.

Williams, M. T., Moran, M. S., & Vorhees, C. V. (2003b). Refining the critical period for methamphetamine-induced spatial deficits in the Morris water maze. *Psychopharmacology, 168*(3), 329–338.

Yamamoto, Y., Yamamoto, K., Fukui, Y., & Kurishita, A. (1992). Teratogenic effects of methamphetamine in mice. *Nihon Hoigaku Zasshi, 46*(2), 126–131.

Zoubková, H., Tomášková, A., Nohejlová, K., Černá, M., & Šlamberová, R. (2019). Prenatal exposure to methamphetamine: Up-regulation of brain receptor genes. *Frontiers in Neuroscience, 13*, 771.

Chapter 17

Alcohol, cannabis, and brain development

Kristen R. Breit
Department of Psychology, West Chester University, West Chester, PA, United States

List of abbreviations

CB1 Cannabinoid-type 1
CBD Cannabidiol
CNS Central nervous system
E-cigarettes Electronic cigarettes
FAS Fetal alcohol syndrome
FASD Fetal alcohol spectrum disorders
THC Delta-9-tetrahydrocannabinol

Mini-dictionary of terms

Cannabidiol a neuroprotective constituent of cannabis
Cannabimimetic synthetic substances that mimic cannabis
Cannabinoids any of a group of closely related compounds that include cannabinol and the active constituents of cannabis
Central nervous system the brain and spinal cord
Choline an essential nutrient necessary for the growth of mammalian cells
Cholinergic systems organized nerve cells that use the neurotransmitter acetylcholine as a chemical messenger
Concurrent use using two substances in close time proximity
Craniofacial abnormalities birth defects of the face or head
Delta-9-tetrahydrocannabinol the primary psychoactive constituent of cannabis
Electronic cigarettes atomization devices that heat a solution of propylene glycol, nicotine, and flavoring agents and deliver this aerosol (vapor) to users
Endocannabinoids naturally occurring cannabinoids within the body
Epicanthal folds eyelid foldings
Fetal alcohol spectrum disorders umbrella term for the range of physical, neurological, and behavioral alterations resulting from prenatal alcohol exposure
Fetal alcohol syndrome most severe diagnosis of FASDs, characterized by behavioral alterations, growth deficiencies, and craniofacial abnormalities
Neurogenesis the process by which nervous system cells, the neurons, are produced by neural stem cells
Neuroimaging produces images of the structure or activity of the nervous system by techniques such as magnetic resonance imaging or computerized tomography
Neuropathology the study of changes in the central nervous system
Neuroplasticity the ability of the brain to change over time
Neuroprotection the preservation of neuronal structure and/or function
Neurotoxicity when exposure to natural or man-made toxic substances alters the normal activity of the nervous system
Palpebral fissures openings between the eyelids
Philtrum indentation in the middle area of the upper lip
Route of administration the way in which a drug is consumed
Teratogen any agent that can disrupt the development of a fetus
Teratology the study of teratogenic effects

Diagnosis, Management and Modeling of Neurodevelopmental Disorders. https://doi.org/10.1016/B978-0-12-817988-8.00017-8

Introduction

A **teratogen** is any agent that can disrupt the development of a fetus, and **teratology** is the study of teratogenic effects. Alcohol is one of the most widely studied teratogens. Prenatal exposure to alcohol can lead to a range of physical, neurological, and behavioral alterations referred to as **fetal alcohol spectrum disorders (FASDs)**. However, alcohol may not be the only drug of abuse consumed by pregnant women. In a recent survey, 5.4% of pregnant women reported using illicit drugs, and younger pregnant women (age 15–17 years) reported higher rates of illicit drug use (14.6%) (Substance Use and Mental Health Services Administration, 2014). The most commonly used illicit drug in this age population is cannabis; however, the effects of prenatal cannabis exposure are still not well understood. Even less is known about the effects of **concurrent use** of alcohol and cannabis during pregnancy, despite the fact that simultaneous use is common among women of childbearing age (Substance Use and Mental Health Services Administration, 2014).

Prenatal alcohol exposure

Prenatal alcohol exposure can reduce birth weight, alter facial and other physical features, and impair brain development (Riley, Infante, & Warren, 2011). FASD pose a serious public health concern, as estimates suggest that approximately 1 in 10 women report some alcohol consumption during pregnancy and recent meta-analyses estimate that the global prevalence of FASD nears 177 per 1000 live births (Roozen et al., 2016). Thus, the prevalence of FASD may be greater than that of autism spectrum disorder (Christensen et al., 2018). In addition to the public health cost, the personal, social, and economic ramifications for the individuals and their families can be overwhelming.

Brain and behavioral alterations associated with FASD

Neuropathology (study of changes in the **central nervous system [CNS]**) associated with prenatal alcohol exposure alters cognitive, emotional, motor, and behavioral functions (Kable et al., 2016) that can lead to serious problems in school and daily life (Mattson, Crocker, & Nguyen, 2011). These behavioral and cognitive dysfunctions may further contribute to secondary disabilities, including mental health issues, problems in work environments, substance abuse, difficulties with independent living, and trouble with the law, among other challenges. In addition, prenatal alcohol exposure may produce lifelong health problems (Pei, Denys, Hughes, & Rasmussen, 2011).

The most severe diagnosis on this spectrum is **fetal alcohol syndrome (FAS)**, which was first described in 1973. In addition to changes in behavioral alterations and growth deficiencies, some of the criteria for FAS include **craniofacial abnormalities** such as short openings between the eyelids (**palpebral fissures**), small eyelid foldings (**epicanthal folds**), and smoothed indentations in the middle area of the upper lip (**philtrum**) (Jones & Smith, 1973) (Fig. 17.1).

Much of what is known about alcohol's effects on the CNS has come from preclinical research using rodent models, which have shown that prenatal alcohol exposure can disrupt cell functioning in a wide variety of ways (Climent, Pascual, Renau-Piqueras, & Guerri, 2002; Lussier et al., 2015; Tyler & Allan, 2014; Wilhelm & Guizzetti, 2015). These reductions can lead to changes in overall brain weight, as well as specific structures, such as the cerebellum (Maier & West, 2001), cerebral cortex (Fabio et al., 2013), corpus callosum (Qiang, Wang, & Elberger, 2002), and hippocampus (Berman & Hannigan, 2000), producing deficits in associated behaviors.

Importantly, FASD-related deficits in learning and memory are linked to alterations in the hippocampus; for example, rodent studies show that developmental alcohol exposure reduces the number of cells in the hippocampus, particularly in the CA1 subfield and dentate gyrus (Livy, Miller, Maier, & West, 2003), alters cell connectivity, and alters neurotransmission (Monk, Leslie, & Thomas, 2012). Such changes can lead to long-lasting dysfunction in the hippocampus, as well as impairments in learning and memory.

It is important to emphasize that the data from both clinical and preclinical research show similar results (Driscoll, Streissguth, & Riley, 1990) (Fig. 17.2). For example, **neuroimaging** studies in humans show both structural and functional neuropathology in various brain regions, including basal ganglia, cerebellum, cortex and hippocampus (Sowell et al., 2002), and white matter tracts (Taylor et al., 2015). Functional imaging studies on children with FASD illustrate dysfunction in these areas (Norman, Crocker, Mattson, & Riley, 2009) resulting from **neurotoxicity**. Individuals with FASD show deficits in executive functioning (including attention, working memory, and planning) (Kodituwakku, 2007), which can severely impact daily living.

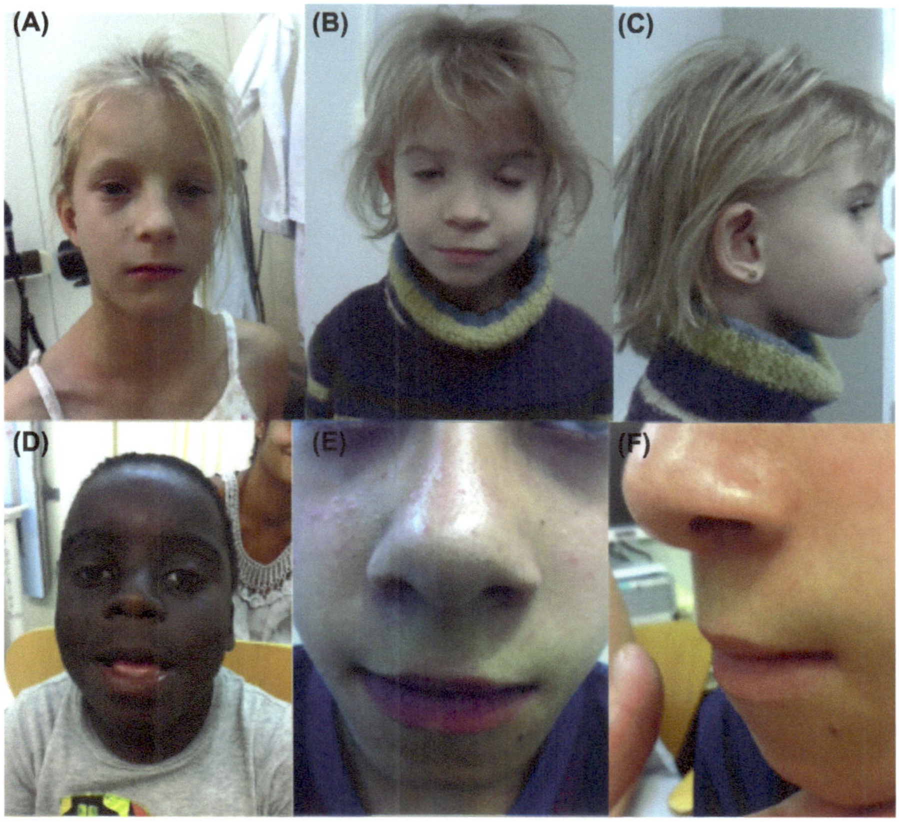

FIGURE 17.1 **Craniofacial abnormalities associated with FASD.** A wide variety of craniofacial abnormalities may result from prenatal alcohol exposure. For example, individuals with FASD may exhibit short openings between the eyelids (palpebral fissures; A, B), small eyelid foldings (epicanthal folds; B), and smooth indentations in the middle area of the upper lip (philtrum; D, E, F). *Used with permission from del Campo and Jones (2017).*

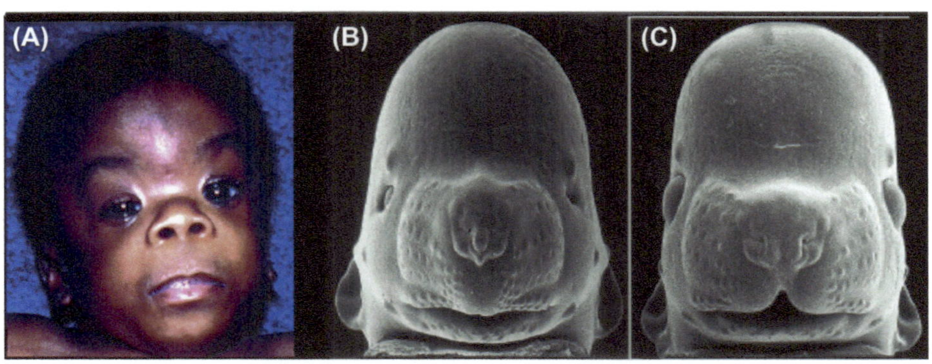

FIGURE 17.2 **Comparison of FASD-related facial abnormalities between clinical and preclinical models.** Similar craniofacial abnormalities can be observed in both children with FASD and rodents exposed to alcohol prenatally. Small palpebral fissures, epicanthal folds, and smooth philtrum can be seen in the child (A) and alcohol-exposed fetal mouse (B) compared to a typically developing fetal mouse (C). *Used with permission from del Campo and Jones (2017); modified from Sulik, Johnston, and Webb (1981). Courtesy of Prof. Kathie Sulik, University of North Carolina—Chapel Hill.*

Possible interventions for FASD

Although FAS was identified over 40 years ago, the identification of treatments for FASD in clinical populations has lagged. Recent clinical studies show some promising interventions for reducing alcohol-related deficits in cognitive, social, and behavioral skills (Petrenko & Alto, 2017). Animal models have suggested a number of potential treatments for FASD (Wellmann, George, Brnouti, & Mooney, 2015); however, the ability to translate many of these treatments is limited. There are a variety of challenges in identifying pregnant women at risk, maintaining compliance, and addressing concerns that pharmacological interventions might potentially harm a developing fetus.

Poor maternal nutrition has long been recognized to increase fetal vulnerability to alcohol because alcohol can interfere with nutrient absorption, utilization (Bode & Bode, 2003), and placental transfer (Syme, Paxton, & Keelan, 2004). Recent studies suggest that children with FASD may not be consuming adequate or recommended levels of a number of nutrients (Nguyen, Risbud, Chambers, & Thomas, 2016), including **choline**. Choline is an essential nutrient necessary for the growth of mammalian cells. Given the critical role of choline in development, recommendations for adequate choline intake levels for pregnant and lactating women, as well as children, have been well established (Food and Nutrition Board, 1998). Numerous preclinical studies illustrate the importance of choline availability for brain and behavioral development and suggest that choline supplementation during prenatal alcohol exposure mitigates alcohol-related birth weight reductions, delays in early reflex development (Thomas, Abou, & Dominguez, 2009), and cognitive impairments (Thomas, Idrus, Monk, & Dominguez, 2010).

But, perhaps more importantly, some preclinical studies show that choline can effectively reduce the severity of FASD-related deficits even after the child is born, including reduced hyperactivity and learning and memory deficits (Thomas, Garrison, & O'Neill, 2004). This is critically important, given that pregnant women may not report alcohol consumption or seek appropriate prenatal care (Lange, Shield, Koren; Rehm, & Popova, 2014). Based on these findings, a recent clinical trial is examining the effects of choline supplementation in children with FASD, aged 2–5, with promising results (Wozniak et al., 2013, 2015). As these infants mature, the effects of prenatal nutritional supplements, including choline, during prenatal alcohol exposure will be better understood.

Other factors may also modify alcohol's effects on **neuroplasticity** (the ability of the brain to change over time) and associated behaviors. For example, preclinical studies suggest that voluntary exercise and an enriched environment can increase neurogenesis (Hamilton et al., 2014) and mitigate hippocampal and memory deficits associated with FASD (Christie et al., 2005). However, it is much more complicated to examine such interventions in clinical populations given various confounding factors. Nonetheless, it is vitally important to remember that while the plasticity of the brain may make it more vulnerable to teratogens compared to a fully developed brain, the brain remains plastic throughout the lifetime, and thus, effective interventions are possible.

Prenatal cannabis exposure

Far less is known about the effects of prenatal cannabis exposure on neurodevelopment. Among pregnant women in Western countries, there has been an increase in cannabis use both among women of childbearing age (Brown et al., 2017) and among pregnant women in the United States (Agrawal et al., 2019), particularly during the first trimester (Volkow, Han, Compton, & McCance-Katz, 2019). From 2014 to 2017, it was estimated that 10% of nonpregnant women and 4.5% of pregnant women consumed cannabis in the past 30 days, with 8% of pregnant women in the first trimester consuming cannabis in the past 30 days. Given the legalization of cannabis for medical and recreational use in several areas, cannabis is becoming increasingly more available and use is increasing.

Pregnant women increasingly use cannabis products to combat undesirable symptoms of pregnancy such as nausea or as an alternative to other prescription medications under the belief that cannabis use during pregnancy is safe (Brown et al., 2017). However, this is a false assumption; recent data suggest that cannabis use during pregnancy can actually provoke recurrent nausea and vomiting rather than combat it (Kim, Moon, Dixon, & Tullar, 2018), and in addition, cannabis can interact negatively with other prescribed medications (Brown & Winterstein, 2019; Ujváry and Hanuš, 2016).

The cannabis plant contains more than 500 chemical compounds, including over 100 naturally occurring **cannabinoids** (Radwan, Wanas, Chandra, & ElSohly, 2017). Of particular concern is the potency of some forms of cannabis that are now available. The mean potency of **delta-9-tetrahydrocannabinol (THC)**, the primary psychoactive constituent, in cannabis products has been continually increasing over time, from 3.4% in 1993 to 55.7% or higher in 2017 (Chandra et al., 2019; Mehmedic et al., 2010), with the average potency being approximately 17.1% (Chandra et al., 2019). Levels are even higher among certain hashish preparations, and **cannabimimetic** substances (synthetic substances that mimic cannabis, e.g., "Spice") may have greater potency than THC. The adverse consequences of prenatal alcohol exposure have been extensively studied over the past 40 years, but less is known about the effects of prenatal cannabis exposure, particularly at the higher THC levels now consumed. Importantly, THC and its metabolites can pass across the placenta (Gómez et al., 2003) and thus directly affect the fetus.

Brain and behavioral alterations associated with prenatal cannabis exposure

There are few prospective clinical studies examining the effects of prenatal cannabis exposure. However, results from these and retrospective studies are mixed, likely due to differences in cannabis exposure levels, prospective versus retrospective approaches, confounds of other drug use (such as tobacco), age and nature of outcome measures, as well as a host of other factors. Unfortunately, the consequences of prenatal exposure to higher THC potencies available today will not be known for years to come. One ongoing clinical prospective study (Generation R) which was initiated in 2001, will provide additional understanding of the effects of prenatal cannabis exposure (Jaddoe et al., 2012). However, it will be years before these data are understood, since THC levels have increased substantially since the initiation of the Generation R study. Thus, preclinical studies utilizing exposure to THC levels that are now publicly consumed are desperately needed.

Exposure to THC in particular can be neurotoxic (Chan, Hinds, Impey, & Storm, 1998). Some preclinical evidence suggests that prenatal THC exposure can alter cell proliferation, migration, and synaptogenesis (Gómez et al., 2003). Since significant cannabinoid receptor binding has been observed in the hippocampus, cerebellum, and caudate–putamen by midpregnancy (Berrendero et al., 1998), effects of prenatal THC exposure may be seen across many behavioral domains. For example, some evidence suggests that children exposed to prenatal cannabis have altered emotional, behavioral, and cognitive development, particularly in executive functioning (Huizink, 2014).

Although THC is the primary psychoactive component, it is by no means the only one. For example, **cannabidiol (CBD)** has both cannabinoid- and non-cannabinoid-receptor mechanisms of action and can elicit **neurogenesis** and **neuroprotection** instead of psychoactive effects (Boggs, Nguyen, Morgenson, Taffe, & Ranganathan, 2018; Madras, 2019). The data on the teratogenic effects of CBD are also mixed. Although some report no teratogenic effects (Valim Brigante et al., 2018), others report that prenatal CBD exposure can lead to future infertility, miscarriage, and several other pregnancy-related disorders, effects not consistently seen in other cannabinoids such as THC (Almada et al., 2020; Neradugomma, Drafton, Mor, & Mao, 2019). On the other hand, CBD is believed to be protective toward several physiological and neurological disorders due to its direct and indirect antioxidant effects (Atalay, Jarocka-Karpowicz, & Skrzydlewska, 2020). In fact, evidence suggests that prenatal exposure to CBD may promote antiinflammatory effects among offspring after birth (Callejas et al., 2018).

However, cannabis products often include a combination of THC and CBD; CBD can mitigate the psychological and cognitive effects of THC to make them more tolerable, although the mechanism by which this occurs is poorly understood (Boggs et al., 2018; Madras, 2019). The mean ratio of THC:CBD has risen from 23 in 2008 to 104 in 2017 when psychoactive effects are desired, but conversely, higher CBD:THC ratios are taken more often (or on its own) when taken as a medication alterative (Chandra et al., 2019). Thus, the effects of prenatal exposure to cannabis, as a whole, are not well understood and will likely highly vary based on cannabinoid presences in the product.

Electronic cigarette use

Furthermore, the **route of administration** may impact the effects of prenatal drug exposure. While alcohol is typically consumed only orally, cannabis and its constituents can be consumed in a variety of ways. Two of the most popular routes of administration for THC are via edible products and **electronic cigarettes** (e-cigarettes). The popularity and use of e-cigarettes have continually increased since their development in the early 2000s, yet little research has examined the health consequences resulting from e-cigarette use during pregnancy (Suter, Mastrobattista, Sachs, & Aagaard, 2015). Prevalence rates of e-cigarette use among pregnant women in the United States are estimated to be between 5% and 14% (Cardenas, Fischbach, & Chowdhury, 2019). In fact, e-cigarette use during pregnancy is likely to occur among women diagnosed with substance use disorders and/or among women who are trying to quit smoking once pregnancy is confirmed (Oncken et al., 2017).

E-cigarette atomization devices heat a solution of propylene glycol, drugs, and flavoring agents and deliver this aerosol (vapor) to users (Grana, Benowitz, & Glantz, 2014). Originally designed as a cessation and replacement device for traditional cigarettes, e-cigarettes provide tobacco consumers with the ability to gradually decrease drug consumption over time. However, even though e-cigarettes contain fewer toxins than traditional smoking routes, toxins are still present (Goniewicz et al., 2014), and physical and neural development of the unborn fetus can still be affected by the psychoactive constituents. Moreover, given the limited regulations on e-cigarettes and accessories, the e-cigarette design also provides consumers with the ability to more easily customize potencies of the psychoactive constituents nicotine and THC (more or less than is found in tobacco or marijuana cigarettes) (Talih et al., 2014) and to combine substances for coconsumption (Henningfield et al., 2015).

Several reports suggest that e-cigarettes are perceived as safer than traditional smoking, including among pregnant women (Baeza-Loya et al., 2014). However, there has been little research directly examining the effects of prenatal exposure to cannabis constituents via e-cigarette consumption, despite requests made by medical and research professionals. This is a critically important area of research, as pregnant women report using e-cigarettes to consume cannabis for both recreational and continued cessation purposes.

Combined use of alcohol and cannabis during pregnancy

According to the National Household Survey on Drug Abuse, half of pregnant women who report consuming cannabis also report drinking alcohol (Substance Abuse and Mental Health Services Administration, 2015). It is still relatively unclear whether THC exposure could exacerbate alcohol's teratogenic effects, since results from both clinical and preclinical studies have been mixed (Abel & Subramanian, 1990; Goldschmidt, Richardson, Cornelius, & Day, 2004). A limited number of animal studies do indicate that cannabinoids may exacerbate alcohol-related neurotoxicity during development.

For example, recent research suggests that cannabinoid exposure may modulate alcohol's teratogenic effects by blocking cannabinoid-type 1 (CB1) receptors during alcohol exposure and may protect against alcohol's teratogenic effects. CB1 and CB2 receptors, and their endogenous ligands, form the **endocannabinoid system**. CB1 receptors, the main binding sites for THC, are predominantly expressed in the CNS (Matsuda, Bonner, & Lolait, 1993). CB1 receptor antagonists can also block alcohol's effects on hippocampal activity (Basavarajappa, Ninan, & Arancio, 2008). In addition to alcohol's acute action, alcohol may also alter development of endocannabinoid systems, leading to CNS dysfunction (Basavarajappa, 2015).

Moreover, alcohol activates the **cholinergic systems** in the brain, and the cholinergic and cannabinoid systems interact in the brain; thus, activation of both the cholinergic and cannabinoid systems may lead to further changes in brain development and functioning (Domino, 1981). Clinical evidence has shown that coconsumption of alcohol and cannabis exacerbates impairments in cognitive tasks (Bura, Castañé, Ledent, Valverde, & Maldonado, 2007; Miller & Branconnier, 1983), and areas important for these tasks (such as the hippocampus and basal forebrain) utilize both the cholinergic and cannabinoid systems (Kim, Isokawa, Ledent, & Alger, 2002).

Importantly, accurate data reflecting concurrent use and exposure levels among pregnant women are difficult to obtain; women who consume either alcohol or cannabis frequently underreport their usage due to stigma (Lange, Shield, Koren, Rehm, & Popova, 2014) and also due to the rapidly changing legalization, accessibility, and potency of cannabis. Given that 40-50% of all pregnancies are unplanned, understanding the effects of prenatal exposure to both of these drugs has important implications, not only for the lives of affected individuals and families, but also for public health and establishing public policy.

Applications to other areas of development

In this chapter, we have reviewed the effects of prenatal alcohol and cannabis exposure on long-term brain and behavioral development. Although the effects of prenatal alcohol exposure have been extensively studied using both preclinical and clinical models, much less is known about the effects of prenatal cannabis use. While more recent research has been done looking at the effects of prenatal THC exposure in preclinical models, it is important to note that other constituents of cannabis are present, including CBD; varying ratios of cannabis components have not been well researched in terms of their effects on neurodevelopment. Additionally, it will take years to complete prospective longitudinal studies in clinical populations; although there is much similarity between the preclinical and clinical models of prenatal exposure to other drugs, more research should be done to understand the potential consequences of prenatal cannabis exposure. Moreover, very little is known about the potential effects of combined exposure to alcohol and cannabis during pregnancy, despite the fact that these drugs are often consumed simultaneously, including among pregnant women. Lastly, given the growing amount of administration routes for cannabis consumption, and possible concurrent use of other drugs, such as nicotine, more research is needed in this area of neurodevelopmental research.

Summary points

- **Bullet of Summary Point 1:** A teratogen is any agent that can disrupt the development of a fetus.
- **Bullet of Summary Point 2:** Prenatal exposure to alcohol leads to a range of physical, neurological, and behavioral alterations referred to as fetal alcohol spectrum disorders (FASDs).
- **Bullet of Summary Point 3:** The most commonly used illicit drug in this age population is cannabis.
- **Bullet of Summary Point 4:** The possible consequences of prenatal cannabis exposure are unclear, even though the constituents of cannabis can pass across the placenta and directly affect a fetus.
- **Bullet of Summary Point 5:** Even less is known about the effects of concurrent use of alcohol and cannabis during pregnancy, despite the fact that simultaneous use is common among pregnant women and women of childbearing age.

- **Bullet of Summary Point 6:** The route of administration of drugs may impact the effects of prenatal drug exposure, and routes of administration for cannabis consumption are rapidly changing.
- **Bullet of Summary Point 7:** Given that 40-50% of all pregnancies are unplanned, understanding the effects of prenatal exposure to both of these drugs has important implications.

References

Abel, E., & Subramanian, M. (1990). Effects of low doses of alcohol on delta-9-tetrahydrocannabinol's effects in pregnant rats. *Life Sciences, 47*(18), 1677−1682.

Agrawal, A., Rogers, C. E., Lessov-Schlaggar, C. N., Carter, E. B., Lenze, S. N., & Grucza, R. A. (2019). Alcohol, cigarette, and cannabis use between 2002 and 2016 in pregnant women from a nationally representative sample. *Journal of the American Medical Association Pediatrics, 173*(1), 95−96.

Almada, M., Amaral, C., Oliveira, A., Fernandes, P. A., Ramos, M. J., Fonseca, B., ... Teixeira, N. (2020). Cannabidiol (CBD) but not tetrahydrocannabinol (THC) dysregulate in vitro decidualization of human endometrial stromal cells by disruption of estrogen signaling. *Reproductive Toxicology, 93*, 75−82.

Atalay, S., Jarocka-Karpowicz, I., & Skrzydlewska, E. (2020). Antioxidative and anti-inflammatory properties of cannabidiol. *Antioxidants, 9*(1), 21.

Baeza-Loya, S., Viswanath, H., Carter, A., Molfese, D. L., Velasquez, K. M., Baldwin, P. R., ... De La Garza, R. (2014). Perceptions about e-cigarette safety may lead to e-smoking during pregnancy. *Bulletin of the Menninger Clinic, 78*(3), 243−252.

Basavarajappa, B. S. (2015). Fetal alcohol spectrum disorder: Potential role of endocannabinoids signaling. *Brain Sciences, 5*(4), 456.

Basavarajappa, B. S., Ninan, I., & Arancio, O. (2008). Acute ethanol suppresses glutamatergic neurotransmission through endocannabinoids in hippocampal neurons. *Journal of Neurochemistry, 107*(4), 1001−1013.

Berman, R. F., & Hannigan, J. H. (2000). Effects of prenatal alcohol exposure on the hippocampus: Spatial behavior, electrophysiology, and neuroanatomy. *Hippocampus, 10*(1), 94−110.

Berrendero, F., Garcia-Gil, L., Hernandez, M., Romero, J., Cebeira, M., De Miguel, R., ... Fernandez-Ruiz, J. (1998). Localization of mRNA expression and activation of signal transduction mechanisms for cannabinoid receptor in rat brain during fetal development. *Development, 125*(16), 3179−3188.

Bode, C., & Bode, J. C. (2003). Effect of alcohol consumption on the gut. *Best Practice and Research. Clinics in Gastroenterology, 17*(4), 575−592.

Boggs, D. L., Nguyen, J. D., Morgenson, D., Taffe, M. A., & Ranganathan, M. (2018). Clinical and preclinical evidence for functional interactions of cannabidiol and δ 9-tetrahydrocannabinol. *Neuropsychopharmacology, 43*(1), 142−154.

Brown, Q. L., Sarvet, A. L., Shmulewitz, D., Martins, S. S., Wall, M. M., & Hasin, D. S. (2017). Trends in marijuana use among pregnant and nonpregnant reproductive-aged women, 2002-2014. *Journal of the American Medical Association, 317*(2), 207−209.

Brown, J. D., & Winterstein, A. G. (2019). Potential adverse drug events and drug−drug interactions with medical and consumer cannabidiol (CBD) use. *Journal of Clinical Medicine, 8*(7), 989.

Bura, S. A., Castañé, A., Ledent, C., Valverde, O., & Maldonado, R. (2007). Genetic and pharmacological approaches to evaluate the interaction between the cannabinoid and cholinergic systems in cognitive processes. *British Journal of Pharmacology, 150*(6), 758−765.

Callejas, G., Figueira, R., Gonçalves, F., Volpe, F., Zuardi, A., Crippa, J., ... Sbragia, L. (2018). Maternal administration of cannabidiol promotes an anti-inflammatory effect on the intestinal wall in a gastroschisis rat model. *Brazilian Journal of Medical and Biological Research, 51*(5).

Cardenas, V. M., Fischbach, L. A., & Chowdhury, P. (2019). The use of electronic nicotine delivery systems during pregnancy and the reproductive outcomes: A systematic review of the literature. *Tobacco Induced Diseases, 17*.

Chandra, S., Radwan, M. M., Majumdar, C. G., Church, J. C., Freeman, T. P., & ElSohly, M. A. (2019). New trends in cannabis potency in USA and europe during the last decade (2008−2017). *European Archives of Psychiatry and Clinical Neuroscience, 269*(1), 5−15

Chan, G. C., Hinds, T. R., Impey, S., & Storm, D. R. (1998). Hippocampal neurotoxicity of δ9-tetrahydrocannabinol. *Journal of Neuroscience, 18*(14), 5322−5332.

Christensen, D. L., Baio, J., Braun, K. V., Bilder, D., Charles, J., Constantino, J. N., ... Yeargin-Allsopp, M. (2018). Prevalence and characteristics of autism spectrum disorder among children aged 8 years—autism and developmental disabilities monitoring network, 11 sites, United States, 2012. *MMWR Surveillance Summaries, 65*(13), 1.

Christie, B. R., Swann, S. E., Fox, C. J., Froc, D., Lieblich, S. E., Redila, V., & Webber, A. (2005). Voluntary exercise rescues deficits in spatial memory and long-term potentiation in prenatal ethanol-exposed male rats. *European Journal of Neuroscience, 21*(6), 1719−1726.

Climent, E., Pascual, M., Renau-Piqueras, J., & Guerri, C. (2002). Ethanol exposure enhances cell death in the developing cerebral cortex: Role of brain-derived neurotrophic factor and its signaling pathways. *Journal of Neuroscience Research, 68*(2), 213−225.

del Campo, M., & Jones, K. L. (2017). A review of the physical features of the fetal alcohol spectrum disorders. *European Journal of Medical Genetics, 60*(1), 55−64. https://doi.org/10.1016/j.ejmg.2016.10.004.

Domino, E. F. (1981). Cannabinoids and the cholinergic system. *The Journal of Clinical Pharmacology, 21*, 249−255.

Driscoll, C. D., Streissguth, A. P., & Riley, E. P. (1990). Prenatal alcohol exposure: Comparability of effects in humans and animal models. *Neurotoxicology and Teratology, 12*(3), 231−237.

Fabio, M. C., March, S. M., Molina, J. C., Nizhnikov, M. E., Spear, N. E., & Pautassi, R. M. (2013). Prenatal ethanol exposure increases ethanol intake and reduces c-fos expression in infralimbic cortex of adolescent rats. *Pharmacology Biochemistry and Behavior, 103*(4), 842−852.

Food and Nutrition Board. (1998). *Dietary reference intakes for thiamin, riboflavin, niacin, vitamin b6, folate, vitamin b12, pantothenic acid, biotin, and choline*.

Goldschmidt, L., Richardson, G. A., Cornelius, M. D., & Day, N. L. (2004). Prenatal marijuana and alcohol exposure and academic achievement at age 10. *Neurotoxicology and Teratology, 26*(4), 521–532.

Gómez, M. A., Hernández, M., Johansson, B., de Miguel, R., Ramos, J. A., & Fernández-Ruiz, J. (2003). Prenatal cannabinoid exposure and gene expression for neural adhesion molecule L1 in the fetal rat brain. *Developmental Brain Research, 147*(1), 201–207.

Goniewicz, M. L., Knysak, J., Gawron, M., Kosmider, L., Sobczak, A., Kurek, J., … Benowitz, N. (2014). Levels of selected carcinogens and toxicants in vapour from electronic cigarettes. *Tobacco Control, 23*(2), 133–139.

Grana, R., Benowitz, N., & Glantz, S. A. (2014). E-cigarettes. *Circulation, 129*(19), 1972–1986.

Hamilton, G., Jablonski, S., Schiffino, F., Cyr, S. S., Stanton, M., & Klintsova, A. (2014). Exercise and environment as an intervention for neonatal alcohol effects on hippocampal adult neurogenesis and learning. *Neuroscience, 265*, 274–290.

Henningfield, J., Cone, E. J., Buchhalter, A. R., Fant, R. V., Gitchell, J. G., & Pinney, J. M. (2015). Evaluation of electronic nicotine delivery systems: Regulatory precedents from the fda 2013 draft guidance for abuse-deterrent opioids. *Drug and Alcohol Dependence, 156*, 94–95.

Huizink, A. (2014). Prenatal cannabis exposure and infant outcomes: Overview of studies. *Progress in Neuro-Psychopharmacology and Biological Psychiatry, 52*, 45–52.

Jaddoe, V. W., Van Duijn, C. M., Franco, O. H., Van Der Heijden, A. J., Van IIzendoorn, M. H., De Jongste, J. C., … Raat, H. (2012). The generation R study: Design and cohort update 2012. *European Journal of Epidemiology, 27*(9), 739–756.

Jones, K., & Smith, D. (1973). Recognition of the fetal alcohol syndrome in early infancy. *The Lancet, 302*(7836), 999–1001.

Kable, J. A., O'Connor, M. J., Olson, H. C., Paley, B., Mattson, S. N., Anderson, S. M., & Riley, E. P. (2016). Neurobehavioral disorder associated with prenatal alcohol exposure (ND-PAE): Proposed DSM-5 diagnosis. *Child Psychiatry and Human Development, 47*(2), 335–346.

Kim, J., Isokawa, M., Ledent, C., & Alger, B. E. (2002). Activation of muscarinic acetylcholine receptors enhances the release of endogenous cannabinoids in the hippocampus. *The Journal of Neuroscience, 22*(23), 10182–10191.

Kim, H. G., Moon, J., Dixon, H., & Tullar, P. (2018). Recurrent nausea and vomiting in a pregnant woman with chronic marijuana use. *Case Reports in Obstetrics and Gynecology, 2018*, 1–3. https://doi.org/10.1155/2018/9746062.

Kodituwakku, P. W. (2007). Defining the behavioral phenotype in children with fetal alcohol spectrum disorders: A review. *Neuroscience and Biobehavioral Reviews, 31*(2), 192–201.

Lange, S., Shield, K., Koren, G., Rehm, J., & Popova, S. (2014). A comparison of the prevalence of prenatal alcohol exposure obtained via maternal self-reports versus meconium testing: A systematic literature review and meta-analysis. *BMC Pregnancy and Childbirth, 14*(1), 1.

Livy, D. J., Miller, E. K., Maier, S. E., & West, J. R. (2003). Fetal alcohol exposure and temporal vulnerability: Effects of binge-like alcohol exposure on the developing rat hippocampus. *Neurotoxicology and Teratology, 25*(4), 447–458.

Lussier, A. A., Stepien, K. A., Neumann, S. M., Pavlidis, P., Kobor, M. S., & Weinberg, J. (2015). Prenatal alcohol exposure alters steady-state and activated gene expression in the adult rat brain. *Alcoholism: Clinical and Experimental Research, 39*(2), 251–261.

Madras, B. K. (2019). Tinkering with THC-to-CBD ratios in marijuana. *Neuropsychopharmacology, 44*, 215–216.

Maier, S. E., & West, J. R. (2001). Regional differences in cell loss associated with binge-like alcohol exposure during the first two trimesters equivalent in the rat. *Alcohol, 23*(1), 49–57.

Matsuda, L. A., Bonner, T. I., & Lolait, S. J. (1993). Localization of cannabinoid receptor mRNA in rat brain. *Journal of Comparative Neurology, 327*(4), 535–550.

Mattson, S. N., Crocker, N., & Nguyen, T. T. (2011). Fetal alcohol spectrum disorders: Neuropsychological and behavioral features. *Neuropsychology Review, 21*(2), 81–101.

Mehmedic, Z., Chandra, S., Slade, D., Denham, H., Foster, S., Patel, A. S., … ElSohly, M. A. (2010). Potency trends of δ9-THC and other cannabinoids in confiscated cannabis preparations from 1993 to 2008. *Journal of Forensic Sciences, 55*(5), 1209–1217.

Miller, L. L., & Branconnier, R. J. (1983). Cannabis: Effects on memory and the cholinergic limbic system. *Psychological Bulletin, 93*(3), 441–456.

Monk, B. R., Leslie, F. M., & Thomas, J. D. (2012). The effects of perinatal choline supplementation on hippocampal cholinergic development in rats exposed to alcohol during the brain growth spurt. *Hippocampus, 22*(8), 1750–1757.

Neradugomma, N. K., Drafton, K., Mor, G. G., & Mao, Q. (2019). Marijuana-derived cannabinoids inhibit uterine endometrial stromal cell decidualization and compromise trophoblast-endometrium cross-talk. *Reproductive Toxicology, 87*, 100–107.

Nguyen, T. T., Risbud, R. D., Chambers, C. D., & Thomas, J. D. (2016). Dietary nutrient intake in school-aged children with heavy prenatal alcohol exposure. *Alcoholism: Clinical and Experimental Research, 40*(5), 1075–1082.

Norman, A. L., Crocker, N., Mattson, S. N., & Riley, E. P. (2009). Neuroimaging and fetal alcohol spectrum disorders. *Developmental Disabilities Research Reviews, 15*(3), 209–217.

Oncken, C., Ricci, K. A., Kuo, C.-L., Dornelas, E., Kranzler, H. R., & Sankey, H. Z. (2017). Correlates of electronic cigarettes use before and during pregnancy. *Nicotine and Tobacco Research, 19*(5), 585–590.

Pei, J., Denys, K., Hughes, J., & Rasmussen, C. (2011). Mental health issues in fetal alcohol spectrum disorder. *Journal of Mental Health, 20*(5), 438–448.

Petrenko, C. L., & Alto, M. E. (2017). Interventions in fetal alcohol spectrum disorders: An international perspective. *European Journal of Medical Genetics, 60*(1), 79–91.

Qiang, M., Wang, M. W., & Elberger, A. J. (2002). Second trimester prenatal alcohol exposure alters development of rat corpus callosum. *Neurotoxicology and Teratology, 24*(6), 719–732.

Radwan, M. M., Wanas, A. S., Chandra, S., & ElSohly, M. A. (2017). Natural cannabinoids of cannabis and methods of analysis. In *Cannabis sativa l.-botany and biotechnology* (pp. 161–182). Springer.

Riley, E. P., Infante, M. A., & Warren, K. R. (2011). Fetal alcohol spectrum disorders: An overview. *Neuropsychology Review, 21*(2), 73–80.

Roozen, S., Peters, G. J. Y., Kok, G., Townend, D., Nijhuis, J., & Curfs, L. (2016). Worldwide prevalence of fetal alcohol spectrum disorders: A systematic literature review including meta-analysis. *Alcoholism: Clinical and Experimental Research, 40*(1), 18–32.

Sowell, E. R., Thompson, P. M., Mattson, S. N., Tessner, K. D., Jernigan, T. L., Riley, E. P., & Toga, A. W. (2002). Regional brain shape abnormalities persist into adolescence after heavy prenatal alcohol exposure. *Cerebral Cortex, 12*(8), 856–865.

Substance Abuse and Mental Health Services Administration. (2015). *Behavioral health trends in the United States: Results from the 2014 national survey on drug use and health.*

Substance Use and Mental Health Services Administration. (2014). *Results from the 2013 national survey on drug use and health: Summary of national findings.*

Sulik, K. K., Johnston, M. C., & Webb, M. A. (1981). Fetal alcohol syndrome: Embryogenesis in a mouse model. *Science, 214*(4523), 936–938.

Suter, M. A., Mastrobattista, J., Sachs, M., & Aagaard, K. (2015). Is there evidence for potential harm of electronic cigarette use in pregnancy? *Birth Defects Research Part A: Clinical and Molecular Teratology, 103*(3), 186–195.

Syme, M. R., Paxton, J. W., & Keelan, J. A. (2004). Drug transfer and metabolism by the human placenta. *Clinical Pharmacokinetics, 43*(8), 487–514.

Talih, S., Balhas, Z., Eissenberg, T., Salman, R., Karaoghlanian, N., El Hellani, A., … Shihadeh, A. (2014). Effects of user puff topography, device voltage, and liquid nicotine concentration on electronic cigarette nicotine yield: Measurements and model predictions. *Nicotine and Tobacco Research, 17*(2), 150–157.

Taylor, P. A., Jacobson, S. W., van der Kouwe, A., Molteno, C. D., Chen, G., Wintermark, P., … Meintjes, E. M. (2015). A DTI-based tractography study of effects on brain structure associated with prenatal alcohol exposure in newborns. *Human Brain Mapping, 36*(1), 170–186.

Thomas, J. D., Abou, E. J., & Dominguez, H. D. (2009). Prenatal choline supplementation mitigates the adverse effects of prenatal alcohol exposure on development in rats. *Neurotoxicology and Teratology, 31*(5), 303–311.

Thomas, J. D., Garrison, M., & O'Neill, T. M. (2004). Perinatal choline supplementation attenuates behavioral alterations associated with neonatal alcohol exposure in rats. *Neurotoxicology and Teratology, 26*(1), 35–45.

Thomas, J. D., Idrus, N. M., Monk, B. R., & Dominguez, H. D. (2010). Prenatal choline supplementation mitigates behavioral alterations associated with prenatal alcohol exposure in rats. *Birth Defects Research Part A: Clinical and Molecular Teratology, 88*(10), 827–837.

Tyler, C. R., & Allan, A. M. (2014). Prenatal alcohol exposure alters expression of neurogenesis-related genes in an ex vivo cell culture model. *Alcohol, 48*(5), 483–492.

Ujváry, I., & Hanuš, L. (2016). Human metabolites of cannabidiol: A review on their formation, biological activity, and relevance in therapy. *Cannabis and Cannabinoid Research, 1*(1), 90–101.

Valim Brigante, T. A., Abe, F. R., Zuardi, A. W., Hallak, J. E. C., Crippa, J. A. S., & de Oliveira, D. P. (2018). Cannabidiol did not induce teratogenicity or neurotoxicity in exposed zebrafish embryos. *Chemico-Biological Interactions, 291*, 81–86.

Volkow, N. D., Han, B., Compton, W. M., & McCance-Katz, E. F. (2019). Self-reported medical and nonmedical cannabis use among pregnant women in the United States. *Journal of the American Medical Association, 322*(2), 167–169.

Wellmann, K. A., George, F., Brnouti, F., & Mooney, S. M. (2015). Docosahexaenoic acid partially ameliorates deficits in social behavior and ultrasonic vocalizations caused by prenatal ethanol exposure. *Behavioural Brain Research, 286*, 201–211.

Wilhelm, C. J., & Guizzetti, M. (2015). Fetal alcohol spectrum disorders: An overview from the glia perspective. *Frontiers in Integrative Neuroscience, 9*.

Wozniak, J. R., Fuglestad, A. J., Eckerle, J. K., Fink, B. A., Hoecker, H. L., Boys, C. J., … Georgieff, M. K. (2015). Choline supplementation in children with fetal alcohol spectrum disorders: A randomized, double-blind, placebo-controlled trial. *The American Journal of Clinical Nutrition, 102*(5), 1113–1125.

Wozniak, J. R., Fuglestad, A. J., Eckerle, J. K., Kroupina, M. G., Miller, N. C., Boys, C. J., … Georgieff, M. K. (2013). Choline supplementation in children with fetal alcohol spectrum disorders has high feasibility and tolerability. *Nutrition Research, 33*(11), 897–904.

Chapter 18

The environmental pollutant trichloroethylene disrupts key neural pathways during brain development

Briana R. De Miranda[1] and Sarah J. Blossom[2]

[1]*Center for Neurodegeneration and Experimental Therapeutics, Department of Neurology, College of Medicine, University of Alabama at Birmingham, Birmingham, AL, United States;* [2]*Department of Pharmaceutical Sciences, College of Pharmacy, University of New Mexico, New Mexico, United States*

List of abbreviations

ATSDR Agency of Toxic Substances and Disease Registry
DCVC S-(1,2-dichlorovinyl)-l-cysteine
DNMT DNA methyltransferase
GABA gamma-Aminobutyric acid
IARC International Agency for Research on Cancer
TaClo 1-trichloromethyl-1,2,3,4-tetrahydro-beta-carboline
TCE Trichloroethylene
TSCA Toxic Substances Control Act
VSCC voltage-sensitive calcium channel

Mini-dictionary of terms

ATSDR Agency of Toxic Substances and Disease Registry that prioritizes and assesses chemicals based on frequency, toxicity, and potential for human exposure
EPA United States Environmental Protection Agency
In utero Any time period during gestation prior to birth
Neuroinflammation Inflammation within the brain or spinal cord caused by pathogen exposure, toxicant exposure, or cellular damage that is not resolved by normal cellular processes
Neurotoxicity Any cellular damage to the brain or spinal cord caused by exogenous or endogenous pathology.
Perinatal The time period surrounding birth, particularly within the few weeks prior to and after birth
Postnatal The time period following birth, usually within the first days or weeks
Superfund Site Areas within the United States designated by the EPA as significantly contaminated by toxic chemicals from improper storage or waste management
Toxicant Chemical compound of synthetic origin that causes pathology in human tissue
Trichloroethylene (TCE) Organic solvent used in industrial operation as a vapor degreasing agent

Introduction

Trichloroethylene overview

Trichloroethylene (TCE; C_2HCl_3) is a chlorinated halocarbon primarily used for its powerful action on greases, oils, resins, waxes, and other natural and synthetic substances. Currently, 80%−90% of TCE manufactured worldwide is used for the

Diagnosis, Management and Modeling of Neurodevelopmental Disorders. https://doi.org/10.1016/B978-0-12-817988-8.00018-X

industrial degreasing of metals (Bakke, Stewart, & Waters, 2007). TCE has also been used in dry cleaning, textile processing, anesthetics, food processing, as well as other industrial, commercial, and military uses. TCE has been an ingredient in some consumer products including paint thinners, disinfectants, and spot removers.

Sources of TCE exposure

TCE is released into the air, water, and soil at places where it is produced or used. Despite reductions of its release into the environment over the last few decades, TCE exposure remains significant due to inappropriate disposal practices from factories or military sites. TCE continues to be classified as a hazardous waste pollutant, a common groundwater contaminant, and is among the most frequently detected contaminants of US public water supplies and US EPA National Priority List (commonly referred to as Superfund) sites. TCE is among the most frequently detected drinking water contaminants regulated by the EPA. The most recent estimates indicate TCE is detected in 4.5% of groundwater and 15% of surface water, the primary sources of US drinking water supplies (ATSDR, 2014). Although most environmental exposure to TCE occurs through consuming contaminated water, a significant number of people are also exposed to TCE in the workplace through inhalation or dermal contact. Individuals are also exposed within residential buildings via vapor intrusion by living near hazardous waste sites.

Human health risks

TCE is on the list of 90 chemicals selected from the 85,000 chemicals in the TSCA (Toxic Substances Control Act) Inventory as having the highest potential for exposure and hazard (EPA, 2014). It is currently ranked 16th out of 275 chemicals on the Agency of Toxic Substances and Disease Registry (ATSDR) Substance Priority List that prioritizes chemicals based on frequency, toxicity, and potential for human exposure. In line with earlier reports of carcinogenicity in rodents, a working group convened by the International Agency for Research on Cancer (IARC) reclassified TCE as carcinogenic to humans based on "sufficient" epidemiologic and animal evidence and qualitative mechanistic similarities between humans and rodents for kidney cancer (Guha et al., 2012). Consequently, the US Department of Health and Human Services 14th Report on Carcinogens added TCE to the list of 248 substances known to be a human carcinogen in 2016.

There are several human and experimental studies to suggest that TCE can lead to noncancer effects, including neurotoxicity. A comprehensive assessment of adult neurotoxicity documented changes in nerve function (Feldman, Niles, Proctor, & Jabre, 1992; Ruijten, Verberk, & Salle, 1991), reduced hearing and visual impairment, decreased cognitive function, and impaired vestibular function (Burg & Gist, 1999; Kilburn, 2002; Kilburn & Warshaw, 1992; Reif et al., 2003; Vernon & Ferguson, 1969). Although there is a high degree of concordance between studies in rodent models and the reported neurological effects in humans, studies in human populations are inherently limited based on lack of exposure information and duration, as well as the inability to adequately determine effect levels. Overall, these studies suggest that TCE imparts central nervous system (CNS) effects.

Developmental neurotoxicity

TCE exposure during development

Most human epidemiological studies of TCE toxicity are based on adult occupational exposures since they are easier to document. However, human contact with TCE can occur at all stages of life. TCE and its metabolites can cross the placenta (Laham, 1970) and accumulate in the breast milk. TCE was detected in 35% of breast milk samples with the maximum concentration of 6 ng/mL in a heavily TCE-contaminated area in Arizona (Beamer et al., 2012). Within that population, TCE concentration in the breast milk significantly correlated with the concentration in household water, meaning that TCE exposure was also a potential concern for bottle-fed infants, who ingest more water on a bodyweight basis than adults.

Developmental neurotoxicity

Studies of mothers exposed to TCE have documented increased adverse birth outcomes including low birth weight and neural tube defects (Forand, Lewis-Michl, & Gomez, 2012). Although less well studied in human population, the dynamic nature, extensive maturational period, and cellular plasticity of the brain during development lend it as a vulnerable target for disruption by environmental toxicants such as TCE. Consequently, TCE is suggested to be classified as a developmental neurotoxicant even at low doses (Salama, El-Naggar, Abdel-Rahman, & Elhak, 2018). There are few human studies documenting TCE exposure and neurotoxicity outcomes in humans. In an occupational study, children of mothers who worked with TCE and who were exposed both gestationally and postnatally had poorer visual acuity, impaired motor coordination, and behaviors characterized by inattention and hyperactivity (Laslo-Baker et al., 2004; Till, Koren, & Rovet, 2001). TCE exposure through consumption of contaminated drinking water by residents living near the TCE-contaminated Rocky Mountain Arsenal Superfund site was associated with higher scores for depression, lower intelligence scores, and impaired memory recall, as compared to individuals who did not ingest contaminated water (Reif et al., 2003). In another study, subjects who were children at the time of TCE exposure from drinking contaminated well water had enhanced cognitive deficits compared to subjects exposed as adults (White, Feldman, Eviator, Jabre, & Niles, 1997). These studies suggest that developmental exposure to TCE may impart human neurological effects; however, the mechanisms are not well established.

Toxic mechanisms of neuronal pathology

While there are no data from human studies that can elucidate the exact mechanisms of TCE-induced neurotoxicity over the lengthy window of neurodevelopment, there are several proposed mechanisms by which TCE-induced neurotoxicity occurs in model organisms and cellular systems. Here, we discuss evidence of neuronal pathology as a direct result of TCE exposure during development, including mitochondrial dysfunction and oxidative stress, receptor and neurotransmitter disruption, genotoxicity, and epigenetic changes.

Receptor and channel disruption

The predominant neurological response to acute volatile organic exposures, including TCE, is depression of the CNS (Shafer, Bushnell, Benignus, & Woodward, 2005). Indeed, TCE was used historically as an anesthetic beginning in the 1940s, but was phased out for general surgical purposes and replaced by halothane between the 1950s and 1960s. One exception, however, was the use of TCE in obstetrics, which continued until the 1970s in Europe and the United States (Kennedy, 1974). Due to its potent anesthetic properties, TCE was hypothesized to interact with the $GABA_A$ receptor, which was later confirmed in studies within the adult mouse brain (Shelton & Nicholson, 2014). In this context, $GABA_A$ receptors represent a vulnerability in perinatal exposure to TCE, as this neurotransmitter and receptor are especially important in the developing brain. For example, neurons that express GABA (GABAergic cells) are expressed at an early developmental period, embryonic day 12 in rat, and therefore likely play a key role in signaling events to drive CNS development (Wu & Sun, 2015). There is evidence that GABAergic neurons drive proliferation and migration of cells within the developing brain, and positive regulation of $GABA_A$ results in a decrease of DNA synthesis within cortical progenitor cells (LoTurco, Owens, Heath, Davis, & Kriegstein, 1995). In this context, TCE disruption of $GABA_A$ during critical neurodevelopmental windows has the potential to disrupt protein expression and overall proliferation of cells within the CNS.

In addition to GABA receptors, TCE has been shown to modulate CNS signaling through interaction with voltage-sensitive ion channels (VSCC) within ex vivo neuronal cultures. Data from these studies implicates TCE as a modulator of intracellular calcium levels within neurons, which could strongly impact neurotransmitter signaling (Shafer et al., 2005). Indeed, some of the earliest evidence for TCE interaction with VSCCs suggests that the volatile compound affects potassium and calcium influx, which correlated with increased cyclic adenosine monophosphate (cAMP) in the brainstem (Nilsson, 1987). This alteration of brainstem cAMP signaling was postulated to be downstream of changes in noradrenaline release and binding to adrenergic receptors at the postsynapse (Nilsson, 1986). As solvent exposure is associated with increased noradrenaline levels in the rat brain (Andersson et al., 1981), alterations in noradrenaline as a result of VSCC-mediated calcium influx are a putative mechanism for TCE to influence neurotransmitter signaling in the CNS. TCE-induced neurotransmitter alterations are depicted in Fig. 18.1.

Mitochondrial dysfunction

There is a clear link between TCE exposure and mitochondrial dyshomeostasis, which has been demonstrated within multiple cell types and organ systems (Chen, Cai, Anders, Stevens, & Jones, 2001; Gash et al., 2008; Lash, Parker, & Scott, 2000). The disruption of normal mitochondrial function by TCE during fetal development is best characterized within cardiomyocytes, which contributes to fetal heart abnormalities following developmental exposure to TCE (Forand et al., 2012). Like cardiomyocytes, neurons are metabolically active cells with a high energy demand. Neurons require high

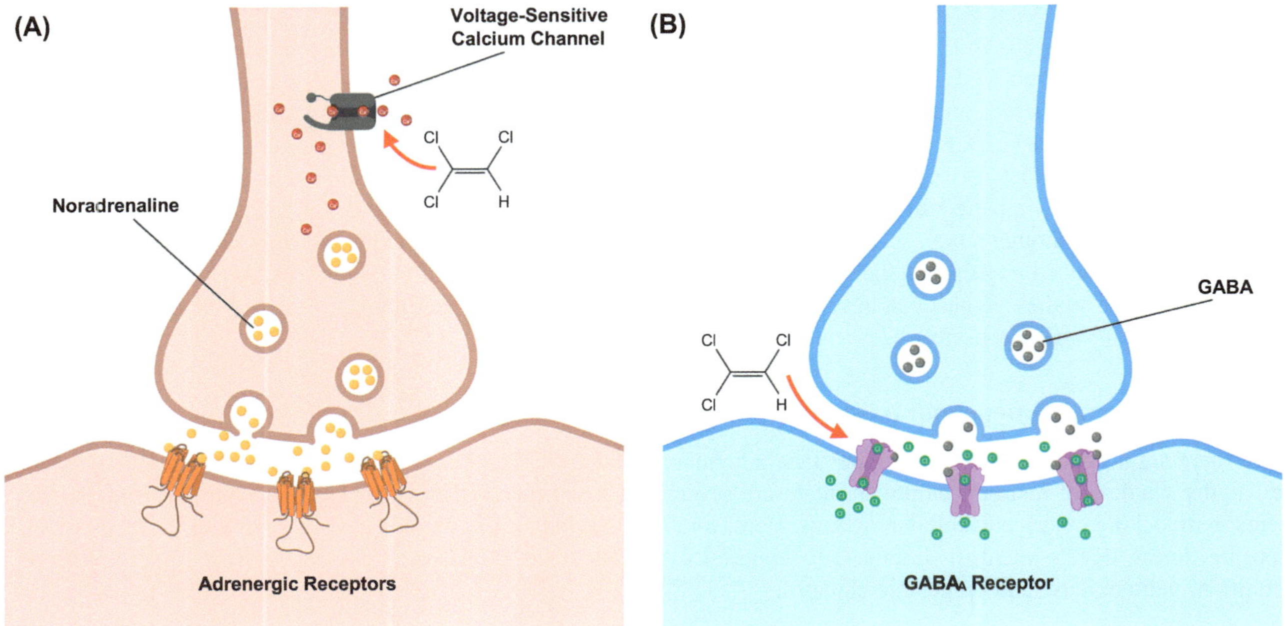

FIGURE 18.1 Putative neurotransmitter alterations associated with TCE exposure. (A) TCE is reported to alter voltage-sensitive calcium channels (VSCC), which increases intracellular calcium, influencing the release of synaptic vesicals containing the neurotransmitter noradrenaline. TCE-induced changes in noradrenaline are postulated to cause postsynaptic cAMP elevations associated with noradrenaline binding to adrenergic receptors. (B) TCE interaction with the $GABA_A$ receptor is reported following exposure, increasing chloride influx into postsynaptic neurons. Alterations in neurotransmitter expression and release caused by TCE are potential mechanisms in neurodevelopmental toxicity following exposure to the solvent. Figure generated with Biorender.com.

ATP levels to maintain resting membrane potential, long-distance axonal trafficking, vesicular sequestration of neurotransmitters, and enzymatic processing of signaling molecules (Bauernfeind et al., 2014). Thus, neuronal mitochondrial function is a key component to cellular health, and even minimal disruption of this process may result in neuronal pathology, dysfunction, or cell death.

TCE exposure results in a decrease in mitochondrial function within neurons of the adult striatum (Sauerbeck, Hunter, Bing, & Sullivan, 2012) and midbrain (Gash et al., 2008) of Fisher 344 rats. As these studies are within adult rodents, there is a limited amount of inference that can be applied upon the immature CNS; however, these data have elucidated that TCE appears to specifically inhibit complex I of the mitochondria, a common target of environmental mitochondrial toxicants (Gash et al., 2008). TCE-induced mitochondrial complex I inhibition results in a cascade of cellular dysfunction and generates reactive oxygen species (ROS), which contributes to the depletion of cellular antioxidant proteins (e.g., glutathione) and the loss of normal ATP production (Boulton, Keane, Morris, McNeil, & Manning, 2012; Gash et al., 2008). Cellular metabolism of TCE to reactive intermediates, such as the halogenated beta-carboline, 1-trichloromethyl-1,2,3,4-tetrahydro-beta-carboline (TaClo), results in neurotoxicity of human neuroblastoma cell lines (SK-N-SH, IMR-32), including alteration of cell cycle progression, and inhibited neurite formation (Sharma et al., 2017). In addition, human neuroblastoma cells treated with TaClo had reduced ATP levels, increased glycolysis dependence, and changes in autophagy (Sharma et al., 2017).

There are other metabolites of TCE, such as S-(1,2-dichlorovinyl)-L-cysteine (DCVC), that cause mitochondrial toxicity; however, these cysteine S-conjugates have predominately been investigated as peripheral organ targets. For example, the proximal tubular kidney cell line LLC-PK1 exposed to DCVC results in cytochrome *c* release, decreased mitochondrial membrane potential, and swollen mitochondria (Chen et al., 2001). While data from neuronal exposure to DCVC are limited, DCVC can be measured in the brain following TCE exposure, either through the generation of DCVC through glutathione conjugation pathways of TCE or by the deacetylation of another TCE metabolite, *N*-acetyl-*S*-1,2-dichlorovinyl-L-cysteine (NA-DCVC, Hsieh et al., 2010). Taken together, there are potentially multiple mechanisms for TCE-induced mitochondrial dysfunction in the developing brain; however, more studies are needed to explore the role of mitochondrial toxicity caused by specific metabolites of TCE.

Genotoxicity and epigenetics

Genotoxicity and epigenetic changes resulting from toxicant exposure have the potential to significantly influence functional effects during brain development (Keverne, Pfaff, & Tabansky, 2015; Lewis & Kroll, 2018). The case for TCE-induced genotoxicity is postulated to involve bioactivation of TCE to reactive metabolites by normal cellular metabolism, which may damage cellular macromolecules such as proteins, DNA, and lipids. While there is strong evidence that genotoxicity occurs within peripheral organs in response to TCE exposure (e.g., liver and kidney, De La Rosa et al., 2017; Lash et al., 2000; Robbiano, Baroni, Carrozzino, Mereto, & Brambilla, 2004), measurement of genotoxicity within the TCE-exposed brain remains poorly understood and represents another important area where further research is needed.

In contrast, increasing evidence suggests that alterations in epigenetic processes including DNA methylation, microRNAs (miRNAs), and histone acetylation in the brain promote neuroinflammation, neurodegeneration, and neurotoxicity (Neal & Richardson, 2018). Among the epigenetic processes, DNA methylation has been the most widely studied in models of TCE exposure designed to investigate immunotoxicity. This was recently shown in a study documenting TCE-induced alteration of DNA methylation of peripheral blood mononuclear cells in pathways important in autoimmunity in occupationally exposed humans (Phillips et al., 2019). In autoimmune-prone mice, exposure to TCE altered the DNA methylation of the promoter region of the proinflammatory cytokine IFN-γ in $CD4^+$ T cells (Gilbert et al., 2016). More recently, developmental TCE exposure altered DNA methylation of regions that regulate differentiation and function in effector/memory CD4s (Byrum et al., 2019). TCE promoted hypermethylation of these regions, which contrasts to other studies showing that TCE promotes hypomethylation in cultured cell lines (Cui, Choudhury, & Irudayaraj, 2016). Thus, the ability of TCE to either hypo- or hypermethylate DNA to either increase or decrease gene expression, respectively, is likely to be cell type/organ-specific.

Aside from the immune system, a number of studies suggest that TCE exposure causes altered DNA methylation in the brain during development. One line of evidence to support TCE-induced epigenetic changes within the postnatal period is the dysregulation of neurotrophic genes within the hippocampus. A mouse model exploring continuous low-level TCE exposure during postnatal days 1−42 (0.01 or 0.1 mg/mL TCE) resulted in a dose-dependent decrease in brain-derived neurotrophic factor (BDNF), neurotrophin-3 (NT-3), and nerve growth factor (NGF) within the hippocampus (Blossom, Melnyk, Cooney, Gilbert, & James, 2012). Postnatal TCE exposure in these mice also induced changes in the transmethylation and transsulfuration pathways, which affect glutathione synthesis and metabolism, but may also have implications in epigenetic regulation of neurotrophic gene expression. Furthermore, these measures suggest lower methylation capacity. Additionally, these mice exposed to TCE during developmental periods also had lower plasma levels of SAM + SAH and homocysteine and methionine levels suggesting that more homocysteine is being converted to cysteine, presumably in response to oxidative stress. These changes are all in the direction that would be expected to compromise cellular methylation reactions including DNA Methylation. In cerebellum, TCE-treated male mice had lower methionine levels and lower total 5-methylcytosine than untreated controls (Blossom, 2013). Overall, the data indicates that TCE produces oxidative stress and lower methylation capacity in brain regions during early postnatal exposure that may impart significant impacts on the immature CNS. Cellular mechanisms of toxicity are summarized in Fig. 18.2.

Neurodevelopmental pathology

Exposure to TCE during early development is associated with behavioral anomalies in both human and animal studies, implicating the neurotoxicity produced by TCE has long-term impact on neurobehavior. Over 30 years ago, Taylor and colleagues reported that gestational TCE exposure through postnatal day 21 in male rats resulted in increased exploratory and locomotor behavior, suggesting that hyperactivity may occur in exposed offspring (Taylor, Lagory, Zaccaro, Pfohl, & Laurie, 1985). These data were in contrast to the well-known effects of acute TCE exposure in adults, in which CNS depression (e.g., fatigue, dizziness, and drowsiness) is the predominant outcome (Annau, 1981; Lowry, Vandervort, & Polakoff, 1974). In more recent years, there has also been evidence that long-term environmental exposure to high levels of TCE results in poor neurobehavioral test performance (Reif et al., 2003). While the precise mechanisms behind these observations are still poorly understood, there are clear neuropathological changes associated with gestational and nonadult exposure to TCE, including specific brain regions with selective sensitivity, such as the hippocampus and cerebellum.

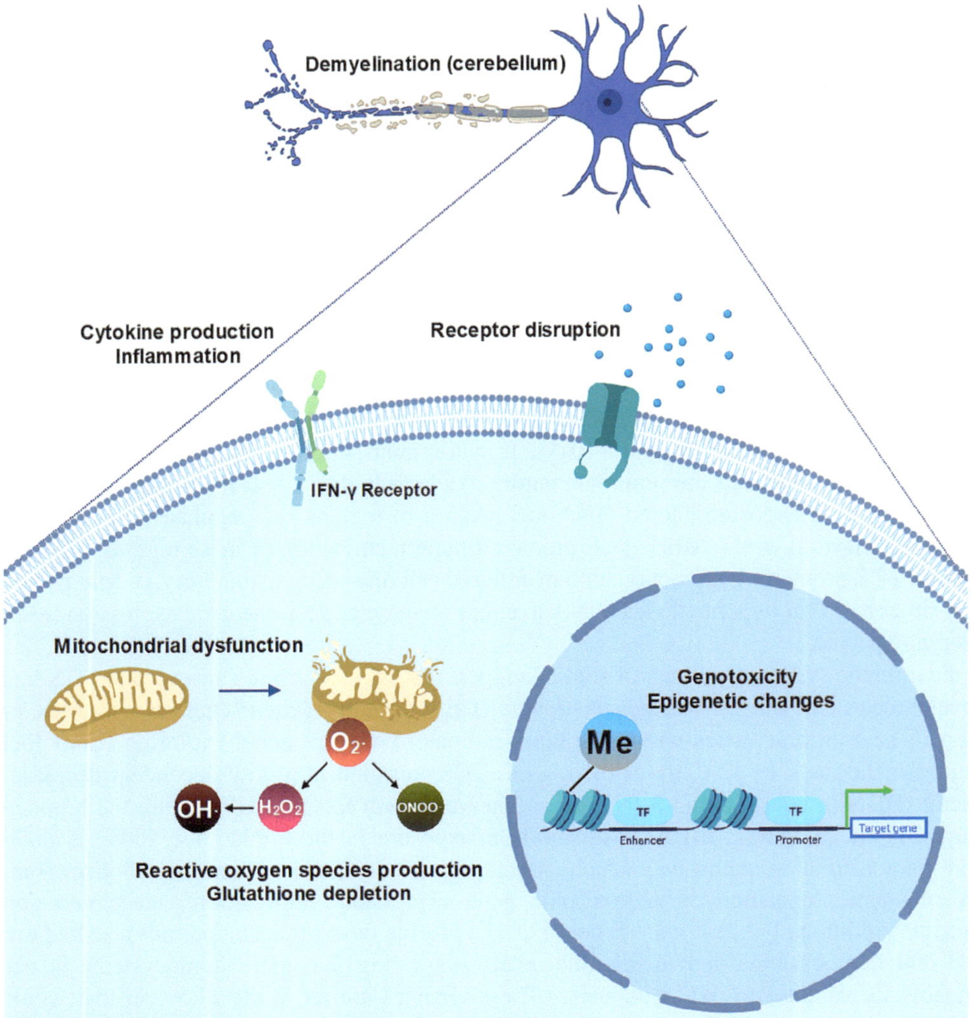

FIGURE 18.2 **Observed mechanisms of neurotoxicity by TCE and its metabolites.** Developmental exposure to TCE is associated with loss of myelin within cerebellar neurons. Mechanisms of intracellular dysfunction caused by TCE involve receptor disruption (voltage-sensitive calcium channels), inflammatory signaling changes involving interferon (IFN)-γ, mitochondrial toxicity and resulting ROS production, reduction of cellular glutathione through epigenetic modification, and genotoxicity from DNA damage. Figure generated with Biorender.com.

Hippocampal toxicity

There is evidence that the hippocampal brain region may be selectively vulnerable to perinatal TCE exposure, as prenatal and neonatal TCE exposure in drinking water reportedly resulted in decreased myelination of hippocampal neurons of the CA1 region (Isaacson, Spohler, & Taylor, 1990; Isaacson & Taylor, 1989). Similarly, TCE treatment in mice appears to affect long-term potentiation (LTP) in hippocampal slices, possibly as a mechanism that may lead to learning and/or memory impairment associated with TCE exposure (Ohta, 1998; Ohta, Saito, Saito, Kurasaki, & Hosokawa, 2001). Underlying this damage may be an increase in oxidant stress within the hippocampus, as postnatal exposure to TCE (postnatal days 0–42) resulted in elevated levels of oxidized glutathione, the major cellular ROS scavenging enzyme, as well as disruption within the glutathione synthesis pathways (Blossom et al., 2012). Additionally, TCE exposure during this postnatal time periods also resulted in increased protein markers for oxidative damage, measured by elevated 3-nitrotyrosine levels, a protein adduct of the peroxynitrite radical (Blossom et al., 2012).

In addition to oxidative damage, a limited amount of evidence implicates epigenetic changes in the hippocampus related to TCE exposure (Blossom et al., 2012). DNA methylation and histone acetylation play important roles in animal behavior; therefore, TCE-induced epigenetic changes in the hippocampus could affect the long-term behavioral consequences observed following gestational or early-life TCE exposure. In rodents, maternal behavior toward their pups in the first postnatal week had lifelong effects on pup behavior (Weaver et al., 2005) attributable in part to epigenetic changes in the hippocampal glucocorticoid receptor promoter (Curley, Davidson, Bateson, & Champagne, 2009). Epigenetics and DNA methylation are essential for memory formation where hippocampal RNA levels for the DNA methyltransferases (3a and 3b) were increased with fear conditioning in rats (Miller, Campbell, & Sweatt, 2008). Subsequent studies deleting either DNMT 1 or 3a in forebrain excitatory neurons allowed memory formation, but when both genes were deleted, synaptic plasticity, learning, and memory were inhibited (Feng et al., 2010). More recently, it was shown that methylation of histone residues in hippocampal neurons may facilitate memory formation (Collins, Sweatt, & Greer, 2019). Together, these mechanisms may underlie the observed experimental change in learning behavior from perinatal exposure to TCE; however, more data is needed to relate these pathological mechanisms to human TCE exposure.

Cerebellar abnormalities

Within the developing brain, the cerebellum appears to be specifically vulnerable to perinatal toxicant exposure, as this brain region is critical for motor coordination, social approach behavior, and exploration. There is evidence from animal models that early-life TCE exposure results in increased locomotor activity (Taylor et al., 1985; Waseem, Ali, Dogra, Dutta, & Kaw, 2001). These behavioral changes correlated with DNA hypomethylation within the same brain region, indicating that epigenetic changes induced by TCE within the cerebellum could be partially responsible for the observed motor behavioral hyperactivity within rodent models (Blossom et al., 2013). Additionally, postnatal exposure to TCE resulted in increased markers of inflammatory damage within the cerebellum, suggesting that TCE-induced oxidative stress could also influence the development of specific brain regions (Blossom, Melnyk, Li, Wessinger, & Cooney, 2017).

Immune modulation

There is strong evidence that TCE exposure during development alters the peripheral immune system, even when TCE exposure occurred solely during in utero (Blossom et al., 2017). Similarly, elevated levels of cytokines have been measured in the brain of mice exposed to low levels of TCE during development, suggesting that neuroinflammation may be either a primary or secondary outcome to the toxicity of TCE (Blossom et al., 2017; Meadows, Parker, Gilbert, Blossom, & DeWitt, 2017). In addition, there is evidence that TCE exposure produces a neuroinflammatory response in the adult rat brain, which results in the activation of microglia, the resident myeloid-derived immune cells of the brain (Liu et al., 2010).

Disruption to normal microglial activity during brain development is implicated in the development of schizophrenia (De Picker et al., 2017), with evidence that increased microglial activation may alter neuronal plasticity by pruning synapses (Bilbo & Schwarz, 2012). While there are currently no studies that have examined the longitudinal outcome of gestational TCE exposure on schizophrenia risk, there is evidence that the structurally related compound tetrachloroethylene (PERC) increases the rate of mental illness following prenatal exposure. Like TCE, PERC is a volatile organic solvent, which has been extensively used by the dry cleaning industry. In a retrospective cohort study, Aschengrau and colleagues report that prenatal and early childhood exposure to PERC in drinking water during resulted in an increased likelihood for schizophrenia, as well as bipolar disorder and posttraumatic stress disorder (Aschengrau et al., 2012). Similarly, in a population birth cohort study, children born to individuals within the dry cleaning industry had an increased incidence of schizophrenia development (Perrin et al., 2007). Whether this increased risk is due to microglial activation within the developing brain is unclear, but warrants further investigation.

Conclusions

Much of the current literature exploring TCE-induced damage to the brain has been collected from correlative studies, animal models utilizing adult exposures, or inferred through mechanisms of cellular damage from peripheral organs. The few studies directly linking TCE exposure during critical perinatal windows have corroborated these hypotheses, showing a clear link between TCE and neurodevelopmental pathology, which range from gross behavioral abnormalities to cellular and molecular dysfunction (Fig. 18.3). TCE remains an important environmental exposure to all populations; however, as much of the world continues to grow in use of TCE despite its known deleterious effects, there is still a strong need to investigate its toxicity throughout all stages of life, particularly during vulnerable CNS development periods.

Applications to other areas of development

TCE is toxic to multiple organ systems. One of the major noncancer developmental pathologies associated with TCE exposure during fetal development is cardiac toxicity, resulting in several types of structural malformations including

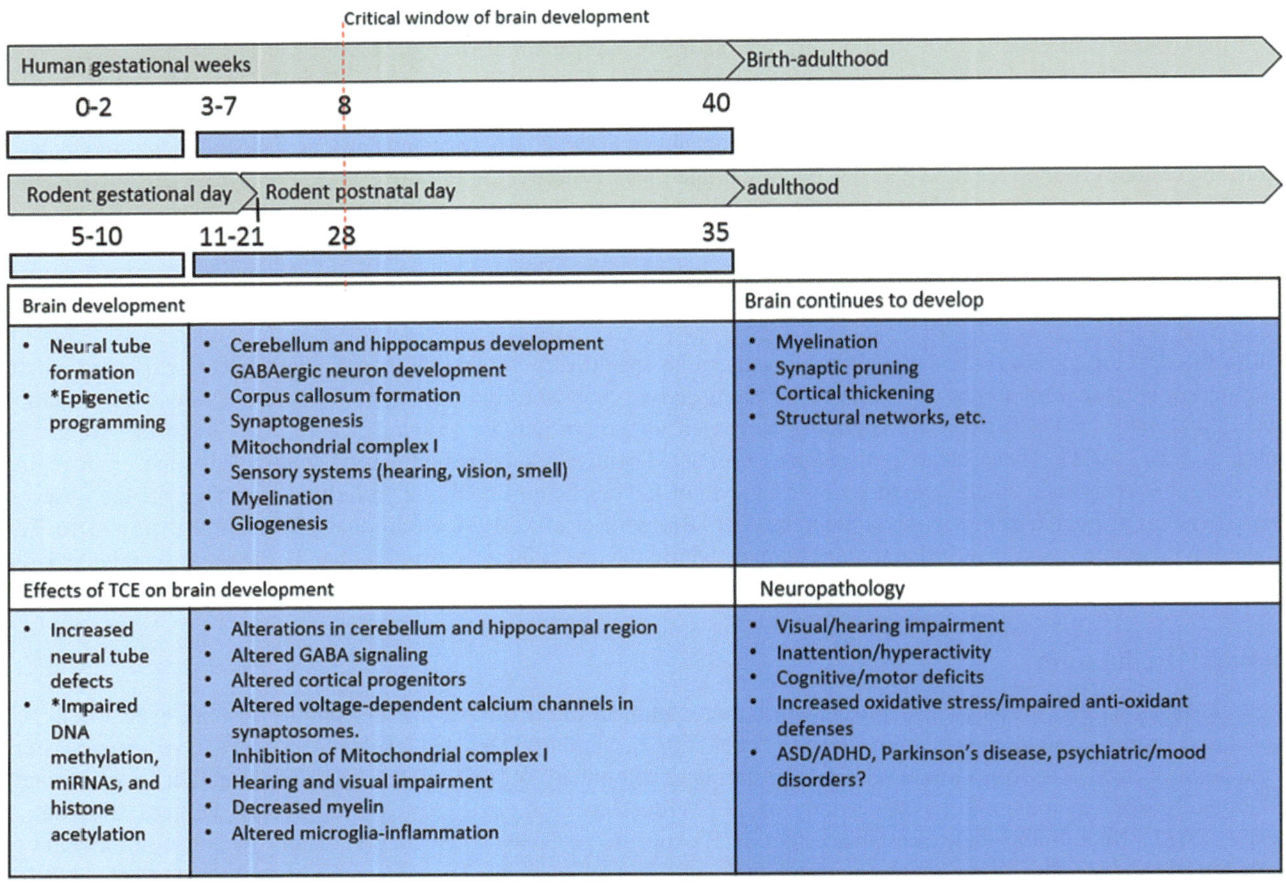

FIGURE 18.3 **Summary of key human/rodent brain developmental checkpoints that correspond with known or possible TCE-induced changes.** The broad steps of brain development span from gestation to adulthood. The period of development represents a sensitive window for perturbation due to environmental insults in both rodents and humans that may later manifest in neuropathology and/or disease. *Epigenetic programming spans throughout life with the potential to impact many neurological processes.

valvuloseptal, atrial septal, and ventricular septal defects, as well as aortic stenosis (Makris et al., 2016). There is evidence that gestational exposure to TCE affects calcium signaling within cardiac myocytes (Hoffmann, Heinroth, Richards, Plews, & Toraason, 1994; Selmin, Thorne, Caldwell, & Taylor, 2008) and influences gene expression through epigenetic modification (Makwana et al., 2010; Palbykin et al., 2011). The similarities of these mechanisms with those implicated in brain development suggest that perinatal exposure to TCE may cause pleiotropic damage within multiple organ systems. The resulting developmental defects, whether in the heart or brain, are likely influenced by a number of complex factors, including genetic variability, dosage and length of TCE exposure, gestational window, and other environmental interactions. Together, this evidence suggests that limiting TCE exposure is critical during the perinatal and developmental time period.

Key facts of developmental TCE exposure

- Exposure to TCE during the period when the brain is developing is associated with pathology in behavior and neurological health.
- TCE is one of the most abundant environmental contaminants and is the most commonly sampled organic contaminant within drinking water within the United States.
- Certain areas of the world have high levels of TCE contamination, such as EPA-designated Superfund Sites within the United States.
- TCE exposure during gestation or perinatal periods is of particular concern, as TCE can readily cross the placenta and affect the developing nervous system of the fetus.

- Areas of the brain such as the hippocampus and cerebellum appear to be the most vulnerable to TCE exposure during development.
- Several mechanisms of toxicity from developmental TCE exposure have been identified, including epigenetic reprogramming and elevated oxidative stress.
- Identifying the impact of TCE on neurodevelopmental toxicity is poorly understood and requires further research.

Summary points

- Perinatal TCE exposure is implicated in neurobehavioral deficits.
- Oxidative metabolism of TCE results in neurotoxic metabolites.
- Mitochondrial dysfunction, oxidative stress, neuroinflammation, and epigenetic modification are among the neurotoxic mechanisms implicated in TCE exposure during brain development.
- Direct action of TCE upon voltage-sensitive calcium channels may influence neurotransmitter signaling and alter synapse development.
- Certain brain regions, such as the hippocampus and cerebellum, appear to be selectively vulnerable to perinatal TCE exposure.

References

Andersson, K., Fuxe, K., Nilsen, O. G., Toftgard, R., Eneroth, P., & Gustafsson, J. A. (1981). Production of discrete changes in dopamine and noradrenaline levels and turnover in various parts of the rat brain following exposure to xylene, ortho-, meta-, and para-xylene, and ethylbenzene. *Toxicology and Applied Pharmacology, 60*, 535–548.

Annau, Z. (1981). The neurobehavioral toxicity of trichloroethylene. *Neurobehavioral Toxicology and Teratology, 3*, 417–424.

Aschengrau, A., Weinberg, J. M., Janulewicz, P. A., Romano, M. E., Gallagher, L. G., Winter, M. R., … Ozonoff, D. M. (2012). Occurrence of mental illness following prenatal and early childhood exposure to tetrachloroethylene (PCE)-contaminated drinking water: A retrospective cohort study. *Environmental Health, 11*, 2.

ATSDR. (2014). *Toxicological profile for trichloroethylene (TCE) draft for public comment.* Atlanta, GA: US Department of Health and Human Services.

Bakke, B., Stewart, P. A., & Waters, M. A. (2007). Uses of and exposure to trichloroethylene in U.S. industry: A systematic literature review. *Journal of Occupational and Environmental Hygiene, 4*, 375–390.

Bauernfeind, A. L., Barks, S. K., Duka, T., Grossman, L. I., Hof, P. R., & Sherwood, C. C. (2014). Aerobic glycolysis in the primate brain: Reconsidering the implications for growth and maintenance. *Brain Structure and Function, 219*, 1149–1167.

Beamer, P. I., Luik, C. E., Abrell, L., Campos, S., Martinez, M. E., & Saez, A. E. (2012). Concentration of trichloroethylene in breast milk and household water from Nogales, Arizona. *Environmental Science and Technology, 46*, 9055–9061.

Bilbo, S. D., & Schwarz, J. M. (2012). The immune system and developmental programming of brain and behavior. *Frontiers in Neuroendocrinology, 33*, 267–286.

Blossom, S. J., Cooney, C. A., Melnyk, S. B., Rau, J. L., Swearingen, C. J., & Wessinger, W. D. (2013). Metabolic changes and DNA hypomethylation in cerebellum are associated with behavioral alterations in mice exposed to trichloroethylene postnatally. *Toxicology and Applied Pharmacology, 269*, 263–269.

Blossom, S. J., Melnyk, S., Cooney, C. A., Gilbert, K. M., & James, S. J. (2012). Postnatal exposure to trichloroethylene alters glutathione redox homeostasis, methylation potential, and neurotrophin expression in the mouse hippocampus. *Neurotoxicology, 33*, 1518–1527.

Blossom, S. J., Melnyk, S. B., Li, M., Wessinger, W. D., & Cooney, C. A. (2017). Inflammatory and oxidative stress-related effects associated with neurotoxicity are maintained after exclusively prenatal trichloroethylene exposure. *Neurotoxicology, 59*, 164–174.

Boulton, S. J., Keane, P. C., Morris, C. M., McNeil, C. J., & Manning, P. (2012). Real-time monitoring of superoxide generation and cytotoxicity in neuroblastoma mitochondria induced by 1-trichloromethyl-1,2,3,4-tetrahydro-beta-carboline. *Redox Report, 17*, 108–114.

Burg, J. R., & Gist, G. L. (1999). Health effects of environmental contaminant exposure: An intrafile comparison of the trichloroethylene subregistry. *Archives of Environmental Health: An International Journal, 54*, 231–241.

Byrum, S. D., Washam, C. L., Patterson, J. D., Vyas, K. K., Gilbert, K. M., & Blossom, S. J. (2019). Continuous developmental and early life trichloroethylene exposure promoted DNA methylation alterations in polycomb protein binding sites in effector/memory CD4(+) T cells. *Frontiers in Immunology, 10*, 2016.

Chen, Y., Cai, J., Anders, M. W., Stevens, J. L., & Jones, D. P. (2001). Role of mitochondrial dysfunction in S-(1,2-dichlorovinyl)-l-cysteine-induced apoptosis. *Toxicology and Applied Pharmacology, 170*, 172–180.

Collins, B. E., Sweatt, J. D., & Greer, C. B. (2019). Broad domains of histone 3 lysine 4 trimethylation are associated with transcriptional activation in CA1 neurons of the hippocampus during memory formation. *Neurobiology of Learning and Memory, 161*, 149–157.

Cui, Y., Choudhury, S. R., & Irudayaraj, J. (2016). Epigenetic toxicity of trichloroethylene: A single-molecule perspective. *Toxicology Research, 5*, 641–650.

Curley, J. P., Davidson, S., Bateson, P., & Champagne, F. A. (2009). Social enrichment during postnatal development induces transgenerational effects on emotional and reproductive behavior in mice. *Frontiers in Behavioral Neuroscience, 3*, 25.

De La Rosa, V. Y., Asfaha, J., Fasullo, M., Loguinov, A., Li, P., Moore, L. E., ... Vulpe, C. D. (2017). Editor's highlight: High-throughput functional genomics identifies modulators of TCE metabolite genotoxicity and candidate susceptibility genes. *Toxicological Sciences, 160*, 111–120.

De Picker, L. J., Morrens, M., Chance, S. A., & Boche, D. (2017). Microglia and brain plasticity in acute psychosis and schizophrenia illness course: A meta-review. *Frontiers in Psychiatry, 8*, 238.

EPA. (2014) *TSCA work plan for chemical assessments: 2014 update*. US EPA).

Feldman, R. G., Niles, C., Proctor, S. P., & Jabre, J. (1992). Blink reflex measurement of effects of trichloroethylene exposure on the trigeminal nerve. *Muscle and Nerve, 15*, 490–495.

Feng, J., Zhou, Y., Campbell, S. L., Le, T., Li, E., Sweatt, J. D., ... Fan, G. (2010). Dnmt1 and Dnmt3a maintain DNA methylation and regulate synaptic function in adult forebrain neurons. *Nature Neuroscience, 13*, 423–430.

Forand, S. P., Lewis-Michl, E. L., & Gomez, M. I. (2012). Adverse birth outcomes and maternal exposure to trichloroethylene and tetrachloroethylene through soil vapor intrusion in New York state. *Environmental Health Perspectives, 120*, 616–621.

Gash, D. M., Rutland, K., Hudson, N. L., Sullivan, P. G., Bing, G., Cass, W. A., ... Prince, T. S. (2008). Trichloroethylene: Parkinsonism and complex 1 mitochondrial neurotoxicity. *Annals of Neurology, 63*, 184–192.

Gilbert, K. M., Blossom, S. J., Erickson, S. W., Broadfoot, B., West, K., Bai, S., ... Cooney, C. A. (2016). Chronic exposure to trichloroethylene increases DNA methylation of the Ifng promoter in CD4(+) T cells. *Toxicology Letters, 260*, 1–7.

Guha, N., Loomis, D., Grosse, Y., Lauby-Secretan, B., El Ghissassi, F., Bouvard, V., ... Straif, K. (2012). Carcinogenicity of trichloroethylene, tetrachloroethylene, some other chlorinated solvents, and their metabolites. *The Lancet Oncology, 13*, 1192–1193.

Hoffmann, P., Heinroth, K., Richards, D., Plews, P., & Toraason, M. (1994). Depression of calcium dynamics in cardiac myocytes–a common mechanism of halogenated hydrocarbon anesthetics and solvents. *Journal of Molecular and Cellular Cardiology, 26*, 579–589.

Hsieh, J. M., Tsirulnikov, K., Sawaya, M. R., Magilnick, N., Abuladze, N., Kurtz, I., ... Pushkin, A. (2010). Structures of aminoacylase 3 in complex with acetylated substrates. *Proceedings of the National Academy of Sciences of the United States of America, 107*, 17962–17967.

Isaacson, L. G., Spohler, S. A., & Taylor, D. H. (1990). Trichloroethylene affects learning and decreases myelin in the rat hippocampus. *Neurotoxicology and Teratology, 12*, 375–381.

Isaacson, L. G., & Taylor, D. H. (1989). Maternal exposure to 1,1,2-trichloroethylene affects myelin in the hippocampal formation of the developing rat. *Brain Research, 488*, 403–407.

Kennedy, R. L. (1974). General analgesia and anesthesia in obstetrics. *Clinical Obstetrics and Gynecology, 17*, 227–239.

Keverne, E. B., Pfaff, D. W., & Tabansky, I. (2015). Epigenetic changes in the developing brain: Effects on behavior. *Proceedings of the National Academy of Sciences of the United States of America, 112*, 6789–6795.

Kilburn, K. H. (2002). Is neurotoxicity associated with environmental trichloroethylene (TCE)? *Archives of Environmental Health: An International Journal, 57*, 113–120.

Kilburn, K. H., & Warshaw, R. H. (1992). Prevalence of symptoms of systemic lupus erythematosus (SLE) and of fluorescent antinuclear antibodies associated with chronic exposure to trichloroethylene and other chemicals in well water. *Environmental Research, 57*, 1–9.

Laham, S. (1970). Studies on placental transfer. Trichlorethylene. *IMS Industrial Medicine and Surgery, 39*, 46–49.

Lash, L. H., Parker, J. C., & Scott, C. S. (2000). Modes of action of trichloroethylene for kidney tumorigenesis. *Environmental Health Perspectives, 108*(Suppl. 2), 225–240.

Laslo-Baker, D., Barrera, M., Knittel-Keren, D., Kozer, E., Wolpin, J., Khattak, S., ... Koren, G. (2004). Child neurodevelopmental outcome and maternal occupational exposure to solvents. *Archives of Pediatrics and Adolescent Medicine, 158*, 956–961.

Lewis, E. M., & Kroll, K. L. (2018). Development and disease in a dish: The epigenetics of neurodevelopmental disorders. *Epigenomics, 10*, 219–231.

Liu, M., Choi, D. Y., Hunter, R. L., Pandya, J. D., Cass, W. A., Sullivan, P. G., ... Bing, G. (2010). Trichloroethylene induces dopaminergic neurodegeneration in Fisher 344 rats. *Journal of Neurochemistry, 112*, 773–783.

LoTurco, J. J., Owens, D. F., Heath, M. J., Davis, M. B., & Kriegstein, A. R. (1995). GABA and glutamate depolarize cortical progenitor cells and inhibit DNA synthesis. *Neuron, 15*, 1287–1298.

Lowry, L. K., Vandervort, R., & Polakoff, P. L. (1974). Biological indicators of occupational exposure to trichloroethylene. *Journal of Occupational Medicine, 16*, 98–101.

Makris, S. L., Scott, C. S., Fox, J., Knudsen, T. B., Hotchkiss, A. K., Arzuaga, X., ... Narotsky, M. G. (2016). A systematic evaluation of the potential effects of trichloroethylene exposure on cardiac development. *Reproductive Toxicology, 65*, 321–358.

Makwana, O., King, N. M., Ahles, L., Selmin, O., Granzier, H. L., & Runyan, R. B. (2010). Exposure to low-dose trichloroethylene alters shear stress gene expression and function in the developing chick heart. *Cardiovascular Toxicology, 10*, 100–107.

Meadows, J. R., Parker, C., Gilbert, K. M., Blossom, S. J., & DeWitt, J. C. (2017). A single dose of trichloroethylene given during development does not substantially alter markers of neuroinflammation in brains of adult mice. *Journal of Immunotoxicology, 14*, 95–102.

Miller, C. A., Campbell, S. L., & Sweatt, J. D. (2008). DNA methylation and histone acetylation work in concert to regulate memory formation and synaptic plasticity. *Neurobiology of Learning and Memory, 89*, 599–603.

Neal, M., & Richardson, J. R. (2018). Epigenetic regulation of astrocyte function in neuroinflammation and neurodegeneration. *Biochimica et Biophysica Acta - Molecular Basis of Disease, 1864*, 432–443.

Nilsson, K. B. (1986). Actions of 1,1,1-trichloroethane on the cAMP metabolism in mouse brain. *Acta Pharmacologica et Toxicologica, 59*, 362–369.

Nilsson, K. B. (1987). Effects of 1,1,1-trichloroethane on synaptosomal calcium accumulation in mouse brain. *Pharmacology and Toxicology, 61*, 215–219.

Ohta, M. (1998). [Changes in spatio-temporal patterns after long-term potentiation (LTP) in mouse hippocampal slices and effects of trichloroethylene on LTP]. *Hokkaido Igaku Zasshi, 73*, 365–378.

Ohta, M., Saito, T., Saito, K., Kurasaki, M., & Hosokawa, T. (2001). Effect of trichloroethylene on spatiotemporal pattern of LTP in mouse hippocampal slices. *International Journal of Neuroscience, 111*, 257–271.

Palbykin, B., Borg, J., Caldwell, P. T., Rowles, J., Papoutsis, A. J., Romagnolo, D. F., & Selmin, O. I. (2011). Trichloroethylene induces methylation of the Serca2 promoter in H9c2 cells and embryonic heart. *Cardiovascular Toxicology, 11*, 204–214.

Perrin, M. C., Opler, M. G., Harlap, S., Harkavy-Friedman, J., Kleinhaus, K., Nahon, D., … Malaspina, D. (2007). Tetrachloroethylene exposure and risk of schizophrenia: Offspring of dry cleaners in a population birth cohort, preliminary findings. *Schizophrenia Research, 90*, 251–254.

Phillips, R. V., Rieswijk, L., Hubbard, A. E., Vermeulen, R., Zhang, J., Hu, W., … Lan, Q. (2019). Human exposure to trichloroethylene is associated with increased variability of blood DNA methylation that is enriched in genes and pathways related to autoimmune disease and cancer. *Epigenetics, 14*, 1112–1124.

Reif, J. S., Burch, J. B., Nuckols, J. R., Metzger, L., Ellington, D., & Anger, W. K. (2003). Neurobehavioral effects of exposure to trichloroethylene through a municipal water supply. *Environmental Research, 93*, 248–258.

Robbiano, L., Baroni, D., Carrozzino, R., Mereto, E., & Brambilla, G. (2004). DNA damage and micronuclei induced in rat and human kidney cells by six chemicals carcinogenic to the rat kidney. *Toxicology, 204*, 187–195.

Ruijten, M. W., Verberk, M. M., & Salle, H. J. (1991). Nerve function in workers with long term exposure to trichloroethene. *British Journal of Industrial Medicine, 48*, 87–92.

Salama, M. M., El-Naggar, D. A., Abdel-Rahman, R. H., & Elhak, S. A. G. (2018). Toxic effects of trichloroethylene on rat neuroprogenitor cells. *Frontiers in Pharmacology, 9*, 741.

Sauerbeck, A., Hunter, R., Bing, G., & Sullivan, P. G. (2012). Traumatic brain injury and trichloroethylene exposure interact and produce functional, histological, and mitochondrial deficits. *Experimental Neurology, 234*, 85–94.

Selmin, O. I., Thorne, P. A., Caldwell, P. T., & Taylor, M. R. (2008). Trichloroethylene and trichloroacetic acid regulate calcium signaling pathways in murine embryonal carcinoma cells p19. *Cardiovascular Toxicology, 8*, 47–56.

Shafer, T. J., Bushnell, P. J., Benignus, V. A., & Woodward, J. J. (2005). Perturbation of voltage-sensitive Ca^{2+} channel function by volatile organic solvents. *Journal of Pharmacology and Experimental Therapeutics, 315*, 1109–1118.

Sharma, R. K., Candelario-Jalil, E., Feineis, D., Bringmann, G., Fiebich, B. L., & Akundi, R. S. (2017). 1-Trichloromethyl-1,2,3,4-tetrahydro-beta-carboline (TaClo) alters cell cycle progression in human neuroblastoma cell lines. *Neurotoxicity Research, 32*, 649–660.

Shelton, K. L., & Nicholson, K. L. (2014). Pharmacological classification of the abuse-related discriminative stimulus effects of trichloroethylene vapor. *Journal of Drug and Alcohol Research, 3*, 235839.

Taylor, D. H., Lagory, K. E., Zaccaro, D. J., Pfohl, R. J., & Laurie, R. D. (1985). Effect of trichloroethylene on the exploratory and locomotor activity of rats exposed during development. *The Science of the Total Environment, 47*, 415–420.

Till, C., Koren, G., & Rovet, J. F. (2001). Prenatal exposure to organic solvents and child neurobehavioral performance. *Neurotoxicology and Teratology, 23*, 235–245.

Vernon, R. J., & Ferguson, R. K. (1969). Effects of trichloroethylene on visual-motor performance. *Archives of Environmental Health: An International Journal, 18*, 894–900.

Waseem, M., Ali, M., Dogra, S., Dutta, K. K., & Kaw, J. L. (2001). Toxicity of trichloroethylene following inhalation and drinking contaminated water. *Journal of Applied Toxicology, 21*, 441–444.

Weaver, I. C., Champagne, F. A., Brown, S. E., Dymov, S., Sharma, S., Meaney, M. J., & Szyf, M. (2005). Reversal of maternal programming of stress responses in adult offspring through methyl supplementation: Altering epigenetic marking later in life. *Journal of Neuroscience, 25*, 11045–11054.

White, R. F., Feldman, R. G., Eviator, II, Jabre, J. F., & Niles, C. A. (1997). Hazardous waste and neurobehavioral effects: A developmental perspective. *Environmental Research, 73*, 113–124.

Wu, C., & Sun, D. (2015). GABA receptors in brain development, function, and injury. *Metabolic Brain Disease, 30*, 367–379.

Chapter 19

Developmental neurotoxicity of the herbicide atrazine

Janiel Ahkin Chin Tai and Jennifer L. Freeman

School of Health Sciences, Purdue University, West Lafayette, IN, United States

List of abbreviations

5-HIAA 5-Hydroxyindoleacetic
ACTH Adrenocorticotropic hormone
ATZ Atrazine
CORT Corticosterone
CRH Corticotropin-releasing hormone
DACT Diaminochlorotriazine
DEA Deethyl-atrazine
DIA Deisopropyl-atrazine
DOPAC 3,4-Dihydroxyphenylacetic acid
EDC Endocrine disrupting chemical
HPA Hypothalamic–pituitary–adrenal
HPG Hypothalamic–pituitary–gonadal
HPT Hypothalamic–pituitary–thyroid
HVA Homovanillic acid
Kiss1r g-Protein-coupled receptor 54
KNDy Kisspeptin–neurokinin B–dynorphin
LH Luteinizing hormone
MCL Maximum contaminant level
PND Postnatal day
ppb Parts per billion
T_3 Thyroxine
T_4 Triiodothyronine
TSH Thyroid-stimulating hormone

Mini-dictionary of terms

Dopamine Catecholamine neurotransmitter that regulates motor movement, learning capabilities, and addiction to substances.
Endocrine Disrupting Chemical (EDC) Chemicals that interfere with the body's endogenous endocrine system that do not always act in a dose-dependent manner.
Epidemiology Science of using statistics relating to human health parameters such as incidence to understand human health.
Hypothalamic–pituitary–adrenal axis (HPA) System composed of the hypothalamus and pituitary in the brain and the adrenal gland, which regulates the body's stress response and glucocorticoid production.
Hypothalamic–pituitary–gonadal axis (HPG) System composed of the hypothalamus and pituitary in the brain and the gonads that regulate reproduction and sex steroid hormone production.
Serotonin Monoamine neurotransmitter that regulates many systems in the body such as vasoconstriction and mood. Alterations can cause mood disorders such as depression.

Diagnosis, Management and Modeling of Neurodevelopmental Disorders. https://doi.org/10.1016/B978-0-12-817988-8.00019-1

Introduction

Atrazine (ATZ) is a synthetic agricultural herbicide used to control broad leafy weeds. ATZ is commonly applied to corn and other crop fields, though it can be used on lawns as well (British Crop Protection Council, 1977; Li, Li, Bi, & Li, 2019; Russart & Rhen, 2016; USGS, 2016). ATZ works by interfering with the electron transport chain to interrupt photosynthesis. Resistant plants can absorb and metabolize ATZ, but are not vulnerable to impeded photosynthesis (Cheremisinoff & Rosenfeld, 2011; Eldridge, Wetzel, & Tyrey, 1999). ATZ was introduced in 1958 and is the second most commonly applied herbicide in the United States (Cheremisinoff & Rosenfeld, 2011; Ochoa-Acuña, Frankenberger, Hahn, & Carbajo, 2009; USGS, 2016). In 2003, the European Union banned its use due to contamination concerns (Bethsass & Colangelo, 2006). The triazine herbicides are potential carcinogens and endocrine disrupting chemicals (EDC).

Toxicant profile

Triazine herbicides consist of nitrogenous heterocyclic rings. These herbicides work by binding to the Q_B protein in photosystem II of broad leafy weeds. This protein binding blocks the movement of electrons through the photosynthetic electron transport chain. Resistant crop plants have rapid metabolic detoxification through glutathione S-transferase and dealkylation through cytochrome P_{450} (LeBaron, McFarland, & Burnside, 2008). ATZ is classified as a triazine herbicide and is the most commonly used in its class. ATZ is composed of a triazine ring with an ethylamine, propylamine, and chlorine as its functional groups (Fig. 19.1). ATZ is a white, odorless powder that is moderately soluble in water.

Toxicokinetics

ATZ can break down into over 40 metabolites (Fig. 19.1). ATZ metabolites appear to have toxicities comparable to or lower than ATZ. Syngenta, the manufacturer of ATZ, selected the metabolites diaminochlorotriazine (DACT), deethyl-atrazine (DEA), and deisopropyl-atrazine (DIA) for monitoring through the Syngenta Monitoring Program. These daughter compounds can be generated through degradation of ATZ in surface water and soil via photolysis of microorganisms and have been found in animal tissue. ATZ is the most commonly detected, followed by DEA, DACT, and DIA in rural wells in the United States. Another metabolite, hydroxyatrazine, is also commonly found in the environment. This metabolite is not water-soluble, causing crystal formation and is considered a nephrotoxicant. Although the half-life of ATZ is 125 days, the half-life of its metabolites is around 20–50 days (WHO, 2011).

ATZ and its metabolites can be absorbed gastrointestinally and minimally through the skin. ATZ is reported to be rapidly eliminated through urine and feces. Circulating in the system, ATZ and its metabolites have an affinity for protein, binding readily to erythrocytes, liver, spleen, and kidney. In humans, ATZ is primarily metabolized via dealkylation (Barr, Panuwet, Nguyen, Udunka, & Needham, 2007; WHO, 2011).

Occupational exposure

Chemical pesticide applicators and factory workers are the occupational population of concern. Although applicators may only get exposure seasonally when the herbicide is applied, factory workers who manufacture the chemical have the

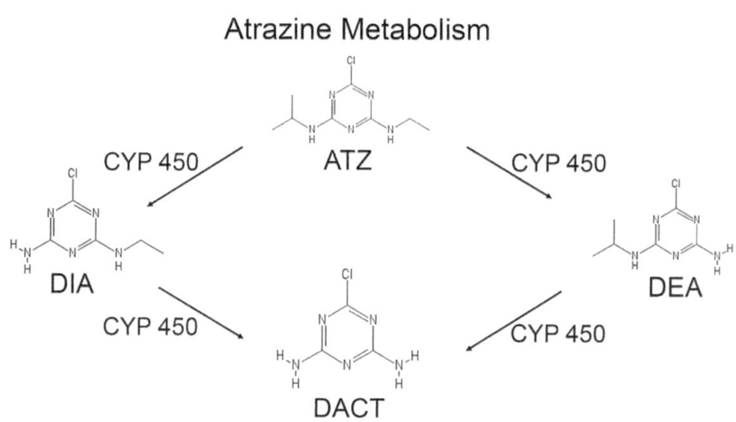

FIGURE 19.1 Atrazine metabolism. The parent chemical ATZ can break down into over 40 different metabolites, but many are not regulated or monitored. ATZ detection is most common, followed by DEA, DACT, and DIA in rural wells of the United States, which are the primary ATZ metabolites.

possibility of chronic exposure. A triazine factory in Switzerland in a matched cohort study found an association between increased incidence of gastritis and gastroenteritis (Ware, 1987). In response to health concerns, the Occupational Safety and Health Administration set a limit of 5 g ATZ/m^3 of exposure in an 8-hour workday. ATZ is also registered as a restricted use pesticide to prevent uncertified applicators from applying (ATSDR, 2003).

Environmental exposure

Exposure routes via atmosphere, food, and groundwater are reported; however, water contamination is the main route of environmental exposure for several reasons. Many areas of heavy ATZ application have clay soils or other low filtration soils where the herbicide not absorbed by crops or weeds is only partially absorbed by the soil. ATZ also does not bind efficiently to soil particles. Combined with heavy application in the Midwestern United States, these factors contribute to the herbicide being highly mobile in runoff. These factors and ATZ's persistence in the environment make chemical exposure a concern (Gammon, Aldous, Carr, Sanborn, & Pfeifer, 2005; Meffe & de Bustamante, 2014; Wang & Xie, 2012; WHO, 2011). ATZ contaminates potable water, and although the US EPA has set the maximum contaminant level (MCL) at 3 μg/L, concentrations can greatly exceed reaching up to 30 μg/L or even 100 μg/L (Blanchard & Lerch, 2000; WHO, 2011). Since ATZ frequently contaminates potable water supplies, is applied heavily in the Midwestern United States, and can be found at concentrations above the MCL, understanding the health risks from environmental exposures and mechanisms of toxicity is also important.

Health risks

Populations of concern are centered around occupational exposures and residents who live in areas near crop fields where ATZ is applied. Runoff from crop fields has the ability to infiltrate lakes, rivers, streams, and groundwater. Epidemiological studies link ATZ contaminated potable water to developmental toxicity, birth defects, and risks for cancer and neurodegenerative disease. Maternal exposure to ATZ during gestation is linked with increased risk of preterm birth as well as birth defects (Agopian, Lupo, Canfield, & Langois, 2013; Migeot et al., 2013; Stayner et al., 2017; Waller, Paul, Peterson, & Hitti, 2010). Although a clear association with ATZ exposure and cancer is controversial, trends for increased risk of certain types of tumors and thyroid cancer have been suggested (Freeman et al., 2011; Rusiecki et al., 2004). In addition, a few studies suggest increased incidence in Parkinson's disease (PD) with pesticide exposure with ATZ as a potential pesticide of interest (Li et al., 2014a,b; Priyadarshi, Khuder, Schaub, & Shrivastava, 2000; Semchuk, Love, & Lee, 1992). Overall, these outcomes suggest concern for developmental exposure and neurological disorders. Although there is some controversy and mostly associative data in epidemiological studies, investigations have been conducted in various model organisms to further define toxicity outcomes.

Toxicity studies

Studies in various model organisms at a range of concentrations have assessed health concerns related to ATZ exposure. Chronic exposure to ATZ at high doses caused earlier onset of mammary gland tumors in Sprague Dawley rats (Wetzel et al., 1994). The mechanism of tumor induction is suggested to be related to ovarian function. Although this exposure level was above biologically relevant exposure levels, this study raised questions regarding atrazine's safety (WHO, 2011). This study also helped spur restricted personnel exposure levels. A follow-up study noted a reduction of luteinizing hormone (LH), which led to disrupted estrous cycle with 100, 200, or 300 mg/kg body weight of ATZ/day (Cooper, Stoker, Tyrey, Goldman, & McElroy, 2000). LH and prolactin levels are regulated by the hypothalamus, and there is speculation about ATZ disruption occurring at the level of the hypothalamus or the kisspeptin neurons above the hypothalamus. Tumor formation in Sprague Dawley rats occurs by altering catecholamine function and gonadotropin-releasing hormone in the hypothalamus. This ultimately affects the HPG axis (WHO, 2011). Studies in other model organisms also support ATZ's role as an EDC along the HPG axis.

As an EDC, ATZ does not necessarily act in a dose-dependent manner, and several studies have displayed sex differences. A study exposing embryonic snapping turtles to 2 or 40 μg/L ATZ determined that androgen receptors and kisspeptin 1 receptors were upregulated in both sexes at hatching, but female-specific upregulation of aromatase was observed. The gene prodynorphin, which is associated with the HPA axis, was also upregulated (Russart and Rhen, 2016). Six months later, at 2 μg/L, aromatase and prolactin-releasing hormone mRNA expression was increased in both sexes.

In vitro studies utilizing the H295R adrenocortical carcinoma cell line had a 1.5–2-fold increase in aromatase mRNA activity with doses ranging from 0 to 30 μM ATZ, with a follow-up study determining that DEA and DIA also upregulated aromatase in the same cell line, as well as breast and placental cells (Suzawa & Ingraham, 2008). Building on this study, Caron-Beaudoin, Denison, and Sanderson (2016) found promoter regions PIII and I.3 transcription levels to be increased for aromatase.

Neuroendocrine system toxicity

Many studies support the hypothalamus as the site of action for ATZ (ATSDR, 2003; WHO, 2011). The HPG, HPA, and HPT axes in particular work together to regulate synthesis and sensitivity to stress, sex steroid hormones, growth, and reproductive function.

Hypothalamic–pituitary–gonadal axis

The specific site within the hypothalamus where regulation of the HPG axis starts is the infundibular nucleus in humans and arcuate nucleus in rats. To understand how ATZ disrupts hypothalamic function, it is important to understand how the HPG axis works (Fig. 19.2). *KISS1* gene produces kisspeptin, which is important for normal reproductive function as well as its G-protein-coupled receptor 54 (*KISS1R*). Kisspeptin and *KISS1R* are expressed in several places throughout the body, but hypothalamic *KISS1* is responsible for regulating metabolic information to the HPG axis. This peptide hormone is also important for signaling GnRH neurons for the onset of puberty. Kisspeptin–neurokinin B–dynorphin (KNDy) neurons coordinate pulsatile GnRH secretions in GnRH neurons via KISS1R receptors. GnRH is then released into the anterior pituitary gonadotropins. The anterior pituitary creates follicle-stimulating hormone (FSH) and LH, which are carried throughout the body in the bloodstream. LH secretion then stimulates theca cell development in the secondary follicle of the ovaries by binding to an LH receptor. Theca cells synthesize estradiol, which binds to KNDy neurons to stop pulsatile secretion of GnRH to regulate the system (Karapanou & Papadimitriou, 2010; Magoffin, 2005; Marques, Skorupskaite, George, & Anderson, 2000).

A handful of developmental studies delve into ATZ exposure and its effects on the HPG axis (Table 19.1). For example, an embryonic ATZ exposure in snapping turtles reported changes in genes associated with the HPG axis (Russart & Rhen, 2016). Sex- and age-dependent changes in *Kiss1* and its receptor *Kiss1R* were found. Young female Japanese quails exposed developmentally to ATZ had decreases in hypothalamic *GnRH* and underdeveloped reproductive organs (Qin et al., 2015). Many studies focus on adult ATZ exposure in various rat models along the HPG axis and report decreases in *Kiss1* expression, decreases in GnRH pulse frequency and amplitude, as well as a surge in LH (Fig. 19.3). These effects have the potential to alter downstream gonadal development with several studies noting apoptosis in female

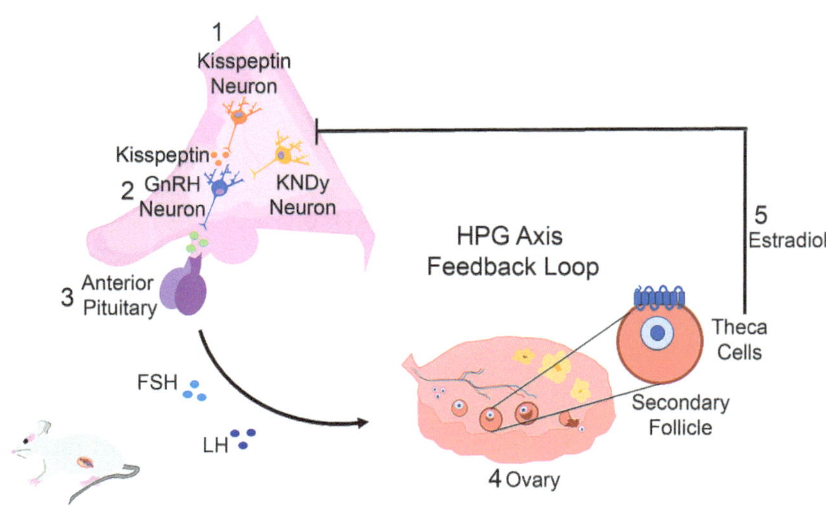

FIGURE 19.2 **HPG axis in rats.** Kisspeptin neurons secrete kisspeptin (1), which binds to receptors on the GnRH neuron (2). The GnRH neurons send pulsatile secretion of GnRH to the anterior pituitary (3). GnRH binds to receptors on gonadotrophs in the anterior pituitary releasing FSH and LH. LH stimulates thecal cell growth in the secondary follicle of the ovary (4). Theca cells generate estradiol, which acts on KNDy neurons to inhibit GnRH activation (5).

TABLE 19.1 ATZ neurotoxicity along the HPG axis.

References	Model organism	Results	ATZ concentration
Russart and Rhen (2016)	Developing snapping turtles	Males had higher *Kiss1* and *GnRH* mRNA in embryos; *Kiss1R* higher in hatchling females	2, 40 μg/L during embryogenesis
Foradori et al. (2009)	Ovariectomized adult female Wistar rats	High dose of ATZ altered LH pulse and amplitude and suppressed GnRH	50, 100, 200 mg/kg
Qin et al. (2015)	Developing female quails	Underdeveloped ovaries/oviduct; decreased hypothalamic *GnRH* mRNA, LH in pituitary, and FSH in ovary; upregulated FSH and prolactin in pituitary	50, 250, 500 mg/kg at 18 days for 45 days
Wilhelms, Fitzpatrick, Scanes, and Anderson (2006)	Developing quails	At 14 days, hatchling and ovarian weight decreased as well as circulating progesterone at highest concentration	123, 246, 504 μg/kg *in ovo*
Goldman et al. (2013)	Adult Long Evans female rats	4 days of ATZ suppress LH surge; a single exposure elevated *Kiss1* mRNA expression	40 days of 100 mg/kg or a single exposure
Kimura et al. (2019)	Ovariectomized female rats	Increased LH surge and decrease in *Kiss1* mRNA in AVPV	100 mg/kg

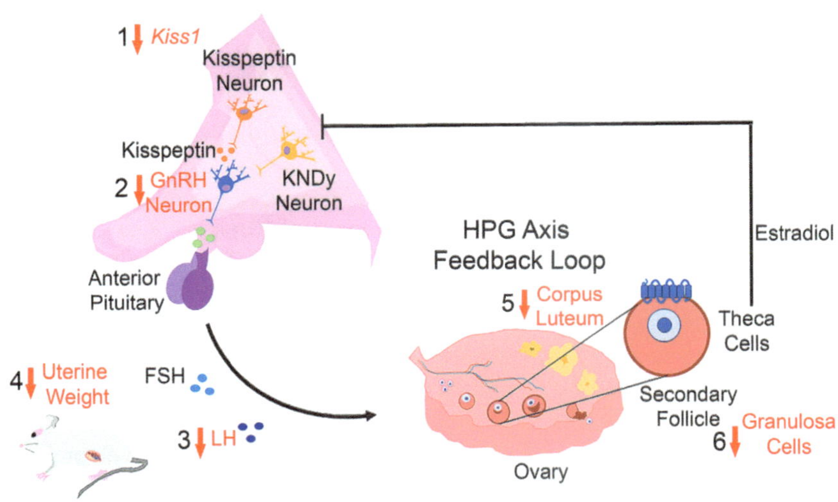

FIGURE 19.3 ATZ neurotoxicity along the HPG axis. ATZ is shown to decrease expression of *Kiss1* (1) and decrease GnRH neurons (2) in the hypothalamus. A decrease in the LH surge is widely documented with ATZ exposure (3). ATZ has been associated with uterine weight loss (4), corpus luteum loss (5), and a decline in granulosa cell quality (6). Integrating the findings from multiple studies in rat models implicates ATZ acting through the neuroendocrine system to influence reproductive health.

reproductive cells and a decrease in uterine and ovarian weight (Goldman, Davis, Murr, & Cooper, 2013; Kimura et al., 2019; Marques et al., 2000). Although no widely accepted conclusion is drawn on what reproductive cells are specifically targeted with ATZ exposure, there is a trend in decreased integrity in female reproductive tissue that can ultimately affect reproductive function. Sex steroids generated from the HPG axis to promote reproductive function can also regulate the HPA axis.

Hypothalamic–pituitary–adrenal axis

The HPA axis regulates stress-related hormones, particularly corticotropin-releasing hormone (CRH), adrenocorticotropic hormone (ACTH), and corticosterone (CORT). CRH is synthesized in the paraventricular nucleus within the hypothalamus to help synthesize ACTH in the pituitary and is considered a neuropeptide mediator of stress and HPA activation. CRH

TABLE 19.2 ATZ neurotoxicity along the HPA axis.

Reference	Model organism	Results	ATZ concentration
Foradori et al. (2018)	Adult female Sprague Dawley rats	Changes in adrenal morphology and indirect adrenal stimulation to generate CORT (CRH receptor important for ATZ modulating corticosterone increase); ATZ, DIA, and DEA elevate plasma corticosterone concentrations	100, 200 mg/kg of ATZ, DEA, DIA, or DACT
Russart and Rhen (2016)	Developing snapping turtles	Increase in hypothalamic POMC in hatchlings and at 6 months in females	2, 40 μg/L during embryogenesis
Fraites et al. (2009)	Adult female Long Evans rats	Increase in ACTH, progesterone, and CORT; DACT had highest plasma levels, but didn't affect HPA hormone release	60.5, 70 mg/kg of ATZ, DIA, or DACT
Riffle et al. (2014)	Castrated male Wistar rats	Increase in adrenal proteins associated with stress; Increase in serum CORT and progesterone	5, 25, 75, 200 mg/kg
Lin et al. (2014)	C57BL/6 mouse dams	Increased anxiety-like behavior, decreased perirhinal cortex serotonin, and perirhinal cortex serotonin homeostasis	1.4 mg/kg/day (3 mg/L)

releases ACTH to stimulate glucocorticoids (e.g., CORT) in response to stress in a negative feedback loop. Stress during proestrus or exposure to estradiol in females causes an increase in plasma levels of ACTH and generally higher levels of CRH, CORT, and progesterone. Conversely in males, testosterone decreases ACTH levels and gonadectomy increases ACTH. Although the HPA axis regulates stress hormones and is also regulated by sex hormones, it can also indirectly regulate the HPG axis by altering levels of GnRH, pituitary sensitivity to GnRH, and LH secretion (Fraites et al., 2009; Toufexis, Rivarola, Lara, & Viau, 2014).

Few studies have addressed developmental ATZ exposure and changes in glucocorticoids or other endpoints for HPA activation from a neurotoxicity standpoint (Table 19.2). A study with embryonic ATZ exposure in snapping turtles looked at immediate and later in life changes in genes associated with stress in the HPA pathway. The authors found upregulated proopiomelanocortin (POMC), a precursor to ACTH, at hatching and at 6 months in females. This implicates developmental ATZ exposure can cause changes in the HPA later in life. Many studies focus on various stress responses in adult rats. These studies support ATZ and some of its metabolites causing increases in blood plasma levels of CORT and ACTH (Foradori et al., 2009, 2018; Riffle et al., 2014). A study in adult female Sprague Dawley rats found that pituitary removal prevented CORT synthesis, implicating that the pituitary and specifically the CRH receptor are necessary for ATZ to modulate increases in CORT levels (Foradori et al., 2018). More studies focusing on developmental ATZ exposure and its effect on the HPA axis will shed light on how ATZ neurotoxicity affects stress responses at the neuroendocrine level.

Serotonin and the HPA axis

Higher circulating levels of CORT and other stress-related hormones and proteins in females not only modulate the sensitivity to sex steroids estrous cyclicity by altering GnRH and LH production through the HPG axis exposure, but also increase risk for developing psychological disorders (Fraites et al., 2009; Toufexis, Rivarola, Lara, & Viau, 2014). The developmental origin of adult disease hypothesis suggests that toxicant exposure during development can increase the risk of certain diseases during adulthood, so developmental exposure could also increase risk for adult neurological disorders. The HPA axis is also important for regulation of psychological disorders such as depression and anxiety, which are higher in human females. A study exposed zebrafish to 0, 0.3, 3, or 30 μg/L during embryogenesis and found a significant reduction in 5-hydroxyindoleacetic acid (5-HIAA), a major metabolite of serotonin in adult females. A subset of genes was downregulated throughout the serotonergic pathway (Wirbisky et al., 2015). Another study in C57BL/6 mice used a combination of behavioral and neurochemical assays to determine ATZ effects on the brain and behavior. Offspring were exposed in utero from gestational day 6 to postnatal day (PND) 23 to 0 or 3 mg/L. The authors found decreases in perirhinal cortex serotonin in adult females and increases in anxiety-like behavior. These results suggest that ATZ exposure during development causes sex-specific alterations in serotonergic systems that can ultimately affect behavior (Lin et al., 2014).

Hypothalamus–pituitary–thyroid (HPT) axis

The HPT axis is important for regulating metabolism, growth, and reproduction. The thyroid produces the hormones thyroxine (T_3) and triiodothyronine (T_4) that regulate activity of thyroid receptor signaling found on Sertoli cells across vertebrates. At the pituitary level, there is speculation that GnRH causes an increase in TSH and T_3 in some amphibians and fish (Castañeda Cortés et al., 2014). Thyroid hormones can affect gonadal development and androgen levels, particularly in males. Effects of ATZ and its metabolites have been noted in thyroid histology and outcomes in gonadal pubertal development, but not many studies to date have linked endpoints in the HPT axis to ATZ neurotoxicity (Stoker, Guidici, Laws, & Cooper, 2002).

Dopamine

ATZ exposure has been shown to affect neurotransmission, particularly with dopamine (DA), which has implications for diseases such as PD (Table 19.3). DA is an important monoaminergic transmitter responsible for facilitating motor movement with the main metabolites being 3,4-dihydroxyphenylacetic acid (DOPAC) and homovanillic acid (HVA) (Li et al., 2014a,b; Meiser, Weindl, & Hiller, 2013). In a study using 25 and 50 mg/kg of ATZ in pregnant mice with exposure from gestational day 0 to PND 1, offspring exposed in utero had decreased levels of DA, its precursors and transporters (DAT, TH), and a decrease in expression of genes and proteins necessary for mesencephalic DA neuron growth (Nurr1, Vmat2) 6 months after developmental exposure ceased in the substantia nigra (Li et al., 2014a). Results from this study (Li et al., 2014a) and a second study with ATZ exposure during juvenile development showed affects to the dopaminergic neurons during these growth periods (Li et al., 2014b). These results are particularly important because Nurr1, which was changed in both studies, is important for regulating development of dopaminergic neurons. Follow-up studies on ATZ exposure in pubertal rats around 20 days old were performed with 0, 50, 100, or 200 mg/kg with the same endpoints and found similar results at the highest ATZ exposure (Li et al., 2015). In contrast, CB57BL/6 mice exposed in utero from gestational day 6 to PND 23 at 3 mg/L ATZ had increases in striatal DA and decreases in DA turnover as juveniles at PND 35. As adults, DA levels were not altered, but increases in metabolites methoxytyramine and 5-HT were detected in both sexes and decreases in the metabolite DOPAC in females were found. Behavioral analysis revealed hyperactivity as juveniles in exposed animals and increases in anxiety-like behavior in females (Lin et al., 2014) (Table 19.2). These findings suggest that developmental exposure to ATZ can have lasting neurological impacts into adulthood.

A study has also shown microglial activation with ATZ exposure, which signals inflammation and brain pathology (Ma, Wu, Zhang, He, & Li, 2015). A study using BV-2 microglia and SH-SY5Y neuroblastoma cells found dose-dependent relationships with SH-SY5Y cell viability, an increase in ROS in BV-2 microglial cells, and proteins associated with inflammation and microglial activation when exposed to 12.5, 25, or 50 μM ATZ.

TABLE 19.3 Summary of ATZ exposure and developmental DA effects.

Reference	Model organism	Results	ATZ concentration
Ma et al. (2015)	SH-SY5Y (neuroblastoma) BV-2 cells (microglia)	Microglia increase ROS, inflammatory cytokines, and decrease cell viability	12.5, 25, 50 μM
Li et al. (2014a)	Sprague Dawley rat pups exposed in utero	Decrease in DA, precursors, and genes for DA neurogenesis in substantia nigra	0, 25, 50 mg/kg gestational day 0-PND 1
Li, He, Ma, Wu, and Li (2015)	Pubertal Sprague Dawley male rats	Decrease in DA and precursors and decrease in genes for mesencephalic DA neurogenesis in substantia nigra at highest ATZ concentration	0, 50, 100, 200 mg/kg from 27 to 54 days old
Li et al. (2014b)	Pubertal Sprague Dawley rats	Decrease in DA, its precursors, and genes for mesencephalic DA neurogenesis in the substantia nigra	0, 25, 50 mg/kg from 22 to 62 days old
Bardullas, Giordano, and Rodriguez (2011)	Male juvenile Sprague Dawley rats	Impaired motor coordination after 10–12 month exposure; Striatal decrease of DA	10 mg/kg at 21 days until rat was 300 g

Conclusions

Although ATZ has been banned by the European Union, it is still a common herbicide used in the Midwestern United States and other global regions. Sex differences with ATZ exposure have been shown, with sex-dependent changes in genes, proteins, and circulating hormone levels. Although some studies show ATZ acting in a dose-dependent manner, other studies have evidence to support ATZ as an EDC that does not act in a dose-dependent manner. Association with cancer is still controversial, but health concerns related to developmental outcomes and neurotoxicity have been documented in several model organisms. As an EDC, ATZ's site of action is proposed to be the neuroendocrine system and in particular the hypothalamus. The hypothalamus is an important site in the brain that modulates development, reproduction, and stress through multiple endocrine axes. Studies support disruption of the HPG axis with alterations in female reproduction and the HPA pathway related to stress with a developmental ATZ exposure. Furthermore, ATZ is reported to alter the dopaminergic system.

Although there is growing evidence to support the hypothalamus as ATZ's site of action, there are still gaps in the literature that need to be explored to further elucidate ATZ-mediated changes and to what extent the HPG, HPA, and HPT axes are affected. In addition, more developmental studies exploring effects of ATZ on all three axes will provide a comprehensive understanding on how ATZ neurotoxicity impairs hypothalamic function. Moreover, there is a need to further our understanding on the influences of a developmental ATZ exposure on neurotransmission as studies to date have primarily focused on the dopaminergic system with a few studies on the serotonergic system.

Application to other areas of development

We summarized the effects of ATZ on the neurological system, focusing on the developing central nervous system. While most studies assessed downstream endocrine health outcomes, it is now generally accepted that ATZ's primary site of action is the neuroendocrine system, specifically the hypothalamus. The hypothalamic target can serve to explain how ATZ exposure influences the number of endocrine-related health outcomes that are reported in the literature along the HPG, HPA, and HPT axes. Furthermore, there are direct connections between the neuroendocrine and neurotransmission systems that are also reported to be altered by ATZ exposure. Current literature supports ATZ neuroendocrine disruption of kisspeptin, GnRH neuronal development, and subsequent alterations in LH surge. These alterations lead to downstream effects on neuronal development and stress responses that affect organismal and gonadal weight, reproductive health, and fecundity. Studies also support ATZ acting on the dopaminergic system by decreasing expression of genes associated with dopaminergic neurogenesis leading to decreases in DA and its precursors in the substantia nigra and impaired motor performance. Several studies support increased incidence of PD with pesticide exposure, but more studies are needed to conclude if ATZ exposure is linked to PD pathogenesis. In addition, a few studies show sex-specific alterations in the serotonergic system linked to increases in anxiety-like behavior; however, more developmental studies are needed. These alterations support ATZ acting as a neuroendocrine toxicant through a hypothalamic target effecting downstream endocrine axes and the dopaminergic and serotonergic systems by altering gene expression, neurogenesis, and hormone levels among other outcomes that ultimately affect reproduction and behavior, but the exact neuroendocrine toxicity mechanism is yet to be elucidated.

Key facts of the developmental neurotoxicity of the herbicide atrazine

- Water monitoring indicates atrazine (ATZ) is generally below 0.1 μg/L, but concentrations can be higher in agricultural areas.
- The guideline value for ATZ in drinking water set by the World Health Organization (WHO) is 100 μg/L.
- The maximum contaminant level in drinking water set by the US Environmental Protection Agency (EPA) is 3 μg/L.
- ATZ was excluded from reregistration in the European Union in 2003 based on risk of contaminating drinking water sources, not health risk.
- According to the US EPA, the concentration of ATZ and its three main metabolites in drinking water taken together with dietary exposure ranges from 12.5 to 68 μg/L seasonally, while the United Kingdom has less than 0.1 μg/L.
- The US EPA reports that the neuroendocrine effects are considered as the primary toxicological effects of regulatory concern.

Summary points

- Atrazine (ATZ) is an endocrine disrupting chemical (EDC) commonly applied to crop fields that is detected in potable water supplies wherever it is applied resulting in risk of environmental exposure.
- ATZ is the second most common agricultural herbicide applied in the United States, but was banned due to water contamination issues in the European Union.
- Developmental exposure during embryogenesis or puberty causes long-lasting alterations at the hypothalamic level, which can ultimately affect reproduction and stress response.
- Developmental ATZ exposure is shown to affect dopamine (DA), by decreasing DA prevalence, dopaminergic neurons in certain areas of the brain, as well as genes associated with DA transport.
- Additional studies indicate serotonin alterations associated with learning and anxiety-like behavior changes with ATZ exposure.

References

Agency for Toxic Substances and Disease Registry. (2003). *FAQ of atrazine*. US Department of Health and Human Services. Retrieved from: https://www.atsdr.cdc.gov/tfacts153.pdf.

Agopian, A. J., Lupo, P. J., Canfield, M. A., & Langlois, P. H. (2013). Case-control study of maternal residential atrazine exposure and male genital malformations. *American Journal of Medical Genetics, Part A, 161*(5), 977–982.

Bardullas, U., Giordano, M., & Rodriguez, V. M. (2011). Chronic atrazine exposure causes disruption of the spontaneous locomotor activity and alters the striatal dopaminergic system of the male Sprague-Dawley rat. *Neurotoxicology and Teratology, 33*(2), 263–272.

Barr, D. B., Panuwet, P., Nguyen, J. V., Udunka, S., & Needham, L. L. (2007). Assessing exposure to atrazine and its metabolites using biomonitoring. *Environmental Health Perspectives, 115*(10), 1474–1478.

Bethsass, J., & Colangelo, A. (2006). European union bans atrazine, while the United States negotiates continued use. *International Journal of Occupational and Environmental Health, 12*(3), 260–267.

Blanchard, P. E., & Lerch, R. N. (2000). Watershed vulnerability to losses of agricultural chemicals: Interactions of chemistry, hydrology and land use. *Environmental Science and Technology, 34*, 3315–3322.

British Crop Protection Council. (1977). In Hubert Martin, & Charles R. Worthing (Eds.), *Pesticide manual: Basic information on the chemicals used as active components of pesticides* (5th ed.) Droitwich, Eng: The Council.

Caron-Beaudoin, É., Denison, M. S., & Sanderson, J. T. (2016). Effects of neonicotinoids on promoter-specific expression and activity of aromatase (CYP19) in human adrenocortical carcinoma (H295R) and primary umbilical vein endothelial (HUVEC) cells. *Toxicological Sciences, 149*(1), 134–144.

Castañeda Cortés, D. C., Langlois, V. S., & Fernandino, J. I. (2014). Crossover of the hypothalamic pituitary-adrenal/interrenal, -thyroid, and -gonadal axes in testicular development. *Frontiers in Endocrinology, 5*, 139.

Cheremisinoff, N. P., & Rosenfeld, P. E. (2011). *Handbook of pollution prevention and cleaner production: Best practices in the agrochemical industry*. In N. P. Cheremisinoff, & P. E. Rosenfeld (Eds.). *3*. Elsevier (Chapters 3,4).

Cooper, R. L., Stoker, T. E., Tyrey, L., Goldman, J. M., & McElroy, W. K. (2000). Atrazine disrupts the hypothalamic control of pituitary-ovarian function. *Toxicological Sciences, 53*(2), 297–307.

Eldridge, J. C., Wetzel, L. T., & Tyrey, L. (1999). Estrous cycle patterns of Sprague-Dawley rats during acute and chronic atrazine administration. *Reproductive Toxicology, 13*(6), 491–499.

Foradori, C. D., Healy, J. E., Zimmerman, A. D., Kemppainen, R. J., Jones, M. A., Read, C. C., … Handa, R. J. (2018). Characterization of activation of the hypothalamic-pituitary-adrenal axis by the herbicide atrazine in the female rat. *Endocrinology, 159*(9), 3378–3388.

Foradori, C. D., Hinds, L. R., Hanneman, W. H., Legare, M. E., Clay, C. M., & Handa, R. J. (2009). Atrazine inhibits pulsatile luteinizing hormone release without altering pituitary sensitivity to a gonadotropin-releasing hormone receptor agonist in female Wistar Rats. *Biology of Reproduction, 81*(1), 40–45.

Fraites, M. J. P., Cooper, R. L., Buckalew, A., Jayaraman, S., Mills, L., & Laws, S. C. (2009). Characterization of the hypothalamic-pituitary-adrenal axis response to atrazine and metabolites in the female rat. *Toxicological Sciences, 112*(1), 88–99.

Freeman, L. E. B., Rusiecki, J. A., Hoppin, J. A., Lubin, J. H., Koutros, S., Andreotti, G., … Alavanja, M. C. R. (2011). Atrazine and cancer incidence among pesticide applicators in the agricultural health study (1994–2007). *Environmental Health Perspectives, 119*(9), 1253–1259.

Gammon, D. W., Aldous, C. N., Carr, W. C., Sanborn, J. R., & Pfeifer, K. F. (2005). A risk assessment of atrazine use in California: Human health and ecological aspects. *Pest Management Science, 61*(4), 331–355.

Goldman, J. M., Davis, L. K., Murr, A. S., & Cooper, R. L. (2013). Atrazine-induced elevation or attenuation of the LH surge in the ovariectomized, estrogen-primed female rat: Role of adrenal progesterone. *Reproduction, 146*(4), 305–314.

Karapanou, O., & Papadimitriou, A. (2010). Determinants of menarche. *Reproductive Biology and Endocrinology, 8*(1), 115.

Kimura, M., Ishii, M. N., Seki, N., Sakai, Y., Yamashita, T., Awatsuji, H., … Matsui, H. (2019). Reduction of *Kiss1* expression in the anteroventral periventricular nucleus is associated with atrazine-induced attenuation of the luteinizing hormone surge in female rats. *Biology of Reproduction, 100*(1), 41–48.

LeBaron, H. M., McFarland, J. E., & Burnside, O. C. (2008). *The triazine herbicides*. Elseiver.

Li, Y. S., He, X., Ma, K., Wu, Y. P., & Li, B. X. (2015). The effect of exposure to atrazine on dopaminergic development in pubertal male SD rats. *Birth Defects Research B: Developmental and Reproductive Toxicology, 104*(5), 184–189.

Li, J., Li, X., Bi, H., & Li, B. (2019). The MEK/ERK/CREB signaling pathway is involved in atrazine induced hippocampal neurotoxicity in Sprague Dawley rats. *Ecotoxicology and Environmental Safety, 170*, 673–681.

Lin, Z., Dodd, C. A., Xiao, S., Krishna, S., Ye, X., & Filipov, N. M. (2014). Gestational and lactational exposure to atrazine via the drinking water causes specific behavioral deficits and selectively alters monoaminergic systems in C57BL/6 mouse dams, juvenile and adult offspring. *Toxicological Sciences. 141*(1), 90–102.

Li, Y., Sun, Y., Yang, J., Wu, Y., Yu, J., & Li, B. (2014a). Age-dependent dopaminergic dysfunction following fetal exposure to atrazine in SD rats. *Environmental Toxicology and Pharmacology, 37*(3), 1275–1282.

Li, Y., Sun, Y., Yang, J., Wu, Y., Yu, J., & Li, B. (2014b). The long-term effects of the herbicide atrazine on the dopaminergic system following exposure during pubertal development. *Mutation Research: Genetic Toxicology and Environmental Mutagenesis, 763*, 23–29.

Magoffin, D. A. (2005). Ovarian theca cell. *The International Journal of Biochemistry and Cell Biology, 37*(7), 1344–1349.

Marques, P., Skorupskaite, K., George, J. T., & Anderson, R. A. (2000). In K. R. Feingold, B. Anawalt, A. Boyce, G. Chrousos, K. Dungan, A. Grossman, & D. P. Wilson (Eds.), *Physiology of GnRH and gonadotropin secretion* (Endotext).

Ma, K., Wu, H.-Y., Zhang, B., He, X., & Li, B.-X. (2015). Neurotoxicity effects of atrazine-induced SH-SY5Y human dopaminergic neuroblastoma cells via microglial activation. *Molecular BioSystems, 11*(11), 2915–2924.

Meffe, R., & de Bustamante, I. (2014). Emerging organic contaminants in surface water and groundwater: A first overview of the situation in Italy. *The Science of the Total Environment, 481*, 280–295.

Meiser, J., Weindl, D., & Hiller, K. (2013). Complexity of dopamine metabolism. *Cell Communication and Signaling: CCS, 11*, 34.

Migeot, V., Albouy-Llaty, M., Carles, C., Limousi, F., Strezlec, S., Dupuis, A., & Rabouan, S. (2013). Drinking-water exposure to a mixture of nitrate and low-dose atrazine metabolites and small-for-gestational age (SGA) babies: A historic cohort study. *Environmental Research, 122*, 58–64.

Ochoa-Acuña, H., Frankenberger, J., Hahn, L., & Carbajo, C. (2009). Drinking-water herbicide exposure in Indiana and prevalence of small-for-gestational-age and preterm delivery. *Environmental Health Perspectives; Research Triangle Park, 117*(10), 1619–1624.

Priyadarshi, A., Khuder, S. A., Schaub, E. A., & Shrivastava, S. (2000). A meta-analysis of Parkinson's disease and exposure to pesticides. *Neurotoxicology, 21*(4), 435–440.

Qin, L., Du, Z. H., Zhu, S. Y., Li, X. N., Li, N., Guo, J. A., … Zhang, Y. (2015). Atrazine triggers developmental abnormality of ovary and oviduct in quails (Coturnix *Coturnix coturnix*) via disruption of hypothalamo-pituitary-ovarian axis. *Environmental Pollution, 207*, 299–307.

Riffle, B. W., Klinefelter, G. R., Cooper, R. L., Winnik, W. M., Swank, A., Jayaraman, S., … Laws, S. C. (2014). Novel molecular events associated with altered steroidogenesis induced by exposure to atrazine in the intact and castrate male rat. *Reproductive Toxicology, 47*, 59–69.

Rusiecki, J. A., De Roos, A., Lee, W. J., Dosemeci, M., Lubin, J. H., Hoppin, J. A., … Alavanja, M. C. R. (2004). Cancer incidence among pesticide applicators exposed to atrazine in the agricultural health study. *JNCI Journal of the National Cancer Institute, 96*(18), 1375–1382.

Russart, K. L. G., & Rhen, T. (2016). Atrazine alters expression of reproductive and stress genes in the developing hypothalamus of the snapping turtle, *Chelydra serpentina. Toxicology, 366–367*, 1–9.

Semchuk, M., Love, J., & Lee, G. (1992). Parkinson's disease and exposure to agricultural work and pesticide chemicals. *Neurology, 42*(7), 1328–1335.

Stayner, L. T., Almberg, K., Jones, R., Graber, J., Pedersen, M., & Turyk, M. (2017). Atrazine and nitrate in drinking water and the risk of preterm delivery and low birth weight in four Midwestern states. *Environmental Research, 152*, 294–303.

Stoker, T. E., Guidici, D. L., Laws, S. C., & Cooper, R. L. (2002). The effects of atrazine metabolites on puberty and thyroid function in the male Wistar rat. *Toxicological Sciences, 67*(2), 198–206.

Suzawa, M., & Ingraham, H. A. (2008). The herbicide atrazine activates endocrine gene networks via non-steroidal NR5A nuclear receptors in fish and mammalian cells. *PLoS One, 3*, e2117.

Toufexis, D., Rivarola, M. A., Lara, H., & Viau, V. (2014). Stress and the reproductive Axis. *Journal of Neuroendocrinology, 26*(9), 573–586.

US Geological Survey. (2016). *Pesticide use maps*. Retrieved from: https://water.usgs.gov/nawqa/pnsp/usage/maps/show_map.php?year=2016&map=ATRAZINE&hilo=L.

Waller, S. A., Paul, K., Peterson, S. E., & Hitti, J. E. (2010). Agricultural-related chemical exposures, season of conception, and risk of gastroschisis in Washington state. *American Journal of Obstetrics and Gynecology, 202*(3), 241.e1-241.e6.

Wang, Q., & Xie, S. (2012). Isolation and characterization of a high-efficiency soil atrazine-degrading *Arthrobacter* sp. Strain. *International Biodeterioration and Biodegradation, 71*, 61–66.

Ware, G. (1987). *Reviews of environmental contamination and toxicology* (Vol. 143, pp. 47–57). New York: Springer-Verlag.

Wetzel, L. T., Luempert, L. G., Breckenridge, C. B., Tisdel, M. O., Stevens, J. T., Thakur, A. K., … Eldridge, J. C. (1994). Chronic effects of atrazine on estrus and mammary tumor formation in female Sprague-Dawley and Fischer 344 rats. *Journal of Toxicology and Environmental Health, 43*(2), 169–182.

Wilhelms, K. W., Fitzpatrick, K. F., Scanes, C. G., & Anderson, L. L. (2006). In ovo exposure to a triazine herbicide: Effects of atrazine on circulating reproductive hormones and gonadal histology in young Japanese quail. *Archives of Environmental Contamination and Toxicology, 51*(1), 117–122.

Wirbisky, S. E., Weber, G. J., Sepúlveda, M. S., Xiao, C., Cannon, J. R., & Freeman, J. L. (2015). Developmental origins of neurotransmitter and transcriptome alterations in adult female zebrafish exposed to atrazine during embryogenesis. *Toxicology, 333*, 156–167.

World Health Organization. (2011). *Atrazine and its metabolites in drinking water* (Geneva).

Chapter 20

Neurodevelopmental delays and in utero exposure to hyperemesis gravidarum

Marlena S. Fejzo and Patrick M. Mullin

Department of Obstetrics and Gynecology, Keck School of Medicine, University of Southern California, Los Angeles, CA, United States

List of abbreviations

HG Hyperemesis gravidarum
NVP Nausea and vomiting of pregnancy
ND Neurodevelopmental delay
ASD Autism spectrum disorder
PTSD Posttraumatic stress disorder

Mini-dictionary of terms

- Hyperemesis gravidarum is severe, prolonged nausea and vomiting of pregnancy, in the absence of other diseases, that is generally accompanied by >5% weight loss.
- Neurodevelopmental delay is a broad term that includes diagnoses of attention deficit disorder/attention deficit hyperactivity disorder, learning difficulties/delays, sensory integration disorder/sensory processing disorder, social development delay/social anxiety, and speech or language impairment/delay.
- Autism spectrum disorder is a neurological and developmental disorder covering a spectrum of personal interaction, communication, learning disorders, and behavioral disorders.
- GDF15 is the placenta and appetite hormone growth and differentiation factor 15 encoded by the gene *GDF15*. Genetic variation around the *GDF15* locus is associated with hyperemesis gravidarum.
- IGFBP7 is insulin-like growth factor-binding protein (IGFBP) 7 encoded by the gene *IGFBP7*. It is a member of the IGFBP family. Genetic variation around the *IGFBP7* locus is associated with hyperemesis gravidarum.
- IL-6 is the protein interleukin-6, a cytokine that plays a role in inflammation, B-cell maturation, and induction of fever.

Introduction

Nausea and vomiting of pregnancy (NVP) affects as many as 90% of pregnant women and 18% take medication to treat their symptoms (Clark, Hughes, & McDonald, 2013; Piwko, Koren, Babashov, Vicente, & Einarson, 2013). Hyperemesis gravidarum (HG) is at the severe end of the clinical spectrum and accounts for almost 400,000 emergency department visits in the United States annually (HCUP, 2014). Estimates of severe NVP vary greatly and range from 0.3% in a Swedish registry to as high as 10.8% in a Chinese registry of pregnant women (Kallen, 1987; Zhang & Cai, 1991). While there is no international consensus on the definition, HG has been defined as persistent, debilitating, unexplained nausea and vomiting usually resulting in more than a 5% weight loss, abnormal fluid and nutritional intake, electrolyte imbalance, dehydration, and ketonuria (Goodwin, 1998). Symptoms often extend beyond the first trimester and can last throughout the entire pregnancy in as many as one-third of cases, leading to extreme weight loss and possibly a state of malnutrition and prolonged dehydration of pregnancy (Fejzo et al., 2009). The condition can cause considerable stress and result in posttraumatic stress disorder (PTSD) (DSM-IV-R), especially in cases where symptoms last until term (Christodoulou-Smith et al., 2011).

Diagnosis, Management and Modeling of Neurodevelopmental Disorders. https://doi.org/10.1016/B978-0-12-817988-8.00020-8

HG is associated with poor perinatal outcomes including preterm birth, fetal growth restriction, low birth weight, and requirement for neonatal care and/or resuscitation (Fiaschi, Nelson-Piercy, Gibson, Szatkowski, & Tata, 2018). HG is also associated with an increased risk of preeclampsia and placental abruption, thyrotoxicosis, gestational anemia, and coagulopathy (Fejzo, Trovik, et al., 2019). The profound nutrient and electrolyte disturbances can result in serious maternal morbidity including Wernicke encephalopathy (WE), cardiac, renal, liver, and neuromuscular conditions (Fejzo, Trovik, et al., 2019). The prolonged and violent vomiting can cause splenic avulsion, esophageal rupture, Valsalva retinopathy, and pneumothorax (Fejzo, Trovik, et al., 2019). Maternal and fetal deaths are still reported secondary to HG in this century (Fejzo, Macgibbon, & Mullin, 2016), and there is an increased risk of perinatal death (Vandraas et al., 2013). While there is less information, on outcomes of children exposed to HG in utero, mounting evidence supports a role for HG and adverse fetal brain development (Koren, Ornoy, & Berkovitch, 2018). Herein, we present the data to date on neurodevelopmental outcomes in offspring exposed to HG in utero. The outcomes reported and approximate ages of diagnosis/reporting are summarized in Table 20.1. We introduce possible mechanisms for adverse neurodevelopmental outcomes, which may include genetics, maternal stress, nutrient/fluid deficiencies, poor maternal—infant bonding, and/or antiemetic medication use in pregnancy.

HG and neurodevelopmental delay in babies and children

In 1970, CJ Roberts set out to determine whether pregnancy complications were associated with neurological abnormalities at 1 year of age (Roberts, 1970). Roberts established levels of development using a neurologic examination scored by reference to a neurologic schedule standardized to English children. Infants were tested between 44 and 56 weeks of age and over 60 observations such as muscle tone, reflex activity, neuromotor patterns, and special sensory function were part of the evaluation. Included in the study population were 42 babies from South Wales exposed in utero to HG. No significant differences were identified in these infants compared with the entire group. Given the small study sample and that the definition of HG was not described, evaluation of larger and more recent studies is necessary to determine if there is an association between HG and an increased risk of detectable neurological impairment for infants.

In 1999, a large prospective study in Helsinki comparing first, second, and third trimester nausea to temperament and behavior in children was published by Martin, Wisenbaker, and Huttunen (1999). Trimesters of nausea in pregnancy and maternal ratings of temperament of 2001 infants at 6 months, 1097 children at age 5, and teacher ratings of classroom behaviors and academic grades on 1289 children at age 12 were studied. Infant temperament was scored using five scales measuring biological irregularity (regularity of eating, sleeping, and voiding), distress to novelty (fussiness in a new environment or with a new person), activity emotional intensity (activity and emotional expression), sensory threshold (sensitivity to/fussiness about wet or soiled diaper), negative emotionality (situations resulting in frustration and irritability). The study found that infants whose mothers reported first trimester nausea were more likely to have increased distress to novelty and biological irregularity. Second trimester nausea was associated with an increase in fussiness. Third trimester nausea was associated with more distress to wet/soiled diapers and was more active and emotionally intense. Overall, nausea in the second and/or third trimester had significantly different scores on two of the five scales-lower sensory thresholds and higher levels of activity and emotional intensity. Children at age 5 were scored using a Finnish version of the Thomas, Chess, and Korn Parent Temperament Questionnaire, which had seven scales-3 relating to negative emotionality, task persistence, sensitivity mainly to temperature, sound, and light, activity, biological irregularity, and adaptivity to change. Five-year old children whose mothers reported second or third trimester vomiting were more likely to have lower task persistence (lower ability to remain attentive during play, instructional situations, and in learning difficult tasks). The Lambert and Harsough Adjustment Difficulty Scale was modified to measure five scales in 12-year-old children learning, distractibility, immaturity, negative emotionality, and externalizing problems. Academic grades for carefulness and mean academic grade were extracted. Maternal nausea at any trimester was associated with more immaturity, distractibility, and negative emotionality. Maternal second trimester nausea was significantly associated with less careful and precise work. Overall, second trimester nausea exposure was associated with worse child temperament and behavioral outcomes than first trimester nausea. The study concludes that exposure to second and third trimester nausea is associated with poor self-regulation of attention and emotion, but not cognitive ability, nor internalization of problems. The results may be confounded by exposure to infections in the second and third trimester, because the authors reported second and third trimester nausea was also associated with fever, muscle pain, or cough. The study does not measure vomiting or severity of nausea. Therefore, this study is unable to conclude whether infection or nutritional deficiencies and other factors associated with prolonged NVP or HG are potential causal factors. The study is also limited by maternal reporting of nausea and child temperament. However, teacher ratings at age 12 did provide additional support for the findings. The study also cannot determine whether genetic factors that may be related to both maternal nausea and to child behavior are at

TABLE 20.1 Summary of adverse neurodevelopmental outcomes reported in offspring.

Approximate age	Outcome
6 Months	
	Distress to novelty
	Biological irregularity
	Increased fussiness
	Distress to wet/soiled diapers
	Active and emotionally intense
	Lower sensory thresholds
1–2 Years	
	Externalizing problems
	Disregulated behaviors
	Negative emotionality
	Sleep disturbances
	social communication problems
2 Years	
	Externalizing problems
	Internalizing problems
	Attention problems
	Sleep problems
	Emotional regulation problems
	Symptoms of attention deficit hyperactivity disorder
	Symptoms of affective disorder
2–17 Years	
	Autism spectrum disorder
3–7 Years	
	Lower median verbal, performance, and Full-Scale IQ
5 Years	
	Lower task persistence
8 Years	
	Attention disorders
	Learning delays
	Sensory disorders
	Social developmental delay or social anxiety
	Speech and language delay
9–11 Years	
	Attention-deficit/hyperactivity disorders
	Depression
	Social problems

Continued

TABLE 20.1 Summary of adverse neurodevelopmental outcomes reported in offspring.—cont'd

Approximate age	Outcome
12 Years	
	More immaturity
	More distractibility
	Negative emotionality
	Less careful and precise work
	Worse child temperament and behavior
	Poor self-regulation of attention and emotion
	Attention disorders
	Anxiety
	Sensory disorders
	Sleep difficulty
	Social developmental delay or social anxiety
<18 Years	
	Behavioral disorders
	Emotional disorders
	Attention-deficit/hyperactivity disorders
	Conduct disorders/oppositional defiant disorders
	Emotional disorders
	Pervasive developmental disorders
	Childhood autism
	Developmental disorders
Adulthood	
	Depression
	Anxiety
	Bipolar disorder

play. However, given the large sample size of this prospective study, the association of nausea with child behavior provided strong support for needing additional larger studies to determine whether prolonged nausea and vomiting and HG are associated with neurodevelopmental delay (ND).

Consistent with the findings by Martin et al., children whose mothers were treated with the antiemetic Diclectin for NVP did not have lower cognitive ability compared to controls, and in fact, scored higher on a number of tests (Nulman et al., 2009). In this study by Nulman et al., published in 2009, IQs of children aged 3–7 were divided into three groups based on NVP in their mothers as follows: 45 with NVP treated with Diclectin, 47 with NVP and no Diclectin treatment, and 29 with no NVP. The children were tested using the WPPSI, NEPSY, PLS, Beery, WRAVMA, and McCarthy tests. Mothers also completed the Child Behavior checklist (CBCL) and Conner's Parent Rating Scale. Mother's Full-Scale IQs were measured using the WASI and were not significantly different between the groups. Children with NVP had significantly higher nonverbal intelligence scores, but there was no difference in verbal intelligence scores compared to children whose mothers did not have NVP. After adjustment for age and sex, Diclectin treatment was a significant predictor for Performance IQ and McCarthy Pictorial Memory. In contrast to Martin et al., this study did not find an increased risk of attention problems using CBCL and Conner's questionnaires. However, this study defined the NVP group as nausea, vomiting, and/or retching prior to a gestational age of 20 weeks without any underlying pathology. The study did not compare scores between trimesters of nausea like in the Martin study, and the majority of participants had nausea in the first trimester only (N = 40) compared to 29 participants in the second trimester and 23 in the third trimester. It is not possible to discern whether participants in the first trimester nausea group would skew results from the second or third trimester, when the results were most profound in the study by Martin et al. In addition, this study was on a different and significantly smaller sample population, there were no teacher reports, and children were only assessed between ages 3 and 7. It is also noteworthy that this study was supported by Duchesnay, the manufacturers of Diclectin, which may contribute to bias. Also, this study did not include any participants with the most severe form of NVP, HG. However, a follow-up study by Nulman, Maltepe, Farine, and Koren (2015) included 22 women hospitalized for HG and concluded that children of hospitalized mothers had significantly lower median scores on verbal, performance IQ and Full-Scale IQ. Duration of hospitalization was a significant predictor. A review of this study by Koren et al. (2018) concluded that HG meets three of Shepard's criteria for a teratogen affecting brain development. Taken together, these studies suggest that while normal NVP may have no effect or even be protective, more severe and/or longer lasting symptoms may impact fetal neurodevelopment.

In 2015, we published a study comparing neurodevelopmental outcomes of 312 children from 203 mothers with HG to neurodevelopmental outcomes from 169 children from 89 unaffected mothers (Fejzo, Magtira, Schoenberg, Macgibbon, & Mullin, 2015). Mothers with HG were all treated with intravenous hydration and the control mothers all had at least two pregnancies with either no nausea or NVP that did not require medical attention. The children were on average 8 years old. Mothers of children exposed in utero to HG had a 3.28-fold increase in odds of reporting a neurodevelopmental diagnosis including attention disorders, learning delay, sensory disorders, social development delay or social anxiety, and speech and language delay ($P < .0005$) in their children. The study was not well powered for more rare ND diagnoses. The study did not find any evidence to support a role for antiemetic medication, treatments, or preterm birth, but suggested an association between early start of HG symptoms in the mother and having a child with ND. The study had several limitations. Maternal smoking, alcohol consumption, and recreational drug use were not assessed. However, maternal smoking is inversely correlated with HG, and the HG cases all required intravenous fluid due to low fluid intake. Therefore, if the cases and controls were not well matched for smoking and alcohol, it would likely bias toward the null. Another limitation was the maternal self-reporting of child diagnoses, which may not be accurate. That being said, the rates of self-reported diagnoses in the control group were consistent with published rates suggesting accurate self-reporting in this group. Also, adjustments were not made for multiple comparisons, and therefore the findings reported in this study should be taken with caution. Finally, the cases in the study were from the United States and primarily recruited through the Hyperemesis Education and Research Foundation website. The results, therefore, may not be generalizable to other populations of HG. For example, 15% of the HG cases reporting a child with ND and 11% of the HG cases whose child did not have ND required a feeding tube during their HG pregnancy, which suggests a population with a higher rate of severe HG.

In 2020, two large studies were published that examined the relationship between severe NVP (SNVP) and offspring neurodevelopment. A study of a multiethnic Asian mother–child cohort by Syn et al. contained 1172 women who reported on first and second trimester nausea and vomiting (Syn et al., 2021). Severe vomiting, defined as regular vomiting with inability to retain meals, was reported in 190 women (16.2%). The offspring were analyzed in early childhood using the 1-year Infant-Toddler Social and Emotional Assessment, the 1.5-year Quantitative Checklist for Autism in toddler, the 2-year Bayley Scales of Infant and Toddler Development, Third Edition, the 2- and 4-year Child Behavior Checklist, and at 4.5 years, the Kaufman Brief Intelligence Test, Second Edition. Offspring exposed in utero to SNVP had more externalizing problems, dysregulated behaviors, negative emotionality, and sleep disturbances at age 1 and more social communication problems at age 1.5. At age 2, children exposed to SNVP had more externalizing and internalizing problems, primarily attention and sleep problems and difficulty with emotional regulation, with more symptoms of attention deficit hyperactivity and affective disorders. At age 4, children exposed to SNVP had more internalizing behaviors, primarily anxiety, depressive, and affective symptoms. No differences in verbal and nonverbal IQ scores were detected at age 4.5 years. Syn et al., note that the adverse neurodevelopmental outcomes appear to shift from primarily externalizing symptoms in the first 2 years of life, to primarily internalizing symptoms after age 2. A limitation to the study is that the classification of SNVP was based on self-reports, but the authors note that there was 90.5% congruence with medical records.

A second large study was published in 2020 by Wang et al. and reported on findings from both a US-based cohort of 10,710 children aged 9–11 years old, and a Danish cohort of 2,092,897 children followed from birth until whichever came first—the first psychiatric diagnosis, emigration, death, 18-year birthday, or December 31, 2016 (Wang et al., 2020). For the US-based cohort, mothers were classified as having SNVP if they answered that they had NVP accompanied by weight loss and/or SNVP extending past 6 months' gestation. For the Danish cohort, HG diagnosis (ICD-8 code:638 and ICD-10 code: O21) was obtained from the Danish National Patient Registry (DNPR). For the US cohort, cognition was measured using the NIH Toolbox (abcd_tbss01), and emotional and behavioral diagnoses were made using the Childhood Behavioral

Checklist. For the Danish cohort, psychiatric ICD-8 and ICD-10 diagnoses were pulled from the DNPR and the Danish Psychiatric Central Research Register. In addition, children were also classified as having attention deficit hyperactivity disorder (ADHD) if two or more prescriptions were redeemed in the National Prescription Register for ADHD. In the US cohort, children exposed to SNVP scored significantly higher on all psychiatric problem measures and included increased ADHD, depression, and social problems. Children exposed to SNVP had an overall increased psychiatric problem score that was on average 25% higher than unexposed children and a total cognitive performance score that was 3.7% lower than unexposed children on average. In the Danish cohort, children exposed to HG in utero were found to have significantly increased diagnoses of behavioral and emotional disorders, ADHDs, conduct disorders/oppositional defiant disorders, emotional disorders, pervasive developmental disorders, childhood autism, and developmental disorders. Of note, this is the first study to measure brain morphology in the context of in utero exposure to SNVP. Total cortical volume and area were measured in the US cohort and found to be significantly lower in the SNVP-exposed children. The reduction in brain regions was found to significantly mediate the effects of SNVP on psychiatric problems and cognition. The authors note the limitation that for the US cohort, SNVP was based on self-reporting, but emphasize that the Danish cohort had similar findings and included only women with a clinical diagnosis of HG. The authors also note that the findings were similar for both cohorts despite including different sources of potential bias and adjusting for different confounding factors. In summary, while the aforementioned reports may have used different criteria to define maternal SNVP and HG, different measures for childhood neurodevelopmental outcomes, adjusted for different confounding factors, and had different strengths and limitations, they all support the conclusion that in utero exposure to SNVP and/or HG results in increased risk of adverse neurodevelopmental outcomes.

HG exposure in utero and autism spectrum disorder

A follow-up study to our earlier work looked at children who were on average 12 years old (Fejzo, Kam, Laguna, Macgibbon, & Mullin, 2019). Included in the study were 267 children exposed to HG in utero and 93 unexposed children. This study has the same limitations as the original study described earlier. Similar to the original study, there was a threefold increase in reporting of children with ND in the HG group compared to the control group. The mothers with HG reported a significant increase in children with attention disorders, anxiety, sensory disorders, sleep difficulty, social development delay, or social anxiety. Differences in speech, language delay, or learning delays were no longer detected among children of cases compared to controls. The study also found increased reporting of autism spectrum disorder (ASD)—8% of women with HG reported a child diagnosed with ASD, but none of the unexposed children were diagnosed with ASD. While the rate of ASD may be inflated by self-selective reporting, the observation was not surprising, as several other studies have also found an association between SNVP and ASD. In 2010, a study on prenatal and perinatal risk factors for autism was conducted in China comparing maternal factors in 95 ASD cases and 95 controls (Zhang et al., 2010). Zhang et al., found that severe hyperemesis was significantly associated with risk for ASD in a univariate analysis—17% of ASD cases had a mother with HG, compared to 4% of children without ASD. However, the study was unadjusted due to the low frequency of severe HG in the control group. Also, a definition for "severe" HG was not provided. In support of the Chinese study, an Australian study published in 2018 showed that increasing severity of ASD symptoms in the child correlated with increasing severity of NVP in the mother (Whitehouse et al., 2018). In this study, social and communication assessments of 287 children with ASD were compared to the level of their mother's NVP severity. The study limitations include that it was based on retrospective recall of NVP using a nonvalidated scale that did not include onset or duration of symptoms. Further support for an association between ASD and HG comes from a US-based retrospective cohort study of 8760 children with ASD and 26,280 controls, which identified HG as a risk factor for ASD (Hisle-Gorman et al., 2018). In this study, Hisle-Gorman et al. showed maternal HG was associated with ASD regardless of whether HG was defined by diagnosis alone or by a diagnosis that included pharmaceutical treatment. The study limitations include reliance on ICD-9 diagnosis codes and medication prescriptions of unknown dose, duration, and timing. The study also was unable to control for maternal education, race, and marital status. While the Lasso method and strategies were used to reduce a false discovery rate, the analysis included multiple potential associations, which puts the study at risk for false positive associations. That being said, recently a retrospective US-based cohort study by Getahun et al. (2021) analyzed medical records of 469,789 women (14,516 with HG) and their children and found that children exposed in utero to HG were at significantly increased risk of an ASD diagnosis. Both the studies by Fejzo, Kam et al. (2019) and Getahun et al. (2021) showed no evidence for an association between ASD diagnosis and maternal antiemetic treatment for HG. The study by Hisle-Gorman et al. (2018) supports this because they found the association with autism was independent of pharmaceutical intervention for HG. The limitations of the study by Getahun et al. include possible bias in diagnostic suspicion between women with HG and controls, self-reported smoking, and missing dosage data for prescribed medication. The authors conclude the Bradford Hill criteria for HG as a risk factor for ASD in terms of biological plausibility and timing have been met by this study. Additionally, the large studies in 2020 summarized in the previous section on ND also found evidence linking in utero exposure to SNVP/HG to increased risk of ASD (Syn et al., 2021; Wang et al., 2020). Briefly, the multiethnic Asian cohort study used social communication skills from the Q-CHAT as a proxy for ASD diagnosis in toddlers at 1.5 years of age, while the Danish cohort study used diagnostic coding for ASD. Given that the association between HG and ASD has been reported in the aforementioned small and large studies, although causality cannot be proven, it is highly likely that HG is associated with an increased risk for ASD.

HG exposure and psychological/behavioral disorders in adulthood

In 2011 and 2012, Mullin et al. published two studies of maternal exposure to HG and risk of psychological and behavioral disorders in adulthood (Mullin et al., 2011, 2012). Women with pregnancies affected by HG reported on whether or not their mothers also had HG and if so, whether or not their siblings were exposed to HG in utero. Fifty-five women with HG reported on psychological and behavioral disorders in 87 HG-exposed adult siblings. Ninety-five women with HG reported on psychological and behavioral disorders in 172 adult siblings who were not exposed in utero to HG. Offspring exposed in utero to HG were 3.6-fold more likely to have a psychological or behavioral disorder in adulthood compared to unexposed siblings of women affected by HG with diagnoses primarily being depression, bipolar disorder, and anxiety. This study was limited to women with HG so it was able to control for potential confounding genetic factors contributing to HG that may also contribute to psychological and behavioral disorders. The limitations to the study included an incomplete survey response rate (although response rates were similar between the groups and responders and nonresponders were well matched). In addition, the study is based on a retrospective component of self-reporting maternal HG status and sibling diagnoses. The 2012 study introduced a second control group, 95 women who had pregnancies with normal or no nausea. They reported on emotional and behavioral disorders in 101 unexposed adult siblings. The second study showed that both control groups (both unexposed in utero to HG, but one group with and the other without a family history of HG) reported the same frequency of emotional/behavioral disorders, which was significantly lower than those who were exposed to HG in utero and had a family history of HG. This supports the suspicion that it is the HG exposure itself, and not genetic or familial factors related to HG, that explains the significant increase in emotional and behavioral disorders in adults exposed in utero to HG.

Potential mechanisms

GDF15 and IGFBP7

It is important to note that aforementioned studies provide evidence for an association between HG and ND/ASD, but do not prove that HG causes ND/ASD. That being said, there are several theoretically plausible mechanisms for how HG may cause abnormal brain development. Studies have linked dysregulation of maternal inflammation and fetal inflammation genes to abnormal fetal brain development (Clark et al., 2010; Goeden et al., 2016). A genome-wide association study (GWAS) of HG showed that the placenta and appetite genes GDF15 (macrophage inhibitory cytokine-1) and IGFBP7 are significantly associated with HG ($P = 2.4 \times 10^{-41}$, $P = 9.2 \times 10^{-24}$) (Fejzo et al., 2018), and the GDF15 and IGFBP7 proteins are significantly overexpressed in women hospitalized for HG (Fejzo, Fasching, et al., 2019). It is therefore of interest that both GDF15 and IGFBP7 may also play a role in signaling immune system changes in early pregnancy, suggesting that abnormal levels of these proteins in HG pregnancies could have an effect on fetal brain development. Of note, IL-6, which is upregulated by GDF15, has a variant linked to ND at age 2 (Clark et al., 2010; Tsui et al., 2012). And, blocking IL-6 in a rodent model of maternal immune activation and autism behaviors results in substantial elimination of autism-like behaviors (Smith, Li, Garbett, Mirnics, & Patterson, 2007). GDF15 has also been shown to enhance mouse neural stem cell proliferation and differentiation and promote synaptic activity, suggesting a role in fetal brain development and functioning (Kim et al., 2015). Lastly, *IGFBP7* was the greatest association signal in a GWAS of genetic variants influencing prenatal brain development, achieving genome-wide significance for gray matter volume measured by infant neuroimaging (Xia et al., 2017), and cortical volume was found to be significantly altered in children exposed to SNVP (Wang et al., 2020).

Maternal stress

Maternal stress and anxiety may play a role, as they are common during HG pregnancies (Pirimoglu et al., 2010; Tan, Vani, Lim, & Omar, 2010). Maternal stress, primarily during the first and second trimesters, is associated with permanent changes in neuroendocrine regulation and behavior in children. Neuroendocrine regulation is an important factor underlying both ADHD and depression. Animal studies have convincingly shown that stress during pregnancy results in offspring with increased anxiety and depressive behavior possibly by altered fetal development of the hypothalamic–pituitary–andrenal (HPA) axis and alterations of regulatory and neurotransmitter systems in the brain (Lazinski, Shea, & Steiner, 2008; Mulder et al., 2002). Finally, prenatal bereavement stress is associated with an increased risk of ASD (Class et al., 2014).

Maternal starvation/dehydration

Maternal undernutrition may also be a contributing factor. In one study, more than a quarter of HG pregnancies resulted in greater than 15% weight loss with symptoms persisting until term in over 20% of pregnancies. This study suggests HG can be a form of prolonged starvation (Fejzo et al., 2009). Studies of the Dutch and Chinese famine reveal that in addition to significant low birth weight, smaller head circumference, and cardiovascular disease, there is an increased risk of schizophrenia spectrum disorders, congenital anomalies of the central nervous system, and antisocial personality disorders among people exposed to famine in the first half of gestation. Stunted brain development has been proposed to underly these associations. People exposed in utero to famine in mid or late gestation are more likely to have an affective disorder, possibly due to abnormal programming of the HPA axis (Painter, Roseboom, & Bleker, 2005). Also, maternal diet during pregnancy has been associated with an increased risk for ASD—this offers another explanation for the association between HG and autism risk (Peretti et al., 2018). In support of this, a rodent model of ASD showed that maternal protein malnutrition, a common feature of HG, caused ASD-like symptoms in offspring (Batista, Giusti-Paiva, & Vilela, 2019).

Severe cases of HG can result in vitamin deficiency syndromes such as maternal WE caused by thiamine deficiency and fetal intracranial hemorrhage and dysmorphology caused by vitamin K deficiency (Eventov-Friedman, Klinger, & Shinwell, 2009; Oudman et al., 2019; Toriello et al., 2013). Since both HG and maternal alcoholism can lead to thiamine deficiency and other nutrient deficiencies, it would be interesting to determine whether there may be overlapping phenotypes between children with fetal-alcohol syndrome and children exposed to HG-induced WE or even milder thiamin deficiency in pregnancy. Of note, both fetal exposure to alcohol and HG are associated with increased risk of attention disorder and ASD (Fejzo, Kam, et al., 2019; Mukherjee, Cook, Norgate, & Price, 2019). Early neonatal vitamin K deficiency has been linked to impaired neuronal migration and cortical dysplasia (Bhoj, Dubbs, McDonald-McGinn, & Zackai, 2013; Toriello et al., 2013). Certain nutritional deficiencies in pregnancy, such as deficits of folate and vitamin B12, are associated with disruptions in myelination and inflammatory processes in infants and a greater risk of depression in adulthood (Black, 2008). In animal models, prenatal vitamin D deficiency is associated with adverse neuropsychiatric outcomes (Harms, Eyles, McGrath, Mackay-Sim, & Burne, 2008).

Women with HG are often dehydrated due to low fluid intake and prolonged vomiting, which may be corrected by intravenous fluid. Fluid imbalance can lead to significant electrolyte disturbances. Studies must be undertaken to determine whether dehydration and electrolyte imbalances affect neurodevelopmental outcomes.

Medication/treatment

Women with HG are often prescribed antiemetics and some self-treat with Cannabis. One US-based study compared 37 treatments and medications for HG in 138 women who reported a child with ND to 174 women who reported no children with ND (Fejzo et al., 2015). The study found no evidence to support an increased risk of ND associated with any medication or treatment. However, the study was small and may be underpowered, especially for the less common medications/treatments. For example, while the study included 213 women who took ondansetron, only 26 women reported using Diclectin. A Canadian study on Diclectin treatment for NVP (Nulman et al., 2009) found no increased risk of ND, except when the patient was hospitalized for HG (Nulman et al., 2015). While the hospitalized patients used higher doses of Diclectin, they also tended to have the most severe symptoms of HG. However, the US-based study did not find any evidence that hospitalization for HG was associated with an increased risk for ND. The only potential association they found was an increased (albeit unadjusted) risk of ND if symptoms started between weeks 1 and 4 of pregnancy. A follow-up US study of HG and autism found no evidence to support an association between early symptoms, nor for hospitalization, or the three most commonly prescribed medications ondansetron, promethazine, or metoclopramide and autism risk (FEJZO, Kam, et al., 2019). While that exploratory study was very small (only 15 cases of autism), a much larger study of 14,526 HG pregnancies associating HG with autism also found no evidence to support a link between antiemetic medication (Diphenydramine, Promethazine, Ondansetron, and Metoclopramide) and autism (Getahun et al., 2021). A separate study compared neurodevelopment in 143 children exposed to ondansetron or promethazine to 407 unexposed children (Larrimer et al., 2014). This study also did not find any evidence to support an association with ND and the two commonly prescribed antiemetics. Unlike the other small US-based study that relied on maternal self-reporting of ND diagnoses in children, this study used the Neonatal Behavioral Assessment Scale administered by certified professionals at 7 days old. Additionally, the mother filled out the Child Behavioral Checklist between 17 and 66 months of age. Importantly, in the multiethnic Asian study, among the 190 women self-reporting SNVP, 67 women were hospitalized and diagnosed with HG (Syn et al., 2021). In a sensitivity analysis, it was found that the increased risk in ND was based on the 123 mothers who self-reported SNVP who were *not* hospitalized for HG, providing evidence that undertreatment rather than medications is responsible for the association. However, the US-based studies and the Asian study did not include antiemetic monotherapy, and larger studies are needed to further evaluate any associations with a single medication.

Cannabis use to treat NVP is on the rise due in part to its expanding legalization. For example, more than 7% of pregnant women in Northern California use Cannabis, based on self-report and/or toxicology screens (Young-Wolff et al., 2017). Cannabis use in pregnancy has been associated with similar poor neurodevelopmental outcomes found to be associated with prolonged NVP and HG such as inattention and depression (Huizink & Mulder, 2006). However, this association is controversial because generally studies are not controlled for important confounding factors including HG, making conclusions about cause and effect difficult.

Maternal postpartum complications

HG can be associated with substantial maternal physical and psychological effects and can also pose a financial burden postpartum (Fejzo et al., 2009). Women with extreme weight loss (>15%) due to HG are more likely to have longer recovery times, postpartum digestive problems, muscle pain, gall bladder dysfunction, and PTSD. A child with a behavioral disorder was reported by 9.3% of women with extreme weight loss (Fejzo et al., 2009). These postpartum conditions may negatively affect maternal–infant bonding, which in turn may contribute to the behavioral abnormalities seen later in life. In support of this theory, rodent studies show maternal care in the first week after birth results in epigenetic modification of genes expressed in the brain that shapes neuroendocrine and behavioral stress responsivity throughout life (Weaver, 2009).

Conclusion

In conclusion, mounting evidence suggests that HG is associated with poor neurodevelopmental outcomes including ASD. The cause for this association is unknown, but may be related to genetics, maternal stress, nutrient deficiencies, or poor maternal–infant bonding. In the future, studies that compare HG risk genes or serum levels of GDF15 and IGFBP7 in mother–child pairs with and without ND may elucidate whether genetics is at play. Studies on women with HG are needed to determine whether interventions to reduce maternal stress, increase maternal/fetal nutrient supplementation, and/or improve maternal–infant bonding reduce the risk of ND and ASD in this population. Thus far, associations between prescribed antiemetic therapies and ND/autism have not been found, but large well-controlled monotherapy studies are needed. As similar ND outcomes have been noted with both HG pregnancies and Cannabis-exposed pregnancies, future studies of outcomes related to cannabis exposure must include HG as a confounding factor. HG has been associated with other long-term adverse outcomes including higher baseline cortisol concentrations, reduction of insulin sensitivity, and greater risk of testicular cancer in adulthood (Fejzo, Trovik, et al., 2019). While we do not know the reason for the association between HG and long-term outcomes, current evidence supports that HG can affect brain development (Koren et al., 2018; Wang et al., 2020) and that, in terms of biological plausibility and timing, HG fulfills the Bradford Hill criteria as a risk factor for autism (Getahun et al., 2021).

Of note, in one study, children born to women reporting SNVP were divided into two groups, those whose mothers were hospitalized and those who were not hospitalized. Those who were hospitalized had children with normal neurodevelopmental outcomes, while the children born to women reporting SNVP that were *not* hospitalized were at increased risk of ND (Syn et al., 2021). This suggests that more aggressive treatment and/or treatment in a hospital setting may mitigate the risk. HG is generally understudied and undertreated in part due to the incorrect belief that it is self-limited. Studies are urgently needed to reduce not only the adverse maternal outcomes, but the poor child outcomes as well. Because early intervention may have the potential to improve neurodevelopmental outcomes in children, pediatricians should include maternal history of SNVP/HG and screen-exposed infants and children.

Applications to other areas of neurodevelopmental delay

The study of HG exposure in utero and adverse neurodevelopmental outcomes has several applications to other areas of ND. HG is defined by a minimum of 5% weight loss and represents a form of caloric restriction in pregnancy. The nutritional deficiencies and associated consequences due to HG are similar to those seen in conditions such as chronic alcohol use, as evidenced by the fact that both HG and alcoholism are risk factors for severe thiamin (B1) deficiency and WE. Poor childhood outcomes including ND are seen with fetal alcohol syndrome. Similarly, HG is associated with ND, attention disorders, and ASD. HG can also lead to poor maternal–infant bonding. Risk genes for HG have been identified as *GDF15* and *IGFBP7*. These genes may also play a role in immunity and inflammation. A GWAS study on NVP and a separate GWAS study on neurodevelopment showed *IGFBP7* were significantly associated with both HG and with gray matter volume in infants. Thus, abnormalities in the expression of this gene may contribute to both HG and abnormal brain development.

Key facts of hyperemesis gravidarum

- Hyperemesis gravidarum (HG) affects approximately 2% of pregnant women.
- HG is defined by at least a 5% weight loss in early pregnancy due to nausea and vomiting.
- HG can cause prolonged nutrient deficiency, dehydration, and maternal stress.
- HG is associated with an increased risk of preterm birth.
- HG is associated with an increased risk of neurodevelopmental delay (ND) in exposed offspring.
- HG is associated with an increased risk of autism spectrum disorder (ASD) in exposed offspring.

Summary points

- HG can affect brain development.
- HG is a risk factor for autism spectrum disorder.
- Screening for ND should be implemented following HG pregnancies.
- Studies should be performed to determine whether nutritional intervention improves outcomes.
- Studies should be performed to determine whether interventions that reduce maternal stress and poor maternal–infant bonding improve outcomes.
- Studies on Cannabis should control for HG as a confounding factor to determine whether Cannabis, HG, or both play a role in adverse ND outcomes.
- Pediatricians should include maternal history of HG and screen-exposed infants and children for ND and ASD.

References

Batista, T. H., Giusti-Paiva, A., & Vilela, F. C. (2019). Maternal protein malnutrition induces autism-like symptoms in rat offspring. *Nutritional Neuroscience, 22*, 655–663.

Bhoj, E., Dubbs, H., McDonald-McGinn, D., & Zackai, E. (2013). Late-onset partial complex seizures secondary to cortical dysplasia in a patient with maternal vitamin K deficient embryopathy: Comments on the article by Toriello et al. [2013] and first report of the natural history. *American Journal of Medical Genetics, 161A*, 2396–2398.

Black, M. M. (2008). Effects of vitamin B12 and folate deficiency on brain development in children. *Food and Nutrition Bulletin, 29*, S126–S131.

Christodoulou-Smith, J., Gold, J. I., Romero, R., Goodwin, T. M., Macgibbon, K. W., Mullin, P. M., & Fejzo, M. S. (2011). Posttraumatic stress symptoms following pregnancy complicated by hyperemesis gravidarum. *Journal of Maternal-Fetal and Neonatal Medicine, 24*, 1307–1311.

Clark, S. S., Hughes, S. B., & McDonald, S. S. (2013). The impact of nausea and vomiting of pregnancy on quality of life: Report of a national consumer survey and recommendations for improving care. *Obstetrical and Gynecological Survey, 68*, S1–S10.

Clark, E. A., Mele, L., Wapner, R. J., Spong, C. Y., Sorokin, Y., Peaceman, A., … Human Development Maternal-Fetal Medicine Units Networks. (2010). Association of fetal inflammation and coagulation pathway gene polymorphisms with neurodevelopmental delay at age 2 years. *American Journal of Obstetrics and Gynecology, 203*, 83 e1–83 e10.

Class, Q. A., Abel, K. M., Khashan, A. S., Rickert, M. E., Dalman, C., Larsson, H., … D'onofrio, B. M. (2014). Offspring psychopathology following preconception, prenatal and postnatal maternal bereavement stress. *Psychological Medicine, 44*, 71–84.

Eventov-Friedman, S., Klinger, G., & Shinwell, E. S. (2009). Third trimester fetal intracranial hemorrhage owing to vitamin K deficiency associated with hyperemesis gravidarum. *Journal Pediatric Hematology and Oncology, 31*, 985–988.

Fejzo, M. S., Fasching, P. A., Schneider, M. O., Schwitulla, J., Beckmann, M. W., Schwenke, E., … Mullin, P. M. (2019). Analysis of GDF15 and IGFBP7 in hyperemesis gravidarum support causality. *Geburtshilfe und Frauenheilkunde, 79*, 382–388.

Fejzo, M., Kam, A., Laguna, A., Macgibbon, K., & Mullin, P. (2019). Analysis of neurodevelopmental delay in children exposed in utero to hyperemesis gravidarum reveals increased reporting of autism spectrum disorder. *Reproductive Toxicology, 84*, 59–64.

Fejzo, M. S., Macgibbon, K. W., & Mullin, P. M. (2016). Why are women still dying from nausea and vomiting of pregnancy? *Gynecology and Obstetrics Case report, 2*.

Fejzo, M. S., Magtira, A., Schoenberg, F. P., Macgibbon, K., & Mullin, P. M. (2015). Neurodevelopmental delay in children exposed in utero to hyperemesis gravidarum. *European Journal of Obstetrics and Gynecology and Reproductive Biology, 189*, 79–84.

Fejzo, M. S., Poursharif, B., Korst, L. M., Munch, S., Macgibbon, K. W., Romero, R., & Goodwin, T. M. (2009). Symptoms and pregnancy outcomes associated with extreme weight loss among women with hyperemesis gravidarum. *Journal of Womens Health (Larchmt), 18*, 1981–1987.

Fejzo, M. S., Sazonova, O. V., Sathirapongsasuti, J. F., Hallgrimsdottir, I. B., Vacic, V., Macgibbon, K. W., … Andme Research, T. (2018). Placenta and appetite genes GDF15 and IGFBP7 are associated with hyperemesis gravidarum. *Nature Communications, 9*, 1178.

Fejzo, M. S., Trovik, J., Grooten, I. J., Sridharan, K., Roseboom, T. J., VIKANES, A., … Mullin, P. M. (2019). Nausea and vomiting of pregnancy and hyperemesis gravidarum. *Nature Reviews Disease Primers, 5*, 62.

Fiaschi, L., Nelson-Piercy, C., Gibson, J., Szatkowski, L., & Tata, L. J. (2018). Adverse maternal and birth outcomes in women admitted to hospital for hyperemesis gravidarum: A population-based cohort study. *Paediatric and Perinatal Epidemiology, 32*, 40–51.

Getahun, D., Fassett, M. J., Jacobsen, S. J., Xiang, A. H., Takhar, H. S., WING, D. A., & Peltier, M. R. (2021). Autism spectrum disorders in children exposed in utero to hyperemesis gravidarum. *American Journal of Perinatology, 38*(3), 265–272. https://doi.org/10.1055/s-0039-1696670, 31581303. Epub 2019 Oct 3.

Goeden, N., Velasquez, J., Arnold, K. A., Chan, Y., Lund, B. T., Anderson, G. M., & Bonnin, A. (2016). Maternal inflammation disrupts fetal neurodevelopment via increased placental output of serotonin to the fetal brain. *Journal of Neuroscience, 36*, 6041–6049.

Goodwin, T. M. (1998). Hyperemesis gravidarum. *Clinical Obstetrics and Gynecology, 41*, 597–605.

Harms, L. R., Eyles, D. W., McGrath, J. J., Mackay-Sim, A., & Burne, T. H. (2008). Developmental vitamin D deficiency alters adult behaviour in 129/SvJ and C57BL/6J mice. *Behavioural Brain Research, 187*, 343–350.

HCUP. (2014). *Healthcare cost and utilization project. [Online]*. Rockville, MD: Agency for Healthcare Research and Quality. Available https://hcupnet.ahrq.gov/. (Accessed 14 August 2019).

Hisle-Gorman, E., Susi, A., Stokes, T., Gorman, G., Erdie-Lalena, C., & Nylund, C. M. (2018). Prenatal, perinatal, and neonatal risk factors of autism spectrum disorder. *Pediatric Research, 84*, 190–198.

Huizink, A. C., & Mulder, E. J. (2006). Maternal smoking, drinking or cannabis use during pregnancy and neurobehavioral and cognitive functioning in human offspring. *Neuroscience and Biobehavioral Reviews, 30*, 24–41.

Kallen, B. (1987). Hyperemesis during pregnancy and delivery outcome: A registry study. *European Journal of Obstetrics and Gynecology and Reproductive Biology, 26*, 291–302.

Kim, D. H., Lee, D., Chang, E. H., Kim, J. H., Hwang, J. W., Kim, J. Y., … Chang, J. W. (2015). GDF-15 secreted from human umbilical cord blood mesenchymal stem cells delivered through the cerebrospinal fluid promotes hippocampal neurogenesis and synaptic activity in an Alzheimer's disease model. *Stem Cells Development, 24*, 2378–2390.

Koren, G., Ornoy, A., & Berkovitch, M. (2018). Hyperemesis gravidarum-Is it a cause of abnormal fetal brain development? *Reproductive Toxicology, 79*, 84–88.

Larrimer, M. B., Dajani, N. K., Siegel, E. R., Eswaran, H., Newport, D. J., & Stowe, Z. N. (2014). Antiemetic medications in pregnancy: A prospective investigation of obstetric and neurobehavioral outcomes. *American Journal of Obstetrics and Gynecology, 210*, 270 e1-7.

Lazinski, M. J., Shea, A. K., & Steiner, M. (2008). Effects of maternal prenatal stress on offspring development: A commentary. *Archives Womens Mental Health, 11*, 363–375.

Martin, R. P., Wisenbaker, J., & Huttunen, M. O. (1999). Nausea during pregnancy: Relation to early childhood temperament and behavior problems at twelve years. *Journal of Abnormal Child Psychology, 27*, 323–329.

Mukherjee, R. A. S., Cook, P. A., Norgate, S. H., & Price, A. D. (2019). Neurodevelopmental outcomes in individuals with fetal alcohol spectrum disorder (FASD) with and without exposure to neglect: Clinical cohort data from a national FASD diagnostic clinic. *Alcohol, 76*, 23–28.

Mulder, E. J., Robles de Medina, P. G., Huizink, A. C., Van Den Bergh, B. R., Buitelaar, J. K., & Visser, G. H. (2002). Prenatal maternal stress: Effects on pregnancy and the (unborn) child. *Early Human Development, 70*, 3–14.

Mullin, P. M., Bray, A., Schoenberg, F., MacGibbon, K. W., Romero, R., Goodwin, T. M., & Fejzo, M. S. (2011). Prenatal exposure to hyperemesis gravidarum linked to increased risk of psychological and behavioral disorders in adulthood. *Journal of Developmental Origins of Health and Disease, 2*, 200–204.

Mullin, P. M., Bray, A., Vu, V., Schoenberg-Paik, F., MacGibbon, K., Romero, R., … Fejzo, M. S. (2012). No increased risk of psychological/behavioral disorders in siblings of women with hyperemesis gravidarum (HG) unless their mother had HG. *Journal of Developmental Origins of Health and Disease, 3*, 375–379.

Nulman, I., Maltepe, C., Farine, D., & Koren, G. (2015). Neurodevelopment of children after maternal hospitalization for nausea and vomiting of pregnancy. *Obstetrics and Gynecology, 125*, 81s-81s.

Nulman, I., Rovet, J., Barrera, M., Knittel-Keren, D., Feldman, B. M., & Koren, G. (2009). Long-term neurodevelopment of children exposed to maternal nausea and vomiting of pregnancy and diclectin. *The Journal of Pediatrics, 155*, 45–50, 50 e1-2.

Oudman, E., Wijnia, J. W., Oey, M., Van Dam, M., Painter, R. C., & Postma, A. (2019). Wernicke's encephalopathy in hyperemesis gravidarum: A systematic review. *European Journal of Obstetrics and Gynecology and Reproductive Biology, 236*, 84–93.

Painter, R. C., Roseboom, T. J., & Bleker, O. P. (2005). Prenatal exposure to the Dutch famine and disease in later life: An overview. *Reproductive Toxicology, 20*, 345–352.

Peretti, S., Mariano, M., Mazzocchetti, C., Mazza, M., Pino, M. C., Verrotti Di Pianella, A., & Valenti, M. (2018). Diet: The keystone of autism spectrum disorder? *Nutritional Neuroscience*, 1–15.

Pirimoglu, Z. M., Guzelmeric, K., Alpay, B., Balcik, O., Unal, O., & Turan, M. C. (2010). Psychological factors of hyperemesis gravidarum by using the SCL-90-R questionnaire. *Clinical and Experimental Obstetrics and Gynecology, 37*, 56–59.

Piwko, C., Koren, G., Babashov, V., Vicente, C., & Einarson, T. R. (2013). Economic burden of nausea and vomiting of pregnancy in the USA. *Journal of Population Therapeutics and Clinical Pharmacology, 20*, e149–e160.

Roberts, C. J. (1970). Developmental and neurological sequelae of the common complications of pregnancy and birth. *British Journal of Preventive and Social Medicine, 24*, 33–38.

Smith, S. E., Li, J., Garbett, K., Mirnics, K., & Patterson, P. H. (2007). Maternal immune activation alters fetal brain development through interleukin-6. *Journal of Neuroscience, 27*, 10695–10702.

Syn, N. L., Chan, S. Y., Chia, E. W. Y., Ong, W. X., Phua, D., Cai, S., Shek, L. P. C., Chong, Y. S., Daniel, L. M., Broekman, B. F. P., Godfrey, K. M., Meaney, M. J., & Law, E. C. (2021 Jan). Severity of nausea and vomiting in pregnancy and early childhood neurobehavioural outcomes: the growing up in Singapore towards healthy outcomes study. *Paediatric and Perinatal Epidemiology, 35*(1), 98–108. https://doi.org/10.1111/ppe.12703. Epub 2020 Jun 23. PMID: 32578237; PMCID: PMC7116637.

Tan, P. C., Vani, S., Lim, B. K., & Omar, S. Z. (2010). Anxiety and depression in hyperemesis gravidarum: Prevalence, risk factors and correlation with clinical severity. *European Journal of Obstetrics and Gynecology and Reproductive Biology, 149*, 153–158.

Toriello, H. V., Erick, M., Alessandri, J. L., Bailey, D., Brunetti-Pierri, N., Cox, H., ... Lin, A. E. (2013). Maternal vitamin K deficient embryopathy: Association with hyperemesis gravidarum and crohn disease. *American Journal of Medical Genetics Part A, 161A*, 417–429.

Tsui, K. H., Chang, Y. L., Feng, T. H., Chung, L. C., Lee, T. Y., Chang, P. L., & Juang, H. H. (2012). Growth differentiation factor-15 upregulates interleukin-6 to promote tumorigenesis of prostate carcinoma PC-3 cells. *Journal of Molecular Endocrinology, 49*, 153–163.

Vandraas, K. F., Vikanes, A. V., Vangen, S., Magnus, P., Støer, N. C., & Grjibovski, A. M. (2013). Hyperemesis gravidarum and birth outcomes-a population-based cohort study of 2.2 million births in the Norwegian Birth Registry. *BJOG, 120*(13), 1654–1660. https://doi.org/10.1111/1471-0528.12429, 24021026. Epub 2013 Sep 10.

Wang, H., Rolls, E. T., Du, X., Du, J., Yang, D., Li, J., ... Feng, J. (2020). Severe nausea and vomiting in pregnancy: Psychiatric and cognitive problems and brain structure in children. *BMC Medicine, 18*, 228.

Weaver, I. C. (2009). Shaping adult phenotypes through early life environments. *Birth Defects Research C Embryo Today, 87*, 314–326.

Whitehouse, A. J. O., Alvares, G. A., Cleary, D., Harun, A., Stojanoska, A., Taylor, L. J., ... Maybery, M. (2018). Symptom severity in autism spectrum disorder is related to the frequency and severity of nausea and vomiting during pregnancy: A retrospective case-control study. *Molecular Autism, 9*, 37.

Xia, K., Zhang, J., Ahn, M., Jha, S., Crowley, J. J., Szatkiewicz, J., ... Knickmeyer, R. C. (2017). Genome-wide association analysis identifies common variants influencing infant brain volumes. *Translational Psychiatry, 7*, e1188.

Young-Wolff, K. C., Tucker, L. Y., Alexeeff, S., Armstrong, M. A., Conway, A., Weisner, C., & Goler, N. (2017). Trends in self-reported and bio-chemically tested marijuana use among pregnant females in California from 2009-2016. *Journal of the American Medical Association, 318*, 2490–2491.

Zhang, J., & Cai, W. W. (1991). Severe vomiting during pregnancy: Antenatal correlates and fetal outcomes. *Epidemiology, 2*, 454–457.

Zhang, X., Lv, C. C., Tian, J. A., Miao, R. J., Xi, W., Hertz-Picciotto, I., & Qi, L. H. (2010). Prenatal and perinatal risk factors for autism in China. *Journal of Autism and Developmental Disorders, 40*, 1311–1321.

Chapter 21

Neurodevelopment and PCDH19-related developmental and epileptic encephalopathy (PCDH19-DDE)

Silvia Bassani[1], Carla Marini[2] and Maria Passafaro[1]

[1]*Neuroscience Institute, Consiglio Nazionale delle Ricerche (CNR), Milan, Italy;* [2]*Pediatric Neurology Unit and Laboratories, Children's Hospital A. Meyer, Florence, Italy*

List of abbreviations

AEDs Antiepileptic drugs
APs Apical progenitors
aRGCs Apical radial glial cells
ASD Autism spectrum disorder
BPs Basal progenitors
bRGCs Basal radial glial cells
CP Cortical plate
DEE Developmental and epileptic encephalopathy
DS Dravet syndrome
EEG Electroencephalogram
EFMR Epilepsy and mental retardation limited to females
GABA Gamma-aminobutyric acid
EIEE9 Early infantile epileptic encephalopathy-9
IPCs Intermediate progenitor cells
GABA$_A$Rs GABA type A receptors
GCE Girls clustering epilepsy
MRI Magnetic resonance imaging
NCAD Neural cadherin
NFLE Nocturnal frontal lobe epilepsy
NPCs Neural precursor cells
NP Neural plate
NSPCs Neural stem and progenitor cells
NT Neural tube
PCDH19 Protocadherin-19
PCDH19-DDE PCDH19-related DDE
RG Radial glia
shRNA Short hairpin RNA
SVZ Subventricular zone
VZ Ventricular zone
WT Wild type

Diagnosis, Management and Modeling of Neurodevelopmental Disorders. https://doi.org/10.1016/B978-0-12-817988-8.00021-X

Mini-dictionary of terms

Autism a disorder that originates during brain development and that affects how children perceive the world around them and how they relate to others, leading to problems in interpersonal relationships and repetitive behaviors.

Cell-interference the phenomenon originating from the coexistence within the same tissue of cell populations with different genetic and phenotypic features that interfere with each other affecting the properties of the whole tissue.

Cortical dysplasia a malformation in the cortex due to defects of brain development, such as cell migration and positioning or proliferation.

De novo mutation a DNA alteration that differs from mutations inherited from parents due to its appearance in the unborn child for the first time.

Encephalopathy a general term denoting a disease affecting the brain.

Intellectual disability a disorder that originates during brain development and is characterized by impaired cognitive abilities leading to problems in learning, memory, and everyday skills.

Neurosteroids steroids that are able to affect the level of neurons excitability, for instance by binding to $GABA_A$Rs receptors and modulating their function.

Protocadherin a transmembrane protein of the cadherin superfamily that mediates adhesion.

Status epilepticus a long-lasting seizure or seizure sequences that put at risk patient's life

X-inactivation the process by which one of the two copies of chromosome X is randomly inactivated in females to prevent gene transcription and keep gene-dosage between females and males, who possess only one chromosome X.

Introduction

Early infantile epileptic encephalopathy-9 (EIEE9; OMIM # 300088) is a developmental and epileptic encephalopathy (DEE) characterized by epilepsy associated with intellectual disability, and autism spectrum disorder (ASD) in about 70% of patients (Dibbens et al., 2008; Juberg & Hellman, 1971).

This pathology is due to inherited or de novo mutations in the X-chromosome gene *PCDH19* (Xq22.1) (Dibbens et al., 2008). PCDH19-related DDE (PCDH19-DDE) inheritance mode is peculiar since healthy carrier fathers transmit the mutations to their daughters who develop the symptoms. For this reason, PCDH19-DDE was defined as epilepsy and mental retardation limited to females (EFMR) or girls clustering epilepsy (GCE) (Dibbens et al., 2008; Juberg & Hellman, 1971; Ryan et al., 1997; Scheffer et al., 2008).

PCDH19 mutations rates are estimated to range between 2% and 20%, and even though definitive data on incidence and prevalence are still lacking, *PCDH19*, with more than 200 patients on record, is considered the second most clinically relevant epilepsy gene after *SCN1A* (Carvil et al., 2013; Depienne et al., 2011; Depienne & LeGuern, 2012; Hynes et al., 2010; Terracciano et al., 2012; Trivisano et al., 2018).

PCDH19 is a 6-exon gene encoding protocadherin-19 (PCDH19) that is a member of the delta-2 protocadherin subclass of the cadherin superfamily (Kahr, Vandepoele, & van Roy, 2013). PCDH19 is a cell-adhesion molecule composed of six extracellular cadherin domains, a transmembrane domain, and an intracellular C-terminus. Protocadherins are emerging as key molecules in neural circuit formation and functioning given their role in neurite arborization, axon outgrowth, and at synapses (reviewed in Peek, Mah, & Weiner, 2017). PCDH19 is mainly expressed in the nervous system, especially in the cortex and limbic areas (reviewed in (Gerosa, Francolini, Bassani, & Passafaro, 2019), and its biological function is under investigation.

Here, we report the clinical aspects of PCDH19-DDE and the biological impact of PCDH19 loss of function on the milestones of brain development as inferred from experimental models.

Clinical features of PCDH19-DDE

PCDH19-DDE core symptoms

PCDH19 belongs to the group of genes whose mutations cause DDE (Table 21.1), and the main clinical features of PCDH19-DEE are summarized in Table 21.2.

The first symptom drawing patients to medical attention is the sudden-onset focal seizures clusters triggered by febrile illness at an average age of 10 months (Depienne et al., 2009; Marini et al., 2010, 2012; Scheffer et al., 2008; Trivisano et al., 2018). Within each cluster, girls exhibit multiple daily seizures for 2–3 days, and the seizure-free interval between clusters ranges usually from 2 to 10 months.

Most patients have focal seizures with variable epileptogenic focuses, suggesting a wide distribution of neuronal hyperexcitability in the brain (Ikeda et al., 2016; Marini et al., 2012; Specchio, Fusco, & Vigevano, 2011; Specchio, Marini, et al., 2011; Trivisano et al., 2018). About 80% of patients manifest prominent affective semiology during onset

TABLE 21.1 DDE syndromes.

Phenotype/Syndrome	Gene	Protein
PCDH19-related DDE	*PCDH19*	Protocadherin-19
Neonatal onset DDE	*KCNQ2*	Potassium voltage-gated channel subfamily Q member 2
Ohtahara syndrome	*STXBP1, ARX*	Syntaxin binding protein 1; Aristaless-related homeobox protein
Migrating partial seizures of infancy	*KCNT1*	Potassium sodium-activated channel subfamily T member 1
Other unspecified DDE with infantile onset	*CDKL5; STXBP1; ARX; SCN2A, SCN8A; GABRB2; GABRB3; HCN1*	Cyclin-dependent kinase like 5; syntaxin binding protein 1; Aristaless-related Homeobox protein; sodium voltage-gated channel alpha subunit 2; sodium voltage-gated channel alpha subunit 8; GABA(A)R beta 2 subunit; GABA(A)R beta 3 subunit; hyperpolarization-activated cyclic nucleotide-gated potassium channel 1
Dravet syndrome	*SCN1A*	Sodium voltage-gated channel alpha subunit 1
Genetic generalized epilepsy with febrile seizure plus (GEFS+) spectrum	*SCN1A; SCN1B; GABRG2; GABRA1; HCN1; STX1B*	Sodium voltage-gated channel alpha subunit 1; sodium voltage-gated channel beta subunit 1; GABA(A)R gamma 2 subunit; GABA(A)R alpha1 subunit; hyperpolarization-activated cyclic nucleotide-gated potassium channel 1; syntaxin 1B

DDE syndromes with their related genes and proteins.

TABLE 21.2 Clinical features of PCDH19-DEE.

Summary of PCDH19-DEE phenotype		
Seizures	Onset	Mean 10 months; range 5–30 months
	Type	80% focal, 20% generalized
	Occurring in clusters	>80% of patients
	Triggered by fever	Most patients
EEG	Ictal	Focal discharge; generalized spike-wave complexes/discharges
	Interictal	Normal background activity
Motor and cognitive development	0–12 months	Normal
	Infancy and childhood	70% mild to moderate to severe delay
		30% normal
Autistic traits		50% of patients
Other features	Behavioral and psychiatric problems	Depression, belligerent, hostile and disruptive conduct with also psychotic traits
		Schizophrenia
Brain MRI		4% cortical dysplasia

Clinical features of PCDH19-DEE with a focus on seizures, EEG, motor and cognitive development, autistic traits, and other features.

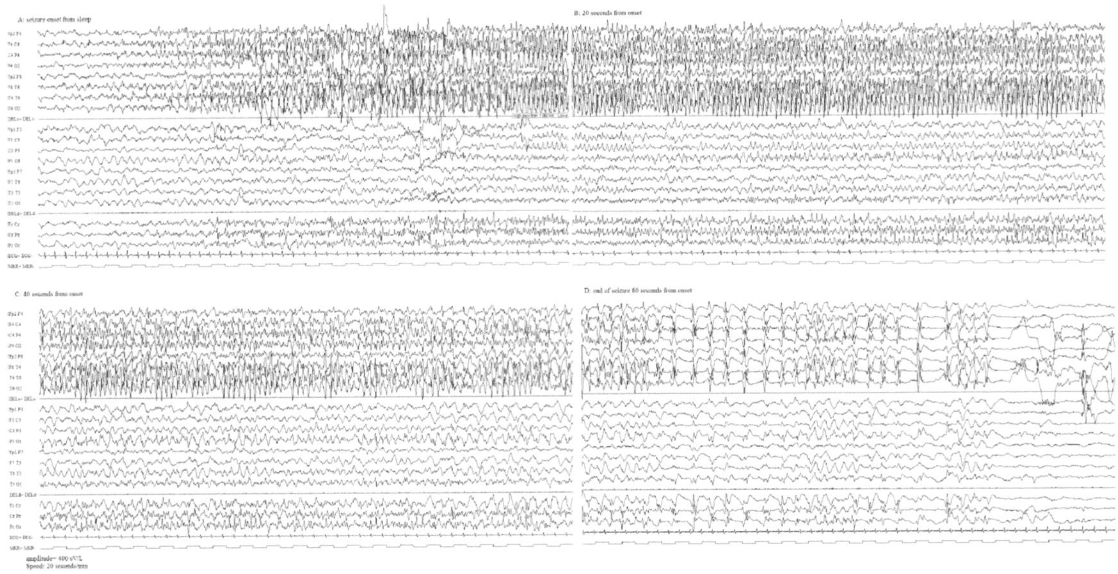

FIGURE 21.1 **PCDH19-DDE focal seizure.** (A) Seizure arises from sleeping with sudden onset of an ictal discharge over the right hemisphere; the epileptiform activity involves also the vertex regions; (B) 20 s from onset, the ictal discharge still ongoing on the right hemisphere; (C) 40 s from onset, ongoing right ictal discharge with minimal spread to the controlateral hemisphere; (D) 80 s from onset, the discharge stops and the EEG is dominated by postictal diffuse flattening of the background activity. From the clinical point of view, the patient has a focal seizure with fearful facial expression, staring, unresponsiveness, vocalization, and hypomotor phase. *DELd*, right deltoid muscle; *DELs*, left deltoid muscle.

seizures, including fearful facial expression and screaming, suggesting an epileptogenic dysfunction involving the limbic system (Marini et al., 2012). A minority of patients manifest generalized seizures (Marini et al., 2012; Specchio, Fusco, & Vigevano, 2011; Specchio, Marini, et al., 2011).

Electroencephalogram (EEG) recording at onset epilepsy during seizures exhibits an alternating pattern with paroxysmal ictal activity shifting from one hemisphere to the other. Frontotemporal areas of both hemispheres are often involved (Fig. 21.1); irregular and generalized spike-wave discharges are typically associated with absences (Marini et al., 2012; Trivisano et al., 2018; van Harssel et al., 2013).

Patients carrying PCDH19 mutations have a normal early psychomotor development followed, in about 70% of girls, by cognitive impairment emerging around age two and half. Although a specific developmental profile has not been recognized (Cappelletti et al., 2015), cognitive impairment ranges from moderate (40%−50%) to severe (10%) (Breuillard et al., 2016; Depienne et al., 2011; Marini et al., 2010; Scheffer et al., 2008; Trivisano et al., 2018; van Harssel et al., 2013).

Behavioral and psychiatric problems are common features of PCDH19-DDE and include depression, belligerent, hostile, disruptive conduct with also psychotic and autistic traits, and schizophrenia (Breuillard et al., 2016; Camacho et al., 2012; Kolk et al., 2018; Vlaskamp et al., 2019). Behavioral problems occur in about half of the patients and more than 44% can be diagnosed with ASD (Camacho et al., 2012). *PCDH19* mutations were associated with psychiatric comorbidities in approximately 60% of females (Kolc et al., 2018). Several patients experience behavioral and mood changes, consisting of episodes of confusion, speech difficulties, or aggressive behavior appearing between the seizures of a cluster (Marini et al., 2012).

Differential diagnosis and neuroimaging

Outside the core features of *PCDH19*-related epilepsy, there is a spectrum of clinical manifestations that might overlap with other epilepsies or epilepsy syndromes including Dravet syndrome (DS).

The occurrence of seizures precipitated by fever, normal early development, and subsequent delay or regression might induce clinicians to formulate a diagnosis of DS (Depienne et al., 2009; Higurashi et al., 2013; van Harssel et al., 2013). Yet, one of the most peculiar PCDH19-related disorders is the occurrence of focal seizures in clusters and the evolution of the disorder, with more than 50% of patients developing autistic symptoms. Patients presenting with acute-febrile onset of seizures might, very rarely, mimic the febrile infection-related epilepsy syndrome—FIRES (Specchio, Fusco, & Vigevano, 2011; Specchio, Marini, et al., 2011) (Fig. 21.2). Considering that seizures often manifest during sleep with motor and affective, limbic semiology, *PCDH19*-related epilepsy might overlap also with focal epilepsies including nocturnal frontal lobe epilepsy (NFLE).

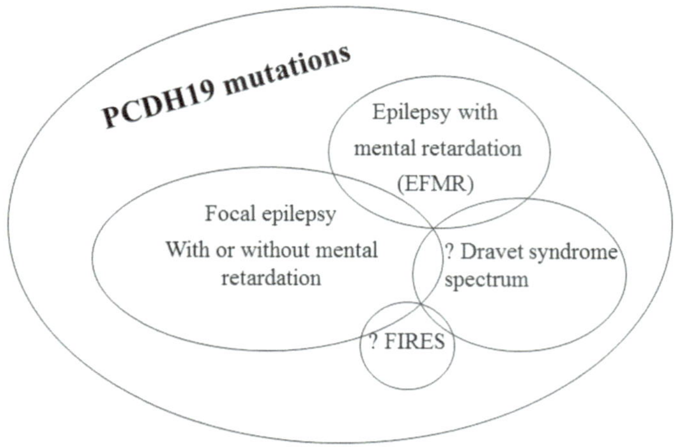

FIGURE 21.2 **PCDH19-DDE phenotype spectrum.** Spectrum of phenotypical manifestations associated with *PCDH19* gene mutations (FIRES = febrile infection-related epilepsy syndrome).

Most studies report normal brain magnetic resonance imaging (MRI) in PCDH19-DDE patients (Depienne et al., 2009; Marini et al., 2010; Scheffer et al., 2008; van Harssel et al., 2013). However, focal cortical dysplasia was reported in some patients (Kurian et al., 2018; Lotte et al., 2016), thus raising the open question on whether malformation of cortical development should be included among the symptoms of PCDH19-related disorder.

PCDH19 mutations

PCDH19 mutations can be missense, frameshift, nonsense, splicing, copy number variations, and, although originally described in families (Juberg & Hellman, 1971; Ryan et al., 1997; Scheffer et al., 2008), most mutations are de novo (Depienne et al., 2009; Ikeda et al., 2016; Marini et al., 2010; Specchio, Fusco, & Vigevano, 2011; Specchio, Marini, et al., 2011; Trivisano et al., 2018; van Harssel et al., 2013; Vincent et al., 2012).

No genotype−phenotype correlation has been observed. At present, the only known prognostic clinical factor remains the evidence that patients with earlier-onset seizures have a more severe outcome in terms of epilepsy, intellectual disability, and psychiatric symptoms (Kolc et al., 2018; Trivisano et al., 2018).

In the original description of this disorder, males carrying hemizygous mutations were reported as unaffected transmitting parents (Juberg & Hellman, 1971; Ryan et al., 1997; Scheffer et al., 2008) even though a detailed clinical analysis revealed that some had moderate cognitive impairment and inflexible, rigid, obsessive controlling personalities (Scheffer et al., 2008; van Harssel et al., 2013).

The selective susceptibility of females to seizures led to the so-called "cell-interference model" (Fig. 21.3), according to which PCDH19-DDE core symptoms would arise from the coexistence of PCDH19-positive and PCDH19-negative cells, as a consequence of the random X-chromosome inactivation in females. This hypothesis was supported in 2009 when the first affected male carrying somatic mosaicism for *PCDH19* deletion was reported (Depienne et al., 2009). Since then, nine additional male patients having comorbid psychiatric features have been reported (de Lange et al., 2017; Depienne et al., 2009; Perez, Hsieh, & Rohena, 2017; Terracciano et al., 2016; Thiffault et al., 2016).

PCDH19-DDE treatments

Treatment of *PCDH19*-related disorder is difficult and no consensus has been reached on how to treat PCDH19-DDE patients.

Especially at onset and during the first years of life, clusters of seizures are resistant to antiepileptic drugs (AEDs). Over time, especially around and after puberty, frequency and pharmacoresistance tend to decrease.

Assessing the efficacy of drugs is difficult, since a possible age-dependent spontaneous seizure remission must be considered. However, among antiepileptic drugs, phenytoin, bromide, and clonazepam were effective in decreasing seizures frequency. Some efficacy was also showed by valproic acid, topiramate, phenobarbital, levetiracetam, lamotrigine, and carbamazepine (Higurashi et al., 2013; Lotte et al., 2016).

Opportunities for targeted therapeutic interventions came from the observation of the deficit of the neurosteroid allopregnanolone in both patients and experimental models (Tan et al. 2015). Some clinical trials already established the efficacy of the synthetic methyl analog of allopregnanolone, ganaxolone, in epilepsy and *status epilepticus* in some patients (Broomall et al., 2014; Kerrigan et al., 2000; Pieribone et al., 2007; Yawno et al., 2017).

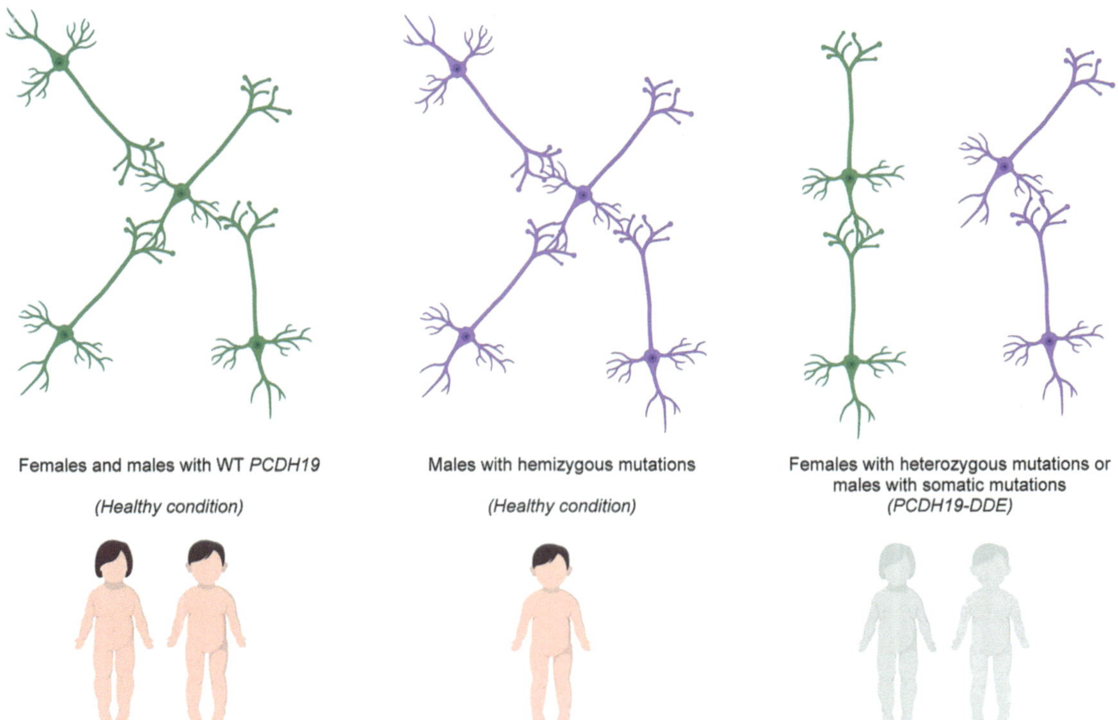

FIGURE 21.3 **The "cell-interference model".** According to the "cell-interference model", PCDH19-DDE symptoms would arise from the coexistence of PCDH19-positive and PCDH19-negative cells, as a consequence of somatic mutations in males or the random X-chromosome inactivation in females with heterozygous *PCDH19* mutations (right panel). By contrast, males with hemizygous *PCDH19* mutations, and therefore a homogeneous population of PCDH19-negative cells, are seizure-free (middle panel) as the children with WT *PCDH19* alleles (left panel).

The contribution of Pcdh19 to the different phases of brain development

The role of PCDH19 in the brain development is starting to emerge. Data on PCDH19 functions came from different models, including zebrafish, KO mice, and human induced neural precursors cells (NPCs). Even though a complete and coherent picture is still missing, PCDH19 is emerging as a protein involved in several core stages of neurodevelopment, from neurulation to neuronal network wiring, passing through cell migration and proliferation (Fig. 21.4).

Neurulation

The first key event in neurodevelopment is the neurulation, the process by which the neural plate transforms into the neural tube, from which the entire brain will originate. Neurulation relies on highly regulated cell migration processes, which are compromised upon Pcdh19 loss in zebrafish. In particular, the movement of Pcdh19 null cells in the anterior neural tube becomes more randomized and less coordinated with that of neighboring cells, thus affecting the medially directed cell migration (Biswas, Emond, & Jontes, 2010; Emond, Biswas, & Jontes, 2009).

Proliferation of NPCs and neurogenesis

The cavity of the neural tube, from which the brain ventricular system will develop, is lined by neural progenitors that proliferate and generate all brain cells. The balance between the expansion of progenitors and their differentiation into postmitotic cells guarantees the generation of the correct brain cell number in a timely manner. Evidence from zebrafish and mouse embryos and induced (i) NPCs from human origin points to a role of PCDH19 in the proliferation of neuronal precursors (Cooper et al., 2015; Fujitani, Zhang, Fujiki, Fujihara, & Yamashita, 2017; Homan et al., 2018). In the optic tectum of zebrafish larvae, pcdh19 loss is associated with increased cell proliferation and concomitant defects in the columnar organization of neurons (Cooper et al., 2015). In mouse embryos, it has been shown that PCDH19 promotes the proliferation of cortical progenitors, and PCDH19 downregulation allows the differentiation of radial glia (RG) into basal progenitors (BPs). During development, the transient PCDH19 downregulation between embryonic day (E) 14.5 and E 16.5, necessary to allow this differentiation step, is accomplished by Mir-484, which is involved in the recently described

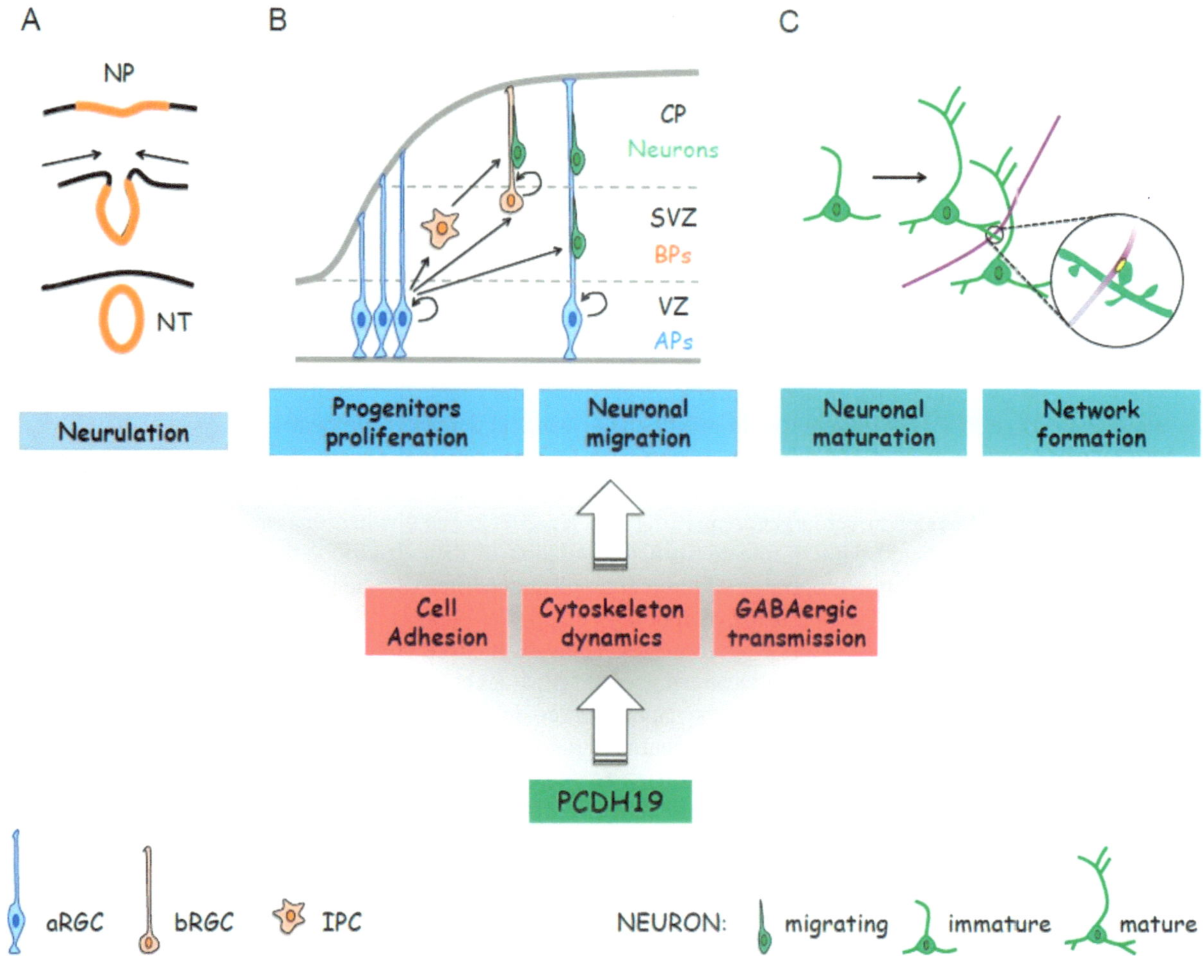

FIGURE 21.4 PCDH19 role in neurodevelopment. This simplified scheme recapitulates the main functions of PCDH19 during brain development (top, A–C), which rely on PCDH19 involvement in the regulation of cell adhesion, cytoskeleton dynamics, and GABAergic transmission (bottom). (A) In zebrafish pcdh19 is required for cell migration during the anterior neurulation, the process by which the neural plate (NP) gives rise to the neural tube (NT). (B) In the developing neocortex, apical progenitors (APs) of the ventricular zone (VZ) divide and generate basal progenitors (BPs) of the subventricular zone (SVZ). Apical radial glial cells (aRGCs) belong to the APs pool, while intermediate progenitor cells (IPCs) and basal radial glial cells (bRGCs) belong to the BPs pool. Both APs and BPs give rise to neurons that migrate radially to their final destination in the cortical plate (CP). PCDH19 promotes the proliferation of neuronal progenitors in mouse cortex, and PCDH19 downregulation allows the differentiation of the radial glia into basal BPs. Furthermore, PCDH19 regulates neuronal migration and sorting. (C) Recent evidence points to PCDH19 as a regulator of neurite outgrowth and arborization and of neuronal network formation.

microduplication syndrome 16p13.11, characterized by hyperactivity. Mir-484 targets *Pcdh19* 3′UTR and reduces mRNA stability and translation and hence PCDH19 expression (Fujitani et al., 2017). Consistently, neural stem and progenitor cells (NSPCs) from reprogrammed fibroblasts of patients with PCDH19 loss-of-function mutations showed reduced cell polarity, premature neurogenesis, and maturation (Homan et al., 2018).

Neuronal migration and sorting

Neurons produced in the ventricular zone (VZ) migrate radially and reach their final destination. PCDH19 has emerged as a regulator of neuronal migration, as inferred from defects in neuronal localization and sorting in models of PCDH19 loss of function (Bassani et al., 2018; Cooper et al., 2015; Pederick et al., 2016, 2018). In vitro experiments show that *Pcdh19* null progenitor neurons from E 15.4 mice embryos are characterized by increased migration when cultured as neurospheres (Pederick et al., 2016). In vivo, rat hippocampal neurons of the CA1 region, in which PCDH19 has been downregulated by in utero electroporation of short hairpin (sh)RNA, showed ectopic localization postnatally, index of improper migration (Bassani et al., 2018). Furthermore, PCDH19 determines the sorting of neuronal populations that are homogeneous with regard of PCDH19 expression, i.e., of PCDH19-positive and PCDH19-negative neurons. This spatial segregation has been reported under physiological conditions in the optic tectum of zebrafish (Cooper et al., 2015) and as a consequence of PCDH19 mosaic deletion in heterozygous PCDH19 KO mice (Hayashi et al., 2017; Pederick et al., 2018). Similarly, PCDH19 overexpression in the cortex of wild-type (WT) mice caused ectopic clustering of PCDH19-positive neurons (Hayashi et al., 2017). In the optic tectum of zebrafish, the neurons originating from a common Pcdh19-positive progenitor formed discrete columns of Pcdh19-expressing neurons interspersed with Pcdh19-negative neurons. This columnar organization was disrupted upon *pcdh19* loss (Cooper et al., 2015). A similar spatial pattern, which was not visible in WT mice, was reported in the cortex of heterozygous PCDH19 KO mice, as a result of the columnar segregation of neurons with the PCDH19 WT or null allele (Pederick et al., 2018). The functional consequence of these distribution patterns and their potential implication for pathological conditions are still unknown.

Neuronal maturation and circuit formation

After migration, neurons acquire their mature appearance and establish synaptic connections. There is some evidence that in this developmental stage PCDH19 plays a role in neurite outgrowth and dendritic arborization (Bassani et al., 2018; Cooper et al., 2015). In the optic tectum of zebrafish larvae, in which Pcdh19-positive neurons segregate within distinct columns, Pcdh19 loss is associated not only with loss of columnar architecture but also with loss of dendritic arbor confinement and causes the dendritic arbor to extend over larger areas (Cooper et al., 2015). The downregulation of PCDH19 in a subset of rat hippocampal neurons is associated with an altered spatial orientation of their dendrites, the overall reduction of dendritic length, and the change of the relative proportion of basal and apical dendrites (Bassani et al., 2018). However, *Pcdh19* KO displayed no abnormalities in dendrite and spine morphology of layer V neurons (Hayashi et al., 2017). Altogether, where and how far neurites extend influence neuronal connections. The synaptic role of PCDH19 is gradually emerging. PCDH19 can be detected at synapses of cortical and hippocampal neurons (Bassani et al., 2018; Hayashi et al., 2017). The observation that PCDH19 affects the expression levels of some synaptic markers in cultured hippocampal neurons suggests a pro-wiring effect of PCDH19. Furthermore, PCDH19 binds gamma-aminobutyric acid (GABA) type A receptors ($GABA_ARs$) regulating their surface expression and currents in primary neurons (Bassani et al., 2018).

Defects in all neurodevelopmental stages up to synapse formation determine neuronal network properties and functioning. With this regard, in vivo calcium imaging in zebrafish *pcdh19* mutants during neuronal network formation revealed a functional pattern consistent with altered synaptogenesis and network topology (Light & Jontes, 2019). Additional evidence of network functional defects comes from abnormalities in electrocorticogram recordings from heterozygous *Pcdh19* KO mice (Pederick et al., 2018) and the higher seizure susceptibility of rats in which *Pcdh19* was downregulated in a subset of hippocampal neurons (Bassani et al., 2018).

Furthermore, loss of *Pcdh19* in hemizygous males and heterozygous females was associated with impaired behaviors including activity defects under stress conditions. Notably, only heterozygous females exhibited decreased fear responses (Hayashi et al., 2017).

The molecular mechanisms underpinning PCDH19 neurodevelopmental role

PCDH19 is in the first place an adhesion molecule able to associate both *in cis* and *in trans* with itself (Cooper, Jontes, & Sotomayor, 2016) and other adhesion molecules such as neural cadherin (NCAD) (Biswas et al., 2010; Emond, Biswas, Blevins, & Jontes, 2011). In addition, the intracellular C-terminus of PCDH19 harbors the binding sites for components of the actin remodeling WAVE regulatory complex (Chen et al., 2014) and for the $GABA_AR$ alpha subunits (Bassani et al., 2018) (Fig. 21.5). Each of these features has the potential to contribute to most of PCDH19 neurodevelopmental roles.

By means of its adhesive properties, PCDH19 might influence cell response to spatial cues that drive migration, and also provide a molecular code for cell—cell recognition on the basis of cell sorting and neuronal network formation. By regulating the cytoskeleton directly, via interaction with the WAVE complex, or indirectly through its binding partner NCAD, PCDH19 is well equipped for regulating processes as cell motility, cell morphology, and polarity, the last being in relationship with neuronal precursors expansion (Homan et al., 2018). Finally, the interplay between PCDH19 and $GABA_ARs$ provides additional means to regulate the migration and morphological maturation of neurons, since GABA exerts a trophic action during the neurodevelopment (Ben-Ari, 2014). The finding that PCDH19 mutations are associated with deficiency of neurosteroids (Tan et al., 2015), which are $GABA_ARs$-positive allosteric modulators with neurodevelopmental functions, further supports this hypothesis.

An additional level of complexity on the multifaceted role of PCDH19 in neurodevelopment is given by the "cell-interference hypothesis" (Fig. 21.3) mentioned earlier that originates from the peculiar inheritance pattern of PCDH19-related epilepsy, which affects exclusively female patients with heterozygous PCDH19 mutations, but spares male

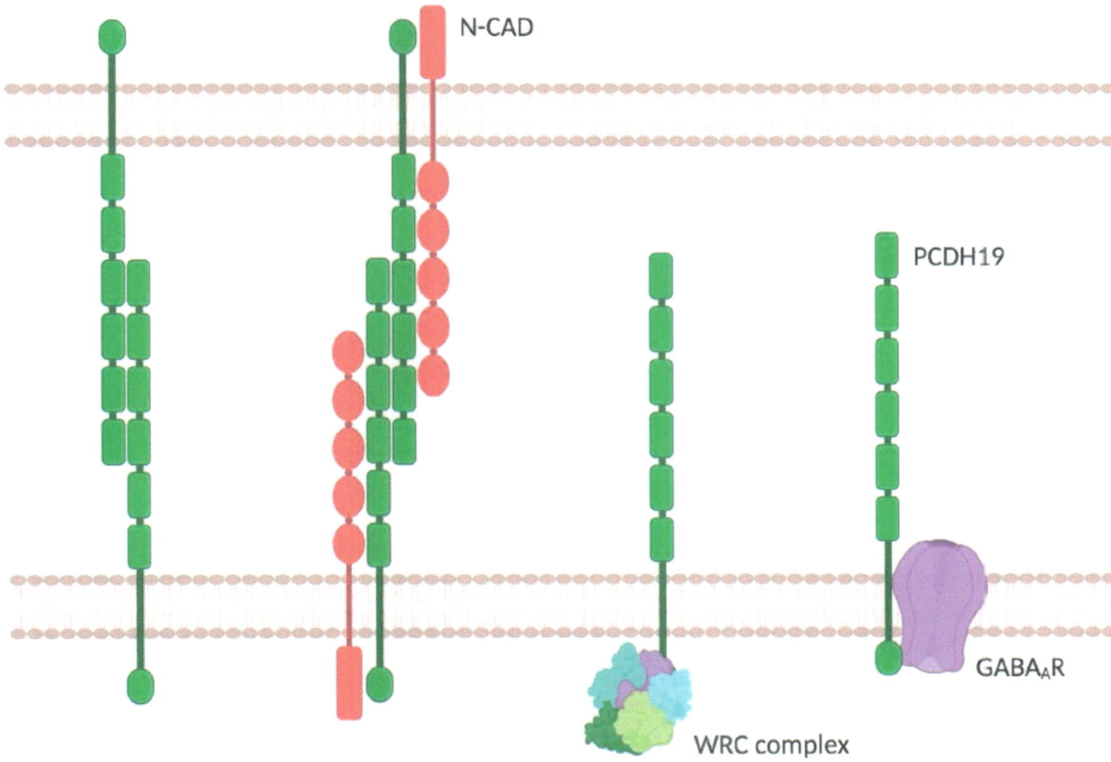

FIGURE 21.5 **PCDH19 interactions.** PCDH19 (in green) is an adhesion molecule able to associate with itself and neural cadherin (NCAD, magenta). The intracellular C-terminus of PCDH19 harbors the binding sites for components of the actin remodeling WAVE regulatory complex (WRC) and for the GABA$_A$R alpha subunits.

patients with one mutant PCDH19 allele (Depienne et al., 2009). Hence, the coexistence of two cell populations, expressing either the WT or the mutant *PCDH19*, would be a critical determinant of the PCDH19-related epilepsy. In particular, the heterogeneous expression of PCDH19 would lead to two cell populations with different and deleterious spatial and temporal patterns of behavior. According to PCDH19 biological roles, this heterogeneity is expected to affect cell migration and recognition, thus leading to spatial segregation, (Hayashi et al., 2017; Pederick et al., 2018) or cause asynchrony in neuronal developmental program (Homan et al., 2018). Given PCDH19 role in GABAergic transmission (Bassani et al., 2018), the spatial patterns of neuronal excitability might be also affected. However, the precise mechanism by which PCDH19 heterogeneous expression translates into structural and functional defects of neuronal networks still awaits elucidation.

Applications to other areas of development

In this chapter, we have reviewed the clinical features of PCDH19-related syndrome and the biological roles of PCDH19 so far investigated.

A significant overlapping exists between the clinical phenotype of PCDH19-related syndrome and other neurodevelopmental diseases, including the Dravet syndrome (Higurashi et al., 2013), and this suggests that some key molecular pathways arising from different gene mutations might be shared or converge downstream.

PCDH19 is a member of the protocadherin family within the cadherin superfamily. Both cadherins and protocadherins have been implicated in neurodevelopmental disorders (Redies, Hertel, & Hübner, 2012), possibly indicating that members of the same family might play similar roles.

In this chapter, we describe the cell-interference model as a putative pathogenic mechanism for PCDH19-related syndrome. This mechanism was previously postulated for another X-linked condition named Craniofrontonasal syndrome (Twigg et al., 2013). This syndrome is caused by mutations in the gene encoding the transmembrane protein Ephrin-B1 (Twigg et al., 2013). It is possible that this mechanism might apply to other diseases involving proteins that regulate cell—cell interaction and communication.

Key facts of PCDH19-related neurodevelopmental disorder

- Among epilepsy-related genes, PCDH19 is second only to *SCN1A* for clinical relevance
- PCDH19-related epilepsy affects almost exclusively females, while generally sparing carrier males
- In 2009, the first male patient carrying somatic mosaicism for *PCDH19* deletion was reported, thus supporting the so-called "cell-interference hypothesis"
- The "cell-interference hypothesis" states that neighboring cells with either WT or mutated PCDH19 interfere with each other and compromise brain physiology
- Ganaxolone, a synthetic analog of the neurosteroid allopregnanolone, provides opportunities for targeted therapeutic interventions

Summary points

- PCDH19-related syndrome is characterized by early-onset epilepsy, intellectual disability, and autism
- In patients carrying PCDH19 mutations, a normal early psychomotor development is generally followed by cognitive impairment
- A positive correlation has been observed between the age of seizure onset and the severity of cognitive impairment
- PCDH19 mutations are either de novo or transmitted through an X-linked pattern from seizure-free fathers to their daughters who will manifest the neurodevelopmental syndrome
- PCDH19 is a cell-adhesion protein involved in neurulation, cell proliferation, cell migration, and neuronal circuit formation
- PCDH19-binding partners include the adhesion molecule NCAD, the actin remodeling WAVE regulatory complex, and $GABA_AR$ alpha subunits

References

Bassani, S., Cwetsch, A. W., Gerosa, L., Serratto, G. M., Folci, A., Hall, I. F., … Passafaro, M. (2018). The female epilepsy protein PCDH19 is a new GABAAR-binding partner that regulates GABAergic transmission as well as migration and morphological maturation of hippocampal neurons. *Human Molecular Genetics, 27*(6), 1027–1038.

Ben-Ari, Y. (2014). The GABA excitatory/inhibitory developmental sequence: A personal journey. *Neuroscience, 279*, 187–219.

Biswas, S., Emond, M. R., & Jontes, J. D. (2010). Protocadherin-19 and N-cadherin interact to control cell movements during anterior neurulation. *The Journal of Cell Biology, 191*(5), 1029–1041.

Breuillard, D., Leunen, D., Chemaly, N., Auclair, L., Pinard, J. M., Kaminska, A., … Nabbout, R. (2016). Autism spectrum disorder phenotype and intellectual disability in females with epilepsy and PCDH-19 mutations. *Epilepsy and Behavior, 60*, 75–80.

Broomall, E., Natale, J. E., Grimason, M., Goldstein, J., Smith, C. M., Chang, C., … Wainwright, M. S. (2014). Pediatric super-refractory status epilepticus treated with allopregnanolone. *Annals of Neurology, 76*(6), 911–915.

Camacho, A., Simón, R., Sanz, R., Viñuela, A., Martínez-Salio, A., & Mateos, F. (2012). Cognitive and behavioral profile in females with epilepsy with PDCH19 mutation: Two novel mutations and review of the literature. *Epilepsy and Behavior, 24*(1), 134–137.

Cappelletti, S., Specchio, N., Moavero, R., Terracciano, A., Trivisano, M., Pontrelli, G., … Cusmai, R. (2015). Cognitive development in females with PCDH19 gene-related epilepsy. *Epilepsy and Behavior, 42*, 36–40.

Carvill, G. L., Heavin, S. B., Yendle, S. C., McMahon, J. M., O'Roak, B. J., Cook, J., … Mefford, H. C. (2013). Targeted resequencing in epileptic encephalopathies identifies de novo mutations in CHD2 and SYNGAP1. *Nature Genetics, 45*(7), 825–830.

Chen, B., Brinkmann, K., Chen, Z., Pak, C. W., Liao, Y., Shi, S., … Rosen, M. K. (2014). The WAVE regulatory complex links diverse receptors to the actin cytoskeleton. *Cell, 156*(1–2), 195–207.

Cooper, S. R., Emond, M. R., Duy, P. Q., Liebau, B. G., Wolman, M. A., & Jontes, J. D. (2015). Protocadherins control the modular assembly of neuronal columns in the zebrafish optic tectum. *The Journal of Cell Biology, 211*(4), 807–814.

Cooper, S. R., Jontes, J. D., & Sotomayor, M. (2016). Structural determinants of adhesion by Protocadherin-19 and implications for its role in epilepsy. *Elife, 5*.

Depienne, C., Bouteiller, D., Keren, B., Cheuret, E., Poirier, K., Trouillard, O., … Leguern, E. (2009). Sporadic infantile epileptic encephalopathy caused by mutations in PCDH19 resembles Dravet syndrome but mainly affects females. *PLoS Genetics, 5*(2), e1000381.

Depienne, C., & LeGuern, E. (2012). PCDH19-related infantile epileptic encephalopathy: An unusual X-linked inheritance disorder. *Human Mutation, 33*(4), 627–634.

Depienne, C., Trouillard, O., Bouteiller, D., Gourfinkel-An, I., Poirier, K., Rivier, F., … LeGuern, E. (2011). Mutations and deletions in PCDH19 account for various familial or isolated epilepsies in females. *Human Mutation, 32*(1), E1959–E1975.

Dibbens, L. M., Tarpey, P. S., Hynes, K., Bayly, M. A., Scheffer, I. E., Smith, R., … Gécz, J. (2008). X-linked protocadherin 19 mutations cause female-limited epilepsy and cognitive impairment. *Nature Genetics, 40*(6), 776–781.

Emond, M. R., Biswas, S., Blevins, C. J., & Jontes, J. D. (2011). A complex of Protocadherin-19 and N-cadherin mediates a novel mechanism of cell adhesion. *The Journal of Cell Biology, 195*(7), 1115–1121.

Emond, M. R., Biswas, S., & Jontes, J. D. (2009). Protocadherin-19 is essential for early steps in brain morphogenesis. *Developmental Biology, 334*(1), 72–83.

Fujitani, M., Zhang, S., Fujiki, R., Fujihara, Y., & Yamashita, T. (2017). A chromosome 16p13.11 microduplication causes hyperactivity through dysregulation of miR- 484/protocadherin-19 signaling. *Molecular Psychiatry, 22*(3), 364–374.

Gerosa, L., Francolini, M., Bassani, S., & Passafaro, M. (2019). The role of protocadherin 19 (PCDH19) in neurodevelopment and in the pathophysiology of early infantile epileptic encephalopathy-9 (EIEE9). *Developmental Neurobiology, 79*(1), 75–84.

van Harssel, J. J., Weckhuysen, S., van Kempen, M. J., Hardies, K., Verbeek, N. E., de Kovel, C. G., … Brilstra, E. H. (2013). Clinical and genetic aspects of PCDH19-related epilepsy syndromes and the possible role of PCDH19 mutations in males with autism spectrum disorders. *Neurogenetics, 14*(1), 23–34.

Hayashi, S., Inoue, Y., Hattori, S., Kaneko, M., Shioi, G., Miyakawa, T., & Takeichi, M. (2017). Loss of X-linked Protocadherin-19 differentially affects the behavior of heterozygous female and hemizygous male mice. *Scientific Reports, 7*(1), 5801.

Higurashi, N., Nakamura, M., Sugai, M., Ohfu, M., Sakauchi, M., Sugawara, Y., … Hirose, S. (2013). PCDH19-related female-limited epilepsy: Further details regarding early clinical features and therapeutic efficacy. *Epilepsy Research, 106*(1–2), 191–199.

Homan, C. C., Pederson, S., To, T. H., Tan, C., Piltz, S., Corbett, M. A., … Gecz, J. (2018). PCDH19 regulation of neural progenitor cell differentiation suggests asynchrony of neurogenesis as a mechanism contributing to PCDH19 Girls Clustering Epilepsy. *Neurobiology of Disease, 116*, 106–119.

Hynes, K., Tarpey, P., Dibbens, L. M., Bayly, M. A., Berkovic, S. F., Smith, R., … Scheffer, I. E. (2010). Epilepsy and mental retardation limited to females with PCDH19 mutations can present de novo or in single generation families. *Journal of Medical Genetics, 47*(3), 211–216.

Ikeda, H., Imai, K., Shigematsu, H., Takahashi, Y., Inoue, Y., Higurashi, N., & Hirose, S. (2016). Characteristic phasic evolution of convulsive seizure in PCDH19-related epilepsy. *Epileptic Disorders, 18*(1), 26–33.

Juberg, R. C., & Hellman, C. D. (1971). A new familial form of convulsive disorder and mental retardation limited to females. *The Journal of Pediatrics, 79*(5), 726–732.

Kahr, I., Vandepoele, K., & van Roy, F. (2013). Delta-protocadherins in health and disease. *Progress in Molecular Biology and Translational Science, 116*, 169–192.

Kerrigan, J. F., Shields, W. D., Nelson, T. Y., Bluestone, D. L., Dodson, W. E., Bourgeois, B. F., … Monaghan, E. P. (2000). Ganaxolone for treating intractable infantile spasms: A multicenter, open- label, add-on trial. *Epilepsy Research, 42*(2–3), 133–139.

Kolc, K. L., Sadleir, L. G., Scheffer, I. E., Ivancevic, A., Roberts, R., Pham, D. H., & Gecz, J. (2018). A systematic review and meta-analysis of 271 PCDH19-variant individuals identifies psychiatric comorbidities, and association of seizure onset and disease severity. *Molecular Psychiatry, 24*(2), 241–251.

Kurian, M., Korff, C. M., Ranza, E., Bernasconi, A., Lübbig, A., Nangia, S., … Bast, T. (2018). Focal cortical malformations in children with early infantile epilepsy and PCDH19 mutations: Case report. *Developmental Medicine and Child Neurology, 60*(1), 100–105.

de Lange, I. M., Rump, P., Neuteboom, R. F., Augustijn, P. B., Hodges, K., Kistemaker, A. I., … Brilstra, E. H. (2017). Male patients affected by mosaic PCDH19 mutations: Five new cases. *Neurogenetics, 18*(3), 147–153.

Light, S. E. W., & Jontes, J. D. (2019). Multiplane calcium imaging reveals disrupted development of network topology in zebrafish. *eNeuro, 6*(3).

Lotte, J., Bast, T., Borusiak, P., Coppola, A., Cross, J. H., Dimova, P., … Kluger, G. (2016). Effectiveness of antiepileptic therapy in patients with PCDH19 mutations. *Seizure, 35*, 106–110.

Marini, C., Darra, F., Specchio, N., Mei, D., Terracciano, A., Parmeggiani, L., … Guerrini, R. (2012). Focal seizures with affective symptoms are a major feature of PCDH19 gene-related epilepsy. *Epilepsia, 53*(12), 2111–2119.

Marini, C., Mei, D., Parmeggiani, L., Norci, V., Calado, E., Ferrari, A., … Guerrini, R. (2010). Protocadherin 19 mutations in girls with infantile-onset epilepsy. *Neurology, 75*(7), 646–653.

Pederick, D. T., Homan, C. C., Jaehne, E. J., Piltz, S. G., Haines, B. P., Baune, B. T., … Thomas, P. Q. (2016). Pcdh19 loss-of-function increases neuronal migration in vitro but is dispensable for brain development in mice. *Scientific Reports, 6*, 26765.

Pederick, D. T., Richards, K. L., Piltz, S. G., Kumar, R., Mincheva-Tasheva, S., Mandelstam, S. A., … Thomas, P. Q. (2018). Abnormal cell sorting underlies the unique X-linked inheritance of PCDH19 epilepsy. *Neuron, 97*(1), 59–66.e55.

Peek, S. L., Mah, K. M., & Weiner, J. A. (2017). Regulation of neural circuit formation by protocadherins. *Cellular and Molecular Life Sciences, 74*(22), 4133–4157.

Perez, D., Hsieh, D. T., & Rohena, L. (2017). Somatic Mosaicism of PCDH19 in a male with early infantile epileptic encephalopathy and review of the literature. *American Journal of Medical Genetics, 173*(6), 1625–1630.

Pieribone, V. A., Tsai, J., Soufflet, C., Rey, E., Shaw, K., Giller, E., & Dulac, O. (2007). Clinical evaluation of ganaxolone in pediatric and adolescent patients with refractory epilepsy. *Epilepsia, 48*(10), 1870–1874.

Redies, C., Hertel, N., & Hübner, C. A. (2012). Cadherins and neuropsychiatric disorders. *Brain Research, 1470*, 130–144.

Ryan, S. G., Chance, P. F., Zou, C. H., Spinner, N. B., Golden, J. A., & Smietana, S. (1997). Epilepsy and mental retardation limited to females: An X-linked dominant disorder with male sparing. *Nature Genetics, 17*(1), 92–95.

Scheffer, I. E., Turner, S. J., Dibbens, L. M., Bayly, M. A., Friend, K., Hodgson, B., … Berkovic, S. F. (2008). Epilepsy and mental retardation limited to females: An under-recognized disorder. *Brain, 131*(Pt 4), 918–927.

Specchio, N., Fusco, L., & Vigevano, F. (2011). Acute-onset epilepsy triggered by fever mimicking FIRES (febrile infection-related epilepsy syndrome): The role of protocadherin 19 (PCDH19) gene mutation. *Epilepsia, 52*(11), e172–175.

Specchio, N., Marini, C., Terracciano, A., Mei, D., Trivisano, M., Sicca, F., ... Vigevano, F. (2011). Spectrum of phenotypes in female patients with epilepsy due to protocadherin 19 mutations. *Epilepsia, 52*(7), 1251–1257.

Tan, C., Shard, C., Ranieri, E., Hynes, K., Pham, D. H., Leach, D., ... Gecz, J. (2015). Mutations of protocadherin 19 in female epilepsy (PCDH19-FE) lead to allopregnanolone deficiency. *Human Molecular Genetics, 24*(18), 5250–5259.

Terracciano, A., Specchio, N., Darra, F., Sferra, A., Bernardina, B. D., Vigevano, F., & Bertini, E. (2012). Somatic mosaicism of PCDH19 mutation in a family with low-penetrance EFMR. *Neurogenetics, 13*(4), 341–345.

Terracciano, A., Trivisano, M., Cusmai, R., De Palma, L., Fusco, L., Compagnucci, C., ... Specchio, N. (2016). PCDH19-related epilepsy in two mosaic male patients. *Epilepsia, 57*(3), e51–55.

Thiffault, I., Farrow, E., Smith, L., Lowry, J., Zellmer, L., Black, B., ... Saunders, C. (2016). PCDH19- related epileptic encephalopathy in a male mosaic for a truncating variant. *American Journal of Medical Genetics, 170*(6), 1585–1589.

Trivisano, M., Pietrafusa, N., Terracciano, A., Marini, C., Mei, D., Darra, F., ... Specchio, N. (2018). Defining the electroclinical phenotype and outcome of PCDH19-related epilepsy: A multicenter study. *Epilepsia, 59*(12), 2260–2271.

Twigg, S. R., Babbs, C., van den Elzen, M. E., Goriely, A., Taylor, S., McGowan, S. J., ... Wilkie, A. O. (2013). Cellular interference in craniofrontonasal syndrome: Males mosaic for mutations in the X-linked EFNB1 gene are more severely affected than true hemizygotes. *Human Molecular Genetics, 22*(8), 1654–1662.

Vincent, A. K., Noor, A., Janson, A., Minassian, B. A., Ayub, M., Vincent, J. B., & Morel, C. F. (2012). Identification of genomic deletions spanning the PCDH19 gene in two unrelated girls with intellectual disability and seizures. *Clinical Genetics, 82*(6), 540–545.

Vlaskamp, D. R. M., Bassett, A. S., Sullivan, J. E., Robblee, J., Sadleir, L. G., Scheffer, I. E., & Andrade, D. M. (2019). Schizophrenia is a later-onset feature of PCDH19 girls clustering epilepsy. *Epilepsia, 60*(3), 429–440.

Yawno, T., Miller, S. L., Bennet, L., Wong, F., Hirst, J. J., Fahey, M., & Walker, D. W. (2017). Ganaxolone: A new treatment for neonatal seizures. *Frontiers in Cellular Neuroscience, 11*, 246.

Chapter 22

Neurodevelopment in turner syndrome

Reid Blanchett and Rebecca C. Knickmeyer
Institute of Quantitative Health Science and Engineering, Bioengineering Facility, Michigan State University, East Lansing, MI, United States

List of abbreviations

ADHD Attention deficit hyperactivity disorder
ASD Autism spectrum disorder
CHD Congenital heart disease
GH Growth hormone
HRT Hormone replacement therapy
ID Intellectual disability
ILF Inferior longitudinal fasciculus
IPS Intraparietal sulcus
MRI Magnetic resonance imaging
SLF Superior longitudinal fasciculus
TS Turner syndrome

Mini-dictionary of terms

Attentional control The ability to allocate one's attention based on current goals. A top-down process.
Dyscalculia A developmental disorder characterized by profound difficulties in numerical understanding in children of normal or above-average intelligence.
Functional disomy Occurs when X chromosome genes are expressed at twice the expected level.
Gonadal dysgenesis Abnormal development of gonads leading to a developmental hormonal milieu that is androgen- and estrogen-deficient. In TS, gonadal dysgenesis occurs when pairing failure during meiosis leads to rapid loss of oocytes.
Growth hormone Also known as somatotropin. A peptide hormone produced by the pituitary gland, which stimulates growth.
Haploinsufficiency Occurs when a dosage-sensitive gene is expressed at approximately half the expected level. May result from a de novo or inherited loss-of-function mutation in one allele or the deletion of one allele.
Imprinted gene A gene that is expressed in a parent-of-origin-specific manner.
Inferior longitudinal fasciculus A bundle of axons connecting occipital and temporal areas. Thought to facilitate many different functions including face processing.
Intellectual disability A developmental disorder characterized by significant limitations in intellectual functioning and adaptive behavior.
Oxandrolone An androgenic anabolic steroid sometimes used to promote growth in children with TS.
Response inhibition The ability to inhibit inappropriate actions or actions that are no longer required, so as to successfully accomplish one's current goals.
Ring chromosome X A structurally abnormal X chromosome whose ends have fused together to form a ring.
Superior longitudinal fasciculus A bundle of axons connecting parietal and frontal areas. Facilitates many different functions including spatial working memory and visually guided action.
Turner syndrome A rare condition of phenotypic females in which all or part of the second X chromosome is missing in some or all of the body's cells.

Introduction

Turner syndrome (TS) is a rare condition of phenotypic females in which all or part of the second sex chromosome is missing in some or all of the body's cells. TS occurs in approximately one in 2000 live female births, making it one of the

Diagnosis, Management and Modeling of Neurodevelopmental Disorders. https://doi.org/10.1016/B978-0-12-817988-8.00022-1

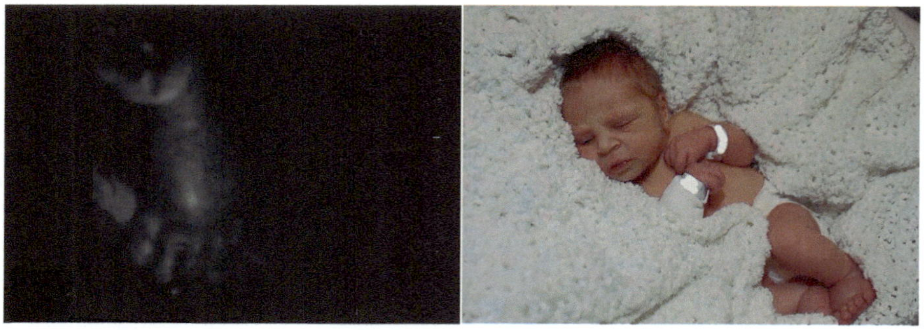

FIGURE 22.1 **A child with Turner syndrome.** Ultrasound image on the left. Infant photograph on the right.

most common human chromosomal abnormalities (Jacobs, Melville, Ratcliff, Keay, & Syme, 1974; Nielsen & Wohlert, 1991). It is a complex and heterogeneous condition, which can affect many different body systems including the reproductive system, cardiovascular system, and the central nervous system (see Fig. 22.2). This chapter will focus primarily on the central nervous system. However, to place this material in its proper context, we will begin with a brief overview of the genetics of TS, associated physical features, and key medical issues that may impinge on neurodevelopment (Fig. 22.1).

Genetics

Three criteria must be met before assigning a diagnosis of TS. First, the individual must be missing either the entire second sex chromosome or the entire distal short arm of the second X-chromosome (Xp22.3); deletions impacting other regions of the X-chromosome and small distal deletions on the short arm of the X chromosome are not included. Second, the

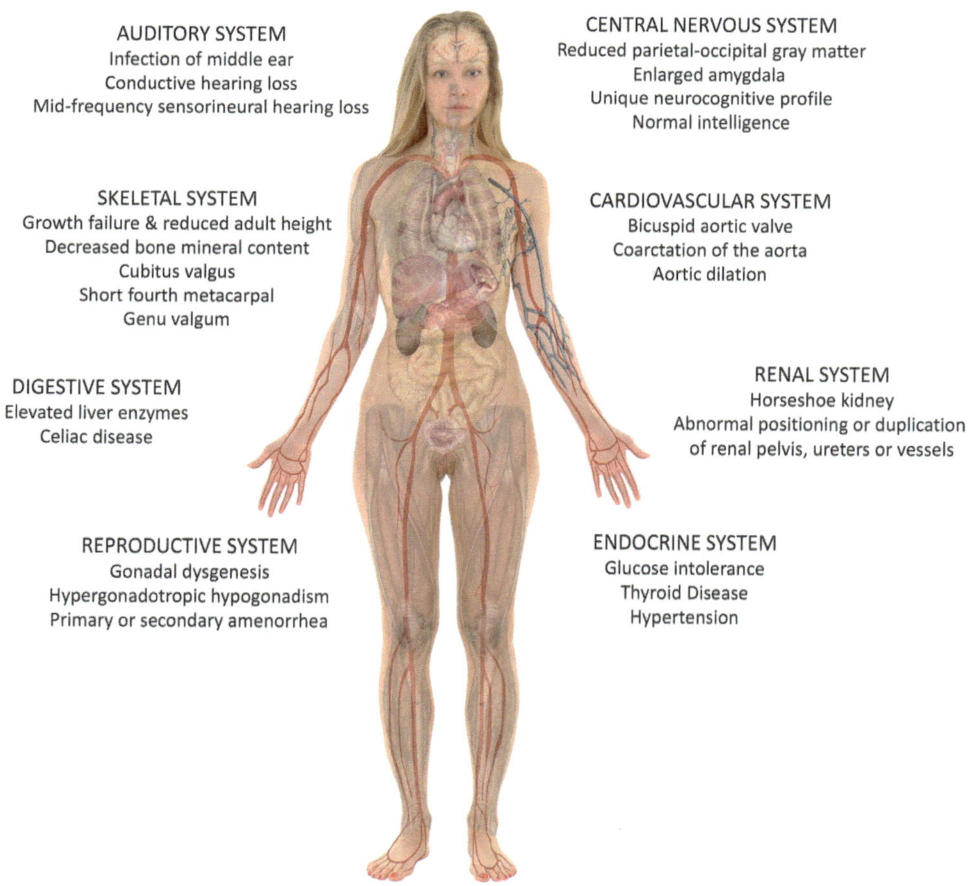

FIGURE 22.2 **Common features of Turner syndrome.** Turner syndrome can affect many different body systems including the skeletal, reproductive, cardiovascular, auditory, endocrine, digestive, renal, and central nervous systems. Female figure with organ systems created by Mikael Häggström (public domain).

individual must exhibit one or more clinical characteristics, as described in the following section. Third, the individual must be a phenotypic female; phenotypic males and individuals with sexual ambiguity are not currently included, even if they harbor some cells that are 45,X. Finally, women over 50 years of age with less than 5% 45,X cells are excluded, as X chromosome loss represents a normal part of female aging (Gravholt et al., 2017).

Although there is variability across studies, current estimates suggest 40%–50% of patients with TS have complete X monosomy (45,X). In most cases, the remaining X is maternal in origin (Davenport, 2010). Approximately 30% of patients have a mosaic karyotype that includes a 45,X cell line and a normal cell line (46,XX or 46,XY). Another 10% of patients carry either an isochromosome Xq or an isodicentric Xp. Additional (rare) karyotypes include deletion of Xp22.3, mosaicism with "Triple X" (45,X/47,XXX or 45,X/46,XX/47,XXX), mosaicism with a ring chromosome X (46,X,r(X)/46,XX), and unbalanced X-autosome translocations (Gravholt et al., 2017). It was once thought that complete X monosomy resulted from a nondisjunction error during meiosis while mosaic karyotypes arose due to postfertilization mitotic errors (Davenport, 2010). However, some scientists have hypothesized that all viable 45,X individuals (those that survive past birth) are cryptic mosaics and that absence of the X chromosome in 45,X individuals is also the result of postmitotic errors in the majority of cases (Hook & Warburton, 2014).

Physical features and key medical issues

Clinical manifestations of TS are extremely wide-ranging. In this section we focus on features that are both common and potentially relevant to neurodevelopment. A comprehensive list of common abnormalities associated with Turner syndrome and their approximate prevalence can be found in the recent clinical care guidelines (Gravholt et al., 2017).

Skeletal system

Growth failure leading to reduced adult height is the most common physical feature of TS, affecting 95%–100% of patients. Consequently, most children with TS are now treated with growth hormone (GH) to increase adult height (Gravholt et al., 2017). Treatment may begin as early as 9 months of age (Davenport et al., 2007). In some cases, concomitant treatment with oxandrolone is recommended in children 10 years of age and above (Gravholt et al., 2017). The impact of growth failure and its treatment upon neurodevelopment is relatively understudied. For many years it was presumed that short stature would negatively affect psychosocial outcomes in TS and that these effects would be ameliorated by GH treatment. However, the vast majority of studies on this question have suffered from a high risk of bias (Gardner, Boshart, Yeguez, Desai, & Sandberg, 2016). Well-designed epidemiological studies suggest GH treatment has little to no effect on psychosocial outcomes in adults with TS (Amundson, Boman, Barrenas, Bryman, & Landin-Wilhelmsen, 2010; Carel et al., 2005). The impact of oxandrolone treatment on neurodevelopment is also unclear and may be age-specific. One study reported that low doses of oxandrolone improved working memory in girls aged 10–14 years (Ross et al., 2003). In contrast, a study of adult women with TS observed higher anxiety and depression in those treated with oxandrolone and no positive neurocognitive effects (Freriks et al., 2015).

Reproductive system

Gonadal dysgenesis occurs in the majority of individuals with TS, leading to hypergonadotropic hypogonadism and primary or secondary amenorrhea. Consequently, most children with TS receive hormone replacement therapy (HRT) for induction of puberty, maintenance of female secondary sex characteristics, and attainment of peak bone mass. HRT typically begins around 12 years of age (Gravholt et al., 2017). Research to date suggests that HRT has minimal or no effect on characteristic neurocognitive profiles (intact verbal abilities and impaired visuospatial skills, see following section for more detail) (Ross et al., 2002), but may improve verbal and nonverbal memory, processing speed, and motor function (Ross, Roeltgen, Feuillan, Kushner, & Cutler, 1998; Ross, Roeltgen, Feuillan, Kushner, & Cutler, 2000). On-time puberty is expected to promote positive psychosocial and psychosexual development (Gravholt et al., 2017).

Cardiovascular system

Congenital heart disease (CHD) is common in TS. Bicuspid aortic valve is present in 15%–30% of patients, coarctation in 7%–18%, and aortic dilation/aneurysm in 3%–42% (Gravholt et al., 2017). Altered cerebral blood flow during fetal and infant development, as a consequence of CHD, could negatively impact neurodevelopment leading to white matter dysmaturation and injury. Corrective heart surgery is also associated with a variety of factors that can influence neurodevelopment including general anesthesia, inflammatory responses, and low blood pressure and low cardiac output during the recovery period (Morton, Ishibashi, & Jonas, 2017).

Auditory system

Hearing problems are present in about 30% of individuals with TS. In younger children, conductive hearing loss is common due to middle-ear effusion, frequent otitis media, and tympanic membrane pathology. Midfrequency sensorineural hearing loss is more common in adulthood, but can occur as early as 6 years. Early hearing difficulties have the potential to negatively impact language development and socialization. Indeed, hearing impairment is one of the few factors associated with lower self-esteem and poorer psychosocial adaptation in women with TS (Gravholt et al., 2017).

Neurocognitive profile

Most individuals with TS have intelligence scores in the normal range. However, they frequently exhibit specific deficits in visuospatial ability, arithmetical abilities, social cognition, attention, and executive function. With specific regard to attention, impaired attentional control has been observed in multiple studies, while results for sustained attention are inconsistent. As regards executive functions, the most frequently affected domains are planning, response inhibition, mental flexibility, and working memory. These challenges typically emerge during the school-age years and may negatively impact academic performance and health-related quality of life. Areas of preserved or enhanced function have also been reported, including preserved orienting attention, enhanced phonological processing, enhanced receptive and expressive vocabulary, and preserved or enhanced semantic language. As in other genetic syndromes, there is significant individual variability in the neurocognitive profile. Factors contributing to this variability remain poorly understood (Gravholt et al., 2017; Knickmeyer & Davenport, 2011; Knickmeyer & Hooper, 2019). Despite these challenges, young women with TS show an equal or higher level of educational attainment and employment status compared to the general population (Gravholt et al., 2017).

Vulnerability to learning disabilities and neuropsychiatric conditions

Intellectual disability (ID) is present in approximately 10% of individuals with TS. ID is most common in individuals carrying a small ring X. Dyscalculia (a specific learning disability affecting the acquisition of mathematical skills in children of normal or above-average intelligence) is present in 50%–75% of individuals with TS (Gravholt et al., 2017). Intriguingly, individuals with TS appear to be at greater risk for several neuropsychiatric conditions that are usually more common in phenotypic males. This includes attention deficit hyperactivity disorder (ADHD) and autism spectrum disorder (ASD). ADHD is observed in approximately 25% of school-aged girls with TS (Green et al., 2015), 15% of school-aged males, and approximately 7% of school-aged girls without TS (NIMH, 2017). Regarding ASD, an unselected sample of 150 individuals with TS yielded five ASD cases, for a prevalence of one in 30 (approximately 3%) (Creswell & Skuse, 1999). To put this in context, in 2014, 2.7% of boys had been diagnosed with ASD, but only 0.7% of girls (NIMH, 2018). Girls and women with TS may also be more vulnerable to developmental coordination disorder (Nijhuis-van der Sanden, Eling, & Otten, 2003) and depression (Cardoso et al., 2004; Reimann, Perman, Ho, Parks, & Comis, 2018). Finally, adolescent girls with TS have higher levels of state anxiety than their female peers (Kilic, Ergur, & Ocal, 2005).

Altered brain structure and function

Visuospatial circuits

The structure and function of the dorsal visual processing stream differ in individuals with TS as compared to 46,XX females (Knickmeyer & Hooper, 2019). The dorsal stream includes projections from early visual areas to the posterior parietal cortex (Galletti et al., 2001) and parietal–frontal, parietal–premotor, and parietal–medial temporal circuits (Kravitz, Saleem, Baker, & Mishkin, 2011). Structural neuroimaging studies consistently reveal decreased parietal–occipital gray matter in children and adults with TS, and diffusion tensor imaging studies reveal altered white matter microstructure in the superior longitudinal fasciculus (SLF) (Knickmeyer & Hooper, 2019), which connects parietal and frontal areas (Schmahmann et al., 2007). The SLF likely facilitates spatial working memory and visually guided action. Functional MRI studies confirm the importance of these circuits (Knickmeyer & Hooper, 2019). For example, individuals with TS showed altered activation of the left and right supramarginal gyrus during a paradigm designed to target spatial working memory (Haberecht et al., 2001).

Social processing circuits

Children and adults with TS also show enlargement of the amygdala and orbitofrontal cortex (Knickmeyer & Hooper, 2019). These areas play important roles in emotion recognition and the attribution of mental states (theory of mind) (Adolphs, 2009; Heberlein & Adolphs, 2004). Interestingly, enlargement of the amygdala and orbitofrontal cortex is also a frequently reported finding in children with ASD (Bernhardt, Di Martino, Valk, & Wallace, 2016; Ecker, 2017). Altered white matter microstructure in the inferior longitudinal fasciculus (ILF) (Villalon-Reina et al., 2013; Yamagata et al., 2012) may contribute to poor facial and emotion recognition. fMRI studies of social cognition in TS are extremely limited, but one study has reported greater—and more persistent—right amygdala activation to fearful faces in females with TS as compared to females without TS (Skuse, Morris, & Dolan, 2005).

Circuits involved in attention and executive functions

Decreased gray matter in dorsal parietal cortex is expected to influence executive network function (Hermans, Henckens, Joels, & Fernandez, 2014), contributing to impaired planning, response inhibition, mental flexibility, and working memory. Reduced gray matter in the superior parietal lobule and inferior parietal sulcus could affect dorsal attention networks (Aboitiz, Ossandon, Zamorano, Palma, & Carrasco, 2014), contributing to impairments in sustained attention, which have sometimes (but not always) been observed. Altered white matter microstructure in the SLF may also contribute to impairments in attention and executive function. Functional MRI studies confirm the importance of these particular neural circuits (Knickmeyer & Hooper, 2019). For example, during a response inhibition task, girls with TS had activated additional regions of the prefrontal cortex when compared to girls without TS, suggesting either altered cortical organization or functional immaturity of the executive function network (Tamm, Menon, & Reiss, 2003).

Circuits involved in arithmetical abilities

Most arithmetical and numerical processes involve the parietal lobes in general and the intraparietal sulcus (IPS) in particular (Butterworth, Varma, & Laurillard, 2011). Consequently, gray matter reductions in the IPS are thought to play an important role in impaired numerical processing and dyscalculia in individuals with TS. Functional MRI studies confirm the importance of these particular circuits (Knickmeyer & Hooper, 2019). For example, a recent resting-state MRI study identified a functional network that exhibited reduced connectivity in girls with TS. This network consisted of three modules. Reduced connectivity in Module 2, which overlaps with the salience network and shows strong parietal involvement, was associated with math-related cognition (Xie et al., 2017).

Early development

Most neuroimaging studies of TS have been carried out in older children and adults, using cross-sectional designs. Consequently, they cannot address when in development these differences arise. To address this question, my research group has been studying infants with TS. We recently published the first quantitative neuroimaging study in this age group. Twenty-six infants with TS were compared to 47 typically developing female and 39 typically developing male infants. Like older children and adults with TS, 1-year-old infants with TS exhibited decreased gray matter volumes in premotor, somatosensory, and parietal–occipital cortex. This suggests that many aspects of the neuroanatomical phenotype of TS are established in the prenatal and/or early postnatal period. Other aspects of the typical neuroanatomical phenotype were not evident in infancy. For example, we did not observe volume reductions in the cuneus or the superior parietal lobule. Our results indicate that volume reductions in primary and secondary visual cortex precede (and may contribute to) reductions in high-order visual–spatial processing areas (Davenport et al., 2020).

Mechanisms

There are several mechanisms by which the absence of the second sex chromosome could alter neurodevelopment. This includes 1. haploinsufficiency of gene expression, 2. failure to express imprinted genes, 3. the uncovering of X-linked mutations, and 4. functional disomy.

Haploinsufficiency of gene expression

The human X chromosome is significantly larger than the Y chromosome (~150 Mb vs. ~50 Mb) and carries many more genes (~1400 vs. ~200) (NCBI, 1998). In order to compensate for the substantial imbalance between XX males and XY females, one X chromosome is inactivated early in female development. However, some genes continue to be expressed from the "silent" X (Balaton, Dixon-McDougall, Peeters, & Brown, 2018). These escapees likely play important roles in TS (Knickmeyer & Davenport, 2011). Indeed, haploinsufficiency of *SHOX* is responsible for short stature in TS (Gravholt et al., 2017), and haploinsufficiency of *TIMP1* has been linked to increased incidence of bicuspid aortic valve (Corbitt et al., 2018). Although no genes have been conclusively linked to the neurocognitive phenotype, many escapees have known roles in neurodevelopment including *NLGN4X*, *STS*, *L1CAM*, *KDM5C*, and *KAL1*. Deletion mapping studies suggest that some aspects of the cognitive phenotype arise due to haploinsufficiency of one or more genes in Xp22.3 (Zinn et al., 2007), while impairments in fear recognition may arise due to haploinsufficiency of one or more genes in Xp11.3 (Good et al., 2003).

It is important to note that some escapees have Y homologues, but others do not (Balaton et al., 2018). Haploinsufficiency for genes without Y homologues may explain the increased risk for male-biased neurodevelopmental disorders observed in TS, including ADHD and ASD. As illustrated in Fig. 22.3, some of these genes may support the development of neural circuits involved in social cognition and executive function, protecting 46,XX females from autosomal variants with adverse effects on these domains. Reduced X chromosome gene dosage in 46,XY males and in females with TS would lower their liability threshold for developing ADHD and ASD. This model would also explain why some individuals with TS develop ASD or ADHD while others do not, as girls with TS and low genetic liability for ASD or ADHD would not cross the diagnostic threshold.

Failure to express imprinted genes

Expression of imprinted genes depends upon their parent of origin. As noted previously, most individuals with TS carry an X chromosome that is maternal in origin (Davenport, 2010). In 1997, David Skuse reported that girls with TS and a maternal X chromosome experienced greater academic and social difficulties compared to girls with TS and a paternal X chromosome. They also had significantly lower verbal IQ and poorer behavioral inhibition (Skuse et al., 1997). The authors hypothesized that a gene expressed on the paternal X enhances verbal and higher-order executive function skills, which in turn promotes positive social interactions and protects against ASD. Parent-of-origin effects have also been reported for some aspects of the neuroanatomical phenotype, but these analyses are typically underpowered (Knickmeyer & Davenport, 2011).

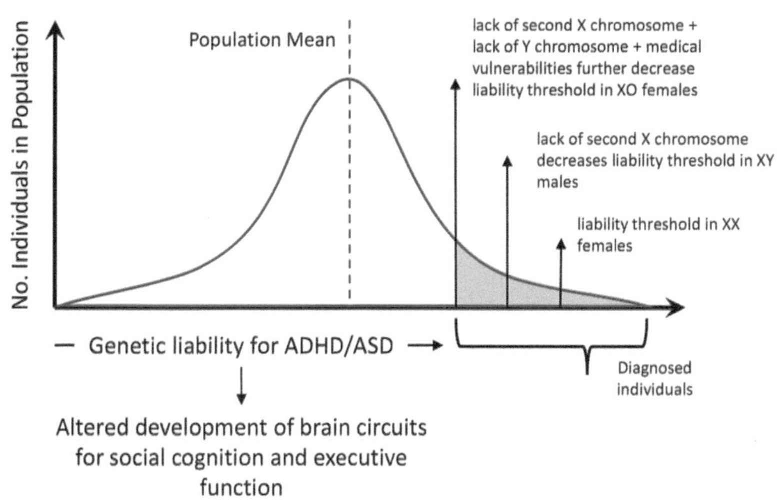

FIGURE 22.3 **Reduced gene dosage in females with TS may lower their liability threshold for developing ADHD and ASD.** Under a multiple threshold liability/female-protective effect model, risk for ASD and ADHD is quantitative and distributed in the population. XX females may be protected by the presence of a second X chromosome, which supports the development of brain circuits involved in social cognition and executive function. Similar to XY males, individuals with TS may have a lower threshold due to the lack of a second X chromosome. Additional factors such as gonadal insufficiency and congenital heart disease may lower their threshold even further.

Uncovering of X-linked mutations

As of 2017, at least 141 X-chromosome genes have been associated with ID (Neri, Schwartz, Lubs, & Stevenson, 2018). Like typical males, females with TS are at an increased risk for X-linked ID as a second X is not present to compensate for the abnormal gene. Interestingly, female carriers of deleterious mutations in X-linked ID genes often show skewed X-chromosome inactivation (Plenge, Stevenson, Lubs, Schwartz, & Willard, 2002). This suggests that female cells can preferentially silent X-chromosomes carrying damaging variants. This protective mechanism would be absent in most females with TS.

Functional disomy

As noted previously, ID is most common in individuals with TS who carry a small ring X. Ring X chromosomes may fail to inactivate, resulting in functional disomy. In contrast to other karyotypes, individuals with a ring X may overexpress X chromosome genes that are typically silenced (Leppig & Disteche, 2001). At present, no specific X-chromosome genes have been linked to the increased incidence of ID in this relatively rare group. However, it is notable that duplications of X-linked ID genes may produce a phenotype identical or similar to that associated with loss-of-function mutations or deletion (Neri et al., 2018).

Clinical recommendations

The most recent clinical guidelines for TS strongly recommend that integrative care for women and girls with TS includes behavioral health specialists who are trained to diagnose the educational and psychosocial difficulties sometimes associated with TS, as well as individuals who can provide appropriate interventions. Given significant clinical heterogeneity and potential changes over the course of development, the guidelines recommend annual developmental and behavioral screenings until adulthood and more comprehensive neuropsychological assessments at key transitional stages in schooling. These stages would include early life (preschool), school entry, transition to high school, and transition to higher education. Children should be referred for occupational, physical, and/or speech therapy in early life or at school entry, if assessments indicate that this support is needed. Academic and occupational adjustments should also be made as appropriate for each individual. The guidelines recommend aiming for on-time puberty and aggressive management of predictors of hearing impairment to promote positive psychosocial outcomes. Finally, the guidelines suggest that evidence-based interventions for cognitive or psychosocial problems, developed for other populations (children with ADHD and ASD, for example), may help meet the needs of girls and women with TS (Gravholt et al., 2017).

Conclusions

TS is a complex condition, which is frequently accompanied by neurocognitive challenges, psychosocial issues, and psychiatric comorbidities. Neuroimaging studies suggest that cognitive and behavioral features reflect altered neural development. Circuits involved in visuospatial processing, social cognition, arithmetical ability, attention, and executive function may be particularly vulnerable to the loss of the second sex chromosome. The mechanisms linking sex chromosome loss to altered neural development remain unclear, but may involve haploinsufficiency of gene expression, failure to express imprinted genes, the uncovering of X-linked mutations, and, in those with a ring X chromosome, functional disomy. Additional research is needed to formulate a comprehensive neurodevelopmental model of TS, which incorporates genetic vulnerabilities, associated clinical features, neurodevelopmental processes, cognitive processes, and psychosocial function (see Fig. 22.4).

Applications to other areas of development

There are striking sex differences in the prevalence of early-onset psychiatric disorders with both ASD and ADHD being 3–5 times more common in males as compared to females. Prior research suggests that females are partially protected from mutations causing early-onset psychiatric disorders, but the mechanisms underlying this protection remain unknown. One compelling hypothesis is that differences in X-chromosome gene dosage contribute to this female protective effect by shaping the development of neural circuits involved in social cognition and executive function. TS, the topic of the current chapter, is a well-defined genetic disorder resulting from the partial or complete loss of one of the sex chromosomes. Affected individuals are haploinsufficient for genes that are normally expressed from both X-chromosomes in females.

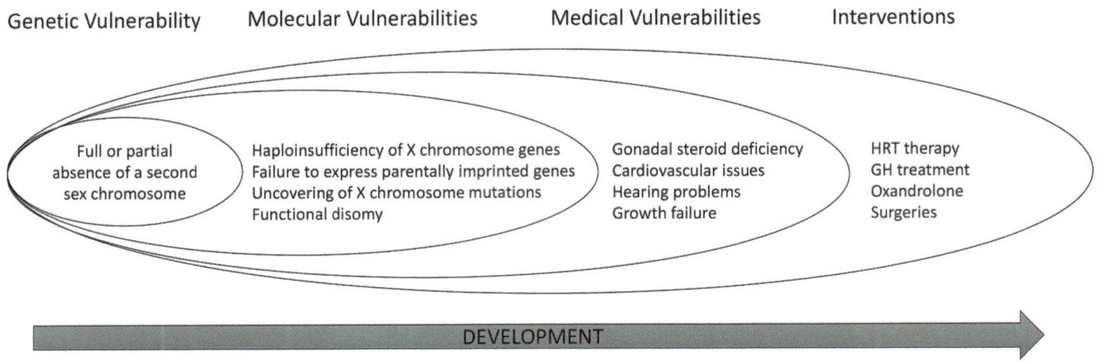

FIGURE 22.4 **Risk factors for adverse neurodevelopmental outcomes in TS.** Neurodevelopment in individuals with TS occurs in the context of genetic, molecular, and medical vulnerabilities, as well as vulnerabilities arising from various medical interventions, which accumulate across the lifespan.

Consequently, TS is an excellent model to investigate whether X-chromosome gene dosage influences neuroimaging phenotypes and cognitive domains relevant to male-biased neurodevelopmental disorders. Individuals with TS are also more likely to be diagnosed with developmental dyscalculia, a disorder characterized by profound difficulties in numerical understanding. Dyscalculia has a prevalence rate between 3% and 6% in the general population, but occurs in 50%–75% of girls with TS. The high occurrence of dyscalculia in TS suggests that X chromosome genes play an important role in shaping neuronal networks necessary for number processing. Dyscalculia is also highly heterogeneous. Individuals with TS may represent a more homogeneous subgroup in which to study genetic contributors to and neuronal correlates of impaired numerical abilities (Table 22.1).

TABLE 22.1 Sex chromosome aneuploidies.

Disorder	Mechanism	Prevalence	Cognitive characteristics	Brain features	Further Resources
Turner syndrome	Complete or partial monosomy of the X chromosome (45,X0) most commonly caused by a sporadic chromosomal nondisjunction.	1 in 2000 female live births	Deficits in attention, executive function, social cognition, visuospatial skills. Relatively preserved language skills. Associated with ADHD, dyscalculia, potentially ASD	Decreased GM in occipital–parietal areas. Increased GM in amygdala, insula, orbitofrontal cortex.	(Green, Flash, & Reiss, 2019)
Klinefelter syndrome	47,XXY karyotype with equal probability of a maternal or paternal chromosomal nondisjunction in meiosis.	1 in 450 male live births	Deficits in attention, executive function, social cognition, language Relatively preserved visuospatial skills. Associated with ASD and ADHD.	Reduced total brain volumes. Reduced volume of cerebellar hemispheres. Decreased GM volume in the frontal and temporal lobes, amygdala, hippocampus	(Green, Flash, & Reiss, 2019; Steinman, Ross, Lai, Reiss, & Hoeft, 2009)
XYY syndrome	47,XYY karyotype caused by an error in paternal meiosis II or during early cleavage divisions	1 in 1000 male live births	Decreased IQ and adaptive behavior. Deficits in language and sociocommunicative functioning. Associated with ASD and ADHD.	Increased total WM and GM. Increased GM in language areas.	(Bryant et al., 2012; Green, Flash, & Reiss, 2019; Joseph et al., 2018)

Continued

TABLE 22.1 Sex chromosome aneuploidies.—cont'd

Disorder	Mechanism	Prevalence	Cognitive characteristics	Brain features	Further Resources
XXX syndrome	47,XXX karyotype usually caused by a maternal meiotic error	1 in 1000 female live births	Deficits in language and visuospatial skills. Associated with ADHD, depression, and anxiety disorders.	Decreased total brain volume. Decreased regional brain volumes with the exception of parietal GM.	(Green, Flash, & Reiss, 2019; Lenroot et al., 2014)

A summary of cognitive and neuroanatomical phenotypes in individuals with conditions characterized by an abnormal number of sex chromosomes, that is, other than XX (female) or XY (male).
Based on information provided in Bryant, D. M., Hoeft, F., Lai, S., Lackey, J., Roeltgen, D., Ross, J., & Reiss, A. L. (2012). Sex chromosomes and the brain: a study of neuroanatomy in XYY syndrome. *Developmental Medicine and Child Neurology, 54*(12), 1149–1156. https://doi.org/10.1111/j.1469-8749.2012.04418.x; Green, T., Flash, S., & Reiss, A. L. (2019). Sex differences in psychiatric disorders: what we can learn from sex chromosome aneuploidies. *Neuropsychopharmacology, 44*(1), 9–21. https://doi.org/10.1038/s41386-018-0153-2; Joseph, L., Farmer, C., Chlebowski, C., Henry, L., Fish, A., Mankiw, C.,... Raznahan, A. (2018). Characterization of autism spectrum disorder and neurodevelopmental profiles in youth with XYY syndrome. *Journal of Neurodevelopmental Disorders, 10*; Lenroot, R. K., Blumenthal, J. D., Wallace, G. L., Clasen, L. S., Lee, N. R., & Giedd, J. N. (2014). A case-control study of brain structure and behavioral characteristics in 47,XXX syndrome. *Genes Brain and Behavior, 13*(8), 841–849. https://doi.org/10.1111/gbb.12180; Steinman, K., Ross, J., Lai, S., Reiss, A. L., & Hoeft, F. (2009). Structural and functional neuroimaging in Klinefelter (47,XXY) syndrome, A review of the literature and preliminary results from a functional magnetic resonance imaging study of language. *Developmental Disabilities Research Reviews, 15*, 295–308.

Key facts of the sex chromosomes

- The sex chromosomes (X and Y) are derived from an ancestral pair of autosomes, but over evolutionary history the Y chromosome has lost large swathes of genetic material.
- Approximately 13% of X chromosome genes consistently escape X inactivation. Another 15% escape inactivation in either a subset of individuals or a subset of tissues.
- Approximately 30% of genes known to completely escape X inactivation are located in two pseudoautosomal regions (PAR1 and PAR2). Identical PARs exist at the distal ends of the Y chromosome allowing the X and Y chromosome to pair up during male meiosis. Genes in PAR1 and PAR2 do not need additional dosage compensation as XX females and XY males have the same copy number.
- X-linked genes play a disproportionate role in intellectual disability.

Summary points

- TS is a rare condition of phenotypic females in which all or part of the second sex chromosome is missing in some or all of the body's cells.
- TS is a complex and heterogeneous condition, which can affect many different body systems.
- TS is often associated with a unique neurocognitive profile that includes specific deficits in visuospatial processing, arithmetical abilities, social cognition, attentional control, and executive function, accompanied by preserved intellectual ability, orienting attention, and semantic language processing.
- TS is often accompanied by decreased gray matter volumes in parietal–occipital cortex and increased volumes of amygdala and orbitofrontal cortex, in comparison to 46,XX females.
- Neuroanatomical features of TS are detectable in infancy.
- Haploinsufficiency of X chromosome genes, failure to express parentally imprinted genes, the uncovering of X chromosome mutations, and functional disomy may all contribute to altered neurodevelopment in individuals with TS.
- Clinical care for individuals with TS ought to include behavioral health specialists.

References

Aboitiz, F., Ossandon, T., Zamorano, F., Palma, B., & Carrasco, X. (2014). Irrelevant stimulus processing in ADHD: catecholamine dynamics and attentional networks. *Frontiers in Psychology, 5*, 183.

Adolphs, R. (2009). The social brain: Neural basis of social knowledge. *Annual Review of Psychology, 60*, 693–716.

Amundson, E., Boman, U. W., Barrenas, M. L., Bryman, I., & Landin-Wilhelmsen, K. (2010). Impact of growth hormone therapy on quality of life in adults with turner syndrome. *Journal of Clinical Endocrinology and Metabolism, 95*(3), 1355–1359. https://doi.org/10.1210/jc.2009-1754.

Balaton, B. P., Dixon-McDougall, T., Peeters, S. B., & Brown, C. J. (2018). The eXceptional nature of the X chromosome. *Human Molecular Genetics, 27*(R2), R242–R249. https://doi.org/10.1093/hmg/ddy148.

Bernhardt, B. C., Di Martino, A., Valk, S. L., & Wallace, G. L. (2016). Neuroimaging-based phenotyping of the autism spectrum. *Current Topics in Behavioral Neurosciences*. https://doi.org/10.1007/7854_2016_438.

Bryant, D. M., Hoeft, F., Lai, S., Lackey, J., Roeltgen, D., Ross, J., & Reiss, A. L. (2012). Sex chromosomes and the brain: a study of neuroanatomy in XYY syndrome. *Developmental Medicine & Child Neurology, 54*(12), 1149–1156.

Butterworth, B., Varma, S., & Laurillard, D. (2011). Dyscalculia: From brain to education. *Science, 332*(6033), 1049–1053.

Cardoso, G., Daly, R., Haq, N. A., Hanton, L., Rubinow, D. R., Bondy, C. A., & Schmidt, P. (2004). Current and lifetime psychiatric illness in women with Turner syndrome. *Gynecological Endocrinology, 19*(6), 313–319.

Carel, J. C., Ecosse, E., Bastie-Sigeac, I., Cabrol, S., Tauber, M., Leger, J., ... Coste, J. (2005). Quality of life determinants in young women with turner's syndrome after growth hormone treatment: Results of the StaTur population-based cohort study. *Journal of Clinical Endocrinology and Metabolism, 90*(4), 1992–1997. https://doi.org/10.1210/jc.2004-1395.

Corbitt, H., Morris, S. A., Gravholt, C. H., Mortensen, K. H., Tippner-Hedges, R., Silberbach, M., ... Investigators, G. R. (2018). TIMP3 and TIMP1 are risk genes for bicuspid aortic valve and aortopathy in Turner syndrome. *PLoS Genetics, 14*(10). https://doi.org/ARTN e100769210.1371/journal.pgen.1007692.

Creswell, C. S., & Skuse, D. H. (1999). Autism in association with turner syndrome: Genetic implications for male vulnerability to pervasive developmental disorders. *Neurocase, 5*, 511–518.

Davenport, M. L. (2010). Turner syndrome. In S. B. Cassidy, & J. E. Allanson (Eds.), *Management of genetic syndromes* (pp. 847–869). Hoboken: John Wiley & Sons, Inc.

Davenport, M. L., Crowe, B. J., Travers, S. H., Rubin, K., Ross, J. L., Fechner, P. Y., ... Quigley, C. A. (2007). Growth hormone treatment of early growth failure in toddlers with turner syndrome: A randomized, controlled, multicenter trial. *Journal of Clinical Endocrinology and Metabolism, 92*(9), 3406–3416.

Davenport, M. L., Cornea, E., Xia, K., Crowley, J. J., Halvorsen, M. W., Goldman, B. D., & Knickmeyer, R. C. (2020). Altered brain structure in infants with turner syndrome. *Cerebral Cortex, 30*(2), 587–596.

Ecker, C. (2017). The neuroanatomy of autism spectrum disorder: An overview of structural neuroimaging findings and their translatability to the clinical setting. *Autism, 21*(1), 18–28. https://doi.org/10.1177/1362361315627136.

Freriks, K., Verhaak, C. M., Sas, T. C., Menke, L. A., Wit, J. M., Otten, B. J., ... Timmers, H. J. (2015). Long-term effects of oxandrolone treatment in childhood on neurocognition, quality of life and social-emotional functioning in young adults with Turner syndrome. *Hormones and Behavior, 69*, 59–67. https://doi.org/10.1016/j.yhbeh.2014.12.008.

Galletti, C., Gamberini, M., Kutz, D. F., Fattori, P., Luppino, G., & Matelli, M. (2001). The cortical connections of area V6: An occipito-parietal network processing visual information. *European Journal of Neuroscience, 13*(8), 1572–1588.

Gardner, M., Boshart, M. L., Yeguez, C. E., Desai, K. M., & Sandberg, D. E. (2016). Coming up short: Risks of bias in assessing psychological outcomes in growth hormone therapy for short stature. *Journal of Clinical Endocrinology and Metabolism, 101*(1), 23–30. https://doi.org/10.1210/jc.2015-3256.

Good, C. D., Lawrence, K., Thomas, N. S., Price, C. J., Ashburner, J., Friston, K. J., ... Skuse, D. H. (2003). Dosage-sensitive X-linked locus influences the development of amygdala and orbitofrontal cortex, and fear recognition in humans. *Brain, 126*(Pt 11), 2431–2446. https://doi.org/10.1093/brain/awg242awg242 (pii).

Gravholt, C. H., Andersen, N. H., Conway, G. S., Dekkers, O. M., Geffner, M. E., Klein, K. O., ... Backeljauw, P. F. (2017). Clinical practice guidelines for the care of girls and women with turner syndrome: Proceedings from the 2016 cincinnati international turner syndrome meeting. *European Journal of Endocrinology/European Federation of Endocrine Societies, 177*(3), G1–G70. https://doi.org/10.1530/EJE-17-0430EJE-17-0430 (pii).

Green, T., Bade Shrestha, S., Chromik, L. C., Rutledge, K., Pennington, B. F., Hong, D. S., & Reiss, A. L. (2015). Elucidating X chromosome influences on attention deficit hyperactivity disorder and executive function. *Journal of Psychiatric Research, 68*, 217–225. https://doi.org/10.1016/j.jpsychires.2015.06.021S0022-3956(15)00192-2 (pii).

Green, T., Flash, S., & Reiss, A. L. (2019). Sex differences in psychiatric disorders: what we can learn from sex chromosome aneuploidies. *Neuropsychopharmacology, 44*(1), 9–21.

Haberecht, M. F., Menon, V., Warsofsky, I. S., White, C. D., Dyer-Friedman, J., Glover, G. H., ... Reiss, A. L. (2001). Functional neuroanatomy of visuo-spatial working memory in Turner syndrome. *Human Brain Mapping, 14*(2), 96–107.

Heberlein, A. S., & Adolphs, R. (2004). Functional anatomy of human social cognition. In A. Easton, & N. Emery (Eds.), *The cognitive neuroscience of social behavior*. London: Psychology Press.

Hermans, E. J., Henckens, M. J., Joels, M., & Fernandez, G. (2014). Dynamic adaptation of large-scale brain networks in response to acute stressors. *Trends in Neurosciences, 37*(6), 304–314. https://doi.org/10.1016/j.tins.2014.03.006.

Hook, E. B., & Warburton, D. (2014). Turner syndrome revisited: Review of new data supports the hypothesis that all viable 45,X cases are cryptic mosaics with a rescue cell line, implying an origin by mitotic loss. *Human Genetics, 133*(4), 417–424. https://doi.org/10.1007/s00439-014-1420-x.

Jacobs, P. A., Melville, M., Ratcliff, S., Keay, A. J., & Syme, J. (1974). Cytogenetic survey of 11,680 newborn-infants. *Annals of Human Genetics, 37*(MAY), 359–376.

Joseph, L., Farmer, C., Chlebowski, C., Henry, L., Fish, A., Mankiw, C., & Raznahan, A. (2018). Characterization of autism spectrum disorder and neurodevelopmental profiles in youth with XYY syndrome. *Journal of Neurodevelopmental Disorders, 10*(1), 30.

Kilic, B. G., Ergur, A. T., & Ocal, G. (2005). Depression, levels of anxiety and self-concept in girls with Turner's syndrome. *Journal of Pediatric Endocrinology and Metabolism, 18*(11), 1111–1117.

Knickmeyer, R. C., & Davenport, M. (2011). Turner syndrome and sexual differentiation of the brain: Implications for understanding male-biased neurodevelopmental disorders. *Journal of Neurodevelopmental Disorders, 3*(4), 293–306. https://doi.org/10.1007/s11689-011-9089-0.

Knickmeyer, R. C., & Hooper, S. R. (2019). The deep biology of cognition: Moving toward a comprehensive neurodevelopmental model of turner syndrome. *American Journal of Medical Genetics Part C-Seminars in Medical Genetics, 181*(1), 91–99. https://doi.org/10.1002/ajmg.c.31679.

Kravitz, D. J., Saleem, K. S., Baker, C. I., & Mishkin, M. (2011). A new neural framework for visuospatial processing. *Nature Reviews Neuroscience, 12*(4), 217–230. https://doi.org/10.1038/nrn3008.

Lenroot, R. K., Blumenthal, J. D., Wallace, G. L., Clasen, L. S., Lee, N. R., & Giedd, J. N. (2014). A case-control study of brain structure and behavioral characteristics in 47,XXX syndrome. *Genes, Brain and Behavior, 13*(8), 841–849.

Leppig, K. A., & Disteche, C. M. (2001). Ring X and other structural X chromosome abnormalities: X inactivation and phenotype. *Seminars in Reproductive Medicine, 19*(2), 147–157. https://doi.org/10.1055/s-2001-15395.

Morton, P. D., Ishibashi, N., & Jonas, R. A. (2017). Neurodevelopmental abnormalities and congenital heart disease: Insights into altered brain maturation. *Circulation Research, 120*(6), 960–977. https://doi.org/10.1161/CIRCRESAHA.116.309048.

NCBI. (1998). Chromosome map. In *Genes and disease*. Bethesda, MD: National Center for Biotechnology Information (US).

Neri, G., Schwartz, C. E., Lubs, H. A., & Stevenson, R. E. (2018). X-linked intellectual disability update 2017. *American Journal of Medical Genetics, Part A, 176*(6), 1375–1388. https://doi.org/10.1002/ajmg.a.38710.

Nielsen, J., & Wohlert, M. (1991). Chromosome-abnormalities found among 34910 newborn children - results from a 13-year incidence study in Arhus, Denmark. *Human Genetics, 87*(1), 81–83.

Nijhuis-van der Sanden, M. W., Eling, P. A., & Otten, B. J. (2003). A review of neuropsychological and motor studies in Turner Syndrome. *Neuroscience and Biobehavioral Reviews, 27*(4), 329–338.

NIMH. (2017). November 2017. Retrieved from https://www.nimh.nih.gov/health/statistics/attention-deficit-hyperactivity-disorder-adhd.shtml.

NIMH. (2018). April 2018. Retrieved from https://www.nimh.nih.gov/health/statistics/autism-spectrum-disorder-asd.shtml.

Plenge, R. M., Stevenson, R. A., Lubs, H. A., Schwartz, C. E., & Willard, H. F. (2002). Skewed X-chromosome inactivation is a common feature of X-linked mental retardation disorders. *The American Journal of Human Genetics, 71*(1), 168–173. https://doi.org/10.1086/341123.

Reimann, G. E., Perman, M. M. B., Ho, P. S., Parks, R. A., & Comis, L. E. (2018). Psychosocial characteristics of women with a delayed diagnosis of turner syndrome. *The Journal of Pediatrics, 199*, 206–211.

Ross, J. L., Roeltgen, D., Feuillan, P., Kushner, H., & Cutler, G. B. (1998). Effects of estrogen on nonverbal processing speed and motor function in girls with Turner's syndrome. *Journal of Clinical Endocrinology and Metabolism, 83*(9), 3198–3204.

Ross, J. L., Roeltgen, D., Feuillan, P., Kushner, H., & Cutler, G. B. (2000). Use of estrogen in young girls with Turner syndrome - effects on memory. *Neurology, 54*(1), 164–170.

Ross, J. L., Roeltgen, D., Stefanatos, G. A., Feuillan, P., Kushner, H., Bondy, C., & Cutler, G. B. (2003). Androgen-responsive aspects of cognition in girls with Turner syndrome. *Journal of Clinical Endocrinology and Metabolism, 88*(1), 292–296.

Ross, J. L., Stefanatos, G. A., Kushner, H., Zinn, A., Bondy, C., & Roeltgen, D. (2002). Persistent cognitive deficits in adult women with Turner syndrome. *Neurology, 58*(2), 218–225. https://doi.org/10.1212/wnl.58.2.218.

Schmahmann, J. D., Pandya, D. N., Wang, R., Dai, G., D'Arceuil, H. E., de Crespigny, A. J., & Wedeen, V. J. (2007). Association fibre pathways of the brain: Parallel observations from diffusion spectrum imaging and autoradiography. *Brain, 130*, 630–653.

Steinman, K., Ross, J., Lai, S., Reiss, A., & Hoeft, F. (2009). Structural and functional neuroimaging in Klinefelter (47,XXY) syndrome: a review of the literature and preliminary results from a functional magnetic resonance imaging study of language. *Developmental Disabilities Research Reviews, 15*(4), 295–308.

Skuse, D. H., James, R. S., Bishop, D. V. M., Coppins, B., Dalton, P., Aamondt-Leeper, G., … Jacobs, P. A. (1997). Evidence from Turner's syndrome of an imprinted X-linked locus affecting cognitive function. *Nature, 387*(12), 705–708.

Skuse, D. H., Morris, J. S., & Dolan, R. J. (2005). Functional dissociation of amygdala-modulated arousal and cognitive appraisal, in Turner syndrome. *Brain, 128*, 2084–2096.

Tamm, L., Menon, V., & Reiss, A. L. (2003). Abnormal prefrontal cortex function during response inhibition in turner syndrome: Functional magnetic resonance imaging evidence. *Biological Psychiatry, 53*(2), 107–111. doi:S0006322302014889 (pii).

Villalon-Reina, J., Jahanshad, N., Beaton, E., Toga, A. W., Thompson, P. M., & Simon, T. J. (2013). White matter microstructural abnormalities in girls with chromosome 22q11.2 deletion syndrome, Fragile X or Turner syndrome as evidenced by diffusion tensor imaging. *NeuroImage, 81*, 441–454. https://doi.org/10.1016/j.neuroimage.2013.04.028S1053-8119(13)00375-3 (pii).

Xie, S., Yang, J. T., Zhang, Z. X., Zhao, C. X., Bi, Y. C., Zhao, Q. L., … Gong, G. L. (2017). The effects of the X chromosome on intrinsic functional connectivity in the human brain: Evidence from turner syndrome patients. *Cerebral Cortex, 27*(1), 474–484.

Yamagata, B., Barnea-Goraly, N., Marzelli, M. J., Park, Y., Hong, D. S., Mimura, M., & Reiss, A. L. (2012). White matter aberrations in prepubertal estrogen-naive girls with monosomic Turner syndrome. *Cerebral Cortex, 22*(12), 2761–2768. https://doi.org/10.1093/cercor/bhr355bhr355 (pii).

Zinn, A. R., Roeltgen, D., Stefanatos, G., Ramos, P., Elder, F. F., Kushner, H., … Ross, J. L. (2007). *A Turner syndrome neurocognitive phenotype maps to Xp22.3* (Vol. 3). Behavioral and Brain Functions.

Chapter 23

Very preterm children and the impact on neurodevelopmental outcomes

Leona Pascoe and Peter J. Anderson

Turner Institute for Brain and Mental Health, School of Psychological Sciences, Monash University, Clayton, VIC, Australia

List of abbreviations

ADHD, Attention deficit hyperactivity disorder
ASD, Autism spectrum disorder
CP, cerebral palsy
DCD, Developmental coordination disorder
ELBW, extremely low birth weight
EP, extremely preterm
IQ, intelligence quotient
NDI, neurodevelopmental impairment
NICU, Neonatal intensive care unit
VLBW, very low birth weight
VP, very preterm

Mini-dictionary of terms

Attention deficit hyperactivity disorder (ADHD) A neurodevelopmental disorder that affects children's behavior and learning and generally includes inattention, impulsivity, and/or overactivity symptoms.
Autism spectrum disorder (ASD) A complex neurodevelopmental disorder that primarily affects children's social and communication skills and includes the presence of restricted and/or repetitive behaviors and interests.
Cerebral palsy A physical disability often caused by prenatal brain injury that primarily affects motor functioning, coordination, and posture.
Developmental Coordination Disorder (DCD) A motor skills disorder that affects children's ability to coordinate their movements and results in a delay in early motor skill development, but generally does not affect their intellectual abilities.
Low birth weight A birth weight of less than 2500 g.
Morbidity The presence of a disease or disorder.
Mortality Survival status; the number of deaths.
Neonatal intensive care unit A specialized hospital unit for neonatal care of sick or premature newborn babies.
Neurodevelopmental impairment Impaired visual, hearing, physical, and cognitive functioning.
Preterm birth A live birth prior to 37 completed weeks of gestation.

Introduction

Preterm birth (birth prior to 37 completed weeks' gestation) is a major worldwide burden and is a leading cause of morbidity and mortality among children under the age of 5. Global estimates of preterm birth were reported to be 10.6% in 2014, which equates to approximately 14.8 million infants born preterm annually (Chawanpaiboon et al., 2019). Of those, approximately 1.7 million live births are very preterm (VP, <32 weeks' gestation) (Chawanpaiboon et al., 2019); however, advances in neonatal intensive care practices have reduced global mortality rates, particularly for those born at earlier gestations (Saigal & Doyle, 2008).

Diagnosis, Management and Modeling of Neurodevelopmental Disorders. https://doi.org/10.1016/B978-0-12-817988-8.00023-3

A range of neonatal complications are commonly faced by VP infants, including respiratory conditions (e.g., respiratory distress syndrome and bronchopulmonary dysplasia), infection (e.g., sepsis, necrotizing enterocolitis), neurological conditions (e.g., periventricular leukomalacia, intraventricular hemorrhage), and feeding difficulties. These complications are more common in the most immature infants, such as those born extremely preterm (EP, <28 completed weeks' gestation). While a reduction in mortality and neonatal morbidity has been reported (Chen, Bajwa, Rimensberger, Posfay-Barbe, & Pfister, 2016; Pierrat et al., 2017; Stoll et al., 2015), this has not been accompanied by an equal reduction in long-term neurodevelopmental impairments (Saigal & Doyle, 2008). The goal of this review is to summarize long-term neurodevelopmental outcomes following VP birth, with a focus on longitudinal research that highlights changes in developmental trajectories across neurodevelopmental domains with increasing age.

Neurodevelopmental impairments following very preterm birth

It is well established that children born VP are at increased risk for neurodevelopmental impairment (NDI) than infants born at term, with approximately 24% of VP infants worldwide having some degree of NDI (Blencowe et al., 2013). While studies differ subtly in the criteria used to classify NDI, it generally includes the presence of one or more of the following: profound hearing loss requiring amplification in both ears, visual impairment with visual acuity <20/200 in both eyes, moderate to severe cognitive delay (>2 SD below the normative mean on a standardized cognitive measure), and moderate to severe cerebral palsy (Adams-Chapman et al., 2018; Hintz, Newman, & Vohr, 2016). While the majority of VP children do not develop severe NDI, it is unclear whether rates of NDI and developmental delay are improving over time (Adams-Chapman et al., 2018; Cheong et al., 2017).

Two years of age is a common developmental time point for assessing NDI in children born VP (Pascal et al., 2018; Rogers & Hintz, 2016). The Bayley Scales of Infant and Toddler Development (third version or earlier) is the most commonly administered standardized measure of early developmental status; however, its ability to predict later cognitive and motor functioning is modest (Hack et al., 2005; Spencer-Smith, Spittle, Lee, Doyle, & Anderson, 2015; Spittle et al., 2013; Sutcliffe, Soo, & Barnes, 2010) and has been reported to underestimate rates of developmental delay in children born VP (Anderson et al., 2010). Moderate to severe cognitive and motor delays are relatively common at preschool age in children born <27 weeks' gestation (11%–30% and 3%–5% for cognitive and motor delays, respectively), while blindness and deafness are rare (0%–4.6% and 0.9%–2.5% for severe visual and hearing impairment, respectively) (Rogers & Hintz, 2016). As such, assessment of cognitive and motor development is the cornerstone for determining NDI.

Up to 50% of VP children are thought to experience milder forms of neurodevelopmental difficulties, particularly in terms of their cognitive and motor functioning (Aarnoudse-Moens, Weisglas-Kuperus, van Goudoever, & Oosterlaan, 2009; Anderson & Doyle, 2003; Marlow, 2004). These mild to moderate cognitive and motor difficulties can become more apparent at older ages as a result of increasing functional demands in daily life and schooling, and the use of more refined and appropriate assessment tools for specific functions. Therefore, VP toddlers with seemingly typical early developmental outcomes can go on to develop cognitive, behavioral, and academic deficits at later ages. These early cognitive and motor delays can later manifest as learning, attention, and memory, behavioral and socioemotional difficulties, as well as motor coordination problems. Concerningly, data from birth cohorts across eras suggest that the long-term neurodevelopmental outcomes for children born VP and EP are not improving (Twilhaar, Wade, et al., 2018) and may be worsening over time (Burnett et al., 2018; Cheong et al., 2017; Spittle, Cameron, Doyle, Cheong, & Victorian Infant Collaborative Study Group, 2018). This highlights the ongoing need to investigate the long-term neurodevelopmental outcomes of children born VP, with a focus on specific cognitive, motor, and behavioral domains as this information helps to inform surveillance and early intervention programs.

Cognitive outcomes

General cognition is typically assessed using measures of general intelligence (IQ), which is an important outcome to monitor in children born VP given its strong associations with adaptive and functional outcomes, such as academic achievement (Aarnoudse-Moens, Weisglas-Kuperus, Duivenvoorden, van Goudoever, & Oosterlaan, 2013; Johnson, Wolke, Hennessy, & Marlow, 2011). A previous meta-analytic review showed that VP children have, on average, an IQ approximately 13 points (0.86 SD) below their term-born peers (Twilhaar, Wade, et al., 2018). Evidence suggests that this discrepancy is not improving with more contemporaneous preterm cohorts (Cheong et al., 2017; Twilhaar, Wade, et al., 2018).

Neuropsychological evaluations are needed to explain the nature of cognitive difficulties observed on measures of IQ. Indeed, there is mounting evidence that children born VP exhibit deficits in attention, language, memory, executive function, processing speed, and visual–spatial abilities (Anderson, 2014). However, it is important to stress that these group-level findings mask the vast heterogeneity observed in VP children; for example, a significant proportion of this population do well and escape moderate to severe cognitive deficits.

Attention and memory

Attention and memory are core cognitive skills that underpin learning, everyday functioning, and knowledge acquisition and have been well investigated in VP children. Attention is often conceptualized as a multifaceted system that involves focusing (selective), sustaining, shifting, and dividing attention. Similarly, memory is a complex system composed of immediate memory stores (short-term memory), processes that hold and manipulate information entering the system (working memory), and long-term storage systems for procedural and declarative information (Anderson, 2014).

Problems across the attention domains are commonly reported in VP infants, children, and adolescents (Breeman, Jaekel, Baumann, Bartmann, & Wolke, 2016; de Kieviet, van Elburg, Lafeber, & Oosterlaan, 2012; Wilson-Ching et al., 2013). It has been shown that EP and/or extremely low birth weight (ELBW, <1000g) children perform 0.3–0.6 SD below their term-born peers on measures of selective, sustained, shifting, and divided attention at 8 years of age and are 2–3 times more likely than term-born controls to have an attention impairment (Anderson et al., 2011). Furthermore, inattention is one of the most common behavioral concerns reported by parents and teachers of VP children (de Kieviet et al., 2012). Memory deficits have also been reported for VP children and adolescents particularly for immediate and working memory domains (Luu, Ment, Allan, Schneider, & Vohr, 2011; Omizzolo et al., 2014). Compared with term born controls, VP children have been reported to be 2–3.5 times more likely to have a memory impairment, with the rate of impairment varying between 19% and 41% for the different memory domains (Omizzolo et al., 2014). In terms of stability, attention difficulties have been noted to remain relatively stable from childhood to adulthood for VP or very low birth weight (VLBW, ≤1500g) individuals compared to their term-born counterparts (Breeman et al., 2016); however, limited work has investigated stability of memory deficits in children born VP.

Executive function

Executive function is an umbrella term that encompasses a range of associated processes required for goal-directed behavior and is fundamental to cognition, social functioning, behavior, and academic achievement. Executive function has been broadly described to include attentional control, cognitive flexibility, working memory, inhibition, planning and organization, and strategic reasoning (Fig. 23.1; Anderson, 2002; Haebich et al., 2018) and has been an area of considerable interest in long-term

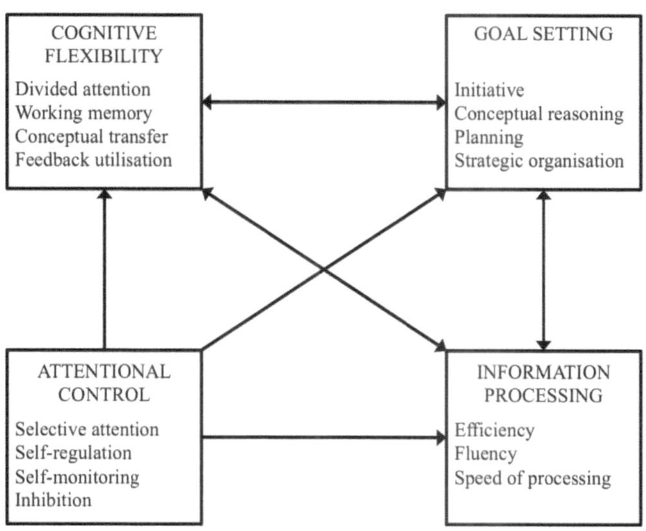

FIGURE 23.1 **A model of executive function.** An executive function model proposed by Anderson (2002) with four distinct domains: cognitive flexibility, goal setting, attentional control, and information processing. These domains involve highly integrated processes and are considered to be discrete functions but operate in an integrative manner to perform tasks. Attentional control is thought to greatly influence the functioning of the other domains, while information processing, cognitive flexibility, and goal setting domains are considered to be interrelated and interdependent. *Model is from Anderson, P. (2002). Assessment and development of executive function (EF) during childhood.* Child Neuropsychology, 8*(2), 71-82, with permission from* Child Neuropsychology.

outcome studies of VP children given suspicions that this domain may be a particular area of vulnerability for these children (Aarnoudse-Moens et al., 2009; Anderson, 2014). Indeed, children born VP and EP have been found to have executive function deficits based on parent and teacher reports as well as standardized assessment (Anderson & Doyle, 2004; Costa et al., 2017; Taylor & Clark, 2016). A meta-analytic review in VP/VLBW children highlighted that these children perform 0.4 SD and 0.5 SD below term-born controls on measures of working memory and cognitive flexibility, respectively (Aarnoudse-Moens et al., 2009). A meta-analysis in preterm children further reported group differences of 0.5 SD and 0.7 SD for inhibition and planning ability, respectively between children born <26 weeks' gestation and those born at term (Mulder, Pitchford, Hagger, & Marlow, 2009). Planning, organization, and strategic reasoning abilities, collectively referred to as goal setting, have also been found to be impaired in VP adolescents using a comprehensive goal setting assessment, with reduced completion times and increased error rates noted (Haebich et al., 2018). When examining the stability of executive dysfunction, longitudinal studies in EP children have reported that 15%–16% of these children have persistent executive function deficits from ages 8 to 18, with 12%–19% having late-onset difficulties in adolescence (Costa et al., 2017). Further, executive functioning in children born EP does not appear to be improving in more recent birth eras, with some aspects of executive functioning appearing to worsen despite advances in medical care (Burnett et al., 2018).

Language

Language functioning is a complex functional domain including interdependent domains of semantics, phonology, grammar, and pragmatics, but is often simplified into expressive and receptive language domains. There has been considerable research demonstrating that these children have poorer language skills across most aforementioned language components (Barre, Morgan, Doyle, & Anderson, 2011; Nguyen et al., 2018; Reidy et al., 2013). A meta-analysis on language abilities in VP/VLBW children reported that these children performed approximately 0.6–0.8 SD below term-born controls on expressive and receptive language, respectively (Barre et al., 2011). In terms of semantics, group differences were reported to be 0.4–0.6 SD in favor of controls (Barre et al., 2011). Findings from a large prospective cohort of VP/VLBW children further reported reduced language functioning across the domains of phonological awareness, semantics, grammar, discourse, and pragmatics at 7 years of age compared to term-born controls (Reidy et al., 2013). Furthermore, the proportion of VP/VLBW children with low language scores (1.25 SD below test mean) was greater in VP/VLBW compared with controls, with reported rates of 20%–30% in VP/VLBW children compared with 1.4%–14.6% in controls across language domains (Reidy et al., 2013). Results from the same prospective, longitudinal cohort of VP/VLBW children, found these individuals continued to perform 0.5–1.0 SD below term-born controls at 13 years of age, being 2–8 times more likely to have an impairment across language domains, particularly for expressive language (Nguyen et al., 2018). In terms of language trajectories, VP children have demonstrated lower language functioning from toddlerhood to adolescence compared with term-born peers, indicating persistent language difficulties over time (Fig. 23.2; Nguyen et al., 2018).

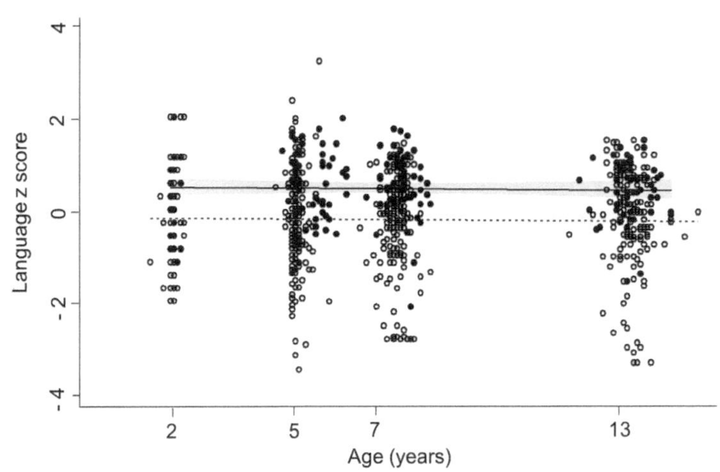

FIGURE 23.2 **Language trajectories from toddlerhood to adolescence in very preterm and term-born children.** Language outcomes from 2 to 13 years of age for children born very preterm (open circles) and at term (solid circles). The regression line of best fit is dotted for the very preterm group and solid for term-born controls. Shadows around the regression lines represent 95% confidence intervals (CIs). *Data are from Nguyen, T. N., Spencer-Smith, M., Zannino, D., Burnett, A., Scratch, S. E., Pascoe, L., … Anderson, P. J. (2018). Developmental trajectory of language from 2 to 13 years in children born very preterm.* Pediatrics, 141*(5), with permission from* Pediatrics.

Academic achievement

Children born VP are more vulnerable to scholastic underachievement compared with their term-born counterparts, with children born at the earliest gestations and smallest birth weights displaying more marked deficits in reading and mathematics and increased rates of specific learning difficulties (Anderson & Doyle, 2003). VP children are approximately three times more likely to receive educational assistance (28% vs. 8%) and twice as likely to repeat grades than term-born controls (30% vs. 13%; Twilhaar, de Kieviet, van Elburg, & Oosterlaan, 2019). Similarly, approximately 57% of EP children require special education services, with 50% of children rated by their teachers as having an academic attainment level below the national average at 11 years of age (Johnson & Marlow, 2016). Difficulties in mathematics are particularly pertinent in this population, with the majority of studies reporting greater discrepancies in math performance between VP and term-born children compared with other academic domains, such as reading (Johnson & Marlow, 2016). These academic difficulties tend to persist even after accounting for IQ (Anderson & Doyle, 2003).

In a longitudinal study of academic functioning in VP children across Dutch primary schooling, VP children performed below term-born controls, on average, 0.5 SD for mathematics, 0.3 SD for reading comprehension, and 0.2 SD for spelling (Twilhaar, de Kieviet, Bergwerff, et al., 2019; Twilhaar, de Kieviet, van Elburg, & Oosterlaan, 2019). The findings demonstrated that academic deficits in VP children remained stable over the primary school period and did not resolve with time. Indeed, similar reports have noted in a population-based cohort of EP/ELBW children in Australia, where the disadvantage in academic performance between EP/ELBW and term-born children remained relatively constant from preschool through to late adolescence, reflecting a persisting academic deficit for EP/ELBW individuals (Doyle et al., 2015). Additionally, the odds of having word reading, spelling, and mathematics difficulties were found to be greater for children born in 2005 than those born in earlier eras (early 1990s and 1997), which could not be explained by differences in perinatal care or sociodemographic factors between birth eras (Fig. 23.3; Cheong et al., 2017).

Motor outcomes

A meta-analysis reported that VP/VLBW children perform 0.6–0.7 SD below their term-born peers on standardized measures of motor functioning (de Kieviet, Piek, Aarnoudse-Moens, & Oosterlaan, 2009). Cerebral palsy (CP) is the most severe form of motor disability following VP birth and encompasses a group of disorders relating to movement and posture development that results in activity limitations (Spittle & Orton, 2014). While there is evidence that the rate of CP may be stabilizing or decreasing (Pascal et al., 2018; Spittle et al., 2018), CP is present in approximately 14.6% of infants born EP and 6.2% in those born VP (Himpens, Van den Broeck, Oostra, Calders, & Vanhaesebrouck, 2008). Bilateral spastic CP is the most prevalent type of CP in VP children (Himpens et al., 2008), defined by "increased tone with a velocity-dependent increase in resistance to passive movement" (Duncan & Matthews, 2018). A diagnosis of CP or being "at risk" of CP can now be provided before 6 months

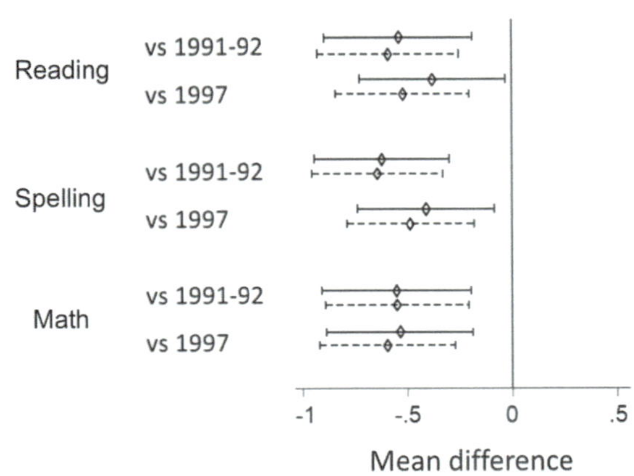

FIGURE 23.3 Changing academic functioning in extremely preterm/extremely low birth weight children across the birth eras. Mean differences on standardized academic scores comparing extremely preterm/extremely low birth weight children born in 2005 with earlier eras (1991–92 and 1997). Comparisons are displayed as regression coefficients (point estimate) and 95% confidence intervals (CIs). If CIs are less than zero, performance is worse in 2005 cohort compared with earlier eras. Solid line is adjusted for age at assessment, age of mother, and sociodemographic variables (e.g., lower maternal education, multilingual households); dashed line is adjusted for age at assessment, age of mother, sociodemographic variables, and perinatal variables (e.g., antenatal corticosteroids, gestational age at birth, sex, grade 3 or 4 intraventricular hemorrhage). *Data are from Cheong, J. L., Anderson, P. J., Burnett, A. C., Roberts, G., Davis, N., Hickey, L., … Victorian Infant Collaborative Study Group (2017). Changing neurodevelopment at 8 years in children born extremely preterm since the 1990s.* Pediatrics, *139(6), e20164086, with permission from* Pediatrics.

corrected age, via a combination of medical history, standardized motor assessments (such as the Prechtl General Movements Assessment and the Hammersmith Infant Neurologic Examination (HINE)), and magnetic resonance imaging at term-equivalent age (Novak et al., 2017). While a diagnosis of CP can usually be made early in development, the severity of CP is thought to be more accurately assessed at or after 2 years corrected age (Novak et al., 2017).

A greater proportion of VP children present with milder, non-CP motor impairments such as poor balance, coordination, and ball skills, which are often referred to as developmental coordination disorder (DCD; Edwards et al., 2011). Reported prevalence rates of mild−moderate motor impairments are suggested to range between 9.5% and 72.2% in VP/VLBW children, with the variability thought to be dependent on the motor outcome utilized and sample size and composition (Edwards et al., 2011). There is some evidence to suggest that the rate of these mild to moderate motor impairments may be on the increase over time (Spittle et al., 2018), which may be due to the reported reduction in CP in VP children. Motor impairments such as DCD are more commonly diagnosed at older ages as children start to display difficulties in hand−eye coordination, motor control, and sensory-motor integration at late preschool and early school age when compared with their term-born peers. It has been reported that in the absence of CP, VP/VLBW children are 6−9 times more likely to have mild−moderate motor impairments compared with their term-born peers (Edwards et al., 2011).

Mild to moderate motor impairments are often overlooked in VP children, but there is considerable evidence to demonstrate that these difficulties are associated with functional challenges in other domains such as cognition, academic achievement, social and peer relations, independent self-care, and physical health (Burns et al., 2009; Davis, Ford, Anderson, Doyle, & Group, 2007; Spittle & Orton, 2014).

Behavioral outcomes

Behavioral challenges for VP children can include difficulties in behavior regulation, attention, activity levels, sensory sensitivity, peer and social relations, and mood and somatic symptoms. VP birth appears to confer an increased risk of a pattern of behavioral problems referred to as the "preterm behavioral phenotype" characterized by inattention, socioemotional difficulties, and increased internalizing problems (e.g., anxiety, social withdrawal) (Johnson & Marlow, 2011). While behavioral and emotional problems have been reported in very young VP children (Spittle et al., 2009), most investigations have been focused on the school-age period or beyond (Arpi & Ferrari, 2013). Limited longitudinal investigations have suggested that these deficits may remain stable from early childhood to middle childhood (Gray, Indurkhya, & McCormick, 2004) and even persist through to early adulthood (Linsell et al., 2019).

Socialemotional outcomes

Socioemotional functioning refers to an individual's ability to successfully interact and communicate within a social setting and to efficiently manage and monitor emotions. It is closely linked to concepts such as "social competence" and encompasses emotion regulation, social cognitive processing, prosocial skills, and positive communication. Children and adolescents born VP are at greater risk of socioemotional difficulties compared to their term-born peers (Montagna & Nosarti, 2016; Ritchie, Bora, & Woodward, 2018; Spittle et al., 2009; Twilhaar, de Kieviet, Bergwerff, et al., 2019; Twilhaar, de Kieviet, van Elburg, & Oosterlaan, 2019). Children and adolescents born VP have been reported to experience difficulties in emotion recognition, emotion regulation, and social cognition, particularly the ability to understand that people's motivations and emotions are different from one's own (theory of mind) (Montagna & Nosarti, 2016; Twilhaar, de Kieviet, Bergwerff, et al., 2019; Twilhaar, de Kieviet, van Elburg, & Oosterlaan, 2019). These children are reported to have more peer social difficulties than term-born children, with 12%−14% of VP children experiencing chronic bullying at school (two or more times a week) compared to 4% of full-term peers (Ritchie et al., 2018). Children born EP are reported to have a 2−4 times increased odd of clinically significant emotional or social problems at 6 years of age, with 4.6 times increased odds of exhibiting mood disorders at 11 years of age (Johnson & Marlow, 2016).

Mental health outcomes

Approximately 24% of VP children have a mental health diagnosis in midchildhood compared with 9% of term-born children, with a three times higher odds of meeting diagnostic criteria for a mental health disorder (Johnson et al., 2010; Treyvaud et al., 2013). For attention deficit hyperactivity disorder (ADHD; any subtype), VP children are four times more likely to have a diagnosis compared to their term-born peers (Johnson et al., 2010; Trevyaud et al., 2013). The rate of autism spectrum disorders (ASD) has also been shown to be greater in VP compared to term-born children (4.5% vs. 0%), with the rate even higher in EP individuals (EP: 8.0% vs. 0%; Treyvaud et al., 2013; Johnson et al., 2010). For internalizing disorders, such as anxiety, there is an almost 50% higher odds of the disorder for VP children compared to term-born children (11% vs. 8%, Treyvaud et al., 2013), which increases to three and a half times the odds for EP children (Johnson et al., 2010).

Factors associated with neurodevelopmental outcomes

The etiology of these cognitive, motor, and behavioral difficulties in VP children is multifactorial, resulting from interrelated genetic, biological, and social–environmental risk factors. Indeed, neurodevelopmental outcomes have been shown to be associated with a range of biological, perinatal, neonatal (Anderson et al., 2017; Doyle et al., 2015; Linsell, Malouf, Morris, Kurinczuk, & Marlow, 2016; Treyvaud et al., 2013), socioeconomic, demographic (Doyle et al., 2015; Spittle et al., 2009; Treyvaud et al., 2013; Wong & Edwards, 2013), and psychosocial factors (Clark, Woodward, Horwood, & Moor, 2008; Treyvaud et al., 2016), wherein the presence of these factors either exacerbates or buffers the adverse effects of preterm birth. Gene–environment interactions may result in an increased risk for poorer neurodevelopmental outcomes following VP birth, with a number of identified genetic factors, inclusive of 43 genes, previously associated with atypical neurodevelopmental outcomes (cognition, perception, behavior, brain structure, CP, and mental health disorders) (Blair, Pickler, & Anderson, 2016). Perinatal brain injury has been identified to be a key biological contributor to poor neurodevelopmental outcomes following VP birth. Cerebral white and gray matter pathologies, as well as cerebellar alterations, have been associated with cognitive, language, and motor deficits in VP children (Anderson et al., 2017). Sociodemographic factors, such as low parental education, and biological factors, such as male sex, have also been identified to be significant predictors of cognition impairment in children born VP (Linsell, Malouf, Morris, Kurinczuk, & Marlow, 2015). Psychosocial variables, such as parenting, maternal stress, parent–child interaction, as well as early stressors such as invasive procedures and the high sensory environment of the neonatal intensive care unit (NICU), have been shown to be important environmental moderators that can alter neurodevelopmental outcomes following VP birth (Treyvaud et al., 2016; Vinall et al., 2014). Taken together, it is clear that there are modifiable risk and protective factors that influence neurodevelopmental outcomes for VP children. Further research into the cumulative effects of these factors using high-quality longitudinal studies will enable their relative contribution and importance on neurodevelopmental outcomes at varying ages to be assessed. Such investigations will assist in the development and implementation of therapies and interventions appropriate for differing outcomes at distinct developmental periods in VP individuals.

Applications to other areas of development

This chapter focused on VP birth as an example of the effects of perinatal adversity on the developing brain by examining long-term neurodevelopmental outcomes across childhood and adolescence. However, these findings have wide applicability to children who are not reaching their developmental potential, be it as a result of early adversity, such as poverty (Barajas, Philipsen, & Brooks-Gunn, 2007), or child maltreatment and neglect (Hildyard & Wolfe, 2002), or exposure to other neonatal conditions (e.g., congenital heart disease, prenatal substance exposure) (Latal, 2016; Shankaran et al., 2007). Collectively, these perinatal and postnatal events and exposures can affect the timing and pattern of gene expression, the formation and maturation of the developing brain, alter brain structure and function and subsequent functional outcomes, such as cognition, language, and socioemotional skills (Walker et al., 2011). These unfavorable early events occur during critical periods of development, which can also impact other developmental markers, such as growth, physiological function, and immune and metabolic response, and augment the risk of behavioral or mental health disorders in later life (Tomalski & Johnson, 2010). Given the impact of early adversity on long-term child development, several streams of focus have emerged. As noted in this chapter, an important area of investigation is to identify risk and protective factors that worsen or negate neurodevelopmental difficulties following early adversity. Sociodemographic factors, such as social status, parental education and occupation, and home and neighborhood environments, are known to have a profound impact on structural and functional brain development and subsequent child development. Additional factors such as parental mental health and early stress have also been considered important moderators of outcomes in children across these different populations. A second area of focus has been to harness these modifiable targets for use in the development of new therapies and interventions. Indeed, the provision of environmental enrichment, parent education around strengthening attachment relationships and positive maternal caregiving behavior, and supporting parental mental health have been promising avenues for intervention and supporting neurodevelopmental outcomes in children at high biological and psychosocial risk (Walker et al., 2011). The findings of this review lend support for the need to monitor and support children at high risk of poor neurodevelopmental outcomes well into childhood, adolescence, and beyond and to identify and account for the cumulative effect of biological and social factors that can augment the risk of adverse long-term outcomes in vulnerable children.

Key facts about prematurity

- There has been growing interest in the complex, subtle neurodevelopmental deficits of preterm children, given increased survival without severe neurosensory or neurodevelopmental impairment.
- Neurodevelopmental outcomes following preterm birth are inversely related to gestational age at birth, with greater risk for difficulties for those born with greater immaturity.
- While preterm children, as a group, have persistent challenges across a range of outcomes from infancy to adolescence, individual children may demonstrate different developmental trajectories.
- Accurate and sensitive early markers of later functional outcomes are needed so that children at greatest long-term risk can be identified earlier.
- Determining the optimal timing for developmental intervention and the appropriate combination of targets for intervention is needed.

Summary points

- **Bullet of Summary Point 1**: Mortality rates of very preterm infants are improving; however, long-term morbidity following very preterm birth continues to be an area of challenge and investigation.
- **Bullet of Summary Point 2**: High rates of mild to moderate neurodevelopmental deficits have been reported in very preterm children across a range of cognitive, motor, and behavioral domains.
- **Bullet of Summary Point 3:** There is growing evidence that neurodevelopmental difficulties faced by very preterm individuals remain stable through childhood and adolescence and may potentially persist into adulthood in some domains.
- **Bullet of Summary Point 4**: Despite group-level deficits, considerable interindividual variability has been reported, highlighting the contribution of risk and resilience factors to child outcomes following very preterm birth.
- **Bullet of Summary Point 5**: A number of genetic, biological, sociodemographic, and psychosocial factors have been identified to influence short- and long-term neurodevelopmental outcomes in very preterm individuals.
- **Bullet of Summary Point 6**: Identification of biological, sociodemographic, and familial factors help to provide opportunities for developing therapies and intervention strategies to promote positive long-term outcomes for children born very preterm.

References

Aarnoudse-Moens, C. S. H., Weisglas-Kuperus, N., Duivenvoorden, H. J., van Goudoever, J. B., & Oosterlaan, J. (2013). Executive function and IQ predict mathematical and attention problems in very preterm children. *PloS One, 8*(2), e55994.

Aarnoudse-Moens, C. S. H., Weisglas-Kuperus, N., van Goudoever, J. B., & Oosterlaan, J. (2009). Meta-analysis of neurobehavioral outcomes in very preterm and/or very low birth weight children. *Pediatrics, 124*(2), 717–728.

Adams-Chapman, I., Heyne, R. J., DeMauro, S. B., Duncan, A. F., Hintz, S. R., Pappas, A., … Newman, J. E. (2018). Neurodevelopmental impairment among extremely preterm infants in the neonatal research network. *Pediatrics, 141*(5), e20173091.

Anderson, P. (2002). Assessment and development of executive function (EF) during childhood. *Child Neuropsychology, 8*(2), 71–82.

Anderson, P. J. (2014). Neuropsychological outcomes of children born very preterm. *Seminars in Fetal and Neonatal Medicine, 19*(2), 90–96.

Anderson, P. J., De Luca, C. R., Hutchinson, E., Roberts, G., Doyle, L. W., & Group, t. V. I. C. (2010). Underestimation of developmental delay by the new Bayley-III scale. *JAMA Pediatrics, 164*(4), 352–356.

Anderson, P. J., De Luca, C. R., Hutchinson, E., Spencer-Smith, M. M., Roberts, G., Doyle, L. W., & Victorian Infant Collaborative Study Group. (2011). Attention problems in a representative sample of extremely preterm/extremely low birth weight children. *Developmental Neuropsychology, 36*(1), 57–73.

Anderson, P., & Doyle, L. W. (2003). Neurobehavioral outcomes of school-age children born extremely low birth weight or very preterm in the 1990s. *The Journal of the American Medical Association, 289*(24), 3264–3272.

Anderson, P. J., & Doyle, L. W. (2004). Executive functioning in school-aged children who were born very preterm or with extremely low birth weight in the 1990s. *Pediatrics, 114*(1), 50–57.

Anderson, P. J., Treyvaud, K., Neil, J. J., Cheong, J. L. Y., Hunt, R. W., Thompson, D. K., … Inder, T. E. (2017). Associations of newborn brain magnetic resonance imaging with long-term neurodevelopmental impairments in very preterm children. *The Journal of Pediatrics, 187*, 58–65.e51.

Arpi, E., & Ferrari, F. (2013). Preterm birth and behaviour problems in infants and preschool-age children: A review of the recent literature. *Developmental Medicine and Child Neurology, 55*(9), 788–796.

Barajas, R. G., Philipsen, N., & Brooks-Gunn, J. (2007). *Cognitive and emotional outcomes for children in poverty. Handbook of families and poverty.* Thousand Oaks, CA: Sage.

Barre, N., Morgan, A., Doyle, L. W., & Anderson, P. J. (2011). language abilities in children who were very preterm and/or very low birth weight: A meta-analysis. *The Journal of Pediatrics, 158*(5), 766–774.e761.

Blair, L. M., Pickler, R. H., & Anderson, C. (2016). Integrative review of genetic factors influencing neurodevelopmental outcomes in preterm infants. *Biological Research for Nursing, 18*(2), 127–137.

Blencowe, H., Lee, A. C., Cousens, S., Bahalim, A., Narwal, R., Zhong, N., … Katz, J. (2013). Preterm birth–associated neurodevelopmental impairment estimates at regional and global levels for 2010. *Pediatric Research, 74*(S1), 17.

Breeman, L. D., Jaekel, J., Baumann, N., Bartmann, P., & Wolke, D. (2016). Attention problems in very preterm children from childhood to adulthood: The Bavarian Longitudinal Study. *Journal of Child Psychology and Psychiatry, 57*(2), 132–140.

Burnett, A. C., Anderson, P. J., Lee, K. J., Roberts, G., Doyle, L. W., Cheong, J. L., & Group, V. I. C. S. (2018). Trends in executive functioning in extremely preterm children across 3 birth eras. *Pediatrics, 141*(1), e20171958.

Burns, Y. R., Danks, M., O'Callaghan, M. J., Gray, P. H., Cooper, D., Poulsen, L., & Watter, P. (2009). Motor coordination difficulties and physical fitness of extremely-low-birthweight children. *Developmental Medicine & Child Neurology, 51*(2), 136–142.

Chawanpaiboon, S., Vogel, J. P., Moller, A.-B., Lumbiganon, P., Petzold, M., Hogan, D., … Gülmezoglu, A. M. (2019). Global, regional, and national estimates of levels of preterm birth in 2014: A systematic review and modelling analysis. *The Lancet Global Health, 7*(1), e37–e46.

Chen, F., Bajwa, N. M., Rimensberger, P. C., Posfay-Barbe, K. M., & Pfister, R. E. (2016). Thirteen-year mortality and morbidity in preterm infants in Switzerland. *Archives of Disease in Childhood – Fetal and Neonatal Edition, 101*(5), F377–F383.

Cheong, J. L., Anderson, P. J., Burnett, A. C., Roberts, G., Davis, N., Hickey, L., … Victorian Infant Collaborative Study Group. (2017). Changing neurodevelopment at 8 years in children born extremely preterm since the 1990s. *Pediatrics, 139*(6), e20164086.

Clark, C. A., Woodward, L. J., Horwood, L. J., & Moor, S. (2008). Development of emotional and behavioral regulation in children born extremely preterm and very preterm: Biological and social influences. *Child Development, 79*(5), 1444–1462.

Costa, D. S., Miranda, D. M., Burnett, A. C., Doyle, L. W., Cheong, J. L., Anderson, P. J., & Group, V. I. C. S. (2017). Executive function and academic outcomes in children who were extremely preterm. *Pediatrics, 140*(3), e20170257.

Davis, N. M., Ford, G. W., Anderson, P. J., Doyle, L. W., & Group, V. I. C. S. (2007). Developmental coordination disorder at 8 years of age in a regional cohort of extremely-low-birthweight or very preterm infants. *Developmental Medicine & Child Neurology, 49*(5), 325–330.

Doyle, L. W., Cheong, J. L., Burnett, A., Roberts, G., Lee, K. J., Anderson, P. J., & Group, V. I. C. S. (2015). Biological and social influences on outcomes of extreme-preterm/low-birth weight adolescents. *Pediatrics, 136*(6), e1513–e1520.

Duncan, A. F., & Matthews, M. A. (2018). Neurodevelopmental outcomes in early childhood. *Clinics in Perinatology, 45*(3), 377–392.

Edwards, J., Berube, M., Erlandson, K., Haug, S., Johnstone, H., Meagher, M., … Zwicker, J. G. (2011). Developmental coordination disorder in school-aged children born very preterm and/or at very low birth weight: A systematic review. *Journal of Developmental & Behavioral Pediatrics, 32*(9), 678–687.

Gray, R. F., Indurkhya, A., & McCormick, M. C. (2004). Prevalence, stability, and predictors of clinically significant behavior problems in low birth weight children at 3, 5, and 8 years of age. *Pediatrics, 114*(3), 736–743. https://doi.org/10.1542/peds.2003-1150-L. In this issue.

Hack, M., Taylor, H. G., Drotar, D., Schluchter, M., Cartar, L., Wilson-Costello, D., … Morrow, M. (2005). Poor predictive validity of the Bayley Scales of Infant Development for cognitive function of extremely low birth weight children at school age. *Pediatrics, 116*(2), 333–341.

Haebich, K. M., Willmott, C., Ellis, R., Burnett, A. C., Scratch, S. E., Pascoe, L., … Anderson, P. J. (2018). Goal setting deficits at 13 years in very preterm born children. *Journal of the International Neuropsychological Society, 24*(4), 372–381.

Hildyard, K. L., & Wolfe, D. A. (2002). Child neglect: Developmental issues and outcomes. *Child Abuse & Neglect, 26*(6), 679–695.

Himpens, E., Van den Broeck, C., Oostra, A., Calders, P., & Vanhaesebrouck, P. (2008). Prevalence, type, distribution, and severity of cerebral palsy in relation to gestational age: A meta-analytic review. *Developmental Medicine & Child Neurology, 50*(5), 334–340.

Hintz, S. R., Newman, J. E., & Vohr, B. R. (2016). Changing definitions of long-term follow-up: Should "long term" be even longer? *Seminars in Perinatology, 40*(6), 398–409.

Johnson, S., Hollis, C., Kochhar, P., Hennessy, E., Wolke, D., & Marlow, N. (2010). Psychiatric disorders in extremely preterm children: Longitudinal finding at age 11 years in the EPICure study. *Journal of the American Academy of Child & Adolescent Psychiatry, 49*(5), 453–463.e451.

Johnson, S., & Marlow, N. (2011). Preterm birth and childhood psychiatric disorders. *Pediatric Research, 69*(5 Pt 2), 11r–18r.

Johnson, S., & Marlow, N. (2016). Charting the survival, health and development of extremely preterm infants: EPICure and beyond. *Paediatrics and Child Health, 26*(11), 498–504.

Johnson, S., Wolke, D., Hennessy, E., & Marlow, N. (2011). Educational outcomes in extremely preterm children: Neuropsychological correlates and predictors of attainment. *Developmental Neuropsychology, 36*(1), 74–95.

de Kieviet, J. F., Piek, J. P., Aarnoudse-Moens, C. S., & Oosterlaan, J. (2009). Motor development in very preterm and very low-birth-weight children from birth to adolescence: A meta-analysis. *The Journal of the American Medical Association, 302*(20), 2235–2242.

de Kieviet, J. F., van Elburg, R. M., Lafeber, H. N., & Oosterlaan, J. (2012). Attention problems of very preterm children compared with age-matched term controls at school-age. *The Journal of Pediatrics, 161*(5), 824–829.e821.

Latal, B. (2016). Neurodevelopmental outcomes of the child with congenital heart disease. *Clinics in Perinatology, 43*(1), 173–185.

Linsell, L., Johnson, S., Wolke, D., Morris, J., Kurinczuk, J. J., & Marlow, N. (2019). Trajectories of behavior, attention, social and emotional problems from childhood to early adulthood following extremely preterm birth: A prospective cohort study. *European Child & Adolescent Psychiatry, 28*(4), 531–542.

Linsell, L., Malouf, R., Morris, J., Kurinczuk, J. J., & Marlow, N. (2015). Prognostic factors for poor cognitive development in children born very preterm or with very low birth weight: A systematic review. *JAMA Pediatrics, 169*(12), 1162–1172.

Linsell, L., Malouf, R., Morris, J., Kurinczuk, J. J., & Marlow, N. (2016). Prognostic factors for cerebral palsy and motor impairment in children born very preterm or very low birthweight: A systematic review. *Developmental Medicine & Child Neurology, 58*(6), 554–569.

Luu, T. M., Ment, L., Allan, W., Schneider, K., & Vohr, B. R. (2011). Executive and memory function in adolescents born very preterm. *Pediatrics, 127*(3), e639–646.

Marlow, N. (2004). Neurocognitive outcome after very preterm birth. *Archives of Disease in Childhood - Fetal and Neonatal Edition, 89*(3), F224–F228.

Montagna, A., & Nosarti, C. (2016). Socio-emotional development following very preterm birth: Pathways to psychopathology. *Frontiers in Psychology, 7*, 80.

Mulder, H., Pitchford, N. J., Hagger, M. S., & Marlow, N. (2009). Development of executive function and attention in preterm children: A systematic review. *Developmental Neuropsychology, 34*(4), 393–421.

Nguyen, T. N., Spencer-Smith, M., Zannino, D., Burnett, A., Scratch, S. E., Pascoe, L., … Anderson, P. J. (2018). Developmental trajectory of language from 2 to 13 years in children born very preterm. *Pediatrics, 141*(5).

Novak, I., Morgan, C., Adde, L., Blackman, J., Boyd, R. N., Brunstrom-Hernandez, J., … Badawi, N. (2017). Early, accurate diagnosis and early intervention in cerebral palsy: Advances in diagnosis and Treatment. *JAMA Pediatrics, 171*(9), 897–907.

Omizzolo, C., Scratch, S. E., Stargatt, R., Kidokoro, H., Thompson, D. K., Lee, K. J., … Anderson, P. J. (2014). Neonatal brain abnormalities and memory and learning outcomes at 7 years in children born very preterm. *Memory, 22*(6), 605–615.

Pascal, A., Govaert, P., Oostra, A., Naulaers, G., Ortibus, E., & Van den Broeck, C. (2018). Neurodevelopmental outcome in very preterm and very-low-birthweight infants born over the past decade: A meta-analytic review. *Developmental Medicine & Child Neurology, 60*(4), 342–355.

Pierrat, V., Marchand-Martin, L., Arnaud, C., Kaminski, M., Resche-Rigon, M., Lebeaux, C., … Ancel, P. Y. (2017). Neurodevelopmental outcome at 2 years for preterm children born at 22 to 34 weeks' gestation in France in 2011: EPIPAGE-2 cohort study. *British Medical Journal, 358*, j3448.

Reidy, N., Morgan, A., Thompson, D. K., Inder, T. E., Doyle, L. W., & Anderson, P. J. (2013). Impaired language abilities and white matter abnormalities in children born very preterm and/or very low birth weight. *The Journal of Pediatrics, 162*(4), 719–724.

Ritchie, K., Bora, S., & Woodward, L. J. (2018). Peer relationship outcomes of school-age children born very preterm. *The Journal of Pediatrics, 201*, 238–244.

Rogers, E. E., & Hintz, S. R. (2016). Early neurodevelopmental outcomes of extremely preterm infants. *Seminars in Perinatology, 40*(8), 497–509.

Saigal, S., & Doyle, L. W. (2008). An overview of mortality and sequelae of preterm birth from infancy to adulthood. *The Lancet, 371*(9608), 261–269.

Shankaran, S., Lester, B. M., Das, A., Bauer, C. R., Bada, H. S., Lagasse, L., & Higgins, R. (2007). Impact of maternal substance use during pregnancy on childhood outcome. *Seminars in Fetal and Neonatal Medicine, 12*(2), 143–150.

Spencer-Smith, M. M., Spittle, A. J., Lee, K. J., Doyle, L. W., & Anderson, P. J. (2015). Bayley-III cognitive and language scales in preterm children. *Pediatrics, 135*(5), e1258–e1265.

Spittle, A. J., Cameron, K., Doyle, L. W., Cheong, J. L., & Victorian Infant Collaborative Study Group. (2018). Motor impairment trends in extremely preterm children: 1991–2005. *Pediatrics, 141*(4), e20173410.

Spittle, A. J., & Orton, J. (2014). Cerebral palsy and developmental coordination disorder in children born preterm. *Seminars in Fetal and Neonatal Medicine, 19*(2), 84–89.

Spittle, A. J., Spencer-Smith, M. M., Eeles, A. L., Lee, K. J., Lorefice, L. E., Anderson, P. J., & Doyle, L. W. (2013). Does the Bayley-III Motor Scale at 2 years predict motor outcome at 4 years in very preterm children? *Developmental Medicine & Child Neurology, 55*(5), 448–452.

Spittle, A. J., Treyvaud, K., Doyle, L. W., Roberts, G., Lee, K. J., Inder, T. E., … Anderson, P. J. (2009). Early emergence of behavior and social-emotional problems in very preterm infants. *Journal of the American Academy of Child & Adolescent Psychiatry, 48*(9), 909–918.

Stoll, B. J., Hansen, N. I., Bell, E. F., Walsh, M. C., Carlo, W. A., Shankaran, S., … Wyckoff, M. (2015). Trends in care practices, morbidity, and mortality of extremely preterm neonates, 1993–2012. *The Journal of the American Medical Association, 314*(10), 1039–1051.

Sutcliffe, A. G., Soo, A., & Barnes, J. (2010). Predictive value of developmental testing in the second year for cognitive development at five years of age. *Pediatric Reports, 2*(2), e15.

Taylor, H. G., & Clark, C. A. C. (2016). Executive function in children born preterm: Risk factors and implications for outcome. *Seminars in Perinatology. 40*(8), 520–529.

Tomalski, P., & Johnson, M. H. (2010). The effects of early adversity on the adult and developing brain. *Current Opinion in Psychiatry, 23*(3), 233–238.

Treyvaud, K., Doyle, L. W., Lee, K. J., Ure, A., Inder, T. E., Hunt, R. W., & Anderson, P. J. (2016). Parenting behavior at 2 years predicts school-age performance at 7 years in very preterm children. *Journal of Child Psychology and Psychiatry, 57*(7), 814–821.

Treyvaud, K., Ure, A., Doyle, L. W., Lee, K. J., Rogers, C. E., Kidokoro, H., … Anderson, P. J. (2013). Psychiatric outcomes at age seven for very preterm children: Rates and predictors. *Journal of Child Psychology and Psychiatry, 54*(7), 772–779.

Twilhaar, E. S., de Kieviet, J. F., Bergwerff, C. E., Finken, M. J. J., van Elburg, R. M., & Oosterlaan, J. (2019). Social adjustment in adolescents born very preterm: Evidence for a cognitive basis of social problems. *The Journal of Pediatrics, 213*.

Twilhaar, E. S., de Kieviet, J. F., van Elburg, R. M., & Oosterlaan, J. (2019). Academic trajectories of very preterm born children at school age. *Archives of Disease in Childhood - Fetal and Neonatal Edition, 104*(4), F419–f423.

Twilhaar, E. S., Wade, R. M., De Kieviet, J. F., Van Goudoever, J. B., Van Elburg, R. M., & Oosterlaan, J. (2018). Cognitive outcomes of children born extremely or very preterm since the 1990s and associated risk factors: A meta-analysis and meta-regression. *JAMA Pediatrics, 172*(4), 361–367.

Vinall, J., Miller, S. P., Bjornson, B. H., Fitzpatrick, K. P., Poskitt, K. J., Brant, R., … Grunau, R. E. (2014). Invasive procedures in preterm children: Brain and cognitive development at school age. *Pediatrics, 133*(3), 412–421.

Walker, S. P., Wachs, T. D., Grantham-McGregor, S., Black, M. M., Nelson, C. A., Huffman, S. L., … Richter, L. (2011). Inequality in early childhood: Risk and protective factors for early child development. *The Lancet, 378*(9799), 1325–1338.

Wilson-Ching, M., Molloy, C. S., Anderson, V. A., Burnett, A., Roberts, G., Cheong, J. L., … Anderson, P. J. (2013). Attention difficulties in a contemporary geographic cohort of adolescents born extremely preterm/extremely low birth weight. *Journal of the International Neuropsychological Society, 19*(10), 1097–1108.

Wong, H. S., & Edwards, P. (2013). Nature or nurture: A systematic review of the effect of socio-economic status on the developmental and cognitive outcomes of children born preterm. *Maternal and Child Health Journal, 17*(9), 1689–1700.

Chapter 24

Linking congenital heart disease and brain functional connectivity in newborns

Josepheen De Asis-Cruz and Catherine Limperopoulos

Developing Brain Institute, Diagnostic Imaging and Radiology, Children's National Health System, Washington, D.C., United States

List of abbreviations

BOLD Blood oxygen-level-dependent signal
CHD Congenital heart disease
HLHS Hypoplastic left heart syndrome
MRI Magnetic resonance imaging
rs-fMRI Resting-state functional connectivity MRI
RSN Resting-state networks

Mini-dictionary of terms

Diffusion tensor imaging An MRI technique that measures the brain's microstructure and anatomical connectivity
Functional connectome A comprehensive network map of the functional connections between different elements of the brain (or nervous system)
Functional MRI The detection of changes in BOLD signal in the brain due to changes in the concentration of deoxygenated blood induced by neural activity and accompanying changes in blood flow
Global efficiency Describes the ability of a network to integrate information across distant nodes
Graph theory A branch of mathematics that studies the relationship between nodes and their connections. In neuroscience, it is used to study the relationship between regions of the brain (i.e., vertices/nodes) and the structural or functional connections between them (i.e., edges)
Hubs Critical nodes in a network, usually defined by a centrality measure
Magnetic resonance spectroscopy An MRI technique that measures the chemical composition of brain tissue
Module A collection of nodes that are highly connected with each other but sparsely connected to other nodes
Resting state fMRI An fMRI approach that measures temporal correlations between spontaneous, low-frequency fluctuations of the BOLD signal from different areas of the brain
Resting state networks Distributed map of brain regions that perform similar or related functions
Rich club organization The propensity of highly connected nodes to densely connect with each other
Small-world topology A network with high local clustering between nodes (segregated) but with short path between any pair of nodes (integrated)
Task-based fMRI An fMRI approach that measures changes in BOLD signal evoked by a sensorimotor or cognitive task

The long-term survival of infants with complex congenital heart disease (CHD) has dramatically improved over the last two decades owing to advances in diagnosis and treatment (Gaynor et al. 2015; Wernovsky & Licht, 2016). This has shifted the focus to understanding the genesis and evolution of neurodevelopmental impairments that have remained prevalent in CHD survivors. These deficits involve a broad range of developmental domains, including motor function, speech and language, social behavior, and cognition (Donofrio & Massaro, 2010; Newburger et al., 2012). Not surprisingly, these pose lifelong medical and economic burden, among others, to children, their families, and society at large (Marelli, Miller, Marino, Jefferson, & Newburger, 2016).

Diagnosis, Management and Modeling of Neurodevelopmental Disorders. https://doi.org/10.1016/B978-0-12-817988-8.00024-5

Initial studies ascribed adverse neurologic outcomes to brain injury acquired from intraoperative procedures, such as the bypass method used and the length of hypothermic arrest (Bellinger et al., 1995; Hovels-Gurich, 2002). While these remain significant risk factors for acquired brain injury in CHD (Gaynor et al., 2015; Wernovsky & Licht, 2016), neurologic issues including microcephaly, motor asymmetries, muscle tone abnormalities, and absent suck were shown to be present in more than 50% of infants with CHD before open heart surgery (Limperopoulos et al., 1999, 2001). These impairments were predictive of neurodevelopmental outcomes and suggested that brain injury in CHD may have its origins in fetal life.

More recently, magnetic resonance image (MRI) of brain structure and function has played a critical role in advancing our understanding of brain injury that may account for the neurodevelopmental deficits seen in CHD in the months and years following surgical repair/palliation (Clouchoux et al., 2013; De Asis-Cruz et al., 2018; Miller et al., 2007; Limperopoulos et al., 2010; Panigrahy et al., 2015; Schmithorst et al., 2016). In this chapter, we describe atypical brain growth, microstructure, and metabolism in CHD during the perinatal period, focusing on how these relate to functional connectivity changes in CHD newborns.

Brain dysmaturity in CHD

Brain development is delayed in CHD (Licht et al., 2009; Peyvandi, Latal, Miller, & McQuillen, 2019). A complex interplay of genetic, hemodynamic, and perioperative factors that compromise oxygen and nutrient delivery to the brain likely account for brain dysmaturation in CHD (Morton, Ishibashi, & Jonas, 2017; Newburger et al., 2012; Sun et al., 2015; Wernovsky & Licht, 2016). Unlike the heart, which is fully formed at 7 weeks' gestation, the brain has a protracted maturational course (Srivastava, 2006). During the third trimester, the brain undergoes rapid cell division and growth making it especially vulnerable to insults (Winick & Rosso, 1969). Impaired blood flow to the developing brain, as seen in CHD, may be especially deleterious as the brain receives about 25% of the cardiac output and accounts for almost 50% oxygen consumption during this period (Peyvandi et al., 2019). Studies in sheep and humans have shown that chronic hypoxemia interferes with critical neural processes that shape brain development, including neurogenesis, interneuron migration, and maturation of neurons (Leonetti, Back, Gallo, & Ishibashi, 2019). Using a composite measure of brain maturity called the Total Maturation Score (Childs et al., 2001) that takes into account myelination, cortical folding, germinal matrix distribution, and glial cell migration, Licht et al. (Licht et al., 2009) have demonstrated delayed brain maturation in infants with d-TGA and HLHS.

Brain injury in newborns with CHD

Focal white matter injury and small focal strokes are the most common types of brain injury in newborns with CHD (Marelli et al., 2016). Conventional MRI reliably detects these, but more sophisticated MRI-based techniques have provided accruing evidence of a more distributed and diffuse type of injury that is more consistent with the neurobehavioral profile of children with CHD. Further, we outline the morphometric, microstructural, and metabolic delays and abnormalities that have been observed in CHD brains.

Morphometry. MRI-based volumetric brain measurements are acquired by using automatic or semiautomatic segmentation algorithms that define the boundaries of different tissue types (i.e., white matter, gray matter, and brainstem) in the brain (Gousias et al., 2013). Volumetric assessments in newborns with CHD requiring corrective surgery have shown global and regional reductions in cerebral and cerebellar volumes (Ortinau et al., 2012; von Rhein et al., 2015), as well as regional gyrification abnormalities (Ortinau et al., 2012, 2013). Important differences in CHD and control brain volumes in vivo have also been reported (Limperopoulos et al., 2010; Clouchoux et al., 2013). These differences become more pronounced with advancing gestational age as the total and regional brain growth trajectories in complex CHD progressively fall off compared to their healthy counterparts (Fig. 24.1). Fetuses with greater circulatory compromise, such as those with hypoplastic left heart syndrome (HLHS) with absent anterograde flow in the aortic arch (i.e., atresia), have the smallest global brain volumes (Limperopoulos et al., 2010). Along with reductions in brain volumes, fetuses with HLHS have also been shown to have delayed cortical folding in the frontal, parietal, calcarine, temporal, and collateral regions (Clouchoux et al., 2013). These findings underscore the potential role of impaired cerebral hemodynamics in abnormal brain development in CHD.

Diffusion tensor imaging. Diffusion tensor imaging (DTI) probes the brain's cellular structure by measuring the restricted diffusion of water molecules in brain tissue. Two commonly reported measures are apparent diffusion coefficient (ADC) and fractional anisotropy (FA). ADC decreases and FA increases with advancing gestational age reflecting the increasing myelination of white matter (Nomura et al., 1994; Sakuma et al., 1991), specifically oligodendrocyte maturation (Drobyshevsky et al., 2005). CHD infants, prior to corrective surgery, have higher ADC and lower FA values compared to

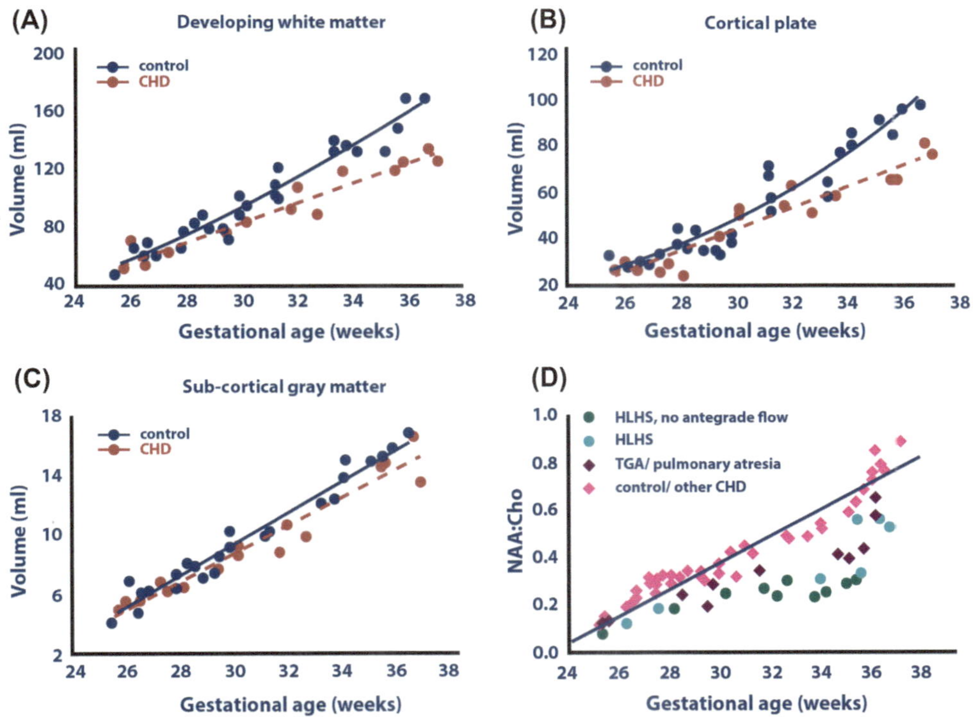

FIGURE 24.1 **Decreased regional brain volumes and altered metabolism in CHD fetuses.** Developing white matter (A), cortical plate (B), and subcortical gray matter volumes in HLHS (red) fetuses are reduced compared to controls (red). NAA:Cho ratio are also decreased in HLHS (D), with more severe CHD types having greater reductions. *Data as previously reported and figures adapted Limperopoulos, C., Tworetzky, W., Mcelhinney, D.B., Newburger, J.W., Brown, D.W., Robertson, R.L., … du Plessis, A. (2010). Brain volume and metabolism in fetuses with congenital heart disease.* Circulation, 121*(1), 26–33; Clouchoux, C., Du Plessis, A.J., Bouyssi-Kobar, M., Tworetzky, W., McElhinney, D.B., Brown, D.W., … Limperopoulos, C. (2013). Delayed cortical development in fetuses with complex congenital heart disease. Cerebral Cortex, 23(12), 2932–2943.*

controls suggesting diffuse white matter injury (Miller et al., 2007) consistent with the wide range of neurodevelopmental issues seen in CHD. A later study by Sethi et al. (Sethi et al., 2013) also showed that the degree of microstructural abnormalities correlated with severity of CHD diagnosis, such that HLHS patients with smaller caliber ascending aorta and aortic valve atresia had higher ADC and lower FA.

Spectroscopy. Magnetic resonance spectroscopy (MRS) uses the signal from hydrogen protons to determine the concentration of different metabolites such as N-acetylaspartate (NAA), choline (Cho), and lactate (Lac) in tissue. Using MRS, Miller et al. (Miller et al., 2007) reported a 10% decrease in NAA:Cho ratio and a 28% increase in Lac:Cho ratio in 41 newborns with CHD compared to controls. These changes suggest a loss of neuronal and axonal integrity and increased anaerobic metabolism. Concentrations of citrate and glutamine, metabolites involved in energy metabolism, have also been shown to be altered in term and preterm CHD newborns, changes that were related to reductions in cerebellar and thalamic volumes (Gertsvolf et al., 2018). In 55 third trimester CHD fetuses, Limperopoulos et al. (Limperopoulos et al., 2010) showed similar trends: lower NAA:Cho ratio and trending lower lactate levels in CHD. In both groups, the ratio of NAA:Cho increased with advancing GA, but trajectories in CHD lagged behind controls. This study also showed that the lowest NAA:Cho ratios were seen in fetuses with more severe CHD.

Morphometry, DTI, and MRS findings highlight the early onset of brain abnormalities in children with CHD. How these growth, microstructural, and metabolic changes relate to neuronal network dysfunction and consequent neurologic deficits in CHD in fetuses and newborns remains unclear, but recent investigations using functional MRI (fMRI) have allowed us to better understand in utero and ex utero brain function in high-risk populations such as CHD (De Asis-Cruz et al., 2018; Thomason et al., 2013).

Resting-state functional connectivity MRI

Functional MRI relies on the natural contrast afforded by different magnetic susceptibilities of oxygenated and deoxygenated blood (Ogawa, Lee, Kay, & Tank, 1990). The latter, because it is paramagnetic, causes signal loss on T2*-weighted images. When there is increased neural activity, local blood flow is also increased to meet the brain's oxygen requirements. The increase in perfusion, however, surpasses oxygen demands such that the concentration of deoxyhemoglobin relative to oxyhemoglobin decreases. This translates to an increase in the blood oxygen-level-dependent

(BOLD) signal in the areas of the brain that are active. As the BOLD signal is a vascular response, it is an indirect measure of brain activity. BOLD signals are associated with perisynaptic local field potentials in the gamma frequency range (30−100 Hz), which reflect neural excitability, as opposed to the spiking rate of individual neurons (Logothetis, 2003).

Functional MRI studies are either task-based or resting-state experiments. In task-based studies, subjects perform motor, sensory, or cognitive tasks that are designed to evoke BOLD responses from the brain. Resting-state fMRI (rs-fMRI), on the other hand, studies intrinsic fluctuations of the BOLD signal. First described by Biswal (Biswal, Zerrin Yetkin, Haughton, & Hyde, 1995) in the motor cortex, rs-fMRI detects temporal correlations between spontaneous, low-frequency (<0.1 Hz) oscillations of the BOLD signal arising from different areas of the brain. Since the motor network was described, multiple networks have now been identified including the visual, auditory, salience, attention, and default mode networks. RSN patterns closely correspond to anatomical organization (Skudlarski et al., 2008) and to task-elicited network activations (Damoiseaux et al. 2006), show reproducibility within and between subjects over time (Damoiseaux et al,. 2006), are present in different mammalian species (Chuang & Nasrallah, 2017; Hutchison et al., 2011), and persist during sleep (Larson-Prior et al., 2009).

In patient populations, especially in pediatric cohorts, rs-fRMI has several advantages over task-based fMRI. While both are noninvasive and confer very good spatial specificity, rs-fMRI imposes no cognitive demands on subjects and thus can be conveniently used even for fetal imaging. In addition, rs-fMRI queries multiple brain systems simultaneously. Lastly, acquisition times for rs-fMRI tend to have shorter duration.

Resting-state networks in newborns

The first whole-brain analyses of resting-state networks in preterm infants at term age revealed five networks (Fransson et al., 2007). These included the primary visual cortex, bilateral sensorimotor cortex, bilateral auditory cortex, a network that included the lateral parietal cortex, precuneus, and cerebellum, and the mediolateral prefrontal cortex. Over the years, an increasing number of studies have applied rs-fMRI to preterm and healthy newborns and have helped elucidate developmental trajectories of RSNs. Studies have shown that most networks, including the somatomotor, visual, and auditory networks, resemble their adult configuration at term age and the foundations of some emerge as early as 26 weeks' PMA as shown in preterm infants (Fransson et al., 2007; Smyser et al., 2010; Doria et al., 2010; Gao et al. 2014). Precursors of higher-order cognitive networks, such as the default mode network, have also been described in the newborn period (Doria et al. 2010; Fransson et al. 2007). Studies have also shown that maturation of RSNs, in terms of both connectivity strength and patterns, proceeds in a hierarchical manner that reflects cortical maturation sequence and neurodevelopmental milestones (Doria et al., 2010; Fransson et al., 2007; Fransson, Åden, Blennow, & Lagercrantz, 2011; Gao et al., 2009; Smyser et al., 2010). Sensorimotor system develops first, followed by temporal and parietal cortices and then systems that involve higher association cortices (Gao et al., 2011; Gao et al., 2014; Lin et al., 2008). The pattern of development also follows a posterior to anterior, inferior to superior, and medial to lateral pattern (Doria et al. 2010; Smyser et al., 2010).

Functional connectome of the newborn brain

Graph theory provides a robust, quantitative framework for understanding the relationship between different areas of the brain (Sporns, 2011). In rs-fMRI, brain regions are modeled as nodes connected by links or edges that represent the statistical correlation between BOLD signals arising from the connected nodes (Fig. 24.2). Recent studies have shown that consistent with the structural connectome described using DTI in healthy infants (Ball et al., 2013) and the adult functional architecture (Bullmore & Sporns, 2012), healthy newborn networks exhibited small-world topology, modular organization, and a right-tailed distribution with a few functional "hubs" occupying a critical role in the network (Table 24.1) (De Asis-Cruz et al., 2015; Gao et al., 2011; Fransson, Åden, Blennow, & Lagercrantz, 2011). While network organization in infants resembled adults, several differences were also noted. In newborns, modules tended to be anatomically contiguous (Fig. 24.3); adjacent regions likely connected with each other compared to distant nodes. Adult modules, in contrast, may span regions that are spatially disconnected. In addition, neonatal hubs were mostly localized in areas that subserve primary functions such as the sensorimotor and visual cortices, although there are a few in association areas such as the insula and posterior cingulate cortex similar to adults (De Asis-Cruz et al., 2015; Gao et al., 2011). A study in a small cohort of 17 preterm infants at term age also suggests that there is significant overlap between structural and functional network topology (van den Heuvel et al., 2014).

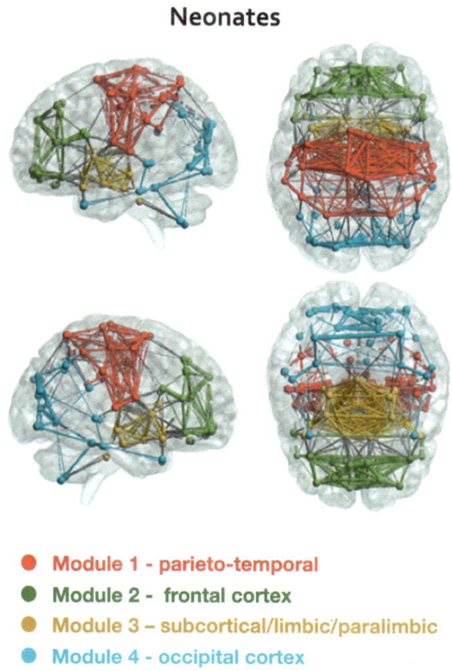

FIGURE 24.2 **Modular organization of the newborn brain.** Newborn brains cluster into four functional modules; spheres are nodes, lines are edges; color represents module membership (Bouyssi-Kobar, De Asis-Cruz, Murnick, Chang, & Limperopoulos, 2019).

TABLE 24.1 Hubs in the newborn brain.

Degree	Betweenness
Insula, L	Olfactory, L
Insula, R	Parahippocampal gyrus, L
Postcentral gyrus, L	Fusiform gyrus, L
Postcentral gyrus, R	Insula, L
Hippocampus, L	Rolandic operculum, L
Rolandic operculum, R	Inferior frontal gyrus (opercular), L
Hippocampus, R	Precuneus, L
Amygdala, L	Postcentral gyrus, R
Amygdala, R	Middle cingulate gyrus, R
Putamen, L	Supplementary motor area, R
Rolandic operculum, L	Superior parietal gyrus, R
Pallidum, L	Anterior cingulate gyrus, R
Paracentral lobule, L	Orbitofrontal cortex (inferior), R
Paracentral lobule, R	Olfactory, R
Putamen, R	
Pallidum, R	
Thalamus, R	

Hubs defined using degree and betweenness measures (De Asis-Cruz et al. 2015; Fransson, Åden, Blennow, & Lagercrantz, 2011; Gao et al. 2011).

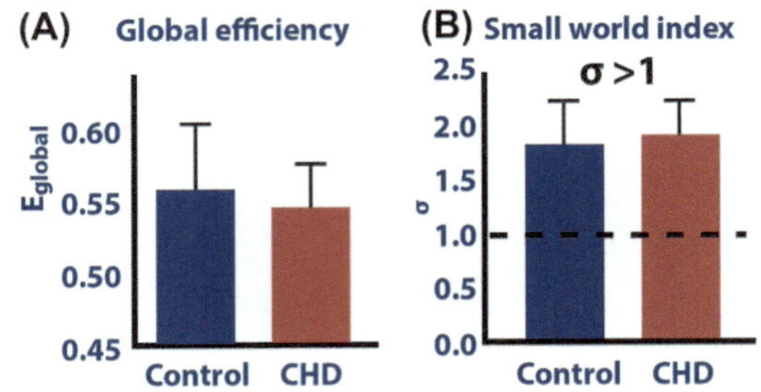

FIGURE 24.3 **Global network topology preserved in CHD neonates.** Global efficiency of CHD (red) and healthy (red) newborn RNSs are comparable (A); both display small-world topology with small-world index > 1 (B). *Figure adapted from De Asis-Cruz, J., Donofrio, M. T., Vezina, G., & Limperopoulos, C. (2018). Aberrant brain functional connectivity in newborns with congenital heart disease before cardiac surgery.* NeuroImage: Clinical, 17, *31–42.*

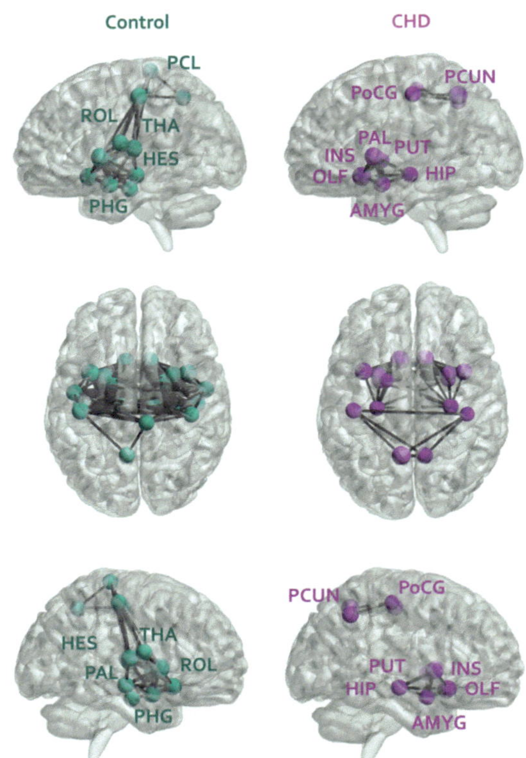

FIGURE 24.4 **Rich-club organization in CHD newborns.** Fewer connections exist between rich-club hubs in CHD (right panel); note density of edges in the control brain (left) versus CHD (right). *All data and analyses as previously described here De Asis-Cruz, J., Donofrio, M. T., Vezina, G., & Limperopoulos, C. (2018). Aberrant brain functional connectivity in newborns with congenital heart disease before cardiac surgery.* NeuroImage: Clinical, 17, *31–42.*

Functional connectome in CHD

Disruption of brain networks has been associated with a wide array of neurodevelopmental disabilities including autism spectrum disorders, ADHD, and schizophrenia (Itahashi et al., 2014; Moseley et al., 2015). To date, however, few studies have examined functional connectivity in newborns with CHD (Birca et al., 2016; De Asis-Cruz et al., 2018), and only one network-based study examining resting-state connectivity (De Asis-Cruz et al., 2018). De Asis-Cruz et al. (2018) compared network organization in 82 healthy neonates and 30 newborns with complex CHD before surgery. Similar to healthy controls, CHD infants showed intact global network topology (i.e., preserved small-world topology) (Fig. 24.4). Possibly, the small-world architecture of the newborn CHD brain confers resilience to insults while in utero. This topology, as shown in other studies in newborns and adults, is more robust against simulated targeted insults to the network (Achard, Salvador, Whitcher, Suckling, & Bullmore, 2006; De Asis-Cruz et al., 2018; Gao et al., 2011). Interestingly, structural connectomes in adolescents who underwent corrective surgery for complex CHD during infancy showed increased "small-worldness"

TABLE 24.2 Edges with reduced connectivity in CHD.

	Node 1	Node 2
1	Insula, L	Parahippocampal gyrus, L
2	Insula, L	Amygdala, L
3	Middle cingulate gyrus, R	Pallidum, L
4	Postcentral gyrus, R	Pallidum, L
5	Precentral gyrus, R	Pallidum, L
6	Postcentral gyrus, R	Pallidum, R
7	Rolandic operculum, R	Brainstem
8	Supplementary motor area, L	Brainstem
9	Inferior parietal lobule, R	Caudate, L
10	Superior frontal gyrus (medial), L	Parahippocampal gyrus, L

Top 10 connections with reduced connectivity strength in CHD newborns compared to controls (De Asis-Cruz et al. 2018).

and reduced global efficiency that correlated with patients' cognitive outcomes (Panigrahy et al., 2015). Specifically, these abnormalities were associated with a higher risk of attention deficit disorder and cognitive dysfunction (Panigrahy et al., 2015; Schmithorst et al., 2016). Together, we hypothesize that these findings suggest a disease continuum, where efficient global processing of information is initially preserved during early childhood, and the findings in adolescence reflect cumulative effects of altered cerebral perfusion and possibly neurologic sequelae associated with surgery.

Regional differences between CHD and controls were also described (De Asis-Cruz et al., 2018). Hubs of complex networks are not only highly connected to other nodes but to each other. This property, called rich-club organization, was preserved in CHD. Rich-club nodes in healthy infants and CHD included bilateral insulae, amygdalae, hippocampus, and bilateral putamen, among others. While there is overlap in these nodes between the two groups, there are more rich-club nodes in controls compared to CHD. In addition, there are denser connections between high-degree nodes in controls (Fig. 24.4). Similar to rich-club nodes, centrality-based hubs, here based on a composite measure of betweenness, degree, and closeness centrality, tend to be shared between the cases and controls, suggesting that while there are regional network differences, critical nodes in healthy infants are also the important nodes in CHD. Using network-based statistics (Zalesky, Fornito, & Bullmore, 2010), this study also showed that connections that were weaker in CHD were those that connected the subcortical regions and the brainstem. The greatest connectivity differences between the two groups were in the edges between the globus pallidus, brainstem, and caudate (Table 24.2). Brain regions with reduced connectivity also were proximal to regions that had delayed sulcation (such as the cingulate sulcus, frontal sulcus, and anterior ascending ramus—sulci that are boundaries of regions that have altered connectivity—such as the middle cingulate, left mid-to-superior frontal gyri, and right rolandic operculum) and gyrification. Most of the affected connections are interhemispheric and most involved rich-club nodes.

Similar to morphological changes in CHD that persist across the life span (von Rhein et al., 2014), differences in functional connectivity likely persist into adulthood (King et al., 2017). King et al. studied 17 adolescents and emerging adult survivors of CHD using a task-based *n*-back task that tests either vigilance or working memory depending on task difficulty. They showed that frontoparietal regions activated similarly in healthy and CHD groups, but the right inferior gyrus and left precuneus, areas that are important in executive functioning, have lesser task-induced deactivation in CHD. The degree to which BOLD activity in these regions was deactivated correlated with CHD subjects' working memory. This highlights the long-term repercussions of early brain injury on the developing brain.

Application to other areas of development

In this chapter, we reviewed alterations in brain growth, metabolism, and microstructure in newborns with CHD and related them to functional connectivity changes. Notably, brain abnormalities were present before term-equivalent age and in utero. Fetal resting-state connectivity in fetuses with congenital heart disease has not been previously described. Recently, rs-fMRI has been successfully used to probe brain function in utero (Schöpf, Kasprian, Brugger, & Prayer, 2012; Thomason et al., 2013; Thomason et al., 2017). These reports describe robust bilateral connectivity between some homologous brain structures in fetuses. In Thomason et al.'s study, they showed significant connection strengths in parts of the cingulate, the dorsal frontal cortex, and the somatosensory association cortex, among others. Consistent with data in preterm infants (Doria et al., 2010; Smyser et al., 2010), majority of the strongly connected regions were medially and posteriorly located and followed a mediolateral connectivity gradient. In about 80% of areas, connectivity strength between homologous counterparts correlated with gestational age, mirroring trends observed in preterm infants (Smyser et al., 2010). Interestingly, abnormalities in connectivity during the fetal period seem to be a barometer of the state of the fetus and the pregnancy. Fetuses who were eventually born before term had diminished connectivity in areas of the left hemisphere that were involved in language processing; the degree to which connectivity at this region was different from controls was linked to how premature the fetus was eventually born (Thomason et al., 2017). These studies provide evidence that rs-fMRI could be an effective way of determining neural connectivity abnormalities in fetuses with complex CHD.

Key facts about resting-state functional MRI (rs-fMRI)

- Resting-state fMRI in increasingly being used for newborn imaging because the technique allows us to noninvasively, rapidly, reliably, and simultaneously assess multiple brain systems.
- Development of resting-state networks progresses in a hierarchical fashion.
- Resting-state networks maturation trajectory corresponds to cortical maturation sequence and reflects developmental milestones.
- In newborns, functional connectomes derived using rs-fMRI correlate with structural connectomes defined using DTI.
- Disruption in resting-state networks has been reported in several neurodevelopmental disorders including autism spectrum disorders. In the newborn period, atypical network topology has been reported in prematurity and CHD.

Summary points

- This chapter relates resting-state functional connectivity to other MRI-based evidence of brain dysmaturity in newborns with complex CHD.
- Despite recent advances in surgical techniques that have greatly reduced mortality in newborns with CHD, the incidence of long-term neurologic disabilities in survivors has remained unchanged.
- Brain volume trajectories in CHD begin to progressively fall off compared to healthy fetuses during the third trimester of gestation.
- Fetuses and newborns with CHD have cortical folding and sulcation delays.
- Newborns and third trimester fetuses with CHD have increased WM average diffusivity and reduced fractional anisotropy compared to controls.
- In CHD newborns and third trimester fetuses, ratio of NAA to Cho is decreased and that of lactate to Cho is increased suggesting loss of neuronal integrity and increase in anaerobic metabolism, respectively.
- Functional connectome of CHD newborns before surgery reveals small-world architecture similar to controls.
- Rich-club organization in CHD infants is preserved; however, the density of connections among rich-club hubs and the number of hubs are reduced in CHD.
- Strength of connectivity for some regional connections, most of which are interhemispheric and connected hubs in the subcortical regions and brainstem, was reduced in CHD.

References

Achard, S., Salvador, R., Whitcher, B., Suckling, J., & Bullmore, E. D. (2006). A resilient, low-frequency, small-world human brain functional network with highly connected association cortical hubs. *Journal of Neuroscience, 26*(1), 63–72.

Ball, G., Srinivasan, L., Aljabar, P., Counsell, S. J., Durighel, G., Hajnal, J. V., ... Edwards, A. D. (2013). Development of cortical microstructure in the preterm human brain. *Proceedings of the National Academy of Sciences, 110*(23), 9541–9546.

Bellinger, D. C., Jonas, R. A., Rappaport, L. A., Wypij, D., Wernovsky, G., Kuban, K. C., ... Walsh, A. Z. (1995). Developmental and neurologic status of children after heart surgery with hypothermic circulatory arrest or low-flow cardiopulmonary bypass. *New England Journal of Medicine, 332*(9), 549–555.

Birca, A., Vakorin, V. A., Porayette, P., Madathil, S., Chau, V., Seed, M., ... Hahn, C. D. (2016). Interplay of brain structure and function in neonatal congenital heart disease. *Annals of Clinical and Translational Neurology, 3*(9), 708–722.

Biswal, B., Zerrin Yetkin, F., Haughton, V. M., & Hyde, J. S. (1995). Functional connectivity in the motor cortex of resting human brain using echo-planar MRI. *Magnetic Resonance in Medicine, 34*(4), 537–541.

Bouyssi-Kobar, M., De Asis-Cruz, J., Murnick, J., Chang, T., & Limperopoulos, C. (2019). Altered functional brain network integration, segregation, and modularity in infants born very preterm at term-equivalent age. *The Journal of Pediatrics, 213*, 13–21.

Bullmore, E., & Sporns, O. (2012). The economy of brain network organization. *Nature Reviews Neuroscience, 13*(5), 336.

Childs, A. M., Ramenghi, L. A., Cornette, L., Tanner, S. F., Arthur, R. J., Martinez, D., & Levene, M. I. (2001). Cerebral maturation in premature infants: Quantitative assessment using MR imaging. *American Journal of Neuroradiology, 22*(8), 1577–1582.

Chuang, K. H., & Nasrallah, F. A. (2017). Functional networks and network perturbations in rodents. *NeuroImage, 163*, 419–436.

Clouchoux, C., Du Plessis, A. J., Bouyssi-Kobar, M., Tworetzky, W., McElhinney, D. B., Brown, D. W., … Limperopoulos, C. (2013). Delayed cortical development in fetuses with complex congenital heart disease. *Cerebral Cortex, 23*(12), 2932–2943.

Damoiseaux, J. S., Rombouts, S. A. R. B., Barkhof, F., Scheltens, P., Stam, C. J., Smith, S. M., & Beckmann, C. F. (2006). Consistent resting-state networks across healthy subjects. *Proceedings of the National Academy of Sciences, 103*(37), 13848–13853.

De Asis-Cruz, J., Bouyssi-Kobar, M., Evangelou, I., Vezina, G., & Limperopoulos, C. (2015). Functional properties of resting state networks in healthy full-term newborns. *Scientific Reports, 5*, 17755.

De Asis-Cruz, J., Donofrio, M. T., Vezina, G., & Limperopoulos, C. (2018). Aberrant brain functional connectivity in newborns with congenital heart disease before cardiac surgery. *NeuroImage: Clinical, 17*, 31–42.

Donofrio, M. T., & Massaro, A. N. (2010). Impact of congenital heart disease on brain development and neurodevelopmental outcome. *International Journal of Pediatrics, 2010*.

Doria, V., Beckmann, C. F., Arichi, T., Merchant, N., Groppo, M., Turkheimer, F. E., … Larkman, D. J. (2010). Emergence of resting state networks in the preterm human brain. *Proceedings of the National Academy of Sciences, 107*(46), 20015–20020.

Drobyshevsky, A., Song, S. K., Gamkrelidze, G., Wyrwicz, A. M., Derrick, M., Meng, F., … Luo, N. L. (2005). Developmental changes in diffusion anisotropy coincide with immature oligodendrocyte progression and maturation of compound action potential. *Journal of Neuroscience, 25*(25), 5988–5997.

Fransson, P., Åden, U., Blennow, M., & Lagercrantz, H. (2011). The functional architecture of the infant brain as revealed by resting-state fMRI. *Cerebral Cortex, 21*(1), 145–154.

Fransson, P., Skiöld, B., Horsch, S., Nordell, A., Blennow, M., Lagercrantz, H., & Åden, U. (2007). Resting-state networks in the infant brain. *Proceedings of the National Academy of Sciences, 104*(39), 15531–15536.

Gao, W., Alcauter, S., Elton, A., Hernandez-Castillo, C. R., Smith, J. K., Ramirez, J., & Lin, W. (2014). Functional network development during the first year: Relative sequence and socioeconomic correlations. *Cerebral Cortex, 25*(9), 2919–2928.

Gao, W., Gilmore, J. H., Giovanello, K. S., Smith, J. K., Shen, D., Zhu, H., & Lin, W. (2011). Temporal and spatial evolution of brain network topology during the first two years of life. *PloS One, 6*(9), e25278.

Gao, W., Zhou, H., Giovanello, K., Smith, J. K., Shen, D., Gilmore, J., & Lin, W. (2014). Brain functional networks in the developing brain using resting BOLD. In *Advanced brain neuroimaging topics in health and disease-methods and applications*. IntechOpen.

Gao, W., Zhu, H., Giovanello, K. S., Smith, J. K., Shen, D., Gilmore, J. H., & Lin, W. (2009). Evidence on the emergence of the brain's default network from 2-week-old to 2-year-old healthy pediatric subjects. *Proceedings of the National Academy of Sciences, 106*(16), 6790–6795.

Gaynor, J. W., Stopp, C., Wypij, D., Andropoulos, D. B., Atallah, J., Atz, A. M., … Goldberg, C. S. (2015). Neurodevelopmental outcomes after cardiac surgery in infancy. *Pediatrics, 135*(5), 816–825.

Gertsvolf, N., Votava-Smith, J. K., Ceschin, R., del Castillo, S., Lee, V., Lai, H. A., … Panigrahy, A. (2018). Association between subcortical morphology and cerebral white matter energy metabolism in neonates with congenital heart disease. *Scientific Reports, 8*.

Gousias, I. S., Hammers, A., Counsell, S. J., Srinivasan, L., Rutherford, M. A., Heckemann, R. A., … Edwards, A. D. (2013). Magnetic resonance imaging of the newborn brain: Automatic segmentation of brain images into 50 anatomical regions. *PloS One, 8*(4), e59990.

van den Heuvel, M. P., Kersbergen, K. J., de Reus, M. A., Keunen, K., Kahn, R. S., Groenendaal, F., … Benders, M. J. (2014). The neonatal connectome during preterm brain development. *Cerebral Cortex, 25*(9), 3000–3013.

Hövels-Gürich, H. H., Konrad, K., Wiesner, M., Minkenberg, R., Herpertz-Dahlmann, B., Messmer, B. J., & Von Bernuth, G. (2002). Long term behavioural outcome after neonatal arterial switch operation for transposition of the great arteries. *Archives of Disease in Childhood, 87*(6), 506–510.

Hutchison, R. M., Leung, L. S., Mirsattari, S. M., Gati, J. S., Menon, R. S., & Everling, S. (2011). Resting-state networks in the macaque at 7 T. *NeuroImage, 56*(3), 1546–1555.

Itahashi, T., Yamada, T., Watanabe, H., Nakamura, M., Jimbo, D., Shioda, S.,., & Hashimoto, R. (2014). Altered network topologies and hub organization in adults with autism: A resting-state fMRI study. *PloS One, 9*(4), e94115.

King, T. Z., Smith, K. M., Burns, T. G., Sun, B., Shin, J., Jones, R. A., … Mahle, W. T. (2017). fMRI investigation of working memory in adolescents with surgically treated congenital heart disease. *Applied Neuropsychology: Child, 6*(1), 7–21.

Larson-Prior, L. J., Zempel, J. M., Nolan, T. S., Prior, F. W., Snyder, A. Z., & Raichle, M. E. (2009). Cortical network functional connectivity in the descent to sleep. *Proceedings of the National Academy of Sciences, 106*(11), 4489–4494.

Leonetti, C., Back, S. A., Gallo, V., & Ishibashi, N. (2019). Cortical dysmaturation in congenital heart disease. *Trends in Neurosciences, 42*(3), 192–204.

Licht, D. J., Shera, D. M., Clancy, R. R., Wernovsky, G., Montenegro, L. M., Nicolson, S. C., … Vossough, A. (2009). Brain maturation is delayed in infants with complex congenital heart defects. *The Journal of Thoracic and Cardiovascular Surgery, 137*(3), 529–537.

Limperopoulos, C., Majnemer, A., Rosenblatt, B., Shevell, M. I., Rohlicek, C., Tchervenkov, C., & Gottesman, R. (2001). Association between electroencephalographic findings and neurologic status in infants with congenital heart defects. *Journal of Child Neurology, 16*(7), 471–476.

Limperopoulos, C., Majnemer, A., Shevell, M. I., Rosenblatt, B., Rohlicek, C., & Tchervenkov, C. (1999). Neurologic status of newborns with congenital heart defects before open heart surgery. *Pediatrics, 103*(2), 402–408.

Limperopoulos, C., Tworetzky, W., Mcelhinney, D. B., Newburger, J. W., Brown, D. W., Robertson, R. L., ... du Plessis, A. (2010). Brain volume and metabolism in fetuses with congenital heart disease. *Circulation, 121*(1), 26–33.

Lin, W., Zhu, Q., Gao, W., Chen, Y., Toh, C. H., Styner, M.,., & Gilmore, J. H. (2008). Functional connectivity MR imaging reveals cortical functional connectivity in the developing brain. *American Journal of Neuroradiology, 29*(10), 1883–1889.

Logothetis, N. K. (2003). The underpinnings of the BOLD functional magnetic resonance imaging signal. *Journal of Neuroscience, 23*(10), 3963–3971.

Marelli, A., Miller, S. P., Marino, B. S., Jefferson, A. L., & Newburger, J. W. (2016). Brain in congenital heart disease across the lifespan: The cumulative burden of injury. *Circulation, 133*(20), 1951–1962.

Miller, S. P., McQuillen, P. S., Hamrick, S., Xu, D., Glidden, D. V., Charlton, N., ... Vigneron, D. B. (2007). Abnormal brain development in newborns with congenital heart disease. *New England Journal of Medicine, 357*(19), 1928–1938.

Morton, P. D., Ishibashi, N., & Jonas, R. A. (2017). Neurodevelopmental abnormalities and congenital heart disease: Insights into altered brain maturation. *Circulation Research, 120*(6), 960–977.

Moseley, R. L., Ypma, R. J. F., Holt, R. J., Floris, D., Chura, L. R., Spencer, M. D., ... Rubinov, M. (2015). Whole-brain functional hypoconnectivity as an endophenotype of autism in adolescents. *NeuroImage: Clinical, 9*, 140–152.

Newburger, J. W., Sleeper, L. A., Bellinger, D. C., Goldberg, C. S., Tabbutt, S., Lu, M., ... Pike, N. (2012). Early developmental outcome in children with hypoplastic left heart syndrome and related anomalies: The single ventricle reconstruction trial. *Circulation, 125*(17), 2081–2091.

Nomura, Y., Sakuma, H., Takeda, K., Tagami, T., Okuda, Y., & Nakagawa, T. (1994). Diffusional anisotropy of the human brain assessed with diffusion-weighted MR: Relation with normal brain development and aging. *American Journal of Neuroradiology, 15*(2), 231–238.

Ogawa, S., Lee, T. M., Kay, A. R., & Tank, D. W. (1990). Brain magnetic resonance imaging with contrast dependent on blood oxygenation. *Proceedings of the National Academy of Sciences, 87*(24), 9868–9872.

Ortinau, C., Alexopoulos, D., Dierker, D., Van Essen, D., Beca, J., & Inder, T. (2013). Cortical folding is altered before surgery in infants with congenital heart disease. *The Journal of Pediatrics, 163*(5), 1507–1510.

Ortinau, C., Inder, T., Lambeth, J., Wallendorf, M., Finucane, K., & Beca, J. (2012). Congenital heart disease affects cerebral size but not brain growth. *Pediatric Cardiology, 33*(7), 1138–1146.

Panigrahy, A., Schmithorst, V. J., Wisnowski, J. L., Watson, C. G., Bellinger, D. C., Newburger, J. W., & Rivkin, M. J. (2015). Relationship of white matter network topology and cognitive outcome in adolescents with d-transposition of the great arteries. *NeuroImage: Clinical, 7*, 438–448.

Peyvandi, S., Latal, B., Miller, S. P., & McQuillen, P. S. (2019). The neonatal brain in critical congenital heart disease: Insights and future directions. *NeuroImage, 185*, 776–782.

von Rhein, M., Buchmann, A., Hagmann, C., Dave, H., Bernet, V., Scheer, I., ... Kretschmar, O. (2015). Severe congenital heart defects are associated with global reduction of neonatal brain volumes. *The Journal of Pediatrics, 167*(6), 1259–1263.

von Rhein, M., Buchmann, A., Hagmann, C., Huber, R., Klaver, P., Knirsch, W., & Latal, B. (2014). Brain volumes predict neurodevelopment in adolescents after surgery for congenital heart disease. *Brain, 137*(1), 268–276.

Sakuma, H., Nomura, Y., Takeda, K., Tagami, T., Nakagawa, T., Tamagawa, Y., ... Tsukamoto, T. (1991). Adult and neonatal human brain: Diffusional anisotropy and myelination with diffusion-weighted MR imaging. *Radiology, 180*(1), 229–233.

Schmithorst, V. J., Panigrahy, A., Gaynor, J. W., Watson, C. G., Lee, V., Bellinger, D. C., ... Newburger, J. W. (2016). Organizational topology of brain and its relationship to ADHD in adolescents with d-transposition of the great arteries. *Brain and Behavior, 6*(8), e00504.

Schöpf, V., Kasprian, G., Brugger, P. C., & Prayer, D. (2012). Watching the fetal brain at 'rest'. *International Journal of Developmental Neuroscience, 30*(1), 11–17.

Sethi, V., Tabbutt, S., Dimitropoulos, A., Harris, K. C., Chau, V., Poskitt, K., ... Miller, S. P. (2013). Single-ventricle anatomy predicts delayed microstructural brain development. *Pediatric Research, 73*(5), 661.

Skudlarski, P., Jagannathan, K., Calhoun, V. D., Hampson, M., Skudlarska, B. A., & Pearlson, G. (2008). Measuring brain connectivity: Diffusion tensor imaging validates resting state temporal correlations. *NeuroImage, 43*(3), 554–561.

Smyser, C. D., Inder, T. E., Shimony, J. S., Hill, J. E., Degnan, A. J., Snyder, A. Z., & Neil, J. J. (2010). Longitudinal analysis of neural network development in preterm infants. *Cerebral Cortex, 20*(12), 2852–2862.

Sporns, O. (2011). The human connectome: A complex network. *Annals of the New York Academy of Sciences, 1224*(1), 109–125.

Srivastava, D. (2006). Making or breaking the heart: From lineage determination to morphogenesis. *Cell, 126*(6), 1037–1048.

Sun, L., Macgowan, C. K., Sled, J. G., Yoo, S. J., Manlhiot, C., Porayette, P., ... Hickey, E. (2015). Reduced fetal cerebral oxygen consumption is associated with smaller brain size in fetuses with congenital heart disease. *Circulation, 131*(15), 1313–1323.

Thomason, M. E., Dassanayake, M. T., Shen, S., Katkuri, Y., Alexis, M., Anderson, A. L., ... Studholme, C. (2013). Cross-hemispheric functional connectivity in the human fetal brain. *Science Translational Medicine, 5*(173), 173ra24.

Thomason, M. E., Scheinost, D., Manning, J. H., Grove, L. E., Hect, J., Marshall, N., ... Hassan, S. S. (2017). Weak functional connectivity in the human fetal brain prior to preterm birth. *Scientific Reports, 7*, 39286.

Wernovsky, G., & Licht, D. J. (2016). Neurodevelopmental outcomes in children with congenital heart disease—what can we impact? *Pediatric Critical Care Medicine: A Journal of the Society of Critical Care Medicine and the World Federation of Pediatric Intensive and Critical Care Societies, 17*(8 Suppl. 1), S232.

Winick, M., & Rosso, P. (1969). Head circumference and cellular growth of the brain in normal and marasmic children. *The Journal of Pediatrics, 74*(5), 774–778.

Zalesky, A., Fornito, A., & Bullmore, E. T. (2010). Network-based statistic: Identifying differences in brain networks. *NeuroImage, 53*(4), 1197–1207.

Chapter 25

Brain growth in congenital heart disease from prenatal environment to adulthood

Cynthia Ortinau
Department of Pediatrics, Washington University in St. Louis School of Medicine, St. Louis, MO, United States

Abbreviations

CHD Congenital heart disease
DHCA Deep hypothermic circulatory arrest
MRI Magnetic resonance imaging
Pre-OL Premyelinating oligodendrocytes
SVP Single ventricle physiology
TBV Total brain volume
TGA Transposition of the great arteries

Mini-dictionary of terms

Congenital heart disease An anatomical defect of the heart or great vessels.
Deep hypothermic circulatory arrest Surgical approach that decreases the body temperature and stops circulation to provide optimal operating conditions while maintaining cerebral protection.
Single ventricle physiology A form of congenital heart disease that results in one functioning ventricle of the heart.

Introduction

Congenital heart disease (CHD) is a structural abnormality of the heart, or its related vasculature, that occurs in nearly 1% of the population (van der Linde et al., 2011). CHD encompasses a variety of diagnoses that result in minimal to severe effects on cardiac functioning and hemodynamics. Moderate−severe forms generally require cardiac surgery or catheter-based procedures in the first year of life to correct or palliate the anatomical defect. Advances in surgical and cardiac intensive care over the past several decades have improved survival for moderate−severe CHD, resulting in a growing population of children and adults with this diagnosis (Marelli et al., 2014). This has contributed to increasing recognition that lifelong neurodevelopmental deficits are common in survivors (Marino et al., 2012; Warnes et al., 2008). Brain magnetic resonance imaging (MRI) studies have identified that abnormal brain development, including impaired brain growth, is an important predictor of neurodevelopmental deficits in this population (Beca et al., 2013; Rollins et al., 2017; von Rhein et al., 2014). This chapter will describe the timing, pattern, and trajectory of impaired brain growth, the potential mechanisms impeding brain growth, and the correlation of impaired brain growth with neurodevelopmental impairments in CHD.

Timing, pattern, and trajectory of impaired brain growth

Abnormal brain growth begins prenatally

Fetal biometry is an ultrasound technique applied in routine obstetrical care to measure fetal growth. Studies of clinically acquired biometric measures demonstrate reduced head circumference and biparietal diameter in CHD fetuses by 20−22 weeks' gestation (Fig. 25.1) (Lauridsen et al., 2019; Masoller et al., 2014). Smaller head circumference is even more pronounced at birth

Diagnosis, Management and Modeling of Neurodevelopmental Disorders. https://doi.org/10.1016/B978-0-12-817988-8.00025-7

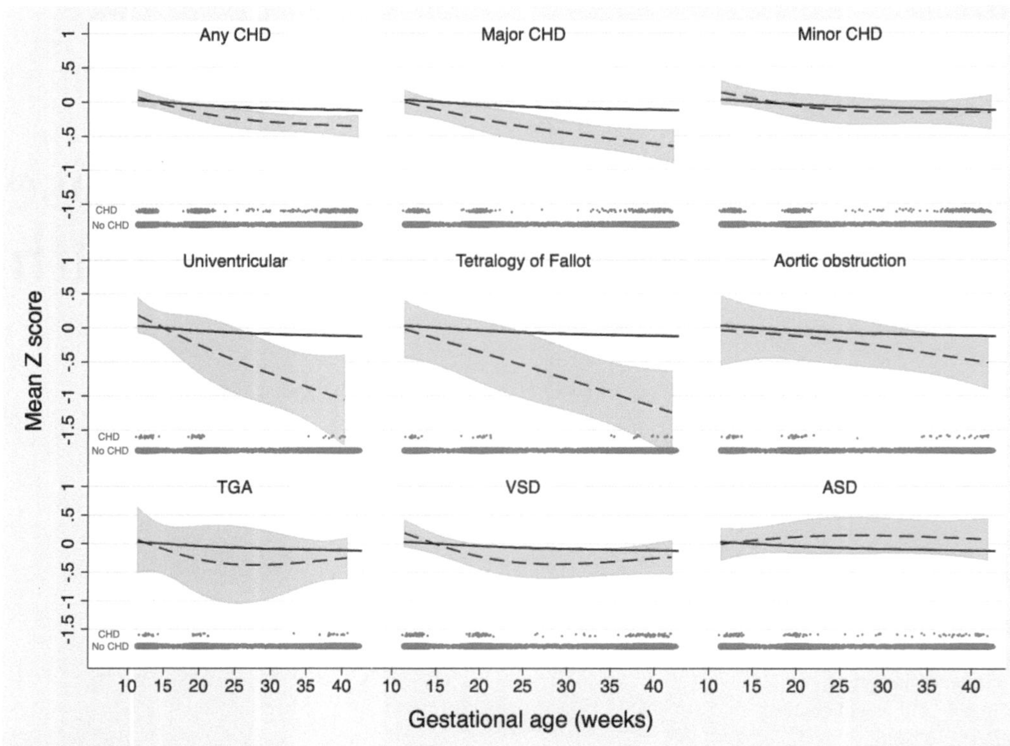

FIGURE 25.1 Biometric measure of brain size. Mean z-scores of biparietal diameter and head circumference (y-axis) plotted by gestational age (x-axis) in fetuses with CHD (*dashed line*) and fetuses without CHD (*straight line*). *Gray areas,* confidence intervals; *Gray dots,* gestational ages with available measurements; *TGA,* transposition of the great arteries; *VSD,* ventricular septal defect; *ASD,* atrial septal defect. Reductions in head size are evident for any CHD and major CHD by 20 weeks' gestation and become more pronounced by birth. *Reprinted with permission from Lauridsen et al., (2010). Fetal heart defects and measures of cerebral size.* The Journal of Pediatrics, 210, *146–153.*

(Lauridsen et al., 2019). Importantly, reductions in head circumference persist after adjusting for placental weight (Matthiesen et al., 2016). Advances in fetal brain MRI techniques now permit greater delineation of abnormal fetal brain growth, beyond that of fetal biometry. Reduced total brain volume (TBV) and higher cerebrospinal fluid volume are evident on MRI even when ultrasound biometrics are normal (Griffiths et al., 2019). Decreased TBV is present in CHD fetuses during the second and third trimesters of pregnancy (Limperopoulos et al., 2010; Schellen et al., 2015; Sun et al., 2015). This finding progresses along a slower rate of growth during the third trimester (Limperopoulos et al., 2010).

Abnormal development of the white matter, cortex, subcortical gray matter, and cerebellum have been identified on fetal MRI and contribute to overall impaired brain growth for CHD fetuses (Clouchoux et al., 2013; Rajagopalan et al., 2018; Schellen et al., 2015). MRI studies performed after birth, but before cardiac surgery, reflect the impact of CHD over the entire pregnancy. These data show reduced brain volume across all lobes of the brain (von Rhein et al., 2015). Although volumetric analyses have not identified region-specific vulnerabilities (von Rhein et al., 2015), two-dimensional measures of brain size show the frontal lobe and brainstem to be most affected in neonates (Ortinau, Beca, et al., 2012). Two small cohorts have identified that the cerebral cortex, subcortical gray matter, and cerebellum have the greatest contributions to a slower rate of brain growth from the prenatal period through early infancy (Ortinau, Mangin-Heimos et al., 2018; Rajagopalan et al., 2018).

In addition to volumetric deficits, MRI cortical morphometric analyses show delayed gyrification and abnormal sulcation in second and third trimester CHD fetuses (Clouchoux et al., 2013; Ortinau, Rollins, et al., 2018). These alterations are present even before cortical volume becomes reduced (Clouchoux et al., 2013). In the first week after birth, impaired gyrification occurs alongside reduced cortical surface area and altered sulcal depth patterns (Fig. 25.2) (Ortinau et al., 2013). A lower gyrification index also correlates with reduced cortical orientation dispersion index, a microstructural measure reflective of dendritic arborization (Kelly et al., 2019).

Abnormal microstructural development has also been reported in the developing white matter and thalamus of CHD fetuses (Berman et al., 2011; Khan, 2018). By the time CHD infants are born, they display a 12% lower white matter fractional anisotropy (Miller et al., 2007). Furthermore, abnormal white matter microstructural measures correlate with

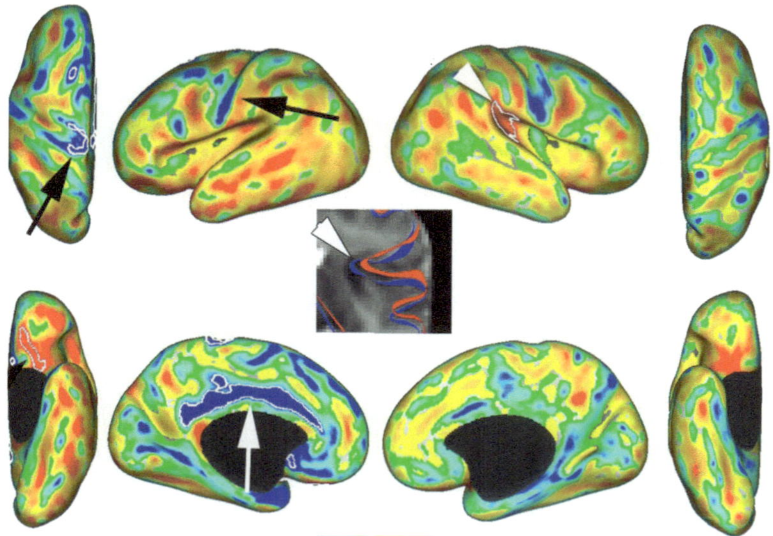

FIGURE 25.2 Sulcal depth patterns. Images represent the mean sulcal depth differences between CHD and control groups. White contours, significant group differences; black arrowhead, orbitofrontal region; black arrow, central sulcus; white arrowhead, superior ascending limb of the lateral sulcus; white arrow, cingulate gyrus. The central T2-weighted image displays the contour of the lateral sulcus for CHD (red) and controls (blue) and highlights that greater depth can also reflect deeper and more open sulci. *Adapted with permission from Ortinau et al. Cortical folding is altered before surgery in infants with congenital heart disease.* The Journal of Pediatrics, 163, *1507–1510.*

smaller brain size (Ortinau, Beca, et al., 2012). Abnormal brain metabolism, reflected by a lower *N*-acetylaspartate to choline ratio, is present in CHD fetuses and progresses along a slower rate of rise compared to controls (Limperopoulos et al., 2010). Neonatal, preoperative imaging has identified that *N*-acetylaspartate to choline ratio is reduced by 10% in CHD infants and occurs alongside microstructural abnormalities (Miller et al., 2007). Collectively, these data support in utero disruption of axonal/neuronal development and dendritic arborization that likely underlie macrostructural abnormalities in brain development and contribute to impaired fetal brain growth. The impact of these processes on brain connectivity in CHD is described in Chapter 24.

Perioperative and infant brain growth

Infancy is a vulnerable time for the developing CHD brain due to the hemodynamic changes and critical illness that occur after birth and surrounding neonatal cardiac surgery. A slower rate of brain growth is seen in CHD infants from the fetal environment through 3 months of age (Ortinau, Mangin-Heimos et al., 2018). Brain injury is one clinical factor that plays a role in this trajectory (Fig. 25.3). Prenatal reductions in white matter, cortical, and cerebellar volumes increase the risk of developing perioperative brain injury (Claessens et al., 2019). Infant data also demonstrates an association between more immature brains and perioperative white matter injury (Beca et al., 2013). While disrupted prenatal brain development leads to increased risk of brain injury, moderate–severe brain injury also contributes to poorer perioperative brain growth (Peyvandi et al., 2018). Despite this, brain injury does not fully explain the altered trajectory of perioperative brain development seen in CHD (Ortinau, Mangin-Heimos et al., 2018; Peyvandi et al., 2018). Other clinical factors associated with a slower rate of perioperative brain growth include single ventricle physiology (SVP), lower infant weight, and longer hospital stay (Meuwly et al., 2019; Ortinau, Inder, et al., 2012; Ortinau, Mangin-Heimos et al., 2018; Peyvandi et al., 2018).

Limited data exist regarding infant brain growth beyond the period of cardiac surgery. At 3 months of age, infants with CHD display smaller frontal lobe, parietal lobe, and brainstem measures (Ortinau, Inder, et al., 2012). Gray matter deficits appear to have the greatest contributions to reduced brain growth through 3 months (Ortinau, Mangin-Heimos et al., 2018). Children with CHD who were imaged approximately 8 months after cardiac surgery also demonstrate reduced gray matter volume without significant white matter deficits (Watanabe et al., 2009). However, at 1 year of age, reduced TBV is primarily due to smaller white matter and brainstem volumes in infants with biventricular physiology (Rollins et al., 2017).

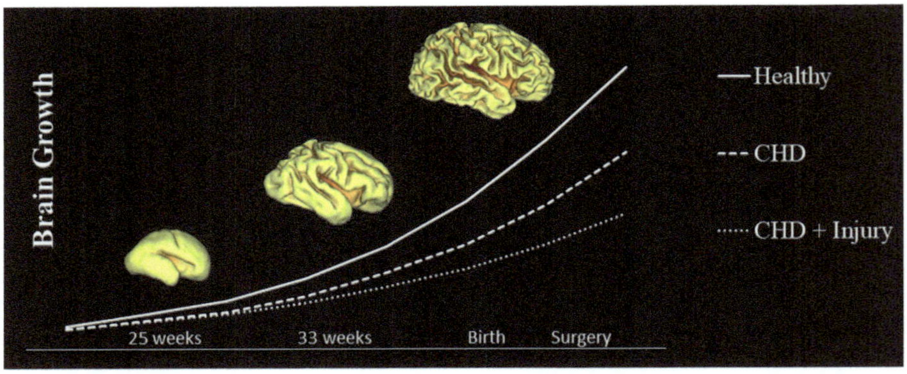

FIGURE 25.3 Schematic representation of brain growth in CHD with and without brain injury. Altered brain growth begins in the second trimester and continues at a slower rate of growth in CHD. Those CHD fetuses with more abnormal brain growth are at higher risk of perioperative brain injury after birth. Brain injury also further alters the trajectory of brain growth around the time of cardiac surgery.

Persistent impairments in childhood, adolescence, and adulthood

Deficits in brain growth continue to be present across the life span for children with CHD. Toddlers with SVP have smaller white and gray matter volumes compared to controls (Heye et al., 2018). A separate cohort of SVP children identified that impaired brain growth at 3 years of age was primarily related to white matter deficits, but gray matter structures also contributed by age 9 (Hiraiwa et al., 2019). Adolescents with SVP display reduced gray matter density in cortical, deep gray matter, hippocampal, and cerebellar regions (Singh et al., 2018). Children and adolescents with transposition of the great arteries (TGA) demonstrate impaired hippocampal brain growth (Munoz-Lopez et al., 2017), smaller cortical and subcortical volumes, and cortical thickness abnormalities (Watson et al., 2016). In a heterogeneous cohort of complex CHD, adolescents with cyanotic lesions display greater impairments in the white matter, thalami, hippocampi, and corpus callosum compared to those with acyanotic lesions (von Rhein et al., 2014). Adults with CHD continue to display volumetric deficits specifically in the hippocampi and white matter (Fontes et al., 2019).

Potential mechanisms impeding brain growth

Fetal cardiac physiology

The third trimester of pregnancy is marked by a dramatic increase in synapse formation, gyrification, and sulcation and an almost threefold increase in brain weight (Clouchoux et al., 2012; du Plessis, 2009; Hansen et al., 2003). An exponential increase in cerebral blood flow and vascularity also occurs, presumably related to the oxygen requirement needed to facilitate third trimester brain development (du Plessis, 2009). Delivery of oxygen- and nutrient-rich blood to the fetal brain is dependent on normal placental-cardiac hemodynamics (Fig. 25.4A) (McQuillen, Goff, & Licht, 2010). For the CHD fetus, abnormal cardiac anatomy can lead to impaired cerebral oxygenation, altered cerebral blood flow, or both, depending on the underlying defect (Fig. 25.4B−C) (McQuillen et al., 2010; Rudolph, 2016). Studies of cerebrovascular resistance in CHD fetuses support hypoxia-related cerebral vasodilation and/or altered cerebral blood flow patterns that likely contribute to abnormal brain growth (Rudolph, 2016). Further, novel fetal MRI studies show impaired cerebral oxygenation across multiple types of CHD during the third trimester (Lauridsen et al., 2017; Sun et al., 2015) and decreased cerebral oxygen consumption is associated with reduced brain size (Sun et al., 2015).

Because CHD generally develops early in gestation, the aforementioned cerebral hemodynamic alterations are chronic in nature. Exposure to chronic in utero hypoxia leads to decreased neuronal density in the deep white matter, impaired myelination, and abnormal cortical folding (Lawrence et al., 2019). A perinatal chronic hypoxia model has also identified reduced neural progenitor cells and neurogenesis in the subventricular zone, which result in abnormal cortical development, mimicking neuropathology findings in CHD infants (Morton et al., 2017). Data in preterm infants (at similar gestational ages to the CHD fetus) have identified an exquisite vulnerability of premyelinating oligodendrocytes (Pre-OLs) to hypoxia/ischemia-mediated cell injury via free radical production, excitotoxicity, and microglial activation (Ortinau & Neil, 2015). This injury upregulates precursor cells but an arrested maturation occurs that leads to decreased mature oligodendrocytes and hypomyelination (Billiards et al., 2008; Buser et al., 2012; Segovia et al., 2008). Myelin deficits disrupt thalamocortical and corticocortical connectivity and ultimately affect cortical and subcortical gray matter development. Direct injury to developing white matter axons, subplate neurons, and migratory GABA-ergic neurons, which are all critical for cortical and subcortical gray matter development, can also occur with hypoxia/ischemia (Ortinau & Neil, 2015).

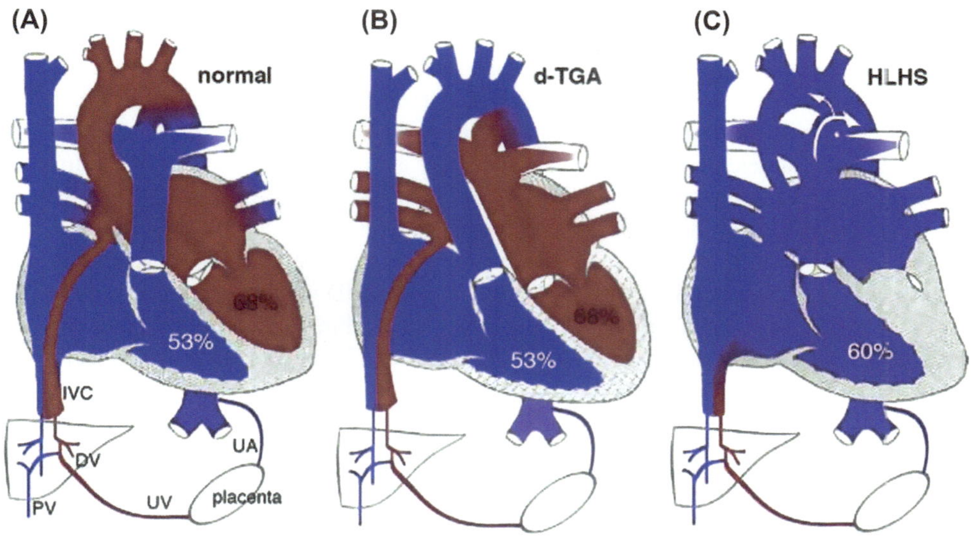

FIGURE 25.4 Fetal cardiac physiology. With normal fetal cardiac anatomy (A), oxygenated (red) blood is streamed from the placenta to the fetal heart, is preferentially shunted through the patent foramen ovale, and is pumped out the aortic arch to the head/neck vessels so that blood with the highest oxygen content reaches the developing brain. With dextro-transposition of the great arteries (d-TGA) (B), the aorta and pulmonary artery are switched and deoxygenated (blue) blood is supplied to the brain from the right side of the heart. With hypoplastic left heart syndrome (HLHS) (C), underdevelopment of the left ventricle leads to mixing of oxygenated and deoxygenated (purple) blood that is often supplied to the brain via retrograde flow through the ductus arteriosus from the right side of the heart. *Reprinted with permission from McQuillen et al., (2010). Effects of congenital heart disease on brain development.* Progress in Pediatric Cardiology, 29*(2), 79–85.*

Effects of transitional physiology

Several physiological changes occur at birth that can affect the CHD infant: 1) separation from the placental leads to increased systemic vascular resistance, 2) spontaneous breathing and a rise in arterial oxygen saturation facilitate decreased pulmonary vascular resistance, and 3) closure of the foramen ovale and ductus arteriosus occur (Fanaroff, Martin, & Walsh, 2006). This transitional physiology may result in clinical instability for the CHD infant that occurs in the delivery room with placental separation, in the first few days of life as the ductus arteriosus closes, or over several weeks as pulmonary vascular resistance continues to fall (Donofrio et al., 2014). In the first few days after birth, cerebral oxygen delivery also progressively declines in CHD infants, despite an increase in oxygen demand (Lynch et al., 2014). In some infants, prostaglandin E1 infusion is used to maintain patency of the ductus arteriosus and prevent hemodynamic compromise until surgery. However, a greater combined diameter of the foramen ovale and ductus arteriosus has been associated with decreased cerebral oxygen saturation (Baik et al., 2016). Therefore, the use of prostaglandin E1 may have implications for maintaining cerebral hemodynamics. This is particularly relevant due to the increased risk of postoperative white matter injury as cerebral oxygenation declines (Lynch et al., 2014) and the association of reduced cerebral oxygenation with smaller cortical gray matter volume (Kelly et al., 2017).

Cardiac surgery

Cardiopulmonary bypass with deep hypothermic circulatory arrest (DHCA) is a common surgical approach for CHD. DHCA is associated with specific vulnerability of Pre-OLs that, similar to preterm data, leads to maturational arrest of oligodendrocyte development and hypomyelination (Ishibashi et al., 2012). Repeated hypoxic exposure and hypoxia prior to ischemia reperfusion (i.e., cardiac surgery) are common clinical scenarios that further increase the susceptibility of Pre-OLs in animal models (Agematsu et al., 2014; Segovia et al., 2008). A porcine model has also demonstrated an effect of bypass on oligodendrocytes, altering white matter microstructure well after cardiac surgery (Stinnett et al., 2017). Human studies have identified an association of postoperative white matter injury with lower cerebral oxygenation and use of DHCA (Beca et al., 2013; Lynch et al., 2014), thus supporting preclinical data. The full impact of perioperative white matter vulnerability on subsequent global brain growth is not extensively studied. However, a slower trajectory of perioperative brain growth occurs in infants with white matter injury (Peyvandi et al., 2018).

Genetic variants and brain growth

Ongoing advances in genetic and genomic technologies have provided a growing body of evidence that genetic variants likely contribute to abnormal brain development in CHD. Whole exome sequencing has identified enrichment of damaging de novo variants in genes highly expressed in the heart and brain in CHD patients with neurodevelopmental disabilities (Homsy et al., 2015). Using a targeted gene panel, deleterious rare and novel variants occur 4–5 times more frequently in children with CHD who have neurodevelopmental deficits (Blue et al., 2018). Furthermore, CHD infants with pathogenic proangiogenic variants display smaller head circumference at birth (Russell et al., 2019). Specific to brain imaging, CHD fetuses display altered sulcal patterns in early emerging sulci, which are thought to be under tighter genetic control than later emerging sulci (Ortinau, Rollins, et al., 2018). Adolescents with CHD who demonstrate more atypical sulcal patterns are more likely to have damaging de novo variants (Morton et al., 2019). Finally, a cohort of patients with 22q11.2 deletion syndrome identified a CHD-specific brain vulnerability in the cortex and hippocampus (Fountain et al., 2014). These initial data suggest that genetic and epigenetic factors are important for altered brain growth.

Brain growth and neurodevelopment

A growing body of evidence has begun to describe the structure–function relationships of altered brain growth in CHD. One of the first studies linking brain imaging with outcome identified that 3-month-old infants with more dysmature brains had worse cognitive, language, and motor outcomes at age 2 (Beca et al., 2013). Specific to volumetric analyses, reduced white matter and cerebellar volumes are associated with worse language development at 1 year of age (Rollins et al., 2017). Toddlers with SVP prior to Fontan operation display an association between increasing cerebrospinal fluid volume and lower cognitive, language, and motor scores (Heye et al., 2018).

Data in adolescents and adults demonstrate continued effects of abnormal brain growth on neurodevelopmental functioning. Qualitative structural abnormalities are associated with poorer neurocognitive outcomes across several cohorts (Bolduc, Lambert, Ganeshamoorthy, & Brossard-Racine, 2018). Abnormal hippocampal development has been identified as an important predictor of lower intelligence quotient, poorer verbal comprehension, and memory impairment (Fontes et al., 2019; Latal et al., 2016; Munoz-Lopez et al., 2017). Adolescents with CHD also display an association of smaller white matter and cerebellar volumes with poorer cognitive and motor abilities (Table 25.1) (von Rhein et al., 2014). Abnormalities in microstructural brain development continue to be present in adolescents and adults with CHD and are linked with poorer performance in academic achievement, executive function, social cognition, memory, attention, motor function, and visuospatial skills (Bolduc et al., 2018).

TABLE 25.1 Association of brain volumes with neurodevelopment in adolescents with CHD. Corrected model included total brain volume, white matter, cortical gray matter, cerebellum, corpus callosum, hippocampus, sex, and age at MRI.

	Predictor	Corrected R^2	Beta	*P*-value
Cognitive abilities				
Total IQ	Hippocampal volume	0.18	0.45	0.004
Verbal comprehension	Total white matter	0.17	0.44	0.006
Perceptual reasoning	Hippocampal volume	0.11	0.36	0.03
Working memory	Cerebellar volume	0.16	0.42	0.008
Processing speed	–	–	–	–
Motor abilities				
Pure motor	Total white matter	0.15	0.41	0.01
Adaptive fine motor	–	–	–	–
Dynamic balance	–	–	–	–
Static balance	Cerebellar volume	0.11	0.36	0.03

Reprinted with permission from von Rhein et al., (2014). Brain volumes predict neurodevelopment in adolescents after surgery for congenital heart disease. *Brain, 137*, 268–76.

Applications to other areas of development

In this chapter, the impact of CHD on brain growth over the life span has been reviewed. Multiple studies have identified disrupted brain growth as early as the second and third trimesters of pregnancy. The MRI methods used to study fetal brain development in CHD could be readily applied to other high-risk populations to understand in utero brain growth for both fetal and maternal complications. The multifactorial contributions of pre- and postnatal physiologic, medical, and genetic pathways to abnormal brain development emphasize the complexity of altered brain growth in the CHD population. This is directly relevant for other birth defects or pregnancy complications (such as congenital diaphragmatic hernia or twin–twin transfusion syndrome) that compromise fetal hemodynamics, lead to clinical instability postnatally, require surgical intervention or intensive care, and/or result from genetic variants. These data highlight the need to extend research investigations to approaches that combine genetics, brain imaging, and neurodevelopment into a single model. This is directly relevant to the CHD population, as well as other birth defects. Furthermore, the growing body of literature linking MRI abnormalities with neurodevelopmental outcome in CHD provides a foundation for utilizing neuroimaging as a biomarker for parental counseling and for testing the effects of neuroprotective strategies. This is particularly relevant because neuroprotection trials are already underway.

Key facts

- CHD is the most common birth defect and occurs in 9/1000 live births.
- Moderate–severe CHD requires procedural intervention to correct or palliate the anatomical defect in the first year of life.
- Medical and surgical advances have improved survival for moderate–severe CHD and there are now more adults than children living with this diagnosis.
- Improved survival for CHD has shifted clinical and research focus to reducing the long-term neurodevelopmental sequelae of this diagnosis.
- Long-term neurodevelopmental deficits are present in 50% of children with moderate–severe CHD, persist into adulthood, and affect quality of life, educational achievement, and employment.

Summary points

- Abnormal brain growth in CHD begins in utero and persists throughout childhood, adolescence, and adulthood.
- Disturbances in white and gray matter structures contribute to abnormal brain growth.
- Altered fetal physiology, postnatal hemodynamic and surgical factors, and genetic variants are all mechanistic pathways of disrupted brain development and growth.
- White matter injury is common and is associated with poorer brain growth.
- Abnormal brain growth is associated with neurocognitive and motor deficits.

References

Agematsu, K., Korotcova, L., Scafidi, J., Gallo, V., Jonas, R. A., & Ishibashi, N. (2014). Effects of preoperative hypoxia on white matter injury associated with cardiopulmonary bypass in a rodent hypoxic and brain slice model. *Pediatric Research, 75*(5), 618–625. https://doi.org/10.1038/pr.2014.9.

Baik, N., Urlesberger, B., Schwaberger, B., Schmolzer, G. M., Kostenberger, M., Avian, A., & Pichler, G. (2016). Foramen ovale (FO) - the under-estimated sibling of ductus arteriosus (DA): Relevance during neonatal transition. *Early Human Development, 103*, 137–140. https://doi.org/10.1016/j.earlhumdev.2016.08.009.

Beca, J., Gunn, J. K., Coleman, L., Hope, A., Reed, P. W., Hunt, R. W., … Shekerdemian, L. S. (2013). New white matter brain injury after infant heart surgery is associated with diagnostic group and the use of circulatory arrest. *Circulation, 127*(9), 971–979. https://doi.org/10.1161/CIRCULATIONAHA.112.001089.

Berman, J. I., Hamrick, S. E., McQuillen, P. S., Studholme, C., Xu, D., Henry, R. G., … Glenn, O. A. (2011). Diffusion-weighted imaging in fetuses with severe congenital heart defects. *American Journal of Neuroradiology, 32*(2), E21–E22. https://doi.org/10.3174/ajnr.A1975.

Billiards, S. S., Haynes, R. L., Folkerth, R. D., Borenstein, N. S., Trachtenberg, F. L., Rowitch, D. H., … Kinney, H. C. (2008). Myelin abnormalities without oligodendrocyte loss in periventricular leukomalacia. *Brain Pathology, 18*(2), 153–163. https://doi.org/10.1111/j.1750-3639.2007.00107.x.

Blue, G. M., Ip, E., Walker, K., Kirk, E. P., Loughran-Fowlds, A., Sholler, G. F., … Winlaw, D. S. (2018). Genetic burden and associations with adverse neurodevelopment in neonates with congenital heart disease. *American Heart Journal, 201*, 33–39. https://doi.org/10.1016/j.ahj.2018.03.021.

Bolduc, M. E., Lambert, H., Ganeshamoorthy, S., & Brossard-Racine, M. (2018). Structural brain abnormalities in adolescents and young adults with congenital heart defect: A systematic review. *Developmental Medicine and Child Neurology, 60*(12), 1209–1224. https://doi.org/10.1111/dmcn.13975.

Buser, J. R., Maire, J., Riddle, A., Gong, X., Nguyen, T., Nelson, K., … Back, S. A. (2012). Arrested preoligodendrocyte maturation contributes to myelination failure in premature infants. *Annals of Neurology, 71*(1), 93–109. https://doi.org/10.1002/ana.22627.

Claessens, N. H. P., Khalili, N., Isgum, I., Ter Heide, H., Steenhuis, T. J., Turk, E., … Benders, M. (2019). Brain and CSF volumes in fetuses and neonates with antenatal diagnosis of critical congenital heart disease: A longitudinal MRI study. *American Journal of Neuroradiology, 40*(5), 885–891. https://doi.org/10.3174/ajnr.A6021.

Clouchoux, C., du Plessis, A. J., Bouyssi-Kobar, M., Tworetzky, W., McElhinney, D. B., Brown, D. W., … Limperopoulos, C. (2013). Delayed cortical development in fetuses with complex congenital heart disease. *Cerebral Cortex, 23*(12), 2932–2943. https://doi.org/10.1093/cercor/bhs281.

Clouchoux, C., Kudelski, D., Gholipour, A., Warfield, S. K., Viseur, S., Bouyssi-Kobar, M., … Limperopoulos, C. (2012). Quantitative in vivo MRI measurement of cortical development in the fetus. *Brain Structure and Function, 217*(1), 127–139. https://doi.org/10.1007/s00429-011-0325-x.

Donofrio, M. T., Moon-Grady, A. J., Hornberger, L. K., Copel, J. A., Sklansky, M. S., Abuhamad, A., … Stroke, N. (2014). Diagnosis and treatment of fetal cardiac disease: A scientific statement from the American Heart Association. *Circulation, 129*(21), 2183–2242. https://doi.org/10.1161/01.cir.0000437597.44550.5d.

Fanaroff, A. A., Martin, R. J., & Walsh, M. C. (2006). *Fanaroff and Martin's neonatal-perinatal medicine: Diseases of the fetus and infant* (8th ed.). Philadelphia, Pa: Mosby Elsevier.

Fontes, K., Rohlicek, C. V., Saint-Martin, C., Gilbert, G., Easson, K., Majnemer, A., … Brossard-Racine, M. (2019). Hippocampal alterations and functional correlates in adolescents and young adults with congenital heart disease. *Human Brain Mapping, 40*(12), 3548–3560. https://doi.org/10.1002/hbm.24615.

Fountain, D. M., Schaer, M., Mutlu, A. K., Schneider, M., Debbane, M., & Eliez, S. (2014). Congenital heart disease is associated with reduced cortical and hippocampal volume in patients with 22q11.2 deletion syndrome. *Cortex, 57*, 128–142. https://doi.org/10.1016/j.cortex.2014.04.004.

Griffiths, P. D., Mousa, H. A., Finney, C., Mooney, C., Mandefield, L., Chico, T. J. A., & Jarvis, D. (2019). An integrated in utero MR method for assessing structural brain abnormalities and measuring intracranial volumes in fetuses with congenital heart disease: Results of a prospective case-control feasibility study. *Neuroradiology, 61*(5), 603–611. https://doi.org/10.1007/s00234-019-02184-2.

Hansen, K., Sung, C. J., Huang, C., Pinar, H., Singer, D. B., & Oyer, C. E. (2003). Reference values for second trimester fetal and neonatal organ weights and measurements. *Pediatric and Developmental Pathology, 6*(2), 160–167. https://doi.org/10.1007/s10024-002-1117-3.

Heye, K. N., Knirsch, W., Latal, B., Scheer, I., Wetterling, K., Hahn, A., … Reich, B. (2018). Reduction of brain volumes after neonatal cardiopulmonary bypass surgery in single-ventricle congenital heart disease before Fontan completion. *Pediatric Research, 83*(1–1), 63–70. https://doi.org/10.1038/pr.2017.203.

Hiraiwa, A., Kawasaki, Y., Ibuki, K., Hirono, K., Matsui, M., Yoshimura, N., … Ichida, F. (2019). Brain development of children with single ventricle physiology or transposition of the great arteries: A longitudinal observation study. *Seminars in Thoracic and Cardiovascular Surgery*. https://doi.org/10.1053/j.semtcvs.2019.06.013.

Homsy, J., Zaidi, S., Shen, Y., Ware, J. S., Samocha, K. E., Karczewski, K. J., … Chung, W. K. (2015). De novo mutations in congenital heart disease with neurodevelopmental and other congenital anomalies. *Science, 350*(6265), 1262–1266. https://doi.org/10.1126/science.aac9396.

Ishibashi, N., Scafidi, J., Murata, A., Korotcova, L., Zurakowski, D., Gallo, V., & Jonas, R. A. (2012). White matter protection in congenital heart surgery. *Circulation, 125*(7), 859–871. https://doi.org/10.1161/CIRCULATIONAHA.111.048215.

Kelly, C. J., Christiaens, D., Batalle, D., Makropoulos, A., Cordero-Grande, L., Steinweg, J. K., … Counsell, S. J. (2019). Abnormal microstructural development of the cerebral cortex in neonates with congenital heart disease is associated with impaired cerebral oxygen delivery. *J Am Heart Assoc, 8*(5), Article e009893. https://doi.org/10.1161/JAHA.118.009893.

Kelly, C. J., Makropoulos, A., Cordero-Grande, L., Hutter, J., Price, A., Hughes, E., … Counsell, S. J. (2017). Impaired development of the cerebral cortex in infants with congenital heart disease is correlated to reduced cerebral oxygen delivery. *Scientific Reports, 7*(1), 15088. https://doi.org/10.1038/s41598-017-14939-z.

Khan, S., Rollins, C., Ortinau, C., Afacan, O., Warfield, S., & Gholipour, A. (2018). *Tract-specific group analysis in fetal cohorts using in utero diffusion tensor imaging*. https://doi.org/10.1007/978-3-030-00931-1_4.

Latal, B., Patel, P., Liamlahi, R., Knirsch, W., O'Gorman Tuura, R., & von Rhein, M. (2016). Hippocampal volume reduction is associated with intellectual functions in adolescents with congenital heart disease. *Pediatric Research, 80*(4), 531–537. https://doi.org/10.1038/pr.2016.122.

Lauridsen, M. H., Uldbjerg, N., Henriksen, T. B., Petersen, O. B., Stausbol-Gron, B., Matthiesen, N. B., … Hjortdal, V. E. (2017). Cerebral oxygenation measurements by magnetic resonance imaging in fetuses with and without heart defects. *Circulation: Cardiovascular Imaging, 10*(11), e006459. https://doi.org/10.1161/CIRCIMAGING.117.006459.

Lauridsen, M. H., Uldbjerg, N., Petersen, O. B., Vestergaard, E. M., Matthiesen, N. B., Henriksen, T. B., … Hjortdal, V. E. (2019). Fetal heart defects and measures of cerebral size. *The Journal of Pediatrics, 210*, 146–153. https://doi.org/10.1016/j.jpeds.2019.02.042.

Lawrence, K. M., McGovern, P. E., Mejaddam, A., Rossidis, A. C., Baumgarten, H., Kim, A., … Gaynor, J. W. (2019). Chronic intrauterine hypoxia alters neurodevelopment in fetal sheep. *The Journal of Thoracic and Cardiovascular Surgery*. https://doi.org/10.1016/j.jtcvs.2018.12.093.

Limperopoulos, C., Tworetzky, W., McElhinney, D. B., Newburger, J. W., Brown, D. W., Robertson, R. L., Jr., … du Plessis, A. J. (2010). Brain volume and metabolism in fetuses with congenital heart disease: Evaluation with quantitative magnetic resonance imaging and spectroscopy. *Circulation, 121*(1), 26–33. https://doi.org/10.1161/CIRCULATIONAHA.109.865568.

van der Linde, D., Konings, E. E., Slager, M. A., Witsenburg, M., Helbing, W. A., Takkenberg, J. J., & Roos-Hesselink, J. W. (2011). Birth prevalence of congenital heart disease worldwide: A systematic review and meta-analysis. *Journal of the American College of Cardiology, 58*(21), 2241–2247. https://doi.org/10.1016/j.jacc.2011.08.025.

Lynch, J. M., Buckley, E. M., Schwab, P. J., McCarthy, A. L., Winters, M. E., Busch, D. R., … Licht, D. J. (2014). Time to surgery and preoperative cerebral hemodynamics predict postoperative white matter injury in neonates with hypoplastic left heart syndrome. *The Journal of Thoracic and Cardiovascular Surgery, 148*(5), 2181–2188. https://doi.org/10.1016/j.jtcvs.2014.05.081.

Marelli, A. J., Ionescu-Ittu, R., Mackie, A. S., Guo, L., Dendukuri, N., & Kaouache, M. (2014). Lifetime prevalence of congenital heart disease in the general population from 2000 to 2010. *Circulation, 130*(9), 749–756. https://doi.org/10.1161/CIRCULATIONAHA.113.008396.

Marino, B. S., Lipkin, P. H., Newburger, J. W., Peacock, G., Gerdes, M., Gaynor, J. W., … Stroke, C. (2012). Neurodevelopmental outcomes in children with congenital heart disease: Evaluation and management: A scientific statement from the American Heart Association. *Circulation, 126*(9), 1143–1172. https://doi.org/10.1161/CIR.0b013e318265ee8a.

Masoller, N., Martinez, J. M., Gomez, O., Bennasar, M., Crispi, F., Sanz-Cortes, M., … Gratacos, E. (2014). Evidence of second-trimester changes in head biometry and brain perfusion in fetuses with congenital heart disease. *Ultrasound in Obstetrics and Gynecology, 44*(2), 182–187. https://doi.org/10.1002/uog.13373.

Matthiesen, N. B., Henriksen, T. B., Agergaard, P., Gaynor, J. W., Bach, C. C., Hjortdal, V. E., & Ostergaard, J. R. (2016). Congenital heart defects and indices of placental and fetal growth in a nationwide study of 924 422 liveborn infants. *Circulation, 134*(20), 1546–1556. https://doi.org/10.1161/CIRCULATIONAHA.116.021793.

McQuillen, P. S., Goff, D. A., & Licht, D. J. (2010). Effects of congenital heart disease on brain development. *Progress in Pediatric Cardiology, 29*(2), 79–85. https://doi.org/10.1016/j.ppedcard.2010.06.011.

Meuwly, E., Feldmann, M., Knirsch, W., von Rhein, M., Payette, K., Dave, H., … On Behalf of the Research Group Heart and Brain. (2019). Postoperative brain volumes are associated with one-year neurodevelopmental outcome in children with severe congenital heart disease. *Scientific Reports, 9*(1), 10885. https://doi.org/10.1038/s41598-019-47328-9.

Miller, S. P., McQuillen, P. S., Hamrick, S., Xu, D., Glidden, D. V., Charlton, N., … Vigneron, D. B. (2007). Abnormal brain development in newborns with congenital heart disease. *New England Journal of Medicine, 357*(19), 1928–1938. https://doi.org/10.1056/NEJMoa067393.

Morton, P. D., Korotcova, L., Lewis, B. K., Bhuvanendran, S., Ramachandra, S. D., Zurakowski, D., … Ishibashi, N. (2017). Abnormal neurogenesis and cortical growth in congenital heart disease. *Science Translational Medicine, 9*(374). https://doi.org/10.1126/scitranslmed.aah7029.

Morton, S. U., Maleyeff, L., Wypij, D., Yun, H. J., Newburger, J. W., Bellinger, D. C., … Im, K. (2019). Abnormal left-hemispheric sulcal patterns correlate with neurodevelopmental outcomes in subjects with single ventricular congenital heart disease. *Cerebral Cortex*. https://doi.org/10.1093/cercor/bhz101.

Munoz-Lopez, M., Hoskote, A., Chadwick, M. J., Dzieciol, A. M., Gadian, D. G., Chong, K., … Vargha-Khadem, F. (2017). Hippocampal damage and memory impairment in congenital cyanotic heart disease. *Hippocampus, 27*(4), 417–424. https://doi.org/10.1002/hipo.22700.

Ortinau, C., Alexopoulos, D., Dierker, D., Van Essen, D., Beca, J., & Inder, T. (2013). Cortical folding is altered before surgery in infants with congenital heart disease. *The Journal of Pediatrics, 163*(5), 1507–1510. https://doi.org/10.1016/j.jpeds.2013.06.045.

Ortinau, C., Beca, J., Lambeth, J., Ferdman, B., Alexopoulos, D., Shimony, J. S., … Inder, T. (2012). Regional alterations in cerebral growth exist preoperatively in infants with congenital heart disease. *The Journal of Thoracic and Cardiovascular Surgery, 143*(6), 1264–1270. https://doi.org/10.1016/j.jtcvs.2011.10.039.

Ortinau, C., Inder, T., Lambeth, J., Wallendorf, M., Finucane, K., & Beca, J. (2012). Congenital heart disease affects cerebral size but not brain growth. *Pediatric Cardiology, 33*(7), 1138–1146. https://doi.org/10.1007/s00246-012-0269-9.

Ortinau, C. M., Mangin-Heimos, K., Moen, J., Alexopoulos, D., Inder, T. E., Gholipour, A., … Smyser, C. D. (2018). Prenatal to postnatal trajectory of brain growth in complex congenital heart disease. *NeuroImage: Clinical, 20*, 913–922. https://doi.org/10.1016/j.nicl.2018.09.029.

Ortinau, C., & Neil, J. (2015). The neuroanatomy of prematurity: Normal brain development and the impact of preterm birth. *Clinical Anatomy, 28*(2), 168–183. https://doi.org/10.1002/ca.22430.

Ortinau, C. M., Rollins, C. K., Gholipour, A., Yun, H. J., Marshall, M., Gagoski, B., … Im, K. (2018). Early-emerging sulcal patterns are atypical in fetuses with congenital heart disease. *Cerebral Cortex*. https://doi.org/10.1093/cercor/bhy235.

Peyvandi, S., Kim, H., Lau, J., Barkovich, A. J., Campbell, A., Miller, S., … McQuillen, P. (2018). The association between cardiac physiology, acquired brain injury, and postnatal brain growth in critical congenital heart disease. *The Journal of Thoracic and Cardiovascular Surgery, 155*(1), 291–300 e293. https://doi.org/10.1016/j.jtcvs.2017.08.019.

du Plessis, A. J. (2009). Cerebral blood flow and metabolism in the developing fetus. *Clinics in Perinatology, 36*(3), 531–548. https://doi.org/10.1016/j.clp.2009.07.002.

Rajagopalan, V., Votava-Smith, J. K., Zhuang, X., Brian, J., Marshall, L., Panigrahy, A., & Paquette, L. (2018). Fetuses with single ventricle congenital heart disease manifest impairment of regional brain growth. *Prenatal Diagnosis, 38*(13), 1042–1048. https://doi.org/10.1002/pd.5374.

von Rhein, M., Buchmann, A., Hagmann, C., Dave, H., Bernet, V., Scheer, I., … Heart and Brain Research Group. (2015). Severe congenital heart defects are associated with global reduction of neonatal brain volumes. *The Journal of Pediatrics, 167*(6), 1259–1263 e1251. https://doi.org/10.1016/j.jpeds.2015.07.006.

von Rhein, M., Buchmann, A., Hagmann, C., Huber, R., Klaver, P., Knirsch, W., & Latal, B. (2014). Brain volumes predict neurodevelopment in adolescents after surgery for congenital heart disease. *Brain, 137*(Pt 1), 268–276. https://doi.org/10.1093/brain/awt322

Rollins, C. K., Asaro, L. A., Akhondi-Asl, A., Kussman, B. D., Rivkin, M. J., Bellinger, D. C., … Soul, J. S. (2017). White matter volume predicts language development in congenital heart disease. *The Journal of Pediatrics, 181*, 42–48 e42. https://doi.org/10.1016/j.jpeds.2016.09.070.

Rudolph, A. M. (2016). Impaired cerebral development in fetuses with congenital cardiovascular malformations: Is it the result of inadequate glucose supply? *Pediatric Research, 80*(2), 172–177. https://doi.org/10.1038/pr.2016.65.

Russell, M. W., Moldenhauer, J. S., Rychik, J., Burnham, N. B., Zullo, E., Parry, S. I., ... Gaynor, J. W. (2019). Damaging variants in proangiogenic genes impair growth in fetuses with cardiac defects. *The Journal of Pediatrics, 213*, 103–109. https://doi.org/10.1016/j.jpeds.2019.05.013.

Schellen, C., Ernst, S., Gruber, G. M., Mlczoch, E., Weber, M., Brugger, P. C., ... Kasprian, G. (2015). Fetal MRI detects early alterations of brain development in tetralogy of fallot. *American Journal of Obstetrics and Gynecology, 213*(3), 392 e391–397. https://doi.org/10.1016/j.ajog.2015.05.046.

Segovia, K. N., McClure, M., Moravec, M., Luo, N. L., Wan, Y., Gong, X., ... Back, S. A. (2008). Arrested oligodendrocyte lineage maturation in chronic perinatal white matter injury. *Annals of Neurology, 63*(4), 520–530. https://doi.org/10.1002/ana.21359.

Singh, S., Kumar, R., Roy, B., Woo, M. A., Lewis, A., Halnon, N., & Pike, N. (2018). Regional brain grey matter changes in adolescents with single ventricle heart disease. *Neuroscience Letters, 665*, 156–162. https://doi.org/10.1016/j.neulet.2017.12.011.

Stinnett, G. R., Lin, S., Korotcov, A. V., Korotcova, L., Morton, P. D., Ramachandra, S. D., ... Ishibashi, N. (2017). Microstructural alterations and oligodendrocyte dysmaturation in white matter after cardiopulmonary bypass in a Juvenile Porcine model. *J Am Heart Assoc, 6*(8). https://doi.org/10.1161/JAHA.117.005997.

Sun, L., Macgowan, C. K., Sled, J. G., Yoo, S. J., Manlhiot, C., Porayette, P., ... Seed, M. (2015). Reduced fetal cerebral oxygen consumption is associated with smaller brain size in fetuses with congenital heart disease. *Circulation, 131*(15), 1313–1323. https://doi.org/10.1161/CIRCULATIONAHA.114.013051.

Warnes, C. A., Williams, R. G., Bashore, T. M., Child, J. S., Connolly, H. M., Dearani, J. A., ... Webb, G. D. (2008). ACC/AHA 2008 guidelines for the management of adults with congenital heart disease: A report of the American College of Cardiology/American Heart Association Task Force on Practice Guidelines (Writing Committee to develop guidelines on the management of adults with congenital heart disease). Developed in collaboration with the American Society of Echocardiography, Heart Rhythm Society, International Society for Adult Congenital Heart Disease, Society for Cardiovascular Angiography and Interventions, and Society of Thoracic Surgeons. *Journal of the American College of Cardiology, 52*(23), e143–e263. https://doi.org/10.1016/j.jacc.2008.10.001.

Watanabe, K., Matsui, M., Matsuzawa, J., Tanaka, C., Noguchi, K., Yoshimura, N., ... Gur, R. C. (2009). Impaired neuroanatomic development in infants with congenital heart disease. *The Journal of Thoracic and Cardiovascular Surgery, 137*(1), 146–153. https://doi.org/10.1016/j.jtcvs.2008.06.036.

Watson, C. G., Asaro, L. A., Wypij, D., Robertson, R. L., Jr., Newburger, J. W., & Rivkin, M. J. (2016). Altered grey matter in adolescents with d-transposition of the great arteries. *The Journal of Pediatrics, 169*, 36–43.e1. https://doi.org/10.1016/j.jpeds.2015.09.084.

Chapter 26

Adolescent dopamine development: connecting experience with vulnerability or resilience to psychiatric disease

Lauren M. Reynolds[1,2,3] and Cecilia Flores[3]

[1]*McGill University, Integrated Program in Neuroscience, Montreal, QC, Canada;* [2]*Sorbonne Université, CNRS UMR 8246, INSERM U1130, Neuroscience Paris Seine - Institut de Biologie Paris Seine, Paris, France;* [3]*McGill University, Department of Psychiatry, Douglas Mental Health University Institute, Montreal, QC, Canada*

List of abbreviations

DCC Deleted in colorectal cancer
GWAS Genome-wide association studies
MRI Magnetic resonance imaging
mRNA Messenger RNA
NAc Nucleus accumbens
PFC Prefrontal cortex
PND Postnatal day
SNP Single nucleotide polymorphism
VTA Ventral tegmental area

Mini-dictionary of terms

Adolescence A developmental period marking the transition from a juvenile state to independence/adulthood.
Axon Long neurite extending from the soma of one neuron to its signaling partner.
Dopamine A chemical neurotransmitter with a modulatory effect over postsynaptic neurons.
Guidance cue Secreted or cell-bound molecules that exist in the extracellular environment. Axons navigate to their destination by interpreting the guidance cues in their environment, a process that depends on the guidance cue receptors each axon expresses.
Mesocortical dopamine projection Dopaminergic neurons with cell bodies in the ventral tegmental area and axons that project to the prefrontal cortex.
Mesolimbic dopamine projection Dopaminergic neurons with cell bodies in the ventral tegmental area and axons that project to limbic regions, including the nucleus accumbens.
microRNA A small noncoding RNA molecule that represses the expression of specific target genes.
Psychostimulant drugs A class of drugs that produce a temporary increase in motor activity, mental processing, attention, and/or euphoria. Psychostimulant drugs are used medically or recreationally, and the effects of this class of drugs generally depend on the dose taken.

Adolescence is a critical period for prefrontal cortex development

Psychiatric diseases are more likely to emerge during adolescence than during any other period of life (Fig. 26.1; Kessler et al., 2007; Lee et al., 2014; Paus, Keshavan, & Giedd, 2008). A common denominator of disorders that have an adolescent onset, particularly drug addiction, depression, and schizophrenia, are deficits in the function of the prefrontal cortex (PFC) and cognitive impairments (Goldstein & Volkow, 2011; Rice et al., 2019; Tan, Callicott, & Weinberger, 2009). The PFC and cognitive behaviors continue to mature across adolescence (Liston et al., 2006;

Diagnosis, Management and Modeling of Neurodevelopmental Disorders. https://doi.org/10.1016/B978-0-12-817988-8.00026-9

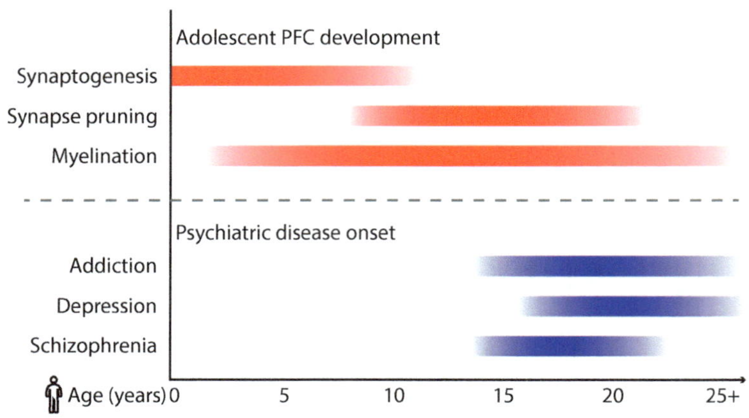

FIGURE 26.1 Psychiatric disease onset coincides with adolescent prefrontal cortex development. During adolescence, the prefrontal cortex undergoes a dynamic period of development including synaptogenesis, synaptic pruning, and myelination. This critical developmental period coincides with the onset of psychiatric disorders marked by prefrontal cortex dysfunction, including addiction, depression, and schizophrenia (Kessler et al., 2007; Lee et al., 2014; Paus et al., 2008).

Somerville, Hare, & Casey, 2011), and variations in their developmental trajectories likely contribute to individual differences in vulnerability to psychopathology. Environmental factors, such as drug use, are known to increase the risk of later psychiatric disease and may do so by interfering with PFC development (Andréasson, Engström, Allebeck, & Rydberg, 1987; Anthony & Petronis, 1995; Brook, Brook, Zhang, Cohen, & Whiteman, 2002; French et al., 2015; Hallfors, Waller, Bauer, Ford, & Halpern, 2005; Weiser, Knobler, Noy, & Kaplan, 2002).

The PFC is among the final brain regions to fully develop (Fig. 26.1). Longitudinal and cross-sectional magnetic resonance imaging (MRI) studies have mapped the trajectory of human brain development and found dynamic changes in the PFC during adolescence. Gray matter thickness in the PFC decreases across adolescence before stabilizing in adulthood, while white matter volume increases (Gogtay et al., 2004; Sowell et al., 2003). Postmortem human brain analysis further suggests that these macroscale changes likely result from modifications at the cellular level, with dramatic developmental changes in neuronal architecture, synapse density, and neurotransmitter concentration occurring across the adolescent period (Catts et al., 2013; Fung et al., 2010; Petanjek et al., 2011). Since the PFC is essential for cognitive function, including behavioral inhibition and flexibility, motivation, and reward processing, cognitive ability matures in parallel to PFC structural development (Larsen & Luna, 2018).

Sex differences in brain structure emerge across human adolescence (Ingalhalikar et al., 2014; Lenroot & Giedd, 2010; Sowell et al., 2007); however, this dimorphism does not translate into pronounced cognitive performance disparities between adult men and women. Instead, research indicates that men and women engage different cognitive strategies in order to produce similarly advantageous behavioral outcomes (Grissom & Reyes, 2018; Lenroot & Giedd, 2010). It is therefore critical to understand adolescent neurodevelopment in both sexes, as exposure to the same risk factors for psychopathologies targeting PFC function may lead to divergent outcomes.

Developmental rodent models for psychiatric-like traits are valuable for research. However, defining the adolescent period in rodents can prove challenging, as there are not clear-cut boundaries. While puberty and adolescence necessarily overlap temporally, the terms are not interchangeable. Puberty is defined by the onset of sexual maturity, while adolescence is a more diffuse period representing the gradual transition from a juvenile state to independence. Work from our group and others suggests that adolescence extends from the age of weaning (postnatal day (PND) 21) until adulthood (PND 60) and can be further subdivided into early, middle, and late periods that represent discrete developmental timepoints (Fig. 26.2;

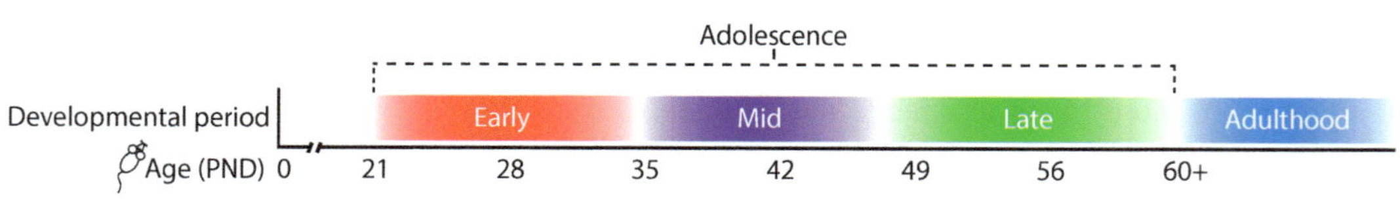

FIGURE 26.2 Defining adolescence in rodent models. Adolescence in rodents can be considered to extend from postnatal day (PND) 21 (weaning) until PND 60 (the start of adulthood) and can be further subdivided into early (PND 21 − 34), mid (PND 35 − 46) and late (PND 47 − 59) adolescent periods.

Hoops & Flores, 2017; Reynolds, Pokinko, et al., 2018; Reynolds, Yetnikoff, et al., 2018; Schneider, 2013; Spear, 2000; Tirelli, Laviola, & Adriani, 2003).

The maturational trajectory of the PFC in rodents is similar to humans, with cortical thickness peaking around 30 days of age and stabilizing after PND 60 in both rats and mice (Hammelrath et al., 2016; Mengler et al., 2014). These macroscale changes are also linked to alterations to the structure of PFC neurons, synaptic density, and myelination of axons (Caballero, Granberg, & Tseng, 2016; Delevich, Thomas, & Wilbrecht, 2019; Juraska & Willing, 2017). Cognitive and goal-directed behaviors in rodents are likewise refined across adolescence, mirroring what has been observed in humans (Andrzejewski et al., 2011; Naneix, Marchand, Scala, Pape, & Coutureau, 2012).

Mesocortical dopamine development drives prefrontal cortex maturation

Dopaminergic input to the PFC is a driver of PFC development. In contrast to other neurotransmitter systems innervating this region, dopamine axons continue to grow from the ventral tegmental area (VTA) to the PFC during adolescence (Reynolds, Pokinko, et al., 2018), resulting in a progressive increase in innervation between PNDs 21 and 60 (Fig. 26.3; Hoops & Flores, 2017). During this protracted process, growing dopamine axons undergo substantial modifications in fiber density, shape, and organization (Hoops & Flores, 2017). Prolonged development of PFC dopamine connectivity extrapolates to primates, including humans (Rosenberg & Lewis, 1995; Rothmond, Weickert, & Webster, 2012; Weickert et al., 2007).

PFC neurons receiving dopamine input, pyramidal neurons in particular, also undergo a dynamic period of structural changes during adolescence in both rodents and primates, including humans (Juraska & Willing, 2017; Lambe, Krimer, & Goldman-Rakic, 2000; Manitt et al., 2011; Petanjek et al., 2011). This structural development is linked to the maturation of their intrinsic electrophysiological properties (Manitt et al., 2013). The expression and functionality of dopamine receptors in the PFC change across adolescence in rats (Fig. 26.3; Caballero et al., 2016; Jordan & Andersen, 2017; Naneix, Marchand, Scala, Pape, & Coutureau, 2012), although this age-dependent expression pattern is less apparent in mice (Cullity, Madsen, Perry, & Kim, 2018; Pokinko et al., 2017). The changing influence of dopamine over PFC neurons is proposed to establish the precise balance of excitation and inhibition necessary for information processing (Caballero et al., 2016). In fact, we have recently confirmed that dopamine development in the PFC drives the establishment of local networks because alterations to the extent and/or organization of dopamine input to the PFC result in significant reorganization of postsynaptic pyramidal neuron morphology and function, ultimately manifesting as changes in the performance of PFC-dependent cognitive tasks (Reynolds, Pokinko, et al., 2018).

The same developmental pattern of PFC dopamine receptor expression seen in rodents has been observed in humans (Rothmond, Weickert, & Webster, 2012; Weickert et al., 2007), suggesting that the role of dopamine as a driver of PFC development is conserved. The maturation of dopamine input to the PFC has indeed been proposed to drive both dendritic spine pruning from pyramidal neurons across human adolescence (Petanjek et al., 2011) and adolescent cognitive development (Galvan, 2017; Luciana, Wahlstrom, Porter, & Collins, 2012; Wahlstrom, White, & Luciana, 2010).

While an adolescent increase in PFC dopamine innervation density has been observed in both male and female rodents (Willing, Cortes, Brodsky, Kim, & Juraska, 2017), the extent of sex differences in this process remains understudied.

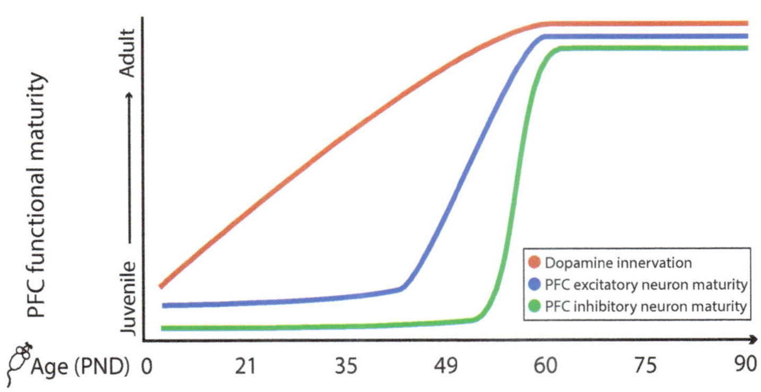

FIGURE 26.3 **Dopaminergic maturation of the prefrontal cortex in adolescence.** In primates and in rodents (highlighted here), dopamine innervation density increases across adolescence before stabilizing in adulthood (>PND 60). Postsynaptic neurons mature across the same timeline, with mature responses to dopaminergic modulation emerging over the adolescent period. Together, these maturation changes are thought to calibrate cognitive function in adolescence (Caballero et al., 2016).

Given that sex hormone receptor expression by dopamine neurons differs significantly between male and female rodents (Kritzer & Creutz, 2008), whether the adolescent development of PFC dopamine innervation is a puberty-dependent or puberty-independent process remains a key question (Juraska & Willing, 2017; Walker et al., 2017).

The Netrin-1/DCC guidance cue system is a molecular regulator of PFC dopamine development

In the search for molecular mechanisms controlling adolescent brain development, the Netrin-1/DCC guidance cue system has emerged as a critical regulator of the establishment of dopamine input to the PFC in adolescence. Guidance cues are secreted soluble or cell-bound molecules in the extracellular environment that act as a signal for growing axons. Dopamine neurons highly and conspicuously express DCC receptors from embryonic life to adulthood across species including humans (Flores, 2011), whereas its ligand Netrin-1 is expressed in the terminal regions that dopamine axons innervate, including nucleus accumbens (NAc) and PFC (Manitt et al., 2011). Expression levels of DCC and Netrin-1 are high during embryonic and early postnatal development and wane across adolescence until reaching a plateau in adulthood (Cuesta et al., 2018; Manitt et al., 2010). Their function also seems to change from early life to adulthood: once axon growth is completed, they contribute to synaptic plasticity and maintenance (Glasgow et al., 2018; Horn et al., 2013; Yetnikoff, Labelle-Dumais, & Flores, 2007).

Netrin-1 signaling through DCC receptors mediates guidance "decisions" that axons make at intermediate choice points along their pathfinding trajectory or once they reach their target (Deiner et al., 1997; Nelson & Colón-Ramos, 2013). During adolescence, DCC signaling within dopamine neurons actively segregates between axons that will innervate the NAc (mesolimbic) and those that will grow to the PFC (mesocortical) (Fig. 26.4; Manitt et al., 2013, 2011; Reynolds, Pokinko, et al., 2018). Reduced DCC function within dopamine neurons during adolescence induces targeting errors in mesolimbic dopamine axons, leading to their ectopic growth to the PFC (Reynolds, Pokinko, et al., 2018). Shifting the balance of DCC-dependent dopamine axon growth to the PFC in adolescence leads to profound changes in the structure and function of postsynaptic neurons, altered sensitivity to psychostimulant drugs of abuse, and enduring changes to performance on PFC-dependent cognitive tasks that persist into adulthood (Manitt et al., 2013, 2011; Pokinko, Moquin, Torres-Berrío, Gratton, & Flores, 2015; Reynolds, Pokinko, et al., 2018). *DCC* haploinsufficiency also occurs in humans and results in altered VTA-PFC connectivity and behavioral phenotypes related to addiction (Vosberg et al., 2018).

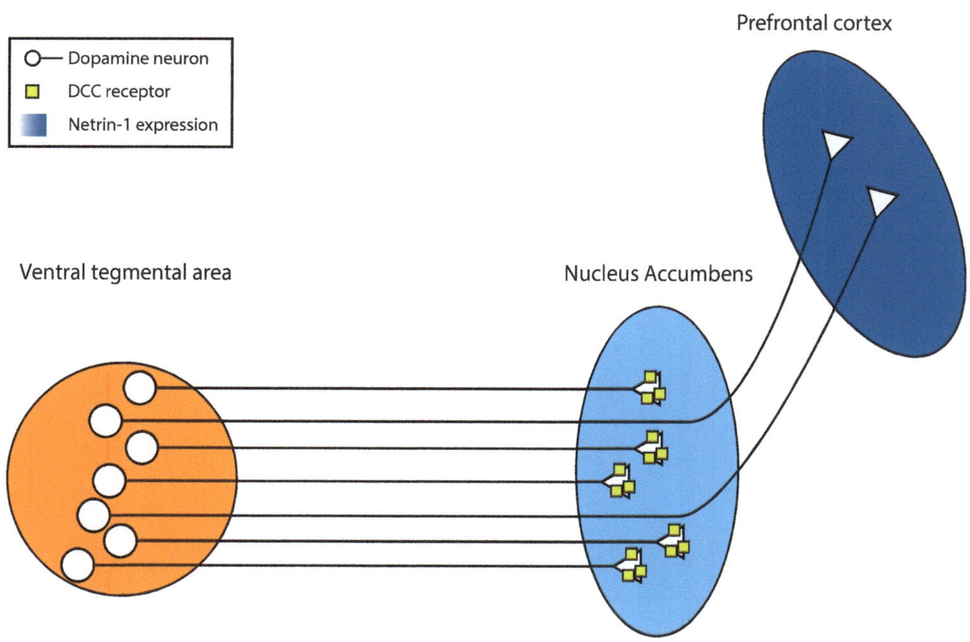

FIGURE 26.4 **Netrin-1/DCC signaling in dopamine neurons determines their targeting in adolescence.** Dopamine neurons with high DCC-expressing axons innervate the nucleus accumbens (NAc), a region of low Netrin-1 expression. Dopamine neurons that express little or no DCC in their axons do not recognize the NAc as their final target and instead continue to grow toward the prefrontal cortex (PFC). DCC receptors actively segregate axons into the mesocortical (PFC-projecting) and mesolimbic (NAc- projecting) pathways during adolescence.

Evidence for the involvement of DCC signaling in psychopathologies is rapidly exploding (Vosberg, Leyton, & Flores, 2019). Multiple independent genome-wide association studies (GWAS) have linked polymorphisms in the human *DCC* gene to psychopathologies marked by PFC dysfunction and adolescent onset. Single nucleotide polymorphisms (SNPs) in *DCC* have been found in patients suffering from major depressive disorder (Vosberg et al., 2019), and both postmortem and translational studies suggest a causal role for PFC *DCC* in depression-like behaviors (Manitt et al., 2013; Torres-Berrío et al., 2017). SNPs in the *DCC* gene have also been identified in schizophrenia, Parkinson's disease, and addiction patient populations (Grant, Fathalli, Rouleau, Joober, & Flores, 2012; Kim et al., 2011; Li et al., 2016; Smeland et al., 2017).

Drugs of abuse alter adolescent dopamine development via DCC signaling

During adolescence organisms are often navigating complex environments alone for the first time, setting the occasion for a unique period of sensitivity to environmental influence over PFC development (Kolb et al., 2012). Among environmental factors, drug use in adolescence is common and is known to increase vulnerability to later psychiatric diseases, particularly those involving PFC impairments such as addiction (Anthony & Petronis, 1995; Grant & Dawson, 2003; Robins & Przybeck, 1985), schizophrenia (Andréasson et al., 1987; French et al., 2015; Weiser, Knobler, Noy, & Kaplan, 2002), and depression (Brook et al., 2002; Hallfors et al., 2005). Despite this risk, adolescence remains a period of drug experimentation, and over half of drug use initiates in any given year are under the age of 18 years old (SAMHSA, 2012; Swendsen et al., 2012).

Because this time of high likelihood to begin taking drugs overlaps with ongoing PFC development, drug use in adolescence has been suggested to confer later addiction risk by disrupting ongoing cortical and cognitive maturation. Individuals struggling with addiction in adulthood show deficits in cognitive behaviors and PFC function (Goldstein & Volkow, 2011), and PFC activation during a cognitive task in adolescence is associated with later substance use level, particularly when that individual is already a drug user (Mahmood et al., 2013). However, the causal relationship between cognitive function, adolescent drug use, and later addiction cannot be determined from clinical studies. Preclinical studies have shed light into this matter, indicating that adolescent exposure to drugs alters the development and function of the mesocorticolimbic dopamine system and the PFC in particular (Gulley & Juraska, 2013; Jordan & Andersen, 2017).

While many classes of drugs may impact adolescent development, the stimulant drug amphetamine is of particular translational relevance: it is taken for therapeutic purposes, but also prescription drug misuse by adolescents is an increasingly common pathway to drug abuse (McCabe, West, Morales, Cranford, & Boyd, 2007). Exposure to amphetamine at recreational-like doses in adolescent rats has been shown to alter PFC dopamine receptor levels, change the electrophysiological properties of pyramidal neurons, and lead to cognitive impairments (Gulley & Juraska, 2013).

Until recently, the molecular mechanisms by which amphetamine interferes with adolescent PFC development remained unknown. We have found that repeated exposure to a recreational-like dose of amphetamine downregulates the expression of DCC only during early adolescence (Cuesta et al., 2019, 2018; Yetnikoff, Almey, Arvanitogiannis, & Flores, 2011; Yetnikoff et al., 2007; Yetnikoff, Pokinko, Arvanitogiannis, & Flores, 2014). We also showed that early adolescence is a critical period where exposure to amphetamine leads to long-term impairments in cognitive function and dopamine function in the PFC of mice (Reynolds & Flores, 2019; Reynolds et al., 2015; Reynolds, Yetnikoff, et al., 2018). These negative long-term effects of exposure to amphetamine in early adolescence are limited to doses corresponding to those taken recreationally by humans, as exposure to a dose that reaches a therapeutic-like plasma concentration upregulates DCC protein expression in the VTA (Cuesta et al., 2019). In turn, therapeutic-like exposure in adolescence does not disrupt PFC dopamine organization and leads to improved cognitive performance in adulthood (Cuesta et al., 2019), in line with studies in nonhuman primates (Soto et al., 2012). Thus, the *type* of the experience, positive or negative, can bidirectionally regulate DCC expression, suggesting that *DCC* functions as a plasticity gene rather than a vulnerability gene (Barth, Portella, Dubé, Meaney, & Silveira, 2019; Belsky, Jonassaint, Pluess, antonBrummett, & Williams, 2009).

MicroRNAs: linking experience, DCC expression, and adolescent dopamine development

Experiences in adolescence can regulate *Dcc* messenger RNA (mRNA) and protein expression by recruiting the microRNA, miR-218 (Cuesta et al., 2018; Torres-Berrío et al., 2017, 2019). MicroRNAs are potent regulators of mRNA stability and translation and can link experience to alterations in protein expression, neuronal structure, and function across development (Hollins & Cairns, 2016). We identified miR-218 as a powerful repressor of DCC expression in mouse and human adult PFC pyramidal neurons and showed that it mediates stress-induced dysregulation of DCC expression and

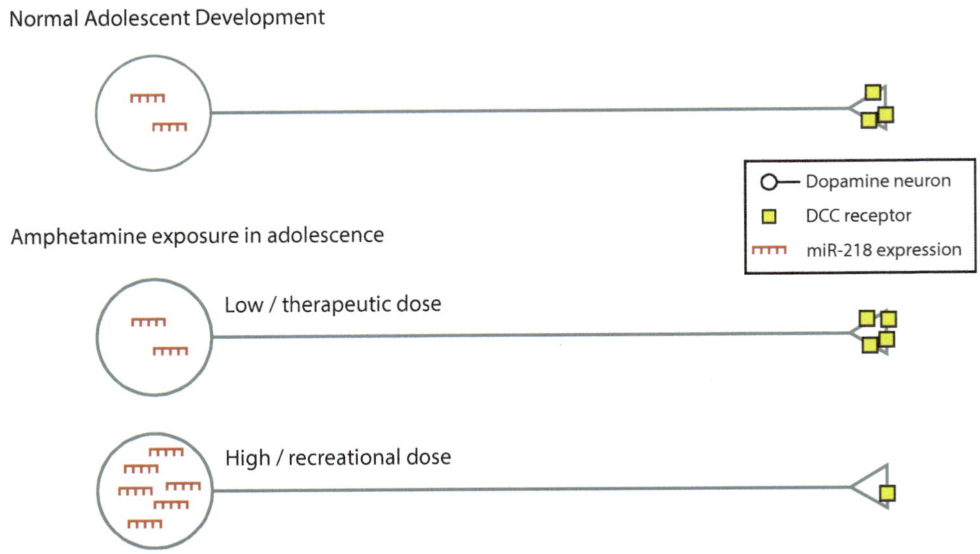

FIGURE 26.5 **Amphetamine in adolescence regulates DCC expression in dopamine neurons via microRNA miR-218.** The microRNA miR-218 is a posttranscriptional repressor of *Dcc* and is differentially expressed across the lifetime. Amphetamine in adolescence regulates DCC expression in a miR-218-dependent manner as a function of drug dose. Low doses that produce a therapeutic-like plasma amphetamine concentration do not regulate miR-218 expression, but increase DCC protein expression most likely via posttranslational mechanisms. In contrast, high doses that reach a recreational use-like plasma amphetamine concentration upregulate miR-218 expression and, in turn, decrease DCC mRNA and protein levels. Thus, positive and negative experiences can bidirectionally regulate DCC expression via miR-218.

depression-like outcomes (Torres-Berrío et al., 2017), acting as a molecular switch of susceptibility and resilience (Torres-Berrío et al., 2019).

MiR-218 is expressed in dopamine neurons at levels inversely correlated with DCC expression across the lifetime (Cuesta et al., 2018), and amphetamine in adolescence recruits miR-218 to decrease DCC levels in dopamine neurons in an age- and dose-dependent manner (Fig. 26.5; Cuesta et al., 2019). Thus, miR-218/DCC interaction may be a molecular mechanism by which adolescent experiences program adult brain structure, regulating vulnerability or resilience to psychiatric disease.

Conclusion

Adolescence is a dynamic developmental period defined in part by a changing PFC. The extended maturation of the PFC is a unique window of plasticity for environmental factors to confer risk for, or protection from, psychiatric disorders. The gradual unfolding of dopamine connectivity dictates PFC development and is controlled by Netrin-1/DCC guidance cue signaling. Susceptibility factors, such as abused drugs, regulate Netrin-1/DCC function via posttranscriptional mechanisms, influencing PFC and cognitive trajectories. Micro-RNA regulation of *Dcc* expression in adolescence controls PFC development allowing experience to construct adult networks. Emerging evidence that both positive and negative interventions regulate DCC levels, albeit in an opposite manner, suggests that *DCC* is a plasticity gene rather than a vulnerable gene. This molecular pathway represents an opportunity to develop interventions aimed at reducing the incidence of psychiatric disease by promoting healthy brain development and cognitive maturation in adolescence.

Applications to other areas of development

Here we reviewed the role of dopamine innervation in organizing the maturation of the prefrontal cortex, identified a key molecular pathway regulating adolescent dopamine development, and described how environmental factors impact the molecular, anatomical, and functional development of the adolescent brain conferring either susceptibility or resilience to psychiatric disease. We focus on the mesocortical dopamine pathway; however, recent evidence also suggests that later postnatal axon growth may extend to other pathways (Arruda-Carvalho, Wu, Cummings, & Clem, 2017; Yetnikoff, Reichard et al., 2014).

Key facts

Key facts of axon growth

- Axons are the projections of neurons that conduct action potentials away from the soma toward downstream signaling partners.
- Axons grow long distances to reach their targets using guidance cues to navigate their path.
- Most long-distance axonal growth occurs during embryonic and early postnatal development.
- Mesocortical dopamine axons continue to grow long distances across adolescence, undergoing targeting "decisions" along their path.
- Adolescent axon growth is particularly vulnerable to environmental insult.

Summary points

- Adolescence is a period of dynamic physical, hormonal, and behavioral changes.
- Psychiatric conditions, including major depression, addiction, and schizophrenia, are most likely to emerge in adolescence.
- The prefrontal cortex continues to develop in adolescence and is dysfunctional in psychiatric disease.
- Experience in adolescence shapes prefrontal cortex development.
- Dopamine input to the prefrontal cortex drives its maturation.
- Guidance cue genes organize the targeting and growth of dopamine pathways, are regulated by experience in adolescence, and are altered in psychiatric disorders.
- The mesocortical system in adolescence remains sensitive to both positive and negative environmental influences.
- Guidance cues may be a promising site to promote healthy brain development in adolescence.

References

Andréasson, S., Engström, A., Allebeck, P., & Rydberg, U. (1987). Cannabis and schizophrenia. A longitudinal study of Swedish conscripts. *The Lancet, 330*(8574), 1483–1486.

Andrzejewski, M. E., Schochet, T. L., Feit, E. C., Harris, R., McKee, B. L., & Kelley, A. E. (2011). A comparison of adult and adolescent rat behavior in operant learning, extinction, and behavioral inhibition paradigms. *Behavioral Neuroscience, 125*(1), 93–105.

Anthony, J. C., & Petronis, K. R. (1995). Early-onset drug use and risk of later drug problems. *Drug and Alcohol Dependence, 40*(1), 9–15.

Arruda-Carvalho, M., Wu, W.-C., Cummings, K. A., & Clem, R. L. (2017). Optogenetic examination of prefrontal-amygdala synaptic development. *Journal of Neuroscience, 37*(11), 2976–2985.

Barth, B., Portella, A. K., Dubé, L., Meaney, M. J., & Silveira, P. (2019). *Early life origins of ageing and longevity* (pp. 121–140). https://doi.org/10.1007/978-3-030-24958-8_7.

Belsky, J., Jonassaint, C., Pluess, M., anton, Brummett, B., & Williams, R. (2009). Vulnerability genes or plasticity genes? *Molecular Psychiatry, 14*(8). mp200944.

Brook, D. W., Brook, J. S., Zhang, C., Cohen, P., & Whiteman, M. (2002). Drug use and the risk of major depressive disorder, alcohol dependence, and substance use disorders. *Archives of General Psychiatry, 59*(11), 1039–1044.

Caballero, A., Granberg, R., & Tseng, K. Y. (2016). Mechanisms contributing to prefrontal cortex maturation during adolescence. *Neuroscience and Biobehavioral Reviews, 70*.

Catts, V. S., Fung, S. J., Long, L. E., Joshi, D., Vercammen, A., Allen, K. M., … Weickert, C. (2013). Rethinking schizophrenia in the context of normal neurodevelopment. *Frontiers in Cellular Neuroscience, 7*, 60.

Cuesta, S., Restrepo-Lozano, J., Popescu, C., He, S., Reynolds, L. M., Israel, S., … Flores, C. (2019). DCC-related developmental effects of abused-versus therapeutic-like amphetamine doses in adolescence. *Addiction Biology*, e12791.

Cuesta, S., Restrepo-Lozano, J., Silvestrin, S., Nouel, D., Torres-Berrío, A., Reynolds, L. M., … Flores, C. (2018). Non-contingent exposure to amphetamine in adolescence recruits miR-218 to regulate Dcc expression in the VTA. *Neuropsychopharmacology, 43*(4), 900–911.

Cullity, E., Madsen, H., Perry, C., & Kim, J. (2018). Postnatal developmental trajectory of dopamine receptor 1 and 2 expression in cortical and striatal brain regions. *The Journal of Comparative Neurology, 527*(6).

Deiner, M. S., Kennedy, T. E., Fazeli, A., Serafini, T., Tessier-Lavigne, M., & Sretavan, D. W. (1997). Netrin-1 and DCC mediate axon guidance locally at the optic disc: Loss of function leads to optic nerve hypoplasia. *Neuron, 19*(3), 575–589.

Delevich, K., Thomas, W. A., & Wilbrecht, L. (2019). Adolescence and "late blooming" synapses of the prefrontal cortex. *Cold Spring Harbor Symposia on Quantitative Biology*, 037507.

Flores, C. (2011). Role of netrin-1 in the organization and function of the mesocorticolimbic dopamine system. *Journal of Psychiatry and Neuroscience: Journal of Psychiatry and Neuroscience, 36*(5), 296–310.

French, L., Gray, C., Leonard, G., Perron, M., Pike, B. G., Richer, L., ... Paus, T. (2015). Early cannabis use, polygenic risk score for schizophrenia and brain maturation in adolescence. *JAMA Psychiatry, 72*(10), 1002–1011.

Fung, S. J., Webster, M. J., Sivagnanasundaram, S., Duncan, C., Elashoff, M., & Weickert, C. (2010). Expression of interneuron markers in the dorsolateral prefrontal cortex of the developing human and in schizophrenia. *American Journal of Psychiatry, 167*(12), 1479–1488.

Galvan, A. (2017). Adolescence, brain maturation and mental health. *Nature Neuroscience, 20*(4), 503–504.

Glasgow, S. D., Labrecque, S., Beamish, I. V., Aufmkolk, S., Gibon, J., Han, D., ... Kennedy, T. E. (2018). Activity-dependent netrin-1 secretion drives synaptic insertion of GluA1-containing AMPA receptors in the hippocampus. *Cell Reports, 25*(1), 168 182.e6.

Gogtay, N., Giedd, J. N., Lusk, L., Hayashi, K. M., Greenstein, D., Vaituzis, C. A., ... Thompson, P. M. (2004). Dynamic mapping of human cortical development during childhood through early adulthood. *Proceedings of the National Academy of Sciences of the United States of America, 101*(21), 8174–8179.

Goldstein, R. Z., & Volkow, N. D. (2011). Dysfunction of the prefrontal cortex in addiction: Neuroimaging findings and clinical implications. *Nature Reviews Neuroscience, 12*(11), 652–669.

Grant, B., & Dawson, D. (2003). Age of onset of drug use and its association with DSM-IV drug abuse and dependence: Results from the national longitudinal alcohol epidemiologic survey. *Journal of Substance Abuse, 10*(2), 163–173.

Grant, A., Fathalli, F., Rouleau, G., Joober, R., & Flores, C. (2012). Association between schizophrenia and genetic variation in DCC: A case-control study. *Schizophrenia Research, 137*(1–3), 26–31.

Grissom, N. M., & Reyes, T. M. (2018). Let's call the whole thing off: Evaluating gender and sex differences in executive function. *Neuropsychopharmacology, 44*(1), 1.

Gulley, J. M., & Juraska, J. M. (2013). The effects of abused drugs on adolescent development of corticolimbic circuitry and behavior. *Neuroscience, 249*.

Hallfors, D. D., Waller, M. W., Bauer, D., Ford, C. A., & Halpern, C. T. (2005). Which comes first in adolescence—sex and drugs or depression? *American Journal of Preventive Medicine, 29*(3), 163–170.

Hammelrath, L., Škokić, S., Khmelinskii, A., Hess, A., van der Knaap, N., Staring, M., ... Hoehn, M. (2016). Morphological maturation of the mouse brain: An in vivo MRI and histology investigation. *NeuroImage, 125*, 144–152.

Hollins, S. L., & Cairns, M. J. (2016). MicroRNA: Small RNA mediators of the brains genomic response to environmental stress. *Progress in Neurobiology, 143*, 61–81 (Dev. Biol. 283 2005).

Hoops, D., & Flores, C. (2017). Making dopamine connections in adolescence. *Trends in Neurosciences, 40*(12).

Horn, K. E., Glasgow, S. D., Gobert, D., Bull, S.-J., Luk, T., Girgis, J., ... Kennedy, T. E. (2013). DCC expression by neurons regulates synaptic plasticity in the adult brain. *Cell Reports, 3*(1), 173–185.

Ingalhalikar, M., Smith, A., Parker, D., Satterthwaite, T. D., Elliott, M. A., Ruparel, K., ... Verma, R. (2014). Sex differences in the structural connectome of the human brain. *Proceedings of the National Academy of Sciences of the United States of America, 111*(2), 823–828.

Jordan, C. J., & Andersen, S. L. (2017). Sensitive periods of substance abuse: Early risk for the transition to dependence. *Developmental Cognitive Neuroscience, 25*, 29–44.

Juraska, J. M., & Willing, J. (2017). Pubertal onset as a critical transition for neural development and cognition. *Brain Research, 1654*(Pt B), 87–94.

Kessler, R. C., Amminger, P. G., Aguilar-Gaxiola, S., Alonso, J., Lee, S., & Üstün, B. T. (2007). Age of onset of mental disorders: A review of recent literature. *Current Opinion in Psychiatry, 20*(4), 359–364.

Kim, J.-M., Park, S. K., Yang, J., Shin, E.-S., Lee, J.-Y., Yun, J., ... Jeon, B. S. (2011). SNPs in axon guidance pathway genes and susceptibility for Parkinson's disease in the Korean population. *Journal of Human Genetics, 56*(2), 125–129.

Kolb, B., Mychasiuk, R., Muhammad, A., Li, Y., Frost, D. O., & Gibb, R. (2012). Experience and the developing prefrontal cortex. *Proceedings of the National Academy of Sciences of the United States of America, 109*(Suppl. 2), 17186–17193.

Kritzer, M. F., & Creutz, L. M. (2008). Region and sex differences in constituent dopamine neurons and immunoreactivity for intracellular estrogen and androgen receptors in mesocortical projections in rats. *Journal of Neuroscience, 28*(38), 9525–9535.

Lambe, E. K., Krimer, L. S., & Goldman-Rakic, P. S. (2000). Differential postnatal development of catecholamine and serotonin inputs to identified neurons in prefrontal cortex of Rhesus monkey. *Journal of Neuroscience, 20*(23), 8780–8787.

Larsen, B., & Luna, B. (2018). Adolescence as a neurobiological critical period for the development of higher-order cognition. *Neuroscience and Biobehavioral Reviews, 94*. Behav.Neurosci. 112 1998.

Lee, F. S., Heimer, H., Giedd, J. N., Lein, E. S., Šestan, N., Weinberger, D. R., & Casey, B. (2014). Mental health. Adolescent mental health–opportunity and obligation. *Science, 346*(6209), 547–549.

Lenroot, R. K., & Giedd, J. N. (2010). Sex differences in the adolescent brain. *Brain and Cognition, 72*(1), 46–55.

Li, Y., Qiao, X., Yin, F., Guo, H., Huang, X., Lai, J., & Wei, S. (2016). A population-based study of four genes associated with heroin addiction in han Chinese. *PLoS One, 11*(9), e0163668.

Liston, C., Watts, R., Tottenham, N., Davidson, M. C., Niogi, S., Ulug, A. M., & Casey, B. (2006). Frontostriatal microstructure modulates efficient recruitment of cognitive control. *Cerebral Cortex, 16*(4), 553–560.

Luciana, M., Wahlstrom, D., Porter, J. N., & Collins, P. F. (2012). Dopaminergic modulation of incentive motivation in adolescence: Age-related changes in signaling, individual differences, and implications for the development of self-regulation. *Developmental Psychology, 48*(3), 844–861.

Mahmood, O., Goldenberg, D., Thayer, R., Migliorini, R., Simmons, A., & Tapert, S. (2013). Adolescents' fMRI activation to a response inhibition task predicts future substance use. *Addictive Behaviors, 38*(1), 1435–1441.

Manitt, C., Eng, C., Pokinko, M., Ryan, R., Torres-Berrío, A., Lopez, J., ... Flores, C. (2013). Dcc orchestrates the development of the prefrontal cortex during adolescence and is altered in psychiatric patients. *Translational Psychiatry, 3*(12), e338.

Manitt, C., Labelle-Dumais, C., Eng, C., Grant, A., Mimee, A., Stroh, T., & Flores, C. (2010). Peri-pubertal emergence of UNC-5 homologue expression by dopamine neurons in rodents. *PLoS One, 5*(7), e11463.

Manitt, C., Mimee, A., Eng, C., Pokinko, M., Stroh, T., Cooper, H. M., … Flores, C. (2011). The netrin receptor DCC is required in the pubertal organization of mesocortical dopamine circuitry. *Journal of Neuroscience, 31*(23), 8381–8394.

McCabe, S. E., West, B. T., Morales, M., Cranford, J. A., & Boyd, C. J. (2007). Does early onset of non-medical use of prescription drugs predict subsequent prescription drug abuse and dependence? Results from a national study. *Addiction, 102*(12), 1920–1930.

Mengler, L., Khmelinskii, A., Diedenhofen, M., Po, C., Staring, M., Lelieveldt, B. P., & Hoehn, M. (2014). Brain maturation of the adolescent rat cortex and striatum: Changes in volume and myelination. *NeuroImage, 84*, 35–44.

Naneix, F., Marchand, A. R., Scala, G., Pape, J.-R., & Coutureau, E. (2012). Parallel maturation of goal-directed behavior and dopaminergic systems during adolescence. *Journal of Neuroscience, 32*(46), 16223–16232.

Nelson, J. C., & Colón-Ramos, D. A. (2013). Serotonergic neurosecretory synapse targeting is controlled by netrin-releasing guidepost neurons in *Caenorhabditis elegans*. *Journal of Neuroscience: The Official Journal of the Society for Neuroscience, 33*(4), 1366–1376.

Paus, T., Keshavan, M., & Giedd, J. N. (2008). Why do many psychiatric disorders emerge during adolescence? *Nature Reviews Neuroscience, 9*(12), 947–957.

Petanjek, Z., Judaš, M., Šimic, G., Rasin, M., Uylings, H. B., Rakic, P., & Kostovic, I. (2011). Extraordinary neoteny of synaptic spines in the human prefrontal cortex. *Proceedings of the National Academy of Sciences of the United States of America, 108*(32), 13281–13286.

Pokinko, M., Grant, A., Shahabi, F., Dumont, Y., Manitt, C., & Flores, C. (2017). Dcc haploinsufficiency regulates dopamine receptor expression across postnatal lifespan. *Neuroscience, 346*, 182–189.

Pokinko, M., Moquin, L., Torres-Berrío, A., Gratton, A., & Flores, C. (2015). Resilience to amphetamine in mouse models of netrin-1 haploinsufficiency: Role of mesocortical dopamine. *Psychopharmacology, 232*(20).

Reynolds, L. M., & Flores, C. (2019). Guidance cues: Linking drug use in adolescence with psychiatric disorders. *Neuropsychopharmacology, 44*(1), 225–226.

Reynolds, L. M., Makowski, C. S., Yogendran, S. V., Kiessling, S., Cermakian, N., & Flores, C. (2015). Amphetamine in adolescence disrupts the development of medial prefrontal cortex dopamine connectivity in a Dcc-dependent manner. *Neuropsychopharmacology, 40*(5), 1101–1112.

Reynolds, L. M., Pokinko, M., Torres-Berrío, A., Cuesta, S., Lambert, L. C., Cid-Pellitero, E., … Flores, C. (2018). DCC receptors drive prefrontal cortex maturation by determining dopamine axon targeting in adolescence. *Biological Psychiatry, 83*(2), 181–192.

Reynolds, L. M., Yetnikoff, L., Pokinko, M., Wodzinski, M., Epelbaum, J. G., Lambert, L. C., … Flores, C. (2018). Early adolescence is a critical period for the maturation of inhibitory behavior. *Cerebral Cortex, 11*(2), 214.

Rice, F., Riglin, L., Thapar, A. K., Heron, J., Anney, R., O'Donovan, M. C., & Thapar, A. (2019). Characterizing developmental trajectories and the role of neuropsychiatric genetic risk variants in early-onset depression. *JAMA Psychiatry, 76*(3), 306–313.

Robins, L. N., & Przybeck, T. R. (1985). Age of onset of drug use as a factor in drug and other disorders. *NIDA Research Monograph, 56*, 178–192.

Rosenberg, D., & Lewis, D. (1995). Postnatal maturation of the dopaminergic innervation of monkey prefrontal and motor cortices: A tyrosine hydroxylase immunohistochemical analysis. *The Journal of Comparative Neurology, 358*(3), 383–400.

Rothmond, D. A., Weickert, C. S., & Webster, M. J. (2012). Developmental changes in human dopamine neurotransmission: Cortical receptors and terminators. *BMC Neuroscience, 13*(1), 18.

SAMHSA. (2012). *Results from the 2011 national survey on drug use and health: Summary of national findings. NSDUH series H-44, HHS publication No. (SMA) 12-4713.* Rockville, MD: Substance Abuse and Mental Health Services Administration.

Schneider, M. (2013). Adolescence as a vulnerable period to alter rodent behavior. *Cell and Tissue Research, 354*(1).

Smeland, O. B., Wang, Y., Frei, O., Li, W., Hibar, D. P., Franke, B., … Andreassen, O. A. (2017). Genetic overlap between schizophrenia and volumes of hippocampus, putamen, and intracranial volume indicates shared molecular genetic mechanisms. *Schizophrenia Bulletin, 44*(4).

Somerville, L. H., Hare, T., & Casey, B. (2011). Frontostriatal maturation predicts cognitive control failure to appetitive cues in adolescents. *Journal of Cognitive Neuroscience, 23*(9), 2123–2134.

Soto, P. L., Wilcox, K. M., Zhou, Y., Ator, N. A., Riddle, M. A., Wong, D. F., & Weed, M. R. (2012). Long-term exposure to oral methylphenidate or dl-amphetamine mixture in peri-adolescent Rhesus monkeys: Effects on physiology, behavior, and dopamine system development. *Neuropsychopharmacology, 37*(12), 2566–2579.

Sowell, E. R., Peterson, B. S., Kan, E., Woods, R. P., Yoshii, J., Bansal, R., … Toga, A. W. (2007). Sex differences in cortical thickness mapped in 176 healthy individuals between 7 and 87 years of age. *Cerebral Cortex, 17*(7), 1550–1560.

Sowell, E. R., Peterson, B. S., Thompson, P. M., Welcome, S. E., Henkenius, A. L., & Toga, A. W. (2003). Mapping cortical change across the human life span. *Nature Neuroscience, 6*(3), 309–315.

Spear, L. (2000). The adolescent brain and age-related behavioral manifestations. *Neuroscience and Biobehavioral Reviews 24*(4), 417–463.

Swendsen, J., Burstein, M., Case, B., Conway, K. P., Dierker, L., He, J., & Merikangas, K. R. (2012). Use and abuse of alcohol and illicit drugs in US adolescents: Results of the national comorbidity survey-adolescent supplement. *Archives of General Psychiatry, 69*(4), 390–398.

Tan, H.-Y., Callicott, J. H., & Weinberger, D. R. (2009). Prefrontal cognitive systems in schizophrenia: Towards human genetic brain mechanisms. *Cognitive Neuropsychiatry, 14*(4–5), 277–298.

Tirelli, E., Laviola, G., & Adriani, W. (2003). Ontogenesis of behavioral sensitization and conditioned place preference induced by psychostimulants in laboratory rodents. *Neuroscience and Biobehavioral Reviews, 27*(1–2), 163–178.

Torres-Berrío, A., Lopez, J., Bagot, R. C., Nouel, D., Bo, G., Cuesta, S., … Flores, C. (2017). DCC confers susceptibility to depression-like behaviors in humans and mice and is regulated by miR-218. *Biological Psychiatry, 81*(4), 306–315.

Torres-Berrío, A., Nouel, D., Cuesta, S., Parise, E. M., Restrepo-Lozano, J., Larochelle, P., … Flores, C. (2019). MiR-218: A molecular switch and potential biomarker of susceptibility to stress. *Molecular Psychiatry, 1*.

Vosberg, D. E., Leyton, M., & Flores, C. (2019). The netrin-1/DCC guidance system: Dopamine pathway maturation and psychiatric disorders emerging in adolescence. *Molecular Psychiatry*.

Vosberg, D. E., Zhang, Y., Menegaux, A., Chalupa, A., Manitt, C., Zehntner, S., … Leyton, M. (2018). Mesocorticolimbic connectivity and volumetric alterations in DCC mutation carriers. *Journal of Neuroscience: The Official Journal of the Society for Neuroscience, 38*(20), 4655–4665.

Wahlstrom, D., White, T., & Luciana, M. (2010). Neurobehavioral evidence for changes in dopamine system activity during adolescence. *Neuroscience and Biobehavioral Reviews, 34*(5), 631–648.

Walker, D. M., Bell, M. R., Flores, C., Gulley, J. M., Willing, J., & Paul, M. J. (2017). Adolescence and reward: Making sense of neural and behavioral changes amid the chaos. *Journal of Neuroscience: The Official Journal of the Society for Neuroscience, 37*(45), 10855–10866.

Weickert, C., Webster, M., Gondipalli, P., Rothmond, D., Fatula, R., Herman, M., … Akil, M. (2007). Postnatal alterations in dopaminergic markers in the human prefrontal cortex. *Neuroscience, 144*(3), 1109–1119.

Weiser, M., Knobler, H. Y., Noy, S., & Kaplan, Z. (2002). Clinical characteristics of adolescents later hospitalized for schizophrenia. *American Journal of Medical Genetics, 114*(8), 949–955.

Willing, J., Cortes, L. R., Brodsky, J. M., Kim, T., & Juraska, J. M. (2017). Innervation of the medial prefrontal cortex by tyrosine hydroxylase immunoreactive fibers during adolescence in male and female rats. *Developmental Psychobiology, 59*(5).

Yetnikoff, L., Almey, A., Arvanitogiannis, A., & Flores, C. (2011). Abolition of the behavioral phenotype of adult netrin-1 receptor deficient mice by exposure to amphetamine during the juvenile period. *Psychopharmacology, 217*(4), 505–514.

Yetnikoff, L., Labelle-Dumais, C., & Flores, C. (2007). Regulation of netrin-1 receptors by amphetamine in the adult brain. *Neuroscience, 150*(4), 764–773.

Yetnikoff, L., Pokinko, M., Arvanitogiannis, A., & Flores, C. (2014). Adolescence: A time of transition for the phenotype of dcc heterozygous mice. *Psychopharmacology, 231*(8), 1705–1714.

Yetnikoff, L., Reichard, R. A., Schwartz, Z., Parsely, K. P., & Zahm, D. S. (2014). Protracted maturation of forebrain afferent connections of the ventral tegmental area in the rat. *The Journal of Comparative Neurology, 522*(5), 1031–1047.

Chapter 27

Developmental coordination disorder

Jill G. Zwicker

Department of Occupational Science & Occupational Therapy, University of British Columbia, Vancouver, BC, Canada; Brain, Behaviour, & Development, BC Children's Hospital Research Institute, Vancouver, BC, Canada

List of abbreviations

ADHD attention deficit hyperactivity disorder
ASD autism spectrum disorder
CNV copy number variation
CO-OP Cognitive Orientation to daily Occupational Performance
DCD developmental coordination disorder
DSM Diagnostic and Statistical Manual
IMD internal modeling deficit
MNS mirror neuron system
MRI magnetic resonance imaging
NTT Neuromotor Task Training

Mini-dictionary of terms

Cognitive Orientation to daily Occupational Performance a cognitive treatment approach developed by Canadian occupational therapists specifically for children with DCD. The approach is grounded in theories of learning and problem-solving, strategy use, motor learning, and child-centered practice (Missiuna, Mandich, Polatajko, & Malloy-Miller, 2001). The therapist analyzes the child's motor performance and then guides the child to discover strategies to enable motor skill acquisition (Polatajko et al., 2001).

Diffusion Tensor Imaging measures the amount of water molecule displacement (mean diffusivity) and relative direction of water diffusion (fractional anisotropy or FA) in axons; lower mean diffusivity and higher FA indicate greater structural connectivity (Brown-Lum & Zwicker, 2015).

Dynamic Systems Theory a theory of motor development that proposes that movement is the result of the interaction between the person, the task, and the environment (Wilson, Smits-Engelsman, Caeyenberghs, & Steenbergen, 2017; Wilson, Smits-Engelsman, Caeyenberghs, Steenbergen, Sugden, et al., 2017).

Electroencephalography (EEG) measures electrical activity of neurons by placing electrodes on the scalp; measures neuronal events occurring at the millisecond level (Brown-Lum & Zwicker, 2015).

Environmental Stress Hypothesis a conceptual model that illustrates the complex relationship between motor skills and internalizing problems; DCD acts as a primary stressor that exposes the individual to secondary psychosocial stressors (e.g., physical inactivity, low-self-esteem), which then creates internalizing problems, such as anxiety and depression (Mancini, Rigoli, Cairney, Roberts, & Piek, 2016).

Glucocorticoids corticosteroid medications, such as dexamethasone and hydrocortisone, given to the mother when preterm birth is suspected to accelerate lung development of the infant (Tam, 2018).

Internal Modeling Deficit Hypothesis contends that the motor deficits associated with DCD are related to deficits in predictive control when planning and executing movements (Wilson, Smits-Engelsman, Caeyenberghs, and Steenbergen, 2017; Wilson, Smits-Engelsman, Caeyenberghs, Steenbergen, Sugden, et al., 2017).

Mirror Neuron System a network of neurons within the inferior frontal gyrus, ventral premotor cortex, and inferior parietal lobule that is active during the production and observation of actions (Reynolds et al., 2015).

Neuromotor Task Training a task-specific treatment approach developed by physical therapists in the Netherlands for children with DCD. It is based on motor learning theory and uses verbal guidance and feedback to teach motor skills (Schoemaker & Smits-Engelman, 2005).

Resting-state MRI a functional MRI sequence that identifies blood-oxygen-level-dependent (BOLD) signals that are correlated when the brain is at rest; brain regions that are functionally linked in resting state networks are anatomically connected (Brown-Lum & Zwicker, 2015).

Transcranial Magnetic Stimulation (TMS) a noninvasive stimulation of the brain that creates a magnetic pulse to cause an electrical current to excite or inhibit brain regions (Hyde, Fuelscher, & Williams, 2019).

Diagnosis, Management and Modeling of Neurodevelopmental Disorders. https://doi.org/10.1016/B978-0-12-817988-8.00027-0

Introduction

In this chapter, I describe and define a common neurodevelopmental disorder known as developmental coordination disorder (DCD). I outline the diagnostic criteria of DCD, common co-occurring conditions, and prognosis of the disorder. I then highlight the neural basis, genetic contributions, environmental exposures, and hypothesized mechanisms underlying the disorder. I conclude with information on evidence-based interventions for DCD. Throughout this chapter, I cite review articles that have summarized the literature; the reader is directed to these articles (and the original studies cited within the reviews) to supplement the brief overview presented in this chapter.

What is DCD?

DCD is a chronic motor disorder of unknown etiology that affects 5%–6% of school-aged children (American Psychiatric Association (APA), 2013). DCD significantly interferes with a child's ability to learn motor skills and perform everyday activities, such as tying shoes, using cutlery, printing, riding a bicycle, or playing sports. The motor deficits have a significant impact on daily life and may affect school, work, and leisure activities (Zwicker, Missiuna, Harris, & Boyd, 2012).

DCD is described in the Diagnostic and Statistical Manual—5th ed. (DSM-5; APA, 2013) under *Neurodevelopmental Disorders*, in a subsection called *Motor Disorders*. While DCD is a primarily a motor disorder, you will see later in this chapter that DCD also has a profound effect on one's mental health, well-being, and quality of life. The term DCD is not new, but may be known by other terms, such as dyspraxia, clumsy child syndrome, or perceptual motor dysfunction, to name a few (Zwicker et al., 2012). DCD first appeared in DSM-III and is the preferred term to use to describe these children as per an international consensus meeting in 1994 (Blank et al., 2019).

DCD is a heterogeneous condition, ranging in severity, presentation, and presence of co-occurring conditions. The cardinal feature of DCD is difficulty learning fine and/or gross motor skills, with motor performance being slower, less accurate, and more variable than typically developing peers (Zwicker et al., 2012). While difficulty with postural control, motor coordination, and motor learning are commonly described (Wilson, Ruddock, Smits-Engelsman, Polatajko, & Blank, 2013; Zwicker, Missiuna, & Boyd, 2009), additional features of DCD may include deficits in executive function and visual–spatial skills (Ferguson, Jelsma, Versfeld, & Smits-Engelsman, 2014; Leonard & Hill, 2015; Wilson et al., 2013).

Diagnosis of DCD

The DSM-5 outlines four diagnostic criteria for DCD (APA, 2013):

A. The acquisition and execution of coordinated motor skills are substantially below that expected given the individual's chronological age and opportunity for skill learning and use.
B. The motor skills deficit significantly and persistently interferes with activities of daily living appropriate to chronological age and impacts academic/school productivity, prevocational and vocational activities, leisure, and play.
C. Onset of symptoms is in the early developmental period.
D. The motor skills deficit is not better explained by intellectual disability, visual impairment, or a neurological condition affecting movement (e.g., cerebral palsy).

Despite its prevalence, DCD is under-recognized and under-diagnosed (Blank, Smits-Engelsman, Polatajko, & Wilson, 2012). To help bridge this gap, several articles have been published to outline how to diagnose the disorder (Blank et al., 2019; Harris, Mickelson, & Zwicker, 2015; Kirby, Sugden, & Purcell, 2014). In addition, a DCD Toolkit for Pediatricians has been made available online that highlights the importance of diagnosis and summarizes how to operationalize the diagnostic criteria (Fig. 27.1; Zwicker, Montgomery, & Mickelson, 2018: http://www.childdevelopment.ca/DCDAdvocacyToolkit/DCDAdvocacyToolkitIntro.aspx).

Common co-occurring conditions

DCD often co-occurs with other neurodevelopmental disorders, most commonly attention deficit attention deficit hyperactivity disorder (ADHD) with a 50% overlap of the two disorders (Goulardins et al., 2015; Licari et al., 2019; Zwicker et al., 2012). Other common co-occurring conditions include learning disabilities and specific language impairment (Zwicker et al., 2012). Since the publication of DSM-5 in 2013, a co-occurring diagnosis of DCD and autism spectrum disorder (ASD) has been permitted; an estimated 50%–80% of children with ASD may meet the diagnostic criteria for DCD (Blank et al., 2019; Dewey, 2018; Licari et al., 2019). The high degree of overlap between neurodevelopmental

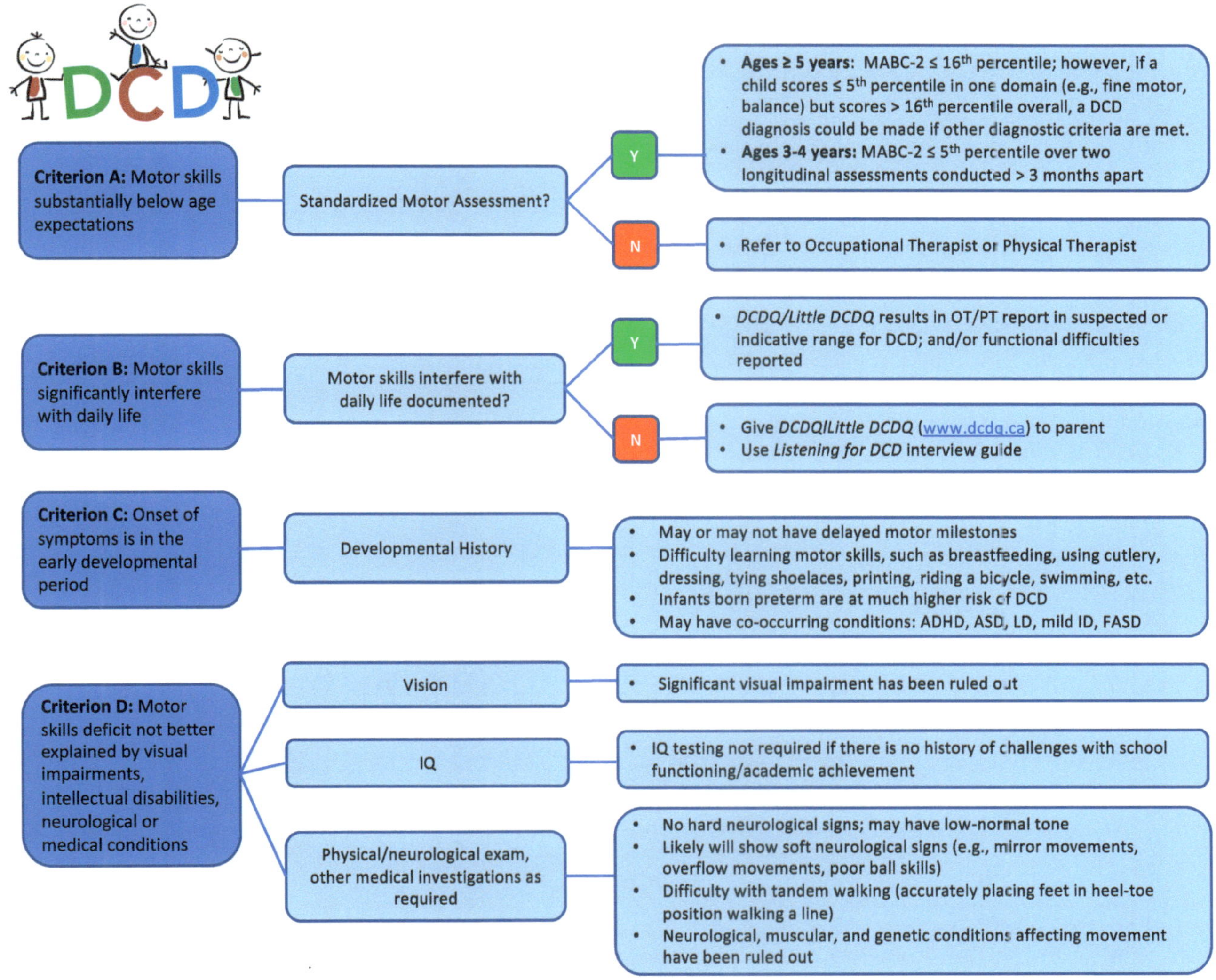

FIGURE 27.1 **Application of DSM-5 diagnostic criteria for DCD.** *ADHD*, attention deficit hyperactivity disorder; *ASD*, autism spectrum disorder; *DCD*, developmental coordination disorder; *DCDQ*, Developmental Coordination Disorder Questionnaire; *FASD*, fetal alcohol spectrum disorder; *ID*, intellectual disability; *IQ*, intelligence quotient; *LD*, learning disability; *MABC-2*, Movement Assessment Battery for Children—2nd ed.; *N*, no; *Y*, yes. *DCD Toolkit for Pediatricians available from: http://www.childdevelopment.ca/Libraries/DCD/PediatriciansDCDToolkitforPediatricians.sflb.ashx and is based current evidence from the American Psychiatric Association. (2013).* Diagnostic and statistical manual of mental disorders - 5th ed. *(DSM-5). Washington, DC: Author and Blank, R., Barnett, A. L., Cairney, J., Green, D., Kirby, A., Polatajko, H., … Vinçon, S. (2019). International clinical practice recommendations on the definition, diagnosis, intervention, and psychosocial aspects of developmental coordination disorder.* Developmental Medicine and Child Neurology, 61, *242–285. https://doi.org/10.1111/dmcn.14132.*

disorders has prompted researchers to consider shared etiology as a result of atypical brain development (Dewey, 2018; Dewey & Bernier, 2016; Licari et al., 2019); however, other researchers contend that DCD is distinct from other disorders (Blank et al., 2019), including ADHD (Goulardins et al., 2015) and ASD (Caçola, Miller, & Williamson, 2017; Kilroy, Cermak, & Aziz-Zadeh, 2019).

Another high co-occurrence with DCD is prematurity (Spittle & Orton, 2014). Infants who are born very preterm are six to eight times more likely to develop DCD compared to their full-term peers (Edwards et al., 2011).

Prognosis of DCD

It was once believed that children would outgrow their motor difficulties, but longitudinal studies have shown that the motor difficulties tend to persist into adolescence and adulthood (Zwicker et al., 2012). As nearly 75% of children will continue to have difficulties as adults, DCD should be considered a life-long disorder (Kirby et al., 2014). Individuals with co-occurring conditions tend to have poorer outcomes than those with DCD alone (Blank et al., 2019).

Missiuna, Gaines, and Soucie (2006) outline a trajectory of DCD from the preschool years to adolescence (Fig. 27.2). Young children may first present with difficulty in learning fine and gross motor skills, such as catching a ball or riding

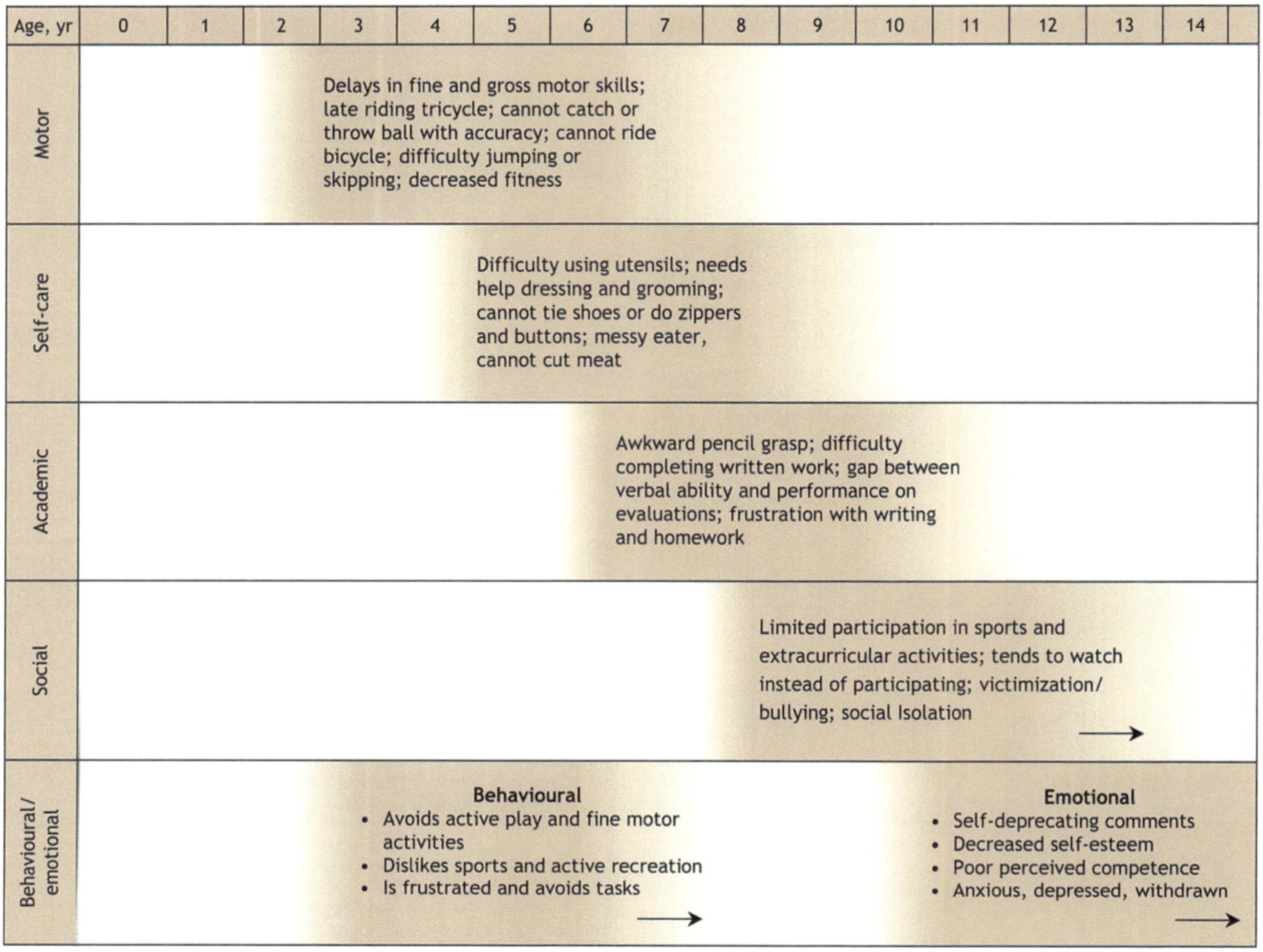

FIGURE 27.2 **Developmental trajectory of developmental coordination disorder.** *Reprinted with permission from Missiuna, C., Gaines, R., & Soucie, H. Why every office needs a tennis ball.* Canadian Medical Association Journal, *175, 471. doi: https://doi.org/10.1503/cmaj.051202.*

a tricycle. Early in the school years, the motor difficulties begin to impact self-care and academic skills, and by middle childhood, children are beginning to withdraw from sports and physical activities. In late childhood and the teen years, the emotional toll becomes evident, with self-deprecating comments, poor self-esteem, and mental health concerns (e.g., anxiety and depression). Others have highlighted the impact of DCD in adulthood, including difficulty in transitioning from school to higher education and/or employment, reduced participation in physical activities, poorer mental health, and lower life satisfaction (Blank et al., 2019; Tal Saban & Kirby, 2018).

The impact of DCD on psychosocial well-being and mental health has been well documented, particularly in relation to internalizing symptoms of anxiety and depression (Cairney, Veldhuizen, & Szatmari, 2010; Caçola, 2016; Draghi, Cavalcante Neto, Rohr, Jelsma, & Tudella, 2020; Omer, Jijon, & Leonard, 2019; Zwicker, Harris, & Klassen, 2013). An environmental stress hypothesis has been proposed, which suggests that poor motor skills in childhood may lead to the development of internalizing symptoms later in life (Mancini et al., 2016). In addition to adverse mental health, DCD has been associated with poorer physical health, including reduced physical activity and fitness, greater risk for overweight and obesity, and higher risk for cardiovascular disease (Cacola, 2016; Hendrix, Prins, & Dekkers, 2014; Rivilis et al., 2011). What starts as a seemingly "minor" motor impairment in childhood can lead to chronic physical and mental health concerns in adulthood.

Neural basis of DCD

While altered brain development or brain function has been suspected in DCD, it has only been in the last decade or so that brain imaging studies have been conducted to support this hypothesis. Several reviews have summarized the findings to date, showing that children with DCD compared to typically developing peers have altered brain structure and brain function (Biotteau et al., 2016; Brown-Lum & Zwicker, 2015; Peters, Maathuis, & Hadders-Algra, 2013; Wilson, Smits-Engelsman, Caeyenberghs, Steenbergen, Sugden, et al., 2017). While there is no consensus on the specific neural correlates of DCD, findings have consistently implicated the cerebellum, basal ganglia, parietal lobes, and parts of the frontal lobes (Biotteau et al., 2016; Bo & Lee, 2013; Brown-Lum & Zwicker, 2015; Dewey & Bernier, 2016; Wilson, Smits-Engelsman, Caeyenberghs, Steenbergen, Sugden, et al., 2017; Zwicker et al., 2009) (Fig. 27.3). As noted by Biotteau et al. (2016), no two studies observed the same brain differences, which may be due to the heterogeneity of the disorder, how cases of DCD were defined, the presence of co-occurring disorders, small sample sizes, and differences in MRI acquisition, processing, and analyses. Despite these limitations, it is clear that children with DCD are neurobiologically different from their typically developing peers (Brown-Lum & Zwicker, 2015).

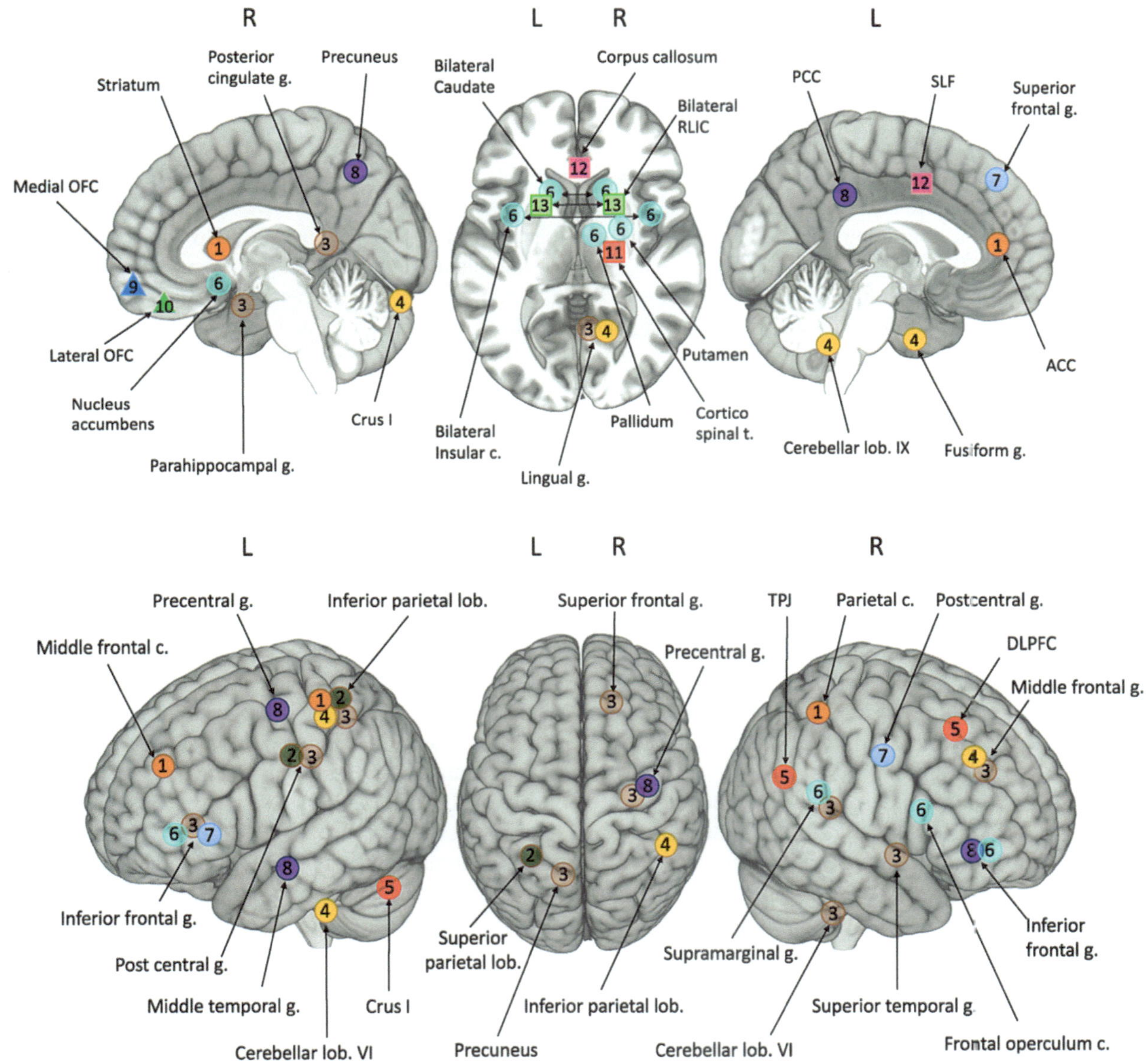

FIGURE 27.3 **Foci of atypical brain involvement reported from neuroimaging studies of children with DCD compared to TD children.** *Studies: circles* 1–8: functional MRI; *triangles* 9–10: structural MRI; *squares* 11–13: diffusion tensor imaging. Each number/color represents one study. [1] Querne et al. https://doi.org/10.1016/j.brainres.2008.07.066; [2] Kashiwagi et al. https://doi.org/10.1097/WNR.0b013e32832f4d87; [3] Zwicker et al. https://doi.org/10.1542/peds.2010-0059; [4] Zwicker et al. https://doi.org/10.1016/j.ijdevneu.2010.12.002; [5] Debrabant et al. https://doi.org/10.1016/j.ridd.2013.02.008; [6] McLeod et al. https://doi.org/10.1016/j.nicl.2014.03.010; [7] Licari et al. https://doi.org/10.1007/s00221-015-4243-7; [8] Reynolds et al. https://doi.org/10.1016/j.ijdevneu.2015.10.003; [9] Langevin et al. https://doi.org/10.1111/dmcn.12561; [10] Caeyenberghs et al. https://doi.org/10.1111/desc.12424; [11] Zwicker et al. https://doi.org/10.1016/j.pediatrneurol.2011.12.007; [12] Langevin et al. https://doi.org/10.1016/j.jpeds.2014.01.018; [13] Debrabant et al. https://doi.org/10.1016/j.jpeds.2015.09.069. *Brain areas*: *ACC*, anterior cingulate cortex; *c*, cortex; *DLPFC*, dorsolateral prefrontal cortex; *g*, gyrus; *lob.*, lobule; *OFC*, orbitofrontal cortex; *PCC*, posterior cingulate cortex; *RLIC*, retrolenticular part of the internal capsule; *SLF*, superior longitudinal fasciculus; *t*, tract; *TPJ*, temporoparietal junction. *Reprinted with permission from: Biotteau, M., Chaix, Y., Blais, M., Tallet, J., Péran, P., & Albaret, J. M. (2016). Neural signature of DCD: A critical review of MRI neuroimaging studies.* Frontiers in Neurology, 7, *Article 227. https://doi.org/10.3389/fneur.2016.00227.*

In terms of brain structure, children with DCD show differences in both gray matter and white matter. Gray matter alterations include cortical thinning in the temporal region, reduction in gray matter volume in premotor regions, and altered structural covariance networks in frontal regions associated with motor planning (Biotteau et al., 2016; Dewey & Bernier, 2016; Licari & Reynolds, 2017; Wilson, Smits-Engelsman, Caeyenberghs, Steenbergen, Sugden, et al., 2017). White matter alterations have been noted at the microstructural level in sensorimotor regions, including the corticospinal tract, posterior thalamic radiations, internal capsule, superior longitudinal fasciculus, and parietal regions of the corpus callosum (Biotteau et al., 2016; Brown-Lum & Zwicker, 2015; Dewey & Bernier, 2016; Licari et al., 2019; Wilson, Smits-Engelsman, Caeyenberghs, Steenbergen, Sugden, et al., 2017).

Children with DCD also show differences in brain activation compared to typically developing peers. While results vary between studies, it is clear that children with DCD are using different networks to perform motor skills, and they show under-activation of cerebellar, parietal, and prefrontal regions during motor learning (Brown-Lum & Zwicker, 2015; Wilson, Smits-Engelsman, Caeyenberghs, Steenbergen, Sugden, et al., 2017). It has been widely accepted that the cerebellum is implicated in DCD (Bo & Lee, 2013; Zwicker et al., 2009, 2012), but neuroimaging evidence suggests that the connections to and from the cerebellum are also likely involved (e.g., parietal—cerebellar and cerebellar—frontal tracts) (Bo & Lee, 2013; Brown-Lum & Zwicker, 2015; Dewey & Bernier, 2016). Altered functional connectivity of the primary motor cortex to other motor and sensory regions has also been noted in children with DCD (Biotteau et al., 2016; Brown-Lum & Zwicker, 2015; Dewey & Bernier, 2016; Licari et al., 2019).

Neurophysiology is also altered in DCD. For example, electroencephalography studies have shown different patterns of cortical activation and compromised interhemispheric communication in children with DCD (Brown-Lum & Zwicker, 2015; Hyde et al., 2019). Findings from transcranial magnetic stimulation studies in adults with DCD show reduced interhemispheric cortical inhibition and corticospinal excitability within the motor cortices (Hyde et al., 2019). Similarly, children with DCD show differential neural activation patterns on functional near-infrared spectroscopy during fine motor tasks in the premotor and supplementary motor areas, as well as the dorsolateral prefrontal cortex (Hyde et al., 2019).

In summary, evidence is accumulating to determine the neural basis of DCD. While individual studies vary in their results, it is apparent that findings from structural MRI, diffusion tensor imaging, task-based and resting-state functional MRI, and neurophysiological measures are converging. Children with DCD show a host of brain differences compared to typically developing children, many of which correlate with motor function. These findings may help to explain why children with DCD show variable motor performance and struggle to learn motor skills (Brown-Lum & Zwicker, 2017).

Genetic contributions to DCD

DCD is estimated to be 70% heritable, yet few studies have examined genetic and genomic variations in this clinical population (Dewey & Bernier, 2016; Licari & Reynolds, 2017). Recent advances in genomic techniques, such as high-resolution microarrays and next-generation sequencing, have allowed investigation of structural variation in chromosomes; one such structural variation is copy number variation (CNV), which is the result of duplications or deletions in the genome (Dewey & Bernier, 2016). Only one study to date has investigated CNV in children with DCD, which found enrichment for brain-expressed genes and genes previously associated with other neurodevelopmental disorders, including duplications in SHANK3 and 16p11.2 and deletions in VIPR2, 22q11, GAP43, and LSAMP (summarized in Dewey & Bernier, 2016; Licari & Reynolds, 2017). Evidence suggests that, at least in some cases, DCD may have a genetic basis, but further investigations are necessary to further elucidate genetic or genomic contributions to the disorder. It is likely that the etiology of DCD is multifactorial and involves the complex interplay of genetic, environmental, and developmental factors associated with altered brain development (Gomez & Sirigu, 2015; Licari et al., 2019).

Environmental exposures

Environmental contributions have not been well studied in DCD. Potential exposures include teratogens, infections, poor nutrition, and perinatal brain injury that affect the developing brain (Dewey & Bernier, 2016). Infants who are born preterm are at much higher risk of DCD (Edwards et al., 2011), but it may not be the degree prematurity that influences outcomes, but rather exposures in neonatal intensive care. For example, postnatal infection, early illness severity, repeated exposure to noxious stimuli (e.g., heel lances for blood collection, intubation) have been associated with atypical brain development and poor neurodevelopmental outcomes (Duerden, Taylor, & Miller, 2013). The cerebellum, which has been implicated in DCD (Zwicker et al., 2009), is rapidly developing during the period of prematurity. Exposure to glucocorticoids, opioids, pain-related stress, and infection during this period is associated with altered cerebellar growth and poorer cognitive and motor outcomes (Ranger et al., 2015; Tam, 2018; Zwicker et al., 2016). The association between environmental exposures and DCD in term-born children has not been thoroughly investigated, with the exception of prenatal exposure to alcohol (APA, 2013; Doney et al., 2014; Lucas et al., 2014).

Another important environmental consideration is motor experience (Dewey & Bernier, 2016; Licari & Reynolds, 2017). One of the diagnostic criteria for DCD is ensuring that the motor skill deficit is not related to lack of opportunity for skill learning and use (APA, 2013). Thus, it is important to ensure that individuals have had ample opportunity to practice and learn motor skills before a DCD diagnosis is considered (Blank et al., 2012, 2019). Lower socioeconomic status is also a risk factor for DCD (Gomez & Sirigu, 2015).

Hypothesized mechanisms underlying DCD

Several hypotheses have emerged to explain the mechanism underlying DCD. One hypothesized mechanism is the mirror neuron system (MNS) (Reynolds et al., 2015; Wilson, Smits-Engelsman, Caeyenberghs, Steenbergen, Sugden, et al., 2017). The MNS typically fires when observing another person perform an action. It was initially thought that children with DCD may have deficits in the MNS given their difficulties in mental imagery and imitation of movement, but neuroimaging evidence suggests that there is no difference in MNS activation between children with and without DCD (Blank et al., 2019).

Another potential explanation is the automatization deficit hypothesis, which suggests that children with DCD have difficulty in making motor skills automatic; this explanation is consistent with the cerebellar hypothesis (Brown-Lum & Zwicker, 2015; Zwicker et al., 2009, 2012). Also aligned with involvement of the cerebellum and parietal−cerebellar networks in DCD is the internal modeling deficit (IMD) hypothesis. The internal model is involved in predictive motor control by providing rapid online correction to a motor command; an efferent copy of the motor command is sent to the cerebellum to compare the predicted movement with actual movement, allowing for online adjustments (Wilson, Smits-Engelsman, Caeyenberghs, Steenbergen, Sugden, et al., 2017; Zwicker et al., 2012). Accumulating evidence from studies of visuospatial attention and oculomotor control, manual control, motor imagery, and postural control support the IMD hypothesis by showing that children with DCD have deficits in predictive control of movement (Adams, Lust, Wilson, & Steenbergen, 2014; Barhoun et al., 2019; Gabbard & Bobbio, 2011; Wilson, Smits-Engelsman, Caeyenberghs, and Steenbergen (2017); Wilson, Smits-Engelsman, Caeyenberghs, Steenbergen, Sugden, et al., 2017).

Others argue strongly against the IMD hypothesis and assert that poor perception−action coupling is a more likely explanation for DCD (Wade & Kazeck, 2018). This perspective is based in dynamic systems theory and contends that perception−action coupling results from the interaction of the individual and their environment. Wade and Kazeck (2018) posit that "the deficits in the motor ability of children with DCD may reside in poor perceptual-motor ability that is a consequence of reduced sensitivity to the action opportunities present in the environment, and a limited capacity to fit their own motor abilities to the opportunities for action (affordances) present in the environment" (p. 496).

Recently a hybrid model has been proposed, which combines dynamic systems theory with cognitive neuroscience, demonstrating the link between the brain, cognition, and motor behavior (Blank et al., 2019; Wilson, Smits-Engelsman, Caeyenberghs, and Steenbergen, 2017). Others have extended this model to include a transactional element, by considering interaction of the person, the activity, and the environment (Green & Payne, 2018).

In short, there continues to be controversy and debate regarding the underlying mechanism(s) of DCD. We have only just begun to discover potential neural, genetic, and environmental factors contributing to DCD, let alone the complex interactions between them. A better understanding of the mechanism(s) underlying DCD may inform development or refinement of interventions to improve outcomes of those affected.

Intervention

Several systematic reviews of intervention for DCD have been conducted (Miyahara, Lagisz, Nakagawa, & Henderson, 2017; Preston et al., 2017; Smits-Engelsman et al., 2013, 2018; Yu, Sit, & Burnett, 2018). While each review had different inclusion criteria and slightly different conclusions, evidence has been summarized in the recently published international clinical practice guidelines for DCD (Blank et al., 2019). In short, interventions can be broadly categorized as process-oriented, where the focus of treatment is on the underlying motor deficit (e.g., balance), or task-oriented, where the focus is on functional motor difficulties (e.g., riding a bicycle). The bulk of the evidence favors the task-oriented approach and includes such interventions as task-specific training, Neuromotor Task Training (NTT), and Cognitive Orientation to daily Occupational Performance (CO-OP) (Blank et al., 2012, 2019). While these approaches vary, the common element is that the treatment is focused on specific tasks aimed to improve activity and participation of the child. NTT and CO-OP also use verbal mediation and cognitive strategies to overcome the child's functional motor difficulties. While CO-OP is traditionally an individual-based intervention, recent evidence shows that this treatment can also be effective when delivered in a group format (Anderson, Wilson, & Williams, 2017). Other interventions offered in groups have been difficult to evaluate, but seem to offer some promise (de Hóra, Larkin, & Connell, 2019; Smits-Engelman et al., 2018). Previous recommendations suggested that group intervention may not be effective for young children or those with more severe DCD (Blank et al., 2012).

Intervention using virtual reality appeared to show promising results (Smits-Engelman et al., 2018); however, more recent reviews have found limited evidence to support virtual reality interventions for improving motor skills of children with DCD (Cavalcante Neto et al., 2019; Mentiplay et al., 2019).

The ideal dose, frequency, and intensity of intervention have not been determined (Blank et al., 2019), but evidence suggests that a large training dose and high frequency of practice are more effective (Blank et al., 2019; Yu et al., 2018). To date, most interventions have focused on improving motor skills, but have not directly addressed the psychosocial implications of the disorder. Best practice guidelines suggest that intervention needs to focus on fostering function and participation as well as preventing secondary physical and mental health consequences (Camden, Wilson, Kirby, Sugden, & Missiuna, 2015). In addition, future research needs to examine the long-term outcomes of intervention and whether intervention in childhood can prevent issues in adulthood.

Summary

In summary, DCD is a common neurodevelopmental disorder that is present from early in life but often persists into adulthood. While the core features of the disorder are related to difficulties in motor coordination and motor learning, DCD can have a profound effect on physical health, mental health, psychosocial well-being, and quality of life. The etiology of DCD is likely a complex interplay of neural, genetic, and environmental factors, which may also explain the high co-occurrence with other neurodevelopmental disorders. Effective interventions exist for improving motor skills, but more holistic treatment options are needed to improve overall quality of life and long-term outcomes.

Applications to other areas of development

Cases of "pure" DCD are rare. Estimates suggest that up to 70% of children with DCD meet criteria for at least one other neurodevelopmental disorder (Dewey & Bernier, 2016; Zwicker et al., 2012). The most common co-occurring disorder is ADHD, with more than 50% of children with DCD having ADHD; the reverse is also the case, with at least 50% of children with ADHD having DCD (Blank et al., 2019; Dewey, 2018; Goulardins et al., 2015; Licari et al., 2019; Zwicker et al., 2012). More than half of children with ASD (50%–80%) show signs of DCD (Dewey, 2018; Licari et al., 2019) but only 10%–30% of children with DCD may have ASD (Blank et al., 2019; Licari et al., 2019). In total, 30% to 70% of children with language problems and more than 50% of children with dyslexia will have motor difficulties consistent with DCD (Blank et al., 2019; Zwicker et al., 2012). The high degree of co-occurrence of DCD with other neurodevelopmental disorders has prompted some researchers to suggest that these are not separate disorders, but are reflective of atypical brain development (Dewey & Bernier, 2016; Licari et al., 2019). On the contrary, other researchers provide evidence to support that DCD is a distinct disorder (Blank et al., 2019; Caçola et al., 2017; Goulardins et al., 2015; Kilroy et al., 2019).

There is a strong association of DCD and prematurity; infants who are born very preterm are six to eight times more likely to develop DCD compared to term-born peers (Edwards et al., 2011). While low birthweight or low gestational age has been associated with DCD, recent evidence suggests that exposures in neonatal intensive care (e.g., severe illness, infection, pain-related stress, and medications, such as glucocorticoids and opioids) are associated with altered brain development in the neonatal period and poorer neurodevelopmental outcomes in childhood (Duerden et al., 2013; Ranger et al., 2015; Tam, 2018; Zwicker et al., 2016). These findings provide encouraging evidence that there may be modifiable risk factors to decrease the incidence of DCD in this high-risk population.

Key facts

- DCD affects 5%–6% of school-age children.
- Infants born very preterm are at high risk of developing DCD.
- DCD extends into adulthood in almost 75% of cases.
- Co-occurring conditions are common, with up to 70% of children with DCD meeting criteria for at least one other neurodevelopmental disorder.
- Compared to typically developing children, children with DCD show differences in brain structure and function, most commonly in the cerebellum, basal ganglia, parietal lobe, and frontal lobe.
- Heritability of DCD is estimated at 70%, but few genetic studies have been conducted.

Summary points

- This chapter focuses on a common but under-recognized neurodevelopmental disorder known as developmental coordination disorder (DCD).
- DCD affects the ability to learn motor skills and significantly interferes with daily life, school achievement, vocational choices, and leisure pursuits.
- DCD often occurs with other neurodevelopmental disorders, such as attention deficit hyperactivity disorder, autism spectrum disorder, learning disabilities, and specific language impairment.
- While considered a developmental disorder, DCD can persist into adulthood and have long-term consequences on physical health (e.g., obesity, cardiovascular disease), mental health (e.g., anxiety, depression), and overall quality of life.
- The cause of DCD is largely unknown, but is likely a complex interplay of neural, genetic, and environmental factors.
- Task-specific rehabilitation interventions can improve specific motor skills, but more comprehensive treatments are required to address or prevent the secondary consequences of DCD.

References

Adams, I. L., Lust, J. M., Wilson, P. H., & Steenbergen, B. (2014). Compromised motor control in children with DCD: A deficit in the internal model? A systematic review. *Neuroscience and Biobehavioral Reviews, 47*, 225–244. https://doi.org/10.1016/j.neubiorev.2014.08.011.

American Psychiatric Association. (2013). *Diagnostic and statistical manual of mental disorders - 5th ed. (DSM-5)*. Washington, DC: Author.

Anderson, L., Wilson, J., & Williams, G. (2017). Cognitive Orientation to daily Occupational Performance (CO-OP) as group therapy for children living with motor coordination difficulties: An integrated literature review. *Australian Occupational Therapy Journal, 64*, 170–184. https://doi.org/10.1111/1440-1630.12333.

Barhoun, P., Fuelscher, I., Kothe, E. J., He, J. L., Youssef, G. J., Enticott, P. G., … Hyde, C. (2019). Motor imagery in children with DCD: A systematic and meta-analytic review of hand-rotation task performance. *Neuroscience and Biobehavioral Reviews, 99*, 282–297. https://doi.org/10.1016/j.neubiorev.2019.02.002.

Biotteau, M., Chaix, Y., Blais, M., Tallet, J., Péran, P., & Albaret, J. M. (2016). Neural signature of DCD: A critical review of MRI neuroimaging studies. *Frontiers in Neurology, 7*. https://doi.org/10.3389/fneur.2016.00227. Article 227.

Blank, R., Barnett, A. L., Cairney, J., Green, D., Kirby, A., Polatajko, H., … Vinçon, S. (2019). International clinical practice recommendations on the definition, diagnosis, intervention, and psychosocial aspects of developmental coordination disorder. *Developmental Medicine and Child Neurology, 61*, 242–285. https://doi.org/10.1111/dmcn.14132.

Blank, R., Smits-Engelsman, B., Polatajko, H., & Wilson, P. (2012). European Academy of Childhood Disability (EACD): Recommendations on the definition, diagnosis, and intervention of developmental coordination disorder (long version). *Developmental Medicine and Child Neurology, 54*, 54–93. https://doi.org/10.1111/j.1469-8749.2011.04171.x.

Bo, J., & Lee, C. M. (2013). Motor skill learning in children with developmental coordination disorder. *Research in Developmental Disabilities, 34*, 2047–2055. https://doi.org/10.1016/j.ridd.2013.03.012.

Brown-Lum, M., & Zwicker, J. G. (2015). Brain imaging increases our understanding of developmental coordination disorder: A review of literature and future directions. *Current Developmental Disorders Reports, 2*(2), 131–140. https://doi.org/10.1007/s40474-015-0046-6.

Brown-Lum, M., & Zwicker, J. G. (2017). Neuroimaging and occupational therapy: Bridging the gap to advance rehabilitation in developmental coordination disorder. *Journal of Motor Behavior, 49*, 98–110. https://doi.org/10.1080/00222895.2016.1271295.

Cairney, J., Veldhuizen, S., & Szatmari, P. (2010). Motor coordination and emotional-behavioral problems in children. *Current Opinion in Psychiatry, 23*, 324–329. https://doi.org/10.1097/YCO.0b013e32833aa0aa.

Camden, C., Wilson, B., Kirby, A., Sugden, D., & Missiuna, C. (2015). Best practice principles for management of children with developmental coordination disorder: Results of a scoping review. *Child: Care, Health and Development, 41*, 147–159. https://doi.org/10.1111/cch.12128.

Cavalcante Neto, J. L., de Oliveira, C. C., Greco, A. L., Zamunér, A. R., Moreira, R. C., & Tudella, E. (2019). Is virtual reality effective in improving the motor performance of children with developmental coordination disorder? A systematic review. *European Journal of Physical and Rehabilitation Medicine, 55*, 291–300. https://doi.org/10.23736/s1973-9087.18.05427-8.

Caçola, P. (2016). Physical and mental health of children with developmental coordination disorder. *Frontiers in Public Health, 4*. https://doi.org/10.3389/fpubh.2016.00224. Article 224.

Caçola, P., Miller, H. L., & Williamson, P. O. (June 2017). Behavioral comparisons in autism spectrum disorder and developmental coordination disorder: A systematic literature review. *Research in Autism Spectrum Disorders, 38*, 6–18. https://doi.org/10.1016/j.rasd.2017.03.004.

Dewey, D. (2018). What is comorbidity and why does it matter in neurodevelopmental disorders? *Current Developmental Disorders Reports, 5*(4), 235–242. https://doi.org/10.1007/s40474-016-0086-6.

Dewey, D., & Bernier, F. P. (2016). The concept of atypical brain development in developmental coordination disorder (DCD) — a new look. *Current Developmental Disorders Reports, 3*(2), 161–169. https://doi.org/10.1007/s40474-016-0086-6.

Doney, R., Lucas, B. R., Jones, T., Howat, P., Sauer, K., & Elliott, E. J. (2014). Fine motor skills in children with prenatal alcohol exposure or fetal alcohol spectrum disorder. *Journal of Developmental and Behavioral Pediatrics, 35*, 598–609. https://doi.org/10.1097/DBP.0000000000000107.

Draghi, T. T. G., Cavalcante Neto, J. L., Rohr, L. A., Jelsma, L. D., & Tudella, E. (2020). Symptoms of anxiety and depression in children with developmental coordination disorder: A systematic review. *Jornal de Pediatria, 96*(1), 8–19. https://doi.org/10.1016/j.jped.2019.03.002.

Duerden, E. G., Taylor, M. J., & Miller, S. P. (2013). Brain development in infants born preterm: Looking beyond injury. *Seminars in Pediatric Neurology, 20*, 65–74. https://doi.org/10.1016/j.spen.2013.06.007.

Edwards, J., Berube, M., Erlandson, K., Haug, S., Johnstone, H., Meagher, M., … Zwicker, J. G. (2011). Developmental coordination disorder in school-aged children born very preterm and/or with very low birth weight: A systematic review. *Journal of Developmental and Behavioral Pediatrics, 32*, 678–687. https://doi.org/10.1097/DBP.0b013e31822a396a.

Ferguson, G. D., Jelsma, J., Versfeld, P., & Smits-Engelsman, B. C. M. (2014). Using the ICF framework to explore the multiple interacting factors associated with developmental coordination disorder. *Current Developmental Disorders Reports, 1*(2), 86–101. https://doi.org/10.1007/s40474-014-0013-7.

Gabbard, C., & Bobbio, T. (2011). The inability to mentally represent action may be associated with performance deficits in children with developmental coordination disorder. *International Journal of Neuroscience, 121*, 113–120. https://doi.org/10.3109/00207454.2010.535936.

Gomez, A., & Sirigu, A. (2015). Developmental coordination disorder: Core sensori-motor deficits, neurobiology, and etiology. *Neuropsychologia, 79*(Pt B), 272–287. https://doi.org/10.1016/j.neuropsychologia.2015.09.032.

Goulardins, J. B., Rigoli, D., Licari, M., Piek, J., Hasue, R. H., Oosterlaan, J., & Oliveira, J. A. (2015). Attention deficit hyperactivity disorder and developmental coordination disorder: Two separate disorders or do they share a common etiology. *Behavioural Brain Research, 292*, 484–492. https://doi.org/10.1016/j.bbr.2015.07.009.

Green, D., & Payne, S. (2018). Understanding organisational ability and self-regulation in children with developmental coordination disorder. *Current Developmental Disorders Reports, 5*(1), 34–42. https://doi.org/10.1007/s40474-018-0129-2.

Harris, S. R., Mickelson, E. C., & Zwicker, J. G. (2015). Diagnosis and management of developmental coordination disorder. *Canadian Medical Association Journal, 187*, 659–665. https://doi.org/10.1503/cmaj.140994.

Hendrix, C. G., Prins, M. R., & Dekkers, H. (2014). Developmental coordination disorder and overweight and obesity in children: A systematic review. *Obesity Reviews, 15*, 408–423. https://doi.org/10.1111/obr.12137.

de Hóra, N., Larkin, L., & Connell, A. (2019). Is group-based or individual-based intervention more effective for quality of life outcomes in children with developmental coordination disorder? A systematic review. *Physical and Occupational Therapy in Pediatrics, 39*, 353–371. https://doi.org/10.1080/01942638.2018.1490847.

Hyde, C., Fuelscher, I., & Williams, J. (2019). Neurophysiological approaches to understanding motor control in DCD: Current trends and future directions. *Current Developmental Disorders Reports, 6*(2), 78–86. https://doi.org/10.1007/s40474-019-00161-1.

Kilroy, E., Cermak, S. A., & Aziz-Zadeh, L. (2019). A review of functional and structural neurobiology of the action observation network in autism spectrum disorder and developmental coordination disorder. *Brain Sciences, 9*. https://doi.org/10.3390/brainsci9040075. Article 75.

Kirby, A., Sugden, D., & Purcell, C. (2014). Diagnosing developmental coordination disorders. *Archives of Disease in Childhood, 99*, 292–296. https://doi.org/10.1136/archdischild-2012-303569.

Leonard, H. C., & Hill, E. L. (2015). Executive difficulties in developmental coordination disorder: Methodological issues and future directions. *Current Developmental Disorders Reports, 2*(2), 141–149. https://doi.org/10.1007/s40474-015-0044-8.

Licari, M. K., Finlay-Jones, A., Reynolds, J. E., Alvares, G. A., Spittle, A. J., Downs, J., … Varcin, K. (2019). The brain basis of comorbidity in neurodevelopmental disorders. *Current Developmental Disorders Reports, 6*(1), 9–18. https://doi.org/10.1007/s40474-019-0156-7.

Licari, M. K., & Reynolds, J. E. (2017). Understanding performance variability in developmental coordination disorder: What does it all mean? *Current Developmental Disorders Reports, 4*(2), 53–59. https://doi.org/10.1007/s40474-017-0112-3.

Lucas, B. R., Latimer, J., Pinto, R. Z., Ferreira, M. L., Doney, R., Lau, M., … Elliott, E. J. (2014). Gross motor deficits in children prenatally exposed to alcohol: A meta-analysis. *Pediatrics, 134*, e192–209. https://doi.org/10.1542/peds.2013-3733.

Mancini, V., Rigoli, D., Cairney, J., Roberts, L. D., & Piek, J. (2016). The elaborated environmental stress hypothesis as a framework for understanding the association between motor skills and internalizing problems: A mini-review. *Frontiers in Psychology, 7*. https://doi.org/10.3389/fpsyg.2016.00239. Article 239.

Mentiplay, B. F., FitzGerald, T. L., Clark, R. A., Bower, K. J., Denehy, L., & Spittle, A. J. (2019). Do video game interventions improve motor outcomes in children with developmental coordination disorder? A systematic review using the ICF framework. *BMC Pediatrics, 19*. https://doi.org/10.1186/s12887-018-1381-7. Article 22.

Missiuna, C., Gaines, R., & Soucie, H. (2006). Why every office needs a tennis ball: A new approach to assessing the clumsy child. *Canadian Medical Association Journal, 175*, 471–473. https://doi.org/10.1503/cmaj.051202.

Missiuna, C., Mandich, A., Polatajko, H. J., & Malloy-Miller, T. (2001). Cognitive Orientation to daily Occupational Performance (CO-OP): Part I — theoretical foundations. *Physical and Occupational Therapy in Pediatrics, 20*, 69–81.

Miyahara, M., Lagisz, M., Nakagawa, S., & Henderson, S. E. (2017). A narrative meta-review of a series of systematic and meta-analytic reviews on the intervention outcome for children with developmental co-ordination disorder. *Child: Care, Health and Development, 43*, 733–742. https://doi.org/10.1111/cch.12437.

Omer, S., Jijon, A. M., & Leonard, H. C. (2019). Research review: Internalising symptoms in developmental coordination disorder: A systematic review and meta-analysis. *Journal of Child Psychology and Psychiatry, 60*, 606–621. https://doi.org/10.1111/jcpp.13001.

Peters, L. H. J., Maathuis, C. G. B., & Hadders-Algra, M. (2013). Neural correlates of developmental coordination disorder. *Developmental Medicine and Child Neurology, 55*(Suppl. 4), 59–64. https://doi.org/10.1111/dmcn.12309.

Polatajko, H. J., Mandich, A. D., Missiuna, C., Miller, L. T., Macnab, J. J., Malloy-Miller, T., & Kinsella, E. A. (2001). Cognitive Orientation to daily Occupational Performance (CO-OP): Part III − the protocol in brief. *Physical and Occupational Therapy in Pediatrics, 20*, 107−123.

Preston, N., Magallón, S., Hill, L. J., Andrews, E., Ahern, S. M., & Mon-Williams, M. (2017). A systematic review of high quality randomized controlled trials investigating motor skill programmes for children with developmental coordination disorder. *Clinical Rehabilitation, 31*, 857−870. https://doi.org/10.1177/0269215516661014.

Ranger, M., Zwicker, J. G., Chau, C. M. Y., Park, M. T. M., Chakravarty, M. M., Poskitt, K. J., … Grunau, R. E. (2015). Neonatal pain and infection relate to smaller cerebellum in very preterm children at school age. *The Journal of Pediatrics, 167*, 292−298. https://doi.org/10.1016/j.jpeds.2015.04.055. e1.

Reynolds, J. E., Thornton, A. L., Elliott, C., Williams, J., Lay, B. S., & Licari, M. K. (2015). A systematic review of mirror neuron system function in developmental coordination disorder: Imitation, motor imagery, and neuroimaging evidence. *Research in Developmental Disabilities, 47*, 234−283. https://doi.org/10.1016/j.ridd.2015.09.015.

Rivilis, I., Hay, J., Cairney, J., Klentrou, P., Liu, J., & Faught, B. E. (2011). Physical activity and fitness in children with developmental coordination disorder: A systematic review. *Research in Developmental Disabilities, 32*, 894−910. https://doi.org/10.1016/j.ridd.2011.01.017.

Schoemaker, M. M., & Smits-Engelman, B. M. C. (2005). Neuromotor task training: A new approach to treat children with DCD. In D. A. Sugden, & M. Chambers (Eds.), *Children with developmental coordination disorder* (pp. 212−227). London: Whurr.

Smits-Engelsman, B. C., Blank, R., van der Kaay, A. C., Mosterd-van der Meijs, R., Vlugt-van den Brand, E., … Wilson, P. H. (2013). Efficacy of interventions to improve motor performance in children with developmental coordination disorder: A combined systematic review and meta-analysis. *Developmental Medicine and Child Neurology, 55*, 229−237. https://doi.org/10.1111/dmcn.12008.

Smits-Engelsman, B., Vinçon, S., Blank, R., Quadrado, V., Polatajko, H., & Wilson, P. H. (2018). Evaluating the evidence for motor-based interventions in developmental coordination disorder: A systematic review and meta-analysis. *Research in Developmental Disabilities, 74*, 72−102. https://doi.org/10.1016/j.ridd.2018.01.002.

Spittle, A. J., & Orton, J. (2014). Cerebral palsy and developmental coordination disorder in children born preterm. *Seminars in Fetal and Neonatal Medicine, 19*, 84−89. https://doi.org/10.1016/j.siny.2013.11.005.

Tal Saban, M., & Kirby, A. (2018). Adulthood in developmental coordination disorder (DCD): A review of current literature based on ICF perspective. *Current Developmental Disorders Reports, 5*(1), 9−17. https://doi.org/10.1007/s40474-017-0115-0.

Tam, E. (2018). Cerebellar injury in preterm infants. *Handbook of Clinical Neurology, 155*, 49−59. https://doi.org/10.1016/B978-0-444-64189-2.00003-2.

Wade, M. G., & Kazeck, M. (2018). Developmental coordination disorder and its cause: The road less travelled. *Human Movement Science, 57*, 489−500. https://doi.org/10.1016/j.humov.2016.08.004.

Wilson, P. H., Ruddock, S., Smits-Engelsman, B., Polatajko, H., & Blank, R. (2013). Understanding performance deficits in developmental coordination disorder: A meta-analysis of recent research. *Developmental Medicine and Child Neurology, 55*, 217−228. https://doi.org/10.1111/j.1469-8749.2012.04436.x.

Wilson, P. H., Smits-Engelsman, B., Caeyenberghs, K., & Steenbergen, B. (2017). Towards a hybrid model of developmental coordination disorder. *Current Developmental Disorders Reports, 4*(3), 64−71. https://doi.org/10.1007/s40474-017-0115-0.

Wilson, P. H., Smits-Engelsman, B., Caeyenberghs, K., Steenbergen, B., Sugden, D., Clark, J., … Blank, R. (2017). Cognitive and neuroimaging findings in developmental coordination disorder: New insights from a systematic review of recent research. *Developmental Medicine and Child Neurology, 59*, 1117−1129. https://doi.org/10.1111/dmcn.13530.

Yu, J. J., Burnett, A. F., & Sit, C. H. (2018). Motor skill interventions in children with developmental coordination disorder: A systematic review and meta-analysis. *Archives of Physical Medicine and Rehabilitation, 99*, 2076−2099. https://doi.org/10.1016/j.apmr.2017.12.009.

Zwicker, J. G., Harris, S. R., & Klassen, A. F. (2013). Quality of life domains affected in children with developmental coordination disorder: A systematic review. *Child: Care, Health and Development, 39*, 562−580. https://doi.org/10.1111/j.1365-2214.2012.01379.x.

Zwicker, J. G., Miller, S. P., Grunau, R. E., Chau, V., Brant, R., Studholme, C., … Tam, E. W. (2016). Smaller cerebellar growth and poorer neurodevelopmental outcomes in very preterm infants exposed to morphine. *Journal of Pediatrics, 172*, 81−87. https://doi.org/10.1016/j.jpeds.2015.12.024. e2.

Zwicker, J. G., Missiuna, C., & Boyd, L. A. (2009). Neural correlates of developmental coordination disorder: A review of hypotheses. *Journal of Child Neurology, 24*, 1273−1281. https://doi.org/10.1177/0883073809333537.

Zwicker, J. G., Missiuna, C., Harris, S. R., & Boyd, L. A. (2012). Developmental coordination disorder: A review and update. *European Journal of Paediatric Neurology, 16*, 573−581. https://doi.org/10.1016/j.ejpn.2012.05.005.

Zwicker, J. G., Montgomery, I., & Mickelson, E. C. R. (2018). *DCD toolkit for pediatricians.* Available from: http://www.childdevelopment.ca/DCDAdvocacyToolkit/DCDAdvocacyToolkitIntro.aspx.

Part III

Biomarkers, screening, methods and diagnosis

Chapter 28

Neurodevelopment and the Ages and Stages Questionnaire, third edition (ASQ-3)

N. Granana[1] and A.M. Romero Otalvaro[2]

[1]Department of Pediatrics, Hospital C Durand, Buenos Aires, Argentina; [2]Universidad Pontificia Bolivariana, Monteria, Cordoba, Colombia

List of abbreviations

ASQ-3 Ages and Stages Questionnaire, third edition
EI Early intervention
LMIC Low-middle income countries

Mini-dictionary of terms

Autism Difficulty in social interaction and communication and by restricted or repetitive patterns of thought and behavior.
Autistic disorder Neurodevelopmental disorder with social communication deficits and repetitive and restrictive behaviors, which begin during development.
Daycare centers Also called day nursery, nursery school, institution that provides supervision and care of infants and young children during the daytime, particularly so that their parents can hold jobs.
Early detection During development, intervention is considered before a pathological diagnosis is confirmed.
Global development Development skills in mainly five areas: communication, fine motor, gross motor, social−personal, and problem-solving skills.
Language delay Language development below capacities for a certain age and is one of the language disorders in which the child fails to develop language abilities at the usual age appropriate for their developmental.
Neurodevelopmental skills Skills emerging during child maturation.
Primary care provider Is a health care practitioner who sees children that have common medical problems. This person is most often a doctor, but may be a physician assistant or a nurse practitioner.
Psychoeducation Refers to the process of providing education and information to those seeking or receiving mental health services, such as people diagnosed with mental health conditions (or life-threatening/terminal illnesses) and their family members.
Systematized screening Systematized assessment during well-health control in children during the first 5 years at pediatric controls.
Well-child visits A well-child visit is a routine visit scheduled by your child primary care provider (PCP) for routine checkup. This visit is different form sick visit because more is done for your child in a well-child visit also referred to as preventive visit.

Introduction

There is a 3%−8% estimated risk for neurodevelopmental disorders in children in advanced countries (Weiss & Landrigan, 2000). The etiologies are neurobiological injuries and socio environmental adversity, with higher prevalences in developing countries. The California Adverse Childhood Experiences Study followed 17,000 children at risk from childhood. Without early intervention (EI), the study found they developed poor academic achievement at school and increased the frequency of behavior problems, criminal conduct, violence and mental disorders (Anda et al. 2006; Perry, 2002). They reached less employment and social inclusion. The scientific rationale for EI is overwhelming (Sinclair, 2007).

Diagnosis, Management and Modeling of Neurodevelopmental Disorders. https://doi.org/10.1016/B978-0-12-817988-8.00028-2

Experienced early life stress is associated with a deficit in the development of nucleus accumbens at fMRI, mostly known as a reward center. Hypoactivation during adolescence is followed by anxiety, depression, less resilience skills, and a higher risk of addictive behaviors (Goff et al., 2013).

Moreover, early detection and intervention are followed by a reduction in criminal aggression, parenting stress, and improvement in parenting competencies (AAP, American Academy of Pediatrics, 2006; Allen, 2011). For parents, there are fewer subsequent pregnancies, months on welfare, reports of child abuse, and arrests of teenagers (DCSF, 2007).

Long-term benefits of EI predicted higher academic achievements (Vandell, Burchinal, & Pierce, 2016).

Systemized screening of neurodevelopment at well-child controls in children is useful for early detection of global developmental delays from 1 to 66 months of age. It covers also specific language and autistic traits. Use of questionnaires in well-child visits helps psychoeducation of parents on development and let to improve skills in the long-term follow-up. It is effective for daycare centers for detection but also EI for primary care providers.

Systemized neurodevelopmental screening

The systemized assessment of neurodevelopment during the first 60 months of life constitute a fundamental part of well-child visits. The American Academy of Pediatrics recommends using a standardized instrument at least at 9, 18, and 30 (or 24) months of age covered by legislation (AAP, 2006); in the United States, the identification and intervention are mandatory through the Individuals with Disabilities Education Act (IDEA, 2004). Questionnaires improve rates of detection (Table 28.1), and early intervention rates. Furthermore, underdetection is followed by even less access to treatment.

The policy statement showed that primary care providers tend to under detect developmental disorders. Considerable evidence demostrated clinical guidelines are insufficient to ensure their widespread adoption (King et al., 2010). King et al. study found that no pediatrician in the community settings had been implementing developmental screening defined by the AAP policy statement. Some questionnaires like Denver II are used in an unstructured way; or only in some patients whose surveillance had already raised suspicions for developmental delays, for clinical evident concerns.

It is feasible to use a standardized instrument during well-child visits by pediatricians and residents of pediatrics. During our validation of Ages & Stages Questionnaires (ASQ-3), residents employed it more consistently (Otalvaro et al., 2018). Senior staff is untrained to manipulate structured instruments at regular follow-up. It can be implemented also by psychological and educational teams.

Types of instruments

There are different options to decide what instrument can be implemented:

- Type of instrument:
  - Structured interviews: with activities to be held with children, increase the reliability and validity of traditional child diagnostic procedures.
  - Questionnaires: reports from parents, caregivers or professionals. These are cost-effective instruments and have become increasingly used for screening surveillance at epidemiological studies.
  - Combined questionnaires: questionnaires with activities, like ASQ-3; improve confidence. ASQ-3 includes activity sheets at every range of age, which additionally serve to provide tools to the family for EI, as a unique opportunity for psychoeducation. Combinations between questionnaires and playing activities increase sensitivity and specificity. Most parents have reported ASQ as very easy to use, not too time-consuming, and the preferred screening instrument for community health primary care providers (Hix-Small, Marks, Squires, & Nickel, 2007).

TABLE 28.1 Early detection of developmental disorders.

Early detection	Without scales	With scales
Developmental Delay	30%	70%–80%
Author	Palfrey et al. (1987)	Squires, Nickel, & Eisert, 1996

%: percentages.

Range of application: most questionnaires cover 0–3 years of age assessment and do not include social skills. Those including 0–5 years of age give a better amplitude of range on the assessment.

- Domains of development:
  - General surveillance instruments should include motor, language, socioemotional and cognitive domains.
  - Specific instruments include certain areas: a separate measure, the ASQ: SE-2, focus on behavioral and social-emotional/mental health issues; the Checklist of Autism in Toddlers (CHAT) covers autism detection. Adding specific areas increases the length of the interview; it is significant to decide what to prioritize on a given schedule.
  - Time: screeners routinely can take from 10 to 40 minutes to be filled. ASQ-3 is easy to administer, it took approximately 10–13 to complete and 2–3 minutes to score during our validation process.
  - Label of certification: Some instruments require certification to be administered by specialists and others can be filled by parents or caregivers without specific training.

ASQ-3 does not require too much training, and it is practical for professionals or families far away from training centers, using appropriate cutoff scores (Alvik & Grøholt, 2011). It should be stated that these reports rely on the level of reading. In low-middle income countries (LMIC) we consider that is more adequate to interview parents by the professional or caregiver to increase confidence (Otalvaro et al., 2018).

- Psychometric properties: Validation on sensitivity, specificity, positive predicted value, and language adaptation is required for the closest population. ASQ-3 was standardized originally in 15,438 children and has many translations and validation all over the world.
- Results: ASQ-3 scores are depicted simply by a numerical output that is characterized by cutoff points on normal, at-risk, or below-normal areas. However, each domain also is scored separately to provide more informative results. This would allow to monitor general progress over time and perform comparisons between areas. Children scoring at or below the cutoff on one or more areas should be considered for referral for further assessment.

Table 28.2 provides a comparison between instruments. A review of child development measures showed that ASQ-3 and Parents' Evaluation of Developmental Status (PEDS) scales met most criteria from 35 scales. It doesn't measure the same areas as ASQ-3. The authors could not locate formal peer-reviewed evaluations and only limited evidence of parental acceptability for PEDS (Bedford, Walton, & Ahn, 2013).

TABLE 28.2 Developmental screening instruments.

Scale	Type	Age rank	Domains	Time	Psychometrics
Ages and Stages Questionnaire (ASQ-3) (Squires et al., 1996)	Combination of parent report and direct elicitation.	1 –66 months	Communication, fine motor, gross motor, problem-solving, social personal	10 –15 min	Sensitivity ranged 70%–90% specificity 76%–91% reliability test-retest; 44%–83% internal consistency; 76%–91% concurrent validity
DDST: Denver II-(Glascoe, 1993).	Combination of parent report, direct elicitation, and observation.	0 month- 3 years	Personal–social, fine motor, gross motor language	15 –20 min	22% sensitivity, 86% specificitiviy; positive predictive value 28%
CAT/CLAMSClinical Adaptive Test/Clinical Linguistic Auditory Milestones Scale (Capute, 1996)	Observational and parent questionnaire	1–3 years	Motor, language, problem-solving	15 –20 min	66% sensitivity; 79% specificity, positive predictive value 80%, negative predictive value 65%

Comparison between psychometric properties within scales.

ASQ-3

ASQ-3 focuses on early detection for delays in five developmental domains: communication, gross and fine motor, problem-solving, and personal–social skills. Questionnaires include a set of 21 structured questionnaires to screen the development of children from 1 to 66 months of age. It can be addressed with preterm infants correcting the gestational age.

Detection may be employed during the regular childcare from physician and parents reports. Written for a fourth–sixth-grade level of reading skills, it can also be applied in mass mail-outs for Child Find programs. The Manual contains extensive instructions. There are many training options and can be administered by parents at home or in waiting rooms. The kit of materials can be assembled from the list provided in the manual. Cutoff scores (set at 1 ½ or two SDs below the mean) indicate the need for referral or monitoring.

Each questionnaire consists of 30 items scoring "yes," "sometimes," or "not yet" depending on whether the child can perform the activity. It has been validated and translated from English to many languages with other translations underway. We describe the validation of ASQ-3 in Argentine Spanish at an LMIC population and its applications.

Validations of ASQ-3

The ASQ has been validated in many countries, in diverse languages. It has been translated and used in a number of European settings as well as in North and South America, Asia, and Australia, but not validated in all of them (Kerstjens et al., 2009). It possesses good psychometric properties and clinical utility as a cross-cultural measure of neurodevelopment, as is presented in a comparative revision of 53 articles written in English (Small, Hix-Small, Vargas-Baron, & Marks, 2019).

Validation requires an assessment of a large representative sample and production of scores. Hamilton has suggested that, for developmental screening tests, sensitivity of 70%–80% and specificity close to 80% are acceptable (Hamilton, 2006).

As an example, Argentine validation was conducted in 630 children between 1 and 66 months of age, with a homogeneous distribution by sex. It was taken randomly from the well-child visits, whose parents agreed to participate. Thirty-five children who consulted due to delayed acquisition of development were excluded. For local adaptation, the combined sample obtained was established. It was brief to administer and quick to rate. It was comprehensible for the hospital team to be trained. The administrator only required pencil and paper for scoring and a booklet with the EI guidelines to deliver to the parents who handled the survey.

The socioeconomic level of the families was determined according to Graffar Mendez scale (Mendez Castellano, 1994). Sociodemographic factors include maternal schooling degree, labor activity of the head of the family, and housing facilities. There are five grades: I, being the highest to V, the lowest in social status (Mendez Castellano, 1994). Most subjects were at IV (low-level: 27%) and V strata (Very low income level: 33%). The distribution of levels was consistent with the local population rates.

Interviews were conducted by members of the developmental team, which included pediatricians, psychologists and pedagogues, taking account of diverse levels of skills (interrater reliability). Questionnaires were returned by the mothers in 95% of the sample. Most assessments completed at low-middle income countries were obtained by professional interviewers to parents or caregivers or by direct assessment to the child (Small et al., 2019). Our experience is that some parents do not possess enough reading skills to complete a reliable answer. The questions need to be adapted to the level of comprehension. Before conducting the interview, it was explained to parents that the purpose was to monitor their child's development and to plan education.

The empirical validity was acceptable. Both the sensitivity of 88% and specificity of 94%, as the positive predictive value of 88% and negative of 96% compared to the local version of Denver Scale-II (PRUNAPE). The results show a Chi-square statistical value, with $P < 0.01$; therefore, there was evidence of a highly significant dependence between the results. The correlation of the tests for the overall analysis between the identification of the deficit and the standard subjects was significant (r: 0.81; *P*: .00).

The sample showed that 19.5% of the children were cataloged at risk (Fig. 28.1). Children, from 36 months of age reported the most considerable frequency of the deficit. The risk sample presented neurodevelopmental disorders in 12.5%, as depicted in Table 28.3. Frequent domains were language and global delay. Both were the most consistent diagnoses if the risk was persistent at the communicational and global level. Autism spectrum disorder (ASD) obtained a consistent syndrome when a inadequacy in language and personal–social skills was persistent at follow-up. The motor compromise was less consistent with gross and fine motor deficits at the follow-up visits and problem-solving alone did not match any specific syndrome.

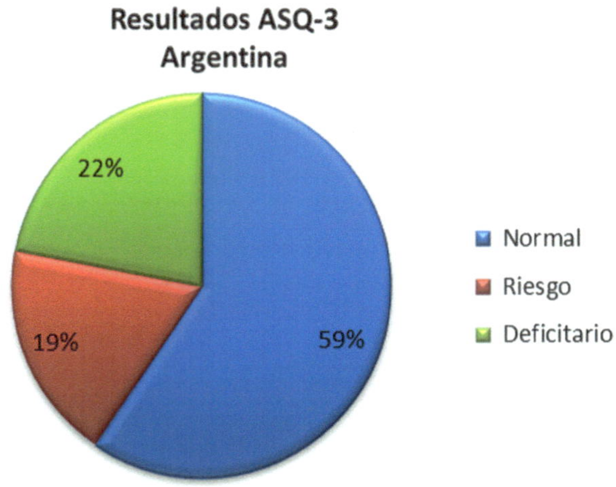

FIGURE 28.1 Argentine sample ASQ-3. Blue: normal; Red: at risk; Green: below risk.

TABLE 28.3 Frequency of risk of neurodevelopmental deficit in a sample of 123 out of 630 Latin children.

Deficit	N	Frequency (%)	Clinical diagnoses	N	Frequency (%)
Communication	29	4,6	Language delay	27	4,3
Gross motor	25	4			
Fine motor	24	3,8	Motor delay	16	2,5
Problem-solving	23	3,6			
Personal social	22	3,5			
Communication plus personal–social	7	1,1	Autism spectrum disorder	7	1,1
Global	31	4,9	Global delay	30	4,7
Total	123	19,5	Total	80	12,6

From Otalvaro, A. M. R., Grañana, N., Gaeto, N., Torres, M. D. L. Á., Zamblera, M. N., Vasconez, M. A., ... & Squires, J. (2018). ASQ-3: Validation of the Ages and Stages Questionnaire for the detection of neurodevelopmental disorders in Argentine children. *Archivos Argentinos de Pediatria, 116*(1), 7–13.

Comparison validations of ASQ-3 between different populations

Kerstjens (2009) made a comparison between Dutch and global ASQ validations. At least 10 out of 15 mean scores with other countries yielded statistically significant differences, but remarkably small. Problem-solving percentiles were higher in the US sample. Fine motor ratings were higher in the Dutch, Norwegian, and Korean samples. They considered that there was no real explanation for these characteristics. The small effect size of most of the disagreement and the absence of more clinically relevant divergences supported the cross-continental usefulness of the ASQ. Even though there are few cross-cultural variances, there still is the need for careful adaptation and validation of developmental screeners for specific cultural settings and languages.

A comparison between US 24 months-old lactants between Latin validations from Colombia, Ecuador, and Argentina showed that communication, personal–social and fine motor cutoffs were significantly lower at Argentine and Ecuadorian children compared with US samples. Colombian sample showed differences in fine motor only (Table 28.4).

Analysis of gender differences showed that girls scored higher on all domains in Dutch validation, which reached statistical significance for fine motor functioning, personal–social, problem-solving, and total results. These differences are consistent with the Norwegian findings (Kerstjens et al., 2009). The "gold standard" neurodevelopmental tests have identical cutoff points for boys and girls in this age group (Hintz et al., 2006). It could be debated whether separate cutoff points are required for girls and boys, as it is the case with behavioral measures like the Child Behavior Checklist.

ASQ performance resulted accurate in detecting legitimate problems in apparently healthy and with biological risk factors such as prematurity (Klamer, Lando, Pinborg, & Greisen, 2005), single-parent families (Kerstjens et al., 2009), with low income and maternal education (Otalvaro et al., 2018).

TABLE 28.4 Comparison of Argentine mean scores with US, Colombian, Ecuadorian, and Argentine scores on the ASQ 24 months form.

	US	Colombia	Argentine	Ecuador
N	30?	30	30	30
Commun cation	51 (13)	53 (8.7)	44 (11)p***#	42 (13)p***#
Fine motor	53(9)	44.7(13.1)p***#	45 (12)p***#	48 (7)p*
Gross motor	54 (8)	49.5(10.6)p**	53 (12)	50 (7)p*
Problem-solving	48 (9)	52 (8.9)p*	47 (9)	42 (8)p**#
Personal social	50,5 (9.5)	53 (9.2)	45 (9)p**#	40 (8)p***#

p*** (<0.005); p** (<0.01); # difference >5 points.

Reports comparing ASQ, Second Edition, and Bayley Scales of Infant Development, Second Edition, in low-risk term children (Gollenberg, Lynch, Jackson, McGuinness, & Msall, 2010) and Bayley III (Schonhaut, Armijo, Schönstedt, Alvarez, & Cordero, 2013) showed agreement between results. Studies with no agreement such as from Rubio-Codina, Araujo, Attanasio, Muñoz, and Grantham-McGregor (2016) had been made without validation on the population screened. Canadian Task Force considered the scales were not confident only because of a leak of adult follow-up studies (Canadian Task Force 2016; McPheeters et al., 2016). This information could provide valuable data to determine where more (not less) resources are required to reduce inequalities. It would also help to include baseline data to monitor the effects of interventions.

Screening and assessment of global development and specific developmental problems: the case of language delay and autism

The increased prevalence of autistic spectrum disorders has fostered research efforts on the development and validation of specific instruments in young children. In fact, there are several autism-specific tools meant to be used with young children in various stages of development; they include the Socio-Emotional Ages and Stages Questionnaire (ASQ-SE), diverse versions of Checklist for Autism in Toddlers (CHAT) and Pervasive Developmental Disorders Screening Test (PDDST). The AAP recommends using at 18, 24 months and less than 3 years of well-child visits.

A revision of assessment with specific scales showed that CHAT checklists were disrupted with high false-positive rates. MCHAT, the version that has been translated and adapted to diverse languages, maintained a lot of reasons for false-positive findings, including developmental concerns, that may resolve with maturation. Behaviors in typically developing toddlers can overlap with ASD deficits, such as repetitive symptoms at a young age. They are complex to translate into clinical practice, so screening rates in many regions remain low (Zwaigenbaum et al., 2015).

Screening in younger children is critically important to early referral to treatment. The long-term stability of ASD diagnosis in children aged 24 months is firmly established. The evidence also demonstrates that siblings are at elevated risk for autism spectrum and other developmental disorders and therefore should receive intensified surveillance (Zwaigenbaum et al., 2015). Implementing Early Screening for Autistic Traits (ESAT) has been associated with higher false-negative rates. The Screening Tool for Autism in Toddlers and Young Children (STAT) has been validated in clinical samples referred for suspected ASD, with better sensitivity and specificity. It requires a certified level of expertise to administer. It can help to observe autistic behaviors as a second-stage screener between ages of 14 and 48 months and would be feasible for autism early detection for primary care providers (Mazurek, Curran, Burnette, & Sohl, 2019).

With regard to potential barriers to ASD specific screening, there is not enough time to cover autism-specific features; lack of familiarity with tools; and lack of office-based systems for requiring referrals and monitoring outcomes (Zwaigenbaum et al., 2015).

Marlow, Servili, and Towson (2019) identified a substantial number of screening tests from LMIC. These tools are brief, low-cost, and can be implemented by paraprofessionals or lay in community health workers. Routine controls provide a significant first step toward addressing the need for services in LMIC, but high-quality strategies take time to be conceptualized, developed, piloted, and validated before implementation can happen. A focus on improving the scientific rigor of early detection approaches and enhancing the reach to underserved populations should be prioritized. A barrier to primary identification through surveillance pivots around the selection process of the instruments themselves. Screening tools may be general, encompassing multiple domains (e.g., the ASQ), or specific as M-CHAT. ASQ-3 is copyright licensed, but it is adapted to many languages and countries.

False-negatives mCHAT at 18 months demonstrated significantly greater deficits in the developmental domains of the ASQ compared with those with a negative MCHAT screen. Children without impairment at ASQ did not receive a subsequent ASD diagnosis (true negatives) (Øien et al., 2018).

Although ASQ-3 refers to all the developmental domains of interest, it focuses on personal-social rather than social emotional skills, it should be significant to recognize the sensitivity for autism screening.

We analyzed the validity and predictive value comparing ASQ-3 to MCHAT as a screening for autism. An analytical cross-sectional study was carried out with 100 children aged 12–48 months at well-child visits referred for language and/or communication concerns. The ASQ-3, Denver II locally validated (National Test of Screening: PRUNAPE) and M-CHAT developmental screening scales were administered. The specificity and sensitivity for the detection for ASD were compared. We found out that ASQ-3 was the most sensitive and specific instrument to detect cases for suspected ASD compared to MCHAT, and Denver II (Prunape). It showed good sensitivity, but lower specificity (Rouvier et al., unpublished data).

Data from a single pediatric practice report showed that around 75% of children with positive results on the autism-specific screening tool (the M-CHAT) were missed by the PEDS, a standardized general developmental screening questionnaire (Pinto-Martin et al., 2008).

It would be critical to continue examining screening strategies that additionally include the measurement of biomarkers as EEG. This techniques can capture potential risks of developing autism at 12 months of age (Bosl, Tager-Flusberg, & Nelson, 2018).

Outcomes from longitudinal well-child visits in young children with psychoeducative intervention sheets for parents and teachers with ASQ-3

Maturational development represent an essential aspect of periodic well-child visits in children. The pediatrician performs an integral assessment of development, recognizing risk factors and investigating its alterations systematically, to intervene promptly and in a hasty manner. ASQ-3 offers intervention sheets for every age range for parents and caregivers. We revised the outcomes of children with early and systematic detection and intervention through well-child visits. A prospective study was conducted with 188 of 415 children from 0 to 60 months in whom one or more areas of neurodevelopment delays were found. Parents received a session about EI practices with the activity sheets from ASQ-3 and rescheduled visits every 3 months during a year. We analyzed referrals required, adherence to medical indications and clinical outcomes.

Of the combined initial sample of 415 children, 188 (45%) were found at risk, 93% improved simply with rapid detection and orientations to parents. Only 7% persisted with a neurodevelopmental disorder in the final trial. Communication proved to remain the most afflicted area, both in the first and in the second test.

In addition, the antecedent of socio environmental adversity in children who failed at the second trial was more frequent, in 24% of them. More than half of the parents of children who failed in the second assessment had incomplete elementary education level and several adversity factors. Early detection and pediatric psychoeducative measures in neurodevelopment allowed improvement of youngsters in the scope of well-child visits (Carlos-Oliva, Vitale, Grañana, Rouvier, & Zeltman, 2019).

Assessment of neurodevelopment at daycare centers from adverse communities

Social-communication development can represent the first sign of alarm for autism in young children. Although autism may be equally frequent in children from different economic status, those at low-income communities tend to be underrepresented in research.

Macy, Glascoe, Ji, Macy, and Zhang (2017) analyzed barriers for early detection of childhood delay or disability. Cross-sector communication is a hurdle for many initial identification programs: it is not uncommon for problematic communication issues to arise due to multiple agencies and disciplines involved in the advance identification process. Another issue is with the screening assessment itself. Data is often mismanaged from initial screening results or from developmental surveillance activities. The third barrier is the management of the process. Children may be incorrectly referred to or not for a comprehensive evaluation when screening results are interpreted. Tipically, there are problems getting referral information to the place where it needs to go. The last barrier is the follow-up. Information may get lost or incorrectly/insufficiently acted upon during the follow-up stage of the process.

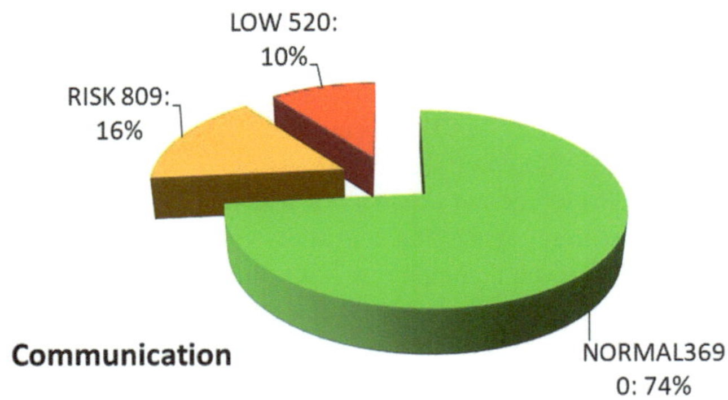

FIGURE 28.2 Communication domain ASQ-3 daycare center. Green: normal; Orange: at risk; Red: below risk.

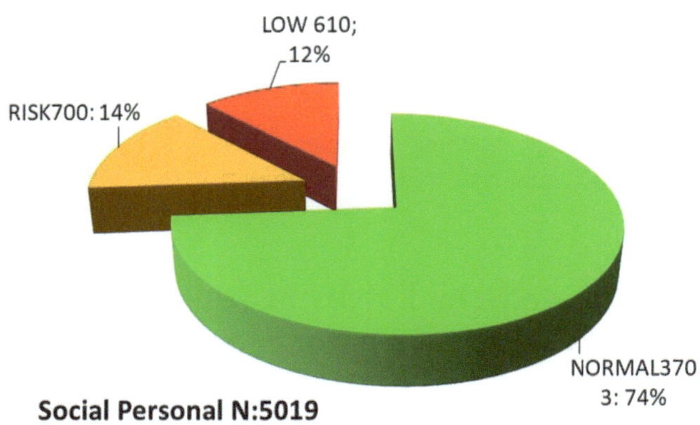

FIGURE 28.3 Social personal domain ASQ-3 daycare center. Green: normal; Orange: at risk; Red: below risk.

There is an increasing interest in EI programs in community settings. It may help reduce the gap in the detection of developmental disorders and ASDs in the adverse population. Daycare centers are a potential setting where detection of at-risk children can be conducted. We analyzed the frequency of communication deficits and social personal skills in toddlers with ASQ-3 in a low-income community sample. From 5.019 children screened, 73.5% scored at or above the risk cutoff on the communication domain (Fig. 28.2) and 74% in the social–personal area (Fig. 28.3). Around a quarter of children were at risk on both subdomains, and 12% at language and 10% were below social personal cutoffs (Granana et al. unpublished). Universal developmental screening is feasible in adverse populations at daycare centers with ASQ-3. Children identified as at-risk can be tracked and effectively addressed to initial assessment and intervention.

Applications to other areas of development

In this chapter, we have reviewed neurodevelopmental assessment with ASQ-3, the Ages and Stages Questionnaire, third version (2009).

We explored the value of systemized screening in young children, which has been proven to increase health professionals' recognition of appropriate developmental skills during well-child visits. Other areas to consider would be detection of global delay and particular developmental problems such as language impairment or autism warnings, compared with adding specific questionnaires. It would reduce time and would increase the areas of detection. During well-child visits, we explored ASQ-3 utility for psychoeducation on EI for parents through working sheets with activity guidelines. The outcome demonstrated a reduction in developmental problems in most children when their parents practiced play activities as a first-line EI.

In conclusion, we tried ASQ-3 as a systematic screening for caregivers at daycare centers for vulnerable communities. The Questionnaires proved to be sensitive to ensure early detection and interventions at routine care for adverse socio-economic groups, ASQ-3 helped in the context of initial referral to well-child centers, to create activities in an ordinary program and to train families on areas of development.

Key facts

- Neurodevelopmental disorders: conditions with onset early in development within the International Classification of Diseases from World Health Organization (WHO).
- Early intervention: Is the term employed to describe the services and supports that are available to babies and young children with potential disabilities and their families (CDC, 2019).
- Developmental monitoring: represent a flexible and ongoing process whereby health care professionals ask about children's progress and obtain informed clinical judgments based on their education and experience.
- Developmental screening: professionals use validated tests to help identify the presence of concerns. Evidence indicates that monitoring and screening should be used in tandem to maximize sensitivity to potential delays (Barger, Rice, Wolf, & Roach, 2018).

Summary points

- The use of standardized instruments is essential for rapid (AAP, 2006) detection and intervention.
- Parent reports increase sensitivity and specificity on surveillance.
- ASQ-3 fills out most developmental scales requirements.
- ASQ-3 has been validated for screening at well-child visits in children at LMIC.
- Global development and specific developmental problems can be detected with one tool: the case of language delay and autism detection with ASQ-3.
- Children improve outcome with psychoeducative intervention at activity sheets for parents and teachers with ASQ-3.
- Assessment of neurodevelopment at daycare centers in adverse socioeconomic communities can detect populations at risk.

References

AAP, American Academy of Pediatrics. (2006). Council on children with disabilities. Identifying infants and young children with developmental disorders in the medical home: An algorithm for developmental surveillance and screening. *Pediatrics, 118*(1), 405–420.

Allen, G. (2011). *Early intervention: The next steps, an independent report to her majesty's government by Graham Allen MP*. The Stationery Office.

Alvik, A., & Grøholt, B. (2011). Examination of the cut-off scores determined by the Ages and Stages Questionnaire in a population-based sample of 6 month-old Norwegian infants. *BMC Pediatrics, 11*(1), 117.

Anda, R. F., Felitti, V. J., Bremner, J. D., Walker, J. D., Whitfield, C. H., Perry, B. D., … Giles, W. H. (2006). The enduring effects of abuse and related adverse experiences in childhood. *European Archives of Psychiatry and Clinical Neuroscience, 256*(3), 174–186, 256(3): 174–86.

Barger, B., Rice, C., Wolf, R., & Roach, A. (2018). Better together: Developmental screening and monitoring best identify children who need early intervention. *Disability and Health Journal, 11*(3), 420–426.

Bedford, H., Walton, S., & Ahn, J. (2013). *Measures of child development: A review. Policy research unit in the health of children, young people and families*. Centre for Paediatric Epidemiology and Biostatistics, UCL Institute of Child Health. Available from: http://www.ucl.ac.uk/cpru/publications.

Bosl, W. J., Tager-Flusberg, H., & Nelson, C. A. (2018). EEG analytics for early detection of autism spectrum disorder: A data-driven approach. *Scientific Reports, 8*(1), 6828.

Canadian Task Force on Preventive Health Care. (2016). Recommendations on screening for developmental delay. *Canadian Medical Association Journal, 188*(8), 579–587.

Capute, A. J. (1996). *The capute scales: CAT/CLAMS*. Baltimore: Kennedy Fellows Association.

Carlos-Oliva, D., Vitale, M. P., Grañana, N., Rouvier, M. E., & Zeltman, C. (2019). *Evolución del neurodesarrollo con el uso del cuestionario de edades y etapas ASQ-3 en el control de salud de niños*. Rev Neurol (in press).

CDC. Center for Disease Control and Prevention. From: https://www.cdc.gov/ncbddd/actearly/parents/states.html. Accessed October, 2019.

DCSF. (2007). *Department for children, schools and families. Cost–Benefit Analysis of Interventions with parents*. Research Report DCSF-RW008.

Glascoe, F. P., & Byrne, K. E. (1993). The accuracy of three developmental screening tests. *Journal of Early Intervention, 17*(4), 368–379.

Goff, B., Gee, D. G., Telzer, E. H., Humphreys, K. L., Gabard-Durnam, L., Flannery, J., … Tottenham, N. (2013). Reduced nucleus accumbens reactivity and adolescent depression following early-life stress. *Neuroscience, 249*, 129–138.

Gollenberg, A. L., Lynch, C. D., Jackson, L. W., McGuinness, B. M., & Msall, M. E. (2010). Concurrent validity of the parent-completed Ages and Stages Questionnaires, with the Bayley scales of infant development II in a low-risk sample. *Child: Care, Health and Development, 36*(4), 485–490.

Hamilton, S. (2006). Screening for developmental delay: Reliable, easy-to-use tools: Win-win solutions for children at risk and busy practitioners. *Journal of Family Practice, 55*(5), 415–423.

Hintz, S. R., Kendrick, D. E., Vohr, B. R., Poole, W. K., Higgins, R. D., & Nichd Neonatal Research Network. (2006). Gender differences in neurodevelopmental outcomes among extremely preterm, extremely-low-birthweight infants. *Acta Paediatrica, 95*(10), 1239–1248.

Hix-Small, H., Marks, K., Squires, J., & Nickel, R. (2007). Impact of implementing developmental screening at 12 and 24 months in a pediatric practice. *Pediatrics, 120*(2), 381–389.

IDEA. (2004). *Individuals with disabilities education improvement Act of 2004* (pp. 108–446). Pub. L. No.

Kerstjens, J. M., Bos, A. F., ten Vergert, E. M., de Meer, G., Butcher, P. R., & Reijneveld, S. A. (2009). Support for the global feasibility of the Ages and Stages Questionnaire as developmental screener. *Early Human Development, 85*(7), 443–447.

King, T. M., Tandon, S. D., Macias, M. M., Healy, J. A., Duncan, P. M., Swigonski, N. L., … Lipkin, P. H. (2010). Implementing developmental screening and referrals: Lessons learned from a national project. *Pediatrics, 125*(2), 350–360.

Klamer, A., Lando, A., Pinborg, A., & Greisen, G. (2005). Ages and Stages Questionnaire used to measure cognitive deficit in children born extremely preterm. *Acta Paediatrica, 94*(9), 1327–1329.

Macy, M., Glascoe, F. P., Ji, Y., Macy, R., & Zhang, B. (2017). Management of an early detection program: Supporting helpful features. *Current Developmental Disorders Reports, 4*(4), 118–129.

Marlow, M., Servili, C., & Tomlinson, M. (2019). A review of screening tools for the identification of autism spectrum disorders and developmental delay in infants and young children: Recommendations for use in low-and middle-income countries. *Autism Research, 12*(2), 176–199.

Mazurek, M. O., Curran, A., Burnette, C., & Sohl, K. (2019). ECHO Autism STAT: Accelerating early access to autism diagnosis. *Journal of Autism and Developmental Disorders, 49*(1), 127–137.

McPheeters, M. L., Weitlauf, A., Vehorn, A., Taylor, C., Sathe, N. A., Krishnaswami, S., … Warren, Z. E. (2016). *Screening for autism spectrum disorder in young children*.

Méndez Castellano, H., & De Méndez, M. C. (1994). Sociedad y estratificación: Método Graffar Mendez Castellano. In *Sociedad y estratificación: Método Graffar Mendez Castellano*. Mexico DF: Fundacredesa.

Øien, R. A., Schjølberg, S., Volkmar, F. R., Shic, F., Cicchetti, D. V., Nordahl-Hansen, A., … Ventola, P. (2018). Clinical features of children with autism who passed 18-month screening. *Pediatrics, 141*(6), e20173596.

Otalvaro, A. M. R., Grañana, N., Gaeto, N., Torres, M. D. L.Á., Zamblera, M. N., Vasconez, M. A., … Squires, J. (2018). ASQ-3: Validation of the Ages and Stages Questionnaire for the detection of neurodevelopmental disorders in Argentine children. *Archivos Argentinos de Pediatria, 116*(1), 7–13.

Palfrey, J. S., Singer, J. D., Walker, D. K., & Butler, J. A. (1987). Early identification of children's special needs: A study in five metropolitan communities. *The Journal of Pediatrics, 111*(5), 651–659.

Perry, B. D. (2002). Childhood experience and the expression of genetic potential: What childhood neglect tells us about nature and nurture. *Brain and Mind, 3*(1), 79–100.

Pinto-Martin, J. A., Young, L. M., Mandell, D. S., Poghosyan, L., Giarelli, E., & Levy, S. E. (2008). Screening strategies for autism spectrum disorders in pediatric primary care. *Journal of Developmental and Behavioral Pediatrics, 29*(5), 345–350.

Rubio-Codina, M., Araujo, M. C., Attanasio, O., Muñoz, P., & Grantham-McGregor, S. (2016). Concurrent validity and feasibility of short tests currently used to measure early childhood development in large scale studies. *PloS One, 11*(8), e0160962.

Schonhaut, L., Armijo, I., Schönstedt, M., Alvarez, J., & Cordero, M. (2013). Validity of the Ages and Stages Questionnaires in term and preterm infants. *Pediatrics, 131*(5), e1468–e1474.

Sinclair, A. (2007). *0–5: How small children make a big difference*. London: The Work Foundation. Provocation series 3, No 1.

Small, J. W., Hix-Small, H., Vargas-Baron, E., & Marks, K. P. (2019). Comparative use of the Ages and Stages Questionnaires in low-and middle-income countries. *Developmental Medicine and Child Neurology, 61*(4), 431–443.

Squires, J., Nickel, R. E., & Eisert, D. (1996). Early detection of developmental problems: Strategies for monitoring young children in the practice setting. *Journal of Developmental and Behavioral Pediatrics, 17*(6), 420–427.

Vandell, D. L., Burchinal, M., & Pierce, K. M. (2016). Early childcare and adolescent functioning at the end of high school: Results from the NICHD study of early child care and youth development. *Developmental Psychology, 52*(10), 1634.

Weiss, B., & Landrigan, P. J. (2000). The developing brain and the environment: An introduction. *Environmental Health Perspectives, 108*(Suppl. 3), 373–374.

Zwaigenbaum, L., Bauman, M. L., Fein, D., Pierce, K., Buie, T., Davis, P. A., … Kasari, C. (2015). Early screening of autism spectrum disorder: Recommendations for practice and research. *Pediatrics, 136*(Suppl. 1), S41–S59.

Chapter 29

Attention deficit hyperactivity disorder screening in childhood and adolescence

Steven J. Korzeniewski
Department of Family Medicine and Public Health Sciences, Wayne State University School of Medicine, Detroit, MI, United States

List of abbreviations

4th Revision, SNAP-IV Swanson, Nolan, and Pelham Questionnaire
A-TAC Autism − tics, attention deficit hyperactivity disorder and other comorbidities
ADHD Attention deficit hyperactivity disorder
ARS ADHD Rating Scale
ASQ Conners Abbreviated Symptom Questionnaire
AUC Area under the receiver operating characteristic curve
BSFQ Before School Functioning Questionnaire
CABI Child and Adolescent Behavior Inventory
CBCL Child Behavior Checklist
DAWBA Development and Well-Being Assessment
DSM Diagnostic and Statistical Manual of Mental Disorders
ECI-4 Early Childhood Inventory-4
HI Hyperactivity−inattention
LR ± Likelihood ratio positive and negative
PPV and NPV Positive and negative predictive values
SDQ Strengths and Difficulties Questionnaire
PREMB-R Morning (AM) and Evening (PM) Subscales Parent Rating of Evening and Morning Behavior Scale Revised

Mini-dictionary of terms adapted from last's dictionary of epidemiology (Last, 2002)

Area under the receiver operating characteristic curve average sensitivity across all specificities.
Diagnostic odds ratio the ratio of the odds of a positive screen in subjects who have the disorder relative to the odds of a positive screen in subject who do not have the disorder.
Likelihood ratio negative (1 − sensitivity)/specificity.
Likelihood ratio positive the ratio sensitivity/(1 − specificity) or sensitivity/false-positive rate.
Negative predictive value the probability that a person with a negative test result does not have the disease.
Positive predictive value the probability that a person with a positive test does have the disease.
Sensitivity the probability that a diseased person (case) in the population will be identified as diseased (syn: true positive probability).
Specificity the probability that a person without the disease (noncase) will be correctly identified as nondiseased.

Introduction

Attention deficit hyperactivity disorder (ADHD) is a common and burdensome neurodevelopmental disorder that affects both children and adults (Posner, Polanczyk, & Sonuga-Barke, 2020). The Diagnostic and Statistical Manual of Mental Disorders (DSM)-5 currently defines ADHD in children below 17 years of age as the presence of $\geq$ six symptoms under

Diagnosis, Management and Modeling of Neurodevelopmental Disorders. https://doi.org/10.1016/B978-0-12-817988-8.00029-4

the inattentive or hyperactive and impulsive presentations (or both) prior to age 12 (American Psychiatric Association, 2013). The DSM-5 further requires that ADHD symptoms be persistent for ≥ 6 months; have negative impact on social, academic, or occupational activities; occur in two or more settings; and not be attributed to other behavioral or mental disorders. Previous editions required symptom presentation (DSM-3) and impairment (DSM-4-TR) by age 7. The 11th edition of the International Classification of Diseases similarly requires a persistent patter of inattention (IA), hyperactivity—impulsivity, or both, for at least 6 months, and the symptoms must interfere with academic, occupational, or social functioning to warrant ADHD diagnosis (World Health Organization, 2018).

Early efforts to establish ADHD prevalence in childhood yielded a wide range of estimates that was later attributed to methodological differences across studies, mostly in definitions (Posner et al., 2020). Meta-analysis of 143 studies of children and adolescents supports a global prevalence of 5.3% with little international variation and generally stable time trends (Polanczyk, de Lima, Horta, Biederman, & Rohde, 2007), although not all studies agree (Rydell, Lundstrom, Gillberg, Lichtenstein, & Larsson, 2018; Safer, 2018).

In contrast to the reliance on rigorous diagnosis when establishing prevalence, ADHD screening instruments are used to identify persons who probably have the disorder and those who probably do not (Porta, 2014). This chapter describes ADHD screening performance in children and adolescents. Large (N ≥ 500) and representative studies of community samples are emphasized, and the focus is on screening performance metrics that are comparable across studies (e.g., area under the receiver operating characteristic curve (AUC), likelihood ratio positive and negative (LR±), and positive and negative predictive values (PPV and NPV).

Screening performance among children and adolescents

Large (N ≥ 500) and/or population representative samples of school-age children and adolescents

Strengths and Difficulties Questionnaire (SDQ) (Goodman, 1997) (Fig. 29.1; Table 29.1)

Multiple studies have used the SDQ to identify children who probably have ADHD and those who probably do not. This 25-item parent (or caregiver), teacher (and less often self) completed questionnaire assesses emotional symptoms, conduct problems, hyperactivity—inattention (HI), peer problems, and prosocial behavior (each subscale score ranges from 0 to 10). In 2016, investigators developed SDQ scoring weights for clinical use by leveraging the British Child and Adolescent Mental Health Survey of nationally representative youths aged 5—15 (N = 18,232) (Algorta, Dodd, Stringaris, & Youngstrom, 2016). Parent, teacher, and eligible 11—15-year-old SDQ HI responses were used to identify ADHD (n = 264) established by Development and Well-Being Assessment (DAWBA), a validated semistructured child and parent interview accompanied by a teacher questionnaire (Goodman, Ford, Richards, Gatward, & Meltzer, 2000). The AUC ranged from 0.81 to 0.96 across age (<10, 11—13, or 14—16 years) and sex groups. By contrast, screening performance was reduced in discriminating ADHD from other diagnoses; accuracy ranged from 0.70 to 0.88 and the diagnostic odds ratio was 4.5 using a 10+ HI scale score cutoff.

Another study (N = 2988) of 12—17-year-old Dutch adolescent- and parent-reported SDQ scores recently reported similar ADHD (n = 872) screening performance for the HI subscale (AUC corrected for optimism, 0.80), although sensitivity was somewhat lower using cutoffs with equivalent specificity (Table 29.1) (Vugteveen, de Bildt, Theunissen, Reijneveld, & Timmerman, 2019). Because parent ratings are the easiest to implement, Danish investigators examined only parent-reported SDQ subscale scores among participants in the Danish National Birth Cohort who were assessed at age 8—15 (Overgaard, Madsen, Oerbeck; Friis, & Obel, 2019). The results question the utility of relying only on parent ratings, as AUC ranged from 0.71 to 0.83 across SDQ subscales. Moreover, fewer than 4% of positive screens met diagnostic criteria when half or more children with ADHD were identified, and LR+ ≥ 10 occurred only with sensitivity <30%.

An earlier study in Bergen, Norway, used parent- and teacher-completed SDQ ratings and the Swanson, Nolan, and Pelham Questionnaire 4th Revision (SNAP-IV) scores to identify 7—9-year-olds with ADHD that was confirmed by DSM-4 criteria assessed by both informants (Ullebo, Posserud, Heiervang, Gillberg, & Obel, 2011). Parent SDQ HI scores were slightly, but statistically significantly, less accurate than teacher ratings; AUC 0.91 (95% CI: 0.90—0.92) versus 0.95 (95% CI: 0.94—0.95). By contrast, sensitivity and specificity were equal or greater comparing teacher to parent scores at each SDQ HI cutoff; LR+ was 14.8 with 59% sensitivity using teacher ratings, while LR+ was 12.3 for parent ratings with 49% sensitivity using the same cutoff score (≥7). ADHD presentation-specific analyses revealed decreased sensitivity in detecting the inattentive subtype.

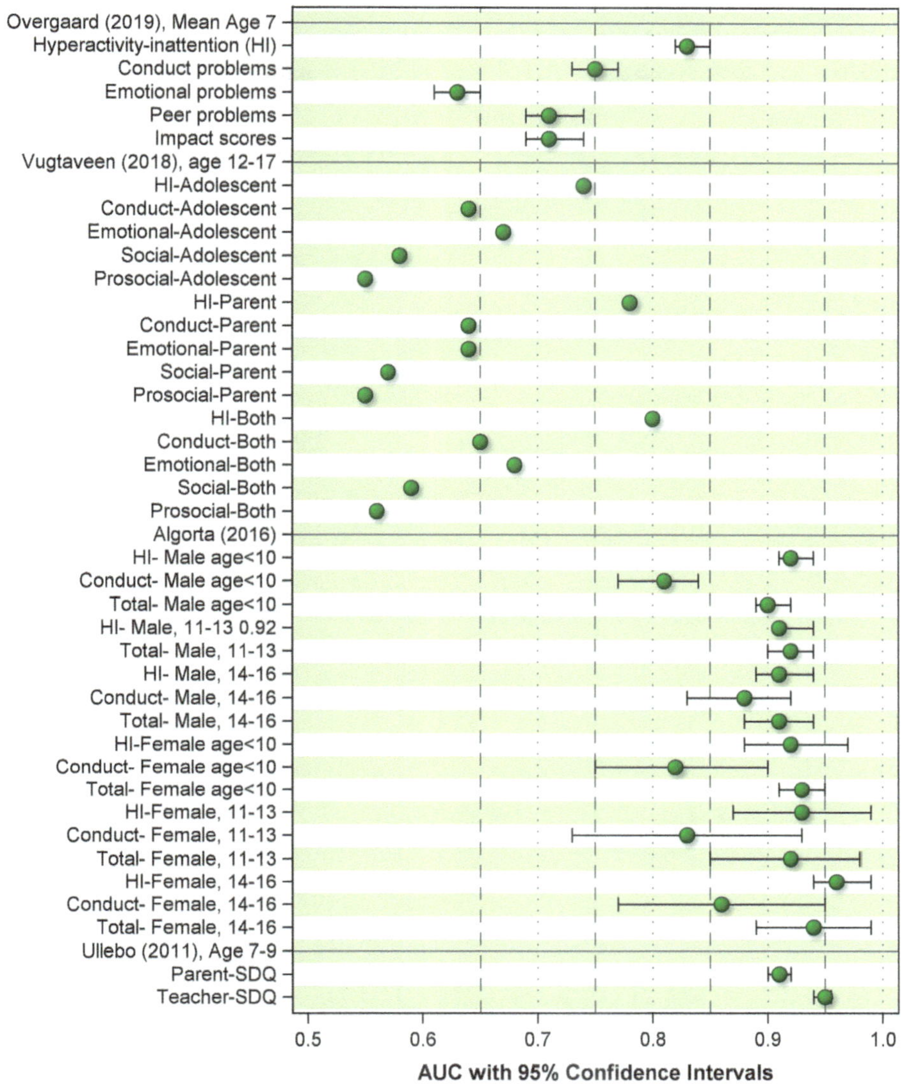

FIGURE 29.1 **Area under the receiver operating characteristic curve (AUC) using the Strengths & Difficulties Questionnaire (SDQ) to identify children and adolescents who have ADHD.** Screening performance estimates for attention deficit hyperactivity disorder instruments (and subscales) completed by adolescents, parents, and/or teachers. *Black filled circles* represent area under the receiver operating characteristic curve (i.e., the average proportion (%) of children with ADHD who are detected across all possible cutoffs; higher values equal better screening performance). *Horizontal lines* represent 95% confidence intervals around the estimated AUC.

Child Behavior Checklist (CBCL) (Achenbach, Dumenci, & Rescorla, 2003; Achenbach & Edelbrock, 1983), *ADHD Rating Scale (ARS)* (Dupaul, 1991) Attention Deficit Disorders Evaluation Scale (ADDES) (McCarney, 1989), and Conners Abbreviated Symptom Questionnaire (ASQ) (Conners, 1990) (Fig. 29.2; Table 29.2).

The CBCL, a parent/caregiver questionnaire, is one of the most commonly used instruments to screen for ADHD in children aged 6–11 and 12–18. In one of the largest studies to date, n = 49,088 first and fourth graders were screened using the Korean version of the ARS (K-ARS), which measures IA (nine items) and hyperactivity-impulsivity (nine items) using a four-point Likert scale. During a second screening phase, K-CBCL assessments were completed by parents to identify ADHD that was confirmed by interviewers and psychiatric specialist raters using the Diagnostic Interview Schedule for Children Version IV (Shaffer, Fisher, Lucas, Dulcan, & Schwab-Stone, 2000) in children who were assigned top decile K-ARS scores by teachers or parents (Park et al., 2014). The AUC was significantly higher using K-ARS parent scores than teacher scores in each of the three study groups (AUC 0.80–0.85 vs. 0.51–0.5.6, respectively, among first graders in 2010, first graders in 2011, and fourth graders in 2011). A 90th percentile cutoff for K-ARS teacher, parent, and combined scores across the three study groups resulted in LR+ of 0.92–0.97, 2.5–2.9 and 2.4–3.1, respectively, due to low specificity, particularly with parent ratings (Table 29.2). The AUC ranged from 0.69 to 75 across the three study groups using the K-CBCL; the attention problems and externalizing problems subscales respectively identified half or more of children who met ADHD criteria, but specificity was limited, resulting in LR+ values of 2.8–3.9, 2.9–3.3, and 2.6–4.5 among first graders in 2010, first graders in 2011, and fourth graders in 2011.

TABLE 29.1 Strengths and Difficulties Questionnaire hyperactivity inattention subscale ADHD screening performance in children and adolescents.

Name (Year)	Sample source	Country	Participants (ADHD)	Age (years)	Cutoff	Sens.	Spec.	PPV	LR+	DOR
Overgaard et al. (2019)	Com	Denmark	51,096 (943)	7 avg.	1	0.98	0.21	–	1.3	
					2	0.94	0.43	0.01	1.7	
					3	0.86	0.61	0.02	2.2	
					4	0.76	0.76	0.02	3.2	
					5	0.64	0.86	0.04	4.5	
					6	0.50	0.93	0.06	6.7	
					7	0.39	0.96	0.09	9.5	
					8	0.29	0.98	0.14	13.2	
					9	0.19	0.99	0.21	17.6	
					10	0.09	1.00	0.31	23.4	
Vugteveen, De Bildt, Hartman, & Timmerman (2018)	Com	Neth.	2753 (872)	12–17	90th (P)	0.63	0.77	–	2.7	5.8
					90th (A)	0.59	0.76	–	2.5	4.6
Carballo, Rodriguez-Blanco, Garcia-Nieto, & Baca-Garcia (2018)	Clinic Referral	Spain	523 (283)	3–17	0	1.00	0.00		1.0	
					1	0.99	0.11		1.1	
					2	0.99	0.17		1.2	
					3	0.97	0.32		1.4	
					4	0.93	0.49		1.8	
					5	0.84	0.67		2.6	
					6	0.70	0.84		4.5	
					7	0.57	0.92		6.7	
					8	0.38	0.99		27.4	
					9	0.20	1.00		–	
					10	0.00	1.00		–	
Ullebo et al. (2011)	Com	Norway	5997/236	7–9						
		Parent	403 (34)		4	0.95	0.82	0.22	5.3	
			347 (42)		5	0.84	0.88	0.28	7.0	
			239 (41)		6	0.72	0.93	0.37	10.3	
			129 (42)		7	0.59	0.96	0.48	14.8	
			103 (43)		8	0.46	0.98	0.56	23.0	
			58 (36)		9	0.33	0.99	0.65	33.0	
			107 (71)		10	0.22	0.99	0.66	22.0	
		Teacher	605 (28)		4	0.88	0.76	0.17	3.7	
			458 (44)		5	0.79	0.86	0.24	5.6	

Continued

TABLE 29.1 Strengths and Difficulties Questionnaire hyperactivity inattention subscale ADHD screening performance in children and adolescents.—cont'd

Name (Year)	Sample source	Country	Participants (ADHD)	Age (years)	Cutoff	Sens.	Spec.	PPV	LR+	DOR
			277 (54)		6	0.66	0.93	0.34	9.4	
			143 (39)		7	0.49	0.96	0.44	12.3	
			101 (43)		8	0.37	0.98	0.55	18.5	
			62 (31)		9	0.24	0.99	0.62	24.0	
			65 (48)		10	0.15	1.00	0.74	–	

Study characteristics, instrument score cutoffs, and screening performance metric estimates are described. Sensitivity: the probability that a diseased person (case) in the population will be identified as diseased (syn: true positive probability). Specificity: the probability that a person without the disease (noncase) will be correctly identified as nondiseased. Positive predictive value: the probability that a person with a positive test does have the disease. Likelihood ratio positive: the ratio sensitivity/(1 − specificity) or sensitivity/false-positive rate. Diagnostic odds ratio: the ratio of the odds of a positive screen in subjects who have the disorder relative to the odds of a positive screen in subject who do not have the disorder. *Com.*, community; *DOR*, diagnostic odds ratio; *F*, female; *LR+*, Likelihood Ratio Positive; *M*, male; *PPV*, Positive Predictive Value; *Sens.*, sensitivity; *Spec.*, Specificity.

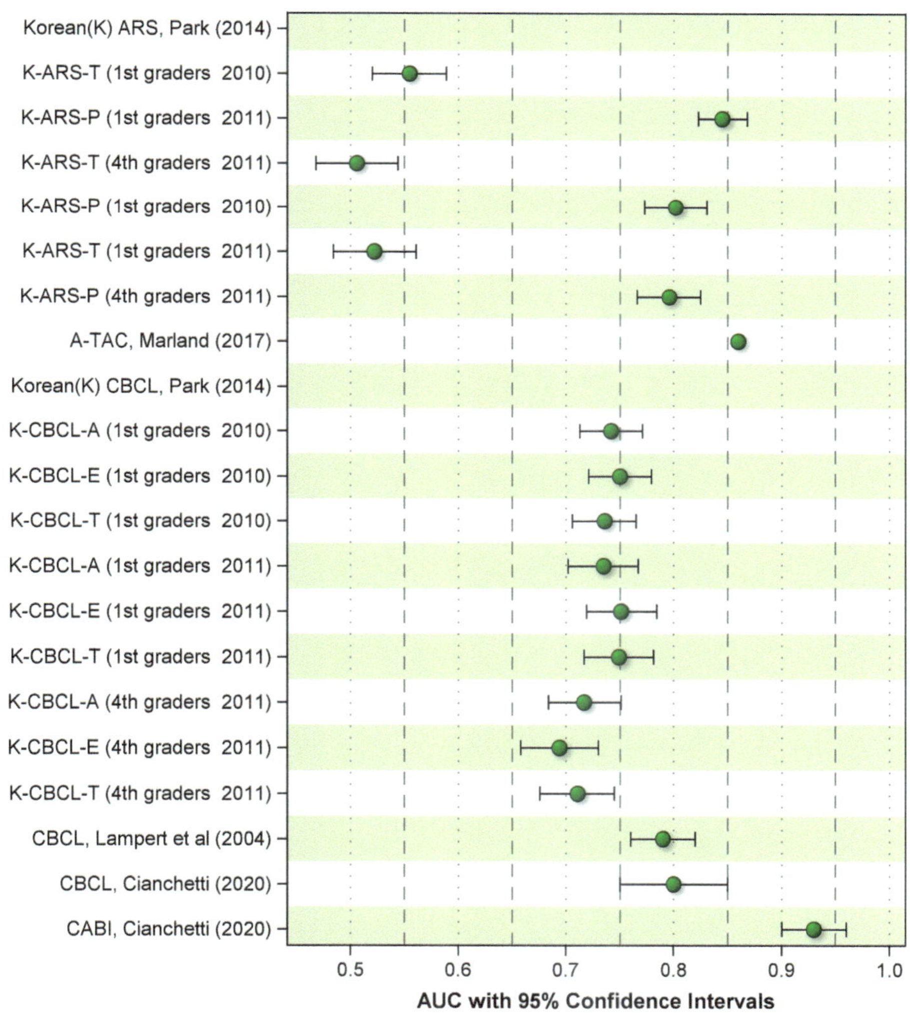

FIGURE 29.2 **Area under the receiver operating characteristic curve (AUC) using the ADHD Rating Scale (ARS), Autism – tics, attention deficit hyperactivity disorder and other comorbidities (A-TAC), Child Behavior Checklist (CBCL), and Child and Adolescent Behavior Inventory (CABI) to identify children and adolescents with ADHD.** Screening performance estimates for attention deficit hyperactivity disorder instruments (and subscales) completed by adolescents, parents, and/or teachers. *Black filled circles* represent area under the receiver operating characteristic curve (i.e., the average proportion (%) of children with ADHD who are detected across all possible cutoffs; higher values equal better screening performance). *Horizontal lines* represent 95% confidence intervals around the estimated AUC.

TABLE 29.2 ADHD Rating Scale (ARS), conners Abbreviated Symptom Questionnaire (ASQ), Attention Deficit Disorders Evaluation Scale (ADDES), Child Behavior Checklist (CBCL), and Child and Adolescent Behavior Inventory (CABI) screening performance in children and adolescents.

Instrument	Name (Year)	Sample source	Country	Participants (ADHD)	Age	Cutoff Value	Sens.	Spec.	PPV	LR+
ARS	Park (2014)	Com.	South Korea	49,0888 (1215)	1st graders 2010	K-ARS-T ≥ 90th	0.64	0.34	0.36	1.0
						K-ARS-P ≥ 90th	0.78	0.73	0.64	2.9
					1st graders 2011	K-ARS-T ≥ 90th	0.66	0.29	0.37	0.9
						K-ARS-P ≥ 90th	0.76	0.7	0.61	2.5
					4th graders 2011	K-ARS-T ≥ 90th	0.61	0.35	0.42	0.9
						K-ARS-P ≥ 90th	0.71	0.72	0.67	2.5
ASQ	Bussing et al. (1998)	School	US	318 (83)	9.7	T ≥ 60	0.64	0.73	0.59	2.4
Parent						T ≥ 65	0.58	0.80	0.64	2.9
						T ≥ 70	0.50	0.85	0.66	3.3
Teacher	Holmberg et al. (2013)	Com.	Sweden	398 (21)	9.5	10	0.63	0.91		7
ADDES	Bussing et al. (1998)	School	US	318 (83)	9.7	15th	0.58	0.82	0.64	3.2
						10th	0.48	0.89	0.70	4.4
						5th	0.36	0.93	0.77	5.1
CBCL	Park et al. (2014)	Com.	South Korea	49,0888 (1215)	4th graders 2010	K-CBCL-A ≥ 60T	0.38	0.87	0.62	2.8
						K-CBCL-E				
						K-CBCL-T ≥ 63T	0.3	0.92	0.69	3.9
					1st graders 2011	K-CBCL-A ≥ 60T	0.39	0.87	0.65	3.0
						K-CBCL-T ≥ 63T	0.26	0.92	0.68	3.3
					4th graders 2011	K-CBCL-A ≥ 60T	0.43	0.84	0.68	2.6
						K-CBCL-T ≥ 63T	0.28	0.94	0.79	4.5

Continued

TABLE 29.2 ADHD Rating Scale (ARS), conners Abbreviated Symptom Questionnaire (ASQ), Attention Deficit Disorders Evaluation Scale (ADDES), Child Behavior Checklist (CBCL), and Child and Adolescent Behavior Inventory (CABI) screening performance in children and adolescents.—cont'd

Instrument	Name (Year)	Sample source	Country	Participants (ADHD)	Age	Cutoff Value	Sens.	Spec.	PPV	LR+
	Lampert et al. (2004)	School and clinic	Brazil	763(235)	10.9	Raw score = 9	0.75	0.7		2.5
	Cianchetti et al. (2020)	Com.	Italy	462 (145)	6–18	T ≥ 70	0.59	0.89		5.4
CABI	Cianchetti et al. (2020)	Com.	Italy	462 (145)	6–18	T ≥ 70	0.83	0.92		10.4

Study characteristics, instrument score cutoffs, and screening performance metric estimates are described. Sensitivity: the probability that a diseased person (case) in the population will be identified as diseased (syn: true positive probability). Specificity: the probability that a person without the disease (noncase) will be correctly identified as nondiseased. Positive predictive value: the probability that a person with a positive test does have the disease. Likelihood ratio positive: the ratio sensitivity/(1 − specificity) or sensitivity/false-positive rate. *Com.*,community; *LR+*, Likelihood Ratio Positive; *PPV*, Positive Predictive Value; *Sens.*, sensitivity; *Spec.*, Specificity.

A smaller study conducted in four Italian cities located in different regions of the country recently examined CBCL screening performance compared to the Child and Adolescent Behavior Inventory (CABI) of problem areas defined by DSM-oriented scales for ADHD and other developmental disorders (Cianchetti et al., 2013). Parent or caregiver assessments were followed by standard clinical diagnosis according to DSM-5 criteria to confirm ADHD diagnosis (Cianchetti et al., 2020). The AUC using the CBCL was comparable to that observed in previous studies, 0.80 (95% CI 0.76−0.83) and LR+ was 5.4. The CABI instrument had greater sensitivity and equivalent specificity, resulting in a 9.7 LR+ and increased AUC (0.93; 95% CI 0.90−0.96). An earlier Brazilian study that used the CBCL to identify ADHD that was confirmed by psychiatrists using DSM-4 criteria reported similar screening performance in a community sample (N = 763; AUC 0.79 and LR+ 2.5, respectively) (Lampert, Polanczyk, Tramontina, Mardini, & Rohde, 2004)

A study of N = 422 first graders in one municipality in Stockholm County used the ASQ scale to screen for ADHD that was confirmed at follow-up in fourth grade (Holmberg, Sundelin, & Hjern, 2013). Teacher-reported scores ≥10 identified 63% (95% CI: 43%−79%) of children diagnosed with pervasive ADHD, PPV was 29% and LR+ was 6.7. Parent-reported scores ≥10 were less accurate (sensitivity (29%, 95% CI: 15−49, PPV 20% and LR+ 4.2); however, the combination of both respondents' scores increased PPV to 50% and LR+ to 16.6. Sensitivity and specificity were reduced for detection of situational ADHD (i.e., LR+ 3.0).

In contrast to the aforementioned findings, an earlier study that used the 10-item ASQ and the ADDES in a district population of special education students and in a random sample of low-risk children reported poorer performance identifying ADHD confirmed by parental diagnostic interviews based on DSM-IV- criteria (Bussing, Schuhmann, Belin, Widawski, & Perwien, 1998). The LR+ ranged from 2.4 to 3.4 using the ASQ and from 3.2 to 5.1 using ADDES.

Screening of childhood autism − tics, attention deficit hyperactivity disorder and other comorbidities (A-TAC) (Larson et al., 2010)

The Child and Adolescent Twin Study in Sweden has surveyed parents of all 9-year-olds by telephone since 2004 using the A-TAC to identify ADHD diagnoses recorded in the National Patient Register (Marland et al., 2017). Among n = 25,828 twins, the AUC for identifying ADHD (n = 723) recorded at age 9 or 12 was 0.86; use of a low threshold (6) for concentration and attention resulted in 64% sensitivity, 17% PPV, and LR+ of 18, while use of a high threshold for impulsiveness and activity (12.5) resulted in 28% sensitivity, 38% PPV, and an LR+ of 29. The balance of sensitivity and specificity resulted in increased LR+ relative to a previous Child and Adolescent Twin Study report of LR+ of 3.4 and 8.0 using the A-TAC concentration and attention and impulsiveness scores and activity scores to identify ADHD, respectively (Anckarsater et al., 2011).

Before School Functioning Questionnaire (BSFQ) (Faraone, Hammerness, & Wilens, 2018) *and the Parent Rating of Evening and Morning Behavior Scale Revised (PREMB-R) Morning (AM) and Evening (PM) Subscales* (Faraone, Childress, et al., 2018) (Table 29.3).

TABLE 29.3 ADHD screening performance in children and adolescents using the screening of childhood autism – tics, attention deficit hyperactivity disorder and other comorbidities (A-TAC), Before School Functioning Questionnaire (BSFQ), and the Parent Rating of Evening and Morning Behavior Scale Revised (PREMB-R) Morning (AM) and Evening (PM) Subscales.

Instrument	Name (Year)	Sample source	Country	Participants (ADHD)	Age (years)	Cutoff Value	Sensitivity	Specificity	PPV	LR+
A-TAC	Marland et al. (2017)	Com	Sweden	25,828 twins (723)	9–12					
Concentration and attention						6 [low]	0.64	0.91	0.17	18*
Impulsiveness and activity						12.5 [high]	0.28	0.99	0.38	29*
BSFQ	Faraone et al. (2019)	Com.	US	1200 (121)	6–17	80th	0.53	0.87		4.1
		Without comorbidities				90th	0.39	0.94		6.5
						93rd	0.26	0.96		6.5
						98th	0.08	0.99		8.0
		With comorbidities				80th	0.78	0.56		1.8
						90th	0.56	0.79		2.7
						93rd	0.4	0.88		3.3
						98th	0.13	0.94		2.2
PREMB-R AM						80th	0.44	0.89		4.0
		Without comorbidities				90th	0.17	0.97		5.7
						93rd	0.17	0.97		5.7
						98th	0.03	0.99		3.0
		With comorbidities				80th	0.51	0.66		1.5
						90th	0.18	0.83		1.1
						93rd	0.18	0.83		1.1
						98th	0.11	0.96		2.8
PREMB-R PM						80th	0.48	0.88		4.0
		Without comorbidities				90th	0.32	0.95		6.4
						93rd	0.18	0.97		6.0
						98th	0.05	0.99		5.0
		With comorbidities				80th	0.80	0.63		2.2
						90th	0.53	0.81		2.8
						93rd	0.40	0.91		4.4
						98th	0.22	0.98		11.0

Study characteristics, instrument score cutoffs, and screening performance metric estimates are described. Sensitivity: the probability that a diseased person (case) in the population will be identified as diseased (syn: true positive probability). Specificity: the probability that a person without the disease (noncase) will be correctly identified as nondiseased. Positive predictive value: the probability that a person with a positive test does have the disease. Likelihood ratio positive: the ratio sensitivity/(1 − specificity) or sensitivity/false-positive rate. Diagnostic odds ratio: the ratio of the odds of a positive screen in subjects who have the disorder relative to the odds of a positive screen in subject who do not have the disorder. *, Diagnostic Odds Ratio; *Com.*, community; *LR+*, Likelihood Ratio Positive; *PPV*, Positive Predictive Value; *Sens.*, sensitivity; *Spec.*, Specificity.

The 20-item BSFQ assesses ADHD-related symptoms prior to attending school, while the PREMB-R subscales measure functional impairments in the early morning before school or late afternoon after school. US investigators recently performed a normative online survey of a representative sample of 1200 primary caregivers of children and adolescents aged 6–17 years with 50 male and 50 females randomly selected in each of 12 age groups. The BSFQ and PREMB-R subscales were used to identify caregiver report of ADHD diagnosed by a health professional separately in youths with or without comorbidities (Faraone et al., 2019). Using each of the three instruments, sensitivity was reduced while specificity was increased among children and adolescents without comorbidities, whereas the opposite was observed in children with comorbidities, using 80th, 90th, 93rd, and 98th percentile thresholds to define a positive screen. Performance was comparable in each group using the three instruments to assess children and adolescents with or without comorbidities when sensitivity was above 25% (with comorbidities, LR+ 1.1–4.4; without comorbidities, LR+ 4–6.5).

Swanson, Nolan, and Pelham Questionnaire (4th Revision, SNAP-IV) (Swanson et al., 2001)

A population representative case-control study recruited 2000 pediatricians from across Spain and asked them to assess three children with a clinical impression of ADHD and three children without a clinical impression of ADHD defined according to the Clinical Global Impression Scales (CGI) (Alda & Serrano-Troncoso, 2013). The CGI is a seven-point scale that evaluates the severity of disease (1 = normal, not at all ill; 2 = borderline mentally ill; 3 = mildly ill; 4 = moderately ill; 5 = markedly ill; 6 = severely ill; or 7 = extremely ill) (Guy, 1976); scores of three or more were considered to have met DSM-IV ADHD criteria. The LR+ was 4.7 using the SNAP-IV to identify CGI-defined ADHD and estimated PPV was 12.6% at 3% prevalence and 34% at 10% prevalence.

Screening performance in large (N ≥ 500) studies of pre-school age children (Fig. 29.3; Table 29.4)

A northeast Iranian study used the ASQ parent and teacher questionnaires to screen n = 4143 5–6-year-olds in 10 different urban areas, and in a second stage n = 150 children were randomly selected and clinical interviews were performed among children with positive scores to confirm ADHD diagnosis (Abdekhodaie, Tabatabaei, & Gholizadeh, 2012). The estimated sensitivity and specificity were 90% and 81%, respectively, resulting in an LR+ of 10.

Another study used the Japanese versions of the home and school form of parent and teacher versions of the ARS-IV and the SDQ to screen children (N = 838) for ADHD diagnoses that was established after an age-5 community health checkup by diagnostic interview according to DSM-5 criteria (Takayanagi et al., 2016). The AUC was significantly lower using the parent compared to the teacher ARS-IV (0.95 vs. 0.66, respectively), as sensitivity was reduced while specificity was comparable using either the 80th or 90th percentile cutoffs. Inspection of the receiver operating characteristic curve revealed equivalent sensitivity when comparing parent and teacher ARS-IV ratings to parent and teacher SDQ scores with a fixed specificity of ~80%.

More recently, parent and teacher reports of the Early Childhood Inventory-4 (ECI-4) (Gadow, Sprafkin, & Nolan, 2001) hyperactivity–impulsivity and IA subscales were used in the Norwegian Mother and Child Cohort study to establish the validity of US norms around age 3.5 (Overgaard, Oerbeck, et al., 2019). The AUC was higher for ECI-4 HI than IA parent ratings (0.85 vs. 0.73) and teacher ECI-4 HI was significantly reduced compared to parent ratings (0.70 vs. 0.85) among boys but not girls. While specificity was equivalent, sensitivity was reduced in girls compared to boys using parent and teacher ECI-4 HI ratings, such that a cutoff of 13 produced LR+ 5.7 with 49% specificity in boys and LR+ 3.6 with 31% sensitivity in girls.

Meta-analysis of small studies of children and adolescents

A meta-analysis and systematic review described the results of 25 studies published from 1991 to 2015 that used CBCL attention problem subscale (n = 16), CRS-R parent (n = 4) or teacher (n = 5) ratings, or the ASQ (n = 5) to identify children aged 3–18 who probably had ADHD and those who probably did not (Chang, Wang, & Tsai, 2016); all but one study had a sample size below N = 500. The 95% confidence intervals overlapped using the three instruments, but the ASQ had the highest pooled estimated sensitivity (range: 0.75–0.83), specificity (range: 0.73–0.84), LR+ (range: 2.9–5.3), LR- (range: 0.2–0.3), and diagnostic odds ratios (range: 9–27). US studies (n = 8) that used the CBCL attention subscale tended to report nonsignificantly lower sensitivity and significantly higher specificity than non-US studies. Otherwise, specificity, unlike sensitivity, was increased with higher fractions of females and participants aged ≥11.

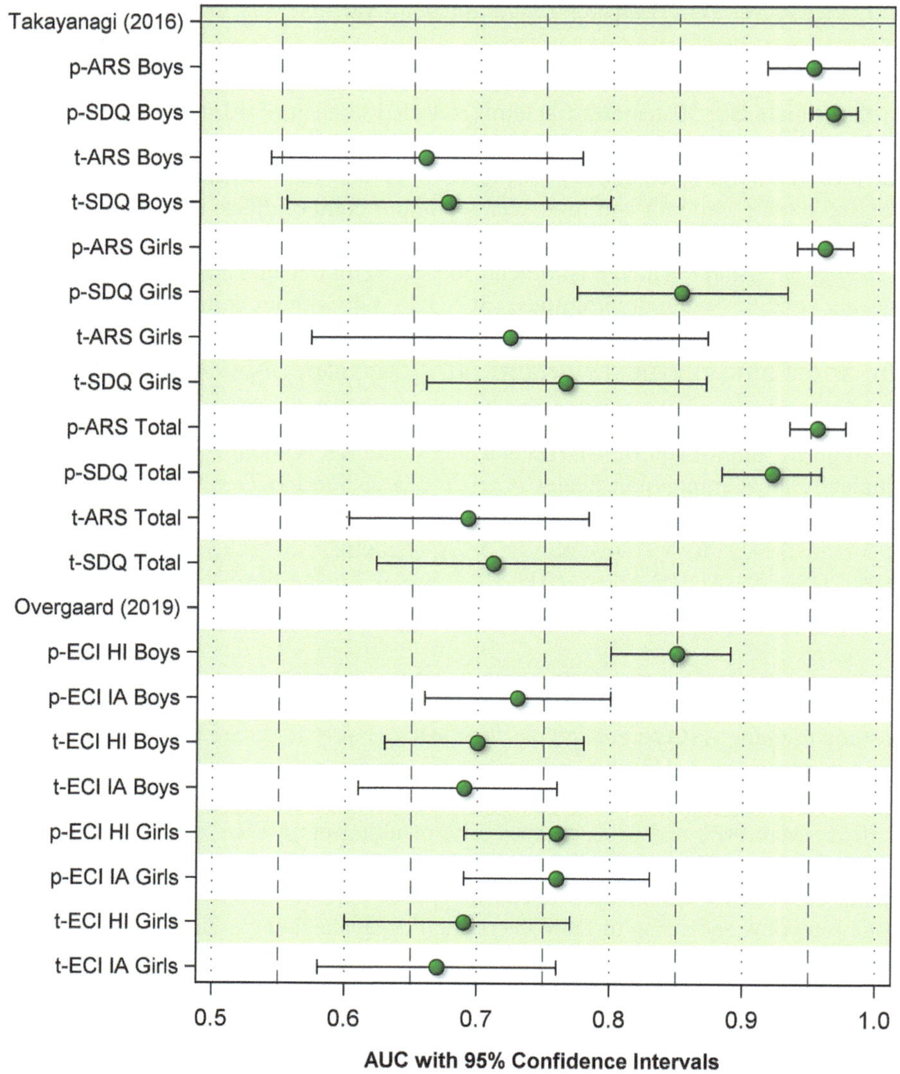

FIGURE 29.3 **Area under the receiver operating characteristic curve (AUC) using the ADHD Rating Scale (ARS), Strengths & Difficulties Questionnaire (SDQ), and the Early Childhood Inventory-4 (ECI) to identify ADHD in preschool-age children.** Screening performance estimates for attention deficit hyperactivity disorder instruments (and subscales) completed by adolescents, parents, and/or teachers. *Black filled circles* represent area under the receiver operating characteristic curve (i.e., the average proportion (%) of children with ADHD who are detected across all possible cutoffs; higher values equal better screening performance). *Horizontal lines* represent 95% confidence intervals around the estimated AUC.

Applications to other areas of development

This chapter described multiple instruments that can identify children and adolescents who probably have ADHD and those who probably do not; however, discriminative performance is typically reduced in the setting of multiple comorbidities. This is not surprising as symptom-based case definitions inherently lack specificity (Hyams, 1998), and attention problems tend to cooccur with multiple other neurodevelopmental and behavioral deficits (Gordon-Lipkin, Marvin, Law, & Lipkin, 2018; Mohammadi et al., 2019; Pahlman, Gillberg, Wentz, & Himmelmann, 2020), especially in high risk populations (e.g., children born extremely preterm (Johnson, Hollis, Marlow, Simms, & Wolke, 2014; Scott et al., 2017)). Recognition that hyperactivity, distractibility, and impulsivity are sensitive but nonspecific brain functions and behavioral patterns has prompted the inference that these abnormalities should be viewed as a spectrum that reflects underlying developmental processes rather than components of discrete nonoverlapping disorders (Berger, Remington, Leitner, & Leviton, 2015). Nevertheless, current DSM-5 and ICD-11 diagnostic criteria describe ADHD as categorically independent (Posner et al., 2020).

TABLE 29.4 Screening performance in large studies of preschool-age children using the ADHD Rating Scale (ARS), Strengths & Difficulties Questionnaire (SDQ), and Early Childhood Inventory-4 (ECI).

Instrument	Name (Year)	Sample source	Country	N (ADHD)	Age (years)	Cut-off	Sen.	Spec.	PPV	LR+
ARS	Takayanagi et al. (2016)	Com.	Japan	838 (46)	5–6	80th	0.94	0.83	0.24	5.5
Parent						90th	0.89	0.94	0.47	14.8
Teacher						80th	0.51	0.81	0.13	2.7
						90th	0.3	0.91	0.16	3.3
SDQ Parent						[a]	0.90	0.80		4.5
SDQ Teacher						[a]	0.50	0.80		2.5
ASQ Teacher	Abdekhodaie et al. (2012)	Com.	Iran	54 (31)	5.5	15	0.90	0.81		10
P and T ECI HI	Overgaard et al. (2019)	Com.	Norway	651 (247)	3.5 avg.	10	0.72	0.78	0.48	3.3
(among boys)						11 12	0.63 0.56	0.82 0.87	0.49 0.54	3.5 4.2
						13	0.49	0.91	0.61	5.7
						14	0.42	0.94	0.65	6.7
						15	0.35	0.96	0.74	9.9
						16	0.25	0.97	0.69	8.0
(among girls)						10	0.44	0.84	0.43	2.8
						11	0.4	0.87	0.45	3.0
						12	0.35	0.89	0.47	3.2
						13	0.31	0.91	0.50	3.6
						14	0.24	0.95	0.55	4.5
						15	0.15	0.95	0.45	3.0
						16	0.15	0.96	0.50	3.6

Study characteristics, instrument score cutoffs, and screening performance metric estimates are described. Sensitivity: the probability that a diseased person (case) in the population will be identified as diseased (syn: true positive probability). Specificity: the probability that a person without the disease (noncase) will be correctly identified as nondiseased. Positive predictive value: the probability that a person with a positive test does have the disease. Likelihood ratio positive: the ratio sensitivity/(1 − specificity) or sensitivity/false-positive rate. *Com.*, community; *HI*, Hyperactivity–Impulsivity and IA inattention; *LR+*, Likelihood Ratio Positive; *p*, parent; *PPV*, Positive Predictive Value; *Sens.*, sensitivity; *Spec.*, Specificity; *t*, teacher.
[a]*visual inspection of receiver operating characteristic curve used to establish fixed specificity cutoff at 80%.*

ADHD screening instruments, diagnostic criteria, presentation subtypes, and treatment modalities are bound to change as investigators increasingly conduct transdiagnostic research (McTeague et al., 2020) and identify underlying pathophysiology shared by multiple neurobehavioral disorders (e.g., the National Institute of Mental Health Research Domain Criteria initiative (Cuthbert, 2014; Morris & Cuthbert, 2012)). Indeed, emerging evidence supports the possibility that dimensionality may convey information about ADHD treatment response in some children (Ros & Graziano, 2019).That said, progress will require much effort and novel approaches to overcome the significant challenges in the integration of psychological and biological measures *and* constructs (Lake, Yee, & Miller, 2017).

Efforts to build consensus on pathophysiological dimensions that are most clinically useful in caring for children with ADHD are likely to bear fruit (Posner et al., 2020). Similar appreciation for dimensionality is needed in studies that seek to identify ADHD antecedents, as it is possible that specific risk factors (or mosaics) will convey information about elevated risks and trajectories of multiple disorders; especially ADHD, autism, and nonautistic social impairment (Rints, McAuley, & Nilsen, 2015; Sokolova et al., 2017; Stergiakouli et al., 2017). For example, hypertensive pregnancy disorders (and their correlates including preterm delivery) are associated with elevated risks of ADHD (Bohm et al., 2019; Dachew, Scott, Mamun, & Alati, 2019), autism, and other neuropsychiatric disorders (Dachew, Mamun, Maravilla, & Alati, 2018; Maher et al., 2018; Walker et al., 2015; Xu et al., 2018). This has important study design implications in investigations of etiology (Weiss & Liff, 1983).

It is possible that new software applications and analytic techniques including machine learning models may improve screening performance in children with ADHD (Amado-Caballero et al., 2020; Duda, Haber, Daniels, & Wall, 2017; Mwamba, Fourie, & den Heever, 2019, pp. 1444–1447). Emerging evidence also supports the possibility that molecular biomarkers (Mooney et al., 2020) and neuroimaging (Guo et al., 2020) may provide useful information about ADHD in children. However, additional research in large unselected populations is needed (Uddin, Dajani, Voorhies, Bednarz, & Kana, 2017).

Conclusion

Multiple assessment instruments can identify preschool and school-age children and adolescents who probably have ADHD and those who probably do not with moderate-to-high sensitivity and moderate specificity in community samples. The ASQ is lauded for its brevity, while the CBCL questionnaire is among the most commonly used comprehensive assessment tools (Chang et al., 2016). Variation in ADHD screening performance across instruments often appears to be explained by differences in diagnostic methods (e.g., clinical examination vs. parent-report), sample characteristics (e.g., age, sex, and comorbidities), and/or respondent ratings (e.g., parent vs. teacher). Specificity tends to be moderate because ADHD symptoms commonly occur in children who meet diagnostic criteria for other neurodevelopmental and behavioral disorders. Transdiagnostic research and appreciation for dimensionality might accelerate progress in developing new tools to assess, diagnose, subtype, and treat ADHD in children. Focus on early assessment and intervention to reduce risk or improve trajectories across different dimensions of brain functioning and behavior may yield benefits. Success will require rigorous study designs, appreciation for an ADHD spectrum, and support for large studies of unselected populations.

Key facts of attention deficit hyperactivity disorder (ADHD) in children

- ADHD in children below 17 years of age is defined in the Diagnostic and Statistical Manual of Mental Disorders –5 as the presence of $\geq$ six symptoms under the inattentive or hyperactive and impulsive presentations (or both) prior to age 12; symptoms must be persistent for $\geq$ 6 months; have negative impact on social, academic, or occupational activities; occur in two or more settings; and not be attributed to other behavioral or mental disorders.
- The 11th edition of the International Classification of Diseases similarly requires a persistent patter of inattention, hyperactivity–impulsivity, or both, for at least 6 months, and the symptoms must interfere with academic, occupational, or social functioning to warrant ADHD diagnosis.
- Children with ADHD are more likely than their typically developing peers to meet diagnostic criteria for other neurodevelopmental and behavioral disorders.
- The three subtypes of ADHD are: predominantly inattentive, predominantly hyperactive/impulsive, and combined.
- Meta-analysis of 143 studies of children and adolescents supports a global prevalence of 53 per 1000 with little international variation and generally stable time trends.

Summary points

- Multiple assessment instruments can identify preschool and school-age children and adolescents who probably have attention deficit hyperactivity disorder (ADHD) and those who probably do not with moderate-to-high sensitivity and moderate specificity in community samples.
- Specificity tends to be moderate because ADHD symptoms commonly occur in children who meet diagnostic criteria for other neurodevelopmental, behavioral, and neuropsychiatric disorders.
- Variation in ADHD screening performance across instruments appears largely attributable to differences in diagnostic methods (e.g., clinical examination vs. parent report), sample characteristics (e.g., age, sex, and comorbidities) and/or respondents (e.g., parent vs. teacher).

- Multiple types of respondents (i.e., parents, teachers or children) tend to convey additional information about ADHD compared to relying only on parent ratings.
- Recognition that hyperactivity, distractibility, and impulsivity are sensitive but nonspecific brain functions and behavioral patterns has prompted the inference that attention problems should be viewed as a spectrum that reflects underlying developmental processes rather than components of a discrete disorder.

References

Abdekhodaie, Z., Tabatabaei, S. M., & Gholizadeh, M. (2012). The investigation of ADHD prevalence in kindergarten children in northeast Iran and a determination of the criterion validity of Conners' questionnaire via clinical interview. *Research in Developmental Disabilities, 33*(2), 357–361. https://doi.org/10.1016/j.ridd.2011.10.006.

Achenbach, T. M., Dumenci, L., & Rescorla, L. A. (2003). DSM-oriented and empirically based approaches to constructing scales from the same item pools. *Journal of Clinical Child and Adolescent Psychology, 32*(3), 328–340. https://doi.org/10.1207/S15374424JCCP3203_02.

Achenbach, T. M., & Edelbrock, C. S. (1983). *Manual for the child behavior checklist and revised child behavior profile.*

Alda, J. A., & Serrano-Troncoso, E. (2013). Attention-deficit hyperactivity disorder: Agreement between clinical impression and the SNAP-IV screening tool. *Actas Españolas de Psiquiatría, 41*(2), 76–83. Retrieved from: https://www.ncbi.nlm.nih.gov/pubmed/23592067.

Algorta, G. P., Dodd, A. L., Stringaris, A., & Youngstrom, E. A. (2016). Diagnostic efficiency of the SDQ for parents to identify ADHD in the UK: A ROC analysis. *European Child and Adolescent Psychiatry, 25*(9), 949–957. https://doi.org/10.1007/s00787-015-0815-0.

Amado-Caballero, P., Casaseca-de-la-Higuera, P., Alberola-Lopez, S., Andres-de-Llano, J. M., Lopez-Villalobos, J. A., Garmendia-Leiza, J. R., & Alberola-Lopez, C. (2020). Objective ADHD diagnosis using convolutional neural networks over daily-life activity records. *IEEE Journal of Biomedical and Health Informatics.* https://doi.org/10.1109/JBHI.2020.2964072.

American Psychiatric Association. (2013). *Diagnostic and statistical manual of mental disorders (DSM-5®).* American Psychiatric Pub.

Anckarsater, H., Lundstrom, S., Kollberg, L., Kerekes, N., Palm, C., Carlstrom, E., … Lichtenstein, P. (2011). The child and adolescent twin study in Sweden (CATSS). *Twin Research and Human Genetics, 14*(6), 495–508. https://doi.org/10.1375/twin.14.6.495.

Berger, I., Remington, A., Leitner, Y., & Leviton, A. (2015). Brain development and the attention spectrum. *Frontiers in Human Neuroscience, 9*, 23. https://doi.org/10.3389/fnhum.2015.00023.

Bohm, S., Curran, E. A., Kenny, L. C., O'Keeffe, G. W., Murray, D., & Khashan, A. S. (2019). The effect of hypertensive disorders of pregnancy on the risk of ADHD in the offspring. *Journal of Attention Disorders, 23*(7), 692–701. https://doi.org/10.1177/1087054717690230.

Bussing, R., Schuhmann, E., Belin, T. R., Widawski, M., & Perwien, A. R. (1998). Diagnostic utility of two commonly used ADHD screening measures among special education students. *Journal of the American Academy of Child and Adolescent Psychiatry, 37*(1), 74–82. https://doi.org/10.1097/00004583-199801000-00020.

Carballo, J. J., Rodriguez-Blanco, L., Garcia-Nieto, R., & Baca-Garcia, E. (2018). Screening for the ADHD Phenotype Using the Strengths and Difficulties Questionnaire in a Clinical Sample of Newly Referred Children and Adolescents. *Journal of Attention Disorders, 22*(11), 1032–1039.

Chang, L. Y., Wang, M. Y., & Tsai, P. S. (2016). Diagnostic accuracy of rating scales for attention-deficit/hyperactivity disorder: A meta-analysis. *Pediatrics, 137*(3). https://doi.org/10.1542/peds.2015-2749. e20152749.

Cianchetti, C., Faedda, N., Pasculli, M., Ledda, M. G., Diaz, G., Peschechera, A., … Margari, L. (2020). Predictive validity for the clinical diagnosis of a new parent questionnaire, the CABI, compared with CBCL. *Clinical Child Psychology and Psychiatry.* https://doi.org/10.1177/1359104519895056.

Cianchetti, C., Pittau, A., Carta, V., Campus, G., Littarru, R., Ledda, M. G., … Fancello, G. S. (2013). Child and adolescent behavior inventory (CABI): A new instrument for epidemiological studies and pre-clinical evaluation. *Clinical Practice and Epidemiology in Mental Health, 9*, 51–61. https://doi.org/10.2174/1745017901309010051.

Conners, C. K. (1990). *Conners' abbreviated symptom questionnaire.* Multi-Health Systems.

Cuthbert, B. N. (2014). The RDoC framework: Facilitating transition from ICD/DSM to dimensional approaches that integrate neuroscience and psychopathology. *World Psychiatry, 13*(1), 28–35. https://doi.org/10.1002/wps.20087.

Dachew, B. A., Mamun, A., Maravilla, J. C., & Alati, R. (2018). Pre-eclampsia and the risk of autism-spectrum disorder in offspring: Meta-analysis. *British Journal of Psychiatry, 212*(3), 142–147. https://doi.org/10.1192/bjp.2017.27.

Dachew, B. A., Scott, J. G., Mamun, A., & Alati, R. (2019). Pre-eclampsia and the risk of attention-deficit/hyperactivity disorder in offspring: Findings from the ALSPAC birth cohort study. *Psychiatry Research, 272*, 392–397. https://doi.org/10.1016/j.psychres.2018.12.123.

Duda, M., Haber, N., Daniels, J., & Wall, D. P. (2017). Crowdsourced validation of a machine-learning classification system for autism and ADHD. *Translational Psychiatry, 7*(5), e1133. https://doi.org/10.1038/tp.2017.86.

Dupaul, G. J. (1991). Parent and teacher ratings of ADHD symptoms - psychometric properties in a community-based sample. *Journal of Clinical Child Psychology, 20*(3), 245–253. https://doi.org/10.1207/s15374424jccp2003_3.

Faraone, S. V., Childress, A., Wigal, S. B., Kollins, S. H., McDonnell, M. A., DeSousa, N. J., & Sallee, F. R. (2018). Reliability and validity of the daily parent rating of evening and morning behavior scale, revised. *Journal of Attention Disorders, 22*(11), 1066–1073. https://doi.org/10.1177/1087054715619009.

Faraone, S. V., DeSousa, N. J., Komolova, M., Sallee, F. R., Incledon, B., & Wilens, T. E. (2019). Functional impairment in youth with ADHD: Normative data and norm-referenced cutoff points for the before school functioning questionnaire and the parent rating of evening and morning behavior scale, revised. *Journal of Clinical Psychiatry, 81*(1). https://doi.org/10.4088/JCP.19m12956.

Faraone, S. V., Hammerness, P. G., & Wilens, T. E. (2018). Reliability and validity of the before-school functioning scale in children with ADHD. *Journal of Attention Disorders, 22*(11), 1040–1048. https://doi.org/10.1177/1087054714564623.

Gadow, K. D., Sprafkin, J., & Nolan, E. E. (2001). DSM-IVSymptoms in community and clinic preschool children. *Journal of the American Academy of Child and Adolescent Psychiatry, 40*(12), 1383–1392. https://doi.org/10.1097/00004583-200112000-00008.

Goodman, R. (1997). The strengths and difficulties questionnaire: A research note. *Journal of Child Psychology and Psychiatry, 38*(5), 581–586. https://doi.org/10.1111/j.1469-7610.1997.tb01545.x.

Goodman, R., Ford, T., Richards, H., Gatward, R., & Meltzer, H. (2000). The development and well-being assessment: Description and initial validation of an integrated assessment of child and adolescent psychopathology. *Journal of Child Psychology and Psychiatry, 41*(5), 645–655. Retrieved from: https://www.ncbi.nlm.nih.gov/pubmed/10946756.

Gordon-Lipkin, E., Marvin, A. R., Law, J. K., & Lipkin, P. H. (2018). Anxiety and mood disorder in children with autism spectrum disorder and ADHD. *Pediatrics, 141*(4). https://doi.org/10.1542/peds.2017-1377.

Guo, X., Yao, D., Cao, Q., Liu, L., Zhao, Q., Li, H., … Sun, L. (2020). Shared and distinct resting functional connectivity in children and adults with attention-deficit/hyperactivity disorder. *Translational Psychiatry, 10*(1), 65. https://doi.org/10.1038/s41398-020-0740-y.

Guy, W. (1976). *ECDEU assessment manual for psychopharmacology*. US Department of Health, Education, and Welfare, Public Health Service.

Holmberg, K., Sundelin, C., & Hjern, A. (2013). Screening for attention-deficit/hyperactivity disorder (ADHD): Can high-risk children be identified in first grade? *Child: Care, Health and Development, 39*(2), 268–276. https://doi.org/10.1111/j.1365-2214.2012.01382.x.

Hyams, K. C. (1998). Developing case definitions for symptom-based conditions: The problem of specificity. *Epidemiological Review, 20*(2), 148–156. https://doi.org/10.1093/oxfordjournals.epirev.a017977.

Johnson, S., Hollis, C., Marlow, N., Simms, V., & Wolke, D. (2014). Screening for childhood mental health disorders using the Strengths and difficulties questionnaire: The validity of multi-informant reports. *Developmental Medicine and Child Neurology, 56*(5), 453–459. https://doi.org/10.1111/dmcn.12360.

Lake, J. I., Yee, C. M., & Miller, G. A. (2017). Misunderstanding RDoC. *Zeitschrift für Psychologie, 225*(3), 170–174. https://doi.org/10.1027/2151-2604/a000301.

Lampert, T. L., Polanczyk, G., Tramontina, S., Mardini, V., & Rohde, L. A. (2004). Diagnostic performance of the CBCL-attention problem scale as a screening measure in a sample of Brazilian children with ADHD. *Journal of Attention Disorders, 8*(2), 63–71. https://doi.org/10.1177/108705470400800204.

Larson, T., Anckarsater, H., Gillberg, C., Stahlberg, O., Carlstrom, E., Kadesjo, B., … Gillberg, C. (2010). The autism-tics, AD/HD and other comorbidities inventory (A-TAC): Further validation of a telephone interview for epidemiological research. *BMC Psychiatry, 10*(1), 1. https://doi.org/10.1186/1471-244x-10-1. Artn 1.

Last, J. (2002). *A dictionary of epidemiology* (3rd ed.). Oxford: Oxford University Press.

Maher, G. M., O'Keeffe, G. W., Kearney, P. M., Kenny, L. C., Dinan, T. G., Mattsson, M., & Khashan, A. S. (2018). Association of hypertensive disorders of pregnancy with risk of neurodevelopmental disorders in offspring: A systematic review and meta-analysis. *JAMA Psychiatry, 75*(8), 809–819. https://doi.org/10.1001/jamapsychiatry.2018.0854.

Marland, C., Lichtenstein, P., Degl'Innocenti, A., Larson, T., Rastam, M., Anckarsater, H., … Lundstrom, S. (2017). The autism-tics, ADHD and other comorbidities inventory (A-TAC): Previous and predictive validity. *BMC Psychiatry, 17*(1), 403. https://doi.org/10.1186/s12888-017-1563-0.

McCarney, S. (1989). *The attention–deficit disorders evaluation scale home version technical manual*. Columbia, MO: Hawthorne Educational Services.

McTeague, L. M., Rosenberg, B. M., Lopez, J. W., Carreon, D. M., Huemer, J., Jiang, Y., … Etkin, A. (2020). Identification of common neural circuit disruptions in emotional processing across psychiatric disorders. *American Journal of Psychiatry*. https://doi.org/10.1176/appi.ajp.2019.18111271.

Mohammadi, M. R., Zarafshan, H., Khaleghi, A., Ahmadi, N., Hooshyari, Z., Mostafavi, S. A., … Salmanian, M. (2019). Prevalence of ADHD and its comorbidities in a population-based sample. *Journal of Attention Disorders*. https://doi.org/10.1177/1087054719886372, 1087054719886372.

Mooney, M. A., Ryabinin, P., Wilmot, B., Bhatt, P., Mill, J., & Nigg, J. T. (2020). Large epigenome-wide association study of childhood ADHD identifies peripheral DNA methylation associated with disease and polygenic risk burden. *Translational Psychiatry, 10*(1), 8. https://doi.org/10.1038/s41398-020-0710-4.

Morris, S. E., & Cuthbert, B. N. (2012). Research Domain criteria: Cognitive systems, neural circuits, and dimensions of behavior. *Dialogues in Clinical Neuroscience, 14*(1), 29–37. Retrieved from: https://www.ncbi.nlm.nih.gov/pubmed/22577302.

Mwamba, H. M., Fourie, P. R., & den Heever, D. V. (2019). PANDAS: Paediatric attention-deficit/hyperactivity disorder application software. In *Conf Proc IEEE Eng Med Biol Soc* (Vol. 2019) https://doi.org/10.1109/EMBC.2019.8857357.

Overgaard, K. R., Madsen, K. B., Oerbeck, B., Friis, S., & Obel, C. (2019). The predictive validity of the strengths and difficulties questionnaire for child attention-deficit/hyperactivity disorder. *European Child and Adolescent Psychiatry, 28*(5), 625–633. https://doi.org/10.1007/s00787-018-1226-9.

Overgaard, K. R., Oerbeck, B., Friis, S., Biele, G., Pripp, A. H., Aase, H., & Zeiner, P. (2019). Screening with an ADHD-specific rating scale in preschoolers: A cross-cultural comparison of the early childhood inventory-4. *Psychological Assessment, 31*(8), 985–994. https://doi.org/10.1037/pas0000722.

Pahlman, M., Gillberg, C., Wentz, E., & Himmelmann, K. (2020). Autism spectrum disorder and attention-deficit/hyperactivity disorder in children with cerebral palsy: Results from screening in a population-based group. *European Child and Adolescent Psychiatry*. https://doi.org/10.1007/s00787-020-01471-1.

Park, J. I., Shim, S. H., Lee, M., Jung, Y. E., Park, T. W., Park, S. H., … Chung, S. K. (2014). The validities and efficiencies of Korean ADHD rating scale and Korean child behavior checklist for screening children with ADHD in the community. *Psychiatry Investigation, 11*(3), 258–265. https://doi.org/10.4306/pi.2014.11.3.258.

Polanczyk, G., de Lima, M. S., Horta, B. L., Biederman, J., & Rohde, L. A. (2007). The worldwide prevalence of ADHD: A systematic review and metaregression analysis. *American Journal of Psychiatry, 164*(6), 942–948. https://doi.org/10.1176/ajp.2007.164.6.942.

Porta, M. (2014). *A dictionary of epidemiology*. Oxford University Press.

Posner, J., Polanczyk, G. V., & Sonuga-Barke, E. (2020). Attention-deficit hyperactivity disorder. *Lancet, 395*(10222), 450–452. https://doi.org/10.1016/S0140-6736(19)33004-1.

Rints, A., McAuley, T., & Nilsen, E. S. (2015). Social communication is predicted by inhibitory ability and ADHD traits in preschool-aged children: A mediation model. *Journal of Attention Disorders, 19*(10), 901–911. https://doi.org/10.1177/1087054714558873.

Ros, R., & Graziano, P. A. (2019). A transdiagnostic examination of self-regulation: Comparisons across preschoolers with ASD, ADHD, and typically developing children. *Journal of Clinical Child and Adolescent Psychology*, 1–16. https://doi.org/10.1080/15374416.2019.1591280.

Rydell, M., Lundstrom, S., Gillberg, C., Lichtenstein, P., & Larsson, H. (2018). Has the attention deficit hyperactivity disorder phenotype become more common in children between 2004 and 2014? Trends over 10 years from a Swedish general population sample. *Journal of Child Psychology and Psychiatry, 59*(8), 863–871. https://doi.org/10.1111/jcpp.12882.

Safer, D. J. (2018). Is ADHD really increasing in youth? *Journal of Attention Disorders, 22*(2), 107–115. https://doi.org/10.1177/1087054715586571.

Scott, M. N., Hunter, S. J., Joseph, R. M., O'Shea, T. M., Hooper, S. R., Allred, E. N., … Kuban, K. (2017). Neurocognitive correlates of attention-deficit hyperactivity disorder symptoms in children born at extremely low gestational age. *Journal of Developmental and Behavioral Pediatrics, 38*(4), 249–259. https://doi.org/10.1097/DBP.0000000000000436.

Shaffer, D., Fisher, P., Lucas, C. P., Dulcan, M. K., & Schwab-Stone, M. E. (2000). NIMH diagnostic interview Schedule for children version IV (NIMH DISC-IV): Description, differences from previous versions, and reliability of some common diagnoses. *Journal of the American Academy of Child and Adolescent Psychiatry, 39*(1), 28–38. https://doi.org/10.1097/00004583-200001000-00014.

Sokolova, E., Oerlemans, A. M., Rommelse, N. N., Groot, P., Hartman, C. A., Glennon, J. C., … Buitelaar, J. K. (2017). A causal and mediation analysis of the comorbidity between attention deficit hyperactivity disorder (ADHD) and autism spectrum disorder (ASD). *Journal of Autism and Developmental Disorders, 47*(6), 1595–1604. https://doi.org/10.1007/s10803-017-3083-7.

Stergiakouli, E., Davey Smith, G., Martin, J., Skuse, D. H., Viechtbauer, W., Ring, S. M., … St Pourcain, B. (2017). Shared genetic influences between dimensional ASD and ADHD symptoms during child and adolescent development. *Molecular Autism, 8*, 18. https://doi.org/10.1186/s13229-017-0131-2.

Swanson, J. M., Kraemer, H. C., Hinshaw, S. P., Arnold, L. E., Conners, C. K., Abikoff, H. B., … Wu, M. (2001). Clinical relevance of the primary findings of the MTA: Success rates based on severity of ADHD and ODD symptoms at the end of treatment. *Journal of the American Academy of Child and Adolescent Psychiatry, 40*(2), 168–179. https://doi.org/10.1097/00004583-200102000-00011.

Takayanagi, N., Yoshida, S., Yasuda, S., Adachi, M., Kaneda-Osato, A., Tanaka, M., … Nakamura, K. (2016). Psychometric properties of the Japanese ADHD-RS in preschool children. *Research in Developmental Disabilities, 55*, 268–278. https://doi.org/10.1016/j.ridd.2016.05.002.

Uddin, L. Q., Dajani, D. R., Voorhies, W., Bednarz, H., & Kana, R. K. (2017). Progress and roadblocks in the search for brain-based biomarkers of autism and attention-deficit/hyperactivity disorder. *Translational Psychiatry, 7*(8), e1218. https://doi.org/10.1038/tp.2017.164.

Ullebo, A. K., Posserud, M. B., Heiervang, E., Gillberg, C., & Obel, C. (2011). Screening for the attention deficit hyperactivity disorder phenotype using the strength and difficulties questionnaire. *European Child and Adolescent Psychiatry, 20*(9), 451–458. https://doi.org/10.1007/s00787-011-0198-9.

Vugteveen, J., De Bildt, A., Hartman, C. A., & Timmerman, M. E. (2018). Using the Dutch multi-informant Strengths and Difficulties Questionnaire (SDQ) to predict adolescent psychiatric diagnoses. *European Child & Adolescent Psychiatry.*

Vugteveen, J., de Bildt, A., Theunissen, M., Reijneveld, M., & Timmerman, M. (2019). Validity aspects of the strengths and difficulties questionnaire (SDQ) adolescent self-report and parent-report versions among Dutch adolescents. *Assessment*. https://doi.org/10.1177/1073191119858416, 1073191119858416.

Walker, C. K., Krakowiak, P., Baker, A., Hansen, R. L., Ozonoff, S., & Hertz-Picciotto, I. (2015). Preeclampsia, placental insufficiency, and autism spectrum disorder or developmental delay. *JAMA Pediatrics, 169*(2), 154–162. https://doi.org/10.1001/jamapediatrics.2014.2645.

Weiss, N. S., & Liff, J. M. (1983). Accounting for the multicausal nature of disease in the design and analysis of epidemiologic studies. *American Journal of Epidemiology, 117*(1), 14–18. https://doi.org/10.1093/oxfordjournals.aje.a113509.

World Health Organization. (2018). *International classification of diseases, 11th revision (ICD-11). ICD-11 is here*. Geneva: World Health Organization.

Xu, R. T., Chang, Q. X., Wang, Q. Q., Zhang, J., Xia, L. X., Zhong, N., … Huang, Q. T. (2018). Association between hypertensive disorders of pregnancy and risk of autism in offspring: A systematic review and meta-analysis of observational studies. *Oncotarget, 9*(1), 1291–1301. https://doi.org/10.18632/oncotarget.23030.

Chapter 30

The Mini-Mental State Pediatric Examination: a cognitive screening test for Italian children

Valeria Peviani[1,2,3], Pina Scarpa[1,4] and Gabriella Bottini[1,2,4]

[1]*Cognitive Neuropsychology Center, ASST Grande Ospedale Metropolitano Niguarda, Milan, Italy;* [2]*Department of Brain and Behavioural Sciences, University of Pavia, Pavia, Italy;* [3]*Department of Neuroscience, Max Planck Institute for Empirical Aesthetics, Frankfurt am Main, Germany;* [4]*NeuroMI, Milan Center for Neuroscience, Italy*

List of abbreviations

GLM General linear model
GzLM Generalized linear model
IQ Intelligence quotient
MMSE Mini-Mental State Examination
MMSPE Mini-Mental State Pediatric Examination
SYSTEMS School-Years Screening Test for the Evaluation of Mental State

Mini-dictionary of terms

Attention: Ability to actively select, focus on, and orient cognitive resources toward discrete aspects of information. Sustained attention is the capacity to maintain the attentional focus over time, whereas divided attention is the ability to select and focus on two or more sources of information, or to meet two or more task demands, simultaneously.

Body representation: Mental representation of the body. It is built upon the integration of sensory and motor inputs with stored information about functional and structural properties about the body.

Cutoff score: It represents the threshold between normality and pathology. In clinical neuropsychology, the 5th (or 95th, if, for example, scores represent time units and not accuracy) percentile is often used as a cutoff score. When the cutoff is represented by the 5th percentile, it allows to detect a performance that is very unlikely (less than 5%) to be observed in the normative population. and therefore labelled as 'pathological'.

Executive functions: Set of complex cognitive functions involved in planning, guiding, supervising, and evaluating behavior. For instance, working memory, logical reasoning, cognitive flexibility, inhibition of the irrelevant or distracting information, and task switching are some of the most important executive functions.

General and generalized linear models: The general linear model (GLM) is a statistical linear model expressed by the generic formula $Y = A + XB + E$, where Y represents a set of measurements (dependent variable), X represents the independent variable (e.g., experimental condition, age, etc.), A and B are parameters that modulate the relationship between Y and X, and E represents the residual error. The GLM underlies several statistical models, such as ANOVA, MANOVA, linear regression, and t-test. The generalized linear model (GzLM) is a generalization of the GLM, useful when error distributions are not-Guassian.

IQ and mental age: The IQ (intelligence quotient) is a score used to represent the intellectual ability of a person. IQ scores are normally distributed, with mean 100 and standard deviation 15. The cutoff score representing the threshold between normality and pathology is usually set at 2 standard deviations below the mean ($IQ < 70$). The cutoff score representing the threshold between normality and intellectual giftedness is usually set at 2 standard deviations above the mean ($IQ > 130$). The mental age is calculated by multiplying IQ and chronological age and dividing by 100.

Diagnosis, Management and Modeling of Neurodevelopmental Disorders. https://doi.org/10.1016/B978-0-12-817988-8.00030-0

Lexical access: Ability to efficiently access the mental lexicon, i.e., a systematic organization of vocabulary. Neuropsychological tests for lexical access usually require the subject to name an object depicted in a drawing (picture naming), to name as many words as possible beginning with a certain letter (phonemic fluency), or to name as many objects belonging to a certain semantic category (semantic fluency).

Long-term memory: Ability to store information over a long period of time (from minutes to years). Episodic memory (retaining memories of temporally- and spatially-defined situations happened in the past) and semantic memory (information that has been abstracted from the context; e.g., meaning of a concept) are examples of long-term memory components.

Short-term memory: Ability to retain a small amount of information (7 ± 2 chunks, or pieces, of information) for a short period of time, in the order of seconds. In the verbal domain, it is usually assessed by asking the subject to repeat increasing sequences of digits.

Visuoconstructive skills: Set of skills involving the coordination of spatial, visual, and motor abilities. Traditionally, it is tested by asking the subject to copy a geometrical figure.

Working memory: Ability to retain and manipulate information for a short period of time, in the order of seconds. It differs from short-term memcry, as working memory is characterized by an active, i.e., executive, component, which allows to mentally manipulate the information retained. In the verbal domain, it is usually assessed by asking the subject to repeat increasing sequences of digits backward.

Introduction

This chapter focuses on pediatric cognitive screening. Before addressing how cognitive screening can be carried out in pediatric settings, the chapter proposes a general introduction to neuropsychological assessment and screening.

After such brief introduction, we discuss the adaptation to the developmental age of the Mini-Mental State Examination (MMSE; Folstein, Folstein, & McHugh, 1975), a widely-used screening tool for the adult age.

Specifically, we describe the research carried on by our group (Peviani, Scarpa, Vedovelli, & Bottini, 2018; Scarpa, Toraldo, Peviani, & Bottini, 2017), which led to the development and standardization of a pediatric version of the MMSE, i.e., the MMSPE (Mini-Mental State Pediatric Examination), now available to Italian clinicians.

Neuropsychological assessment and cognitive screening

The neuropsychological assessment (Fig. 30.1) is a crucial step within the clinical evaluation of neurological and psychiatric conditions. The main goals of neuropsychological assessment are: (i) to detect neurological dysfunction in order to guide differential diagnosis and prognosis; (ii) to characterize cognitive strengths and weaknesses; (iii) to describe cognitive changes over time; (iv) to assess the efficacy of treatments; (v) to provide recommendations regarding treatment and educational planning, as well as everyday life management (Casaletto & Heaton, 2017; Marshall, 2010).

The most important tool for a clinical neuropsychologist is the standardized test, consisting in one or more tasks designed to assess specific cognitive functions. Neuropsychological tests are administered in a standardized fashion, which must be clearly described in their test manual (e.g., Raven Colored Progressive Matrices; Raven, Court, & Raven, 1990), or

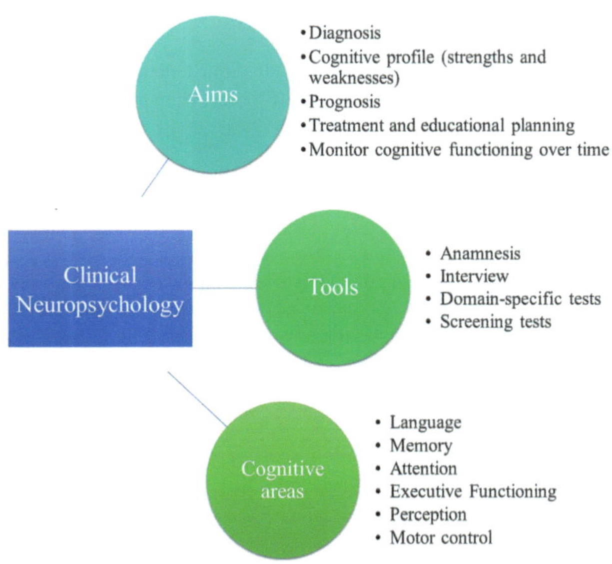

FIGURE 30.1 **Clinical neuropsychology: aims, tools, and areas.** Aims, tools, and main cognitive areas of investigation in clinical neuropsychology are summarized.

published paper (e.g., Short Story test for children; Scarpa et al., 2006). Normative scores (or "norms") are collected by administering the test to a normative sample, i.e., a reference group representing the general population, in terms of age, education, and gender.

A test can be devised to provide an extensive evaluation of a specific cognitive function (e.g., Test of Early Language Development, assessing receptive and expressive language; Hresko, Hammill, & Reid, 1991), or specific cognitive subdomain (e.g., spatial attention in the Bells test, Gauthier, Dehaut, & Joanette, 1989; face recognition in the Camden memory test, Warrington, 1996). These tests are usually referred to as domain-specific tests.

Other tests, instead, are designed to provide a global cognitive score, by means of a brief screening assessment of diverse cognitive functions. These tests are usually referred to as cognitive screening tests. Their purpose is facilitating early detection of cognitive change or decline, and guiding the selection of domain-specific tests to formulate an accurate diagnosis (Porta, 2008). Indeed, cognitive screening tests do not—and are not intended to—substitute a full neuropsychological assessment. For instance, cognitive screening tests are used by general practitioners in order to detect symptoms of alleged cognitive impairment, especially in the context of aging (Cullen, O'Neill, Evans, Coen, & Lawlor, 2007).

The Mini-Mental State Examination (MMSE)

The MMSE (Folstein et al., 1975) is one of the most used screening tests in clinical practice. Although a number of issues can be raised regarding the MMSE, it is characterized by several positive features, such as its reliability and easiness of administration (Creavin et al., 2016).

The MMSE is composed of 11 items, assessing key cognitive functions (Folstein et al., 1975). The first item addresses spatial and temporal orientation (10 points); then, short- and long-term memory (repetition and recall of three words) are assessed (6 points). Attention and working memory are tested by means of a backward calculation task or a backward spelling task (5 points). In the linguistic domain, lexical access and naming (2 points), repetition (1 point), oral and written comprehension (4 points), and writing (1 point) are assessed. Finally, visuoconstructive skills are tested by asking the subject to copy a geometrical figure (1 point).

It takes about 7 minutes to be administered to a patient with cognitive impairment, and only 5 minutes to a person without cognitive impairment (Borson, Scanlan, Brush, Vitaliano, & Dokmak, 2000). Scores range from 1 to 30, and the cutoff is usually set at 24, with a score comprised between 18 and 23 indicating mild cognitive impairment, and 0–17 indicating severe cognitive impairment, although other cutoff scores have been proposed (Crum, Anthony, Bassett, & Folstein, 1993; Kukull et al., 1994).

Cognitive screening in developmental age

When the objects of neuropsychology assessment are pediatric behavior and cognitive functioning, it is crucial to consider the brain as a fast-developing system, with different functions developing (often non-linearly) at different speed (Obrzut, 1986).

Several domain-specific tests and test batteries, usually stratified by age, exist to detect specific neuropsychological impairments, and to formulate a diagnosis (e.g., tests of visuomotor integration, Beery, 2004; tests of memory and learning, Adams & Sheslow, 1990). However, as previously mentioned, the screening approach might be more convenient and efficient compared to a full neuropsychological evaluation in certain pediatric clinical contexts. For instance, cognitive screening assessments can be useful to indicate whether a full neuropsychological assessment is necessary, and to guide the selection of domain-specific tests for further examinations. Such approach has additional beneficial effects on the child and on economic resources (Bornholt, Ajersch, Fisher, Markham, & Ouvrier, 2010; Scarpa et al., 2017).

Primary care pediatricians would take great advantage of a systematic and standardized screening tool, addressing the cognitive global functioning of their patients, to identify children at risk, and monitor the evolution of their cognitive functions and dysfunctions over time (Scarpa et al., 2017). In this context, the identification of children with developmental difficulties and delays is possible by frequently and systematically monitoring their development. Care providers need to effectively identify children whose performance falls below the normative cutoff, and refer them for further assessment and treatment. For instance, the Bayley Infant Neurodevelopmental Screener (BINS; Aylward, 1995) serves such purpose in early development (3–24 months of age), by briefly assessing basic neurological functions (e.g., posture, muscle tone, movement), expressive (e.g., gross and fine motor, oral motor, and verbal), receptive (visual, auditory, and verbal), and cognitive achievements (e.g., object permanence and problem-solving).

However, only few attempts have been made to provide clinicians with a cognitive screening test targeting the pediatric age range. Despite the need of a pediatric cognitive screening tool has been often pointed out (Ouvrier, Goldsmith, Ouvrier, & Williams, 1993; Ouvrier, Hendy, Bornholt, & Black, 1999; Rubial-Álvarez et al., 2007; Scarpa et al., 2017), at present there is no cognitive screening test that serves as a the gold standard for the developmental neuropsychology community.

MMSE and developmental age: different approaches compared

In light of the need of a pediatric cognitive screening test, and of the success of the MMSE as a screening tool in adulthood, the MMSE has been put forward as a candidate to perform cognitive screening in the developmental age (Fig. 30.2). For instance, Ouvrier et al. (1993) administered a slightly-adapted version of the MMSE to 117 Australian children aged between 4 and 15. Some minor modifications to the items were devised (e.g., inclusion of practice trials, simplification of the drawing to be copied), to facilitate children performing the test. Significant correlations with children's mental age and reading skills have been reported. As another example, Shoji et al. (2002) administered the MMSE to 1046 Japanese children aged 3-11, with the purpose of comparing cognitive functioning of patients with Alzheimer disease with that of preschool and school-aged children. Additionally, the MMSE was administered to a sample of 181 Spanish children aged 4−12 (Rubial-Álvarez et al., 2007). They found that scores increased gradually from 4 to 10 years of age. Furthermore, authors reported a significant correlation between the MMSE score and verbal intelligence.

Despite those promising results, one of the main drawbacks of using the original MMSE (Folstein et al., 1975) in developmental settings is that certain neuropsychological functions, critical to cognitive development (e.g., executive functioning, numerical knowledge), are disregarded (Obrzut, 1986; Telzrow, 1983). Furthermore, item complexity is not appropriate for some age ranges (Ouvrier et al., 1993; 1999).

An efficient solution to the aforementioned issue turned out to be the adaptation of the MMSE to the developmental age. Ouvrier et al. (1999) introduced the SYSTEMS (School-Years Screening Test for the Evaluation of Mental State), which includes 46 items, scored dichotomously, covering the MMSE key cognitive domains. Authors collected normative data on a gender-balanced sample of Australian children, aged 5−11, and reported satisfactory interrater and test-retest reliability, as well as concurrent validity, assessed by investigating the correlation between the test score and the mental age. As another example, Jain and Passi (2005) adapted the MMSE to the developmental age and administered it to a sample of 100 Indian children, aged 3−14. The sample included 50 neurologically impaired children and 50 controls. While the MMSE original 11-question structure was maintained, some items were modified or substituted to make the test more appropriate for children. Authors highlighted the high potential of their screening test in clinical practice.

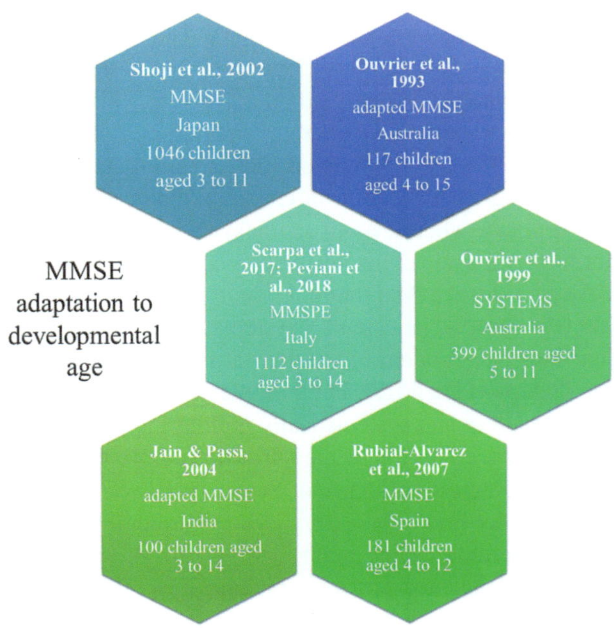

FIGURE 30.2 **MMSE adaptation to the developmental age.** Published studies on MMSE application and adaptation to the developmental age are summarized. The sample size and age range, along with the country where the study has been conducted, are reported.

Structure and flexibility of the MMSPE (Mini-Mental State Pediatric Examination)

Since no cognitive screening test for the developmental age was available in Italy, our research group developed a pediatric version of the MMSE for the Italian population (Peviani et al., 2018; Scarpa et al., 2017). The basic structure of the MMSE has been preserved. In detail, the MMSPE (Mini-Mental State Pediatric Examination) investigates spatial and temporal orientation, memory, attention, verbal production, verbal comprehension, and visuoconstructive skills. Critically, additional items have been included to briefly assess writing, reading, and calculation age-related skills (or prerequisites), given their critical role in cognitive and social development. Finally, we have included items exploring shape and color recognition, body representation, and executive functioning.

Furthermore, we adopted an age-stratification approach in order to tailor items' difficulty to the selected age ranges, and optimize MMSPE adaptation to a large portion of the developmental spectrum. More specifically, we developed four versions of the MMSPE, targeting: (i) children attending the kindergarten; (ii) children attending the first and second grades (early primary school) ; (iii) children attending the third, fourth, and fifth grades (late primary school); and (iv) children attending the sixth, seventh, and eighth grades (middle school). This approach allows to assess the cognitive skills of children belonging to different age groups. A summary of the standardization procedure is depicted in Fig. 30.3, while a summary of the cognitive skills and items is reported in Table 30.1. For a more detailed description of the tests, refer to Peviani et al. (2018) (MMSPE for preschoolers) and Scarpa et al. (2017) (for children aged 6−14). Both tests are fully available as supplementary materials to the articles.

In order to collect normative data, the MMSPE was administered to a total of 1114 children (526 males), speaking Italian as a first language, aged 36 months to 14 years and 11 months, recruited from kindergartens, primary, and middle schools, situated in Northern and Central Italy, in either rural or urban areas. Children were included after having ascertained the absence of any neuropsychiatric disease, auditory, or visual impairment. The time of administration ranged from 8 to 12 minutes. Demographic details of the normative samples are reported in Table 30.2.

Child's age and parental education predict the MMSPE global cognitive score

Importantly, besides children's age and school class, the age and level of education of their parents (years of completed schooling) were collected in our studies. While parental age did not significantly predict the overall MMSPE score in any of the age groups, parental education had a significant effect on the global cognitive score in all age groups. Parental level of education is known to have an influence on the cognitive development of their children, being a moderator of genetic and environmental factors (Neiss & Rowe, 2000; Rowe, Jacobson, & Van Den Oord, 1999). Parental education is significantly correlated with cognitive performance already in early infancy, as shown by research testing the relationship between parental education with the Mental Development Index, assessed through the Bayley Scales of Infant

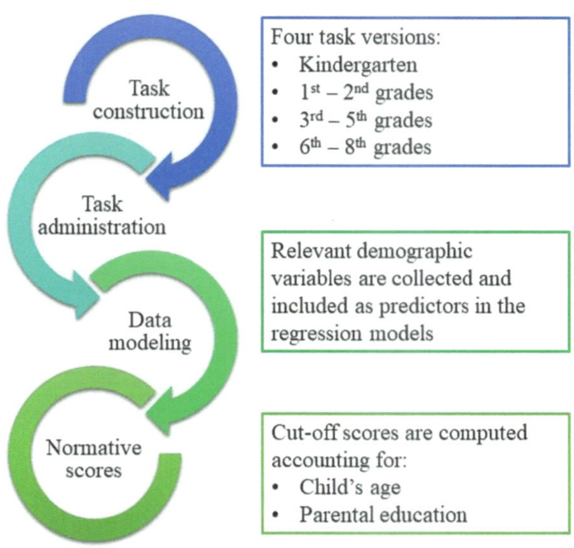

FIGURE 30.3 **MMSPE standardization steps.** Four versions of the MMSPE have been created and administered to a large normative sample of children aged 3−14. Cutoff scores have been obtained by considering relevant demographic variables (child's age and mean parental education) as predictors of the total MMSPE score.

TABLE 30.1 Cognitive domains and tasks of the MMSPE.

	Preschool-age		School-age	
Cognitive domain	**Tasks**	**Score**	**Task**	**Score**
Spatial and temporal orientation	Name and surname Age Address Date	6	Name and surname Age Address Date	11
Productive language	Naming	3	Naming	3
Praxis and body representation	Action execution Figure copying Drawing a human Body parts location	10	Action execution Figure copying Drawing a human	6
Reading and writing prerequisite skills	Phonological fusion Phonological segmentation One's own name writing	3	Phonological fusion Phonological segmentation Spoonerism task Word reading Sentence writing	5
Numerical knowledge	Counting forward Object counting Numerical judgment	3	Counting forward Counting backward Time tables	3
Short and long term memory	Word repetition Word delayed recall	6	Word repetition Word delayed recall	6
Visual memory and visual recognition	Picture memory	3		
Executive functions	Backward word repetition Go–no go task[a]	6	Spelling backward Go–no go task[a]	4
		Tot. 40		Tot. 38

Cognitive domains and tasks assessed by the MMSPE for preschool (Peviani et al., 2018) and school (Scarpa et al., 2017) aged children.
[a]*The Go–no go task has been adapted from Scarpa et al. (2006).*

TABLE 30.2 Demographic details of the normative samples.

	Scarpa et al., 2017	Peviani et al., 2018
Sample	807 (412 M)	305 (151 M)
Child's age (mean ± SD)	9.77 ± 2.32 years	4.59 ± 0.84 years
Father's age (mean ± SD)	43.90 ± 5.99 years	40.82 ± 5.69 years
Mean mother's age (mean ± SD)	40.80 ± 5.25 years	37.89 ± 4.58 years
Mean father's education (mean ± SD)	11.88 ± 3.75 years	14.21 ± 3.38 years
Mean mother's education (mean ± SD)	12.78 ± 3.49 years	15.41 ± 3.27 years

Mean and standard deviations (SD) referred to the main demographic variables of the normative samples for the MMSPE are reported.

Development (Bayley, 1993), at 27 months, as well as with other cognitive and linguistic indexes (Roberts, Bornstein, Slater, & Barrett, 1999). Some of those indexes are correlated with the parental education as early as 9 months of age. The effect of parental education persists in later years, and affects measures of intelligence and of neuropsychological functioning. In detail, Stanford-Binet and Wechsler IQs are modulated by parental education (Mercy & Steelman, 1982; Zhou, Baghurst, Gibson, & Makrides, 2007), as well as a number of neuropsychological measures, such as sustained and divided attention, visual integration skills, episodic memory, verbal learning, math and reading skills, vocabulary, and verbal production (Gibbs & Forste, 2014; Richels, Johnson, Walden, & Conture, 2013; Scarpa et al., 2006; Schady, 2011).

The cutoff scores for the MMSPE (5th percentiles) were computed by fitting a GLM or GzLM, which included child's age and mean parental education as predictors. The mean parental education was computed as the average level of education in years of the parents. In Fig. 30.4, the global MMSPE score is plotted against age.

Applications to other areas of development

So far, we have discussed the utility of a cognitive screening test for the developmental age in the context of pediatric primary care, neurological, neuropsychiatric, and neuropsychological assessments. Indeed, a brief but systematic evaluation of cognitive functioning may be helpful to general pediatricians to detect the risk of delayed cognitive development, or of impaired cognitive abilities. This may be informative to appropriately refer a child for further assessment and treatment. Furthermore, in neurology and neuropsychiatry wards, a cognitive screening test would be extremely useful to monitor the child's cognitive profile, especially when a full neuropsychological test battery cannot be easily administered (e.g., when the child manifests severe attentional difficulties that hamper the reliability of a full neuropsychological evaluation).

However, the utility of a pediatric cognitive screening test extends beyond the clinical settings. For instance, it is useful in the context of empirical research on behavioral, psychological, and neural development. It can be adopted to attest the global cognitive level of a control sample, monitor the cognitive profile of a clinical group in longitudinal studies, or modeled as a relevant variable in cohort studies.

Finally, a cognitive screening test could be used in educational contexts, as part of screening evaluations for learning disabilities. This measure could provide relevant information to be read in light of the evaluation of academic skills, behavioral traits, and contextual factors, in detecting learning difficulties or impairments.

Key facts of pediatric neuropsychological assessment

- The ultimate purpose of pediatric neuropsychological assessment is to optimize child's adaptation, i.e., functional and adaptive interaction with the environment.
- Collecting anamnestic information (birth, medical, family, social, and school history, as well as developmental milestones) is crucial to appropriately select neuropsychological tests and to interpret cognitive scores.
- Behavioral observation may focus on the child's interaction with caregivers, examiners, and care providers, as well as on motivation, attention, and inhibition of the irrelevant environmental stimuli, verbal and nonverbal communication, and response style.

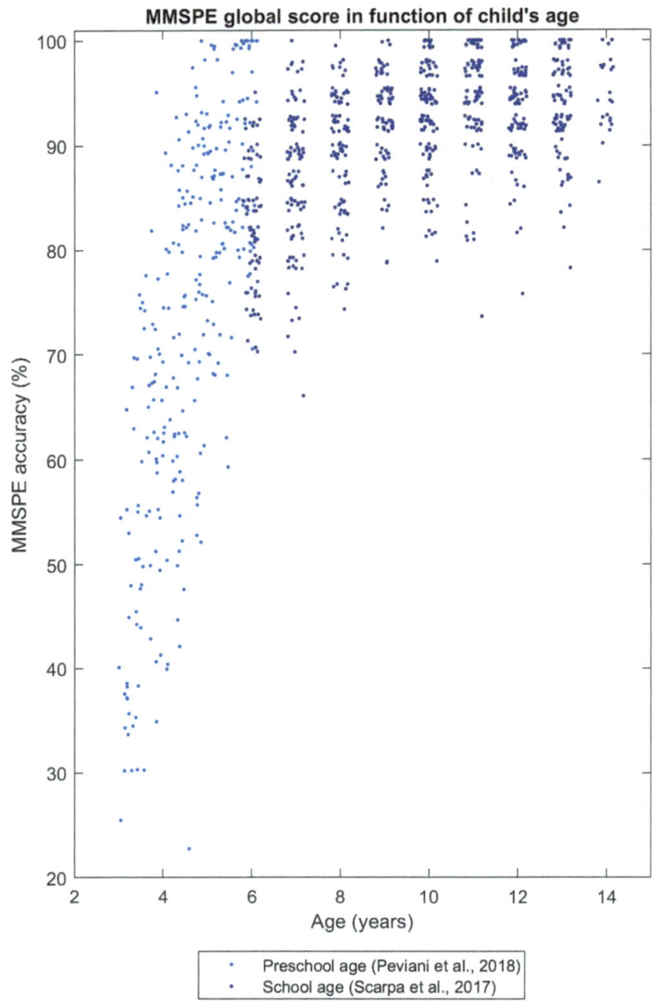

FIGURE 30.4 **MMSPE global score and child age.** MMSPE global score (transformed in percentage) is plotted against child's age.

- Psychological, personological, and emotional factors strongly affect child's test performance and symptoms' expression. Including measures of emotional state and personality in the assessment may be beneficial for a more accurate interpretation of the test outcomes.
- Besides evaluating the performance at each individual task, the clinician should observe and document children' strategies, errors, and performance patterns across tasks.
- The outcome of the assessment needs to be evaluated by integrating different sources of information, gained through the collection of clinical history, behavioral observation, emotional and neuropsychological testing.

Summary points

- Cognitive screening tests are devised to briefly evaluate key cognitive functions.
- Cognitive screening tests cannot substitute the neuropsychological assessment, but can be useful in early detection of cognitive difficulties and in monitoring the individual's congitive profile over time.
- The MMSE is the most used cognitive screening tool for adult and elderly populations.
- The need of a cognitive screening test for the pediatric age has been increasingly pointed out by researchers and clinicians.
- The adaptation of the MMSE to the developmental age has been carried out within specific national contexts.
- For the Italian pediatric population, the MMSPE (Mini-Mental State Pediatric Examination), an adapted version of the MMSE, has been developed and standardized.
- The MMSPE cutoff scores, available for children aged 3–14, have been obtained by accounting for the age of the child and the educational level of the parents.
- Besides the clinical setting, cognitive screening tools could be useful in educational and research contexts.

References

Adams, W., & Sheslow, D. (1990). *The wide range assessment of memory and learning*. Wilmington: DE: Jastak Assessments.

Aylward, G. P. (1995). *Bayley infant neurodevelopmental screener. San Antonio.*

Bayley, N. (1993). *Bayley scales of infant development* (2nd ed.). San Antonio, TX: The Psychological Corporation.

Beery, K. E. (2004). *Beery VMI: The Beery-Buktenica developmental test of visual-motor integration*. Minneapolis, MN: Pearson.

Bornholt, L. J., Ajersch, S., Fisher, I. H., Markham, R. H., & Ouvrier, R. A. (2010). Cognitive screening for children and adolescents: General limits or ceiling effects? *Journal of Child Neurology, 25*(5), 567–571. https://doi.org/10.1177/0883073809352686.

Borson, S., Scanlan, J., Brush, M., Vitaliano, P., & Dokmak, A. (2000). The mini-cog: A cognitive? vital signs? Measure for dementia screening in multilingual elderly. *International Journal of Geriatric Psychiatry, 15*(11), 1021–1027. https://doi.org/10.1002/1099-1166(200011)15:11<1021::AID-GPS234>3.0.CO, 2-6.

Casaletto, K. B., & Heaton, R. K. (2017). Neuropsychological assessment: Past and future. *Journal of the International Neuropsychological Society, 23*(9–10 Special Issue), 778–790. https://doi.org/10.1017/S1355617717001060.

Creavin, S. T., Wisniewski, S., Noel-Storr, A. H., Trevelyan, C. M., Hampton, T., Rayment, D., … Cullum, S. (2016). Mini-Mental State Examination (MMSE) for the detection of dementia in clinically unevaluated people aged 65 and over in community and primary care populations. *Cochrane Database of Systematic Reviews*, (1)https://doi.org/10.1002/14651858.CD011145.pub2.

Crum, R. M., Anthony, J. C., Bassett, S. S., & Folstein, M. F. (1993). Population-based norms for the mini-mental state examination by age and educational level. *JAMA: The Journal of the American Medical Association, 269*(18), 2386. https://doi.org/10.1001/jama.1993.03500180078038.

Cullen, B., O'Neill, B., Evans, J. J., Coen, R. F., & Lawlor, B. A. (2007). A review of screening tests for cognitive impairment. *Journal of Neurology, Neurosurgery and Psychiatry, 78*(8), 790–799. https://doi.org/10.1136/jnnp.2006.095414.

Folstein, M. F., Folstein, S. E., & McHugh, P. R. (1975). "Mini-mental state". A practical method for grading the cognitive state of patients for the clinician. *Journal of Psychiatric Research, 12*(3), 189–198. https://doi.org/10.1016/0022-3956(75)90026-6.

Gauthier, L., Dehaut, F., & Joanette, Y. (1989). The bells test: A quantitative and qualitative test for visual neglect. *International Journal of Clinical Neuropsychology, 11*(2), 49–54.

Gibbs, B. G., & Forste, R. (2014). Breastfeeding, parenting, and early cognitive development. *The Journal of Pediatrics, 164*(3), 487–493. https://doi.org/10.1016/J.JPEDS.2013.10.015.

Hresko, W. P., Hammill, D. D., & Reid, D. K. (1991). *Test of early language development: TELD-2*. Austin, TX: Pro-ed.

Jain, M., & Passi, G. R. (2005). Assessment of a modified mini-mental scale for cognitive functions in children. *Indian Pediatrics, 42*(9), 907–912.

Kukull, W. A., Larson, E. B., Teri, L., Bowen, J., McCormick, W., & Pfanschmidt, M. L. (1994). The mini-mental state examination score and the clinical diagnosis of dementia. *Journal of Clinical Epidemiology, 47*(9), 1061–1067. https://doi.org/10.1016/0895-4356(94)90122-8.

Marshall, J. (2010). *The handbook of clinical neuropsychology*. In J. Gurd, & U. Kischka (Eds.). Oxford University Press. https://doi.org/10.1093/acprof:oso/9780199234110.001.0001.

Mercy, J. A., & Steelman, L. C. (1982). Familial influence on the intellectual attainment of children. *American Sociological Review, 47*(4), 532. https://doi.org/10.2307/2095197.

Neiss, M., & Rowe, D. C. (2000). Parental education and child's verbal IQ in adoptive and biological families in the National Longitudinal Study of Adolescent Health. *Behavior Genetics, 30*(6), 487–495. https://doi.org/10.1023/A:1010254918997.

Obrzut, J. H. G. (1986). *Child neuropsychology*. Orlando: Academic Press.

Ouvrier, R. A., Goldsmith, R. F., Ouvrier, S., & Williams, I. C. (1993). The value of the mini-mental state examination in childhood: A preliminary study. *Journal of Child Neurology, 8*(2), 145–148. https://doi.org/10.1177/088307389300800206.

Ouvrier, R., Hendy, J., Bornholt, L., & Black, F. H. (1999). SYSTEMS: School-years screening test for the evaluation of mental status. *Journal of Child Neurology, 14*(12), 772–780. https://doi.org/10.1177/088307389901401202.

Peviani, V., Scarpa, P., Vedovelli, S., & Bottini, G. (2018). Mini-Mental State Pediatric Examination (MMSPE) standardization and normative data on Italian children aged 36 to 72 months. *Applied Neuropsychology: Child*, 1–5. https://doi.org/10.1080/21622965.2018.1522590.

Porta, M. A. (2008). *Dictionary of epidemiology* (5th ed.). Oxford: Oxford University Press.

Raven, J. C., Court, J. H., & Raven, J. (1990). Section 2: Coloured progressive matrices (1990 edition, with US norms). *Manual for the Raven's progressive Matrices and vocabulary Scales*. Oxford: Oxford Psychologist Press.

Richels, C. G., Johnson, K. N., Walden, T. A., & Conture, E. G. (2013). Socioeconomic status, parental education, vocabulary and language skills of children who stutter. *Journal of Communication Disorders, 46*(4), 361–374. https://doi.org/10.1016/J.JCOMDIS.2013.07.002.

Roberts, E., Bornstein, M. H., Slater, A. M., & Barrett, J. (1999). Early cognitive development and parental education. *Infant and Child Development, 8*(1), 49–62. https://doi.org/10.1002/(SICI)1522-7219(199903)8:1<49::AID-ICD188>3.0.CO, 2-1.

Rowe, D. C., Jacobson, K. C., & Van Den Oord, E. J. C. G. (1999). Genetic and environmental influences on vocabulary IQ: Parental education level as moderator. *Child Development, 70*(5), 1151–1162. https://doi.org/10.1111/1467-8624.00084.

Rubial-Álvarez, S., Machado, M. C., Sintas, E., de Sola, S., Böhm, P., & Peña-Casanova, J. (2007). A preliminary study of the mini-mental state examination in a Spanish child population. *Journal of Child Neurology, 22*(11), 1269–1273. https://doi.org/10.1177/0883073807307098.

Scarpa, P., Piazzini, A., Pesenti, G., Brovedani, P., Toraldo, A., Turner, K., … Bottini, G. (2006). Italian neuropsychological instruments to assess memory, attention and frontal functions for developmental age. *Neurological Sciences, 27*(6), 381–396. https://doi.org/10.1007/s10072-006-0717-5.

Scarpa, P., Toraldo, A., Peviani, V., & Bottini, G. (2017). Let's cut it short: Italian standardization of the MMSPE (Mini-Mental State Pediatric Examination), a brief cognitive screening tool for school-age children. *Neurological Sciences, 38*(1). https://doi.org/10.1007/s10072-016-2743-2.

Schady, N. (2011). Parents' education, mothers' vocabulary, and cognitive development in early childhood: Longitudinal evidence from Ecuador. *American Journal of Public Health, 101*(12), 2299–2307. https://doi.org/10.2105/AJPH.2011.300253.

Shoji, M., Fukushima, K., Wakayama, M., Shizuka-Ikeda, M., Ikeda, Y., Kawakami, A., … Abe, K. (2002). Intellectual faculties in patients with Alzheimer's disease regress to the level of a 4-5-year-old child. *Geriatrics and Gerontology International, 2*(3), 143–147. https://doi.org/10.1046/j.1444-1586.2002.00040.x.

Telzrow, C. F. (1983). Making child neuropsychological appraisal appropriate for children: Alternative to downward extension of adult batteries. *Clinical Neuropsychology, 5*, 136–141. ISSN: 0197-3681.

Warrington, E. K. (1996). *The Camden memory tests manual* (Vol. 1). Psychology Press.

Zhou, S. J., Baghurst, P., Gibson, R. A., & Makrides, M. (2007). Home environment, not duration of breast-feeding, predicts intelligence quotient of children at four years. *Nutrition, 23*(3), 236–241. https://doi.org/10.1016/j.nut.2006.12.011.

Chapter 31

Prediction of neurodevelopment in preterm infants: from the neonatal intensive care unit into adulthood

Pilar Medina–Alva

Department of Neonatology, Instituto Materno Perinatal, Department of Pediatrics, Universidad Peruana Cayetano Heredia, Lima, Peru

List of abbreviations

ASD Autism spectrum disorder
CGMI Cortical gray matter injury
CP Cerebral palsy
CUS Cranial ultrasound
DCD Developmental coordination disorder
ELBW Extremely low birth weight
EP Extremely preterm
GM General movements
IQ Intelligence quotient
IVH Intraventricular hemorrhage
M-CHAT Modified Checklist for Autism in Toddlers
MDI Mental development index
MRI Magnetic resonance imaging
NDI Neurodevelopmental impairment
NPV Negative predictive value
PDI Psychomotor index
PHH Post hemorrhagic hydrocephalus
PPV Positive predictive value
PVHI Periventricular hemorrhagic infarction
PVL Periventricular leukomalacia
VLBW Very low birth weight
VP Very preterm
WMI White matter injury

Mini-dictionary of terms

Corrected age A preterm baby's chronological age minus the number of weeks or months he/she was born early.
Extreme preterm birth Any birth completed before 27 weeks and 6 days' gestation.
Extremely low birth weight infant Infant whose birth weight is less than 1000 g.
Microcephaly Head circumference measurement below −2 standard deviations in the distribution for general population. It correlates with lower brain size and sometimes with impaired development.
Negative predictive value The quotient between the total number of true negatives and the total number of negative tests. The ideal value is 100%. The value is independent from the prevalence of the disease in the population.
Neurodevelopmental impairment A definition used for neurodevelopment studies. It includes low scores in developmental tests (<2SD) (in one or more subscores), moderate or severe cerebral palsy, hearing impairment that needs amplification or bilateral blindness.

Diagnosis, Management and Modeling of Neurodevelopmental Disorders. https://doi.org/10.1016/B978-0-12-817988-8.00031-2

Positive predictive value The quotient between the total number of true positives and the total number of positive tests. The ideal value is 100%. The value is independent from the prevalence of the disease in the population.
Preterm birth Any birth completed before 37 weeks' gestation.
Very low birth weight infant Infant whose birth weight is less than 1500 g.
Very preterm birth Any birth completed before 28 and 31 weeks and 6 days' gestation.

Introduction

An estimated number of 15 million preterm births occur each year, with marked disparities in survival among and within different settings. For survivors, preterm birth can have effects on neurodevelopment, with increased risks of cerebral palsy (CP), sensory and behavioral disorders, and impaired learning, which contribute to its burden of chronic disease in older age. Extremely preterm infants are at higher risk of abnormal neurodevelopment with a shifting spectrum of outcomes, showing decreasing rates of severe CP and increasing rates of other functional disabilities and a specific behavioral phenotype (Adams-Chapman et al., 2018; Church, Banihani, Luther, Maddalena, & Asztalos, 2018; McGowan & Vohr, 2019). Risk factors for neurodevelopmental impairment (NDI) in preterms include gender, gestational age, birth weight, use of antenatal steroids, and brain lesions such as intraventricular hemorrhage (IVH) and periventricular leukomalacia (PVL), among others (Ambalavanan et al., 2012). Identification of risk factors has contributed in reduction of preterm mortality, while medical, economic, and social costs for survivors make prematurity a priority in global health care (Fig. 31.1) (Blencowe et al., 2013).

Neurodevelopmental impairment at 2 years

24 months of corrected age is an estimated frontier for childrens' first developmental phase, also called sensory-motor stage, thus it seems reasonable to evaluate development at this point. NDI is defined as < 2 DS score in developmental tests (motor and/or cognitive), moderate or severe CP, bilateral blindness, or bilateral hearing loss requiring amplification (Aylward, 2018). The incidence of NDI ranges in very low birth weight (VLBW) preterms from 6% to 45% depending on gestational age, period of time, and outcome (Church et al., 2018).

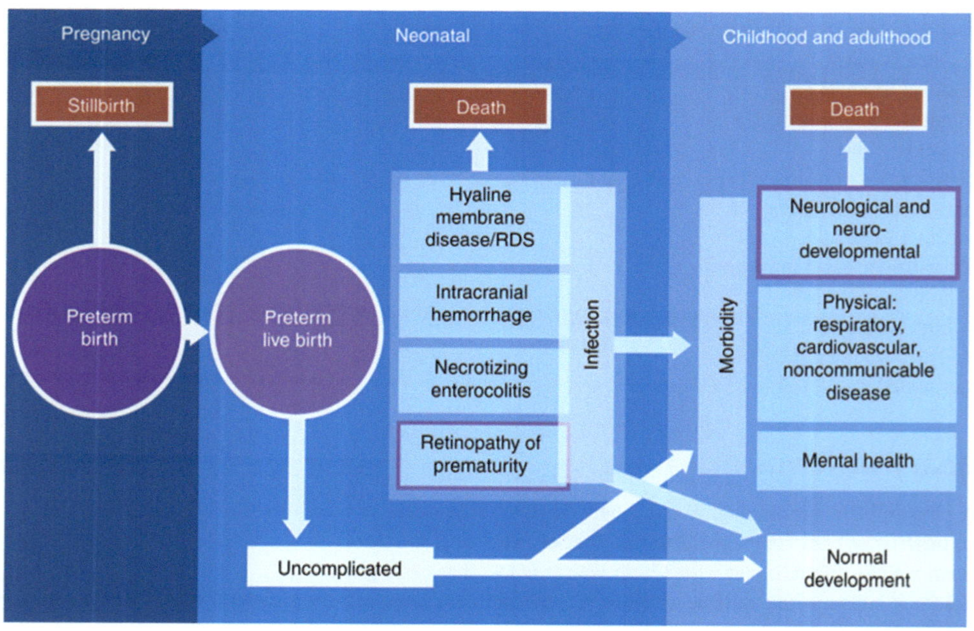

FIGURE 31.1 **Disease schema for sequelae of preterm birth.** The spectrum of prematurity-related lifelong sequelae is shown in this figure. Risk is increased with lower gestational ages, lower birth weights, and certain neonatal conditions. A considerable proportion of preterms survive with mild morbidity with lifelong consequences. *Reproduced with permission from Blencowe, H., Lee, A. C., Cousens, S., Bahalim, A., Narwal, R., Zhong, N., … Lawn, J. E. (2013). Preterm birth—associated neurodevelopmental impairment estimates at regional and global levels for 2010.* Pediatric Research, *74(S1), 17–34.*

Early assessments offer several advantages. Results can help determine which children should receive early intervention, they also can be used as outcome measures in research protocols and also can be used to inform clinicians and families about the appropriateness of care for certain groups of infants.

There are some concerns about some tests underestimating or overestimating children who are at the extreme ends of the normal distribution and about tests taken in different periods of time, as average tests' scores tend to increase (Vohr et al., 2012). The differences between Bayley-II and Bayley-III scores are particularly problematic at the lower end. These disparities are important for long-term follow-up studies in which previous results using the earlier scales are combined with more contemporary data. Some strategies have been made to make scores more comparable, such as creating another category for those babies in the 70–85 range, changing cutoff points, or creating composite scores (Aylward, 2018; Lowe, Erickson, Schrader, & Duncan, 2012).

Another issue is the use of corrected age rather than chronologic age for assessments. Correction for gestational age should be employed for preterm children at 2 years for functional levels or cognitive outcome, but some authors suggest that scores should continue to be adjusted for research evaluations throughout childhood, specially in extremely preterm babies. For clinical use, however, many developmental experts adjust age only up to 24 months (Doyle & Anderson, 2016).

Baby/infant examinations

Neurological examination

The ability of a systematic neurological examination to predict neurodevelopmental abnormalities in preterms has been widely explored with different assessment tools with variable results. Most studies show that the majority of preterm babies with nonoptimal evaluations at term age have further normal development. These findings might suggest that exam abnormalities might be due to transient postural alterations or delayed maturation patterns in otherwise healthy preterms and can explain the very low positive predictive and high negative predictive values found for abnormal neurological examination at term (Amess, McFerran, Khan, & Rabe, 2009; Simard, Lambert, Lachance, Audibert, & Gosselin 2011). Serial neurological examinations may be more useful to predict gross motor outcome at 2 years in very preterm babies. However, more prevalent is delayed motor development (such as transient dystonia of the preterm) that can be present even in babies with optimal neurological examination early on infancy. Other disabilities in cognition, language, and behavior cannot be predicted by neurological evaluation (Church et al., 2018).

Head circumference

Microcephaly (defined as a head circumference z-score below −2 SD for gestational age) is associated with a reduction of brain tissue volumes, especially deep nuclear gray matter. On the contrary, larger total brain tissue, white matter, and cerebellar volumes at term-equivalent age are associated with better neurodevelopment (Cheong et al., 2016). Brain volumes may be an important marker for neurodevelopmental deficits described in preterm children. In large cohorts of VLBW infants, proportions of babies with microcephaly or suboptimal head circumference were associated with impaired developmental outcomes (Kuban et al., 2009; Neubauer, Griesmaier, Pehböck-Walser, Pupp-Peglow, & Kiechl-Kohlendorfer, 2013). Greater head growth in preterms from birth to term age is associated with higher cognitive and motor scores at 18 months; however, some studies suggest that head growth at postdischarge might be a better predictor of cognitive and motor outcomes as far as 16–36 months of age (Belfort et al., 2011; Raghuram et al., 2017).

General movements (GMs)

Preterm infants with abnormal GMs are more likely to have abnormal 2-year outcome. Abnormal GMs were associated with reduced MDI scores and identified all children with CP in a cohort of 122 children born less than 33 weeks when performed at 3 months (De Bock et al., 2017). A pilot study found that it is common for EP infants to have abnormal writhing movements at term-corrected age. While poor repertoire movements were common, all infants with poor repertoire in this study had normal subsequent fidgety movements, implying that their risk of CP will be low. The fidgety age assessment could be more helpful to look at future risk for CP in this population (Sharp, Coenen, & Amery, 2018).

Neuroimaging

There is a variety of neuroimaging modalities for the detection and delimitation of brain injury in preterm babies. Prematurity-related brain injury includes intraventricular, parenchymal and cerebellar hemorrhages, posthemorrhagic ventricular dilatation (PHVD), and white matter injury (WMI) and PVL. Some of these lesions represent primary injury that can occur in the first days of life and others appear as a consequence of the disruption of brain development. Thus, predicting neurodevelopment with neuroimaging depends both on the mode of neuroimaging and the timing of assessment. Knowledge on the benefits and limitations of these studies is key to give appropriate counseling regarding neurodevelopmental outcomes (Sewell & Andescavage, 2018).

Early imaging at the bedside consists in cranial ultrasound (CUS), aimed to detect IVH and its complications (periventricular hemorrhagic infarction (PVHI) and/or PHVD), which may occur mainly in the first 2 weeks of life. Late CUS (between 35 and 40 weeks' corrected age) improves outcome prediction. MRI is the main study for detection of white matter injury, but there is no support for recommending routine MRI for all VLBW preterm babies at term age (Hinojosa-Rodríguez et al., 2017).

Cranial ultrasound

CUS is the preferred neuroimaging study at neonatal intensive care units because of its feasibility and availability. Early CUS findings are classified as major and minor (see Table 31.1). Major or severe lesions are more associated with CP than minor or no lesions at all (17%–25% vs. 4%) (Ancel, 2006). Cerebellar lesions are difficult to identify and appear in less than 10% of preterm brain sonograms. White matter abnormalities are difficult to recognize as increased or persisting echogenic lesions. PVL is easier to identify in serial scans that initially show increased echogenicity and after 2–4 weeks can identify cysts of variable size (de Vries, Benders, & Groenendaal, 2015). Severe ventriculomegaly is considered a major CUS finding, though most studies do not characterize it as mild, moderate, and severe with specific measurements of ventricle size. There is also a difficulty inherent in attempting to analyze only PHVD in the setting of additional injuries (Dorner, Burton, Allen, Robinson, & Soares, 2018).

The predictive value of CUS has been addressed by several studies. Sensitivity and specificity were 43%–44% and 82%–87% respectively to predict CP when sonograms were performed within the first 5 days of life. An abnormal early CUS is a good predictor of CP, motor developmental delay, and cognitive developmental delay. It is remarkable to mention that negative likelihood ratios are poor, thus special attention must be paid therefore when counseling families based only on early normal CUS without follow-up studies (Franckx, Hasaerts, Huysentruyt, & Cools, 2018). Among infants with normal ultrasound at term age (late CUS), it is very infrequent to find moderate or severe WMI or abnormal gray matter on

TABLE 31.1 Major and minor findings in early and late cranial ultrasound.

Major CUS findings	Minor CUS findings
III grade intraventricular hemorrhage	I-II grade intraventricular hemorrhage
Post hemorrhagic ventricular dilatation	Transient parenchymal echodensities
Periventricular cysts	Germinal cysts
Severe ventriculomegaly[a]	Paraventricular cysts[b]
Periventricular cysts	Lenticulostriate vasculopathy[b]
Persistent parenchymal echodensities	Germinolysis[b]
Cerebellar lesions	Calcifications[b]
	Chorid plexus cysts[b]

Major findings in early and/or late CUS are strong predictors of NDI. Minor findings are frequent but do not correlate with neurodevelopment. Serial CUS scans are recommended for higher accuracy of prediction.
[a]*Definitions vary between authors.*
[b]*As exclusive findings.*
Prepared with data from Setänen, S., Lahti, K., Lehtonen, L., Parkkola, R., Maunu, J., Saarinen, K., & Haataja, L. (2014). Neurological examination combined with brain MRI or cranial US improves prediction of neurological outcome in preterm infants. *Early Human Development, 90*(12), 851–856 and Hintz, S. R., Barnes, P. D., Bulas, D., Slovis, T. L., Finer, N. N., Wrage, L. A., ... SUPPORT Study Group of the Eunice Kennedy Shriver National Institute of Child Health and Human Development Neonatal Research Network. (2015). Neuroimaging and neurodevelopmental outcome in extremely preterm infants. *Pediatrics, 135*(1), e32–e42.

MRI (Horsch et al., 2010). Late CUS findings seem therefore to be more useful for prognostication of outcome. In full multivariable models, both late CUS and MRI, but not early CUS, remained independently associated with NDI (Burkitt, Kang, Jyoti, Mohamed, & Chaudhari, 2019; Hintz et al., 2015; de Vries et al., 2015).

Magnetic resonance imaging (MRI)

MRI is a valuable tool for prediction of outcome of neurodevelopment in preterm infants. To maximize its utility in the clinical context, interpretation of neonatal MRI needs to be done with knowledge of complementary information (Anderson, Cheong, & Thompson, 2015). MRI allows the evaluation of the contrast between encephalic and cerebellar white and gray matter (T1 sequence), myelination process (T1 and T2), evolution of ischemic pathologies of arterial origin (DWI sequence), vascular disorders of the Circle of Willis and its branches (MR angiography, time of flight), evolution of hemorrhages (SWI sequence), macroscopic focal lesions (>1 mm), and diffuse WMI (FLAIR and T2) (Hinojosa-Rodríguez et al., 2017) (Table 31.2). MRI is more sensitive than CUS for detecting subtle and/or diffuse neonatal brain disorders and is considered the gold standard for clinical diagnosis of diffuse WMI in the absence of histological information (de Vries et al., 2015). Early MRI yields important information about the pathogenesis of preterm brain injury and therefore is indispensable in research on preterm brain injury. Structural brain abnormalities are related to long-term neurodevelopment; hence, MRI should preferably be performed at term-equivalent age (Plaisier, Govaert, Lequin, & Dudink, 2014).

WMI is the most common pathology seen in preterm infants with conventional MRI, including cystic lesions, punctate lesions, delayed myelination, volume loss, thinning of the corpus callosum, and T2-weighted diffuse excessive high signal intensity (DEHSI). Lesions can vary according to timing of the brain scan and brain maturation sequence. Scores have been designed to classify the severity of WMI and higher scores correlate with NDI (Kidokoro, Neil, & Inder, 2013; Woodward, Anderson, Austin, Howard, & Inder, 2006). Moderate and severe cerebral WMI at term age is predictive of cognitive and motor delay, CP, and neurosensory impairment. Gray matter abnormalities were less strongly associated with cognitive

TABLE 31.2 Variables to assess by clinical MRI at term-equivalent age.

Cerebral white matter
Presence or absence of cysts
Signal abnormality
Myelination of the posterior limb of the internal capsule and corona radiate
Size and morphology of the corpus callosum
Size and morphology of the lateral ventricles
Volume of the periventricular white matter
Cortical gray matter
Signal abnormality
Cortical fold maturation
Size of the extracerebral space
Subcortical gray matter
Signal abnormality
Symmetry and size of the basal ganglia
Symmetry and size of the thalamus
Cerebellum
Signal abnormality
Symmetry and size of the hemispheres

These brain features must be evaluated in MRI performed at term age. Scores have been designed to classify the severity of white and gray matter lesions. Higher scores correlate with NDI.
Reproduced with permission from Hinojosa-Rodríguez, M., Harmony, T., Carrillo-Prado, C., Van Horn, J. D., Irimia, A., Torgerson, C., & Jacokes, Z. (2017). Clinical neuroimaging in the preterm infant: Diagnosis and prognosis. *NeuroImage Clinical, 16*, 355–368.

delay, motor delay, and CP. DEHSI is a common finding in T2 sequences of VP infants. Its underlying pathology is not clear, but to this date there is no evidence that DEHSI is associated with early development or later IQ (Anderson et al., 2015).

Cortical gray matter injury (CGMI) is not frequent in VLBW infants; however, delayed cortical folding, wide interhemispheric fissure, and enlarged extracerebral space are relatively common and are associated with cortical gray matter abnormalities. Babies with CGMI have lower cognitive and motor development scores and higher rate of CP at 24 months of age. MRI can detect cerebellar lesions with higher sensitivity, while CUS mastoid views can detect as half as MRI. Cerebellar hemorrhagic infarction is associated with delayed motor and language development (Anderson et al., 2015).

Although MRI can identify more subtle abnormalities that may be missed on CUS, the predictive value on NDI is poor. For some studies, the predictive value for NDI was only marginally improved with the addition of a term-age MRI compared with that of an early and late CUS examination combined (Burkitt et al., 2019). Other authors propose performing MRI when feasible for all asymptomatic EP infants, while other VP preterms should be considered if clinical indications exist such as abnormal physical examination, seizures, or major CUS findings (Ibrahim, Mir, & Chalak, 2018). The combination of clinical data (neurological examination and head circumference) and neuroimaging (serial CUS or MRI at term) provides another valuable tool for predicting NDI at 2 years age in preterm infants (Medina-Alva et al., 2019; Setänen et al., 2014; Vesoulis, El Ters, Herco, Whitehead, & Mathur, 2018).

Advanced neuroimaging techniques such as diffusion tensor imaging, cortical surface area and cerebral volumetric measures may improve prognostic abilities in preterms. However, they are not widely available or easily interpretable, and hence, their use should be limited to research settings only (Parikh, 2016).

Other prediction studies

Electroencephalography/amplitude integrated encephalography (EEG/aEEG)

Early aEEG patterns, such as absent sleep—wake cycles and low-voltage tracing, can be predictive of adverse neurological outcome in the preterm newborn. Brief seizures are common in patients with IVH but are not associated with outcome (Wikström et al., 2012). A recent systematic review evaluated the prognostic utility of EEG/aEEG in the first 7 days of life of preterm infants, with high risk of bias and high heterogeneity as main findings. There is a need of higher quality studies to confirm the previous observations (Fogtmann, Plomgaard, Greisen, & Gluud, 2017).

Neurodevelopment at school age

Neurodevelopment at school age

The duration of follow-up of preterm babies is still in debate, as is the use of corrected age beyond 2 years. There is considerable concern on the direct effects of an early biological risk factor such as prematurity, which can be modified for better or worse by subsequent experiences and postnatal environmental factors determining long-term cognitive and academic outcomes. It is also evident that subtle, high-frequency/low-severity dysfunctions such as attention-deficit disorder, learning disabilities, behavioral problems, or executive dysfunction may not be apparent earlier and demand longer follow-up to be detected and delimited. Early developmental assessment has poor sensitivity but good specificity and negative predictive value for school-age cognitive deficit (Aylward, 2018; Wong, Santhakumaran, Cowan, Modi, & Medicines for Neonates Investigator Group, 2016).

After the first milestone is reached at 2 years, school age is expected to become the second end point for research studies. At 3—4 years of age, intelligence quotient (IQ), executive function, school readiness, visual motor integrative skills, and verbal/nonverbal language achievements can be assessed. When compared with children born at term, VP and EP 4-year-old children have increased risks of CP, cognitive delay, language delay, and emotional/behavior problems, and it is likely for them to have multiple domain impairments (Woodward et al., 2009) (Figs. 31.2 and 31.3). As CP rates lower, former preterm infants are still at risk of a range of motor abnormalities, including delayed motor milestones, balance and coordination abnormalities, and challenges with manual dexterity, known as developmental coordination disorder (DCD). Specific tests can be applied at 4 years age to predict DCD at 8 years with variable predictive values. At school age, DCD is associated with lower cognitive and academic test scores and greater behavior problems (McGowan & Vohr, 2019). When followed for longer periods, preterms show lower IQs and school grades, primarily in math (Abel et al., 2016; Mu, Lin, Chen, Chang, & Tsou, 2008). By age 6, even VP children without severe neurodevelopmental impairment are 2—3 times more likely to show delays in math (43% vs. 19%), written language (36% vs. 22%), language comprehension (26% vs. 14%), handwriting (36% vs. 17%), spelling (38% vs. 30%), and physical education (33% vs. 11%) when compared to full-term infants (Pritchard et al., 2009).

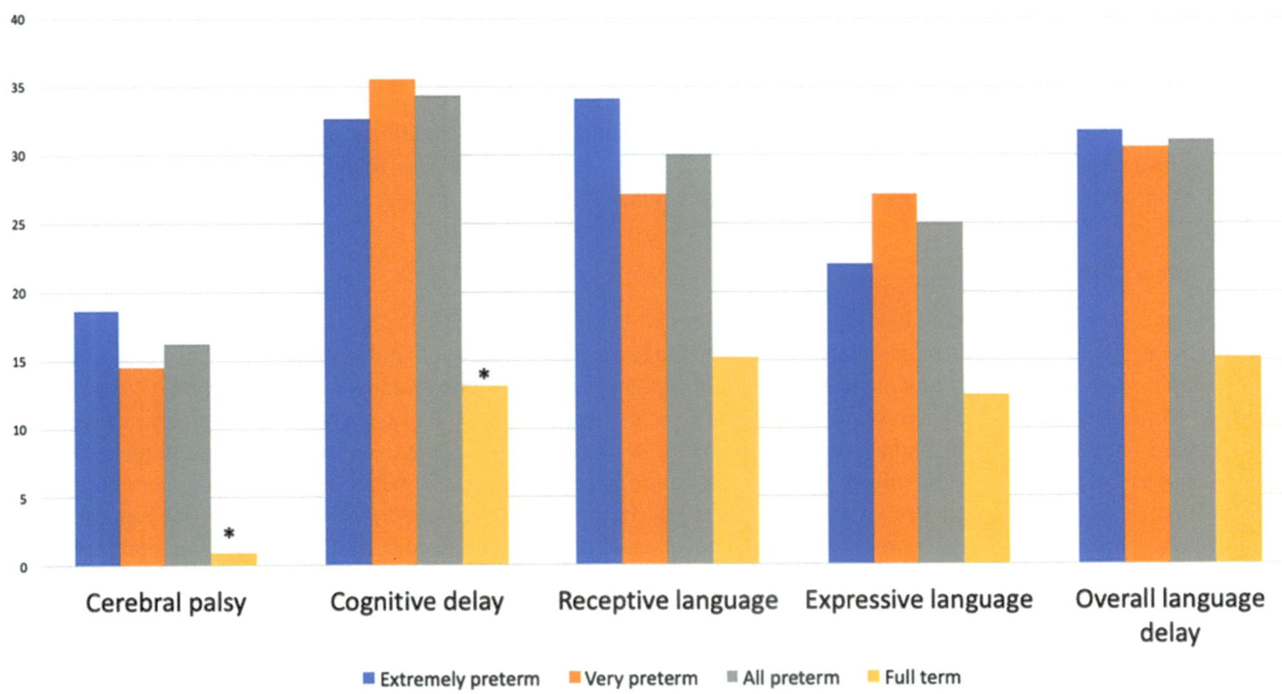

FIGURE 31.2 **Neuromotor, cognitive, and language outcomes at age 4 years in extremely preterm, very preterm, and full-term babies.** Results express percentages of outcomes present in study participants (N = 212, 105 preterm and 107 full-term infants). VP and EP infants always show higher incidence of poor neuromotor, cognitive, and language outcomes when compared to full-term counterparts. Only CP and cognitive delay (measured with Weschler Preschool and Primary Scales of Intelligence, WPPSI-R) were statistically significant (* = $P < .001$). *Adapted with permission from Woodward, L. J., Moor, S., Hood, K. M., Champion, P. R., Foster-Cohen, S., Inder, T. E., & Austin, N. C. (2009). Very preterm children show impairments across multiple neurodevelopmental domains by age 4 years.* Archives of Disease in Childhood - Fetal and Neonatal Edition, 94*(5), 339–344.*

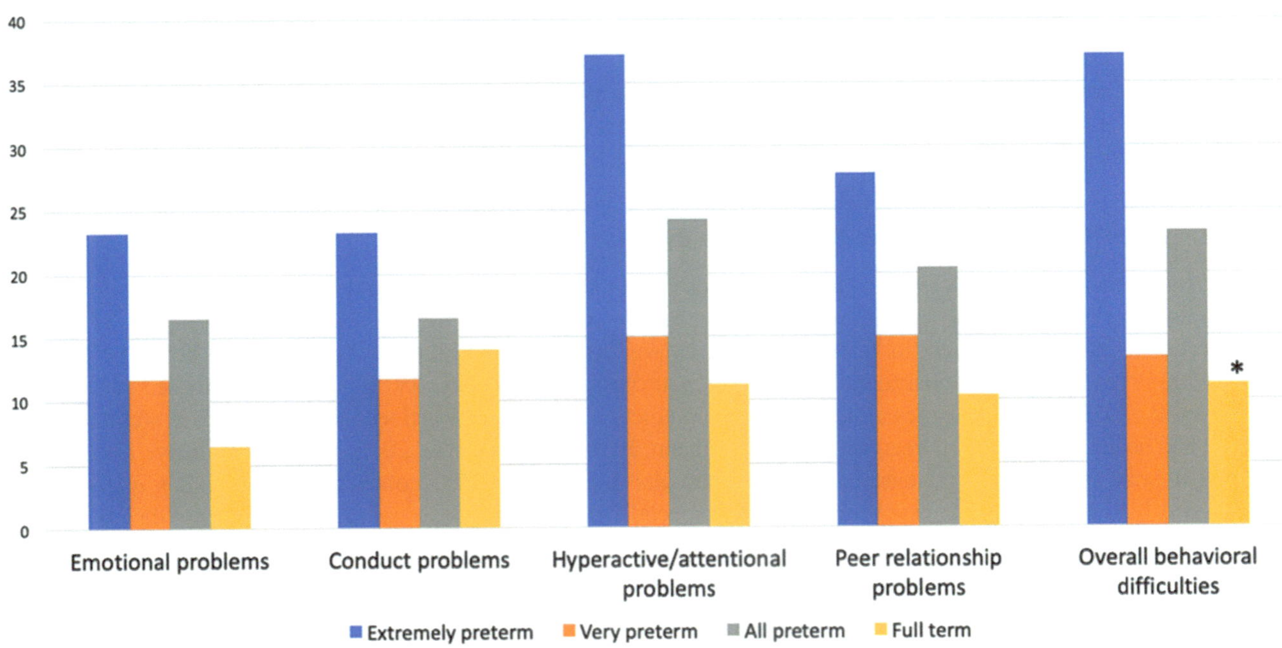

FIGURE 31.3 **Behavioral outcomes at age 4 years in extremely preterm, very preterm, and full-term babies.** Results express percentages of outcomes present in study participants (N = 212, 105 preterm and 107 full-term infants). VP and EP infants always show higher incidence of emotional and behavioral outcomes when compared to full-term babies. Percentages of overall behavioral outcomes were statistically significant (* = $P < .001$) from full-term counterparts. *Modified from Woodward, L. J., Moor, S., Hood, K. M., Champion, P. R., Foster-Cohen, S., Inder, T. E., & Austin, N. C. (2009). Very preterm children show impairments across multiple neurodevelopmental domains by age 4 years.* Archives of Disease in Childhood - Fetal and Neonatal Edition, 94*(5), 339–344.*

Neuroimaging

Severe late CUS findings can be associated with cognitive impairment and disability at school age (Hintz et al., 2018). Preterms with mild to severe WMI findings in term MRI are at much higher risk of motor impairment. Some cohorts have associated WMI to language development, learning capacity, attention, and processing speed at school age. Significant cerebellar lesions on MRI are also associated with disability at school age. These findings highlight the importance of cerebral white matter and cerebellar connectivity for later intact cognitive functioning among children born very preterm.

Prediction of language skills

Receptive and expressive language delays were common in a large cohort of ELBW infants. Children who could not feed themselves at 30 months were more likely to present language delay. Longitudinal cohorts may help understand the relationship between feeding patterns, language delay, and cognitive dysfunction (Adams-Chapman, Bann, Carter, & Stoll, 2015). In addition to brain injury, environmental factors, including both race and low socioeconomic status, are known to be associated with alterations in the language environment, decreased early language exposure, and subsequent language delay. Language acquisition is influenced by early interactive language exposure, which is related to maternal education levels. There is increasing evidence of the impact of psychologic and socioeconomic adversities on preterm neurodevelopmental and behavioral outcomes. Identification of adverse maternal mental health provides an opportunity for intervention in preterm infants and their families (McGowan & Vohr, 2019).

Prediction of sensory disorders

Research on methods for prediction of visual impairment is on the way. They could be especially useful in preterm babies with pulvinar lesions and hydrocephalus, who are more at risk of visual dysfunction (Chau, Taylor, & Miller, 2013). Certain techniques such as tractography could be more useful to predict visual abilities based on the features of the optic radiations (Groppo et al., 2014).

Prediction of autism spectrum disorder (ASD)

There is a concern over major motor, cognitive, visual, and hearing impairments appearing to account for more than half of the positive M-CHAT screens among preterm infants. This makes it difficult to determine whether the ASD diagnosis is inappropriate, is comorbid with developmental disabilities, or is both. Therefore, caution is necessary because there may be a high false-positive rate due to coexisting neurodevelopmental disabilities (Aylward, 2018). Some studies have shown that after correction of these factors, the expected rate of positive tests in the EP and VP population is lower than early reported, but remains higher than the general population (Stephens et al., 2012).

Neurodevelopment in adolescents and adults

A significant proportion of VP/VLBW adults are diagnosed with mild cognitive impairment and show low cognitive scores when evaluated beyond 19 years of age, with no evidence of recovery or deterioration in the trajectory (Fig. 31.4). Boys and gestational-age preterms seem to be at higher risk, as well as those with brain injury. More than 50% of VP/VLBW adults with severe cognitive impairment have severe neurologic or neurosensory impairment diagnosed in childhood (Linsell et al., 2018). Low IQ scores at school age are predictive of low adult socioeconomical status as well as reduced survival and health into old age (Breeman, Jaekel, Baumann, Bartmann, & Wolke, 2015). Preterm adults are also at higher risk of metabolic syndrome, hypertension, and cardiovascular disease, which can contribute to reduce their quality of life, as well as a higher incidence of psychiatric disease, especially anxiety and depression, and personality disorders (Raju, Buist, Blaisdell, Moxey-Mims, & Saigal 2017).

Applications to other areas of development

Increased survival of very preterm and extreme preterm infants raises concern about the growing burden of disabilities across the life span that survivors deal with. The combination of lesions in the immature brain and the distortion of the normal maturation processes may result in a reduction of the brain and cerebellar volumes, which can affect some specific regions of white and gray matter and generate an altered connectivity of such regions. The immaturity of other organs, such as lungs and kidneys, and the occurrence of certain disorders, such as necrotizing enterocolitis or retinopathy of

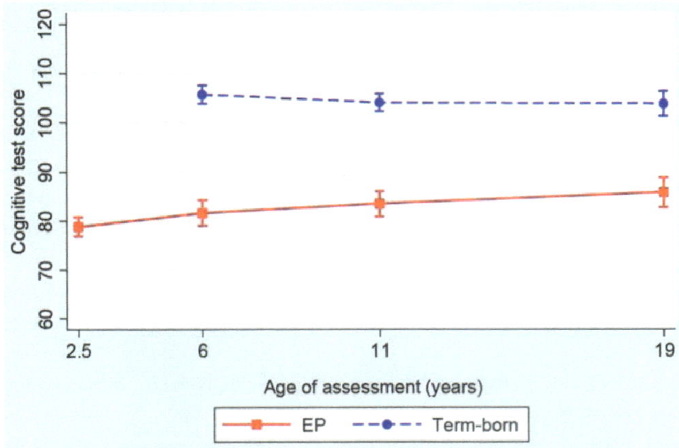

FIGURE 31.4 **Observed and predicted mean cognitive scores plus 95% confidence intervals in EP participants and full-term controls at ages 2.5, 6, 11, and 19.** When compared with full-term counterparts, EP children show cognitive trajectories with low scores that persist into early adulthood, with no evidence of recovery or deterioration in time. *Reproduced with permission from Linsell, L., Johnson, S., Wolke, D., O'Reilly, H., Morris, J. K., Kurinczuk, J. J., & Marlow, N. (2018). Cognitive trajectories from infancy to early adulthood following birth before 26 weeks of gestation: A prospective, population-based cohort study.* Archives of Disease in Childhood, 103*(4), 363–370.*

prematurity, can also have lifelong consequences. The degree of these abnormalities correlates with neurodevelopmental impairment at early age. Later in life these abnormalities correlate with lower IQs, impaired executive functions, working and visual memory, and social abilities, which can be detected at school age and confirmed later on. Long-term studies have shown that a significant fraction of very preterm or VLBW adults have social disadvantages and remain at higher risk for psychiatric and behavioral disorders, as well as cardiovascular and metabolic disorders. Therefore, prematurity must be considered a chronic condition, and studies should focus in its lasting repercussions in life.

Key facts of prediction of neurodevelopment in preterm infants

- Neurodevelopment is altered in preterm infants. Abnormalities are more frequent and severe with lower gestational ages.
- Neuroimaging appears to be of more utility for prediction of neurodevelopment when performed around term age and when combined with clinical data.
- Neurodevelopment and IQ at school age cannot be accurately predicted by tests performed at 2 years.
- There is a need for studies to fully evaluate understand development in some specific areas of development at school age (e.g., language, autism, behavior) and later into adolescence and adulthood.

Summary points

- Prematurity is a chronic condition with consequences extending from birth into adulthood.
- The direct effects of prematurity can be modified by subsequent experiences and postnatal environmental factors determining long-term cognitive and academic outcomes.
- Most studies are focused in prediction of neurodevelopmental impairment at 2 years of age using neurological examination, neuroimaging, and other techniques with variable accuracy of results.
- The majority of preterms survive with only mild to moderate disability, with lower intelligence quotients and impaired executive functions, working/visual memory, and social skills.
- Impaired neurodevelopment at school age in preterms is likely to affect multiple domains and correlates with poor outcome in adulthood.
- Adolescent and adulthood implications of prematurity are not fully determined and include a higher risk of psychiatric and behavioral disorders.

References

Abel, K., Heuvelman, H., Wicks, S., Rai, D., Emsley, R., Gardner, R., & Dalman, C. (2016). Gestational age at birth and academic performance: Population-based cohort study. *International Journal of Epidemiology*, 1–12.

Adams-Chapman, I., Bann, C., Carter, S. L., & Stoll, B. J. (2015). Language outcomes among ELBW infants in early childhood. *Early Human Development, 91*(6), 373–379.

Adams-Chapman, I., Heyne, R. J., DeMauro, S. B., Duncan, A. F., Hintz, S. R., Pappas, A., … Follow-Up Study of the Eunice Kennedy Shriver National Institute of Child Health and Human Development Neonatal Research Network.. (2018). Neurodevelopmental impairment among extremely preterm infants in the neonatal research network. *Pediatrics, 141*(5).

Ambalavanan, N., Carlo, W. A., Tyson, J. E., Langer, J. C., Walsh, M. C., Parikh, N. A., … Follow-Up Subcommittees of the Eunice Kennedy Shriver National Institute of Child Health and Human Development Neonatal Research Network. (2012). Outcome trajectories in extremely preterm infants. *Pediatrics, 130*(1), e115–e125.

Amess, P., McFerran, C., Khan, Y., & Rabe, H. (2009). Early prediction of neurological outcome by term neurological examination and cranial ultrasound in very preterm infants. *Acta Paediatrica, 98*(3), 448–453.

Ancel, P.-Y. (2006). Cerebral palsy among very preterm children in relation to gestational age and neonatal ultrasound abnormalities: The EPIPAGE cohort study. *Pediatrics, 117*(3), 828–835.

Anderson, P. J., Cheong, J. L. Y., & Thompson, D. K. (2015). The predictive validity of neonatal MRI for neurodevelopmental outcome in very preterm children. *Seminars in Perinatology, 39*(2), 147–158.

Aylward, G. P. (2018). Issues in neurodevelopmental testing of infants born prematurely: The Bayley scales of infant development third edition and other tools. In H. Needelman, & B. J. Jackson (Eds.), *Follow-up for NICU graduates* (pp. 241–253).

Belfort, M. B., Rifas-Shiman, S. L., Sullivan, T., Collins, C. T., McPhee, A. J., Ryan, P., … Makrides, M. (2011). Infant growth before and after term: Effects on neurodevelopment in preterm infants. *Pediatrics, 128*(4), e899–e906.

Blencowe, H., Lee, A. C., Cousens, S., Bahalim, A., Narwal, R., Zhong, N., … Lawn, J. E. (2013). Preterm birth–associated neurodevelopmental impairment estimates at regional and global levels for 2010. *Pediatric Research, 74*(S1), 17–34.

Breeman, L. D., Jaekel, J., Baumann, N., Bartmann, P., & Wolke, D. (2015). Preterm cognitive function into adulthood. *Pediatrics, 136*(3), 415–423.

Burkitt, K., Kang, O., Jyoti, R., Mohamed, A.-L., & Chaudhari, T. (2019). Comparison of cranial ultrasound and MRI for detecting brain injury in extremely preterm infants and correlation with neurological outcomes at 1 and 3 years. *European Journal of Pediatrics, 178*(7), 1053–1061.

Chau, V., Taylor, M. J., & Miller, S. P. (2013). Visual function in preterm infants: Visualizing the brain to improve prognosis. *Documenta Ophthalmologica, 127*(1), 41–55.

Cheong, J. L. Y., Thompson, D. K., Spittle, A. J., Potter, C. R., Walsh, J. M., Burnett, A. C., … Doyle, L. W. (2016). Brain volumes at term-equivalent age are associated with 2-year neurodevelopment in moderate and late preterm children. *The Journal of Pediatrics, 174*, 91–97.

Church, P. T., Banihani, R., Luther, M., Maddalena, P., & Asztalos, E. (2018). Premature infants: The behavioral phenotype of the preterm survivor. In H. Needelman, & B. J. Jackson (Eds.), *Follow-up for NICU graduates* (pp. 111–126).

De Bock, F., Will, H., Behrenbeck, U., Jarczok, M. N., Hadders-Algra, M., & Philippi, H. (2017). Predictive value of general movement assessment for preterm infants' development at 2 years – implementation in clinical routine in a non-academic setting. *Research in Developmental Disabilities, 62*, 69–80.

Dorner, R A., Burton, V. J., Allen, M. C., Robinson, S., & Soares, B. P. (2018). Preterm neuroimaging and neurodevelopmental outcome: A focus on intraventricular hemorrhage, post-hemorrhagic hydrocephalus, and associated brain injury. *Journal of Perinatology, 38*(11), 1431–1443.

Doyle, L. W., & Anderson, P. J. (2016). Do we need to correct age for prematurity when assessing children? *The Journal of Pediatrics, 173*, 11–12.

Fogtmann, E. P., Plomgaard, A. M., Greisen, G., & Gluud, C. (2017). Prognostic accuracy of electroencephalograms in preterm infants: A systematic review. *Pediatrics, 139*(2). e20161951.

Franckx, H., Hasaerts, D., Huysentruyt, K., & Cools, F. (2018). Cranial ultrasound and neurophysiological testing to predict neurological outcome in infants born very preterm. *Developmental Medicine and Child Neurology, 60*(12), 1232–1238.

Groppo, M., Ricci, D., Bassi, L., Merchant, N., Doria, V., Arichi, T., … Edwards, A. D. (2014). Development of the optic radiations and visual function after premature birth. *Cortex, 56*, 30–37.

Hinojosa-Rodríguez, M., Harmony, T., Carrillo-Prado, C., Van Horn, J. D., Irimia, A., Torgerson, C., & Jacokes, Z. (2017). Clinical neuroimaging in the preterm infant: Diagnosis and prognosis. *NeuroImage Clinical, 16*, 355–368.

Hintz, S. R., Barnes, P. D., Bulas, D., Slovis, T. L., Finer, N. N., Wrage, L. A., … SUPPORT Study Group of the Eunice Kennedy Shriver National Institute of Child Health and Human Development Neonatal Research Network. (2015). Neuroimaging and neurodevelopmental outcome in extremely preterm infants. *Pediatrics, 135*(1), e32–e42.

Hintz, S. R., Vohr, B. R., Bann, C. M., Taylor, H. G., Das, A., Gustafson, K. E., … SUPPORT study group of the Eunice Kennedy Shriver National Institute of Child Health and Human Development Neonatal Research Network. (2018). Preterm neuroimaging and school-age cognitive outcomes. *Pediatrics, 142*(1). e20174058.

Horsch, S., Skiold, B., Hallberg, B., Nordell, B., Nordell, A., Mosskin, M., … Blennow, M. (2010). Cranial ultrasound and MRI at term age in extremely preterm infants. *Archives of Disease in Childhood - Fetal and Neonatal Edition, 95*(5), F310–F314.

Ibrahim, J., Mir, I., & Chalak, L. (2018). Brain imaging in preterm infants <32 weeks gestation: A clinical review and algorithm for the use of cranial ultrasound and qualitative brain MRI. *Pediatric Research, 84*(6), 799–806.

Kidokoro, H., Neil, J. J., & Inder, T. E. (2013). New MR imaging assessment tool to define brain abnormalities in very preterm infants at term. *American Journal of Neuroradiology, 34*(11), 2208–2214.

Kuban, K. C. K., Allred, E. N., O'Shea, T. M., Paneth, N., Westra, S., Miller, C., … Leviton, A. (2009). Developmental correlates of head circumference at birth and two years in a cohort of extremely low gestational age newborns. *The Journal of Pediatrics, 155*(3), 344–349. e1-3.

Linsell, L., Johnson, S., Wolke, D., O'Reilly, H., Morris, J. K., Kurinczuk, J. J., & Marlow, N. (2018). Cognitive trajectories from infancy to early adulthood following birth before 26 weeks of gestation: A prospective, population-based cohort study. *Archives of Disease in Childhood, 103*(4), 363–370.

Lowe, J. R., Erickson, S. J., Schrader, R., & Duncan, A. F. (2012). Comparison of the Bayley II mental developmental index and the Bayley III cognitive scale: Are we measuring the same thing?: Comparison of the Bayley scales. *Acta Paediatrica, 101*(2), e55–e58.

McGowan, E. C., & Vohr, B. R. (2019). Neurodevelopmental follow-up of preterm infants. *Pediatric Clinics of North America, 66*(2), 509–523.

Medina-Alva, P., Duque, K. R., Zea-Vera, A., Bellomo, S., Cárcamo, C., Guillen-Pinto, D., … Ochoa, T. J. (2019). Combined predictors of neurodevelopment in very low birth weight preterm infants. *Early Human Development, 130*, 109–115.

Mu, S.-C., Lin, C.-H., Chen, Y.-L., Chang, C.-H., & Tsou, K.-I. (2008). Relationship between perinatal and neonatal indices and intelligence quotient in very low birth weight infants at the age of 6 or 8 years. *Pediatrics and Neonatology, 49*(2), 13–18.

Neubauer, V., Griesmaier, E., Pehböck-Walser, N., Pupp-Peglow, U., & Kiechl-Kohlendorfer, U. (2013). Poor postnatal head growth in very preterm infants is associated with impaired neurodevelopment outcome. *Acta Paediatrica, 102*(9), 883–888.

Parikh, N. A. (2016). Advanced neuroimaging and its role in predicting neurodevelopmental outcomes in very preterm infants. *Seminars in Perinatology, 40*(8), 530–541.

Plaisier, A., Govaert, P., Lequin, M. H., & Dudink, J. (2014). Optimal timing of cerebral MRI in preterm infants to predict long-term neurodevelopmental outcome: A systematic review. *American Journal of Neuroradiology, 35*(5), 841–847.

Pritchard, V. E., Clark, C. A. C., Liberty, K., Champion, P. R., Wilson, K., & Woodward, L. J. (2009). Early school-based learning difficulties in children born very preterm. *Early Human Development, 85*(4), 215–224.

Raghuram, K., Yang, J., Church, P. T., Cieslak, Z., Synnes, A., Mukerji, A., … Canadian Neonatal Follow-Up Network Investigators. (2017). Head growth trajectory and neurodevelopmental outcomes in preterm neonates. *Pediatrics, 140*(1). e20170216.

Raju, T. N. K., Buist, A. S., Blaisdell, C. J., Moxey-Mims, M., & Saigal, S. (2017). Adults born preterm: A review of general health and system-specific outcomes. *Acta Paediatrica, 106*(9), 1409–1437.

Setänen, S., Lahti, K., Lehtonen, L., Parkkola, R., Maunu, J., Saarinen, K., & Haataja, L. (2014). Neurological examination combined with brain MRI or cranial US improves prediction of neurological outcome in preterm infants. *Early Human Development, 90*(12), 851–856.

Sewell, E. K., & Andescavage, N. N. (2018). Neuroimaging for neurodevelopmental prognostication in high-risk neonates. *Clinics in Perinatology, 45*(3), 421–437.

Sharp, M., Coenen, A., & Amery, N. (2018). General movement assessment and motor optimality score in extremely preterm infants. *Early Human Development, 124*, 38–41.

Simard, M.-N., Lambert, J., Lachance, C., Audibert, F., & Gosselin, J. (2011). Prediction of developmental performance in preterm infants at two years of corrected age: Contribution of the neurological assessment at term age. *Early Human Development, 87*(12), 799–804.

Stephens, B. E., Bann, C. M., Watson, V. E., Sheinkopf, S. J., Peralta-Carcelen, M., Bodnar, A., … Vohr, B. R. (2012). Screening for autism spectrum disorders in extremely preterm infants. *Journal of Developmental and Behavioral Pediatrics, 33*(7), 535–541.

Vesoulis, Z., El Ters, N., Herco, M., Whitehead, H., & Mathur, A. (2018). A web-based calculator for the prediction of severe neurodevelopmental impairment in preterm infants using clinical and imaging characteristics. *Children, 5*(11), 151.

Vohr, B. R., Stephens, B. E., Higgins, R. D., Bann, C. M., Hintz, S. R., Das, A., … Fuller, J. (2012). Are outcomes of extremely preterm infants improving? Impact of Bayley assessment on outcomes. *The Journal of Pediatrics, 161*(2), 222–228. e3.

de Vries, L. S., Benders, M. J. N. L., & Groenendaal, F. (2015). Progress in neonatal neurology with a focus on neuroimaging in the preterm infant. *Neuropediatrics, 46*(4), 234–241.

Wikström, S., Pupp, I. H., Rosén, I., Norman, E., Fellman, V., Ley, D., & Hellström-Westas, L. (2012). Early single-channel aEEG/EEG predicts outcome in very preterm infants: Early aEEG/EEG predicts outcome in preterms. *Acta Paediatrica, 101*(7), 719–726.

Wong, H. S., Santhakumaran, S., Cowan, F. M., Modi, N., & Medicines for Neonates Investigator Group. (2016). Developmental assessments in preterm children: A meta-analysis. *Pediatrics, 138*(2). e20160251.

Woodward, L. J., Anderson, P. J., Austin, N. C., Howard, K., & Inder, T. E. (2006). Neonatal MRI to predict neurodevelopmental outcomes in preterm infants. *New England Journal of Medicine, 355*(7), 685–694.

Woodward, L. J., Moor, S., Hood, K. M., Champion, P. R., Foster-Cohen, S., Inder, T. E., & Austin, N. C. (2009). Very preterm children show impairments across multiple neurodevelopmental domains by age 4 years. *Archives of Disease in Childhood - Fetal and Neonatal Edition, 94*(5), 339–344.

Chapter 32

Antenatal fetal neurodevelopment test: KANET assessment

Toshiyuki Hata[1,2] and Mohamed Ahmed AboEllail[1]

[1]*Department of Perinatology and Gynecology, Kagawa University, Graduate School of Medicine, Miki, Kagawa, Japan;* [2]*Department of Obstetrics and Gynecology, Miyake Clinic, Okayama, Japan*

Mini-dictionary of terms

ATNAT is Amiel-Tison Neurological Assessment at Term, a neurological test with the score applied to neonates at risk of developing neurological abnormalities to assess their neurological development.

Cerebral palsy is a group of disorders affecting muscle tone and subsequently movement and posture arising due to brain damage during early development, most often before birth.

Ethnic difference is the difference between populations from certain races or areas that result in differences in morphological characters of the face and body.

Fetal head anteflexion is forward movement of the fetal head

Fetal neurobehavior is any observable fetal movement or facial expression that can be visualized or recorded by means of ultrasound or other diagnostic modalities to assess development of the fetal nervous system.

Neurodevelopment is the development of the nervous system

Introduction

The introduction of new technology has expanded perspectives on fetal ultrasound. Ascertaining the proper anatomical development of the fetus was the initial target. However, assessing the functional integrity of these organs was another goal. The new ultrasound technology did not just provide clearer images compared with what was previously seen and diagnosed. It added another dimension to fetal and perinatal ultrasound. Four-dimensional (4D) ultrasound enabled real-time visualization of fetal movements and clear identification of fetal facial details. Therefore, it led to a new era of neurodevelopmental studies analyzing fetal facial expressions and the quality of movements (Kurjak et al., 2002; Merz & Abramowicz, 2012). It was also confirmed that qualitative changes in general fetal and neonatal movements reveal quantitative changes when the integrity of the nervous system is affected. The observation of such movements and facial expressions can be used as an indicator of fetal brain and central nervous system (CNS) functional development to complement the anatomical development assessed in the casual fetal scans performed in everyday practice (Einspieler & Prechtl, 2005; Salihagic-Kadic, Kurjak, Medic, Andonotopo, & Azumendi, 2005). The importance of such studies is their potential application to explore the hidden aspects of many conditions such as autism spectrum disorders and cerebral palsy. The latter has been found to be linked to unfavorable intrauterine conditions rather than delivery accidents (MacLennan, Thompson, & Grecz, 2015). Moreover, 60%−70% of neurodevelopmental disabilities develop prenatally (Kurjak et al., 2004). Therefore, such an observation of fetal behavior might facilitate an important distinction between normal and abnormal brain development, and demand for the ability to observe development of the fetal nervous system has grown markedly.

Kurjak antenatal neurodevelopment test (KANET)

Using 4D ultrasound, Kurjak and his team in Zagreb (Kurjak et al., 2008) proposed a new scoring system that initially consisted of 10 parameters. They was a combination of some parameters derived from postnatal Amiel-Tison Neurological

Diagnosis, Management and Modeling of Neurodevelopmental Disorders. https://doi.org/10.1016/B978-0-12-817988-8.00032-4

Assessment at Term (ATNAT) (Gosselin, Gahagan, & Amiel-Tison, 2005), which can be detected prenatally by 4D ultrasound, plus some parameters of general fetal movement assessment. The basis of this test was the assumption that fetal behavioral patterns directly reflect the function of the fetal brain and CNS. The test should be performed in the third trimester from 28 to 38 weeks of gestation for a period of 15–20 min when the fetus is awake (Kurjak et al., 2008). Several studies were conducted to validate it (Hata, Dai, & Marumo, 2010; Hata, Kanenishi, & Sasaki, 2010; Kurjak et al., 2005, 2010; Miskovic et al., 2010; Sasaki, Yanagihara, Naitoh, & Hata, 2010), and 3 years later, an international standardization of the modified KANET was approved in the Osaka Consensus meeting in order to increase the reproducibility of the KANET score (Stanojevic et al., 2011). Two parameters were omitted, and so the modified KANET included eight parameters (Fig. 32.1). Therefore, the total score was also modified (Fig. 32.2). They also recommended that follow-up should be done for infants until at least the age of 24 months when a diagnosis of disabling or nondisabling cerebral palsy is made. Another consensus was reached in 2015 in Bucharest, emphasizing the importance of routine clinical application of the KANET test and addressing the importance of the postnatal follow-up of cases till the age of at least 3 years and longer whenever possible to improve the correlation between KANET scores and postnatal neurodevelopmental outcomes (Stanojevic et al., 2015).

Applications of KANET

I. Ethnic diversity

Variation in genetics among humans results in various ethnicities with characteristic physical appearances (Farkas, Katic, & Forrest, 2005). However, the effect of such diversity on fetal behavior has not been ascertained. KANET scores were compared between Japanese and Croatian pregnancies as respective representatives of Asians and Caucasians (Hanaoka et al., 2016). Comparison of the total KANET scores revealed significant differences between Japanese (median, 14; range, 10–16) and Croatians (median, 12; range, 10–15) ($P < .0001$), although the total KANET score was normal in both groups. Moreover, a significant difference was also found in four individual parameters when compared between the two groups (isolated head anteflexion, isolated eye blinking, facial alteration or mouth opening, and isolated leg movement). The remaining four parameters were not different on comparing the two groups. The diversity in such four parameters related to facial expressions might be due to the variation in facial morphological characteristics between the two races. Therefore, ethnicity is a key point to be considered when fetal behavior is assessed using KANET.

II. Sex differences

Various biological, genetic, and hormonal factors contribute to anatomical and functional development of the brain (Choleris, Galea, Sohrabji, & Frick, 2018). Whether these parameters can cause differences in fetal neurobehavior between male and female fetuses was the question addressed in a study done by a Kagawa team (Hata et al., 2016b). KANET scores were compared between male and female fetuses. No significant differences were found in either the total or individual KANET scores between the two groups. Therefore, the fetal sex may not be an influential factor in fetal neurobehavior studies.

III. Parity and maternal emotional status

KANET was used to detect whether fetal behavior shows a difference on comparing primiparous and multiparous mothers (Hata et al., 2016a). The total KANET scores were normal in both groups except in one multiparous mother, and no significant difference was found in the total score between the two groups. As for individual parameters, the rate of isolated eye blinking was significantly higher in primiparous (median, 2; range, 0–2) than in multiparous (median, 1; range, 0–2) ($P < .04$) mothers. The more related state of primiparous mothers due to their lighter load and less stress compared with multiparous mothers might promote maturation of the fetal brain, especially the central dopamine system, which controls fetal eye blinking (Bodfish, Powell, Golden, & Lewis, 1995). Therefore, the maternal emotional state should be considered during fetal neurobehavior studies implementing the use of 4D ultrasound.

IV. Psychotropic drugs during pregnancy

Studies examining the effect of the maternal use of psychotropic drugs during pregnancy on fetal behavior showed inconsistent results when two-dimensional (2D) sonography was used (Dieter, Emory, Johnson, & Raynor, 2008; Dieter et al., 2001; Emory & Dieter, 2006; Kean, Gargari, Suwanrath, Sahota, & James, 2001). Therefore, a KANET-based study was conducted to examine the effect of the maternal use of psychotropic drugs on fetal neurobehavior in the third trimester of pregnancy (Hata, Kanenishi, AboEllail, et al., 2019). The total KANET score was normal in both groups, and

Sign	*Score*		
	0	*1*	*2*
Isolated head anteflexion 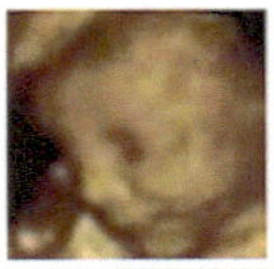	Abrupt	Small range (0-3 times of movements)	Variable in full range, many alteration (>3 times of movements)
Cranial sutures and head circumference 	Overlapping of cranial sutures	Normal cranial sutures with measurement of HC below or above the normal limit (–2 SD) according to GA	Normal cranial sutures with normal measurement of HC according to GA
Isolated eye blinking 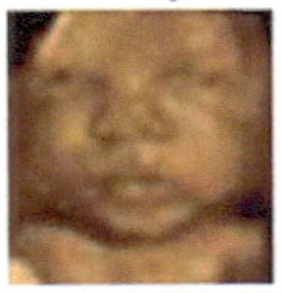	Not present	Not fluent (1-5 times of blinking)	Fluency (>5 times of blinking)
Facial alteration (grimace or tongue expulsion) 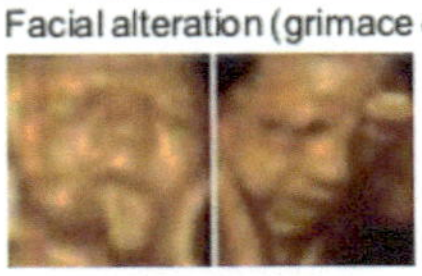 or mouth opening (yawning or mouthing) 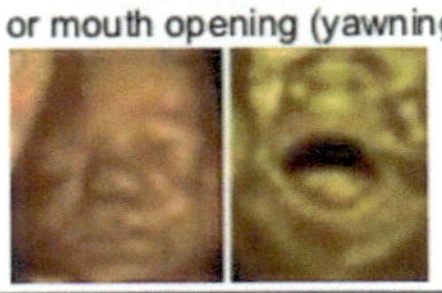	Not present	Not fluent (1-5 times of alteration)	Fluency (>5 times of alteration)
Isolated leg movement 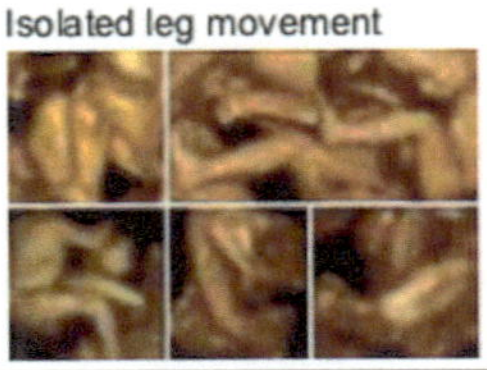	Cramped	Poor repertoire or small in range (0-5 times of movement)	Variable in full range, many alternation (>5 times of movement)
Isolated hand movement 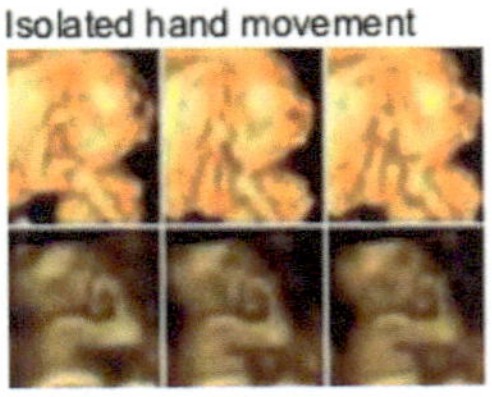 or hand to face movements 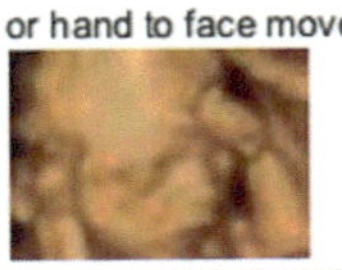	Cramped or abrupt	Poor repertoire or small in range (0-5 times of movement)	Variable in full range, many alternation (>5 times of movements)
Fingers movements 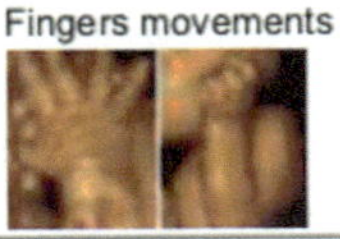	Unilateral or bilateral clenched fist, (neurological thumb)	Cramped invariable finger movements	Smooth and complex, variable finger movements
Gestalt perception of GMS	Definitely abnormal	Borderline	Normal

FIGURE 32.1 **(Modified KANET score).** Note that there are eight parameters, with each parameter having a score of 0–2 according to the movements evaluated. *Reprinted with permission from Stanojevic, M., Talic, A., Miskovic, B., Vasilj, O., Shaddad, A. N., Ahmed, B., … Al-Noobi, M. (2011). An attempt to standardize Kurjak's antenatal neurodevelopmental test: Osaka Consensus Statement.* Donald School Journal of Ultrasound in Obstetrics and Gynecology, *5(4), 317–329.*

TOTAL SCORE	INTERPRETATION
0-5	Abnormal
6-9	Borderline
10-16	Normal

FIGURE 32.2 **(Interpretation of the modified KANET score).** Interpretation of the total score: normal when 10–16, borderline when 6–9, and abnormal when 0–5. *Reprinted with permission from Stanojevic, M., Talic, A., Miskovic, B., Vasilj, O., Shaddad, A. N., Ahmed, B., ... Al-Noobi, M. (2011). An attempt to standardize Kurjak's antenatal neurodevelopmental test: Osaka Consensus Statement.* Donald School Journal of Ultrasound in Obstetrics and Gynecology, *5(4), 317–329.*

comparison of the total and individual scores showed no differences between psychotropic administered pregnancies and healthy controls. Therefore, psychotropic drugs may not affect fetal neurobehavior.

V. Prediction of postnatal developmental disabilities

The ability of KANET to detect postnatal developmental disabilities was assessed, which included 353 fetuses in the third trimester between 28 and 38 weeks of gestation (Hata et al., 2018). Using 4D ultrasound, KANET was performed with follow-up for at least 2 years for all fetuses to detect the development of postnatal disabilities. KANET results showed 337 normal fetuses (score was 10–16) representing 95.47% of the total fetuses, and 16 fetuses (4.53%) showed a borderline score (6–9) with no abnormal score fetuses. Postnatal follow-up revealed eight fetuses with developmental disabilities. Five fetuses were in the normal KANET score group (1.48%). They comprised one fetus with Werdnig–Hoffmann disease detected just after birth, one with autism spectrum disorder diagnosed at 24 months, one with Ullrich congenital muscular dystrophy diagnosed at 9 months, and two with developmental disorders diagnosed at age 3 and 18 months. The borderline KANET score group comprised three fetuses with developmental disabilities (one with motor development delay diagnosed at 6 months, one with Duchenne muscular dystrophy diagnosed at 18 months, and one with autism spectrum disorder diagnosed at age 30 months). The rate of detecting postnatal developmental disabilities was significantly different between the normal and borderline groups ($P < .001$), indicating the usefulness of KANET to predict postnatal disabilities.

VI. KANET in diabetic pregnancies

Edelberg and coworkers showed that persistent material hyperglycemia is associated with decreased fetal movements (Edelberg, Dierker, Kalhan, & Rosen, 1987). Another study reported that diabetes could affect fetal behavior irrespective of glycemic control, probably via the immature development of the sensory-motor response system (Cosmi, Anceschi, Cosmi, Piazze, & La Torre, 2003). KANET was also conducted to assess fetal neurobehavior in diabetic pregnancies (Antsaklis, Porovic, Daskalakis, & Kurjak, 2017). In that study, KANET was applied to 40 pregnancies with preexisting or gestational diabetes mellitus on insulin therapy versus 40 nondiabetic low-risk pregnancies (Antsaklis et al., 2017). The total KANET score was significantly higher in fetuses of nondiabetic mothers (median, 16; range, 6–16) compared with those of diabetic mothers (median, 14; range, 4–16). The number of abnormal scores in the diabetic group (two cases, representing 1.3%) was significantly higher than in the nondiabetic group, which did not have any abnormal scores ($P < .001$). Comparison of the individual parameters revealed that the main differences were in isolated eye blinking, facial alteration, and finger movements. These results indicate the value of applying KANET to assess diabetic pregnancies, as they are vulnerable to impairment of fetal neurodevelopment.

VII. KANET and thyroid disorders

Thyroid disorders, especially hypothyroidism, have been linked to postnatal neurodevelopment and cognitive functions (Henriches et al., 2010). Therefore, a study was conducted to assess fetal neurobehavior in pregnancies with hypothyroidism using KANET, involving 25 euthyroid versus 25 hypothyroid pregnancies (five of them were controlled while 20 were in an uncontrolled hypothyroid state) (Dieb, Salam, Shaheen, & Shaeer, 2019). The results showed that the hypothyroid group demonstrated significantly lower scores compared with the euthyroid group. Moreover, the incidence of the abnormal KANET score was significantly higher in the hypothyroid group (9.36%) than in the euthyroid group (1.4%) ($P < .005$). The uncontrolled hypothyroid fetuses displayed lower mean KANET scores compared with controlled fetuses, with no significant difference between the two groups. This study lacked a postnatal follow-up; however, KANET suggested a difference in neurobehavior between hypothyroid and normal pregnancies. Therefore, it may be useful to add KANET for the clinical assessment of fetal neurodevelopment in mothers with an abnormal thyroid function.

VIII. KANET in high-risk pregnancies

Following the first introduction of KANET and before the Osaka consensus, which promoted use of the modified KANET, there were several studies comparing KANET in high- versus low-risk pregnancies. The largest study was conducted by Talic and her coworkers (Talic et al., 2011); it was a multicenter study involving 620 singletons of low- and high-risk pregnancies. The results of the study demonstrated a significant difference in KANET scores between the two groups. Twenty-four percent of pregnancies involving mothers with a previous history of cerebral palsy showed an abnormal KANET score in the current pregnancy. Moreover, a low KANET score was predictive of a poor outcome, as indicated by two intrauterine and one neonatal death in the low KANET group. Ten out of 36 pregnancies with low KANET scores displayed abnormal postnatal outcomes, with four of them showing severe generalized spasticity. Honemeyer and his team (Honemeyer, Tali, Therwat, Paulose & Patidar, 2013) performed another study of high-risk pregnancies involving mothers with antiphospholipid antibody syndrome, uncontrolled hyperthyroidism, hemoglobinopathies, systemic lupus, hypertension, cyanotic heart disease, renal disease, and diabetes. The study was associated with early neurological evaluation after birth for all the examined cases. No abnormal KANET results were found; however, two-thirds of borderline scores were noted in the high-risk group. The third study compared high- to low-risk pregnancies (Athannasiadis et al., 2013) in which 74 high-risk were compared with 78 low-risk cases. Their results showed significantly lower total KANET scores in high-risk fetuses. The score of the diabetic subgroup was significantly higher than those in preeclampsia and fetal growth restriction subgroups. This study confirmed the feasibility of KANET for prenatal neurological evaluation. A more recent study was done in Brazil involving high-risk pregnancies, and the abnormal scores were found in the high-risk group, emphasizing the importance of KANET application in high-risk pregnancies for antenatal prediction of abnormal neurological development (Neto, 2015).

IX. KANET in twins

Twin pregnancy is a special condition where the intrauterine space and environment are shared by two fetuses. The neurobehavior of twins has always attracted the interest of scientists. KANET was used to investigate fetal neurobehavioral differences between singleton and twin fetuses, comparing singleton versus twin neurobehavior in all three trimesters (Kurjak et al., 2013). The number of abnormal, borderline, and normal KANET scores between singletons and twins did not show any significant differences. However, the individual scores for isolated eye blinking, mouthing, grimacing, hand to head movement, finger movement, gestalt perception, and general movements differed significantly in twins and singletons. They found that the twins showed less activity than singletons. These results were also confirmed in another study and probably due to restricted space in utero (AboEllail et al., 2018).

Another KANET-based study on twins addressed the temperament after birth (Honemeyer & Kurjak, 2012). It found that higher KANET scores in one twin at 29w2d, 31w4d, and 34w6d revealed the postnatal temperament, indicating that the temperament is decided during intrauterine life. Similar results were obtained by Degani and his coworkers (Degani, Leibovitz, Shapiro, & Ohel, 2009).

Limitations of KANET

The aforementioned studies confirmed the value of KANET; however, there are some limitations. First, it is time-consuming. Second, it involves a learning curve, and so might not be familiar for beginners. Also, it requires experience with 4D construction of fetal behaviors, necessitating expertise with handling 4D ultrasound. A detailed familiarity with fetal anatomy is also an important requirement for this test. These factors raise the question of whether KANET can be a suitable screening test for fetal neurobehavioral developmental assessment.

Mini KANET

Due to the aforementioned limitations of KANET, a simplification of the KANET was proposed by (Hata, Kanenishi, Mori, et al., 2019). A 3-year study involving 353 pregnancies with no fetal structural or chromosomal anomalies was performed. Eight out of the 353 fetuses were with postnatal developmental disabilities. After the application of KANET, the score of each parameter in these fetuses was reviewed. Choice of three parameters was done (isolated eye blinking, facial alteration or mouth opening, and isolated leg movement) to create the new Mini KANET (Fig. 32.3). The normal score is 2–6, while 0–1 is considered abnormal (Fig. 32.4). Comparison between KANET and Mini KANET revealed significantly higher sensitivity of Mini KANET (50%) compared with KANET (37.5%) ($P = .001$). Mini KANET also showed higher specificity, accuracy, and positive and negative predictive values; however, there were no significant

Sign	Score			Sign Score
	0	1	2	
Isolated eye blinking 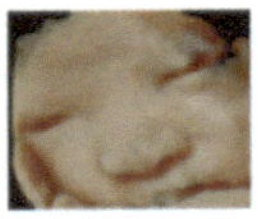	Not present	Not fluent (blinking 1-5 times)	Fluency (blining > 5 times)	
Facial alteration (grimace or tongue expulsion)  or Mouth opening (yawning or mouthing) 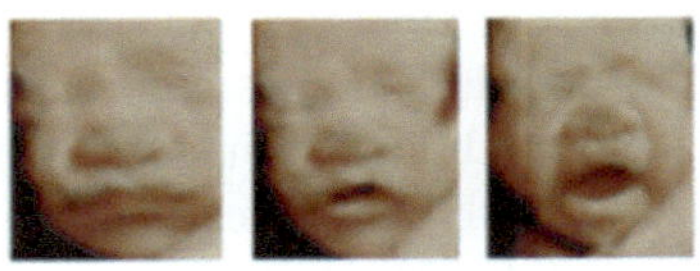	Not present	Not fluent (alteration 1-5 times)	Fluency (alteration > 5 times)	
Isolated leg movement 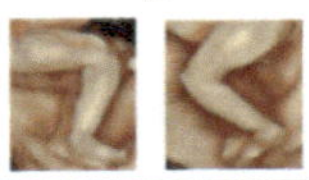	Cramped	Poor repertoire or small in range (movement 0-5 times)	Variable in full range, many alterations (movement > 5 times)	
			Total score	

FIGURE 32.3 **(Mini KANET).** Mini KANET assessment containing the three parameters with each parameter score of 0–2. *Reprinted with permission from Hata, T., Kanenishi, K., Mori, N., AboEllail, M. A. M., Hanaoka, U., Koyano, K., … Kurjak, A. (2019). Mini KANET: Simple fetal antenatal neurodevelopmental test.* Donald School Journal of Ultrasound in Obstetrics and Gynecology, *13(2), 59–63.*

TOTAL SCORE	INTERPRETATION
0-1	Abnormal
2-6	Normal

FIGURE 32.4 **(Interpretation of Mini KANET).** The total score of Mini KANET and its interpretation to differentiate abnormal (score 0–1) from normal (2–6) pregnancies. *Hata, T., Kanenishi, K., Mori, N., AboEllail, M. A. M., Hanaoka, U., Koyano, K., … Kurjak, A. (2019). Mini KANET: Simple fetal antenatal neurodevelopmental test.* Donald School Journal of Ultrasound in Obstetrics and Gynecology, *13(2), 59–63.*

differences between KANET and Mini KANET. Reducing the number of parameters of KANET from eight to three in Mini KANET might render it time-saving and more applicable with a shorter learning curve. The deleted parameters were found to not affect the overall detection rate of KANET; therefore, they were deleted to create the new Mini KANET. Despite its potential merits over KANET, both of them still have limitations. First, there is difficulty performing these tests under certain conditions, such as with a covered fetal face or occipito-anterior position, oligohydramnios, and deep head engagement. Another criticism is the subjective judgment, especially of facial expressions, which characterizes all neurobehavioral studies. Trials were done to devise a coding system for such facial movements and expressions (Reissland, Francis, & Mason, 2012; Reissland, Francis, Mason, & Lincoln, 2011). However, it still lacks universal standardization.

Conclusions

KANET might be a useful diagnostic tool enabling the prediction of postnatal developmental disabilities in healthy as well as high-risk fetuses in utero. Mini KANET assessment might replace the KANET test in healthy fetuses as being simpler and more time-saving. Further studies involving a larger size are needed to confirm the validity and clinical relevance of these tests.

Applications to other areas of development

In this chapter, we reviewed the origin and development of the antenatal fetal neurodevelopmental test (KANET) and standardized modified KANET. We demonstrated several studies showing the application of this test to observe fetal neurodevelopment and behavior. We also described variables that might affect the outcome of this test, such as the maternal emotional state, fetal sex, and ethnicity. Studies on KANET illustrated in this chapter compared fetal behavior in normal fetuses versus fetuses of high-risk pregnancies such as mothers with diabetes, thyroid disorders, twins, and those on psychotropic medications. KANET showed the ability to predict postnatal neurodevelopmental disabilities (Hata et al., 2018). At the end of the chapter, we discussed limitations of the KANET test and our proposed Mini KANET, which has shown diagnostic accuracies comparable to those of the original KANET with less time and efforts (Hata, Kanenishi, AboEllail, et al., 2019; Hata, Kanenishi, Mori, et al., 2019). Finally, we can say that KANET is a useful test to predict unfavorable intrauterine conditions affecting the neurological development of fetal behavior and might cause neurological disabilities such as cerebral palsy later in life. Larger multicenter studies are needed to apply this test using standardized coding for each fetal movement and facial expression to reveal the hidden aspects of fetal neurobehavioral development and the mechanisms of development of such neurological abnormalities.

Key facts regarding KANET

- KANET is an essential tool for fetal neurobehavioral evaluation.
- Standardization of this test was achieved in the Osaka Consensus Meeting, and many multicenter studies were conducted.
- The maternal emotional state and parity might affect fetal behavior.
- Fetal ethnicity should be considered when KANET is applied; however, the fetal sex does not influence the test.
- The intake of psychotropic drugs during pregnancy may not affect fetal neurobehavior, as indicated by the results of KANET.
- KANET is particularly beneficial in high-risk pregnancies to observe fetal neurodevelopment and predict postnatal disabilities.
- KANET use requires a learning curve, sonographic skill, and knowledge of fetal anatomy.
- Mini KANET assessment is more simple and time-saving than KANET and has the same diagnostic indices.

Summary points

- This chapter focuses on KANET, which is an antenatal fetal neurodevelopmental test.
- KANET was proposed by Prof. Asim Kurjak and his team using four-dimensional ultrasound to devise a test combining elements of general movements plus some from the ATANAT score.
- KANET can discriminate normal from abnormal fetal behavior and hence can be used to elucidate the origin of neurological disabilities such as cerebral palsy.
- KANET was successfully applied to high-risk pregnancies such as diabetics or those with thyroid disorders and twins and showed differences in behavior compared with normal fetuses.
- Maternal stress and the emotional state as well as ethnicity should be considered during fetal neurobehavior studies using KANET.
- Infants with borderline and abnormal KANET scores should be followed up for at least 2 years after birth until a diagnosis of cerebral palsy or other neurological abnormalities is ruled out.
- KANET use involves a learning curve, skill in 4D ultrasound, and knowledge of fetal anatomy.
- Mini KANET is simpler and more time-saving than KANET but retains accurate diagnostic indices.

References

AboEllail, M. A. M., Kanenishi, K., Mori, N., Noguchi, J., Marumo, G., & Hata, T. (2018). Ultrasound study of fetal movements in singleton and twin pregnancies at 12-19 weeks. *Journal of Perinatal Medicine, 46*(8), 832–838.

Antsaklis, P., Porovic, S., Daskalakis, G., & Kurjak, A. (2017). 4D assessment of fetal brain function in diabetic patients. *Journal of Perinatal Medicine, 45*(6), 711–715.

Athanasiadis, A. P., Mikos, T., Tambakoudis, G. P., Theodoridis, T. D., Papastergiou, M., Assimakopoulos, E., & Tarlatzis, B. C. (2013). Neurodevelopmental fetal assessment using KANET scoring system in low and high-risk pregnancies. *Journal of Maternal-Fetal and Neonatal Medicine, 26*(4), 363–368.

Bodfish, J. W., Powell, S. B., Golden, R. N., & Lewis, M. H. (1995). Blink rate as an index of dopamine function in adults with mental retardation and repetitive behavior disorders. *American Journal on Mental Retardation, 99*(4), 335–344.

Choleris, E., Galea, L. A. M., Sohrabji, F., & Frick, K. M. (2018). Sex differences in the brain: Implications for behavioral and biomedical research. *Neuroscience and Biobehavoral Reviews, 85*, 126–145.

Cosmi, E. V., Anceschi, M. M., Cosmi, E., Piazze, J. J., & La Torre, R. (2003). Ultrasonographic patterns of fetal breathing movements in normal pregnancy. *International Journal of Gynecology and Obstetrics, 80*(3), 285–290.

Degani, S., Leibovitz, Z., Shapiro, I., & Ohel, G. (2009). Twins' temperament: Early prenatal sonographic assessment and postnatal correlation. *Journal of Perinatology, 29*(5), 337–342.

Dieb, A., Salam, R., Shaheen, D., & Shaeer, E. (2019). Evaluation of fetal neurological behavior in hypothyroid pregnant females - a pilot study. *Journal of Maternal-Fetal and Neonatal Medicine, 32*(16), 2617–2621.

Dieter, J. N., Emory, E. K., Johnson, K. C., & Raynor, B. D. (2008). Maternal depression and anxiety effects on the human fetus: Preliminary findings and clinical implications. *Infant Mental Health Journal, 29*(5), 420–441.

Dieter, J. N., Field, T., Hernandez-Reif, M., Jones, N., Lecanuet, J. P., Salman, F. A., & Rezepi, M. (2001). Maternal depression and increased fetal activity. *Journal of Obstetrics and Gynaecology, 21*(5), 468–473.

Edelberg, S. C., Dierker, L., Kalhan, S., & Rosen, M. G. (1987). Decreased fetal movements with sustained maternal hyperglycemia using the glucose clamp technique. *American Journal of Obstetrics and Gynecology, 156*(5), 1101–1105.

Einspieler, C., & Prechtl, H. F. (2005). Prechtl's assessment of general movements: A diagnostic tool for the functional assessment of the young nervous system. *Mental Retardation and Developmental Disabilities Research Reviews, 11*(1), 61–67.

Emory, E. K., & Dieter, J. N. (2006). Maternal depression and psychotropic medication effects on the human fetus. *Annals of the New York Academy of Sciences, 1094*(1), 287–291.

Farkas, L. G., Katic, M. J., & Forrest, C. R. (2005). International anthropometric study of facial morphology in various ethnic groups/races. *Journal of Craniofacial Surgery, 16*(4), 615–646.

Gosselin, J., Gahagan, S., & Amiel-Tison, C. (2005). The Amiel-Tison neurological assessment at term: Conceptual and methodological continuity in the course of follow-up. *Mental Retardation and Developmental Disabilities Research Reviews, 11*(1), 34–51.

Hanaoka, U., Hata, T., Kanenishi, K., AboEllail, M. A. M., Uematsu, R., Konishi, Y., … Kurjak, A. (2016). Does ethnicity have an effect on fetal behavior? A comparison of Asian and Caucasian populations. *Journal of Perinatal Medicine, 44*(2), 217–221.

Hata, T., Dai, S. Y., & Marumo, G. (2010). Ultrasound for evaluation of fetal neurobehavioural development: From 2D to 4D ultrasound. *Infant and Child Development, 19*(1), 99–118.

Hata, T., Hanaoka, U., AboEllail, M. A. M., Uematsu, R., Noguchi, J., Kusaka, T., & Kurjak, A. (2016a). Does parity have an effect on fetal behavior? A comparison between primi and multiparas. *Donald School Journal of Ultrasound in Obstetrics and Gynecology, 10*(1), 99–102.

Hata, T., Hanaoka, U., AboEllail, M. A. M., Uematsu, R., Noguchi, J., Kusaka, T., & Kurjak, A. (2016b). Is there a sex difference in fetal behavior? A comparison of the KANET test between male and female fetuses. *Journal of Perinatal Medicine, 44*(5), 585–588.

Hata, T., Kanenishi, K., AboEllail, M. A. M., Mori, N., Koyano, K., Kato, I., & Kusaka, T. (2019). Effect of psychotropic drugs on fetal behavior in the third trimester of pregnancy. *Journal of Perinatal Medicine, 47*(2), 207–211.

Hata, T., Kanenishi, K., Mori, N., AboEllail, M. A. M., Hanaoka, U., Koyano, K., … Kusaka, T. (2018). Prediction of postnatal developmental disabilities using the antenatal fetal neurodevelopmental test: KANET assessment. *Journal of Perinatal Medicine, 47*(1), 77–81.

Hata, T., Kanenishi, K., Mori, N., AboEllail, M. A. M., Hanaoka, U., Koyano, K., … Kurjak, A. (2019). Mini KANET: Simple fetal antenatal neurodevelopmental test. *Donald School Journal of Ultrasound in Obstetrics and Gynecology, 13*(2), 59–63.

Hata, T., Kanenishi, K., & Sasaki, M. (2010). Four-dimensional sonographic assessment of fetal movement in the late first trimester. *International Journal of Gynaecology and Obstetrics, 109*(3), 190–193.

Henrichs, J., Bongers-Schokking, J. J., Schenk, J. J., Ghassabian, A., Schmidt, H. G., Visser, T. J., … Tiemeier, H. (2010). Maternal thyroid function during early pregnancy and cognitive functioning in early childhood: The generation R study. *Journal of Clinical Endocrinology and Metabolism, 95*(9), 4227–4234.

Honemeyer, U., & Kurjak, A. (2012). A prenatal beginning for temperament formation- myth or reality? Case study of a twin pregnancy. *Donald School Journal of Ultrasound in Obstetrics and Gynecology, 6*(2), 148–153.

Honemeyer, U., Talic, A., Therwat, A., Paulose, L., & Patidar, R. (2013). The clinical value of KANET in studying fetal neurobehavior in normal and at-risk pregnancies. *Journal of Perinatal Medicine, 41*(2), 187–197.

Kean, L. H., Gargari, S. S., Suwanrath, C., Sahota, D. C., & James, D. K. (2001). A comparison of fetal behaviour in term fetuses exposed to anticonvulsant medication with unexposed controls. *British Journal of Obstetrics and Gynaecology, 108*(11), 1159–1163.

Kurjak, A., Abo-Yaqoub, S., Stanojevic, M., Yigiter, A. B., Vasilj, O., Lebit, D., … Kadic, A. S. (2010). The potential of 4D sonography in the assessment of fetal neurobehavior—multicentric study in high-risk pregnancies. *Journal of Perinatal Medicine, 38*(1), 77–82.

Kurjak, A., Carrera, J., Medic, M., Azumendi, G., Andonotopo, W., & Stanojevic, M. (2005). The antenatal development of fetal behavioral patterns assessed by four-dimensional sonography. *Journal of Maternal-Fetal and Neonatal Medicine, 17*(6), 401–416.

Kurjak, A., Miskovic, B., Stanojevic, M., Amiel-Tison, C., Ahmed, B., Azumendi, G., … Salihagic-Kadic, A. (2008). New scoring system for fetal neurobehavior assessed by three- and four-dimensional sonography. *Journal of Perinatal Medicine, 36*(1), 73–81.

Kurjak, A., Stanojevic, M., Andonotopo, W., Salihagic-Kadic, A., Carrera, J. M., & Azumendi, G. (2004). Behavioral pattern continuity from prenatal to postnatal life – a study by four-dimensional (4D) ultrasonography. *Journal of Perinatal Medicine, 32*(4), 346–353.

Kurjak, A., Talic, A., Stanojevic, M., Honemeyer, U., Serra, B., Prats, P., & Di Renzo, G. C. (2013). The study of fetal neurobehavior in twins in all three trimesters of pregnancy. *Journal of Maternal-Fetal and Neonatal Medicine, 26*(12), 1186–1195.

Kurjak, A., Vecek, N., Hafner, T., Bozek, T., Funduk-Kurjak, B., & Ujevic, B. (2002). Prenatal diagnosis: What does four-dimensional ultrasound add? *Journal of Perinatal Medicine, 30*(1), 57–62.

MacLennan, A. H., Thompson, S. C., & Grecz, J. (2015). Cerebral palsy: Causes, pathways, and the role of genetic variants. *American Journal of Obstetrics and Gynecology, 213*(6), 779–788.

Merz, E., & Abramowicz, J. S. (2012). 3D/4D ultrasound in prenatal diagnosis: Is it time for routine use? *Clinical Obstetrics and Gynecology, 55*(1), 336–351.

Miskovic, B., Vasilj, O., Stanojevic, M., Ivanković, D., Kerner, M., & Tikvica, A. (2010). The comparison of fetal behavior in high risk and normal pregnancies assessed by four dimensional ultrasound. *Journal of Maternal-Fetal and Neonatal Medicine, 23*(12), 1461–1467.

Neto, R. M. (2015). KANET in Brazil: First experience. *Donald School Journal of Ultrasound in Obstetrics and Gynecology, 9*(1), 1–5.

Reissland, N., Francis, B., & Mason, J. (2012). Development of fetal yawn compared with non-yawn mouth openings from 24–36 weeks gestation. *PLoS One, 7*(11), e50569.

Reissland, N., Francis, B., Mason, J., & Lincoln, K. (2011). Do facial expressions develop before birth? *PLoS One, 6*(8), e24081.

Salihagic-Kadic, A., Kurjak, A., Medic, M., Andonotopo, W., & Azumendi, G. (2005). New data about embryonic and fetal neurodevelopment and behavior obtained by 3D and 4D sonography. *Journal of Perinatal Medicine, 33*(6), 478–490.

Sasaki, M., Yanagihara, T., Naitoh, N., & Hata, T. (2010). Four-dimensional sonographic assessment of inter-twin contact late in the first trimester. *International Journal of Gynaecology and Obstetrics, 108*(2), 104–107.

Stanojevic, M., Antsaklis, P., Kadic, A. S., Predojevic, M., Vladareanu, R., Vladareanu, S., & Neto, M. R. (2015). Is Kurjak antenatal neurodevelopmental test ready for routine clinical application? Bucharest consensus statement. *Donald School Journal of Ultrasound in Obstetrics and Gynecology, 9*(3), 260–265.

Stanojevic, M., Talic, A., Miskovic, B., Vasilj, O., Shaddad, A. N., Ahmed, B., … Al-Noobi, M. (2011). An attempt to standardize Kurjak's antenatal neurodevelopmental test: Osaka Consensus Statement. *Donald School Journal of Ultrasound in Obstetrics and Gynecology, 5*(4), 317–329.

Talic, A., Kurjak, A., Ahmed, B., Stanojevic, M., Predojevic, M., Kadic, A. S., & Di Renzo, G. C. (2011). The potential of 4D sonography in the assessment of fetal behavior in high-risk pregnancies. *Journal of Maternal-Fetal and Neonatal Medicine, 24*(7), 948–954.

Part IV

Management and treatments

Chapter 33

Pain evaluation and management in neurodevelopmental conditions

Marlon L. Wong[1] and Eva Widerstrom-Noga[2]

[1]*Department of Physical Therapy, Miller School of Medicine University of Miami, Miami, FL, United States;* [2]*Miami Project to Cure Paralysis University of Miami, Miller School of Medicine, Miami, FL, United States*

List of abbreviations

ADHD attention deficit/hyperactivity disorder
ASDs autism spectrum disorders
CP cerebral palsy
GERD gastroesophageal reflux disease
NDDs neurodevelopmental disorders

Mini-dictionary of terms

Alexithymia Lack of emotional awareness or difficulty in identifying and describing feelings
Neuropathic pain Pain caused by a lesion or disease of the somatosensory nervous system
Nociceptive Pain Pain that arises from actual or threatened damage to nonneural tissue and is due to the activation of nociceptors.

Introduction

It is often overlooked that people with neurodevelopmental disorders (NDDs) face increased risk for both acute and chronic pain (Breau, Camfield, McGrath, & Finley, 2003). Acutely, they are exposed to more medical procedures, accidents, and in some instances self-injurious behavior (Cong et al., 2017; Nigg, 2013; Symons, 2011), which significantly increases the risk for developing persistent pain conditions. Indeed, chronic pain is common in children with NDDs, and over 40% of children with moderate to profound cognitive impairment have been reported to experience daily pain (Breau et al., 2003). In fact, pain is one of the three most common symptoms reported by parents of children with central nervous system impairment (Steele et al., 2014). Improving pain assessment and management will likely have a significant positive impact on quality of life in this population.

The increased risk for acute pain starts early in life as NDDs are associated with preterm birth (Mwaniki, Atieno, Lawn, & Newton, 2012), and preterm-born neonates often undergo hundreds of invasive procedures during intensive care (Carbajal et al., 2008). The increased exposure to procedural pain early in the neonatal period appears to be associated with alterations in thalamic macrostructural and metabolic growth (Duerden et al., 2018) and may prime the basal microglial phenotype toward a proinflammatory state later in life (Burke, Fan, & Trang, 2016). The thalamus is an integral part of the brain's pain network, and it functions as a relay station between the spinal cord and cortex for nociceptive input, while microglia are key cellular modulators for pain processing. Thus, altered thalamic and microglial functions associated with exposure to procedural pain in the neonatal period are proposed to be significant contributors to the observed abnormal somatosensory processing and increased risk for chronic pain later in life in people with NDDs (Burke et al., 2016; Duerden et al., 2018; Low & Schweinhardt, 2012). It has been suggested that children with neurological impairments often

Diagnosis, Management and Modeling of Neurodevelopmental Disorders. https://doi.org/10.1016/B978-0-12-817988-8.00033-6

experience pain more intensely than children with normal neurological function (Solodiuk et al., 2014). Moreover, pain exposure in the neonatal period also contributes to adverse neurobehavioral outcomes, specifically impaired habituation and increased stress responses, in infants with NDDs (Cong et al., 2017).

Impaired ability to self-identify negative feelings, or alexithymia, may complicate the evaluation of chronic pain in people with NDDs. Emotion identification is an essential step in the emotion regulation process, and poor regulation of negative emotions is a risk factor for poor pain outcomes (Di Tella et al., 2017; Fietz, Valencia, & Silani, 2018). Alexithymia is elevated in adults and youth with chronic pain, and it is also associated with autism spectrum disorders (ASDs) (Griffin, Lombardo, & Auyeung, 2016). Furthermore, people with NDDs may be limited in their ability to communicate their pain and/or may have atypical behaviors in response to pain that are misinterpreted or unrecognized by caregivers. This communication barrier may cause additional distress and complicate the management of pain in people with NDDs (Warlow & Hain, 2018).

NDDs are often associated with self-injurious behavior. Self-injurious behavior can be episodic or chronic, and it involves repeated and forceful stimulation, such as head banging, self-biting, self-scratching or hitting, gouging, and other similar behaviors. For some people with NDDs, the self-injurious behavior is believed to be an atypical response to pain (Symons, Harper, McGrath, Breau, & Bodfish, 2009). However, the repetitive and high forces associated with the self-injurious behavior cause local tissue damage and further contribute to their pain, thus creating a vicious cycle (Symons, 2011) (Fig. 33.1).

Given the long-term physiological and behavioral consequences of pain in people with NDDs, early detection and effective pain management are important. However, pain assessment and management are further complicated if there are communication barriers and/or self-injurious behavior as these can amplify the pain experience and cause tissue damage, respectively. The remainder of this chapter will describe concepts for pain assessment in people with NDDs and review current evidence pertaining to pain in NDD conditions. Specifically, we will review pain in cerebral palsy (CP), ASDs, and attention deficit/hyperactivity disorder (ADHD).

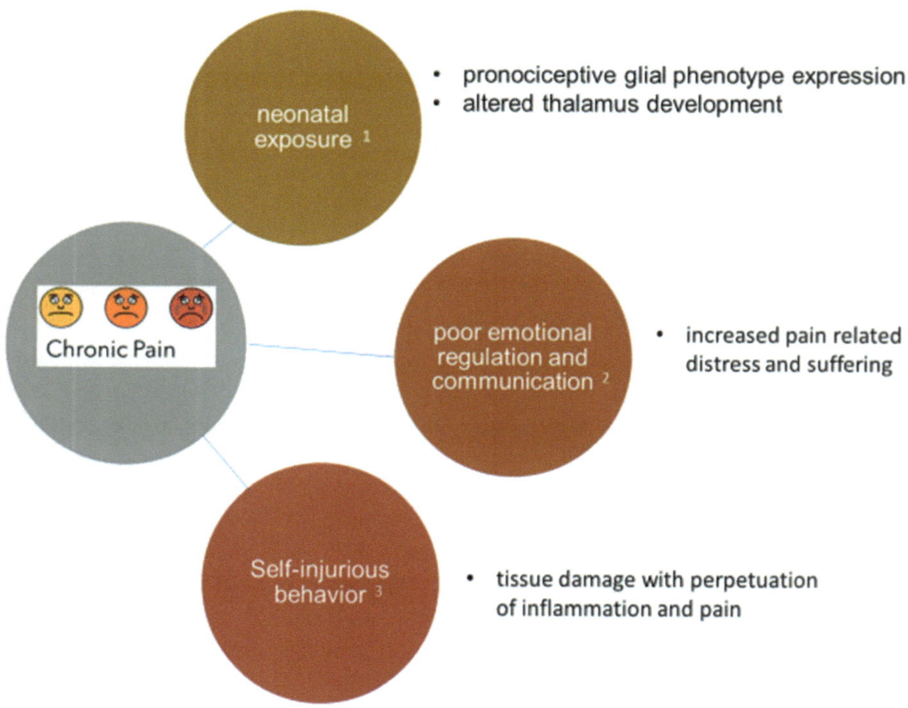

FIGURE 33.1 Factors that contribute to increased prevalence of chronic and persistent pain in people with neurodevelopmental disorders. *Based on Duerden, E. G., Grunau, R. E., Guo, T., Foong, J., Pearson, A., Au-Young, S., … Miller, S. P. (2018). Early procedural pain is associated with regionally-specific alterations in thalamic development in preterm neonates.* Journal of Neuroscience, *38(4), 878–886. https://ds.doi.org/10.1523/JNEUROSCI.0867-17.2017; Griffin, C., Lombardo, M. V., & Auyeung, B. (2016). Alexithymia in children with and without autism spectrum disorders.* Autism Res, *9(7), 773–780. https://ds.doi.org/10.1002/aur.1569; Symons, F. J. (2011). Self-injurious behavior in neurodevelopmental disorders: relevance of nociceptive and immune mechanisms.* Neuroscience and Biobehavioral Reviews, *35(5), 1266–1274. https://ds.doi.org/10.1016/j.neubiorev.2011.01.002.*

Mechanisms of pain

In order to manage pain optimally in these as in all other populations, attempts should always be made to identify both the type and if applicable the source of pain. This evaluation includes the classification of pain into neuropathic or nociceptive pain as these types of pain require principally different treatment approaches (Hauer, Houtrow, Section On, & Palliative Medicine, 2017; Warlow & Hain, 2018). In nociceptive pain types, pain is generated by the activation of nociceptors (or pain receptors) located throughout the body in the skin, musculoskeletal structures, and visceral structures. This type of pain is caused by inflammatory, mechanical, or ischemic stimuli. Prolonged nociceptive stimuli may result in both peripheral and central sensitization (Latremoliere & Woolf, 2009). Nociceptive pain in people with NDDs is often associated with other complications of chronic disease including scoliosis, spasticity, dystonia, and hip dislocation/subluxation in nonambulatory children (Hauer, Houtrow, Section On, & Palliative Medicine, 2017). Nociceptive pains can also be generated in visceral organs, especially in the organs of the digestive tract. Indeed, constipation and gastroesophageal reflux disease (GERD) are common in people with NDDs. In contrast, neuropathic pain is a consequence of injury to the nervous system, and this type of pain can be central (associated with damage to or abnormal processing within the central nervous system) or peripheral (associated with damage to the peripheral nerves).

Pain in specific neurodevelopmental conditions

Cerebral palsy (CP)

A recent review reported prevalence rates of pain in CP to vary widely, between 14% and 76%, with higher rates in females and in older ages (McKinnon, Meehan, Harvey, Antolovich, & Morgan, 2019). Similar to observations in other chronic pain populations, persistent pain is associated with reduced quality of life, participation, and psychological functioning in people with CP (Colver et al., 2015; Parkes et al., 2008). In a systematic review based on the International Classification of Functioning, Disability and Health (ICF) framework, investigators found that "Sensation of pain" was the most frequently reported outcome (37.6%) under the category *body function* (Benner et al., 2019). Further, in a survey study in 406 adults with CP (Jahnsen, Villien, Aamodt, Stanghelle, & Holm, 2004), one-third of the participants had chronic musculoskeletal pain. These data emphasize the critical need to evaluate and treat these consequences of NDDs.

Common causes of nociceptive pain in this population include hip luxation, hypertonia, muscle spasms, constipation, and gastroesophageal reflux (McKearnan, Kieckhefer, Engel, Jensen, & Labyak, 2004). However, neuropathic pain may also be a common type of pain in people with CP. For example, by using quantitative sensory testing (QST), Blankenburg et al. found that 50% of patients with CP had a combination of impaired mechanical and thermal sensation and mechanical hyperalgesia (Blankenburg et al., 2018). These authors proposed that the abnormal spinothalamic and dorsal column medial lemniscus mediated functions were suggestive of neuronal dysfunction in the thalamus region, likely caused by periventricular leukomalacia, and that this was one of the underlying causes of pain. These study results are consistent with findings in a wide range of other chronic pain populations where abnormal somatosensory function involving ascending somatosensory pathways is associated with neuropathic pain (Rolke et al., 2006).

Autism spectrum disorder (ASD)

Sensory abnormalities are a diagnostic criterion for ASDs (Association, 2013), and both hypersensitivity and hyposensitivity to painful stimuli have been reported in this population (Allely, 2013; Moore, 2015). In two separate studies (Klintwall et al., 2011; Militerni et al., 2000), almost half of the parents of children with ASDs reported that their children were hyposensitive to pain (43% and 40% of parents, respectively). Similarly, clinicians in autism treatment centers have reported a high prevalence of children with hyposensitivity to pain (Gillberg, Terenius, & Lonnerholm, 1985). It has been postulated that hyperfunction of the endogenous opioid system underlies the apparent hyposensitivity to painful stimuli in this population (Allely, 2013). However, the reporting of pain insensitivity by parents and clinicians may also stem from the atypical pain responses among children with ASDs, which often lead to poor recognition of their pain by caregivers and clinicians, and these findings are not necessarily evidence of decreased pain reactivity. In fact, studies using facial action coding or physiological responses (i.e., elevated heart rate and plasma b-endorphin levels) to measure pain response during medical procedures have found that people with ASDs have normal or increased pain responses compared to neurotypical people (Nader, Oberlander, Chambers, & Craig, 2004; Tordjman et al., 2009). Moreover, studies of high-functioning adolescents and adults with ASDs found that pain intensity ratings and pain thresholds were either normal or indicative of hypersensitivity for hypothetical pain and experimentally induced pain (Bandstra, Johnson, Filliter, & Chambers, 2012; Cascio et al., 2008; Fan, Chen, Chen, Decety, & Cheng, 2014).

Perseverative behavior is a well-known tendency among individuals with ASDs, and pain can become the perseverative focus of attention for these individuals. Inability or difficulty with shifting attentional focus away from the pain may amplify the pain symptoms and the associated distress. Thus, it is widely believed that the heightened anxiety often experienced by this population contributes to their pain experience through psychological mechanisms as well as physical mechanisms, for instance, prolonged muscle tension (Allely, 2013). One study found that children with ASDs demonstrated diminished or absent pain reactions during venipuncture, but they had elevated heart rate, plasma beta-endorphins, and worsening of negative behaviors including social withdrawal, self-injurious behavior, or aggressiveness (Tordjman et al., 2009). These findings support the importance of evaluating distress and perseveration to better manage pain in this population.

In conclusion, there are likely subgroups within the ASD population of differing pain sensitivity and reactivity. Furthermore, one individual can be hyposensitive to some noxious stimuli and hypersensitive to others. For example, an individual with ASD may demonstrate hyposensitivity to thermal noxious stimuli (e.g., touching a hot stove), but hypersensitivity to noxious stimuli associated with medical procedures (e.g., venipuncture). For individuals who demonstrate diminished sensitivity and reactivity to pain, care should be taken to closely monitor physiological (i.e., heart rate and blood pressure) and behavioral signs of distress, as abnormal pain behaviors are common in people with ASDs.

Attention deficit/hyperactivity disorder (ADHD)

Numerous studies report a higher prevalence of chronic pain conditions in children and adults with ADHD when compared to the general population. Children with ADHD are found to have higher rates of headache (Jameson et al., 2016; Kutuk et al., 2018), with 1.46–3 times the normal risk, and recurrent or severe stomach pain (Holmberg & Hjern, 2006; Jameson et al., 2016). Similarly, adults with ADHD report significantly more pain and lower quality of life than age-matched healthy controls, with one study reporting that 80% of the participants had widespread pain (Kessler, Lane, Stang, & Van Brunt, 2009; Lensing, Zeiner, Sandvik, & Opjordsmoen, 2015; Stickley, Koyanagi, Takahashi, & Kamio, 2016; Stray et al., 2013). There also appears to be an association between ADHD and fibromyalgia syndrome, with studies of fibromyalgia patients reporting rates of ADHD between 25% and 45%, which is significantly higher than the estimated 3% –16% prevalence rate for ADHD in the general population (Derksen, Vreeling, & Tchetverikov, 2015; Reyero et al., 2011; van Rensburg, Meyer, Hitchcock, & Schuler, 2018). Although pain sensitivity appears to be in the normative range as assessed with quantitative sensory testing in people with ADHD, findings do suggest that children with ADHD may be prone to impaired temperature sensation and sharp/dull discrimination (Scherder, Rommelse, Broring, Faraone, & Sergeant, 2008), and adults with ADHD may have lower cold pain thresholds than the general population (Treister, Eisenberg, Demeter, & Pud, 2015). Moreover, Ritalin medication, commonly prescribed for ADHD, appears to have antinociceptive properties in this population (Treister et al., 2015).

Pain assessment and management

Pain evaluation

Optimal pain assessment requires a comprehensive evaluation of pain severity and pain interference with daily life activities (e.g., physical, emotional, social) as well as other pain characteristics (e.g., location, exacerbating and alleviating factors). Across individuals with NDDs, there are wide-ranging intellectual disabilities, somatosensory abnormalities, communication skills/impairments, physical impairments, and self-injurious behaviors. Thus, the different degrees and presence of impairments and consequences associated with NDDs create unique challenges for the assessment and management of pain. Due to this heterogeneity, individualized approaches with respect to both pain assessments and management may be required.

Pain severity, or pain intensity, is the most commonly measured pain assessment domain in both clinical research and practice, but this domain is dependent on self-report. While measures of pain intensity such as the numeric rating scale (0–10), visual analog scale (0–10, or 0–100), color analog scale, or faces scales may be appropriate for individuals without severe cognitive impairments (Tsze, von Baeyer, Bulloch, & Dayan, 2013; Von Baeyer, 2006), they are not useful for infants, toddlers, or individuals with profound communication or cognitive impairment. In the latter cases, pain assessment will be based on observed pain behaviors and parent/caregiver perceptions.

Another domain, pain interference, may be of particular importance in this population as it can be evaluated in individuals with or without cognitive impairments. Pain interference is the extent to which pain hinders or decreases engagement with life activities and enjoyment of life. Pain interference encompasses the physical, emotional, cognitive, and physiological (i.e., sleep) impact of pain (Amtmann et al., 2010).

A toolbox of validated instruments for the assessment of pain in children with CP, published by an interdisciplinary team (Kingsnorth et al., 2015), identified five instruments as having moderate or strong clinical utility for assessment of pain (Table 33.1). These tools are also applicable across a wide spectrum of disorders for assessment of pain-related disability in children. The ***Child Activity Limitations Interview*** (Palermo, Witherspoon, Valenzuela, & Drotar, 2004) is intended to assess the impact of pain on participation in daily activities and can be reported either by the child or by the parent/caregiver. The ***Pediatric Pain Interference Scale*** (Varni et al., 2010) assesses pain-related behaviors and is reported by the child or by the parent. The ***Varni/Thompson Pediatric Pain Questionnaire*** (Varni, Thompson, & Hanson, 1987) has been adapted to CP and assesses verbal and nonverbal cues regarding the presence and quality of pain reported by the child or the parent/caregiver. The ***Noncommunicating Children's Pain Checklist***—Revised (Breau, McGrath, Camfield, & Finley, 2002) and the ***Pediatric Pain Profile*** (Hunt et al., 2004) were developed specifically for children who lack verbal communication skills and/or have severe neurological impairment. These tools require the provider to observe and score the frequency of specific pain behaviors. The revised ***Face, Legs, Activity, Cry Consolability scale*** (r-FLACC) is another observational tool worth mentioning; nurses have identified the r-FLACC as a preferred tool for pain assessment of children with NDDs in acute care settings due to its ease of use and pragmatic qualities (Crosta, Ward, Walker, & Peters, 2014). The r-FLACC is also designed to include pain behaviors specific to each child, which is an important feature for children with atypical pain behaviors.

TABLE 33.1 Tools for assessing pain interference in children with neurodevelopmental disorders.

Tool	Type of assessment	Method of delivery	Population	Unique characteristics
Child activity limitations interview[a]	Interview/ self-report/ observational	Provider reads a list of activities and asks the child to list how much difficulty the child has with these activities over the last 4 weeks due to pain	Verbal children or care giver report	Allows children and parents to select from a list of activities that have personal meaning
Noncommunicating children's pain checklist—Revised[b]	Observational	The provider observes the child for a 2-hour period, then scores the frequency of 30 pain behaviors across 7 domains (vocal, social, facial, activity, body and limbs, physiological, and eating/ sleeping)	Developed to be used with nonverbal children	Addresses gap for children with severe cognitive, motor, and physical impairments
Pediatric pain interference scale[c]	Self-report/ observational	Child answers questions on sleep, emotional, and physical pain interference	Verbal children only or caregiver report	Short questionnaire that covers the major domains of pain interference
Pediatric pain profile[d]	Observational	The provider scores the frequency of 20 pain related behaviors	Developed to be used with nonverbal children	Addresses gap for children with severe cognitive, motor, and physical impairments
Varni/Thompson pediatric pain questionnaire[e]	Self-report/ observational)	The child identifies relevant pain descriptors (both self-generated descriptors and chosen from a list), pain intensity, and completes a body diagram in a structured interview	Adapted and validated for use in children with GMFCS levels I to III	Provides information regarding the intensity of pain, the sensory, affective, and evaluative qualities of pain, and the location of pain

GMFCS, Gross Motor Function Classification System.
[a]*Breau et al., 2002.*
[b]*Hunt et al., 2004.*
[c]*Palermo et al., 2004*
[d]*Varni et al., 2010.*
[e]*Varni et al., 1987.*

In general, tools that capture the child's own perspective should be used when possible. The reliance on observed pain behaviors and parent/caregiver perception has significant drawbacks as these measures will not capture the subjective nature of pain. Furthermore, it may be difficult to interpret typical pain behaviors, such as laughing, a blunted facial expression, sudden stillness (freezing phenomenon), or self-injurious behavior (Hauer et al., 2017). Because pain responses and behaviors in people with NDDs are heterogeneous, it was long thought that they experienced less pain than the general population. Recent research, however, suggests that the opposite may be true (Allely, 2013; Symons, 2011). Atypical responses to pain can make it difficult even for caregivers to determine when individuals with NDDs are experiencing pain, and therefore caregiver perceptions may not be reliable for identifying pain in this population (Nader et al., 2004) (Fig. 33.2).

Pain management

Pain management in individuals with NDDs should follow the same principles as in other chronic pain populations. The pain type and the potential sources of pain should guide a tailored treatment approach (see Table 33.2). It is important to note that individuals with NDDs often have numerous concurrent sources of pain and may therefore experience pain in multiple locations.

Nociceptive pain

Although there are limited reports concerning pain management in NDD populations, a recent systematic review (Ostojic, Paget, & Morrow, 2019) examined various pain management approaches in children and adolescents with CP. When pain is secondary to or associated with hypertonia, common treatments included Botulinum Toxin A injections and intrathecal baclofen. In pain conditions secondary to spastic hip disease, the main treatment was surgery. Other authors (Vogtle, 2009) have reported that in addition to pharmacological treatments (e.g., antispasmodics, antacids, laxatives, and antidepressants), nonpharmacological approaches are sometimes used to relieve pain. Nonpharmacological strategies that may be beneficial include exercise, biofeedback, repositioning and use of supportive pillows, swaddling/cuddling or rocking, warm baths, weighted blankets, and music or audiotherapy (Hauer et al., 2017).

Neuropathic pain

It is suspected that the central nervous system dysfunction inherent to NDDs also makes these individuals more susceptible to neuropathic pain. However, neuropathic pain in NDD populations has not been well documented. Nevertheless, neuropathic pain can occur in this population, and like the general population, it can be associated with e.g., diabetes neuropathy, nerve and brain injuries. When a neuropathic pain condition is suspected in a patient, anticonvulsants and antidepressants in combination with other nonpharmacological approaches may be helpful.

Overall, research regarding pain classification and management is sparse in NDD populations. Because chronic pain affects a significant number of individuals with NDDs, more research on pain evaluation and management is needed in these populations.

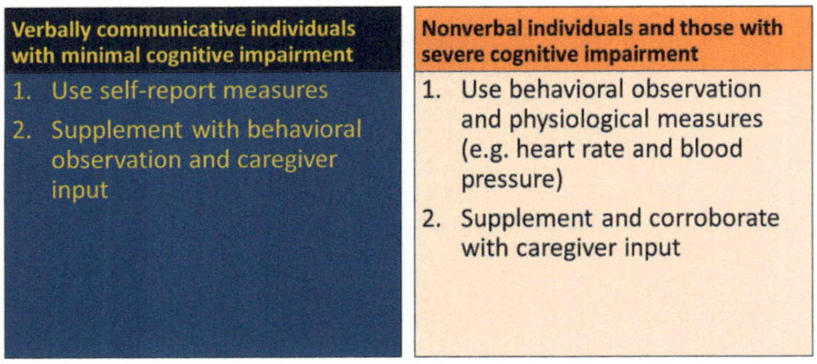

FIGURE 33.2 Recommendations for pain assessment in people with neurodevelopmental disorders.

TABLE 33.2 Potential sources of pain in people with neurodevelopmental disorders.

Neuropathic pain	Nociceptive pain	Nociceptive visceral pain
Nerve compression or ischemia from immobility or sustained postures	Musculoskeletal conditions including scoliosis, hip subluxation, and pathological fractures	Constipation
Widespread pain associated with altered excitability in the central nervous system	Muscle pain from hypertonia, fatigue, or immobility	Gastroesophageal reflux
	Otitis media	Renal stones and urinary tract infections
	Dental caries	Cholecystitis (tube feeding)
	Iatrogenic from medical procedures	Dysmenorrhea

Applications to other areas of development

As discussed earlier, excessive pain exposure in the neonatal period may adversely affect neurodevelopment, and poorly managed pain in children and adults with NDDs may contribute to self-injurious behavior and increased distress. Thus, advancements in pain research may have widespread implications for other areas of research and clinical practice in this population. Specifically, better understanding of the relationships between impairment of the central nervous system and pain sources attributable to the central nervous system in people with NDDs may help to elucidate the underlying mechanisms for both phenomena. Further, advancements in and optimization of pain assessment and management in this population may have beneficial effects on neurobehavioral outcomes and quality of life for people with NDDs.

Key facts

- Pain is one of the three most common reported symptoms by the parents of children with NDDs, and over 40% of children with moderate to profound cognitive impairment have been reported to experience daily pain
- One third of adults with CP have chronic musculoskeletal pain
- Abnormal pain behaviors are common in people with ASDs
- People with ADHD have higher rates of headache, recurrent or severe stomach pain, and widespread pain

Summary points

- People with NDDs have increased risk for acute and chronic pain, and many factors contribute to this increased risk, including increased exposure to painful stimuli in the neonatal period, brain changes susceptive to the development of neuropathic pain, complications of chronic neuromuscular impairment, and self-injurious behavior
- Impaired ability to self-identify negative feelings, impaired ability to communicate suffering, and/or atypical pain behaviors in people with NDDs complicate the assessment and management of pain
- When possible, it is important for clinicians to evaluate pain intensity and pain interference ratings directly from the patient with an NDD, as there may be a lack of concordance with caretaker perceptions
- When self-report is not possible, observational tools such as the Noncommunicating Children's Pain Checklist—Revised, the Pediatric Pain Profile, or the r-FLACC can be used to assess pain interference
- Attempts should be made to identify the origin and type of pain (nociceptive or neuropathic).

References

Allely, C. S. (2013). Pain sensitivity and observer perception of pain in individuals with autistic spectrum disorder. *ScientificWorldJournal, 2013*, 916178. https://doi.org/10.1155/2013/916178.

Amtmann, D., Cook, K. F., Jensen, M. P., Chen, W. H., Choi, S., Revicki, D., … Lai, J. S. (2010). Development of a PROMIS item bank to measure pain interference. *Pain, 150*(1), 173–182. https://doi.org/10.1016/j.pain.2010.04.025.

Association, A. P. (2013). In *Diagnostic and statistical manual of mental disorders* (5th ed.). Arlington, VA: American Psychiatric Publishing.

Bandstra, N. F., Johnson, S. A., Filliter, J. H., & Chambers, C. T. (2012). Self-reported and parent-reported pain for common painful events in high-functioning children and adolescents with autism spectrum disorder. *The Clinical Journal of Pain, 28*(8), 715–721. https://doi.org/10.1097/AJP.0b013e318243ecf6.

Benner, J. L., Noten, S., Limsakul, C., Van Der Slot, W. M. A., Stam, H. J., Selb, M., … Roebroeck, M. E. (2019). Outcomes in adults with cerebral palsy: Systematic review using the international classification of functioning, disability and health. *Developmental Medicine and Child Neurology, 61*(10), 1153–1161. https://doi.org/10.1111/dmcn.14247.

Blankenburg, M., Junker, J., Hirschfeld, G., Michel, E., Aksu, F., Wager, J., & Zernikow, B. (2018). Quantitative sensory testing profiles in children, adolescents and young adults (6-20 years) with cerebral palsy: Hints for a neuropathic genesis of pain syndromes. *European Journal of Paediatric Neurology, 22*(3), 470–481. https://doi.org/10.1016/j.ejpn.2017.12.015.

Breau, L. M., Camfield, C. S., McGrath, P. J., & Finley, G. A. (2003). The incidence of pain in children with severe cognitive impairments. *Archives of Pediatrics and Adolescent Medicine, 157*(12), 1219–1226. https://doi.org/10.1001/archpedi.157.12.1219.

Breau, L. M., McGrath, P. J., Camfield, C. S., & Finley, G. A. (2002). Psychometric properties of the non-communicating children's pain checklist-revised. *Pain, 99*(1–2), 349–357. https://doi.org/10.1016/s0304-3959(02)00179-3.

Burke, N. N., Fan, C. Y., & Trang, T. (2016). Microglia in health and pain: Impact of noxious early life events. *Experimental Physiology, 101*(8), 1003–1021. https://doi.org/10.1113/EP085714.

Carbajal, R., Rousset, A., Danan, C., Coquery, S., Nolent, P., Ducrocq, S., … Breart, G. (2008). Epidemiology and treatment of painful procedures in neonates in intensive care units. *Journal of the American Medical Association, 300*(1), 60–70. https://doi.org/10.1001/jama.300.1.60.

Cascio, C., McGlone, F., Folger, S., Tannan, V., Baranek, G., Pelphrey, K. A., & Essick, G. (2008). Tactile perception in adults with autism: A multidimensional psychophysical study. *Journal of Autism and Developmental Disorders, 38*(1), 127–137. https://doi.org/10.1007/s10803-007-0370-8.

Colver, A., Rapp, M., Eisemann, N., Ehlinger, V., Thyen, U., Dickinson, H. O., … Arnaud, C. (2015). Self-reported quality of life of adolescents with cerebral palsy: A cross-sectional and longitudinal analysis. *Lancet, 385*(9969), 705–716. https://doi.org/10.1016/S0140-6736(14)61229-0.

Cong, X., Wu, J., Vittner, D., Xu, W., Hussain, N., Galvin, S., … Henderson, W. A. (2017). The impact of cumulative pain/stress on neurobehavioral development of preterm infants in the NICU. *Early Human Development, 108*, 9–16. https://doi.org/10.1016/j.earlhumdev.2017.03.003.

Crosta, Q. R., Ward, T. M., Walker, A. J., & Peters, L. M. (2014). A review of pain measures for hospitalized children with cognitive impairment. *Journal for Specialists in Pediatric Nursing, 19*(2), 109–118. https://doi.org/10.1111/jspn.12069.

Derksen, M. T., Vreeling, M. J., & Tchetverikov, I. (2015). High frequency of adult attention deficit hyperactivity disorder among fibromyalgia patients in The Netherlands: Should a systematic collaboration between rheumatologists and psychiatrists be sought? *Clinical and Experimental Rheumatology, 33*(1 Suppl. 88), S141.

Di Tella, M., Ghiggia, A., Tesio, V., Romeo, A., Colonna, F., Fusaro, E., … Castelli, L. (2017). Pain experience in Fibromyalgia Syndrome: The role of alexithymia and psychological distress. *Journal of Affective Disorders, 208*, 87–93. https://doi.org/10.1016/j.jad.2016.08.080.

Duerden, E. G., Grunau, R. E., Guo, T., Foong, J., Pearson, A., Au-Young, S., … Miller, S. P. (2018). Early procedural pain is associated with regionally-specific alterations in thalamic development in preterm neonates. *Journal of Neuroscience, 38*(4), 878–886. https://doi.org/10.1523/JNEUROSCI.0867-17.2017.

Fan, Y. T., Chen, C., Chen, S. C., Decety, J., & Cheng, Y. (2014). Empathic arousal and social understanding in individuals with autism: Evidence from fMRI and ERP measurements. *Social Cognitive and Affective Neuroscience, 9*(8), 1203–1213. https://doi.org/10.1093/scan/nst101.

Fietz, J., Valencia, N., & Silani, G. (2018). Alexithymia and autistic traits as possible predictors for traits related to depression, anxiety, and stress: A multivariate statistical approach. *Journal of Evaluation in Clinical Practice, 24*(4), 901–908. https://doi.org/10.1111/jep.12961.

Gillberg, C., Terenius, L., & Lonnerholm, G. (1985). Endorphin activity in childhood psychosis. Spinal fluid levels in 24 cases. *Archives of General Psychiatry, 42*(8), 780–783. https://doi.org/10.1001/archpsyc.1985.01790310042005.

Griffin, C., Lombardo, M. V., & Auyeung, B. (2016). Alexithymia in children with and without autism spectrum disorders. *Autism Research, 9*(7), 773–780. https://doi.org/10.1002/aur.1569.

Hauer, J., Houtrow, A. J., Section On, H., & Palliative Medicine, C. O. C. W. D. (2017). Pain assessment and treatment in children with significant impairment of the central nervous system. *Pediatrics, 139*(6). https://doi.org/10.1542/peds.2017-1002.

Holmberg, K., & Hjern, A. (2006). Health complaints in children with attention-deficit/hyperactivity disorder. *Acta Paediatrica, 95*(6), 664–670. https://doi.org/10.1080/08035250600717121.

Hunt, A., Goldman, A., Seers, K., Crichton, N., Mastroyannopoulou, K., Moffat, V., … Brady, M. (2004). Clinical validation of the paediatric pain profile. *Developmental Medicine and Child Neurology, 46*(1), 9–18. https://doi.org/10.1017/s0012162204000039.

Jahnsen, R., Villien, L., Aamodt, G., Stanghelle, J. K., & Holm, I. (2004). Musculoskeletal pain in adults with cerebral palsy compared with the general population. *Journal of Rehabilitation Medicine, 36*(2), 78–84.

Jameson, N. D., Sheppard, B. K., Lateef, T. M., Vande Voort, J. L., He, J. P., & Merikangas, K. R. (2016). Medical comorbidity of attention-deficit/hyperactivity disorder in US adolescents. *Journal of Child Neurology, 31*(11), 1282–1289. https://doi.org/10.1177/0883073816653782.

Kessler, R. C., Lane, M., Stang, P. E., & Van Brunt, D. L. (2009). The prevalence and workplace costs of adult attention deficit hyperactivity disorder in a large manufacturing firm. *Psychological Medicine, 39*(1), 137–147. https://doi.org/10.1017/S0033291708003309.

Kingsnorth, S., Orava, T., Provvidenza, C., Adler, E., Ami, N., Gresley-Jones, T., … Fehlings, D. (2015). Chronic pain assessment tools for cerebral palsy: A systematic review. *Pediatrics, 136*(4), e947–960. https://doi.org/10.1542/peds.2015-0273.

Klintwall, L., Holm, A., Eriksson, M., Carlsson, L. H., Olsson, M. B., Hedvall, A., … Fernell, E. (2011). Sensory abnormalities in autism. A brief report. *Research in Developmental Disabilities, 32*(2), 795–800. https://doi.org/10.1016/j.ridd.2010.10.021.

Kutuk, M. O., Tufan, A. E., Guler, G., Yalin, O. O., Altintas, E., Bag, H. G., ... Ozge, A. (2018). Migraine and associated comorbidities are three times more frequent in children with ADHD and their mothers. *Brain and Development, 40*(10), 857–864. https://doi.org/10.1016/j.braindev.2018.06.001

Latremoliere, A., & Woolf, C. J. (2009). Central sensitization: A generator of pain hypersensitivity by central neural plasticity. *The Journal of Pain, 10*(9), 895–926. https://doi.org/10.1016/j.jpain.2009.06.012.

Lensing, M. B., Zeiner, P., Sandvik, L., & Opjordsmoen, S. (2015). Quality of life in adults aged 50+ with ADHD. *Journal of Attention Disorders, 19*(5), 405–413. https://doi.org/10.1177/1087054713480035.

Low, L. A., & Schweinhardt, P. (2012). Early life adversity as a risk factor for fibromyalgia in later life. *Pain Research and Treatment, 2012*, 140832. https://doi.org/10.1155/2012/140832

McKearnan, K. A., Kieckhefer, G. M., Engel, J. M., Jensen, M. P., & Labyak, S. (2004). Pain in children with cerebral palsy: A review. *Journal of Neuroscience Nursing, 36*(5), 252–259. https://doi.org/10.1097/01376517-200410000-00004.

McKinnon, C. T., Meehan, E. M., Harvey, A. R., Antolovich, G. C., & Morgan, P. E. (2019). Prevalence and characteristics of pain in children and young adults with cerebral palsy: A systematic review. *Developmental Medicine and Child Neurology, 61*(3), 305–314. https://doi.org/10.1111/dmcn.14111

Militerni, R., Bravaccio, C., Falco, C., Puglisi-Allegra, S., Pascucci, T., & Fico, C. (2000). Pain reactivity in children with autistic disorder. *The Journal of Headche and Pain, 1*, 53–56.

Moore, D. J. (2015). Acute pain experience in individuals with autism spectrum disorders: A review. *Autism, 19*(4), 387–399. https://doi.org/10.1177/1362361314527839.

Mwaniki, M. K., Atieno, M., Lawn, J. E., & Newton, C. R. (2012). Long-term neurodevelopmental outcomes after intrauterine and neonatal insults: A systematic review. *Lancet, 379*(9814), 445–452. https://doi.org/10.1016/S0140-6736(11)61577-8.

Nader, R., Oberlander, T. F., Chambers, C. T., & Craig, K. D. (2004). Expression of pain in children with autism. *The Clinical Journal of Pain, 20*(2), 88–97.

Nigg, J. T. (2013). Attention-deficit/hyperactivity disorder and adverse health outcomes. *Clinical Psychology Review, 33*(2), 215–228. https://doi.org/10.1016/j.cpr.2012.11.005.

Ostojic, K., Paget, S. P., & Morrow, A. M. (2019). Management of pain in children and adolescents with cerebral palsy: A systematic review. *Developmental Medicine and Child Neurology, 61*(3), 315–321. https://doi.org/10.1111/dmcn.14088.

Palermo, T. M., Witherspoon, D., Valenzuela, D., & Drotar, D. D. (2004). Development and validation of the child activity limitations Interview: A measure of pain-related functional impairment in school-age children and adolescents. *Pain, 109*(3), 461–470. https://doi.org/10.1016/j.pain.2004.02.023.

Parkes, J., White-Koning, M., Dickinson, H. O., Thyen, U., Arnaud, C., Beckung, E., ... Colver, A. (2008). Psychological problems in children with cerebral palsy: A cross-sectional European study. *Journal of Child Psychology and Psychiatry, 49*(4), 405–413. https://doi.org/10.1111/j.1469-7610.2007.01845.x.

van Rensburg, R., Meyer, H. P., Hitchcock, S. A., & Schuler, C. E. (2018). Screening for adult ADHD in patients with fibromyalgia syndrome. *Pain Medicine, 19*(9), 1825–1831. https://doi.org/10.1093/pm/pnx275.

Reyero, F., Ponce, G., Rodriguez-Jimenez, R., Fernandez-Dapica, P., Taboada, D., Martin, V., ... Palomo, T. (2011). High frequency of childhood ADHD history in women with fibromyalgia. *European Psychiatry, 26*(8), 482–483. https://doi.org/10.1016/j.eurpsy.2010.03.012.

Rolke, R., Baron, R., Maier, C., Tolle, T. R., Treede, R. D., Beyer, A., ... Wasserka, B. (2006). Quantitative sensory testing in the German research network on neuropathic pain (DFNS): Standardized protocol and reference values. *Pain, 123*(3), 231–243. https://doi.org/10.1016/j.pain.2006.01.041.

Scherder, E. J., Rommelse, N. N., Broring, T., Faraone, S. V., & Sergeant, J. A. (2008). Somatosensory functioning and experienced pain in ADHD-families: A pilot study. *European Journal of Paediatric Neurology, 12*(6), 461–469. https://doi.org/10.1016/j.ejpn.2007.11.004.

Solodiuk, J. C., Brighton, H., McHale, J., LoChiatto, J., Logan, D. E., Sager, S., ... Berde, C. B. (2014). Documented electronic medical record-based pain intensity scores at a tertiary pediatric medical center: A cohort analysis. *Journal of Pain and Symptom Management, 48*(5), 924–933. https://doi.org/10.1016/j.jpainsymman.2014.01.006.

Steele, R., Siden, H., Cadell, S., Davies, B., Andrews, G., Feichtinger, L., & Singh, M. (2014). Charting the territory: Symptoms and functional assessment in children with progressive, non-curable conditions. *Archives of Disease in Childhood, 99*(8), 754–762. https://doi.org/10.1136/archdischild-2013-305246.

Stickley, A., Koyanagi, A., Takahashi, H., & Kamio, Y. (2016). ADHD symptoms and pain among adults in England. *Psychiatry Research, 246*, 326–331. https://doi.org/10.1016/j.psychres.2016.10.004.

Stray, L. L., Kristensen, O., Lomeland, M., Skorstad, M., Stray, T., & Tonnessen, F. E. (2013). Motor regulation problems and pain in adults diagnosed with ADHD. *Behavioral and Brain Functions, 9*, 18. https://doi.org/10.1186/1744-9081-9-18.

Symons, F. J. (2011). Self-injurious behavior in neurodevelopmental disorders: Relevance of nociceptive and immune mechanisms. *Neuroscience and Biobehavioral Reviews, 35*(5), 1266–1274. https://doi.org/10.1016/j.neubiorev.2011.01.002.

Symons, F. J., Harper, V. N., McGrath, P. J., Breau, L. M., & Bodfish, J. W. (2009). Evidence of increased non-verbal behavioral signs of pain in adults with neurodevelopmental disorders and chronic self-injury. *Research in Developmental Disabilities, 30*(3), 521–528. https://doi.org/10.1016/j.ridd.2008.07.012.

Tordjman, S., Anderson, G. M., Botbol, M., Brailly-Tabard, S., Perez-Diaz, F., Graignic, R., ... Bronsard, G. (2009). Pain reactivity and plasma beta-endorphin in children and adolescents with autistic disorder. *PLoS One, 4*(8), e5289. https://doi.org/10.1371/journal.pone.0005289.

Treister, R., Eisenberg, E., Demeter, N., & Pud, D. (2015). Alterations in pain response are partially reversed by methylphenidate (Ritalin) in adults with attention deficit hyperactivity disorder (ADHD). *Pain Practice, 15*(1), 4–11. https://doi.org/10.1111/papr.12129.

Tsze, D. S., von Baeyer, C. L., Bulloch, B., & Dayan, P. S. (2013). Validation of self-report pain scales in children. *Pediatrics, 132*(4), e971–979. https://doi.org/10.1542/peds.2013-1509.

Varni, J. W., Stucky, B. D., Thissen, D., Dewitt, E. M., Irwin, D. E., Lai, J. S., … Dewalt, D. A. (2010). PROMIS pediatric pain interference scale: An item response theory analysis of the pediatric pain item bank. *The Journal of Pain, 11*(11), 1109–1119. https://doi.org/10.1016/j.jpain.2010.02.005.

Varni, J. W., Thompson, K. L., & Hanson, V. (1987). The Varni/Thompson pediatric pain questionnaire. I. Chronic musculoskeletal pain in juvenile rheumatoid arthritis. *Pain, 28*(1), 27–38. https://doi.org/10.1016/0304-3959(87)91056-6.

Vogtle, L. K. (2009). Pain in adults with cerebral palsy: Impact and solutions. *Developmental Medicine and Child Neurology, 51*(Suppl. 4), 113–121. https://doi.org/10.1111/j.1469-8749.2009.03423.x.

Von Baeyer, C. L. (2006). Children's self-reports of pain intensity: Scale selection, limitations and interpretation. *Pain Research and Management, 11*(3), 157–162.

Warlow, T. A., & Hain, R. D. W. (2018). Total pain' in children with severe neurological impairment. *Children, 5*(1). https://doi.org/10.3390/children5010013.

Chapter 34

Use of levetiracetam in neonates: features and applications (a narrative review)

Shannon Pollock and Louis N. Manganas
Epilepsy Fellow, Department of Neurology, Stony Brook University Hospital, Stony Brook, NY, United States

List of abbreviations

ADHD Attention deficit hyperactivity disorder
AED Antiepileptic drug
ASD Autism spectrum disorder
Ca^{2+} Calcium
Cl^- Chloride
EKG Electrocardiogram
FDA U.S. Food and Drug Administration
GABA Gamma-aminobutyric acid
GFR Glomerular filtration rate
K^+ Potassium
LEV Levetiracetam
PHB Phenobarbital
SV2A Synaptic vesicle protein 2A
VD Volume of distribution

Mini-dictionary of terms

Seizure Abnormal and uncontrollable electrical activity in the brain
Anti-epileptic drugs (AEDs) Drugs used to stop seizures
Efficacy The ability of a drug to produce a beneficial change
Side effects Negative effects produced by a drug
Clinical trial Volunteers to test new treatments to prevent disease

Introduction

Seizures are common in the neonatal period. Although several epilepsy syndromes have been recognized during this period, seizures are most often a result of hypoxia, brain bleed, metabolic derangement, or structural abnormality (Ahrens, Ream, & Slaughter, 2019). Phenobarbital (PHB) remains the standard of care for the medical treatment of newborn seizures. Although alternatives are highly desired as a result of PHB side effects, little progress has been made due the challenges in the design of clinical trials in this patient population. Levetiracetam (LEV) is emerging as a commonly used and potentially safer antiepileptic drug (AED) to PHB as monotherapy, in the treatment of neonatal seizures (Mchugh, Lancaster, & Manganas, 2017). In addition to the lack of high-quality safety and efficacy studies in this population, there are several factors that must also be reviewed when considering seizure management in neonates (Agrawal &

Diagnosis, Management and Modeling of Neurodevelopmental Disorders. https://doi.org/10.1016/B978-0-12-817988-8.00034-8

Banergee, 2017). In this review we will discuss the pharmacologic and pharmacokinetic advantages of LEV as well as favorable side effect profile and efficacy when compared to other AEDs. Furthermore, the benefits of LEV will be discussed under specific contexts.

Pharmacology of LEV

The exact mechanism of LEV is not understood; however, it is known to mediate its effect through the synaptic vesicle protein 2A (SV2A). This protein is specific to synaptic vesicles, which are present throughout the brain and regulate exocytosis (Agrawal & Banergee, 2017). LEV receptors begin to develop as early as 26 weeks' gestation and are present by 37 weeks' gestation (Agrawal & Banergee, 2017). Given this, LEV avoids some of the challenges AEDs with traditional mechanisms must overcome in the neonate brain, such as the relative overexpression of glutamate receptors, immaturity of gamma-aminobutyric acid (GABA) receptors, and inversion of chloride (Cl^-) gradient (Agrawal & Banergee, 2017). Some data also suggests that LEV may exert its effect through potassium (K^+) channels, calcium (Ca^{2+}) channels, and GABA-A.

Pharmacokinetics of LEV and considerations in neonates

LEV was initially approved by the U.S. Food and Drug Administration (FDA) for use as adjunctive treatment for partial seizures in adults. Later it was approved in adult and pediatric populations and has become a more favorable choice due to its relative lack of significant side effects, drug–drug interactions, and availability in both liquid and tablet formulations (Agrawal & Banergee, 2017). Despite its popularity and increased use in neonates, few studies have been conducted to determine safety and efficacy. Trials are hindered by the inherent difficulties in studying this patient population such as: LEV use as add-on treatment, length of follow-up required assessing neurodevelopmental side effects, and comorbid neurologic conditions that may also complicate neurodevelopmental outcomes.

LEV is rapidly absorbed in neonates as well as older infants and adults. LEV displays linear pharmacokinetics with little drug–drug interactions, given minimal protein binding and nonreliance on hepatic pathways (Agrawal & Banergee, 2017). As a result, plasma levels of LEV tend to remain unchanged if other AEDs are added (Agrawal & Banergee, 2017).

In general, drug doses in neonates are usually similar to those in older children (on an mg/kg basis); however, there are two neonate-specific factors that need to be considered when administering LEV: volume of distribution (VD) and renal function. Due to the water-soluble nature of LEV and the higher VD in neonates, a higher loading dose may be needed (Agrawal & Banergee, 2017). Approximately two-thirds of the dose is excreted unchanged via urine, and the remaining third is hydrolyzed in the blood and other tissues into three inactive metabolites, which in turn are also renally excreted (Agrawal & Banergee, 2017). Although elimination half-life is shorter, and excretion shown to be up to 40% higher in neonates and children compared to adults (Agrawal & Banergee, 2017), the need for more frequent dosing may not similarly apply to neonates. The little data that exists is confounded by the inclusion of both term and preterm neonates in some studies. Nevertheless, decreased clearance occurs in neonates compared to older infants and children due to immature renal function: decreased glomerular filtration rate (GFR) and decreased esterase activity (Agrawal & Banergee, 2017). As a result, lower plasma clearance and longer half-life of LEV are expected (Agrawal & Banergee, 2017).

To summarize, a higher loading dose (given higher VD) and more frequent maintenance LEV may be needed for the first few weeks of life. However, as serum reference ranges for LEV are not well established, therapeutic efficacy may be difficult to quantify in this respect (Agrawal & Banergee, 2017).

Acute and chronic effects of traditional AEDs and LEV on the immature brain

Exposure to AEDs in utero and during the neonatal period can have detrimental implications for long-term cognitive and behavioral development (Meador & Loring, 2015). The neonatal period is a critical time during brain development and the specific effects of medication exposures are dependent on a variety of factors, including the time of exposure during gestation, drug type, dosage, and duration of drug use. For example, neural tube defects linked to valproic acid are most associated with first trimester exposure, while cognitive behavioral effects of other AEDs are more linked to third trimester exposure (Kellogg & Meador, 2017). Multiple mechanisms for neurodevelopment effects have been described including folate deficiency, reactive species, ischemia, synaptic changes, and neuronal death or apoptosis, which is a major factor in cognitive and behavioral long-term effects (Kellogg & Meador, 2017; Kim, Kondratyev, & Gale, 2007).

Several AEDs cause neuronal cell death including: PHB, phenytoin, valproic acid (Kim et al., 2007) other benzodiazepines, vigabatrin, carbamazepine, and topiramate especially in combination (Kim et al., 2007). Rat pup brain development in the first 2 weeks of life correlates with the human third trimester (Kim et al., 2007). Studies of rat pup brains during this critical time have shown apoptosis to occur when therapeutic doses of traditional AEDs described earlier are administered (Kim et al., 2007). In contrast, LEV showed no increase in cell death at supratherapeutic doses (250−1000 mg/kg) as monotherapy. Polytherapy use of LEV was also studied in combination with phenytoin. These studies showed no worsening of phenytoin-induced cell death, which was an effect that was seen with phenytoin/carbamazepine polytherapy combination. In contrast, the LEV/carbamazepine combination at LEV doses of 250 mg/kg showed no increase in neuronal cell death. Interestingly, thalamic cell death was seen with LEV (500 mg/kg) but not in combination with phenytoin, suggesting that AED mechanisms are important determinants of polytherapy effect. Additional experiments in rat brains have studied the effect of LEV on other mechanisms of neuronal damage: for example, Trollmann et al. (2008) described a particular hypoxia-induced neuroprotective mechanism regulated by protein HIF-1 (hypoxia-inducible transcription factor) and found that LEV did not interfere with this pathway.

Neuronal death and adverse effects of early exposure to LEV appear to be less severe when compared to traditional AEDs (Kellogg & Meador, 2017; Kim et al., 2007). Alternatively, some studies have described the neuroprotective effects of LEV use (Shetty, 2013; Zou et al., 2013). Pretreatment of animal brains (suffering traumatic brain injury, stroke, or seizures) with LEV did not show a difference in severity or number of seizures, but did demonstrate less neuronal injury to the hippocampus and other limbic structures compared to those animals that were not treated (Shetty, 2013). In another study, pretreatment with LEV prior to induced hypoxic-ischemic injury prevented seizures (compared to no pretreatment) and those animals who received treatment were less likely to develop epilepsy in the future (Shetty, 2013). Depending on the model used in each study, mechanisms proposed for LEV include: effect on epileptogenesis, modulation of Ca^{2+} currents, and effect on antioxidants (Shetty, 2013; Zou et al., 2013). Traumatic brain injury (TBI) animal models have suggested beneficial effects on neurologic outcomes with LEV treatment possibly related to glutamate transporter effects (Zou et al., 2013). Despite these positive studies, controversy remain as other studies, such as a model of amygdala kindling, showed no neuroprotective or antiepileptogenic effects (Shetty, 2013).

In humans, prenatal exposure to AEDs has been linked to congenital malformations, risk of cognitive impairment, and raised concerns for autism spectrum disorder (ASD) and attention deficit hyperactivity disorder (ADHD) (Meador & Loring, 2015). Specifically, exposure of a fetus to valproic acid has the highest association with both major congenital malformations and risk of cognitive impairment (Meador & Loring, 2015). Multiple other AEDs including PHB, topiramate, carbamazepine, and phenytoin have also been linked to various neurodevelopmental problems including neural tube defects, congenital malformations, and cognitive impairment (Meador & Loring, 2015). Specific risks vary depending on which agent the fetus is exposed to, as well as exposure to monotherapy versus polytherapy, and timing of exposure (i.e., first vs. third trimester) (Meador & Loring, 2015). In general, clear high-quality data regarding use in pregnancy is lacking for many AEDs (Meador & Loring, 2015). Although lamotrigine and LEV are considered lowest-risk in pregnancy, in animal studies LEV was associated with increased risk of skeletal deformities, although association with the more traditional malformations such as neural tube defects, cleft lip/palate, and cardiac or renal abnormalities such as those seen with PHB has not been seen (Meador & Loring, 2015). Postnatally, retrospective comparisons of neurodevelopmental outcomes in patients treated with PHB and LEV have led to the conclusion that LEV is significantly less detrimental to neurodevelopment than PHB, specifically in cognitive and motor domains—those domains in which the effects of PHB are most prominent (Maitre, Smolinsky, Slaughter, & Stark, 2013). In addition, the cumulative dose of AED appears to play a role in the degree of later associated impairments (Maitre et al., 2013).

Overall, several lines of evidence demonstrate the use of LEV in treating seizures, lack of LEV-induced neuronal damage, and possible, though controversial, evidence for neuronal protection and reduction in epileptogenesis (Agrawal & Banergee, 2017; Kahle & Staley, 2012; Kellogg & Meador, 2017; Kim et al., 2007; Maitre et al., 2013; Shetty, 2013; Trollmann et al., 2008; Zou et al., 2013). Still, the overall effects of early exposure to LEV on neurodevelopment and epileptogenesis, while clearly felt to be improved compared to traditional AEDs such as PHB (Kellogg & Meador, 2017; Kim et al., 2007; Maitre et al., 2013), require further study in humans.

Safety and efficacy of LEV in neonates

Thus far, data on safety and efficacy of LEV for neonatal seizures is derived from retrospective studies typically in small cohorts (Mchugh et al., 2017). Outcome measures are variable, including the definition of efficacy as is variation of EEG monitoring (Mchugh et al., 2017). Additionally, complicating interpretation is the use of other AEDs along with LEV contributing to difficulty in efficacy measures (Mchugh et al., 2017).

Efficacy of LEV (with complete or near-complete seizure cessation) across studies included in one review and meta-analysis (Mchugh et al., 2017) ranged from 25% to 100% within studies or 77% across all studies. For comparison, in studies evaluating PHB, efficacy varied between 33% and 77%. Side effects of LEV reported include drowsiness, irritability, and, in one study, transient conjugated hyperbilirubinemia (Mchugh et al., 2017) see Table 34.1. Aside from control of intermittent seizures, LEV in older children and adults is commonly used as a second-line therapy in treatment of status epilepticus. By extension, many practitioners have begun using LEV for treatment of status in the neonatal population, as well (Ahrens et al., 2019). However, there are no large studies and in this clinical scenario, LEV is most often used as a second-line agent after failure of PHB. Thus, LEV's efficacy for treatment of status epilepticus in neonates remains unclear (Ahrens et al., 2019).

TABLE 34.1 Results of individual studies.

	Primary levetiracetam			
Source	**Complete/near-complete cessation**	**≥50% reduction**	**Minimal/no improvement or unable to judge**	**Total sample size**
Abend, Gutierrez-Colina, Monk, Dlugos, & Clancy (2011)	1 (0.25)	0 (0.0)	3 (0.75)	4
Ramantani, Ikonomidou, Walter, Rating, & Dinger (2011)	30 (0.79)	0 (0.0)	8 (0.21)	38
Khan et al. (2011)	3 (1.0)	0 (0.0)	0 (0.0)	3
Khan, Cipriani, Wright, Crisp, & Kirmani (2C13)	3 (1.0)	0 (0.0)	0 (0.0)	3
Rakshasbhuvankar, Rao, Kohan, Simmer, & Nagarajan (2013)	N/A	N/A	N/A	N/A
Total	37 (0.77)	0 (0.0)	11 (0.23)	48
Source	**Secondary levetiracetam**			
	Complete/near-complete cessation	**≥50% reduction**	**Minimal/no improvement or unable to judge**	**Total sample size**
Abend, Gutierrez-Colina, Monk, Dlugos, & Clancy (2011)	6 (0.32)	5 (0.26)	8 (0.42)	19
Ramantani, Ikonomidou, Walter, Rating, & Dinger (2011)	N/A	N/A	N/A	N/A
Khan et al. (2011)	16 (0.84)	0 (0.0)	3 (0.16)	19
Khan, Cipriani, Wright, Crisp, & Kirmani (2013)	6 (0.75)	0 (0.0)	2 (0.25)	8
Rakshasbhuvankar, Rao, Kohan, Simmer, & Nagarajan (2013)	6 (0.75)	1 (0.13)	1 (0.13)	8
Total	34 (0.63)	6 (0.11)	14 (0.26)	54
Source	**Primary phenobarbital**			
	Complete/near-complete cessation	**≥50% reduction**	**Minimal/no improvement or unable to judge**	**Total sample size**
Painter et al. (1999)	13 (0.43)	N/A	17 (0.57)	30
Boylan et al. (2004)	11 (0.50)	N/A	11 (0.50)	22
Total	24 (0.46)	N/A	28 (54)	52
Source	**Secondary phenobarbital**			
	Complete/near-complete cessation	**≥50% reduction**	**Minimal/no improvement or unable to judge**	**Total sample size**
Painter et al. (1999)	5 (0.38)	N/A	8 (0.62)	13
Boylan et al. (2004)	N/A	N/A	N/A	N/A
Total	5 (0.38)	N/A	8 (0.62)	13

Gestational age is another consideration when evaluating efficacy. Preterm infants are less likely to experience status epilepticus than term infants, due to immature neuronal networks; instead they experience recurrent seizures (Ahrens et al., 2019). As stated, the SV2A protein is present as early as 26 weeks' gestation (Agrawal & Banergee, 2017). Still, some ability to control seizures in the preterm population has been described. In preterm infants (26–36 weeks' gestational age), one study found that LEV given as first-line therapy led to seizure cessation in 57% of patients with no recurrence for a minimum of 48 h. PHB was added in 43%, and the remainder required a third-line drug for seizure control (Han, Moon, Youn, Sung, & Lee, 2018). Clinical symptoms, laboratory, vital signs, and electrocardiogram (EKG) were monitored and no adverse drug effects were reported (Han et al., 2018).

Although information regarding safety and efficacy of LEV is the most clinically desired aspect for its use in neonates, information regarding both remains limited. Prospective trials of LEV monotherapy (Arican, Gencpinar, Cavusoglu, & Dundar, 2018) are underway (Levetiracetam Treatment of Neonatal Seizures Safety and Efficacy Phase II Study (LEVNEONAT-1)) but are not yet published at the time of this writing (Favrais et al., 2019).

Recently, data assessing short-term neurodevelopmental outcomes in neonates treated with levetiracetam compared to those treated with phenobarbital showed improved outcomes for those treated with levetiracetam 1 month after treatment (Falsaperla et al., 2019).

Application to other areas of development

In this chapter we have reviewed the pharmacokinetics, efficacy, and safety of LEV. Relevant aspects of other AEDs have been included for context. As the neonatal period is a unique and rapidly changing time of neurodevelopment, the effects of brain injury and uncontrolled seizures have repercussions, which can be felt throughout the lifetime. The desire for an antiseizure medication that avoids the adverse developmental effects attributed to traditional AEDs has led to great interest in LEV for treatment of neonatal seizures; although efficacy is hinted at in small heterogeneous studies, clear superiority, or even equality, compared to PHB has yet to be demonstrated. Additionally, LEV has promising properties with regard to neuroprotection in the setting of acute injury (Kim et al., 2007; Shetty, 2013; Trollmann et al., 2008), which has downstream favorable implications for neurodevelopment. Even prior to birth, children exposed to LEV in utero have been shown to have more favorable neurodevelopmental outcomes (Kellogg & Meador, 2017). Although behavioral side effects are more common in children than in adults (Arican et al., 2018), LEV in general has emerged as a safe choice for treatment of seizures across a broad age range, with minimal adverse effects on neurodevelopment (Kellogg & Meador, 2017; Maitre et al., 2013).

Key facts of neonatal seizures

- A neonate is an infant less than 30 days old.
- Although any age group can experience seizures, the underlying etiology of the seizure varies by age. In neonates, it is most commonly due to an injury to the brain (stroke, bleed, infection).
- A seizure is a burst of abnormal electrical activity in the brain. Symptoms of a seizure vary depending on what part of the brain is involved: the larger the area of brain involved, the more pronounced the clinical seizure will be.
- Myelin is the insulation around axons that allows nerves to transmit signals rapidly. Because neonates' brains are not fully myelinated, they are more likely to have focal seizures that only affect a small part of the brain.
- Seizures in neonates may include facial twitching, extremity shaking, or more subtle eye movements, sucking motions, "bicycling" of the legs, or apnea.

Summary points

- Seizures are a common neurologic condition in neonates and are most commonly symptomatic of an underlying neuronal injury or structural abnormality (Ahrens et al., 2019).
- LEV affects a synaptic protein SV2A, which is present as early as 26 weeks' gestation and at near-adult concentration by 37 weeks' gestation (Agrawal & Banergee, 2017).
- LEV has linear pharmacokinetics and no clinically relevant drug–drug interactions (Agrawal & Banergee, 2017).
- Dosing considerations in neonates include higher total body water content leading to higher VD and immature renal function (Maitre et al., 2013).
- Higher loading dose of LEV may be needed due to higher VD, and dose adjustment may be needed especially in the first week of life due to immature and developing renal function (Maitre et al., 2013).

- Neurotoxicity at supratherapeutic doses of LEV (250−1000 mg/kg) was not demonstrated with monotherapy, and increased cell death was also not seen with polytherapy in combination with phenytoin (Kim et al., 2007).
- Human studies, while typically small and heterogeneous in population, seizure type, and etiology, report few adverse side effects of LEV, most commonly drowsiness or irritability (Mchugh et al., 2017).
- Efficacy data is limited and ranges from 25% to 100% control or near-complete control of seizures (Mchugh et al., 2017).
- Clinical trials focused on efficacy of LEV as monotherapy or in comparison to PHB are ongoing, but additional data will likely still be needed to clearly establish the role of LEV in treatment of neonatal seizures (Ahrens et al., 2019; Mchugh et al., 2017).

References

Abend, N. S., Gutierrez-Colina, A. M., Monk, H. M., Dlugos, D. J., & Clancy, R. R. (2011). Levetiracetam for treatment of neonatal seizures. *Journal of Child Neurology, 26*(04), 465−470.

Agrawal, A., & Banergee, A. (2017). A review on pharmacokinetics of levetiracetam in neonates. *Current Drug Metabolism, 18*(8). https://doi.org/10.2174/1389200218666170607100054.

Ahrens, S., Ream, M. A., & Slaughter, L. A. (2019). Status epilepticus in the neonate: Updates in treatment strategies. *Current Treatment Options in Neurology, 21*(2). https://doi.org/10.1007/s11940-019-0546-5.

Arican, P., Gencpinar, P., Cavusoglu, D., & Dundar, N. O. (2018). Levetiracetam monotherapy for the treatment of infants with epilepsy. *Seizure, 56*, 73−77. https://doi.org/10.1016/j.seizure.2018.02.006.

Boylan, G. B., Rennie, J. M., Chorley, G., Pressler, R. M., Fox, G. F., Farrer, K., … Binnie, C. D. (2004). Second-line anticonvulsant treatment of neonatal seizures: a video-EEG monitoring study. *Neurology, 62*(3), 486−488.

Falsaperla, R., Mauceri, L., Pavone, P., Barbagallo, M., Vitaliti, G., Ruggieri, M., … Corsello, G. (2019). Short-term neurodevelopmental outcome in term neonates treated with phenobarbital versus levetiracetam: A single-center experience. *Behavioural Neurology, 2019*, 3683548. https://doi.org/10.1155/2019/3683548.

Favrais, G., Ursino, M., Mouchel, C., Boivin, E., Jullien, V., Zohar, S., & Saliba, E. (2019). Levetiracetam optimal dose-finding as first-line treatment for neonatal seizures occurring in the context of hypoxic-ischaemic encephalopathy (LEVNEONAT-1): Study protocol of a phase II trial. *BMJ Open, 9*(1). https://doi.org/10.1136/bmjopen-2018-022739.

Han, J. Y., Moon, C. J., Youn, Y. A., Sung, I. K., & Lee, I. G. (2018). Efficacy of levetiracetam for neonatal seizures in preterm infants. *BMC Pediatrics, 18*(1). https://doi.org/10.1186/s12887-018-1103-1.

Kahle, K. T., & Staley, K. J. (2012). Neonatal seizures and neuronal transmembrane ion transport. *Jaspers Basic Mechanisms of the Epilepsies*, 1066−1076. https://doi.org/10.1093/med/9780199746545.003.0082.

Kellogg, M., & Meador, K. J. (2017). Neurodevelopmental effects of antiepileptic drugs. *Neurochemical Research, 42*(7), 2065−2070. https://doi.org/10.1007/s11064-017-2262-4.

Khan, O., Chang, E., Cipriani, C., Wright, C., Crisp, E., & Kirmani, B. (2011). Use of intravenous levetiracetam for management of acute seizures in neonates. *Pediatric Neurology, 44*(04), 265−269.

Khan, O., Cipriani, C., Wright, C., Crisp, E., & Kirmani, B. (2013). Role of intravenous levetiracetam for acute seizure management in preterm neonates. *Pediatric Neurology, 49*(05), 340−343.

Kim, J., Kondratyev, A., & Gale, K. (2007). Antiepileptic drug-induced neuronal cell death in the immature brain: Effects of carbamazepine, topiramate, and levetiracetam as monotherapy versus polytherapy. *Journal of Pharmacology and Experimental Therapeutics, 323*(1), 165−173. https://doi.org/10.1124/jpet.107.126250.

Maitre, N. L., Smolinsky, C., Slaughter, J. C., & Stark, A. R. (2013). Adverse neurodevelopmental outcomes after exposure to phenobarbital and levetiracetam for the treatment of neonatal seizures. *Journal of Perinatology, 33*(11), 841−846. https://doi.org/10.1038/jp.2013.116.

Mchugh, D., Lancaster, S., & Manganas, L. (2017). A systematic review of the efficacy of levetiracetam in neonatal seizures. *Neuropediatrics, 49*(01), 012−017. https://doi.org/10.1055/s-0037-1608653.

Meador, K. J., & Loring, D. W. (2015). Developmental effects of antiepileptic drugs and the need for improved regulations. *Neurology, 86*(3), 297−306. https://doi.org/10.1212/wnl.0000000000002119.

Painter, M. J., Scher, M. S., Stein, A. D., Armatti, S., Wang, Z., Gardiner, J. C., … Alvin, J. (1999). Phenobarbital compared with phenytoin for the treatment of neonatal seizures. *New England Journal of Medicine, 341*(7), 485−489.

Rakshasbhuvankar, A., Rao, S., Kohan, R., Simmer, K., & Nagarajan, L. (2013). Intravenous levetiracetam for treatment of neonatal seizures. *Journal of Clinical Neuroscience, 20*(08), 1165−1167.

Ramantani, G., Ikonomidou, C., Walter, B., Rating, D., & Dinger, J. (2011). Levetiracetam: safety and efficacy in neonatal seizures. *European Journal of Paediatric Neurology, 15*(01), 1−7.

Shetty, A. K. (2013). Prospects of levetiracetam as a neuroprotective drug against status epilepticus, traumatic brain injury, and stroke. *Frontiers in Neurology, 4*. https://doi.org/10.3389/fneur.2013.00172.

Trollmann, R., Schneider, J., Keller, S., Strasser, K., Wenzel, D., Rascher, W., … Gassmann, M. (2008). HIF-1-regulated vasoactive systems are differentially involved in acute hypoxic stress responses of the developing brain of newborn mice and are not affected by levetiracetam. *Brain Research, 1199*, 27−36. https://doi.org/10.1016/j.brainres.2007.12.069.

Zou, H., Brayer, S. W., Hurwitz, M., Niyonkuru, C., Fowler, L. E., & Wagner, A. K. (2013). Neuroprotective, neuroplastic, and neurobehavioral effects of daily treatment with levetiracetam in experimental traumatic brain injury. *Neurorehabilitation and Neural Repair, 27*(9), 878−888. https://doi.org/10.1177/1545968313491007.

Chapter 35

Quetiapine treatment in pediatric scenarios

Barbara Plemeniti Tololeski[1] and Maja Drobnič Radobuljac[2,3]

[1]Unit for Adolescent Psychiatry, Center for Mental Health, University Psychiatric Clinic Ljubljana, Ljubljana, Slovenia; [2]Unit for Intensive Child and Adolescent Psychiatry, Center for Mental Health, University Psychiatric Clinic Ljubljana, Slovenia; [3]Department of Psychiatry, Medical Faculty, University of Ljubljana, Ljubljana, Slovenia

List of abbreviations

5HT 5-Hydroxytryptamine
ADHD Attention deficit hyperactivity disorder
CD Cluster of differentiation
CNS Central nervous system
D Dopamine
EMA European Medicines Agency
EPS Extrapyramidal symptoms
FDA U.S. Food and Drug Administration
H Histamine
IL Interleukin
M Muscarinic
MS Multiple sclerosis
NMDAR N-methyl-D-aspartate receptor
NMS Neuroleptic malignant syndrome
PCP Phencyclidine
SOD1 Superoxide dismutase
SSRI Selective serotonin reuptake inhibitor
TNF Tumor necrosis factor
α1 Alpha 1 adrenergic

Mini-dictionary of terms

(Binding) affinity In neuropharmacology, this refers to the "strength" of the connections between the drug and specific neurotransmitter receptors.

Abuse, addiction To misuse a drug for effects different to those it is officially prescribed for (e.g., to induce euphoria, sedation); addiction (dependence) syndrome is a mental disorder with core symptoms: strong desire to take the substance, difficulties in controlling substance-taking behavior, withdrawal symptoms, tolerance, progressive decline in overall functioning and health due to addiction.

Agitation A serious, acute state of psychomotor restlessness, in which the patient may harm himself or others, with numerous etiologies, such as various mental disorders, intoxication, physical disorders, etc.

Autism spectrum disorder A cluster of neurodevelopmental disorders with core symptoms of impaired communication and social interactions and stereotyped behaviors; common comorbid disorders can be intellectual disability, anxiety, and affective disorders, obsessive–compulsive disorder, psychotic decompensation, behavioral, as well as sleep and feeding disturbances.

Bipolar mania A type of episode in a mental disorder called bipolar disorder (American Psychiatric Association, 2013), in which patient experiences elevated or irritable mood, inflated self-esteem, decreased need for sleep, talkativeness, distractibility, and psychomotor agitation.

Diagnosis, Management and Modeling of Neurodevelopmental Disorders. https://doi.org/10.1016/B978-0-12-817988-8.00035-X

Delirium A serious acute state with impaired quality of consciousness, attention, memory, and orientation, with possible psychotic symptoms such as optical hallucinations, delusions, and agitation; relatively frequent in small children when seriously ill.

Extrapyramidal symptoms A group of neurological symptoms of movement disorders, such as: tremor, dystonia, parkinsonism, akathisia, tardive dyskinesia, bradykinesia, which are drug-induced in the case of antipsychotics (side effect).

Multiple sclerosis A chronic autoimmune disease of the central nervous system in myelin (the covering that envelopes the nerve cell's axons) is the main target of a pathological immune response of the body accompanied by numerous symptoms, e.g., muscle spasms, stiffness, weakness, vision problems, fatigue, pain, bladder and bowl problems, cognitive problems, depression, anxiety, psychosis, speech and swallowing problems.

Off-label Use of a drug for an officially unapproved condition or unapproved age of the patient.

Personality disorders Different types of problematic personality with maladaptive thinking, emotional responses and behaviors, which impair functioning.

Psychosis Is a psychiatric disorder, in which the patient loses control of reality; it usually includes positive symptoms (e.g., auditory hallucinations, delusions of different types), negative symptoms (e.g., blunted affect, social withdrawal), and cognitive symptoms (memory and concentration impairments).

Schizophrenia Is a psychiatric disorder characterized by untreated psychosis persisting for at least 1 month and continuous signs of the disturbance (including psychosis) must persist for a period of at least 6 months. Usually it has a relapsing–remitting course (American Psychiatric Association, 2013)

Introduction

Psychosis, especially due to schizophrenia, is a serious neurodevelopmental and neurodegenerative disorder, which rarely occurs in childhood, although its frequency progressively increases in adolescence and young adulthood. According to all relevant guidelines, patients with schizophrenia need to be treated with antipsychotics (Keating et al., 2017). In the past two decades, treatment with atypical antipsychotics such as risperidone, paliperidone, olanzapine, and aripiprazole has prevailed over the older, more typical antipsychotics such as haloperidol and zuclopenthixol. Due to the broader spectrum of mechanisms of actions and high ratio of 5HT2/D2 receptor affinity, the atypical antipsychotics are more successful in the treatment of negative, affective, and cognitive symptoms of schizophrenia, which is very important for disease prognosis and the patient's future functioning. These drugs also have a lower risk of side effects such as extrapyramidal symptoms (EPS) and prolactin elevation, although one needs to be careful as they may nevertheless cause metabolic changes and weight gain, which may be significant, especially in youth.

One of the atypical antipsychotics (Table 35.1), quetiapine, is a dibenzothiazepine derivative with affinity for multiple neurotransmitter receptors. As a weak antagonist of dopamine 2 (D2) receptors, it reduces positive symptoms of psychosis and stabilizes affective symptoms, without causing EPS. As an antagonist of 5HT2 receptors, it causes dopamine release in certain brain regions and consequently improves cognitive and affective symptoms. Its partial antagonism of 5HT1 receptors and the actions of the active metabolite norquetiapine (norepinephrine reuptake blockade and 5HT2C and 5HT7 antagonism) result in antidepressant effects (Stahl, 2017). By blocking H1, $\alpha 1$, and M1 receptors, it has sedating effects and other side effects (see Fig. 35.1 and Table 35.2).

Quetiapine and neuroscience

In recent years, serious psychiatric disorders such as schizophrenia have been associated with neuroinflammation and reactive oxidative/nitrosative stress in the central nervous system (CNS), and future treatment is likely to involve agents with antiimmunoinflammatory, antioxidative, and neuroplasticity effects (Nasrallah, 2013).

Quetiapine can be neuroprotective and plays a role in neurogenesis. For example, it was shown to prevent the oxidative-stress-related cytotoxic effect of amyloid β (Ab)25–35 in cultured cells (Xu et al., 2008) and to decrease the level of nitrotyrosine, a protein marker of oxidative stress, in the hippocampi of PCP-treated rats (a pharmacological model of schizophrenia with cognitive symptoms) (He et al., 2018). The antioxidative effect of quetiapine was also demonstrated in vitro on human blood samples, where it decreased levels of lipid peroxidation (Dietrich-Muszalska, Kontek, & Rabe-Jabłońska, 2011).

Wang and colleagues showed that application of quetiapine could also protect myelin and reduce schizophrenia-like behavior in a mouse model of cuprizone-induced demyelination (Wang, Liu, Zhang et al., 2015). They speculated that the mechanism of action of quetiapine was through a complex cascade inhibiting the release of nitric oxide and TNF-α, resulting in the neutralization of lipopolysaccharide-induced abnormal intercellular calcium homeostasis, with no subsequent microglial activation (Wang, Liu, Tian, et al. 2015; Wang, Liu, Zhang, et al., 2015).

TABLE 35.1 Atypical antipsychotics used in children and adolescents.

Atypical antipsychotics used in children and adolescents	European medicines agency (Europe) indications	Food and drug administration (United States) indications
Risperidone	Short-term symptomatic treatment (up to 6 weeks) of persistent aggression in conduct disorder in children from the age of 5 years and adolescents with subaverage intellectual functioning or mental retardation diagnosed according to DSM-IV criteria, in whom the severity of aggressive or other disruptive behaviors require pharmacologic treatment	• Treatment of schizophrenia in adolescents aged 13–17 years • Short-term treatment of acute manic or mixed episodes associated with bipolar I disorder in children and adolescents aged 10–17 years • Treatment of irritability associated with autistic disorder in children and adolescents aged 5–16 years
Aripiprazole	• Treatment of schizophrenia in adolescents aged 15–17 years • Up to 12 weeks to treat manic episodes in adolescents aged 13–17 years	• Treatment of manic or mixed episodes associated with bipolar I disorder in pediatric patients 10–17 years of age • Treatment of schizophrenia in adolescents aged 13–17 years • Treatment of irritability associated with autistic patients 6–17 years of age • Treatment of Tourette's disorder in pediatric patients 6–18 years of age
Olanzapine	Not approved for use in youth	• Treatment of schizophrenia in adolescents aged 13–17 years • Acute treatment of manic or mixed episodes associated with bipolar I disorder and maintenance treatment of bipolar I disorder in adolescents aged 13–17 years
Paliperidone	Treatment of schizophrenia in adolescents aged 15–17 years	Treatment of schizophrenia in adolescents aged 12–17 years
Quetiapine	Not approved for use in youth	• Treatment of schizophrenia in adolescents aged 13–17 years • Treatment of bipolar mania in adolescents aged 10–17 years
Ziprasidone	Treatment of manic or mixed episodes of moderate severity in bipolar disorder in children and adolescents aged 10–17 years	Not approved for use in youth

The table shows atypical antipsychotics and their indications for use in children and adolescents. There are differences between approved indications by EMA and FDA.

Quetiapine's myelin protecting role was additionally demonstrated in experimental autoimmune encephalomyelitis (a mouse model of multiple sclerosis (MS)), where quetiapine modulated the immune response, for example, by decreasing CD4+/CD8+ T cell infiltration into the spinal cord and inhibiting IL-2 production of effector T cells (Mei et al., 2012). Therefore, it is understandable that quetiapine with its remyelinating and neuroprotective properties in inflammatory and noninflammatory models of demyelination (Zhornitsky et al., 2013) is currently being studied as a possible candidate for remyelinating treatment of MS (Clinicaltrials.gov, NCT02087631). The effectiveness of the medication could be a consequence of the related etiology of MS and schizophrenia in children, namely that children with MS have a 10 times higher risk of psychosis than healthy children (Pakpoor et al., 2017). Currently there are no guidelines for the treatment of psychosis associated with MS in children and adolescents, but quetiapine seems to be a possible candidate, because of its low risk of EPS and its antipsychotic activity, as well as neuroprotective and remyelinating properties.

Quetiapine use in early-onset schizophrenia and bipolar mania

The use of quetiapine in those younger than 18 is officially approved by the U.S. Food and Drug Administration (FDA) for only two indications: early onset schizophrenia, where it is registered for use in patients between 13 and 17 years of age, and bipolar mania, for patients between 10 and 17 years of age. Its safety and effectiveness in youth with schizophrenia

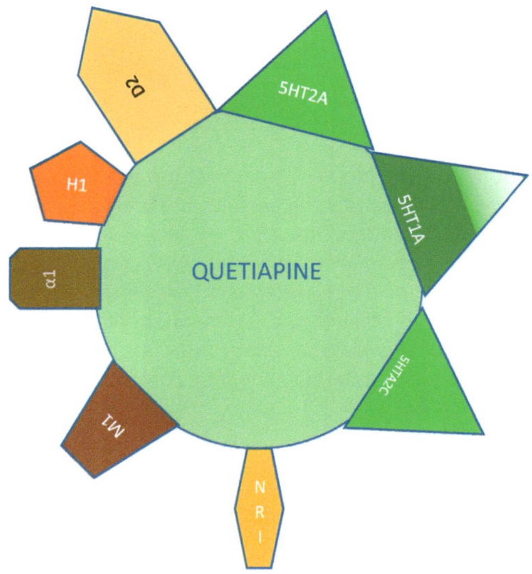

FIGURE 35.1 Simplified outline of quetiapine's binding properties. In this figure we can see some of quetiapine's most clinically important binding properties. Antagonism on D2 receptors produces antipsychotic effect, antagonism on H1 receptors contributes to the ability to enhance sleep or to induce daytime sedation, especially combined with M1-antimuscarinic and α-1 adrenergic antagonism. Antagonism on 5HT2A causes dopamine release in certain brain regions (improves cognitive and affective symptoms). Its partial antagonism on 5HT1A (together with partial antagonism of its active metabolite norquetiapine), norquetiapine's inhibition of norepinephrine reuptake (NRI), and antagonist actions at 5HT2C receptors may contribute to the antidepressant effect and cognitive improvement. *Adapted from Stahl, S. M. (2008).* Stahl's essential psychopharmacology: Neuroscientific basis and practical applications*(3rd ed.). New York, NY, US: Cambridge University Press.*

TABLE 35.2 Quetiapine effects and binding profile (Stahl, 2008).

Action on receptor	Clinical effects	Side effects
D2 antagonism	Antipsychotic	/
5HT2A antagonism	Antidepressant	/
5HT1A partial antagonism	Antidepressant, cognitive improvement	/
H1 antagonism	Sedation, induction of sleep	Weight gain, metabolic changes
M1 antagonism	Sedation	Weight gain, metabolic changes
α1 antagonism	Sedation	Orthostatic hypotension
5HT2C antagonism	Antidepressant	Weight gain
NRI	Mood improvement	/

The table shows how possible actions on different neuroreceptors result in different clinical effects and side effects. Binding properties vary greatly with technique and from one laboratory to another.
Data are from Stahl, S. M. (2008). *Stahl's essential psychopharmacology: Neuroscientific basis and practical applications*(3rd ed.). New York, NY, US: Cambridge University Press.

were proven in a relatively large randomized, placebo-controlled study (Findling, McKenna, Earley, Stankowski, & Pathak, 2012). However, quetiapine still has not received approval for use in this age group in Europe, since the European Medicines Agency (EMA) requires more data on efficiency, as well as long-term effects on growth, maturation, cognitive and behavioral development (Table 35.3).

The aforementioned study involved 220 adolescents (13–17 years old) with schizophrenia, treated with quetiapine for 6 weeks (400 or 800 mg/day), who showed increased body weight, total cholesterol, and triglycerides at the conclusion of the study. More studies on children and adolescents have also compared the use of quetiapine to other atypical antipsychotics. In a small open-label study of 32 adolescents undergoing their first psychosis, its effectiveness was comparable to olanzapine, although the adolescents using quetiapine gained 10 kg less after 6 months of treatment than the ones using olanzapine (Arango et al., 2009). Compared to risperidone use in this population, quetiapine was less effective in treating certain domains (total positive symptoms, negative symptoms, clinical global severity, and depression), although risperidone caused more EPS and prolactin elevation, which are common reasons for treatment discontinuation (Swadi et al., 2010).

TABLE 35.3 Quetiapine's official use in youth.

Official use of quetiapine in youth	United States	Europe
Immediate and extended release	FDA-approved for: schizophrenia in 13–17 year olds bipolar mania in 10–17 year olds	EMA neither approved nor recommended: due to lack of data to support use in youth Long-term safety implications on growth and maturation have not been studied beyond 26 weeks Long-term implications for cognitive and behavioral development are not known

The table shows the differences in approvals for using quetiapine in schizophrenia and bipolar mania in children and adolescents in the United States and Europe. FDA approved it in 2009, while EMA at the time of publication still didn't.

Similarly, quetiapine showed effectiveness in treating bipolar mania in children and adolescents in a large randomized placebo-controlled study by Pathak et al. (2013). Its effectiveness was shown to be as good as that of divalproex, although the improvements were quicker and the response and remission rates in adolescents with bipolar mania better with quetiapine (DelBello et al., 2006). Despite increasingly good results in the adult population, the use of quetiapine as a monotherapy in youth with bipolar depression is still questionable, since two studies on this population did not show a superior effectiveness compared to placebo (DelBello et al., 2009; Findling, Pathak, Earley, Liu, & DelBello, 2014). That said, when neurofunctional correlates of the response to quetiapine in adolescents with bipolar depression were explored, it was shown that quetiapine indeed changed the activation of areas in the prefrontal cortex critical for the emotional regulation, whereas placebo did not (Chang et al., 2018). However, more research is needed to support these suggestive observations.

The use of quetiapine in other disorders in youth

Quetiapine has been quite successfully used off-label in different organic disorders with either psychotic or behavioral symptoms. Its low risk of EPS is probably the reason why it is often chosen. The short-term off-label use of quetiapine (median daily dose of 1.3 mg/kg/day) proved successful in the treatment of 55 cases of pediatric delirium (Joyce et al., 2015), although more extensive controlled trials are needed. Despite the lack of relevant studies, the consensus guidelines of the American Association for Emergency Psychiatry and the American Academy of Child and Adolescent Psychiatry Emergency Child Psychiatry Committee advise the use of quetiapine for emergency department treatment of acute agitation due to delirium in youth (Gerson et al., 2019).

Quetiapine has also been successfully used for the treatment of agitation secondary to anti-NMDAR encephalitis in pediatric male patients without causing any EPS or neuroleptic malignant syndrome (NMS) (Schumacher, Mann, & MacKenzie, 2016). It was also safe and without side effects when used in the treatment of psychosis related to primary brain calcification in an adolescent female (Plemeniti Tololeski et al., 2019).

It has shown good efficacy in treating tics and Tourette's syndrome in children and adolescents in case reports (Paraga, Parraga, Woodward & Fenning, 2001) and small-scale studies (Copur, Arpaci, Demir, & Narin, 2007; Mukaddes & Abali, 2003). However, risperidone and aripiprazole are still the first-line treatment choices and are the most commonly used, with efficacy proven in several studies (Ganos, Martino, & Pringsheim, 2017).

Quetiapine has also shown potential for treatment of behavioral and emotional symptoms in youth with emerging personality disorders. It proved better than placebo in a small controlled study in aggressive youth with conduct disorder (Conner, McLaughlin, & Jeffers-Terry, 2008), causing no extrapyramidal symptoms and was effective even after 18 weeks of treatment (Findling et al., 2007). It was also successful as an add-on therapy to methylphenidate in aggressive children with ADHD (Kronenberger et al., 2007). Special attention is needed when using the drug in this population due to the potential for abuse (see next subheading). Quetiapine has also been successful as an add-on therapy (average dose 390 mg/d) with various selective serotonin reuptake inhibitors (SSRI) in adolescents with developing borderline personality disorder resistant to previous treatments (Podobnik, Foller Podobnik, Grgic, Marcinko, & Pivac, 2012). They were less irritable, emotionally unstable, aggressive, suicidal, and depressed, and the improvement persisted over the 24 weeks of the study.

TABLE 35.4 Quetiapine use in child and adolescent psychiatry.

Disorder, symptom	Dose	Study/case report
Early-onset schizophrenia	400–800 mg/day	Findling et al. (2012)
Bipolar mania	400–600 mg/day	Pathak et al. (2013)
Pediatric delirium	1,3 mg/kg/day	Joyce et al. (2015)
Acute agitation	1–1,5 mg/kg/dose	Gerson et al. (2019)
Agitation in anti-NMDAR encephalitis	25–250 mg/day	Schumacher et al. (2016)
Psychosis due to primary brain calcification	200–900 mg/day	Plemeniti Tololeski et al. (2019)
Tics and Tourette's syndrome	25–100 mg/day	Mukadess and Abali (2003)
Aggressive behavior in adolescent with conduct disorder	200–600 mg/day	Conner et al. (2008)
Different symptoms in adolescent with developing borderline personality disorder	average 390 mg/day	Podobnik et al. (2012)
Autism spectrum disorder	50–150 mg/day	Golubochik et al. (2011)

A few studies have explored the efficacy of quetiapine in youth with autism spectrum disorder (only risperidone and aripiprazole are approved for the treatment of irritability associated with autism). Although the majority failed to show the efficacy of quetiapine (Corson, Barkenbus, Posey, Stigler, & McDougle, 2004; Findling et al., 2004), it was found to be helpful in one retrospective study (Hardan, Jou, & Handen, 2005) where improvements were noticed in conduct, inattention, and hyperactivity in more than half of children and adolescents with pervasive developmental disorders, as well as in a preliminary open-label small-scale observational study (Golubchik, Sever, & Weizman, 2011) where low-dose quetiapine (50–150 mg/d) decreased the intensity of aggressive behavior and sleep disturbances in high-functioning adolescents with autism (Table 35.4).

Quetiapine metabolic side effects and potential for abuse

Apart from sedation in the first weeks of titration and some rare adverse events in youth treated with quetiapine, it is the metabolic side effects that deserve most of the attention (Table 35.5). Although the results of the different studies are inconsistent, quetiapine clearly causes unwanted metabolic changes in children and adolescents. When Cohen, Bonnot, Bodeau, Consoli, and Laurent (2012) reviewed controlled studies of youth treated with quetiapine for 3–12 weeks, they found serum cholesterol (10.77 ± 2.14 mg/dL) and triglyceride levels (19.5 ± 3.92 mg/dL) even higher than with olanzapine and the patients' weight increased (1.74 ± 0.38 kg). Thyroid function was found to be disturbed after treatment with quetiapine; elevation of thyroid stimulating hormone (Findling et al., 2014) and decreased free thyroxin (Fragas et al.,

TABLE 35.5 Metabolic side effects of atypical antipsychotics in children and adolescents.

	Number of studies	Mean weight gain (kg)	Cholesterol (mg/dL)	Triglycerides (mg/dL)	Glucose (mg/dL)
Quetiapine	10	1.74	10.77	19.5	1.64
Olanzapine	14	3.99	4.46	20.18	2.09
Risperidone	25	2.02	0.91	5.57	3.7
Aripiprazole	10	0.89	2.65	0.03	2.21
Clozapine	5	2.38	/	/	/
Ziprasidone	4	−0.1	0.47	−1.52	−1.24

The table presents different metabolic side effects of atypical antipsychotic in youth compared to placebo. Data are reviewed from different studies. Quetiapine's most worrisome side effects seem to be increased levels of cholesterol and triglycerides (Cohen et al., 2012).

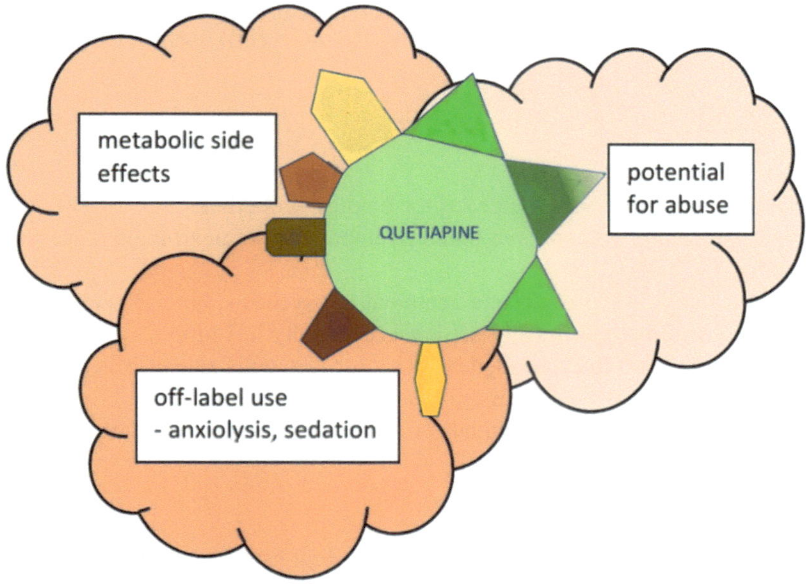

FIGURE 35.2 **Outline of important issues in quetiapine off-label use.** Quetiapine is the leading atypical antipsychotic in off-label prescribing. The most common symptoms treated this way are sleep disturbances and anxiety (Carton et al., 2015) in low dosing regiments. But low dosing cannot be assumed to equate increased safety (Chow et al., 2017), which is worrisome especially in long-term use. Additional potential for abuse doesn't ease the issue.

2008) have been observed in various studies. Long-term effects in the youth population still need to be studied, and weight gain, lipid profile, and thyroid function have to be monitored regularly (Masi et al., 2015).

The use of atypical antipsychotics for off-label indications has increased in recent years. Quetiapine has been the leading medication in adults, where it is used particularly often for anxiety and insomnia (Carton et al., 2015). Insomnia is also commonly treated with quetiapine in in-patient youth (Chow et al., 2017). With two-thirds of insomnia-treated in-patient youth continuing with quetiapine after discharge, quetiapine's potential for misuse and addiction should also be kept in mind. In a retrospective study in the United States, it was found that quetiapine is the most commonly abused drug among atypical antipsychotics (Klein, Bangh, & Cole, 2016). Its potential for misuse/abuse, proven in rodents (Cha et al., 2013), is probably connected to its antihistaminic activity and depressant effects on the CNS (Fisher & Boggs, 2010), which may be exaggerated with nonoral routes such as intranasal and intravenous, since hepatic metabolism is skipped and higher doses of the drug quickly reach the CNS (Piróg-Balcerzak, Habrat, & Mierzejewski, 2015). All this may explain why quetiapine already has a number of street names and an assigned street value (Sansone & Sansone, 2010). Its misuse/abuse in youth should be a concern, even though only solitary cases have been described in adolescents and only one so far with a full addiction syndrome (Kolli, Mary, Garcia-Delgar & Coffey, 2016). Clinicians should therefore be cautious when prescribing quetiapine to youth, especially to those with a higher risk of abuse (e.g., those with ADHD, conduct disorders, a positive family history of addiction, etc.) (Fig. 35.2).

Applications to other areas of development

The present chapter reviews the use of the atypical antipsychotic quetiapine in child and adolescent psychiatry. Recent in vitro animal studies have explored quetiapine's other activities in the CNS. In psychiatry it is well-known as one of the atypical antipsychotics with a tolerable side-effect profile, especially due to its low risk of neurological side effects. Recent studies have also shown its antioxidative, neuroprotective, and remyelinating properties in inflammatory and non-inflammatory models of demyelination (Zhornitsky et al., 2013). Presumably these effects are produced among others by decreasing demyelination, decreasing loss of oligodendrocytes, decreasing number of activated astrocytes and microglia, and increasing SOD1 activity (Zhang et al., 2008). These mechanisms of action could be especially important in the treatment of multiple sclerosis, where demyelination and neural loss produce progressive neurological and psychiatric symptoms. Even though only 5% of patients with multiple sclerosis are children, many of them have psychiatric co-morbidity, most commonly anxiety disorders, attention deficit hyperactivity disorder, mood disorders, and some degree of cognitive impairment (Weisbrot et al., 2014) as well as 10 times higher risk for psychosis (Pakpoor et al., 2017). While for these young patients, quetiapine could be a valuable treatment option in the future, its safety and tolerability are only recently being tested in adults with multiple sclerosis (ClinicalTrials.gov, NCT02087631). Therefore, this area offers many openings for future research.

Key facts about quetiapine

- Quetiapine is an atypical antipsychotic with multireceptor binding properties.
- In vitro it has shown neuroprotective properties of potential importance to neurology, namely in the treatment of multiple sclerosis in the future.
- In child and adolescent psychiatry, it is used for the treatment of psychosis, bipolar mania, agitation, delirium, behavioral disturbances, and sleep disturbances (mostly off-label, especially in Europe).
- It has low EPS and prolactin elevation risk, but moderate risk for metabolic side effects, especially in youth, so careful and frequent monitoring and preventive strategies are essential.
- Professionals must be aware of quetiapine's potential for abuse, particularly in certain subpopulations.

Summary points

- Quetiapine is an atypical antipsychotic with a multireceptor binding profile.
- In vitro studies show that it has antioxidative and antiinflammatory properties as well, which may be additionally beneficial in neuroinflammatory, neurodevelopmental, and neurodegenerative disorders.
- In youth, it is most effective for use in schizophrenia and bipolar mania (although used off-label in Europe).
- It has been effective in children and adolescents with delirium and organic psychosis and as an add-on treatment in emerging borderline personality disorder, conduct disorder, and sleep disturbances (all off-label).
- It has a low risk of EPS and prolactin elevation.
- Careful monitoring for side effects, especially metabolic, is necessary and should be even more frequent with children and adolescents, since they are more sensitive than adults.

References

American Psychiatric Association. (2013). *Diagnostic and statistical manual of mental disorders* (5th ed.). Washington, DC: Author.

Arango, C., Robles, O., Parellada, M., Fraguas, D., Ruiz-Sancho, A., Medina, O., … Moreno, D. (2009). Olanzapine compared to quetiapine in adolescents with a first psychotic episode. *European Child and Adolescent Psychiatry, 18*(7), 418–428.

Carton, L., Cottencin, O., Lapeyre-Mestre, M., Geoffroy, P. A., Favre, J., Simon, N., … Rolland, B. (2015). Off-label prescribing of antipsychotics in adults, children and elderly individuals: A systematic review of recent prescription trends. *Current Pharmaceutical Design, 21*(23), 3280–3297.

Cha, H. J., Lee, H. A., Ahn, J. I., Jeon, S. H., Kim, E. J., & Jeong, H. S. (2013). Dependence potential of quetiapine: Behavioral pharmacology in rodents. *Biomolecules and therapeutics, 21*(4), 307–312.

Chang, K., DelBello, M., Garrett, A., Kelley, R., Howe, M., Adler, C., … Singh, M. (2018). Neurofunctional correlates of response to quetiapine in adolescents with bipolar depression. *Journal of Child and Adolescent Psychopharmacology, 28*(6), 379–386.

Chow, E. S., Zangeneh-Kazemi, A., Akintan, O., Chow-Tung, E., Eppel, A., & Boylan, K. (2017). Prescribing practices of quetiapine for insomnia at a tertiary care inpatient child and adolescent psychiatry unit: A continuous quality improvement project. *Journal of the Canadian Academy of Child and Adolescent Psychiatry, 26*(2), 98–103.

Cohen, D., Bonnot, O., Bodeau, N., Consoli, A., & Laurent, C. (2012). Adverse effects of second-generation antipsychotics in children and adolescents: A bayesian meta-analysis. *Journal of Clinical Psychopharmacology, 32*, 309–316.

Connor, D. F., McLaughlin, T. J., & Jeffers-Terry, M. (2008). Randomized controlled pilot study of quetiapine in the treatment of adolescent conduct disorder. *Journal of Child and Adolescent Psychopharmacology, 18*(2), 140–156.

Copur, M., Arpaci, B., Demir, T., & Narin, H. (2007). Clinical effectiveness of quetiapine in children and adolescents with tourette's syndrome: A retrospective case-note survey. *Clinical Drug Investigation, 27*(2), 123–130.

Corson, A. H., Barkenbus, J. E., Posey, D. J., Stigler, K. A., & McDougle, C. J. (2004). A retrospective analysis of quetiapine in the treatment of pervasive developmental disorders. *Journal of Clinical Psychiatry, 65*(11), 1531–1536.

DelBello, M. P., Chang, K., Welge, J. A., Adler, C. M., Rana, M., Howe, M., … Strakowski, S. M. (2009). A double-blind, placebo-controlled pilot study of quetiapine for depressed adolescents with bipolar disorder. *Bipolar Disorders, 11*, 483–493.

DelBello, M. P., Kowatch, R. A., Adler, C. M., Stanford, K. E., Welge, J. A., Barzman, D. H., … Strakowski, S. M. (2006). A double-blind randomized pilot study comparing quetiapine and divalproex for adolescent mania. *Journal of the American Academy for Child and Adolescent Psychiatry, 45*, 305–313.

Dietrich-Muszalska, A., Kontek, B., & Rabe-Jabłońska, J. (2011). Quetiapine, olanzapine and haloperidol affect human plasma lipid peroxidation in vitro. *Neuropsychobiology, 63*(4), 197–201.

Findling, R. L., McKenna, K., Earley, W. R., Stankowski, J., & Pathak, S. (2012). Efficacy and safety of quetiapine in adolescents with schizophrenia investigated in a 6-week, double-blind, placebo-controlled trial. *Journal of Child and Adolescent Psychopharmacology, 22*(5), 327–342.

Findling, R. L., McNamara, N. K., Gracious, B. L., O'Riordan, M. A., Reed, M. D., Demeter, C., & Blumer, J. L. (2004). Quetiapine in nine youths with autistic disorder. *Journal of Child and Adolescent Psychopharmacology, 14*(2), 287–294.

Findling, R. L., Pathak, S., Earley, W. R., Liu, S., & DelBello, M. P. (2014). Efficacy and safety of extended-release quetiapine fumarate in youth with bipolar depression: An 8-week, double-blind, placebo-controlled trial. *Journal of Child and Adolesentc Psychopharmacology, 24*, 325–335.

Findling, R. L., Reed, M. D., O'Riordan, M. A., Demeter, C. A., Stansbrey, R. J., & McNamara, N. K. (2007). A 26-week open-label study of quetiapine in children with conduct disorder. *Journal of Child and Adolescent Psychopharmacology, 17*(1), 1–9.

Fisher, B. A., & Boggs, D. L. (2010). The role of antihistaminic effects in the misuse of quetiapine: A case report and review of the literature. *Neuroscience and Biobehavioral Reviews, 34*(4), 555–558.

Fraguas, D., Merchán-Naranjo, J., Laita, P., Parellada, M., Moreno, D., Ruiz-Sancho, A., … Arango, C. (2008). Metabolic and hormonal side effects in children and adolescents treated with second-generation antipsychotics. *Journal of Clinical Psychiatry, 69*, 1166–1175.

Ganos, C., Martino, D., & Pringsheim, T. (2017). Tics in the pediatric population: Pragmatic management. *Movement Disorders Clinical Practice, 4*(2), 160–172.

Gerson, R., Malas, N., Feuer, V., Silver, G. H., Prasad, R., & Mroczkowski, M. M. (2019). Best practices for evaluation and treatment of agitated children and adolescents (BETA) in the emergency department: Consensus statement of the American Association for Emergency Psychiatry. *Western Journal of Emergency Medicine, 20*(2), 409–418.

Golubchik, P., Sever, J., & Weizman, A. (2011). Low-dose quetiapine for adolescents with autistic spectrum disorder and aggressive behavior: Open-label trial. *Clinical Neuropharmacology, 34*(6), 216–219.

Hardan, A. Y., Jou, R. J., & Handen, B. L. (2005). Retrospective study of quetiapine in children and adolescents with pervasive developmental disorders. *Journal of Autism and Developmental Disorders, 35*(3), 387–391.

He, J., Liu, F., Zu, Q., Xu, Z., Zheng, H., Li, X., & Wang, W. (2018). Chronic administration of quetiapine attenuates the phencyclidine-induced recognition memory impairment and hippocampal oxidative stress in rats. *NeuroReport, 29*, 1099–1103.

Joyce, C., Witcher, R., Herrup, E., Kaur, S., Mendez-Rico, E., Silver, G., … Traube, C. (2015). Evaluation of the safety of quetiapine in treating delirium in critically ill children: A retrospective review. *Journal of Child and Adolescent Psychopharmacology, 25*(9), 666–670.

Keating, D., McWilliams, S., Schneider, I., Hynes, C., Cousins, G., Strawbridge, J., & Clarke, M. (2017). Pharmacological guidelines for schizophrenia: A systematic review and comparison of recommendations for the first episode. *BMJ Open, 7*. e013881.

Klein, L., Bangh, S., & Cole, J. (2016). Intentional recreational abuse of quetiapine compared to other second-generation antipsychotics. *Western Journal of Emergency Medicine, 18*(2), 243–250.

Kolli, V., Mary, H., Garcia-Delgar, B., & Coffey, B. J. (2016). Quetiapine addiction in an adolescent. *Journal of Child and Adolescent Psychopharmacology, 26*(2), 174–176.

Kronenberger, W. G., Giauque, A. L., Lafata, D. E., Bohnstedt, B. N., Maxey, L. E., & Dunn, D. W. (2007). Quetiapine addition in methylphenidate treatment-resistant adolescents with comorbid ADHD, conduct/oppositional-defiant disorder, and aggression: A prospective, open-label study. *Journal of Child and Adolescent Psychopharmacology, 17*(3), 334–347.

Masi, G., Milone, A., Veltri, S., Iuliano, R., Pfanner, C., & Pisano, S. (2015). Use of quetiapine in children and adolescents. *Pediatric Drugs, 17*(2), 125–140.

Mei, F., Guo, S., He, Y., Wang, L., Wang, H., Niu, J., … Xiao, L. (2012). Quetiapine, an atypical antipsychotic, is protective against autoimmune-mediated demyelination by inhibiting effector T cell proliferation. *PLoS One, 7*(8). e42746.

Mukaddes, N. M., & Abali, O. (2003). Quetiapine treatment of children and adolescents with tourette's disorder. *Journal of Child and Adolescent Psychopharmacology, 13*(3), 295–299.

Nasrallah, H. A. (June 2013). Beyond dopamine: The 'other' effects of antipsychotics (Editorial) *Current Psychiatry, 12*, 6.

Pakpoor, J., Goldacre, R., Schmierer, K., Giovannoni, G., Waubant, E., & Goldacre, M. J. (2017). Psychiatric disorders in children with demyelinating diseases of the central nervous system. *Multiple Sclerosis Journal, 24*(9), 1243–1250.

Párraga, H. C., Párraga, M. I., Woodward, R. L., & Fenning, P. A. (2001). Quetiapine treatment of children with tourette's syndrome: Report of two cases. *Journal of Child and Adolescent Psychopharmacology, 11*(2), 187–191.

Pathak, S., Findling, R. L., Earley, W. R., Acevedo, L. D., Stankowski, J., & DelBello, M. P. (2013). Efficacy and safety of quetiapine in children and adolescents with mania associated with bipolar I disorder: A 3-week, double-blind, placebo-controlled trial. *Journal of Clinical Psychiatry, 74*, 100–109.

Piróg-Balcerzak, A., Habrat, B., & Mierzejewski, P. (2015). Misuse and abuse of quetiapine. *Psychiatria Polska, 49*(1), 81–93.

Plemeniti Tololeski, B., Debeljak, M., Perkovič Benedik, M., Rigler, T., Kyriakopoulos, M., Kotnik, P., … Drobnič Radobuljac, M. (2019). The use of quetiapine in treatment of acute psychotic symptoms in an adolescent patient with primary brain calcification: A case report *BMC Psychiatry, 19*(1), 67.

Podobnik, J., Foller Podobnik, I., Grgic, N., Marcinko, D., & Pivac, N. (2012). The effect of add-on treatment with quetiapine on measures of depression, aggression, irritability and suicidal tendencies in children and adolescents. *Psychopharmacology, 220*(3), 639–641.

Sansone, R. A., & Sansone, L. A. (2010). Is seroquel developing an illicit reputation for misuse/abuse? *Psychiatry, 7*(1), 13–16.

Schumacher, L. T., Mann, A. P., & MacKenzie, J. G. (2016). Agitation management in pediatric males with anti-N-Methyl-D-aspartate receptor encephalitis. *Journal of Child and Adolescent Psychopharmacology, 26*(10), 939–943.

Stahl, S. M. (2008). *Stahl's essential psychopharmacology: Neuroscientific basis and practical applications* (3rd ed.). New York, NY, US: Cambridge University Press.

Stahl, S. M. (2017). *Prescriber's guide stahl's essential psychopharmacology*. Quetiapine (6th ed.). Cambirdge: Clays.

Swadi, H. S., Craig, B. J., Pirwani, N. Z., Black, V. C., Buchan, J. C., & Bobier, C. M. (2010). A trial of quetiapine compared with risperidone in the treatment of first onset psychosis among 15- to 18-year-old adolescents. *International Clinical Psychopharmacology, 25*, 1−6.

Wang, H., Liu, S., Tian, Y., Wu, X., He, Y., Li, C., ... Xiao, L. (2015). Quetiapine inhibits microglial activation by neutralizing abnormal STIM1-mediated intercellular calcium homeostasis and promotes myelin repair in a cuprizone-induced mouse model of demyelination. *Frontiers in Cellular Neuroscience, 9*, 492.

Wang, H. N., Liu, G. H., Zhang, R. G., Xue, F., Wu, D., Chen, Y. C., ... Tan, Q. R. (2015). Quetiapine ameliorates schizophrenia-like behaviors and protects myelin integrity in cuprizone intoxicated mice: The involvement of notch signaling pathway. *The International Journal of Neuropsychopharmacology, 19*(2), pyv088.

Weisbrot, D., Charvet, L., Serafin, D., Milazzo, M., Preston, T., Cleary, R., ... Krupp, L. (2014). Psychiatric diagnoses and cognitive impairment in pediatric multiple sclerosis. *Multiple Sclerosis Journal, 20*(5), 588−593.

Xu, H., Wang, H., Zhuang, L., Yan, B., Yu, Y., Wei, Z., ... Li, X. M. (2008). Demonstration of an anti-oxidative stress mechanism of quetiapine implications for the treatment of Alzheimer's disease. *FEBS Journal, 275*, 3718−3728.

Zhang, Y., Xu, H., Jiang, W., Xiao, L., Yan, B., He, J., ... Li, X. M. (2008). Quetiapine alleviates the cuprizone-induced white matter pathology in the brain of C57BL/6 mouse. *Schizophrenia Research, 106*, 182−191.

Zhornitsky, S., Wee Yong, V., Koch, M. W., Mackie, A., Potvin, S., Patten, S. B., & Metz, L. M. (2013). Quetiapine fumarate for the treatment of multiple sclerosis: Focus on myelin repair. *CNS Neuroscience and Therapeutics, 19*, 737−744.

Chapter 36

Treatments with cannabinoids and brain development

P. Schonhofen[1,2,3], D.M. Vargas[1,2,3], R.B. Parsons[4] and F. Klamt[1,2,3]

[1]*Laboratory of Cellular Biochemistry, ICBS/UFRGS, Porto Alegre, Rio Grande do Sul, Brazil;* [2]*Programa de Pós-Graduação em Ciências Biológicas: Bioquímica, ICBS/UFRGS, Porto Alegre, Rio Grande do Sul, Brazil;* [3]*National Institutes of Science and Technology—Translational Medicine (INCT-TM), Porto Alegre, Rio Grande do Sul, Brazil;* [4]*Institute of Pharmaceutical Science, School of Cancer and Pharmaceutical Studies, King's College London (KCL), London, United Kingdom*

List of abbreviations

2-AG 2-Arachidonoylglycerol
AA Arachidonic acid
AC Adenylyl cyclase
ADHD Attention deficit hyperactivity disorder
AEA N-arachidonoylethanolamine
AIDS Acquired Immunodeficiency Syndrome
ASD Autism spectrum disorder
BD Bipolar disorder
cAMP Cyclic adenosine monophosphate
CB1 Cannabinoid receptors type-1
CB2 Cannabinoid receptors type-2
CBD Cannabidiol
DAG Diacylglycerol
DSM-5 Diagnostic and Statistical Manual of Mental Disorders, Fifth Edition
ECS Endocannabinoid system
ET Ethanolamine
FAAH Fatty acid amide hydrolase
FDA Food and Drug Administration
GPR55 G-protein-coupled receptor 55
MAGL Monoacylglycerol lipase
MAPK Mitogen-activated protein kinase
NAPE N-arachidonoylphosphatidylethanolamine
NDDs Neurodevelopmental disorders
NHIE Neonatal hypoxic-ischemic encephalopathy
PIP2 Phosphatidylinositol biphosphate
PKA Protein kinase A
TNF-α Tumor necrosis factor alpha
TRPV1 Transient receptor potential of vanilloid type-1
VDAC1 Voltage-dependent anion-selective channel protein
Δ^9- THC delta-9-tetrahydrocannabinol

Diagnosis, Management and Modeling of Neurodevelopmental Disorders. https://doi.org/10.1016/B978-0-12-817988-8.00036-1

Mini-dictionary of terms

Agonist A drug that binds to and activates a receptor to produce a biological response.
Antagonist A drug that binds to the same site as the agonist but does not activate it, blocking the agonist's action.
Cannabidiol Enriched *Cannabis* – with higher levels.
Cannabinoids Plant-derived, endogenous, or synthetic *Cannabis* isolated compounds.
Cannabis Smoked or plant extracts of *Cannabis sativa.*
***Cannabis*-derived product** Cannabinoids in commercial drug formulation.
Case report Descriptive study of a single individual.
Parent surveys/chart Surveys used to explore the experience of the parents of children with a determined disease or condition.
Psychoactive Substance that affects how the brain works and causes changes in mood, awareness, thoughts, feelings, or behavior.
Synthetic cannabinoids Human-made mind-altering chemicals.

Introduction

Neurodevelopmental disorders (NDDs) are a group of complex illnesses with typical onset during infancy or childhood associated with brain development issues and promoting deficits on cognition, communication, adaptive behavior, and psychomotor skills (Cardoso et al., 2019). The Diagnostic and Statistical Manual of Mental Disorders, Fifth Edition (DSM-5) classifies the following as NDDs: attention deficit hyperactivity disorder (ADHD), autism spectrum disorder (ASD), communication disorders, intellectual disabilities, motor disorders, and specific learning disorders, among others.

This chapter will discuss the current knowledge about treatments of NDDs based on isolated cannabinoids and *Cannabis* during infancy and adolescence. We will address their benefits and possible adverse effects, as well as their predicted mechanisms of action in the main NDDs. Due to its relevance and age of onset/treatment, infantile epilepsy was also included in our analysis.

Cannabinoids and the endocannabinoid signaling

Cannabis is the most widely used recreational drug, and its recent medical and recreational use legalization in the United States has increased its accessibility (Wong & Wilens, 2017). Among the *Cannabis* extract, delta-9-tetrahydrocannabinol (Δ^9-THC) and cannabidiol (CBD) are the most abundant and the most studied cannabinoids. Their main target are the constituents of the endocannabinoid system (ECS), the G-protein-coupled cannabinoid receptors type-1 (CB1), widely expressed in the nervous system, and type-2 (CB2), mainly expressed in immune cells (Di Marzo & De Petrocellis, 2012), the endogenous cannabinoids homologues (endocannabinoids), such as N-arachidonoylethanolamine (anandamide, AEA) and 2-arachidonoylglycerol (2-AG), and their related enzymes (Pertwee et al., 2010).

In neurons, endocannabinoids are produced as a response to neuronal activity and control neurotransmitters release (Fig. 36.1). This process is initiated by an increase in the Ca^{2+} influx, triggered by neurotransmission in the postsynaptic neuron, which induces endocannabinoids synthesis from its plasma membrane precursor lipids. AEA is synthesized from phospholipase-D-mediated hydrolysis of N-arachidonoylphosphatidylethanolamine (NAPE), while 2-AG originates from the diacylglycerol lipase-mediated hydrolysis of diacylglycerol (DAG), derived mainly from phosphatidylinositol biphosphate (PIP_2). AEA and 2-AG diffuse toward the presynaptic terminals where they bind and activate the presynaptic CB1 receptors. This binding triggers the activation and release of Gi/Go proteins from the CB1 and inhibits adenylyl cyclase (AC), thus decreasing cyclic AMP (AMPc) formation and protein kinase A (PKA) activity. These events reduce presynaptic intracellular K^+ and Ca^{2+}, arresting the release of stored neurotransmitters. Finally, AEA and 2-AG reenter the post- and presynaptic terminals, where they are catabolized by fatty acid amide hydrolase (FAAH) and monoacylglycerol lipase (MAGL), to yield arachidonic acid (AA) and ethanolamine (ET) and AA and glycerol, respectively.

Additionally, CB1 receptor activation stimulates the mitogen-activated protein kinase (MAPK) activity, a mechanism by which cannabinoids affect synaptic plasticity, cell migration, and neuronal growth (Mechoulam & Parker, 2013).

ECS in neurodevelopment

In the central nervous system, CB1 is particularly enriched in the cortex, hippocampus, amygdala, basal ganglia outflow tracts, and cerebellum. This distribution corresponds to the most prominent behavioral effects of *Cannabis* and helps to predict neurological and psychological effects of ECS manipulation (Mackie, 2005). CB1 is also in intracellular compartments, such as the mitochondrial surface, where they modulate intracellular levels of ATP, Ca^{2+}, and reactive oxygen species (Djeungoue-Petga & Hebert-Chatelain, 2017).

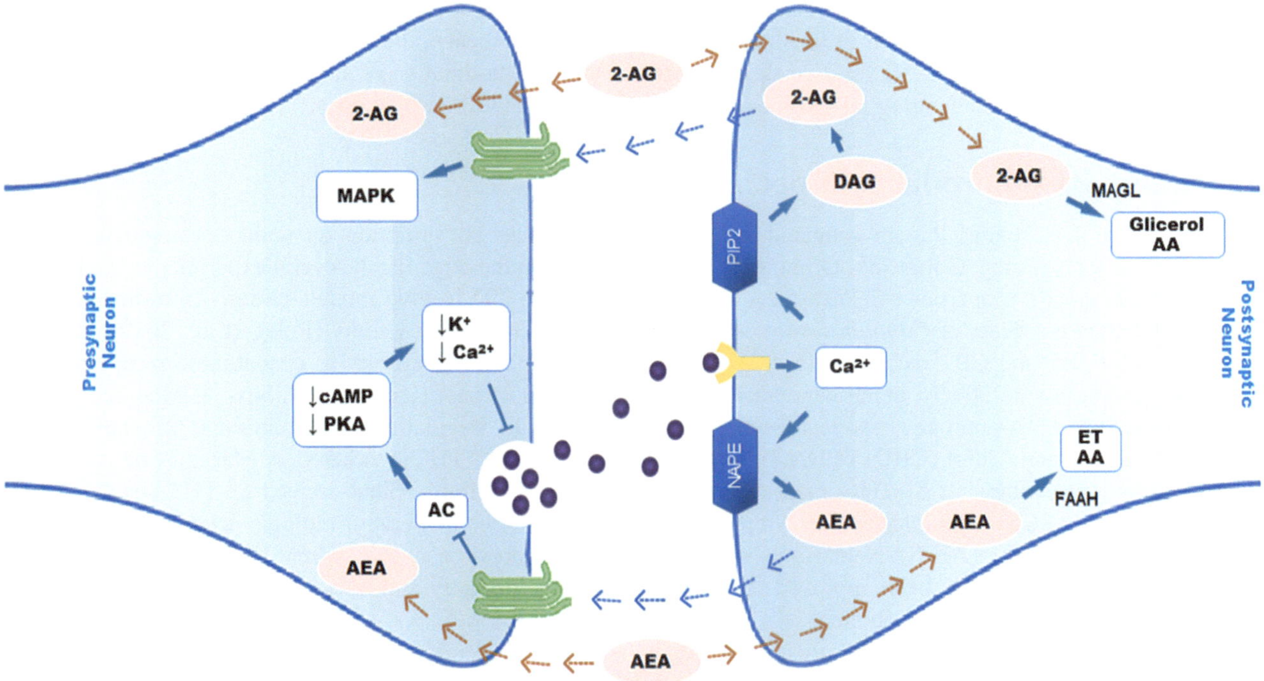

FIGURE 36.1 **Retrograde endocannabinoid signaling.** Endocannabinoids are produced on demand in the postsynaptic neuron, released in the synaptic cleft, and activate CB1 receptor in the presynaptic neuron. *2-AG*, 2-arachydonoilglycerol; *AA*, arachidonic acid; *AC*, adenylyl cyclase; *AEA*, Anandamide; *cAMP*, cyclic adenosine monophosphate; *CB1*, Cannabinoid receptor 1; *DAG*, lipase, diacylglycerol lipase; *DAG*, diacylglycerol; *ET*, ethanolamine; *FAAH*, fatty acid amide hydrolase; *MAGL*, monoacylglycerol lipase; *MAPK*, mitogen-activated protein kinase; *NAPE*, N-arachidonoylphosphatidylethanolamine; PIP_2, phosphatidylinositol biphosphate; *PKA*, protein kinase A; *PLD*, phospholipase D.

The ECS is active from early brain development periods, undergoing neurodevelopment until adolescence. This activity pattern may promote an increased vulnerability to cannabinoid exposure throughout early life stages (Mechoulam & Parker, 2013). In the developing nervous system and in the neurogenic areas of adult brain (hippocampal subgranular zone and subventricular zone), the ECS exerts a regulatory role on neural progenitor cells survival, proliferation, differentiation, and migration via CB1 (Díaz-Alonso et al., 2012) (Fig. 36.2).

The ECS maintains the homeostasis of prefrontal cortex interactions with the amygdala and hippocampus, which are responsible for behaviors such as emotional memory and anxiety-related behaviors, and endocannabinoids are required for

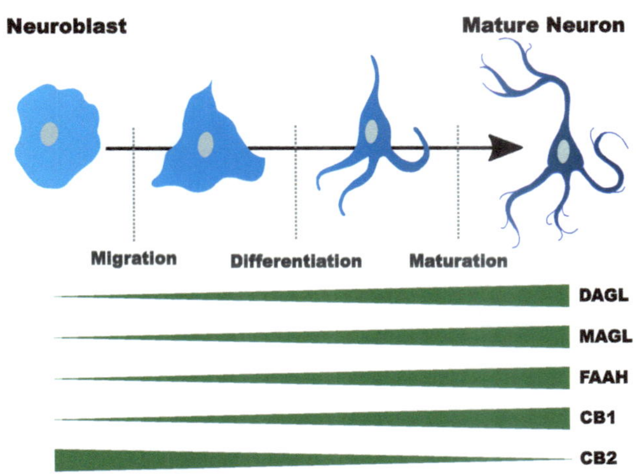

FIGURE 36.2 **The endocannabinoid system in neurodevelopment.** The endocannabinoid system participates in the regulation of neural progenitor cell proliferation, migration, differentiation, and survival. Its main constituents increase their expression during neuronal differentiation, with CB1 and CB2 demonstrating opposite patterns of expression. *CB1*, cannabinoid receptor 1; *CB2*, cannabinoid receptor 2; *DAGL*, Diacylglycerol lipase; *FAAH*, fatty acid amide hydrolase; *MAGL*, monoacylglycerol lipase.

the normal stress response, a process that matures during adolescence (Dow-Edwards & Silva, 2017). Besides, as the prefrontal cortex is the last brain region to finish development after adolescence, the abundance of CB1 may explain the effects of *Cannabis* use in this age (Meyer, Lee, & Gee, 2017). Thus, the modulation of the ECS as a therapeutic approach in developmental stages is challenging, as its unbalance could lead to undesired outcomes.

Cannabinoid-based products in pediatrics

Preclinical and clinical studies that investigated the effects of *Cannabis* consumption on brain development reported several harmful effects (Poleg, Golubchik, Offen, & Weizman, 2019). On the other hand, several experimental studies have demonstrated therapeutic properties of isolated cannabinoids in in vitro and in vivo models (Kaur, Ambwani, & Singh, 2016). The deleterious effects of *Cannabis* are mostly related to Δ^9-THC psychoactivity (Poleg et al., 2019), due to its agonism for both CB1 and CB2 receptors. In contrast, CBD binds with lower affinity to cannabinoid receptors and is devoid of psychotomimetic effects, being beneficial on several brain functions (Campbell, Phillips, & Manasco, 2017).

Cannabis-derived FDA-approved products are already available ("FDA Regulation of Cannabis and Cannabis-Derived Products, Including Cannabidiol (CBD) | FDA,"). Dronabinol (synthetic Δ^9-THC; marketed as Marinol), nabilone (synthetic dimethylheptyl analogue of Δ^9-THC; marketed as Cesamet), and nabiximols (plant-derived Δ^9-THC and CBD, trade name Sativex) (Table 36.1) are indicated for symptoms associated with chemotherapy, multiple sclerosis, and acquired immunodeficiency syndrome (AIDS). Nabiximols is also on trial for cerebral palsy in children. Epidiolex, a plant-derived cannabinoid product, has already been approved for refractory infantile epilepsy and is in clinical trial for other children disorders, including ASD and schizophrenia.

Until recently, the beneficial outcomes of cannabinoids in pediatrics were mainly based upon case reports and parents' surveys (Table 36.2). These surveys reported improvements in neurodevelopmental features, such as cognitive and motor function, and reduced seizure frequency in epileptic children. On the other hand, not all studies reported favorable results and rarely described adverse effects and drug administration issues. For these reasons, clinical trials are indispensable for both the investigation of therapeutic and toxicological aspects of cannabinoid therapy use during neurodevelopmental stages and the standardization of drug administration protocols.

TABLE 36.1 *Cannabis*-derived products indications for pediatric patients.

Cannabinoid	*Cannabis*-derived products	FDA-approved indication	Clinical trials for neurodevelopmental pediatric conditions
CBD (Plant-derived)	Epidiolex	Seizures associated with Lennox–Gastaut or Dravet syndromes in patients 2 years of age and older;	Autism Spectrum Disorder Infantile Spasms; Epilepsy Childhood Absence Epilepsy Prader–Willi Syndrome Seizures Lennox–Gastaut Syndrome Dravet Syndrome Intractable Epilepsy Sturge–Weber Syndrome Rett Syndrome Schizophrenia
Dronabinol (Synthetic Δ^9-THC)	Marinol, Syndros	Anorexia, weight loss in patients with AIDS; nausea and vomiting associated with cancer chemotherapy;	
Nabilone (Synthetic Δ^9-THC)	Cesamet	Nausea and vomiting associated with cancer chemotherapy;	
Nabiximols (1:1 - Plant-derived CBD and Δ^9-THC)	Sativex[a]	Moderate to severe spasticity due to multiple sclerosis[a].	Cerebral Palsy

CBD, Cannabidiol; *Δ^9-THC*, delta-9-tetrahydrocannabinol; *FDA*, Food and Drug Administration; *EMA*, European Medicines Agency.
[a]*EMA approval and FDA accepted for investigation. Available on www.fda.gov.*

TABLE 36.2 Pure CBD and relevant *Cannabis* extract case reports, parent surveys, and retrospective chart reviews published until October 2019 with children and young adults.

Author/year	Design	Condition	Cannabinoid	Outcomes
Rajaraman, Sankar, and Hussain (2018)	Case report	Refractory status epilepticus	CBD (Epidiolex)	Resolution of status epilepticus
Knupp et al. (2019)	Prospective Observational study	Epilepsy	Artisanal *Cannabis* extracts	Seizures improvement Adverse effects: Increased seizures
Hausman-Kedem, Menascu, and Kramer (2018)	Observational study	Refractory epilepsy	*Cannabis* oil extract	Seizures improvement Mild adverse reactions
Treat, Chapman, Colborn, and Knupp (2017)	Retrospective chart review	Epilepsy	*Cannabis* extract	Seizures improvement Adverse effects: worsening seizures, somnolence, and gastrointestinal symptoms
Rosemergy, Adler, and Psirides (2016)	Case report	Refractory status epilepticus	*Cannabis* extract	Not effective
Crippa, Crippa, Hallak, Martín-Santos, and Zuardi (2016)	Case report	Refractory epilepsy	CBD	Complete seizure remission Improvement of previous Δ^9-THC intoxication
Aguirre-Velázquez (2017)	Parent survey	Refractory epilepsy	CBD-based homemade *Cannabis* extracts	Improvement in seizures, emotional state, attention, communication, sleep, and diet Adverse effects: increased appetite, changes in sleep patterns exacerbation of seizures
Shannon and Opila-Lehman (2016)	Case report	Anxiety, post-traumatic stress disorder	CBD oil	Increase in sleep quality and quantity Decrease in anxiety No side effects
Tzadok et al. (2016)	Retrospective chart review	Refractory epilepsy	CBD-enriched *Cannabis* extract	Improvements in seizure, behavior and alertness, language, communication, motor skills, and sleep Adverse effects: somnolence, fatigue, gastrointestinal disturbances and irritability, aggravation of seizures
Press, Knupp, and Chapman (2015)	Parental retrospective chart	Refractory epilepsy	Any oral *Cannabis* (mostly high CBD extract)	Improvement in seizures, behavior/alertness, language, and motor skills Adverse effects: worsening of seizures, somnolence, gastrointestinal symptoms
Saade and Joshi (2015)	Case report	Malignant migrating partial seizures	CBD	Seizures improvement Developmental gains
Hussain et al. (2015)	Parent survey	Refractory epilepsy	CBD-enriched *Cannabis*	Seizures improvement Increased appetite Weight gain
Maa and Figi (2014)	Case report	Dravet syndrome	CBD-enriched *Cannabis* extract	Seizures improvement Weaned from other antiepileptic drugs Improved autistic behaviors
Porter and Jacobson (2013)	Parent survey	Refractory epilepsy	CBD-enriched *Cannabis* extract	Improvement in seizures, mood, alertness, sleep Adverse effects: drowsiness, fatigue, decreased appetite

Search terms: "Cannabidiol" OR "Cannabis". *CBD*, Cannabidiol.

A search on the clinicaltrials.gov website identified 15 completed protocols for the use of cannabinoids in pediatric patients (0–17 years), as well as 27 ongoing trials (Table 36.3) ("Home - ClinicalTrials.gov,"). There has been a recent increase in clinical trials for children involving CBD, mainly for infantile epilepsy treatment, which along with several ongoing trials for cannabinoids use for NDDs and other children and adolescents' diseases will aid further enlightenment on cannabinoid's therapeutic and adverse effects.

TABLE 36.3 Completed and ongoing clinical trials with cannabinoids until October 2019.

Condition	Study ID	Cannabinoid
Epilepsy syndromes, seizures (*completed trials*)	NCT02447198 NCT02324673 NCT02987114 NCT02224703 NCT02224690 NCT02318602 NCT02091206 NCT02544763 NCT02224560 NCT02091375	CBD, GWP42003-P (CBD), PLT101 (CBD), Epidiolex (CBD)
Autism spectrum disorder (*completed trial*)	NCT02956226	CBD-enriched extract
Chronic pain (*completed trial*)	NCT00153192	Δ^9-THC (Marinol)
Multiple sclerosis/diabetes/cancer/neuropathic pain (*completed trials*)	NCT02073474 NCT02073474	Sativex (CBD: Δ^9-THC)
Acne vulgaris (*completed trial*)	NCT03573518	BTX 1503 (CBD)
Autism spectrum disorder (*ongoing trials*)	NCT03900923	CBD
	NCT03202303	CBDV
	NCT03699527	Medical *Cannabis*
Infantile spasms (*ongoing trials*)	NCT02953548 NCT02954887 NCT03421496	GWP42003-P (CBD)
Childhood absence epilepsy (*ongoing trials*)	NCT03336242 NCT03355300	CBD
Fragile X syndrome (*ongoing trial*)	NCT03614663	ZYN002 (CBD)
Schizophrenia (*ongoing trial*)	NCT02932605	CBD
Infantile tumor (*ongoing trial*)	NCT03052738	Medical *Cannabis*
Atopic dermatitis (*ongoing trial*)	NCT03824405	BTX 1204 (CBD)
Rett syndrome (*ongoing trial*)	NCT03848832	GWP42003-P (CBD)
Epilepsy syndromes, seizures (*ongoing trials*)	NCT02844933 NCT03458416 NCT02332655 NCT02286986 NCT02783092 NCT02224573 NCT02695537 NCT02397863 NCT04133480 NCT02544750	CBD, GWP42003-P (CBD), Epidiolex (CBD)
	NCT03848481	CBDV
	NCT02523183	Medical *Cannabis*
	NCT02983695	TIL-TC150 (CBD: Δ^9-THC)
	NCT03024827	CanniMed (CBD: Δ^9-THC)

Search on clinicaltrials.gov with the terms: "Cannabidiol" OR "THC" OR "Cannabinoids", child: birth–17 years. *CBD*, Cannabidiol; *CBDV*, Cannabidivarin; *Δ^9-THC*, delta-9-tetrahydrocannabinol.

However, the use of cannabinoid products in pediatrics remains a controversial subject, since *Cannabis*, and its derived compounds, may promote adverse effects on the development of children and adolescents brain through the ECS (Campbell et al., 2017). The current knowledge about isolated cannabinoids or *Cannabis* for the main NDDs as well as the ECS′ role for such disorders is presented as follows.

Attention deficit/hyperactivity disorder (ADHD)

ADHD is the most diagnosed behavioral disorder in children, with increasing prevalence over the last years (Felt, Biermann, Christner, Kochhar, & Van Harrison, 2014). Despite the lack of scientific evidence, some adults with the disease self-medicate with *Cannabis* for ameliorating symptoms. A pilot randomized placebo-controlled trial of nabiximols (Δ^9-THC:CBD) in 30 adults with ADHD found improvements in hyperactivity/impulsivity and cognitive inhibition measures, but not in inattention or emotional lability (Cooper et al., 2017). A possible mechanism of cannabinoids action in ADHD is via alterations in striatal dopamine, as Δ^9-THC administration increase dopamine in brain regions implicated in this disease (Stokes et al., 2010). Additionally, a genetic investigation in adolescents with ADHD found enriched expression of the ECS pathway components (Khadka et al., 2016).

On the other hand, structural brain imaging studies in adult ADHD patients have shown that *Cannabis* users have decreased cortical thickness in regions related to motor control, working memory, and inhibitory control, compared to nonusers (Lisdahl et al., 2016). A study with adolescent synthetic cannabinoids users, with and without ADHD, found that these compounds also exert negative effects on brain morphology and comorbidity of ADHD (Colak, Cakmak Celik, Zorlu, Kitis, & Yuncu, 2019).

A recent parent survey with ASD children treated with CBD reported substantial improvements in hyperactivity symptoms, a common comorbid symptom of ASD also related to ADHD (Barchel et al., 2019), meaning a possible use of this cannabinoid for ADHD too.

Thus, data regarding *Cannabis* and cannabinoid use in ADHD suggest that, although it can improve symptoms, it can also cause potentially deleterious structural changes of important brain areas.

Autism spectrum disorder (ASD)

ASD is a group of neurodevelopmental disorders frequently associated with general cognitive deficits (Zamberletti, Gabaglio, & Parolaro, 2017). The ECS is often affected in ASD patients with comorbidities such as seizures, anxiety, cognitive impairments, and sleep disturbances (Zamberletti et al., 2017). ASD patients present lower plasma levels of AEA, and an interaction between endocannabinoid signaling and oxytocin-mediated social reward has also been demonstrated (Karhson, Hardan, & Parker, 2016). Thus, deficits in AEA signaling may contribute to social impairment in ASD (Wei et al., 2015).

CBD is suggested as a candidate for treatment of ASD, since this cannabinoid is able to modulate AEA plasma levels (Poleg et al., 2019). Indeed, a recent survey with parents of 53 children (4–22 years old) reported improvements in the comorbidity symptoms after CBD treatment in children with ASD (Barchel et al., 2019). Self-injury, rage attacks, hyperactivity symptoms, sleep problems, and anxiety improved considerably. Adverse effects, mostly somnolence and change in appetite, were mild; however, the long-term effects should be evaluated in large-scale studies (Barchel et al., 2019). Further preclinical, together with the ongoing clinical, studies (Table 36.3) are needed to establish the use of CBD and other cannabinoids for ASD symptoms and comorbidities.

Motor and tic disorders

Given the benefits of cannabinoids reported on behavior and motor skills and the ECS's function in motor control (Hoch et al., 2019), they might be of interest for neurodevelopmental motor disorders.

A recent study evaluated cannabinoid (CBD:Δ^9-THC 20:1 or 6:1) in children with complex motor disorders. Significant improvement in spasticity and dystonia, sleep, pain severity, and quality of life was observed. Rare adverse effects included worsening of seizures, behavioral changes, and somnolence (Libzon et al., 2018).

Tourette syndrome, characterized by motor and vocal tics, presents comorbidities similar to ADHD and obsessive–compulsive behaviors (Gill & Kompoliti, 2020). As discussed earlier, cannabinoids seem to be effective in the attenuation of ADHD-related symptoms.

In a case report, Δ^9-THC decreased tic severity and improved life quality of a 15-year-old with refractory Tourette syndrome, by modulating neurotransmitters via ECS (Hasan et al., 2010). In a more recent case report, a female patient showed a rapid and highly significant improvement in the severity of symptoms following administration of a daily dosage of Δ^9-THC:CBD (1:2). As the patient had no improvements with regular medicines or Δ^9-THC alone, these beneficial effects may be due solely to CBD (Pichler, Kawohl, Seifritz, & Roser, 2019).

Although anecdotal case reports have described patients having improvement in tics after using *Cannabis*, only two small randomized, placebo-controlled trials of Δ^9-THC in adult patients have been published (Artukoglu & Bloch, 2019). Δ^9-THC was effective and safe in treating tics and obsessive–compulsive behavior (K R Müller-Vahl et al., 2002, 2003). Thus, although there is insufficient evidence to support its use for children with Tourette syndrome, cannabinoid-based treatments seem to be a promising approach for new medications for motor and tics disorders (Wong & Wilens, 2017).

Schizophrenia

Scientific evidence associates *Cannabis* use with increased risk of developing psychotic disorders, including schizophrenia, at earlier ages. Patients with schizophrenia and a history of chronic *Cannabis* use present lower levels of AEA (Manseau & Goff, 2015). This endocannabinoid may be part of a physiological compensatory mechanism for psychosis in the brain, since it is elevated in response to onset of psychosis. Chronic *Cannabis* users with schizophrenia do not have elevated AEA levels, suggesting that *Cannabis* may reduce the brain's ability to balance a psychotic process, pointing toward a mechanism underlying the *Cannabis*–psychosis link in vulnerable individuals (Manseau & Goff, 2015).

Most of the effects of *Cannabis* on psychotic symptoms are due to Δ^9-THC, and CBD can counter many of them (Manseau & Goff, 2015). Indeed, adjuvant treatment with isolated CBD for schizophrenia promotes significant improvements in both psychosis and cognition, with only mild side effects (Hoch et al., 2019). Other *Cannabis*-based medicines, Rimonabant (Acomplia or Redufast) and Drinabant, showed no benefits and presented serious adverse effects, resulting with their withdrawal from the market (Hoch et al., 2019).

Therefore, despite the positive outcomes obtained with CBD for the treatment of schizophrenia and its comorbidities, this disease may represent an example of how *Cannabis* use can induce impairments in brain development through modulation of ECS. A clinical trial of CBD is currently ongoing for schizophrenia in patients including adolescents (Table 36.3). Together with completed and ongoing trials in adult patients, these trials will certainly clarify cannabinoids' effects in this disease.

Bipolar disorder

Bipolar disorder (BD) is a disorder genetically and clinically related to schizophrenia, characterized by unsteady mood states, which switch from (hypo)mania to depression (De Hert et al., 2011). A study with 90 BD patients found that *Cannabis* use is associated with a decrease in the age of onset of BD symptoms, which was more pronounced than the decrease in the age of onset of schizophrenia (De Hert et al., 2011). Some BD patients use it to alleviate symptoms of both mania and depression (Connor, Ford, Pearson, Scranton, & Dusad, 2017). Early-onset BD patients present increased comorbid diagnoses, symptom severity, parental stress, and child high-risk behaviors (Connor et al., 2017). These findings suggest that there is a partly shared, preexisting genetic liability for BD and schizophrenia, involving the ECS during neurodevelopment (De Hert et al., 2011).

Both serotonergic and dopaminergic neurotransmission play a prominent role in the pathophysiology of neuropsychiatric disorders. Since these systems are modulated by the ECS, it is suggested as a target to control mental illnesses including BD. Besides, AA and inflammatory pathways can play a part in the pathophysiology of BD, and a link from the ECS to inflammatory pathways has been strongly established (Arjmand et al., 2019).

Very little is known about cannabinoid treatments for BD. The efficacy and safety of CBD in two adult patients showed symptomatic improvement while on regular treatment plus CBD, but not during CBD monotherapy, suggesting that CBD may be ineffective for the manic episode (Zuardi et al., 2010). In early-onset BD patients, no data is available.

A double-blind, randomized placebo-controlled clinical trial is currently recruiting adult patients to investigate CBD on depressive symptoms of BD (www.clinicaltrials.gov, NCT03310593).

Infantile epilepsy

Cannabis-derived products are used mainly for childhood severe epilepsies including Dravet and Lennox–Gastaut syndromes. These early-onset disorders are characterized by frequent, refractory seizures and neurodevelopmental delays, which lead to impaired quality of life (Devinsky et al., 2018).

In epilepsy, ECS's retrograde signaling regulates seizure activity and neuronal hyperexcitability (Gaston & Friedman, 2017). There is also increasing evidence for ECS′ importance for the epileptogenesis process (Gaston & Szaflarski, 2018). Agonist activity at cannabinoid receptors has demonstrated an anticonvulsant effect, but antagonism can exacerbate seizures (Braakman, van Oostenbrugge, van Kranen-Mastenbroek, & de Krom, 2009).

Studies have demonstrated anticonvulsant properties of both Δ^9-THC and CBD (Gaston & Szaflarski, 2018). However, the use of CB1 agonists such as Δ^9-THC, present in *Cannabis* extracts, is unfeasible in children due to its psychoactive effects (Blair, Deshpande, & DeLorenzo, 2015). Alternatively, CBD has low affinity for cannabinoid receptors, acting indirectly on the ECS increasing the availability AEA, which activates CB1 (Gaston & Szaflarski, 2018). Many other actions of CBD could contribute to its anticonvulsant activity, such as activation of the transient receptor potential of vanilloid type-1 (TRPV1), antagonist action at G-protein-coupled receptor 55 (GPR55), targeting of sodium channels with abnormal activities (mutated in Dravet syndrome) and modulation of tumor necrosis factor alpha release (TNF-α) (Gaston & Szaflarski, 2018).

The effects of CBD-enriched *Cannabis* extracts in infantile epilepsy were first presented in case reports or parents' surveys (Table 36.2). The most famous case, a child with Dravet syndrome who received a CBD-enriched *Cannabis* extract (Charlotte's Web case), reported a remarkable improvement in seizure and cognition (Maa & Figi, 2014).

Published studies of artisanal *Cannabis* products, and open-label, randomized controlled trials of CBD, indicate it is well tolerated despite some adverse effects, such as diarrhea and sedation, and interaction with clobazam. Also, CBD promoted a significant improvement in seizure frequency in a large number of patients with various types of treatment-refractory epilepsy (Gaston & Szaflarski, 2018; Schonhofen et al., 2018).

Recently, in an open-label extension study evaluating the long-term safety and efficacy of CBD, patients with Lennox−Gastaut syndrome who completed one of 2 randomized, double-blind, placebo-controlled trials had improvements in drop seizure frequency and in overall conditions, along with mostly mild or moderate adverse events (Thiele et al., 2019).

Thus, CBD demonstrates sustained reductions in seizures with an acceptable long-term safety profile. Among children with treatment-resistant epilepsy, those suffering from early-onset and severe epilepsies present the greatest neurodevelopmental problems, including intellectual disability and ASD (Gaston & Szaflarski, 2018). As discussed earlier, studies with *Cannabis* products showed improvements for such problems.

Despite the large number of studies, they vary in the dosage and duration of treatment, and many lack long-term follow-up (Wong & Wilens, 2017), what makes it difficult to evaluate the real benefits and risks of CBD-based therapies. Current ongoing trials will improve the quality of evidence (Table 36.2).

Hypoxic-ischemic encephalopathy

Neonatal hypoxic-ischemic encephalopathy (NHIE) is the most common cause of cerebral palsy. Many characteristic actions of cannabinoids suggest a neuroprotective effect after an acute hypoxic-ischemic insult, and the ECS is considered as an endogenous neuroprotective system. In fact, enhanced endocannabinoid levels have been observed in newborn's brains after acute excitotoxic injury or hypoxic-ischemic insults in animal models. Either Δ^9-THC or AEA was able to reduce brain damage in a CB1 activation-dependent manner. There are also evidence of neuroprotective actions of CBD by reducing brain damage, restoring motor function and cognitive damage in NHIE (Sagredo et al., 2018).

Other neurodevelopmental conditions: behavioral, cognitive, and mood disorders

The ECS participates in the control of emotional states, hence cannabinoids could be useful for related conditions. Indeed, positive anxiolytic effects of CBD have been demonstrated in adult patients with generalized social anxiety, with no adverse effects (Hoch et al., 2019). Anxiety and sleep disorders are also often the result of posttraumatic stress disorder. In a case report of a 10-year-old girl with such symptoms, CBD decreased anxiety and improved quality and quantity of sleep (Table 36.2) (Shannon & Opila-Lehman, 2016).

CBD promoted improvements in hyperactivity, irritability, lethargy, and speech in infantile ASD patients and in self-injurious behavior in intellectually disabled adolescents. Some of its benefits on behavior and motor skills in other conditions may be transferable to this population as well (Campbell et al., 2017).

Finally, as discussed earlier, early *Cannabis* use has detrimental effects on cognitive function (e.g., executive function, learning, memory, and attention) and enhances the occurrence of mental illnesses (Manseau & Goff, 2015).

Conclusions

Effects of *Cannabis* and isolated cannabinoids in the main NDDs and the ECS mechanisms are summarized in Fig. 36.3. The ECS might be part of the pathophysiology of NDDs or acts as a protective mechanism. Thus, as cannabinoids can interact with this system, they are relevant as drug candidates. However, as the ECS participates in normal neurodevelopment from the embryo to adolescence, cannabinoid exposure can potentially disrupt this process.

Despite the association of *Cannabis* with neurodevelopmental impairments in BD and schizophrenia, CBD presents a general safe profile. This cannabinoid is effective for some infantile epileptic syndromes, including Dravet and Lennox–Gastaut syndromes, and for the most common comorbidities of NDDs such as sleep disorder, anxiety, and seizures. In other comorbidities such as psychosis, addictive behavior, mood or cognitive disorders, and aggressiveness, the level of evidence is still low (Poleg et al., 2019). Furthermore, side effects occur, though severe ones are rare. To provide reliable treatment recommendations, more high-quality clinical trials are required.

Applications to other areas of development

This chapter is focused upon cannabinoid treatment of NDDs in children (0–17 years) and its beneficial effects, mainly CBD, for most of the NDDs reviewed here. Moreover, they have many other capabilities, such as antioxidant and anti-inflammatory effects, potentially neuroprotective for the underdeveloped brain, which is more susceptible to neurotoxic damage. The ECS participates both in normal development and in the pathogenic processes involved with diseases during this period. ECS is also important in the mature brain's synapses regulation and is active from early embryogenesis. Thus, it is important to consider maternal exposure to *Cannabis* and the implications of cannabinoids treatment upon fetal development. The neurodevelopmental data in human and animal models suggest that prenatal exposure to *Cannabis* or Δ^9-THC may lead to changes in aspects of learning, memory, and decision-making, psychological well-being, and motor development in the offspring. These effects are mediated by the CB1 receptor (Grant, Petroff, Isoherranen, Stella, & Burbacher, 2018). CBD is less associated with such impairments, but less studied in prenatal period.

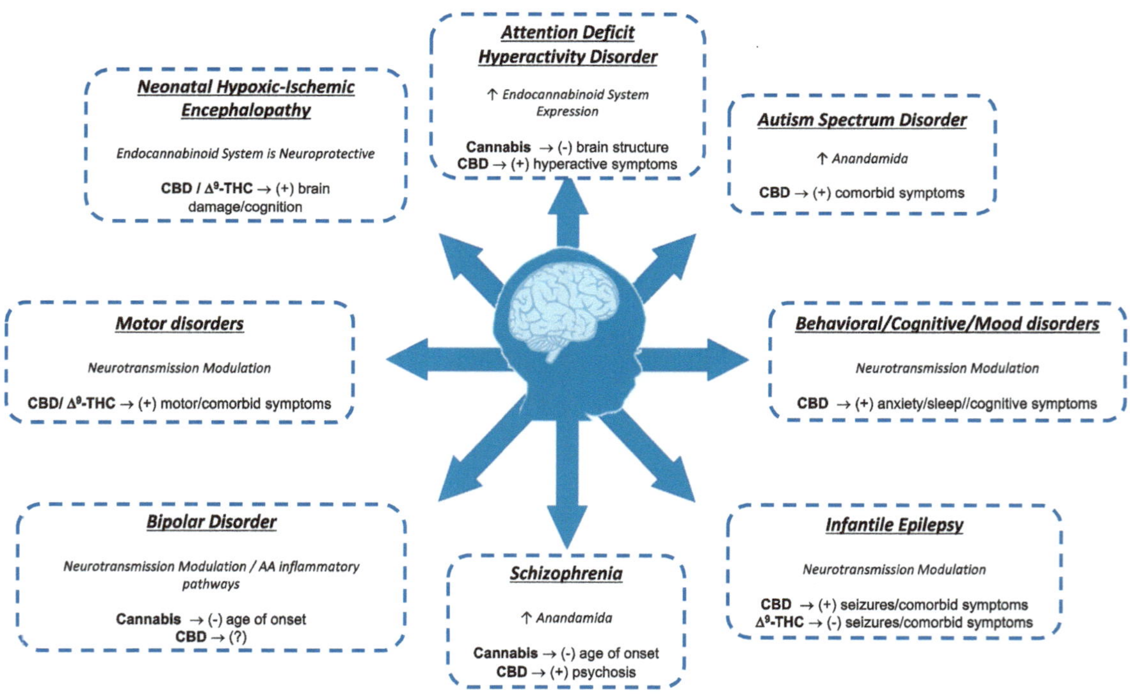

FIGURE 36.3 **Endocannabinoid signaling in neurodevelopmental disorders.** Proposed participation of the endocannabinoid system, and the proposed therapeutic/deleterious effects of cannabinoids. *CBD*, cannabidiol; *Δ^9-THC*, delta-9-tetrahydrocannabinol. Improved (+) and worsened (−) effects.

Key facts of cannabinoids in neurodevelopmental disorders

Early *Cannabis* use induces psychosis and emotional/mood disorders in susceptible people.
Delta-9-tetrahydrocannabinol presents psychoactive effects through cannabinoid receptor 1 agonism.
Cannabidiol is safer in children and is protective for various conditions.
Drug-resistant refractory infantile epileptic syndromes are the most relevant target of cannabidiol.
The endocannabinoid system might be a therapeutic target.

Summary points

- The endocannabinoid system modulates neurodevelopment since embryo until the adolescence.
- The endocannabinoid system might be involved either in the pathophysiology of neurodevelopmental diseases and in protective mechanisms.
- Cannabinoids have shown both improvements and adverse effects depending on the condition.
- *Cannabis* is associated with neurodevelopmental impairments in bipolar disorder and schizophrenia.
- Cannabidiol presents a general safer profile and is effective for some infantile epileptic syndromes and for the most common comorbidities of neurodevelopmental diseases.

References

Aguirre-Velázquez, C. G. (2017). Report from a survey of parents regarding the use of cannabidiol (*Medicinal cannabis*) in Mexican children with refractory epilepsy. *Neurology Research International, 2017*, 2985729. https://doi.org/10.1155/2017/2985729.

Arjmand, S., Behzadi, M., Kohlmeier, K. A., Mazhari, S., Sabahi, A., & Shabani, M. (2019). Bipolar disorder and the endocannabinoid system. *Acta Neuropsychiatrica, 31*, 193–201. https://doi.org/10.1017/neu.2019.21.

Artukoglu, B. B., & Bloch, M. H. (2019). The potential of cannabinoid-based treatments in tourette syndrome. *CNS Drugs, 33*(5), 417–430. https://doi.org/10.1007/s40263-019-00627-1.

Barchel, D., Stolar, O., De-Haan, T., Ziv-Baran, T., Saban, N., Fuchs, D. O., … Berkovitch, M. (January 2019). Oral cannabidiol use in children with autism spectrum disorder to treat related symptoms and co-morbidities. *Frontiers in Pharmacology, 9*. https://doi.org/10.3389/fphar.2018.01521.

Blair, R. E., Deshpande, L. S., & DeLorenzo, R. J. (2015). Cannabinoids: Is there a potential treatment role in epilepsy? *Expert Opinion on Pharmacotherapy, 16*(13), 1911–1914. https://doi.org/10.1517/14656566.2015.1074181.

Braakman, H. M. H., van Oostenbrugge, R. J., van Kranen-Mastenbroek, V. H. J. M., & de Krom, M. C. T. F. M. (2009). Rimonabant induces partial seizures in a patient with a history of generalized epilepsy. *Epilepsia, 50*(9), 2171–2172. https://doi.org/10.1111/j.1528-1167.2009.02203.x.

Campbell, C. T., Phillips, M. S., & Manasco, K. (2017). Cannabinoids in pediatrics. *Journal of Pediatric Pharmacology and Therapeutics: JPPT: The Official Journal of PPAG, 22*(3), 176–185. https://doi.org/10.5863/1551-6776-22.3.176.

Cardoso, A. R., Lopes-Marques, M., Silva, R. M., Serrano, C., Amorim, A., Prata, M. J., & Azevedo, L. (2019). Essential genetic findings in neurodevelopmental disorders. *Human Genomics, 13*(1). https://doi.org/10.1186/s40246-019-0216-4.

Colak, C., Cakmak Celik, Z., Zorlu, N., Kitis, O., & Yuncu, Z. (2019). Cortical thickness and subcortical volume in adolescent synthetic cannabinoid users with or without ADHD: A preliminary study. *Archives of Neuropsychiatry, 56*(3), 167–172. https://doi.org/10.29399/npa.23495.

Connor, D. F., Ford, J. D., Pearson, G. S., Scranton, V. L., & Dusad, A. (2017). Early-onset bipolar disorder: Characteristics and outcomes in the clinic. *Journal of Child and Adolescent Psychopharmacology, 27*(10), 875–883. https://doi.org/10.1089/cap.2017.0058.

Cooper, R. E., Williams, E., Seegobin, S., Tye, C., Kuntsi, J., & Asherson, P. (2017). Cannabinoids in attention-deficit/hyperactivity disorder: A randomised-controlled trial. *European Neuropsychopharmacology: The Journal of the European College of Neuropsychopharmacology, 27*(8), 795–808. https://doi.org/10.1016/j.euroneuro.2017.05.005.

Crippa, J. A. S., Crippa, A. C. S., Hallak, J. E. C., Martín-Santos, R., & Zuardi, A. W. (2016). Δ9-THC intoxication by cannabidiol-enriched *cannabis* extract in two children with refractory epilepsy: Full remission after switching to purified cannabidiol. *Frontiers in Pharmacology, 7*, 359. https://doi.org/10.3389/fphar.2016.00359.

De Hert, M., Wampers, M., Jendricko, T., Franic, T., Vidovic, D., De Vriendt, N., … van Winkel, R. (2011). Effects of *cannabis* use on age at onset in schizophrenia and bipolar disorder. *Schizophrenia Research, 126*(1–3), 270–276. https://doi.org/10.1016/j.schres.2010.07.003.

Devinsky, O., Patel, A. D., Thiele, E. A., Wong, M. H., Appleton, R., Harden, C. L., … GWPCARE1 Part A Study Group, O. behalf of the G. P. A. S. (2018). Randomized, dose-ranging safety trial of cannabidiol in Dravet syndrome. *Neurology, 90*(14), e1204–e1211. https://doi.org/10.1212/WNL.0000000000005254.

Di Marzo, V., & De Petrocellis, L. (2012). Why do cannabinoid receptors have more than one endogenous ligand? *Philosophical Transactions of the Royal Society B: Biological Sciences, 367*(1607), 3216–3228. https://doi.org/10.1098/rstb.2011.0382.

Díaz-Alonso, J., Aguado, T., Wu, C.-S., Palazuelos, J., Hofmann, C., Garcez, P., … Galve-Roperh, I. (2012). The CB_1 cannabinoid receptor drives corticospinal motor neuron differentiation through the Ctip2/Satb2 transcriptional regulation axis. *Journal of Neuroscience: The Official Journal of the Society for Neuroscience, 32*(47), 16651–16665. https://doi.org/10.1523/JNEUROSCI.0681-12.2012.

Djeungoue-Petga, M.-A., & Hebert-Chatelain, E. (2017). Linking mitochondria and synaptic transmission: The CB1 receptor. *BioEssays, 39*(12), 1700126. https://doi.org/10.1002/bies.201700126.

Dow-Edwards, D., & Silva, L. (January 1, 2017). Endocannabinoids in brain plasticity: Cortical maturation, HPA axis function and behavior. *Brain Research*. https://doi.org/10.1016/j.brainres.2016.08.037. Elsevier.

FDA Regulation of Cannabis and Cannabis-Derived Products, Including cannabidiol (CBD) | FDA. (n.d.). Retrieved November 2, 2019 from https://www.fda.gov/news-events/public-health-focus/fda-regulation-cannabis-and-cannabis-derived-products-including-cannabidiol-cbd.

Felt, B. T., Biermann, B., Christner, J. G., Kochhar, P., & Van Harrison, R. (2014). Diagnosis and management of ADHD in children. *American Family Physician, 90*(7), 456–464. https://doi.org/10.1002/psb.1710.

Gaston, T. E., & Friedman, D. (2017). Pharmacology of cannabinoids in the treatment of epilepsy. *Epilepsy and Behavior, 70*, 313–318. https://doi.org/10.1016/j.yebeh.2016.11.016.

Gaston, T. E., & Szaflarski, J. P. (November 1, 2018). *Cannabis* for the treatment of epilepsy: An update. *Current Neurology and Neuroscience Reports, 1*. https://doi.org/10.1007/s11910-018-0882-y. Current Medicine Group LLC.

Gill, C. E., & Kompoliti, K. (2020). Clinical features of tourette syndrome. *Journal of Child Neurology, 35*(2), 166–174. https://doi.org/10.1177/0883073819877335.

Grant, K. S , Petroff, R., Isoherranen, N., Stella, N., & Burbacher, T. M. (2018). *Cannabis* use during pregnancy: Pharmacokinetics and effects on child development. *Pharmacology and Therapeutics, 182*(206), 133–151. https://doi.org/10.1016/j.pharmthera.2017.08.014.

Hasan, A., Rothenberger, A., Münchau, A., Wobrock, T., Falkai, P., & Roessner, V. (2010). Oral Δ9-tetrahydrocannabinol improved refractory gilles de la Tourette syndrome in an adolescent by increasing intracortical inhibition: A case report. *Journal of Clinical Psychopharmacology, 30*(2), 190–192. https://doi.org/10.1097/JCP.0b013e3181d236ec.

Hausman-Kedem, M., Menascu, S., & Kramer, U. (2018). Efficacy of CBD-enriched medical *cannabis* for treatment of refractory epilepsy in children and adolescents - an observational, longitudinal study. *Brain and Development, 40*(7), 544–551. https://doi.org/10.1016/j.braindev.2018.03.013.

Hoch, E., Niemann, D., von Keller, R., Schneider, M., Friemel, C. M., Preuss, U. W., … Pogarell, O. (February 1, 2019). How effective and safe is medical *cannabis* as a treatment of mental disorders? A systematic review. *European Archives of Psychiatry and Clinical Neuroscience*. https://doi.org/10.1007/s00406-019-00984-4. Dr. Dietrich Steinkopff Verlag GmbH and Co. KG.

Home - ClinicalTrials.gov. (n.d.). Retrieved November 4, 2019 from https://clinicaltrials.gov/ct2/home.

Hussain, S. A., Zhou, R., Jacobson, C., Weng, J., Cheng, E., Lay, J., … Sankar, R. (2015). Perceived efficacy of cannabidiol-enriched *cannabis* extracts for treatment of pediatric epilepsy: A potential role for infantile spasms and Lennox-Gastaut syndrome. *Epilepsy and Behavior, 47*, 138–141. https://doi.org/10.1016/j.yebeh.2015.04.009.

Karhson, D. S., Hardan, A. Y., & Parker, K. J. (2016). Endocannabinoid signaling in social functioning: An RDoC perspective. *Translational Psychiatry, 6*(9). https://doi.org/10.1038/tp.2016.169.

Kaur, R.,R., Ambwani, S., & Singh, S. (2016). Endocannabinoid system: A multi-facet therapeutic target. *Current Clinical Pharmacology, 11*(2), 110–117. https://doi.org/10.2174/1574884711666160418105339.

Khadka, S., Pearlson, G. D., Calhoun, V. D., Liu, J., Gelernter, J., Bessette, K. L., & Stevens, M. C. (July 2016). Multivariate imaging genetics study of MRI gray matter volume and SNPs reveals biological pathways correlated with brain structural differences in attention deficit hyperactivity disorder. *Frontiers in Psychiatry, 7*. https://doi.org/10.3389/fpsyt.2016.00128.

Knupp, K. G., Rice, J. D., Helmkamp, L. J., Galinkin, J., Sempio, C., Jost, K., & Chapman, K. E. (2019). Prospective evaluation of oral *cannabis* extracts in children with epilepsy. *Seizure, 72*, 23–27. https://doi.org/10.1016/j.seizure.2019.09.007.

Libzon, S., Schleider, L. B. L., Saban, N., Levit, L., Tamari, Y., Linder, I., … Blumkin, L. (2018). Medical *cannabis* for pediatric moderate to severe complex motor disorders. *Journal of Child Neurology, 33*(9), 565–571. https://doi.org/10.1177/0883073818773028.

Lisdahl, K. M., Tamm, L., Epstein, J. N., Jernigan, T., Molina, B. S. G., Hinshaw, S. P., … MTA Neuroimaging Group. (2016). The impact of ADHD persistence, recent *cannabis* use, and age of regular *cannabis* use onset on subcortical volume and cortical thickness in young adults. *Drug and Alcohol Dependence, 161*, 135–146. https://doi.org/10.1016/j.drugalcdep.2016.01.032.

Maa, E., & Figi, P. (2014). The case for medical marijuana in epilepsy. *Epilepsia, 55*(6), 783–786. https://doi.org/10.1111/epi.12610.

Mackie, K. (2005). Distribution of cannabinoid receptors in the central and peripheral nervous system. In *Handbook of Experimental Pharmacology* (Vol. 168, pp. 299–325). Retrieved from: http://www.ncbi.nlm.nih.gov/pubmed/16596779.

Manseau, M. W., & Goff, D. C. (October 1, 2015). Cannabinoids and schizophrenia: Risks and therapeutic potential. *Neurotherapeutics*. https://doi.org/10.1007/s13311-015-0382-6. Springer New York LLC.

Mechoulam, R., & Parker, L. A. (2013). The endocannabinoid system and the brain. *Annual Review of Psychology, 64*, 21–47. https://doi.org/10.1146/annurev-psych-113011-143739.

Meyer, H. C., Lee, F. S., & Gee, D. G. (July 2017). The role of the endocannabinoid system and genetic variation in adolescent brain development. *Neuropsychopharmacology*, 1–47. https://doi.org/10.1038/npp.2017.143.

Müller-Vahl, Kirsten, R., Schneider, U., Prevedel, H., Theloe, K., Kolbe, H., … Emrich, H. M. (2003). Δ9-tetrahydrocannabinol (THC) is effective in the treatment of tics in tourette syndrome: A 6-week randomized trial. *Journal of Clinical Psychiatry, 64*(4), 459–465. https://doi.org/10.4088/JCP.v64n0417.

Müller-Vahl, K. R., Schneider, U., Koblenz, A., Jöbges, M., Kolbe, H., Daldrup, T., & Emrich, H. M. (2002). Treatment of tourette's syndrome with delta 9-tetrahydrocannabinol (THC): A randomized crossover trial. *Pharmacopsychiatry, 35*(2), 57–61. https://doi.org/10.1055/s-2002-25028.

Pertwee, R. G., Howlett, A. C., Abood, M. E., Alexander, S. P. H., Marzo, V. Di, Elphick, M. R., … Kunos, G. (2010). International union of basic and clinical pharmacology. LXXIX. Cannabinoid receptors and their ligands: Beyond CB_1 and CB_2. *Pharmacological Reviews, 62*, 588–631. https://doi.org/10.1124/pr.110.003004.588.

Pichler, E.-M., Kawohl, W., Seifritz, E., & Roser, P. (2019). Pure delta-9-tetrahydrocannabinol and its combination with cannabidiol in treatment-resistant tourette syndrome: A case report. *International Journal of Psychiatry in Medicine, 54*(2), 150–156. https://doi.org/10.1177/0091217418791455.

Poleg, S., Golubchik, P., Offen, D., & Weizman, A. (2019). Cannabidiol as a suggested candidate for treatment of autism spectrum disorder. *Progress in Neuro-Psychopharmacology and Biological Psychiatry, 89*, 90–96. https://doi.org/10.1016/j.pnpbp.2018.08.030. Elsevier Inc.

Porter, B. E., & Jacobson, C. (2013). Report of a parent survey of cannabidiol-enriched *cannabis* use in pediatric treatment-resistant epilepsy. *Epilepsy and Behavior: E&B, 29*(3), 574–577. https://doi.org/10.1016/j.yebeh.2013.08.037.

Press, C. A., Knupp, K. G., & Chapman, K. E. (2015). Parental reporting of response to oral *cannabis* extracts for treatment of refractory epilepsy. *Epilepsy and Behavior, 45*, 49–52. https://doi.org/10.1016/j.yebeh.2015.02.043. July 2014.

Rajaraman, R. R., Sankar, R., & Hussain, S. A. (2018). Successful use of pure cannabidiol for the treatment of super-refractory status epilepticus. *Epilepsy and Behavior Case Reports, 10*, 141–144. https://doi.org/10.1016/j.ebcr.2018.07.004.

Rosemergy, I., Adler, J., & Psirides, A. (2016). Cannabidiol oil in the treatment of super refractory status epilepticus. A case report. *Seizure, 35*, 56–58. https://doi.org/10.1016/j.seizure.2016.01.009.

Saade, D., & Joshi, C. (2015). Pure cannabidiol in the treatment of malignant migrating partial seizures in infancy: A case report. *Pediatric Neurology, 52*(5), 544–547. https://doi.org/10.1016/j.pediatrneurol.2015.02.008.

Sagredo, O., Palazuelos, J., Gutierrez-Rodriguez, A., Satta, V., Galve-Roperh, I., & Martínez-Orgado, J. (2018). Cannabinoid signalling in the immature brain: Encephalopathies and neurodevelopmental disorders. *Biochemical Pharmacology, 157*, 85–96. https://doi.org/10.1016/j.bcp.2018.08.014. Elsevier Inc.

Schonhofen, P., Bristot, I. J., Crippa, J. A., Hallak, J. E. C., Zuardi, A. W., Parsons, R. B., & Klamt, F. (2018). Cannabinoid-based therapies and brain development: Potential harmful effect of early modulation of the endocannabinoid system. *CNS Drugs*, 1–16. https://doi.org/10.1007/s40263-018-0550-4.

Shannon, S., & Opila-Lehman, J. (2016). Effectiveness of cannabidiol oil for pediatric anxiety and insomnia as part of posttraumatic stress disorder: A case report. *The Permanente Journal, 20*(4), 108–111. https://doi.org/10.7812/TPP/16-005.

Stokes, P. R. A., Egerton, A., Watson, B., Reid, A., Breen, G., Lingford-Hughes, A., … Mehta, M. A. (2010). Significant decreases in frontal and temporal [11C]-raclopride binding after THC challenge. *NeuroImage, 52*(4), 1521–1527. https://doi.org/10.1016/j.neuroimage.2010.04.274.

Thiele, E., Marsh, E., Mazurkiewicz-Beldzinska, M., Halford, J. J., Gunning, B., Devinsky, O., … Roberts, C. (2019). Cannabidiol in patients with Lennox-Gastaut syndrome: Interim analysis of an open-label extension study. *Epilepsia, 60*(3), 419–428. https://doi.org/10.1111/epi.14670.

Treat, L., Chapman, K. E., Colborn, K. L., & Knupp, K. G. (2017). Duration of use of oral *cannabis* extract in a cohort of pediatric epilepsy patients. *Epilepsia, 58*(1), 123–127. https://doi.org/10.1111/epi.13617.

Tzadok, M., Uliel-Siboni, S., Linder, I., Kramer, U., Epstein, O., Menascu, S., … Ben-Zeev, B. (2016). CBD-enriched medical *cannabis* for intractable pediatric epilepsy. *Seizure, 35*, 41–44. https://doi.org/10.1016/j.seizure.2016.01.004.

Wei, D., Lee, D. Y., Cox, C. D., Karsten, C. A., Penagarikano, O., Geschwind, D. H., … Piomelli, D. (2015). Endocannabinoid signaling mediates oxytocin-driven social reward. *Proceedings of the National Academy of Sciences of the United States of America, 112*(45), 14084–14089. https://doi.org/10.1073/pnas.1509795112.

Wong, S. S., & Wilens, T. E. (2017). Medical cannabinoids in children and adolescents: A systematic review. *Pediatrics, 140*(5), e20171818. https://doi.org/10.1542/peds.2017-1818.

Zamberletti, E., Gabaglio, M., & Parolaro, D. (2017). The endocannabinoid system and autism spectrum disorders: Insights from animal models. *International Journal of Molecular Sciences, 18*(9). https://doi.org/10.3390/ijms18091916.

Zuardi, A., Crippa, J., Dursun, S., Morais, S., Vilela, J., Sanches, R., & Hallak, J. (2010). Cannabidiol was ineffective for manic episode of bipolar affective disorder. *Journal of Psychopharmacology, 24*(1), 135–137. https://doi.org/10.1177/0269881108096521.

Chapter 37

Deep brain stimulation for pediatric dystonia: neurophysiological, behavioral, and functional changes and their clinical relevance

Lior M. Elkaim[1] and George M. Ibrahim[2]

[1]*Division of Neurosurgery, McGill University, McGill University Hospital Center, Montreal, QC, Canada;* [2]*Division of Neurosurgery, Department of Surgery, University of Toronto, Toronto, ON, Canada*

List of abbreviations

AD Alzheimer's disease
BFMDRS
Burke–Fahn–Marsden Dystonia Rating Scale
CP Cerebral Palsy
DBS Deep brain stimulation
ER Endoplasmic reticulum
GA1 Glutaric aciduria type 1
LND Lesch–Nyhan disease
MD Myoclonus dystonia
PD Parkinson's disease
PKAN Pantothenate kinase-associated neurodegeneration

Mini-dictionary of terms

Deep brain stimulation A surgical procedure involving implantation of an implantable pulse generator (IPG), leads, and electrode(s) to deliver chronic electrical stimulation of brain region(s).

Dystonia A neurological movement disorder characterized by sustained, repetitive, and involuntary muscle contractions resulting in twisting, unwanted and abnormal movements, and/or fixed posture.

Introduction

Dystonia is one of the most common pediatric movement disorders (Cif & Coubes, 2017). Clinically, dystonia exists on a spectrum of severity that can range from mild to persistent debilitating symptoms. Historically, dystonia etiologies have been divided into two groups: primary and secondary dystonia. In 2013, Albanese et al. published a more detailed and descriptive classification for dystonia. Succinctly, the second axis of the Albanese classification divides dystonia into four main groups: inherited dystonia without nervous system pathology, inherited dystonia with nervous system pathology, acquired dystonia, and idiopathic dystonia. For a more detailed presentation of these groups and the subtypes included in each group, see Table 37.1.

The exact pathophysiology of dystonia remains unclear (Lozano et al., 2019). Evidence suggests that each etiology has its own unique neuropathological and genetic profile (Elkaim, De Vloo, Kalia, Lozano, & Ibrahim, 2018), which may partially explain the heterogeneity among dystonia subtypes. These differences include, among others, differences in baseline severity,

Diagnosis, Management and Modeling of Neurodevelopmental Disorders. https://doi.org/10.1016/B978-0-12-817988-8.00037-3

TABLE 37.1 Classification of patients by dystonia etiology.

Group	Subgroups included	Previous classifications[a]
Inherited dystonia, without nervous system pathology	DYT1 DYT6 Myoclonus dystonia	Primary dystonia, dystonia-plus
Inherited dystonia, with nervous system pathology	Pantothenate kinase-associated neurodegeneration Lesch–Nyhan disease Glutaric aciduria type 1 Mohr–Tranebjaerg syndrome Batten's disease Methylmalonic acidemia	Secondary dystonia or heredodegenerative dystonia
Acquired dystonia	Cerebral palsy Kernicterus Stroke Infectious Nigrostriatal necrosis Metabolic Posttraumatic	Secondary dystonia
Idiopathic		Mostly primary dystonia

Table 37.1 showing the distribution of patients into groups based on dystonia etiology as per the consensus Albanese classification. When available, the table also shows how the patients would have been classified prior to publication of the Albanese classification.
[a]*Previous classification: Where applicable, describes how these patients were categorized in their respective studies before the updated classification system proposed by Albanese et al. (2013).*
Table 37.1 has been adapted with permission from the following paper written by the authors of this book chapter: Elkaim LM, Alotaibi N, Sigal AJ, Ibrahim GM. Deep Brain Stimulation for Pediatric Dystonia: A Meta-Analysis with Individual Participant Data. *Dev Med Child Neurol.* 2019; 61(1):49–56.

natural history, and epidemiological and clinical presentations (Elkaim et al., 2018, Elkaim, Alotaibi, Sigal, & Ibrahim, 2019). Another essential distinction between dystonia etiologies is susceptibility to treatment, particularly to deep brain stimulation (DBS) (Vidailhet, Jutras, Grabli, & Roze, 2013). Some patients, especially those with *DYT1-TOR1A* (Starr et al., 2014), myoclonus dystonia (MD) (Rughani & Lozano, 2013), and idiopathic dystonia (Elkaim et al., 2018), tend to respond particularly well to DBS, while patients with acquired dystonia tend to have poorer clinical outcomes (evaluated with the Burke–Fahn–Marsden Dystonia Rating Scale, or BFMDRS), after DBS (Elkaim et al., 2019). Given the significant contrast between subgroups, some authors have begun to describe dystonia as a "network disorder" (Schirinzi, Sciamanna, Mercuri, & Pisani, 2018), comprised of a heterogeneous group of dystonias. Thus, further understanding of each neuropathological portrait is required to optimize specialized treatments for each subgroup.

The method by which DBS influences dysfunctional neural circuitry in dystonic patients is not fully understood and likely differs depending on underlying dystonia etiology. A recent review by Lozano et al. cites direct inhibition (or excitation) of neural activity, information lesions (or jamming), and synaptic filtering as some of the potential mechanisms by which DBS exerts its effects (Lozano et al., 2019). Regardless, many children and youth have shown significant improvements in functionality and quality of life after the procedure, in both the short and long term. DBS has thus become a standard of care for carefully selected patients with medically refractory dystonia (Lozano et al., 2019). Further efforts are required to optimize patient selection and refine DBS parameters and brain region targets. DBS in pediatric patients has its own set of challenges; differences in epidemiology compared with adults (Lin, Lumsden, Gimeno, & Kaminska, 2014), age-specific ethical concerns (Davidson, Elkaim, Lipsman, & Ibrahim, 2018), and different technical surgical considerations all require particular consideration before surgery. Further, dystonia may affect children at crucial times of physical and neurocognitive growth, yet it is unclear how DBS shapes the developing brain and musculoskeletal system.

The need for further direction regarding the role of DBS in dystonic children and youth is therefore underlined. The objective of the following chapter is to describe, through a neurodevelopmental lens, the neuropathology and genetics of different dystonia etiologies and assess (when the evidence permits) the molecular, functional, and behavioral impacts of DBS thereon. This chapter begins by briefly describing an overview of dystonia pathogenesis, before analyzing more specific pathogenetic profiles within dystonia subgroups. Subsequently, the role of DBS in dystonia, with a focus on proposed mechanisms and clinical impacts, is discussed. The chapter concludes by analyzing the applicability of DBS in other areas of circuit dysfunction, including psychiatric conditions, chronic pain, epilepsy, and Alzheimer's disease (AD).

Dystonia pathogenesis: a brief overview

The pathophysiology of dystonia has been the subject of extensive research (Jinnah & Hess, 2018; Neychev, Gross, Lehéricy, Hess, & Jinnah, 2011; Schirinzi et al., 2018). To avoid oversimplification of complex physiological phenomena and to keep the following chapter succinct, only more widely recognized pathological drivers of dystonia will be discussed. Specifically, the role of genetics, motor networks, and neurophysiology will be examined. Further, this section describes the main mechanisms seen across the entire dystonia spectrum; more specific characteristics unique to each etiology will be discussed later in the corresponding subsections.

The past decade has shown a considerable advance in the genetics of dystonia, in part due to the identification of many novel dystonia genes but also progress as to the understanding of dystonic gene roles (Lohmann & Klein, 2017). Recently, Jinnah and colleagues identified four possible biological pathways through which genetic anomalies lead to dystonia: dopamine signaling anomalies, cation transporters abnormalities, heavy metal accumulation, and mitochondrial dysfunction (Jinnah & Hess, 2018). While the classification is not necessarily comprehensive, it provides a framework for understanding the link between genotype and phenotype expression in dystonia and facilitates the study and identification of shared molecular pathways. In keeping with the theme of this book chapter, the etiology-specific differences in genetics will be discussed in the upcoming respective etiology subsections.

Historically, dystonia has been understood as a disorder of the basal ganglia, specifically the striatum. Some of the current evidence for basal ganglia involvement includes defects in dopamine synthesis in the basal ganglia in dystonic patients (Wijemanne & Jankovic, 2015), GPi overactivity (Lozano et al., 2019), and loss of synaptic downscaling at corticostriatal synapses (Calabresi et al., 2016), which may occur during neurodevelopment (Schirinzi et al., 2018). Other brain regions, including the cerebral cortex, thalamus, and the cerebellum, may also be implicated (Jinnah & Hess, 2018; Jinnah, Neychev, & Hess, 2017; Neychev et al., 2011; Schirinzi et al., 2018). Thus, as Jinnah and colleagues first eluded to in 2006 (Jinnah & Hess, 2006), it may be more appropriate to consider that dystonia arises as a result of multiple faulty system inputs; dystonia should, therefore, be considered a motor network (and not strictly basal ganglia) disorder. This understanding is especially relevant regarding DBS; the network model may partially explain the disparity in DBS efficacy between dystonia subgroups, as most studies focus on basal ganglia (especially GPi) stimulation. Future studies may, therefore, demonstrate optimized DBS targets outside the basal ganglia based on individual pathogenetic profiles.

The neurophysiology of dystonia has been extensively studied and reviewed (Hallett, 1998; 2011; Phukan, Albanese, Gasser, & Warner, 2011), with a comprehensive summary of findings extending beyond the scope of this chapter. Briefly, three general mechanisms seem to underscore many, but not all, dystonias: loss of inhibition, sensory dysfunction, and abnormal neuroplasticity (Quartarone & Hallett, 2013). In dystonia, the loss of inhibition hypothesis stipulates that there exists an imbalance between the direct and indirect pathways responsible for voluntary movement, with one of the pathways either being hyper or hypoactive (Hallett, 2006). Further, despite being previously dubbed a pure-motor disorder, dystonia is now characterized by multiple areas of sensorial circuit dysfunction (Antelmi et al., 2017). These include (among others) deficits in temporal sensory processing (Quartarone & Hallett, 2013), spatial awareness including fingertip sensitivity, and impaired lateral somatosensory inhibition (Antelmi et al., 2017).

Inherited dystonia without nervous system pathology

Largely due to advances in genetic sequencing technologies, a significant number of inherited conditions have recently been added to this subgroup (Lohmann & Klein, 2017). Relatively thorough research (especially for *DYT1-TOR1A)* on the genetics, pathophysiology, and outcomes after DBS exists for two of the etiologies included in this subgroup: *DYT-TOR1A* (or *DYT1*) dystonia or *DYT-SGCE* (or MD). The following sections will focus on these two etiologies.

DYT-TOR1A: pathophysiology and genetics

Among all specific dystonia etiologies, *DYT-TOR1A* dystonia may be the most thoroughly studied. Previously part of the primary dystonia group, *DYT1* dystonia is caused by a mutation in the *TOR1A* gene, identified by Ozelius et al. in 1997. The gene encodes for the protein torsinA, an endoplasmic reticulum (ER) ATPase. Molecularly, torsinA is part of the AAA + superfamily (ATPases associated with various cellular activities), although little is known as to the specific substrates affected by torsinA (Cascalho, Jacquemyn, & Goodchild, 2017). On a cellular level, evidence suggests that torsinA influences a variety of cellular actions, including protein quality control, transcription, protein processing, degradation, and membrane fusion (Elkaim et al., 2018). Among these, three main areas of dysfunction are consistently reported: those pertaining to the nuclear envelope, membrane biology, and ER homeostasis (Cascalho et al., 2017).

Further, recent data has shown that torsinA may be implicated in cellular lipid production and membrane composition (Grillet et al., 2016). The specific mechanisms by which these cellular changes affect underlying neurocircuitry in dystonia remain unclear; one hypothesis stipulates that the presence of larger more complex endoplasmic reticulum (i.e., in striatal anticholinergic neurons) may partially explain the increased susceptibility in these cells (Cascalho et al., 2017). This hypothesis, however, does not explain the more widespread functional connectivity network anomalies (including the sensorimotor network and dorsal attention network) present in *DYT1* dystonia (Premi et al., 2016).

Another important pathophysiological consideration in *DYT1* dystonia is maladaptive neuroplasticity, especially in pediatrics, where plasticity is a crucial component of brain development (Cramer et al., 2011). Through exaggerated responsiveness to rTMS conditioning protocols relative to healthy controls, several studies have demonstrated excessive plasticity in the sensory cortex, motor cortex, and even brainstem in patients affected with *DYT1* dystonia (Quartarone & Pisani, 2011). Further evidence of abnormal plasticity in *DYT1* dystonia includes reduced eye-blink classical conditioning (Kojovic et al., 2013), the loss of synaptic downscaling at corticostriatal synapses (Calabresi et al., 2016), and abnormal firing patterns in the GPi (Quartarone & Pisani, 2011). Moreover, evidence in mouse models suggests that a critical neuroplastic development period for dystonia exists for carriers of the *DYT-TOR1A* mutation, since carriers that do not develop symptoms within a certain time frame never develop dystonic symptoms (Maltese et al., 2018). Some may argue that, because of the presence of a specific timeframe for the development of dystonia, early DBS may be more effective. While some studies have indeed shown this correlation (Lumsden et al., 2013), a more recent meta-analysis including only pediatric patients showed that older age at onset, and not the proportion of life lived with dystonia, predicts DBS efficacy (Elkaim et al., 2019). Nonetheless, because the study did not show an advantage to delaying surgical intervention, DBS should remain an early consideration.

Overall, and despite significant progress, it remains unclear as to how the *TOR1A* mutation and subsequent cellular dysfunction lead to early-onset generalized dystonia in an autosomal dominant fashion. In the upcoming years, studies linking the several distinct pathological changes seen in *DYT1* dystonia should further the understanding of the disorder.

Myoclonus dystonia (SGCE/DYT11): pathophysiology and genetics

MD is an autosomal dominant disorder characterized by a combination of dystonia and myoclonus. While dystonic symptoms are often present, myoclonic jerks, most often affecting the neck and upper limbs, typically dominate the clinical portrait (Charlesworth, Bhatia, & Wood, 2013). Mutation or deletion of the ε-sarcoglycan (SGCE) gene accounts for 30%–50% of MD cases (Popa et al., 2014), although more significant genetic heterogeneity has been described (Elkaim et al., 2018). The SGCE gene encodes the 438 amino acid protein ε-sarcoglycan; when the gene is mutated, the current hypothesis stipulates that the protein is misplaced in the ER (instead of plasma membrane), thus promoting its destruction by cellular proteasomes (Charlesworth et al., 2013). The specific role of ε-sarcoglycan, and how its dysfunction leads to MD, remains unclear. The significant positive outcomes of MD patients after DBS point to underlying basal ganglia dysfunction (Elkaim et al., 2019), while more recent morphological, imaging, and functional studies strongly suggest cerebellothalamic pathway anomalies (Popa et al., 2014). Roze and colleagues summarize many anomalies seen in MD neurophysiological studies in a recent, thorough review of MD pathogenesis. These anomalies include (among others) abnormal saccade adaptation (Popa et al., 2014) and eye-blink conditioning, suggestive of olivopontocerebellar circuit dysfunction, abnormal motor learning improved by alcohol consumption, suggestive of GABAergic deficits and underlying Purkinje cell dysfunction, and abnormal neuronal firing activity in the Gpi, suggestive of basal ganglia dysfunction (Roze, Lang, & Vidailhet, 2018). Notably, MD is also associated with increased rates of comorbid psychiatric disorders, including obsessive–compulsive and anxiety disorder. These findings suggest that, as with other forms of dystonia, MD is more likely a network disorder arising from multiple faulty system inputs.

Inherited dystonia with nervous system pathology

According to the Albanese classification, etiologies are considered to have nervous system pathology if they have evidence of neurodegeneration, at the gross, microscopic, or molecular level. Many disorders, including pantothenate kinase-associated neurodegeneration (PKAN), glutaric aciduria type 1 (GA1), and Lesch–Nyhan disease (LND) are included in the inherited dystonia with nervous system pathology subgroup. Children and youth included within this subgroup represent a significant portion (roughly 16% (Elkaim et al., 2019)) of dystonic patients receiving DBS in the known literature; despite this, research regarding the pathology, genetics, and neurophysiology of the underlying etiologies as they pertain to dystonia is scarce, especially when compared with *DYT1*, MD, and cerebral palsy (CP). The following section will focus on the two etiologies within the inherited nervous system pathology group that are most commonly treated with DBS: PKAN and GA1.

Pantothenate kinase-associated neurodegeneration (PKAN): pathophysiology and genetics

PKAN is a rare autosomal recessive disorder caused by mutations in the *PANK2* gene, which usually codes for protein isoforms that catalyze the first step of coenzyme A (CoA) biosynthesis (Aoun et al., 2017). PANK2, which in humans is highly expressed in the brain (Hörtnagel, Prokisch, & Meitinger, 2003), plays a key role in energy and lipid metabolism. PANK2 gene mutations and resulting CoA dysfunction lead to cysteine, and eventually, iron accumulation in the brain. In patients with PKAN, iron accumulation occurs primarily in the basal ganglia, which may manifest as the characteristic "eye of the tiger" sign on T2-weighted MRI. Ultimately, the excessive iron and resulting damage lead to dystonia, parkinsonism, dysarthria, spasticity, and mental retardation (Aoun et al., 2017).

Glutaric aciduria type 1 (GA1)

GA1 is caused by mutations in the GCDH gene encoding glutaryl-CoA dehydrogenase. The dysfunction of this enzyme, which is typically involved in the degradation of lysine, hydroxylysine, and tryptophan, leads to an accumulation of glutaric acid and 3-hydroxyglutaric acid, especially in the basal ganglia. Diagnosis can be made by detecting glutaconic acid and glutaryl carnitine using mass spectrometry and tandem mass spectrometry (Boy et al., 2017). Ultimately, many patients with untreated GA1 experience encephalopathic crises during their first 6 years of life, resulting in lifelong neurological dysfunction including cognitive disabilities, CP with hypo- or hypertonia, dystonia, and choreoathetosis.

Acquired dystonia

The acquired dystonia subgroup (previously secondary dystonia) includes several dystonia etiologies, including stroke, infectious, posttraumatic, and CP. Dystonia in these patients tends to be more severe when compared with other dystonia subgroups (Lin et al., 2014). CP is by far the most common etiology included in acquired dystonia and is also the most common indication for DBS within the subgroup (Elkaim et al., 2019); thus, the following section will only discuss CP.

Cerebral palsy (CP): pathophysiology

CP remains one of the most significant causes of global morbidity; it is the most common cause of severe early childhood disability, affecting roughly one in 500 neonates, and often causes lifelong physical, developmental, psychological, and social limitations (Graham et al., 2016). CP is a term used to describe a clinical portrait caused by nonprogressive central nervous system lesions that occur during fetal or infant brain development. While CP is typically considered a disorder of movement and posture, over half of patients exhibit disturbances of sensation, perception, cognition, and communication (Monbaliu et al., 2017). Further, although nonprogressive lesions cause CP, the musculoskeletal and neurological consequences of CP often inhibit healthy childhood development, thus making CP progressive in nature.

Many movement disorders may be evident in patients with CP, including spasticity, ataxia, and dyskinesia (which further separates into dystonic and chorea-athetotic forms). For dystonia, abnormal sensory thalamocortical pathways (McClelland et al., 2018), excessive plasticity (Monbaliu et al., 2017), and lack of selective muscle control may be involved. Further, abnormally prolonged embryonic physiologic processes such as cocontraction and overflow muscle contraction may contribute to dystonic symptoms (Graham et al., 2016). During healthy development, the fetus is seen in different flexor and extensor postural stages. Cocontractions, which precede the development of voluntary movements, characterize these positions. In patients with early brain injury, such as those with CP, cocontraction may continue after the first few years of life, leading to dystonic symptoms (Graham et al., 2016). As highlighted by a recent review on CP by Graham et al. (2016), other potential explanations for dystonia in children with CP are exaggerated tonic labyrinth reflex postures (which, in normal developing children, stabilizes children using a "distal to proximal" muscle activation) and abnormal timing and patterns of muscle activation. Finally, over 70% of dystonic patients with CP show anomalies in the brainstem or thalamus (Monbaliu et al., 2017); however, a study in dystonic patients with CP did not show any significant correlation between the severity of imaging lesions and clinical severity of dystonia (Monbaliu et al., 2016).

Idiopathic dystonia

Idiopathic dystonia refers to dystonia with an unknown cause, showing no nervous system pathology and without any associated genetic anomalies. Previously, patients with idiopathic dystonia were (along with inherited dystonia without nervous system pathology) included in the primary dystonia group. Patients with idiopathic dystonia tend to have worse clinical severity before DBS when compared with patients with inherited dystonia, although the clinical improvements after DBS are similar (Tables 37.2 and 37.5) (Elkaim et al., 2019). Overall, minimal research exists surrounding the pathophysiology of idiopathic dystonia, despite these children and youth representing over 22% of patients undergoing DBS (Elkaim et al., 2019). It is likely that in the future, idiopathic dystonias will be reclassified as inherited dystonia as novel genes are recognized.

TABLE 37.2 Dystonia outcomes for children and youth with inherited dystonia without degenerative pathology.

Heritability	Nervous system pathology	Subtype	*n*	Preoperative severity (mot), median, (25/75 centiles)	BFMDRS-M % improvement, median, (25/75 centiles)	Follow-up (mo), median (25/75 centiles)	Proportion with clinically significant (<20%) improvement
Inherited	None						
		DYT1 DYT6	102	45.5 (33.6/64.7)	78.1 (53.7/94.0)	15.0 (12.0/42.0)	88%
		Myoclonus dystonia	9	18.5 (9.7/31.5)	68.3 (35.8/84.6)	–	100%

Table 37.2 showing the outcomes using the BFMDRS after DBS in dystonic children and youth without inherited dystonia with degenerative pathology.

TABLE 37.3 Dystonia outcomes for children and youth with inherited dystonia with degenerative pathology.

Heritability	Nervous system pathology	Subtype	*n*	Preoperative severity (mot), median, (25/75 centiles)	BFMDRS-M % improvement, median, (25/75 centiles)	Follow-up (mo), median (25/75 centiles)	Proportion with clinically significant (<20%) improvement
Inherited	Degenerative						
		PKAN	36	83.7 (64.0/96.5)	27.7 (0.6/65.4)	12.0 (7.1/30.1)	57%
		GA1	5	111.2 (64.5/113.6)	6.4 (1.3/13.1)	12.0 (2.9/12.0)	0

Table 37.2 showing the outcomes using the BFMDRS after DBS in dystonic children and youth with inherited dystonia with degenerative pathology.

TABLE 37.4 Dystonia outcomes for children and youth with acquired dystonia.

Heritability	Nervous system pathology	Subtype	*n*	Preoperative severity (mot), median, (25/75 centiles)	BFMDRS-M % improvement, median, (25/75 centiles)	Follow-up (mo), median (25/75 centiles)	Proportion with clinically significant (<20%) improvement
Acquired	Static lesions						
		Cerebral palsy	59	86.2 (66.7/100.3)	11.1 (0.0/21.4)	12.0 (6.0/16.5)	26.9%

Table 37.2 showing the outcomes using the BFMDRS after DBS in dystonic children and youth with acquired dystonia.

TABLE 37.5 Dystonia outcomes for children and youth with idiopathic dystonia.

Heritability	Nervous system pathology	Subtype	*n*	Preoperative severity (mot), median, (25/75 centiles)	BFMDRS-M % improvement, median, (25/75 centiles)	Follow-up (mo), median (25/75 centiles)	Proportion with clinically significant (<20%) improvement
Idiopathic			72	62.0 (42.5/89.2)	50.5 (24.3/73.1)	20.0 (12.0/36.0)	73%

Table 37.2 showing the outcomes using the BFMDRS after DBS in dystonic children and youth with idiopathic dystonia.
Tables 37.2–37.5 have been adapted and adjusted with permission from the following paper written by the authors of this book chapter: Elkaim LM, Alotaibi N, Sigal AJ, Ibrahim GM. Deep Brain Stimulation for Pediatric Dystonia: A Meta-Analysis with Individual Participant Data. *Dev Med Child Neurol.* 2019; 61(1):49–56.

Deep brain stimulation

DBS clinical impact

There is consensus as to the efficacy and safety of DBS for dystonia in many adult settings (including *DYT1*, MD, and tardive dystonia). Until recently, data pertaining to DBS efficacy in dystonic children and youth were limited to case reports and small case series. This year, the authors of this book chapter published the first meta-analysis analyzing DBS efficacy for dystonic children and youth (Elkaim et al., 2019). The paper included 321 patients and showed varying degrees of efficacy depending on underlying pathology (Tables 37.2–37.5). Narratively, at last follow-up (median 12 months), 66.1% of patients showed clinically significant (defined as >20% (Vidailhet et al., 2009)) improvement on BFMDRS motor score. On multivariable regression analysis, older age at dystonia onset, inherited dystonia without nervous system pathology, idiopathic dystonia, and truncal involvement portended better outcome after DBS. Of note, although the 20% threshold for clinical significance is mentioned in numerous studies, no formal analysis has determined clinically significant change on the BFMDRS. For a more detailed presentation of the results (stratified by pathology), see Tables 37.2–37.5.

The best responders to DBS among children and youth are those with inherited dystonia without nervous system pathology (*DYT1* and MD), with over 88% of patients showing clinically significant improvement postoperatively (Elkaim et al., 2019). Patients with idiopathic dystonia respond similarly well. Among the worst responders are patients with CP (median improvement BFMDRS motor of 11%) and patients with GA1 (median improvement BFMDRS motor of 6%) (Elkaim et al., 2019). Importantly, many authors have criticized the use of the BFMDRS to measure improvement in patients with CP (Pagliano et al., 2018); they argue that it fails to evaluate DBS efficacy in other areas, such as pain and quality of life, which are sometimes mentioned as the most common concerns in patients with CP (Lumsden, Gimeno, Tustin, Kaminska, & Lin, 2015). A recent study reviewing treatment options for children with CP showed that no current studies are analyzing the effects of DBS on pain (Fehlings et al., 2018); thus, along with the BFMDRS, future studies should include alternative measures of outcome after DBS.

Applications in other areas of development

DBS has become one of the foremost medical developments of the last century, with widespread and demonstrated success in patients with Parkinson's disease (PD) and dystonia. Since its inception, over 160,000 patients have been treated with DBS for various medical conditions (Lozano & Lipsman, 2013). The precise mechanisms by which DBS affects the brain remain unclear; however, it is clear that DBS significantly improves many circuit pathologies, leading to decreased patient and caretaker disease burden, increased functionality and mobility, and increased quality of life. Since its widespread success in patients with PD, DBS has been used for many pathologies, with varying success. These include (among others) chronic pain, epilepsy, psychiatric disorders, and Tourette syndrome. More recently, DBS has been used in a double-blind, open-label study (11 months) evaluating AD (Kuhn et al., 2015)). After 12 months, four of six patients demonstrated a response to treatment. However, another randomized, double-blind study for patients with AD showed no improvement after DBS (Lozano et al., 2016). The vast spectrum of neurological diseases being treated with DBS highlights its potential as a powerful therapeutic tool. Future applications for DBS will hopefully lead to a diversification and increase in its use; in doing so, DBS will help deepen our understanding of the human brain, while helping many people in the process.

Key facts (5–10 in bulleted format)

- Deep brain stimulation (DBS) is increasingly used to treat children and youth with refractory dystonia.
- DBS efficacy for treatment of dystonia depends on underlying dystonia pathology.
- Optimizing treatment patterns and guiding surgical candidate selection require proper understanding of underlying pathophysiology-specific mechanisms.
- DBS is effective in dystonic children and youth with *DYT1* dystonia, myoclonus dystonia, and idiopathic dystonia.
- Further studies (evaluating alternative outcome measures in addition to BFMDRS) are required to properly evaluate DBS efficacy for other dystonias.
- Potential future indications for DBS include psychiatric conditions, epilepsy, chronic pain, and Alzheimer's disease.

Summary points (5–10 in bulleted format)

- Deep brain stimulation (DBS) is increasingly used to treat children and youth with refractory dystonia.
- DBS efficacy for treatment of dystonia depends on underlying dystonia pathology.
- Optimizing treatment patterns and guiding surgical candidate selection require proper understanding of underlying pathophysiology-specific mechanisms.
- DBS is effective in dystonia children and youth with *DYT1* dystonia, MD, and idiopathic dystonia.
- Further studies (evaluating alternative outcome measures in addition to BFMDRS) are required to properly evaluate DBS efficacy in other forms of dystonia.
- Potential future indications for DBS include psychiatric conditions, epilepsy, chronic pain, and Alzheimer's disease.

Appendix

Table 37.1 showing the distribution of patients into groups based on dystonia etiology.

References

Albanese, A., Bhatia, K., Bressman, S. B., Delong, M. R., Fahn, S., Fung, V. S. C., … Teller, J. K. (2013). Phenomenology and classification of dystonia: A consensus update. *Movement Disorders, 28*(7), 863–873. https://doi.org/10.1002/mds.25475.

Antelmi, E., Erro, R., Rocchi, L., Liguori, R., Tinazzi, M., Di Stasio, F., … Bhatia, K. P. (2017). Neurophysiological correlates of abnormal somatosensory temporal discrimination in dystonia. *Movement Disorders, 32*(1), 141–148. https://doi.org/10.1002/mds.26804.

Aoun, M., Corsetto, P. A., Nugue, G., Montorfano, G., Ciusani, E., Crouzier, D., … Tiranti, V. (2017). Changes in red blood cell membrane lipid composition: A new perspective into the pathogenesis of PKAN. *Molecular Genetics and Metabolism, 121*(2), 180–189. https://doi.org/10.1016/j.ymgme.2017.04.006.

Boy, N., Mühlhausen, C., Maier, E. M., Heringer, J., Assmann, B., Burgard, P., … vom Dahl, S. (2017). Proposed recommendations for diagnosing and managing individuals with glutaric aciduria type I: Second revision. *Journal of Inherited Metabolic Disease, 40*(1), 75–101. https://doi.org/10.1007/s10545-016-9999-9.

Calabresi, P., Pisani, A., Rothwell, J., Ghiglieri, V., Obeso, J. A., & Picconi, B. (2016). Hyperkinetic disorders and loss of synaptic downscaling. *Nature Neuroscience, 19*(7), 868–875. https://doi.org/10.1038/nn.4306.

Cascalho, A., Jacquemyn, J., & Goodchild, R. E. (2017). Membrane defects and genetic redundancy: Are we at a turning point for DYT1 dystonia? *Movement Disorders, 32*(3), 371–381. https://doi.org/10.1002/mds.26880.

Charlesworth, G., Bhatia, K. P., & Wood, N. W. (2013). The genetics of dystonia: New twists in an old tale. *Brain, 136*(7), 2017–2037. https://doi.org/10.1093/brain/awt138.

Cif, L., & Coubes, P. (2017). Historical developments in children's deep brain stimulation. *European Journal of Paediatric Neurology, 21*(1), 109–117. https://doi.org/10.1016/j.ejpn.2016.08.010.

Cramer, S. C., Sur, M., Dobkin, B. H., O'Brien, C., Sanger, T. D., Trojanowski, J. Q., … Vinogradov, S. (2011). Harnessing neuroplasticity for clinical applications. *Brain, 134*(6), 1591–1609. https://doi.org/10.1093/brain/awr039.

Davidson, B., Elkaim, L. M., Lipsman, N., & Ibrahim, G. M. (2018). Editorial. An ethical framework for deep brain stimulation in children. *Neurosurgical Focus, 45*(3), 11.

Elkaim, L. M., Alotaibi, N., Sigal, A. J., & Ibrahim, G. M. (2019). Deep brain stimulation for pediatric dystonia: A meta-analysis with individual participant data. *Developmental Medicine and Child Neurology, 61*(1), 49–56.

Elkaim, L. M., De Vloo, P., Kalia, S. K., Lozano, A. M., & Ibrahim, G. M. (2018). Deep brain stimulation for childhood dystonia: Current evidence and emerging practice. *Expert Review of Neurotherapeutics, 18*(10), 773–784. https://doi.org/10.1080/14737175.2018.1523721.

Fehlings, D., Brown, L., Harvey, A., Himmelmann, K., Lin, J.-P., Macintosh, A., … Walters, I. (2018). Pharmacological and neurosurgical interventions for managing dystonia in cerebral palsy: A systematic review. *Developmental Medicine and Child Neurology, 60*(4), 356–366. https://doi.org/10.1111/dmcn.13652.

Graham, H.,K., Rosenbaum, P., Paneth, N., Dan, B., Lin, J.-P., Damiano, D. L., … Lieber, R. L. (2016). Cerebral palsy. *Nature Reviews: Disease Primers, 2*(1), 2–24. https://doi.org/10.1016/B978-0-444-63916-5.00020-3.

Grillet, M., Dominguez Gonzalez, B., Sicart, A., Pöttler, M., Cascalho, A., Billion, K., … Goodchild, R. E. (2016). Torsins are essential regulators of cellular lipid metabolism. *Developmental Cell, 38*(3), 235–247. https://doi.org/10.1016/j.devcel.2016.06.017.

Hallett, M. (1998). The neurophsiology of dystonia. *Archives of Neurology, 55*(5), 601.

Hallett, M. (2006). Pathophysiology of dystonia. *Journal of Neural Transmission Supplementum, 70*, 485–488.

Hallett, M. (2011). Neurophysiology of dystonia: The role of inhibition. *Neurobiology of Disease, 42*(2), 177–184. https://doi.org/10.1016/j.nbd.2010.08.025.

Hörtnagel, K., Prokisch, H., & Meitinger, T. (2003). An isoform of hPANK2, deficient in pantothenate kinase-associated neurodegeneration, localizes to mitochondria. *Human Molecular Genetics, 12*(3), 321–327. https://doi.org/10.1093/hmg/ddg026.

Jinnah, H. A., & Hess, E. J. (2006). A new twist on the anatomy of dystonia: The basal ganglia and the cerebellum? *Neurology, 67*(10), 1740–1741. https://doi.org/10.1212/01.wnl.0000246112.19504.61.

Jinnah, H. A., & Hess, E. J. (2018). Evolving concepts in the pathogenesis of dystonia. *Parkinsonism and Related Disorders, 46*, S62–S65. https://doi.org/10.1016/j.parkreldis.2017.08.001.

Jinnah, H. A., Neychev, V., & Hess, E. J. (2017). The anatomical basis for dystonia: The motor network model. *Tremor and Other Hyperkinetic Movements (New York, N.Y.), 7*, 506. https://doi.org/10.7916/D8V69X3S.

Kojovic, M., Pareés, I., Kassavetis, P., Palomar, F. J., Mir, P., Teo, J. T., … Edwards, M. J. (2013). Secondary and primary dystonia: Pathophysiological differences. *Brain, 136*(7), 2038–2049. https://doi.org/10.1093/brain/awt150.

Kuhn, J., Hardenacke, K., Lenartz, D., Gruendler, T., Ullsperger, M., Bartsch, C., … Sturm, V. (2015). Deep brain stimulation of the nucleus basalis of Meynert in Alzheimer's dementia. *Molecular Psychiatry, 20*(3), 353–360. https://doi.org/10.1038/mp.2014.32.

Lin, J. P., Lumsden, D. E., Gimeno, H., & Kaminska, M. (2014). The impact and prognosis for dystonia in childhood including dystonic cerebral palsy: A clinical and demographic tertiary cohort study. *Journal of Neurology Neurosurgery and Psychiatry, 85*(11), 1239–1244. https://doi.org/10.1136/jnnp-2013-307041.

Lohmann, K., & Klein, C. (2017). Update on the genetics of dystonia. *Current Neurology and Neuroscience Reports, 17*(3). https://doi.org/10.1007/s11910-017-0735-0.

Lozano, A. M., Fosdick, L., Chakravarty, M. M., Leoutsakos, J. M., Munro, C., Oh, E., … Smith, G. S. (2016). A phase II study of fornix deep brain stimulation in mild Alzheimer's disease. *Journal of Alzheimer's Disease, 54*(2), 777–787. https://doi.org/10.3233/JAD-160017.

Lozano, A. M., & Lipsman, N. (2013). Probing and regulating dysfunctional circuits using deep brain stimulation. *Neuron, 77*(3), 406–424. https://doi.org/10.1016/j.neuron.2013.01.020.

Lozano, A. M., Lipsman, N., Bergman, H., Brown, P., Chabardes, S., Chang, J. W., … Krauss, J. K. (2019). Deep brain stimulation: Current challenges and future directions. *Nature Reviews Neurology, 15*(3), 148–160. https://doi.org/10.1038/s41582-018-0128-2.

Lumsden, D. E., Gimeno, H., Tustin, K., Kaminska, M., & Lin, J. P. (2015). Interventional studies in childhood dystonia do not address the concerns of children and their carers. *European Journal of Paediatric Neurology, 19*(3), 327–336. https://doi.org/10.1016/j.ejpn.2015.01.003.

Lumsden, D. E., Kaminska, M., Gimeno, H., Tustin, K., Baker, L., Perides, S., … Lin, J. P. (2013). Proportion of life lived with dystonia inversely correlates with response to pallidal deep brain stimulation in both primary and secondary childhood dystonia. *Developmental Medicine and Child Neurology, 55*(6), 567–574. https://doi.org/10.1111/dmcn.12117.

Maltese, M., Tassone, A., Sciamanna, G., Ponterio, G., Vanni, V., Martella, G., … Gardoni, F. (2018). Early structural and functional plasticity alterations in a susceptibility period of DYT1 dystonia mouse striatum. *eLife, 7*, 1–23. https://doi.org/10.7554/eLife.33331.

McClelland, V. M., Fialho, D., Flexney-Briscoe, D., Holder, G. E., Elze, M. C., Gimeno, H., … Lin, J. P. (2018). Somatosensory evoked potentials and central motor conduction times in children with dystonia and their correlation with outcomes from deep brain stimulation of the globus pallidus internus. *Clinical Neurophysiology, 129*(2), 473–486. https://doi.org/10.1016/j.clinph.2017.11.017.

Monbaliu, E., de Cock, P., Ortibus, E., Heyrman, L., Klingels, K., & Feys, H. (2016). Clinical patterns of dystonia and choreoathetosis in participants with dyskinetic cerebral palsy. *Developmental Medicine and Child Neurology, 58*(2), 138–144. https://doi.org/10.1111/dmcn.12846.

Monbaliu, E., Himmelmann, K., Lin, J.-P., Ortibus, E., Bonouvrié, L., Feys, H., … Dan, B. (2017). Clinical presentation and management of dyskinetic cerebral palsy. *The Lancet Neurology, 16*(9), 741–749. https://doi.org/10.1016/S1474-4422(17)30252-1.

Neychev, V. K., Gross, R. E., Lehéricy, S., Hess, E. J., & Jinnah, H. A. (2011). The functional neuroanatomy of dystonia. *Neurobiology of Disease, 42*(2), 185–201. https://doi.org/10.1016/j.nbd.2011.01.026.

Ozelius, L. J., Hewett, J. W., Page, C. E., Bressman, S. B., Kramer, P. L., Shalish, C., … Breakefield, X. O. (1997). The early-onset torsion dystonia gene (DYT1) encodes an ATP-binding protein. *Nature Genetics, 17*, 40–48.

Pagliano, E., Baranello, G., Masson, R., Foscan, M., Arnoldi, M. T., Marchi, A., … Pantaleoni, C. (2018). Outcome measures for children with movement disorders. *European Journal of Paediatric Neurology*, 1–8. https://doi.org/10.1016/j.ejpn.2018.01.014.

Phukan, J., Albanese, A., Gasser, T., & Warner, T. (2011). Primary dystonia and dystonia-plus syndromes: Clinical characteristics, diagnosis, and pathogenesis. *The Lancet Neurology, 10*(12), 1074–1085. https://doi.org/10.1016/S1474-4422(11)70232-0.

Popa, T., Milani, P., Richard, A., Hubsch, C., Brochard, V., Tranchant, C., … Roze, E. (2014). The neurophysiological features of myoclonus-dystonia and differentiation from other dystonias. *JAMA Neurology, 71*(5), 612–619. https://doi.org/10.1001/jamaneurol.2014.99.

Premi, E., Diano, M., Gazzina, S., Cauda, F., Gualeni, V., Tinazzi, M., … Borroni, B. (2016). Functional connectivity networks in asymptomatic and symptomatic DYT1 carriers. *Movement Disorders, 31*(11), 1739–1743. https://doi.org/10.1002/mds.26725.

Quartarone, A., & Hallett, M. (2013). Emerging concepts in the physiological basis of dystonia. *Movement Disorders, 28*(7), 958–967. https://doi.org/10.1002/mds.25532.

Quartarone, A., & Pisani, A. (2011). Abnormal plasticity in dystonia: Disruption of synaptic homeostasis. *Neurobiology of Disease, 42*(2), 162–170. https://doi.org/10.1016/j.nbd.2010.12.011.

Roze, E., Lang, A. E., & Vidailhet, M. (2018). Myoclonus-dystonia: Classification, phenomenology, pathogenesis, and treatment. *Current Opinion in Neurology, 31*(4), 484–490. https://doi.org/10.1097/WCO.0000000000000577.

Rughani, A. I., & Lozano, A. M. (2013). Surgical treatment of myoclonus dystonia syndrome. *Movement Disorders, 28*(3), 282–287. https://doi.org/10.1002/mds.25326.

Schirinzi, T., Sciamanna, G., Mercuri, N. B., & Pisani, A. (2018). Dystonia as a network disorder. *Current Opinion in Neurology, 31*(4), 1. https://doi.org/10.1097/wco.0000000000000580.

Starr, P. A., Markun, L. C., Larson, P. S., Volz, M. M., Martin, A. J., & Ostrem, J. L. (2014). Interventional MRI–guided deep brain stimulation in pediatric dystonia: First experience with the ClearPoint system. *Journal of Neurosurgery: Pediatrics, 14*(4), 400–408. https://doi.org/10.3171/2014.6.PEDS13605.

Vidailhet, M., Jutras, M.-F., Grabli, D., & Roze, E. (2013). Deep brain stimulation for dystonia. *Journal of Neurology, Neurosurgery and Psychiatry, 84*(9), 1029–1042. https://doi.org/10.1136/jnnp-2011-301714.

Vidailhet, M., Yelnik, J., Lagrange, C., Fraix, V., Grabli, D., Thobois, S., … Pollak, P. (2009). Bilateral pallidal deep brain stimulation for the treatment of patients with dystonia-choreoathetosis cerebral palsy: A prospective pilot study. *The Lancet Neurology, 8*(8), 709–717. https://doi.org/10.1016/S1474-4422(09)70151-6.

Wijemanne, S., & Jankovic, J. (2015). Dopa-responsive dystonia - clinical and genetic heterogeneity. *Nature Reviews Neurology, 11*(7), 414–424. https://doi.org/10.1038/nrneurol.2015.86.

Chapter 38

Use of repetitive transcranial magnetic stimulation in neurodevelopment: a narrative review

Yoshihiro Noda
Multidisciplinary Translational Research Lab, Department of Neuropsychiatry, Keio University School of Medicine, Shinjuku-ku, Tokyo, Japan

List of abbreviations

ADHD Attention deficit hyperactivity disorder
ASD Autism spectrum disorder
LTD Long-term depression
LTP Long-term potentiation
TBS Theta-burst stimulation
TMS Transcranial magnetic stimulation

Mini-dictionary of terms

ADHD Attention deficit hyperactivity disorder is one of the most common mental disorders affecting children. Symptoms of ADHD include inattention, hyperactivity, and impulsivity.

ASD Autism spectrum disorder is a neurodevelopmental disorder that affects communication and behavior. Symptoms of ASD generally appear in the first 2 years of life.

LTD Long-term depression is a persistent decrease in synaptic efficacy that occurs in an activity-dependent manner. LTD works synergistically with LTP to generate a balanced and stable memory mechanism.

LTP Long-term potentiation means a strengthening of synapses, which makes communication between neurons more likely. LTP is thought to be a crucial mechanism involved in learning and memory formation.

TBS Theta-burst stimulation is a newer form of brain stimulation that more closely mimics the natural rhythms of activity in the neurons of the brain. Theta bursts are short bursts of stimulation at high frequencies. TBS has been shown to produce stronger and longer-lasting effects in brain activity, while requiring much less time to administer.

TMS Transcranial magnetic stimulation is a neuromodulatory technique that applies magnetic pulses to the brain via a special coil. When an electric current is delivered to the coil, the magnetic field is generated according to Faraday's law of induction. Subsequently, the generated magnetic field induces an electrical current in the brain. TMS is typically applied at a sufficient intensity to cause action potentials in proximate neurons.

Introduction

In this chapter, we introduce the efforts for the clinical application of repetitive transcranial magnetic stimulation (rTMS) for neurodevelopmental disorders. Although neurodevelopmental disorders include a variety of conditions, this chapter outlines the focus on "autism spectrum disorders (ASD)," which have the most clinical needs and are being studied so far.

ASD is one of the common developmental disorders characterized by the presence of persistent impairments in social communication and interaction, restricted and repetitive patterns of behaviors, interests, and activities (see Fig. 38.1). The symptoms of ASD cause clinically significant impairment in social, occupational, or other areas of functioning (Durkin et al., 2017; Stahlberg, Soderstrom, Rastam, & Gillberg, 2004). In addition, individuals of ASD often have the symptom of

Diagnosis, Management and Modeling of Neurodevelopmental Disorders. https://doi.org/10.1016/B978-0-12-817988-8.00038-5

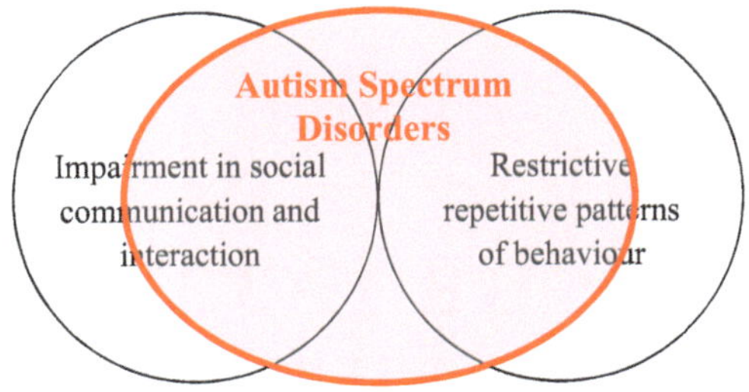

FIGURE 38.1 **Diagnostic criteria for ASD (DSM-5).** Core features of ASD symptoms according to the Diagnostic and Statistical Manual of Mental Disorders, fifth ed. (DSM-5). The clinical diagnosis of ASD is based on two core symptoms, impairment in social communication and interaction as well as restrictive, repetitive patterns of behavior.

sensory abnormalities (Posar & Visconti, 2018), difficulties in mentalizing processes, and executive dysfunction. Furthermore, the following features are often associated with the core symptoms of ASD: cognitive delay, epilepsy, and motor clumsiness, and psychiatric comorbidity such as obsessive–compulsive disorder, affective disorders, and attention deficit and hyperactivity disorder (ADHD) (Hutton, Goode, Murphy, Le Couteur, & Rutter, 2008; Mouridsn, Rich, & Isager, 2008; Stahlberg et al., 2004).

Although various therapeutic strategies have been proposed, including pharmacological, behavioral, educational, and medical interventions for individuals with ASD, there is no consensus for which approach is the most effective for the treatment in this population. This is mainly because the pathophysiology of ASD itself is thought to be highly heterogeneous (Yenkoyan, Grigoryan, Fereshetyan, & Yepremyan, 2017). Thus, it is necessary to stratify individuals with ASD based on clinical features and biomarkers.

For the pathophysiology of ASD, the excitatory and inhibitory imbalance is thought to be a final common pathway (Port, Oberman, & Roberts, 2019). However, this excitatory and inhibitory imbalance appears differently depending on the experimental method, ASD subtype, brain region, and stage of brain development. In addition, postmortem studies in individuals with ASD indicated the neuroanatomical microstructure abnormalities in the neuropil space of cortical mini-columns (Ecker, Schmeisser, Loth, & Murphy, 2017).

Since ASD is a lifelong disorder and most patients remain severely impaired in terms of psychosocial functioning throughout their adult lives (Billstedt, Gillberg, & Gillberg, 2005), it is essential to invent the new therapeutic strategy for this disorder. In this context, in recent years rTMS, a technique for noninvasive neuromodulation, has been considered in the treatment for individuals with ASD due to its therapeutic potential.

Applications of rTMS in individuals of ASD

Transcranial magnetic stimulation (TMS)

TMS is a noninvasive tool with applications in neurophysiological and clinical research (see Fig. 38.2). Although the majority of psychiatric studies have so far involved adult patients with depression, clinical scientists are recently applying

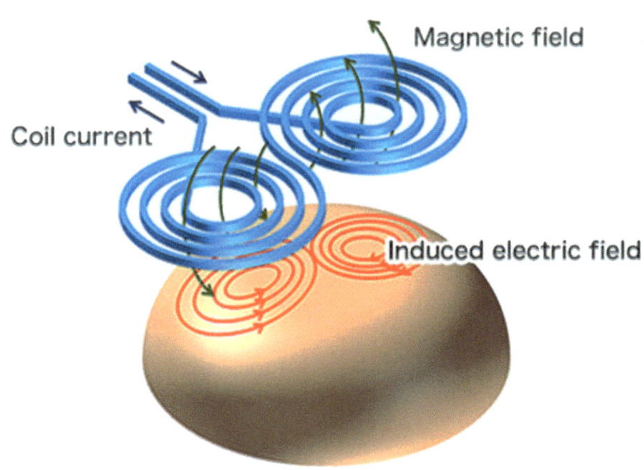

FIGURE 38.2 **Induced current to cerebral cortex by figure-of-eight coil.** TMS is a neuromodulatory technique that applies magnetic pulses to the brain via a special coil. When an electric current is delivered to the coil, the magnetic field is generated according to Faraday's law of induction. Subsequently, the generated magnetic field induces an electrical current in the brain.

this modality in pediatric psychiatric patients. Thus, TMS is a rapidly emerging field that has a big potential to develop new therapeutic modalities in children and adolescents with neurodevelopmental disorders such as ASD (Oberman et al., 2016). Noninvasive brain stimulation represented by repetitive TMS (rTMS) has emerged as novel techniques to promote plasticity and alleviate symptoms in neuropsychiatric disorders (Casanova, Sokhadze; Opris, Wang, & Li, 2015).

Mechanism of action of TMS

rTMS can modulate the cortical excitability in the scale of minutes to hours and further it has clinical effects in a range of weeks to months (Cantone et al., 2017). Effects on neurophysiology as well as clinical outcomes by rTMS depend on the stimulation protocol and the number of pulses and frequency of rTMS sessions, likely by both modulating GABA receptor-mediated activity (Cantone et al., 2017; Enticott et al., 2013) and inducing long-term potentiation (LTP)-like or long-term depression (LTD)-like synaptic plasticity (Hoogendam, Ramakers, & Di Lazzaro, 2010). Indeed, for example, one study noted that rTMS applied to the left dorsolateral prefrontal cortex in children (10Hz, 120% resting motor threshold, 15 sessions) has induced the regional increase in glutamate levels from baseline as measured by magnetic resonance spectroscopy in children with depression, associated with clinical improvement (Yang et al., 2014). Thus, rTMS has potential effects on both cortical neurophysiology (e.g., excitability and plasticity) as well as clinical treatments (Casanova et al., 2015).

In general, high-frequency rTMS (5–20Hz) and intermittent theta-burst stimulation (TBS) induce cortical facilitation, whereas low-frequency rTMS (≤1Hz) and continuous TBS induce cortical suppression based on the presumed mechanisms of use-dependent synaptic plasticity (Maeda, Keenan, Tormos, Topka, & Pascual-Leone, 2000a; 2000b). Since it is thought that individuals with ASD have altered excitation/inhibition balance and aberrant synaptic plasticity of specific neural circuits (Port, Oberman, & Roberts, 2019), neuromodulatory therapeutics by rTMS would be useful and helpful to improve the pathophysiological neural basis of this population. Specifically, when rTMS is applied to the prefrontal cortex of individuals with ASD, it is supposed to restore excitation and inhibition imbalance and to modulate gamma oscillations and neuroplasticity (Uzunova, Pallanti, & Hollander, 2016). Further, for the stimulation parameters such as targeting location of stimulation (e.g., dorsolateral prefrontal cortex, inferior prefrontal cortex, supplementary motor area, etc.) and frequency of stimulation (high-frequency or intermittent theta-burst stimulation (iTBS), low-frequency or continuous TBS), various rTMS protocols have been investigated in clinical trials in individuals with ASD so far (Barahona-Correa, Velosa, Chainho, Lopes, & Oliveira-Maia, 2018; Oberman, Rotenberg, & Pascual-Leone, 2015). However, a certain consensus has not yet been obtained in terms of the optimal brain area to target, frequency of stimulation, and the rhythm that sends the TMS stimulus.

Clinical effectiveness of rTMS in ASD

To date, there have been several published studies that examined the clinical effectiveness of rTMS for ASD (Masuda et al., 2019a; Oberman et al., 2015) (see Table 38.1). Specifically, rTMS treatment for ASD appears to be effective for symptoms such as repetitive behaviors, irritability, and specific items of attention (Barahona-Correa et al., 2018; Hauer et al., 2019). Collectively, based on the previous findings of these studies, the therapeutic effect of rTMS in individuals with ASD would be promising but still inconclusive due to the small sample sizes of these studies and the heterogeneity in clinical symptoms and outcomes and various rTMS protocols and targets of stimulation among the previous studies (Baio et al., 2018; Cole et al., 2019). Thus, at present, the clinical effectiveness of rTMS in ASD would be likely classified as Level C (possibly effective) (Lefaucheur et al., 2020). Thus, considering the fact that currently available medications do not work for the core symptoms of ASD, well-designed and controlled clinical trials are warranted to appropriately evaluate the therapeutic potential of rTMS in ASD.

A new type of TMS treatment, theta-burst stimulation

A new form of rTMS called intermittent theta-burst stimulation (iTBS) can be delivered in 3 min and has demonstrated comparable effectiveness to the conventional 37.5 min 10Hz rTMS protocol in patients with depression (Blumberger et al., 2018). TBS is a patterned form of TMS that can be applied using intermittent or continuous protocols (Di Lazzaro et al., 2008; Huang, Edwards, Rounis, Bhatia, & Rothwell, 2005) that is widely believed to represent cellular learning in a Hebbian form of long-term synaptic plasticity (Larson, Wong, & Lynch, 1986). TBS protocol has characteristics that mimic endogenous theta rhythms of the brain, which can lead to the improvement of the induction of synaptic LTP

TABLE 38.1 rTMS for ASD.

Authors, year	# Of subjects	Age (mean)	Type of rTMS	Parameters
Gomez et al. (2017)	24	10–13	LF-rTMS (left DLPFC)	90%RMT, 1Hz, 4 trains, 1500 pulses, 20 sessions
Wang et al. (2016)	30	7–21	LF-rTMS (left/right DLPFC)	90%RMT, 0.5Hz, 8 trains, 160 pulses, 12 sessions
Sokhadze et al. (2014)	20	10–21	LF-rTMS (left/right/bilateral DLPFC)	90%RMT, 1Hz, 9 trains, 180 pulses, 18 sessions
Casanova et al. (2014)	18	(13.1)	LF-rTMS (left/right/bilateral DLPFC)	90%RMT, 0.5Hz, 8 trains, 160 pulses, 18 sessions
Sokhadze et al. (2014)	27	9–21	LF-rTMS (left/right/bilateral DLPFC)	90%RMT, 1Hz, 9 trains, 180 pulses, 18sessions
Panerai et al. (2014)	17	(13.5)	HF-rTMS/LF-rTMS (left/right PrMC)	8 Hz/1Hz, 30 trains, 900 pulses, 10 sessions
Sokhadze et al. (2012)	20	9-21 (13.5)	LF-rTMS (left/right DLPFC)	90%RMT, 1Hz, 15 trains, 150 pulses, 12 sessions
Casanova et al. (2012)	25	9-19 (12.9)	LF-rTMS (left/right DLPFC)	90%RMT, 1Hz, 15 trains, 150 pulses, 12 sessions

DLPFC, dcrsolateral prefrontal cortex; *HF-rTMS*, high-frequency rTMS; *LF-rTMS*, low-frequency repetitive transcranial magnetic stimulation; *PrMC*, premotor cortex; *RMT*, resting motor threshold

(Suppa et al., 2016). When applied in an intermittent TBS, LTP-like potentiation of cortical activity will be induced, whereas when applied in a continuous TBS, LTD-like depression of cortical activity will be induced.

To date, very few TBS clinical studies have been conducted on individuals with ASD (see Table 38.2). Abujadi et al. conducted an open-label, pilot study of iTBS to target executive function deficits and restricted repetitive behaviors in male children and adolescents with ASD. Ten individuals with ASD participated in this study and 15 sessions of neuronavigated iTBS targeting the right dorsolateral prefrontal cortex were delivered 5 days a week for 3 weeks. The intensity of iTBS was 100% of motor threshold with three pulses of 50 Hz bursts delivered at 5 Hz in 30 trains of 10 bursts with 8 s on and 2 s off over 300 s. As a result, following iTBS intervention, parent report scores on the Repetitive Behavior Scale-Revised and the Yale–Brown Obsessive–Compulsive Scale were improved. Further, individuals with ASD showed improvements in perseverative errors on the Wisconsin Card Sorting Test and total time for the Stroop test. Moreover, the intervention of iTBS was well tolerated with no serious adverse effects. Thus, these preliminary results warrant further sham-controlled iTBS trials with larger sample sizes for ASD (Abujadi, Croarkin, Bellini, Brentani, & Marcolin, 2018).

Safety and tolerability concerns of rTMS in ASD

Of note, the following are often reported as common side effects: headache, scalp discomfort, twitching, mood changes, fatigue, and tinnitus. Further, the most serious side effect is a seizure. Overall the risk of seizure is considered to be less than 0.01% across all kinds of patients and all TMS paradigms (Rossi, Hallett, Rossini, Pascual-Leone, & Safety of, 2009). Indeed, no seizures have been reported during single or paired-pulse TMS paradigms in neurologically healthy individuals. Further, there have been no reports of adverse events on cognitive function so far. On the other hand, although effects of rTMS are usually the desired effects of the stimulation and currently no ASD-specific risk factors are identified for TMS-associated adverse effects, it should still be noted that certain TMS protocols can have both positive and negative impacts,

TABLE 38.2 TBS for ASD.

Authors, year	# Of subjects	Age (mean)	Type of rTMS	Parameters
Abujadi et al., 2018	10	9–17	iTBS (right DLPFC)	100%RMT, 50Hz, 30 trains, 900 pulses, 15 sessions

TBS, theta-burst stimulation.

especially for individuals with neuropsychiatric disorders including ASD. In addition, there may be tolerability concerns in rTMS for individuals with ASD. For example, individuals with ASD often have hyperreactivity to sensory input as well as comorbid hyperactivity. Thus, the feasible and tolerable rTMS protocols should be established for these individuals because rTMS treatment often requires a long treatment session and has a unique uncomfortable sensation including muscle contraction during the session (Rossi et al., 2009).

Ethical concerns of rTMS in ASD

Moreover, there are also ethical concerns when considering the application of rTMS in a vulnerable population such as children and adolescents with ASD (Croarkin, Wall, & Lee, 2011). For example, an important issue that has not yet been fully clarified is whether neuromodulation by rTMS on excitability in a certain brain area may cause compensatory neuromodulations in other areas of the brain in the opposite direction since the pathophysiology underlying the clinical symptoms of ASD is thought to be a complex balance within specific brain regions as well as across multiple neural networks. Another ethical concern relates to the application of rTMS for children and adolescents whose brains are still in development, since there may be a possibility that therapeutic interventions by rTMS can induce unpredictable and potentially long-lasting effects on neurodevelopment (Jotterand & Giordano, 2011). Thus, it is crucial to systematically and broadly evaluate the behavioral and cognitive outcomes and side effects of rTMS over time (George et al., 2007).

Applications to other areas of neurodevelopmental disorders

The major neurodevelopmental disorders other than ASD include ADHD and tic disorders. Individuals with ADHD often struggle with hyperactivity, impulsivity, and inattentiveness, while tic disorders, including Tourette syndrome, manifest as sudden, rapid, recurrent, and nonrhythmic motor movements and/or vocalizations. ADHD affects approximately 10% of the population and typical treatment strategies include pharmacotherapy such as stimulants (e.g., methylphenidate) (Cortese et al., 2018), which is usually effective for improving symptoms in individuals with ADHD in most cases. Tourette syndrome is thought to involve hyperexcitability of the basal ganglia and motor cortex, and pharmacotherapy with antipsychotics has a certain effect on Tourette syndrome. Although pharmacotherapy has a certain effect on these neurodevelopmental disorders, in recent years, neuromodulation therapy has been gradually attempted as a therapeutic strategy with fewer adverse effects.

For ADHD, two rTMS clinical trials were conducted so far in individuals with ADHD (Gomez et al., 2014; Weaver et al., 2012). In both studies, the behavioral rating scores improved compared with the baseline. Particularly, one study without controls demonstrated that five sessions of low-frequency 1-Hz over the left dorsolateral prefrontal cortex improved the scores on symptoms checklist in individuals with ADHD (Gomez et al., 2014). Weaver et al. applied 10 sessions of high-frequency 10-Hz rTMS over the right dorsolateral prefrontal cortex during an active rTMS phase in a crossover design. The scores on the clinical global impressions and ADHD-IV scale significantly improved compared with the baseline, although no differences were observed between active and sham groups (Weaver et al., 2012). For tic disorders, two clinical trials have been conducted so far (Bersani et al., 2013; Kwon et al., 2011). Bersani et al. applied 20 sessions of 1-Hz rTMS over the supplementary motor area with 1200 pulses per session while Kwon et al. used the same rTMS protocol with 10 sessions. In these studies, scores on both the Yale global tic severity scale and the clinical global impressions were improved. However, the rating scales related to attention deficit or anxiety did not change following rTMS treatment (Kwon et al., 2011). No adverse effects were observed in these studies. Furthermore, Wu et al. applied two sessions of cTBS over the supplementary motor area for individuals with chronic tic disorders, and they assessed the clinical effects with the rating scales of the Yale global tic severity scale and Gilles de la Tourette syndrome—quality of life. However, no significant differences were found between active and sham TBS groups (Wu et al., 2014).

Conclusions

Finally, future work of single and paired-pulse TMS neurophysiology research will contribute to the development of more optimal and efficient rTMS protocols as a clinical intervention in children and adolescents with ASD (Masuda, Nakajima, Miyazaki, Yoshida, et al., 2019b) as well as a probe for developmental neuroscience (Noda, 2020).

Key facts of TMS

- TMS is a noninvasive relatively new therapeutic technology that applies magnetic fields to stimulate neurons in the brain to improve symptoms of depression. TMS is typically used when other standard treatments such as pharmacotherapy and/or psychotherapy for depression have not worked well. Treatment for depression involves delivering repetitive magnetic pulses. Thus, this mode of TMS called repetitive TMS (rTMS). During an rTMS session, a TMS coil is placed over the scalp near the forehead and the electromagnetic pulse penetrates the skull painlessly that can stimulate a targeted neuronal population of the brain involved in mood control and depression. TMS is thought to activate the brain regions whose activities are reduced by depression. At present, the therapeutic mechanism of rTMS for depression has not yet been fully elucidated and is under investigation. Moreover, there are various methods for rTMS treatment. Various protocols and techniques will be developed depending on the target disorders and target symptoms.

Key facts of ASD

- Autism spectrum disorder (ASD) refers to a wide range of conditions characterized by some disabilities in social skills, repetitive behaviors, speech, and nonverbal communication. There are many subtypes of ASD, and it is most influenced by a combination of genetic and environmental factors. Due to the nature of spectrum disorder, each individual with ASD has a distinct set of strengths and disabilities, and further the degree of their characteristics varies from person to person. Some people with ASD need significant support in their daily lives, while others do not. Several factors may affect the lives of individuals with ASD, and it is often accompanied by sensory sensitivities and medical issues such as gastrointestinal disorders, seizures or sleep disorders, as well as mental health problems such as anxiety, depression, and attention issues. Autism symptoms usually appear by the age of 2. Previous clinical studies of ASD have shown that early intervention can lead to positive outcomes later in their lives for individuals with ASD.

Summary points

- Currently, effective pharmacotherapy and psychotherapy for core symptoms of ASD have not been established yet.
- Excitatory and inhibitory imbalance in the neural circuits has been suggested as a final common pathway of the pathophysiology of ASD.
- Neuromodulation by rTMS is expected to be a promising treatment targeting the underlying pathophysiology of the symptoms of ASD.
- Although rTMS for ASD has been suggested to be a promising treatment, it is still at the preliminary stage and further research is needed.
- In particular, since children with ASD are in the developmental stage of their brain, neuromodulation for this vulnerable population requires careful assessment of behavior and cognitive function.

References

Abujadi, C., Croarkin, P. E., Bellini, B. B., Brentani, H., & Marcolin, M. A. (2018). Intermittent theta-burst transcranial magnetic stimulation for autism spectrum disorder: An open-label pilot study. Version 2. *Brazilian Journal of Psychiatry, 40*, 309–311. https://doi.org/10.1590/1516-4446-2017-2279.

Baio, J., Wiggins, L., Christensen, D. L., Maenner, M. J., Daniels, J., Warren, Z., … Dowling, N. F. (2018). Prevalence of autism spectrum disorder among children aged 8 Years - autism and developmental disabilities monitoring network, 11 sites, United States, 2014. *MMWR Surveillance Summaries, 67*(6), 1–23. https://doi.org/10.15585/mmwr.ss6706a1.

Barahona-Correa, J. B., Velosa, A., Chainho, A., Lopes, R., & Oliveira-Maia, A. J. (2018). Repetitive transcranial magnetic stimulation for treatment of autism spectrum disorder: A systematic review and meta-analysis. *Frontiers in Integrative Neuroscience, 12*, 27. https://doi.org/10.3389/fnint.2018.00027.

Bersani, F. S., Minichino, A., Enticott, P. G., Mazzarini, L., Khan, N., Antonacci, G., … Biondi, M. (2013). Deep transcranial magnetic stimulation as a treatment for psychiatric disorders: A comprehensive review. *European Psychiatry, 28*(1), 30–39. https://doi.org/10.1016/j.eurpsy.2012.02.006.

Billstedt, E., Gillberg, I. C., & Gillberg, C. (2005). Autism after adolescence: Population-based 13- to 22-year follow-up study of 120 individuals with autism diagnosed in childhood. *Journal of Autism and Developmental Disorders, 35*(3), 351–360. https://doi.org/10.1007/s10803-005-3302-5.

Blumberger, D. M., Vila-Rodriguez, F., Thorpe, K. E., Feffer, K., Noda, Y., Giacobbe, P., … Downar, J. (2018). Effectiveness of theta burst versus high-frequency repetitive transcranial magnetic stimulation in patients with depression (THREE-D): A randomised non-inferiority trial. *Lancet, 391*(10131), 1683–1692. https://doi.org/10.1016/S0140-6736(18)30295-2.

Cantone, M., Bramanti, A., Lanza, G., Pennisi, M., Bramanti, P., Pennisi, G., & Bella, R. (2017). Cortical plasticity in depression. *ASN Neuro, 9*(3). https://doi.org/10.1177/1759091417711512.

Casanova, M. F., Baruth, J. M., El-Baz, A., Tasman, A., Sears, L., & Sokhadze, E. (2012). Repetitive transcranial magnetic stimulation (rTMS) modulates event-related potential (ERP) indices of attention in autism. *Translational Neuroscience, 3*, 170–180. https://doi.org/10.2478/s13380-012-0022-0.

Casanova, M. F., Hensley, M. K., Sokhadze, E. M., El-Baz, A. S., Wang, Y., Li, X., & Sears, L. (2014). Effects of weekly low-frequency rTMS on autonomic measures in children with autism spectrum disorder. *Frontiers in Human Neuroscience, 8*, 851. https://doi.org/10.3389/fnhum.2014.00851.

Casanova, M. F., Sokhadze, E., Opris, I., Wang, Y., & Li, X. (2015). Autism spectrum disorders: Linking neuropathological findings to treatment with transcranial magnetic stimulation. *Acta Paediatrica, 104*(4), 346–355. https://doi.org/10.1111/apa.12943.

Cole, E. J., Enticott, P. G., Oberman, L. M., Gwynette, M. F., Casanova, M. F., Jackson, S. L. J., … r, T. M. S.i. A. S. D. C. G. (2019). The potential of repetitive transcranial magnetic stimulation for autism spectrum disorder: A consensus statement. *Biological Psychiatry, 85*(4), e21–e22. https://doi.org/10.1016/j.biopsych.2018.06.003.

Cortese, S., Adamo, N., Del Giovane, C., Mohr-Jensen, C., Hayes, A. J., Carucci, S., … Cipriani, A. (2018). Comparative efficacy and tolerability of medications for attention-deficit hyperactivity disorder in children, adolescents, and adults: A systematic review and network meta-analysis. *Lancet Psychiatry, 5*(9), 727–738. https://doi.org/10.1016/S2215-0366(18)30269-4.

Croarkin, P. E., Wall, C. A., & Lee, J. (2011). Applications of transcranial magnetic stimulation (TMS) in child and adolescent psychiatry. *International Review of Psychiatry, 23*(5), 445–453. https://doi.org/10.3109/09540261.2011.623688.

Di Lazzaro, V., Pilato, F., Dileone, M., Profice, P., Oliviero, A., Mazzone, P., … Rothwell, J. C. (2008). The physiological basis of the effects of intermittent theta burst stimulation of the human motor cortex. *The Journal of Physiology, 586*(16), 3871–3879. https://doi.org/10.1113/jphysiol.2008.152736.

Durkin, M. S., Maenner, M. J., Baio, J., Christensen, D., Daniels, J., Fitzgerald, R., … Yeargin-Allsopp, M. (2017). Autism spectrum disorder among US children (2002-2010): Socioeconomic, racial, and ethnic disparities. *American Journal of Public Health, 107*(11), 1318–1826. https://doi.org/10.2105/AJPH.2017.304032.

Ecker, C., Schmeisser, M. J., Loth, E., & Murphy, D. G. (2017). Neuroanatomy and neuropathology of autism spectrum disorder in humans. *Advances in Anatomy, Embryology, and Cell Biology, 224*, 27–48. https://doi.org/10.1007/978-3-319-52498-6_2.

Enticott, P. G., Kennedy, H. A., Rinehart, N. J., Tonge, B. J., Bradshaw, J. L., & Fitzgerald, P. B. (2013). GABAergic activity in autism spectrum disorders: An investigation of cortical inhibition via transcranial magnetic stimulation. *Neuropharmacology, 68*, 202–209. https://doi.org/10.1016/j.neuropharm.2012.06.017.

George, M. S., Nahas, Z., Borckardt, J. J., Anderson, B., Foust, M. J., Burns, C., … Short, E. B. (2007). Brain stimulation for the treatment of psychiatric disorders. *Current Opinion in Psychiatry, 20*(3), 250–254. https://doi.org/10.1097/YCO.0b013e3280ad4698. discussion 247-259.

Gómez, L., Vidal, B., Maragoto, C., Morales, L. M., Berrillo, S., Vera Cuesta, H., & Robinson, M. (2017). Non-invasive brain stimulation for children with autism spectrum disorders: A short-term outcome study. *Behavioral Sciences, 7*, 63. https://doi.org/10.3390/bs7030063.

Gomez, L., Vidal, B., Morales, L., Baez, M., Maragoto, C., Galvizu, R., … Sanchez, A. (2014). Low frequency repetitive transcranial magnetic stimulation in children with attention deficit/hyperactivity disorder. Preliminary results. *Brain Stimul, 7*(5), 760–762. https://doi.org/10.1016/j.brs.2014.06.001.

Hauer, L., Sellner, J., Brigo, F., Trinka, E., Sebastianelli, L., Saltuari, L., … Nardone, R. (2019). Effects of repetitive transcranial magnetic stimulation over prefrontal cortex on attention in psychiatric disorders: A systematic review. *Journal of Clinical Medicine, 8*(4). https://doi.org/10.3390/jcm8040416.

Hoogendam, J. M., Ramakers, G. M., & Di Lazzaro, V. (2010). Physiology of repetitive transcranial magnetic stimulation of the human brain. *Brain Stimulation, 3*(2), 95–118. https://doi.org/10.1016/j.brs.2009.10.005.

Huang, Y. Z., Edwards, M. J., Rounis, E., Bhatia, K. P., & Rothwell, J. C. (2005). Theta burst stimulation of the human motor cortex. *Neuron, 45*(2), 201–206. https://doi.org/10.1016/j.neuron.2004.12.033.

Hutton, J., Goode, S., Murphy, M., Le Couteur, A., & Rutter, M. (2008). New-onset psychiatric disorders in individuals with autism. *Autism, 12*(4), 373–390. https://doi.org/10.1177/1362361308091650.

Jotterand, F., & Giordano, J. (2011). Transcranial magnetic stimulation, deep brain stimulation and personal identity: Ethical questions, and neuroethical approaches for medical practice. *International Review of Psychiatry, 23*(5), 476–485. https://doi.org/10.3109/09540261.2011.616189.2011.616189.

Kwon, H. J., Lim, W. S., Lim, M. H., Lee, S. J., Hyun, J. K., Chae, J. H., & Paik, K. C. (2011). 1-Hz low frequency repetitive transcranial magnetic stimulation in children with Tourette's syndrome. *Neuroscience Letters, 492*(1), 1–4. https://doi.org/10.1016/j.neulet.2011.01.007.

Larson, J., Wong, D., & Lynch, G. (1986). Patterned stimulation at the theta frequency is optimal for the induction of hippocampal long-term potentiation. *Brain Research, 368*(2), 347–350. https://doi.org/10.1016/0006-8993(86)90579-2.

Lefaucheur, J. P., Aleman, A., Baeken, C., Benninger, D. H., Brunelin, J., Di Lazzaro, V., … Ziemann, U. (2020). Evidence-based guidelines on the therapeutic use of repetitive transcranial magnetic stimulation (rTMS): An update (2014-2018). *Clinical Neurophysiology, 131*(2), 474–528. https://doi.org/10.1016/j.clinph.2019.11.002.

Maeda, F., Keenan, J. P., Tormos, J. M., Topka, H., & Pascual-Leone, A. (2000a). Interindividual variability of the modulatory effects of repetitive transcranial magnetic stimulation on cortical excitability. *Experimental Brain Research, 133*(4), 425–430. https://doi.org/10.1007/s002210000432.

Maeda, F., Keenan, J. P., Tormos, J. M., Topka, H., & Pascual-Leone, A. (2000b). Modulation of corticospinal excitability by repetitive transcranial magnetic stimulation. *Clinical Neurophysiology, 111*(5), 800–805. https://doi.org/10.1016/s1388-2457(99)00323-5.

Masuda, F., Nakajima, S., Miyazaki, T., Tarumi, R., Ogyu, K., Wada, M., … Noda, Y. (2019a). Clinical effectiveness of repetitive transcranial magnetic stimulation treatment in children and adolescents with neurodevelopmental disorders: A systematic review. *Autism, 23*(7), 1614–1629. https://doi.org/10.1177/1362361318822502.

Masuda, F., Nakajima, S., Miyazaki, T., Yoshida, K., Tsugawa, S., Wada, M., … Noda, Y. (2019b). Motor cortex excitability and inhibitory imbalance in autism spectrum disorder assessed with transcranial magnetic stimulation: A systematic review. *Translational Psychiatry, 9*(1), 110. https://doi.org/10.1038/s41398-019-0444-3.

Mouridsen, S. E., Rich, B., & Isager, T. (2008). Psychiatric disorders in adults diagnosed as children with atypical autism. A case control study. *Journal of Neural Transmission, 115*(1), 135–138. https://doi.org/10.1007/s00702-007-0798-1.

Noda, Y. (2020). Toward the establishment of neurophysiological indicators for neuropsychiatric disorders using transcranial magnetic stimulation-evoked potentials: A systematic review. *Psychiatry and Clinical Neurosciences, 74*(1), 12–34. https://doi.org/10.1111/pcn.12936.

Oberman, L. M., Enticott, P. G., Casanova, M. F., Rotenberg, A., Pascual-Leone, A., McCracken, J. T., & Group, T. M. S.i. A. C. (2016). Transcranial magnetic stimulation in autism spectrum disorder: Challenges, promise, and roadmap for future research. *Autism Research, 9*(2), 184–203. https://doi.org/10.1002/aur.1567.

Oberman, L. M., Rotenberg, A., & Pascual-Leone, A. (2015). Use of transcranial magnetic stimulation in autism spectrum disorders. *Journal of Autism and Developmental Disorders, 45*(2), 524–536. https://doi.org/10.1007/s10803-013-1960-2.

Panerai, S., Tasca, D., Lanuzza, B., Trubia, G., Ferri, R., Musso, S., & Elia, M. (2014). Effects of repetitive transcranial magnetic stimulation in performing eye-hand integration tasks: Four preliminary studies with children showing low-functioning autism. *Autism, 18*, 638–650. https://doi.org/10.1177/1362361313495717.

Port, R. G., Oberman, L. M., & Roberts, T. P. (2019). Revisiting the excitation/inhibition imbalance hypothesis of ASD through a clinical lens. *British Journal of Radiology, 92*(1101), 20180944. https://doi.org/10.1259/bjr.20180944.

Posar, A., & Visconti, P. (2018). Sensory abnormalities in children with autism spectrum disorder. *The Journal of Pediatrics, 94*(4), 342–350. https://doi.org/10.1016/j.jped.2017.08.008.

Rossi, S., Hallett, M., Rossini, P. M., Pascual-Leone, A., & Safety of, T. M. S. C. G. (2009). Safety, ethical considerations, and application guidelines for the use of transcranial magnetic stimulation in clinical practice and research. *Clinical Neurophysiology, 120*(12), 2008–2039. https://doi.org/10.1016/j.clinph.2009.08.016.

Sokhadze, E. M., Baruth, J. M., Sears, L., Sokhadze, G. E., El-Baz, A. S., & Casanova, M. F. (2012). Prefrontal neuromodulation using rTMS improves error monitoring and correction function in autism. *Applied Psychophysiology and Biofeedback, 37*, 91–102. https://doi.org/10.1007/s10484-012-9182-5.

Sokhadze, E. M., El-Baz, A. S., Sears, L. L., Opris, I., & Casanova, M. F. (2014). rTMS neuromodulation improves electrocortical functional measures of information processing and behavioral responses in autism. *Frontiers in Systems Neuroscience, 8*, 134. https://doi.org/10.3389/fnsys.2014.00134.

Stahlberg, O., Soderstrom, H., Rastam, M., & Gillberg, C. (2004). Bipolar disorder, schizophrenia, and other psychotic disorders in adults with childhood onset AD/HD and/or autism spectrum disorders. *Journal of Neural Transmission, 111*(7), 891–902. https://doi.org/10.1007/s00702-004-0115-1.

Suppa, A., Huang, Y. Z., Funke, K., Ridding, M. C., Cheeran, B., Di Lazzaro, V., … Rothwell, J. C. (2016). Ten years of theta burst stimulation in humans: Established knowledge, unknowns and prospects. *Brain Stimulation, 9*(3), 323–335. https://doi.org/10.1016/j.brs.2016.01.006.

Uzunova, G., Pallanti, S., & Hollander, E. (2016). Excitatory/inhibitory imbalance in autism spectrum disorders: Implications for interventions and therapeutics. *World Journal of Biological Psychiatry, 17*(3), 174–186. https://doi.org/10.3109/15622975.2015.1085597.

Wang, Y., Hensley, M. K., Tasman, A., Sears, L., Casanova, M. F., & Sokhadze, E. M. (2016). Heart rate variability and skin conductance during repetitive TMS course in children with autism. *Applied Psychophysiology and Biofeedback, 41*, 47–60. https://doi.org/10.1007/s10484-015-9311-z.

Weaver, L., Rostain, A. L., Mace, W., Akhtar, U., Moss, E., & O'Reardon, J. P. (2012). Transcranial magnetic stimulation (TMS) in the treatment of attention-deficit/hyperactivity disorder in adolescents and young adults: A pilot study. *The Journal of ECT, 28*(2), 98–103. https://doi.org/10.1097/YCT.0b013e31824532c8.

Wu, S. W., Maloney, T., Gilbert, D. L., Dixon, S. G., Horn, P. S., Huddleston, D. A., … Vannest, J. (2014). Functional MRI-navigated repetitive transcranial magnetic stimulation over supplementary motor area in chronic tic disorders. *Brain Stimulation, 7*(2), 212–218. https://doi.org/10.1016/j.brs.2013.10.005.

Yang, X. R., Kirton, A., Wilkes, T. C., Pradhan, S., Liu, I., Jaworska, N., … MacMaster, F. P. (2014). Glutamate alterations associated with transcranial magnetic stimulation in youth depression: A case series. *The Journal of ECT, 30*(3), 242–247. https://doi.org/10.1097/YCT.0000000000000094.

Yenkoyan, K., Grigoryan, A., Fereshetyan, K., & Yepremyan, D. (2017). Advances in understanding the pathophysiology of autism spectrum disorders. *Behavioural Brain Research, 331*, 92–101. https://doi.org/10.1016/j.bbr.2017.04.038.

Chapter 39

Cognitive behavioral therapy in children with anxiety disorders

Gabrielle Gloston[1], Michael A. Sustaíta[1], Ana C. Ramirez[1], Saira A. Weinzimmer[1], Andres G. Viana[2], Eric A. Storch[1] and Sophie C. Schneider[1]

[1]*Menninger Department of Psychiatry and Behavioral Sciences, Baylor College of Medicine, Houston, TX, United States;* [2]*Department of Psychology, University of Houston, Houston, TX, United States*

List of abbreviations

ADHD Attention deficit/hyperactivity disorder
CBT Cognitive behavioral therapy
DSM-5 Diagnostic and Statistical Manual of Mental Disorders, Fifth Edition
ICD-11 The International Classification of Diseases and Related Health Problems, 11th Edition
OCD Obsessive–compulsive disorder

Mini-dictionary of terms

Agoraphobia Fear of having difficulty escaping a situation in which something anxiety provoking, embarrassing, or incapacitating could occur.
Cognitive behavioral therapy (CBT) Therapist-led interventions to modify maladaptive thoughts and behaviors.
Cognitive restructuring CBT technique to identify maladaptive thoughts and replace them with more realistic or helpful thoughts.
Exposure CBT technique where the patient is gradually exposed to a feared object or situation until the anxiety response decreases.
Generalized anxiety disorder Chronic worry that is difficult to control and is accompanied by at least one physical symptom.
Manualized treatment protocol A formalized treatment format with training manuals and resources for therapists and, often, resources for patients.
Panic disorder Fear of the prospect of unexpected panic attacks, with compensatory behavior intended to mitigate the negative effects of panic attacks.
Psychoeducation Information given to patients to help them understand the causes, nature, and treatment of their anxiety symptoms.
Selective mutism Persistent refusal to speak in social situations despite the ability to communicate effectively.
Separation anxiety disorder Excessive anxiety concerning separation from an attachment figure.
Social anxiety disorder Fear of negative evaluation by others that drives anxiety when in specific social situations.
Specific phobias Excessive fear in response to specific objects of situations.

Introduction

Anxiety disorders are among the most common psychiatric disorders in both adult and pediatric populations and often begin in mid-to-late childhood (Bandelow & Michaelis, 2015; Rapee, Schniering, & Hudson, 2009). While feelings of anxiety are developmentally normal, persistent elevated anxiety that is accompanied by substantial distress or functional impairment may indicate psychopathology that requires clinical attention. Anxiety disorders are diagnosed according to criteria outlined in the Diagnostic and Statistical Manual of Mental Disorders, Fifth Edition (DSM-5; American Psychiatric Association, 2013) or the International Classification of Diseases and Related Health Problems 11th Edition (ICD-11; World Health Organization, 2018). Obsessive–compulsive disorder (OCD) and posttraumatic stress disorder were previously classified as anxiety disorders, but their classifications changed in DSM-5, so they will not be considered in this chapter. DSM-5 anxiety disorders and their core features are summarized in Table 39.1.

Diagnosis, Management and Modeling of Neurodevelopmental Disorders. https://doi.org/10.1016/B978-0-12-817988-8.00039-7

TABLE 39.1 DSM-5 anxiety disorders and key features in youth.

Anxiety disorder	Key features
Agoraphobia	Fear about potential difficulty escaping specific scenarios in the event of experiencing something anxiety provoking, embarrassing, or incapacitating, such as a panic attack or uncontrollable physical symptom. Specific feared situations: public transportation, open spaces, enclosed spaces, lines or crowds, and/or being outside (of home) alone.
Generalized anxiety disorder	Inappropriate worry that is difficult to control. Worry is accompanied by at least one physical symptom: feeling on edge, easily fatigued, difficulty concentrating, irritability, muscle tension, or sleep disturbances.
Panic disorder	Recurrent unexpected panic attacks; that is, not triggered by known anxiety-provoking situations or stimuli. Panic attacks followed by avoidant behavior and/or persistent worry.
Selective mutism	Failure to speak in certain social settings, despite the ability to produce speech (that is, there are no known deficiencies in production of speech).
Separation anxiety disorder	Fear or anxiety related to separation from an attachment figure (e.g., parent or other caregiver) that is developmentally inappropriate.
Social anxiety disorder	Significant inappropriate anxiety or fear in social situations that subject an individual to potential scrutiny from peers or adults. Relevant social settings consistently trigger anxiety and/or are avoided Child or adolescent fears negative evaluation from others.
Specific phobias	Intense fear or anxiety in response to a specific stimulus, which may manifest as crying, tantrums, freezing, or clinging. The stimulus nearly always elicits an immediate fear response and is actively avoided. Subtypes: Animal, natural environment, blood injection injury, situational, or other.

DSM-5: Diagnostic and Statistical Manual of Mental Disorders, Fifth Edition (American Psychiatric Association, 2013).
This table describes the key presenting features of the primary pediatric anxiety disorders.

Similar diagnostic criteria are used for pediatric and adult patients, though some diagnostic criteria are modified for pediatric presentations; they may be required to display fewer core symptoms of the disorder or display symptoms for a shorter period of time (American Psychiatric Association, 2013). Further, separation anxiety disorder and selective mutism are considered to be largely disorders of childhood (Viana, Beidel, & Rabian, 2009). The presentation of specific anxiety symptoms can be affected by developmental factors. For example, children may respond to anxiety by crying, throwing tantrums, freezing, or clinging to caregivers. Anxiety-related physical symptoms are particularly common in youth due to their inability to effectively identify and communicate feelings of intense anxiety (Wehry, Beesdo-Baum, Hennelly, Connolly, & Strawn, 2015).

Prevalence and onset of pediatric anxiety disorders

The lifetime prevalence of anxiety disorders among US adolescents is approximately 32.4%, with higher rates among females (38.3%) compared to males (26.8%; Kessler, Petukhova, Sampson, Zaslavsky, & Wittchen, 2012). See Table 39.2 for an overview of lifetime disorder prevalence rates. Selective mutism was only added to the anxiety disorders chapter in DSM-5, and it has not yet been included in large-scale prevalence studies. Specific phobias are the most common pediatric anxiety disorders, and generalized anxiety disorder is the least common. Anxiety disorders have an earlier age of onset than most other mental health disorders (Kessler et al., 2005), and a meta-analysis of studies found that separation anxiety disorder, specific phobias, and social anxiety disorder have an average age of onset before 15 years old (Lijster et al., 2017). Pediatric anxiety disorders are associated with poor health outcomes and interference in educational, interpersonal, and daily functioning (Rapee, Schniering, & Hudson, 2009), elevated comorbidity, and increased risk of suicide (Balázs et al., 2013; Kessler, Avenevoli, et al., 2012a; Olfson et al., 2017). If left untreated, pediatric anxiety disorders tend to follow a chronic course (Wehry et al., 2015). The high prevalence of pediatric anxiety disorders, negative impacts on functioning, and chronic disorder course necessitate effective early intervention.

TABLE 39.2 Lifetime prevalence of specific anxiety disorders among male and female adolescents.

Anxiety disorder	Total	Females	Males
Agoraphobia	2.7%	3.7%	1.7%
Generalized anxiety disorder	2.2%	2.8%	1.6%
Panic disorder	2.3%	2.5%	2.1%
Selective mutism[a]	–	–	–
Separation anxiety disorder	7.7%	9.5%	5.9%
Social anxiety disorder	8.6%	11.2%	6.2%
Specific phobias	20.0%	23.0%	17.1%

Note: Prevalence data is from Kessler, Petukhova, et al. (2012b).
[a]*Prevalence data is not available for selective mutism.*
Table 39.2 displays the lifetime prevalence rates of specific anxiety disorders in male and female adolescents from the United States with data from Kessler, R. C., Petukhova, M., Sampson, N. A., Zaslavsky, A. M., & Wittchen, H.-U. (2012b). Twelve-month and lifetime prevalence and lifetime morbid risk of anxiety and mood disorders in the United States. *International Journal of Methods in Psychiatric Research, 21*(3), 169-184. doi:10.1002/mpr.1359.

Cognitive behavioral therapy for pediatric anxiety disorders

Cognitive behavioral therapy (CBT) is a well-established, evidence-supported treatment for pediatric anxiety disorders (James, James, Cowdrey, Soler, & Choke, 2013; Wang et al., 2017). Further, CBT has lower rates of treatment dropout and adverse events compared to pharmacologic or placebo interventions (Wang et al., 2017). CBT is typically short-term and goal-oriented, designed to affect change by employing interventions at the cognitive and behavioral level (Kaczkurkin & Foa, 2015). Treatment packages for pediatric (and adult) anxiety are often manualized, with therapist training materials and materials to use in collaboration with the child and caregivers.

In the cognitive behavioral model of anxiety, psychopathology develops and is maintained from a dysfunctional relationship between thoughts, behaviors, and emotions (Rapee et al., 2009). Maladaptive thoughts can affect emotional states and behavior; accordingly, anxiety disorders are conceptualized as the result of stimuli being evaluated as a potential threat (cognition), triggering fear or anxiety (emotion), resulting in efforts to avoid the fear- or anxiety-provoking stimulus in the future (avoidant behavior). Avoidance negatively reinforces the maladaptive cognitions about the stimuli and gradually heightens the individual's fear or anxiety response for future encounters with the stimuli. CBT for anxiety therefore aims to correct maladaptive responses using core components outlined in Table 39.3. Initially, psychoeducation is used to educate the child and caregivers about the nature and causes of anxiety and the plan for treatment. Cognitive

TABLE 39.3 Core components of cognitive behavioral therapy.

Core component	Examples
Cognitive restructuring	Identifying and replacing distorted cognitions (e.g., tendency to catastrophize potential outcomes of a situation) with more adaptive thoughts relevant to feared objects or situations.
Exposure	Anxiety hierarchy construction, gradual exposure to feared stimuli (increasing in intensity) in the absence of avoidant behaviors, continued exposure practice until anxiety response is extinguished.
Parenting strategies	Modeling effective coping skills, reinforcing appropriate (nonavoidant) behavior, reducing parental accommodation of anxiety.
Psychoeducation	Normalization of anxiety symptoms, understanding the role of avoidance in maintaining anxiety, rationale for CBT, and treatment overview.
Relapse prevention	Creation of contingency plan to maintain treatment effects, problem-solving to address potential for relapse, booster sessions, or check-in calls if clinically indicated.
Somatic management skills	Relaxation techniques, diaphragmatic breathing, progressive muscle relaxation.

The table outlines several core components of CBT that are commonly used in the treatment of pediatric anxiety.

TABLE 39.4 Example fear hierarchy.

Item	Scenario	SUDS
1	Playing alone in bedroom while parents are in the next room	10
2	Sleeping in bedroom alone all night	20
3	Going to school for a whole day	30
4	Parents leave child alone in therapy session for 5 min	40
5	Staying at an after-school program for 30 min	50
6	Waiting for a ride from school and parents are 10 min late	60
7	Getting 'lost' in a grocery store for 5 min	70
8	Attending a sleepover with an acquaintance	80
9	Being supervised by a family friend for 2 h while parents go out to dinner	90
10	Attending a 1-week summer camp	100

SUDS, subjective units of distress, as rated at the beginning of treatment from 0 (no distress) to 100 (worst possible level).
Table 39.4 provides an example of a fear hierarchy created for an 8-year-old patient with separation anxiety disorder. Anxiety-provoking scenarios are presented in increasing order of difficulty.

restructuring first teaches the child to detect cognitive distortions relevant to his or her anxiety symptoms, then challenge them by assessing the validity of the thought, and finally to replace maladaptive thoughts with more adaptive cognitions (Seligman & Ollendick, 2011). The major behavioral component of CBT for anxiety is exposure; children are required to gradually confront an anxiety-provoking stimulus without avoidance in order to modify their established fear beliefs with new, corrective information (Kaczkurkin & Foa, 2015). Graded exposure allows the child to either tolerate or completely extinguish his or her fear response to specific cues (Craske, Liao, Brown, & Vervliet, 2012). An example of graded exposure for a child with separation anxiety disorder is presented in Table 39.4. Exposure can occur in different forms; real-world engagement with feared objects or situations (in vivo exposure) is typically the foundation of CBT, though mental exposure tasks can be helpful (imaginal exposure), and individuals with strong physical sensations of anxiety (e.g., panic attacks) can benefit from exposure to feared internal cues (interoceptive exposure); examples of each are given in Table 39.5.

TABLE 39.5 Examples of different forms of exposure.

Type of exposure	Description	Symptom	Example exercise
In-vivo exposure	Patient confronts a feared stimulus in the world	Fear of dogs	Patient is asked to pet a dog
		Fear of talking to people at parties, leading to avoidance of all parties	Patient is asked to go to a party and introduce himself or herself to one person
Imaginal exposure	Patient imagines a situation where they experience a feared stimulus	Fear that parent has gotten into car crash	Patient is asked to imagine their parent getting into a car crash and the consequences the patient would experience afterward
		Fear of being bitten by venomous snake	Patient is asked to imagine what it would be like to be bitten by a snake
Interoceptive exposure	Patient is exposed to physical sensations that accompany a panic attack or other anxiety-related physical symptoms	Rapid heartbeat	Patient is asked to run on a treadmill until their heart is beating rapidly
		Digestive discomfort	Patient is asked to consume something that will cause them mild stomach discomfort, e.g., carbonated beverage

Table 39.5 provides examples of three different types of exposure: real-word (in vivo) tasks, interoceptive exposure, and imaginal exposure.

Evidence supporting CBT for pediatric anxiety disorders

A recent comprehensive meta-analysis of 115 studies (including 88 CBT studies with 6978 participants) demonstrated that CBT is an effective intervention to treat anxiety in children and adolescents (Wang et al., 2017). Although the combination of CBT with pharmacotherapy (e.g., selective serotonin reuptake inhibitors) provided better treatment outcomes than monotherapy overall, combined CBT and pharmacotherapy may only be needed when anxiety symptoms are severe (Taylor et al., 2018). CBT is associated with fewer adverse events and lower attrition rates compared with medication or placebo interventions, supporting the acceptability and tolerability of treatment (Wang et al., 2017). Previous literature highlights the central role of exposure within CBT (Ale, McCarthy, Rothschild, & Whiteside, 2015); treatments with higher amounts of exposure practice are associated with better treatment outcomes (Tiwari, Kendall, Hoff, Harrison, & Fizur, 2013). Similarly, length of treatment, parental involvement, and comorbid disorders have been found to be potential mediators of treatment outcome (Knight, McLellan, Jones, & Hudson, 2014; Öst & Ollendick, 2017).

Manualized treatment protocols

A number of CBT protocols have been formalized and published to help therapists meet the needs of children with anxiety disorders. These manualized treatment protocols involve resources to guide therapists in implementing treatments, as well as handouts, workbooks, or other resources for children and/or parents undergoing treatment. Examples of well-studied interventions include Coping Cat and Cool Kids, which have shown to be effective in the reduction of anxiety symptoms (Arendt, Thastum, & Hougaard, 2016; Villabø, Narayanan, Compton, Kendall, & Neumer, 2018). See Table 39.6 for an example of a treatment plan informed by manualized treatment approaches.

TABLE 39.6 Example treatment plan.

Session	Week	Focus of session
1	1	Psychoeducation: What is anxiety, linking thoughts, feelings, and behaviors
2	2	Psychoeducation: Learning to identify anxiety, introducing the cognitive behavioral model of anxiety
3	3	Cognitive restructuring: Learning to identify anxious thoughts, differentiating helpful responses to anxiety versus unhelpful responses
4	4	Cognitive restructuring: Practicing identifying anxious thoughts and challenging their content, replacing the thought with a calm thought
5	5	Exposure: Learning about the concept of graded exposure, developing fear hierarchies, developing a reward system for exposure practice
6	6	Exposure: Reviewing plans for graded exposure, practice of relatively easy exposure steps, applications of rewards. Addition of practice of social skills to improve experience of socially focused exposure tasks
7	7	Exposure: Practicing exposure steps, learning how to adapt exposure practice based on difficulty and response to the situation
8	8	Exposure: Troubleshooting difficulties during exposure or unwillingness to engage in planned tasks, how to remove avoidance Addition of relaxation skills to improve ability to tolerate distress during exposure tasks
9	10	Exposure: Continued practice, developing new hierarchies, revising existing hierarchies Interoceptive exposure: Development of exposure tasks to desensitize child to internal physical sensations of anxiety, such as rapid heartbeat
10	12	Exposure: Continued practice, review of progress and remaining challenges to treatment
11	16	Relapse prevention: Planning for continued skills practice, how to manage increases in anxiety in the long term

Example of a treatment plan for a child with anxiety; detail is provided about the number of sessions, weekly schedule, and the types of skills that the therapist would be applying with the child and/or parent each session.

Parent involvement

Parents are often involved in the delivery of CBT, though their direct involvement in therapy sessions varies according to the needs of the family and the treatment protocol used. Parental involvement can include reinforcing brave behavior and minimizing child avoidance to feared stimuli, modeling appropriate anxiety management skills, completing treatment-related homework assignments, and promoting the generalization of learned coping skills (Seligman & Ollendick, 2011). Even though CBT can be effective with or without active parental involvement, active parental involvement has been associated with longer maintenance of treatment effects (Manassis, 2014). It may be particularly important to involve parents when there are high levels of parental accommodation of anxiety, such as allowing children to avoid anxiety-provoking situations (La Buissonnière-Ariza et al., 2018).

Treatment timing and intensity

Standard CBT usually consists of 11−18 weekly sessions lasting 1−1.5 h; however, the timing and intensity of delivery can be flexible. Meta-analytic findings support the use of alternative treatment formats, including brief (reduced total number of sessions), concentrated (standard number of sessions completed over a shorter period), and intensive formats (total number of sessions and weeks of treatment reduced). CBT formats with a reduced number of sessions and/or more frequent sessions are successful in treating anxiety symptoms and have slightly lower attrition rates than standard CBT in clinical trials (2% vs. 6%; Öst & Ollendick, 2017). Moreover, an intensive 5-day outpatient CBT model for pediatric anxiety produced moderate-to-large anxiety reduction effects (Whiteside, Dammann, Tiede, Biggs, & Jensen, 2018; Whiteside et al., 2014). In cases of highly severe or complex disorder presentations, CBT can be offered in intensive residential settings (Schneider et al., 2017).

Treatment delivery formats

CBT can be successfully delivered individually, in groups, or with therapist-supported self-guided formats (Hudson et al., 2015; James et al., 2013; McKinnon et al., 2018; Wergeland et al., 2014). Therapist-supported, self-guided interventions are an efficacious, convenient, and cost-effective treatment option for anxiety in children and adolescents (Lyneham & Rapee, 2006; Rapee, Abbott, & Lyneham, 2006). "Teletherapy" is the term used to describe therapy delivered via digital formats, such as telephone, video conferencing, and website delivery (American Psychological Association, 2013). Teletherapy can increase access to services (Salloum, Johnco, Lewin, McBride, & Storch, 2016) while maintaining treatment effects (Ebert et al., 2015; Vigerland et al., 2016).

Disorder-specific treatment

Most CBT treatment protocols for anxiety are intended to treat a range of anxiety disorders in youth. However, clinicians may prefer to use disorder-specific treatments in some circumstances. For example, specific phobias may respond well to a specialized one-session treatment protocol (Oar, Farrell, & Ollendick, 2015). People with social anxiety disorder have demonstrated poorer treatment outcomes to CBT compared to those with other anxiety disorders (Hudson et al., 2015), and there is promising evidence for social anxiety disorder-specific protocols (Scaini, Belotti, Ogliari, & Battaglia, 2016). However, it is not yet clear whether disorder-specific treatments generally improve outcomes compared to generic anxiety treatments (Oldham-Cooper and Loades, 2017), so further research is needed to determine under what circumstances a particular type of intervention is needed.

Impact of comorbid disorders

Children with anxiety disorders often have comorbid depression, OCD, conduct disorders, or attention deficit/hyperactivity disorder (ADHD; Garber & Weersing, 2010; Ollendick, Jarrett, Grills-Taquechel, Hovey, & Wolff, 2008). Comorbid disorders have been associated with poorer CBT outcomes (Knight et al., 2014). Comorbid disorders can impact treatment outcomes due to the impact of core disorder symptoms, such as a lack of motivation in depression or difficulty concentrating in ADHD (Walczak, Ollendick, Ryan, & Esbjørn, 2018). The study with the largest sample of treated youth with anxiety found that depressive (major depressive disorder or dysthymia) and externalizing disorders (oppositional defiant disorder, conduct disorder, or ADHD) were both associated with poorer treatment response (Hudson et al., 2015).

Effectiveness studies

Most studies examining the outcomes of CBT for pediatric anxiety are conducted in specialized treatment environments with narrowly selected participant groups (efficacy research), so it is important to establish how well CBT performs in more "real world" treatment environments such as community clinics (effectiveness research). A number of studies have supported CBT as an effective treatment option for youth anxiety, whether delivered in a group, individual, or computer-assisted treatment format (Djurhuus & Bikic, 2019; Storch et al., 2015; Villabø et al., 2018; Wergeland et al., 2014).

Conclusion

In this chapter we have reviewed the efficacy of CBT for pediatric anxiety disorders, one of the most common psychiatric disorders in youth (Kessler, Avenevoli, et al., 2012a). Pediatric anxiety disorders have been associated with poor outcomes in adulthood (Rapee et al., 2009) and an increased risk for chronic mental illness, comorbidity, and suicide (Balázs et al., 2013; Kessler, Avenevoli, et al., 2012a; Olfson et al., 2017; Wehry et al., 2015). Empirical evidence supports the use of CBT for youth anxiety with efficacy comparable to pharmacological treatments (Wang et al., 2017) and superior outcomes regarding dropout and adverse effects. CBT is flexible in terms of delivery format and selection of specific protocols. Overall, CBT is highly effective, acceptable, and useful for the treatment of pediatric anxiety disorders.

Applications to other areas of development

The core elements of CBT can be adapted to treat symptoms of many other conditions in youth. CBT is an empirically supported treatment for pediatric depression (Crowe & McKay, 2017), trauma-related conditions such as posttraumatic stress disorder (Leenarts, Diehle, Doreleijers, Jansma, & Lindauer, 2013), externalizing disorders and related behavioral problems (Battagliese et al., 2015), and substance abuse (Winters et al., 2018). Moreover, CBT-based approaches have been used to address core concerns in pediatric chronic pain (Lonergan, 2016) and tic disorders (Pringsheim et al., 2019). Further, it can be used to successfully treat psychiatric problems in youth with neuropsychiatric problems such as autism spectrum disorder (for a review, see the following Chapter 40).

Mental health disorders are projected to be a leading factor contributing to problems among youth by 2020, including mortality and overall dysfunction (Fusar-Poli, 2019). Given its established efficacy, providing CBT to treat anxiety and other mental health problems in youth has great potential as a corrective intervention that can potentially restore normative development in many youth and reduce the burden of mental illness across the population (Fusar-Poli, 2019).

Key facts about cognitive behavioral therapy for pediatric anxiety disorders

- Pediatric anxiety disorders affect approximately one third of youth.
- CBT is a first-line treatment for pediatric anxiety and has comparable efficacy to current pharmacologic interventions.
- CBT for youth anxiety is often delivered using manualized protocols that emphasize core components including psychoeducation, cognitive restructuring, and exposure.
- The evidence base supporting CBT is strong, including meta-analyses and reviews of over 7000 participants.
- CBT can be delivered with significant flexibility in the timing, format, and focus of treatment sessions.
- Further research is required to understand the precise predictors of treatment outcome across studies.

Summary points

- Anxiety disorders are among the most common psychiatric disorders in child and adolescent populations.
- Pediatric anxiety disorders are marked by inappropriate fear or worry, accompanied by distress and/or impairment in daily functioning.
- CBT is recognized as the gold standard behavioral intervention to treat pediatric anxiety disorders.
- Common components of CBT treatment for pediatric anxiety disorders include cognitive restructuring, psychoeducation, exposure, and parenting strategies.
- A large body of empirical evidence suggests that CBT reduces anxiety symptomology with similar efficacy to pharmacologic interventions, accompanied by less attrition and fewer adverse events compared to pharmacologic interventions.
- Empirical evidence suggests that several parameters of CBT for anxiety can be modified with little to no significant difference in efficacy including treatment length, delivery format, and treatment intensity.

- There is still debate regarding the impact of comorbidity on treatment response; however, there is a body of evidence to suggest that comorbid disorders (i.e., ADHD, depression, OCD, or conduct disorders) are associated with poorer treatment outcomes.
- CBT is also effective for a wide range of pediatric mental health conditions.

References

Ale, C. M., McCarthy, D. M., Rothschild, L. M., & Whiteside, S. P. H. (2015). Components of cognitive behavioral therapy related to outcome in childhood anxiety disorders. *Clinical Child and Family Psychology Review, 18*(3), 240–251. https://doi.org/10.1007/s10567-015-0184-8.

American Psychiatric Association. (2013). In *Diagnostic and statistical manual of mental disorders* (5th ed.). Arlington, VA: American Psychiatric Publishing.

American Psychological Association. (2013). *Guidelines for the practice of telepsychology*. Retrieved from; http://www.apa.org/practice/guidelines/telepsychology.aspx.

Arendt, K., Thastum, M., & Hougaard, E. (2016). Efficacy of a Danish version of the Cool Kids program: A randomized wait-list controlled trial. *Acta Psychiatrica Scandinavica, 133*(2), 109–121. https://doi.org/10.1111/acps.12448.

Balázs, J., Miklósi, M., Keresztény, A., Hoven, C. W., Carli, V., Wasserman, C., … Wasserman, D. (2013). Adolescent subthreshold-depression and anxiety: psychopathology, functional impairment and increased suicide risk. *Journal of Child Psychology and Psychiatry, 54*(6), 670–677. https://doi.org/10.1111/jcpp.12016.

Bandelow, B., & Michaelis, S. (2015). Epidemiology of anxiety disorders in the 21st century. *Dialogues in Clinical Neuroscience, 17*(3), 327–335. https://doi.org/10.31887/DCNS.2015.17.3/bbandelow.

Battagliese, G., Caccetta, M., Luppino, O. I., Baglioni, C., Cardi, V., Mancini, F., & Buonanno, C. (2015). Cognitive-behavioral therapy for externalizing disorders: A meta-analysis of treatment effectiveness. *Behaviour Research and Therapy, 75*, 60–71. https://doi.org/10.1016/j.brat.2015.10.008.

Craske, M. G., Liao, B., Brown, L., & Vervliet, B. (2012). Role of inhibition in exposure therapy. *Journal of Experimental Psychopathology, 3*(3), 322–345. https://doi.org/10.5127/jep.026511.

Crowe, K., & McKay, D. (2017). Efficacy of cognitive-behavioral therapy for childhood anxiety and depression. *Journal of Anxiety Disorders, 49*, 76–87. https://doi.org/10.1016/j.janxdis.2017.04.001.

Djurhuus, I. D., & Bikic, A. (2019). Is the Cool Kids programme working in outpatient psychiatric clinics? a Danish naturalistic effectiveness study. *Nordic Journal of Psychiatry, 73*(2), 141–148. https://doi.org/10.1080/08039488.2019.1574893.

Ebert, D. D., Zarski, A.-C., Christensen, H., Stikkelbroek, Y., Cuijpers, P., Berking, M., & Riper, H. (2015). Internet and computer-based cognitive behavioral therapy for anxiety and depression in youth: A meta-analysis of randomized controlled outcome trials. *PloS One, 10*(3), e0119895. https://doi.org/10.1371/journal.pone.0119895.

Fusar-Poli, P., & The Healthy London Partnership. (2019). Integrated mental health services for the developmental period (0 to 25 Years): A critical review of the evidence. *Frontiers in Psychiatry, 10*. https://doi.org/10.3389/fpsyt.2019.00355, 355-355.

Garber, J., & Weersing, V. R. (2010). Comorbidity of anxiety and depression in youth: Implications for treatment and prevention. *Clinical Psychology: Science and Practice, 17*(4), 293–306. https://doi.org/10.1111/j.1468-2850.2010.01221.x.

Hudson, J. L., Keers, R., Roberts, S., Coleman, J. R. I., Breen, G., Arendt, K., … Eley, T. C. (2015). Clinical predictors of response to cognitive-behavioral therapy in pediatric anxiety disorders: the genes for treatment (GxT) study. *Journal of the American Academy of Child and Adolescent Psychiatry, 54*(6), 454–463. https://doi.org/10.1016/j.jaac.2015.03.018.

James, A.,C., James, G., Cowdrey, F.,A., Soler, A., & Choke, A. (2013). Cognitive behavioural therapy for anxiety disorders in children and adolescents. *Cochrane Database of Systematic Reviews*. https://doi.org/10.1002/14651858.CD004690.pub3.

Kaczkurkin, A. N., & Foa, E. B. (2015). Cognitive-behavioral therapy for anxiety disorders: an update on the empirical evidence. *Dialogues in Clinical Neuroscience, 17*(3), 337–346. https://doi.org/10.31887/DCNS.2015.17.3/akaczkurkin.

Kessler, R. C., Avenevoli, S., McLaughlin, K. A., Green, J. G., Lakoma, M. D., Petukhova, M., … Merikangas, K. R. (2012a). Lifetime co-morbidity of DSM-IV disorders in the US National Comorbidity Survey Replication Adolescent Supplement (NCS-A). *Psychological Medicine, 42*(09), 1997–2010. https://doi.org/10.1017/S0033291712000025.

Kessler, R. C., Berglund, P., Demler, O., Jin, R., Merikangas, K. R., & Walters, E. E. (2005). Lifetime prevalence and age-of-onset distributions of dsm-iv disorders in the national comorbidity survey replication. *Archives of General Psychiatry, 62*(6), 593–602. https://doi.org/10.1001/archpsyc.62.6.593.

Kessler, R. C., Petukhova, M., Sampson, N. A., Zaslavsky, A. M., & Wittchen, H.-U. (2012b). Twelve-month and lifetime prevalence and lifetime morbid risk of anxiety and mood disorders in the United States. *International Journal of Methods in Psychiatric Research, 21*(3), 169–184. https://doi.org/10.1002/mpr.1359.

Knight, A., McLellan, L., Jones, M., & Hudson, J. (2014). Pre-treatment predictors of outcome in childhood anxiety disorders: A systematic review. *Psychopathology Review, 1*(1), 77–129. https://doi.org/10.5127/pr.034613.

La Buissonnière-Ariza, V., Schneider, S. C., Højgaard, D. R. M. A., Kay, B., Riemann, B. C., Eken, S., … Storch, E. A. (2018). Family accommodation of anxiety symptoms in youth undergoing intensive multimodal treatment for anxiety disorders and obsessive-compulsive disorder: nature, clinical correlates, and treatment response. *Comprehensive Psychiatry, 80*, 1–13. https://doi.org/10.1016/j.comppsych.2017.07.012.

Leenarts, L. E. W., Diehle, J., Doreleijers, T. A. H., Jansma, E. P., & Lindauer, R. J. L. (2013). Evidence-based treatments for children with trauma-related psychopathology as a result of childhood maltreatment: A systematic review. *European Child and Adolescent Psychiatry, 22*(5), 269–283. https://doi.org/10.1007/s00787-012-0367-5.

Lijster, J. M.d., Dierckx, B., Utens, E. M. W. J., Verhulst, F. C., Zieldorff, C., Dieleman, G. C., & Legerstee, J. S. (2017). The age of onset of anxiety disorders: A meta-analysis. *The Canadian Journal of Psychiatry, 62*(4), 237–246. https://doi.org/10.1177/0706743716640757.

Lonergan, A. (2016). The effectiveness of cognitive behavioural therapy for pain in childhood and adolescence: A meta-analytic review. *Irish Journal of Psychological Medicine, 33*(4), 251–264. https://doi.org/10.1017/ipm.2015.59.

Lyneham, H. J., & Rapee, R. M. (2006). Evaluation of therapist-supported parent-implemented CBT for anxiety disorders in rural children. *Behaviour Research and Therapy, 44*(9), 1287–1300. https://doi.org/10.1016/j.brat.2005.09.009.

Manassis, K., Changgun Le, T., Bennett, K., Zhao, X. Y., Mendlowitz, S., Duda, S., … Wood, J. J. (2014). Types of parental involvement in CBT with anxious youth: a preliminary meta-analysis. *Journal of Consulting and Clinical Psychology, 82*(6), 10.1037.a0036969.

McKinnon, A., Keers, R., Coleman, J. R. I., Lester, K. J., Roberts, S., Arendt, K., … Hudson, J. L. (2018). The impact of treatment delivery format on response to cognitive behaviour therapy for preadolescent children with anxiety disorders. *Journal of Child Psychology and Psychiatry, 59*(7), 763–772. https://doi.org/10.1111/jcpp.12872.

Oar, E. L., Farrell, L. J., & Ollendick, T. H. (2015). One session treatment for specific phobias: An adaptation for paediatric blood–injection–injury phobia in youth. *Clinical Child and Family Psychology Review, 18*(4), 370–394. https://doi.org/10.1007/s10567-015-0189-3.

Oldham-Cooper, R., & Loades, M. (2017). Disorder-specific versus generic cognitive-behavioral treatment of anxiety disorders in children and young people: A systematic narrative review of evidence for the effectiveness of disorder-specific CBT compared with the disorder-generic treatment, coping Cat. *Journal of Child and Adolescent Psychiatric Nursing, 30*(1), 6–17. https://doi.org/10.1111/jcap.12165.

Olfson, M., Blanco, C., Wall, M., Liu, S.-M., Saha, T. D., Pickering, R. P., & Grant, B. F. (2017). National trends in suicide attempts among adults in the United States. *JAMA Psychiatry, 74*(11), 1095–1103. https://doi.org/10.1001/jamapsychiatry.2017.2582.

Ollendick, T. H., Jarrett, M. A., Grills-Taquechel, A. E., Hovey, L. D., & Wolff, J. C. (2008). Comorbidity as a predictor and moderator of treatment outcome in youth with anxiety, affective, attention deficit/hyperactivity disorder, and oppositional/conduct disorders. *Clinical Psychology Review, 28*(8), 1447–1471. https://doi.org/10.1016/j.cpr.2008.09.003.

Öst, L.-G., & Ollendick, T. H. (2017). Brief, intensive and concentrated cognitive behavioral treatments for anxiety disorders in children: A systematic review and meta-analysis. *Behaviour Research and Therapy, 97*, 134–145. https://doi.org/10.1016/j.brat.2017.07.008.

Pringsheim, T., Okun, M. S., Müller-Vahl, K., Martino, D., Jankovic, J., Cavanna, A. E., … Piacentini, J. (2019). Practice guideline recommendations summary: treatment of tics in people with Tourette syndrome and chronic tic disorders. *Neurology, 92*(19), 896–906. https://doi.org/10.1212/wnl.0000000000007466.

Rapee, R. M., Abbott, M. J., & Lyneham, H. J. (2006). Bibliotherapy for children with anxiety disorders using written materials for parents: A randomized controlled trial. *Journal of Consulting and Clinical Psychology, 74*(3), 436–444. https://doi.org/10.1037/0022-006x.74.3.436.

Rapee, R. M., Schniering, C. A., & Hudson, J. L. (2009). Anxiety disorders during childhood and adolescence: Origins and treatment. *Annual Review of Clinical Psychology, 5*(1), 311–341. https://doi.org/10.1146/annurev.clinpsy.032408.153628.

Salloum, A., Johnco, C., Lewin, A. B., McBride, N. M., & Storch, E. A. (2016). Barriers to access and participation in community mental health treatment for anxious children. *Journal of Affective Disorders, 196*, 54–61. https://doi.org/10.1016/j.jad.2016.02.026.

Scaini, S., Belotti, R., Ogliari, A., & Battaglia, M. (2016). A comprehensive meta-analysis of cognitive-behavioral interventions for social anxiety disorder in children and adolescents. *Journal of Anxiety Disorders, 42*, 105–112. https://doi.org/10.1016/j.janxdis.2016.05.008.

Schneider, S. C., Højgaard, D. R. M. A., La Buissonnière-Ariza, V., Kay, B., Riemann, B. C., Eken, S., … Storch, E. A. (2017). Multimodal residential treatment for adolescent anxiety: outcome and associations with pre-treatment variables. *Child Psychiatry and Human Development, 49*(3), 434–442. https://doi.org/10.1007/s10578-017-0762-8.

Seligman, L. D., & Ollendick, T. H. (2011). Cognitive-behavioral therapy for anxiety disorders in youth. *Child and Adolescent Psychiatric Clinics, 20*(2), 217–238. https://doi.org/10.1016/j.chc.2011.01.003.

Storch, E. A., Salloum, A., King, M. A., Crawford, E. A., Andel, R., McBride, N. M., & Lewin, A. B. (2015). A randomized controlled trial in community mental health centers of computer-assisted cognitive behavioral therapy versus treatment as usual for children with anxiety. *Depression and Anxiety, 32*(11), 843–852. https://doi.org/10.1002/da.22399.

Taylor, J. H., Lebowitz, E. R., Jakubovski, E., Coughlin, C. G., Silverman, W. K., & Bloch, M. H. (2018). Monotherapy insufficient in severe anxiety? Predictors and moderators in the child/adolescent anxiety multimodal study. *Journal of Clinical Child and Adolescent Psychology, 47*(2), 266–281. https://doi.org/10.1080/15374416.2017.1371028.

Tiwari, S., Kendall, P. C., Hoff, A. L., Harrison, J. P., & Fizur, P. (2013). Characteristics of exposure sessions as predictors of treatment response in anxious youth. *Journal of Clinical Child and Adolescent Psychology, 42*(1), 34–43. https://doi.org/10.1080/15374416.2012.738454.

Viana, A. G., Beidel, D. C., & Rabian, B. (2009). Selective mutism: A review and integration of the last 15 years. *Clinical Psychology Review, 29*(1), 57–67. https://doi.org/10.1016/j.cpr.2008.09.009.

Vigerland, S., Lenhard, F., Bonnert, M., Lalouni, M., Hedman, E., Ahlen, J., … Ljótsson, B. (2016). Internet-delivered cognitive behavior therapy for children and adolescents: a systematic review and meta-analysis. *Clinical Psychology Review, 50*, 1–10. https://doi.org/10.1016/j.cpr.2016.09.005.

Villabø, M. A., Narayanan, M., Compton, S. N., Kendall, P. C., & Neumer, S.-P. (2018). Cognitive–behavioral therapy for youth anxiety: An effectiveness evaluation in community practice. *Journal of Consulting and Clinical Psychology, 86*(9), 751–764. https://doi.org/10.1037/ccp0000326.

Walczak, M., Ollendick, T., Ryan, S., & Esbjørn, B. H. (2018). Does comorbidity predict poorer treatment outcome in pediatric anxiety disorders? An updated 10-year review. *Clinical Psychology Review, 60*, 45–61. https://doi.org/10.1016/j.cpr.2017.12.005.

Wang, Z., Whiteside, S. P. H., Sim, L., Farah, W., Morrow, A. S., Alsawas, M., … Murad, M. H. (2017). Comparative effectiveness and safety of cognitive behavioral therapy and pharmacotherapy for childhood anxiety disorders: a systematic review and meta-analysis. *JAMA Pediatrics, 171*(11), 1049–1056. https://doi.org/10.1001/jamapediatrics.2017.3036.

Wehry, A. M., Beesdo-Baum, K., Hennelly, M. M., Connolly, S. D., & Strawn, J. R. (2015). Assessment and treatment of anxiety disorders in children and adolescents. *Current Psychiatry Reports, 17*(7). https://doi.org/10.1007/s11920-015-0591-z, 591-591.

Wergeland, G. J. H., Fjermestad, K. W., Marin, C. E., Haugland, B. S.-M., Bjaastad, J. F., Oeding, K., … Heiervang, E. R. (2014). An effectiveness study of individual vs. group cognitive behavioral therapy for anxiety disorders in youth. *Behaviour Research and Therapy, 57*, 1–12. https://doi.org/10.1016/j.brat.2014.03.007.

Whiteside, S. P. H., Dammann, J. E., Tiede, M. S., Biggs, B. K., & Jensen, A. H. (2018). Increasing availability of exposure therapy through intensive group treatment for childhood anxiety and OCD. *Behavior Modification, 42*(5), 707–728. https://doi.org/10.1177/0145445517730831.

Whiteside, S. P. H., McKay, D., De Nadai, A. S., Tiede, M. S., Ale, C. M., & Storch, E. A. (2014). A baseline controlled examination of a 5-day intensive treatment for pediatric obsessive-compulsive disorder. *Psychiatry Research, 220*(1–2), 441–446. https://doi.org/10.1016/j.psychres.2014.07.006.

Winters, K. C., Botzet, A. M., Stinchfield, R., Gonzales-Castaneda, R., Finch, A. J., Piehler, T. F., … Hemze, A. (2018). Adolescent substance abuse treatment: a review of evidence-based research. In C. G. Leukefeld, & T. P. Gullotta (Eds.), *Adolescent Substance Abuse: Evidence-Based Approaches to Prevention and Treatment* (pp. 141–171). Cham: Springer International Publishing.

World Health Organization. (2018). In *The international classification of Diseases and related health problems* (11th ed. Draft). Retrieved from;. https://icd.who.int/browse11/l-m/en.

Chapter 40

Cognitive behavioral therapy for anxiety in children with autism spectrum disorders

Ana C. Ramirez, Mayra I. Perez, Danica Limon, Eric A. Storch and Sophie C. Schneider
Menninger Department of Psychiatry and Behavioral Sciences, Baylor College of Medicine, Houston, TX, United States

List of abbreviations

ASD Autism spectrum disorder
CBT Cognitive-behavioral therapy
DSM-5 Diagnostic and Statistical Manual of Mental Disorders, Fifth Edition
HFA High-functioning autism

Mini-dictionary of terms

Agoraphobia Fear of being in specific situations in which escape would be difficult in the case of a panic attack or other embarrassing physical symptoms.
Autism spectrum disorder Neurodevelopmental disorder characterized by persistent deficits in communication and social interaction across multiple contexts and restricted and repetitive behaviors.
Cognitive behavioral therapy Psychosocial intervention that challenges and changes unhelpful cognitive distortions and behaviors to improve mental health.
Generalized anxiety disorder Excessive chronic uncontrollable worry about a number of events or activities.
High-functioning autism Although not an official diagnosis, it is often used to refer to people with autism spectrum disorder who read, write, speak, and manage life skills without much assistance.
Neurotypical Not displaying autistic or other neurologically atypical patterns of thought or behavior.
Panic disorder Excessive fear regarding the possibility of experiencing unexpected panic attacks.
Selective mutism Consistent failure to speak in specific social situations in which there is an expectation for speaking, despite speaking in other situations.
Separation anxiety Developmentally inappropriate and excessive fear or anxiety concerning separation from those to whom the individual is attached.
Social anxiety disorder Marked fear or anxiety about one or more social situations in which the individual is exposed to possible scrutiny by others.
Specific phobia Marked fear or anxiety about a specific object or situation. In children the fear or anxiety may be expressed by crying, tantrums, freezing, or clinging.

Introduction

Autism spectrum disorder (ASD) is a common neurodevelopmental condition that has been shown to impact one out of 59 children and adolescents in the United States (Baio et al., 2018). According to the Diagnostic and Statistical Manual of Mental Disorders, Fifth Edition (DSM-5; American Psychiatric Association, 2013), ASD is characterized by reduced verbal and/or nonverbal communication abilities leading to impaired interactive skills and restricted interests and repetitive behaviors. ASD symptoms are evaluated dimensionally and are classified according to the level of support that the

Diagnosis, Management and Modeling of Neurodevelopmental Disorders. https://doi.org/10.1016/B978-0-12-817988-8.00040-3

individual requires for daily functioning; requiring support, requiring substantial support, or requiring very substantial support (American Psychiatric Association, 2013; Ousley & Cermak, 2014). Likewise, severity levels are often described as a range between high-functioning ASD (HFA) and severe ASD.

Anxiety disorders are highly prevalent in children with ASD, affecting approximately 50%, with prevalence estimates ranging from 42% to 79% (Kent & Simonoff, 2017). Cognitive behavioral therapy (CBT) is a well-supported treatment for anxious children in neurotypical populations (see previous chapter for a review). There is good evidence for the efficacy of anxiety-focused CBT in children with HFA when treatment is adapted to suit the needs of their ASD presentation (Kester & Lucyshyn, 2018; Perihan et al., 2020). This chapter will provide an overview of anxiety disorders in HFA pediatric populations, the evidence base for the application of CBT for anxiety in youth with ASD, and techniques used to adapt CBT for ASD populations.

Presentation of anxiety disorders in ASD

ASD individuals can experience both traditional symptoms of anxiety consistent with standard anxiety diagnostic criteria (American Psychiatric Association, 2013) and atypical symptoms of anxiety that reflect the impact of core ASD characteristics (Kerns, Renno, Kendall, Wood, & Storch, 2017). In youth with ASD, common anxiety-provoking situations include changes in routine, novel situations, difficulties with transitions, social rejection, sensory overload, and restrictions to special interests or repetitive behaviors (Kent & Simonoff, 2017; Kerns et al., 2014; Ozsivadjian, Knott, & Magiati, 2012; South, Rodgers, & Van Hecke, 2017). Core features of DSM-5 anxiety disorders and ASD-related symptoms are presented in Table 40.1. It is important to recognize and treat atypical anxiety symptoms as they are found in 61% of youth

TABLE 40.1 Core symptoms of anxiety disorders and their presentation in ASD.

Anxiety disorders and core symptoms	Common anxiety features in children with ASD
Social anxiety disorder: Anxiety about negative evaluation, being judged, fear of embarrassment, or rejection in social situations.	Anxiety triggers can include situations that require understanding of social context and predicting the use of appropriate social skills for interaction. May experience increased social distress or withdrawal due to awareness over developmental differences, social deficits, and/or worry about potential rejection.
Specific phobias: Extreme fear and distress caused by exposure to a specific object or situation. Subtypes: Animal, natural environment, blood injection injury, situational, other.	Idiosyncratic fears (e.g., fear of beards). Fear of loud noises or specific sounds, though must differentiate between phobias and sensory sensitivity related to ASD. Fear of toilets. Fear of dentists is common.
Separation anxiety disorder: Anxiety about separation from core attachment figure and/or from the home.	Excessive rigidity about family routines.
Generalized anxiety disorder: Persistent and uncontrollable worry about a range of areas. Somatic symptoms (headaches, stomachaches, sleep problems) are often paired with cognitive worries.	Daily worries often involve adherence to schedules, with excessive distress about changes to routine and new situations. Worries can be tied to special interests.
Panic disorder: Child experiences recurrent unexpected and becomes anxious about the possibility of future attacks.	Children with ASD may have unique presenting features due to their sensory sensitivities or body-related self-soothing behaviors.
Agoraphobia: Fear about situations where the individual cannot easily escape or get help if needed, such as for a panic attack.	Youth with ASD may find it difficult to articulate the feared outcome of being in specific situations. May present with meltdowns/tantrums rather than typical fear responses.
Selective Mutism: Failure to speak and respond to others in certain settings in which the individual is expected to speak. The child is able to speak, but refuses.	It is important to differentiate ASD-related communication challenges versus selective mutism. The child might lack social interest.

A summary of the core features of anxiety disorders in DSM-5, and examples of how these disorders may present in youth with ASD.
Information regarding core symptoms of DSM-5 anxiety disorders is adapted from American Psychiatric Association. (2013). *Diagnostic and statistical manual of mental disorders*(5th ed.). Arlington, VA: American Psychiatric Publishing. Some information on the presentation of anxiety in ASD is adapted from Kerns, C. M., Kendall, P. C., Berry, L., Souders, M. C., Franklin, M. E., Schultz, R. T., ... Herrington, J. (2014). Traditional and atypical presentations of anxiety in youth with autism spectrum disorder. *Journal of Autism and Developmental Disorders, 44*(11), 2851–2861. https://doi.org/10.1007/s10803-014-2141-7.

TABLE 40.2 Clinician-rated assessment tools for anxiety in ASD.

Measure	Age range	Administration time	Description
Anxiety Disorders Interview Schedule for Children (ADIS)[a]	7–18	1–3 h	Assesses level of severity/impairment associated with each DSM anxiety disorder
Autism Spectrum Addendum for the ADIS[b]	7–18	15–30 min	Designed to aid clinicians in the differential diagnosis of ASD and anxiety symptoms when using the ADIS.
Autism Comorbidities Interview Present and Lifetime[c]	5–17	1–3 h	Modified version of the KSADS for use in ASD.
Children's Interview for Psychiatric Syndromes Parent Version[d]	6–17	60 min	Assesses 20 DSM disorders, including phobias, separation, social, and generalized anxiety disorders.

Table 40.2 presents information about clinician-rated measures to assess anxiety in children and adolescents with ASD.
[a]*Silverman and Albano (1996).*
[b]*Kerns et al. (2014).*
[c]*Leyfer et al. (2006).*
[d]*Witwer, Lecavalier, and Norris (2012).*

TABLE 40.3 Parent and child-rated questionnaire measures for anxiety in ASD.

Measure	Age range	Number of items
Multidimensional Anxiety Scale for Children[a]	8–19	39
Screen for Child Anxiety and Related Emotional Disorders[b]	9–18	41
Spence Children's Anxiety Scale[c]	7–14	44

Table 40.3 displays a list of parent and child-rated measures to assess anxiety in children and adolescents with ASD.
[a]*March, Parker, Sullivan, Stallings, and Conners (1997).*
[b]*Birmaher et al. (1997).*
[c]*Spence (1998).*

with ASD and anxiety (Kerns et al., 2014). These findings underline the need for ASD-adapted treatments in order to address the full range of anxiety symptoms found in this population.

Children and adolescents with ASD may not have the emotional awareness or expressive skills to clearly communicate their anxiety; thus, behavioral indications and proxy symptom reports (e.g., parents and teachers) are particularly important for establishing an anxiety diagnosis (Kent & Simonoff, 2017). Fortunately, there is an increasing availability of anxiety interviews and questionnaires that have been validated for the assessment of anxiety in HFA youth; see Table 40.2 for examples of clinician-administered diagnostic interviews and Table 40.3 for examples of useful questionnaires.

Treating anxiety in children with autism spectrum disorder

If left untreated, anxiety disorders in children with ASD can be debilitating and lead to elevated risk of future internalizing disorders, gastrointestinal issues, and self-injurious behaviors (Bellini, 2004; Teh, Chan, Tan, & Magiati, 2017). Anxiety disorders add a substantial burden to the autistic individual and their family by hindering independent living skills; negatively affecting educational attainment, vocational achievement, and causing strain on familial relationships (Kerns et al., 2014; South et al., 2017). Treatment intervention at early stages of development is crucial given that untreated anxiety symptoms are likely to persist and continue to interfere with many aspects of functioning (Teh et al., 2017).

As outlined in the previous chapter, CBT and pharmacotherapy are the recommended treatments for pediatric anxiety disorders in neurotypical populations. However, there is insufficient evidence to support the use of pharmacotherapy for anxiety in youth with ASD (Stepanova, Dowling, Phelps, & Findling, 2017), and youth with ASD may be at higher risk of serious side effects relating to the use of selective serotonin reuptake inhibitors (Johnco & Storch, 2015; Vasa & Mazurek, 2015). CBT is therefore the only recommended treatment for anxiety in children with ASD.

Given the unique presenting features and needs of youth with ASD, modifications to CBT protocols have been recommended to better accommodate features of the disorder that may interfere with treatment (Fujii et al., 2013).

TABLE 40.4 CBT treatment adaptations for treating anxiety in children with ASD.

Treatment	Session number and duration	Examples of treatment components	Parent involvement in sessions
Behavioral Interventions for Anxiety in Children with Autism[a]	16 sessions, 60–90 min.	Coping skills Exposure.	60 out of 90 min including parents in session.
Building Confidence Adapted for ASD[b]	16 sessions, 90 min.	Coping skills Building independence in self-help skills Exposure.	60 of 90 min are spent with the parents/family.
Cool Kids Adapted for ASD[c]	12 sessions, 90–120 min.	Psychoeducation Exposure Optional relaxation content.	Parents typically involved in some part of each session, parent treatment materials available.
Coping Cat Adapted for ASD[d]	16 sessions, 60–90 min.	Psychoeducation Skills training Exposure.	Two parent-only sessions. 10–15 min spent reviewing the content covered.
Multimodal Anxiety and Social Skills Intervention[e]	12 sessions, 50–75 min.	Psychoeducation Exposure Problem-solving Coping Social skills Optional group therapy.	Parents join the individual therapy sessions for the last 10–20 min

Examples of CBT treatments that have been specifically adapted to treat anxiety in youth with ASD.
[a]*Wood et al. (2015).*
[b]*Wood et al. (2009).*
[c]*Chalfant et al. (2007).*
[d]*McNally Keehn, Lincoln, Brown, and Chavira (2013).*
[e]*White et al. (2010).*

Modifications driven by ASD symptomology include following a more structured and concrete approach, simplifying language, discussing preferred topics, psychoeducation about emotions, visual cues, and social skills training (Cooper, Loades, & Russell, 2018). Multiple CBT protocols have been successfully modified (see Table 40.4), demonstrating that CBT is an efficacious treatment for anxiety in pediatric populations (Chalfant, Rapee, & Carroll, 2007; Kester & Lucyshyn, 2018; Storch et al., 2013; Wood et al., 2015).

Examples of adapting CBT for youth with ASD

Parental Involvement. Although parental (or other caregiver) involvement is often encouraged or required in standard CBT protocols for anxious children, parental involvement is particularly important when providing CBT to youth with ASD (Nadeau et al., 2011). Parents are typically involved in all core therapy activities, and parents are taught how to encourage independence, provide rewards to maintain motivation, and work with schools when needed (Storch et al., 2013). Further, CBT often involves identifying and modifying unhelpful or maladaptive parental responses to child anxiety, such as overprotection or high levels of accommodation of anxiety (Johnco & Storch, 2015; Storch et al., 2015). Parental involvement is particularly important outside of the therapy session in order to motivate completion of homework tasks and real-world practice of CBT skills. Completion of CBT skills outside of therapy is crucial for the success of CBT, and parental involvement is particularly vital in encouraging this practice given the reduced motivation and restricted interests typical of ASD.

Tailoring Information to Youth with ASD. CBT provides the child and parent with an understanding of the connections between emotions, thoughts, and behaviors and is vital to explain the rationale behind later treatment stages (Johnco & Storch, 2015; Nadeau et al., 2011; Ung, Selles, Small, & Storch, 2015). Given that language and communication impairment is common in ASD, it is important to adapt and simplify the information provided to youth with ASD, such as when providing psychoeducation (Chalfant et al., 2007). Identifying anxious feelings and the connection between emotions, behaviors, and thoughts is vital for the success of CBT. Given that difficulties in emotion recognition are

TABLE 40.5 Strategies to tailor information for youth with ASD.

Strategy	Example
Increasing visual presentation of information	Use pictures of favorite cartoon character to show examples of different emotions.
Repetition of ideas	Explain core concepts several times using slightly different examples, ask follow-up questions to assess comprehension of information presented.
Normalization of anxiety	Use social stories about children facing anxiety to improve emotion recognition and enhance engagement.
Present information in concrete terms	Reduce reliance on metaphors and abstract thinking. Use literal language to describe concepts, provide examples using acting, pictures, and puppets.
Reduce reliance on verbal communication of ideas	Use structured worksheets for important activities and concepts, provide summaries of information to parents to enhance learning.

Table 40.5 describes examples of specific strategies that clinicians can use to explain CBT-related information to youth with ASD.

TABLE 40.6 Example exposure hierarchy for social anxiety.

Step	Goal: Have two new friends by the end of school year
1	Show child modeling video, in session, of how to approach peers.
2	Role play, in session, child approaching.
3	Approach a familiar peer who is by himself on the school playground and tell him about your interests, while holding preferred item.
4	Approach a familiar peer who is by himself on the school playground and talk about his interests, while holding preferred item.
5	Approach a familiar peer who is by himself on the school playground and talk about his interests, without holding preferred item.
6	Approach a familiar peer who is with someone else on the school playground and tell them about your interests, without holding preferred item.
7	Approach a familiar peer who is with someone else on the school playground and ask them about their interests, without holding preferred item.
8	Approach an unfamiliar peer who is by himself on the school playground and ask them about his interests, without holding preferred item.
9	Have mom set a play date with familiar peer, at home.
10	Have mom set a play date with two familiar peers, at home.
11	Have play date outside of home with two familiar peers.

Table 40.6 describes an example exposure hierarchy that may be applied for a child with ASD and social anxiety.

common in ASD (Law Smith, Montagne, Perrett, Gill, & Gallagher, 2010), it is helpful to include concrete examples of key concepts and repeat key concepts to ensure comprehension. Examples of specific strategies to tailor information are presented in Table 40.5.

Graded Exposure. Exposure is the core behavioral technique used in CBT; the therapist and patient work together to form a hierarchy of feared situations that gradually increase in difficulty. For an example hierarchy, see Table 40.6. The intent of exposure is for the child to learn that their feared outcome will not happen, or if it happens, that they are able to cope with the outcomes. Given that core symptoms of ASD may interfere with the ability to successfully complete some types of exposure (for example, to have a positive social interaction with a strange child), it is of high importance to address social skills deficits and independent living skills when planning exposure tasks by rehearsing social skills in real-world settings (Wood et al., 2009). For example, approaching a familiar peer who is alone increases the chance of a positive response compared to more complex social interactions (Fujii et al., 2013). Providing rewards immediately after exposure tasks is important to increase child motivation, and the special interests of the child can be recruited to ensure that rewards are particularly salient (Storch et al., 2013). In order to facilitate exposure therapy, additional strategies such as relaxation techniques have been proposed as one way to reduce anxiety by allowing approach behaviors and reducing avoidance (Johnco & Storch, 2015). Further, ensuring that children are not distressed by sensory aspects of the situation is important in ensuring the success of exposure tasks.

Review of key supporting evidence

The most recent meta-analysis of studies examining the efficacy of CBT for anxiety in youth with ASD included 23 studies and 745 participants, and it supported CBT as an efficacious anxiety-management treatment with a moderate effect size (Perihan et al., 2020). This is consistent with earlier meta-analyses (Ung et al., 2015; van Steensel, Zegers, & Bögels, 2017). These and other studies have examined treatment and patient-related variables that may be associated with treatment outcome.

Treatment Dose. The number of CBT sessions specified in available treatment protocols vary widely, e.g., from 6 to 32 sessions (Fujii et al., 2013; Sofronoff, Attwood, & Hinton, 2005), with sessions usually lasting around 90 min (Wood & McLeod, 2008). Evidence shows that treatment effects in short-term interventions (shorter than 12 weeks) were significantly weaker than those associated with standard-term (12–15 weeks) and long-term (more than 16 weeks) interventions (Perihan et al., 2020).

Group or Individual Sessions. Individual and group CBT formats are similarly efficacious in the treatment of anxiety disorders in HFA youth (Ung et al., 2015), so selection of treatment format will depend on family preference and treatment availability. Benefits of group CBT include increasing access to services while also providing a cost-effective treatment option, normalization of anxiety, social support, increased accountability, among others (Ishikawa, Okajima, Matsuoka, & Sakano, 2007). Alternately, individual CBT may be a better option for children that require additional individualization and tailoring to their specific needs and interests (Ung et al., 2015).

Social Skills Training. Social skills training can be a helpful component when treating anxiety in HFA youth as it can address social skills deficits that contribute to poor social interactions, such as poor eye contact, difficulty initiating and maintaining conversations, and poor social awareness (Ung et al., 2015). Similarly, social skills training can be quite useful in order to help children with ASD build self-confidence, improve social competence (Wood et al., 2009), and ameliorate symptoms of social anxiety (Ung et al., 2013).

Child Age and Gender. van Steensel et al. (2017) found that child gender and age are not associated with anxiety severity or treatment effectiveness, suggesting that CBT is an intervention that can be used broadly in high-functioning children with ASD. Parent and clinician ratings of anxiety are reliable tools to detect treatment change (Sukhodolsky, Bloch, Panza, & Reichow, 2013).

Family Factors. Parental involvement in CBT interventions has been associated with better treatment outcomes; protocols incorporating parental involvement resulted in large treatment effects, while interventions that focused on the child only produced small effects (Perihan et al., 2020). Family involvement during treatment is also relevant, as high levels of accommodation of anxiety are common in parents of children and adolescents with ASD and other neurodevelopmental disorders (Johnco & Storch, 2015; Storch et al., 2015) and may lead to poorer outcomes. Another potential treatment mediator is parental psychopathology, with children of anxious fathers presenting less severe anxiety symptoms than children of nonanxious fathers (van Steensel et al., 2017). Lastly, higher parental stress has also been identified as a potential outcome mediator (Weiss, Viecili, & Bohr, 2014).

Limitations to available evidence

To date the evidence base for CBT for anxious youth with ASD is largely limited to high-functioning and verbally fluent individuals. Although there is promising evidence that minimally verbal youth with ASD may respond to a modified version of CBT focusing on behavioral techniques including graded exposure paired with reinforcement (Hagopian, Lilly, & Davis, 2017), further research is needed to establish the efficacy of CBT in youth with a wider range of functioning. Another important limitation to consider is the underrepresentation of diverse groups in CBT research to treat comorbid anxiety and ASD. A review of 14 studies over a 9-year span concluded that CBT research tends to have an overrepresentation of Caucasian and highly educated families (Pickard, Reyes, & Reaven, 2019). Noninclusive representation in research samples leads to issues with treatment generalizability, questionable treatment applicability in community settings, and increased racial and ethnic disparities (Pickard et al., 2019). Engaging low SES and diverse families in CBT research may reduce mental health stigma, increase treatment accessibility in health care settings, and help bridge gaps in understanding cooccurring anxiety in ASD populations.

Applications to other areas of development

In this chapter we have reviewed the efficacy of CBT for anxiety disorders in children and adolescents with ASD. If left untreated, anxiety disorders can be extremely debilitating and lead to future internalizing disorders and/or concomitant disorders such as depression, other anxiety disorders, and substance use in later years (de Vries et al., 2019). CBT can also be used to improve functioning in other important domains for youth with ASD and other neurodevelopmental disorders. Youth with ASD are at high risk of severe emotional dysregulation (Joshi et al., 2018), which can be manifested as tantrums, uncontrolled outbursts, aggression, and self-injurious behavior (Mazefsky & White, 2014). CBT has been used to improve emotion regulation in youth with ASD (Scarpa & Reyes, 2011; Weiss et al., 2018), with associated improvements in emotionality, adaptive behaviors, and reductions in psychopathology and aggressive behaviors. As deficit in emotion regulation is a transdiagnostic construct (Weiss et al., 2018), utilizing CBT to improve emotion regulation has the potential to greatly improve emotional and behavioral functioning in youth with ASD.

CBT has also been shown to be beneficial in the treatment of insomnia in children with ASD (Cortesi, Gianotti, Sebastiani, Panunzi, & Valente, 2012). Sleep disorders are common among children with neurodevelopmental disorders, including ASD, epilepsy, Down syndrome, Rett syndrome, Tourette syndrome, and attention deficit hyperactivity disorder (Gregory & Sadeh, 2016). Sleep problems are also predictive of later psychopathology, so utilizing CBT to improve sleep quality among youth with a wide range of neurodevelopmental disorders has significant potential both to improve current functioning and to reduce the risk of later psychiatric problems.

Key facts about cognitive behavioral therapy for anxiety in youth with autism spectrum disorder

- Autism spectrum disorder (ASD) affects one in 59 children in the United States.
- Approximately 50% of youth with ASD meet criteria for an anxiety disorder.
- Cognitive behavioral therapy (CBT) can be modified to meet the needs of youth with ASD.
- CBT is an effective treatment for anxiety in high-functioning and verbally fluent youth with ASD.
- Further research is needed to establish the efficacy of CBT in less verbal youth with ASD and in more diverse populations.

Summary points

- Anxiety disorders affect approximately 50% of children and adolescents with autism spectrum disorder (ASD).
- If left untreated, anxiety in children with ASD is associated with interference in different aspects of functioning.
- Cognitive behavioral therapy (CBT) is the only evidence-supported treatment for anxiety in children with ASD.
- Adaptations to standard CBT such as following a more structured approach, simplifying language and abstract constructs, incorporating visual aids, etc., have been used to maximize treatment outcomes.
- CBT can be effectively implemented in both group and individual formats.
- Future research should focus on treatment mediators, predictors of treatment outcomes, and adaptations.

References

American Psychiatric Association. (2013). *Diagnostic and statistical manual of mental disorders* (5th ed.). Arlington, VA: American Psychiatric Publishing.

Baio, J., Wiggins, L., Christensen, D. L., Maenner, M. J., Daniels, J., Warren, Z., … White, T. (2018). Prevalence of autism spectrum disorder among children aged 8 years—autism and developmental disabilities monitoring network, 11 Sites, United States, 2014. *MMWR Surveillance Summaries, 67*(6), 1. https://doi.org/10.15585/mmwr.ss6706a1.

Bellini, S. (2004). Social skill deficits and anxiety in high-functioning adolescents with autism spectrum disorders. *Focus on Autism and Other Developmental Disabilities, 19*, 78–86. https://doi.org/10.1177/10883576040190020201.

Birmaher, B., Khetarpal, S., Brent, D., Cully, M., Balach, L., Kaufman, J., & Neer, S. M. (1997). The screen for child anxiety related emotional disorders (SCARED): Scale construction and psychometric characteristics. *Journal of the American Academy of Child and Adolescent Psychiatry, 36*(4), 545–553. https://doi.org/10.1097/00004583-199704000-00018.

Chalfant, A. M., Rapee, R., & Carroll, L. (2007). Treating anxiety disorders in children with high functioning autism spectrum disorders: A controlled trial. *Journal of Autism and Developmental Disorders, 37*(10), 1842–1857. https://doi.org/10.1007/s10803-006-0318-4.

Cooper, K., Loades, M. E., & Russell, A. (2018). Adapting psychological therapies for autism. *Research in Autism Spectrum Disorders, 45*, 43–50. https://doi.org/10.1016/j.rasd.2017.11.002.

Cortesi, F., Gianotti, F., Sebastiani, T., Panunzi, S., & Valente, D. (2012). Controlled-release melatonin, singly and combined with cognitive behavioural therapy, for persistent insomnia in children with autism spectrum disorders: A randomized placebo-controlled trial. *Journal of Sleep Research, 21*(6), 700–709. https://doi.org/10.1111/j.1365-2869.2012.01021.x.

Fujii, C., Renno, P., McLeod, B. D., Lin, C. E., Decker, K., Zielinski, K., & Wood, J. J. (2013). Intensive cognitive behavioral therapy for anxiety disorders in school-aged children with autism: A preliminary comparison with treatment-as-usual. *School Mental Health, 5*(1), 25–37. https://doi.org/10.1007/s12310-012-9090-0.

Gregory, A. M., & Sadeh, A. (2016). Annual Research Review: Sleep problems in childhood psychiatric disorders – a review of the latest science. *Journal of Child Psychology and Psychiatry, 57*(3), 296–317. https://doi.org/10.1111/jcpp.12469.

Hagopian, L. P., Lilly, M., & Davis, T. E. (2017). Behavioral treatments for anxiety in minimally verbal children with ASD. In C. M. Kerns, P. Renno, E. A. Storch, P. C. Kendall, & J. J. Wood (Eds.), *Anxiety in children and adolescents with autism spectrum disorder* (pp. 193–210). Academic Press.

Ishikawa, S.-i., Okajima, I., Matsuoka, H., & Sakano, Y. (2007). Cognitive behavioural therapy for anxiety disorders in children and adolescents: A meta-analysis. *Child and Adolescent Mental Health, 12*(4), 164–172. https://doi.org/10.1111/j.1475-3588.2006.00433.x.

Johnco, C., & Storch, E. A. (2015). Anxiety in youth with autism spectrum disorders: Implications for treatment. *Expert Review of Neurotherapeutics, 15*(11), 1343–1352. https://doi.org/10.1586/14737175.2015.1111762.

Joshi, G., Wozniak, J., Fitzgerald, M., Faraone, S., Fried, R., Galdo, M., … Biederman, J. (2018). High risk for severe emotional dysregulation in psychiatrically referred youth with autism spectrum disorder: a controlled study. *Journal of Autism and Developmental Disorders, 48*(9), 3101–3115. https://doi.org/10.1007/s10803-018-3542-9.

Kent, R., & Simonoff, E. (2017). Prevalence of anxiety in autism spectrum disorders. In C. M. Kerns, P. Renno, E. A. Storch, P. C. Kendall, & J. J. Wood (Eds.), *Anxiety in children and adolescents with autism spectrum disorder* (pp. 5–32). Academic Press.

Kerns, C. M., Kendall, P. C., Berry, L., Souders, M. C., Franklin, M. E., Schultz, R. T., … Herrington, J. (2014). Traditional and atypical presentations of anxiety in youth with autism spectrum disorder. *Journal of Autism and Developmental Disorders, 44*(11), 2851–2861. https://doi.org/10.1007/s10803-014-2141-7.

Kerns, C. M., Renno, P., Kendall, P. C., Wood, J. J., & Storch, E. A. (2017). Anxiety disorders interview schedule – autism addendum: Reliability and validity in children with autism spectrum disorder. *Journal of Clinical Child and Adolescent Psychology, 46*(1), 88–100. https://doi.org/10.1080/15374416.2016.1233501.

Kester, K. R., & Lucyshyn, J. M. (2018). Cognitive behavior therapy to treat anxiety among children with autism spectrum disorders: A systematic review. *Research in Autism Spectrum Disorders, 52*, 37–50. https://doi.org/10.1016/j.rasd.2018.05.002.

Law Smith, M. J., Montagne, B., Perrett, D. I., Gill, M., & Gallagher, L. (2010). Detecting subtle facial emotion recognition deficits in high-functioning autism using dynamic stimuli of varying intensities. *Neuropsychologia, 48*(9), 2777–2781. https://doi.org/10.1016/j.neuropsychologia.2010.03.008.

Leyfer, O. T., Folstein, S. E., Bacalman, S., Davis, N. O., Dinh, E., Morgan, J., … Lainhart, J. E. (2006). Comorbid psychiatric disorders in children with autism: interview development and rates of disorders. *Journal of Autism and Developmental Disorders, 36*(7), 849–861. https://doi.org/10.1007/s10803-006-0123-0.

March, J. S., Parker, J. D. A., Sullivan, K., Stallings, P., & Conners, C. K. (1997). The multidimensional anxiety scale for children (MASC): Factor structure, reliability, and validity. *Journal of the American Academy of Child and Adolescent Psychiatry, 36*(4), 554–565. https://doi.org/10.1097/00004583-199704000-00019.

Mazefsky, C. A., & White, S. W. (2014). Emotion regulation: Concepts & practice in autism spectrum disorder. *Child and Adolescent Psychiatric Clinics of North America, 23*(1), 15–24. https://doi.org/10.1016/j.chc.2013.07.002.

McNally Keehn, R. H., Lincoln, A. J., Brown, M. Z., & Chavira, D. A. (2013). The coping cat program for children with anxiety and autism spectrum disorder: A pilot randomized controlled trial. *Journal of Autism and Developmental Disorders, 43*(1), 57–67. https://doi.org/10.1007/s10803-012-1541-9.

Nadeau, J., Sulkowski, M. L., Ung, D., Wood, J. J., Lewin, A. B., Murphy, T. K., … Storch, E. A. (2011). Treatment of comorbid anxiety and autism spectrum disorders. *Neuropsychiatry, 1*(6), 567–578. https://doi.org/10.2217/NPY.11.62.

Ousley, O., & Cermak, T. (2014). Autism spectrum disorder: Defining dimensions and subgroups. *Current Developmental Disorders Reports, 1*(1), 20–28. https://doi.org/10.1007/s40474-013-0003-1.

Ozsivadjian, A., Knott, F., & Magiati, I. (2012). Parent and child perspectives on the nature of anxiety in children and young people with autism spectrum disorders: A focus group study. *Autism, 16*(2), 107–121. https://doi.org/10.1177/1362361311431703.

Perihan, C., Burke, M., Bowman-Perrott, L., Bicer, A., Gallup, J., Thompson, J., & Sallese, M. (2020). Effects of cognitive behavioral therapy for reducing anxiety in children with high functioning ASD: A systematic review and meta-analysis. *Journal of Autism and Developmental Disorders, 50*(6), 1958–1972. https://doi.org/10.1007/s10803-019-03949-7.

Pickard, K., Reyes, N., & Reaven, J. (2019). Examining the inclusion of diverse participants in cognitive behavior therapy research for youth with autism spectrum disorder and anxiety. *Autism, 23*(4), 1057–1064. https://doi.org/10.1177/1362361318795678.

Scarpa, A., & Reyes, N. M. (2011). Improving emotion regulation with CBT in young children with high functioning autism spectrum disorders: A pilot study. *Behavioural and Cognitive Psychotherapy, 39*(4), 495–500. https://doi.org/10.1017/S1352465811000063.

Silverman, W. K., & Albano, A. M. (1996). *Anxiety disorders interview schedule for children for DSM–IV: (Child and parent versions)*. San Antonio,TX: Psychological Corporation/Graywind.

Sofronoff, K., Attwood, T., & Hinton, S. (2005). A randomised controlled trial of a CBT intervention for anxiety in children with Asperger syndrome. *Journal of Child Psychology and Psychiatry, 46*(11), 1152–1160. https://doi.org/10.1111/j.1469-7610.2005.00411.x.

South, M., Rodgers, J., & Van Hecke, A. (2017). Anxiety and ASD: Current progress and ongoing challenges. *Journal of Autism and Developmental Disorders, 47*(12), 3679–3681. https://doi.org/10.1007/s10803-017-3322-y.

Spence, S. H. (1998). A measure of anxiety symptoms among children. *Behaviour Research and Therapy, 36*(5), 545–566. https://doi.org/10.1016/S0005-7967(98)00034-5.

van Steensel, F. J. A., Zegers, V. M., & Bögels, S. M. (2017). Predictors of treatment effectiveness for youth with ASD and comorbid anxiety disorders: It all depends on the family? *Journal of Autism and Developmental Disorders, 47*(3), 636–645. https://doi.org/10.1007/s10803-016-2956-5.

Stepanova, E., Dowling, S., Phelps, M., & Findling, R. L. (2017). Pharmacotherapy of emotional and behavioral symptoms associated with autism spectrum disorder in children and adolescents. *Dialogues in Clinical Neuroscience, 19*(4), 395–402.

Storch, E. A., Arnold, E. B., Lewin, A. B., Nadeau, J. M., Jones, A. M., De Nadai, A. S., … Murphy, T. K. (2013). The effect of cognitive-behavioral therapy versus treatment as usual for anxiety in children with autism spectrum disorders: a randomized, controlled trial. *Journal of the American Academy of Child and Adolescent Psychiatry, 52*(2), 132–142. e132. https://doi.org/10.1016/j.jaac.2012.11.007.

Storch, E. A., Zavrou, S., Collier, A. B., Ung, D., Arnold, E. B., Mutch, P. J., … Murphy, T. K. (2015). Preliminary study of family accommodation in youth with autism spectrum disorders and anxiety: incidence, clinical correlates, and behavioral treatment response. *Journal of Anxiety Disorders, 34*, 94–99. https://doi.org/10.1016/j.janxdis.2015.06.007.

Sukhodolsky, D. G., Bloch, M. H., Panza, K. E., & Reichow, B. (2013). Cognitive-behavioral therapy for anxiety in children with high-functioning autism: A meta-analysis. *Pediatrics, 132*. https://doi.org/10.1542/peds.2013-1193.

Teh, E. J., Chan, D. M.-E., Tan, G. K. J., & Magiati, I. (2017). Continuity and change in, and child predictors of, caregiver reported anxiety symptoms in young people with autism spectrum disorder: A follow-up study. *Journal of Autism and Developmental Disorders, 47*(12), 3857–3871. https://doi.org/10.1007/s10803-017-3136-y.

Ung, D., Selles, R., Small, B. J., & Storch, E. A. (2015). A systematic review and meta-analysis of cognitive-behavioral therapy for anxiety in youth with high-functioning autism spectrum disorders. *Child Psychiatry and Human Development, 46*(4), 533–547. https://doi.org/10.1007/s10578-014-0494-y.

Ung, D., Wood, J. J., Ehrenreich-May, J., Arnold, E. B., Fuji, C., Renno, P., … Storch, E. A. (2013). Clinical characteristics of high-functioning youth with autism spectrum disorder and anxiety. *Neuropsychiatry, 3*(2), 10.2217/npy.2213.2219. https://doi.org/10.2217/npy.13.9.

Vasa, R. A., & Mazurek, M. O. (2015). An update on anxiety in youth with autism spectrum disorders. *Current Opinion in Psychiatry, 28*(2), 83–90. https://doi.org/10.1097/yco.0000000000000133.

de Vries, Y.A., Al-Hamzawi, A., Alonso, J., Borges, G., Bruffaerts, R., Bunting, B., . . . on behalf of the WHO World Mental Health Survey Collaborators. (2019). Childhood generalized specific phobia as an early marker of internalizing psychopathology across the lifespan: results from the World Mental Health Surveys. BMC Medicine, 17(1), 101. https://doi.org/10.1186/s12916-019-1328-3.

Weiss, J. A., Thomson, K., Burnham Riosa, P., Albaum, C., Chan, V., Maughan, A., … Black, K. (2018). A randomized waitlist-controlled trial of cognitive behavior therapy to improve emotion regulation in children with autism. *Journal of Child Psychology and Psychiatry, 59*(11), 1180–1191. https://doi.org/10.1111/jcpp.12915.

Weiss, J. A., Viecili, M. A., & Bohr, Y. (2014). Parenting stress as a correlate of cognitive behavior therapy responsiveness in children with autism spectrum disorders and anxiety. *Focus on Autism and Other Developmental Disabilities, 30*(3), 154–164. https://doi.org/10.1177/1088357614547808.

White, S. W., Albano, A. M., Johnson, C. R., Kasari, C., Ollendick, T., Klin, A., … Scahill, L. (2010). Development of a cognitive-behavioral intervention program to treat anxiety and social deficits in teens with high-functioning autism. *Clinical Child and Family Psychology Review, 13*(1), 77–90. https://doi.org/10.1007/s10567-009-0062-3.

Witwer, A. N., Lecavalier, L., & Norris, M. (2012). Reliability and validity of the children's interview for psychiatric syndromes-parent version in autism spectrum disorders. *Journal of Autism and Developmental Disorders, 42*(9), 1949–1958. https://doi.org/10.1007/s10803-012-1442-y.

Wood, J. J., Drahota, A., Sze, K., Har, K., Chiu, A., & Langer, D. A. (2009). Cognitive behavioral therapy for anxiety in children with autism spectrum disorders: A randomized, controlled trial. *Journal of Child Psychology and Psychiatry, 50*(3), 224–234. https://doi.org/10.1111/j.1469-7610.2008.01948.x.

Wood, J. J., Ehrenreich-May, J., Alessandri, M., Fujii, C., Renno, P., Laugeson, E., … Storch, E. A. (2015). Cognitive behavioral therapy for early adolescents with autism spectrum disorders and clinical anxiety: a randomized, controlled trial. *Behavior Therapy, 46*(1), 7–19. https://doi.org/10.1016/j.beth.2014.01.002.

Wood, J. J., & McLeod, B. D. (2008). *Child anxiety disorders: A family-based treatment manual for practitioners*. New York, NY, US: W W Norton & Co.

Chapter 41

Nutritional treatment in fetal alcohol spectrum disorder (FASD)

Oscar García-Algar[1], Giorgia Sebastiani[1], Vicente Andreu- Fernández[2,3] and Sebastian Sailer[1]

[1]Neonatal Unit, Hospital Clinic-Maternitat, ICGON, BCNatal-Barcelona Center for Maternal-Fetal and Neonatal Medicine, Grup de Recerca Infancia i Entorn (GRIE), Institut d'investigacions Biomèdiques August Pi i Sunyer (IDIBAPS), Barcelona, Spain; [2]Grup de Recerca Infancia i Entorn (GRIE), Institut d'investigacions Biomèdiques August Pi i Sunyer (IDIBAPS), Barcelona, Spain; [3]Valencian International University (VIU), Valencia, Spain

List of abbreviations

ARBD Alcohol-related birth defects
ARND Alcohol-related neurodevelopmental disorders
BDNF Brain-derived neurotrophic factor
CNS Central nervous system
CYP2E1 Cytochrome P450 2E1
DHA Docosahexaenoic acid
DS Down syndrome
EGCG Epigallocatechin gallate
FASD Fetal alcohol spectrum disorders
GD Gestational days
GDNF Glial-cell-derived neurotrophic factor
GGT Gamma-glutamyltransferase
H2O2 Hydrogen peroxide
IOM The Institute of Medicine
LTP Long-term potentiation
MDA Malondialdehyde
NADH Nicotinamide adenosine dinucleotide phosphate
NO Nitric oxide
Nrf Nuclear factor
PD Postnatal days
ROS Reactive oxygen species
SOD Superoxide dismutase
WHO World Health Organization

Mini-dictionary of terms

Apoptosis A mechanism of programmed cell death.
Brain-derived neurotrophic factor (BDNF) A protein released in the brain to support the survival of neurons, it plays a role in the growth, differentiation, and maintenance of these cells.
Cerebellum A critical area of the brain playing a role in coordinating motor function, learning, and attention.
Choline An essential nutrient involved in an important number of metabolic pathways related to neurotransmission, structural integrity of cell plasma membranes, and cell signaling.
Corpus callosum A bundle of fibers that links the left and right hemispheres of the brain.

Diagnosis, Management and Modeling of Neurodevelopmental Disorders. https://doi.org/10.1016/B978-0-12-817988-8.00041-5

Cortex The superficial layer of the brain that is composed of folded gray matter, and it is associated with perception, voluntary movement, and connection of information to support cognitive functions such as memory, language, and abstract thinking.
DHA (docosahexaenoic acid) An essential polyunsaturated fatty omega-3 acid. It plays a critical role in neurocognitive and visual development, as well as the development of the CNS.
EGCG (epigallocatechin gallate) A natural antioxidant with the highest antioxidant and free-radical-scavenging properties. It is known to produce the majority of the health benefits generated by green tea intake.
Epigenetics The study of factors that change gene expression without directly altering the DNA sequence.
Hippocampus The elongated edges on the floor of each lateral ventricle of the brain, being the center of emotion, memory, and the autonomic nervous system.
Microcephaly A condition where a baby is born with a small head or the head stops growing after birth.
NADH (nicotinamide adenosine dinucleotide phosphate) Cofactors involved in several redox reactions in all forms of cellular life. NAD is utilized as a cosubstrate in a number of nonredox reactions playing an important role in signaling and regulatory pathways.
Nitric oxide (NO) A compound that plays a number of critical roles in the body: it acts as a vasodilator (blood vessel relaxant), it influences blood flow to tissues, it regulates the binding and release of oxygen to hemoglobin.
Nuclear factor (Nrf) An essential transcription factor that regulates the antioxidant enzymes induction and antioxidant genes.
Oxidative stress A redox imbalance between the body's production of reactive oxygen species (free radicals) and antioxidants, which protect against reactive oxygen species.
Philtrum The vertical groove between the upper lip and nose.
Resveratrol A polyphenol derived from red grapes, berries, and peanuts, with antioxidant, antiapoptotic, free-radical-scavenging and antilipoprotein peroxidation properties.
Teratogen A substance that impairs fetal development and causes birth defects.
Vitamin C A natural radical scavenger, acting as an antioxidant.
Vitamin E A fat-soluble vitamin, which prevents lipid peroxidation in the cell membrane.

Introduction

Drinking alcohol is a common practice among pregnant women and the incidence of this type of addiction has increased worldwide in the last decades.

Ethanol is one of the major preventable teratogens in humans, and its intake might lead to a broad range of disorders in developing fetus. The most severe disease caused by maternal alcohol consumption is denominated fetal alcohol spectrum disorders (FASD), comprehending facial abnormalities, growth delay, and impaired neurodevelopment such as microcephaly, lifelong cognitive disabilities, and behavioral disturbances (Dorrie, Focker, Freunscht, & Hebebrand, 2014). FASD is a spectrum of disorders that may be influenced by maternal nutritional profile, ethanol consumption patterns, genetic predisposition, and environmental factors. Malnutrition is one of the most important factors that worsen the toxic effects of alcohol on fetal development (Shankar, Ronis, & Badger, 2007). Updated evidence shows that the five countries with the highest prevalence of FASD are South Africa (58.5 per 1000), Croatia (11.5 per 1000), Ireland (8.9 per 1000), Italy (8. per 1000), and Belarus (6.9 per 1000). Recent estimates of the global prevalence of FASD among the general population indicate 1.4 per 1000 (Popova, Lange, Probst, Gmel, & Rehm, 2017).

According to the World Health Organization (WHO), the harmful effects of alcohol depend on the volume and the quality of the alcohol intake and the patterns of drinking. Chronic heavy alcohol consumption of more than six drinks a day or frequent intermittent alcohol use in high quantity known as "binge drinking" causes more toxicity for the fetus than acute, moderate alcohol consumption (Abel et al., 1995). Furthermore, the most sensitive period of embryonic growth is during the first trimester when alcohol intake may disrupt fetal development (Young, Giesbrecht, Eskin, Aliani, & Suh, 2014).

Fetal alcohol spectrum disorders (FASD)

Exposure to alcohol during pregnancy leads to the injury of the brain structures. The most common brain damages are microcephaly, reduced volume of white and gray matter, malformations or agenesis of the corpus callosum, reduction of the volumes of the caudate nucleus, hippocampus, and frontal, temporal, and parietal lobes (Label, Roussotte, & Sowell, 2011). According to the Institute of Medicine (IOM), FASD can be classified into four specific disorders: FAS, partial FAS, alcohol-related birth defects (ARBD), alcohol-related neurodevelopmental disorders (ARND) (Hoyme et al., 2016). FAS is the most severe type of FASD, and it includes growth delay, short palpebral fissures, broad philtrum thin upper lip, and impaired development of the central nervous system (CNS), which include microcephaly and other brain alterations. The most common neuropsychological disabilities are intellectual delay, memory deficiency, language impairment, behavioral abnormalities, and altered development of social, mental, and motor functions (Mattson, Crocker, & Nguyen, 2011).

Mechanisms of alcohol toxicity in FASD

When ethanol is ingested during pregnancy, it arrives quickly at the fetal circulation, and it reaches 40% of the concentration in amniotic fluids in relation to the concentration achieved in maternal blood. The clearance of ethanol is more gradual in amniotic fluid than in maternal blood. As a consequence, maternal ethanol intake produces a long interval of alcohol exposure in the fetal compartment, leading to a severe injury of developing structures (Mancinelli et al., 2014).

The most described teratogenic effects are caused directly by ethanol itself and also by its metabolic products or indirectly by alcohol-related oxidative stress (ROS). Oxidative stress from alcohol is caused by an excess production of NADH (nicotinamide adenosine dinucleotide phosphate) during alcohol oxidation and the increase of the reactive oxygen species (ROS). Ethanol not only enhances ROS species but also decreases the levels of antioxidant pathways and molecules. ROS are extremely reactive molecules that influence the integrity of cellular membranes, impair RNA and protein functions, lipid peroxidation, and promote the activation of cell death pathways as apoptosis. Alcohol-related ROS can affect lipids and proteins and can produce DNA alteration and epigenetic modifications since DNA is very sensitive to the ROS effect (Brocardo, Gil-Mohapel, & Christie, 2011).

Alcohol impairs placental circulation leading to a diminution of placental blood flow. Moreover, ethanol decreases nitric oxide (NO) availability, which plays an important role in the regulation of fetal-placental circulation. Alcohol prompts apoptosis in retinoic acid (a type of vitamin A) deficient cells from the neural crest and also decreases the levels of antioxidant compounds such as free radical scavengers, leading to neurological injury. The metabolic products of ethanol such as acetaldehyde are the principal cause of the direct alcohol toxicity (Bosco & Diaz, 2012).

On the other hand, poor maternal nutritional status may exacerbate the toxic effects of alcohol. Heavy alcohol intake may alter the absorption of critical nutrients into the body and interfere with their uptake from the bowel. These micronutrients include essential amino acids and vitamins, particularly B1 (thiamine), B2 (riboflavin), B6 (pyridoxine), vitamin A and C, omega-3 fatty acids, and folic acid. The severity of these deficiencies correlates with the degree of alcohol consumption. Furthermore, alcoholics with malnutrition break down the alcohol more slowly and therefore enhance high blood alcohol levels that worsen their condition (Sebastiani et al., 2018). All these mechanisms compromise the adequate exchange of nutrients across the placenta causing fetal malnutrition and disabilities involved in FASD. Choline, a micronutrient that is a component of the cell membrane (phospholipids, phosphatidylcholine), is decreased in FASD and has been related to the impairment of brain development and cognitive dysfunction (Zeisel et al., 2011). In this context some authors tried to treat pregnant women with choline, minerals, and antioxidants to reverse the inadequate nutritional status usually seen in FASD mothers (Kable et al., 2015).

Postnatal nutritional treatment in FASD

Several studies have identified feeding problems in children with FASD and impaired dietary habits such as compulsive intake, lack of satiety, or conversely a poor appetite (Murawski, Moore, Thomas, & Riley, 2015). Many children with FASD don't reach the average intake of micronutrients such as omega-3 fatty acids, vitamins C, D, E, and K, magnesium, potassium, zinc, calcium, and choline (Nguyen, Risbud, Chambers, & Thomas, 2016; Werts, Van Calcar, Wargowski, & Smith, 2014). Recent evidence shows that nutritional deficits and consequent neurocognitive disorders may be ameliorated with nutritional supplements in the postnatal period. Target nutritional therapy might change the epigenetic profile disrupted by alcohol (Bastons-Compta, Astals, Andreu-Fernandez, Navarro-Tapia, & Garcia-Algar, 2018). There is a lack of research in humans as the majority of studies have been assessed in animal models. It is critical to determine how a nutritional treatment may reduce the neurological disabilities as a result of prenatal alcohol insult.

Choline

Choline is an essential nutrient involved in an important number of metabolic pathways related to neurotransmission, structural integrity of cell plasma membranes, and cell signaling.

Animal experiments demonstrated that choline supplements attenuate ethanol's teratogenic effects. Ethanol exposure produces severe neurological disabilities such as impairment of memory function or coordination, which are mitigated by choline treatment, regardless of the period of administration (Bearer, 2015; Ryan, 2008). In experimental models, the administration of choline during adolescence didn't attenuate alcohol-related overactivity but mitigated alcohol-related deficits in working memory, which may suggest that choline administration at this later developmental time acts on functioning of the prefrontal cortex (*Schneider* et al.). A study conducted in Sprague Dawley rat pups examined the effects of developmental alcohol exposure and perinatal choline supplementation on hippocampal M(1) and M(2/4) muscarinic receptors, which indicate alcohol-related behavioral effects. Developmental alcohol exposure significantly increased M(2/4) receptor density, an effect mitigated by choline supplementation. Nevertheless, alcohol decreased the density of muscarinic M(1) receptors in the dorsal hippocampus, an effect that was not altered by choline supplementation. This data suggests that developmental alcohol exposure may impair hippocampal cholinergic system and that perinatal choline supplementation may attenuate alcohol-related behavioral problems by influencing cholinergic systems (Monk, Leslie, & Thomas, 2012).

In the literature, there is a lack of studies about the effects of choline supplements in human subjects prenatally exposed to alcohol. A double-blind, randomized, placebo-controlled pilot trial was conducted in children with FASD aged 2.5–5 years at enrollment, to determine if choline supplementation improved neurocognitive functioning, in particular hippocampal-dependent memory. The authors found that the effects of choline supplements on elicited imitation items recalled were significant in the younger participants, and there was an inverse relation between choline dose and memory improvement, suggesting that there might be a potential sensitive period for nutritional interventions (Wozniak et al., 2015). Another study assessed the efficacy of a choline intervention in children with FASD aged 5–10 years. The authors observed that choline group did not improve cognitive functioning in any domain, probably due to the late stage of intervention (Nguyen, Risbud, Mattson, Chambers, & Thomas, 2016).

Vitamin E (a-tocopherol) and vitamin C

Vitamin E is a fat-soluble vitamin, which prevents lipid peroxidation in the cell membrane. In animal models, vitamin E normalized fetal development in rats exposed to ethanol acting as antioxidant treatment (Wentzel, Rydberg, & Eriksson, 2006). Nevertheless, in humans, the supplements with vitamin E in pregnant women consuming alcohol may be associated with a reduction in birth weight (Boskovic, Gargaun, Oren, Djulus, & Koren, 2005) even if this effect has not been clearly demonstrated.

During the postnatal period, vitamin E decreased oxidative stress in offspring rat pups exposed prenatally to ethanol. Vitamin E ameliorated apoptosis-related protein and protected against oxidative stress levels in neonatal rats' cerebellar granule cells and hippocampus (Heaton, Madorsky, Paiva, & Siler-Marsiglio, 2004; Marino, Aksenov, & Kelly, 2004; Siler-Marsiglio, Shaw, & Heaton, 2004; Siler-Marsiglio et al., 2005). Moreover, vitamin E-treated rats experimented a significant amelioration of brain atrophy and DNA damage and the reinstatement of the elevated level of homocysteine if compared to controls (Shirpoor, Salami, Khadem-Ansari, Minassian, & Yegiazarian, 2009). In neonatal rats pups treated with vitamin E supplements, the reduction and damage of the number of Purkinje cells in the cerebellum were attenuated (Heaton, Mitchell, & Paiva, 2000; Heaton, Mitchell, Paiva, & Walker, 2000), although in another study this benefit was not found (Tran, Jackson, Horn, & Goodlett, 2005).

A study conducted in Guinea pigs demonstrated that Vitamin C, a natural radical scavenger, reduced hepatic lipid peroxidation, enhanced glutathione peroxidase and reductase activity, and decreased GGT (gamma-glutamyltransferase) activity acting as antioxidant (Suresh, Sreeranjit kumar, Lal, & Indira, 1999).

Omega-3

DHA (docosahexaenoic acid) is an essential polyunsaturated fatty omega-3 acid. DHA is of relevant importance during fetal development because it plays a critical role in neurocognitive and visual development, as well as the development of the CNS (Innis, 2007).

A study conducted in rat models found that prenatal ethanol exposure decreased glutathione concentrations in the brain and the long-term potentiation (LTP), leading to increased lipid peroxidation and consequent oxidative stress. The supplementation with DHA augmented glutathione's concentrations, reversed the deficits of LTP, and reduced lipid peroxidation, thereby partially reversing the negative effects of prenatal ethanol exposure. This study observed that DHA supplementation could decrease oxidative stress and potentiate antioxidant capacity in fetus exposed to ethanol (Patten, Brocardo, & Christie (2013); Patten, Sickmann, Dyer, Innis, & Christie, 2013). A recent study conducted in animal models demonstrated that betaine and omega-3 alone or together may inhibit neurodegeneration in offsprings (Kuşat, Kanbak, Oğlakcõ Ilhan, Burukoglu, & Yücel, 2016). There is a lack of studies about DHA supplementation in pregnant mothers who consume alcohol.

Resveratrol

Resveratrol is a polyphenol derived from red grapes, berries, and peanuts, with antioxidant, antiapoptotic, free-radical-scavenging, and antilipoprotein peroxidation properties (Shakibaei, Harikumar, & Aggarwal, 2009). According to studies conducted in rat models of FASD, the treatment with resveratrol before ethanol exposure decreased ROS levels because of its antioxidant characteristics and also reestablished nuclear factor (Nrf), which has been described as an essential transcription factor that regulates the antioxidant enzymes induction and antioxidant genes (Kumar, Singh, Lavoie, Dipette, & Singh, 2011; Nguyen, Nioi, & Pickett, 2009; Zhang, 2006). Resveratrol also protected and improved Swann cell viability and enhanced brain-derived neurotrophic factor (BDNF) and glial-cell-derived neurotrophic factor (GDNF) expression (Yuan, Zhang, Liu, & Li, 2013).

Epigallocatechin gallate (EGCG)

EGCG, epigallocatechin, epicatechin gallate, and epicatechin are poliphenols known as natural antioxidants. EGCG is the essential element, and it is also the component with the highest antioxidant and free-radical-scavenging property. It is known to produce the majority of the health benefits generated by green tea intake. In experimental models of FASD, antioxidant therapy may mitigate the unfavorable effects of oxidative stress. EGCG reduces lipid peroxidation, prevents the CYP2E1 induction by ethanol (which generates ROS), and decreases free radical protein adducts generation (Cohen-Kerem & Koren, 2003; Joya, Garcia-Algar, Salat-Batlle, Pujades, & Vall, 2015; Oliva, Bardag-Gorce, Tillman, & French, 2011). Furthermore, in animal studies, EGCG enhances adult hippocampal neurogenesis so it has been proposed for the prevention and treatment of a wide range of neurodegenerative diseases and Down syndrome (DS) (De la Torre et al., 2014; Wang et al., 2012). Recent evidence shows a common origin at cellular and molecular levels that are disrupted by trisomy or alcohol intake during pregnancy, leading to craniofacial and neurological phenotypes, which characterized DS or FASD (Solzak, Liang, Zhou, & Roper, 2013). In a murine model, EGCG could prevent some of the embryonic injuries caused by ethanol such as growth delay and the inhibition of hydrogen peroxide (H2O2) and malondialdehyde (MDA) (Long et al., 2010).

The only pilot study in humans assessing the efficacy of EGCG as a therapy to improve cognitive performance in children diagnosed with FASD was conducted by García-Algar (Garcia-Algar, 2014). In a cohort of 40 FASD children, between 7 and 14 years old, an oral dose of 9 mg/kg/day was administered during 1 year, with six control visits until 6 months after finishing the treatment. Preliminary results after 6 months of treatment showed that EGCG improved the recent and delayed memory and also working memory of treated children.

Other antioxidant therapies

Recently, promising results about the antioxidant properties of curcumin, alpha-lipoic acid, and selenium administered with folic acid have been described to improve cognitive deficits and neuronal inflammation in animal studies (Ojeda et al., 2009; Shirpoor at al., 2008; Tiwari & Chopra., 2013).

Applications to other areas of development

In this chapter we have reviewed the nutritional therapy of neurological disabilities of FASD The most described teratogenic effect of alcohol includes the increase of the ROS (Brocardo et al., 2011) that leads to the most common neuropsychological problems such as intellectual delay, memory deficiency, language impairment, behavioral abnormalities, and altered development of social, mental, and motor functions (Mattson et al., 2011).

Recent evidence showed that a target intervention with nutritional supplements characterized by antioxidant, antiapoptotic, free-radical-scavenging, and antilipoprotein peroxidation properties might attenuate the neurological impairment caused by ethanol consumption. The most studied nutritional supplements are choline, Vitamin E, DHA, resveratrol, and EGCG. Most of them could improve neurocognitive functioning, in particular hippocampal-dependent memory assessed in animal models, even if there is a lack of research in humans (Bastons-Compta et al., 2018) (Figs. 41.1−41.5)

In the past years, particular attention has been focused on EGCG, the component of green tea with the highest antioxidant and free-radical-scavenging properties. EGCG reduces lipid peroxidation, prevents the CYP2E1 induction by ethanol (which generates ROS), and decreases free radical protein adducts generation (Cohen-Kerem & Koren, 2003; Joya et al., 2015; Oliva et al., 2011). Moreover, in animal studies, EGCG enhances adult hippocampal neurogenesis so it has been proposed for the prevention and treatment of a wide range of neurodegenerative diseases (Wang et al., 2012). For example, DS and FASD share similar mechanisms, which disrupt fetal development during pregnancy, leading to the typical craniofacial and neurological abnormalities of both phenotypes (Solzak et al., 2013). Experimental studies conducted in animal models of DS observed that EGCG reversed cognitive deficits with effects on memory recognition, working memory, and quality of life (De la Torre et al., 2014). Moreover, due to its neuroprotective properties, EGCG limits the inflammation of the brain and neuronal

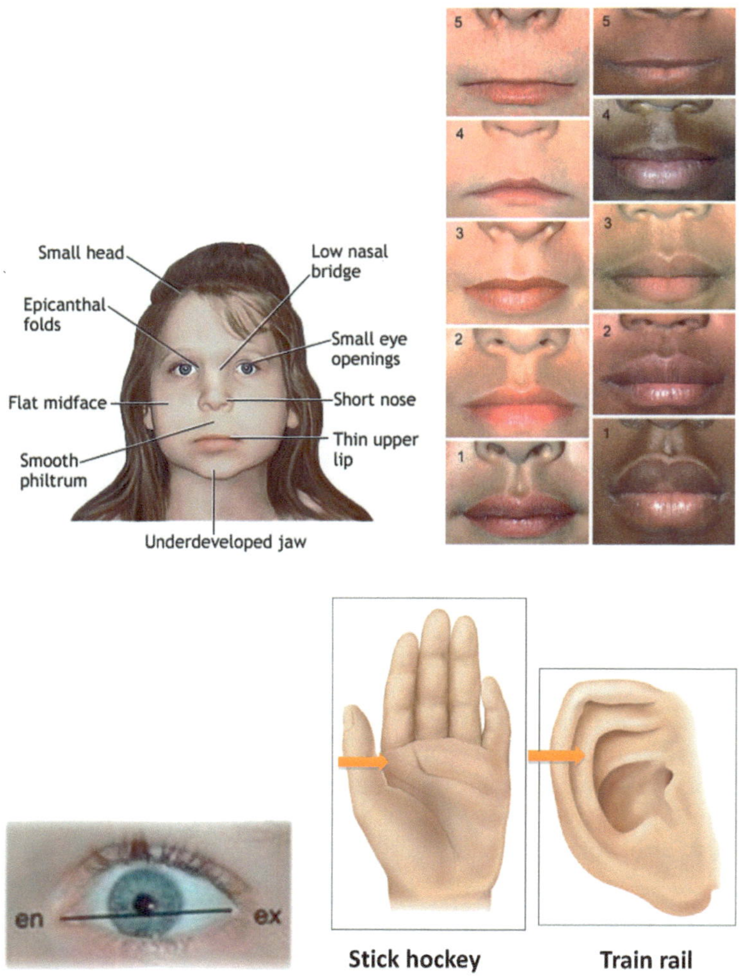

FIGURE 41.1 **Morphological characteristics of FASD.**

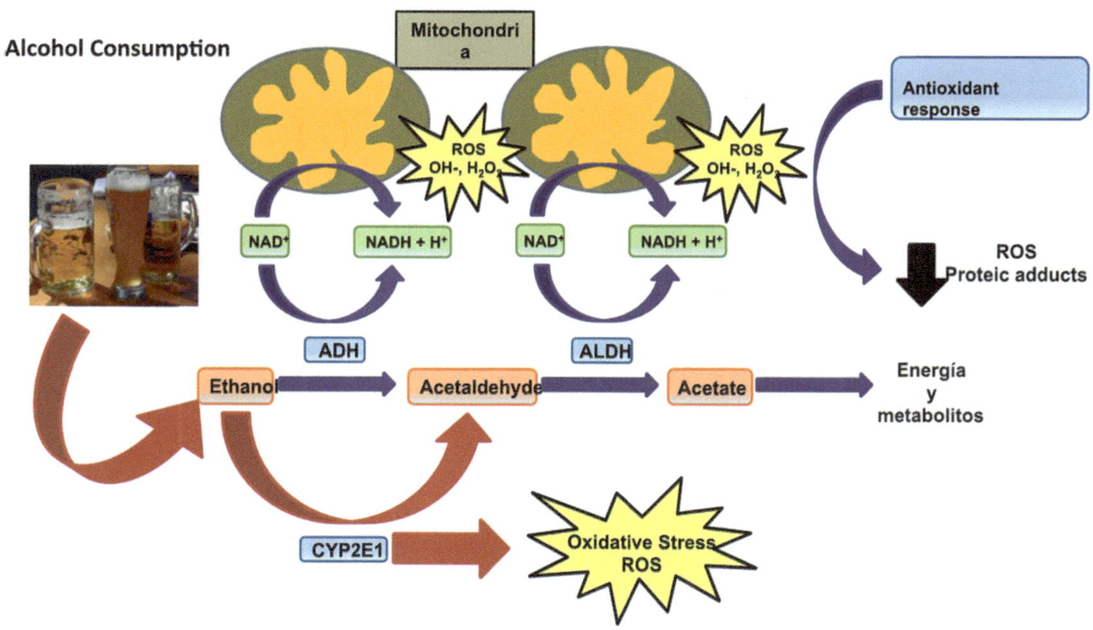

FIGURE 41.2 **Alcohol metabolism.** *ADH*, alcohol dehydrogenases; *ALDH*, aldehyde dehydrogenases.

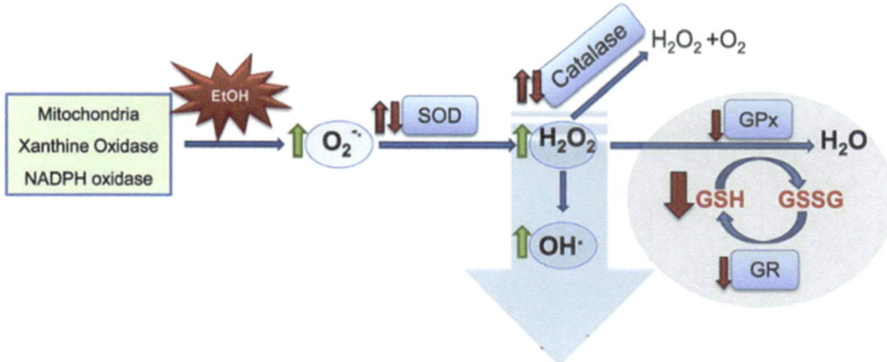

FIGURE 41.3 **Alcohol metabolism.** Ethanol metabolism involves the mitochondrial respiratory chain, the xanthine oxidase, and the NADPH oxidase pathways producing oxidant agents as acetaldehyde and ROS. Superoxide (O_2^-), hydrogen peroxide (H_2O_2), and hydroxyl radical (OH-) increase inside the cell. Ethanol exposure also decreases antioxidant enzymes such as superoxide dismutase (SOD), catalase, glutathione peroxidase (SPx), and glutathione reductase (GR) as well as the intracellular levels of reduced glutathione (GSH). *With permission from: Bastons-Compta, A., Astals, M., Andreu-Fernandez, V., Navarro-Tapia, E., Garcia-Algar, O. (2018). Postnatal nutritional treatment of neurocognitive deficits in fetal alcohol spectrum disorder.* Biochemistry and Cell Biology, 96, *213–221.*

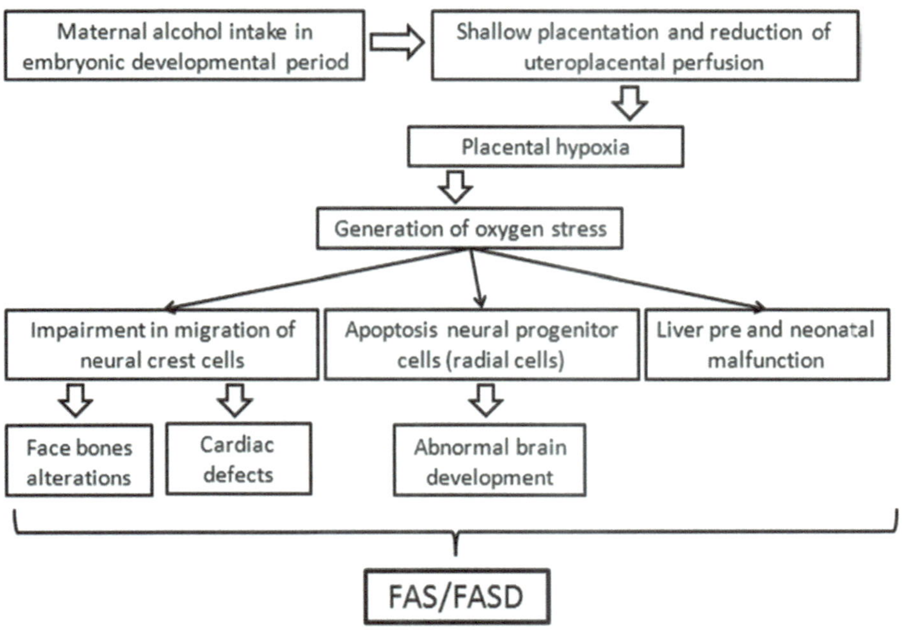

FIGURE 41.4 **Effect of maternal alcohol consumption on the placental development.** Alcohol generates ROS promoting a shallow placentation and a diminished uterine irrigation of the placenta. ROS alters the normal development of the embryo by impairment in migration of neural crest cells (which can cause face bones alterations and cardiac defects) and apoptosis of neural progenitor cells (causing abnormal brain development). *With permission from: Bastons-Compta, A., Astals, M., Andreu-Fernandez, V., Navarro-Tapia, E., Garcia-Algar, O. (2018). Postnatal nutritional treatment of neurocognitive deficits in fetal alcohol spectrum disorder.* Biochemistry and Cell Biology, 96, *213–221.*

damage; therefore, it may be a good candidate for therapeutic agent in neurodegenerative disease such as Alzheimer's disease and Parkinson's disease (Singh, Mandal, & Khan, 2016) (Table 41.1).

Key facts of postnatal nutritional treatment of FASD

- Key fact of prevalence of FASD: Recent estimates of the global prevalence of FASD among the general population indicate 1.4 per 1000.
- Key fact of the World Health Organization (WHO): chronic heavy alcohol consumption of more than six drinks a day or frequent intermittent alcohol use in high quantity known as "binge drinking" causes more toxicity for the fetus than acute, moderate alcohol consumption.

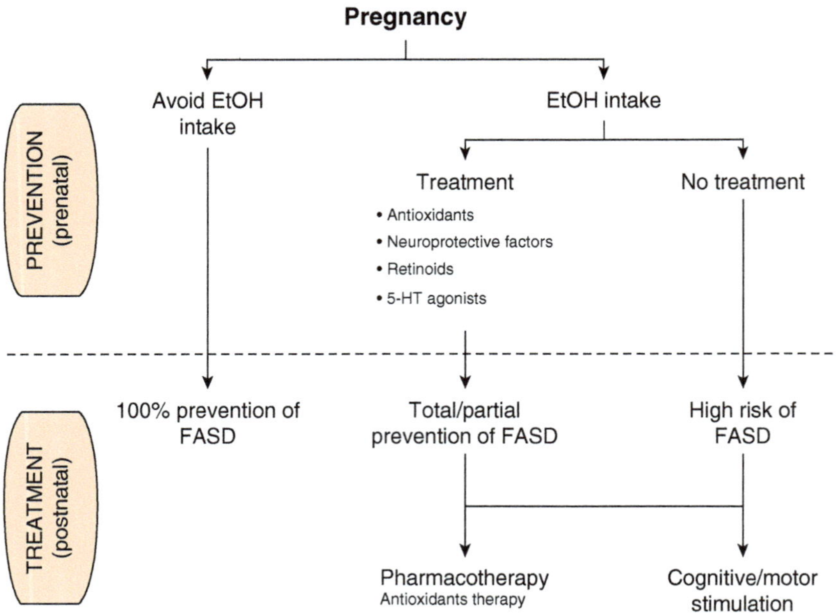

FIGURE 41.5 Current strategies for the prevention and treatment of FASD. *With permission from: Bastons-Compta, A., Astals, M., Andreu-Fernandez, V., Navarro-Tapia, E., Garcia-Algar, O. (2018). Postnatal nutritional treatment of neurocognitive deficits in fetal alcohol spectrum disorder.* Biochemistry and Cell Biology, 96, *213–221.*

TABLE 41.1 Studies on nutrient supplementation to improve postnatal ethanol exposure effects.

Nutrient/Drug	Animal model	Drug dose/exposure period	EtOH dose/exposure period	Results	Reference
Vitamin C	Guinea pigs	250 mg/kg	9 g/kg/day	Reduces hepatic lipid peroxidation Enhances glutathione peroxidase and reductase activity Reduces GGT activity	Suresh et al. (1999)
Vitamin E	Long–Evans rats	30–60 IU/100 mL PD 4–5	12% PD 4–5	At high dose complete prevention of *Purkinje Cell* damage and loss	Heaton, Mitchell, Paiva, & Walker (2000), Heaton, Mitchell, & Paiva (2000)
	Long–Evans rats	0.05mM/24h PD8	86.7–346.6mM/24h PD8	Restores expression of NTFs (BDNF and neurotrophin-3) Reduces cellular disturbances in oxidative processes	Heaton et al. (2004)
	Long–Evans rats	2000 g/kg PD6	5.25 g/kg PD 7–9	Reduces increase in protein carbonyls No improvement of spatial learning in the ethanol-exposed animals	Marino et al. (2004)
	Long–Evans rats	12.26 mg/kg/day PD 4–9	2.625 g/kg/day PD 4–9	Fails to protect against reduction of cerebellar *Purkinje cells*	Tran et al. (2005)

Continued

TABLE 41.1 Studies on nutrient supplementation to improve postnatal ethanol exposure effects.—cont'd

Nutrient/Drug	Animal model	Drug dose/exposure period	EtOH dose/exposure period	Results	Reference
	Long–Evans rats	0.05mM/24h PD9	86.7–346.6mM/24h PD9	Amplifies gGCS and total GSH protein expression levels	Siler-Marsiglio et al. (2005)
	Wistar rats	300 mg/day GD7–PD21	4.5 g/kg/day GD7–PD21	Reduces DNA damage Restores elevated Hcy levels	Shirpoor et al. (2009)
Vitamin E + Pycnogenol	Long–Evans rats	25–100 μg/mL/5s-24h PD9	86.7–346.6mM/5s24h PD9	Decreases cell death and reduced caspase-3 activation	Siler-Marsiglio et al. (2004)
Resveratrol	Long–Evans rats	2–100 mg/kg/1-24h before EtOH treatment PD7	80 mM/5h PD7	Decreases ROS levels Restores Nrf expression Retains the expression and activity of NADPH quinine oxidoreductase 1 and SOD	Kumar et al. (2011)
	Wistar rats	30mM/96h PD3	1500 mg/dL/96h	Recovers cell viability Increases BDNF and GDNF expression	Yuan et al. (2013)
Curcumin	Wistar rats	30–60 mg/kg PD6–28	5 g/kg PD7–9	Improves neuroinflammation (oxidative nitrosative stress, TNFα, IL-1β, and TGF-β1). Decreases neuronal apoptosis (NF-κβ and caspase 3) in cerebral cortex and hippocampus	Tiwari and Chopra (2013)
Alpha-lipoic acid	Wistar rats	100 mg/kg GD7–PD21	4.5 g/kg GD7–PD21	Decreases DNA damage Restores elevated protein carbonyl and lipid hydroperoxide levels	Shirpoor at al. (2008)
Selenium + folic acid	Wistar rats	0.5 μg/g of selenium and 8 μg/g of folic acid PD21	5% w1, 10%w2, 15% w3, 20% from w8 before pregnancy until end of lactation period	Reduces selenium loss, improved balance among oxidative enzymes	Ojeda et al. (2009)
Omega-3	Sprague Dawley rats	34.2% Dams: GD21–PD22 Pups: PD22–60	35.5% GD1–21	Increases glutathione concentrations Decreases lipid peroxidation Reduces oxidative stress Enhances antioxidant capacity	Patten, Brocardo, & Christie (2013)
	Sprague Dawley rats	34.2% PD1–65	36% GD1–22	Reduction in long-term potentiation (LTP) Omega-3 reverses LTP deficit	Patten, Sickmann, et al. (2013)
Omega-3 & Betaine	Sprague Dawley rats	Omega-3 and betaine supplementation: 30 mg/100 g/day and 1% respectively (GD3–PD21)	25–75 mLethanol (96.5%)	Betaine and omega-3 alone or together may inhibit neurodegeneration in offsprings.	Kuşat, Kanbak, Oğlakcı Ilhan, Burukoglu, & Yücel (2016)

Continued

TABLE 41.1 Studies on nutrient supplementation to improve postnatal ethanol exposure effects.—cont'd

Nutrient/Drug	Animal model	Drug dose/exposure period	EtOH dose/ exposure period	Results	Reference
Choline	Sprague Dawley rats	100 mg/kg/day PD11–20, 21–30 or 11–30; measured after PD45	5.25 g/kg/day PD4–9	Mitigates spatial memory deficits	Ryan, Williams, & Thomas (2008)
	Sprague Dawley rats	100 mg/kg/day PD4–30; measured at PD30–33	5.25 g/kg/day PD4–9	Attenuates hyperactivity, mitigates increase in M2/4 receptor density No decrease in muscarinic M1 receptor density in dorsal hippocampus	Monk et al. (2012)
	C57B16/ J mice	18.8 mg/mL PD1–5, PD6–20	13.6% (6 g/kg) P5	Increases distance crossed No effect in non-alcohol-exposed mice	Bearer, Wellmann, Tang, He and Mooney (2015)
	Sprague Dawley rats	100 mg/kg/day PD40–60	5.25 g/kg/day PD4–9	Mitigates alcohol-related deficits in working memory No improvement of alcohol-related hyperactivity	Schneider and Thomas (2016)
	Children with FASD aged 2.5 –5 years	500 mg/day during 9 month	–	Treatment effect on elicited imitation items recalled was significant in the younger participants (2.5 to ≤4.0 years) Inverse relation between choline dose (in mg/kg) and memory improvement	Wozniak et al. (2015)
	Children with FASD aged 5 –10 years	625 mg/day during 6 weeks	–	No improvement in cognitive performance	Nguyen, Risbud, Mattson, et al. (2016)
EGCG	C57B16/ J mice	200, 300, or 400 mg/kg/day GD7–8	0.005–0.02 mL/g GD8	EGCG ameliorated ethanol-induced growth retardation EGCG inhibited the increase in H2O2 and MDA	Long et al. (2010)
	Children with FASD aged 7–15 years	9 mg/kg/day during 12 months	–	After 6 months of treatment with EGCG, there was an improvement in the recent and delayed memory and working memory	Garcia-Algar (2014)

EGCG, Epigallocatechin-3-gallate; *FASD*, fetal alcohol spectrum disorder; *GD*, gestational days; *GGT*, Gamma-glutamyltransfere; *MDA*, malondialdehyde; *Nrf*, nuclear factor; *PD*, postnatal days; *SOD*, superoxide dismutase.

- Key fact of definition of FASD: FASD includes growth delay, short palpebral fissures, broad philtrum, thin upper lip, and impaired development of CNS, which include microcephaly and other brain alterations.
- Key fact of neurocognitive functioning: the most common neuropsychological disabilities of FASD are intellectual delay, memory deficiency, language impairment, behavioral abnormalities, and altered development of social, mental, and motor functions.
- Key fact of nutritional therapy: A target intervention with nutritional supplements characterized by antioxidant properties might attenuate the neurological impairment caused by ethanol consumption. The most studied nutritional supplements are choline, Vitamin E, DHA, resveratrol, and EGCG.
- The most described teratogenic effect of alcohol includes an excess production of NADH during alcohol oxidation and the increase of the reactive oxygen species (ROS). Ethanol also decreases the levels of antioxidant pathways and molecules. Alcohol-related ROS can affect lipids and proteins and can produce DNA alteration and epigenetic modifications.
- Ethanol decreases nitric oxide (NO) availability, which plays an important role in the regulation of fetal-placental circulation.
- The metabolic products of ethanol such as acetaldehyde are the principal cause of the direct alcohol toxicity.
- Poor maternal nutritional status may exacerbate the toxic effects of alcohol. Pregnant women with malnutrition break down the alcohol more slowly and therefore enhance high blood alcohol levels that compromise the adequate exchange of nutrients across the placenta causing fetal malnutrition and disabilities involved in FASD
- Nutritional supplements characterized by antioxidant, antiapoptotic, free-radical-scavenging, and antilipoprotein peroxidation properties might attenuate the neurological impairment typical of FASD.
  - The most studied nutritional supplements are choline, Vitamin E, DHA, resveratrol, and EGCG.
  - In animal models, nutritional supplements could improve neurocognitive functioning, in particular hippocampal-dependent memory.

References

Abel, E. L., & Hannigan, J. H. (1995). Maternal risk factors in fetal alcohol syndrome: Provocative and permissive influences. *Neurotoxicology and Teratology, 17*, 445–462.

Bastons-Compta, A., Astals, M., Andreu-Fernandez, V., Navarro-Tapia, E., & Garcia-Algar, O. (2018). Postnatal nutritional treatment of neurocognitive deficits in fetal alcohol spectrum disorder. *Biochemistry and Cell Biology, 96*, 213–221.

Bearer, C. F., Wellmann, K. A., Tang, N., He, M., & Mooney, S. M. (2015). Choline ameliorates deficits in balance caused by acute neonatal ethanol exposure. *The Cerebellum, 14*(4), 413–420.

Bosco, C., & Diaz, E. (2012). Placental hypoxia and fetal development versus alcohol exposure in pregnancy. *Alcohol, 47*(2), 109–117.

Boskovic, R., Gargaun, L., Oren, D., Djulus, J., & Koren, G. (2005). Pregnancy outcome following high doses of Vitamin E supplementation. *Reproductive Toxicology, 20*(1), 85–88.

Brocardo, P., Gil-Mohapel, J., & Christie, B. (2011). The role of oxidative stress in fetal alcohol spectrum disorders. *Brain Research Reviews, 67*(1–2), 209–225.

Cohen-Kerem, R., & Koren, G. (2003). Antioxidants and fetal protection against ethanol teratogenicity. I. Review of the experimental data and implications to humans. *Neurotoxicology and Teratology, 25*(1), 1–9.

De la Torre, R., de Sola, S., Pons, M., Duchon, A., de Lagran, M. M., Farr, M., … Dierssen, M. (2014). Epigallocatechin-3-gallate, a DYRK1A inhibitor, rescues cognitive deficits in Down síndrome mouse models and in humans. *Molecular Nutrition & Food Research, 58*(2), 278–288.

Dorrie, N., Focker, M., Freunscht, I., & Hebebrand, J. (2014). Fetal alcohol spectrum disorders. *European Child & Adolescent Psychiatry, 23*(10), 863–875.

Garcia-Algar, O. (2014). *Epigallocatechin gallate (EGCG) as therapeutic tool to improve cognitive performance in fetal alcohol syndrome (FAS) children*. Protocol ID: 2014/5553 https://clinicaltrials.gov/ct2/show/NCT025558933.

Heaton, M. B., Madorsky, I., Paiva, M., & Siler-Marsiglio, K. I. (2004). Vitamin E amelioration of ethanol neurotoxicity involves modulation of apoptosis-related protein levels in neonatal rat cerebellar granule cells. *Brain Research Developmental Brain Research, 150*(2), 117–124.

Heaton, M. B., Mitchell, J. J., & Paiva, M. (2000). Amelioration of ethanol induced neurotoxicity in the neonatal rat central nervous system by antioxidant therapy. *Alcoholism: Clinical and Experimental Research, 24*(4), 512–518.

Heaton, M. B., Mitchell, J. J., Paiva, M., & Walker, D. W. (2000). Ethanol induced alterations in the expression of neurotrophic factors in the developing rat central nervous system. *Brain Research Developmental Brain Research, 121*(1), 97–107.

Hoyme, H. E., Kalberg, W. O., Elliott, A. J., Blankenship, J., Buckley, D., Marais, A. S., … May, P. A. (2016). Updated clinical guidelines for diagnosing fetal alcohol spectrum disorders. *Pediatrics, 138*(2).

Innis, S. M. (2007). Dietary (n-3) fatty acids and brain development. *Journal of Nutrition, 137*, 855–859.

Joya, X., Garcia-Algar, O., Salat-Batlle, J., Pujades, C., & Vall, O. (2015). Advances in the development of novel antioxidant therapies as an approach for fetal alcohol síndrome prevention. *Birth Defects Research Part A: Clinical and Molecular Teratology, 103*(3), 163–177.

Kable, J. A., Coles, C. D., Keen, C. L., Uriu-Adams, J. Y., Jones, K. L., Yevtushok, L., … CIFASD. (2015). The impact of micronutrient supplementation in alcohol exposed pregnancies on information processing skills in Ukrainian infants. *Alcohol, 49*(7), 647–656.

Kumar, A., Singh, C. K., Lavoie, H. A., Dipette, D. J., & Singh, U. S. (2011). Resveratrol restores Nrf2 level and prevents ethanol-induced toxic effects in the cerebellum of a rodent model of fetal alcohol spectrum disorders. *Molecular Pharmacology, 80*(3), 446–457.

Kuşat, K., Kanbak, G., Oğlakcõ Ilhan, A., Burukoglu, D., & Yücel, F. (2016). The investigation of the prenatal and postnatal alcohol exposure-induced neurodegeneration in rat brain: Protection by betaine and/or omega-3. *Child's Nervous System, 32*(3), 467–474.

Label, C., Roussotte, F., & Sowell, E. R. (2011). Imaging the impact of prenatal alcohol exposure on the structure of the developing human brain. *Neuropsychology Review, 21*(2), 102–118.

Long, L., Li, Y., Wang, Y. D., He, Q. Y., Li, M., Cai, X. D., … Peng, Y. (2010). The preventive effect of oral EGCG in a fetal alcohol spectrum disorder mouse model Alcoholism. *Clinical and Experimental Research, 34*(11), 1929–1936.

Mancinelli, R. (2014). Fetal alcohol spectrum disorders: Can alcohol-related oxidative stress concur to the prenatal damage? *Alcohol, 2*, 9.

Marino, M. D., Aksenov, M. Y., & Kelly, S. J. (2004). Vitamin E protects against alcohol-induced cell loss and oxidative stress in the neonatal rat hippocampus. *International Journal of Developmental Neuroscience, 2*(5–6), 363–377.

Mattson, S. N., Crocker, N., & Nguyen, T. T. (2011). Fetal alcohol spectrum disorders: Neuropsychological and behavioral features. *Neuropsychology Review, 21*(2), 81–101.

Monk, B. R., Leslie, F. M., & Thomas, J. D. (2012). The effects of perinatal choline supplementation on hippocampal cholinergic development in rats exposed to alcohol during the brain growth spurt. *Hippocampus, 22*(8), 1750–1757.

Murawski, N. J., Moore, E. M., Thomas, J. D., & Riley, E. P. (2015). Advances in diagnosis and treatment of fetal alcohol spectrum disorders: From animal models to human studies. *Alcohol Research, 37*(1), 97–108.

Nguyen, T., Nioi, P., & Pickett, C. B. (2009). The Nrf2-antioxidant response element signaling pathway and its activation by oxidative stress. *Journal of Biological Chemistry, 284*(20), 13291–13295.

Nguyen, T. T., Risbud, R. D., Chambers, C. D., & Thomas, J. D. (2016). Dietary nutrient intake in school-aged children with heavy prenatal alcohol exposure. *Alcoholism: Clinical and Experimental Research, 40*(5), 1075–1082.

Nguyen, T. T., Risbud, R. D., Mattson, S. N., Chambers, C. D., & Thomas, J. D. (2016). Randomized, double-blind, placebo-controlled clinical trial of choline supplementation in school aged children with fetal alcohol spectrum disorders. *American Journal of Clinical Nutrition, 104*(6), 1683–1692.

Ojeda, M. L., Nogales, F., Jotty, K., Barrero, M. J., Murillo, M. L., & Carreras, O. (2009). Dietary selenium plus folic acid as an antioxidant therapy for ethanol exposed pups. *Birth Defects Research Part B: Developmental and Reproductive Toxicology, 86*(6), 490–495.

Oliva, J., Bardag-Gorce, F., Tillman, B., & French, S. W. (2011). Protective effect of quercetin, EGCG, catechin and betaine against oxidative stress induced by ethanol in vitro. *Experimental and Molecular Pathology, 90*(3), 295–299.

Patten, A. R., Brocardo, P. S., & Christie, B. R. (2013). Omega-3 supplementation can restore glutathione levels and prevent oxidative damage caused by prenatal ethanol exposure. *The Journal of Nutritional Biochemistry, 24*(5), 760–769.

Patten, A. R , Sickmann, H. M., Dyer, R. A., Innis, S. M., & Christie, B. R. (2013). Omega-3 fatty acids can reverse the long term deficits in hippocampal synaptic plasticity caused by prenatal ethanol exposure. *Neuroscience Letters, 13*(551), 7–11.

Popova, S., Lange, S., Probst, C., Gmel, G., & Rehm, J. (2017). Estimation of national, regional, and global prevalence of alcohol use during pregnancy and fetal alcohol syndrome: A systematic review and meta-analysis. *Lancet Global Health, 5*(3), e290–e299.

Ryan, S. H., Williams, J. K., & Thomas, J. D. (2008). Choline supplementation attenuates learning deficits associated with neonatal alcohol exposure in the rat: Effects of varying the timing of choline administration. *Brain Research, 27*(1237), 91–100.

Schneider, R. D., & Thomas, J. D. (2016). Adolescent choline supplementation attenuates working memory deficits in rats exposed to alcohol during the third trimester equivalent. *Alcoholism: Clinical and Experimental Research, 40*(4), 897–905.

Sebastiani, G., Borrás-Novell, C., Casanova, M. A., Pascual Tutusaus, M., Ferrero Martínez, S., Gómez Roig, M. D., & García-Algar, O. (2018). The effects of alcohol and drugs of abuse on maternal nutritional profile during pregnancy. *Nutrients, 10*(8). pii: E1008.

Shakibaei, M., Harikumar, K. B., & Aggarwal, B. B. (2009). Resveratrol addiction: To die or not to die. *Molecular Nutrition & Food Research, 53*(1), 115–128.

Shankar, K., Ronis, M. J., & Badger, T.,M. (2007). Effects of pregnancy and nutritional status on alcohol metabolism. *Alcohol Research & Health, 30*(1), 55–59.

Shirpoor, A., Minassian, S., Salami, S., Khadem-Ansari, M. H., & Yeghiazaryan, M. (2008). Alpha–lipoic acid decreases DNA damage and oxidative stress induced by alcohol in the developing hippocampus and cerebellum of rat. *Cellular Physiology and Biochemistry, 22*(5–6), 769–776.

Shirpoor, A., Salami, S., Khadem-Ansari, M. H., Minassian, S., & Yegiazarian, M. (2009). Protective effect of vitamin E against ethanol-induced hyperhomocysteinemia, DNA damage, and atrophy in the developing male rat brain. *Alcoholism: Clinical and Experimental Research, 33*(7), 1181–1186.

Siler-Marsiglio, K. I., Pan, Q., Paiva, M., Madorsky, I., Khurana, N. C., & Heaton, M. B. (2005). Mitochondrially targeted vitamin E and vitamin E mitigate ethanol-mediated effects on cerebellar granule cell antioxidant defense systems. *Brain Research, 1052*, 202–211.

Siler-Marsiglio, K. I., Shaw, G., & Heaton, M. B. (2004). Pycnogenol and vitamin E inhibit ethanol-induced apoptosis in rat cerebellar granule cells. *Journal of Neurobiology, 59*(3), 261–271.

Singh, N. A., Mandal, A. K., & Khan, Z. A. (2016). Potential neuroprotective properties of epigallocatechin-3-gallate (EGCG). *Nutrition Journal, 15*(1), 60.

Solzak, J. P., Liang, Y., Zhou, F. C., & Roper, R. J. (2013). Commonality in Down and fetal alcohol syndromes. *Birth Defects Research Part A: Clinical and Molecular Teratology, 97*(4), 187–197.

Suresh, M. V., Sreeranjit kumar, C. V., Lal, J. J., & Indira, M. (1999). Impact of massive ascorbic acid supplementation on alcohol induced oxidative stress in Guinea pigs. *Toxicology Letters, 104*(3), 221–229.

Tiwari, V., & Chopra, K. (2013). Protective effect of curcumin against chronic alcohol-induced cognitive deficits and neuroinflammation in the adult rat brain. *Neuroscience, 244*, 147–158.

Tran, T. D., Jackson, H. D., Horn, K. H., & Goodlett, C. R. (2005). Vitamin E does not protect against neonatal ethanol-induced cerebellar damage or deficits in eye blink classical conditioning in rats. *Alcoholism: Clinical and Experimental Research, 29*(1), 117–129.

Wang, Y., Li, M., Xu, X., Song, M., Tao, H., & Bai, Y. (2012). Green tea epigallocatechin-3- gallate (EGCG) promotes neural progenitor cell proliferation and sonic hedgehog pathway activation during adult hippocampal neurogenesis. *Molecular Nutrition & Food Research, 56*(8), 1292–1303.

Wentzel, P., Rydberg, U., & Eriksson, U. J. (2006). Antioxidative treatment diminishes ethanol induced congenital malformations in the rat. *Alcoholism: Clinical and Experimental Research, 30*(10), 1752–1760.

Werts, R. L., Van Calcar, S. C., Wargowski, D. S., & Smith, S. M. (2014). Inappropriate feeding behaviors and dietary intakes in children with fetal alcohol spectrum disorder or probable prenatal alcohol exposure. *Alcoholism: Clinical and Experimental Research, 38*(3), 871–878.

Wozniak, J. R., Fuglestad, A. J., Eckerle, J. K., Fink, B. A., Hoecker, H. L., Boys, C. J., … Georgieff, M. K. (2015). Choline supplementation in children with fetal alcohol spectrum disorders: A randomized, double-blind, placebo-controlled trial. *American Journal of Clinical Nutrition, 102*(5), 1113–1125.

Young, J., Giesbrecht, H. E., Eskin, M. N., Aliani, M., & Suh, M. (2014). Nutrition implications for fetal alcohol spectrum disorder. *Advances in Nutrition, 5*, 675–692.

Yuan, H., Zhang, J., Liu, H., & Li, Z. (2013). The protective effects of resveratrol on Schwann cells with toxicity induced by ethanol in vitro. *Neurochemistry International, 63*(3), 146–153.

Zeisel, S. H. (2011). What choline metabolism can tell us about the underlying mechanisms of fetal alcohol spectrum disorders. *Molecular Neurobiology, 44*, 185–191.

Zhang, D. D. (2006). Mechanistic studies of the Nrf2-Keap1 signaling. *Drug metabolism reviews, 38*(4), 769–789.

Chapter 42

Nurturing the developing brain to reduce neurological delay

Kelly A. Vaughn[1], Bartlett D. Moore, IV[2] and Dana DeMaster[1]
[1]*Department of Pediatrics, University of Texas Health Science Center at Houston, Children's Learning Institute, Houston, TX, United States;*
[2]*Department of Neurosurgery, University of Texas Health Science Center at Houston, McGovern Medical School, Houston, TX, United States*

List of abbreviations

DLLs Dual language learners
FA Fractional anisotropy
FrS Frontostriatal
NICU Neonatal intensive care unit
PrLS Prefrontal limbic system

Mini-dictionary of terms

Diathesis stress model This psychological theory predicts outcomes for vulnerable populations based on exposure to environmental stressors.
Differential susceptibility hypothesis Characterizes vulnerability as increasing plasticity factors—predicting that better-than-average outcomes will result from high-quality environments but worse-than-average outcomes will result from low-quality environments.
Dual language learners Children who are educated in a language that is different from the home language.
Home language environment The quantity and quality of verbal communication within the household.
Rich language exposure High-quality language interactions, including use of descriptive words, between a child and their caregiver.
Sensitive and highly responsive parenting Parents who consistently attend to child's signals, appropriately interpret them, and respond sensitively.
Verbal scaffolding Using language to provide a framework for thought or to facilitate learning of a new skill.

Introduction

This chapter will focus on nurturance factors in the *early* caregiving environment. Within the context of neurological delays, this chapter provides an overview of the effects of caregiving and language environment on brain development. Caregiving and language environment show great variability within the normal range, yet they can be modified through intervention. This discussion of development is focused on the infant and early childhood years because this period represents a window of rapid brain development and intense interaction with caregivers. However, the interaction between nurturance and neurological health extends beyond childhood, and therapeutic tools for preventing and resolving neurological delay may have lasting effects across the life span.

A large body of research has pointed to the diathesis stress model of vulnerability to explain atypical neurological development resulting from highly prevalent genetic traits (e.g., temperament) or acquired characteristics (e.g., prematurity; see Fig. 42.1). The diathesis stress model suggests that specific characteristics increase an individual's vulnerability to environmental stressors, resulting in increased risk for disrupted neurological development (Monroe & Simons, 1991). For example, the diathesis stress model would predict that in identical poor rearing environments (e.g., low-quality parenting), children at risk (e.g., children born preterm—before 37 weeks) will be more vulnerable to negative outcomes compared to others who do not possess that particular characteristic (e.g., children born full-term).

Diagnosis, Management and Modeling of Neurodevelopmental Disorders. https://doi.org/10.1016/B978-0-12-817988-8.00042-7

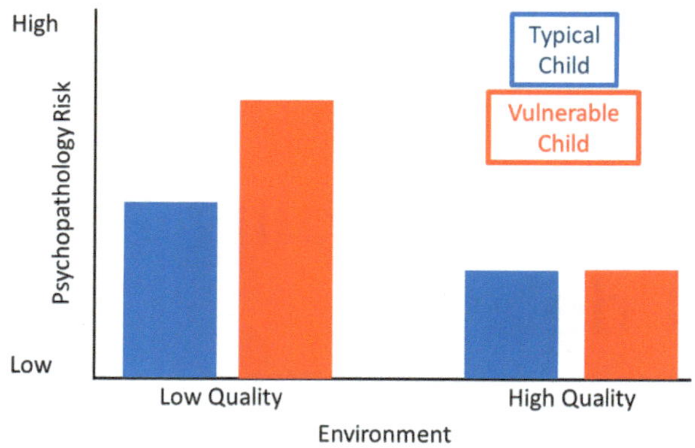

FIGURE 42.1 Predicted psychological health outcomes from the diathesis–stress model. Predictions of outcomes corresponding to the diathesis–stress model (red) compared to what might be expected for a healthy child (blue).

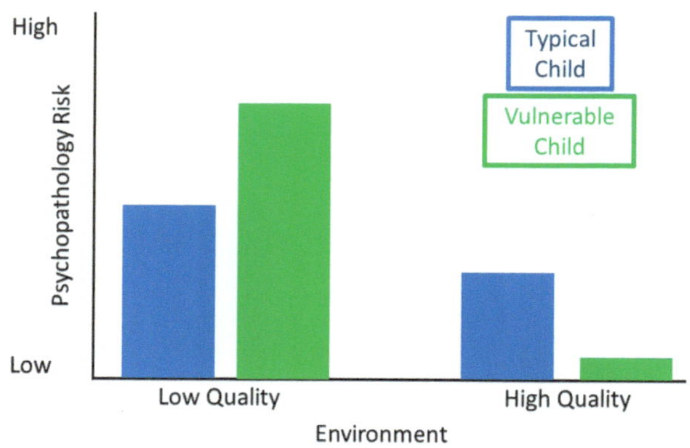

FIGURE 42.2 Predicted psychological health outcomes from the differential susceptibility hypothesis. Predictions of outcomes corresponding to the differential susceptibility hypothesis (green) compared to a healthy child (blue).

Research guided by the diathesis stress model focuses on environmental factors that *negatively* impact developmental outcomes in high-risk populations. However, there is growing awareness of the potential benefit of identifying specific characteristics that result in differential susceptibility to *positive* environmental factors. Originated by Belsky and colleagues, the differential susceptibility hypothesis has emerged as an alternate characterization of the interplay between neurological risk and subsequent child outcome (Belsky & de Haan, 2011). The differential susceptibility model reconceptualizes vulnerability factors such as prematurity or difficult temperament as plasticity factors that *improve* outcomes in high-quality environments and *worsen* outcomes in poor-quality environments (Belsky, 2016; see Fig. 42.2). The diathesis stress model and the differential susceptibility model share their prediction of outcome in negative rearing environments—those who are at risk will fare worse. In contrast, the differential susceptibility model extends the expected capacity of the rearing environment further by proposing that characteristics that increase risk for adverse neurological outcome in negative rearing environments might also increase probability of positive outcome with higher quality of care (Belsky, 2016). From this perspective, interventions that target improving caregiving behavior possess the capacity to leverage neural plasticity to improve outcomes for children at risk for developmental delay.

Considering differential susceptibility within prematurity

Converging evidence suggests that children born at elevated birth risk (i.e., very low birth weight and/or preterm) are at increased risk for emotional dysregulation and academic difficulties (see the chapter "Very preterm children and the impact on neurodevelopment" by Pascoe and Anderson for more information). An adverse caregiving environment is reliably linked to higher rates of psychopathology and reduced school performance for children born premature (Jaekel, 2016; Treyvaud et al., 2016). In line with the differential susceptibility model, studies indicate that interventions targeting parental sensitivity facilitate children's positive development, particularly in areas of social development and early self-regulation (Landry, Smith, & Swank, 2006; Landry, Smith, Swank, Assel, & Vellet, 2001; Landry, Smith, Swank, &

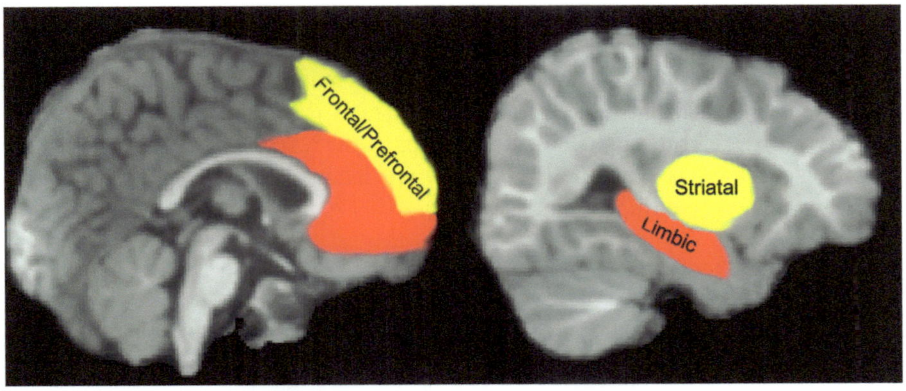

FIGURE 42.3 **Prefrontal limbic system and frontostriatal regions.** The prefrontal limbic system (PrLS; in red) and frontostriatal regions (FS; in yellow). These regions of the brain respond to stress and develop across the life span. Because of experience-dependent plasticity, they may be altered by interventions occurring early in life, such as interventions to promote sensitive and responsive caregiving.

Guttentag, 2008; Landry, Taylor, Guttentag, & Smith, 2008). Results also indicate that children born at high birth risk benefit more from high levels of warm and sensitive responsive parenting than term-born children. There is growing awareness that sensitive and highly responsive parenting supports and improves the parent–child bond. Parents learn how to effectively become the child's "co-regulator" and gradually help the child self-regulate as they mature (Landry, Taylor, et al., 2008). With respect to academic performance, 5-year-olds born at mild birth risk (born 33–36 gestational weeks or at a low birth weight) show reduced preliteracy skills compared to children without birth risk. However, children with mild birth risk show greater gains in literacy skill following intervention compared to children enrolled in the same intervention but with no birth risk (Van der Kooy-Hofland, Van der Kooy, Bus, van IJzendoorn, & Bonsel, 2012).

Relative to the behavioral research discussed earlier, there is a dearth of neuroimaging research targeting the potential for parenting to prevent or remediate neurodevelopmental atypicality. The prefrontal limbic system (PrLS), including frontostriatal (FrS) connections (see Fig. 42.3), rapidly develops in the first three postnatal years. PrLS connections supporting complex cognition and emotion regulation skills include frontal brain regions that are connected to limbic regions (e.g., hippocampus and amygdala) through white matter tracts (e.g., uncinate fasciculus and cingulum bundle). Neural circuitry has the most plasticity when it is rapidly developing, so the protracted developmental course of PrLS makes the network heavily influenced by experience. Early in life, experiences will primarily stem from caregiver interaction (Belsky & de Haan, 2011; Luby, Belden, Harms, Tillman, & Barch, 2016). However, the mechanism driving the improved outcome is unknown. To date, there have been very few investigations on the developing brain as a potential mediator between the environment and behavioral outcomes. The remainder of this chapter will review and integrate the available literature targeting nurturing in the context of brain development.

Effects of nurturing within prematurity

Interventions after the perinatal period target experience-dependent development with the expectation that the resulting change in environment will endure over a longer period of time. A lesser considered implication of intervention is that sustained environmental change may affect maturation of the neurocircuitry that has a more protracted developmental window. Specifically, neurocircuitry with a longer developmental period is shaped by experience-dependent plasticity, a process active throughout the lifetime, as evidenced by interventions that show neurological effects throughout adulthood (Bennett, Kirby, & Finnerty, 2018). As a rule of thumb, interventions that target experience-dependent plasticity are most effective for neurodevelopment if risk for neurodevelopmental deficit is identified early when neural plasticity is highest.

In line with the idea of experience-dependent plasticity, the FrS and PrLS networks are heavily influenced by experience, particularly experiences with caregivers in early childhood, because of the prolonged development of these networks (Belsky & de Haan, 2011; Braun, 2011; Luby et al., 2016). Drawing from animal literature, findings converge to indicate experience-driven synaptic reorganization in the PrLS of offspring constantly exposed to maternal nurturing behaviors (e.g., licking/grooming) (Bock & Braun, 2011; Bock, Rether, Groger, Xie, & Braun, 2014). Effects of maternal nurturing behaviors were also evident for rat pups born preterm with licking and grooming maternal behavior showing possible protective effects to early stress exposure (Francis, Young, Meaney, & Insel,

2002; Walker, Xu, Rochford, & Johnston, 2008). The benefit of nurturing and sensitive caregiver interaction for normalizing neurological development is, to a lesser extent, supported in humans. When evaluated at term-equivalent age, infants born preterm showed improved development of white matter resulting from an intervention targeting increased early parental sensitivity (Milgrom et al., 2010). In later development, cortical thickness and cortical asymmetry in adolescents born preterm were moderated by parental responsiveness such that worse neurological outcomes were evident for adolescents with mothers who were inconsistently responsive or unresponsive during childhood (Frye, Malmberg, Swank, Smith, & Landry, 2010).

Prematurity: mechanism of change

Although findings suggest an intimate link between the parent–child relationship and PrLS development, we do not yet understand the mechanism of change—*why* does highly responsive caregiving shape development of this neurocircuitry? One possibility is that highly responsive parents modulate infant and toddler response to psychological stress through coregulation by reducing repeated stress responses that adversely impact brain development in limbic and cortical regions. In support of this idea, results converge to indicate that limbic development is stress-sensitive. Specifically, limbic development is affected by the stress hormone cortisol, and maturational changes in limbic regions are consistently linked to early life adversity, including negative parent–child relationships (Rao et al., 2010; Vythilingam et al., 2002). Premature birth increases an infant's risk for atypical PrLS neurocircuitry, including regional volume differences as well as disruption to structural and functional connections. Thus, infants born preterm are highly susceptible to environmental stress. Therefore, preterm infants and toddlers benefit from minimizing environmental stress through parental coregulation. In the infant and toddler years, responsive parenting behaviors facilitate children's positive development by regulating the brain's stress response, thus encouraging healthy development of the neural network ultimately charged with regulating response to environmental stressors. Considering that the first 3 years of life represent a period of rapid brain development, evidence that parenting behavior might be a *modifiable* psychosocial factor with significant potential to normalize atypical neurodevelopment strongly suggests the potential advantage of broadening therapeutic intervention beyond the neonatal period.

Nurturing child language development

Many lines of research converge on the finding that the home language environment shapes a child's long-term language and literacy outcomes. Furthermore, findings indicate that home language environment is highly variable, yet also responsive to family-based intervention. Some of the earliest findings suggested large gaps in child vocabulary associated with the quality of the home language environment—specifically that children in the United States raised in poverty heard 30 million fewer words by age 4 than peers from higher socioeconomic status (SES) families (Hart & Risley, 1992). More recent findings suggest that the difference is considerably smaller, perhaps around 4 million words (Gilkerson et al., 2017); nevertheless, the quality of the language children experience at home has an impact on children's language development.

The number of different words that children hear at home is important. When parents use more varied vocabulary, their children develop larger vocabularies (Hoff, 2003; Rowe, 2012), which in turn directly supports their later literacy development (Hemphill & Tivnan, 2008; Lee, 2011). One efficient way to enrich the vocabulary environment at home is to increase the number of shared parent–child book reading sessions as part of routine activities (Fletcher & Reese, 2005; Frijters, Barron, & Brunello, 2000). *Quantity* of language is not the only way parents contribute to their child's language environment—the *quality* of language interactions, often referred to as rich language exposure, matters as well (Hockenberger, Goldstein, & Sirianni Haas, 1999; Taylor, Donovan, Miles, & Leavitt, 2009). When parents use language to provide a framework or scaffold for thought, they help their children to develop both language and cognitive skills (Bernier, Carlson, & Whipple, 2010; Smith, Landry, & Swank, 2000; see Table 42.1). Similarly, parents can use language to maintain their child's attention by elaborating on a single topic (Landry, Smith, Swank, & Miller-Loncar, 2000). These interactions can help children to develop self-regulation skills that allow them to control their own attention in other situations (Landry et al., 2000).

Although there are differences in the quantity and quality of language interactions across families, there are effective interventions for improving the way parents use language (Landry et al., 2012, 2017; Landry, Smith, et al., 2008). These interventions, which encourage parents to engage in verbal scaffolding, use language to maintain the child's attention, and engage in shared book reading, have been effective in improving child language outcomes for low SES families (Landry et al., 2012, 2017; Landry, Smith, et al., 2008).

TABLE 42.1 Comparing high-quality and low-quality home language environments.

	High-quality home language environment	Low-quality home language environment
Verbal scaffolding	Mom: "See how this missing piece looks like the shape of a cow? I wonder if there is a puzzle piece that looks like a cow to put there."	Mom: "Come on, do the puzzle."
Maintaining attention	Child: "Cow"Mom: "Yes that's a cow. Where might that cow live?"	Child: "Cow"Mom: "Yes, that's a cow. This is a pig. This is a horse."
Shared book reading	Mom: "This is a cow. What do you think the cow is going to do?"	Mom: "This is a cow."
Conversational turn-taking	Mom: "Can you find the cow?"Child: "Cow."Mom: "Yes, that's the cow. Where might that cow live?"Child: "Farm."	Mom: "Can you find the cow?"Child: "Cow."Mom: "Yes."

Examples of high-quality and low-quality interactions between parents and children. In high-quality home language environments, parents are more responsive to their children and make efforts to extend the conversation. In low-quality home environments, parents may respond to their children briefly and without efforts to continue the conversation.

Language: clinical implications

In addition to verbal scaffolding and rich language exposure, conversational turn-taking is an important predictor of language development (Zimmerman et al., 2009). This finding highlights the fact that the relationship between the home language environment and a child's language development is reciprocal (Huttenlocher, Waterfall, Vasilyeva, Vevea, & Hedges, 2010). That is, when children have better language skills, they may elicit more (and longer) conversations with their parents, and when children have more conversations with their parents, they will improve their language skills (Hoff-Ginsberg, 1994). Unfortunately, the opposite is also true—when children are struggling with language, for example, when children have expressive language delays or are "late-talkers," they speak less to their parents, so their parents have fewer conversations with them, which leaves children with fewer opportunities to improve their language skills (Paul & Elwood, 1991). Interventions that target the home language environment may be particularly important for children who are struggling with language, as these interventions can encourage parents to speak to their children more, even when their children are not prompting conversations. In particular, parenting interventions that highlight responsiveness can train parents to create language opportunities in response to their child (Landry et al., 2006).

Indeed, interventions that focus on responsive parenting have been effective in improving the quality of the home language environment and ultimately improving child language development. One population that has benefited from these interventions is children born preterm at a very low birth weight, who are at risk for language difficulties (Landry et al., 2006). Infants who are born preterm are born before prenatal brain development is complete, and they spend the early part of their development in an NICU environment, where their language environment is quite different from a home language environment (Caskey, Stephens, Tucker, & Vohr, 2011; McMahon, Wintermark, & Lahav, 2012). The combination of biological and environmental factors surrounding their premature birth puts these children at risk for long-term language difficulties (McMahon et al., 2012; Reidy et al., 2013). When parents of children born preterm receive an intervention focused on responsive parenting, they improve the language input they provide to their child, and their child's language outcomes improve (Landry et al., 2012; Landry, Smith, et al., 2008).

Unique challenges for dual language learners

The nature of bilingualism in the United States highlights the importance of the home language environment. One-fifth of school-aged children in the United States speak a language other than English at home (Kids Count Data Center, 2018). If the home language environment is lacking in vocabulary, book reading, and high-quality conversations, as described earlier, children in these homes may have the same difficulties with acquiring their native language as children in monolingual English homes (Reese, Garnier, Gallimore, & Goldenberg, 2000; Weisleder & Fernald, 2013). Even for monolingual children, a high-quality school environment is often not enough to offset a poor-quality home language environment (Weiland, Ulvestad, Sachs, & Yoshikawa, 2013). For dual language learners (DLLs), the school environment often does not support development of the home language, meaning that most of the home language support must come from the home environment (Fillmore, 1991; Hammer, Davison, Lawrence, & Miccio, 2009). When DLLs start school in the United States, their parents may reduce the use of their

TABLE 42.2 Common misconceptions about dual language learners (DLLs).

Misconception	Explanation
DLLs have smaller vocabularies	DLLs know fewer words in each language, but their total vocabulary size (summed across both languages) is similar to that of monolinguals
DLLs confuse their two languages	From birth, DLLs can distinguish between their two languages. If DLLs switch languages during a conversation, it is likely because they are making the most of the vocabulary that they have in each language
Parents should use only English at home	The quality of the language environment is important. If parents do not have strong English skills, they may only be able to provide a low-quality English environment. When DLLs hear high-quality input in a language other than English, they can build early language skills that transfer to English.

This table describes some common misconceptions about DLLs and explains why they are incorrect.

home language and increase their use of English (Hammer et al., 2009). When this occurs, children receive little input in their home language and are at risk of losing it (Fillmore, 1991).

Parent attempts to support their child's English development are justified. When US DLLs enter school, they are expected to use English, a language with which they have much less experience than monolingual children (Castro, Páez, Dickinson, & Frede, 2011). Fortunately, DLLs can use their home language as a base when learning English in school, but when the home language environment was of low quality, these children may not have built the fundamental language skills in their native language that can help them to acquire a new language (Cummins, 1979). Enhancing the home language environment for US DLLs can give them the language foundation they need to successfully learn English in school and to maintain their home language skills.

Notably, aside from the quantity and quality of language in the home, DLLs have a different language environment than monolingual children. This different language environment has consequences for their language development. The vocabulary of DLLs may be divided across their two languages; for example, children may know more words that refer to home activities in their home language and more words that refer to school activities in English (Hoff & Core, 2013). This may lead to code switching—inserting a word from one language within the context of the other language. Contrary to what some adults believe, this is not an indication that the DLLs are confusing their two languages (Genesee, 1989; see Table 42.2).

Hearing one language versus two languages also influences brain function. DLLs have neural responses to each language that look slightly different than neural responses monolingual children have to their native language (Ferjan Ramírez, Ramírez, Clarke, Taulu, & Kuhl, 2017). Over time, these different experiences and patterns of brain activity may alter the structure of the language network in the brain (Mohades et al., 2015).

Language: mechanism of change

Just as hearing two languages can alter the language network in the brain, so can differences in the home language environment. A robust definition of the language network includes left hemisphere frontal and temporal brain regions (Friederici & Gierhan, 2013; see Fig. 42.4). White matter (measured as fractional anisotropy, FA) connecting the left temporal and frontal regions (i.e., arcuate and superior longitudinal fasciculi) is increased for 4–6 year-old children who experience more adult–child conversational turns throughout the day (Romeo, Segaran, et al., 2018). These children also have higher verbal skills than their peers, and their higher verbal skills can be explained by increased activity in a left frontal brain region (Broca's area) when listening to a story (Romeo, Leonard, et al., 2018). These findings suggest that the mechanism of change underlying home language effects on a child's language development is through the language network. Specifically, improving home language environments leads to more neural activity during language exposure and to structural brain changes that support efficient processing of language, which leads to enhanced language skills for the child.

Applications to other areas of development

Here we have discussed nurturing in infants and young children when the brain is highly plastic and rapidly developing in response to environmental stimulation. However, these principles are generalizable across the life span. First the

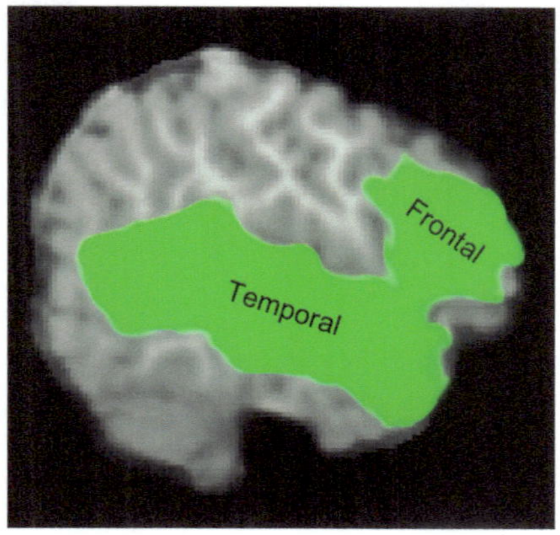

FIGURE 42.4 **Brain regions for language.** The regions in green, including frontal and temporal regions in the left hemisphere, reflect a robust definition of the language network. Interventions that improve the home language environment can shape the development of these brain regions through experience-dependent plasticity.

interventions we have discussed are grounded in increasing and improving social interaction, which is important for brain development and is consistently reported to be important for brain health over the life span. For example, in young adults and older adults, lack of social interaction is reported to increase risk for mental health problems. Although interventions often target parents of young children, interventions that utilize social interaction to change neurological function can still leverage the specialized functioning of the human "social brain" to achieve positive outcome in adult populations; for more on this, see Feldman (2015) and Hari, Henriksson, Malinen, and Parkkonen (2015). Second, the interventions we have reviewed are family-based and target caregivers. Family-based interventions are highly efficient because they empower caregivers to become agents of change in improving their loved one's long-term outcomes. Family-based interventions are highly successful in areas other than parenting including reducing obesity and improving self-esteem (Danielsen, Nordhus, Júlíusson, Mæhle, & Pallesen, 2013; St Jeor, Perumean-Chaney, Sigman-Grant, Williams, & Foreyt, 2002). Although interventions most often target amelioration of negative outcomes of children, family-based interventions are a lesser, although promising, tool for improving outcome across the life span (Biegel & Schulz, 1999; Myers, 2003).

Key facts of nurturing child language development

- Research suggests that children raised in poverty hear at least 4 million fewer words by age 4 than children from higher socioeconomic backgrounds.
- Parents can increase the number of words their children hear by reading books with their children.
- When parents are trained to follow their children's lead, they can learn to use language to help children think through problems and stay focused on a single topic.
- Even if the child is a late-talker, parents can create conversations in response to their child's behaviors.

Key facts about dual language learners

- The home language environment is especially important for dual language learners (DLLs), who must learn a new language when they start school.
- If parents of DLLs provide a good home language environment, their children can build off of their home language to learn a new language in school.
- Because DLLs are learning two languages during childhood, they may not know as many words in each language.
- If DLLs switch languages in the middle of a sentence, it may be because they don't know the word in both languages, not because they are confused.

Summary points

- The human brain has a lengthy window of neural plasticity, particularly in the prefrontal limbic system and frontostriatal network.
- Interventions can use social interaction to support healthy development of vulnerable neural circuitry.

- Different theories describe how experience shapes neurological development in at-risk populations.
- The weight of evidence indicates that neurological deficits can be normalized through intensive intervention.
- The home language environment—which can be changed through parent intervention—is key to early language and literacy development.
- Regions of the brain in the left frontal and temporal cortex change in response to the home language environment.

References

Belsky, J. (2016). The differential susceptibility hypothesis: Sensitivity to the environment for better and for worse. *JAMA Pediatrics, 170*(4), 321–322. https://doi.org/10.1001/jamapediatrics.2015.4263.

Belsky, J., & de Haan, M. (2011). Annual research review: Parenting and children's brain development: The end of the beginning. *Journal of Child Psychology and Psychiatry, 52*(4), 409–428. https://doi.org/10.1111/j.1469-7610.2010.02281.x.

Bennett, S. H., Kirby, A. J., & Finnerty, G. T. (2018). Rewiring the connectome: Evidence and effects. *Neuroscience and Biobehavioral Reviews, 88*, 51–62. https://doi.org/10.1016/j.neubiorev.2018.03.001.

Bernier, A., Carlson, S. M., & Whipple, N. (2010). From external regulation to self-regulation: Early parenting precursors of young children's executive functioning. *Child Development, 81*(1), 326–339. https://doi.org/10.1111/j.1467-8624.2009.01397.x.

Biegel, D. E., & Schulz, R. (1999). Caregiving and caregiver interventions in aging and mental illness. *Family Relations, 48*(4), 345–354.

Bock, J., & Braun, K. (2011). The impact of perinatal stress on the functional maturation of prefronto-cortical synaptic circuits: Implications for the pathophysiology of ADHD? *Progress in Brain Research, 189*, 155–169. https://doi.org/10.1016/B978-0-444-53884-0.00023-3.

Bock, J., Rether, K., Groger, N., Xie, L., & Braun, K. (2014). Perinatal programming of emotional brain circuits: An integrative view from systems to molecules. *Frontiers in Neuroscience, 8*, 11. https://doi.org/10.3389/fnins.2014.00011.

Braun, K. (2011). The prefrontal-limbic system: Development, neuroanatomy, function, and implications for socioemotional development. *Clinics in Perinatology, 38*(4), 685–702. https://doi.org/10.1016/j.clp.2011.08.013.

Caskey, M., Stephens, B., Tucker, R., & Vohr, B. (2011). Importance of parent talk on the development of preterm infant vocalizations. *Pediatrics, 128*(5), 910–916. https://doi.org/10.1542/peds.2011-0609.

Castro, D. C., Páez, M. M., Dickinson, D. K., & Frede, E. (2011). Promoting language and literacy in young dual language learners: Research, practice, and policy. *Child Development Perspectives, 5*(1), 15–21. https://doi.org/10.1111/j.1750-8606.2010.00142.x.

Cummins, J. (1979). Linguistic interdependence and the educational development of bilingual children. *Review of Educational Research, 49*(2), 222–251. https://doi.org/10.3102/00346543049002222.

Danielsen, Y. S., Nordhus, I. H., Júlíusson, P. B., Mæhle, M., & Pallesen, S. (2013). Effect of a family-based cognitive behavioural intervention on body mass index, self-esteem and symptoms of depression in children with obesity (aged 7–13): A randomised waiting list controlled trial. *Obesity Research and Clinical Practice, 7*(2), e116–e128. https://doi.org/10.1016/j.orcp.2012.06.003.

Feldman, R. (2015). The adaptive human parental brain: Implications for children's social development. *Trends in Neurosciences, 38*(6), 387–399. https://doi.org/10.1016/j.tins.2015.04.004.

Ferjan Ramírez, N., Ramírez, R. R., Clarke, M., Taulu, S., & Kuhl, P. K. (2017). Speech discrimination in 11-month-old bilingual and monolingual infants: A magnetoencephalography study. *Developmental Science, 20*(1), e12427. https://doi.org/10.1111/desc.12427.

Fillmore, L. W. (1991). When learning a second language means losing the first. *Early Childhood Research Quarterly, 6*(3), 323–346. https://doi.org/10.1016/S0885-2006(05)80059-6.

Fletcher, K. L., & Reese, E. (2005). Picture book reading with young children: A conceptual framework. *Developmental Review, 25*(1), 64–103. https://doi.org/10.1016/j.dr.2004.08.009.

Francis, D. D., Young, L. J., Meaney, M. J., & Insel, T. R. (2002). Naturally occurring differences in maternal care are associated with the expression of oxytocin and vasopressin (V1a) receptors: Gender differences. *Journal of Neuroendocrinology, 14*(5), 349–353. https://doi.org/10.1046/j.0007-1331.2002.00776.x.

Friederici, A. D., & Gierhan, S. M. (2013). The language network. *Current Opinion in Neurobiology, 23*(2), 250–254. https://doi.org/10.1016/j.conb.2012.10.002.

Frijters, J. C., Barron, R. W., & Brunello, M. (2000). Direct and mediated influences of home literacy and literacy interest on prereaders' oral vocabulary and early written language skill. *Journal of Educational Psychology, 92*(3), 466. https://doi.org/10.1037/0022-0663.92.3.466.

Frye, R. E., Malmberg, B., Swank, P., Smith, K., & Landry, S. (2010). Preterm birth and maternal responsiveness during childhood are associated with brain morphology in adolescence. *Journal of the International Neuropsychological Society, 16*(5), 784–794. https://doi.org/10.1017/S1355617710000585.

Genesee, F. (1989). Early bilingual development: One language or two? *Journal of Child Language, 16*(1), 161–179. https://doi.org/10.1017/S0305000900013490.

Gilkerson, J., Richards, J. A., Warren, S. F., Montgomery, J. K., Greenwood, C. R., Kimbrough Oller, D., … Paul, T. D. (2017). Mapping the early language environment using all-day recordings and automated analysis. *American Journal of Speech-Language Pathology, 26*(2), 248–265. https://doi.org/10.1044/2016_AJSLP-15-0169.

Hammer, C. S., Davison, M. D., Lawrence, F. R., & Miccio, A. W. (2009). The effect of maternal language on bilingual children's vocabulary and emergent literacy development during Head Start and kindergarten. *Scientific Studies of Reading, 13*(2), 99–121. https://doi.org/10.1080/10888430902769541.

Hari, R., Henriksson, L., Malinen, S., & Parkkonen, L. (2015). Centrality of social interaction in human brain function. *Neuron, 88*(1), 181–193. https://doi.org/10.1016/j.neuron.2015.09.022.

Hart, B., & Risley, T. R. (1992). American parenting of language-learning children: Persisting differences in family-child interactions observed in natural home environments. *Developmental Psychology, 28*(6), 1096. https://doi.org/10.1037/0012-1649.28.6.1096.

Hemphill, L., & Tivnan, T. (2008). The importance of early vocabulary for literacy achievement in high-poverty schools. *Journal of Education for Students Placed at Risk, 13*(4), 426–451. https://doi.org/10.1080/10824660802427710.

Hockenberger, E. H., Goldstein, H., & Sirianni Haas, L. (1999). Effects of commenting during joint book reading by mothers with low SES. *Topics in Early Childhood Special Education, 19*(1), 15–27. https://doi.org/10.1177/027112149901900102.

Hoff, E. (2003). The specificity of environmental influence: Socioeconomic status affects early vocabulary development via maternal speech. *Child Development, 74*(5), 1368–1378. https://doi.org/10.1111/1467-8624.00612.

Hoff, E., & Core, C. (2013). Input and language development in bilingually developing children. *Seminars in Speech and Language, 34*(4), 215–226. https://doi.org/10.1055/s-0033-1353448.

Hoff-Ginsberg, E. (1994). Influences of mother and child on maternal talkativeness. *Discourse Processes, 18*(1), 105–117. https://doi.org/10.1080/01638539409544886.

Huttenlocher, J., Waterfall, H., Vasilyeva, M., Vevea, J., & Hedges, L. V. (2010). Sources of variability in children's language growth. *Cognitive Psychology, 61*(4), 343–365. https://doi.org/10.1016/j.cogpsych.2010.08.002.

Jaekel, J. (2016). Commentary: Supporting preterm children's parents matters—a reflection on Treyvaud et al. (2016). *Journal of Child Psychology and Psychiatry, 57*(7), 822–823. https://doi.org/10.1111/jcpp.12525.

Kids Count Data Center. (2018). *Children who speak a language other than English at home in the United States.*

Landry, S. H., Smith, K. E., & Swank, P. R. (2006). Responsive parenting: Establishing early foundations for social, communication, and independent problem-solving skills. *Developmental Psychology, 42*(4), 627–642. https://doi.org/10.1037/0012-1649.42.4.627.

Landry, S. H., Smith, K. E., Swank, P. R., Assel, M. A., & Vellet, S. (2001). Does early responsive parenting have a special importance for children's development or is consistency across early childhood necessary? *Developmental Psychology, 37*(3), 387–403. https://doi.org/10.1037//0012-1649.37.3.387.

Landry, S. H., Smith, K. E., Swank, P. R., & Guttentag, C. (2008). A responsive parenting intervention: The optimal timing across early childhood for impacting maternal behaviors and child outcomes. *Developmental Psychology, 44*(5), 1335. https://doi.org/10.1037/a0013030.

Landry, S. H., Smith, K. E., Swank, P. R., & Miller-Loncar, C. L. (2000). Early maternal and child influences on children's later independent cognitive and social functioning. *Child Development, 71*(2), 358–375. https://doi.org/10.1111/1467-8624.00150.

Landry, S. H., Smith, K. E., Swank, P. R., Zucker, T., Crawford, A. D., & Solari, E. F. (2012). The effects of a responsive parenting intervention on parent—child interactions during shared book reading. *Developmental Psychology, 48*(4), 969. https://doi.org/10.1037/a0026400.

Landry, S. H., Taylor, H. B., Guttentag, C., & Smith, K. E. (2008). Responsive parenting: Closing the learning gap for children with early developmental problems. *International Review of Research in Mental Retardation, 36*, 27–60.

Landry, S. H., Zucker, T. A., Williams, J. M., Merz, E. C., Guttentag, C. L., & Taylor, H. B. (2017). Improving school readiness of high-risk preschoolers: Combining high quality instructional strategies with responsive training for teachers and parents. *Early Childhood Research Quarterly, 40*, 38–51. https://doi.org/10.1016/j.ecresq.2016.12.001.

Lee, J. (2011). Size matters: Early vocabulary as a predictor of language and literacy competence. *Applied Psycholinguistics, 32*(1), 69–92. https://doi.org/10.1017/S0142716410000299.

Luby, J. L., Belden, A., Harms, M. P., Tillman, R., & Barch, D. M. (2016). Preschool is a sensitive period for the influence of maternal support on the trajectory of hippocampal development. *Proceedings of the National Academy of Sciences USA, 113*(20), 5742–5747. https://doi.org/10.1073/pnas.1601443113.

McMahon, E., Wintermark, P., & Lahav, A. (2012). Auditory brain development in premature infants: The importance of early experience. *Annals of the New York Academy of Sciences, 1252*(1), 17–24. https://doi.org/10.1111/j.1749-6632.2012.06445.x.

Milgrom, J., Newnham, C., Anderson, P. J., Doyle, L. W., Gemmill, A. W., Lee, K., … Inder, T. (2010). Early sensitivity training for parents of preterm infants: Impact on the developing brain. *Pediatric Research, 67*(3), 330–335. https://doi.org/10.1203/PDR.0b013e3181cb8e2f.

Mohades, S. G., Van Schuerbeek, P., Rosseel, Y., Van De Craen, P., Luypaert, R., & Baeken, C. (2015). White-matter development is different in bilingual and monolingual children: A longitudinal DTI study. *PLoS One, 10*(2), e0117968. https://doi.org/10.1371/journal.pone.0117968.

Monroe, S. M., & Simons, A. D. (1991). Diathesis-stress theories in the context of life stress research: Implications for the depressive disorders. *Psychological Bulletin, 110*(3), 406–425. https://doi.org/10.1037/0033-2909.110.3.406.

Myers, J. E. (2003). Coping with caregiving stress: A wellness-oriented, strengths-based approach for family counselors. *The Family Journal, 11*(2), 153–161. https://doi.org/10.1177/1066480702250162.

Paul, R., & Elwood, T. J. (1991). Maternal linguistic input to toddlers with slow expressive language development. *Journal of Speech and Hearing Research, 34*, 982–988.

Rao, H., Betancourt, L., Giannetta, J. M., Brodsky, N. L., Korczykowski, M., Avants, B. B., … Farah, M. J. (2010). Early parental care is important for hippocampal maturation: Evidence from brain morphology in humans. *NeuroImage, 49*(1), 1144–1150. https://doi.org/10.1016/j.neuroimage.2009.07.003.

Reese, L., Garnier, H., Gallimore, R., & Goldenberg, C. (2000). Longitudinal analysis of the antecedents of emergent Spanish literacy and middle-school English reading achievement of Spanish-speaking students. *American Educational Research Journal, 37*(3), 633–662. https://doi.org/10.3102/00028312037003633.

Reidy, N., Morgan, A., Thompson, D. K., Inder, T. E., Doyle, L. W., & Anderson, P. J. (2013). Impaired language abilities and white matter abnormalities in children born very preterm and/or very low birth weight. *The Journal of Pediatrics, 162*(4), 719–724. https://doi.org/10.1016/j.jpeds.2012.10.017.

Romeo, R. R., Leonard, J. A., Robinson, S. T., West, M. R., Mackey, A. P., Rowe, M. L., & Gabrieli, J. D. (2018). Beyond the 30-million-word gap: Children's conversational exposure is associated with language-related brain function. *Psychological Science, 29*(5), 700–710. https://doi.org/10.1177/0956797617742725.

Romeo, R. R., Segaran, J., Leonard, J. A., Robinson, S. T., West, M. R., Mackey, A. P., … Gabrieli, J. D. (2018). Language exposure relates to structural neural connectivity in childhood. *Journal of Neuroscience, 38*(36), 7870–7877. https://doi.org/10.1523/JNEUROSCI.0484-18.2018.

Rowe, M. L. (2012). A longitudinal investigation of the role of quantity and quality of child-directed speech in vocabulary development. *Child Development, 83*(5), 1762–1774. https://doi.org/10.1111/j.1467-8624.2012.01805.x.

Smith, K. E., Landry, S. H., & Swank, P. R. (2000). Does the content of mothers' verbal stimulation explain differences in children's development of verbal and nonverbal cognitive skills? *Journal of School Psychology, 38*(1), 27–49. https://doi.org/10.1016/S0022-4405(99)00035-7.

St Jeor, S. T., Perumean-Chaney, S., Sigman-Grant, M., Williams, C., & Foreyt, J. (2002). Family-based interventions for the treatment of childhood obesity. *Journal of the Academy of Nutrition and Dietetics, 102*(5), 640. https://doi.org/10.1016/S0002-8223(02)90146-X.

Taylor, N., Donovan, W., Miles, S., & Leavitt, L. (2009). Maternal control strategies, maternal language usage and children's language usage at two years. *Journal of Child Language, 36*(2), 381–404. https://doi.org/10.1017/S0305000908008969.

Treyvaud, K., Doyle, L. W., Lee, K. J., Ure, A., Inder, T. E., Hunt, R. W., & Anderson, P. J. (2016). Parenting behavior at 2 years predicts school-age performance at 7 years in very preterm children. *Journal of Child Psychology and Psychiatry, 57*(7), 814–821. https://doi.org/10.1111/jcpp.12489.

Van der Kooy-Hofland, V. A., Van der Kooy, J., Bus, A. G., van IJzendoorn, M. H., & Bonsel, G. J. (2012). Differential susceptibility to early literacy intervention in children with mild perinatal adversities: Short-and long-term effects of a randomized control trial. *Journal of Educational Psychology, 104*(2), 337. https://doi.org/10.1007/2288-6729-7-2-15.

Vythilingam, M., Heim, C., Newport, J., Miller, A. H., Anderson, E., Bronen, R., … Bremner, J. D. (2002). Childhood trauma associated with smaller hippocampal volume in women with major depression. *American Journal of Psychiatry, 159*(12), 2072–2080. https://doi.org/10.1176/appi.ajp.159.12.2072.

Walker, C. D., Xu, Z., Rochford, J., & Johnston, C. C. (2008). Naturally occurring variations in maternal care modulate the effects of repeated neonatal pain on behavioral sensitivity to thermal pain in the adult offspring. *Pain, 140*(1), 167–176. https://doi.org/10.1016/j.pain.2008.08.004.

Weiland, C., Ulvestad, K., Sachs, J., & Yoshikawa, H. (2013). Associations between classroom quality and children's vocabulary and executive function skills in an urban public prekindergarten program. *Early Childhood Research Quarterly, 28*(2), 199–209. https://doi.org/10.1016/j.ecresq.2012.12.002.

Weisleder, A., & Fernald, A. (2013). Talking to children matters: Early language experience strengthens processing and builds vocabulary. *Psychological Science, 24*(11), 2143–2152. https://doi.org/10.1177/0956797613488145.

Zimmerman, F. J., Gilkerson, J., Richards, J. A., Christakis, D. A., Xu, D., Gray, S., & Yapanel, U. (2009). Teaching by listening: The importance of adult-child conversations to language development. *Pediatrics, 124*(1), 342–349. https://doi.org/10.1542/peds.2008-2267.

Chapter 43

Normalizing perinatal neurological development via intervention

Dana DeMaster[1], Kelly A. Vaughn[1] and Bartlett D. Moore, IV[2,3]
[1]*Department of Pediatrics, University of Texas Health Science Center at Houston, Children's Learning Institute, Houston, TX, United States;*
[2]*Department of Neurosurgery, University of Texas Health Science Center at Houston, McGovern Medical School, Houston, TX, United States;*
[3]*Program in Cognitive Sciences, School of Social Science, Rice University, Houston, Texas, United States*

List of abbreviations

MRI Magnetic resonance imaging
NICU Neonatal intensive care unit
TBI Traumatic brain injury

Mini-dictionary of terms

Brain network A network of brain regions that are structurally connected through white matter and show high functional connectivity.
Neonatal intensive care unit (NICU) Neonatal refers to the first 28 days after birth. NICU refers to a clinic specializing in infant care immediately after birth. Patients receiving long-term NICU are most often born preterm.
Neuroplasticity The process through which the brain dynamically adapts to the environment.
Nurture External environmental factors, or life experiences, that shape developmental outcome.
Perinatal period Broadly referring to the period immediately preceding birth and a short window after birth. Although the specific definition can vary, here perinatal period refers to the end of the second trimester of pregnancy—lasting through the first month after full-term birth.
Prematurity and preterm birth Birth prior to 36 gestational weeks.
Term equivalent Point at which a preterm infant reaches their expected due date roughly corresponding to the 40th gestational week.

Introduction

Ultimately resulting in a complex system of integrated neural networks, human brain development begins in the third week after conception and progresses over the next 30 years (DeMaster et al., 2019). This extended neurodevelopmental window is largely considered to be adaptive for the human species (Bjorklund, 1997; Tottenham, 2014). Fig. 43.1 illustrates the development of the brain of a toddler born extremely preterm. Comparison of the image shows growth of gray and white matter over the course of 4 months. Containing most of the brains cell bodies and connections, gray matter exhibits regional increase until the onset of puberty resulting in a U-shaped curve of growth (Paus, 2005). Brain development is shaped by nurturance through the developmental processes of synaptogenesis and myelination as the architecture of the brain responds to environmental demand. The next section provides a brief overview of normative human brain development to guide later discussion of the role of nurturance in shaping the human brain's developmental course.

Normative brain development in the perinatal period

During the perinatal period, normative brain development involves the following processes: neurogenesis, neuronal migration, synaptogenesis, apoptosis, synaptic pruning, and myelination (Fig. 43.2). Each of these processes is important for the development of neurocircuitry that supports healthy cognition. In the later months of gestation, the process of neuronal migration of cells remains active as neurons continue to relocate to predestined brain regions (Fig. 43.2). During this period, neurons branch and undergo synaptogenesis, forming new neuronal connections. In the weeks before full-term

Diagnosis, Management and Modeling of Neurodevelopmental Disorders. https://doi.org/10.1016/B978-0-12-817988-8.00043-9

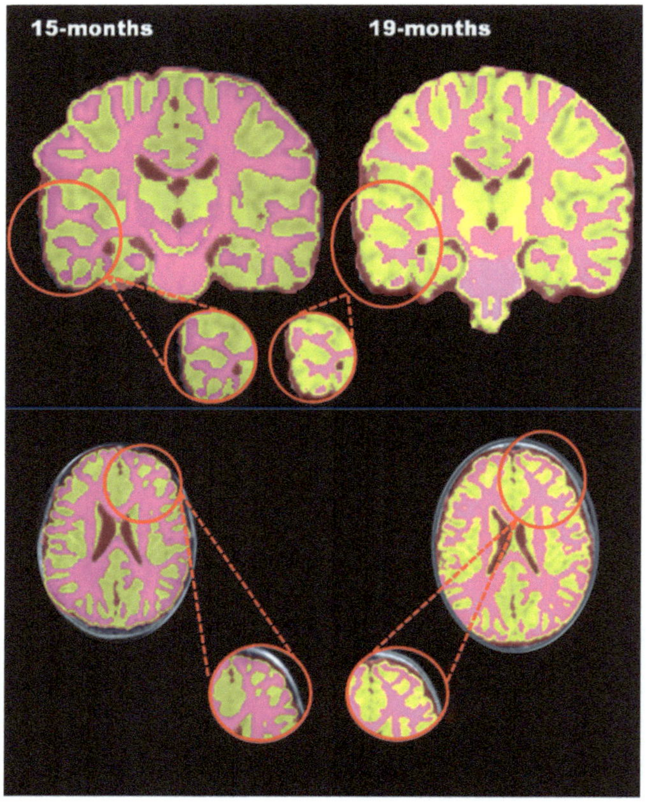

FIGURE 43.1 **White matter and gray matter development in a toddler born extremely preterm.** This figure illustrates brain development in early childhood. Here gray matter is lime green and white matter is pink. Images on the left and the right show increased gray matter in a single infant born at 25 gestational week. T1 images were collected at 15 months (left) and 19 months (right) of age, after correcting for prematurity. Circles capture areas of gray matter increase and are placed adjacent to facilitate comparison.

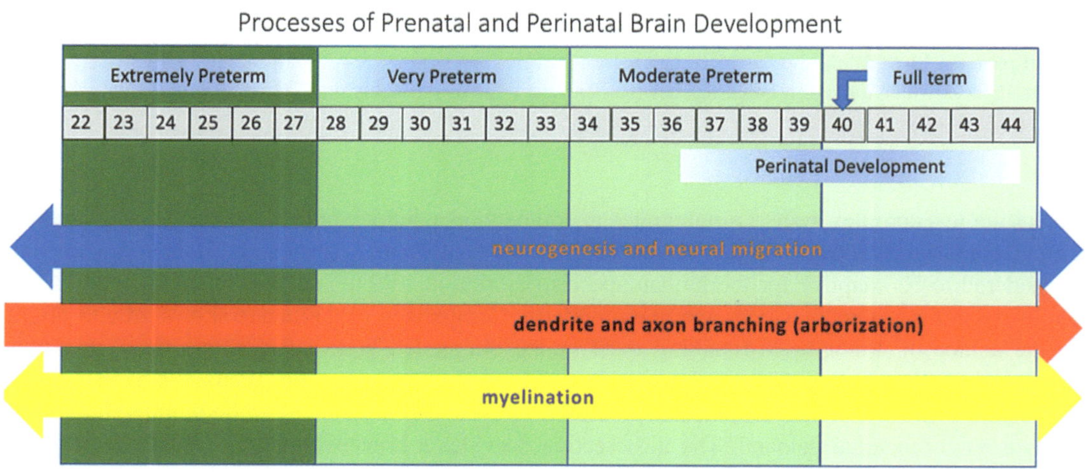

FIGURE 43.2 **Factors influencing neurological development.** The time course of neural development spans pre- and postnatal periods. Following neurogenesis, cells must migrate, branch, and form synapses to support normal cognition. Furthermore, neuroglial cells, including oligodendrocytes, must develop in tandem to support proper brain health. Oligodendrocytes participate in axonal myelination that is critical for neuronal plasticity resulting from the precise timing of input and output action potentials.

birth, synaptogenesis produces a sharp increase in synaptic density throughout the brain resulting in production of more synapses than are needed or even helpful for normal brain function (Kostovic & Vasung, 2009; Stiles & Jernigan, 2010). Synaptogenesis is rapid in the first 2 years of life and is evident in the thickening of gray matter in Fig. 43.1. Pruning of the resulting early neurocircuitry progresses through mediated apoptosis, also called programmed cell death. For example,

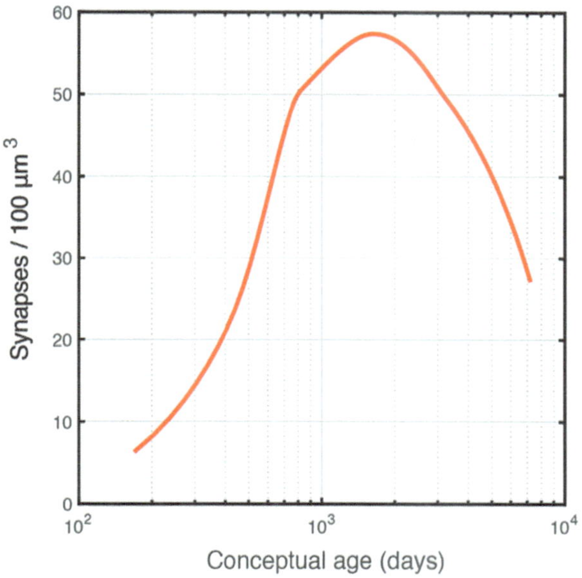

FIGURE 43.3 **Schematic illustration showing synaptogenesis and pruning over time.** In normal human neural development, a sharp rise in synaptogenesis occurs in the cerebral cortex during the perinatal period. Following this drastic increase, weaker synapses are pruned leaving a set of circuits capable of normal cognitive conduct. *Adapted from Stiles, J., & Jernigan, T. L. (2010). The basics of brain development.* Neuropsychology Review, 20*(4), 327–348. https://doi.org/10.1007/s11065-010-9148-4. Synapses were counted in serial electron micrographs of human cortex.*

programmed cell death is required for neural tube closure, healthy development of neurocircuitry and contributes to sex-specific regional neuronal cell numbers (Forger, 2009; Kuan, Roth, Flavell, & Rakic, 2000). However, synaptic pruning following birth is largely guided by environmental experience because synapses that are rarely engaged are more likely to recede (Fig. 43.3). It is through this experience-based synaptic pruning process that environmental factors shape neurological development as the structural connections in the brain adapt to environmental demands (Tottenham, 2014). Experience-based synaptic pruning is most active in infants and young children but extends into childhood (Stiles & Jernigan, 2010). The final process of neurodevelopment, myelination, is the most protracted. Myelin is fatty-white lipid sheath formed by oligodendrocyte neuroglial cells that surrounds long axons (Sampaio & Truwit, 2001). Although axon myelination represents a lengthy developmental process, it plays a critical role in mature brain function by enhancing the efficiency and temporal precision of axonal signaling (Fan et al., 2011; Richards, 2010).

The temporal window of these developmental processes is not homogeneous throughout the brain. Instead, onset, rate of development, and attenuation show regional specificity (Fan et al., 2011). Broadly speaking, synaptic production in primitive brain structures, such as the brain stem and diencephalon, peaks early—roughly corresponding to the perinatal period. In more complex structures such as the prefrontal cortex, synaptogenesis will peak in early childhood around the second year of life (Bourgeois, Goldman-Rakic, & Rakic, 1994; Hodel, 2018; Huttenlocher & Dabholkar, 1997). With regard to pruning, each brain region's temporal window for experience-based synaptic pruning depends on the timing of peak synaptogenesis, with delayed onset of pruning evident for complex brain structures that support later cognition (e.g., prefrontal cortex and hippocampus) and show a late peak of synaptogenesis (Huttenlocher & Dabholkar, 1997).

In contrast to the relatively narrow maturational window of other neurodevelopmental processes, myelination occurs throughout childhood and into adolescence with attenuation in the years approaching adulthood. Similar to synaptogenesis and pruning, the temporal window of myelination is highly variable between brain structures (Brody, Kinney, Kloman, & Gilles, 1987). Myelination of circuits regulating more basic functions such as sensory processing is highly limited after early childhood. However, myelination of white matter connecting the neurocircuitry underlying human cognition, such as the prefrontal cortex, and emotion regulation including some limbic structures, continues beyond childhood (Brody et al., 1987; Jabès & Nelson, 2015; Stiles & Jernigan, 2010). The developmental trajectories of brain structures and the white matter structurally connecting brain regions affect the periods when specific neural circuitry is most sensitive to environmental influence, resulting in different sensitive periods for different cognitive attributes (Hodel, 2018).

Neuroplasticity

Due to the human brain's lengthy developmental timeframe, the resultant long period of plasticity is a double-edged sword with favorable and unfavorable consequences (Fig. 43.4). On one side, relatively slow brain development is beneficial because it facilitates neurocircuitry that dynamically adapts in response to important environmental cues (Hodel, 2018). On the other side, this long period of plasticity also extends the period of vulnerability such that neural circuitry is more susceptible to adverse pre- and postnatal events (DeMaster et al., 2019). This section will discuss the neural circuitry underlying complex cognition within the context of plasticity. Specific attention will be paid to findings indicating that the neural circuitry underlying cognition is highly sensitive to adverse events but also heavily influenced by developmental environment and experience, particularly experiences with caregivers.

To date, research of human neuroplasticity has targeted adverse environmental factors that result in cascading neurological problems such that the weight of neurodevelopmental research is dedicated to identifying factors in the environment that push an otherwise healthy individual off the course of normal neurological development or worsen atypical neurological development in populations at risk (Bick & Nelson, 2016). Thus, historically the link between brain development and parental nurturing environment was investigated by evaluating children who spent part of their early childhood years in institutional settings (Bick & Nelson, 2016; Sheridan, Fox, Zeanah, McLaughlin, & Nelson, 2012; Sheridan, Sarsour, Jutte, D'Esposito, & Boyce, 2012). These investigations target children reared in institutions where most of their physical needs are met but where caregivers did not spend time in social or warm interactions that are typical of a home rearing environment. Sheridan, Fox, et al. (2012) and Sheridan, Sarsour, et al. (2012) collected structural magnetic resonance images (MRIs) of children raised in institutions compared to children reared entirely in a home setting and children who were initially reared in institutions but were then transferred to high-quality home-based foster care by age 3. Results indicated that children with a history of institutional rearing showed reduced white and gray matter compared to age-matched children from exclusively family-based rearing environments. However, children initially reared in an institution who were then transferred to foster care showed white matter volume that was not significantly different to children who were never institutionalized. This finding suggests that moving children from a deprivation setting to an environment with sensitive and responsive caregiving may normalize white matter because the process of myelination is still active. However, caregiving environment did not change gray matter development because the window of high plasticity was closed. In line with findings from other deprivation studies (Bick & Nelson, 2016), results from Sheridan, Fox, et al. (2012) and Sheridan, Sarsour, et al. (2012) provide an excellent example of how adverse caregiving environments shape brain development. These results also demonstrate the potential for treatments involving rearing environment to correct the course of brain development during sensitive periods—when the brain is rapidly developing (Fig. 43.5).

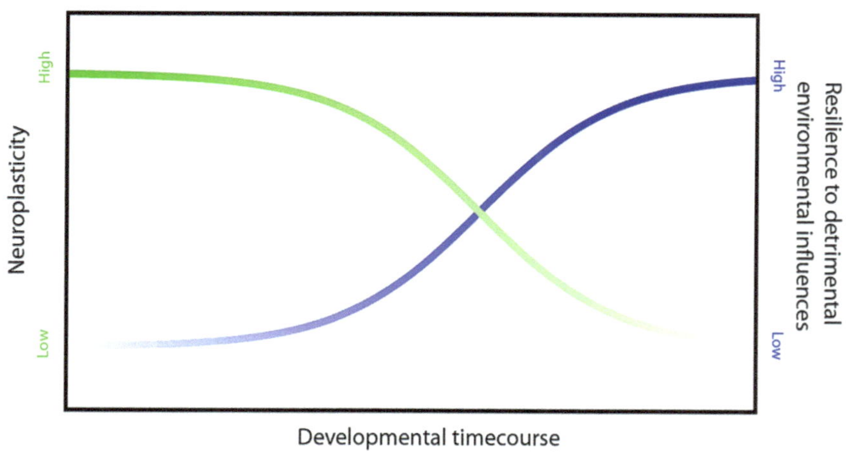

FIGURE 43.4 **Neuroplasticity as related to resilience to detrimental environments.** The relatively slow development of the human brain has both advantages and drawbacks. An extended period of neuroplasticity is beneficial because it facilitates neurocircuitry that can dynamically adapt to environmental cues. However, because the human brain has such a long developmental window, there is an extended window of vulnerability such that neural circuitry is more susceptible to adverse pre- and postnatal events. As neurocircuitry crystalizes, the brain becomes more resilient and is able to recover from detrimental environmental influences. This schematic illustration shows a general decrease in neuroplasticity in green and an increase in resilience in blue.

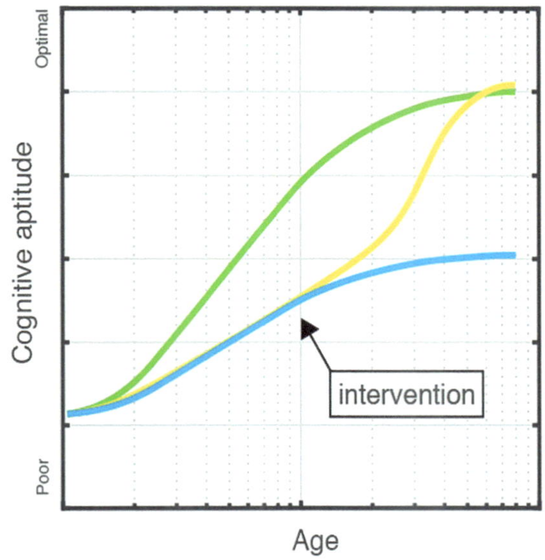

FIGURE 43.5 **Theoretical schematic of differential influence of intervention cognitive attributes.** Green: Normal developmental trajectory of a cognitive capability. Blue: Developmental trajectory of a cognitive capability, such as language acquisition, in an infant who suffered an early detrimental influence. Note that cognitive aptitude is not recovered in this case due to the lasting effects of an earlier detrimental influence. Yellow: Developmental trajectory of a cognitive capability in an infant that suffered an early detrimental influence followed by a positive intervention. Note that the aptitude of this cognitive capability is eventually recovered following the intervention.

For detailed information on the impact of preterm birth of neurological development, see the chapter "Very preterm children and the impact on neurodevelopment" by Pascoe and Anderson in this volume. The next section builds on research identifying risk and protective factors that predict developmental outcome, within the context of prematurity, with a discussion of the multidimensional impact of the rearing environment.

Nurturance in the NICU—pain

There is growing awareness that early environmental stress impacts development of the brain's stress response system through repeated activation of the hypothalamic–pituitary–adrenal (HPA) axis (Dedovic, Duchesne, Andrews, Engert, & Pruessne, 2009) during stressful events. A consequence of low gestational age is that early brain development occurs in a context that is rich in environmental stressors. Premature birth (i.e., birth prior to 36 gestational weeks) results in an immediate demand to engage life-sustaining functions to maintain physiological homeostasis related to breathing, nutrition, temperature regulation, and regulation of response to stimuli. Engaging these systems in the perinatal period places extreme stress on underdeveloped physiology. MRIs acquired at term-equivalent age and those that follow infants over a longer period of time consistently report differences in postnatal brain development between infants born preterm compared to term-born infants. One possibility is in order to sustain life after premature birth, divergent neural network development occurs that prioritizes development of neural network topology of life-sustaining brain circuitry. However, far less is understood about the implication of divergent neural development on neural systems implicated in higher-order cognition and language development.

Beyond differences in brain development in response to life-sustaining physiology, there is also evidence that the caregiving environment of the NICU contributes to differences in brain development between preterm and full-term infants. First, results converge to indicate that repeated exposure to pain intensifies overall reactivity to environmental stressors, including physiological response to pain (Valeri et al., 2016; Valeri, Holsti, & Linhares, 2015). For example, increased HPA access reactivity was evident in a group of infants born preterm compared to term-born infants (Provenzi et al., 2016). Furthermore, in this study the increase in HPA reactivity in the preterm group was modulated by the extent of pain exposure occurring during NICU stay (Provenzi et al., 2016). MRI investigations also show that repeated pain exposure during the sensitive perinatal period can alter neurocircuitry underlying pain processing and emotion regulation (Valeri et al., 2015, 2016). Specifically, increased NICU pain exposure resulted in a delayed development of white matter connecting the cerebral cortex to spinal cord (i.e., corticospinal tract) at term-equivalent age and reduced cortical gray matter thickness in childhood (Zwicker et al., 2013).

Nurturance in the NICU—caregiver interaction

In addition to exposure to pain, the NICU environment may restrict other important factors required for healthy brain development—most notably parental contact. Investigations in rodents indicate that maternal deprivation, or pup separation from the dam, activates the HPA axis of developing pups (Levine & Wiener, 1988). Animal research also indicates that early stress has lasting effects on the brain's stress response system such that adult rats exposed early stress including periods of maternal deprivation showed altered HPA hormone expression linked to maladaptive behaviors (Anisman, Zaharia, Meaney, & Merali, 1998). Stress-induced changes to HPA axis functioning are also evident in nonhuman primates when infants are exposed to negative caregiving environments. Specifically, infant rhesus monkeys repeatedly separated from their mother showed increased cortisol reactivity in response to repeated separation. Moreover, repeated separation also resulted in a lasting increase in reactivity to startling sounds and changed maternal and infant behavior during dyadic interaction (Sanchez et al., 2005). In humans, the consequence of maternal separation on neurological development is less well understood. However, it is clear that HPA axis reactivity is modulated through parent–infant interactions such that human infants react to a mild stressor with a spike in cortisol that is more quickly reduced or soothed by sensitive and responsive parenting (Albers, Riksen-Walraven, Sweep, & de Weerth, 2008).

Interventions in the NICU context

As discussed earlier, there are several factors associated with preterm birth, particularly for those born at the lowest gestational ages, that result in early brain development occurring in a highly stressful context. First, there is the increased demand required to maintain physiological homeostasis in a system that is not yet ready to regulate life-sustaining physiology. Second, there is repeated exposure to pain and deprivation of maternal contact. There is growing awareness that interventions occurring during the NICU stays that reduce environmental stressors through nurturance might have lasting benefit for neurological development of individuals born preterm (Grunau et al., 2009; Milgrom et al., 2010; Mitchell, Yates, Williams, Chang, & Hall, 2013).

Generally, the difference between interventions targeting nurturance during clinical NICU care and those targeting care of toddlers and children is that NICU interventions have the goal of improving developmental outcome by delivering experience-expectant experiences; whereas interventions targeting later development provide experience-dependent experiences (Fig. 43.6). See Table 43.1 for examples. Experiences within the NICU environment influence the somatosensory scaffolding of cognitive, perceptual, and social–emotional development (Maitre et al., 2017). Though this concept remains

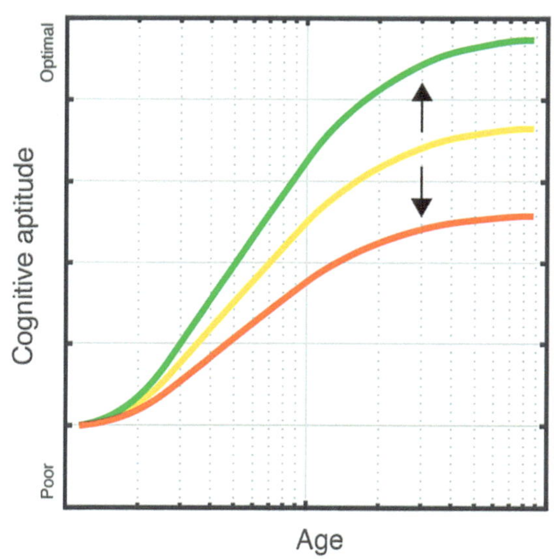

FIGURE 43.6 **Experience-expectant and experience-dependent cognitive development.** Yellow: Changes in cognitive aptitude due to experience-expectant plasticity. Experience-expectant plasticity requires environmental factors that the brain *expects* to be present in a typical nurturing environment. Green: Changes in cognitive aptitude due to positive experience-dependent plasticity. Positive experience-dependent plasticity follows from environmental factors including enriched developmental environments with stimulating and engaging features. Red: Changes in cognitive aptitude due to negative experience-dependent plasticity. Detrimental or deprived environments can negatively affect neural development.

TABLE 43.1 This table gives examples of experience-expectant and experience-dependent plasticity in relation to different skills. Experience-expectant plasticity involves deviations from typical development, whereas experience-dependent plasticity involves experiences that may be common or uncommon. Note that experience-expectant and experience-dependent plasticity are not mutually exclusive, and in fact, experience-dependent plasticity may allow a child to compensate for experience-expectant plasticity (i.e., if a child is born deaf but exposed to sign language, he or she may experience typical language development).

Skill	Experience-expectant plasticity example	Experience-dependent plasticity example
Language	Expected: Child is born hearing Unexpected: Child is born deaf	Child is exposed to sign language Child is exposed to spoken language Child is exposed to multiple languages
Mental rotation	Expected: Child is born sighted Unexpected: Child is born blind	Child frequently plays with blocks Child frequently plays with puzzles
Gross motor skills	Expected: Child is born with two arms and two legs Unexpected: Child is born without a limb	Child has frequent "tummy time" Child is often held/carried

controversial, there are numerous studies suggesting that environmental stressors in the NICU such as loud noises and bright lights may also adversely impact brain development. Conversely, decreased auditory input—maternal voice in particular—may adversely affect language development (Jarjour, 2015; Pineda et al., 2017; Vohr et al., 2017). Over the past decade, there has been a shift in clinical practices to minimize premature infant neonatal care that adversely affects experience-expectant brain development during NICU care by regulating sensory stimulation associated with medical care (e.g., loud sounds, bright light, and pain) (Als et al., 2004; Ranger & Grunau, 2014) and by minimizing maternal separation (Welch et al., 2015, 2016). Results indicate that painful experiences resulting from medically necessary tests and treatments are associated with reduced brain response to touch; whereas warm nurturing experiences provided through skin-to-skin contact with caregivers are associated with stronger brain responses to identical touch stimuli (Maitre et al., 2017). Although there are few neuroimaging investigations of NICU nurturing interventions, the available findings indicate that effects of interventions that modify NICU care to provide a developmentally supportive environment are long-lasting. Specifically, when evaluated at 8 years, improved brain function and structural development, as well as cognitive performance, were evident children at elevated birth risk who received an enhanced NICU experience compared to those receiving standard care (McAnulty et al., 2013).

Applications to other areas of development

Together, findings suggest that clinical implementation of nurturing practices—particularly those aimed at reducing stress during NICU care—is a promising tool for improving neurodevelopmental outcomes in infants and children born preterm, even when measured at age 8 (Welch et al., 2014, 2017). However, research investigating treatments to mitigate birth-related impairment is largely focused on the application of technological advances and medical treatment occurring during the NICU stay (Lester et al., 2016; Pickler, Reyna, Wetzel, & Lewis, 2015; Pickler, Wetzel, Meinzen-Derr, Tubbs-Cooley, & Moore, 2015). One clear benefit of many interventions targeting reduction of NICU stressors is inclusion of parents and parental contact. Parents often report feelings of helplessness in the NICU, family-based interventions improve parent–infant bonding and empower parents to become agents of change with the capacity to directly affect their child's positive neurological development. Moreover, inclusion of parents in NICU interventions also prepares the family to fully assume the role as nurturers after NICU release. Considering that the first 3 years of life represent a period of profound brain development, there is also an urgent need to broaden therapeutic intervention beyond the neonatal period (Fig. 43.5). This idea is further explored in the next chapter.

Findings indicate that in other pediatric populations with similar cognitive deficits to children born preterm, parental intervention has proven highly successful for improving outcomes, suggesting nurturance is an important factor for normalizing neurological development in other at-risk pediatric populations. For example, the available literature converges to indicate that an early model of clinical intervention reduces impact of traumatic brain injury (TBI) on developmental, neuropsychological, and health outcomes (Crowe, Catroppa, Babl, & Anderson, 2012; Crowe, Catroppa, Babl, Rosenfeld, & Anderson, 2012; Goddings, Beltz, Peper, Crone, & Braams, 2019), such that family-based interventions are found to improve both child and family outcome following TBI (Brown, Whittingham, Boyd, & Sofronoff, 2013). First, results indicate that interventions that reduce exposure to environmental stress following pediatric TBI increase parental self-efficacy, reduce parenting anxiety, and improve parent—child interactions (Wade, Oberjohn, Burkhardt, & Greenberg, 2009; Wade et al., 2012). Moreover, there is also evidence that targeting parental factors contributing to the home environment through intervention in turn improves cognitive outcome following pediatric TBI (Durish, Yeates, Stancin, Taylor, Walz, & Wade 2018). However, the majority of children who sustain a TBI do not receive outpatient care or community-based supportive services (Di Scala, Gallagher, & Schneps, 1997; Laatsch et al., 2007).

Together, findings point to reduction of environmental stress as the mechanism through which interventions normalize brain development resulting in improved developmental outcome. Identification of nurturance factors that result in improved neurological outcome in at-risk populations, including children with history of prematurity and pediatric TBI, is a promising area of research. If identified, nurturance factors may be leveraged to improved existing family-based interventions and justify implementation of new interventions to improve neurodevelopment following injury.

Key facts

Key facts about human brain development

- The primary processes of neural development are neural migration, synaptogenesis, and myelination.
- Neural development processes are heterogeneous in the brain such that development onset and attenuation vary by region and brain network.

Key facts about nurturance and brain development

- The brain is most susceptible to environmental influence, nurturance, when it is rapidly developing.
- Neural plasticity of the developing brain can be leveraged through nurturance to improve neurodevelopmental outcome.
- Family-based intervention is an underutilized tool in clinical treatment of pediatric neurological injury.

Key facts about experience-expectant versus experience-dependent plasticity

- Together, these terms characterize how environmental factors shape brain development.
- Experience-expectant plasticity refers to environmental factors that are present for virtually all humans during typical development or the types of experiences that characterize typical human development (see Table 43.1). These environmental factors are *expected* to be present in the nurturing environment and required for normal brain development to occur.
- Experience-expectant plasticity occurs in response to experiences *resulting* from ubiquitous environmental factors such as visual stimulation and the sound of people talking to each other.
- Experience-dependent plasticity refers to the way experiences shape brain development through the process of synaptogenesis and neural network formation.

Summary points

- The perinatal brain is vulnerable to preterm birth as evidenced by divergent neural network development at term-equivalent age.
- Environmental stress disrupts brain development and this may be particularly true following neurological injury resulting from preterm birth.
- Caregiving that is responsive, warm, and sensitive can help to minimize effects of stress on brain development
- Gray matter exhibits regional increase until the onset of puberty resulting in a U-shaped curve of growth. Much of this growth is shaped by environmental factors.
- To support healthy brain development, clinical care environment should be structured to provide ubiquitous environmental experiences such as frequent and warm interactions with a caregiver.
- There is emerging evidence that atypical neurological development may be normalized through family-based intervention targeting parent—infant or parent—child interactions.

References

Albers, E. M., Riksen-Walraven, J. M., Sweep, F. C., & de Weerth, C. (2008). Maternal behavior predicts infant cortisol recovery from a mild everyday stressor. *Journal of Child Psychology and Psychiatry, 49*(1), 97–103. https://doi.org/10.1111/j.1469-7610.2007.01818.x.

Als, H., Duffy, F. H., McAnulty, G. B., Rivkin, M. J., Vajapeyam, S., Mulkern, R. V., … Eichenwald, E. C. (2004). Early experience alters brain function and structure. *Pediatrics, 113*(4), 846–857.

Anisman, H., Zaharia, M. D., Meaney, M. J., & Merali, Z. (1998). Do early-life events permanently alter behavioral and hormonal responses to stressors? *International Journal of Developmental Neuroscience, 16*(3–4), 149–164. https://doi.org/10.1016/s0736-5748(98)00025-2.

Bick, J., & Nelson, C. A. (2016). Early adverse experiences and the developing brain. *Neuropsychopharmacology, 41*(1), 177–196. https://doi.org/10.1038/npp.2015.252.

Bjorklund, D. F. (1997). The role of immaturity in human development. *Psychological Bulletin, 122*(2), 153–169. https://doi.org/10.1037/0033-2909.122.2.153.

Bourgeois, J. P., Goldman-Rakic, P. S., & Rakic, P. (1994). Synaptogenesis in the prefrontal cortex of rhesus monkeys. *Cerebral Cortex, 4*(1), 78–96. https://doi.org/10.1093/cercor/4.1.78.

Brody, B. A., Kinney, H. C., Kloman, A. S., & Gilles, F. H. (1987). Sequence of central nervous system myelination in human infancy. I. An autopsy study of myelination. *Journal of Neuropathology and Experimental Neurology, 46*(3), 283–301. https://doi.org/10.1097/00005072-198705000-00005.

Brown, F. L., Whittingham, K., Boyd, R., & Sofronoff, K. (2013). A systematic review of parenting interventions for traumatic brain injury: Child and parent outcomes. *The Journal of Head Trauma Rehabilitation, 28*(5), 349–360. https://doi.org/10.1097/HTR.0b013e318245fed5.

Crowe, L. M., Catroppa, C., Babl, F. E., & Anderson, V. (2012). Intellectual, behavioral, and social outcomes of accidental traumatic brain injury in early childhood. *Pediatrics, 129*(2), e262–e268. https://doi.org/10.1542/peds.2011-0438.

Crowe, L. M., Catroppa, C., Babl, F. E., Rosenfeld, J. V., & Anderson, V. (2012). Timing of traumatic brain injury in childhood and intellectual outcome. *Journal of pediatric psychology, 37*(7), 745–754. https://doi.org/10.1093/jpepsy/jss070.

Dedovic, K., Duchesne, A., Andrews, J., Engert, V., & Pruessner, J. C. (2009). The brain and the stress axis: the neural correlates of cortisol regulation in response to stress. *Neuroimage, 47*(3), 864–871. https://doi.org/10.1016/j.neuroimage.2009.05.074.

DeMaster, D., Bick, J., Johnson, U., Montroy, J. J., Landry, S., & Duncan, A. F. (2019). Nurturing the preterm infant brain: Leveraging neuroplasticity to improve neurobehavioral outcomes. *Pediatric Research, 85*(2), 166–175. https://doi.org/10.1038/s41390-018-0203-9.

Di Scala, C., Gallagher, S. S., & Schneps, S. E. (1997). Causes and outcomes of pediatric injuries occurring at school. *Journal of school health, 67*(9), 384–389. https://doi.org/10.1111/j.1746-1561.1997.tb07182.x.

Durish, C. L., Yeates, K. O., Stancin, T., Taylor, H. G., Walz, N. C., & Wade, S. L. (2018). Home environment as a predictor of long-term executive functioning following early childhood traumatic brain injury. *Journal of the International Neuropsychological Society 24*(1), 11. https://doi.org/10.1017/S1355617718000437.

Goddings, A. L., Beltz, A., Peper, J. S., Crone, E. A., & Braams, B. R. (2019). Understanding the role of puberty in structural and functional development of the adolescent brain. *Journal of Research on Adolescence, 29*(1), 32–53. https://doi.org/10.1111/jora.12408.

Fan, Y., Shi, F., Smith, J. K., Lin, W., Gilmore, J. H., & Shen, D. (2011). Brain anatomical networks in early human brain development. *NeuroImage, 54*(3), 1862–1871.

Forger, N. G. (2009). Control of cell number in the sexually dimorphic brain and spinal cord. *Journal of Neuroendocrinology, 21*(4), 393–399. https://doi.org/10.1111/j.1365-2826.2009.01825.x.

Grunau, R. E., Whitfield, M. F., Petrie-Thomas, J., Synnes, A. R., Cepeda, I. L., Keidar, A., … Johannesen, D. (2009). Neonatal pain, parenting stress and interaction, in relation to cognitive and motor development at 8 and 18 months in preterm infants. *Pain, 143*(1–2), 138–146. https://doi.org/10.1016/j.pain.2009.02.014.

Hodel, A. S. (2018). Rapid infant prefrontal cortex development and sensitivity to early environmental experience. *Developmental Review, 48*, 113–144. https://doi.org/10.1016/j.dr.2018.02.003.

Huttenlocher, P. R., & Dabholkar, A. S. (1997). Regional differences in synaptogenesis in human cerebral cortex. *Journal of Comparative Neurology, 387*(2), 167–178. https://doi.org/10.1002/(sici)1096-9861(19971020)387:2<167::aid-cne1>3.0.co;2-z.

Jabès, A., & Nelson, C. A. (2015). 20 years after "the ontogeny of human memory: A cognitive neuroscience perspective," where are we? *International Journal of Behavioral Development, 39*(4), 293–303.

Jarjour, I. T. (2015). Neurodevelopmental outcome after extreme prematurity: A review of the literature. *Pediatric Neurology, 52*(2), 143–152. https://doi.org/10.1016/j.pediatrneurol.2014.10.027.

Kostovic, I., & Vasung, L. (2009). Insights from in vitro fetal magnetic resonance imaging of cerebral development. *Seminars in Perinatology, 33*(4), 220–233. https://doi.org/10.1053/j.semperi.2009.04.003.

Kuan, C. Y., Roth, K. A., Flavell, R. A., & Rakic, P. (2000). Mechanisms of programmed cell death in the developing brain. *Trends in Neurosciences, 23*(7), 291–297. https://doi.org/10.1016/s0166-2236(00)01581-2.

Laatsch, L., Harrington, D., Hotz, G., Marcantuono, J., Mozzoni, M. P., Walsh, V., & Hersey, K. P. (2007). An evidence-based review of cognitive and behavioral rehabilitation treatment studies in children with acquired brain injury. *The Journal of head trauma rehabilitation, 22*(4), 248–256. https://doi.org/10.1097/01.HTR.0000281841.92720.0a.

Lester, B. M., Salisbury, A. L., Hawes, K., Dansereau, L. M., Bigsby, R., Laptook, A., … Padbury, J. F. (2016). 18-Month follow-up of infants cared for in a single-family room neonatal intensive care unit. *The Journal of Pediatrics, 177*, 84–89. https://doi.org/10.1016/j.jpeds.2016.06.069.

Levine, S., & Wiener, S. G. (1988). Psychoendocrine aspects of mother-infant relationships in nonhuman primates. *Psychoneuroendocrinology, 13*(1–2), 143–154. https://doi.org/10.1016/0306-4530(88)90011-x.

Maitre, N. L., Key, A. P., Chorna, O. D., Slaughter, J. C., Matusz, P. J., Wallace, M. T., & Murray, M. M. (2017). The dual nature of early-life experience on somatosensory processing in the human infant brain. *Current Biology, 27*(7), 1048–1054. https://doi.org/10.1016/j.cub.2017.02.036.

McAnulty, G., Duffy, F. H., Kosta, S., Weisenfeld, N. I., Warfield, S. K., Butler, S. C., … Als, H. (2013). School-age effects of the newborn individualized developmental care and assessment program for preterm infants with intrauterine growth restriction: Preliminary findings. *BMC Pediatrics, 13*, 25. https://doi.org/10.1186/1471-2431-13-25.

Milgrom, J., Newnham, C., Anderson, P. J., Doyle, L. W., Gemmill, A. W., Lee, K., … Inder, T. (2010). Early sensitivity training for parents of preterm infants: Impact on the developing brain. *Pediatric Research, 67*(3), 330–335. https://doi.org/10.1203/PDR.0b013e3181cb8e2f.

Mitchell, A. J., Yates, C. C., Williams, D. K., Chang, J. Y., & Hall, R. W. (2013). Does daily kangaroo care provide sustained pain and stress relief in preterm infants? *Journal of Neonatal-Perinatal Medicine, 6*(1), 45–52. https://doi.org/10.3233/NPM-1364212.

Paus, T. (2005). Mapping brain maturation and cognitive development during adolescence. *Trends in Cognitive Sciences, 9*(2), 60–68. https://doi.org/10.1016/j.tics.2004.12.008.

Pickler, R. H., Reyna, B. A., Wetzel, P. A., & Lewis, M. (2015). Effect of four approaches to oral feeding progression on clinical outcomes in preterm infants. *Nursing Research and Practice, 2015*, 716828. https://doi.org/10.1155/2015/716828.

Pickler, R. H., Wetzel, P. A., Meinzen-Derr, J., Tubbs-Cooley, H. L., & Moore, M. (2015). Patterned feeding experience for preterm infants: Study protocol for a randomized controlled trial. *Trials, 16*, 255. https://doi.org/10.1186/s13063-015-0781-3.

Pineda, R., Durant, P., Mathur, A., Inder, T., Wallendorf, M., & Schlaggar, B. L. (2017). Auditory exposure in the neonatal intensive care unit: Room type and other predictors. *The Journal of Pediatrics, 183*. https://doi.org/10.1016/j.jpeds.2016.12.072, 56–66 e53.

Provenzi, L., Giusti, L., Fumagalli, M., Tasca, H., Ciceri, F., Menozzi, G., … Montirosso, R. (2016). Pain-related stress in the neonatal intensive care unit and salivary cortisol reactivity to socio-emotional stress in 3-month-old very preterm infants. *Psychoneuroendocrinology, 72*, 161–165. https://doi.org/10.1016/j.psyneuen.2016.07.010.

Ranger, M., & Grunau, R. E. (2014). Early repetitive pain in preterm infants in relation to the developing brain. *Pain Management, 4*(1), 57–67. https://doi.org/10.2217/pmt.13.61.

Richards, J. E. (2010). Attention in the brain and early infancy. *Neoconstructivism: The New Science of Cognitive Development, 1*.

Sampaio, R. C., & Truwit, C. L. (2001). Myelination in the developing human brain. In *Handbook of developmental cognitive neuroscience* (pp. 35–44).

Sanchez, M. M., Noble, P. M., Lyon, C. K., Plotsky, P. M., Davis, M., Nemeroff, C. B., & Winslow, J. T. (2005). Alterations in diurnal cortisol rhythm and acoustic startle response in nonhuman primates with adverse rearing. *Biological Psychiatry, 57*(4), 373–381. https://doi.org/10.1016/j.biopsych.2004.11.032.

Sheridan, M. A., Fox, N. A., Zeanah, C. H., McLaughlin, K. A., & Nelson, C. A., 3rd (2012). Variation in neural development as a result of exposure to institutionalization early in childhood. *Proceedings of the National Academy of Sciences of the United States of America, 109*(32), 12927–12932. https://doi.org/10.1073/pnas.1200041109.

Sheridan, M. A., Sarsour, K., Jutte, D., D'Esposito, M., & Boyce, W. T. (2012). The impact of social disparity on prefrontal function in childhood. *PLoS One, 7*(4), e35744. https://doi.org/10.1371/journal.pone.0035744.

Stiles, J., & Jernigan, T. L. (2010). The basics of brain development. *Neuropsychology Review, 20*(4), 327–348. https://doi.org/10.1007/s11065-010-9148-4.

Tottenham, N. (2014). The importance of early experiences for neuro-affective development. *Current Topics in Behavioral Neurosciences, 16*, 109–129. https://doi.org/10.1007/7854_2013_254.

Valeri, B. O., Holsti, L., & Linhares, M. B. (2015). Neonatal pain and developmental outcomes in children born preterm: A systematic review. *The Clinical Journal of Pain, 31*(4), 355–362. https://doi.org/10.1097/AJP.0000000000000114.

Valeri, B. O., Ranger, M., Chau, C. M., Cepeda, I. L., Synnes, A., Linhares, M. B., & Grunau, R. E. (2016). Neonatal invasive procedures predict pain intensity at school age in children born very preterm. *The Clinical Journal of Pain, 32*(12), 1086–1093. https://doi.org/10.1097/AJP.0000000000000353.

Vohr, B., McGowan, E., McKinley, L., Tucker, R., Keszler, L., & Alksninis, B. (2017). Differential effects of the single-family room neonatal intensive care unit on 18- to 24-month Bayley scores of preterm infants. *The Journal of Pediatrics, 185*. https://doi.org/10.1016/j.jpeds.2017.01.056, 42–48 e41.

Wade, S. L., Oberjohn, K., Burkhardt, A., & Greenberg, I. (2009). Feasibility and preliminary efficacy of a web-based parenting skills program for young children with traumatic brain injury. *The Journal of head trauma rehabilitation, 24*(4), 239–247. https://doi.org/10.1097/HTR.0b013e3181ad6680.

Wade, S. L., Walz, N. C., Carey, J., McMullen, K. M., Cass, J., Mark, E., & Yeates, K. O. (2012). A randomized trial of teen online problem solving: Efficacy in improving caregiver outcomes after brain injury. *Health psychology, 31*(6), 767. https://doi.org/10.1037/a0028440.

Welch, M. G., Firestein, M. R., Austin, J., Hane, A. A., Stark, R. I., Hofer, M. A., … Myers, M. M. (2015). Family nurture intervention in the neonatal intensive care unit improves social-relatedness, attention, and neurodevelopment of preterm infants at 18 months in a randomized controlled trial. *The Journal of Child Psychology and Psychiatry and Allied Disciplines, 56*(11), 1202–1211. https://doi.org/10.1111/jcpp.12405.

Welch, M. G., Halperin, M. S., Austin, J., Stark, R. I., Hofer, M. A., Hane, A. A., & Myers, M. M. (2016). Depression and anxiety symptoms of mothers of preterm infants are decreased at 4 months corrected age with Family Nurture Intervention in the NICU. *Archives of Women's Mental Health, 19*(1), 51–61. https://doi.org/10.1007/s00737-015-0502-7.

Welch, M. G., Myers, M. M., Grieve, P. G., Isler, J. R., Fifer, W. P., Sahni, R., … FNI Trial Group. (2014). Electroencephalographic activity of preterm infants is increased by family nurture intervention: A randomized controlled trial in the NICU. *Clinical Neurophysiology, 125*(4), 675–684. https://doi.org/10.1016/j.clinph.2013.08.021.

Welch, M. G., Stark, R. I., Grieve, P. G., Ludwig, R. J., Isler, J. R., Barone, J. L., & Myers, M. M. (2017). Family nurture intervention in preterm infants increases early development of cortical activity and independence of regional power trajectories. *Acta Paediatrica, 106*(12), 1952–1960. https://doi.org/10.1111/apa.14050.

Zwicker, J. G., Grunau, R. E., Adams, E., Chau, V., Brant, R., Poskitt, K. J., … Miller, S. P. (2013). Score for neonatal acute physiology-II and neonatal pain predict corticospinal tract development in premature newborns. *Pediatric Neurology, 48*(2). https://doi.org/10.1016/j.pediatrneurol.2012.10.016, 123–129 e121.

Part V

Models and modelling

Chapter 44

Microplatforms as a model for neurological conditions

George Deeb[1], Zeina Bou Diab[1], Jessica Saliba[2], Arij Daou[1] and Rami Mhanna[1]
[1]*Biomedical Engineering Program, Maroun Semaan Faculty of Engineering and Architecture, American University of Beirut, Beirut, Lebanon;*
[2]*Department of Biology, Faculty of Science, Lebanese University, Beirut, Lebanon*

List of abbreviations

A-β Amyloid-beta
AD Alzheimer's disease
ASD Autism spectrum disorder
BBB Blood–brain barrier
CNS Central nervous system
CSD Cortical spreading depression
CSF Cerebrospinal fluid
DAI Diffuse axonal injury
MEA Microelectrode array
MS Multiple sclerosis
NSC Neural stem cell
PD Parkinson's disease
TLR Toll-like receptors
TRIF TIR-domain-containing adapter-inducing interferon-β
μ-flow Microflow
α-syn Alpha-synuclein

Mini-dictionary of terms

Hydrogel Water insoluble, three-dimensional network of polymer chains capable of holding a large amount of water. The polymer chains could be of natural or synthetic material.
Lab-on-a-chip (LOC) A device which incorporates one or many laboratory functions onto a chip of a few millimeter-centimeter square in size. LOCs frequently require an understanding of microfluidics when fluid flow is essential in the required analyses and function of the chip.
Microfluidics The realm of science that deals with fluid flows at the micrometer and submicrometer level.
Neural network A circuit of a few or many neurons that are chemically connected by synapses or functionally associated.
Reynolds number A dimensionless number that is used to characterize the type of fluid flow (laminar or turbulent) in a certain geometry.
Three-dimensional cell culture The practice of growing cells in a laboratory setting, which mimics the *in vivo* environment, whereby there are materials that mimic the extra-cellular matrix so that the cells grow in a radial fashion along the 3 spatial axes (x, y and z).
Two-dimensional cell culture The practice of growing cells *in vitro*, whereby they are grown on a flat surface, hence easing the cell-culture process and analysis.

Introduction

Modeling the nervous system has advanced in the previous decades from two-dimensional (2D) cultures of animal cells, to three-dimensional (3D) cultures of human neurons, and even further to cocultures of several types of cells. Recent technological advances and the immense developments in the nano/microworld have brought new opportunities to create

Diagnosis, Management and Modeling of Neurodevelopmental Disorders. https://doi.org/10.1016/B978-0-12-817988-8.00044-0

biomimetic systems with intricate details. The scientific progress has enabled the development of numerous models to mimic a variety of neurodevelopmental conditions leading to a better understanding of normal or disease conditions at a faster rate.

Neural network development

Neural networks begin forming as early as 18 weeks after zygote formation (Ranson, 1920) (Fig. 44.1). These networks generally begin between the multiple primary brain regions such as the basal forebrain, brainstem, and thalamus (Hartlaub, McElroy, Maitre, & Hester, 2019). Between 24 and 28 weeks after fertilization, connections within the cerebral cortex, as well as between the cortex and the thalamus, begin to form. This time is very delicate for fetal brain development whereby drugs may have significant detrimental effects on the formed primary neural circuits. After birth, more networks form among cortical neurons, and other networks connect the various parts of the brain, in addition to the strengthening of existing networks (Fig. 44.2) (Tau & Peterson, 2010). The neural networks that govern motivation and reward and are associated with addiction and mood disorders develop after birth (Russo & Nestler, 2013). Understanding biological neural networks' function and dysfunction would lead to identifying underlying causes of neurodevelopmental diseases and even inspire the fabrication of electronic circuits that integrate multiple signals (Hahnloser, Sarpeshkar, Mahowald, Douglas, & Seung, 2000; Hartlaub, McElroy, Maitre, & Hester, 2019).

The nervous system in research

A variety of models have been proposed to study the nervous system including *in vitro, in vivo, in situ,* and *in silico*. For the most part, nervous research was either done on live subjects, especially for experiments pertaining to understanding the concepts of learning and consciousness, or *in vitro* using cells cultured on tissue culture plastic. As technology improved and we have gained the access to the micro and nanorealms, microfluidic devices have helped to push the boundaries of experimental capabilities in laboratories around the globe. This section will discuss the different *in vitro* models and their various advantages and drawbacks.

a *In vitro* models: pros and cons
   i 2D and 3D *in vitro* models

Briefly, 2D culture models are when cells are grown on plasticware forming monolayers. The 2D culture approach drives cells to proliferate at a rate greater than *in vivo* and to express different genes and gene levels. Most cells are at the

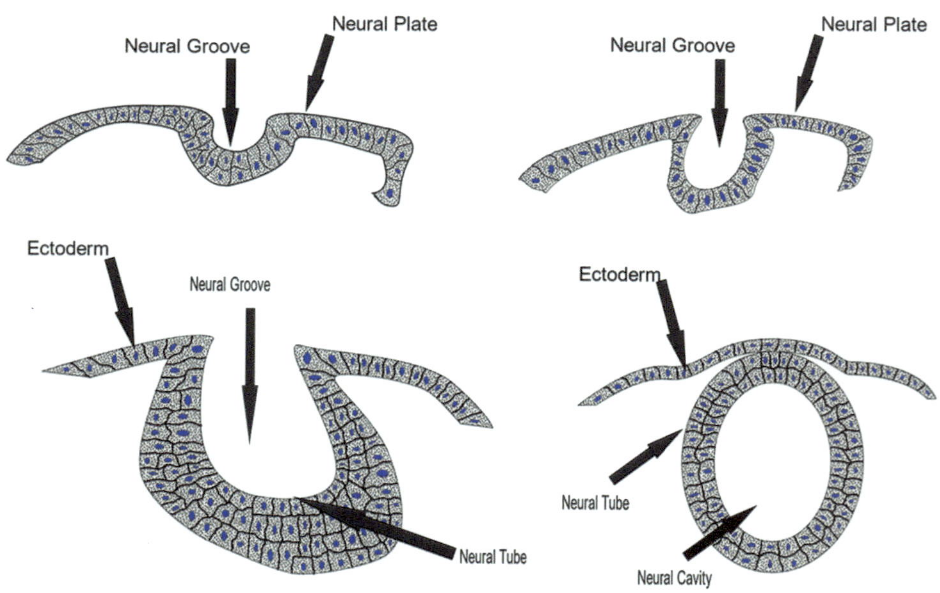

FIGURE 44.1 **Neurological development in a human embryo.** Morphogenesis and the change in the cell shape: development of the neural tube in human embryos. (A) An early embryo (B) At 2 mm. (C) At 2 mm with prominent neural tube formation (D) At 2.7 mm. *Adapted from Ranson, S. W. (1920). The anatomy of the nervous system from the standpoint of development and function. Saunders.*

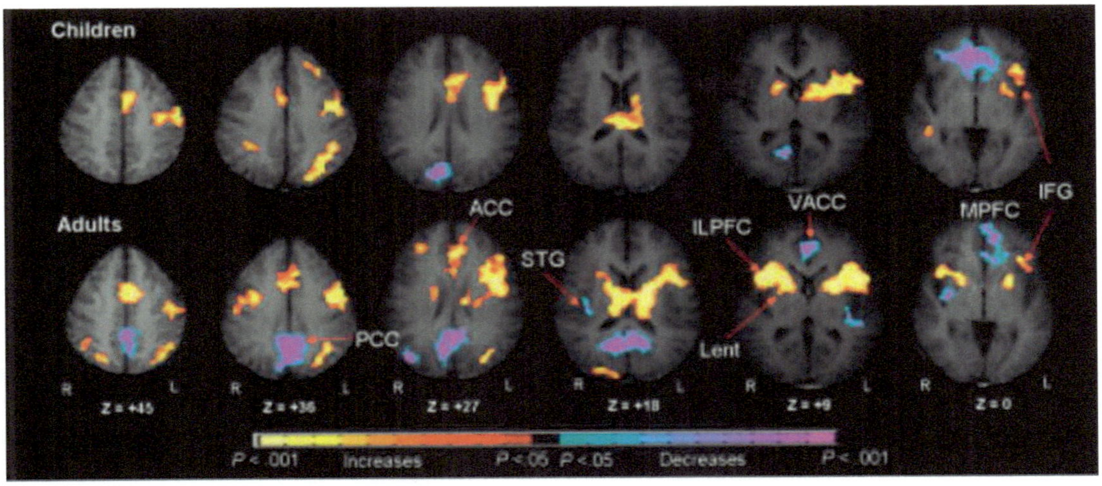

FIGURE 44.2 **Neural networks of a child and adult.** MRI scans of a child's brain and an adult brain showing how different neurons are excited and inhibited when different colors are named. The images clearly show how neural networks change and evolve continuously throughout life. It is worthy to note that areas of active neurons increase and areas of inhibition are developed or increase in adults (Tau & Peterson, 2010).

same stage of the cell cycle, cell–cell contact is at the edges, and the environment lacks a diffusion gradient of oxygen, nutrients, and waste (Saliba et al., 2018). On the other hand, 3D cell cultures enable scientists to provide a better biomimetic environment for cell cultures, achieved by suspending cells in a hydrogel, on porous scaffolds or decellularized tissues, among others (Fig. 44.3A). This cell culture method encourages cells to grow in a natural-like spheroidal shape and make significant cell–matrix contacts, to exhibit cells at different cell cycle stages, and to have a diffusion gradient of oxygen, nutrients, and waste (Saliba et al., 2018). In fact, in neuronal cultures, 2D models do not allow neurons to produce ramified processes over long distances as observed *in vivo*, neither do they provide adequate scaffolding to support cell adhesion (Sasaki, Takagi, Soma, & Yoshida, 2002). Brännvall et al. showed that neuronal stem cells/progenitor cells grown in 3D were able to proliferate and differentiate, as suggested by an increase in markers associated with neurons, astrocytes, and oligodendrocytes (Brännvall et al., 2007).

ii Microfluidic system cultures

Given the significant technological advancements since the 1980s, cell and microenvironment manipulation has become possible with the engineering of microfluidic systems. Tissues *in vivo* are heavily vascularized, which is not mimicked in 2D and 3D cultures. This is a major shortcoming of laboratory engineered tissues and a major challenge of

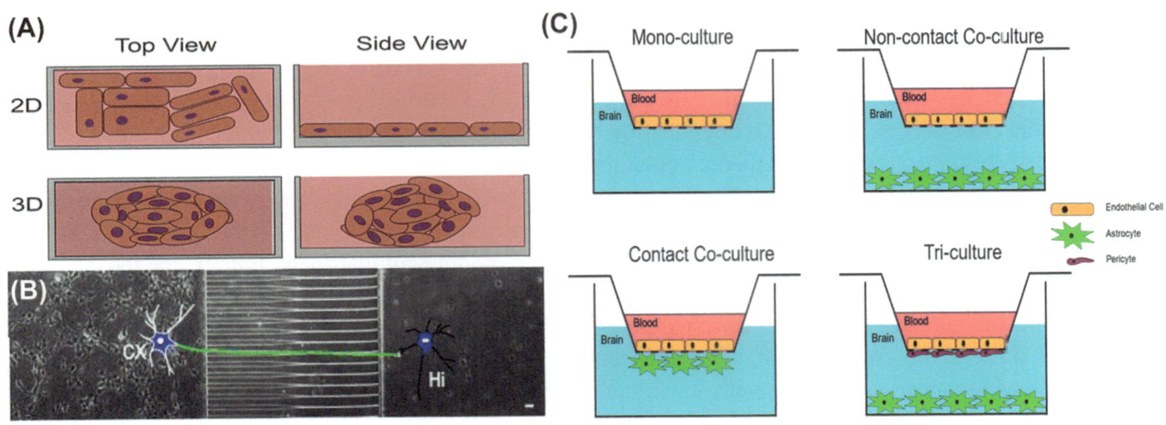

FIGURE 44.3 **Various cell cultures used in neuronal modeling.** (A) A schematic showing the difference between 2D and 3D cell cultures. (B) Cortical and hippocampal neurons cultured in the left and right chambers respectively of a microplatform, whereby the axons extended into the microchannel to form a cortico-hippocampal connection. Schematics of neuronal cells inserted for clarity. Scale bar is 10 μm (Deleglise et al., 2014). (C) Various models of the BBB.

growing larger 3D organs *in vitro*. However, microfluidics enable researchers to culture cells alongside channels with flowing media, whereby the exchange between cells and capillaries is imitated *in vitro* (Paoli & Samitier, 2016). Moreover, microfluidic devices require very little raw material to function. This is a very advantageous aspect when working with proteins, antibodies, and other high-cost materials. Several microfluidic devices have been tailored to be a lab-on-a-chip (LOC). LOCs incorporate several laboratory practices on a single chip from media flow, to thermal control, to pH and chemical monitoring (Fomina et al., 2016). Designing an LOC or any microfluidic device that contains many wells or channels could result in carrying out several experiments under almost identical conditions, giving high repeatability and throughput.

b Neuron cell culture in practice (mono- and cocultures)

i Challenges in growing neural cells in the laboratory

Growing cells onto 2D substrates varies considerably from the vigorous 3D *in vivo* setting as cells are devoid of tissue-specific polarity and have restricted contact with adjacent cells, which change the way cells recognize and react to their microenvironment (Antoni, Burckel, Josset, & Noel, 2015). However, such limitations can be diminished by certain approaches that simulate the 3D configuration, copious vasculature, and the cerebrospinal fluid (CSF) of the brain (van der Meer & van den Berg, 2012). Current advances in microfluidics and microelectromechanical systems (MEMS) have led to the development of distinctive fluidic and patterning techniques for generating *in vitro* central nervous system (CNS) models that best approximate the in vivo cell morphology and habitat Fig. 44.3B. This permits the compliant control of the immediate environment, coculturing, and carrying real-time cell analysis (Whitesides, 2006).

ii Control of neurons on microplatforms

The CNS comprises neurons and several types of supporting glial cells that communicate with the extracellular matrix (McRae & Porter, 2012). Hence, in order to accurately mimic the *in vivo* state of the brain, it is critical to maintain the appropriate ratio of the diverse cell types and control the physical and chemical signaling environment. Microfluidic platforms have been established with proper spatial arrangement of neurons through channels that confine the motion of cell bodies and the production of chemical gradients (Tehranirokh, Kouzani, Francis, & Kanwar, 2013). For the control of surface topology, different patterning techniques are used such as microcontact printing (Vogt, Brewer, & Offenhäusser, 2005), photolithography, and laser ablation (Wheeler & Brewer, 2010). The modulation of cell adhesion and growth is established through various surface-modification materials including poly-L-lysine (Romero-Oliva, Jacob, & Allende, 2003), laminin (Matsuzawa, Tabata, Knoll, & Kano, 2000), polyethylene glycol (Branch, Wheeler, Brewer, & Leckband, 2001), and membrane proteins such as halo rhodopsin and bacteriorhodopsin (Chiu et al., 2000). Neuron−glia interactions can be studied through regulating cell ratios to mimic the *in vivo* setting of a normal functioning brain as well as various atypical coculture conditions, mimicking distinct brain disorders in the laboratory (Park, Koito, Li, & Han, 2012).

c *In vitro* blood−brain barrier models

The blood−brain barrier (BBB) is an integral component of the CNS, which controls the trafficking of materials between the blood stream and the neural tissues of the brain. Several models have been created to mimic this critical structure Fig. 44.3C. Since the BBB is formed of endothelial cells, astrocytes, and pericytes, a worthy model would incorporate these three cell types (Helms et al., 2016). Cultures to model the BBB were very primitive at first. The first attempts were achieved by seeding endothelial cells on one side of a compartment's permeable membrane and then immersing it in media (Rubin et al., 1991). An upgraded model involved a coculture of endothelial cells on one side of the membrane with astrocytes or pericytes in the media-containing compartment (Gaillard et al., 2001). The coculture could be either a contact or a noncontact coculture. In contact coculture, the two types of cells are grown on opposite sides of the permeable membrane. In noncontact coculture, endothelial cells are grown on one side of the membrane, as in monoculture, while the other cells are grown on the surface of the media-containing compartment. Finally, the most complex, yet the most biologically faithful model is a triculture of endothelial cells on one side of the permeable membrane, pericytes on the other side of the membrane, and astrocytes on the surface of the media-containing well (Nakagawa et al., 2009). Contact cocultures present the challenge of separating the cells for different experimental analyses, which requires considerable effort and technical skills, especially for triculture models.

Microfluidic systems

Due to the microscale governing the phenomena at play, microfluidic devices result in high heat transfer speeds, high mass transfer rates, and extremely low reaction time and diffusion rates (Saliba et al., 2018). When studying fluids, the flow is characterized by a nondimensional value known as the Reynold's number (Re). This number is dependent on the channel and the fluid flowing in it. Re is given by the following formula:

$$Re = \frac{\rho V D_h}{\mu}$$

ρ: density of fluid
V: velocity of the fluid flow
D_h: hydraulic diameter of the channel (this is dependent on the geometry of the channel)
μ: viscosity of the fluid

Fluid flow could be either laminar or turbulent. In laminar flow, the fluid flows in streamlines, whereby a velocity profile of the flow in the channel can be drawn, and lateral mixing along the radial axes is virtually nil. Turbulent flow, on the other hand, is chaotic, and there are no streamlines of which particles follow inside the flow. Turbulence causes mixing of the fluid along all axes and results in the chaotic motion of particles within the channel. Flows are deemed laminar when their Re is below 2300 and turbulent when Re is above 2900 (Fig. 44.4). In microfluidic devices and LOCs, owing to dimensions of the flow being in the micrometers, the flow is most frequently laminar (Squires & Quake, 2005).

Microfluidic devices also present very quick heat transfer and diffusion times due to the small dimensions of distance at play. In these devices, the volume-to-surface ratio is very high, resulting in low times to achieve thermal homogeneity within the flow, as well as concentration homogeneity when injecting new substances. The high thermal transfer rate is seen very clearly in microfluidic PCR devices whereby one device has several regions of different temperatures pertaining to the different stages of a PCR workflow. In one microfluidic PCR chip, the solution flows through a channel which loops between the several regions of the chip, each loop corresponding to a complete PCR cycle. Microfluidic chips utilized in PCR reactions demonstrate the high thermal transfer rate and the high throughput of microfluidic chips where one single run in the chip enables several thermal cycles (Zhang & Xing, 2007).

Microfluidic models for brain disease

a Microfluidics in brain cancer and metastasis research

The most common and fatal type of brain cancer is glioma (Kievit et al., 2010) which is a primary tumor originating from glial cells. The brain can also be invaded by metastasizing cells from distant primary tumors such as the breast, colon and lungs among others. Limitations of 2D and transwell-based models in studying brain cancer and cancer metastasis to the brain have prompted the development of brain-on-chip and microfluidic models that better mimic the physiological microenvironment of the nervous tissue, given the critical role of neurons and their microenvironment in cancers of the brain (Denkins et al., 2004; Schmid et al., 2007). In recent years, microfluidic systems have emerged as valuable tools used in cancer drug discovery (Ma et al., 2012) and the study of brain tissue permeability to chemotherapeutic drugs (Terrell-Hall, Ammer, Griffith, & Lockman, 2017). Microfluidics were also used to study glioma cell alignment and migration

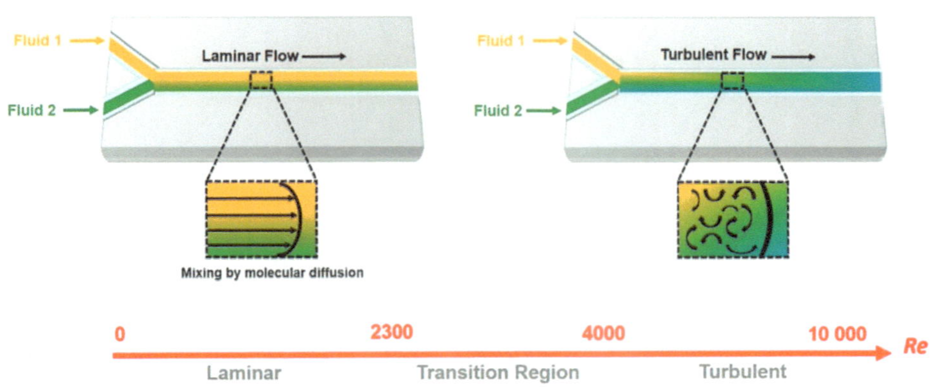

FIGURE 44.4 **Microfluidic flows and Reynold's number.** Schematic showing the difference between laminar and turbulent flows, as well as the variation of flow with respect to Reynold's number (Saliba et al., 2018).

(Lee et al., 2014), to model angiogenesis and the BBB complexity (Tourovskaia, Fauver, Kramer, Simonson, & Neumann, 2014), to better describe cancer cell migration and extravasation (Logan, 1990), to understand the interaction between cortical neurons and different cancer cell lines (Lei et al., 2016), as well as organ preference of metastasis (Shumakovich et al., 2017; Xu et al., 2016).

b Models for neurodegenerative diseases

i Alzheimer's disease

The features of Alzheimer's disease (AD) modeled in microplatforms allows the study of plaques and tangles propagation through neural networks (Deleglise et al., 2013; Dujardin et al., 2014; Song et al., 2014; Stoothoff et al., 2009), amyloid-beta (A-β) and tau protein toxicity (Choi et al., 2013; Jahnke et al., 2012; Kunze, Meissner, Brando, & Renaud, 2011; Poon et al., 2011), glial role in neuroinflammation (Bianco et al., 2012), and the impact of synaptic impairment (Deleglise et al., 2013). A-β deposition in the brain is the result of an imbalance between the peptide's generation and demolition. Numerous contemporary studies have examined the propagation of A-β using innovative microfluidic devices, establishing A-β dissemination throughout neural connections (Song et al., 2014), the stress of the peptide accumulation on those networks (Deleglise et al., 2014), and the recruitment and localization of microglia in response to different A-β gradients (Choi et al., 2013). The microfluidic platforms used by Cho et al. can enable efficient monitoring of microglial migration and its variation in response to therapy targeting neuroinflammation in AD. The platforms' usage can also be extended to the study of the relocation of further cells involved in development of other neurodegenerative diseases such as cerebral ischemia, BBB damage, Parkinson's disease (PD), and amyotrophic lateral sclerosis (Franciosi, Choi, Kim, & McLarnon, 2005; Fuhrmann et al., 2010).

Deleglise et al. used multichamber devices with funnel-shaped microchannels that drive unidirectional axon growth to create an *in vitro* oriented cortico-hippocampal pathway in order to examine the effect of A-β accumulation at the synaptic level (Fig. 44.5A). The study has revealed that somatodendritic A-β deposits on cortical neurons lead to a prompt cortical presynaptic loss that is part of the initial steps characterizing the pathogenic mechanism of AD (Deleglise et al., 2014). Clustering of A-β peptides thus distorts neurotransmitter and neurotrophin signaling leading to synaptic impairment. Dujardin et al. used a microfluidic device with microchannel chambers to demonstrate the neuron-to-neuron transfer of tau proteins initiating the earliest steps of tau pathology in secondary neurons (Dujardin et al., 2014). In almost all tau pathologies, the protein becomes abnormally hyperphosphorylated with irregular spatial arrangements and ultimately accumulates forming intraneuronal neurofibrillary tangles, which is a neuropathological hallmark of AD pathogenesis (Crespo-Biel, Theunis, & Van Leuven, 2012). Kunze et al. used a microfluidic-based model with connected cell culture compartments, hence producing copathological states of phosphorylated tau proteins. This novel model permits the study

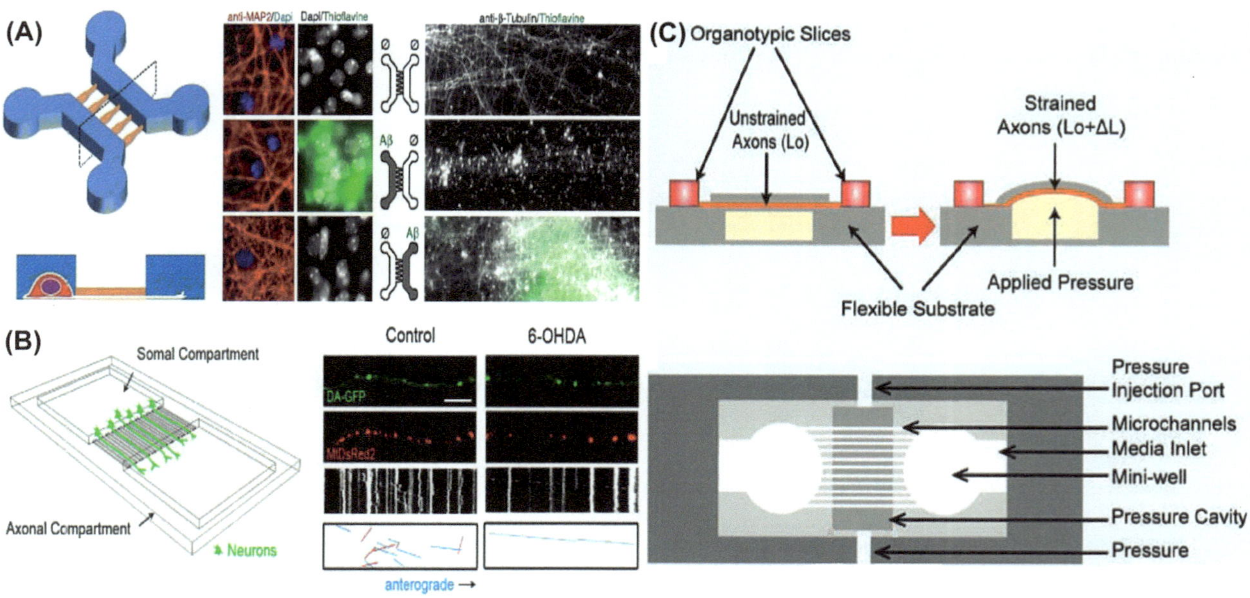

FIGURE 44.5 **Microplatform modeling of AD, PD, and diffuse axonal injuries.** Various microplatforms used to model (A) Alzheimer's disease (Deleglise et al., 2014), (B) Parkinson's disease (Lu et al., 2014), and (C) diffuse axonal injury (Dollé et al., 2013).

of the mechanism of neurodegenerative propagation as well as conducting drug efficiency experiments while assessing side effects between physiological and pathological cell compartments (Kunze et al., 2011). Through a compartmented microfluidic device with a microchannel, Stoothoff et al. revealed that those proteins induce variable changes in retrograde and anterograde axonal transport, with the 3-repeat exhibiting a marginally greater influence on the transport dynamics (Stoothoff et al., 2009). In their model also, inhibiting tau overexpression reestablished neural function despite the persistent neurofibrillary lesions, indicating that dysfunction is the result of tau isoform specific expression and its interaction with microtubules rather than the sole formation of neurofibrillary tangles.

Bianco et al. used overflow microfluidic networks to compare cortical and hippocampal astrocytic activities after introduction of A-β fibrils (Bianco et al., 2012). Results showed that astrocytes derived from the hippocampus, not the cortex, have a deleterious effect on neurons. This gives insight into the mechanism of memory decline in AD. Exploring the communications between different cell populations using an adaptable microfluidic platform and diverse signal readouts is a potent approach for inspecting the precise dynamics of molecular mechanisms governing these interactions during neuroinflammatory events (Bianco et al., 2012)

ii Parkinson's disease

Moreno et al. used a phase-guided 3D microfluidic cell culture bioreactor to differentiate human iPSC-derived neuroepithelial cells into nigrostriatal dopaminergic neurons (DNs). The efficient generation of DNs *in vitro* is vital to model injury as a hallmark of PD. The biological reliability of this model was established through assessing the electrophysiological activity of DNs (Moreno et al., 2015).

A recent study used RBL-2H3 mast cells as a model system to investigate human α-synuclein (α-syn) interactions. Ramezani et al. showed that α-syn colocalizes with mitochondria and upsurges following mitochondrial and cellular stress (Ramezani et al., 2019). Experimental evidence suggests the contribution of mitochondrial transport defects to abnormal trafficking of dopamine axonal mitochondria and the loss of their function (Kim-Han, Antenor-Dorsey, & O'Malley, 2011). However, the latter experiments were hindered by the unavailability of a standardized device that allows fast and efficient separation of axons from cell bodies. Lu et al. designed a novel microplatform that makes monitoring the transport of mitochondria along single dopaminergic axons possible (Lu, Kim-Han, Harmon, Sakiyama-Elbert, & O'Malley, 2014); consisting of an open chamber together with an array of parallel microchannels that permits culturing dopaminergic neurons, isolating, orienting, and imaging their axons in order to ultimately record the motion of axonal mitochondria over time. The developed devices are sufficiently small to allow the installation of a couple of them on one Petri dish, hence facilitating experimentation with more than one condition simultaneously Fig. 44.5B (Lu et al., 2014).

iii Multiple and amyotrophic lateral sclerosis models

The loss of myelin in demyelinating disorders such as multiple sclerosis (MS) leads to the interruption of the nerve signal, injury to the axon, and ultimately, neurodegeneration. It is essential to understand not only the destructive processes but also the repair mechanisms, and MS is a pertinent disease for exploration of such mechanisms. Microglia are the main cells involved in pathogen defense phagocytosis and inflammatory reactions. A microfluidic axon−microglia coculture device, created by Hosmane et al. established that the clearance of unmyelinated axon remains by microglia is mediated through a Toll/interleukin-1 receptor domain-containing adapter that stimulates interferon-b (TRIF) (Hosmane et al., 2012). Notably, these results were relevant in vivo, emphasizing the prospective efficacy of this platform to enhance our understanding of microglial−axon interactions. This observation has various significant consequences. For instance, modifying TRIF activity could be an innovative therapeutic approach to stimulate axon debris removal and boost neuronal regeneration. This would be essentially significant given the evidence that in MS, axonal damage happens early and may even precede demyelination. Many antagonists to Toll-like receptor (TLR)/TRIF pathways are now under clinical studies for possible therapeutic functions in inflammatory and autoimmune disorders (Gambuzza et al., 2011). Up till now, however, management of MS has been concentrated only on relapsing−remitting states with narrow efficacy. Microplatforms could promote the development of actual therapy through detecting real-time disease progression. The exact spatial and temporal regulation of microenvironments that is feasible by microfluidic devices permits the study of several features of neuronal behavior, such as growth, synapse development, neuropharmacology, and electrophysiology (Taylor & Jeon, 2010).

Amyotrophic lateral sclerosis is another progressive neurodegenerative disease of motor neurons characterized by progressive weakness and muscle atrophy resulting in the death of motor neurons and eventual fatal effect (Boillée, Vande Velde, & Cleveland, 2006). *In vitro* models have recently emerged to help understand and better tackle the disease. One important study in this context was that of Kunze et al., who used microfluidic devices to assess neuronal responses to astrocytic regulations by coculturing neurons and genetically altered astrocytes that overexpress mutations in superoxide dismutase (SOD1) (Kunze et al., 2013). This allowed them to study *in vitro* the noncell autonomous conditions in ALS that ignite the regulatory interactions between neurons and astrocytes. Microfluidic devices had been also used *in vitro* to mimic neuromuscular junctions and revealed that this could actually replicate ALS conditions in a dish (Park, Liu, McDonald, Thakor, & Yang, 2013; Uzel et al., 2016).

b Models for traumatic brain and nerve injury

i i- Diffuse axonal injury

The hallmark of diffuse axonal injury (DAI) is enlarged and separated axons with several spheroids developing on discrete axons (Siddique & Thakor, 2014). Mechanoporation of the plasma membranes resulting from this axonal injury produces axon protrusions and focal microtubule disruption (Kilinc, Gallo, & Barbee, 2008). In order to enhance axonal repair and cognitive recovery, it is essential to reduce neuronal inflammation directly after the insult. Proper therapy warrants models that can accurately simulate DAI, several of which were established on microfluidic platforms, with CNS injury being replicated by mechanical means such as stretch (Dollé, Morrison, Schloss, & Yarmush, 2013; Tang-Schomer, Patel, Baas, & Smith, 2010), compression (Hosmane et al., 2011), shear force (Maneshi, Ziegler, Sachs, Hua, & Gottlieb, 2018), and laser (Hellman et al., 2010). Stretch injury to axons replicates instant rupture of microtubules and their gradual loss (Tang-Schomer et al., 2010). A uniaxial strain injury model proposed by Dollé et al. Fig. 44.5C permits connections between two slices and therefore enables extended real-time inspection of injury reactions (Dollé et al., 2013). Hosmane et al. created a valve-based microfluidic axonal injury microcompression platform that allows examination of axonal distortion before, during, and instantly after directed mechanical insult (Hosmane et al., 2011). All of the aforementioned models are based on inspection of neuronal responses to injuries of adjustable levels. A new study, however, monitored intracellular calcium ion levels in cultured astrocytes after exposure to shear forces (Mohammad Mehdi Maneshi, Sachs, & Hua, 2015) and revealed Ca^{2+} influx into astrocytes following injury. Hellman et al. combined pulsed laser microbeams and microfluidic cultures creating more defined areas of neuronal injury compared to other approaches (Hellman et al., 2010). Moreover, using cultured hippocampal neurons and microfluidic devices, Sampson et al. were able to replicate the spread of neurotoxicity observed after traumatic brain injury (Samson, Robertson, Zagnoni, & Connolly, 2016).

ii BBB disease models

Disorders of the BBB have been widely associated with several neurological conditions such as brain tumors, epilepsy, cerebral ischemia, and neurodegenerative syndromes (Akaishi, Takahashi, & Nakashima, 2018; Merali, Huang, Mikulis, Silver, & Kassner, 2017; Miranda, Cova, Sousa, Vitorino, & Pais, 2017; Rüber et al., 2018; Sweeney, Sagare, & Zlokovic, 2018). In order to acquire complete knowledge of the BBB physical and functional assets and to allow delivery of drugs and other compounds to the brain, studies of the BBB are gaining great attention from researchers as well as pharmaceutical firms. Due to its complex organization and difficulty of access, it is very imperative to create appropriate BBB models in research (Alluri, Shaji, Davis, & Tharakan, 2018; Galla, 2018).

New progresses in MEMS and microfabrication technology allowed the manufacture of microfluidic techniques that simulate the biological microenvironments (Michna, Gadde, Ozkan, DeWitt, & Rylander, 2018; Shang, Soon, Lim, Khoo, & Han, 2019; Virumbrales-Muñoz et al., 2019). Microfluidic chips consisting of microchannel networks allow performing accurate and complex operations and hence have been used to model the BBB microenvironment. Brown examined the effect of inflammatory factors such as cytokines on the BBB and revealed that inflammatory stimulus enhances penetrability by diminishing the number of tight junctions (Brown et al., 2016). On the contrary, an astrocyte-rich setting supported the BBB function through increasing tight junctions and as such reducing permeability (Prabhakarpandian et al., 2013).

Microfluidic models can also be used to assess the permeability of different compounds, including free forms, bound to nanomaterials, or functionalized (Marino et al., 2018). Falanga et al. used a microfluidic chip to assess an innovative nano drug delivery vector for the brain and showed that nanoparticles can better cross the BBB when merged with the membranotropic peptide GH625 (Falanga et al., 2017). Wevers and colleagues investigated the permeability of therapeutic antibodies and found that antibodies of human transferrin receptor (MEM-189) can move through the BBB easier than their control counterparts (Wevers et al., 2018). Bonakdar also found that subelectroporation pulsed electric field can disturb the coherence of the BBB rendering it more penetrable (Bonakdar, Graybill, & Davalos, 2017).

c Models for migraine and epilepsy

Cortical spreading depression (CSD) is a wave of depolarization followed by suppression commonly observed in migraine and epilepsy (Charles & Baca, 2013). It is an important pattern of CNS signaling that offers a chance for better understanding of neural functioning as well as the identification of novel treatments for migraine, epilepsy, stroke, and other brain conditions. Better understanding of CSD is possible when chemical stimuli are focally administered, and the information transferred to the adjacent neurons is only through network signaling. Based on this notion, Tang et al. have established a CSD model with a microfluidic platform that offers accurate focal control of chemical stimuli (Tang et al., 2014; Tang, Kim, López-Valdés, Brennan, & Ju, 2011). The experiments confirmed that CSD could be provoked under various settings, including those encountered during migraine aura. This result can hence be used competently in migraine modeling and drug screening.

Epilepsy is a neurological syndrome characterized by extreme coordinated neural activity. Commonly, microelectrode arrays (MEAs) are more convenient than extracellular electrode recording methods. Brain cuts are frequently used in *in vitro* epilepsy simulations to investigate the core epileptic propagation properties (Trevelyan, Baldeweg, Drongelen, Yuste, & Whittington, 2007). Cultured hippocampal cells develop chronic epilepsy few days after slicing and are hence utilized as an *in vitro* model of posttraumatic epilepsy. Liu et al. created a hybrid microfluidic-microelectrode platform that integrates a perfusion setup and electrodes into a fine device (Liu et al., 2018).This epilepsy-on-a-chip system offers a powerful technique for prompt analysis of complex signaling pathways in epilepsy, paving the way for efficient antiepileptic drug discovery (Liu et al., 2018).

d Models for neurodevelopmental disorders

Neural circuits are the fundamental functional entities of the brain responsible for directing complex behavior and cognition. Disturbances in circuitry have been associated with the pathogenesis of several neurodevelopmental disorders such as autism spectrum disorder (ASD) and attention deficit hyperactivity disorder (Hartlaub et al., 2019). Several studies have revealed that there is a discrepancy in oxytocin (Green et al., 2001; Modahl et al., 1998) levels between patients with ASD and healthy counterparts. Thus, oxytocin is used as a biomarker for autism. Currently, oxytocin level is measured through enzyme- and radioimmunoassay in body fluids including plasma, saliva, urine, and CSF (Szeto et al., 2011). However, further investigation revealed the existence of many immunoreactive products interfering with oxytocin and as such compromising the reliability of these methods and hence the need to establish effective procedures to measure oxytocin concentrations (Szeto et al., 2011). Sharma et al. have developed a diagnostic microfluidic platform with channels for oxytocin determination that is selective to common interferences, including oxytocin equivalents and other compounds (Sharma, Gonda, & Tarazi, 2018). The platform also permits the use of less sample volume and reduces the analysis time.

The hope and hype of using microplatforms in brain pathology

The low success rate of animal models being able to faithfully replicate brain pathology could be rescued nowadays by microfluidic devices that allow efficient manipulation of neuronal environments. Microplatforms involved in CNS modeling have succeeded in explaining identified or assumed mechanisms or confirming unique experimental observations and have emerged as a powerful platform for reverse engineering of the brain physiology ex vivo. However, it remains difficult to completely reconstruct the *in vivo* CNS environment. For example, microfluidic approaches exhibit numerous inherent limitations in modeling neural circuitry. Technically, shear stress produced by the platform can damage cells and the microfluidic channels can get filled with cellular remains and air bubbles (Huang, Kao, Curry, & Durbin, 2012). Moreover, the flow rate of these devices is significantly low throughput, thus preventing the study of large sample volumes tests that entail prompt reaction kinetics such as exposing neurons to metabolites that modify their electrophysiological reactions (Huang et al., 2012). Besides, the implementation of a coculture model is affected by numerous considerations, such as cell type and origin, cell number and ratio, and flow rates. Hence, these factors should be accounted for and optimized.

The definitive purpose of these models is to find novel treatment plans. Therefore, it is vital that they precisely simulate the human CNS. Future improvements in 3D structural systems and personalized models of CNS disorders together with the emergence of novel microfluidic-based designs and perfected chip technologies will pave the way to innovative drug delivery systems and treatment strategies. Brain-on-chip has already captured the attention of several pharmaceutical organizations in the world as well as governmental agencies including the FDA. Combining stem cell research with biomaterials and microfabrication technology could foster the development of more realistic *in vivo*—like systems that replicate brain pathophysiology, thereby bringing forward new routes to tackle various brain diseases with a stronger grip of control.

Application to other areas of development

In this chapter we have discussed in brief the embryological development of the nervous system and its development during growth (Chao, Haxby, Lalonde, Ungerleider, & Martin, 1998; Sadler & Langman, 2012). The various models currently used in understanding the nervous system were then presented. The difference between 2D and 3D tissue culturing was illustrated, and the advantages of 3D culture over 2D culture, especially for neurological tissues, were put forward (Brännvall et al., 2007; Saliba et al., 2018). The customary practices of neuronal cell cultures were expressed as well as some challenges faced in growing neurons in a laboratory setting. It is important to balance the ratio of the different neural cells since different ratios would correspond to different physiological and pathophysiological conditions. There are different mechanisms of controlling neurons in microfluidics platforms, such as patterning surface topology and influencing cell-surface adhesion modification. The various model types of the BBB were explained along with the advantages and disadvantages. Finally, the concept of microfluidics and its advantageous aspect concerning laboratory experiments, namely little material expenses and high throughput aspects, were highlighted and expanded upon. The concept of microfluidics for neural development could be expanded to various areas of biomedicine. Various cocultures of different tissues could be mimicked in microfluidic devices. Biological neural networks could be constructed on microplatforms and studied (Aebersold et al., 2016). The effect of different drugs on neural network development and deterioration could be modeled, helping scientists and neurologists comprehend and possibly find cures to various psychological ailments.

Key facts of neural development modeling on microplatforms

- Neural development begins very early in life, only a couple of weeks after zygote formation.
- Neural networks begin to form in the embryo and develop throughout gestation.
- After birth, new neural networks are formed, and old ones are reinforced.
- This process continues throughout life as new experiences are associated with abstract emotions, physical memory is integrated with motion and balance.
- Strengthening of neural networks occurs by strengthening certain neural connections and inhibiting others.
- Microplatforms enable the modeling of the cellular environment, whether physiological or pathological, to conduct experiments and studies that allow researchers to better understand diseases and developmental malformations+.

Summary points

- Neural development begins in the embryo and persists throughout life with the rewiring of various neural networks as a person grows and learns.
- Three-dimensional cell cultures present cells with a more biomimetic environment for growth than two-dimension cultures and significantly affect cell proliferation and maturation especially neural cells.
- Microfluidics is a field that involves conducting experiments in platforms engineered at the microscale, thus enabling researchers to conduct high-throughput experiments with little material cost.
- Several models of various neurodevelopmental diseases have been implemented on microfluidic platforms and have enabled scientists to better understand Alzheimer's, Parkinson's, multiple sclerosis, blood–brain barrier diseases, diffuse axonal injuries, migraine, epilepsy, and attention deficit hyperactivity disorder (Tables 44.1 and 44.2).
- Microplatform devices give researchers the potential to easily gauge the efficacy of several drugs in a cell culture environment, which is very similar to natural physiological or pathological conditions.
- They also allow researchers to better evaluate the efficacy of novel drugs and better understand neural interactions and development due to the more biomimetic environment achieved when compared to 2D and 3D cell cultures.

TABLE 44.1 Summary of major microplatforms for some nervous system disorders.

Disease	Pathology	Microplatform	References
Alzheimer's disease	Accumulation of amyloid beta	Chemotaxis-based microfluidic device	Cho et al., 2016
		Gradient-based microfluidic device	Choi et al., 2013
	Accumulation of tau	Compartmented microchanneled platform	Stoothoff et al., 2009
	Glial inflammation	Overflow microfluidic network	Bianco et al., 2012
Parkinson's disease	DA mitochondrial transport defects	Open multicompartment microchanneled device	Lu et al., 2014
Multiple sclerosis	Axonal degeneration	Microfluidic axon–microglia coculture	Hosmane et al., 2012
Amyotrophic lateral sclerosis	Neuronal response to astrocytic mutations	Microfluidic neuron–astrocyte coculture	Kunze et al., 2013
Diffuse axonal injury	Axonal protrusions and microtubule distortions	Microfluidic compartmented device with controlled air pulses	Tang-Schomer et al., 2010
Migraine	Cortical spreading depression	Perfusion chamber with fluid injection ports	Tang et al., 2014; Tang et al., 2011
Epilepsy	Cortical spreading depression	Hybrid microfluidic-microelectrode pathway	Liu et al., 2018
Autism spectrum disorder	Oxytocin deficiency	Microfluidic platform with selective oxytocin channels	Sharma et al., 2018

TABLE 44.2 Various biological techniques that have been conducted on neural cells.

	Cell growth	References
2D	Culturing neurons on tissue culture plates/dishes	
3D	Cell culture on nonadherent plates	Eitoku, Suganuma, & Kiyosawa, 2016; Jana, Jana, Pal, & Pahan, 2007
	Cell culture on or within gel-like substances (agar, hydrogels)	Martin et al., 2018; Yoo, Kim, Lee, & Nam, 2011
	Cell growth in engineered ECM (electrospun scaffolds, protein fibers, etc.)	Lee, Bashur, Gomez, Goldstein, & Schmidt, 2010
Biochemical assays		
Protein identification and quantization	ELISA	Pellett, Tepp, Johnson, & Sesardic, 2017
	Western blot	Koob et al., 2012
Molecular biology		
Protein sequencing	Next-generation sequencing	Gasparini, Tosatto, Murgia, & Leonardi, 2017

References

Aebersold, M. J., Dermutz, H., Forró, C., Weydert, S., Thompson-Steckel, G., Vörös, J., & Demkó, L. (2016). "Brains on a chip": Towards engineered neural networks. *TRAC Trends in Analytical Chemistry, 78*, 60–69. https://doi.org/10.1016/j.trac.2016.01.025

Akaishi, T., Takahashi, T., & Nakashima, I. (2018). Oligoclonal bands and periventricular lesions in multiple sclerosis will not increase blood-brain barrier permeability. *Journal of the Neurological Sciences, 387*, 129–133. https://doi.org/10.1016/j.jns.2018.02.020

Alluri, H., Shaji, C. A., Davis, M. L., & Tharakan, B. (2018). A mouse controlled cortical impact model of traumatic brain injury for studying blood-brain barrier dysfunctions. *Methods in Molecular Biology, 1717*, 37–52. https://doi.org/10.1007/978-1-4939-7526-6_4

Antoni, D., Burckel, H., Josset, E., & Noel, G. (2015). Three-dimensional cell culture: A breakthrough in vivo. *International Journal of Molecular Sciences, 16*(3), 5517–5527. https://doi.org/10.3390/ijms16035517

Bianco, F., Tonna, N., Lovchik, R. D., Mastrangelo, R., Morini, R., Ruiz, A., … Matteoli, M. (2012). Overflow microfluidic networks: Application to the biochemical analysis of brain cell interactions in complex neuroinflammatory scenarios. *Analytical Chemistry, 84*(22), 9833–9840. https://doi.org/10.1021/ac302094z

Boillée, S., Vande Velde, C., & Cleveland, D. W. (2006). Als: A disease of motor neurons and their nonneuronal neighbors. *Neuron, 52*(1), 39–59. https://doi.org/10.1016/j.neuron.2006.09.018

Bonakdar, M., Graybill, P. M., & Davalos, R. V. (2017). A microfluidic model of the blood–brain barrier to study permeabilization by pulsed electric fields. *RSC Advances, 7*(68), 42811–42818. https://doi.org/10.1039/C7RA07603G

Branch, D. W., Wheeler, B. C., Brewer, G. J., & Leckband, D. E. (2001). Long-term stability of grafted polyethylene glycol surfaces for use with microstamped substrates in neuronal cell culture. *Biomaterials, 22*(10), 1035–1047. https://doi.org/10.1016/S0142-9612(00)00343-4

Brännvall, K., Bergman, K., Wallenquist, U., Svahn, S., Bowden, T., Hilborn, J., & Forsberg-Nilsson, K. (2007). Enhanced neuronal differentiation in a three-dimensional collagen-hyaluronan matrix. *Journal of Neuroscience Research, 85*(10), 2138–2146. https://doi.org/10.1002/jnr.21358

Brown, J. A., Codreanu, S. G., Shi, M., Sherrod, S. D., Markov, D. A., Neely, M. D., … Wikswo, J. P. (2016). Metabolic consequences of inflammatory disruption of the blood-brain barrier in an organ-on-chip model of the human neurovascular unit. *Journal of Neuroinflammation, 13*. https://doi.org/10.1186/s12974-016-0760-y

Chao, L., Haxby, J., Lalonde, F., Ungerleider, L., & Martin, A. (1998). Pictures of animals and tools differentially engage object-related and motion-related brain regions. *Abstracts Society for Neuroscience, 24*, 1507.

Charles, A. C., & Baca, S. M. (2013). Cortical spreading depression and migraine. *Nature Reviews Neurology, 9*(11), 637–644. https://doi.org/10.1038/nrneurol.2013.192

Chiu, M. L., Nollert, P., Loewen, M. C., Belrhali, H., Pebay-Peyroula, E., Rosenbusch, J. P., & Landau, E. M. (2000). Crystallization in cubo: General applicability to membrane proteins. *Acta Crystallographica Section D Biological Crystallography, 56*(6), 781–784. https://doi.org/10.1107/S0907444900004716

Choi, S.-H., Aid, S., Caracciolo, L., Minami, S. S., Niikura, T., Matsuoka, Y., … Bosetti, F. (2013). Cyclooxygenase-1 inhibition reduces amyloid pathology and improves memory deficits in a mouse model of Alzheimer's disease. *Journal of Neurochemistry, 124*(1), 59–68. https://doi.org/10.1111/jnc.12059

Cho, S. M., Lee, S., Yang, S.-H., Kim, H. Y., Lee, M. J., Kim, H. V., … Kim, Y. (2016). Age-dependent inverse correlations in CSF and plasma amyloid-β(1–42) concentrations prior to amyloid plaque deposition in the brain of 3xTg-AD mice. *Scientific Reports, 6*, 20185. https://doi.org/10.1038/srep20185

Crespo-Biel, N., Theunis, C., & Van Leuven, F. (2012). *Protein tau: Prime cause of synaptic and neuronal degeneration in Alzheimer's disease*. https://doi.org/10.1155/2012/251426 [Research article].

Deleglise, B., Lassus, B., Soubeyre, V., Alleaume-Butaux, A., Hjorth, J. J., Vignes, M., … Peyrin, J.-M. (2013). Synapto-protective drugs evaluation in reconstructed neuronal network. *PloS One, 8*(8). https://doi.org/10.1371/journal.pone.0071103

Deleglise, B., Magnifico, S., Duplus, E., Vaur, P., Soubeyre, V., Belle, M., … Brugg, B. (2014). β-amyloid induces a dying-back process and remote trans-synaptic alterations in a microfluidic-based reconstructed neuronal network. *Acta Neuropathologica Communications, 2*. https://doi.org/10.1186/s40478-014-0145-3

Denkins, Y., Reiland, J., Roy, M., Sinnappah-Kang, N. D., Galjour, J., Murry, B. P., … Marchetti, D. (2004). Brain metastases in melanoma: Roles of neurotrophins. *Neuro-Oncology, 6*(2), 154–165. https://doi.org/10.1215/S115285170300067X

Dollé, J.-P., Morrison, B., Schloss, R. R., & Yarmush, M. L. (2013). AN organotypic uniaxial strain model using microfluidics. *Lab on a Chip, 13*(3). https://doi.org/10.1039/c2lc41063j

Dujardin, S., Bégard, S., Caillierez, R., Lachaud, C., Delattre, L., Carrier, S., … Colin, M. (2014). Ectosomes: A new mechanism for non-exosomal secretion of tau protein. *PloS One, 9*(6). https://doi.org/10.1371/journal.pone.0100760

Eitoku, M., Suganuma, N., & Kiyosawa, H. (2016). Comparison of two types of non-adherent plate for neuronal differentiation of mouse embryonic stem cells. *Cytotechnology, 68*(6), 2761–2768. https://doi.org/10.1007/s10616-016-9968-z

Falanga, A. P., Pitingolo, G., Celentano, M., Cosentino, A., Melone, P., Vecchione, R., … Netti, P. A. (2017). Shuttle-mediated nanoparticle transport across an in vitro brain endothelium under flow conditions. *Biotechnology and Bioengineering, 114*(5), 1087–1095. https://doi.org/10.1002/bit.26221

Fomina, N., Johnson, C. A., Maruniak, A., Bahrampour, S., Lang, C., Davis, R. W., … Ahmad, H. (2016). An electrochemical platform for localized pH control on demand. *Lab on a Chip, 16*(12), 2236–2244. https://doi.org/10.1039/C6LC00421K

Franciosi, S., Choi, H. B., Kim, S. U., & McLarnon, J. G. (2005). IL-8 enhancement of amyloid-beta (Aβ1-42)-induced expression and production of pro-inflammatory cytokines and COX-2 in cultured human microglia. *Journal of Neuroimmunology, 159*(1), 66–74. https://doi.org/10.1016/j.jneuroim.2004.10.006

Fuhrmann, M., Bittner, T., Jung, C. K. E., Burgold, S., Page, R. M., Mitteregger, G., … Herms, J. (2010). Microglial Cx3cr1 knockout prevents neuron loss in a mouse model of Alzheimer's disease. *Nature Neuroscience, 13*(4), 411–413. https://doi.org/10.1038/nn.2511

Gaillard, P. J., Voorwinden, L. H., Nielsen, J. L., Ivanov, A., Atsumi, R., Engman, H., … Breimer, D. D. (2001). Establishment and functional characterization of an in vitro model of the blood–brain barrier, comprising a co-culture of brain capillary endothelial cells and astrocytes. *European Journal of Pharmaceutical Sciences, 12*(3), 215–222. https://doi.org/10.1016/S0928-0987(00)00123-8

Galla, H.-J. (2018). Monocultures of primary porcine brain capillary endothelial cells: Still a functional in vitro model for the blood-brain-barrier. *Journal of Controlled Release, 285*, 172–177. https://doi.org/10.1016/j.jconrel.2018.07.016

Gambuzza, M., Licata, N., Palella, E., Celi, D., Foti Cuzzola, V., Italiano, D., … Bramanti, P. (2011). Targeting Toll-like receptors: Emerging therapeutics for multiple sclerosis management. *Journal of Neuroimmunology, 239*(1), 1–12. https://doi.org/10.1016/j.jneuroim.2011.08.010

Gasparini, A., Tosatto, S. C. E., Murgia, A., & Leonardi, E. (2017). Dynamic scaffolds for neuronal signaling: In silico analysis of the TANC protein family. *Scientific Reports, 7*(1), 1–14. https://doi.org/10.1038/s41598-017-05748-5

Green, L., Fein, D., Modahl, C., Feinstein, C., Waterhouse, L., & Morris, M. (2001). Oxytocin and autistic disorder: Alterations in peptide forms. *Biological Psychiatry, 50*(8), 609–613. https://doi.org/10.1016/S0006-3223(01)01139-8

Hahnloser, R. H. R., Sarpeshkar, R., Mahowald, M. A., Douglas, R. J., & Seung, H. S. (2000). Digital selection and analogue amplification coexist in a cortex-inspired silicon circuit. *Nature, 405*(6789), 947. https://doi.org/10.1038/35016072

Hartlaub, A. M., McElroy, C. A., Maitre, N. L., & Hester, M. E. (2019). Modeling human brain circuitry using pluripotent stem cell platforms. *Frontiers in Pediatrics, 7*. https://doi.org/10.3389/fped.2019.00057

Hellman, A. N., Vahidi, B., Kim, H. J., Mismar, W., Steward, O., Jeon, N. L., & Venugopalan, V. (2010). Examination of axonal injury and regeneration in microfluidic neuronal culture using pulsed laser microbeam dissection. *Lab on a Chip, 10*(16), 2083–2092. https://doi.org/10.1039/b927153h

Helms, H. C., Abbott, N. J., Burek, M., Cecchelli, R., Couraud, P.-O., Deli, M. A., … Brodin, B. (2016). In vitro models of the blood-brain barrier: An overview of commonly used brain endothelial cell culture models and guidelines for their use. *Journal of Cerebral Blood Flow and Metabolism: Official Journal of the International Society of Cerebral Blood Flow and Metabolism, 36*(5), 862–890. https://doi.org/10.1177/0271678X16630991

Hosmane, S., Fournier, A., Wright, R., Rajbhandari, L., Siddique, R., Yang, I. H., … Thakor, N. (2011). Valve-based microfluidic compression platform: Single axon injury and regrowth. *Lab on a Chip, 11*(22), 3888–3895. https://doi.org/10.1039/C1LC20549H

Hosmane, S., Tegenge, M. A., Rajbhandari, L., Uapinyoying, P., Kumar, N. G., Thakor, N., & Venkatesan, A. (2012). TRIF mediates microglial phagocytosis of degenerating axons. *Journal of Neuroscience, 32*(22), 7745–7757. https://doi.org/10.1523/JNEUROSCI.0203-12.2012

Huang, P., Kao, T., Curry, A., & Durbin, D. (2012). Factors associated with driving in teens with autism spectrum disorders *Journal of Developmental and Behavioral Pediatrics, 33*(1), 70–74. https://doi.org/10.1097/DBP.0b013e31823a43b7

Jahnke, H. G., Braesigk, A., Mack, T. G., Pönick, S., Striggow, F., & Robitzki, A. A. (2012). Impedance spectroscopy based measurement system for quantitative and label-free real-time monitoring of tauopathy in hippocampal slice cultures. *Biosensors and Bioelectronics, 32*(1), 250–258. https://doi.org/10.1016/j.bios.2011.12.026

Jana, M., Jana, A., Pal, U., & Pahan, K. (2007). A simplified method for isolating highly purified neurons, oligodendrocytes, astrocytes, and microglia from the same human fetal brain tissue. *Neurochemical Research, 32*(12), 2015–2022. https://doi.org/10.1007/s11064-007-9340-y

Kievit, F. M., Florczyk, S. J., Leung, M., Veiseh, O., Park, J. O., Disis, M. L., & Zhang, M. (2010). Chitosan-alginate 3D scaffolds as a mimic of the glioma tumor microenvironment. *Biomaterials, 31*(22), 5903–5910. https://doi.org/10.1016/j.biomaterials.2010.03.062

Kilinc, D., Gallo, G., & Barbee, K. A. (2008). Mechanically-induced membrane poration causes axonal beading and localized cytoskeletal damage. *Experimental Neurology, 212*(2), 422–430. https://doi.org/10.1016/j.expneurol.2008.04.025

Kim-Han, J. S., Antenor-Dorsey, J. A., & O'Malley, K. L. (2011). The Parkinsonian mimetic, MPP+, specifically impairs mitochondrial transport in dopamine axons. *Journal of Neuroscience: The Official Journal of the Society for Neuroscience, 31*(19), 7212–7221. https://doi.org/10.1523/JNEUROSCI.0711-11.2011

Koob, A. O., Bruns, L., Prassler, C., Masliah, E., Klopstock, T., & Bender, A. (2012). Protein analysis through Western blot of cells excised individually from human brain and muscle tissue. *Analytical Biochemistry, 425*(2), 120–124. https://doi.org/10.1016/j.ab.2012.02.034

Kunze, A., Lengacher, S., Dirren, E., Aebischer, P., Magistretti, P. J., & Renaud, P. (2013). Astrocyte–neuron co-culture on microchips based on the model of SOD mutation to mimic ALS. *Integrative Biology, 5*(7), 964–975. https://doi.org/10.1039/c3ib40022k

Kunze, A., Meissner, R., Brando, S., & Renaud, P. (2011). Co-pathological connected primary neurons in a microfluidic device for alzheimer studies. *Biotechnology and Bioengineering, 108*(9), 2241–2245. https://doi.org/10.1002/bit.23128

Lee, J. Y., Bashur, C. A., Gomez, N., Goldstein, A. S., & Schmidt, C. E. (2010). Enhanced polarization of embryonic hippocampal neurons on micron scale electrospun fibers. *Journal of Biomedical Materials Research Part A, 92A*(4), 1398–1406. https://doi.org/10.1002/jbm.a.32471

Lee, K. H., Lee, K. H., Lee, J., Choi, H., Lee, D., Park, Y., & Lee, S.-H. (2014). Integration of microfluidic chip with biomimetic hydrogel for 3D controlling and monitoring of cell alignment and migration: Biomimetic Hydrogel. *Journal of Biomedical Materials Research Part A, 102*(4), 1164–1172. https://doi.org/10.1002/jbm.a.34772

Lei, Y., Li, J., Wang, N., Yang, X., Hamada, Y., Li, Q., … Jiang, X. (2016). An on-chip model for investigating the interaction between neurons and cancer cells. *Integrative Biology, 8*(3), 359–367. https://doi.org/10.1039/c5ib00309a

Liu, W., Zhong, Z., Hu, N., Zhou, Y., Maggio, L., Miri, A. K., … Zhang, Y. S. (2018). Coaxial extrusion bioprinting of 3D microfibrous constructs with cell-favorable gelatin methacryloyl microenvironments. *Biofabrication, 10*(2), 024102. https://doi.org/10.1088/1758-5090/aa9d44

Logan, A. (1990). CNS growth factors. *British Journal of Hospital Medicine, 43*(6), 428–437.

Lu, X., Kim-Han, J. S., Harmon, S., Sakiyama-Elbert, S. E., & O'Malley, K. L. (2014). The Parkinsonian mimetic, 6-OHDA, impairs axonal transport in dopaminergic axons. *Molecular Neurodegeneration, 9*, 17. https://doi.org/10.1186/1750-1326-9-17

Ma, L., Barker, J., Zhou, C., Li, W., Zhang, J., Lin, B., … Honkakoski, P. (2012). Towards personalized medicine with a three-dimensional micro-scale perfusion-based two-chamber tissue model system. *Biomaterials, 33*(17), 4353–4361. https://doi.org/10.1016/j.biomaterials.2012.02.054

Maneshi, Mehdi, Mohammad, Sachs, F., & Hua, S. Z. (2015). A threshold shear force for calcium influx in an astrocyte model of traumatic brain injury. *Journal of Neurotrauma, 32*(13), 1020–1029. https://doi.org/10.1089/neu.2014.3677

Maneshi, Mohammad M., Ziegler, L., Sachs, F., Hua, S. Z., & Gottlieb, P. A. (2018). Enantiomeric Aβ peptides inhibit the fluid shear stress response of PIEZO1. *Scientific Reports, 8*(1), 14267. https://doi.org/10.1038/s41598-018-32572-2

Marino, A., Tricinci, O., Battaglini, M., Filippeschi, C., Mattoli, V., Sinibaldi, E., & Ciofani, G. (2018). A 3D real-scale, biomimetic, and biohybrid model of the blood-brain barrier fabricated through two-photon lithography. *Small, 14*(6), 1702959. https://doi.org/10.1002/smll.201702959

Martin, A. D., Chua, S. W., Au, C. G., Stefen, H., Przybyla, M., Lin, Y., … Ittner, L. M. (2018). Peptide nanofiber substrates for long-term culturing of primary neurons. *ACS Applied Materials and Interfaces, 10*(30), 25127–25134. https://doi.org/10.1021/acsami.8b07560

Matsuzawa, M., Tabata, T., Knoll, W., & Kano, M. (2000). Formation of hippocampal synapses on patterned substrates of a laminin-derived synthetic peptide. *European Journal of Neuroscience, 12*(3), 903–910. https://doi.org/10.1046/j.1460-9568.2000.00977.x

McRae, P. A., & Porter, B. E. (2012). The perineuronal net component of the extracellular matrix in plasticity and epilepsy. *Neurochemistry International, 61*(7), 963–972. https://doi.org/10.1016/j.neuint.2012.08.007

van der Meer, A. D., & van den Berg, A. (2012). Organs-on-chips: Breaking the in vitro impasse. *Integrative Biology, 4*(5), 461–470. https://doi.org/10.1039/c2ib00176d

Merali, Z., Huang, K., Mikulis, D., Silver, F., & Kassner, A. (2017). Evolution of blood-brain-barrier permeability after acute ischemic stroke. *PloS One, 12*(2). https://doi.org/10.1371/journal.pone.0171558

Michna, R., Gadde, M., Ozkan, A., DeWitt, M., & Rylander, M. (2018). Vascularized microfluidic platforms to mimic the tumor microenvironment. *Biotechnology and Bioengineering, 115*(11), 2793–2806. https://doi.org/10.1002/bit.26778

Miranda, A., Cova, T., Sousa, J., Vitorino, C., & Pais, A. (2017). Computational modeling in glioblastoma: From the prediction of blood–brain barrier permeability to the simulation of tumor behavior. *Future Medicinal Chemistry, 10*(1), 121–131. https://doi.org/10.4155/fmc-2017-0128

Modahl, C., Green, L. A., Fein, D., Morris, M., Waterhouse, L., Feinstein, C., & Levin, H. (1998). Plasma oxytocin levels in autistic children. *Biological Psychiatry, 43*(4), 270–277. https://doi.org/10.1016/S0006-3223(97)00439-3

Moreno, E. L., Hachi, S., Hemmer, K., Trietsch, S. J., Baumuratov, A. S., Hankemeier, T., … Fleming, R. M. T. (2015). Differentiation of neuroepithelial stem cells into functional dopaminergic neurons in 3D microfluidic cell culture. *Lab on a Chip, 15*(11), 2419–2428. https://doi.org/10.1039/C5LC00180C

Nakagawa, S., Deli, M. A., Kawaguchi, H., Shimizudani, T., Shimono, T., Kittel, Á., … Niwa, M. (2009). A new blood–brain barrier model using primary rat brain endothelial cells, pericytes and astrocytes. *Neurochemistry International, 54*(3), 253–263. https://doi.org/10.1016/j.neuint.2008.12.002

Paoli, R., & Samitier, J. (2016). Mimicking the kidney: A key role in organ-on-chip development. *Micromachines, 7*(7), 126. https://doi.org/10.3390/mi7070126

Park, J., Koito, H., Li, J., & Han, A. (2012). Multi-compartment neuron–glia co-culture platform for localized CNS axon–glia interaction study. *Lab on a Chip, 12*(18), 3296–3304. https://doi.org/10.1039/C2LC40303J

Park, H. S., Liu, S., McDonald, J., Thakor, N., & Yang, I. H. (2013). Neuromuscular junction in a microfluidic device. *2013 35th Annual International Conference of the IEEE Engineering in Medicine and Biology Society (EMBC)*, 2833–2835. https://doi.org/10.1109/EMBC.2013.6610130

Pellett, S., Tepp, W. H., Johnson, E. A., & Sesardic, D. (2017). Assessment of ELISA as endpoint in neuronal cell-based assay for BoNT detection using hiPSC derived neurons. *Journal of Pharmacological and Toxicological Methods, 88*(Pt 1), 1–6. https://doi.org/10.1016/j.vascn.2017.04.013

Poon, W. W., Blurton-Jones, M., Tu, C. H., Feinberg, L. M., Chabrier, M. A., Harris, J. W., … Cotman, C. W. (2011). β-Amyloid impairs axonal BDNF retrograde trafficking. *Neurobiology of Aging, 32*(5), 821–833. https://doi.org/10.1016/j.neurobiolaging.2009.05.012

Prabhakarpandian, B., Shen, M.-C., Nichols, J. B., Mills, I. R., Sidoryk-Wegrzynowicz, M., Aschner, M., & Pant, K. (2013). SyM-BBB: A microfluidic blood brain barrier model. *Lab on a Chip, 13*(6), 1093–1101. https://doi.org/10.1039/C2LC41208J

Ramezani, M., Wilkes, M. M., Das, T., Holowka, D., Eliezer, D., & Baird, B. (2019). Regulation of exocytosis and mitochondrial relocalization by Alpha-synuclein in a mammalian cell model. *Npj Parkinson's Disease, 5*(1), 12. https://doi.org/10.1038/s41531-019-0084-6

Ranson, S. W. (1920). *The anatomy of the nervous system from the standpoint of development and function*. Saunders.

Romero-Oliva, F., Jacob, G., & Allende, J. E. (2003). Dual effect of lysine-rich polypeptides on the activity of protein kinase CK2. *Journal of Cellular Biochemistry, 89*(2), 348–355. https://doi.org/10.1002/jcb.10493

Rüber, T., David, B., Lüchters, G., Nass, R. D., Friedman, A., Surges, R., … Elger, C. E. (2018). Evidence for peri-ictal blood–brain barrier dysfunction in patients with epilepsy. *Brain, 141*(10), 2952–2965. https://doi.org/10.1093/brain/awy242

Rubin, L. L., Hall, D. E., Porter, S., Barbu, K., Cannon, C., Horner, H. C., … Morales, J. (1991). A cell culture model of the blood-brain barrier. *The Journal of Cell Biology, 115*(6), 1725–1735. https://doi.org/10.1083/jcb.115.6.1725

Russo, S. J., & Nestler, E. J. (2013). The brain reward circuitry in mood disorders. *Nature Reviews Neuroscience, 14*(9). https://doi.org/10.1038/nrn3381

Sadler, T. W., & Langman, J. (2012). *Langman's medical embryology* (12th ed.). Philadelphia: Wolters Kluwer Health/Lippincott Williams & Wilkins.

Saliba, J., Daou, A., Damiati, S., Saliba, J., El-Sabban, M., & Mhanna, R. (2018). Development of microplatforms to mimic the in vivo architecture of CNS and PNS physiology and their diseases. *Genes, 9*(6), 285. https://doi.org/10.3390/genes9060285

Samson, A. J., Robertson, G., Zagnoni, M., & Connolly, C. N. (2016). Neuronal networks provide rapid neuroprotection against spreading toxicity. *Scientific Reports, 6*. https://doi.org/10.1038/srep33746

Sasaki, T., Takagi, M., Soma, T., & Yoshida, T. (2002). 3D culture of murine hematopoietic cells with spatial development of stromal cells in nonwoven fabrics. *Cytotherapy, 4*(3), 285–291. https://doi.org/10.1080/146532402320219808

Schmid, B. C., Rezniczek, G. A., Fabjani, G., Yoneda, T., Leodolter, S., & Zeillinger, R. (2007). The neuronal guidance cue Slit2 induces targeted migration and may play a role in brain metastasis of breast cancer cells. *Breast Cancer Research and Treatment, 106*(3), 333–342. https://doi.org/10.1007/s10549-007-9504-0

Shang, M., Soon, R. H., Lim, C. T., Khoo, B. L., & Han, J. (2019). Microfluidic modelling of the tumor microenvironment for anti-cancer drug development. *Lab on a Chip, 19*(3), 369–386. https://doi.org/10.1039/C8LC00970H

Sharma, S. R., Gonda, X., & Tarazi, F. I. (2018). Autism spectrum disorder: Classification, diagnosis and therapy. *Pharmacology and Therapeutics, 190*, 91–104. https://doi.org/10.1016/j.pharmthera.2018.05.007

Shumakovich, M. A., Mencio, C. P., Siglin, J. S., Moriarty, R. A., Geller, H. M., & Stroka, K. M. (2017). Astrocytes from the brain microenvironment alter migration and morphology of metastatic breast cancer cells. *The FASEB Journal, 31*(11), 5049–5067. https://doi.org/10.1096/fj.201700254R

Siddique, R., & Thakor, N. (2014). Investigation of nerve injury through microfluidic devices. *Journal of The Royal Society Interface, 11*(90). https://doi.org/10.1098/rsif.2013.0676

Song, H.-L., Shim, S., Kim, D.-H., Won, S.-H., Joo, S., Kim, S., … Yoon, S.-Y. (2014). β-Amyloid is transmitted via neuronal connections along axonal membranes. *Annals of Neurology, 75*(1), 88–97. https://doi.org/10.1002/ana.24029

Squires, T. M., & Quake, S. R. (2005). Microfluidics: Fluid physics at the nanoliter scale. *Reviews of Modern Physics, 77*(3), 977–1026. https://doi.org/10.1103/RevModPhys.77.977

Stoothoff, W., Jones, P. B., Spires-Jones, T. L., Joyner, D., Chhabra, E., Bercury, K., … Hyman, B. T. (2009). Differential effect of three-repeat and four-repeat tau on mitochondrial axonal transport. *Journal of Neurochemistry, 111*(2), 417–427. https://doi.org/10.1111/j.1471-4159.2009.06316.x

Sweeney, M. D., Sagare, A. P., & Zlokovic, B. V. (2018). Blood–brain barrier breakdown in Alzheimer disease and other neurodegenerative disorders. *Nature Reviews Neurology, 14*(3), 133–150. https://doi.org/10.1038/nrneurol.2017.188

Szeto, A., McCabe, P. M., Nation, D. A., Tabak, B. A., Rossetti, M. A., McCullough, M. E., … Mendez, A. J. (2011). Evaluation of enzyme immunoassay and radioimmunoassay methods for the measurement of plasma oxytocin. *Psychosomatic Medicine, 73*(5), 393–400. https://doi.org/10.1097/PSY.0b013e31821df0c2

Tang-Schomer, M. D., Patel, A. R., Baas, P. W., & Smith, D. H. (2010). Mechanical breaking of microtubules in axons during dynamic stretch injury underlies delayed elasticity, microtubule disassembly, and axon degeneration. *Federation of American Societies for Experimental Biology Journal: Official Publication of the Federation of American Societies for Experimental Biology, 24*(5), 1401–1410. https://doi.org/10.1096/fj.09-142844

Tang, Yujie Tanye, Kim, J., López-Valdés, H. E., Brennan, K. C., & Ju, Y. S. (2011). Development and characterization of a microfluidic chamber incorporating fluid ports with active suction for localized chemical stimulation of brain slices. *Lab on a Chip, 11*(13), 2247–2254. https://doi.org/10.1039/C1LC20197B

Tang, Yujie T., Mendez, J. M., Theriot, J. J., Sawant, P. M., López-Valdés, H. E., Ju, Y. S., & Brennan, K. C. (2014). Minimum conditions for the induction of cortical spreading depression in brain slices. *Journal of Neurophysiology, 112*(10), 2572–2579. https://doi.org/10.1152/jn.00205.2014

Tau, G. Z., & Peterson, B. S. (2010). Normal development of brain circuits. *Neuropsychopharmacology, 35*(1), 147–168. https://doi.org/10.1038/npp.2009.115

Taylor, A. M., & Jeon, N. L. (2010). Micro-scale and microfluidic devices for neurobiology. *Current Opinion in Neurobiology, 20*(5), 640–647. https://doi.org/10.1016/j.conb.2010.07.011

Tehranirokh, M., Kouzani, A. Z., Francis, P. S., & Kanwar, J. R. (2013). Microfluidic devices for cell cultivation and proliferation. *Biomicrofluidics, 7*(5). https://doi.org/10.1063/1.4826935

Terrell-Hall, T. B., Ammer, A. G., Griffith, J. I. G., & Lockman, P. R. (2017). Permeability across a novel microfluidic blood-tumor barrier model. *Fluids and Barriers of the CNS, 14*. https://doi.org/10.1186/s12987-017-0050-9

Tourovskaia, A., Fauver, M., Kramer, G., Simonson, S., & Neumann, T. (2014). Brief communication: Tissue-engineered microenvironment systems for modeling human vasculature. *Experimental Biology and Medicine, 239*(9), 1264–1271. https://doi.org/10.1177/1535370214539228

Trevelyan, A. J., Baldeweg, T., Drongelen, W. van, Yuste, R., & Whittington, M. (2007). The source of afterdischarge activity in neocortical tonic–clonic epilepsy. *Journal of Neuroscience, 27*(49), 13513–13519. https://doi.org/10.1523/JNEUROSCI.3005-07.2007

Uzel, S. G. M., Platt, R. J., Subramanian, V., Pearl, T. M., Rowlands, C. J., Chan, V., … Kamm, R. D. (2016). Microfluidic device for the formation of optically excitable, three-dimensional, compartmentalized motor units. *Science Advances, 2*(8). https://doi.org/10.1126/sciadv.1501429

Virumbrales-Muñoz, M., Ayuso, J. M., Lacueva, A., Randelovic, T., Livingston, M. K., Beebe, D. J., … Ochoa, I. (2019). Enabling cell recovery from 3D cell culture microfluidic devices for tumour microenvironment biomarker profiling. *Scientific Reports, 9*(1), 6199. https://doi.org/10.1038/s41598-019-42529-8

Vogt, A.k., Brewer, G.j., & Offenhäusser, A. (2005). Connectivity patterns in neuronal networks of experimentally defined geometry. *Tissue Engineering, 11*(11–12), 1757–1767. https://doi.org/10.1089/ten.2005.11.1757

Wevers, N. R., Kasi, D. G., Gray, T., Wilschut, K. J., Smith, B., van Vught, R., … Obermeier, B. (2018). A perfused human blood–brain barrier on-a-chip for high-throughput assessment of barrier function and antibody transport. *Fluids and Barriers of the CNS, 15*. https://doi.org/10.1186/s12987-018-0108-3

Wheeler, B. C., & Brewer, G. J. (2010). Designing neural networks in culture. *Proceedings of the IEEE. Institute of Electrical and Electronics Engineers, 98*(3), 398–406. https://doi.org/10.1109/JPROC.2009.2039029

Whitesides, G. M. (2006). The origins and the future of microfluidics. *Nature, 442*(7101), 368–373. https://doi.org/10.1038/nature05058

Xu, Z., Li, E., Guo, Z., Yu, R., Hao, H., Xu, Y., … Wang, Q. (2016). Design and construction of a multi-organ microfluidic chip mimicking the in vivo microenvironment of lung cancer metastasis. *ACS Applied Materials & Interfaces, 8*(39), 25840–25847. https://doi.org/10.1021/acsami.6b08746

Yoo, S. J., Kim, J., Lee, C.-S., & Nam, Y. (2011). Simple and novel three dimensional neuronal cell culture using a micro mesh scaffold. *Experimental Neurobiology, 20*(2), 110–115. https://doi.org/10.5607/en.2011.20.2.110

Zhang, C., & Xing, D. (2007). Miniaturized PCR chips for nucleic acid amplification and analysis: Latest advances and future trends. *Nucleic Acids Research, 35*(13), 4223–4237. https://doi.org/10.1093/nar/gkm389

Chapter 45

Disease models in neurodevelopmental disorders

Pietro Fazzari[1], Angela Rodríguez-Prieto[1] and Máximo Ibo Galindo[2,3]

[1]*Laboratory of Cortical Circuits, Centro de Investigacion Príncipe Felipe, Valencia, Spain;* [2]*Instituto Interuniversitario de Investigacion de Reconocimiento Molecular y Desarrollo Tecnologico (IDM), Universitat Politecnica de Valencia, Universitat de Valencia, Valencia, Spain;* [3]*UPV-CIPF Joint Research Unit "Disease Mechanisms and Nanomedicine", Centro de Investigacion Príncipe Felipe, Valencia, Spain*

List of abbreviations

CNS central nervous system
CX cerebral cortex
EGF epidermal growth factor
GABA γ-aminobutyric acid
GE ganglionic eminence
GF giant fiber
NRG1 human neuregulin 1
Nrg1 mouse neuregulin 1
PNS peripheral nervous system
PV parvalbumin
RELN reelin
RGC radial glia cells
RTK receptor tyrosine kinase
SZ schizophrenia

Mini-dictionary of terms

Central nervous system: Part of the nervous system of all bilateral animals that receives information gathered from the environment or the body and processes it.

Cortical interneurons: A class on neurons that emerge from the ganglionic eminence in the subcortical telencephalon and are crucial for the formation and function of neural circuits, enabling communication. In the cortex γ-aminobutyric acid (GABA) inhibitory interneurons control locally the output of pyramidal neurons.

Mitochondrial dynamics: Changes in the mitochondrial network involving mainly fusion and fission (division) and that are required for mitochondrial function and integrity.

Peripheral nervous system: Part of the nervous system that gathers the information to be transmitted to the central nervous system and receives orders from it to trigger actions.

***phiC31* phage:** Virus that infects *Streptomyces* bacteria, whose sequences for integration in the host genome have been adapted for genome editing in *Drosophila* and other animals.

Pyramidal neurons: The main type of cortical neurons, the brain cortex (80% of the total neurons). Pyramidal neurons are glutamatergic cells that transfer information across cortical regions.

Radial glial cells: Bipolar-shaped cells that extend a long fiber onto the cortical wall. RGCs divide either symmetrically to expand the pool of new neuronal progenitors or asymmetrically to generate neurons.

Schizophrenia: A neurodevelopmental disorder with a strong genetic component that interferes with neuronal circuits required for cognition and normal social behavior. Is characterized by psychotic symptoms, emotional distress, and cognitive deficits.

Transposable element: Short DNA sequence that can move within the genome, similar to an integrative virus, but it never leaves the nucleus.

Diagnosis, Management and Modeling of Neurodevelopmental Disorders. https://doi.org/10.1016/B978-0-12-817988-8.00045-2

Introduction

Biomedical research relies heavily on disease models in order to advance in unraveling the mechanisms of disease and in the search for clinical developments; in neurobiology, scientists have used a variety of vertebrate and invertebrate models. If we focus on the study of disease models of neurodevelopmental defects, among the range of model organisms we need to choose those that are most relevant, economically viable, and experimentally accessible. Among vertebrate models, mice and rats are by far the most common. In comparison to other vertebrates, they are smaller and cheaper and have a shorter life cycle, which means they also age faster. Compared to invertebrate models, they are more similar to humans, mainly at the anatomical and physiological levels. Still, invertebrate models have been gaining ground in the last decades, especially in those pathologies that involved conserved genes/proteins and cellular mechanisms. The two most common invertebrates are the fruit fly *Drosophila melanogaster* and the nematode work *Caenorhabditis elegans*. Different kinds of models are complementary, each one displaying specific and useful features. In this chapter, we will use examples of *Drosophila* and mouse models of neurodevelopmental diseases involving synaptic function in the central and peripheral nervous systems (CNS and PNS respectively) to illustrate the range of experimental approaches at our disposal.

Features of the *Drosophila* nervous system

Like all animals, *Drosophila* has a CNS for information processing and a PNS for sensing and controlling movement (Fig. 45.1) (Hartenstein, 1993). A main difference is that invertebrates do not produce myelin; therefore, invertebrate models are not appropriate to study pathologies primarily caused by demyelination. The most anterior part of the CNS is the brain, with two large optic lobes and a central brain with different ganglions and neuropils. Posterior to the brain is the thoracico-abdominal ganglion, formed by three thoracic neuromeres, which are connected by the PNS to the structures of each one of the three thoracic segments, and a much reduced group of abdominal neuromeres fused in a single structure attached to the third thoracic neuromere. As an example of the organization of the PNS, we will focus on the thorax. Each one of the three thoracic segments has a pair of legs, and the PNS neurons connect with or stem from the corresponding thoracic neuromere. The axons of sensory neurons present all along the leg gather in a single leg nerve that connects to the paired thoracic neuromeres. These neuromeres also house the soma of the motor neurons that extend their axons through the four most proximal leg segments, which contain intrinsic muscles: coxa, trochanter, femur, and tibia. Anatomically, the wings are part of the second thoracic segment; therefore, the two tracts that collect the sensory axons run along the first (wing margin) and third wing veins to reach the second thoracic neuromere. The muscles that drive wing beating are not intrinsic to this organ, instead they attach to the thoracic cuticle, and it is the change of shape of this box that drives the beating; hence the name of indirect flight muscles. The motor neurons that innervate the flight muscles also have their somas at the second thoracic neuromere.

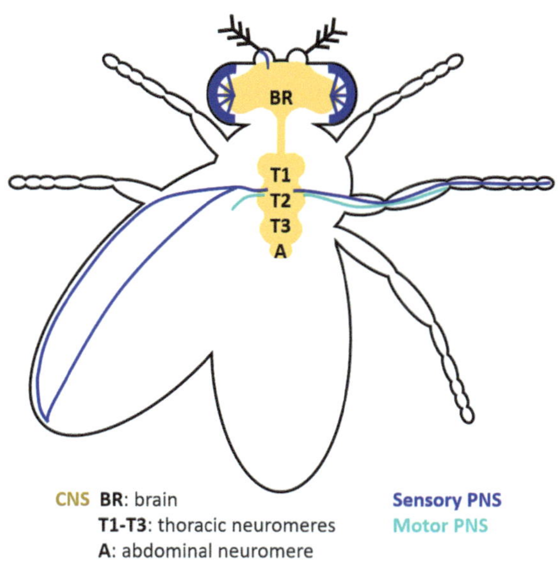

FIGURE 45.1 **Outline of the *Drosophila* nervous system.** The *Drosophila* central nervous system comprises the brain inside the head capsule and the thoracico-abdominal ganglion in the thorax. The image depicts selected peripheral nerves, sensory nerves from the eye and antenna in the head, and the main sensory and motor nerves in the second thoracic segment, corresponding to the second pair of legs and the wings.

The genetic toolkit and resources in *Drosophila*

The *Drosophila* genetic toolkit includes numerous techniques to alter gene function. Transposable element, mainly the *P element*, has been used for mutagenesis, enhancer trapping and as a vector for transgenesis (Ryder & Russell, 2003). *P-element*-mediated generation of transgenics is highly efficient and economical, but in addition to this classic random insertion transgenesis, it is possible to generate site-directed insertions by the use of sequences derived from the *phiC31* phage, thus avoiding insertional mutations and position effects (Groth, Fish, Nusse, & Calos, 2004). The Gal4/UAS system is used for directed expression of transgenes (Brand & Perrimon, 1993). There are hundreds of Gal4 drivers, which express the *Saccharomyces* Gal4 transcription factor under specific promoters, allowing tightly regulated control of expression. Regarding our field of interest, there are, for example, Gal4 lines specific of neurons (*elav-Gal4*), glia (*repo-Gal4*), even specific subpopulations such as motor neurons (*D42-Gal4*) or dopaminergic neurons (*TH-Gal4*). We can combine each Gal4 driver with a UAS construct, which contains the Gal4-responsive UAS promoter upstream of the sequence we want to express, usually a whole gene (wild type, mutant, human, tagged) or an RNAi against a gene we want to downregulate. Finally, genome editing techniques are quite sophisticated, from the widely used CRISPR/Cas9 to *Drosophila*-specific techniques that achieve higher efficiency and complexity (Baena-Lopez, Alexandre, Mitchell, Pasakarnis, & Vincent, 2013).

Drosophila models of Charcot–Marie–Tooth peripheral neuropathy

Charcot–Marie–Tooth (CMT) disease is a rare inherited peripheral neuropathy that can affect both motor and sensory neurons (Barisic et al., 2008). It is a neurodegenerative disease that affects primarily the longest nerves in the body, it usually manifests first in the legs and later on in the arms. The most disabling feature is the muscular degeneration, which results in muscle wasting, movement disability, and foot and hand deformities. It is a very heterogeneous condition regarding age of onset, severity, clinical manifestation, and mode of inheritance (there are dominant, recessive, and X-linked forms). This heterogeneity is mostly due to the underlying genetic heterogeneity, to date mutations in nearly 100 different genes have been shown to cause some form of CMT (Pareyson, Saveri, & Pisciotta, 2017). Some forms of CMT are primarily demyelinating; therefore, due to the fact that invertebrate organisms do not form a myelin sheath around the peripheral nerve axons, these forms cannot be successfully modeled in *Drosophila*. This is not the case for axonal forms of CMT, since axonal function and preservation, and the genes required for them, are well conserved. The genes involved in axonal CMT encode proteins involved in a wide variety of cellular events that are required for the function and survival of the neuron. Further, we give examples of CMT models developed in *Drosophila* involving genes that affect distinct cellular functions.

Aminoacyl-tRNA synthetase genes

To date, six different aminoacyl-tRNA synthetase (aaRS) genes harbor dominant mutations (Wei, Zhang, & Yang, 2019). Although there is still some controversy regarding the mechanism of disease, at least some mutations do not affect the catalytic function, and it is possible that in addition to their primary role as synthetases, they can participate in cell signaling. The first gene of this family to be modeled in *Drosophila* was the tyrosyl-tRNA synthetase (*YARS*) (Storkebaum et al., 2009). Three different clinical mutations were introduced in both human *YARS* and *Drosophila dYARS* cDNAs and expressed in flies by the Gal4/UAS method explained earlier, either ubiquitously or specifically in neurons. In all cases, motor neuron impairment was evident in a negative geotaxis assay (Fig. 45.2A). Negative geotaxis is the tendency of the flies to climb back to the top of the vial after being knocked down to the bottom with a vigorous tap. The experiment is recorded on video, and this is then analyzed to determine parameters such as average time to climb or number of flies reaching a predetermined distance in a time period. Flies expressing the mutant alleles performed much worse than those expressing the human or fly wild-type alleles. In addition to this motor defect, neuronal dysfunction was evident in electrophysiology of the giant fiber (GF) system (Fig. 45.2B). The GF system mediates the reflex escape response: when a possible threat is detected by the visual system, the GF system triggers activation of the tergotrochanteral motor neuron (TTMn) and the tergotrochanteral muscle (TTM) in the leg and the indirect flight muscles (IFM) to jump and fly away. Experimentally, this reaction can be mimicked by electrode stimulation of the brain and measuring the evoked potential by inserting a recording electrode in the muscle. In flies expressing mutant *YARS* or *dYARS*, the muscle response was absent or abnormal. Similar results were obtained upon expression of two dominant mutations in conserved residues of the *Drosophila* glycyl-tRNA synthetase (*dGARS*), in both negative geotaxis assay and GF electrophysiology (Ermanoska et al., 2014). Interestingly, the authors show that mutations in *GARS* and *YARS* share genetic modifiers, which indicates that they also have a common disease mechanism.

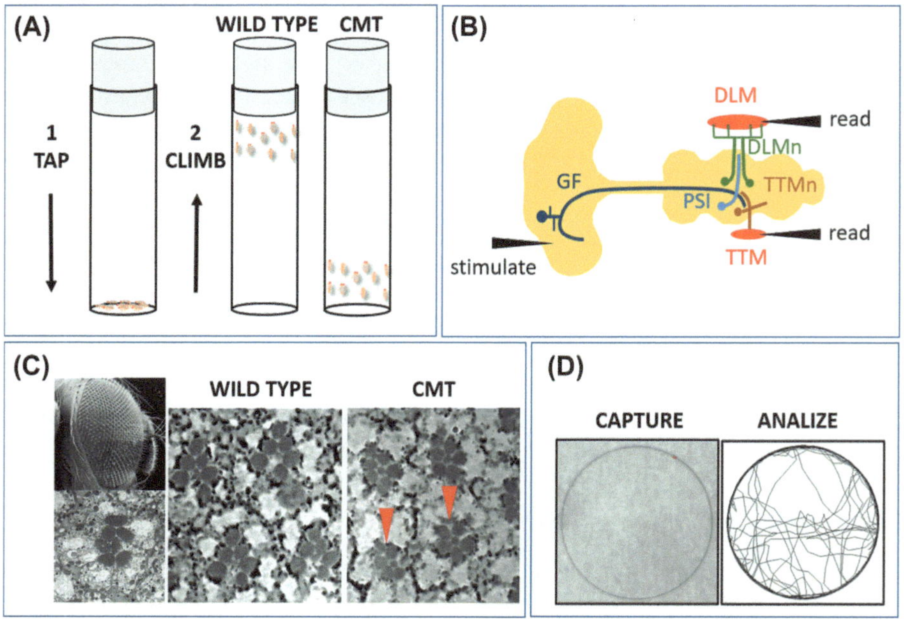

FIGURE 45.2 **Techniques used to characterize neurodegeneration in *Drosophila*.** (A) Negative geotaxis assay, the flies are tapped down to the bottom and filmed to evaluate their climbing ability. (B) Giant fiber system electrophysiology, the stimulating electrode in the brain activates the giant fiber (GF) neuron, which connects with the tergotrochanteral motor neuron (TMNn) to activate the tergotrochanteral muscle (TTM), and the dorsolateral motor neurons (DLMn) via the peripherally synapsing interneuron (PSI) to activate the dorsolateral muscles (DLM). (C) Neurodegeneration can be quantitated by counting the retinal neuron rhabdomeres. In a cross section of the retina, each ommatidium displays seven rhabdomeres, one per neuron; in neurodegeneration there is a time-dependent loss of neurons. (D) To evaluate locomotion activity of a larva or an adult, its movements are recorded in video for a period of time, usually 5–10 min, each photogram is thresholded to identify the individual, and the total trajectory is analyzed to extract locomotion parameters.

Mitochondrial dynamics genes

Mitochondrial dynamics, reshaping of the mitochondrial network due to events of fusion and fission, is intimately related to many neurodegenerative diseases (Itoh, Nakamura, Iijima, & Sesaki, 2013). Mutations in two mitochondrial dynamics genes cause axonal CMT, *Mitofusin-2* (*MFN2*), and *ganglioside-induced differentiation-associated protein 1* (*GDAP1*) (Baxter et al., 2002; Cuesta et al., 2002; Zuchner et al., 2004). MFN2 is involved in mitochondrial fusion, while GDAP1 causes mitochondrial fragmentation, suggesting that the displacement of mitochondrial dynamics in either direction can have similar phenotypes, at least regarding neuronal survival. These functions are represented in *Drosophila* by the *Marf* and *Gdap1* genes, respectively. Expression of mutant human MFN2 in the *Drosophila* eye causes retinal degeneration and dilated cardiomyopathy when expressed in cardiomyocytes (Eschenbacher et al., 2012). Neuron-specific expression of selected clinical mutations had distinct phenotypes, while two of the mutations induce aggregation, the other two produce an accumulation of smaller mitochondria (El Fissi et al., 2018). Thus, it seems that both dominant negative and gain of function of MFN2 can cause CMT. Down- and upregulation of *Drosophila Gdap1* with the Gal4/UAS technique display the expected mitochondrial phenotype, aggregation and fragmentation of the mitochondrial, respectively (Lopez Del Amo et al., 2015). In both conditions, there is an age-dependent neuronal and muscular degeneration, again pointing to the possibility that opposite effects on the mitochondrial network can have similar phenotypes. Neuronal degeneration is normally assessed in the retina, a neural tissue dispensable for survival (Fig. 45.2C). In sections of the retina, each ommatidium of the composite eye contains seven photoreceptor neurons, which can be easily identified by the light-collecting and electron-dense rhabdomere. Neuronal death results in the disorganization of the rhabdomeres, so by counting the average number of rhabdomeres in a field of tens of ommatidia, we can perform accurate statistics of the neurodegeneration. Further work with up- and downregulation of *Gdap1*, followed by nuclear magnetic resonance metabolomics, revealed that there is a dysregulation of the energetic metabolism and the insulin signaling pathway (Lopez Del Amo et al., 2017). This observation is interesting due to the long-known relationship of human and mouse mitofusins with energy, insulin, and obesity (Zorzano, Hernandez-Alvarez, Sebastian, & Munoz, 2015).

Endolysosomal pathway

CMT can also be caused by mutations in genes encoding proteins involved in the endolysosomal pathway, such as the *Factor-Induced-Gene 4* (*FIG4*) and the small GTPase Rab7 (Chow et al., 2007; Verhoeven et al., 2003). FIG4, proteins, including the *Drosophila* dFig4, participate in the synthesis of phosphatidylinositol phosphates required in the early and late endosomes. *Drosophila dFig4* was downregulated in the nervous system using the Gal4/UAS technique to drive expression of two different RNAi constructs (Kyotani et al., 2016). Pan-neuronal expression of these RNAis against *dFig4* results in reduced life span and negative geotaxis of adult flies and in shortened branches in the larval neuromuscular synapse. *dFig4* RNAi in the postmitotic neurons of the retina prevents the targeting of the axons to the neuropiles in the CNS. Expression of *dFig4* carrying clinical mutations in conserved residues is able to partially rescue the phenotype in a null *dFig4* background, confirming the hypothesis that these mutations are hypomorphic (Bharadwaj, Cunningham, Zhang, & Lloyd, 2016). Modeling of the *Rab7* mutations in *Drosophila* has yielded contradictory results. Mutations in *Rab7* causing CMT are dominant, so a dominant negative effect has been proposed. On the one hand, replacement of *Drosophila Rab7* with clinical mutations partially restored the mutant phenotype, pointing to a haploinsufficiency rather than a dominant negative effect (Cherry et al., 2013). In contrast, overexpression of a different CMT mutation in Rab7 in motor neurons in a wild-type background reduced larval mobility, but not overexpression of the wild-type human or *Drosophila* proteins (Janssens et al., 2014). To test mobility, individual larvae are placed in an arena, filmed for a period of time, and the footage is analyzed by specialist software to extract relevant parameters such as distance, average speed, or central exploration (Fig. 45.2D). These mutant flies also displayed abnormal vesicle transport along the axons.

Mouse as a model for neurodevelopmental disorders: from genes to cells, circuits, and behavior

Neurodevelopmental disorders are a heterogeneous group of pathologies that affect neuronal development and result in abnormal brain function. These disorders include rare pathologies and relatively common disorders such as schizophrenia (SZ), autism spectrum disorders, and attention deficit hyperactivity disorder. In general, these different pathologies have in common a high heritability (60%–80%). The study of neuronal disorders requires experimental models that are amenable to genetic, molecular, cellular, and behavioral analysis. The mouse is a very potent model since the current technologies allow to introduce in the mouse genome disease-linked mutations and to investigate the behavioral and cellular alterations that are relevant for the disorders. Therefore, mice became a common mammalian model in neurobiology to investigate the functional relevance of genetic and cellular alterations of neuronal development. The use of mice as an experimental model was particularly driven by advances of developmental neurobiology that allowed a better understanding of the basic biology of neuronal development in mice. Notably, this progress was accompanied by the generation of novel genetic tools and techniques that allow us to dissect out the specific role of different cell types and circuits in the development and function of the brain.

Here, we will first discuss the use of a general mice model for neurodevelopmental disorders. Next, we will review the paradigmatic example of Neuregulin (NRG1) mutant mice as a model for SZ.

Brain development, a stepwise process

It was proposed that multiple genetic and molecular alterations may contribute in different ways to impair one or more facets of neuronal development. Thus, to understand how these genetic lesions may result in functional deficits, we need to understand the complexity of the development of neuronal circuits (Desikan & Barkovich, 2016; Marin, 2016; Washbourne, 2015).

The brain is by far the most complex structure of the body. It arises from a long stepwise process that begins with the differentiation and proliferation of neuronal progenitors in early embryonic development and finishes well after birth with the synaptic remodeling that characterizes the adolescence. Although the pathological symptoms may appear years after birth, the embryonic development and the early postnatal period are critical periods of vulnerability. Genetic and/or environmental stressors during these early stages affect the assembly of neural circuits. Alterations of cortical circuits and of basal ganglia were consistently involved in psychiatric disorders. The early stages of development involve (i) the proliferation and differentiation of neuronal progenitors, (ii) the migration of neurons from the site of origin to their final destination, and (iii) the elongation and guidance of the axons to their putative target (Desikan & Barkovich, 2016). The assembly of neural circuits is completed by the formation of synaptic connections. Next, synapses are remodeled and refined according to the specific developmental program of the different neurons. Neuronal activity is particularly important in this process that determines a continuous refinement of neuronal circuits from late embryonic development throughout childhood until late adolescence (Marin, 2016; Washbourne, 2015). Here, we will primarily focus on the development of the cortex given the importance of this structure in cognitive functions.

Major milestones in cortical development

The proliferation of cortical progenitors. Most cortical neurons arise from undifferentiated progenitors named radial glia cells (RGC) (Fig. 45.3). Mutations in genes that affect the balance between proliferation and differentiation in this process alter the number of neurons that are generated and result in cognitive deficits. For instance, mutations in genes that impair proliferation such as *MCPH1* or *CDK5RAP2* cause microcephaly (Buchman et al., 2010; Desikan & Barkovich, 2016). Conversely, gene mutations that increase proliferation such as in *PIK3CA* provoke megalencephaly (Roy et al., 2015).

Neuronal migration. The migration of neurons in the cortex occurs in two modalities. Principal excitatory neurons are generated close to the ventricular zone and migrate radially toward the outer cortex. This migration follows the guide provided by the radial fibers of the RGCs (Fig. 45.3). Conversely, inhibitory interneurons originate in the ganglionic eminence and migrate tangentially into the cortex (Fig. 45.4). Neuronal migration can be disrupted either by mutations that affect the intrinsic motility of the neurons or by mutations that impair the capacity of neurons to respond to guidance cues that regulate the movement toward and inside the cortex. Mutations in the microtubule-associated protein *DCX* and in the *Reelin* (*RELN*) gene are well-characterized mouse models for human lissencephalies that show alterations of brain development (Desikan & Barkovich, 2016).

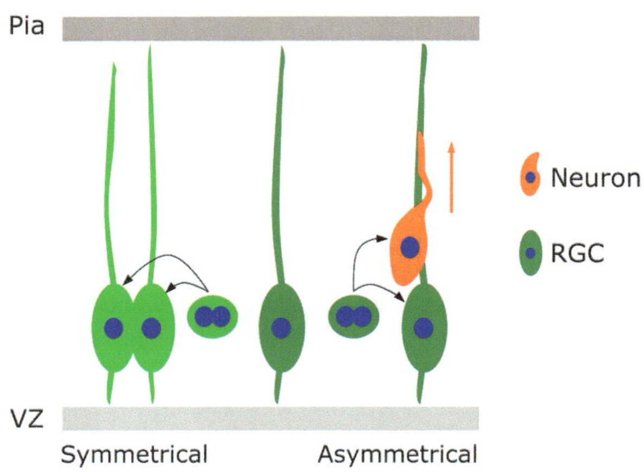

FIGURE 45.3 **Proliferation and radial migration of glutamatergic neurons.** RGCs (green) of the developing neocortex undergo cell cycle. Symmetric and asymmetric cell divisions are schematized in the cartoon. Neurons migrate outward. *Pia*, pia mater; *VZ*, ventricular zone.

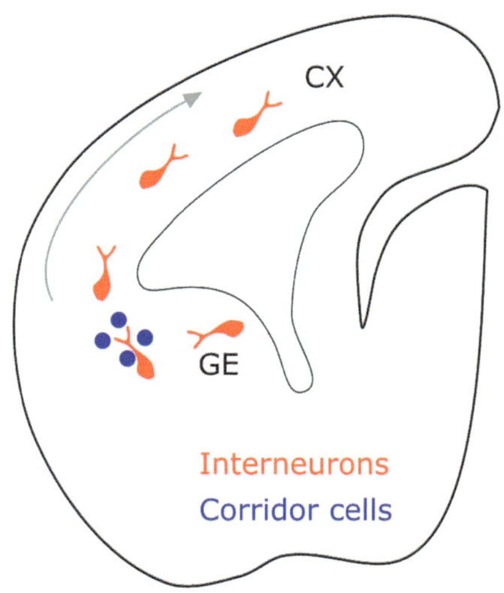

FIGURE 45.4 **Tangential migration of GABA-ergic interneurons.** Inhibitory interneurons originate in the ganglionic eminence and migrate tangentially into the cortex, moving through the corridor cells expressing NRG1. *CX*, cerebral cortex; *GE*, ganglionic eminence.

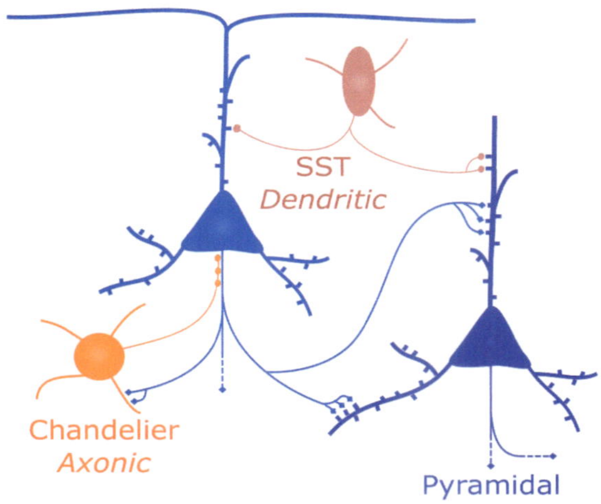

FIGURE 45.5 Circuitry and Nrg1/ErbB4 control of the excitatory input and inhibitory output. Schematic representation of cortical circuits. The output of pyramidal neurons (blue) in the cortex is regulated by excitatory glutamatergic and inhibitory GABA-ergic interneurons, such as somatostatin-expressing neurons (light red) and chandelier cells (orange).

Axon guidance and connections. The assembly of neural circuits relies on the precise navigation of axons to their specific target. This process is particularly delicate since the axonal targets may be very distant from the soma and the guidance depends on the appropriate response to multiple molecular cues (Engle, 2010). L1 syndrome is a paradigmatic example of a neurodevelopmental disorder that involves guidance errors. L1 loss of function in mice mimics different aspects of human pathologies including defects in the corpus callosum, the main nerve tract that connects cortical hemispheres (Dahme et al., 1997).

Synapse formation and remodeling. This final milestone of neuronal wiring is arguably the most complex to study and less understood for different reasons. The timeline of synaptogenesis spans from midembryonic development (~25 gestation weeks in human) to young adulthood, and it is in itself a complex and dynamic stepwise process of assembly, pruning, and refinement (Fig. 45.5). For some pathologies such as SZ or bipolar disorders, the age of onset is after the adolescence, making it difficult to establish a sharp causal link with the developmental alterations (Marin, 2016). Moreover, the structural alterations of synaptic circuits are subtle and their analysis requires very refined methods such as molecular labeling, confocal and electron microscopy, and electrophysiological experiments. Nonetheless, the last decades showed amazing progress in the genetic modeling of human pathologies in mice with the generation of different mouse models for neurodevelopmental disorders such as autism, bipolar disorder, and SZ. Mutant mice for the synaptogenic cues neuroligins, neurexins, and SynCAMs are among the most characterized models (Washbourne, 2015). Hereafter, we will focus on the SZ-risk gene *NRG1*.

NRG1, a major SZ-risk gene

SZ, which affects about 1% of the global population, has a genetic component estimated at 60%–80%. Genome-wide association studies identified *NRG1* as a major SZ-risk gene. Consistently, *Nrg1* mutant mice displayed phenotypes that are relevant for the disease and shed light on the possible role of NRG1 in SZ (Stefansson et al., 2002).

The NRG1 protein is a trophic factor belonging to family of epithelial growth factor (EGF). NRG1 activates ErbB4, a receptor tyrosine kinase (RTK), promoting its dimerization (Chen, Hancock, Role, & Talmage, 2010; Mei & Xiong, 2008). In canonical or forward NRG1-ErbB4 signaling, the activation of the tyrosine kinase domain results in the activation the MAPK/ERK and PI3K pathways. Concurrently, Nrg1/ErbB4 binding drives the noncanonical signaling, also known as intracellular signaling. Briefly, γ-secretase processing of NRG1 releases the intracellular domain of NRG1 that relocates to the nucleus and regulates gene expression (Mei & Xiong, 2008). This intracellular pathway is still largely unknown. However, some studies suggested that the Aph1bc subunit of γ-secretase might have an essential role in this path (Fazzari et al., 2014).

By means of the two mentioned pathways, NRG1 regulates different aspects of neuronal development including neuronal migration, excitatory–inhibitory homeostasis and neurite extension in cortical circuits (Chen et al., 2010; Stefansson et al., 2002).

Roles of NRG1 in neural development: migration

Cortical interneurons emerge from the ganglionic eminence in the subcortical telencephalon (Fig. 45.4). The molecules that control their migration to the cortex are poorly characterized. Nonetheless, studies suggested that NRG1 and ErbB4 are involved in this migratory process.

Interneurons expressing ErbB4 migrate toward the cortex, moving through a corridor expressing a membrane-bound form of NRG1. Moreover, the developing cortex expresses a diffusible isoform of NRG1 acting in vitro as an attractant for migrating interneurons (Flames et al., 2004; López-Bendito et al., 2006). As a result, perturbing ErbB4 function decreases the number of interneurons that migrate to the cortex in vivo (Flames et al., 2004). Notably, in the prefrontal cortex on schizophrenic patients, the number of PV + interneurons that originate in the GE is reduced. Consistently the relevant balance of excitation and inhibition (E/I) is impaired (Ferguson & Gao, 2018; Flames et al., 2004).

Roles of NRG1 in neural development: circuitry

Nrg1/ErbB4 controls the excitatory input and inhibitory output in fast-spiking parvalbumin-expressing (PV+) cells, such as chandelier and basket interneurons. Researches over the past years point to excitatory/inhibitory affected circuitries in the cortex as a risk factor contributing to SZ. The anomalies observed in NRG1/ErB4 mutants support this hypothesis (Fazzari et al., 2010, 2014; Mei & Nave, 2014).

Chandelier and basket cells have a peculiar morphology and uniquely innervate the axon initial segment and soma, respectively, of pyramidal neurons. As a consequence, they have potent control over the cortical network (Cruz, Weaver, Lovallo, Melchitzky, & Lewis, 2009; Tai, Janas, Wang, & Van Aelst, 2014). Collectively, the data indicates that ErbB4 regulates the excitatory input received by interneurons (Fazzari et al., 2010). Moreover, emerging studies indicate how ErbB4 is necessary for the normal integration of basket cells. As a result, conditional ErbB4 mutants show a reduction in the number of inhibitory synapses, so the inhibitory drive they apply on pyramidal cells is also diminished (Del Pino et al., 2017; Fazzari et al., 2010). These functional deficits are linked to increased locomotor activity, reduced levels of anxiety in aversive environment, aberrant social behavior and social withdrawal, poor planning behavior, and spatial learning and working memory dysfunctions (Del Pino et al., 2013).

Moreover, Nrg1 is necessary for excitatory transmission between pyramidal neurons (Fazzari et al., 2014). Using processing-defective mutants of NRG1 Type III, it is demonstrated that the intracellular domain signaling γ-secretase-dependent is specifically required for NRG1 regulation of the growth and branching of dendritic arbors in pyramidal neurons. Also the glutamatergic circuitry is impaired in single-cell Aph1bc-γ-secretase loss of function. In addition, NRG1 intracellular signaling enhances dendritic spine formation. In Aph1BC−/− mice, NRG1-ICD does not translocate to the nucleus and the process of spine formation is disrupted. As a consequence, Aph1BC−/− mutant mice have deficits in synaptic transmission, long-term synaptic plasticity and show behavioral phenotypes SZ-like, such as impaired working memory (Dejaegere et al., 2008; Fazzari et al., 2014).

Applications to other areas of development

In this chapter we describe two representative animal models in neurobiology, the insect *Drosophila melanogaster* and the mammal *Mus musculus*. For each one we explain their main experimental features and basic concepts in nervous system development and function. Although we concentrate on particular examples, modeling the CMT pathophysiology in *Drosophila* and deciphering the roles of mouse Nrg1 in cerebral cortex development and disease, very similar approaches can be employed to study other neurodevelopmental issues. The same techniques we describe in *Drosophila* and mouse have been used to study the disease mechanisms in other pathologies such as neurodegenerative diseases (Alzheimer, Parkinson, and Huntington diseases), epilepsy, behavioral disorders such as autism, and neural tumors.

Key facts of disease models in neurodevelopmental disorders

- Invertebrate animals have central and peripheral nervous systems, but their neurons do not have a myelin sheath.
- Invertebrate models have been used for decades in neurobiological research to understand basic mechanisms such as synapse formation, maintenance, and function.
- The main invertebrate species in classical neurobiology studies are squid, snail, leech, fruit fly, locust, and nematode worm.

- Schizophrenia symptoms are grouped as positive, negative, and cognitive. From a behavioral point of view, the positive symptoms such as hallucinations or delusions are arduous to reproduce in animals models.
- The negative symptoms of schizophrenia such as depression are often present in other neurodevelopmental disorders.
- The cognitive symptoms, such as working memory and sensory-motor gating, are modeled robustly with well-defined mice behavior test.

Summary points

- Invertebrate and vertebrate disease models offer complementary features in the research of disease mechanisms.
- *Drosophila melanogaster* shares nervous system cell types and molecular pathways with vertebrate animals.
- *Drosophila melanogaster* can be used to make relevant models of CMT disease caused by alterations in aminoacyl-tRNA transferases, mitochondrial dynamics, and the endolysosomal pathway.
- The mouse is a potent model for neurodevelopmental disorders.
- Nrg1 mutant mice are a paradigmatic model of schizophrenia.

References

Baena-Lopez, L. A., Alexandre, C., Mitchell, A., Pasakarnis, L., & Vincent, J. P. (2013). Accelerated homologous recombination and subsequent genome modification in *Drosophila. Development, 140*(23), 4818–4825.

Barisic, N., Claeys, K. G., Sirotkovic-Skerlev, M., Lofgren, A., Nelis, E., De Jonghe, P., & Timmerman, V. (2008). Charcot-Marie-Tooth disease: A clinico-genetic confrontation. *Annals of Human Genetics, 72*(Pt 3), 416–441.

Baxter, R. V., Ben Othmane, K., Rochelle, J. M., Stajich, J. E., Hulette, C., Dew-Knight, S., … Vance, J. M. (2002). Ganglioside-induced differentiation-associated protein-1 is mutant in Charcot-Marie-Tooth disease type 4A/8q21. *Nature Genetics, 30*(1), 21–22.

Bharadwaj, R., Cunningham, K. M., Zhang, K., & Lloyd, T. E. (2016). FIG4 regulates lysosome membrane homeostasis independent of phosphatase function. *Human Molecular Genetics, 25*(4), 681–692.

Brand, A. H., & Perrimon, N. (1993). Targeted gene expression as a means of altering cell fates and generating dominant phenotypes. *Development, 118*(2), 401–415.

Buchman, J. J., Tseng, H.-C., Zhou, Y., Frank, C. L., Xie, Z., & Tsai, L.-H. (2010). Cdk5rap2 interacts with pericentrin to maintain the neural progenitor pool in the developing neocortex. *Neuron, 66*(3), 386–402.

Chen, Y., Hancock, M. L., Role, L. W., & Talmage, D. A. (2010). Intramembranous valine linked to schizophrenia is required for neuregulin 1 regulation of the morphological development of cortical neurons. *Journal of Neuroscience: The Official Journal of the Society for Neuroscience, 30*(27), 9199–9208.

Cherry, S., Jin, E. J., Ozel, M. N., Lu, Z., Agi, E., Wang, D., … Hiesinger, P. R. (2013). Charcot-Marie-Tooth 2B mutations in rab7 cause dosage-dependent neurodegeneration due to partial loss of function. *eLife, 2*. e01064.

Chow, C. Y., Zhang, Y., Dowling, J. J., Jin, N., Adamska, M., Shiga, K., … Meisler, M. H. (2007). Mutation of FIG4 causes neurodegeneration in the pale tremor mouse and patients with CMT4J. *Nature, 448*(7149), 68–72.

Cruz, D. A., Weaver, C. L., Lovallo, E. M., Melchitzky, D. S., & Lewis, D. A. (2009). Selective alterations in postsynaptic markers of chandelier cell inputs to cortical pyramidal neurons in subjects with schizophrenia. *Neuropsychopharmacology, 34*(9), 2112–2124.

Cuesta, A., Pedrola, L., Sevilla, T., Garcia-Planells, J., Chumillas, M. J., Mayordomo, F., … Palau, F. (2002). The gene encoding ganglioside-induced differentiation-associated protein 1 is mutated in axonal Charcot-Marie-Tooth type 4A disease. *Nature Genetics, 30*(1), 22–25.

Dahme, M., Bartsch, U., Martini, R., Anliker, B., Schachner, M., & Mantei, N. (1997). Disruption of the mouse L1 gene leads to malformations of the nervous system. *Nature Genetics, 17*(3), 346–349.

Dejaegere, T., Serneels, L., Schäfer, M. K., Van Biervliet, J., Horré, K., Depboylu, C., & De Strooper, B. (2008). Deficiency of Aph1B/C-gamma-secretase disturbs Nrg1 cleavage and sensorimotor gating that can be reversed with antipsychotic treatment. *Proceedings of the National Academy of Sciences of the United States of America, 105*(28), 9775–9780.

Del Pino, I., Brotons-Mas, J. R., Marques-Smith, A., Marighetto, A., Frick, A., Marin, O., & Rico, B. (2017). Abnormal wiring of CCK$^+$ basket cells disrupts spatial information coding. *Nature Neuroscience, 20*(6), 784–792.

Del Pino, I., García-Frigola, C., Dehorter, N., Brotons-Mas, J. R., Alvarez-Salvado, E., Martínez de Lagrán, M., … Rico, B. (2013). Erbb4 deletion from fast-spiking interneurons causes schizophrenia-like phenotypes. *Neuron, 79*(6), 1152–1168.

Desikan, R. S., & Barkovich, A. J. (2016). Malformations of cortical development. *Annals of Neurology, 80*(6), 797–810.

El Fissi, N., Rojo, M., Aouane, A., Karatas, E., Poliacikova, G., David, C., … Rival, T. (2018). Mitofusin gain and loss of function drive pathogenesis in Drosophila models of CMT2A neuropathy. *EMBO Reports, 19*(8).

Engle, E. C. (2010). Human genetic disorders of axon guidance. *Cold Spring Harbor Perspectives in Biology, 2*(3), a001784.

Ermanoska, B., Motley, W. W., Leitao-Goncalves, R., Asselbergh, B., Lee, L. H., De Rijk, P., … Jordanova, A. (2014). CMT-associated mutations in glycyl- and tyrosyl-tRNA synthetases exhibit similar pattern of toxicity and share common genetic modifiers in Drosophila. *Neurobiology of Disease, 68C*, 180–189.

Eschenbacher, W. H., Song, M., Chen, Y., Bhandari, P., Zhao, P., Jowdy, C. C., … Dorn, G. W. (2012). 2nd Two rare human mitofusin 2 mutations alter mitochondrial dynamics and induce retinal and cardiac pathology in Drosophila. *PLoS One, 7*(9). e44296.

Fazzari, P., Paternain, A. V., Valiente, M., Pla, R., Luján, R., Lloyd, K., … Rico, B. (2010). Control of cortical GABA circuitry development by Nrg1 and ErbB4 signalling. *Nature, 464*, 1376.

Fazzari, P., Snellinx, A., Sabanov, V., Ahmed, T., Serneels, L., Gartner, A., … De Strooper, B. (2014). Cell autonomous regulation of hippocampal circuitry via Aph1b-gamma-secretase/neuregulin 1 signalling. *eLife, 3*. e02196.

Ferguson, B. R., & Gao, W.-J. (2018). PV interneurons: Critical regulators of E/I balance for prefrontal cortex-dependent behavior and psychiatric disorders. *Frontiers in Neural Circuits, 12*, 1–13.

Flames, N., Long, J. E., Garratt, A. N., Fischer, T. M., Gassmann, M., Birchmeier, C., … Marín, O. (2004). Short- and long-range attraction of cortical GABAergic interneurons by neuregulin-1. *Neuron, 44*(2), 251–261.

Groth, A. C., Fish, M., Nusse, R., & Calos, M. P. (2004). Construction of transgenic *Drosophila* by using the site-specific integrase from phage phiC31. *Genetics, 166*(4), 1775–1782.

Hartenstein, V. (1993). *Atlas of* Drosophila *development*. New York: Cold Spring Harbor Laboratory Press.

Itoh, K., Nakamura, K., Iijima, M., & Sesaki, H. (2013). Mitochondrial dynamics in neurodegeneration. *Trends in Cell Biology, 23*(2), 64–71.

Janssens, K., Goethals, S., Atkinson, D., Ermanoska, B., Fransen, E., Jordanova, A., … Timmerman, V. (2014). Human Rab7 mutation mimics features of Charcot-Marie-Tooth neuropathy type 2B in Drosophila. *Neurobiology of Disease, 65*, 211–219.

Kyotani, A., Azuma, Y., Yamamoto, I., Yoshida, H., Mizuta, I., Mizuno, T., … Yamaguchi, M. (2016). Knockdown of the Drosophila FIG4 induces deficient locomotive behavior, shortening of motor neuron, axonal targeting aberration, reduction of life span and defects in eye development. *Experimental Neurology, 277*, 86–95.

Lopez Del Amo, V., Palomino-Schatzlein, M., Seco-Cervera, M., Garcia-Gimenez, J. L., Pallardo, F. V., Pineda-Lucena, A., & Galindo, M. I. (2017). A *Drosophila* model of GDAP1 function reveals the involvement of insulin signalling in the mitochondria-dependent neuromuscular degeneration. *Biochimica et Biophysica Acta, 1863*(3), 801–809.

Lopez Del Amo, V., Seco-Cervera, M., Garcia-Gimenez, J. L., Whitworth, A. J., Pallardo, F. V., & Galindo, M. I. (2015). Mitochondrial defects and neuromuscular degeneration caused by altered expression of *Drosophila* Gdap1: Implications for the Charcot-Marie-Tooth neuropathy. *Human Molecular Genetics, 24*(1), 21–36.

López-Bendito, G., Cautinat, A., Sánchez, J. A., Bielle, F., Flames, N., Garratt, A. N., … Garel, S. (2006). Tangential neuronal migration controls axon guidance: a role for neuregulin-1 in thalamocortical axon navigation. *Cell, 125*(1), 127–142.

Marin, O. (2016). Developmental timing and critical windows for the treatment of psychiatric disorders. *Nature Medicine, 22*(11), 1229–1238.

Mei, L., & Nave, K. A. (2014). Neuregulin-ERBB signaling in the nervous system and neuropsychiatric diseases. *Neuron, 83*(1), 27–49.

Mei, L., & Xiong, W. C. (2008). Neuregulin 1 in neural development, synaptic plasticity and schizophrenia. *Nature Reviews Neuroscience, 9*, 437–452.

Pareyson, D., Saveri, P., & Pisciotta, C. (2017). New developments in Charcot-Marie-Tooth neuropathy and related diseases. *Current Opinion in Neurology, 30*(5), 471–480.

Roy, A., Skibo, J., Kalume, F., Ni, J., Rankin, S., Lu, Y., … Millen, K. J. (2015). Mouse models of human PIK3CA-related brain overgrowth have acutely treatable epilepsy. *eLife, 4*. e12703.

Ryder, E., & Russell, S. (2003). Transposable elements as tools for genomics and genetics in *Drosophila*. *Briefings in Functional Genomics and Proteomics, 2*(1), 57–71.

Stefansson, H., Sigurdsson, E., Steinthorsdottir, V., Bjornsdottir, S., Sigmundsson, T., Ghosh, S., … Stefansson, K. (2002). Neuregulin 1 and susceptibility to schizophrenia. *The American Journal of Human Genetics, 71*(4), 877–892.

Storkebaum, E., Leitao-Goncalves, R., Godenschwege, T., Nangle, L., Mejia, M., Bosmans, I., … Jordanova, A. (2009). Dominant mutations in the tyrosyl-tRNA synthetase gene recapitulate in Drosophila features of human Charcot-Marie-Tooth neuropathy. *Proceedings of the National Academy of Sciences of the United States of America, 106*(28), 11782–11787.

Tai, Y., Janas, J. A., Wang, C. L., & Van Aelst, L. (2014). Regulation of chandelier cell cartridge and bouton development via DOCK7-mediated ErbB4 activation. *Cell Reports, 6*(2), 254–263.

Verhoeven, K., De Jonghe, P., Coen, K., Verpoorten, N., Auer-Grumbach, M., Kwon, J. M., … Timmerman, V. (2003). Mutations in the small GTP-ase late endosomal protein RAB7 cause Charcot-Marie-Tooth type 2B neuropathy. *The American Journal of Human Genetics, 72*(3), 722–727.

Washbourne, P. (2015). Synapse assembly and neurodevelopmental disorders. *Neuropsychopharmacology, 40*(1), 4.

Wei, N., Zhang, Q., & Yang, X. L. (2019). Neurodegenerative Charcot-Marie-Tooth disease as a case study to decipher novel functions of aminoacyl-tRNA synthetases. *Journal of Biological Chemistry, 294*(14), 5321–5339.

Zorzano, A., Hernandez-Alvarez, M. I., Sebastian, D., & Munoz, J. P. (2015). Mitofusin 2 as a driver that controls energy metabolism and insulin signaling. *Antioxidants and Redox Signaling, 22*(12), 1020–1031.

Zuchner, S., Mersiyanova, I. V., Muglia, M., Bissar-Tadmouri, N., Rochelle, J., Dadali, E. L., … Vance, J. M. (2004). Mutations in the mitochondrial GTPase mitofusin 2 cause Charcot-Marie-Tooth neuropathy type 2A. *Nature Genetics, 36*(5), 449–451.

Chapter 46

Developmental neurotoxicity of endosulfan: modeling with zebrafish

Karina Dale[1], Josef Daniel Rasinger[2] and Ståle Ellingsen[1]

[1]*University of Bergen, Department of Biological Sciences, Bergen, Norway;* [2]*Institute of Marine Research, Bergen, Norway*

List of abbreviations

a-TOC alpha-Tocopherol
ADHD Attention deficit/hyperactivity disorder
CNS Central nervous system
DNT Developmental neurotoxicity
dpf Days postfertilization
ESF Endosulfan
GABA γ-Amino-butyric acid
GABAr γ-Amino-butyric acid receptor
hpf Hours postfertilization
OC Organochlorine
POP Persistent organic pollutant
ROS Reactive oxygen species

Mini-dictionary of terms

Adverse effects Undesired, harmful, and nonreversible impairment of a functional biological capacity, e.g., from xenobiotics.
Animal development The process leading to the formation of a new mature animal, beginning with the fertilization of the egg.
Antioxidant genes Genes that code for proteins (enzymes) that are responsible for neutralizing reactive oxygen species (ROS; reactive chemicalscontaining oxygen).
Developmental neurotoxicity Effect of xenobiotics on the developing nervous system, including changes to structure and function.
Insecticides A chemical used to kill insects.
Locomotor activity A complex behavioral trait sensitive to environmental stimuli that can be used as an endpoint to measure neurological effect outcomes following exposure to chemicals.
Oxidative stress The abnormal build-up of ROS within living cells. This can cause damage of cell structures and function.
Pesticides Chemicals used to prevent detrimental pest infection of food, crops, and livestock.
Xenobiotic Chemical substance not naturally present in an organism. Can also be used in relation with naturally occurring chemicals if concentrations are much higher than normal.

Introduction

Developmental neurotoxicity: definitions and well-known risk factors.

Developmental stages are in general highly susceptible to perturbations from both intrinsic and extrinsic stimuli. In particular, the nervous system during its development, i.e., when the main structures of the central nervous system (CNS) are formed and organized, is subject to irreversible damages if exposed to xenobiotics. The development of the nervous system differs from formation of other organs in that it changes throughout all developmental stages, instead of being formed at a specific time point (Legradi, el Abdellaoui, van Pomeren, & Legler, 2015); thus damages introduced at an early life stage are highly susceptible to cause negative effects later in life. Developmental neurotoxicity (DNT) is defined as an

Diagnosis, Management and Modeling of Neurodevelopmental Disorders. https://doi.org/10.1016/B978-0-12-817988-8.00046-4

adverse change of the chemistry, structure, or function of the developing nervous system caused by chemical or physical influences (Costa, 1998). The phenotypes of such effects can be assessed at several levels, including anatomical, neurochemical, behavioral, and neurophysiological.

Many of the contaminants recognized to be neurotoxic were historically identified through linking high-dose exposures with clinical symptoms in humans, where exposures often originated from accidents or occupational hazard (Rock & Patisaul, 2018). The association between low-dose chemical exposure and human neurotoxicity was recognized several decades later (Lanphear, 2017), and neurodevelopmental disorders including autism and attention deficit/hyperactivity disorder (ADHD) have increasingly been associated with exposure to environmental contaminants (Grandjean & Landrigan, 2006). Regardless of these findings, the number of chemicals produced commercially has been ever-increasing and the bulk of compounds released each year are not being tested for their DNT potential (Behl et al., 2019). This leads to an unacceptable environmentally mediated hazard to children's learning and development (Trasande & Liu, 2011). There is therefore a growing interest, both scientific and regulatory, for efficient hazard identification and characterization of potential DNT agents (Bal-Price et al., 2018). Identifying and understanding the underlying mechanisms of DNT is key in these processes and an essential criterion for the efficient and timely human and environmental risk assessment of chemicals.

Animal models in testing of chemicals with DNT potential

Traditionally, animal studies have been performed to define lethal doses and to characterize the toxic potential of chemicals. However, there are limitations in current test protocols, including sensitivity, reproducibility, and relevance to human conditions such as autism and ADHD. Recently, a growing toolbox including animal and cell-based models, as well as *in silico* approaches, has led to a better understanding of chemical risk and mechanisms of toxicity at environmentally relevant exposure levels (Rasinger et al., 2018; Rasinger, Carroll, Lundebye, & Hogstrand, 2014; Refatto et al., 2018). In particular, zebrafish (*Danio rerio*) has become a valuable *in vivo* system for studying the impact of exposure to environmental contaminants, in terms of both neurotoxicity and neurodevelopmental outcomes (Dale, Rasinger, Thorstensen, Penglase, & Ellingsen, 2017; Quevedo et al., 2019; Rasinger, Lundebye, Penglase, Ellingsen, & Amlund, 2017). In zebrafish, the main parts of the nervous system appear during the first 24 h; at 3 days postfertilization (dpf), the larvae hatches and most of its morphogenesis has been completed (Kimmel, Ballard, Kimmel, Ullmann, & Schilling, 1995). The combination of external fertilization, ex utero development, transparent embryos, and the large numbers of eggs available from each female fish result in a system well-suited for high-throughput developmental exposure studies.

General mechanisms in developmental neurotoxicity

On a cellular level, effects of DNT can emerge in different ways. One main mechanism is the occurrence of increased levels of reactive oxygen species (ROS) such as peroxides and hydroxyl radicals. ROS form during normal metabolism of oxygen and play important roles in a variety of natural cell functions. However, exposure to contaminants can cause an abnormal build-up of ROS that leads to damage of cell structures known as oxidative stress, which may severely and irreversibly impede neurodevelopment (Rock & Patisaul, 2018). Prior to the development of synapses, neurotransmitters regulate many processes in the developing nervous system and are therefore sensitive targets for chemical exposure. Many contaminants have been associated with neurotransmitter disruption during early brain development (Rock & Patisaul, 2018). The organophosphate insecticide chlorpyrifos, for example, was found to target neurotransmitter pathways including acetylcholine and the three monoamines, serotonin, dopamine, and norepinephrine (Slotkin & Seidler, 2007). Lastly, DNT also can result from incorrect hormonal signaling. In the developing nervous system, hormones play an integral role; for example, in the hypothalamic—pituitary—thyroid axis. Endocrine disruption can occur when androgen, estrogen, or thyroid hormone signaling is affected by environmental contaminants (Rock & Patisaul, 2018).

Developmental neurotoxicity of environmental contaminants

In general, chemicals that cause DNT can be divided into four broad classes; metals, solvents, persistent organic pollutants (POP), and pesticides Table 46.1. The DNT properties of metals including lead, mercury, and arsenic have been known for decades, but a variety of other metals and inorganic compounds may also disrupt neurodevelopment (Rock & Patisaul, 2018). Furthermore, solvents such as toluene, chloroform, and methanol have shown to cause DNT effects in humans (Grandjean & Landrigan, 2006). In addition, POPs were found to affect neurodevelopmental and behavioral outcomes in juveniles in several domains, including attention, motor development, and mental development (Berghuis, Bos, Sauer, &

TABLE 46.1 Selected examples of contaminants known to cause DNT effects.

Contaminant class	Contaminant name	Developmental exposure stage	Exposure level	Reported effects	References
Metals	Lead	0–24 hpf	10–30 nM Pb^{2+}	Decreased larval response to tapping stimulus (reduced startle response)	Rice, Ghorai, Zalewski, and Weber (2011)
	Mercury (methylmercury)	Maternal transmission	12 mg/kg in feed	Larval locomotor hypoactivity Downregulation of genes in antioxidant pathways	Penglase, Hamre, and Ellingsen (2014)
	Arsenic	4–120 hpf	50–150 μg/L AsIII	Oxidative stress, reduced larval survival	Sun HJ et al. (2019)
Solvents	Toluene	4–120 hpf	5–25 μg/mL	Inhibition of hatching, lethality	George S et al. (2011)
	Chloroform	4–72 hpf	0.7–2.1 mM	Malformation of eyes and tail, cardiac edema, decreased hatching frequency, DNA damage	Teixidó, Piqué, Gonzalez-Linares, Llobet, and Gómez-Catalán (2015)
	Methanol	6–24 hpf	2%–4%	Abnormal retinal patterning, reduced eye size, reduced body size, paracardiac and abdominal edema, irregular jaw development, body axis deformities	Fu, Jiao, Weng, Yu and Li (2017)
POPs[a]	TCDD[b]	0–168 hpf	≥100 ppt	Reduction in total neuronal number in the larval brain and decreased gene expression	Hill, Howard, Strahle, and Cossins (2003)
	PBDE[c]	0–6 dpf	500 μg/L	Pronounced hypoactivity at 5 dpf during dark periods	Zhao, Xu, and Yin (2014)
	PCB[d]	4–24 hpf	0.3–1.2 nM	Impaired short- and long-term habituation to a novel environment in adult zebrafish	Glazer, Hahn, and Aluru (2016)
Pesticides	DDT[e]	0–7 dpf	0.1–10 μg/L	Morphological alterations, reduced survival, thyroid disruption	Wu et al. (2019)
		6–30 hpf	1–100 μM	Morphological alterations, increase in chemical-induced seizure activity	Tiedeken and Ramsdell (2009)
	Chlorpyrifos	0–96 hpf	10–300 μg/L	Morphological alterations decreased hatching, hypoactivity. Increased immune gene expression, oxidative stress.	Jin, Liu, Peng, and Fu (2015)
		0–7 dpf	0.001–1 μM	Morphological alterations, hypoactivity	Richendrfer, Pelkowski, Colwill, and Créton (2012)
	Endosulfan[e]	6–72 hpf	0.1–50 mg/L	Morphological alterations, downregulation of antioxidant genes, and hyperactivity	Dale et al. (2017)
		2–96 hpf	0–10 mg/L	Decreased hatchability and induction of vitellogenin (vtg) transcripts	Chow et al. (2013)

This table gives an overview from the zebrafish literature, and examples of effects, on 4 different classes of chemicals highlighted in this chapter that are known to cause developmental neurotoxicity (DNT). For each chemical listed in this table, there is also published studies confirming DNT in other species, e.g rodent models and humans.
[a]*Persistent organic pollutants listed in the Stockholm Convention*
[b]*Tetrachlorodibenzo-p-dioxin*
[c]*Polybrominated diphenyl ethers*
[d]*Polychlorinated biphenyls*
[e]*Endosulfan and Dichlorodiphenyltrichloroethane (DDT) are pesticides listed as POP under the Stockholm Convention.*

Roze, 2015; Rasinger, Carroll, Lundebye, & Hogstrand, 2014; Rasinger, Carroll, et al., 2018; Rasinger, Frenzel, et al., 2018). Finally, an increasing number of agricultural pesticides are recognized to cause DNT in various species (Abreu-Villaça & Levin, 2017; Richardson, Fitsanakis, Westerink, & Kanthasamy, 2019; Rock & Patisaul, 2018). Pesticides are utilized to prevent destruction of food crops and/or infection of livestock by various pests and are classified in different ways; for example, by the pests they control, such as herbicides and insecticides, or by their chemical structure (e.g., organophosphates, organochlorines). Some legacy pesticides are also classified as POPs in the Stockholm Convention (Jennings & Li, 2015). Examples of pesticides with known DNT potential include dichlorodiphenyltrichloroethane (DDT), polychlorinated biphenyls (PCB), and organophosphate compounds such as chlorpyrifos (Grandjean & Landrigan, 2006; Richardson et al., 2019; Saeedi Saravi & Dehpour, 2016).

Developmental neurotoxicity of insecticides

There are several classes of pesticides known to cause adverse effects on neurodevelopment. One important class comprises the organochlorine insecticides. Chemical insecticides are designed to act as neurotoxicants by perturbating the nervous system of the target organisms. However, due to high similarities between insect and mammalian nervous systems, insecticides are not species-selective, and therefore, nontarget organisms are also highly sensitive to DNT effects of insecticides (Costa, 2015). Adverse effects on neurodevelopment are recognized as a prominent outcome of insecticide exposure in nontarget organisms and have therefore been studied in a wide array of animal models. A comprehensive overview of these studies was compiled in 2017 by Abreu-Villaça & Levin. Mainly, insecticides were found to cause neurobehavioral impairments, affect neuronal structures, and were linked to the development of known neurodegenerative diseases such as Parkinson's disease (Abreu-Villaça & Levin, 2017). The mechanisms underlying these detrimental effects were primarily attributed to receptor-mediated disruption, disruption of mechanisms of the endocrine system, and/or disruption of the homeostasis of neurotransmitter and ROS levels (Abreu-Villaça & Levin, 2017; Costa, 2015; Rock & Patisaul, 2018). An example of a prominent insecticide is endosulfan (ESF) whose impact on neurodevelopment we focus on in this chapter.

Neurotoxic properties of endosulfan

ESF acts as a neurotoxicant by perturbating the functions of the γ-amino-butyric acid (GABA) receptor. Under normal conditions, binding of GABA to its receptor (GABAr) causes opening of chloride channels and an inflow of chloride ions into neurons. The ions migrate across an electrochemical gradient, and the migration hinders neuron excitation (Silva & Beauvais, 2010). Thus, GABArs are inhibitory neuroreceptors and antagonists of GABAergic neurons. ESF acts as a noncompetitive GABAr antagonist. It does not directly associate with the GABA binding site, but instead blocks the GABA-coupled chloride channels, thus hindering the binding of GABA to its receptor (Menezes et al., 2017). By preventing influx of chloride ions, the inhibitory effect of GABA is impeded, resulting in a continuous nerve stimulation (Silva & Gammon, 2009), which ultimately can lead to excitotoxic insults and neurodegeneration.

Developmental neurotoxicity of endosulfan

Human

In humans, ESF poisoning manifests mainly as seizures after hyperstimulation of the CNS (Menezes et al., 2017). It is also shown that ESF can accumulate in breast milk and be transferred to the fetus, possibly affecting neurodevelopment. Since the human brain continues to grow and mature many years after birth, both pre- and postnatal exposure to ESF and similar neurotoxic compounds may alter neurodevelopment and cause long-lasting adverse effects (Saeedi Saravi & Dehpour, 2016).

Rodents

In rodents, developmental exposure to ESF is shown to hamper neurotransmission in the frontal cortex (Wilson, Onyenwe, Bradner, Nennig, & Caudle, 2014), affect dopaminergic neurons (Wilson, Shapiro, Bradner, & Caudle, 2014), reduce learning ability (Gómez-Giménez et al., 2017), and induce motor incoordination (Gómez-Giménez, Felipo, et al., 2018; Gómez-Giménez, Llansola, et al., 2018). The latter studies also observed neurotoxic effects in adult rats, suggesting that neurodevelopmental exposure to ESF causes neurological impairments in adult age.

Fish

In addition to studies in mammals, ESF toxicity has been studied in various fish species. ESF is regarded as extremely toxic to fish (Li & Macdonald, 2005), especially to younger life stages. Acute effects of ESF are primarily exerted in the nervous system and mediated effects, e.g., behavior, whereas long-term effects include endocrine disruption, suppression of the immune system, and genotoxicity, in addition to neurological growth effects (Ballesteros et al., 2009; Gormley & Teather, 2003).

Zebrafish as a model organism for studying ESF-induced developmental neurotoxicity

Among fish species, the zebrafish has become a popular model organism in biomedical and toxicological research. The use of zebrafish is advantageous for many reasons as described above, but importantly the zebrafish genome shows a high degree of homology to mammalian species and humans in particular (Seth, Stemple, & Barroso, 2013). Zebrafish are highly suited for studying neurodevelopmental processes because they, similar to other vertebrates, have cognitive, sensorimotor, and basic motor responses controlled by the CNS (Makhija & Jagtap, 2014). Consequently, early life stages of zebrafish are increasingly acknowledged as appropriate models for studying effects of DNT, and over the past decades zebrafish have been widely utilized to investigate effects of developmental exposure to pesticides (Richardson et al., 2019).

Endosulfan-induced developmental neurotoxicity in zebrafish

We will now present findings from various studies where zebrafish have been used as a model organism to discuss the mechanisms and related adverse effects underlying neurodevelopmental exposure to ESF.

In 2017, we published a study in which we investigated DNT effects of ESF exposure on zebrafish embryos, and we assessed if ESF-induced DNT could be reversed by coexposure with the antioxidative nutrient alpha-tocopherol (a-TOC/Vitamin E) (Dale et al., 2017). In this study, zebrafish embryos were subjected to waterborne ESF exposure (0.1–50 mg/L) from 6 to 72 hpf. To evaluate ESF-induced DNT, we looked for the presence of morphological deformities at 48 hpf (pericardial edema, axis malformation, degree of melanin pigmentation), analyzed gene expression at 72 hpf (genes related to oxidative stress pathways), and assessed locomotor activity at 120 hpf (Fig. 46.1).

Morphology

ESF exposure produced a dose–response related increase in morphological deformities at 48 hpf (Dale et al., 2017). ESF-exposed embryos had an increased percentage of pericardial edema, axis malformations, and reduced melanin pigmentation, compared to control embryos (Fig. 46.2). In addition, ESF reduced hatching frequencies at 72 hpf. All of these endpoints are commonly linked to DNT in zebrafish (He, Gao, Huang, & Li, 2014). Another study also showed that ESF exposure at different concentrations (>0.1 mg/L) induced morphological abnormalities at early life stages (48–120 hpf) (Stanley, Curtis, Simonich, & Tanguay, 2009).

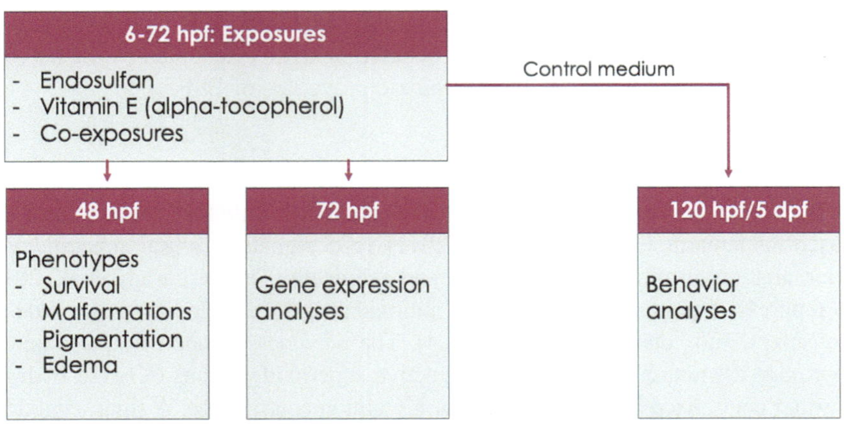

FIGURE 46.1 **Overview of experimental setup.** Duration of exposures presented as hours postfertilization (hpf) and timepoints for various analyses. For behavior analysis, fish larvae were transferred to control medium (clean E3 medium) at 72–120 hpf. *Hpf*, hours postfertilization.

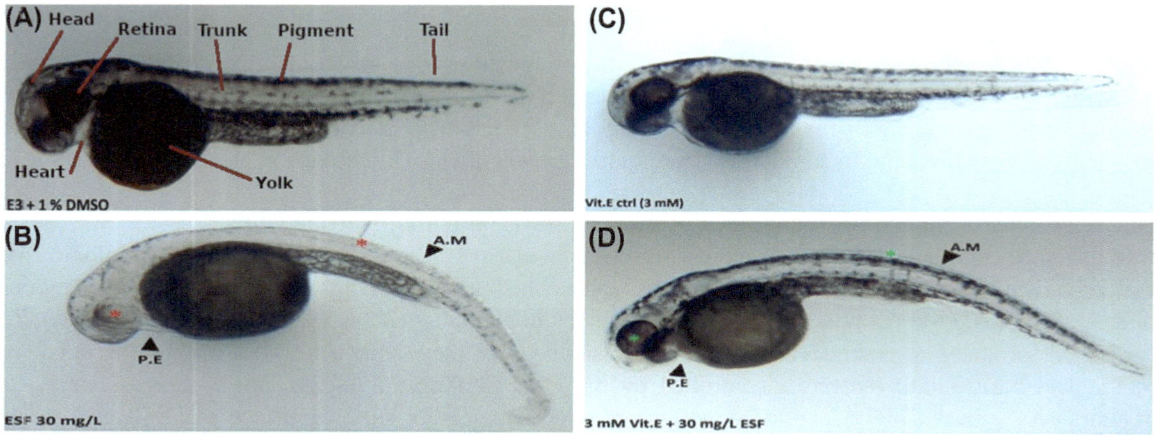

FIGURE 46.2 **Phenotypes of embryos exposed to endosulfan alone or in combination with alpha-tocopherol.** Phenotypes of zebrafish embryos exposed from 6 to 72 hpf to (B) 30 mg/L endosulfan (ESF) or (D) coexposed to 30 mg/L ESF and 3 mM alpha-tocopherol (a-TOC). Controls are (A) E3 medium with 1% DMSO and (C) 3 mM a-TOC. Pictures were captured at 48 hpf. Morphological phenotypes in ESF-exposed embryos (B) and coexposed embryos (D) are shown in terms of reduced pigmentation (red*), improved pigmentation (green*), *P.E*, pericardial edema; *A.M*, axis malformation. *Modified from Dale, K., Rasinger, J. D., Thorstensen, K. L., Penglase, S., & Ellingsen, S. (2017). Vitamin E reduces endosulfan-induced toxic effects on morphology and behavior in early development of zebrafish (*Danio rerio*).* Food and Chemical Toxicology, 101, *84–93. Copyright (2017), with permission from Elsevier.*

Behavior

A common way of assessing neurological effects is by studying the behavior. Although not all behavioral modifications are of neurological origin, all behavior is achieved through control of the nervous system and can be altered by exposure to neurotoxicants (Tierney, 2011). Behavioral assessment can be performed in different ways including measuring number of movements, locomotor activity, and response-to-touch movements. These parameters are usually assessed using video-recording tools (d'Amora & Giordani, 2018), and it has been found that environmental contaminants affect such movement-related behavioral patterns in different ways; some chemicals cause hyperstimulation and increased activity (hyperactivity), whereas others reduce behavioral activity (hypoactivity).

The neurological system of zebrafish is well-studied; zebrafish display behaviors including aggression, fear, and exploration highly comparable to rodents and humans (Miller, Chandrasekaran, Yaghoobi, & Lein, 2018). Several studies have used zebrafish and similar species such as medaka (*Oryzias latipes*) at various life stages including embryos, larvae, and adults, in behavioral assays as a way of studying ESF-induced neurotoxicity (Dale et al., 2017; Gormley & Teather, 2003; Pereira et al., 2012). In our study, we found increased locomotor activity in 120 hpf larvae exposed to 0.5–1.0 mg/L ESF from 6 to 72 hpf, which is in accordance with ESF being a GABAr antagonist causing neuronal hyperstimulation (Fig. 46.3). It was suggested that different concentrations of ESF could have opposite effects on locomotor activity, which is in line with other studies observing both hyper- and hypoactivity following ESF exposure in zebrafish and amphibians (Denoël et al., 2013; Pereira et al., 2012). In either case, as behavior is tightly regulated by the nervous system, behavioral alterations including locomotor activity underline the neurotoxic properties of ESF.

Oxidative stress

Several studies have shown that ESF can increase levels of ROS and cause oxidative stress in zebrafish (Dale et al., 2017; Shao et al., 2012; Velasco-Santamaría, Handy, & Sloman, 2011). ROS signaling is vital in neurodevelopment, particularly for controlling cell differentiation, proliferation, migration, and cell death (Timme-Laragy et al., 2013). In our study, ESF exposure significantly suppressed expression of important antioxidant genes, including superoxide dismutase (*cuzn-sod*), glutathione peroxidase (*gpx1*), and catalase (*cat*) (Fig. 46.4). The enzymes encoded by these genes are responsible for neutralizing ROS; superoxide dismutase converts highly reactive superoxide anions ($\bullet O_2^-$) to hydrogen peroxide (H_2O_2), the latter molecule is further neutralized by either catalase or glutathione peroxidase. A significant downregulation of these genes suggests that ESF affects the antioxidant defense of developing zebrafish larvae.

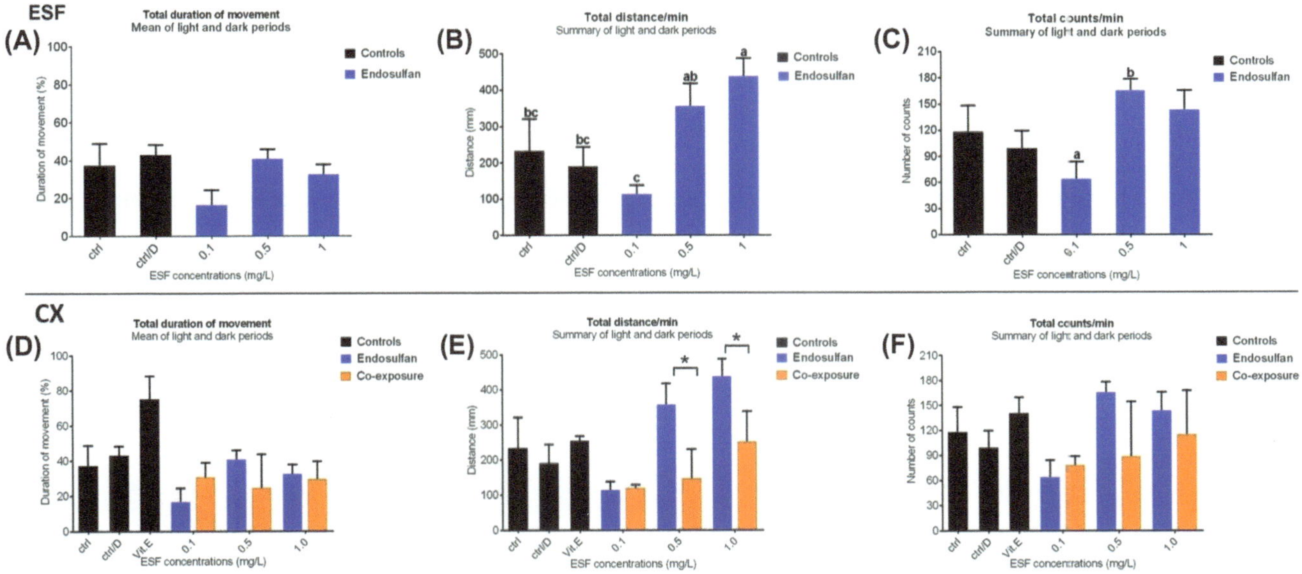

FIGURE 46.3 **Locomotor activity of larvae exposed to endosulfan alone or in combination with alpha-tocopherol.** The percentage of time (A, D), total distance moved (B, E), and number of times movement speed was changed (C, F) in 5 dpf zebrafish larvae that were exposed to endosulfan (ESF; 0.1–1 mg/L) alone or in coexposure (CX) with 3 mM alpha-tocopherol (a-TOC) from 6 to 72 hpf. Data are the average obtained over a 10 min period of alternating light conditions. (A–C) are for ESF-exposed embryos, whereas (D–F) show combined results of ESF (purple) and coexposed (orange) embryos. Controls (black) are E3 medium (ctrl), E3 medium with 1% DMSO (ctrl/D), and 3 mM a-TOC (Vit.E). Data are presented as mean ± SD, n = 12. Different letters indicate statistical differences in mean values between treatments (B + C, One-way ANOVA, $p < 0.05$). *Asterisk* (*) indicates statistical difference between ESF and coexposed embryos (E). *Reprinted from Dale, K., Rasinger, J. D., Thorstensen, K. L., Penglase, S., & Ellingsen, S. (2017). Vitamin E reduces endosulfan-induced toxic effects on morphology and behavior in early development of zebrafish* (Danio rerio). Food and Chemical Toxicology, 101, *84–93. Copyright (2017), with permission from Elsevier.*

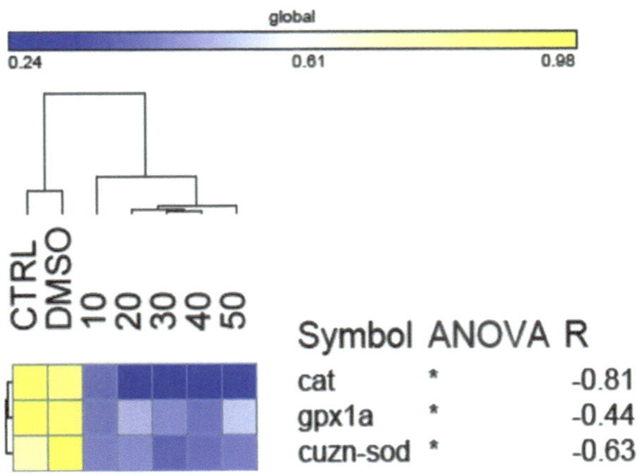

FIGURE 46.4 **Heat map for antioxidant and related gene expression in zebrafish embryos exposed to endosulfan.** The mRNA expression of catalase (*cat*), glutathione peroxidase 1a (*gpx1a*), and superoxide dismutase 1 (*cuzn-sod*) in 72 hpf zebrafish embryos after exposure from 6 to 72 hpf to endosulfan (ESF). Gene expression is color-coded where white, yellow, and blue represent medium, increased, and decreased expression compared to controls, respectively. The heat maps were generated using mean normalized expression (MNE) levels of each gene. The results were clustered according to treatment (above the heat maps) and to gene differences (left of the heat maps). The different treatment groups are listed, indicating controls (CTRL = E3, DMSO = E3 + 1% DMSO), different ESF concentrations in mg/L. Genes with significant changes in expression levels (ANOVA) are marked with *. Rank regression analysis was performed to assess if changes in gene expression were induced in a dose-related manner. The correlation coefficient (R) indicates a positive or a negative dose–response relationship. *Modified from Dale, K., Rasinger, J. D., Thorstensen, K. L., Penglase, S., & Ellingsen, S. (2017). Vitamin E reduces endosulfan-induced toxic effects on morphology and behavior in early development of zebrafish* (Danio rerio). Food and Chemical Toxicology, 101, *84–93. Copyright (2017), with permission from Elsevier.*

Protective effects of vitamin E

Lipid-soluble endogenous antioxidants such as vitamin E may protect against oxidative stress (Dale et al., 2017; Hamre, 2011). In our study, supplementation of a-TOC to ESF exposure significantly reduced the degree of ESF-induced morphological malformations (Fig. 46.2) and also returned ESF-induced hyperactivity within range of control levels (Fig. 46.3). Based on these findings, we suggested that a-TOC had a protective effect against ESF-induced changes of morphology and behavior. Since both these endpoints are linked to neurodevelopment, it is likely that vitamins such as a-TOC may counteract on chemical-induced DNT.

ESF as an endocrine disruptor

In addition to the aforementioned ESF-induced effects, other studies have seen a variety of ESF induced alterations during zebrafish development. Chow, Chan, and Chan (2013) exposed zebrafish embryos to ESF from 2 to 96 hpf and observed induction of the egg precursor protein vitellogenin, suggesting that ESF may act estrogenic and cause male feminization at a later life stage. Another study underlining the possibility of ESF being an endocrine disruptor found that ESF affects gonad structure and function in zebrafish embryos at the 26-somite stage (approx. 24 hpf) (Willey & Krone, 2001).

Applications to other areas of development

Endosulfan—what did we show?

In this chapter, we have reviewed the effects of neurodevelopmental exposure to the organochlorine insecticide ESF, and we have shown that ESF-induced toxic effects have been observed in various organisms including humans, rodents, and fish species; special attention was paid to neurodevelopmental outcomes. Several studies have shown that ESF perturbates the nervous system by acting as a noncompetitive GABAr antagonist, causing continuous nervous stimulation (Menezes et al., 2017; Silva & Gammon, 2009). Effects either related to or independent of this mechanism are disturbance of morphological development and behavior, endocrine disruption, and oxidative stress, among others. We have also shown how zebrafish is a valuable model for studying ESF-induced DNT. Zebrafish are highly comparable to other vertebrates regarding nervous system development; this renders zebrafish embryos and their larvae into valuable tools for high-throughput screening of chemicals that potentially cause DNT (d'Amora & Giordani, 2018; Legradi et al., 2015; Richardson et al., 2019).

Relevance to human disease

There is increasing evidence that environmental contaminants influence neurodevelopmental and neurodegenerative human disorders (Berghuis et al., 2015; Lanphear, 2017; Trasande & Liu, 2011). However, it can be challenging to link these disorders to chemical exposure during early life stages since easily identifiable symptoms may occur only at doses greatly exceeding environmentally relevant doses. In addition, following early onset of low-dose chronic exposure to environmental contaminants, it may take years or decades before any symptoms of disease are manifesting themselves. Irrespective of these challenges, the number of new chemicals released each year is ever-increasing. The estimated costs DNT may cause to society and the environment, if undetected, stress the necessity of focusing research on identifying and characterizing potential health effects early on. The mechanistic information obtained from animal models can be used to derive mode of action (MOA) and adverse outcome pathway (AOP) data required to model exposure outcomes (Bal-Price et al., 2015).

In addition to the toxicity assessment of single chemicals, there also is a growing focus on mixture toxicology and on how specific nutrients may modulate the effects induced by environmental contaminants. Using models such as zebrafish provides an additional, important option for studying how selected nutrients affect the impact of contaminants on neurodevelopment and adult brain function as was recently demonstrated for ESF and alpha-tocopherol (a-TOC) (Dale et al., 2017) and for methylmercury and selenium (Rasinger et al., 2017), respectively.

Relevance to human nutrition

Seafood is an important source of nutrients, including neuroprotective omega-3 polyunsaturated fatty acids, selenium, iodine, and vitamins A and D, among others, and is thus a crucial part of a healthy and balanced diet for humans. Recently, in the aquaculture industry, marine feed resources have been replaced by vegetable feed ingredients (Søfteland et al., 2014), which exposes fish and their consumers to new contaminants including agricultural pesticides (Garlito et al., 2019).

Many of these food contaminants have incomplete DNT profiles and consequently a considerable knowledge gap exists, which hinders comprehensive risk assessments. Thus, modeling DNT impact from contaminants introduced in human food is becoming increasingly challenging and the value of a high-throughput animal model system such as zebrafish in combination with high-throughput omics and in silico computational toxicity tools will become invaluable for regulatory risk assessment of animal and human safety.

Key facts of endosulfan

- Endosulfan (ESF) is a fat-soluble organochlorine insecticide (6,7,8,9,10,10-hexachloro-1,5,5a,6,9,9a-hexahydro-6,9-methano-2,4,3-benzadioxathiepin-3-oxide).
- ESF is known to cause toxicity toward nontarget organisms (including humans).
- For decades, there has been worldwide use of ESF on crops including soy, cotton, vegetables, and coffee.
- In the period from 1950 to 2000, the cumulative global use of ESF was estimated to be 338,000 tons (Li & Macdonald, 2005).
- Technical grade ESF is a mixture composed of two diastereoisomers: α-ESF and β-ESF. Both ESF isomers are semi-volatile and thereby susceptible for atmospheric transport, thus ESF is ubiquitous in the environment (Weber et al., 2010).
- ESF was withdrawn from the global market in 2012, after it was listed as a persistent organic pollutant (POP) under the UNEP Stockholm Convention in 2011 (Jennings & Li, 2015).
- Despite being phased out globally, ESF persists in the environment and will still pose a risk in the years to come.
- ESF is still introduced into the environment as it is being utilized in developing countries due to its beneficial properties including low costs, high efficacy, and environmental stability (Menezes et al., 2017).
- ESF acts as a neurotoxicant by perturbating the functions of the γ-amino-butyric acid (GABA) receptor (Silva & Beauvais, 2010).

Summary points

- Early life stages are particularly sensitive to chemical exposure.
- Current testing of developmental neurotoxicity (DNT)-related adverse health outcomes does not satisfactorily cover new chemicals released to the market each year.
- Zebrafish is a promising model with high-throughput capacity for DNT testing.
- The insecticide endosulfan (ESF) poses a severe concern due to its widespread presence, biological persistance, and DNT potential.
- The DNT potential of ESF has been shown to cause developmental phenotypes in morphology, behavior, and gene expression in zebrafish.

References

d'Amora, M., & Giordani, S. (2018). The utility of zebrafish as a model for screening developmental neurotoxicity. *Frontiers in Neuroscience, 12*, 976.

Abreu-Villaça, Y., & Levin, E. D. (2017). Developmental neurotoxicity of succeeding generations of insecticides. *Environment International, 99*, 55–77.

Bal-Price, A., Crofton, K. M., Leist, M., Allen, S., Arand, M., Buetler, T., … Fritsche, E. (2015). International STakeholder NETwork (ISTNET): creating a developmental neurotoxicity (DNT) testing road map for regulatory purposes. *Archives of Toxicology, 89*(2), 269–287.

Bal-Price, A., Hogberg, H. T., Crofton, K. M., Daneshian, M., FitzGerald, R. E., Fritsche, E., … Leist, M. (2018). Recommendation on test readiness criteria for new approach methods in toxicology: exemplified for developmental neurotoxicity. *ALTEX, 35*(3), 306–352.

Ballesteros, M. L., Durando, P. E., Nores, M. L., Díaz, M. P., Bistoni, M. A., & Wunderlin, D. A. (2009). Endosulfan induces changes in spontaneous swimming activity and acetylcholinesterase activity of *Jenynsia multidentata* (Anablepidae, Cyprinodontiformes). *Environmental Pollution, 157*(5), 1573–1580.

Behl, M., Ryan, K., Hsieh, J.-H., Parham, F., Shapiro, A. J., Collins, B. J., … Tice, R. R. (2019). Screening for developmental neurotoxicity at the national toxicology program: the future is here. *Toxicological Sciences, 167*(1), 6–14.

Berghuis, S. A., Bos, A. F., Sauer, P. J. J., & Roze, E. (2015). Developmental neurotoxicity of persistent organic pollutants: An update on childhood outcome. *Archives of Toxicology, 89*(5), 687–709.

Chow, W. S., Chan, W. K.-L., & Chan, K. M. (2013). Toxicity assessment and vitellogenin expression in zebrafish (*Danio rerio*) embryos and larvae acutely exposed to bisphenol A, endosulfan, heptachlor, methoxychlor and tetrabromobisphenol A. *Journal of Applied Toxicology, 33*(7), 670–678.

Costa, L. G. (1998). Neurotoxicity testing: A discussion of in vitro alternatives. *Environmental Health Perspectives, 106*(Suppl. 2), 505–510.

Costa, L. G. (2015). The neurotoxicity of organochlorine and pyrethroid pesticides. In *Handbook of Clinical Neurology* (Vol. 131, pp. 135–148).

Dale, K., Rasinger, J. D., Thorstensen, K. L., Penglase, S., & Ellingsen, S. (2017). Vitamin E reduces endosulfan-induced toxic effects on morphology and behavior in early development of zebrafish (*Danio rerio*). *Food and Chemical Toxicology, 101*, 84–93.

Denoël, M., Libon, S., Kestemont, P., Brasseur, C., Focant, J.-F., & De Pauw, E. (2013). Effects of a sublethal pesticide exposure on locomotor behavior: A video-tracking analysis in larval amphibians. *Chemosphere, 90*(3), 945–951.

Fu, J., Jiao, J., Weng, K., Yu, D., & Li, R. (2017). Zebrafish methanol exposure causes patterning defects and suppressive cell proliferation in retina. *American Journal of Tourism Research, 9*(6), 2975–2983.

Garlito, B., Ibáñez, M., Portolés, T., Serrano, R., Amlund, H., Lundebye, A.-K., … Hernández, F. (2019). LC-MS/MS method for the determination of organophosphorus pesticides and their metabolites in salmon and zebrafish fed with plant-based feed ingredients. *Analytical and Bioanalytical Chemistry, 1*, 11.

George, S., Xia, T., Rallo, R., Zhao, Y., Ji, Z., Lin, S., … Nel, A. E. (2011). Use of a high-throughput screening approach coupled with in vivo zebrafish embryo screening to develop hazard ranking for engineered nanomaterials. *ACS Nano, 5*(3), 1805–1817.

Glazer, L., Hahn, M. E., & Aluru, N. (2016). Delayed effects of developmental exposure to low levels of the aryl hydrocarbon receptor agonist 3,3′,4,4′,5-pentachlorobiphenyl (PCB126) on adult zebrafish behavior. *NeuroToxicology, 52*, 134–143.

Gómez-Giménez, B., Felipo, V., Cabrera-Pastor, A., Agustí, A., Hernández-Rabaza, V., & Llansola, M. (2018). Developmental exposure to pesticides alters motor activity and coordination in rats: Sex differences and underlying mechanisms. *Neurotoxicity Research, 33*(2), 247–258.

Gómez-Giménez, B., Llansola, M., Cabrera-Pastor, A., Hernández-Rabaza, V., Agustí, A., & Felipo, V. (2018). Endosulfan and cypermethrin pesticide mixture induces synergistic or antagonistic effects on developmental exposed rats depending on the analyzed behavioral or neurochemical end points. *ACS Chemical Neuroscience, 9*(2), 369–380.

Gómez-Giménez, B., Llansola, M., Hernández-Rabaza, V., Cabrera-Pastor, A., Malaguarnera, M., Agusti, A., & Felipo, V. (2017). Sex-dependent effects of developmental exposure to different pesticides on spatial learning. The role of induced neuroinflammation in the hippocampus. *Food and Chemical Toxicology, 99*, 135–148.

Gormley, K. L., & Teather, K. L. (2003). Developmental, behavioral, and reproductive effects experienced by Japanese medaka (*Oryzias latipes*) in response to short-term exposure to endosulfan. *Ecotoxicology and Environmental Safety, 54*(3), 330–338.

Grandjean, P., & Landrigan, P. (2006). Developmental neurotoxicity of industrial chemicals. *The Lancet, 368*(9553), 2167–2178.

Hamre, K. (2011). Metabolism, interactions, requirements and functions of vitamin E in fish. *Aquaculture Nutrition, 17*(1), 98–115.

He, J.-H., Gao, J.-M., Huang, C.-J., & Li, C.-Q. (2014). Zebrafish models for assessing developmental and reproductive toxicity. *Neurotoxicology and Teratology, 42*, 35–42.

Hill, A., Howard, C. V., Strahle, U., & Cossins, A. (2003). Neurodevelopmental defects in zebrafish (*Danio rerio*) at environmentally relevant dioxin (TCDD) concentrations. *Toxicological Sciences, 76*(2), 392–399.

Jennings, A. A., & Li, Z. (2015). Residential surface soil guidance applied worldwide to the pesticides added to the Stockholm Convention in 2009 and 2011. *Journal of Environmental Management, 160*, 226–240.

Jin, Y., Liu, Z., Peng, T., & Fu, Z. (2015). The toxicity of chlorpyrifos on the early life stage of zebrafish: A survey on the endpoints at development, locomotor behavior, oxidative stress and immunotoxicity. *Fish and Shellfish Immunology, 43*(2), 405–414.

Kimmel, C. B., Ballard, W. W., Kimmel, S. R., Ullmann, B., & Schilling, T. F. (1995). Stages of embryonic development of the zebrafish. *Developmental Dynamics: An Official Publication of the American Association of Anatomists, 203*(3), 253–310.

Lanphear, B. P. (2017). Low-level toxicity of chemicals: No acceptable levels? *PLoS Biology, 15*(12), e2003066.

Legradi, J., el Abdellaoui, N., van Pomeren, M., & Legler, J. (2015). Comparability of behavioural assays using zebrafish larvae to assess neurotoxicity. *Environmental Science and Pollution Research, 22*(21), 16277–16289.

Li, Y. F., & Macdonald, R. (2005). Sources and pathways of selected organochlorine pesticides to the arctic and the effect of pathway divergence on HCH trends in biota: A review. *The Science of the Total Environment, 342*(1–3), 87–106.

Menezes, R. G., Qadir, T. F., Moin, A., Fatima, H., Hussain, S. A., Madadin, M., … Senthilkumaran, S. (2017). Endosulfan poisoning: an overview. *Journal of Forensic and Legal Medicine, 51*, 27–33.

Makhija, D. T., & Jagtap, A. G. (2014). Studies on sensitivity of zebrafish as a model organism for Parkinson's disease: comparison with rat model. *Journal of Pharmacology and Pharmacotherapeutics, 5*(1), 39–46. https://doi.org/10.4103/0976-500X.124422.

Miller, G. W., Chandrasekaran, V., Yaghoobi, B., & Lein, P. J. (2018). Opportunities and schallenges for using the zebrafish to study neuronal connectivity as an endpoint of developmental neurotoxicity. *NeuroToxicology, 67*, 102–111.

Penglase, S., Hamre, K., & Ellingsen, S. (2014). Selenium prevents downregulation of antioxidant selenoprotein genes by methylmercury. *Free Radical Biology and Medicine, 75*, 95–104.

Pereira, V. M., Bortolotto, J. W., Kist, L. W., deAzevedo, M. B., Fritsch, R. S., Oliveira, R.daL., … Bogo, M. R. (2012). Endosulfan exposure inhibits brain AChE activity and impairs swimming performance in adult zebrafish (Danio rerio). *Neurotoxicology, 33*(3), 469–475.

Quevedo, C., Behl, M., Ryan, K., Paules, R. S., Alday, A., Muriana, A., & Alzualde, A. (2019). Detection and prioritization of developmentally neurotoxic and/or neurotoxic compounds using zebrafish. *Toxicological Sciences, 168*(1), 225–240.

Rasinger, J. D., Carroll, T. S., Lundebye, A. K., & Hogstrand, C. (2014). Cross-omics gene and protein expression profiling in juvenile female mice highlights disruption of calcium and zinc signalling in the brain following dietary exposure to CB-153, BDE-47, HBCD or TCDD. *Toxicology, 321*, 1–12.

Rasinger, J. D., Carroll, T. S., Maranghi, F., Tassinari, R., Moracci, G., Altieri, I., … Hogstrand, C. (2018). Low dose exposure to HBCD, CB-153 or TCDD induces histopathological and hormonal effects and changes in brain protein and gene expression in juvenile female BALB/c mice. *Reproductive Toxicology, 80*, 105–116.

Rasinger, J. D., Frenzel, F., Braeuning, A., Lampen, A., Braeuning, A., & Lampen, A. (2018). Identification and evaluation of potentially mutagenic and carcinogenic food contaminants. *EFSA Journal, 16* (John Wiley & Sons, Ltd).

Rasinger, J. D., Lundebye, A. K., Penglase, S. J., Ellingsen, S., & Amlund, H. (2017). Methylmercury induced neurotoxicity and the influence of selenium in the brains of adult Zebrafish (*Danio rerio*). *International Journal of Molecular Sciences, 18*(4).

Refatto, V., Rasinger, J. D., Carrol, T. S., Ganay, T., Lundebye, A. K., Sekler, I., ... Hogstrand, C., et al. (2018). Parallel in vivo and in vitro transcriptomics analysis reveals calcium and zinc signalling in the brain as sensitive targets of HBCD neurotoxicity. *Archives of Toxicology, 92*(3), 1189–1203. https://doi.org/10.1007/s00204-017-2119-2.

Rice, C., Ghorai, J. K., Zalewski, K., & Weber, D. N. (2011). Developmental lead exposure causes startle response deficits in zebrafish. *Aquatic Toxicology, 105*(3–4), 600–608.

Richardson, J. R., Fitsanakis, V., Westerink, R. H. S., & Kanthasamy, A. G. (2019). Neurotoxicity of pesticides. *Acta Neuropathologica*, 1–20.

Richendrfer, H., Pelkowski, S. D., Colwill, R. M., & Créton, R. (2012). Developmental sub-chronic exposure to chlorpyrifos reduces anxiety-related behavior in zebrafish larvae. *Neurotoxicology and Teratology, 34*(4), 458–465.

Rock, K. D., & Patisaul, H. B. (2018). Environmental mechanisms of neurodevelopmental toxicity. *Current Environmental Health Reports, 5*(1), 145–157.

Saeedi Saravi, S. S., & Dehpour, A. R. (2016). Potential role of organochlorine pesticides in the pathogenesis of neurodevelopmental, neurodegenerative, and neurobehavioral disorders: A review. *Life Sciences, 145*, 255–264.

Seth, A., Stemple, D. L., & Barroso, I. (2013). The emerging use of zebrafish to model metabolic disease. *Disease Models and Mechanisms, 6*(5), 1080–1088.

Shao, B., Zhu, L., Dong, M., Wang, J. J., Wang, J. J., Xie, H., ... Zhu, S. (2012). DNA damage and oxidative stress induced by endosulfan exposure in zebrafish (Danio rerio). *Ecotoxicology, 21*(5), 1533–1540.

Silva, M. H., & Beauvais, S. L. (2010). Human health risk assessment of endosulfan. I: Toxicology and hazard identification. *Regulatory Toxicology and Pharmacology, 56*(1), 4–17.

Silva, M. H., & Gammon, D. (2009). An assessment of the developmental, reproductive, and neurotoxicity of endosulfan. *Birth Defects Research Part B: Developmental and Reproductive Toxicology, 86*(1), 1–28.

Slotkin, T. A., & Seidler, F. J. (2007). Comparative developmental neurotoxicity of organophosphates in vivo: Transcriptional responses of pathways for brain cell development, cell signaling, cytotoxicity and neurotransmitter systems. *Brain Research Bulletin, 72*(4–6), 232–274.

Søfteland, L., Kirwan, J. A., Hori, T. S. F., Størseth, T. R., Sommer, U., Berntssen, M. H. G., ... Olsvik, P. A. (2014). Toxicological effect of single contaminants and contaminant mixtures associated with plant ingredients in novel salmon feeds. *Food and Chemical Toxicology: An International Journal Published for the British Industrial Biological Research Association, 73*, 157–174.

Stanley, K. A., Curtis, L. R., Simonich, S. L. M., & Tanguay, R. L. (2009). Endosulfan I and endosulfan sulfate disrupts zebrafish embryonic development. *Aquatic Toxicology, 95*(4), 355–361.

Sun, H.-J., Zhang, J.-Y., Wang, Q., Zhu, E., Chen, W., Lin, H., ... Hong, H. (2019). Environmentally relevant concentrations of arsenite induces developmental toxicity and oxidative responses in the early life stage of zebrafish. *Environmental Pollution, 254*, 113022.

Teixidó, E., Piqué, E., Gonzalez-Linares, J., Llobet, J. M., & Gómez-Catalán, J. (2015). Developmental effects and genotoxicity of 10 water disinfection by-products in zebrafish. *Journal of Water and Health, 13*(1), 54–66.

Tiedeken, J. A., & Ramsdell, J. S. (2009). DDT exposure of zebrafish embryos enhances seizure susceptibility: Relationship to fetal p,p′-DDE burden and domoic acid exposure of California sea lions. *Environmental Health Perspectives, 117*(1), 68.

Tierney, K. B. (2011). Behavioural assessments of neurotoxic effects and neurodegeneration in zebrafish. *Biochimica et Biophysica Acta - Molecular Basis of Disease, 1812*(3), 381–389.

Timme-Laragy, A. R., Goldstone, J. V., Imhoff, B. R., Stegeman, J. J., Hahn, M. E., & Hansen, J. M. (2013). Glutathione redox dynamics and expression of glutathione-related genes in the developing embryo. *Free Radical Biology and Medicine, 65*, 89–101.

Trasande, L., & Liu, Y. (2011). Reducing the staggering costs of environmental disease in children, estimated at $76.6 billion in 2008. *Health Affairs, 30*(5), 863–870.

Velasco-Santamaría, Y. M., Handy, R. D., & Sloman, K. A. (2011). Endosulfan affects health variables in adult zebrafish (*Danio rerio*) and induces alterations in larvae development. *Comparative Biochemistry and Physiology - C Toxicology and Pharmacology, 153*(4), 372–380.

Weber, J., Halsall, C. J., Muir, D., Teixeira, C., Small, J., Solomon, K., ... Bidleman, T. (2010). Endosulfan, a global pesticide: a review of its fate in the environment and occurrence in the arctic. *The Science of the Total Environment, 408*(15), 2966–2984.

Willey, J. B., & Krone, P. H. (2001). Effects of endosulfan and nonylphenol on the primordial germ cell population in pre-larval zebrafish embryos. *Aquatic Toxicology, 54*(1–2), 113–123.

Wilson, W. W., Onyenwe, W., Bradner, J. M., Nennig, S. E., & Caudle, W. M. (2014). Developmental exposure to the organochlorine insecticide endosulfan alters expression of proteins associated with neurotransmission in the frontal cortex. *Synapse, 68*(11), 485–497.

Wilson, W. W., Shapiro, L. P., Bradner, J. M., & Caudle, W. M. (2014). Developmental exposure to the organochlorine insecticide endosulfan damages the nigrostriatal dopamine system in male offspring. *NeuroToxicology, 44*, 279–287.

Wu, L., Ru, H., Ni, Z., Zhang, X., Xie, H., Yao, F., ... Zhong, L. (2019). Comparative thyroid disruption by o,p′-DDT and p,p′-DDE in zebrafish embryos/larvae. *Aquatic Toxicology, 216*, 105280.

Zhao, J., Xu, T., & Yin, D.-Q. (2014). Locomotor activity changes on zebrafish larvae with different 2,2′,4,4′-tetrabromodiphenyl ether (PBDE-47) embryonic exposure modes. *Chemosphere, 94*, 53–61.

Chapter 47

Neurodevelopment of the zebrafish spinal serotonin system

Mark A. Masino and Jacob E. Montgomery
Department of Neuroscience, University of Minnesota, Minneapolis, MN, United States

List of abbreviations

5-HT Serotonin (5-hydroxytryptamine)
CNS Central nervous system
CSF Cerebrospinal fluid
CSF-cNs Cerebrospinal fluid–contacting neurons
dpf Days postfertilization
EGFP Enhanced green-fluorescent protein
hpf Hours postfertilization
IRa Inferior raphe
ISN Intraspinal serotonergic neuron
SC Spinal cord
SRa Superior raphe
VLS Ventrolateral stripes

Mini-dictionary of terms

Adult Sexually mature zebrafish (approximately 3 months postfertilization).
Central canal Spinal extension of the brain ventricles containing cerebrospinal fluid.
Cerebrospinal fluid Liquid that surrounds the brain and fills the ventricles of the brain and central canal of the spinal cord.
Embryo Zebrafish developmental stage spanning fertilization through hatching (through 2–3 days postfertilization).
Juvenile Zebrafish developmental stage during which zebrafish obtain adult characteristics (such as dark pigmentation and bone formation), but are not sexually mature (approximately 14 days through 3 months postfertilization).
Larva Zebrafish developmental stage spanning hatching through juvenile stage (approximately 2–3 through 14 days postfertilization).
Raphe nuclei Conserved vertebrate hindbrain regions that contain a large number of serotonergic neurons.
Serotonergic Neurons that produce serotonin, determined by expression of *tryptophan hydroxylase*.
Serotonin (5-hydroxytryptamine; 5-HT) A monoamine neurotransmitter that has a broad range of functions within and outside of the nervous system.
Tryptophan hydroxylase (Tph) The rate-limiting enzyme in serotonin synthesis.
Volume transmission Intracellular signaling that occurs through extracellular space outside of conventional synapses.

Introduction

Although it is most commonly known for its neurotransmitter function, serotonin (5-HT) is an ancient molecule that is believed to predate the evolution of multicellular organisms and nervous tissue (Turlejski, 1996). Consequently, 5-HT retains many functions outside of classical neurotransmission. Indeed, 5-HT was first isolated (and named) not because of its activity within the central nervous system (CNS), but by virtue of its function as a vasoconstrictor (Rapport, Green, & Page, 1948). 5-HT has since been found nearly ubiquitously throughout the nervous system and has been identified as a modulator of a host of functions within the nervous system.

Diagnosis, Management and Modeling of Neurodevelopmental Disorders. https://doi.org/10.1016/B978-0-12-817988-8.00047-6

In vertebrates, the influences of the spinal serotonergic system on functions both related to neurotransmission and development have made it an attractive target for investigating spinal cord (SC) injury and locomotor dysfunction (Ghosh & Pearse, 2014; Sandyk, 2006). Serotonergic signaling modulates, and in some cases initiates, spinal locomotor network activity (Nishimaru, Takizawa, & Kudo, 2000). Although 5-HT's effects on locomotor networks are generally considered excitatory, its mechanisms of action are much more complicated, largely due to the presence of seven families of 5-HT receptor subtypes and the expression and localization of these receptors. Controlling spinal serotonin receptor activity is one approach that investigators are exploiting to improve locomotor function below spinal injuries (Ghosh & Pearse, 2014). 5-HT also influences a number of developmental processes such as neurogenesis (proliferation and differentiation), apoptosis, and neurite growth (Lauder, 1993; Trakhtenberg & Goldberg, 2012), all of which are important for the recovery of the SC from injury. Taken together, our understanding of the spinal 5-HT system has potential to significantly impact the advancement of treatments for SC injuries and disorders.

The use of zebrafish to study vertebrate neurodevelopment

Native to South Asia, the freshwater zebrafish was advanced as an animal model by George Streisinger in the 1970s and 1980s due in large part to their amenability to large-scale genetic approaches (Streisinger, Walker, Dower, Knauber, & Singer, 1981). Single adult mating pairs are capable of producing hundreds of embryos per day. Also, embryos are fertilized externally from the mother and are thus readily accessible for genetic manipulation. The rapid development of embryos and larvae, combined with their nearly complete optical transparency, is ideal for identifying developmental and behavioral mutations. This combination of attributes makes zebrafish more conducive to large-scale forward genetic approaches than other prevalent vertebrate animal models (Table 47.1). The more recent development of technologies to induce targeted mutations, such as transcription activator-like effector nucleases (TALENs) and clustered regularly interspaced short palindromic repeats (CRISPRs), has further increased the versatility of zebrafish as a genetic model system.

The zebrafish CNS develops rapidly, with the majority of neuronal cell types and projections appearing by 3–4 days postfertilization (dpf). Since embryos and larvae develop externally from the mother and are transparent, different aspects of CNS development can be noninvasively visualized in intact, living animals. Progress in the use of transposon-mediated approaches for transgenesis enabled the rapid generation of stable transgenic lines expressing fluorescent reporters to facilitate live imaging of neural tissues (Kawakami, Koga, Hori, & Shima, 1998). Photoconvertible and photoswitchable fluorescent proteins, such as Kaede, continue to play especially prominent roles in cell lineage tracing (Hatta, Tsujii, & Omura, 2006) and examining neurite projection patterns (Sato, Takahoko, & Okamoto, 2006; Scott et al., 2007).

TABLE 47.1 Prevalent neurodevelopmental model organisms.

Model organism	Advantages	Disadvantages
Chicken (*Galus galus*)	Large eggs External development Amniote; development more similar to mammals than aquatic species	Limited genetic tools Low throughput Shell makes embryos less accessible than aquatic species
Frog (*Xenopus laevis* and *Xenopus tropicalis*)	Large eggs High throughput External development Accessible embryos Rapid development	Limited genetic tools Eggs are opaque
Mouse (*Mus musculus*)	Mammal; phylogenetically most similar to humans Relevance to human disease Powerful genetic system	Internal development Inaccessible embryos Low throughput
Zebrafish (*Danio rerio*)	High throughput External development Transparent embryos and larvae Rapid development Abundance of genetic tools	Phylogenetically least similar to humans Genetics complicated by genome duplication

Comparison of advantages and disadvantages of using different vertebrate model organisms to study nervous system development. Information provided by the authors.

Furthermore, the optical accessibility of the larval zebrafish CNS permits intact whole-brain functional imaging of neuronal activity using genetically encoded calcium indicators, as well as light-induced manipulations using optogenetic actuators and photosensitizers. Altogether, the genetic tractability and transparency of zebrafish embryos and larvae permit detailed investigation of an intact and rapidly developing vertebrate nervous system.

Neurodevelopment and ontogeny of swimming behavior

Zebrafish embryonic development encompasses the time of fertilization through hatching, which occurs between 48 and 72 hours postfertilization (hpf). Early zebrafish embryonic development is similar to other vertebrates, although zebrafish embryos develop much more rapidly than mammals. The first cell cleavage occurs approximately 40 min after the egg is fertilized (Kimmel, Ballard, Kimmel, Ullmann, & Schilling, 1995). The neural plate is formed during gastrulation but, unlike other vertebrates, the neural plate of teleost fishes forms an intermediate solid neural rod before eventually hollowing to form the characteristic neural tube (Schmitz, Papan, & Campos-Ortega, 1993). By 24 hpf the major anatomical regions of the brain are distinguishable (Appel, 2000) and several neuronal subtypes have differentiated. At this stage, the SC contains primary motor neurons, Rohon–Beard sensory neurons, and several interneuronal subtypes, including cerebrospinal fluid (CSF) contacting neurons (CSF-cNs; Bernhardt, Chitnis, Lindamer, & Kuwada, 1990; Myers, Eisen, & Westerfield, 1986; Saint-Amant, 2010). Differentiation of these neuronal subtypes coincides with the onset of low-frequency spontaneous tail coiling behavior at about 17 hpf (Saint-Amant, 2010). Although this coiling behavior continues through hatching, embryos that hatch prematurely are capable of producing swimming behavior as early as 28 hpf (Saint-Amant, 2010).

Morphogenesis is considered complete by 3 dpf, after the transition from the embryonic to the larval stage of development. The larval stage is characterized by expression of high-frequency beat and glide swimming behavior and the onset of feeding. During early larval development (3–7 dpf), swimming properties become more refined as individual episodes of swimming activity become shorter in duration and more frequent (Colwill & Creton, 2011; Ingebretson & Masino, 2013). This refinement of locomotor activity is closely tied to the maturation of spinal neuromodulatory systems, such as the serotonergic and dopaminergic systems (Brustein, Chong, Holmqvist, & Drapeau, 2003; Drapeau et al., 2002; Lambert, Bonkowsky, & Masino, 2012). Sex is determined during the juvenile stage (Kossack et al., 2019) and fish are considered adults upon reaching sexual maturity approximately 3 months after fertilization. Body mass increases approximately 5000-fold from embryo to adult (Gore & Burggren, 2012; Tye et al., 2015), imposing substantial changes to hydrodynamic forces during swimming (Muller, Stamhuis, & Videler, 2000). The resultant changes to swimming behavior include a decrease in tail-beat frequency and "coasting" between swimming episodes in adult fish (Kyriakatos et al., 2011). Zebrafish developmental stages are shown in Fig. 47.1.

Development of the serotonergic phenotype

The development of antibodies to 5-HT (Steinbusch, Verhofstad, & Joosten, 1978) enabled researchers to efficiently survey many species and tissues for the presence and location of 5-HT containing cells. However, synthesis of 5-HT is a key criterion for the classification of a cell as serotonergic, since some cell populations express 5-HT transporter proteins and accumulate 5-HT, but do not express the enzymes necessary for 5-HT synthesis (Lebrand et al., 1996). Thus, 5-HT immunoreactivity is not in itself sufficient for identifying serotonergic cells.

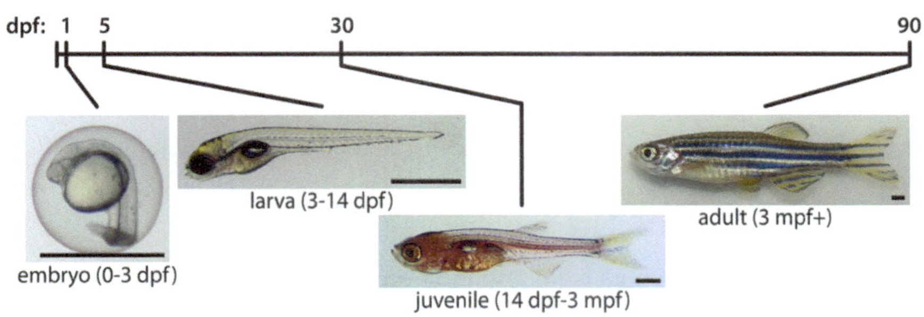

FIGURE 47.1 **Zebrafish developmental stages.** Bright-field images of a zebrafish embryo, larva, juvenile, and adult. The embryonic stage encompasses fertilization through hatching (0 through 2–3 dpf). The larval stage encompasses hatching through approximately 14 dpf. The juvenile stage lasts from approximately 14 dpf until the adult stage, which is determined by sexual maturity (at approximately 3 mpf). *Dpf*, days postfertilization; *mpf*, months postfertilization. Scale bars = 1 mm.

Tryptophan hydroxylase (Tph) is the rate-limiting enzyme in 5-HT synthesis and is specifically found in serotonergic neurons. Whereas humans and other mammals have two genes that encode *tryptophan hydroxylases* (*Tph1* and *Tph2*), zebrafish have three (*tph1a*, *tph1b*, and *tph2*) due to genome duplication (Postlethwait et al., 1998). There is some disagreement about a potential fourth *tph* gene. It was reported that *tyrosine hydroxylase 2* (*th2*), despite having greater sequence homology to *tyrosine hydroxylase 1* (*th1*) than *tph1a*, *tph1b*, or *tph2*, encodes a Tph and is a marker of serotonergic neurons in the hypothalamus (Ren, Li, Zhong, & Lin, 2013). Others have since argued that *th2* is, like *th1*, expressed in dopaminergic neurons and that these cells lie in proximity to, but are distinct from, hypothalamic serotonergic neurons (McPherson et al., 2016). Regardless, *th2*-expressing neurons are not found in the SC and are not believed to innervate the SC; therefore, they will not be described further in this chapter.

Consistent with other vertebrates, fibroblast growth factor and hedgehog signaling are necessary for specification and formation of serotonergic raphe neurons in zebrafish (Teraoka et al., 2004). In mammals, *Nkx2.2* regulates expression of the Ets domain transcription factor *Pet1* (ortholog of human *FEV*) and the *Lim* class homeobox gene *Lmx1b*, both of which are important for sustaining the postmitotic serotonergic phenotype by maintaining expression of genes such as *Tph2* and *5-HT transporter* (Cheng et al., 2003; Deneris & Wyler, 2012). Mammalian *Pet1* expression is restricted to serotonergic raphe neurons and its expression precedes *Tph2* expression and 5-HT production (Hendricks, Francis, Fyodorov, & Deneris, 1999). Comparatively, zebrafish *pet1* expression corresponds to *tph2* expression in neurons in the raphe and SC, but not other *tph2*-expressing populations or neurons that express *tph1a* or *tph1b* (Lillesaar, Tannhauser, Stigloher, Kremmer, & Bally-Cuif, 2007; Montgomery, Wiggin, Rivera-Perez, Lillesaar, & Masino, 2016).

Comparative sources of spinal serotonin

The predominant sources of 5-HT in the mammalian CNS are the superior raphe (SRa) and inferior raphe (IRa) nuclei (Dahlstrom & Fuxe, 1964). These sources are well-conserved among all vertebrates (Jacobs & Azmitia, 1992; Lillesaar, 2011). The SRa sends ascending projections throughout the brain, while the IRa innervates the hindbrain and SC (Jacobs & Azmitia, 1992). The descending IRa projections are considered to represent the largest source of 5-HT to the SC, although several species also possess a relatively small number of serotonergic neurons within the SC itself (Schmidt & Jordan, 2000). Additionally, 5-HT production is conserved in the vertebrate pineal gland, where it is a biosynthetic precursor to melatonin. All three zebrafish *tph* homologs are expressed in the zebrafish pineal gland (Lillesaar, 2011).

Fish possess a comparatively greater number of supraspinal serotonergic cell populations than mammals (Lillesaar, 2011). Most notably in zebrafish is the presence of a large number of *tph1a*-expressing neurons within the hypothalamus and *tph2*-expressing neurons in the pretectal region of the diencephalon (Bellipanni, Rink, & Bally-Cuif, 2002; Teraoka et al., 2004). These serotonergic cell populations are not found in mammals and neither population innervates the SC (Lillesaar, 2011; McLean & Fetcho, 2004a, 2004b).

In the hindbrain, the zebrafish raphe nuclei (Fig. 47.2) bear a strong resemblance to the mammalian raphe nuclei, but with fewer discernible cell clusters within each raphe region (Lillesaar, Stigloher, Tannhauser, Wullimann, & Bally-Cuif, 2009). Similar to the mammalian raphe, serotonergic neurons express *pet1* and *tph2* and the SRa and IRa project to more rostral regions of the brain and to the hindbrain and SC, respectively (Lillesaar et al., 2007, 2009; McLean & Fetcho, 2004b). In addition to the SRa and IRa regions, zebrafish have groups of rostrally projecting serotonergic neurons in the ventrolateral raphe, designated as the ventrolateral stripes (VLS), of which there is no known mammalian equivalent (Lillesaar et al., 2009).

Finally, the zebrafish SC contains a substantial number of intraspinal serotonergic neurons (ISNs) that are distributed along the ventral SC (Brustein et al., 2003; McLean & Fetcho, 2004b; Montgomery et al., 2018; Van Raamsdonk, Bosch, Smit-Onel, & Maslam, 1996). The ISNs are further divided into two, temporally distinct populations that each resides in the ventral SC; one transient population in the embryo and the other that is ostensibly maintained throughout the remaining life of the fish (Montgomery et al., 2016). Altogether, the zebrafish SC is innervated by two serotonergic sources: descending projections from the IRa and two distinct intraspinal neuronal populations (Table 47.2).

Embryonic cerebrospinal fluid–contacting neurons

The first and only known serotonergic cell population to innervate the embryonic SC is a subset of intraspinal CSF-cNs, also known as Kolmer–Agduhr cells, that interact with the central canal (Dale, Roberts, Ottersen, & Storm-Mathisen, 1987). Initially, immunohistochemical labeling demonstrated that the zebrafish SC contains 5-HT immunoreactive cells in the ventral SC, but was unable to discriminate the embryonic and postembryonic populations (Brustein et al., 2003; McLean & Fetcho, 2004a, 2004b). The embryonic CSF-cN ISN population was only later identified by examining serotonergic phenotypes of the ISN populations, such as *pet1* and *tph* gene expression (Bellipanni et al., 2002; Djenoune et al., 2017; Montgomery et al., 2016).

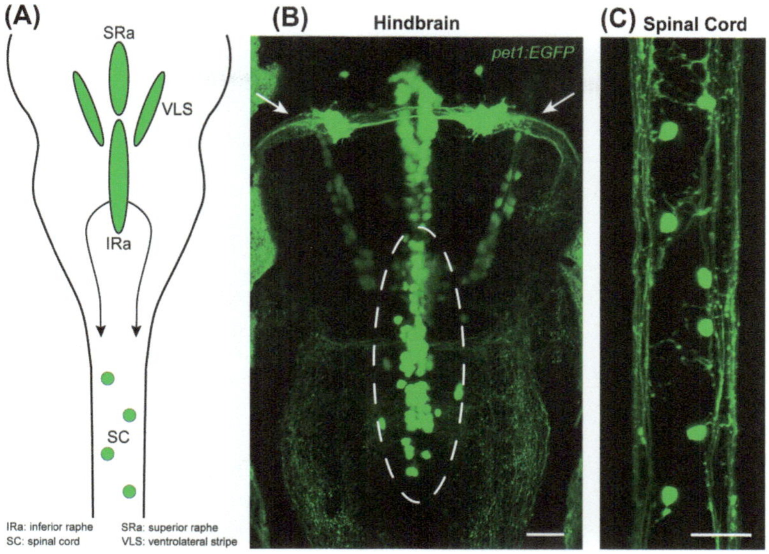

FIGURE 47.2 **IRa and ISNs provide serotonergic innervation to the SC.** (A) Diagram of hindbrain and rostral SC from dorsal perspective (serotonergic cell populations in green; *arrows* represent axonal projections to SC). (B) Dorsal view of larval (5 days postfertilization) *Tg(pet1:EGFP)*^{ne0214} expression in hindbrain. IRa axons project caudally to SC. *Dashed circle* indicates IRa cell bodies, *arrows* indicate ascending IRa axonal projections (full tract not visible in image). (C) Dorsal view of larval *Tg(pet1:EGFP)*^{ne0214} expression in SC. ISNs project laterally. *IRa*, inferior raphe; *ISN*, intraspinal serotonergic neuron; *SC*, spinal cord. Scale bars = 25 μm.

TABLE 47.2 Sources of zebrafish spinal serotonin.

Source	Spinal location	Stages of serotonin immunoreactivity	*tph* gene expression	Cotransmitter
Cerebrospinal fluid −contacting neurons	Ventral; associated with central canal	Embryo (transient; 24–80 hpf)[1]	*tph1a*[1,2]	GABA[1,3]
Inferior raphe projections	Lateral and dorsal	Larva, juvenile, adult (~72 hpf +)[4]	*tph2*[5,6]	None known
Intraspinal serotonergic neurons	Ventral	Larva, juvenile, adult (~64 hpf +)[1]	*tph2*[1]	None known

Summary of the sources of serotonergic innervation of the zebrafish spinal cord. Hpf, *hours postfertilization.* [1]Montgomery et al., 2016; [2]Bellipanni et al., 2002; [3]Djenoune et al., 2017; [4]McLean & Fetcho, 2004a; [5]Teraoka et al., 2004; [6]Lillesaar et al., 2009.

The serotonergic phenotype is restricted to the CSF-cNs that are located in the *nkx2.2a* expression domain below the central canal (known as Kolmer–Agduhr double-prime cells), which indicates that the spinal CSF-cNs are functionally heterogeneous (Djenoune et al., 2014, 2017). The CSF-cNs' primary neurotransmitter phenotype is GABA; however, a subset of these cells begins to express *tph1a* between 20 and 24 hpf (Bellipanni et al., 2002; Montgomery et al., 2016) and become 5-HT immunoreactive between 24 and 32 hpf (McLean & Fetcho, 2004a; Sallinen et al., 2009). *Tph1a* expression indicates that CSF-cNs are serotonergic, synthesizing 5-HT rather than accumulating it from other sources, such as the CSF. Consistent with other known *tph1a*-expressing cell populations (Lillesaar et al., 2007), the intraspinal CSF-cNs do not coexpress *pet1* or *tph2* (Fig. 47.3A; Montgomery et al., 2016). This dual transmitter phenotype (GABA and 5-HT) is transient (Fig. 47.4), as *tph1a* expression is lost by 68 hpf and 5-HT immunoreactivity becomes restricted to a second population of ISNs by approximately 80 hpf (Montgomery et al., 2016). Hence, the CSF-cNs only express a serotonergic phenotype in embryos. Although the function of 5-HT produced in the CSF-cNs is not known, the transience of the serotonergic phenotype during a restricted period of rapid embryonic development suggests a developmental function, possibly through volume transmission of 5-HT through the CSF.

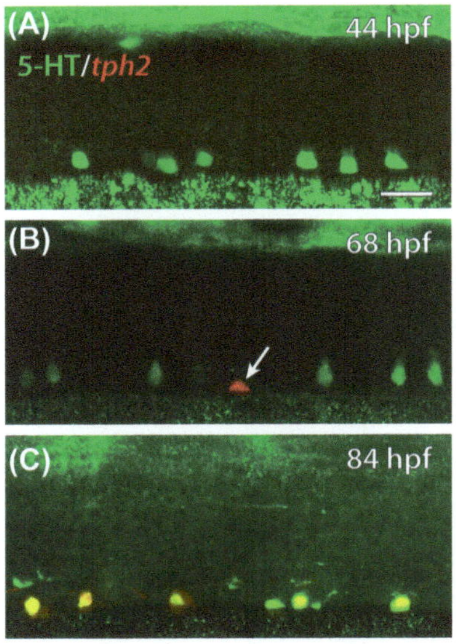

FIGURE 47.3 **Larval, but not embryonic, intraspinal serotonergic neurons express a *tph2* promoter-driven transgene.** Lateral view of overlayed 5-HT antibody labeling (green) and *Tg(tph2:nfsB-mCherry)*^{y226} expression (*tph2*; red). (A) In the embryo (44 hpf), cells in the ventral SC contain 5-HT, but do not express *tph2:mCherry*. (B) During the transition from embryo to larva (68 hpf), *tph2:mCherry*-expressing cells that have not yet synthesized 5-HT are detected (*arrow*). (C) In the larva (84 hpf), 5-HT antibody labeling is restricted to *tph2:mCherry*-expressing cells (yellow). 5-HT, serotonin; *hpf*, hours postfertilization. Scale Bar = 20 μm.

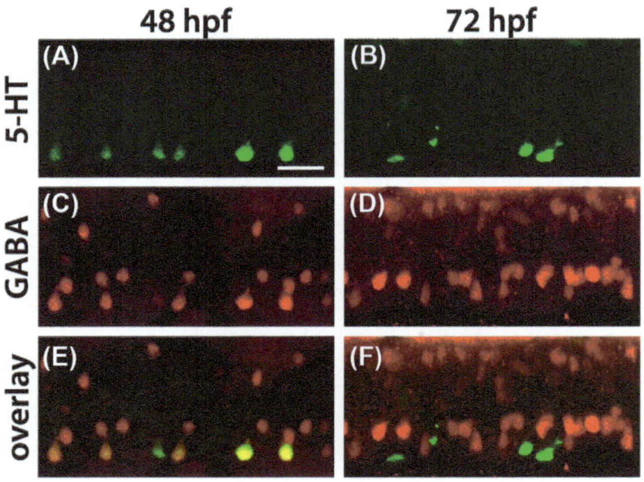

FIGURE 47.4 **Cotransmitter phenotype of embryonic intraspinal serotonergic neurons.** Lateral view of embryonic (48 hpf) and larval (72 hpf) zebrafish SCs double-labeled with antibodies to 5-HT (green; A and B) and GABA (red; C and D), overlayed in (E) and (F). (A, C, and E) In embryos (48 hpf), 5-HT antibody labeling is restricted to GABAergic cells in the ventral SC. (B, D, and F) In larvae (72 hpf), 5-HT antibody labeling does not colocalize with GABA. 5-HT, serotonin; *hpf*, hours postfertilization. Scale bar = 20 μm.

Postembryonic intraspinal serotonergic neurons

At approximately 60 hpf, as the CSF-cNs lose *tph1a* expression, a second population of ISNs in the *Nkx2.2a*-expressing ventral region of the SC begins to express serotonergic markers. In larvae, these cells are arranged bilaterally into two rows and are regularly distributed along the rostrocaudal SC axis (Fig. 47.2C). The ISNs first express *pet1*, then *tph2*, before gradually becoming 5-HT immunoreactive (Fig. 47.3B and C; Montgomery et al., 2016). Expression of these markers progresses in a rostrocaudal direction and virtually all cellular 5-HT antibody labeling in the SC is restricted to approximately 90 *pet1*-expressing ISNs by 96 hpf (Montgomery et al., 2016). The progression of *pet1* and *tph2* gene expression and 5-HT synthesis is consistent with that described in the mouse raphe (Hendricks et al., 1999), but occurs more rapidly. This population of ISNs appears to persist in adults, since all adult spinal 5-HT positive neurons are located in the ventral SC and express *pet1* promoter-driven transgenes (Montgomery et al., 2018; Van Raamsdonk et al., 1996).

Morphologically, ISN neurites grow through 4 or 5 dpf and are found in close proximity to motor neurons (McLean & Fetcho, 2004a, 2004b; Montgomery et al., 2018). ISNs normally have one primary neurite that projects from the soma to the lateral edge of the SC and bifurcates into a rostrally projecting dendritic tree and a caudally projecting axon (Fig. 47.5; Brustein et al., 2003; McLean & Fetcho, 2004a; Montgomery et al., 2018). Together, larval ISN dendrites and axons span approximately 120 μM, or two body segments, rostrocaudally (Montgomery et al., 2018). The larval ISN axons terminate with one or two distinctive enlargements resembling growth cones (Fig. 47.5B′, *arrowheads*). Interestingly, these enlarged structures persist even after axon growth has ceased and contain synaptic vesicles (Montgomery et al., 2018). The distinctive size and shape of the ISN terminal structures suggest that they are functionally specialized, perhaps for extrasynaptic volume transmission.

Inferior raphe axons

The zebrafish raphe is made of up three primary nuclei: the SRa, IRa, and VLS. Serotonergic neurons in the SRa and IRa differentiate simultaneously, with both regions developing in a rostral-to-caudal direction, indicating that the two raphe populations likely originate from distinct progenitor sources (Lillesaar et al., 2007). Although the serotonergic SRa neurons are believed to arise from neuronal progenitor cells near the midbrain–hindbrain boundary, the source of serotonergic IRa neurons is not known (Lillesaar et al., 2007). *Pet1* gene expression is detected in the SRa and IRa between 25 and 36 hpf, which is followed by *tph2* expression approximately 5 h later (Lillesaar et al., 2007; Teraoka et al., 2004). The SRa and IRa are clearly discernible and cells within both regions contain 5-HT by 48 hpf (Lillesaar et al., 2009; McLean & Fetcho, 2004a). The VLS, which was not identified in zebrafish until the generation of *pet1* promoter-driven transgenic lines to facilitate their visualization, is the final serotonergic raphe population to develop, with transgene expression becoming detectable between 2 and 3 dpf (Lillesaar et al., 2009). Interestingly, although the VLS expresses *pet1* and *tph2* transcripts (Lillesaar et al., 2009), studies using *tph2* promoter-driven transgenic zebrafish lines did not describe transgene expression in the VLS (Oikonomou et al., 2019; Yokogawa, Hannan, & Burgess, 2012).

The zebrafish IRa, like the mammalian IRa, is the only serotonergic cell population in the brain that innervates the SC (Lillesaar et al., 2009; McLean & Fetcho, 2004b). Neurites project laterally from *pet1*-expressing cells in the IRa at approximately 36 hpf, and by 48 hpf these neurites clearly extend both rostrally and caudally (Lillesaar et al., 2009). Rostral projections extend dorsally and laterally before forming dense bundles that merge at the midline (Fig. 47.2B, *arrows*) between 2 and 3 dpf (Lillesaar et al., 2009). Meanwhile, caudal projections from the IRa project throughout the hindbrain and into the SC (Lillesaar et al., 2009; McLean & Fetcho, 2004b), entering the rostral SC at approximately 48 hpf (Montgomery et al., 2016). Innervation of the SC by serotonergic IRa axons proceeds rostrocaudally and loosely coincides with the onset of serotonergic marker expression in the larval *tph2*-expressing ISN population (Montgomery et al., 2016). Serotonergic IRa innervation of the entire SC appears complete by 4 or 5 dpf (McLean & Fetcho, 2004a).

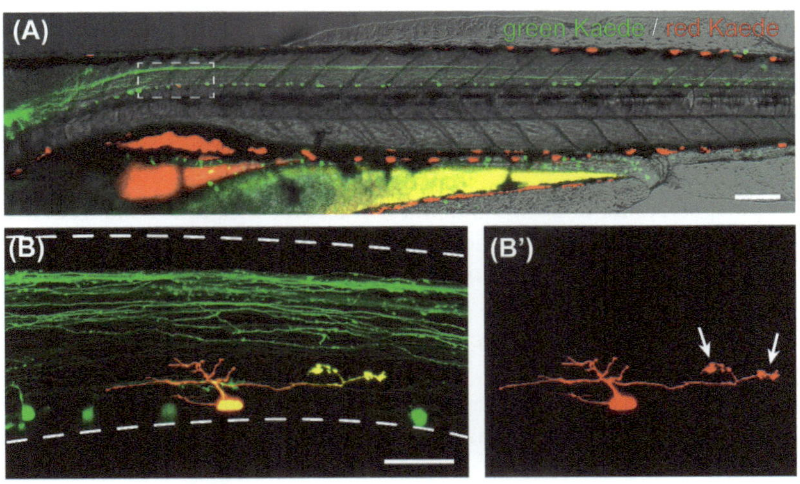

FIGURE 47.5 Larval ISN morphology. A single ISN in a live, 5 days postfertilization *Tg(pet1:Gal4-p2A-EGFP)*^{gy1};*Tg(UAS:Kaede)*^{s1999t} larva was photoconverted with ultraviolet light to reveal its morphological features. (A) *Dashed box* indicates the location of the photoconverted ISN. (B) Higher magnification of the photoconverted ISN showing unconverted (green) and photoconverted (red) Kaede (*dashed lines* represent dorsal and ventral boundaries of the SC). (B′) Photoconverted Kaede reveals ISN morphology. Dendrites project rostrally (left) and the axon projects caudally (right). The axon bifurcates and terminates with two enlarged synaptic structures (*arrows*). *ISN*, intraspinal serotonergic neuron. Scale bar in A = 100 μm; scale bar in B = 25 μm.

Applications to other areas of development

Serotonergic signaling plays important roles in modulating spinal network activity and in the pathogenesis and recovery from neurodegenerative diseases and conditions. 5-HT modulates spinal locomotor network excitability and plays a critical role in shaping activity to produce organized locomotor output. Hence, serotonergic modulation of motor neuronal excitatory responses to glutamate may affect neurodegenerative conditions that result from glutamate excitotoxicity, such as amyotrophic lateral sclerosis (Vermeiren, Janssens, Van Dam, & De Deyn, 2018). 5-HT also influences different aspects of neurogenesis, including cellular proliferation and fate determination (Lauder, 1993) and neurite growth and pathfinding (Trakhtenberg & Goldberg, 2012). This makes spinal 5-HT an attractive target for regenerative studies in the SC (Ghosh & Pearse, 2014). Indeed, transplantation of serotonergic neurons into injured SCs is being explored as an approach to restore locomotor function to injured mammalian SCs (Ghosh & Pearse, 2014). Furthermore, spinal 5-HT was recently demonstrated to promote regeneration of motor neurons following SC injury in adult zebrafish (Barreiro-Iglesias et al., 2015), illustrating the influence of spinal 5-HT on neurogenesis and its potential therapeutic value.

The significance of transient 5-HT production in the ventral zebrafish CSF-cNs is not known. All classes of vertebrates possess spinal CSF-cNs, and although the primary neurotransmitter phenotype of CSF-cNs is GABA, CSF-cN cotransmitter phenotypes (including 5-HT) have also been described in other species (Djenoune & Wyart, 2017) indicating that their functions may be conserved. Developmental signals are transmitted through the CSF, and 5-HT is understood to act as a trophic factor regulating neurodevelopment (Azmitia, 2001). It could be hypothesized that while 5-HT is supplied from the mother to the developing embryo in placental mammals, nonplacental animals would require transient sources of 5-HT during specific periods of embryonic development. Further, serotonergic neurons in the zebrafish hypothalamus also contact the CSF (Reuter et al., 2019) and, like the spinal CSF-cNs, express *tph1a* (Bellipanni et al., 2002). Although this hypothalamic population is permanent and 5-HT synthesis in the spinal CSF-cNs is only transient, functional comparisons between these populations could provide insight into serotonergic volume transmission through the CSF.

Key facts of the zebrafish spinal serotonin system

- Serotonin is an evolutionarily ancient molecule that is found nearly ubiquitously in animals.
- Serotonin is best known as a neurotransmitter, but also has strong developmental effects.
- Serotonergic cells contain Tryptophan hydroxylase, an enzyme required for serotonin synthesis.
- Serotonergic projections from the hindbrain to the spinal cord are conserved among vertebrates.
- The zebrafish spinal cord is innervated by serotonergic axons from the hindbrain and by serotonergic neurons distributed along the ventral boundary of the spinal cord.
- The zebrafish spinal cord becomes fully innervated with serotonergic fibers by 4–5 days postfertilization.

Summary points

- Zebrafish are a powerful model system for visualizing development because they are fertilized externally, develop rapidly, and are nearly transparent.
- Zebrafish development can be classified into embryonic, larval, juvenile, and adult stages.
- Mammals express two tryptophan hydroxylase homologs (*Tph1* and *Tph2*) and zebrafish express three (*tph1a*, *tph1b*, and *tph2*) and possibly a fourth (*th2*).
- The primary source of spinal serotonin in most vertebrates is descending projections from the inferior raphe nuclei.
- In addition to the well-conserved descending inferior raphe (*tph2*-expressing) projections to the spinal cord, zebrafish possess two groups of intraspinal serotonergic neurons that express the serotonergic phenotype at different stages of development.
- A subset of GABAergic cerebrospinal fluid–contacting neurons transiently expresses *tph1a* and produces serotonin in the developing embryonic spinal cord.
- A population of *tph2*-expressing intraspinal serotonergic neurons differentiates during the transition to the larval stage of development and is present throughout subsequent life stages.

References

Appel, B. (2000). Zebrafish neural induction and patterning. *Developmental Dynamics, 219*(2), 155–168. https://doi.org/10.1002/1097-0177(2000)9999:9999<::AID-DVDY1052>3.0.CO;2-K.

Azmitia, E. C. (2001). Modern views on an ancient chemical: Serotonin effects on cell proliferation, maturation, and apoptosis. *Brain Research Bulletin, 56*(5), 413–424. https://doi.org/10.1016/s0361-9230(01)00614-1.

Barreiro-Iglesias, A., Mysiak, K. S., Scott, A. L., Reimer, M. M., Yang, Y., Becker, C. G., & Becker, T. (2015). Serotonin promotes development and regeneration of spinal motor neurons in zebrafish. *Cell Reports, 13*(5), 924–932. https://doi.org/10.1016/j.celrep.2015.09.050.

Bellipanni, G., Rink, E., & Bally-Cuif, L. (2002). Cloning of two tryptophan hydroxylase genes expressed in the diencephalon of the developing zebrafish brain. *Mechanisms of Development, 119*(Suppl. 1), S215–S220. https://doi.org/10.1016/S0925-4773(03)00119-9.

Bernhardt, R. R., Chitnis, A. B., Lindamer, L., & Kuwada, J. Y. (1990). Identification of spinal neurons in the embryonic and larval zebrafish. *Journal of Comparative Neurology, 302*(3), 603–616. https://doi.org/10.1002/cne.903020315.

Brustein, E., Chong, M., Holmqvist, B., & Drapeau, P. (2003). Serotonin patterns locomotor network activity in the developing zebrafish by modulating quiescent periods. *Journal of Neurobiology, 57*(3), 303–322. https://doi.org/10.1002/neu.10292.

Cheng, L., Chen, C. L., Luo, P., Tan, M., Qiu, M., Johnson, R., & Ma, Q. (2003). Lmx1b, Pet-1, and Nkx2.2 coordinately specify serotonergic neurotransmitter phenotype. *Journal of Neuroscience, 23*(31), 9961–9967. https://doi.org/10.1523/JNEUROSCI.23-31-09961.2003.

Colwill, R. M., & Creton, R. (2011). Locomotor behaviors in zebrafish (*Danio rerio*) larvae. *Behavioural Processes, 86*(2), 222–229. https://doi.org/10.1016/j.beproc.2010.12.003.

Dahlstrom, A., & Fuxe, K. (1964). Localization of monoamines in the lower brain stem. *Experientia, 20*(7), 398–399. https://doi.org/10.1007/bf02147990.

Dale, N., Roberts, A., Ottersen, O. P., & Storm-Mathisen, J. (1987). The morphology and distribution of 'Kolmer-Agduhr cells', a class of cerebrospinal-fluid-contacting neurons revealed in the frog embryo spinal cord by GABA immunocytochemistry. *Proceedings of the Royal Society of London Series B Biological Sciences, 232*(1267), 193–203. https://doi.org/10.1098/rspb.1987.0068.

Deneris, E. S., & Wyler, S. C. (2012). Serotonergic transcriptional networks and potential importance to mental health. *Nature Neuroscience, 15*(4), 519–527. https://doi.org/10.1038/nn.3039.

Djenoune, L., Desban, L., Gomez, J., Sternberg, J. R., Prendergast, A., Langui, D., … Wyart, C. (2017). The dual developmental origin of spinal cerebrospinal fluid-contacting neurons gives rise to distinct functional subtypes. *Scientific Reports, 7*(1), 719. https://doi.org/10.1038/s41598-017-00350-1.

Djenoune, L., Khabou, H., Joubert, F., Quan, F. B., Nunes Figueiredo, S., Bodineau, L., … Wyart, C. (2014). Investigation of spinal cerebrospinal fluid-contacting neurons expressing PKD2L1: Evidence for a conserved system from fish to primates. *Frontiers in Neuroanatomy, 8*, 26. https://doi.org/10.3389/fnana.2014.00026.

Djenoune, L., & Wyart, C. (2017). Light on a sensory interface linking the cerebrospinal fluid to motor circuits in vertebrates. *Journal of Neurogenetics, 31*(3), 113–127. https://doi.org/10.1080/01677063.2017.1359833.

Drapeau, P., Saint-Amant, L., Buss, R. R., Chong, M., McDearmid, J. R., & Brustein, E. (2002). Development of the locomotor network in zebrafish. *Progress in Neurobiology, 68*(2), 85–111. https://doi.org/10.1016/S0301008202000758.

Ghosh, M., & Pearse, D. D. (2014). The role of the serotonergic system in locomotor recovery after spinal cord injury. *Frontiers in Neural Circuits, 8*, 151. https://doi.org/10.3389/fncir.2014.00151.

Gore, M., & Burggren, W. W. (2012). Cardiac and metabolic physiology of early larval zebrafish (*Danio rerio*) reflects parental swimming stamina. *Frontiers in Physiology, 3*, 35. https://doi.org/10.3389/fphys.2012.00035.

Hatta, K., Tsujii, H., & Omura, T. (2006). Cell tracking using a photoconvertible fluorescent protein. *Nature Protocols, 1*(2), 960–967. https://doi.org/10.1038/nprot.2006.96.

Hendricks, T., Francis, N., Fyodorov, D., & Deneris, E. S. (1999). The ETS domain factor Pet-1 is an early and precise marker of central serotonin neurons and interacts with a conserved element in serotonergic genes. *Journal of Neuroscience, 19*(23), 10348–10356. https://doi.org/10.1523/JNEUROSCI.19-23-10348.1999.

Ingebretson, J. J., & Masino, M. A. (2013). Quantification of locomotor activity in larval zebrafish: Considerations for the design of high-throughput behavioral studies. *Frontiers in Neural Circuits, 7*, 109. https://doi.org/10.3389/fncir.2013.00109.

Jacobs, B. L., & Azmitia, E. C. (1992). Structure and function of the brain serotonin system. *Physiological Reviews, 72*(1), 165–229. https://doi.org/10.1152/physrev.1992.72.1.165.

Kawakami, K., Koga, A., Hori, H., & Shima, A. (1998). Excision of the tol2 transposable element of the medaka fish, *Oryzias latipes*, in zebrafish, *Danio rerio*. *Gene, 225*(1–2), 17–22. https://doi.org/10.1016/S0378-1119(98)00537-X.

Kimmel, C. B., Ballard, W. W., Kimmel, S. R., Ullmann, B., & Schilling, T. F. (1995). Stages of embryonic development of the zebrafish. *Developmental Dynamics, 203*(3), 253–310. https://doi.org/10.1002/aja.1002030302.

Kossack, M. E., High, S. K., Hopton, R. E., Yan, Y. L., Postlethwait, J. H., & Draper, B. W. (2019). Female sex development and reproductive duct formation depend on Wnt4a in zebrafish. *Genetics, 211*(1), 219–233. https://doi.org/10.1534/genetics.118.301620.

Kyriakatos, A., Mahmood, R., Ausborn, J., Porres, C. P., Buschges, A., & El Manira, A. (2011). Initiation of locomotion in adult zebrafish. *Journal of Neuroscience, 31*(23), 8422–8431. https://doi.org/10.1523/JNEUROSCI.1012-11.2011.

Lambert, A. M., Bonkowsky, J. L., & Masino, M. A. (2012). The conserved dopaminergic diencephalospinal tract mediates vertebrate locomotor development in zebrafish larvae. *Journal of Neuroscience, 32*(39), 13488–13500. https://doi.org/10.1523/JNEUROSCI.1638-12.2012.

Lauder, J. M. (1993). Neurotransmitters as growth regulatory signals: Role of receptors and second messengers. *Trends in Neurosciences, 16*(6), 233–240. https://doi.org/10.1016/0166-2236(93)90162-F.

Lebrand, C., Cases, O., Adelbrecht, C., Doye, A., Alvarez, C., El Mestikawy, S., … Gaspar, P. (1996). Transient uptake and storage of serotonin in developing thalamic neurons. *Neuron, 17*(5), 823–835. https://doi.org/10.1016/S0896-6273(00)80215-9.

Lillesaar, C. (2011). The serotonergic system in fish. *Journal of Chemical Neuroanatomy, 41*(4), 294–308. https://doi.org/10.1016/j.jchemneu.2011.05.009.

Lillesaar, C., Stigloher, C., Tannhauser, B., Wullimann, M. F., & Bally-Cuif, L. (2009). Axonal projections originating from raphe serotonergic neurons in the developing and adult zebrafish, *Danio rerio*, using transgenics to visualize raphe-specific pet1 expression. *Journal of Comparative Neurology, 512*(2), 158–182. https://doi.org/10.1002/cne.21887.

Lillesaar, C., Tannhauser, B., Stigloher, C., Kremmer, E., & Bally-Cuif, L. (2007). The serotonergic phenotype is acquired by converging genetic mechanisms within the zebrafish central nervous system. *Developmental Dynamics, 236*(4), 1072–1084. https://doi.org/10.1002/dvdy.21095.

McLean, D. L., & Fetcho, J. R. (2004a). Ontogeny and innervation patterns of dopaminergic, noradrenergic, and serotonergic neurons in larval zebrafish. *Journal of Comparative Neurology, 480*(1), 38–56. https://doi.org/10.1002/cne.20280.

McLean, D. L., & Fetcho, J. R. (2004b). Relationship of tyrosine hydroxylase and serotonin immunoreactivity to sensorimotor circuitry in larval zebrafish. *Journal of Comparative Neurology, 480*(1), 57–71. https://doi.org/10.1002/cne.20281.

McPherson, A. D., Barrios, J. P., Luks-Morgan, S. J., Manfredi, J. P., Bonkowsky, J. L., Douglass, A. D., & Dorsky, R. I. (2016). Motor behavior mediated by continuously generated dopaminergic neurons in the zebrafish hypothalamus recovers after cell ablation. *Current Biology, 26*(2), 263–269. https://doi.org/10.1016/j.cub.2015.11.064.

Montgomery, J. E., Wahlstrom-Helgren, S., Wiggin, T. D., Corwin, B. M., Lillesaar, C., & Masino, M. A. (2018). Intraspinal serotonergic signaling suppresses locomotor activity in larval zebrafish. *Developmental Neurobiology, 78*(8), 807–827. https://doi.org/10.1002/dneu.22606.

Montgomery, J. E., Wiggin, T. D., Rivera-Perez, L. M., Lillesaar, C., & Masino, M. A. (2016). Intraspinal serotonergic neurons consist of two, temporally distinct populations in developing zebrafish. *Developmental Neurobiology, 76*(6), 673–687. https://doi.org/10.1002/dneu.22352.

Muller, U. K., Stamhuis, E. J., & Videler, J. J. (2000). Hydrodynamics of unsteady fish swimming and the effects of body size: Comparing the flow fields of fish larvae and adults. *Journal of Experimental Biology, 203*(Pt 2), 193–206. Retrieved from http://www.ncbi.nlm.nih.gov/pubmed/10607529.

Myers, P. Z., Eisen, J. S., & Westerfield, M. (1986). Development and axonal outgrowth of identified motoneurons in the zebrafish. *Journal of Neuroscience, 6*(8), 2278–2289. https://doi.org/10.1523/JNEUROSCI.06-08-02278.1986.

Nishimaru, H., Takizawa, H., & Kudo, N. (2000). 5-Hydroxytryptamine-induced locomotor rhythm in the neonatal mouse spinal cord in vitro. *Neuroscience Letters, 280*(3), 187–190. https://doi.org/10.1016/s0304-3940(00)00805-3.

Oikonomou, G., Altermatt, M., Zhang, R. W., Coughlin, G. M., Montz, C., Gradinaru, V., & Prober, D. A. (2019). The serotonergic raphe promote sleep in zebrafish and mice. *Neuron*. https://doi.org/10.1016/j.neuron.2019.05.038.

Postlethwait, J. H., Yan, Y. L., Gates, M. A., Horne, S., Amores, A., Brownlie, A., … Talbot, W. S. (1998). Vertebrate genome evolution and the zebrafish gene map. *Nature Genetics, 18*(4), 345–349. https://doi.org/10.1038/ng0498-345.

Rapport, M. M., Green, A. A., & Page, I. H. (1948). Serum vasoconstrictor, serotonin; isolation and characterization. *Journal of Biological Chemistry, 176*(3), 1243–1251. Retrieved from http://www.ncbi.nlm.nih.gov/pubmed/18100415.

Ren, G., Li, S., Zhong, H., & Lin, S. (2013). Zebrafish tyrosine hydroxylase 2 gene encodes tryptophan hydroxylase. *Journal of Biological Chemistry, 288*(31), 22451–22459. https://doi.org/10.1074/jbc.M113.485227.

Reuter, I., Jackels, J., Kneitz, S., Kuper, J., Lesch, K. P., & Lillesaar, C. (2019). Fgf3 is crucial for the generation of monoaminergic cerebrospinal fluid contacting cells in zebrafish. *Biology Open, 8*(6). https://doi.org/10.1242/bio.040683.

Saint-Amant, L. (2010). Development of motor rhythms in zebrafish embryos. *Progress in Brain Research, 187*, 47–61. https://doi.org/10.1016/B978-0-444-53613-6.00004-6.

Sallinen, V., Sundvik, M., Reenila, I., Peitsaro, N., Khrustalyov, D., Anichtchik, O., … Panula, P. (2009). Hyperserotonergic phenotype after monoamine oxidase inhibition in larval zebrafish. *Journal of Neurochemistry, 109*(2), 403–415. https://doi.org/10.1111/j.1471-4159.2009.05986.x.

Sandyk, R. (2006). Serotonergic mechanisms in amyotrophic lateral sclerosis. *International Journal of Neuroscience, 116*(7), 775–826. https://doi.org/10.1080/00207450600754087.

Sato, T., Takahoko, M., & Okamoto, H. (2006). HuC:Kaede, a useful tool to label neural morphologies in networks in vivo. *Genesis, 44*(3), 136–142. https://doi.org/10.1002/gene.20196.

Schmidt, B. J., & Jordan, L. M. (2000). The role of serotonin in reflex modulation and locomotor rhythm production in the mammalian spinal cord. *Brain Research Bulletin, 53*(5), 689–710. https://doi.org/10.1016/s0361-9230(00)00402-0.

Schmitz, B., Papan, C., & Campos-Ortega, J. A. (1993). Neurulation in the anterior trunk region of the zebrafish *Brachydanio rerio*. *Roux's Archives of Developmental Biology, 202*(5), 250–259. https://doi.org/10.1007/BF00363214.

Scott, E. K., Mason, L., Arrenberg, A. B., Ziv, L., Gosse, N. J., Xiao, T., … Baier, H. (2007). Targeting neural circuitry in zebrafish using GAL4 enhancer trapping. *Nature Methods, 4*(4), 323–326. https://doi.org/10.1038/nmeth1033.

Steinbusch, H. W., Verhofstad, A. A., & Joosten, H. W. (1978). Localization of serotonin in the central nervous system by immunohistochemistry: Description of a specific and sensitive technique and some applications. *Neuroscience, 3*(9), 811–819. https://doi.org/10.1016/0306-4522(78)90033-7.

Streisinger, G., Walker, C., Dower, N., Knauber, D., & Singer, F. (1981). Production of clones of homozygous diploid zebra fish (*Brachydanio rerio*). *Nature, 291*(5813), 293–296. https://doi.org/10.1038/291293a0.

Teraoka, H., Russell, C., Regan, J., Chandrasekhar, A., Concha, M. L., Yokoyama, R., … Wilson, S. W. (2004). Hedgehog and Fgf signaling pathways regulate the development of tphR-expressing serotonergic raphe neurons in zebrafish embryos. *Journal of Neurobiology, 60*(3), 275–288. https://doi.org/10.1002/neu.20023.

Trakhtenberg, E. F., & Goldberg, J. L. (2012). The role of serotonin in axon and dendrite growth. *International Review of Neurobiology, 106*, 105–126. https://doi.org/10.1016/B978-0-12-407178-0.00005-3.

Turlejski, K. (1996). Evolutionary ancient roles of serotonin: Long-lasting regulation of activity and development. *Acta Neurobiologiae Experimentalis, 56*(2), 619–636. Retrieved from http://www.ncbi.nlm.nih.gov/pubmed/8768313.

Tye, M., Rider, D., Duffy, E. A., Seubert, A., Lothert, B., & Schimmenti, L. A. (2015). Nonhatching decapsulated Artemia cysts as a replacement to *Artemia* nauplii in juvenile and adult zebrafish culture. *Zebrafish, 12*(6), 457–461. https://doi.org/10.1089/zeb.2014.1031.

Van Raamsdonk, W., Bosch, T. J., Smit-Onel, M. J., & Maslam, S. (1996). Organisation of the zebrafish spinal cord: Distribution of motoneuron dendrites and 5-HT containing cells. *European Journal of Morphology, 34*(2), 65–77. Retrieved from http://www.ncbi.nlm.nih.gov/pubmed/9090993.

Vermeiren, Y., Janssens, J., Van Dam, D., & De Deyn, P. P. (2018). Serotonergic dysfunction in amyotrophic lateral sclerosis and Parkinson's disease: Similar mechanisms, dissimilar outcomes. *Frontiers in Neuroscience, 12*, 185. https://doi.org/10.3389/fnins.2018.00185

Yokogawa, T., Hannan, M. C., & Burgess, H. A. (2012). The dorsal raphe modulates sensory responsiveness during arousal in zebrafish. *Journal of Neuroscience, 32*(43), 15205–15215. https://doi.org/10.1523/JNEUROSCI.1019-12.2012.

Chapter 48

Inbred mouse model of brain development and intestinal microbiota

Reinaldo B. Oriá[1], Daniel V. Pinto[1], Ronaldo P. Dias[1], Ramon S. Raposo[2], Patricia L. Foley[3], João O. Malva[4] and Richard L. Guerrant[5]

[1]*Laboratory of Biology of Tissue Healing, Federal University of Ceara, Faculty of Medicine, Fortaleza, Ceará, Brazil;* [2]*Experimental Biology Core, University of Fortaleza, Fortaleza, Ceará, Brazil;* [3]*Division of Comparative Medicine, Georgetown University, Washington DC, United States;* [4]*Coimbra Institute for Clinical and Biomedical Research (iCBR), Faculty of Medicine, University of Coimbra, Coimbra, Portugal;* [5]*Center for Global Health, Division of Infectious Diseases and International Health, University of Virginia, School of Medicine, Charlottesville, VA, United States*

List of abbreviations

ADHD Attention deficit hyperactivity disorders
BDNF Brain-derived neurotropic factor
CA1 Cornus ammonis 1
GF Germ-free
LPS Lipopolysaccharide
MUC2 Mucin 2
SPF Specific-pathogen-free

Mini-dictionary of terms

Axenic mice Animals totally free of microbiota.
Enterotypes Clusters of certain microbial communities in the gut.
Germ-free mice Synonymous with axenic mice, i.e., devoid of all microorganisms.
Gnotobiosis Animals with a specific and known microbiota (note this includes axenic).
Inbred mice Animals from at least 20 consecutive generations of mating between siblings with offspring reaching >99% homozygosity. They are also called isogenic.
Intestinal dysbiosis Disruption/impairment of microbial community equilibrium in the gut, favoring disease.
Intestinal microbiota Set of nonpathogenic microorganisms that usually live in the intestinal environment.
Outbred mice Animals intentionally generated from nonrelated mating to maximize genetic diversity (genetic makeup with a high rate of heterozygosity). They are also called heterogeneous.
Specific-pathogen-free mice Animals that are free of specifically identified microorganisms and parasites that have been shown to cause either subclinical or clinical infection.

Introduction

The interaction between the intestinal microbiota and the brain (so-called gut microbiota–brain axis) has been the focus of much research recently (and is expected to substantially grow in the following years). Not only is an unidirectional axis likely playing a role to dynamically shape the brain structure and function during the life span, but a mutual influence is plausibly operating as well, under internal and external environmental cues.

Although much light has been shed on the subject, a more in-depth reflection is needed to untangle the implications of these new discoveries on how we use laboratory animals to efficiently answer research questions and to be translatable to

Diagnosis, Management and Modeling of Neurodevelopmental Disorders. https://doi.org/10.1016/B978-0-12-817988-8.00048-8

the well-being of humans. A controversial issue arises whether current preclinical animal models are overall useful to study brain development, including urgently needed breakthrough research toward new drug discovery to treat the devastating neurodegenerative and other brain-related diseases, including neuropsychiatric disorders (Zhu et al., 2019, 2020), such as worrisome depression (Li et al., 2018; Zheng et al., 2016), in humans in light of recent findings on the role of the intestinal microbiota. Important effects of the intestinal microbiota on multiple sclerosis (Camara-Lemarroy, Metz, Meddings, Sharkey, & Yong, 2018), Parkinson's (Caputi & Giron, 2018), autism spectrum (Adams et al., 2019; Ristori et al., 2019), nervous anorexia (Hata et al., 2019), attention deficit hyperactivity disorder (ADHD) (Bull-Larsen & Mohajeri, 2019), stroke (Xia et al., 2019), autoimmune (De & Shoenfeld, 2019), and Alzheimer's disease (Fujii et al., 2019; Lin et al., 2019; Lin, Zheng, & Zhang, 2018) have been increasingly highlighted.

Appropriate laboratory animals are paramount to understand/solve challenges to human brain health, which has been driving modern neuroscience ever since, including the need to unravel the intricate and complex mechanisms/signaling pathways that govern the brain response to lifelong stressors during normal aging. Poor postnatal brain development may increase the risk for brain neurodegenerative diseases later in life (Wang et al., 2019). Brain neurodegeneration is an inexorable condition of a profound disruption in brain homeostasis that leads to erosion of daily-basis healthy cognitive and social—emotional behaviors, threatening quality of life, that may lead to costly hospitalization and death. As human populations live longer, the need to develop and improve laboratory models to cope with aging-related diseases, in addition, to the better understanding of how early-life events affect brain adaptations to aging, is imperative.

Laboratory animal housing and translational neuroscience research: where are the missing points and challenges?

Current neuroscience research relies mostly on pathogen-free laboratory animal housing conditions and strict adherence to state-of-the-art animal husbandry procedures in clean/barrier-oriented vivaria. Thus, the expectation for a vivarium with good biosecurity control requires highly pristine conditions (almost sterile) in pathogen-free and even sterile in germ-free (GF) housing facilities, the latter utilized to determine the intestinal microbiota's effect during the modeling of diseases in the laboratory-controlled environment. As GF animals do not harbor a native microbiota, if that outcome is not seen after an experimentally driven induction, one can determine the role of intestinal microbiota on that tested condition. Moreover, as transferred intestinal microbiota from a pathogen-free housed donor can be applied to the GF host/recipient, one can reverse that lack of effect for further evidence confirmation.

The preference for inbred animals, isolated from the outside world and maintained in strictly controlled laboratory conditions, has been the caveat of modern research animal facilities with the following goals: reproducibility of the studies, assurance of healthy biological reagents under sanitized conditions, elimination of common rodent pathogens and parasitic vectors, reduction of experimental phenotype variability, standardization of the environment, and the usage of genetically engineered laboratory animals that are constructed from inbred genetic backgrounds (Oria et al., 2018). However, one very important aspect that perhaps has been neglected or not given adequate attention is the translation of the data to humans living in unsanitary environments.

Laboratory housing of unexposed laboratory animals to daily-basis environmental contaminants has led to questions regarding use of these animals to study brain development, as well as neurodegenerative diseases, and whether treatment assessments to these animals are efficiently translated to all human populations. This is an important issue as most of the global population live in poor settings of the world, especially in sub-Sahara Africa, Latin America, and Asia, enduring harsh lives during their life span. Many in this disenfranchised population live and age in deplorable conditions of poor hygiene and sanitation, chronically exposed to environmental toxicants, enteric pathogens and pollutants, and often suffering from a vicious cycle of enteric infections (even subclinically) (Bartelt, Bolick, & Guerrant, 2019) and malnutrition (Oria et al., 2016; Tshala-Katumbay, Mwanza, Rohlman, Maestre, & Oria, 2015). Fig. 48.1 illustrates the exposome one can accumulate during life span due to lifelong environmental exposure to biohazards, toxicants, pathogens, and pollutants, which may even be greater during early life, when the brain is more predisposed to injury (Costa, Aschner, Vitalone, Syversen, & Soldin, 2004).

In such settings, early-life poor environment, unhealthy diets, ubiquitous pathogens, and the lack of continued pro-cognitive stimulants (such as good maternal education, prolonged breastfeeding, schooling, etc.), especially in the first 2 years of life (Murray-Kolb et al., 2014; Pendergast et al., 2018), can jeopardize the full genetic potential for reaching optimal brain development.

Fig. 48.2 shows the vicious cycle of enteric infections and malnutrition in early childhood with lasting impairment in cognitive development and potential ways to ameliorate these effects (Guerrant, DeBoer, Moore, Scharf, & Lima, 2013;

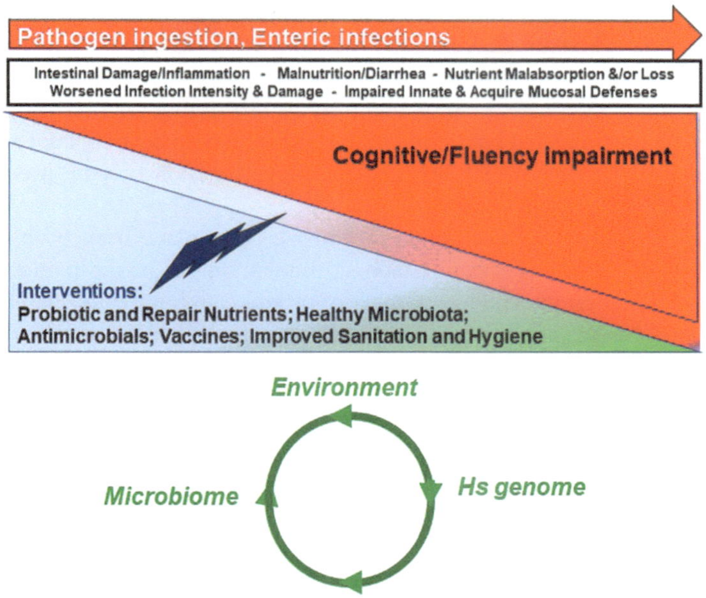

FIGURE 48.1 Diagram of the vicious cycle of diarrhea and malnutrition in the first years of life (one can amplify the other) in children living in poor settings of the developing world. Children chronically exposed to inadequate sanitation and hygiene may have compromised brain development with lasting consequences to cognitive function, including verbal fluency. Nutritional and other interventions may favor a healthy intestinal microbiota to break this vicious cycle. The intestinal microbiome, the environment, and the host genome interact over the life span.

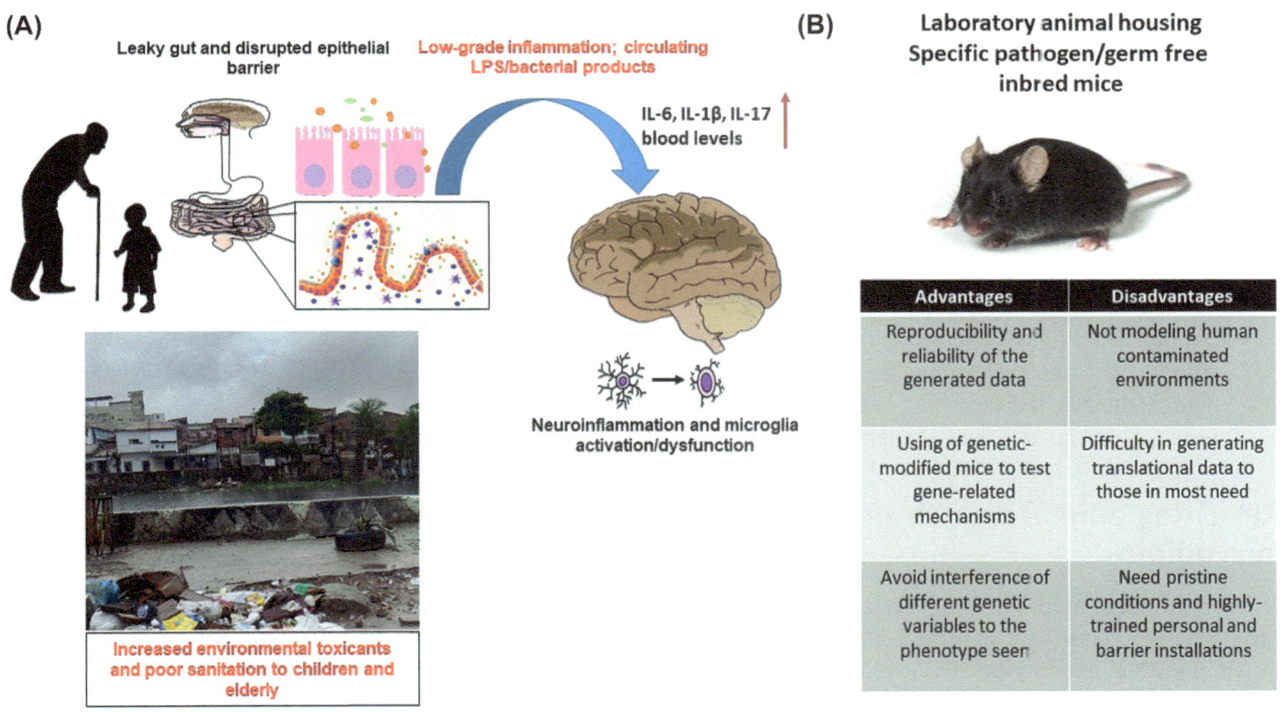

FIGURE 48.2 (A) Schematic model of environmental challenge to children and elderly in the developing world. Environmental toxicants, enteric pathogens, and pollutants may increase the risk of brain neuroinflammation and oxidative stress and reduced cognitive reservoir to deal with neurodegenerative diseases. (B) Box with advantages and disadvantages of the modern laboratory housing and inbred mice for translational neuroscience to benefit disenfranchised populations.

Guerrant, Oria, Moore, Oria, & Lima, 2008; Nataro & Guerrant, 2017). The healthy first years of postnatal life are key for optimal brain development (Ismail, Fatemi, & Johnston, 2017), as rapid myelination of brain neural circuitry, synaptogenesis, and even active neurogenesis in the hippocampus (a brain region strongly associated with memory and learning) occur in this time window. Human brain neurogenesis seems to decline during the first years of life (Sorrells et al., 2018), which may be accelerated by an unstimulated environment. If a full genetic potential is not achieved at this time, it may be forever lost. Unreached optimal language, sociability, and executive function may be irreversible, even after nutritional rehabilitation.

Interestingly, the formation of intestinal microbiota in humans takes place from birth to 3 years of life, as follows: beginning at birth following contact with the vaginal and skin communities of the mother and caregivers, followed by contact with breast milk and the environmental exposure to various symbionts and pathogens. The maturation of the intestinal microbiota (adulthood-like) is achieved in early childhood by the end of the first 3–5 years of life (Rodriguez et al., 2015). This maturation may occur at the same time when children are afflicted with enteric infections in endemic regions of infectious diseases and malnutrition. Improper maturation and diversity of the intestinal microbiota have been found in undernourished children living in poverty and exposed to unsanitary conditions (Subramanian et al., 2014; Yee et al., 2019) and may predispose these children to worse enteric infections (even without overt diarrhea) with lasting health consequences.

In recent decades, animal research has switched primarily from the use of outbred mouse stocks to inbred strains, such as the C57BL/6 mouse strain (most often used). In the latter, the genetic similarity renders these mice amenable for advanced scientific methods requiring greater reliability and reproducibility of the data. In addition to similarity of the genetic background, animals can be classified according to their living biota. The laboratory animal is called gnotobiotic when only certain known strains of bacteria and other microorganisms are native (Bleich & Hansen, 2012). Gnotobiotic animals can be classified as axenic or GF (meaning biota-free) or otherwise gnotobiotic where they have a defined microflora, such as a mixture of eight strains of bacteria, called altered *Schaedler* flora, to form their native biota (Dewhirst et al., 1999). This terminology is also used to describe the needed vivarium conditions in which they live to preserve their microbiota characteristics throughout their experimental life span. Specific-pathogen-free (SPF) rodents, in contrast to gnotobiotic, are tested regularly to ensure that they are free of a specific list of known rodent pathogens; however, their microflora remains undefined.

In order to develop a GF mouse, a cesarean delivery is performed in sterile conditions, without maternal microbiota contamination (from vagina and skin). All housing installations and materials are kept sterile. Interestingly, GF axenic animals show greater motor activity, reduced anxiety, show less social interaction, increased blood–brain barrier permeability (with reduction of the tight junctional proteins, such as occludin and claudin-5) (Braniste et al., 2014), increased neurogenesis, reduced maturation/branching of the microglia (Erny et al., 2015), reduced myelination of the prefrontal cortex, and exhibit less differentiation of effector T cells (Ogbonnaya et al., 2015). Conversely, Crumeyrolle-Arias et al. (2014), found an increase in anxiety behavior in rats raised in GF facility and Arentsen, Raith, Qian, Forssberg, and Diaz (2015) found an increased pattern of social interaction (Arentsen, Raith, Qian, Forssberg, & Diaz, 2015). Different effects on BDNF levels were also found by different authors, with reduced expression in the CA1 region, total hippocampus and exclusive increase in the dentate gyrus (Gareau et al., 2011; Neufeld, Kang, Bienenstock, & Foster, 2011; Sudo et al., 2004). Nevertheless, fecal transplantation has been found to improve anxiety behavior (Schmidt et al., 2020) and Alzheimer's disease-like hallmarks in mice (Kim et al., 2020).

Table 48.1 depicts different sanitary levels and microbiota characteristics of laboratory animal housing.

Revisiting animal models of brain development considering the intestinal microbiota

As the intestinal microbiota have been recognized to affect brain development and neurodegenerative diseases, potential approaches to understand the role of the human intestinal microbiota include the laboratory modeling of "humanized" mouse intestinal microbiota. However, some issues are found with such model and are listed as follows:

Use of humanized mouse microbiota

The major goal is to transfer human intestinal microbiota to mice to find out the role of the human microbiota on different outcomes, especially to reproduce/mimic a human illness condition in mice and to understand the underlying disease mechanisms and how to deal with it.

TABLE 48.1 Comparative environmental and sanitary characterization between laboratory animals (rodents) reared in a vivarium.

Laboratory animals (rodents)	Gnotobiotics (germ-free/defined microbiota)	Specific-pathogen-free (SPF)	Conventional
Pathogen status	Microbiota-free or defined	Defined exclusion list of microbial pathogens	Unknown or presence of nonexcluded microbial pathogens
Housing	Isolators	Microisolators	Open cages or microisolators
Drinking water	Filtered and autoclaved	Filtered and autoclaved, acidified, or chlorinated	Tap water, possibly filtered and/or treated
Diet	Autoclaved or irradiated	Autoclaved or irradiated	Nonsterile
Bedding	Autoclaved/irradiated	Autoclaved/irradiated	Nonsterile
Ventilation	Microenvironment, IVC system	Microenvironment, IVC system	Macroenvironment, room racks
Temperature/humidity control	Mandatory	Mandatory	Mandatory
Dark/Light cycle control	Present	Present	Present
Vector control (pests and vermis)	Mandatory	Mandatory	Ideal
Genetic control	Mandatory	Mandatory	Ideal
Sanitation monitoring	Mandatory	Mandatory	Ideal
Facility sanitary barriers	Rigorous	Rigorous	Absent or minimum

National Research Council. (2011). *Guide for care and use of laboratory animals* (8th ed.). Washington, DC: National Academies Press. 246p.

What is needed to consider

(1) In the human microbiota there are three general enterotypes, whereas in the murine only two (Wang et al., 2014). However, the similarity in the phylum level of these bacteria is relatively high, and the two predominant phyla in humans and murine are the same, *Firmicutes* and *Bacteroidetes*. Even though the phyla are the same, the phylum bacterial species differ around 85% between humans and mice (Ley, Peterson, & Gordon, 2006).

(2) In addition, the gastrointestinal tract of the mouse differs from the human in many aspects, including a lower mean intestinal pH level (e.g., BALB/c mice: pH 5.2 vs. humans: pH ~ 7) (McConnell, Basit, & Murdan, 2008).

(3) Mucus plays an important role in isolating the host from commensal bacteria and even in some cases serving as a stimulus for the growth of some of them. The repetition pattern of the amino acids that make up the mucus PTS (proline, threonine, and serine) domain is distinct from mice and humans. Although the PTS domain of human MUC2 exhibits a similar molecular weight than mice, a difference in sequence is seen (Hugenholtz & de Vos, 2018; Johansson et al., 2011).

(4) When mice with humanized microbiota are compared with those with native microbiota, there is a poor interaction between the new biota and the organism, affecting the immune system. There is a reduction in effector $CD4^+$ and $CD8^+$ lymphocytes and dendritic cell populations, antimicrobial peptides, and protection against Salmonella infection (Chung et al., 2012).

Although important differences exist, there are many general overlapping enterotype group similarities between the enterotypes of microbiota in humans and mice (Wang et al., 2014). More refined studies will likely discriminate potentially important differences that may be critical to better understand pathogen-related human interactions during homeostasis and disease conditions.

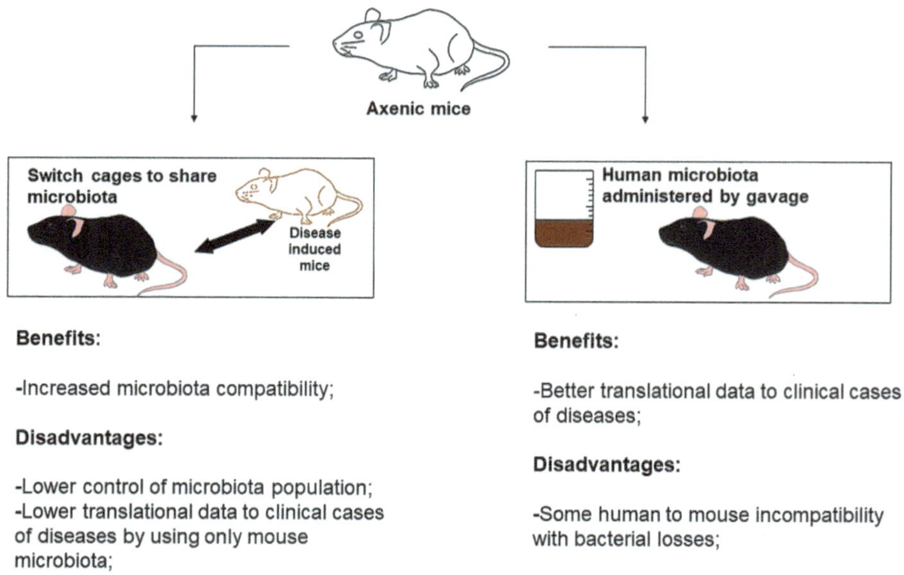

FIGURE 48.3 Diagram illustrating approaches to increase translational neuroscience, utilizing "humanized" intestinal microbiota to the laboratory mice.

A detailed review of differences in murine versus human gastrointestinal tract and microbiota enterotypes can be found elsewhere (Nguyen, Vieira-Silva, Liston, & Raes, 2015).

Pretreatment with antibiotics to knockdown the native biota before donor microbiota transfer may have a high impact on the intestinal mucosal integrity, with reduced production of local baseline cytokines (Hill & Artis, 2010). In addition, GF animals have a larger population of *Bacteroides* and those treated with antibiotics of *Parabacteroides* (Staley et al., 2017).

Interestingly, intestinal microbiota transfer from wild mice (mice living outside the laboratory and exposed to common rodent pathogens) to laboratory mice better approximates the immune system characteristics of humans (Beura et al., 2016) and perhaps that could be a different approach for studying brain development under "real-life" conditions seen in the environment. As new data emerge on integrative roles of the brain and the innate immune system (Kipnis, 2016), peripheral LPS-derived from chronic intestinal dysbiosis and disrupted intestinal barrier could be even more appreciated (Bartelt et al., 2019; Oria et al., 2016).

Fig. 48.3 shows a diagram of the humanized intestinal microbiota model in mice and other protocols in order to take into consideration the environmental factor for laboratory animal models.

Although one would expect that any mouse model data cannot be fully extrapolated to humans, as an intrinsic limitation of all preclinical models, findings from such "humanized" murine model may help to answer important research questions.

An elegant study from Luk and colleagues (Luk et al., 2018), who infected gnotobiotic mice with four types of *Bifidobacteria* from human infants, shed light on how intestinal microbiota affects brain development. The Swiss mouse pups were exposed to *Bifidobacterium* from the first to the 20th day after birth, by gavage. At 6–7 weeks of age, human microbiota–colonized mice showed improved memory and anxiety behaviors, further demonstrating benefits of these bacterial strains even in mice. Interestingly, intestinal microbiota transferred from premature human infants in a neonatal intensive care unit could have different roles in mouse growth (Lu et al., 2018).

Table 48.2 summarizes some studies that have tested humanized intestinal microbiota to assess brain development and behavior in mice.

TABLE 48.2 Summary of selected neuroscience-related studies conducting human intestinal microbiota transfer to mice in laboratory-controlled manner.

References	Objective	Study characteristics	Outcomes
Fujii et al. (2019)	Develop a mouse model useful in investigating the causes of Alzheimer's disease through oral administration of the disease's microbiota	Mice were used to transplant the human intestinal microbiota with fecal samples from a healthy volunteer and an affected patient to reproduce bacterial diversity in donors	- Microbiota of a patient with Alzheimer's disease triggers cognitive decline in the receptor in mice - Object location test and object recognition test have been significantly reduced - There are differences in intestinal metabolites, as a result of differences in the composition of intestinal microorganisms, causing varied behaviors, reinforcing the idea of an intestinal axis - Decrease in metabolites related to the nervous system, including γ-aminobutyrate, taurine, and valine
Hata et al. (2019)	Investigate the role of dysbiosis in specific pathologies of nervous anorexia (AN)	Germ-free mice were reconstituted with the microbiota of AN patients	- Decrease in body weight gain - Reduced food intake - Lower feed efficiency rate - Increased in anxiety related behavior - Lower levels of serotonin in the brain stem
Xia et al. (2019)	Compare the intestinal microbiota of patients with acute ischemic stroke with those of healthy participants and use a stroke dysbiosis Index (SDI)	A model of middle cerebral artery occlusion (MCAO) was used in mice associated with human flora (HFA) to explore the causal relationship between intestinal dysbiosis and the result of stroke	Mice that received fecal transplants from patients with high SDI developed severe brain damage with elevated IL-17^{+} γδ T cells in the intestine
Zhu et al. (2019)	To investigate whether the transplantation of fecal microbiota from patients with drug-free schizophrenia to mice free of specific pathogens could cause schizophrenia-like behavior	- Tissue analysis of transplanted mice: liver, spleen, intestine, prefrontal cortex, striatum, and hippocampus were dissected from the brain - Stimulation of cultured cells and brain slices in vitro with colonic luminal filtrate (CLF) - Behavioral tests	- Behavioral abnormalities (psychomotor hyperactivity, learning difficulties, and memory) - Elevation of the kynurenine–kynurenic acid pathway of tryptophan degradation in the periphery and brain - Increase in basal extracellular dopamine in the prefrontal cortex and 5-hydroxytryptamine in the hippocampus - Increased synthesis of kynurenic acid and kynurenine aminotransferase II in cultured hepatocytes and cortical slices of the forebrain
Li et al. (2018)	Investigate whether the intestinal microbiota may be involved in the pathogenesis of depression through the microbe–gut–brain axis	Germ-free mice underwent fecal microbiota transplantation from patients with major depressive disorder (MDD)	The microbiota influences hepatic metabolism causing changes that may be involved in the pathogenesis of MDD

Continued

TABLE 48.2 Summary of selected neuroscience-related studies conducting human intestinal microbiota transfer to mice in laboratory-controlled manner.—cont'd

References	Objective	Study characteristics	Outcomes
Lu et al. (2018)	Explore the hypothesis that the microbiome influences the growth phenotype and brain development	A germ-free mouse transfer model was used to examine the effects of premature baby microbiota known to induce high or low growth phenotypes on postnatal brain development	- Microbiota that induced low growth phenotype was associated with reductions in neuronal NeuN, neurofilament-L, and in the myelination marker, myelin basic protein (MBP) - It was associated with n euroinflammation marked by an increase in NOS1, a change in the IGF-1 pathway, with a decrease in circulating and brain IGF-1, a decrease in circulating IGFBP3, and an increase in the expression of brain IGFBP3 mRNA
Zheng et al. (2016)	To evaluate how the intestinal microbiota of patients with major depressive disorder (MDD) influences physiology and behavior in mice	- Sequencing of the 16S rRNA gene - Transplant of fecal microbiota from MDD or healthy individuals to mice - Tests to assess depression-like behaviors - Metagenomics and metabolomics in mice that harbored the "microbiota of depression"	- Mice with "depressive-like microbiota" derived from patients with MDD showed depression-like behavior, compared to mice colonized with "healthy microbiota" - Affected mice mainly exhibited disorders of the microbial genes and host metabolites involved in the metabolism of carbohydrates and amino acids

Conclusion

Modern laboratory mouse research has been greatly instrumental for unraveling and addressing key research questions in neuroscience, when experimental conditions are unethical to humans; however, with the increasing appreciation of the critical role of the intestinal microbiota, as a modulatory factor to brain development, renewed interest has been raised in the importance of modeling interactions of specific pathogens with the intestinal microbiota in laboratory mice. This fundamental dilemma of keeping scientific approaches in a reasonable equilibrium between the advantages of reproducibility and reliability of the data offered by modern SPF vivarium housing and those of fulfilling translational relevance to people in real living environments poses difficult challenges. Future studies must take these factors into consideration in order to bring new breakthrough discoveries for the optimal benefit and well-being of human populations.

Applications to other areas of development

In this chapter we review and discuss the advantages and disadvantages of the modern laboratory animal housing to approach the translational neuroscience research to people living in disenfranchised environments, where toxicants, pollutants, and enteric pathogens prevail. Translational neuroscience research is greatly needed to better understand and ameliorate the neurocognitive outcomes of children living in poor sanitary conditions in the developing world, who bear a troubling vicious cycle of enteric infections and malnutrition with lasting consequences for brain and neurocognitive development, jeopardizing the ability to reach their full genetic potential. As accumulating evidence of the importance of the intestinal microbiota to brain development and aging evolves, new strategies to model disease in laboratory mice must consider the role of intestinal pathogenic bacterial populations/gut-to-blood bacterial translocation and often associated systemic low-grade inflammation. We also summarize recent studies conducting human intestinal microbiota transfer to experimental mice as a valuable approach to understand the role of the bacterial populations to brain and behavior research. These novel approaches may help to answer important research questions about how contaminated environments affect the brain and neurocognitive outcomes and to translate breakthrough discoveries, including drug discovery and therapy, to the well-being of human populations still living under unsanitary conditions, which remain tragically common in most of the developing world.

Key facts of intestinal microbiota in translational neuroscience research

- Accumulating exposome of biohazards and pollutants early in life may affect the intestinal microbiota with lasting consequences to human optimal brain development and aging.
- Unhealthy intestinal microbiota may deleteriously affect brain development and behavior in humans.
- Current modern SPF vivarium may not be efficiently translating data to people living in contaminated environments.
- Revisiting inbred mouse models to improve translational data is needed.
- Human microbiota transfer to inbred mice may improve translational data in neuroscience.

Summary points

- Compelling evidence suggests that the intestinal microbiota have a critical role during brain development and aging.
- Children living in poor sanitary conditions in the developing world may suffer a vicious cycle of enteric infections and malnutrition with lasting consequences to brain development.
- Current animal models and housing facilities may not be considering the role of environment contaminants to study human brain development and aging.
- Data generated from animal models to study the brain development and aging may not always be directly translatable to all human populations.
- Animal housing in pathogen-free and germ-free conditions is important to the reproducibility and reliability of the data but may not be fully translational to human populations living in poor sanitary conditions and to dissecting specific causal relationships.
- Revisiting animal models using humanized intestinal microbiota may help to improve translational neuroscience.

Acknowledgments

The authors thank the Brazilian National Council for Scientific and Technological Development (CNPq) especially visiting researcher (n°03/2014, # 400538/2014-8) and the Coordination for the Improvement of Higher Education Personnel (CAPES) Procad (071/2013 # 144494) and CAPES Print (88881.311939/2018-01) funding. RBO was supported in part by the Bill and Melinda Gates Foundation grant # OPP1137923. JOM was funded by National Funds via FCT (Foundation for Science and Technology) through the Strategic Project UIDB/04539/2020 and UIDP/04539/2020 (CIBB): COMPETE-FEDER (POCI-01-0145-FEDER-007440); and CENTRO 2020 Regional Operational Program (CENTRO-01-0145-FEDER-000012: HealthyAgeing 2020; CE NTRO-01-0145-FEDER-000008: BrainHealth 2020). The authors also appreciate Andre Penha Aguiar for aiding with this manuscript.

References

Adams, J. B., Borody, T. J., Kang, D. W., Khoruts, A., Krajmalnik-Brown, R., & Sadowsky, M. J. (2019). Microbiota transplant therapy and autism: Lessons for the clinic. *Expert Review of Gastroenterology and Hepatology, 13*, 1033–1037.

Arentsen, T., Raith, H., Qian, Y., Forssberg, H., & Diaz, H. R. (2015). Host microbiota modulates development of social preference in mice. *Microbial Ecology in Health and Disease, 26*, 29719.

Bartelt, L. A., Bolick, D. T., & Guerrant, R. L. (2019). Disentangling microbial mediators of malnutrition: Modeling environmental enteric dysfunction. *Cellular and Molecular Gastroenterology Hepatology, 7*, 692–707.

Beura, L. K., Hamilton, S. E., Bi, K., Schenkel, J. M., Odumade, O. A., Casey, K. A., … Masopust, D. (2016). Normalizing the environment recapitulates adult human immune traits in laboratory mice. *Nature, 532*, 512–516.

Bleich, A., & Hansen, A. K. (2012). Time to include the gut microbiota in the hygienic standardisation of laboratory rodents. *Comparative Immunology, Microbiology and Infectious Diseases, 35*, 81–92.

Braniste, V., Al-Asmaskh, M., Kowal, C., Anuar, F., Abbaspour, A., Toth, M., … Pettersson, S. (2014). The gut microbiota influences blood-brain permeability in mice. *Sci. Transl. Med., 6*, 263ra158.

Bull-Larsen, S., & Mohajeri, M. H. (2019). The potential influence of the bacterial microbiome on the development and progression of ADHD. *Nutrients, 11*.

Camara-Lemarroy, C. R., Metz, L., Meddings, J. B., Sharkey, K. A., & Yong, V. W. (2018). The intestinal barrier in multiple sclerosis: Implications for pathophysiology and therapeutics. *Brain, 141*, 1900–1916.

Caputi, V., & Giron, M. C. (2018). Microbiome-gut-brain axis and toll-like receptors in Parkinson's disease. *International Journal of Molecular Sciences, 19*.

Chung, H., Pamp, S. J., Hill, J. A., Surana, N. K., Edelman, S. M., Troy, E. B., … Kasper, D. L. (2012). Gut immune maturation depends on colonization with a host-specific microbiota. *Cell, 149*, 1578–1593.

Costa, L. G., Aschner, M., Vitalone, A., Syversen, T., & Soldin, O. P. (2004). Developmental neuropathology of environmental agents. *Annual Review of Pharmacology and Toxicology, 44*, 87–110.

Crumeyrolle-Arias, M., Jaglin, M., Bruneau, A., Vancassel, S., Cardona, A., Dauge, V., … Rabot, S. (2014). Absence of the gut microbiota enhances anxiety-like behavior and neuroendocrine response to acute stress in rats. *Psychoneuroendocrinology, 42*, 207–217.

De, L. F., & Shoenfeld, Y. (2019). The microbiome in autoimmune diseases. *Clinical and Experimental Immunology, 195*, 74–85.

Dewhirst, F. E., Chien, C. C., Paster, B. J., Ericson, R. L., Orcutt, R. P., Schauer, D. B., & Fox, J. G. (1999). Phylogeny of the defined murine microbiota: Altered schaedler flora. *Applied and Environmental Microbiology, 65*, 3287–3292.

Erny, D., Hrabe de Angelis, A. L., Jaitin, D., Wieghofer, P., Staszewski, O., David, E., … Prinz, M. (2015). Host microbiota constantly control maturation and function of microglia in the CNS. *Nature Neuroscience, 18*, 965–977.

Fujii, Y., Nguyen, T. T. T., Fujimura, Y., Kameya, N., Nakamura, S., Arakawa, K., & Morita, H. (2019). Fecal metabolite of a gnotobiotic mouse transplanted with gut microbiota from a patient with Alzheimer's disease. *Bioscience Biotechnology and Biochemistry, 83*, 2144–2152.

Gareau, M. G., Wine, E., Rodrigues, D. M., Cho, J. H., Whary, M. T., Philpott, D. J., … Sherman, P. M. (2011). Bacterial infection causes stress-induced memory dysfunction in mice. *Gut, 60*, 307–317.

Guerrant, R. L., DeBoer, M. D., Moore, S. R., Scharf, R. J., & Lima, A. A. (2013). The impoverished gut—a triple burden of diarrhoea, stunting and chronic disease. *Nature Reviews Gastroenterology and Hepatology, 10*, 220–229.

Guerrant, R. L., Oria, R. B., Moore, S. R., Oria, M. O., & Lima, A. A. (2008). Malnutrition as an enteric infectious disease with long-term effects on child development. *Nutrition Reviews, 66*, 487–505.

Hata, T., Miyata, N., Takakura, S., Yoshihara, K., Asano, Y., Kimura-Todani, T., … Sudo, N. (2019). The gut microbiome derived from anorexia nervosa patients impairs weight gain and behavioral performance in female mice. *Endocrinology, 160*, 2441–2452.

Hill, D. A., & Artis, D. (2010). Intestinal bacteria and the regulation of immune cell homeostasis. *Annual Review of Immunology, 28*, 623–667.

Hugenholtz, F., & de Vos, W. M. (2018). Mouse models for human intestinal microbiota research: A critical evaluation. *Cellular and Molecular Life Sciences, 75*, 149–160.

Ismail, F. Y., Fatemi, A., & Johnston, M. V. (2017). Cerebral plasticity: Windows of opportunity in the developing brain. *European Journal of Paediatric Neurology, 21*, 23–48.

Johansson, M. E., Ambort, D., Pelaseyed, T., Schutte, A., Gustafsson, J. K., Ermund, A., … Hansson, G. C. (2011). Composition and functional role of the mucus layers in the intestine. *Cellular and Molecular Life Sciences, 68*, 3635–3641.

Kim, M. S., Kim, Y., Choi, H., Kim, W., Park, S., Lee, D., … Mook-Jung, I. (2020). Transfer of a healthy microbiota reduces amyloid and tau pathology in an Alzheimer's disease animal model. *Gut, 69*, 283–294.

Kipnis, J. (2016). Multifaceted interactions between adaptive immunity and the central nervous system. *Science, 353*, 766–771.

Ley, R. E., Peterson, D. A., & Gordon, J. I. (2006). Ecological and evolutionary forces shaping microbial diversity in the human intestine. *Cell, 124*, 837–848.

Li, B., Guo, K., Zeng, L., Zeng, B., Huo, R., Luo, Y., … Xie, P. (2018). Metabolite identification in fecal microbiota transplantation mouse livers and combined proteomics with chronic unpredictive mild stress mouse livers. *Translational Psychiatry, 8*, 34.

Lin, C., Zhao, S., Zhu, Y., Fan, Z., Wang, J., Zhang, B., & Chen, Y. (2019). Microbiota-gut-brain axis and toll-like receptors in Alzheimer's disease. *Computational and Structural Biotechnology Journal, 17*, 1309–1317.

Lin, L., Zheng, L. J., & Zhang, L. J. (2018). Neuroinflammation, gut microbiome, and Alzheimer's disease. *Molecular Neurobiology, 55*, 8243–8250.

Luk, B., Veeraragavan, S., Engevik, M., Balderas, M., Major, A., Runge, J., … Versalovic, J. (2018). Postnatal colonization with human "infant-type" Bifidobacterium species alters behavior of adult gnotobiotic mice. *PLoS One, 13*, e0196510.

Lu, J., Lu, L., Yu, Y., Cluette-Brown, J., Martin, C. R., & Claud, E. C. (2018). Effects of intestinal microbiota on brain development in humanized gnotobiotic mice. *Scientific Reports, 8*, 5443.

McConnell, E. L., Basit, A. W., & Murdan, S. (2008). Measurements of rat and mouse gastrointestinal pH, fluid and lymphoid tissue, and implications for in-vivo experiments. *Journal of Pharmacy and Pharmacology, 60*, 63–70.

Murray-Kolb, L. E., Rasmussen, Z. A., Scharf, R. J., Rasheed, M. A., Svensen, E., Seidman, J. C., … Lang, D. (2014). The MAL-ED cohort study: Methods and lessons learned when assessing early child development and caregiving mediators in infants and young children in 8 low- and middle-income countries. *Clinical Infectious Diseases, 59*(Suppl. 4), S261–S272.

Nataro, J. P., & Guerrant, R. L. (2017). Chronic consequences on human health induced by microbial pathogens: Growth faltering among children in developing countries. *Vaccine, 35*, 6807–6812.

Neufeld, K. M., Kang, N., Bienenstock, J., & Foster, J. A. (2011). Reduced anxiety-like behavior and central neurochemical change in germ-free mice. *Neuro-Gastroenterology and Motility, 23*, 255–264. e119.

Nguyen, T. L., Vieira-Silva, S., Liston, A., & Raes, J. (2015). How informative is the mouse for human gut microbiota research? *Disease Models and Mechanisms, 8*, 1–16.

Ogbonnaya, E. S., Clarke, G., Shanahan, F., Dinan, T. G., Cryan, J. F., & O'Leary, O. F. (2015). Adult hippocampal neurogenesis is regulated by the microbiome. *Biological Psychiatry, 78*, e7–e9.

Oria, R. B., Malva, J. O., Foley, P. L., Freitas, R. S., Bolick, D. T., & Guerrant, R. L. (2018). Revisiting inbred mouse models to study the developing brain: The potential role of intestinal microbiota. *Frontiers in Human Neuroscience, 12*, 358.

Oria, R. B., Murray-Kolb, L. E., Scharf, R. J., Pendergast, L. L., Lang, D. R., Kolling, G. L., & Guerrant, R. L. (2016). Early-life enteric infections: Relation between chronic systemic inflammation and poor cognition in children. *Nutrition Reviews, 74*, 374–386.

Pendergast, L. L., Schaefer, B. A., Murray-Kolb, L. E., Svensen, E., Shrestha, R., Rasheed, M. A., … Seidman, J. C. (2018). Assessing development across cultures: Invariance of the Bayley-III scales across seven international MAL-ED sites. *School Psychology Quarterly, 33*, 604–614.

Ristori, M. V., Quagliariello, A., Reddel, S., Ianiro, G., Vicari, S., Gasbarrini, A., & Putignani, L. (2019). Autism, gastrointestinal symptoms and modulation of gut microbiota by nutritional interventions. *Nutrients, 11*.

Rodriguez, J. M., Murphy, K., Stanton, C., Ross, R. P., Kober, O. I., Juge, N., … Collado, M. C. (2015). The composition of the gut microbiota throughout life, with an emphasis on early life. *Microbial Ecology in Health and Disease, 26*, 26050.

Schmidt, E. K. A., Torres-Espin, A., Raposo, P. J. F., Madsen, K. L., Kigerl, K. A., Popovich, P. G., … Fouad, K. (2020). Fecal transplant prevents gut dysbiosis and anxiety-like behaviour after spinal cord injury in rats. *PLoS One, 15*, e0226128.

Sorrells, S. F., Paredes, M. F., Cebrian-Silla, A., Sandoval, K., Qi, D., Kelley, K. W., … Alvarez-Buylla, A. (2018). Human hippocampal neurogenesis drops sharply in children to undetectable levels in adults. *Nature, 555*, 377–381.

Staley, C., Kaiser, T., Beura, L. K., Hamilton, M. J., Weingarden, A. R., Bobr, A., … Khoruts, A. (2017). Stable engraftment of human microbiota into mice with a single oral gavage following antibiotic conditioning. *Microbiome, 5*, 87.

Subramanian, S., Huq, S., Yatsunenko, T., Haque, R., Mahfuz, M., Alam, M. A., … Gordon, J. I. (2014). Persistent gut microbiota immaturity in malnourished Bangladeshi children. *Nature, 510*, 417–421.

Sudo, N., Chida, Y., Aiba, Y., Sonoda, J., Oyama, N., Yu, X. N., … Koga, Y. (2004). Postnatal microbial colonization programs the hypothalamic-pituitary-adrenal system for stress response in mice. *The Journal of Physiology, 558*, 263–275.

Tshala-Katumbay, D., Mwanza, J. C., Rohlman, D. S., Maestre, G., & Oria, R. B. (2015). A global perspective on the influence of environmental exposures on the nervous system. *Nature, 527*, S187–S192.

Wang, J., Linnenbrink, M., Kunzel, S., Fernandes, R., Nadeau, M. J., Rosenstiel, P., & Baines, J. F. (2014). Dietary history contributes to enterotype-like clustering and functional metagenomic content in the intestinal microbiome of wild mice. *Proceedings of the National Academy of Sciences of the United States of America, 111*, E2703–E2710.

Wang, X. J., Xu, W., Li, J. Q., Cao, X. P., Tan, L., & Yu, J. T. (2019). Early-life risk factors for dementia and cognitive impairment in later life: A systematic review and meta-analysis. *Journal of Alzheimer's Disease, 67*, 221–229.

Xia, G. H., You, C., Gao, X. X., Zeng, X. L., Zhu, J. J., Xu, K. Y., … Yin, J. (2019). Stroke dysbiosis index (SDI) in gut microbiome are associated with brain injury and prognosis of stroke. *Frontiers in Neurology, 10*, 397.

Yee, A. L., Miller, E., Dishaw, L. J., Gordon, J. M., Ji, M., Dutra, S., … Groer, M. (2019). Longitudinal microbiome composition and stability correlate with increased weight and length of very-low-birth-weight infantsm. *Systems, 4*.

Zheng, P., Zeng, B., Zhou, C., Liu, M., Fang, Z., Xu, X., … Xie, P. (2016). Gut microbiome remodeling induces depressive-like behaviors through a pathway mediated by the host's metabolism. *Molecular Psychiatry, 21*, 786–796.

Zhu, F., Guo, R., Wang, W., Ju, Y., Wang, Q., Ma, Q., … Ma, X. (2019). Transplantation of microbiota from drug-free patients with schizophrenia causes schizophrenia-like abnormal behaviors and dysregulated kynurenine metabolism in mice. *Molecular Psychiatry*.

Zhu, S., Jiang, Y., Xu, K., Cui, M., Ye, W., Zhao, G., … Chen, X. (2020). The progress of gut microbiome research related to brain disorders. *Journal of Neuroinflammation, 17*, 25.

Chapter 49

The myelin mutant *taiep* rat as a model for developmental brain disorders

Jose R. Eguibar[1], Carmen Cortes[1], Juan C. Ahumada-Juárez[1], Valeria Piazza[2] and Victor Hugo Hernandez Gonzalez[3]

[1]*Behavioral Neurophysiological Laboratory, Institute of Physiology, Benemérita Universidad Autónoma de Puebla, Puebla, Pue, México;* [2]*Center for Research in Optics, Loma Del Bosque, Leon, Mexico;* [3]*Division of Sciences and Engineering, Department of Chemical, Electronic and Biomedical Engineering, Loma del Bosque, León, Mexico*

List of abbreviations

Cas CRISPR-associated
CNP 2′3′ Cyclic nucleotide 3′-phosphodiesterase
CNS Central nervous system
CRISPR Clustered regularly interspaced short palindromic repeats
DYT4
ER Endoplasmic reticulum
H-ABC Hypomyelination with atrophy of the basal ganglia and cerebellum
HLD6 Hypomyelinating leukodystrophy 6
HY High-yawning
jp Jimpy
jp^{rsh} Rumpshaker
jp^{mld} Myelin synthesis deficiency
MAG Myelin-associated glycoprotein
MBP Myelin basic protein
MBP Myelin basic protein
md Myelin-deficient
MOBP Myelin/associated oligodendrocytic basic protein
MRI Magnetic resonance imaging
MT Microtubules
PLP Proteolipid protein
PMD Pelizaeus−Merzbacher disease
PND Postnatal days
PNS Peripheral nervous system
pt Paralytic tremor
qk Quaking
sh Pup shaking pup
SVZ Subventricular zone
TALENs Transcription activator-like effector nucleases
ZFN Zinc finger nucleases

Diagnosis, Management and Modeling of Neurodevelopmental Disorders. https://doi.org/10.1016/B978-0-12-817988-8.00049-X

Mini-dictionary of terms

Adrenoleukodystrophy A central nervous system disease with a progressive loss of the myelin sheath in axons due to an accumulation of long-chain fatty acids.
Animal models An experimental group of laboratory animals that have similarities with actual human conditions and further help to test new therapeutic approaches.
Demyelination A progressive loss of the myelin sheath.
Dysmyelination An abnormal formation of the myelin sheath during development.
Gait The motor organization for alternation during walking patterns.
Krabbe disease Also called as globoid cell leukodystrophy or galactosylceramide lipidosis that produced a progressive damage of the nervous system and early death.
Leukodystrophies A group of diseases with a hypomyelination or demyelination of the central nervous due to different genetic backgrounds.
Multiple sclerosis The most frequent myelin disease characterized by an area or areas in the central nervous system lacking myelin
Myelin An insulation made by specialized cells that covers axons in order to avoid short circuits.
Pelizaeus–Merzbacher disease. A central nervous system disorder that is characterized by alterations in coordination, motor skills, and learning due to an inability to form myelin.
Tubulinopathy A group of genetic diseases that had a mutation in one of the protein subunits that conform microtubules.

Development of central nervous system

Neurodevelopment is a reference for the somatic growth and maturation of neurons with various anatomical and physiological characteristics throughout the life. This is particularly useful when comparing animals with different genetic backgrounds that may affect the growth and neurodevelopment of either the central nervous system (CNS) or the peripheral nervous system (PNS), particularly with respect to myelin sheath formation. Most of the studies about neurodevelopment alterations are directed toward toxicological effects induced by the exposure to metals, contaminants such as pesticides, air pollutants, etc., on neurons, but there are much less studies on evaluating whether glial cells play a critical role during development by offering physical support and guidance for new neurons reaching the target domains (Neman & de Vellis, 2008).

Oligodendrocytes are the myelin-forming cells of the CNS and Schwann cells in the PNS. Myelin, which is essential for the normal functioning of the mammalian CNS, is a fatty insulation composed of modified plasma membrane that surrounds axons and promotes the rapid and efficient conduction of electrical impulses (i.e., action potentials). Many neurological diseases are caused by myelin deficiency, which can be the result of both genetic and environmental factors (Neman & de Vellis, 2008). There are several white matter disorders occurring during development, generally displaying hypomyelination or dysmyelination of the CNS. Deficiency of myelin may also occur after the maturation of the postnatal brain through a degenerative process resulting in demyelination (Neman & de Vellis, 2008). From the development point of view, oligodendrocytes originate in the ventral spinal cord or ventral telencephalon and form the so-called glial-restricted precursor cells that differentiated into both oligodendrocytes and astrocytes. There is a second source of glial cells from the dorsal region of the spinal cord or the telencephalon and importantly the neuroepithelium, which gives rise to the subventricular zone (SVZ) participating in the secondary proliferative population of neuronal and glial cells. Progenitor cells may be specified to become glia, migrate radially from the SVZ into the subcortical white matter in the corpus callosum, striatum, and cerebral cortex, where they differentiate into astrocytes and oligodendrocytes (Marshall, Suzuki, & Goldman, 2003). Additionally, oligodendrocytes continue to be produced in the adult brain and participate in myelin repair (Chari & Blakemore, 2002). Therefore, the possible alterations of myelin during development can affect the normal production of glial cells, oligodendrocyte sheath formation, and its maturation processes.

Animal models of myelin diseases

Until now, there are few available animal models on which specific glial lineage cells are affected, among them the oligodendrocytes that are responsible to form myelin in the CNS. In the case of oligodendrocytes there is a large group of X-linked myelin mutations in humans characterized by the Pelizaeus–Merzbacher disease (PMD), and in animals models in dogs: the shaking pup (sh pup), in pigs: the type III hypomyelinogenesis congenita, in rats: the myelin-deficient (md), in rabbits: the paralytic tremor (pt), and in mouse: jimpy (jp) and its alleles, myelin synthesis deficiency (jp^{mld}) and rumpshaker (jp^{rsh}) mice (Csiza & de Lahunta, 1979; Nussbaum, Neskovis & Kostic, 1969). Among these mutants, the md rat and the jimpy mouse had a short life span due to the severity of myelin loss; however, the shaking pup, pt, and the rumpshaker mice have more myelin amounts and therefore a longer life span (Duncan, 1995; Sidman, Dickie, & Appel, 1964).

Another group of myelin mutants are autosomal recessive, and they are a heterogeneous group including the *shiverer* mouse, which has mutations on the structural gene myelin basic protein (MBP). These mutations result in the absence of MBP in the case of *shiverer* mutant and very low levels in the myelin-deficient mutant *shiverer*^{mld} (Mikoshida, Aruga, Ikenaka, Okano & Maternson, 1992).

The *quaking* (qk) mouse has an autosomal recessive mutation, and after the second week of age, they shake their trunk, accentuated when the animal explore, the tremor is most marked in the caudal part of the trunk and proximal portion of the hind limbs; consequently, the mouse appears to be bouncing on its haunches (Fagg, 1979; Sidman et al., 1964), or the twitcher mouse a model of hereditary leukodystrophy (Duchen, Eicher, Jacobs, Scaravilli, & Teixeira, 1980). The myelin-deficient *Syrian* hamster has an autosomal recessive inheritance with a trembling involving the trunk and the hindquarters and a marked myelin deficiency in the CNS (Nunoya, Takima, Mizutami, & Umezawa, 1985; Redheat & Hood, 1990). Another autosomal recessive mutants are the *taiep* rats, which had a progressive motor syndrome characterized by tremor when reaching the first month of age, followed by ataxia at 4 months, immobility episodes that start at 6 months, and present a peak between 8 and 9 months of age, epilepsy, and hind limb paralysis (Cortes, Gavito, Ita, Valencia, & Eguibar, 2005; Eguibar, Cortes, Ugarte, & León-Chávez, 2014; Holmgren, Urbá-Holmgren, Riboni, & SaenzdeMiera, 1989).

The earlier animal model descriptions are based on the reports about inheritance, morphological characteristics in the CNS, and in some of them, electrophysiological recordings. However, little is known about the associated neurodevelopmental characteristics.

Different paths to obtain myelin mutants

In the case of mice, there are several genetically modified animals, such as: knock-out, knock-in, or those that are modified using Cre-Lox recombination as a site-specific recombinase technology. They may have deletions, insertions, translocations, and inversions at specific sites in the DNA sequence of CNS cells, including oligodendrocytes and other glial cells. Currently there are four major classes of engineered nucleases employed for genome editing purposes: zinc finger nucleases (ZFN), transcription activator-like effector nucleases (TALENs), engineered meganucleases, CRISPR (clustered regularly interspaced short palindromic repeats), and CRISPR-associated (Cas) that allow RNA-guided Cas9 endonuclease to modify the expression of specific genes (Hofker & van Deursen, 2011). However, in most of them there are few studies describing development and neuromaturation of their reflex activities. However, these groups of genetical manipulated mice usually have several alterations, mainly because the modified genes are expressed in different tissues, including neurons and glial cells.

Hence, the heterogeneous group of myelin mutants that only affect the CNS represent better animal models with specific alterations in myelin and lipid proteins, among them, leukodystrophies a heterogeneous group of white matter disorders exclusively affecting the CNS. They mimic equivalent human illness, usually present in infancy or childhood, that frequently leads to early death and rarely in reach adulthood but all them show a progressive disorder associated to myelin dysfunction. Importantly, the genetic basis, clinical manifestations, and pathological landmarks are heterogeneous, including myelinated axons that never get ensheathed or an unbalance between demyelination and different degrees of remyelination (Duncan, Kondo, & Zhang, 2011; Love, 2006; Ohno & Ikenaka, 2019).

Leukodystrophies

In humans there are four groups of leukodystrophies: (1) Pelizaeus-Merzbacher disease (PMD); (2) Krabbe disease; (3) metachromatic leukodystrophy; and (4) adrenoleukodystrophy. The PMD is characterized by a duplication of the *plp1* gene and has two variants: one is a severe disorder with early death, the so-called connatal form, or the classical form in which the myelin deficit is not as severe and therefore life expectancy is longer than the former (Hodes, Woodward, & Spinner, 2000). Krabbe is an autosomal recessive demyelinating disease caused by a genetic deficiency in the activity of the enzyme galactocerebrosidase. Based on that, it is classified as a lysosomal storage disease, where the main cause of demyelination is the accumulation of psychosine, a cytotoxic lipid metabolite that accumulates and appears to selectively kill oligodendrocytes and Schwann cells, leading to rapid demyelination (Wenger, Suzuki, Suzuki, & Suzuki, 2001). Metachromatic leukodystrophy is an autosomal recessive disorder resulting from a mutation of the *arylsulfatase* A gene, which leads to profound demyelination of the central and PNS. There are several clinical forms of the disease with infantile, juvenile, and adult stage onset. The knockout mouse of the *arylsulfatase* gene only shows scattered demyelination and did not reproduce exactly the human disease (Hess et al., 1996). Finally, adrenoleukodystrophy, the most common leukodystrophy in children, results from a mutation in the peroxisomal membrane gene ABCD1. The disease is striking in regard to its phenotypic variation, one of them with a profound subcortical demyelination associated with marked

TABLE 49.1 Animals models of leukodystrophies and the corresponding human disease.

Models	Affected gene	Neuropathology	Longevity	Human disease	References
Jimpy mouse	*plp1*	Dysmyelination	21 days	PMD	Nussbaum et al., 1969
Shiverer mouse	MBP	Dysmyelination	120 days		Mikoshiba et al., 1992
Twitcher mouse	Galactocerebrosidase	Demyelination	35–45 days	Krabbe disease	Duchen et al., 1980
Myelin deficient rat	*plp1*	Dysmyelination	21 days	PMD	Csiza & de Lahunta, 1979
Long–evans shaker rat	MBP	Dysmyelination/ demyelination	6–9 months		Hofker & van Deursen 2011
Tremor rat	Aspartoacetylase	Myelin vacuolation	12 months	Canavan disease	Hofker & van Deursen 2011
taiep rat	β-tubulin 4a	Hypomyelination followed by demyelination	>18 months	Garduno-Robles et al., 2020	Garduno-Robles et al., 2020; Holmgren et al., 1989; Eguibar & Cortes, 2010
Shaking pup canine	*plp1*	Dysmyelination	23–24 days	PMD	Tosic, Matthey, Gow, Lazzarini, & Matthieu, 1997

MBP, Myelin basic protein; *plp*, proteolipid protein gene; *PMD*, Pelizaeus–Merzbacher disease

inflammation and early onset in childhood. In contrast, the adult onset phenotype called adrenomyeloneuropathy has prominent long-tract degeneration in the spinal cord but not a prominent demyelination. Importantly, knockout mouse strategies had failed to yield an accurate model with similar characteristics observed in the human pathology (see Table 49.1; Dubois-Dalcq, Feigenbaum, & Auborug, 1999).

Tubulinopathies and leukodystrophies

Recently, and due to better magnetic resonance imaging (MRI) techniques in association with detail genetic analysis, clinicians have been able to stablish relationships between clinical signs and symptoms in patients with specific alterations in the CNS using MRI and complementary gene analysis. Therefore, during the last decade there are some reports that stablish a clear association of hypomyelination leukodystrophy with several different mutations in the TUBB4A gene (Purnell, Bleyl, & Bonkowsky, 2014; Tonduti et al., 2016); a hypomyelination that mimics PMD (Shimojima et al., 2015); or dystonia-4, an autosomal dystonia that it is characterized by a progressive laryngeal dysphonia associated with two missense mutations of the *tubb4a* gene (Hofker, & van Deursen, 2011).

In human tubulinopathies, there are mutations causing dysfunction of the tubulins and microtubule-associated proteins. Most of them described as cause leading to complex brain malformations, such as dysmorphic basal ganglia (75% of cases), callosal dysgenesis, cerebellar hypoplasia/dysplasia, and cerebral cortex malformations (Gonçalves et al., 2018). Tubulins are structural subunit proteins that form the microtubules and represent a protein superfamily in eukaryotic cells including α, β, γ, δ ξ, and a sixth family called ζ tubulin (McKean et al., 2001). The α, β, and γ tubulins are ubiquitous and present in all nucleated cells organisms. Microtubules are primarily formed from heterodimeric tubulins, consisting of α and β tubulin polypeptides, and they are assembled in a head-to-tail heterodimer organization (McKean et al., 2001). The spectrum of tubulinopathies includes patients with TUBA1A, TUBB2B, TUBB2A, TUBB3, TUB4A, DYNCH1H1, and TUB8ADCX mutations, all them causing either developmental malformations or degenerative changes in different cell types of the CNS (Curiel et al., 2017). These problems include microcephaly, ventriculomegaly, callosal agenesis/ dysgenesis, brainstem hypoplasia, and cerebellar and vermian hypodysplasia, among other neuroimaging findings (Gonçalves et al., 2018).

The case of TUBB4A represents an exception, it being directly associated to hypomyelination and/or a demyelination, hypomyelination with atrophy of the basal ganglia and cerebellum (H-ABC), the main anatomopathological finding. These conditions caused by TUBB4A mutations have the most profound effects on CNS white matter and even myelin degeneration during childhood. H-ABC has been found to be responsible for hypomyelination with atrophy of the basal ganglia and cerebellum (H-ABC). This inborn leukodystrophy is characterized by a combination of defective myelination/cerebellar–neostriatal atrophy, as observed by MRI scans, and a complex neurological phenotype, including delayed motor development, intellectual disability, extrapyramidal movement abnormalities, ataxia, and spasticity (Zech et al., 2015).

Other mutations in the TUBB4A gene also produce hereditary isolated dystonia (DYT4), displaying considerable phenotypic variability, ranging from focal/segmental to severe dystonia, with prominent laryngeal involvement and a missense mutation in the TUBB4A mutation. These pathologies are well in accordance with the prominent role played by tubulins in brain development through neuronal genesis and migration, cortical organization, and in axon guidance (Breuss, Leca, Gstrein, Hansen, & Keays, 2017).

Taiep a myelin mutant model of leukodystrophy

We obtained as a spontaneous mutation in the process of inbreeding to obtain an SD rat with a high-yawning (HY) frequency (Urbá-Holmgren et al., 1990). The myelin mutants develop a progressive motor syndrome characterized by tremor, ataxia, immobility episodes, epilepsy, and paralysis, the acronym of these symptoms given is *taiep* (Holmgren et al., 1989). They had an initial hypomyelination, followed by a progressive demyelination in the CNS (Duncan, Lunn, Holmgren, Urbá-Holmgren, & Brignolo-Holmes, 1992). The myelin alterations are due to an abnormal accumulation of microtubules in oligodendrocytes that disrupt transporting mechanism from endoplasmic reticulum. to the Cis portion of Golgi apparatus (Couve, Cabello, Krsulovic, & Roncagliolo, 1997). So, all myelin proteins MBP, PLP, MAG, MOBP, and CNP decreased with the age as well as lipid content in different structures of the brain and the spinal cord of *taiep* rats (Eguibar, Cortes, Lara-Lozano, & Mendiola, 2012; Möller, Durr, Quarter, & Duncan, 1997).

At the structural level, *taiep* rats have an accumulation of microtubules (MT) physically bound to the endoplasmic reticulum in the cytoplasm of their oligodendrocytes (Couve et al., 1997). It is well stablished that microfilaments and MT are essential for maintaining anterograde transport of proteins and lipids. In fact, stable MT are fundamental to maintain the highly polarized morphology of the oligodendrocytes (Song, Goetz, & Baas, 2001). When the *taiep's* oligodendrocytes were plated, MT-associated proteins MAP1b and *tau* were higher with respect to those levels obtained from oligodendrocytes isolated from control SD rats (Song, O'Connor, Yu, Baas, & Duncan, 1999). In addition, tight bundles of MT and counterclockwise-hooked are statistically higher in cultured oligodendrocytes with respect to those obtained from SD rats (Eguibar & Cortes, 2010; Song et al., 1999), which disrupts the transport of PLP protein or MBP mRNA that can be reversed by nocodazole, an MT disassembling drug, or by inhibiting dynein-activity by specific antibodies (Song, Carson, Barbarese, Li, & Duncan, 2003).

Electrophysiologically, *taiep* rats showed progressive deterioration of auditory brainstem responses along their postnatal development (Roncagliolo, Benítez, & Eguibar, 2000; Lopez-Juarez et al., 2020). Importantly, optic nerves from *taiep* rats showed a significant reduction on the amplitude for compound actions potentials and a diminution of their conduction velocity with the age of the animals (Roncagliolo et al., 2006). Demyelination is concomitant to reactive astrocytosis and the activation of nitric oxide synthase (León-Chávez et al., 2006; León-Chávez et al., 2001), which are able to stimulate lipoperoxidation and activate caspase-3, leading to an increase in the degeneration of myelin, but paradoxically also to activate CNS repair mechanisms (Soto-Rodriguez et al., 2012).

samples and using MRI we corroborate that *taiep* rats have a hypomyelination associated with atrophy of the basal ganglia and cerebellum and in H-ABC patients (Garduño-Robles et al., 2020).

Developmental characteristics of *taiep* rats

The main allometric characteristic of the *taiep* rat is the weight, which normally increases along the first stages in the lactation period. Results in our laboratory showed that there are not significant differences between *taiep* animals with respect to SD rats, with similar weight increments per day of around 1.4 ± 0.2 g between 3 and 15 postnatal days (PND) and around 5.4 ± 0.3 g between 30 and 40 PND, resulting in similar body weights in both groups (see Table 49.2). Comparable results were obtained for the size of the subjects (data not shown) supporting SD rats that are allometrically like *taiep* rats. Development characteristics are based on the day that the pups open their eyes, *taiep* opened their eyes between 13 and 15 PND, in the case of SD rats between 14 and 15 PND (Table 49.3). The age when a superior incisive eruption happened was also similar in both groups of animals varying between 6 and 8 PND (see Table 49.4). These results clearly showed that *taiep* rats, a tubulinopathy model with leukodystrophy, have similar maturational processes as control SD rats during the first 15 days of life.

To measure the brain neurodevelopment process in SD and *taiep* pups, we measure reflex responses, such as surface righting, cliff avoidance, and negative geotaxis. Surface righting reflex is time that the pup spent when the neonate was

TABLE 49.2 Body weight of Sprague Dawley and taiep rats during lactation.

Postnatal days	Sprague dawley	*taiep*
3	6.85 ± 0.15	7.70 ± 0.66
5	8.90 ± 0.10	9.83 ± 1.21
7	11.33 ± 0.49	12.63 ± 1.32
9	14.56 ± 0.63	16.95 ± 0.85
10	15.90 ± 1.10	18.13 ± 1.56
15	25.05 ± 0.25	29.86 ± 1.61

Data are the mean ± SDM on grams.

TABLE 49.3 Percentage of eye opening in Sprague Dawley and taiep pups.

Postnatal days	Spraguedawley (%)	*taiep* (%)
12	0	0
13	4	0
14	64	79
15	100	100

TABLE 49.4 Postnatal eruption of incisive teeth in Sprague Dawley and taiep pups.

Postnatal days	Sprague dawley (%)	*taiep* (%)
6	2	2
7	66	57
8	100	100

placed in a supine position and immediately turns over to restore its normal prone position by touching all the four paws with the plane surface, we measured the time taken to restore the prone position from 3 to 6 PND. The percentage of subjects that executed the reflex successfully were similar in both groups of rats in the time needed to recover the prone position (data not shown). In the case of the cliff avoidance test, the rats displayed risk assessment behaviors by avoiding the cliffs and heights. Cliff avoidance reflex was evaluated in the pups from 3 to 10 PND with the pups placed on the edge of a wooden box, with nose and forepaws touching just over the edge. The time taken by the subjects to avoid the cliff and turn backward was recorded, all rats were capable to do so (see Table 49.5).

Negative geotaxia is a test considered as a diagnostic response for vestibular and/or proprioceptive function, being an automatic postural reaction by turning upright in an inclined plane test. We used a wooden board with rough surface that was inclined 25 degrees, and the pups were placed on it with their snout downward. The time taken to perform 180 degrees rotation was recorded (Naik, Patro, & Patro, 2015). The responsiveness for the execution of the negative geotaxia was similar in both groups of rats between 3 and 15 PND.

Developmental of gait in *taiep* rats

We measured gait since weaning day 21–28 PND in order to measure the ontogeny of locomotion in both groups of rats using the CatWalk system (Noldus Technologies, The Netherlands). The speed of gaiting in *taiep* rats was significantly lower with respect to Sprague Dawley rats (see Fig. 49.1; RMANOVA, $F_{(1,7)} = 57.9$; $P \leq .002$, followed by Bonferroni test $P \leq .05$). These results further imply that since the early stages of development, central pattern generators in the spinal

TABLE 49.5 Postnatal display of complete cliff avoidance in Sprague Dawley and taiep pups.

Postnatal days	Sprague dawley (%)	taiep (%)
3	47.0	25.0
4	52.9	33.3
5	80	64.5
8	100	100
10	100	100

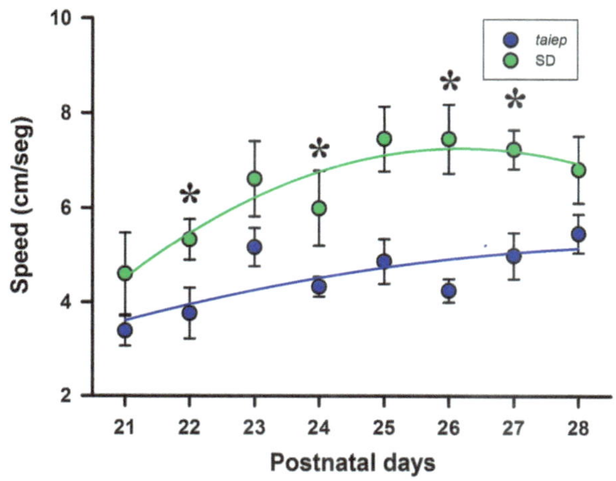

FIGURE 49.1 **Speed of gait in the first week after weaning in *taiep* and Sprague Dawley rats.** Detailed analysis of the velocity of gait in the first week after weaning shows that *taiep* rats have significantly lower speed, measured in the CatWalk, being significant at 22, 24, 26, and 27 days of age with respect to SD rats (RMANOVA, $F_{(1,7)} = 57.9$, $P \leq .002$, followed by Bonferroni test, $P \leq .05$).

cord may be affected by hypomyelination in this structure and corresponds with anatomical data showing that dorsal columns and corticospinal spinal tracts are more affected than ventral columns in the myelin mutant (Lunn, Clayton, & Duncan, 1997). There are only few reports about myelin mutant mice on which authors analyze development characteristics. This is a relevant point because it allows to determine coherent and well-organized body and CNS growth and function. In fact, from the medical point of view, it is well stablished that allometric and reflex activity have prediction value for the future neurodevelopment characteristics.

Future trends in myelin diseases

In conclusion, animal models are fundamental tools for the analysis of normal and altered myelin sheaths, particularly during demyelination associated or not to neuroinflammation because general evolution of the diseases is characterized by progressive deterioration of different sensory, motor, or neurocognitive capabilities until death of the patient. The models are also useful to determine the value of an early intervention through physical therapy, the therapeutic use of different pharmacological, immunological, or even genetic approaches, Gene therapy strategies have been developed to tackle leukodystrophies using hematopoietic stem cells (HCT) or allogeneic HCT produced a mild improvement in patients. Other approaches include the use of human immunodeficiency virus type 1 (HIV)-derived lentiviral vectors that allow more efficient gene transfer into HCT than murine gamma-retroviruses or even intracerebral injections of adeno-associated virus using as a vector to target mainly neurons. Currently there are clinical trials in X-linked adrenoleukodystrophy and several genetic therapies in mouse models (Biffi, Aubourg, & Cartier, 2011). It is clear that soon gene therapy could be an adequate way to treat myelin and other CNS diseases.

Applications to other areas of development

In this chapter, we reviewed some of the most relevant developmental characteristics of myelin mutants and particularly tubulinopathies in humans. Leukodystrophies are a heterogeneous group of diseases with different genetics backgrounds and are grouped in four diseases: (1) PMD; (2) Krabbe disease; (3) metachromatic leukodystrophy; and (4) adrenoleukodystrophy. Recently, there are several reports involving several mutations in the TUBB4A gene, leading to several clinical illnesses such as hypomyelination with atrophy of the basal ganglia and cerebellum (H-ABC), dystonia type 4 (DYT4-isolated dystonia), or hypomyelinating leukodystrophy 6 (HLD6). Recent availability of resonance imaging techniques, associated with genetic analysis and molecular biology techniques, has allowed to detect different causes and provides a detailed clinical characterization of different diseases and further the reproduction of the syndromes in animal models, particularly in genetically modified mice.

So far, the best available model for TBB4A tubulinopathy is the *taiep* rats, a Sprague Dawley—derived animal model with hypomyelination, followed by progressive demyelination in the CNS clinically resembling adrenoleukodystrophy. As previously mentioned, *taiep* is the acronym of the motor syndrome that characterize them: tremor, ataxia, immobility episodes, epilepsy, and paralysis. These mutants do not present delayed maturation of the morphological characteristics such as: body weight, size, incisor eruption, and eye opening. Furthermore, neurologically the *taiep* rats show normal reflex maturation determined by righting reflex, cliff avoidance, and negative geotaxia tests, thus indicating that maturation processes are normal, instead of the hypomyelination of the CNS observed in the human condition. Nonetheless, gait analysis uncovers a significant reduction in the speed of locomotion, suggesting an alteration at the spinal cord level in this leukodystrophy model. This is quite relevant because in the four leukodystrophies reported for gait alterations in humans, such as muscle tone changes and ataxia are remarkable signs and can be of clinical relevance for new therapies. Genetic analysis provides a possibility for early detection and concomitantly gene therapy treatments using different genetic, viral, or stem cell options.

Key facts of myelin *taiep* rats

- *Taiep* rats are an adequate model of myelin diseases because they have an initial hypomyelination followed by a progressive demyelination.
- Myelin disease represents 5% of the total central nervous diseases.
- Adrenoleukodystrophies are a heterogeneous group of myelin diseases characterized by hypomyelination and/or demyelination.
- Leukodystrophies are myelin diseases with different clinical evolutions, and *taiep* rats showed some of their clinical and radiologic characteristics.
- There are not adequate treatments for the disease, and therefore animal models are extremely important for testing new options.
- Human leukodystrophies are characterized by the loss of myelin and associated demyelination.

Summary points

- This chapter focuses on myelin diseases, which are a large group of illnesses that affect the insulation system of the axons.
- *Taiep* rats are an animal model for myelin disease with an initial hypomyelination followed by a progressive demyelination.
- *Taiep* rats develop a locomotor syndrome associated to the progressive demyelination of the central nervous system.
- The therapeutic options for the treatment of myelin diseases are still limited, and animal models are the best option to evaluate new therapies.
- *Taiep* rats is a myelin mutant of TUBB4A gene with similar characteristics of human leukodystrophies.
- Animal models of myelin are fundamental to test new therapeutic options because there are not adequate treatments.
- Gene therapy requires adequate animal models of human diseases.

References

Biffi, A., Aubourg, O., & Cartier, N. (2011). Gene therapy for leukodystrophies. *Human Molecular Genetics, 20*, R42–R53.

Breuss, M. W., Leca, I., Gstrein, T., Hansen, A. H., & Keays, D. A. (2017). Tubulins and brain development—The origins of functional specification. *Molecular and Cellular Neuroscience, 84*, 58–67.

Chari, D. M., & Blakemore, W. F. (2002). Efficient recolonization of progenitor-depleted area in the CNS by adult oligodendrocyte progenitor cells. *Glia, 37*, 307–313.

Cortes, M. C., Gavito, B., Ita, M. L., Valencia, J., & Eguibar, J. R. (2005). Characterization of the spontaneous and gripping-induced immobility episodes on *taiep* rats. *Synapse, 58*, 95–101.

Couve, E., Cabello, F., Krsulovic, J., & Roncagliolo, M. (1997). Binding of microtubules to transitional elements in oligodendrocytes of the myelin mutant *taiep* rat. *Journal of Neuroscience Research, 47*, 573–581.

Csiza, C. K., & de Lahunta, A. (1979). Myelin deficiency (md): A neurologic mutant in the wistar rat. *American Journal of Pathology, 95*, 215–224.

Curiel, J., Rodríguez Bey, G., Takanohashi, A., Bugaighis, M., Fu, X., Wolf, N. I., & Nmezi, B. (2017). TUBB4A mutations result in specific neuronal and oligodendrocytic defects that closely match clinically distinct phenotypes. *Human Molecular Genetics, 26*, 4506–4518.

Dubois-Dalcq, M., Feigenbaum, V., & Auborug, P. (1999). The neurobiology of X-linked adrenoleukodystrophy, a demyelinating peroxisomal disorder. *Trends in Neurosciences, 22*, 4–12.

Duchen, L. W., Eicher, E. M., Jacobs, J. M., Scaravilli, F., & Teixeira, F. (1980). Hereditary leukodystrophy in the mouse: The new mutant twicher. *Brain, 103*, 695–710.

Duncan, I. D. (1995). Inherited disorders of myelination of the central nervous system. In H. Kettenmann, & B. R. Ranson (Eds.), *Neuroglia (990-1009)*. New York: Oxford University Press.

Duncan, I. A., Kondo, Y., & Zhang, S. C. (2011). The myelin mutants as models to study myelin repair in the leukodystrophies. *Neurotherapeutics, 8*, 607–624.

Duncan, I. D., Lunn, K. F., Holmgren, B., Urbá-Holmgren, R., & Brignolo-Holmes, L. (1992). The *taiep* rat: A myelin mutant with an associated oligodendrocyte microtubular defect. *Journal of Neurocytology, 21*, 870–884.

Eguibar, J. R., & Cortes, M. C. (2010). The myelin mutant *taiep* as a model for absence crisis. *Gaceta Médica Mexicana, 146*, 12–18.

Eguibar, R. J., Cortes, M. C., Lara-Lozano, M., & Mendiola, D. M. (2012). Dopaminergic D2-like agonists produce yawning in the myelin mutant *taiep* and Sprague-Dawley rats. *Pharmacology, Biochemistry and Behavior, 102*, 118–123.

Eguibar, R. J., Cortes, M. C., Ugarte, A., & León-Chávez, A. (2014). The myelin mutant rat taiep as a model of neuroimmunological disease. *Advances in Neuroimmune Biology, 5*, 9–17.

Fagg, G. E. (1979). The quaking mouse: Regional variations in the content and protein composition of myelin isolated from the central nervous system. *Neuroscience, 4*, 973–978.

Garduno-Robles, Á., Alata, M., Piazza, V., Cortes, C., Eguibar, J. R., Pantano, S., & Hernandez, V. H. (2020). MRI features in a murine model of TUBB4A tubulinopathy. *Frontiers in Neuroscience, 14*, 12. https://doi.org/10.3389/fnins.2020.00555.

Gonçalves, F. G., Freddi, T. L., Taranath, A., Lakshmanan, R., Goetti, R., Feltrin, F. S., … Arrigoni, F. (2018). Tubulinopathies. *Topics in Magnetic Resonance Imaging, 27*, 395–408.

Hess, B., Saftig, P., Hartmann, D., Coenen, R., Lullmann-Rauch, R., & Goebel, H. H. (1996). Phenotype of arylsulfatase A-deficient mice: Relationship to human metachromatic leukodystrophy. *Proceedings of the National Academy of Sciences USA, 93*, 14821–14826.

Hodes, M. E., Woodward, K., & Spinner, N. B. (2000). Additional copies of the proteolipid protein gene causing Pelizaeus-Merzbacher disease arise by separate integration into the X chromosome. *The American Journal of Human Genetics, 67*, 14–22.

Holmgren, B., Urbá-Holmgren, R., Riboni, L., & Vega-Saenz de Miera, E. C. (1989). Sprague Dawley rat mutant with tremor, ataxia, tonic immobility episodes, epilepsy and paralysis. *Laboratory Animal Science, 39*, 226–228.

Hofker, M. H., & van Deursen, J. M. (2011). Transgenic Mouse Methods and Protocols. In *Myelin: Biology and chemistry* (2nd, p. 365). Springer New York, Dordrecht, Heidelberg, London: Humana Press. https://doi.org/10.1007/978-1-60761-974-1.

Leon-Chavez, B. A., Aguilar-Alonso, P., Gonzalez-Barrios, J. A., Eguibar, J. R., Ugarte, A., Brambila, E., … Martinez-Fong, D. (2006). Increased nitric oxide levels and nitric oxide synthase isoform expression in the cerebellum of the *taiep* rat during its severe demyelination stage. *Brain Research, 1121*, 221–230.

León-Chávez, B. A., Guevara, J., Galindo, S., Luna, J., Ugarte, A., Villegas, O., … Martinez-Fong, D. (2001). Regional and temporal progression of reactive astrocytosis in the brain of the myelin mutant *taiep* rat. *Brain Research, 900*, 152–155.

Lopez-Juarez, A., Vega, A. G., Kleinert-A, A., Piazza, V., Garduno-Robles, A., Alata, M., … Hernández, V. H. (2020). Auditory impairment in H-ABC tubulinopathy. *Journal of Comparative Neurology*, 1–12. https://doi.org/10.1002/cne.24990.

Love, S. (2006). Demyelinating diseases. *Journal of Clinical Pathology, 59*, 1151–1159.

Lunn, K. F., Clayton, M. K., & Duncan, I. D. (1997). The temporal progression of the myelination defect in the *taiep* rat. *Journal of Neurocytology, 26*, 267–281.

Marshall, C. A., Suzuki, S. O., & Goldman, J. E. (2003). Gliogenic and neurogenic progenitors of the subventricular zone: Who are they, where did they come, and where are they going? *Glia, 43*, 52–61.

McKean, P. G., Vaughan, S., & Gull, K. (2001). The extended tubulin superfamily. *Journal of Cell Science, 27*, 395–408.

Mikoshiba, K., Aruga, J., Ikenaka, K., Okano, H., & Martenson, R. E. (1992). Shiverer and allelic mutant mld mice. In R. E. Matterson (Ed.), *Myelin: Biology and chemistry* (pp. 723–744). Boca Raton: CRC Press.

Möller, J. R., Durr, P. G., Quarter, R. H., & Duncan, I. D. (1997). Biochemical analysis of myelin proteins in a novel neurological mutant: The *taiep* rats. *Journal of Neurochemistry, 69*, 773–779.

Naik, A. A., Patro, I. K., & Patro, N. (2015). Slow physical growth, delayed reflex ontogeny, and permanent behavioral as well as cognitive impairments in rats following intra-generational protein malnutrition. *Frontiers in Neuroscience, 9*, 18.

Neman, J., & de Vellis, J. (2008). Myelinating cells in the central nervous system-development, aging, and disease. In R. Pérez-Polo, & R. Steffen (Eds.), *Handbook of neurochemistry and molecular neurobiology. Developmental and aging changes in the nervous system* (pp. 61–75). New York: Springer Press.

Nunoya, T., Takima, M., Mizutani, M., & Umezawa, H. (1985). A new mutant strain of Syrian hamster with myelin deficiency. *Acta Neuropathologica, 65*, 305–312.

Nussbaum, J. L., Neskovic, N., & Kostic, D. (1969). Characterization of the demyelination disorder in the "jimpy" mouse. *Journal of Physiology, 61*, 156.

Ohno, N., & Ikenaka, K. (2019). Axonal and neuronal degeneration in myelin diseases. *Neuroscience Research, 139*, 48–57.

Purnell, S. M., Bleyl, S. B., & Bonkowsky, J. L. (2014). Clinical exome sequencing identifies a novel TUBB4A mutation in a child with static hypomyelinating leukodystrophy. *Pediatric Neurology, 50*, 608–611.

Redheat, C., & Hood, L. (1990). The dysmyelinating mouse mutations Shiverer (shi) and myelin deficient (shi^{mld}). *Behavior Genetics, 20*, 213–234.

Roncagliolo, M., Benítez, J., & Eguibar, J. R. (2000). Progressive deterioration of central components of auditory brainstem responses during postnatal development of the myelin mutant *taiep* rat. *Audiology and Neuro-Otology, 5*, 267–275.

Roncagliolo, M., Schlageter, C., León, C., Couve, E., Bonansco, C., & Eguibar, J. R. (2006). Developmental impairment of compound action potential in the optic nerve of myelin mutant *taiep* rats. *Brain Research, 1067*, 78–84.

Shimojima, K., Okumura, A., Ikeno, M., Nishimura, A., Saito, A., Saitsu, H., Matsumoto, N., et al. (2015). A *de novo* TUBB4A mutation in a patient with hypomyelination mimicking Pelizaeus-Merzbacher disease. *Brain and Development, 37*, 281–285.

Sidman, R. L., Dickie, M. M., & Appel, S. H. (1964). Mutant mice (quaking and jimpy) with deficient myelination in the central nervous system. *Science, 144*, 309–311.

Song, J., Carson, J. H., Barbarese, E., Li, F. Y., & Duncan, I. D. (2003). RNA transport in oligodendrocytes from the *taiep* mutant rat. *Molecular and Cellular Neuroscience, 24*, 926–938.

Song, J., Goetz, B. D., & Baas, P. W. (2001). Cytoskeletal reorganization during the formation of oligodendrocyte processes and branches. *Molecular and Cellular Neuroscience, 17*, 624–636.

Song, J., O'Connor, L. T., Yu, W., Baas, P. W., & Duncan, I. D. (1999). Microtubule alterations in cultured *taiep* rat oligodendrocytes lead to deficits in myelin membrane formation. *Journal of Neurocytology, 28*, 671–683.

Soto-Rodríguez, G., Martínez-Fong, D., Arroyo, R., Aguilar-Alonso, P., Rubio, H., Eguibar, J. R., … León-Chávez, B. A. (2012). Nitric oxide production is associated to increased lipoperoxidation and active caspase-3 in demyelinated brain regions of the taiep rat. *Advances in Bioscience and Biotechnology, 3*, 695–704.

Tonduti, D., Aiello, C., Renaldo, F., Dorboz, I., Saaman, S., & Rodríguez, D. (2016). TUBB4A.related hypomyelinating leukodystrophy: New insights from a series of 12 patients. *European Journal of Paediatric Neurology, 20*, 323–330.

Tosic, M., Matthey, B., Gow, A., Lazzarini, R. A., & Matthieu, J. M. (1997). Intracellular transport of the DM-20 bearing shaking pup (shp) mutation and its possible phenotypic consequences. *Journal of Neuroscience Research, 50*, 844–852.

Urbá-Holmgren, R., Trucios, N., Holmgren, B., Eguibar, J. R., Gavito, A., Cruz, G., & Santos, A. (1990). Genotypic dependency of spontaneous yawning frequency in the rat. *Behavioural Brain Research, 40*, 29–35.

Wenger, D., Suzuki, K., Suzuki, Y., & Suzuki, K. (2001). Galactosylceramide lipidosis. Globoid cell leukodystrophy (Krabbe disease). In C. R. Scriver, A. L. Beaudet, W. S. Sly, D. Valle, B. Childs, W. K. Kinzler, & B. Vogeistein (Eds.), *The metabolic and molecular bases of inherited disease* (8th ed., pp. 3669–3694). New York: McGraw-Hill Inc.

Zech, M., Boesch, S., Jochim, A., Graf, S., Lichtner, P., Peters, A., & Gieger, C. (2015). Large-scale TUBB4A mutational screening in isolated dystonia and controls. *Parkinsonism and Related Disorders, 21*, 1278–1281.

Chapter 50

The MAM-E17 neurodevelopmental model of schizophrenia

Marzena Maćkowiak
Laboratory of Pharmacology and Brain Biostructure, Department of Pharmacology, Maj Institute of Pharmacology, Polish Academy of Sciences, Kraków, Poland

List of abbreviations

EE Environmental enrichment
GAD Glutamic acid decarboxylase
HDAC Histone deacetylase
mPFC Medial prefrontal cortex
MAM Methylazoxymethanol
NMDA N-methyl-D-aspartate acid
GluN2B, Grin2b NMDA receptor subunit 2B
PV Parvalbumin
P Postnatal day
H3K4me3 Trimethylation of histone H3 at lysine 4
GABA γ-Amino butyric acid

Mini-dictionary of terms

Adolescence A period between childhood and adulthood with massive changes in behavioral responses, hormones, and neurotransmission that are related to brain maturation.

Construct validity Comparison of a mechanism that induces a specific phenotype in an animal model with the currently known etiology of the disease in humans.

Environmental enrichment Experimental procedure housing animals in cages containing additional objects for playing: running wheels, tunnels, small balls, blocks of various colors and shapes.

Epigenetic regulation Epigenetic mechanisms regulating gene expression by the chemical modification of DNA and/or histone proteins and by the regulatory action of small nuclear RNAs, microRNAs.

Face validity Comparison of an animal model phenotype with the disease symptoms in humans

MAM-E17 model A neurodevelopmental animal model of schizophrenia based on the prenatal administration of mitotoxin, methylazoxymethanol (MAM) at embryonic day (E) 17.

Predictive validity Evaluation using an animal model to predict unknown aspects of the disease in humans and therapeutic outcomes.

Schizophrenia A neuropsychiatric disorder whose symptoms can be divided into positive (hallucinations, delusions, confused thoughts, and disorganized speech), negative (social withdrawal, flat affect), and cognitive (impairment in attention, memory, decision-making, executive function) disabilities.

Social isolation Experimental procedure housing animals individually in cages, from which they can smell and hear other animals but are unable to make physical contact.

Socioenvironmental factors Elements related to safety, violence, social connections, and influence on human health.

Diagnosis, Management and Modeling of Neurodevelopmental Disorders. https://doi.org/10.1016/B978-0-12-817988-8.00050-6

Introduction

Schizophrenia is a mental disorder occurring in 1% of the human population. The etiology of schizophrenia is difficult to define because the symptoms are observed in late adolescence or early adulthood, but predisposing risk factors act prenatally or during the early postnatal stage of development (Fig. 50.1). The aforementioned trajectory of schizophrenia emergence has been supported by evidence from epidemiological, brain imaging, and neuropathological studies in humans and suggests that schizophrenia is a neurodevelopmental disorder involving the disruption of normal maturational processes in the brain (Lewis & Levitt, 2002). Because of the long latency between exposure to risk factors and the manifestation of disease, early intervention and prevention strategies in schizophrenia are still an open question. One such approach might improve diagnosis and treatment before schizophrenia onset, having a beneficial effect on schizophrenia therapy, since 10%–20% of schizophrenic patients are resistant to antipsychotic therapy (Tamminga & Holcomb, 2005). However, such studies in humans have several diagnostic, practical, and methodological limitations. Therefore, validated animal neurodevelopmental models of schizophrenia are very useful tools for looking at answers to the aforementioned question. This article will focus on an animal model based on the prenatal administration of methylazoxymethanol (MAM), which is widely used to investigate schizophrenia in basic research (Modinos, Allen, Grace, & McGuire, 2015).

Validity of the MAM model of schizophrenia

The value of animal models in investigating schizophrenia abnormalities is based on validity criteria, such as construct validity, face validity, and predictive validity. Although this model does not reproduce all human conditions, it has translational value in schizophrenia research.

Construct validity

Epidemiological data indicate that cycad plant genotoxins impair human brain development and increase the risk for neurodevelopmental disorders, e.g., schizophrenia (Kisby, Moore, & Spencer, 2013). Genotoxins induce DNA damage and may disrupt neuronal proliferation, especially at the prenatal stage of development. It is suggested that interference with cell proliferation via environmental or genetic factors during human embryonic life (especially during the first trimester) might be a common cause for abnormal brain development leading to neurodevelopmental diseases such as schizophrenia (Selemon & Zecevic, 2015). MAM is an active metabolite of the cycad plan toxin cycasin, and although there is no data referring to its direct involvement in human neurodevelopmental diseases, animal studies have shown that MAM induces aberrant DNA methylation that selectively affects neuronal cell proliferation during embryonic development (Cattabeni & Di Luca, 1997).

Face validity

An effect of prenatal MAM exposure on offspring maturation is related to the specific timing of embryonic development. MAM injection between embryonic days (E) 9 and E15 induced morphometric changes in brain structures (i.e., entorhinal cortex, prefrontal cortex, and striatum) (Jongen-Relo et al., 2004; Talamini, Koch, Luiten, Koolhaas, & Korf, 1999), but impairments in schizophrenia-like behavior in adult rats were not always observed (i.e., deficits in sensorimotor gating) (Gourevitch, Rocher, Le Pen, Krebs, & Jay, 2004; Jongen-Relo et al., 2004; Talamini et al., 1999). In contrast, several findings have demonstrated that MAM administration at E17 (MAM-E17) results in schizophrenia-like abnormalities in adult rats that are highly correlated with schizophrenia symptoms. At the behavioral level, adult MAM-E17 rats showed spontaneous hyperactivity that may reflect positive symptoms in schizophrenia (Le Pen et al., 2006), but some findings did

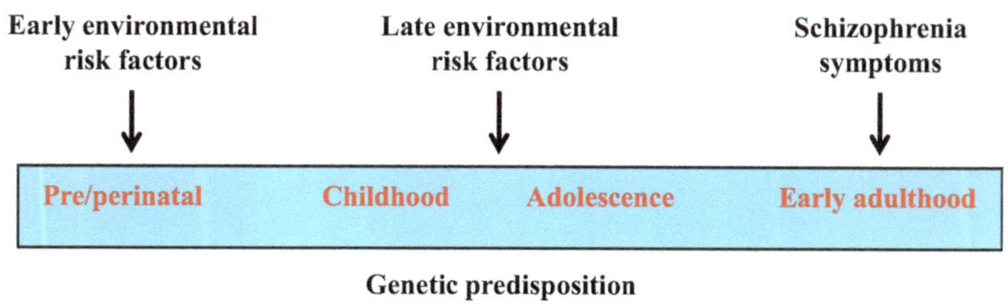

FIGURE 50.1 A schematic depicts schizophrenia development. Schizophrenia appearance is related to genetic predisposition and environmental factors acting at different stages of neurodevelopment.

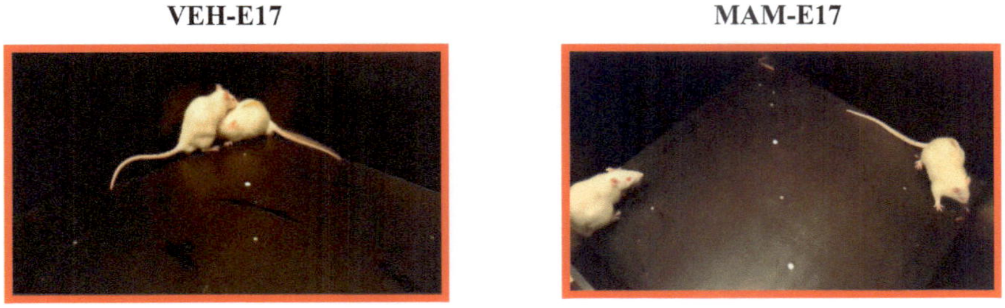

FIGURE 50.2 An example of social behavior of adult rats treated with VEH or MAM at embryonic day 17 (VEH-17 or MAM-E17, respectively) in the social interaction test.

not support these aspects of schizophrenia in the MAM-E17 model (Bator, Latusz, Glowacka, et al., 2018; Flagstad et al., 2004). With respect to the negative symptoms of schizophrenia, such as social withdrawal, MAM-E17 rats showed deficits in the social interaction test (Fig. 50.2) (Bator, Latusz, Glowacka, et al., 2018; Bator, Latusz, Wedzony, & Mackowiak, 2018; Flagstad et al., 2004; Le Pen et al., 2006). Cognitive disabilities related to schizophrenia abnormalities were also observed in MAM-E17 rats (deficits in attention, working memory, recognition memory, sensorimotor gating) (Bator, Latusz, Glowacka, et al., 2018; Bator, Latusz, Wedzony, et al., 2018; Featherstone, Rizos, Nobrega, Kapur, & Fletcher, 2007; Le Pen et al., 2006). Changes in morphological and cytoarchitectural factors related to schizophrenia, i.e., a decrease in the volume of the hippocampus and medial prefrontal cortex (mPFC), were also observed in the brain structures of adult MAM-E17 rats (Featherstone et al., 2007; Moore, Jentsch, Ghajarnia, Geyer, & Grace, 2006). Decreased mPFC volume was correlated with increased neuron density with no neuron loss (Mackowiak, Bator, Latusz, Mordalska, & Wedzony, 2014; Moore et al., 2006), which may reflect the neuropil theory of schizophrenia (Selemon & Goldman-Rakic, 1999). In addition, a decrease in parvalbumin (PV) expression was found in the hippocampus (Gill & Grace, 2014; Lodge, Behrens, & Grace, 2009; Penschuck, Flagstad, Didriksen, Leist, & Michael-Titus, 2006), but in the mPFC, the results are inconsistent (Lodge et al., 2009; Mackowiak, Latusz, Glowacka, Bator, & Bilecki, 2019; Penschuck et al., 2006). Alterations in PV interneurons are also observed in patients with schizophrenia and might be involved in cognitive dysfunction (Gonzalez-Burgos, Cho, & Lewis, 2015). Magnetic resonance imaging also showed the enlargement of the lateral and third ventricles in MAM-E17 rats (Chin et al., 2011). With respect to neurochemical alterations in adult MAM-E17 animals, changes in dopaminergic and glutamatergic transmission were reported in the hippocampus and the mPFC (Hradetzky et al., 2012; Lavin, Moore, & Grace, 2005; Lena, Chessel, Le Pen, Krebs, & Garcia, 2007; Lodge & Grace, 2008), which is consistent with dopaminergic and glutamatergic theories of schizophrenia (Howes, McCutcheon, & Stone, 2015). Although most studies using MAM-E17 were performed in rats, recent data have also reported the same effects in mice with prenatal exposure to MAM (Chalkiadaki et al., 2019; Huo et al., 2018).

Predictive validity

To estimate therapeutic outcomes in the MAM-E17 model, antipsychotic drugs were evaluated because they are the main form of treatment for schizophrenia. The findings showed some effects of typical (haloperidol) and atypical (risperidone, clozapine) neuroleptics on spontaneous hyperactivity induced by MAM (Le Pen, Jay, & Krebs, 2011). Moreover, the antipsychotic drug sulpiride restored impairments in synaptic plasticity in the MAM-E17 model (Belujon, Patton, & Grace, 2013), and the D2 partial agonist aripiprazole prevented hyperdopaminergic activity in the MAM-E17 model (Sonnenschein, Gill, & Grace, 2019). To investigate the glutamatergic pathway in schizophrenia treatment, an mGlu5-positive allosteric modulator was studied in the MAM-E17 model and showed an effect in some cognitive disabilities (Gastambide et al., 2012). Thus, the MAM-E17 model appears to be useful not only for understanding the mechanism of action of clinically used antipsychotic drugs but also for searching for new molecules with antipsychotic ability.

Epigenetic regulation in the MAM-E17 model

Recent findings indicate that the etiology of schizophrenia is related to genetic predisposition and environmental factors acting in the early stages of life (Tsuang, 2000) (Fig. 50.1) Environmental risk factors impair gene expression during brain development by affecting epigenetic regulation and inducing an abnormal trajectory of brain maturation (Fig. 50.3). Thus, the identification of epigenetic mechanisms altered during schizophrenia development might be an important target for schizophrenia prevention. Epigenetic regulation is highly dynamic and using animal models might help to understand the

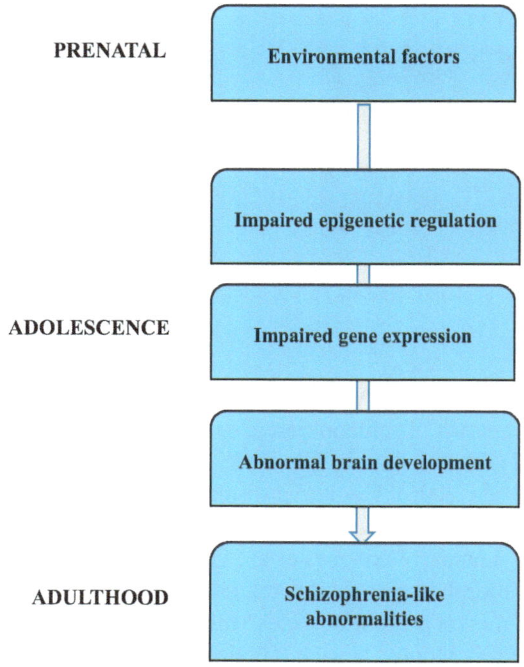

FIGURE 50.3 A schematic depicts the epigenetic regulation in the development of schizophrenia-like abnormalities. Impairments in epigenetic regulation and gene expression induce abnormal brain development and schizophrenia-like malfunctions.

epigenetic control of gene transcription during brain maturation (Millan, 2013). Total histone H3 modifications, i.e., methylation, were analyzed in the mPFC of MAM-E17 rats at different stages of development (Mackowiak et al., 2014). The results showed a different trajectory of changes in the dimethylation of histone H3 at lysine 9 (H3K9me2) compared with the trimethylation of histone H3 at lysine 4 (H3K4me3). Prenatal MAM administration affected H3K9me2 levels in adolescence, and no changes were observed in adulthood. In contrast, global H3K4me3 levels decreased only in adulthood. H3Kme3 proteins are present in γ-amino butyric acid (GABA)-ergic cells positive for glutamic acid decarboxylase 67 (GAD67) protein (Fig. 50.4), and a specific reduction in H3K4me3 was found at the promoter of GAD67 gene (Gad1) and PV in MAM-E17 rats that was related to a decrease in the mRNA levels of these genes (Bator, Latusz, Wedzony, et al., 2018; Mackowiak et al., 2019) (Table 50.1). Histone H3 methylation in the mPFC appears to be sensitive to prenatal risk factor action because an increase in global H3K27me3 was observed in the juvenile mPFC of MAM-E17 rats. The observed effect was correlated with an increase in levels of the trimethylation of histone H3 at lysine 27 (H3K27me3) and the transcriptional repressor REST at the proximal promoter region of Grin2b, which might contribute to the decrease in GluN2B protein, an NMDA receptor subunit 2B, in the early stages of postnatal development in the MAM-E17 model (Gulchina, Xu, Snyder, Elefant, & Gao, 2017). Another mechanism of epigenetic regulation, DNA methylation,

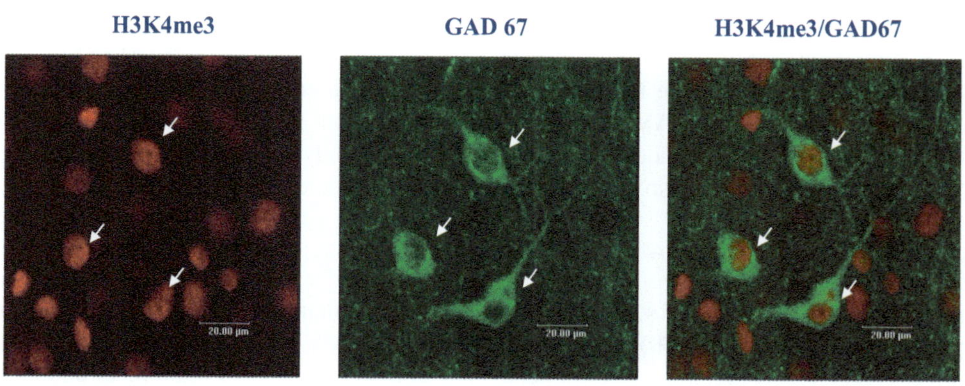

FIGURE 50.4 Representative confocal photomicrographs of the adult medial prefrontal cortex, which was immunoprobed for trimethylation of histone H3 at lysine 4 (H3K4me3) and glutamic acid decarboxylase (GAD) 67. *Arrows* indicate examples of H3K4me3 and GAD67 colocalization. The images are single scans captured with a 63 x objective.

TABLE 50.1 The effects of environmental factors in adolescence on behavioral and neurochemical responses in the medial prefrontal cortex in the MAM-E17 model.

Schizophrenia-like abnormalities	Late adolescence (P45–P49)			Early adulthood (P70–P75)			References
Behavioral response	***MAM-SH***	***MAM-EE***	***MAM-SI***	***MAM-SH***	***MAM-EE***	***MAM-SI***	
Locomotor activity	↔	n.a.	↔	↔	n.a.	↔	Bator, Latusz, Glowacka et al. (2018)
Social interaction	↓	n.a.	↔	↓	↔	↔	Bator, Latusz, Glowacka et al. (2018), Bator, Latusz, Wedzony et al. (2018)
Sensorimotor gating	↔	n.a.	↔	↓	↔	↓	Bator, Latusz, Glowacka et al. (2018), Bator, Latusz, Wedzony et al. (2018)
Recognition memory	↔	n.a.	↓	↓	↔	↓	Bator, Latusz, Glowacka et al. (2018), Bator, Latusz, Wedzony et al. (2018)
Neurochemical response	***MAM-SH***	***MAM-EE***	***MAM-SI***	***MAM-SH***	***MAM-EE***	***MAM-SI***	
GAD67 mRNA	↔	n.a.	n.a.	↓	↔	n.a.	Bator, Latusz, Wedzony et al. (2018), Mackowiak et al. (2014)
Total GAD67 protein	↔	n.a.	↓	↔	↔	↔	Bator, Latusz, Glowacka et al. (2018)
The number of GAD67 cells	n.a	n.a	n.a	↔	↔	n.a.	Bator, Latusz, Wedzony et al. (2018)
Total GAD65 protein	↔	n.a.	↔	↔	n.a.	↔	Bator, Latusz, Glowacka et al. (2018)
PV mRNA	↔	n.a.	↓	↓	n.a.	↔	Maćkowiak et al. (2019)
Total PV protein	↔	n.a.	↑	↓	n.a.	↓	
The number of PV cells	↔	n.a.	↔	↔	n.a.	↔	
Total H3K4me3	↔	n.a.	↓	↓	↔	↔	Bator, Latusz, Wedzony et al. (2018) Maćkowiak et al. (2019), Mackowiak et al. (2014)
H3K4me3 at GAD67 promoter	n.a.	n.a.	n.a.	↓	↔	n.a.	Bator, Latusz, Wedzony et al. (2018)
H3K4me3 at PV promoter	↔	n.a.	↓	↓	n.a.	↔	Maćkowiak et al. (2019)

↑, an increase; ↓, a decrease; ↔,no change; *EE*, enriched environment in early adolescence (P23–P29); *n.a.*, not analyzsed; *SH*, standard housing during postnatal life; *SI*, social isolation in adolescence (P30–P40).

was also investigated in the MAM-E17 model, and the results showed a decrease in DNA methylation at the gene promoter of cannabinoid receptor 1 (CB1) in the adult mPFC, which correlated with an increase in mRNA and protein levels of CB1 at the same age (Stark et al., 2019). The observed alterations in epigenetic regulation reflect the changes found in postmortem studies in schizophrenia subjects (Millan, 2013), and the expression of genes regulated by this modification is also affected in schizophrenia (Tamminga and Holcomb, 2005).

Adolescent abnormalities in the MAM-E17 model

Adolescence is an important stage in human development related to the transition from childhood to adulthood and is associated with several behavioral, hormonal, and neurochemical changes (Spear, 2000). This period is also crucial for schizophrenia emergence because according to the "double-hit" model of schizophrenia, brain development is very vulnerable to environmental factors during adolescence, and this vulnerability might prevent or accelerate schizophrenia onset (Schmitt, Malchow, Hasan, & Falkai, 2014; Selemon & Zecevic, 2015). This suggestion is important in the context of epidemiological and clinical observations that some nonspecific symptoms, i.e., cognitive disability or impairment in social life, might occur during this period (Sommer et al., 2016). However, it is very difficult to predict whether schizophrenia will develop or whether intervention is needed based on these symptoms.

Animal models are very useful for understanding the role of the adolescent period in schizophrenia development. Several behavioral and biochemical deficits are observed in adolescent animals in the MAM-E17 model of schizophrenia. At the behavioral level, impairments in social interactions have been detected, and they were still observed in adulthood (Bator et al., 2018; Le Pen et al., 2006). Moreover, an increased level of anxiety (Du & Grace, 2013) and a higher response to stress (Zimmerman, Bellaire, Ewing, & Grace, 2013) were found in adolescent MAM-E17 animals. At the biochemical level, a decrease in GluN2B protein was detected in the juvenile mPFC of MAM-E17 rats (Gulchina et al., 2017; Li et al., 2017). Moreover, changes in the expression of CB1 receptors in the mPFC (Gomes, Edelson, Volk, & Grace, 2018) or a decrease in PV levels in the hippocampus (Chen, Perez, & Lodge, 2014; Gill & Grace, 2014) starts in adolescent MAM-E17 rats.

To recognize a role of adolescence in aberrant brain maturation, some studies were performed to investigate whether pharmacological interventions during the early stages of life might prevent the development of schizophrenia-like abnormalities in the MAM-E17 model. The findings showed that mGluR2 agonist/mGluR3 antagonist treatment during the preadolescent period prevented changes in GluN2B levels in the mPFC of juvenile MAM-E17 rats and learning deficits in adulthood (Li et al., 2017). Cannabidiol treatment during adolescence not only prevented changes in the CB1 receptor in the adult MAM-E17 mPFC but also abolished deficits in social interactions and recognition memory (Stark et al., 2019). Diazepam administration during adolescence normalized anxiety levels in adolescent MAM-E17 rats and prevented a hyperdopaminergic state and hippocampal PV loss in adult MAM-treated animals (Du & Grace, 2013, 2016). The administration of valproic acid, a nonselective inhibitor of histone deacetylase (HDAC), during early adolescence not only prevented changes in epigenetic markers, i.e., a decrease in H3K4me3 or an increase in HDAC2 in the adult mPFC of MAM-E17 rats, but also abolished deficits in sensorimotor gating in adulthood (Bator, Latusz, Radaszkiewicz, Wedzony, & Mackowiak, 2015; Mackowiak et al., 2014). Thus, critical periods of brain development, particularly childhood and adolescence, appear to be very important for the development of abnormalities observed in adult MAM-E17 animals. This observation is in line with suggestions that increased therapeutic intervention in the early stages of life might prevent social and cognitive impairments before they are established (Sommer et al., 2016).

Adolescent environmental factors in the development of schizophrenia-like abnormalities in the MAM-E17 model

Several epidemiological findings indicate that socioenvironmental factors acting in the critical periods of development, i.e., childhood or adolescence, might prevent or accelerate the transition to schizophrenia, especially in people with a high risk of schizophrenia (Sommer et al., 2016). Environmental interventions such as nutritional, educational, and physical exercise enrichment programs (Raine et al., 2001; Raine, Mellingen, Liu, Venables, & Mednick, 2003) and cognitive behavioral therapy (Fusar-Poli, Bonoldi, et al., 2012; Fusar-Poli, Deste, et al., 2012) might protect against the onset of schizophrenia. In contrast, environmental insults, i.e., victimization, social disadvantages, and urbanity, might increase the risk of schizophrenia (van Os, Kenis, & Rutten, 2010). Thus, the effect of adolescent environmental factors on the onset of schizophrenia-like phenotypes in the MAM-E17 model was also investigated.

Adolescent environmental enrichment

Temporary exposure to environmental enrichment (EE) in early adolescence (P23–P29) prevented the appearance of both behavioral (impairments in social interactions, memory recognition, and sensorimotor gating) and neurochemical (a decrease in GAD67 mRNA in the mPFC) schizophrenia-like symptoms in the MAM-E17 model (Bator, Latusz, Wedzony, et al., 2018) (Table 50.1). The fact that EE action was related to H3K4me3 protein levels, both total and at the GAD67 gene (Gad1) promoter (Bator, Latusz, Wedzony, et al., 2018), supports a role of epigenetic regulation in schizophrenia development (Fig. 50.5, Table 50.1). Thus, the results confirmed the idea that childhood and adolescence might be the most important periods in postnatal life for noninvasive, environmental interventions to help alter the course of schizophrenia (Millan et al., 2016; Sommer et al., 2016). It must be stressed that longer environmental manipulation and cognitive training (up to 5 months) also had a positive effect on schizophrenia-like malfunction in the MAM-E17 model (Jenks et al., 2013). Nevertheless, even short, positive early-life experiences during the critical period of brain maturation might initiate the prevention of schizophrenia.

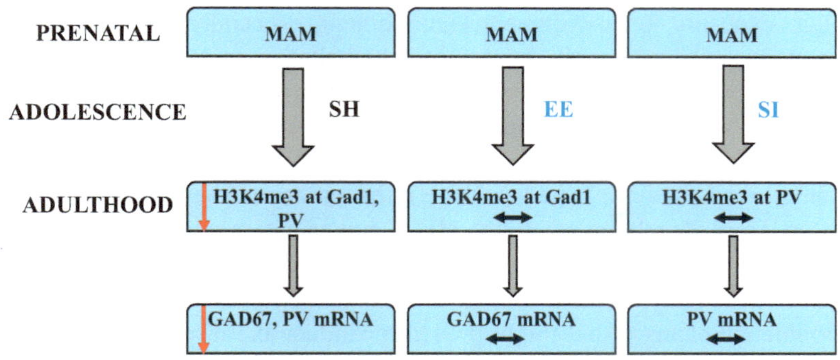

FIGURE 50.5 A schematic depicts the effect of prenatal (MAM) and adolescent (EE, SI) environmental factors in epigenetic regulation (trimethylation of histone H3 at lysine 4; H3K4me3) of parvalbumin (PV) and glutamic acid decarboxylase (GAD67; Gad1) genes expression in adulthood. ↓, a decrease; ↔, no change; *EE*, enriched environment in early adolescence (P23–P29); *SH*, standard housing during postnatal life; *SI*, social isolation in adolescence (P30–P40).

Adolescent environmental risk factors

Social isolation (SI) during adolescence is used in animal studies as a model of social exclusion from peers, which is related to victimization behavior (Buwalda, Geerdink, Vidal, & Koolhaas, 2011). The findings showed that 10-day SI rearing in adolescence (P30–P40) had diverse effects on schizophrenia-like behaviors in adult MAM-E17 rats: it prevented social interaction impairment, accelerated dysfunction in memory recognition, and did not affect deficits in sensorimotor gating (Bator, Latusz, Glowacka, et al., 2018). The effect of adolescent SI on GABA-ergic system markers, i.e., GAD65, GAD67, and PV, in the adult mPFC of MAM-E17 rats was also studied. SI had no effect on total GAD65 and GAD67 protein levels in adult MAM-E17 groups (Bator, Latusz, Glowacka, et al., 2018) (Table 50.1). For PV expression, SI abolished the decrease in PV mRNA levels in MAM-E17 rats but had no effect on the reduction in PV protein levels in this group. However, the effect of SI on PV mRNA was related to alterations in H3K4me3 protein levels, both total and at the PV promoter (Mackowiak et al., 2019) (Fig. 50.5, Table 50.1). Thus, the aforementioned results indicate that MAM-E17 animals are vulnerable to temporary SI exposure during adolescence; however, the impact of SI on the trajectory of schizophrenia-like abnormalities is different and is most likely related to the age when the intervention and the malfunctions began. In the case of impairment in social interactions or PV expression in the adult mPFC of MAM-E17 rats adolescent SI exposure might even have positive effects, preventing their appearance in adulthood. In contrast, it has also been postulated that adolescent stress might influence the development of schizophrenia-like abnormalities in MAM-E17 rats, i.e., PV loss in the hippocampus and a hyperdopaminergic state (Gomes, Zhu, & Grace, 2019). Further studies are needed to better understand the outcomes of exposure to environmental risk factors during adolescence on the development of schizophrenia-like symptoms.

It must be stressed that EE had a beneficial effect on MAM-E17 abnormalities, but SI also showed some positive effects on schizophrenia-like symptoms. Both of them normalized some of GABA-ergic dysfunctions in the adult mPFC of MAM-E17 rats acting via epigenetic mechanisms (H3K4me3) (Fig. 50.5). The aforementioned effects of adolescent environmental factors seem to be important for normal maturation of GABA-ergic system in the prefrontal cortex that is commonly altered in schizophrenic patients (Tamminga & Holcomb, 2005). Further studies are needed to obtain more information about the effect of temporary exposure to environmental factors on the development of abnormalities related to schizophrenia.

Applications to other areas of development

In this article, a neurodevelopmental animal model of schizophrenia, MAM-E17, was reviewed. The validity of the model has been evaluated, and the results suggest that the MAM-E17 model might be useful as a translational tool in animal schizophrenia studies. Moreover, an important role of early-life stages in the development of schizophrenia-like abnormalities has been discussed. The findings showed that in both humans and animal models, i.e., MAM-E17, some nonspecific symptoms related to social and cognitive abilities can already be observed during childhood and adolescence. These observations indicate a crucial time window not only for the prevention and treatment of schizophrenia but also for brain maturation and suggest that pharmacological or environmental interventions during childhood/adolescence periods might alter the trajectory of brain development and the emergence of disease symptoms. These suggestions are supported by the results showing that exposure to environmental enrichment, especially during early adolescence, might be a promising, noninvasive strategy to help alter the course of schizophrenia in subjects with a genetic or environmental predisposition to disease development. However, temporary SI during adolescence might also change the dynamics of schizophrenia development, but these effects are dependent on symptoms. The appearance of recognition memory dysfunction was accelerated in MAM-E17 animals, but impairment in social interactions was abolished. These results suggest that SI might be either a risk factor or have a positive effect on neurodevelopmental disorders. Moreover, an effect of environmental factors on brain development might affect epigenetic regulators that alter gene expression during adolescence and change the trajectory of brain development. A highly dynamic trajectory of epigenetic regulation during adolescence might make individuals very vulnerable to environmental conditions. Thus, the adolescence period appears to be very sensitive to environmental changes that might prevent or induce the disorder's appearance.

Key facts of schizophrenia

- Schizophrenia affects approximately 1% of the human population.
- Males might experience earlier onset of the disease than females.
- There is no cure for schizophrenia.
- Many patients respond well to psychological and pharmacological treatment, but 10%–20% of people suffering from schizophrenia are resistant to therapy.
- People with schizophrenia might have some social problems (unemployment, homelessness, poverty).

Summary points

- MAM-E17 is proposed as an animal model to investigate schizophrenia development.
- The MAM-E17 model has a good construct, face and predictive validity.
- Epigenetic mechanisms are impaired in the MAM-E17 model, suggesting that epigenetic regulation might be an important target for schizophrenia prevention and therapy.
- The adolescence period is crucial for the development of schizophrenia-like abnormalities in the MAM-E17 model.
- Therefore, adolescent environmental conditions might influence the emergence of schizophrenia dysfunction.
- However, temporary exposure to social isolation in adolescence has diverse effects on impairments in the MAM-E17 model.
- Nevertheless, environmental enrichment during adolescence prevents the development of abnormalities in the MAM-E17 model.

References

Bator, E., Latusz, J., Glowacka, U., Radaszkiewicz, A., Mudlaff, K., & Mackowiak, M. (2018). Adolescent social isolation affects schizophrenia-like behavior in the MAM-E17 model of schizophrenia. *Neurotoxicity Research, 34*, 305–323.

Bator, E., Latusz, J., Radaszkiewicz, A., Wedzony, K., & Mackowiak, M. (2015). Valproic acid (VPA) reduces sensorimotor gating deficits and HDAC2 overexpression in the MAM animal model of schizophrenia. *Pharmacological Reports, 67*, 1124–1129.

Bator, E., Latusz, J., Wedzony, K., & Mackowiak, M. (2018). Adolescent environmental enrichment prevents the emergence of schizophrenia-like abnormalities in a neurodevelopmental model of schizophrenia. *European Neuropsychopharmacology, 28*, 97–108.

Belujon, P., Patton, M. H., & Grace, A. A. (2013). Disruption of prefrontal cortical-hippocampal balance in a developmental model of schizophrenia: Reversal by sulpiride. *International Journal of Neuropsychopharmacology, 16*, 507–512.

Buwalda, B., Geerdink, M., Vidal, J., & Koolhaas, J. M. (2011). Social behavior and social stress in adolescence: A focus on animal models. *Neuroscience and Biobehavioral Reviews, 35*, 1713–1721.

Cattabeni, F., & Di Luca, M. (1997). Developmental models of brain dysfunctions induced by targeted cellular ablations with methylazoxymethanol. *Physiological Reviews, 77*, 199–215.

Chalkiadaki, K., Velli, A., Kyriazidis, E., Stavroulaki, V., Vouvoutsis, V., Chatzaki, E., … Sidiropoulou, K. (2019). Development of the MAM model of schizophrenia in mice: Sex similarities and differences of hippocampal and prefrontal cortical function. *Neuropharmacology, 144*, 193–207.

Chen, L., Perez, S. M., & Lodge, D. J. (2014). An augmented dopamine system function is present prior to puberty in the methylazoxymethanol acetate rodent model of schizophrenia. *Developmental Neurobiology, 74*, 907–917.

Chin, C. L., Curzon, P., Schwartz, A. J., O'Connor, E. M., Rueter, L. E., Fox, G. B., … Basso, A. M. (2011). Structural abnormalities revealed by magnetic resonance imaging in rats prenatally exposed to methylazoxymethanol acetate parallel cerebral pathology in schizophrenia. *Synapse, 65*, 393–403.

Du, Y., & Grace, A. A. (2013). Peripubertal diazepam administration prevents the emergence of dopamine system hyperresponsivity in the MAM developmental disruption model of schizophrenia. *Neuropsychopharmacology, 38*, 1881–1888.

Du, Y., & Grace, A. A. (2016). Loss of parvalbumin in the hippocampus of MAM schizophrenia model rats is attenuated by peripubertal diazepam. *The International Journal of Neuropsychopharmacology, 19*.

Featherstone, R. E., Rizos, Z., Nobrega, J. N., Kapur, S., & Fletcher, P. J. (2007). Gestational methylazoxymethanol acetate treatment impairs select cognitive functions: Parallels to schizophrenia. *Neuropsychopharmacology, 32*, 483–492.

Flagstad, P., Mork, A., Glenthoj, B. Y., van Beek, J., Michael-Titus, A. T., & Didriksen, M. (2004). Disruption of neurogenesis on gestational day 17 in the rat causes behavioral changes relevant to positive and negative schizophrenia symptoms and alters amphetamine-induced dopamine release in nucleus accumbens. *Neuropsychopharmacology, 29*, 2052–2064.

Fusar-Poli, P., Bonoldi, I., Yung, A. R., Borgwardt, S., Kempton, M. J., Valmaggia, L., … McGuire, P. (2012). Predicting psychosis: meta-analysis of transition outcomes in individuals at high clinical risk. *Archives of General Psychiatry, 69*, 220–229.

Fusar-Poli, P., Deste, G., Smieskova, R., Barlati, S., Yung, A. R., Howes, O., … Borgwardt, S. (2012). Cognitive functioning in prodromal psychosis: A meta-analysis. *Archives of General Psychiatry, 69*, 562–571.

Gastambide, F., Cotel, M. C., Gilmour, G., O'Neill, M. J., Robbins, T. W., & Tricklebank, M. D. (2012). Selective remediation of reversal learning deficits in the neurodevelopmental MAM model of schizophrenia by a novel mGlu5 positive allosteric modulator. *Neuropsychopharmacology, 37*, 1057–1066.

Gill, K. M., & Grace, A. A. (2014). Corresponding decrease in neuronal markers signals progressive parvalbumin neuron loss in MAM schizophrenia model. *The International Journal of Neuropsychopharmacology, 17*, 1609–1619.

Gomes, F. V., Edelson, J. R., Volk, D. W., & Grace, A. A. (2018). Altered brain cannabinoid 1 receptor mRNA expression across postnatal development in the MAM model of schizophrenia. *Schizophrenia Research, 201*, 254–260.

Gomes, F. V., Zhu, X., & Grace, A. A. (2019). The pathophysiological impact of stress on the dopamine system is dependent on the state of the critical period of vulnerability. *Molecular Psychiatry*. https://doi.org/10.1038/s41380-019-0514-1.

Gonzalez-Burgos, G., Cho, R. Y., & Lewis, D. A. (2015). Alterations in cortical network oscillations and parvalbumin neurons in schizophrenia. *Biological Psychiatry, 77*, 1031–1040.

Gourevitch, R., Rocher, C., Le Pen, G., Krebs, M. O., & Jay, T. M. (2004). Working memory deficits in adult rats after prenatal disruption of neurogenesis. *Behavioural Pharmacology, 15*, 287–292.

Gulchina, Y., Xu, S. J., Snyder, M. A., Elefant, F., & Gao, W. J. (2017). Epigenetic mechanisms underlying NMDA receptor hypofunction in the prefrontal cortex of juvenile animals in the MAM model for schizophrenia. *Journal of Neurochemistry, 143*, 320–333.

Howes, O., McCutcheon, R., & Stone, J. (2015). Glutamate and dopamine in schizophrenia: An update for the 21st century. *Journal of Psychopharmacology, 29*, 97–115.

Hradetzky, E., Sanderson, T. M., Tsang, T. M., Sherwood, J. L., Fitzjohn, S. M., Lakics, V., … Bahn, S. (2012). The methylazoxymethanol acetate (MAM-E17) rat model: Molecular and functional effects in the hippocampus. *Neuropsychopharmacology, 37*, 364–377.

Huo, C., Liu, X., Zhao, J., Zhao, T., Huang, H., & Ye, H. (2018). Abnormalities in behaviour, histology and prefrontal cortical gene expression profiles relevant to schizophrenia in embryonic day 17 MAM-Exposed C57BL/6 mice. *Neuropharmacology, 140*, 287–301.

Jenks, K. R., Lucas, M. M., Duffy, B. A., Robbins, A. A., Gimi, B., Barry, J. M., & Scott, R. C. (2013). Enrichment and training improve cognition in rats with cortical malformations. *PLoS One, 8*, e84492.

Jongen-Relo, A. L., Leng, A., Luber, M., Pothuizen, H. H., Weber, L., & Feldon, J. (2004). The prenatal methylazoxymethanol acetate treatment: A neurodevelopmental animal model for schizophrenia? *Behavioural Brain Research, 149*, 159–181.

Kisby, G. E., Moore, H., & Spencer, P. S. (2013). Animal models of brain maldevelopment induced by cycad plant genotoxins. *Birth Defects Research Part C Embryo Today, 99*, 247–255.

Lavin, A., Moore, H. M., & Grace, A. A. (2005). Prenatal disruption of neocortical development alters prefrontal cortical neuron responses to dopamine in adult rats. *Neuropsychopharmacology, 30*, 1426–1435.

Le Pen, G., Gourevitch, R., Hazane, F., Hoareau, C., Jay, T. M., & Krebs, M. O. (2006). Peri-pubertal maturation after developmental disturbance: A model for psychosis onset in the rat. *Neuroscience, 143*, 395–405.

Le Pen, G., Jay, T. M., & Krebs, M. O. (2011). Effect of antipsychotics on spontaneous hyperactivity and hypersensitivity to MK-801-induced hyperactivity in rats prenatally exposed to methylazoxymethanol. *Journal of Psychopharmacology, 25*, 822–835.

Lena, I., Chessel, A., Le Pen, G., Krebs, M. O., & Garcia, R. (2007). Alterations in prefrontal glutamatergic and noradrenergic systems following MK-801 administration in rats prenatally exposed to methylazoxymethanol at gestational day 17. *Psychopharmacology, 192*, 373–383.

Lewis, D. A., & Levitt, P. (2002). Schizophrenia as a disorder of neurodevelopment. *Annual Review of Neuroscience, 25*, 409–432.

Li, M. L., Gulchina, Y., Monaco, S. A., Xing, B., Ferguson, B. R., Li, Y. C., … Gao, W. J. (2017). Juvenile treatment with a novel mGluR2 agonist/mGluR3 antagonist compound, LY395756, reverses learning deficits and cognitive flexibility impairments in adults in a neurodevelopmental model of schizophrenia. *Neurobiology of Learning and Memory, 140*, 52–61.

Lodge, D. J., Behrens, M. M., & Grace, A. A. (2009). A loss of parvalbumin-containing interneurons is associated with diminished oscillatory activity in an animal model of schizophrenia. *Journal of Neuroscience, 29*, 2344–2354.

Lodge, D. J., & Grace, A. A. (2008). Hippocampal dysfunction and disruption of dopamine system regulation in an animal model of schizophrenia. *Neurotoxicity Research, 14*, 97–104.

Mackowiak, M., Bator, E., Latusz, J., Mordalska, P., & Wedzony, K. (2014). Prenatal MAM administration affects histone H3 methylation in postnatal life in the rat medial prefrontal cortex. *European Neuropsychopharmacology, 24*, 271–289.

Mackowiak, M., Latusz, J., Glowacka, U., Bator, E., & Bilecki, W. (2019). Adolescent social isolation affects parvalbumin expression in the medial prefrontal cortex in the MAM-E17 model of schizophrenia. *Metabolic Brain Disease, 34*, 341–352.

Millan, M. J. (2013). An epigenetic framework for neurodevelopmental disorders: From pathogenesis to potential therapy. *Neuropharmacology, 68*, 2–82.

Millan, M. J., Andrieux, A., Bartzokis, G., Cadenhead, K., Dazzan, P., Fusar-Poli, P., … Weinberger, D. (2016). Altering the course of schizophrenia: Progress and perspectives. *Nature Reviews Drug Discovery, 15*, 485–515.

Modinos, G., Allen, P., Grace, A. A., & McGuire, P. (2015). Translating the MAM model of psychosis to humans. *Trends in Neurosciences, 38*, 129–138.

Moore, H., Jentsch, J. D., Ghajarnia, M., Geyer, M. A., & Grace, A. A. (2006). A neurobehavioral systems analysis of adult rats exposed to methylazoxymethanol acetate on E17: Implications for the neuropathology of schizophrenia. *Biological Psychiatry, 60*, 253–264.

van Os, J., Kenis, G., & Rutten, B. P. (2010). The environment and schizophrenia. *Nature, 468*, 203–212.

Penschuck, S., Flagstad, P., Didriksen, M., Leist, M., & Michael-Titus, A. T. (2006). Decrease in parvalbumin-expressing neurons in the hippocampus and increased phencyclidine-induced locomotor activity in the rat methylazoxymethanol (MAM) model of schizophrenia. *European Journal of Neuroscience, 23*, 279–284.

Raine, A., Mellingen, K., Liu, J., Venables, P., & Mednick, S. A. (2003). Effects of environmental enrichment at ages 3–5 years on schizotypal personality and antisocial behavior at ages 17 and 23 years. *American Journal of Psychiatry, 160*, 1627–1635.

Raine, A., Venables, P. H., Dalais, C., Mellingen, K., Reynolds, C., & Mednick, S. A. (2001). Early educational and health enrichment at age 3–5 years is associated with increased autonomic and central nervous system arousal and orienting at age 11 years: Evidence from the Mauritius Child Health Project. *Psychophysiology, 38*, 254–266.

Schmitt, A., Malchow, B., Hasan, A., & Falkai, P. (2014). The impact of environmental factors in severe psychiatric disorders. *Frontiers in Neuroscience, 8*, 19.

Selemon, L. D., & Goldman-Rakic, P. S. (1999). The reduced neuropil hypothesis: A circuit based model of schizophrenia. *Biological Psychiatry, 45*, 17–25.

Selemon, L. D., & Zecevic, N. (2015). Schizophrenia: A tale of two critical periods for prefrontal cortical development. *Translational Psychiatry, 5*, e623.

Sommer, I. E., Bearden, C. E., van Dellen, E., Breetvelt, E. J., Duijff, S. N., Maijer, K., … Vorstman, J. A. (2016). Early interventions in risk groups for schizophrenia: What are we waiting for? *NPJ Schizophrenia, 2*, 16003.

Sonnenschein, S. F., Gill, K. M., & Grace, A. A. (2019). State-dependent effects of D2 partial agonist aripiprazole on dopamine neuron activity in the MAM neurodevelopmental model of schizophrenia. *Neuropsychopharmacology, 44*, 572–580.

Spear, L. P. (2000). The adolescent brain and age-related behavioral manifestations. *Neuroscience and Biobehavioral Reviews, 24*, 417–463.

Stark, T., Ruda-Kucerova, J., Iannotti, F. A., D'Addario, C., Di Marco, R., Pekarik, V., Drazanova, E., Piscitelli, F., Bari, M., Babinska, Z., Giurdanella, G., Di Bartolomeo, M., Salomone, S., Sulcova, A., Maccarrone, M., Wotjak, C. T., Starcuk, Z., Jr., Drago, F., Mechoulam, R., Di Marzo, V., & Micale, V. (2019). Peripubertal cannabidiol treatment rescues behavioral and neurochemical abnormalities in the MAM model of schizophrenia. *Neuropharmacology, 146*, 212–221.

Talamini, L. M., Koch, T., Luiten, P. G., Koolhaas, J. M., & Korf, J. (1999). Interruptions of early cortical development affect limbic association areas and social behaviour in rats; possible relevance for neurodevelopmental disorders. *Brain Research, 847*, 105–120.

Tamminga, C. A., & Holcomb, H. H. (2005). Phenotype of schizophrenia: A review and formulation. *Molecular Psychiatry, 10*, 27–39.

Tsuang, M. (2000). Schizophrenia: Genes and environment. *Biological Psychiatry, 47*, 210–220.

Zimmerman, E. C., Bellaire, M., Ewing, S. G., & Grace, A. A. (2013). Abnormal stress responsivity in a rodent developmental disruption model of schizophrenia. *Neuropsychopharmacology, 38*, 2131–2139.

Part VI

Resources

Chapter 51

Recommended resources on the neuroscience of development

Rajkumar Rajendram[1,2,3] and Victor R. Preedy[3]

[1]*College of Medicine, King Saud Bin Abdulaziz University for Health Sciences, Riyadh, Central Region, Saudi Arabia;* [2]*Department of Medicine, King Abdulaziz Medical City, King Abdullah International Medical Research Center, Ministry of National Guard Health Affairs, Riyadh, Central Region, Saudi Arabia;* [3]*Diabetes and Nutritional Sciences Research Division, Faculty of Life Science and Medicine, King's College London, London, United Kingdom*

Introduction

Scientists have been fascinated by the nervous system since ancient times. Epileptic seizures, for example, were astutely diagnosed and described as early as the 7th century BCE (Kostopoulos, 2016). Although initially thought to be a curse from any of a range of deities; the pre-Socratic ancient Greeks initiated the concept that this affliction was the result of a defective brain (Kostopoulos, 2016). In Croton in Magna Graecia, Alkmaion (circa 450 BCE) was the first scientist to dissect and study the anatomy of the brain and peripheral nerves (Kostopoulos, 2016). He suggested that the brain was the site of sensation and cognition; believing that the optic nerves were "*paths that bore light to the brain*" (Kostopoulos, 2016) So, Alkmaion was the first to ascribe a specific function to a defined region of the brain (Kostopoulos, 2016). This paradigm shift was a fundamental prerequisite for the birth of developmental neuroscience.

The origins of the comparatively modern field of developmental neuroscience began in earnest with the histological observations Santiago Ramon y Cajal described in the early 1900s (Raman & Cajal, 1911). His accounts of neuronal development and how neurones connect with their targets initiated every analysis of the development of the vertebrate nervous system.

In the 1960s, neuroanatomists, following Ramon y Cajal's footsteps, used electron microscopy (EM) to examine cell structure and synaptic connections (Mason, 2009). New methods for serial section electron microscopy and tomography should allow more detailed analysis of the structure of developing circuitry of simple organisms (Luo, Callaway, & Svoboda, 2008). However, EM is labor-intensive and does not scale well to larger nervous systems (Luo et al., 2008).

In the 1970s, tracers such as horseradish peroxidase were injected into target areas, and retrograde transport of these tracers or fluorescently labeled beads localized the somata of cells that innervated different regions (Katz & Iarovici, 1990; Mason, 2009).

In the late 1980s, tracers that could be applied precisely to fixed tissue in tiny amounts allowed the early phases of axon outgrowth and developing pathways to be defined in embryos (Godement, Vanselow, Thanos, & Bonhoeffer, 1987; Mason, 2009). These tracers led to a resurgence in developmental neuroanatomy (Mason, 2009).

Immunohistochemistry was initially lauded as a technique that could usher in a new understanding of the relationships between cells during neuronal development (Reichardt, 1984). However, many antibodies are nonspecific (Mason, 2009). So, many scientists turned to in situ hybridization to localize messenger ribonucleic acid instead (Mason, 2009).

A huge paradigm shift occurred with the application of molecular genetics to developmental neuroscience (Luo et al., 2008). Astounding achievements have recently arisen from genetic insertion, either intentionally or serendipitously, of fluorescent proteins and other markers into select genes (Luo et al., 2008).

Modifying genes and proteins to analyze molecular mechanisms in "knockouts" is now the norm. Advances in molecular biology have enabled transcription factors to be pinpointed (Luo et al., 2008; Qiu & Ghosh, 2008).

Diagnosis, Management and Modeling of Neurodevelopmental Disorders. https://doi.org/10.1016/B978-0-12-817988-8.00051-8

So hypothesis-driven developmental neuroscientists can investigate the complete pathway of a molecular factor. However, the development of microarray and high-throughput technology has led to a blunderbuss approach to genomics, proteomics, and metabolomics. This has turned the traditional hypothesis-driven approach to neuroscience on its head.

In recent years, an overwhelming number of novel methods and tools for developmental neuroscience have emerged (Luo et al., 2008; Mason, 2009). These are being used to visualize cells, tracts, and pathways, tracking their development (Mason, 2009). Neurones can be isolated and cultivated (Mason, 2009). Molecular factors can be manipulated to elucidate their function (Luo et al. 2008; Mason, 2009; Qiu & Ghosh, 2008).

Genes are sought out (Mason, 2009). So, there has been an explosion of knowledge and understanding in developmental neuroscience. It is now difficult even for experienced scientists to remain up-to-date. To assist colleagues who are interested in understanding more about oxidative stress and aging, we have therefore produced tables containing up-to-date resources in this chapter. The experts who assisted with the compilation of these tables of resources are acknowledged as follows.

Resources

Tables 51.1–51.6 list the most up-to-date information on the regulatory bodies (Table 51.1), professional societies (Table 51.2), journals (Table 51.3), books (Table 51.4), online resources and platforms (Table 51.5), and other resources for health care professionals or patients that are relevant to an evidence-based approach to the neuroscience of development (Table 51.6).

TABLE 51.1 Regulatory bodies and organizations.

1000 Genomes Project	http://www.internationalgenome.org/
Allen Brain Atlas	https://human.brain-map.org/
Allen Institute	http://portal.brain-map.org
American Academy for Cerebral Palsy and Developmental Medicine	https://www.aacpdm.org/
American Bone Health	https://americanbonehealth.org/
Autism Genetic Resource Exchange (AGRE)	https://www.autismspeaks.org/agre
Bio-Atlas	http://bio-atlas.psu.edu/index.php
Cambridge Neuroscience	https://www.neuroscience.cam.ac.uk/research/themes/developmental.php
CanChild Centre for Childhood Disability	https://www.canchild.ca/en/diagnoses/developmental-coordination-disorder
Center for Developmental Neuroscience	https://www.csi.cuny.edu/academics-and-research/research-centers/center-developmental-neuroscience
Developmental Medicine and Child Neurology	https://onlinelibrary.wiley.com/journal/14698749
European Society for Child and Adolescent Psychiatry	https://www.escap.eu/
Federation of European Neuroscience Societies (FENS) Neuroscience	https://www.fens.org/
Harvard University: The Science of Early Childhood Development	https://developingchild.harvard.edu/resources/inbrief-science-of-ecd/
Hyperemesis Education and Research Foundation (HER Foundation)	www.helpher.org
International Association for Child and Adolescent Psychiatry and Allied Professions	https://iacapap.org/
International Osteoporosis Foundation	https://www.iofbonehealth.org/latin-america
Journal of Child Neurology	https://us.sagepub.com/en-us/nam/journal/journal-child-neurology

Continued

TABLE 51.1 Regulatory bodies and organizations.—cont'd

Kyoto Pharmaceutical University (Division of Pathological Sciences—Department of Pathological Biochemistry)	https://www.kyoto-phu.ac.jp/english/education_research/laboratory/?c=laboratory_view&pk=14
Mexican Society Physiological Sciences	www.smcf.org.mx
National Center of Neurology and Psychiatry	https://www.ncnp.go.jp/english/index.html
National Institute of Mental health	https://www.nimh.nih.gov/health/topics/child-and-adolescent-mental-health/index.shtml
NeuroMorpho.org	http://neuromorpho.org/index.jsp
Neuroscientifically Challenged	https://www.youtube.com/watch?v=Tp25wrm-AoA
Pediatric Neurology	https://www.journals.elsevier.com/pediatric-neurology
Researchgate.net Adele Diamond profile	https://www.researchgate.net/profile/Adele_Diamond
Researchgate.net Jane Squires profile	https://www.researchgate.net/profile/Jane_Squires2
Riken Center for Brain Science	https://www.riken.jp/en/research/labs/cbs/
Royal Osteoporosis Society	https://theros.org.uk/
Developmental Cognitive Neuroscience	https://www.sciencedirect.com/topics/medicine-and-dentistry/developmental-cognitive-neuroscience
Simons Foundation for Autism Research (SFARI)	https://www.sfari.org/
Society for Developmental Biology	https://www.sdbonline.org/
Society for Neuroscience	https://www.sfn.org/
Society of Toxicology	https://www.toxicology.org/
Springer Nature Embryology	https://www.nature.com/subjects/embryology
The Zebrafish Information Network	https://zfin.org
Timeline Human Development	https://embryology.med.unsw.edu.au/embryology/index.php/Timeline_human_development
Understood	https://www.understood.org/en
World Health Organization	https://www.who.int/
Xenbase	http://www.xenbase.org/entry/
Z Brain Atlas	https://engertlab.fas.harvard.edu/Z-Brain/home/

This table lists the regulatory bodies and organizations involved with neurological development and associated specialities or interests. The links were accurate at the time of going to press but may move or alter. There are also lists of sites in Tables 51.2, 51.5, and 51.6.

TABLE 51.2 Professional societies.

Allen Institute	https://alleninstitute.org/what-we-do/brain-science
American Academy of Cerebral Palsy and Developmental Medicine	https://www.aacpdm.org
American Academy of Neurology	https://www.aan.com
American Association of Neuropathologists	https://www.neuropath.org
American College of Neuropsychopharmacology	https://acnp.org
American Epilepsy Society	https://www.aesnet.org
American Neurological Association	https://myana.org
American Psychological Association	https://www.apa.org
American Society for Nutrition	https://nutrition.org
American Society of Human Genetics	http://www.ashg.org
Australasian Winter Conference on Brain Development	https://www.otago.ac.nz/awcbr/index.html

Continued

TABLE 51.2 Professional societies.—cont'd

British Neuroscience Association	https://www.bna.org.uk
Child Neurology Society	https://www.childneurologysociety.org
Cognitive Development Society	https://cogdevsoc.org
Dana Foundation	https://www.dana.org
Developmental Origins of Health and Disease (DOHaD) Latin American Chapter	https://dohadsoc.org/associated-societies/ibero-american
Developmental Origins of Health and Disease (DOHaD) African Chapter	https://www.dohadafrica.org.za
Developmental Origins of Health and Disease (DOHaD) Canada	http://www.dohad.ca
Developmental Origins of Health and Disease (DOHaD) Japan Society	http://square.umin.ac.jp/Jp-DOHaD/J-DOHaD.html
DOHaD Society—International Society for Developmental Origins of Health and Disease	https://dohadsoc.org
European College of Neuropsychopharmacology	https://www.ecnp.eu
European Society for Child and Adolescent Psychiatry	https://www.escap.eu
European Teratology Society	https://www.etsoc.com
Federation of European Neuroscience Societies (FENS)	https://www.fens.org
Flux Society	https://fluxsociety.org
International Society of Neurochemistry	https://www.neurochemistry.org
International Association for Child and Adolescent Psychiatry and Allied Professions	https://iacapap.org
International Association for the Study of Affective Touch	www.iasat.org
International Behavioral Neuroscience Society (IBNS)	https://www.ibnsconnect.org
International Behavioural and Neural Genetics Society (IBANGS)	https://www.ibngs.org
International Brain Research Organization	https://ibro.org
International Neuropsychological Society	https://www.the-ins.org
International Society for Autism Research	https://www.autism-insar.org
International Society for Developmental Neuroscience (ISDN)	http://www.isdn-conference.elsevier.com/about-isdn.asp
International Society for Developmental Psychobiology	https://isdp.org
International Zebrafish Society (IZSF)	https://www.izfs.org
Mexican Society Physiological Sciences	www.smcf.org.mx
NeuroImaging Tools and Resources Collaboratory	https://www.nitrc.org/frs/?group_id=383
North American Society for Pediatric Gastroenterology, Hepatology and Nutrition	https://naspghan.org
Organisation of Human Brain Mapping	https://www.humanbrainmapping.org/i4a/pages/index.cfm?pageid=3267&pageid=1
Simons Foundation	https://www.sfari.org/resource/simons-simplex-collection/
Sociedad Argentina de Neurolog Infantil	www.saniargentina.org
Sociedad Argentina de Pediatria	https://www.sap.org.ar
Societitaliana di Neuroscienze	http://www.sins.it/EN/index.xhtml
Society for Developmental Biology	https://www.sdbonline.org
Society for Neuroscience	https://www.sfn.org
Society of Toxicology	https://www.toxicology.org
Spanish Society of Neuroscience	https://www.senc.es/en
The Japan Neuroscience Society	https://www.jnss.org/en

Continued

TABLE 51.2 Professional societies.—cont'd

The Japanese Society for Neurochemistry	http://www.neurochemistry.jp/?lang=english
The Japanese Society of Child Neurology	https://www.childneuro.jp/modules/en/index.php?content_id=1
The Teratology Society	https://teratology.org
The Zebrafish Information Network (ZFIN)	https://zfin.org
Trisomy 21 Research Society	https://www.t21rs.org
World Federation of Neurology	https://wfneurology.org
Zebrafish Disease Models Society (ZDMS)	https://www.zdmsociety.org/home

This table lists some societies and organizations devoted to understanding the neuroscience of development. See also Tables 51.1, 51.5, and 51.6. Please note, occasionally the location of the websites or web address changes. In these cases the use of the "Search" tabs should be explored at the parent address or site.

TABLE 51.3 Journals relevant to neurological development.

Plos One
Neurology
Scientific Reports
Brain and Development
Neuroimage
Nature Communications
Frontiers in Neuroscience
Cerebral Cortex
Journal Of Neuroscience
Neuroimage Clinical
Frontiers in Cellular Neuroscience
Neuroscience
World Neurosurgery
Molecular Neurobiology
International Journal Of Molecular Sciences
Human Brain Mapping
European Journal Of Paediatric Neurology
Frontiers in Neurology
Proceedings Of The National Academy Of Sciences Of The United States Of America
Neuroscience Letters
Brain Research
Nature
Autism Research
Frontiers in Human Neuroscience
Journal Of Alzheimer's Disease
Developmental Medicine and Child Neurology
Environmental Research
Neurotoxicology
Cell Reports
Translational Psychiatry

Journals publishing original research and review articles related to neurological development. Included in this list are the top 30 journals, which have published the most number of articles on neurological development over the past 5 years. Data derived from Scopus.

TABLE 51.4 Relevant books.

Affective Touch and the Neurophysiology of CT Afferents. Hakan Olausson, Johan Wessberg, India Morrison, Francis McGlone. Springer, 2016.
Autism Spectrum Disorders. David Amaral, Daniel Geschwind, Geraldine Dawson. Oxford University Press, 2011.
Casarett and Doulls Toxicology. Curtis D. Klaassen. McGraw-Hill Education, 2019.
Child Neuropsychology: Concepts, Theory, and Practice. Jonathan Reed, Jody Warner—Rogers. Wiley-Blackwell, 2008.
Cognitive Development: An Advanced Textbook. Marc H. Bornstein, Michael E. Lamb. Psychology Press, 2011.
Developmental Biology 12th Edition. Michael J.F. Barresi, Scott F. Gilbert. Oxford University Press, 2019.
Developmental Cognitive Neuroscience. Mark H. Johnson. John Wiley and Sons, 2011.
Developmental Cognitive Neuroscience: An Introduction. Mark H. Johnson, Michelle de Haan. Wiley_Blackwell, 2015.
Developmental Neuroscience: A Concise Introduction. Susan E. Fahrbachh. Princeton University Press, 2013.
Follow-Up for NICU Graduates: Promoting Positive Developmental and Behavioral Outcomes for At-Risk Infants. Howard Needelman, Barbara J. Jackson. Springer, 2018.
Fundamental Neuroscience. Larry R. Squire, Floyd E. Bloom, Darwin Berg, Anirvan Ghosh, Sascha du Lac. Academic Press, 2008.
Handbook of Developmental Cognitive Neuroscience, 2nd Edition. Charles Nelson, Monica Luciana. MIT Press, 2008.
Life Span Human Development: Australian and New Zealand Edition (3rd Ed). Carol Sigelman, Elizabeth Rider. Cengage Learning Australia, 2019.
Netters Atlas of Neuroscience. David L. Felten, M. Kerry O'Banion, Mary Summo Maida. Elsevier, 2015
Neural Tube Defects. Jorge A. Lazareff. World Scientific, 2011.
Neural Tube Defects: From Origin to Treatment. Diego F. Wyszynski. Oxford University Press, 2006.
Neurologpedica. Natalio Fejerman, Nora Grañana. Paidos Argentina. 2017.
Neuropsicologia Infantil. Marisa Arnedo Montoro, Judit Bembibre Serrano, Armando Montes Lozano, Mónica Triviño Mosquera. Editorial Médica Panamericana S.A., 2017.
Neuroscience. George J. Augustine, David Fitzpatrick. Sinauer Associates, 2007.
Neuroscience. Claudia Krebs, Joanne Weinberg, Elizabeth Akesson. Lippincott Williams and Wilkins, 2011.
Neuroscience: Exploring the Brain 4th Edition. Mark F. Bear, Barry W. Connors, Michael A. Paradiso. Lippincott Williams and Wilkins, 2015.
Physical Therapy for Children. Suzann Campbel, Robert Palisano, Margo Orlin. Elsevier, 2012.
Principles of Neural Science. Eric R. Kandel, James H. Schwartz, Thomas M. Jessell, Steven A. Siegelbaum, A.J. Hudspeth. McGraw-Hill Medical, 2013.
Reproductive and Developmental Toxicology. Ramesh C. Gupta. Academic Press, 2017.
Statistical Genetics: Gene Mapping Through Linkage and Association. Benjamin M. Neale, Manuel A.R. Ferreira, Sarah E. Medland, Danielle Posthuma. Taylor and Francis, 2008.
The Development Of The Human Body. J. Playfair McMurrich. Blakiston's Son and Co., 1913.
The Epigenome and Developmental Origins Of Health and Disease. Cheryl S. Rosenfeld. Academic Press, 2015.
The Human Brain Prenatal Development and Structure. Miguel Marin-Padilla. Springer, 2011.
The Neuroscience Of Human Relationships. Louis Cozolino. W. W. Norton and Company, 2014.
Trastornos del desarrollo. Natalio Fejerman, Hugo A. Arroyo. Paidos Ed, 2010.

This table lists books on the neuroscience of development.

TABLE 51.5 Resources and emerging technologies relevant to neurological development.

AIM Biotech Pte. Ltd.	https://www.aimbiotech.com/
Allen Brain Atlas	http://portal.brain-map.org/
Bitbrain	https://www.bitbrain.com/neurotechnology-products/software/bitbrain-software-tools
BrainTranscriptome Database	http://www.cdtdb.neuroinf.jp/CDT/Top.jsp
Centre for Advanced MRI	https://camri.auckland.ac.nz/
CN Bio Innovations	https://cn-bio.com
Comparative Mammalian Brain Collection	http://neurosciencelibrary.org/index.html
GWAS Catalog EMBL-EBI	https://www.ebi.ac.uk/gwas/
Hesperos Inc.	https://hesperosinc.com/
Hub4Organoids (Bank of organoids)	https://hub4organoids.eu/
Laboratories of Cognitive Research	http://www.childrenshospital.org/research/labs/laboratories-of-cognitive-neuroscience
Mcgill University-Neuro	https://www.mcgill.ca/neuro/
MIMETAS B.V. (Organ-on-a-Chip)	https://mimetas.com/
Montreal Neurological Institute	https://en.wikipedia.org/wiki/Montreal_Neurological_Institute_and_Hospital
NeuroNexus	https://neuronexus.com/
NeuronUP (web platform of cognitive rehabilitation)	https://www.neuronup.com/
Nikon	https://www.microscope.healthcare.nikon.com/products/confocal-microscopes
Nortis Inc. (Organ-on-Chip Technology)	https://www.nortisbio.com/
Research gate	www.researchgate.net
SFARI Gene Human Gene Module	https://gene.sfari.org/database/human-gene/
Sooma	https://soomamedical.com/en/
Thorlabs	https://www.thorlabs.com/
Tucker-Davis Technologies	https://www.tdt.com/
UCSC Genome Browser	http://genome-euro.ucsc.edu

This table lists some internet resources and emerging technologies relevant to neurological development. See also Table 51.1. Please note, occasionally the location of the websites or web address changes. In these cases the use of the "Search" tabs should be explored at the parent address or site.

TABLE 51.6 Other resources of interest or relevance for health care professionals or patients in relation to neurological development.

Apadea Parent Association	www.apadea.org.ar
Brincar Parent Association Net	www.brincar.org.ar
Centre for Advanced MRI	www.camri.co.nz
Cerebral palsy news today	https://cerebralpalsynewstoday.com/
Cerebral palsy org	https://www.cerebralpalsy.org/information/nutritional-health
DCD Toolkit for Pediatricians	http://www.childdevelopment.ca/DCDAdvocacyToolkit/DCDAdvocacyToolkitIntro.aspx
ENIGMA ORIGINs (Organization for Imaging Genomics in Infancy)	http://enigma.ini.usc.edu/ongoing/enigma-origins/
European Down Syndrome Association	http://www.edsa.eu

Continued

TABLE 51.6 Other resources of interest or relevance for health care professionals or patients in relation to neurological development.—cont'd

Food Fortification Initiative	http://www.ffinetwork.org/well
Global Down Syndrome Foundation	https://www.globaldownsyndrome.org/
HG Care Application	https://apps.apple.com/us/app/hg-care-pregnancy-wellness/id1148105670
Hyperemesis Education and Research Foundation	www.helpher.org
International Federation for Spina Bifida and Hydrocephalus	https://www.ifglobal.org/
Jerome Lejeune Foundation	https://www.fondationlejeune.org/en/#
JM Reys IACAPAP e-Textbook of Child and Adolescent Mental Health. Rey JM, Martin A (eds). Geneva: International Association for Child and Adolescent Psychiatry and Allied Professions, 2019	https://iacapap.org/iacapap-textbook-of-child-and-adolescent-mental-health/
LuMind Down Syndrome Foundation	https://www.lumindidsc.org/pr-hampus-hillerstrom/
March of Dimes	https://www.marchofdimes.org/
Mcgill University	https://www.mcgill.ca/neuro/
Montreal Neurological Institute	https://en.wikipedia.org/wiki/Montreal_Neurological_Institute_and_Hospital
NIH Genetics Home Reference Help Me Understand Genetics	https://ghr.nlm.nih.gov/primer
Online Mendelian Inheritance in Man (OMIM)	https://www.omim.org/
Primer on Medical and Population Genetics, Broad Institute	https://www.broadinstitute.org/scientific-community/science/programs/medical-and-population-genetics/primers/primer-medical-and-pop
PROTECTEA Professional Net	Sites.google.com/view/protectea
Redea Parent Associations	www.redea.org.ar
Royal College of Paediatrics and Child Health	https://www.rcpch.ac.uk/resources/stroke-childhood-clinical-guideline-diagnosis-management-rehabilitation
Stroke Association—Childhood Stroke	https://www.stroke.org.uk/childhood-stroke
The Feta, Infant and Toddler Neuroimaging Group (FITNG)	http://fitng.org/
The Stroke Foundation	https://strokefoundation.org.au/About-Stroke/Types-of-stroke/Childhood-stroke
Turner Syndrome Society of the United States	https://www.turnersyndrome.org/

This table lists some other resources of interest or relevance for health care professionals or patients in relation to neurological development. See also Tables 51.1, 51.2 and 51.5. Please note, occasionally the location of the websites or web address changes.

Summary points

- Development results in many changes in the central nervous system.
- In recent years, a staggering array of new methods and tools for the study of developmental neuroscience have emerged.
- Microarray technology for genomics, proteomics, and metabolomics has turned the traditional hypothesis-driven approach to developmental neuroscience on its head.
- There has been an explosion of the knowledge and understanding in developmental neuroscience. It is difficult even for experienced scientists to remain up-to-date.
- This chapter lists resources relating to the regulatory and professional bodies, societies, journals, books, and websites that are relevant to an understanding of the neuroscience of development.

Acknowledgments

We would like to thank the following authors for contributing to the development of this resource. Eguibar JR, Bartesaghi R, Guidi S, Stagni F, Bendas J, Cantor R, Lowe J, Drobni Radobuljac M, Fejzo M, Ahkin Chin Tai J, Freeman J, Gerlai R, Gomez C, Granana N, Guadano Ferraz A, Homman-Ludiye J, Bourne J, Katada T, Sakurai H, Knickmeyer R, Ma Z, Manganas L, Matte, Mhanna R, Deeb G, Numakawa T, Odaka H, Peviani V, Ren A, Rodriguez-Martinez E I, Waldie K, Zwicker J, Albert M, Alvarez-Zaragoza C, Ishihara K, Montgomery J, Masino M, OKeeffe F, Takeuchi A.

References

Godement, P., Vanselow, J., Thanos, S., & Bonhoeffer, F. (1987). A study in developing visual systems with a new method of staining neurones and their processes in fixed tissue. *Development, 101*, 697–713.

Katz, L. C., & Iarovici, D. M. (1990). Green fluorescent latex microspheres: A new retrograde tracer. *Neuroscience, 34*, 511–520.

Kostopoulos, G. K. (2016). History of neuroscience in Greece: From alkmaion to austerity. *European Journal of Neuroscience, 44*, 2742–2745.

Luo, L., Callaway, E. M., & Svoboda, K. (2008). Genetic dissection of neural circuits. *Neuron, 57*, 634–660.

Mason, C. (2009). The development of developmental neuroscience. *Journal of Neuroscience, 29*, 12735–12747.

Qiu, Z., & Ghosh, A. (2008). A brief history of neuronal gene expression: Regulatory mechanisms and cellular consequences. *Neuron, 60*, 449–455.

Ramon, y, & Cajal, S. (1911). *Histologie du systeme nerveux de l'homme et des vertebres*. Paris: Maloine.

Reichardt, L. F. (1984). Immunological approaches to the nervous system. *Science, 225*, 1294–1299.

Index

Note: 'Page numbers followed by "*f*" indicate figures and "*t*" indicate tables.'

J

K

L

M